# Creep: Characterization, Damage and Life Assessments

*Proceedings of the*
Fifth International Conference on Creep of Materials
18-21 May 1992
Lake Buena Vista, Florida, USA

*Sponsored by*
ASM International®

*Co-sponsored by the*
Americal Society of Mechanical Engineers
ASTM
The Institute of Mechanical Engineers
Japan Society of Mechanical Engineers
The Japan Society of Materials Science

ASM INTERNATIONAL®
The Materials
Information Society

Library of Congress Catalog Card Number: 92-81173
ISBN: 0-87170-444-7
SAN: 204-7586

Production Manager
Linda Kacprzak

Production Coordinator
Randall Boring

**ASM International®**
Materials Park, Ohio 44073-0002

Printed in the United States of America

## NORTH AMERICAN ORGANIZING COMMITTEE

# MESSAGE FROM THE ORGANIZERS

This international conference addresses the important creep design considerations covered in the previous conferences in this series. However, the scope of coverage has been broadened beyond power plant technology and beyond high-temperature metallic materials. For the first time, papers are included on ceramics and composites, and these contributions are presented in integrated sessions under general topic headings reflecting the generic nature of creep testing and analysis.

Although the emphasis is on applications, papers describing microstructural evaluation and theoretical modeling are also included. High-temperature fatigue in which frequency and hold time effects may be traced to time-dependent deformation and/or environmental damage is covered under the general topic of nonsteady stress and temperature effects. The important aspects of materials degradation or damage evolution are related to creep and creep fracture design. The final group of papers deals with creep of structures and life assessment of components.

This conference will, therefore, span the full range of materials for structural applications at high homologous temperatures. It will also relate to a wide spectrum of high-temperature applications. We hope that the presentations and the published proceedings will provide a useful framework on which to examine the current status of this technology and to prepare for the expanding opportunities and challenges in the future.

# Table Of Contents

## Opening Session



## MATERIALS TESTING AND CHARACTERIZATION

## CREEP DATA ANALYSIS AND CONSTITUTIVE EQUATIONS

## CREEP IN STRUCTURES

## LIFE PREDICTION



### Poster Sessions:

*Proceedings of the Fifth International Conference on Creep of Materials, Lake Buena Vista, Florida, USA, 18-21 May 1992*

# High Temperature Design Procedures—Developments in Europe Since 1986

**C.H.A. Townley**
Berkeley Nuclear Laboratories
Berkeley, Gloucestershire, England

ABSTRACT

The paper reviews the progress towards improved design procedures for High Temperature structures that has been made in Europe since the Fourth International Conference on Creep in 1986. The paper does not claim to be comprehensive and deals principally with the methods associated with fossil-fired and nuclear generating plant.

AT THE FOURTH INTERNATIONAL CONFERENCE in 1986, the author reviewed the then current state of development in the United Kingdom on High Temperature Design Rules for nuclear plant (1). Since then, significant advances have been made in the development of the Nuclear Electric R5 Procedure for the Assessment of the High Temperature Response of Structures (2). There have been equally important contributions arising from the decision to proceed with the joint design of a European Fast Reactor.

At an early stage of the European project, it was decided that Design Procedures would be based on the French RCC-MR Code, but with important modifications and additions. For the past six years, there has been active participation by research organisations and industrial companies in France, Germany and Britain aimed at improving the Code. Limitations of space prevent more than an outline of these developments in the present paper. A more complete description will be found in recent publications by Acker et al (3) and Laue et al (4).

The LMFBR differs from many other types of high temperature plant, where pressure stresses are all important. In the Fast Reactor, primary stresses are generally very low, and the design of components is dictated by the need to accommodate transient thermal stresses which occur under certain conditions. Despite these differences, much of what is now being developed in the context of the Fast Reactor can be applied with advantage to other types of plant.

There is no difficulty in ensuring that high temperature equipment meets the highest standards when it first goes into service. The real question is how to design plant so that it will survive for its full required life without signs of distress. There are severe economic penalties associated with premature replacement or refurbishment.

There are effectively three stages in assessing fitness for purpose throughout life, and the paper will deal with each of these in turn. These comprise

1 Estimation of stresses and strains at critical points of the components throughout life
2 Comparison of these predictions with allowable values for the materials of construction
3 Confirmation that the design is not sensitive to the presence of defects and flaws

## STRESS AND STRAIN ANALYSIS

**Elastic Analysis Route.** With modern computing techniques, it is relatively straightforward to calculate stresses and strains in a component throughout its working life, assuming fully elastic behaviour of the material.

Such elastic analyses are attractive to the designer because of their simplicity and their low cost. However, the computed stresses and strains are not representative of those which will actually occur in the component. Before they can be used to assess the performance of a structure which behaves inelastically, they have to be adjusted to allow for non-linear effects. This approach was first incorporated into the procedures of Code Case N47 and a number of disadvantages have become apparent. Often pessimistic assumptions have to be made, which lead to the rejection of proposed designs which, in reality, would be quite adequate.

The rules given in N47, and largely repeated in RCC-MR, require the classification of the calculated stresses into primary membrane, primary bending, secondary and peak. This procedure dates back to the period before finite element methods became available and stress analysis had to be based on shell theory. When using the output of a finite element computation, it is often not obvious which stresses should be regarded as primary, and which as secondary. This is particularly difficult when large thermal stresses are present. In those cases where classification is not clear, the conservative assumption has to be made that the stresses are primary.

A further difficulty is introduced by the adoption of different formulae to limit primary stresses above and below 425°C. This can lead to a step change in the allowable thickness of components at this temperature.

Below 425°C, the governing inequalities are

$$P_L \leq S_m$$
$$P_L + P_B \leq 1.5S_m$$

where $P_L$ is the primary membrane stress (either local or general)

$P_B$ is the primary bending stress

and $S_m$ is the design stress based on yield

Above 425°C, they become

$$P_L \leq S_t$$
$$P_L + \Phi P_B \leq S_t$$

where $S_t$ is the design stress based on creep rupture

and $\Phi$ is taken as 0.7 or 0.8

Work in the UK has demonstrated that failure under creep conditions is governed by reference stress, (5) as is plastic collapse at low temperatures. The limits derived from the reference stress for a rectangular beam under combined tension and bending are compared with those given by the above inequalities in Figure 1.

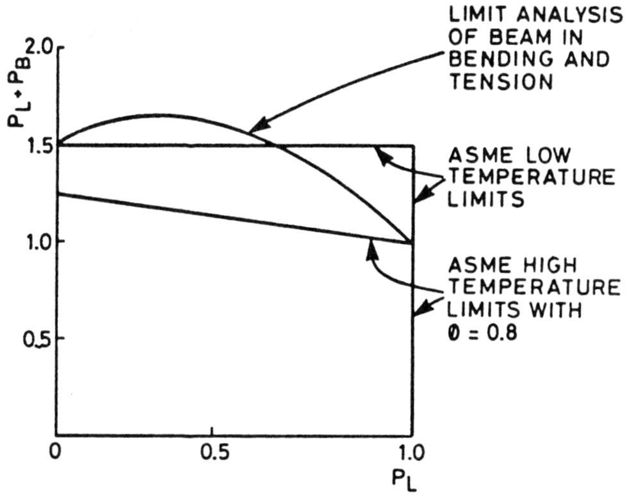


Fig 1 - Comparison of limit states, using the reference stress solution and the ASME inequalities

The gross conservatism of the ASME High Temperature limits on $P_L + P_B$, particularly at low values of primary stress, will be noted. It is also clear from the Figure that, for primary membrane stresses greater than 0.7, the ASME Low Temperature limits on $P_L + P_B$ are unconservative. The maximum error on permitted stress is about 18%. However, it will be noted that the beam in bending represents a "worst case" and that no difficulties have been encountered in service with vessels designed to this criterion.

It, therefore, follows that the inequalities above and below 425°C should be of the same form:

$$P_L \leq \langle S_m, S_t \rangle$$
$$P_L + P_B \leq 1.5 \langle S_m, S_t \rangle$$

where $\langle S_m, S_t \rangle$ denotes the lower value of $S_m$ and $S_t$

These provide a smooth transition from design based on yield properties to design based on creep rupture at about 530°C in the case of type 316 austenitic steel.

The principal corrections which have to be applied to the elastically calculated stresses to allow for inelastic effects are in the areas of strain range enhancement and elastic follow up. The strain range enhancement factor allows for the fact that in regions of a structure where stresses exceed yield, the actual strain range which causes fatigue damage is greater than the elastically calculated value.

Work carried out as part of the European Fast Reactor collaboration has shown that the strain range enhancement factor, $k_e$, as defined in the ASME codes is unconservative for a nominal stress of less than $3S_m$ and is overconservative for higher values. The corresponding RCC-MR procedure is less conservative for higher stresses, but may be unsafe when applied to structures with a large amount of elastic follow up.

A revised method of calculating $k_e$ has been developed, (3) which includes the strain amplification factor due to Neuber and the elastic follow-up factor derived for a cantilever beam. This is illustrated in Fig 2.

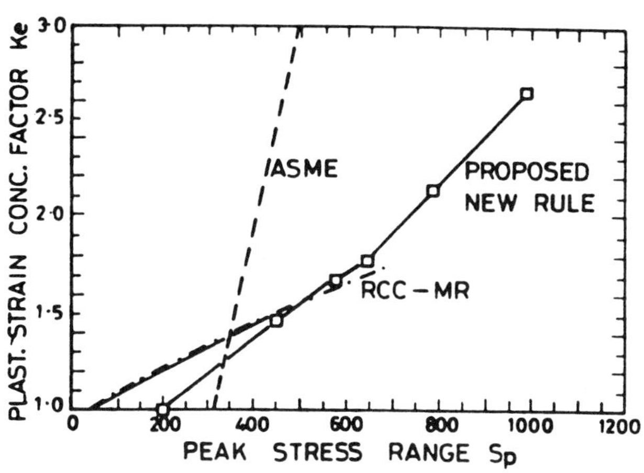


Fig 2 - Comparison of $k_e$ factors

**The Shakedown Method.** Procedures based on shakedown are seen as providing a simplified method of analysis which does not suffer from excessive conservatism of the elastic design route and avoids the difficulties of stress classification. Such methods were the subject of much research in the United Kingdom in the 1970s and have been used extensively for the design of both low and high temperature power plant. Shakedown concepts also underpin the well known inequalities which appear in the ASME and RCC-MR Codes, such as $P_L + P_B + Q \leq 3S_m$

Shakedown analyses may be performed routinely by using a post processor program, such as ADAPT, developed to operate on the output of an elastic finite element analysis of the component (6, 7, 8). If the ADAPT calculations show that shakedown is achieved, then estimation of strain ranges in critical regions of the component is relatively straightforward. Where there are notch effects to be taken into account strain range enhancement rules, such as those mentioned in the previous section, can be applied to the elastically calculated stresses in order to estimate fatigue damage.

In the case of components which operate at high temperature, the ADAPT post processor makes use of a creep modified shakedown limit, similar to that described in a paper by Goodall et al (5). This takes into account the fact that the residual stresses introduced by a transient which causes plasticity will relax by creep during a period of steady operation.

Calculation of strain range proceeds as before, but it is now necessary to take into account creep damage as well. A conservative estimate can be made by assuming that the stress at the beginning of the dwell period remains constant. Where it is necessary to be more precise, simple methods, such as those described in the Nuclear Electric R5 Procedure, (2) are available to estimate the relaxation which takes place, including, where relevant, the effects of local elastic follow up. Consideration of global elastic follow up, such as occurs in piping systems, is the subject of ongoing research.

The importance of the shakedown procedure as a design tool is illustrated in Table 1, where the ADAPT calculations are compared with predictions based on full inelastic analyses. The ADAPT calculations do not allow for stress relaxation during the dwell period, and it will be noted that the creep and fatigue damage predicted by the shakedown procedure is somewhat greater than that given by the inelastic analysis. At the same time, the gross pessimism of the elastic design route is eliminated.

**Full Inelastic Analysis.** Given the adoption of simplified methods of analysis for the majority of high temperature components, full inelastic analysis is likely to be confined to a limited number of cases which need special treatment.

Computer programs are now generally available which are capable of the numerical analysis of inelastic situations. They are relatively straightforward to use where only a simple creep analysis is required. They are much more difficult to apply where a full cyclic elastic-plastic-creep calculation must be performed. In this case, computing costs are very high and, moreover, a considerable amount of skill is needed if the computations are to provide correct estimates of stresses and strains. A series of international benchmarks (9), conducted partly under the auspices of the Commission for European Communities and partly within the European Fast Reactor project, has led to a good appreciation of the issues involved.

Table 1. Comparison of ADAPT calculations with inelastic calculations

| Test Case | Shakedown $\Delta\varepsilon$(%) | Shakedown $D_f$ | Shakedown $D_c$ | Inelastic Analysis $\Delta\varepsilon$(%) | Inelastic Analysis $D_f$ | Inelastic Analysis $D_c$ |
|---|---|---|---|---|---|---|
| 1 | 0.279 | 0.075 | 25 | 0.170 | 0.01 | 5 |
| 2 | 0.508 | 0.536 | 15 | 0.403 | 0.256 | 3.5-6.2 |
| 3 | - | 0.13 | 0.14 | - | 0.05 | 0.1 |
| 4 | 0.3 | 0.1 | 0.25 | 0.135 | 0.002 | 0.043 |

where $\Delta\varepsilon$ is the equivalent strain range
$D_f$ is the fatigue damage fraction
$D_c$ is the creep damage fraction

Test case 1 relates to a T butt configuration
Test case 2 relates to a constricted cylinder problem
Test case 3 relates to a thin plate with double sided thermal shock
Test case 4 relates to a rising fluid situation

Correct choice of mesh, correct choice of load step and correct choice of time step are all essential for reliable results. The accuracy of the calculations also depends critically on the constitutive equation which is used to describe the deformation behaviour of the material. There are significant differences in the predictions due to what seem to be quite minor differences in the constitutive equations.

What has become apparent is that many of the recent proposals for constitutive equations are too complex for use in design calculations. In Fast Reactor components stress levels and strain ranges are restricted to relatively low values by creep and creep-fatigue considerations. This means that many of the deformation phenomena which are observed in accelerated tests are relatively unimportant and can probably be ignored in a practical constitutive equation.

Further work is needed to rationalise the information which has been generated in the past and to provide a simple tool for practical application. It is somewhat unfortunate that much of the present fundamental research seems to be aimed at increasing complexity, rather than reducing it.

The ideal constitutive equation from the point of view of practical application should:

1 Have a minimum number of terms, ideally not more than 3 or 4, to keep computer storage within reasonable bounds and to avoid the need for excessive materials testing.
2 Be confined to strain ranges and stress levels of practical importance, ignoring effects which only become significant under unrealistic conditions.
3 Be of adequate accuracy for engineering calculations without striving for a high degree of precision.

**Interaction Diagrams and Their Use in Preliminary Assessments** From what has been said above, it will be seen that the designer has three different methods of verifying that a proposed design of component is fit for purpose. In order to avoid undue complexity in his analysis he needs to decide if one of the simpler alternatives can be applied.

Interaction Diagrams, such as those originally developed by Bree (10) provide a simple but effective means of doing this and only require an elastic analysis of the stresses. The principles are best illustrated by reference to Figure 3.

The abscissa is the non dimensional primary plus bending stress C, defined as:

$$C = \sigma_p/(K_1 S_y)$$

where $\sigma_p$ is the Tresca primary stress, calculated elastically
$S_y$ is the minimum 0.2% proof stress at the maximum temperature
$K_1$ is a constant to cover the cyclic hardening or softening characteristics of the material

The ordinate, T, is here defined as:

$$T = \sigma_t/(K_1\ S_y)$$

where $\sigma_t$ is the elastically calculated range of secondary stress in the component (excluding F stresses)

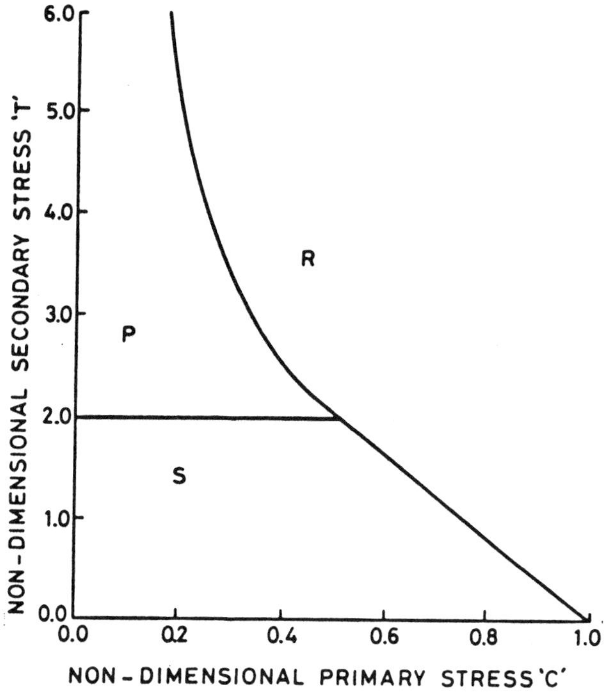


Fig 3 - Interaction diagram for through-wall temperature gradients or secondary bending stresses

The diagram is divided into three regions marked S, P and R where S indicates a region where ratcheting and reversed plastic strains are avoided P indicates a region where no ratcheting takes place, but cyclic plastic straining occurs locally. R indicates a region where freedom from ratcheting cannot be demonstrated.

If the point (C,T) estimated for the component lies within the S region then the component is likely to prove acceptable. Detailed calculations will usually be necessary to confirm this and if the point lies towards the origin an elastic calculation should suffice. Otherwise, a shakedown analysis will be needed.

If the point (C,T) lies outside the S region, a full inelastic analysis will be required. However, it should be noted that if the point is well inside the P or R regions, redesign will probably be necessary as refined calculations are unlikely to demonstrate that the component is satisfacory. In the former case progressive deformation is likely to prove excessive. In the latter, low cycle fatigue is likely to be life limiting.

In high temperature components it is necessary to estimate the creep damage which occurs. This can be done from a knowledge of the effective stress which is relevant to the operating cycle. This may be estimated by a construction described in a paper by Carter et al (11), and illustrated in Figure 4. A straight line is drawn from the origin through the loading point (C,T) and extended to the boundary of the R region. The effective creep stress, $\sigma_r$, is then given by

$$\sigma_r = (T/T^*)/(K_1 S_y)$$

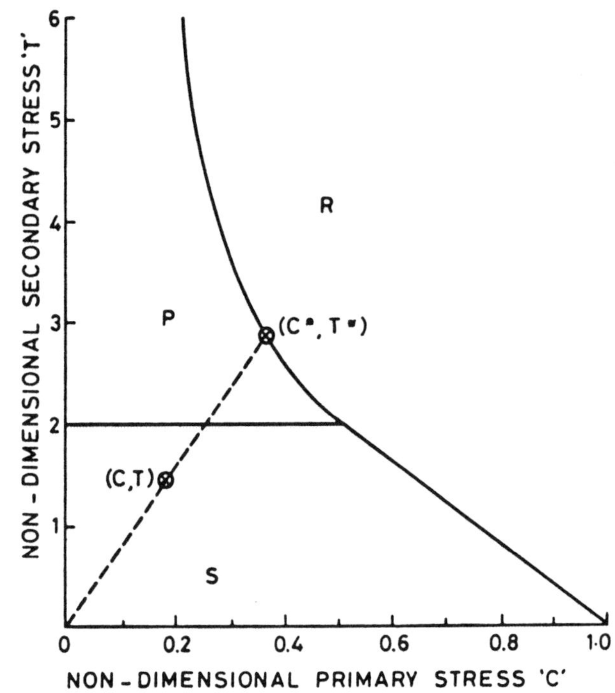


Fig 4 - Construction to estimate effective creep stress

The simple Bree Diagram is only valid for certain types of structure an certain loading systems, for example a plate or a shell subjected to a through thickness temperature gradient or to a bending moment. For other structures, such as cylinder subjected to a moving axial temperature gradient, different criteria apply. The subject has been studied extensively by Professor Ponter and co-workers (12, 13) and additional diagrams have been developed based on their investigations, as shown in Figure 5.

The basic Bree Diagram is modified by a series of "cut-off" lines, which are defined by the parameter F. The value of F depends on the severity of the axial temperature gradient and is determined from a series of auxiliary curves, which are described in the paper by Carter et al (11). It will be noted that many components which would have been deemed to be satisfactory according to the simple Bree type diagrams can potentially undergo ratcheting.

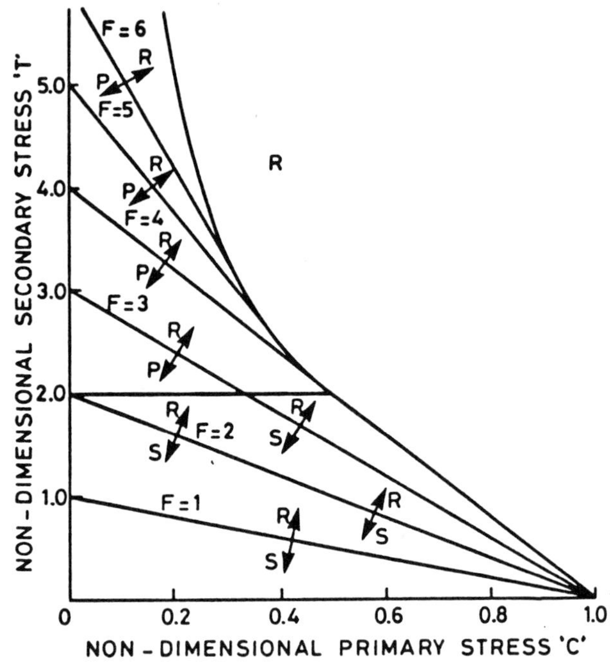


Fig 5 - Modified Bree Diagram, including axial temperature gradients.

## MATERIALS CRITERIA

**Steady Creep Conditions** Many types of high temperature industrial and power plants operate under relatively steady conditions for long periods of time. Provided there are no severe transients and provided that stresses at all times satisfy the creep modified shakedown criteria referred to earlier, then it is relatively straightforward to assess fitness for purpose. Theoretical investigations (5), backed up by experimental validation (14), have shown that failure can be predicted, for creep ductile materials, by reading off the stress to rupture versus time to rupture curve at the reference stress level.

In order to avoid over-conservatism, proposals are now being considered to take advantage of the reduction in creep damage which occurs under compressive stress. The Von Mises and Tresca criteria do not fit the available experimental data for failure under multi axial conditions. The proposal is to replace them with the Hayhurst-Pineau expression, as described by Acker et al (3).

$$\sigma^* = \sigma_{kk} + (1 - \gamma)\sigma_{vm}$$

where $\sigma_{kk}$ is the hydrostatic pressure
$\sigma_{vm}$ is the Von Mises equivalent stress
and $\gamma$ is taken as 0.25 for type 316 steel

The design criterion, which contains the conventional factors of safety, is given by

$$\sigma_{ref} \leq S_t^*$$

where $\sigma_{ref}$ is the reference stress
and $S_t^*$ is the design stress for the material, modified as above to allow for multi axial stressing

The reference stress can be estimated from the conventional limit load of the structure from the formula

$$\sigma_{ref} = P(\sigma_y / P_u)$$

where $P$ is the applied load
$P_u$ is the lower bound limit load
and $\sigma_y$ is the assumed yield stress used to calculate $P_u$

Where the temperature and stressing conditions in the component depend on the output of the plant, it is necessary to divide the proposed operating profile of the component into series of blocks, during which conditions remain sensibly steady. The design requirement then becomes

$$\sum t_i / t_t \leq 1$$

where $t_i$ is the time span of block i
and $t_t$ is read from the $S_t$ versus time curve at the temperature appropriate to the block, and corrected for multi axial stressing if necessary

Where there are local stress concentrations in the component, or where the material is creep-brittle, additional factors have to be included. These are described in the R5 Procedure (2).

**Creep fatigue loadings** In some types of plant, the additional damage caused by thermal shocks and similar transient loads has to be taken into account. This is true, for example of the fast reactor, where primary circuit components are subjected to rapid changes of temperature following a reactor trip.

The materials data used to assess creep fatigue damage are derived from relatively straightforward push-pull or bending tests, carried out under accelerated conditions with reduced hold times. So far, extrapolation to plant conditions has been carried out on an empirical basis and progress is now being made towards assessment methods based on an understanding of the metallurgical mechanisms which are involved.

Observations show that at high strain ranges and short hold times, cracking in the specimen is due to trans-granular fatigue. At lower strain ranges and longer hold times, the damage is entirely due to creep mechanisms (15). As is shown in figure 6, there is only a limited region in which there is a combination of both mechanisms.

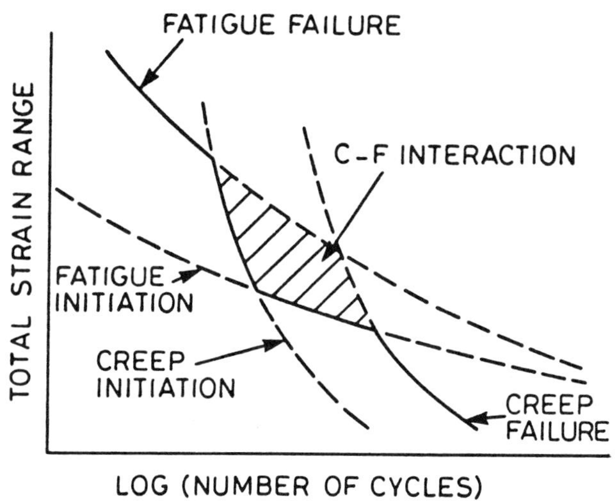


Fig 6 - A mechanism map for a constant lifetime showing different types of failure

Developments have been directed towards a method of estimating creep and fatigue damage which takes these facts into account. The resulting procedure (14, 16) makes use of creep strains which occur during the dwell periods rather than stresses.

Data from strain-controlled tests are used to establish the continuous cycling fatigue curves, in place of the more conventional stress controlled data. Apart from this, the fatigue damage, V, in the component is evaluated in the normal way and is defined as

$$V = \sum(n/N_i)$$

where n is the number of applications of the cycle being considered

and $N_i$ is the number of cycles to cause failure at the strain range relevant to that cycle

The creep damage factor, W, is evaluated from the expression

$$W = \sum(\varepsilon_c/\varepsilon_f)$$

where $\varepsilon_c$ is the effective creep strain relevant to the cycle

and $\varepsilon_f$ is the minimum creep ductility

The full procedure takes into account non damaging creep, which occurs at high rates of strain. At the same time, creep strains which occur during the transients may be included, where necessary, through the introduction of so-called quasi creep terms.

Having evaluated creep and fatigue damages separately, creep-fatigue failure is avoided by limiting the total damage to

$$W + V \leq D$$

where D takes account of any creep-fatigue interaction.

At the low strain ranges relevant to plant, evidence for a number of materials suggest that a linear interaction (D = 1) adequately describes creep-fatigue when ductility exhaustion is used to describe creep damage (17). However, for conservatism the L-shaped diagram is currently proposed as an interim approach for fast reactor components.

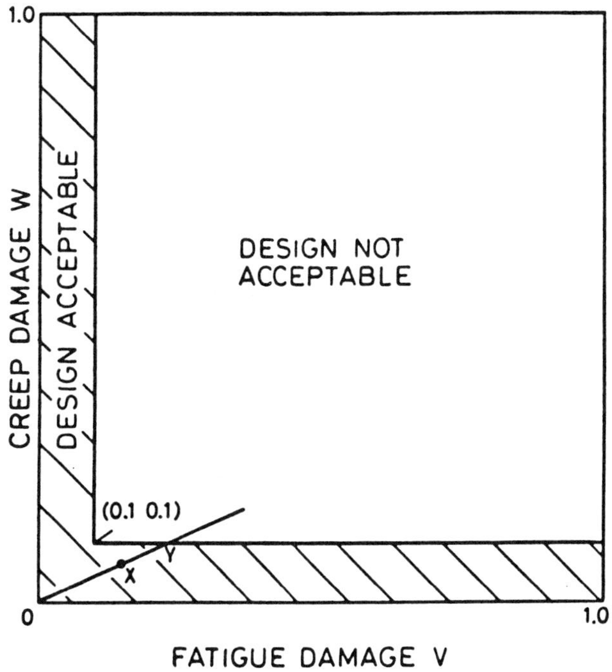


Fig 7 - Creep-Fatigue Interaction Diagram

More recently, the creep-fatigue approach utilising ductility exhaustion has been modified to cover multiaxial effects (18). This requires both stress relaxation under multiaxial stress and the influence of multiaxiality on creep ductility to be described. For pure creep loading, the result is similar to the Hayhurst-Pineau stress-modified criterion described above, leading to much reduced creep damage under compressive stress.

**Special Consideration for Welds** As the majority of high temperature plant components are fabricated structures, the performance of the welded joints has to be taken into account in determining fitness for purpose. This is important because experience shows that the welds are usually the life limiting feature. As papers presented to a recent seminar (19) have shown, components often have to be taken out of service at a point in time where there is little damage in the parent material. When a component is being designed, it is only possible to allow for the presence of the welds indirectly. The stress and strain analysis is carried out as if the component were homogeneous and isotropic and consisted entirely of parent material. Factors are then applied to the permissible strain ranges and stress levels, to take into account the difference of the performance of the welds as compared with the parent material.

Evaluation of these strength reduction factors requires special consideration. Separate tests on the weld metal and the parent material are not satisfactory for this purpose. They do not take into account the stresses built up due to a mismatch in the deformation properties of the various bands of material nor do they allow for the weakening which often occurs in the vicinity of the heat affected zone. The strength reduction factors can thus only be obtained by tests on complete weldments and, even when this is done, such tests must be over a sufficiently long period to show up any degradation of the materials which may occur in service.

Preliminary estimates of creep strength reduction factors for some commonly used steels are given in Table 2, which is taken from Townley (20).

Further information will be found about creep and fatigue strength reduction factors for welds in Type 316 steel in recent papers by Picker et al (21) and Escaravage et al (22). Tests are now in hand to confirm that the tentative values which have been established apply in the long term. They will also investigate whether it is completely valid to obtain a creep-fatigue strength reduction factor by combining the creep and fatigue factors which have been established separately.

Table 2 - Transverse Strength Reduction Factors

| Base Metal | Weld Metal | Welding Process | Temperature | $J_{RT}$ |
|---|---|---|---|---|
| 1/2 CMV | 2 Cr | Probably MMA | 565 | 0.6 |
| P 91 | | Probably MMA | 570 | 0.65 |
| X 20 | | | 600 | 0.5 |
| | | | 575 | 0.8 |
| | | | 550 | 0.95 |
| 316L | 19-12-2 L | TIG | 600 | ~ 1.0 |
| | 19-12-2 L | TIG | 550 | ~ 1.0 |
| 316L | 19-12-2 L | MMA | 550 | 0.85 |
| | 19-12-2 L | SAW | 550 | 0.8 |
| | 19-12-2 L | EBW | 550 | 0.93 to 1.0 |
| | 19-12-2 L | Probably TIG | 550 | 1.0 |
| 316Ti | 19-12-3 Nb | TIG | 700 | 0.75 |
| | 19-12-3 Nb | TIG | 600 | 0.8 |

**The strength reduction factor, $J_{RT}$, is defined as the stress applied to weld in the transverse direction to cause failure in the stated time, divided by the stress which would have to be applied to the parent material to cause failure in the same time.**

## DEFECT ASSESSMENT

The design rules which have been discussed so far assume that a component is free from cracks and similar flaws. After the dimensions have been chosen to satisfy these rules, the next stage is to demonstrate that the component has adequate defect tolerance. This is needed to verify that any defect which has not been detected during pre-service inspection, or which has been generated in service, cannot lead to failure within the design life.

The way in which this is done takes account of the amount of growth which can occur during the life of the plant. The final size of defect is then compared with that which would cause fast fracture, taking changes of material properties into account.

In some circumstances, additional margins of safety can be demonstrated by showing that detectable leakage will occur well in advance of disruptive failure. In others, leakage of the contents of a vessel cannot be tolerated and leak tightness must be assured. In the latter case a demonstration is required that cracks cannot grow sufficiently to penetrate the wall of the component.

The R5 Procedure (2) provides a method of estimating crack growth in high temperature plant, where either creep or creep-fatigue is the controlling factor. It has been adapted for Fast Reactor purposes. There are essentially four steps in the assessment:

1 Before undertaking the main part of the analysis, two checks are carried out. The first is to verify that the initial defect has adequate margins of safety against fast fracture, under the most onerous overload conditions likely to be applied. The second is to verify that the remaining, uncracked portion of the structure is of adequate strength to avoid failure by creep rupture during the plant life.

2 Assuming that these preliminary checks are satisfactory, the assessment then proceeds to the estimation of crack growth. This is characterised by the C* parameter, which can be estimated relatively simply from reference stress calculations using the methods given, for example, by Ainsworth et al (23). The incubation period before the crack starts to grow can also be taken into account in certain circumstances.

3 The final crack size is compared with that which would cause fast fracture. There must be adequate margins against the greatest overload which could occur. It is important to use materials property data which represent the end of life conditions, taking into account any degradation which may have occurred due to environmental effects and irradiation.

4 Finally, a sensitivity study is carried out to investigate the effects of changes in the data which has been used and to check that the conclusions are not affected by assumptions about initial crack size and shape.

Assessment of defects in components subjected to combined creep and fatigue loading follows similar lines, and is described by Ainsworth (17). Crack propagation rates due to the creep and the fatigue are evaluated separately and summed to give the total rate. It is often sufficient to assume that stresses are constant during the dwell periods, but allowance can be made for the relaxation where this is considered significant. Allowance can also be made for the enhanced crack growth rate which may occur before the component has achieved a steady cyclic state.

When considering the additional margins of safety which are provided by demonstrating leak before break, changes in crack shape have to be considered. This is shown diagrammatically in Figure 8, which relates to fatigue crack propagation.

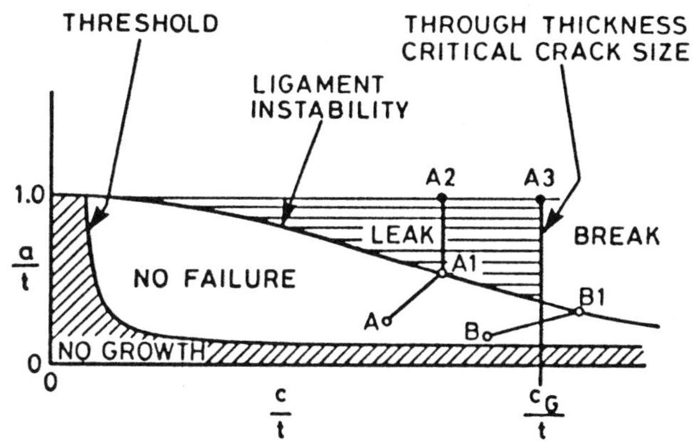


Fig 8 - Leak before Break diagram

A crack with initial dimensions represented by the point A will grow in such a way that, when the remaining ligament snaps through, there will be leakage but no fast fracture. Conversely, a crack with dimensions represented by point B will grow in such a way that fast fracture will occur when the remaining ligament fails. In those situations where leak tightness must be assured, it is necessary to show that the final crack dimensions lie well within the "No Failure" region. This means that the initial defect sizes which can be tolerated are very much smaller than in the case of leak before break.

Evaluation of the way in which the crack grows involves a series of iterations. At each step, the stress intensity factors are calculated at the position of maximum depth and also at the surface. These are then used to estimate the crack growth at both positions and to provide the revised crack dimensions at the end of the step. The iterations continue until the end of design life or until ligament instability is reached. The procedure is now being extended to cover creep and creep-fatigue crack propagation as well as fatigue.

## CONCLUDING REMARKS

As has been described in the paper, significant progress has been made in the past six years in providing better design and assessment procedures.

These developments would not have been possible in the absence of supporting research work, much of it of a speculative nature at the time that it was started. Nor would they have been possible without a close collaboration across the disciplines, particularly in the fields of structural analysis, materials and fracture mechanics.

Design procedures are never static and always capable of further improvement. We can look forward to new developments in the period up to the Sixth International Conference on Creep, based on the results of ongoing fundamental work.

Finally, the author wishes to thank his colleagues in the UK and on the Continent who have provided material for this paper.

## REFERENCES

1 - Rose, R.T., Tomkins B. and Townley, C.H.A., The Development of Design Criteria for Fast Reactors in the U.K., Proceedings Int Conf on Creep, Tokyo, Japan, April 1986.

2 - Nuclear Electric plc, R5: An Assessment Procedure for the High Temperature Response of Structures 1990.

3 - Acker, D., Laue, H. and Townley, C.H.A., Recent research on structural integrity in support of the European Fast Reactor, 11th International Conference on Structural Mechanics in Reactor Technology. Seminar No 11: Construction Codes and Engineering Mechanics, Oarai, Japan, 1991.

4 - Laue, H., Bauermeister, M., Debaene, J.P. and Rose, R.T. Design and Construction Rules Committee for the E F R: Current status and development, Proceedings of 11th Int Conf on Structural Mechanics in Reactor Technology, Tokyo, Japan, 1991.

5 - Goodall, I.W., Leckie, F.A., Ponter, A.R.S. and Townley, C.H.A., The development of high temperature design methods based on reference stress and bounding theorems. ASME Journal of Engineering and Materials Technology, 101, 349 - 355 1979.

6 - Rose, R.T. and White, P.S., Shakedown methods for high temperatures: Seminar on Recent Advances in Design Procedures for High Temperature Plant, Institution of Mechanical Engineers, London, England, 1988.

7 - Toulios, M., White, P.S. and Buckthorpe, D.E., Simplified design methods by shakedown and an algorithm for FE implementation. 4th Applied Solid Mechanics Conference, University of Leicester, England, April, 1991. (Proceedings to be published by Elsevier)

8 - White, P.S., Recent developments of the shakedown design method. 11th Int Conf on Structural Mechanics in Reactor Technology, Tokyo, Japan, 1991.

9 - White, P.S., Elastic benchmark exercise with complex loading and geometry. Transactions of 10th Int Conf on Structural Mechanics in Reactor Technology, Los Angeles, U.S.A. B, 125, 1989.

10- Bree, J., Elastic-plastic behaviour of thin tubes subjected to internal pressure and intermittent high heat fluxes with application to fast nuclear reactor fuel components. Journal of Strain Analysis 1967, 2, 226, 1987.

11- Carter, K.F., Love J.B. and Ponter, A.R.S., Simplified design rules for the prevention of ratcheting. 11th Int Conf on Structural Mechanics in Reactor Technology, Tokyo, Japan. Paper E 07/3, 1991.

12- Ponter, A.R.S. and Carter, K.F., Upper bound methods for use in the design and assessment of axisymmetric thin shells subjected to cyclic thermal loading. Nuclear Engineering and Design, 116, 239, 1989.

13- Ponter, A.R.S., Karadeniz, S. and Carter, K.F., The computation of shakedown limits for structural components subjected to variable thermal loadings: 'Brussels Diagrams' EUR Report 12686 EN, Commission of European Communities, Brussels, Belgium, 1990.

14- Goodall, I.W. and Ainsworth, R., R5: An assessment procedure for the high temperature response of structures. 11th Int Conf on Structural Mechanics in Reactor Technology, Tokyo, Japan, Paper L 04/3, 1991.

15- Hales, R., The application of creep strain as a parameter for estimating creep-fatigue performance. Proceedings of Second Int Conf on Creep 1990 and Fracture of Engineering Materials and Structures, Swansea, Wales, 1984.

16- Picker, C., U K development of a strain based creep-fatigue assessment procedure for fast reactor components. Third Int Post-SMiRT Seminar on Construction Codes and Engineering Mechanics, Los Angeles, U.S.A., 1989.

17- Ainsworth, R.A., Defect Assessment procedure at high temperature, Trans 10th Int Conf on Structural Mechanics in Reactor Technology, Anaheim, U.S.A., L, 77-90, 1989.

18- Hales, R and Ainsworth, R.A., Multiaxial creep-fatigue rule, 8th Int Seminar on Inelastic Analysis, Fracture and Life Prediction, Shizuska, Japan, 1991.

19- Institution of Mechanical Engineers, London, England, Seminar: Life of welds at high temperature, 1990.

20- Townley, C.H.A., Life of welds at high temperature. Int Conf on High Temperature Structural Design, Venice, Italy, October, 1990

21- Picker, C., Bretherton, I. and Bestwick R.D.W., The Performance of Elevated Temperature Stainless Steel Weldments Under Cyclic Loading. Seminar on Life of Welds at High Temperature, Institution of Mechanical Engineers, London, England, June, 1990.

22- Escaravage, M., Bauermeister, M. and Buckthorpe, D., Weldment Design Rules for the European Fast Reactor, Seminar on Life of Welds at High Temperature, Institution of Mechanical Engineers, London, England, June, 1990.

23- Ainsworth, R.A., Chell G.G., Coleman, M.C., Goodall, I.W., Gooch, D.J., Haigh, J.R., Kimmins, S.T. and Neate G.J. CEGB Assessment Procedure for defects in plant operating in the creep range, Jnl Fatigue and Fracture Engineering Materials and Structures, 10, 115-127, 1987.

*Proceedings of the Fifth International Conference on Creep of Materials, Lake Buena Vista, Florida, USA, 18-21 May 1992*

# Design Criteria for Multiaxial Creep-Fatigue

**M. Ohnami and M. Sakane**
Department of Mechanical Engineering
Ritsumeikan University
Kyoto, Japan

## Abstract

This paper describes the multiaxial creep–fatigue lives under multiaxial stress/strain states, focused on the parameter for correlating multiaxial creep–fatigue lives. Multiaxial fatigue lives not only in tension/torsion tests but also in equi–biaxial tension/tension tests were discussed. The equivalent Mises strain was not effective in correlating multiaxial fatigue lives but the equivalent strain based on crack opening displacement and $\Gamma^*$–plane parameter were effective for the correlation. Energy criterion was proposed for the correlation of nonproportional fatigue lives. The creep damage under multiaxial stress states was also discussed.

Practical applications operated at high temperatures receive multiaxial creep–fatigue loading rather than uniaxial loading and multiaxial design is required. For most cases, uniaxial design is sufficient and safety factor covers the life reduction caused by multiaxial loading. However, recent design of high temperature components, like gas turbine and nuclear vessel, multiaxial creep–fatigue design is essential. Sufficient safety factor cannot be taken in the former case and safety must perfectly guaranteed in the latter case.

Von Mises equivalent strain (Mises strain) has been historically used as a multiaxial fatigue design parameter and has been used in the design code of pressure vessel(1). However, recent studies(2,3) have been revealing that the Mises strain is not effective in correlating multiaxial low cycle fatigue lives. Instead, several multiaxial low cycle fatigue parameters have been proposed(4). $\Gamma$–plane parameter(5,6) (BMK strain), $\Gamma^*$–plane parameter(7) (LE strain) and equivalent strain based on crack opening displacement(8–10) (COD strain) are a typical parameter, but their effectiveness has not been fully verified by applying these parameters to the experimental results. The reason for the insufficient verification is resulted from the lack of the multiaxial low cycle fatigue experimental data.

The aim of this paper is to describe multiaxial creep–fatigue criteria which are applicable to the practical design of high temperature components. The multiaxial stress state discussed in this paper is not only the tension/torsion condition but also the equi–biaxial tension/tension condition. Proportional and nonproportional multiaxial creep–fatigue is also discussed.

## Proportional Low Cycle Fatigue Criteria

**Combined Push–pull and Revered Torsion Fatigue.** Figures 1(2) and 2(3) show the correlation of tension/torsion low cycle fatigue(LCF) lives with the Mises strain for type SUS304 stainless and 1Cr–1Mo–1/4V steel hollow cylinder specimens (I.D.9mm, O.D.12mm, G.L.20mm), respectively. The test temperature was 923K for SUS304 stainless steel and 823K for 1Cr–1Mo–1/V steels. The Mises strain cannot correlate the tension/torsion LCF lives for SUS304 steel within a factor of two scatter band. Fatigue data at large $\phi$ are placed conservatively. $\phi$ is the principal strain ratio and is defined as $\phi=\varepsilon_x/\varepsilon_y$; $\varepsilon_x$ and $\varepsilon_y$ are the minimum and maximum principal strains, respectively.

The Mises strain has been used for the multiaxial fatigue design criteria, but the data correlation shown in the two figures indicates that the Mises strain is not an appropriate multiaxial fatigue parameter. It gives too excessive safety factor margin for the data at small $\phi$. Recent studies of multiaxial low cycle fatigue also indicates the ineffectiveness of the Mises strain for the multiaxial fatigue data correlation (2,3,9).

Instead of the Mises strain, many parameters have been proposed(4). The BMK strain(5,6), LE strain(7) and COD strain(8–10) are the typical parameters. The respective parameters are shown by the following equations.

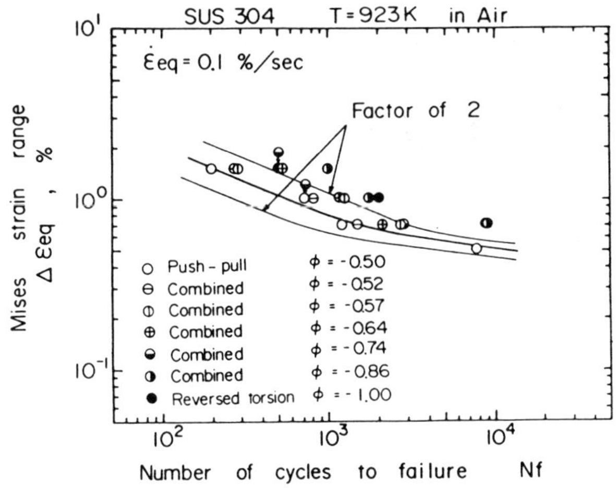


Fig.1 Correlation of the tension/torsion multiaxial low cycle fatigue lives with the Mises strain for SUS 304 stainless steel at 923K.

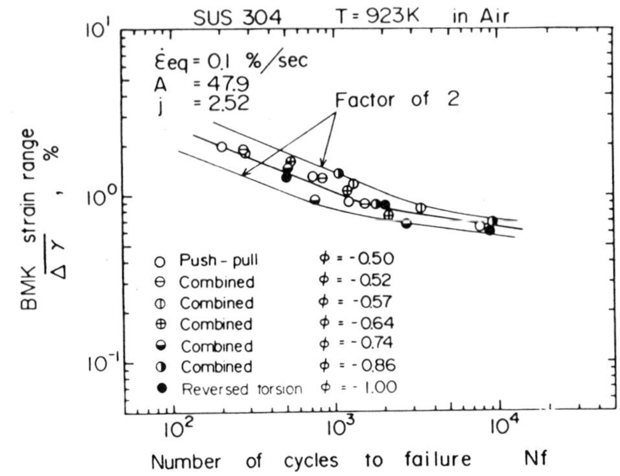


Fig.3 Correlation of the tension/torsion multiaxial low cycle fatigue lives with Γ–plane parameter for SUS 304 stainless steel.

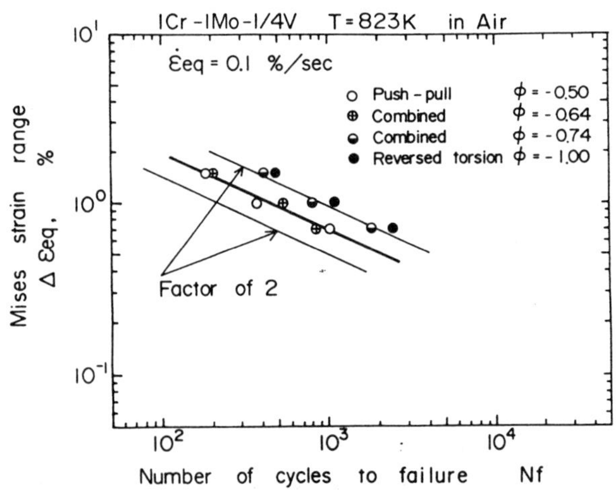


Fig.2 Correlation of the tension/torsion multiaxial low cycle fatigue lives with the Mises strain for 1Cr–1Mo–1/4V steel at 823K.

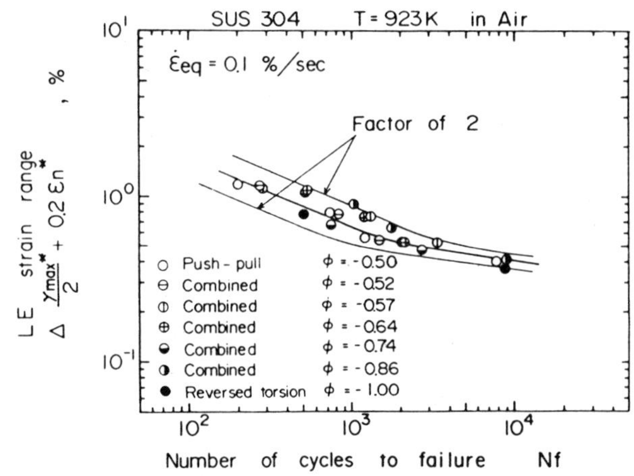


Fig.4 Correlation of the tension/torsion multiaxial low cycle fatigue lives with Γ*–parameter for SUS 304 stainless steel.

BMK strain ;

$$\Delta\bar{\gamma}=\{A(\Delta\varepsilon_n)^j+(\frac{1}{2}\Delta\gamma_{max})^j\}^{\frac{1}{j}} \qquad (1)$$

LE strain ;

$$\frac{\gamma^*_{max}}{2}+K\varepsilon^*_n=C \qquad (K=0.2) \qquad (2)$$

COD strain ;

$$\Delta\varepsilon^*=\beta(2-\phi)^{m'}\Delta\varepsilon_y \qquad (3)$$
$$\beta=1.83,\quad m'=-0.66$$

Figures 3–5(2) represent the tension/torsion multiaxial LCF lives correlated with respective parameters. All the parameters correlate the tension/torsion multiaxial LCF lives within a factor of two scatter band. Among them, the BMK and LE strains have the material constants which presumably

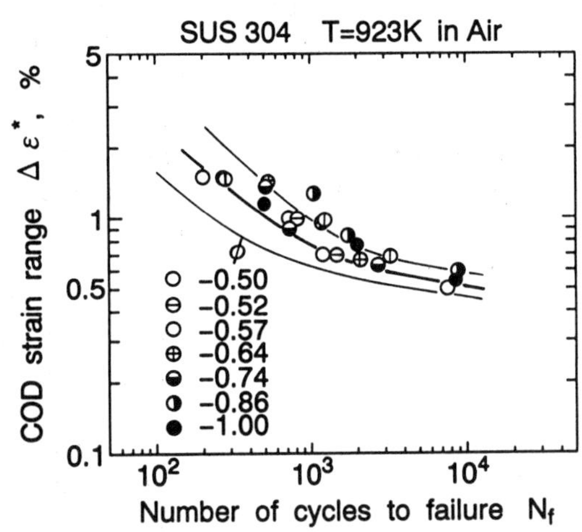


Fig.5 Correlation of the tension/torsion multiaxial low cycle fatigue lives with the COD strain range for SUS 304 stainless steel.

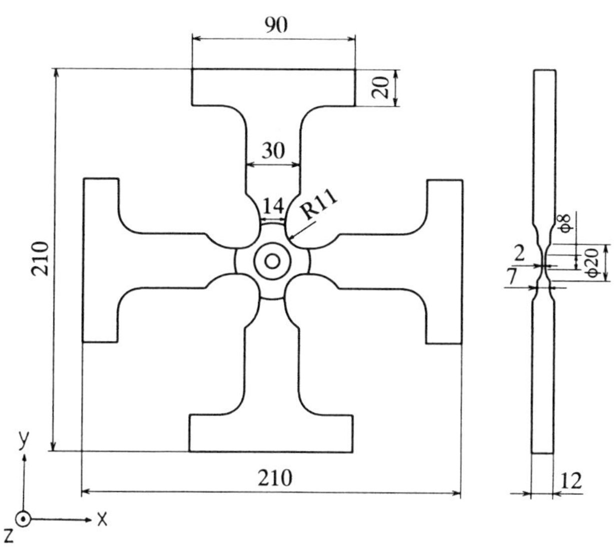


Fig.6 Shape and dimensions of cruciform specimen together with the coordinates employed.

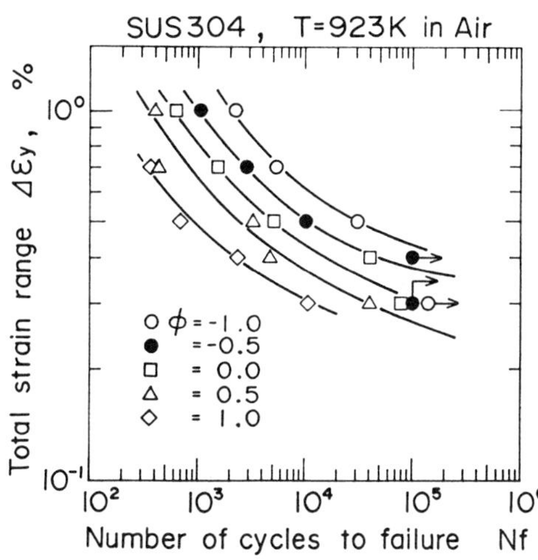


Fig.7 Correlation of multiaxial fatigue lives for SUS 304 cruciform specimens with total strain range in y direction.

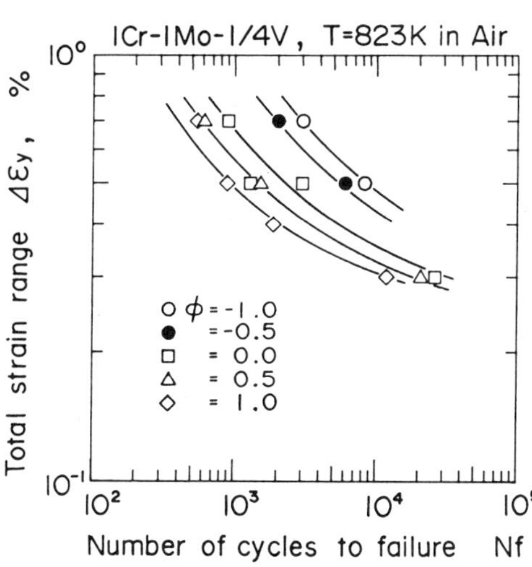


Fig.8 Correlation of multiaxial fatigue lives for 1Cr–1Mo–1/4V cruciform specimens with total strain range in y direction.

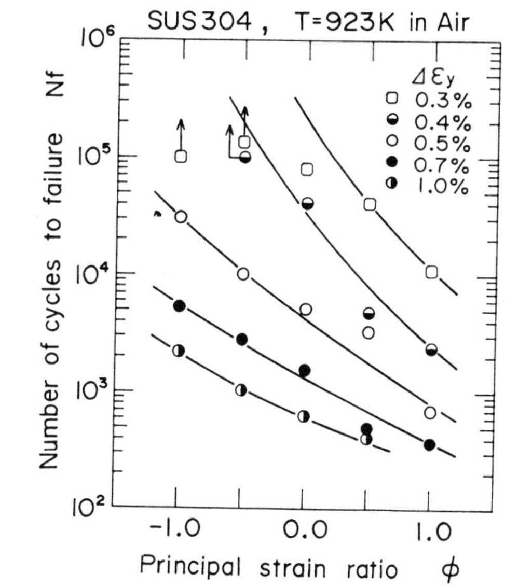


Fig.9 Relationship between multiaxial fatigue life and principal strain ratio for SUS 304 cruciform specimens.

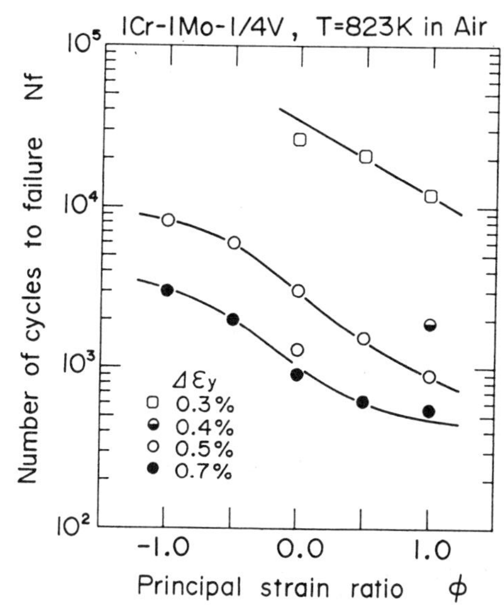


Fig.10 Relationship between multiaxial fatigue life and principal strain ratio for 1Cr–1Mo–1/4V cruciform specimens.

depend on the material. The COD strain has also two material constants but they do not depend on the material, so the COD strain range can apply to almost all the materials unchanging the value of two constants.

**Multiaxial Low Cycle Fatigue Under a Full Range of Multiaxial Stress State.** Combined push–pull and reversed torsion multiaxial fatigue tests have been carried out for studying multiaxial low cycle fatigue lives. This kind of test method, however, can perform the multiaxial LCF tests under limited multiaxial stress/strain states. The possible principal strain ratio $\phi$ in this test method is $-1 \le \phi \le -0.5$. More widely ranged multiaxial stress/strain states sometimes occur in

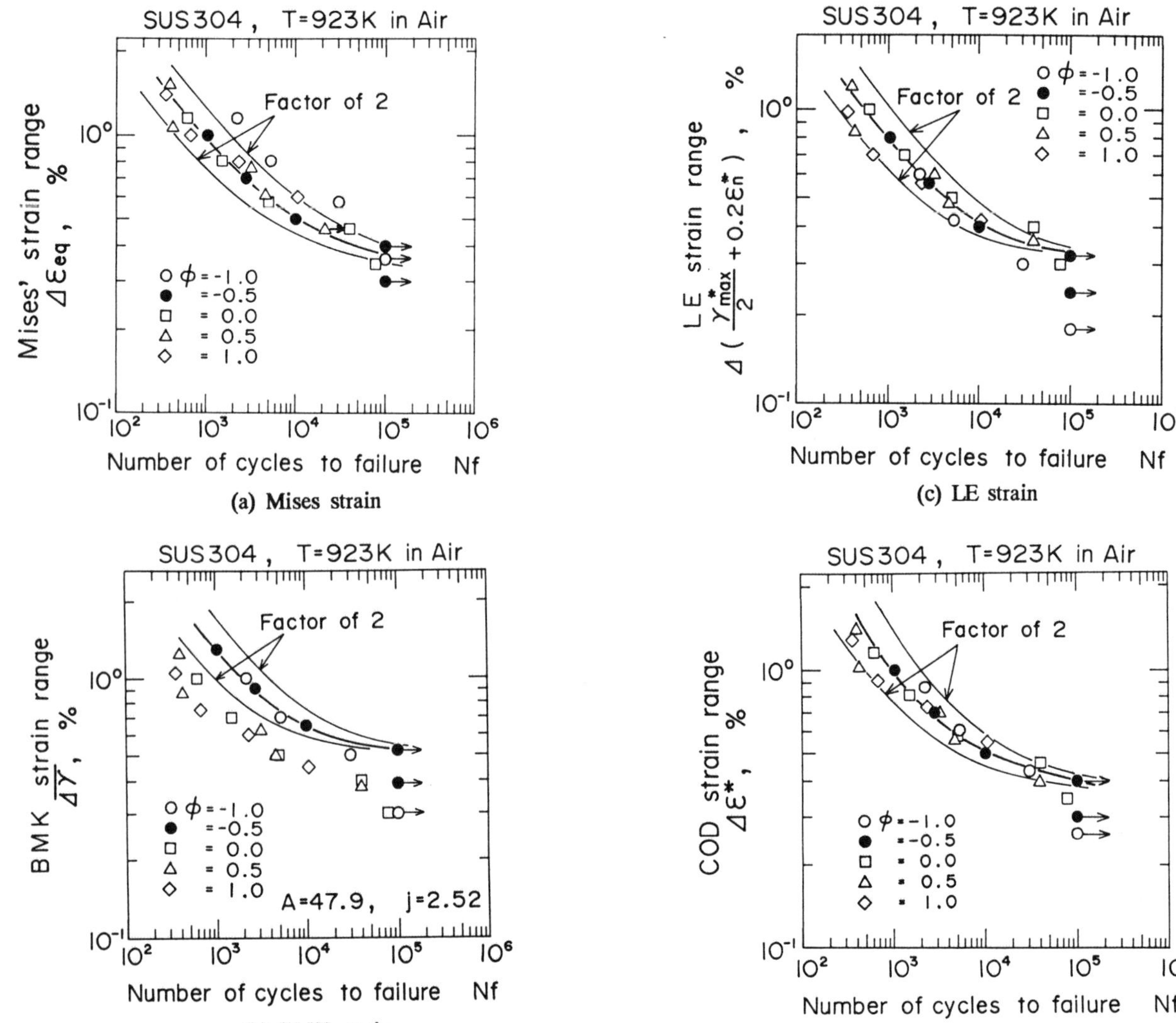


(a) Mises strain

(c) LE strain

(b) BMK strain

(d) COD strain

Fig.11 Correlation of multiaxial fatigue lives of SUS304 cruciform specimens with (a) Mises strain, (b) BMK strain, (c) LE strain and (d) COD strain.

practical components, so the applicability of the multiaxial fatigue parameter to widely ranged multiaxial stress/strain states should be confirmed.

The authors have developed a multiaxial low cycle fatigue machine for the cruciform specimen and have performed multiaxial low cycle fatigue tests in a full range of multiaxial strain state(11-15). The shape of the cruciform specimen is shown in Fig.6(12). Experiments in the multiaxial strain state of $-1 \leq \phi \leq 1$ are possible using the test specimen shown in Fig.6.

Figures 7 and 8(11) show the multiaxial low cycle fatigue data in $\varepsilon_y$ constant tests for SUS304 stainless steel and 1Cr–1Mo–1/4V steel cruciform specimens, respectively. These figures indicate that the strain biaxiality has a significant effect on LCF lives. Fatigue lives decrease with increasing $\phi$. Figures 9 and 10(11)are a replot of Figs.7 and 8, and they show the effect of principal strain ratio on multiaxial fatigue lives. Fatigue lives decrease monotonously with increasing $\phi$ and the strain biaxiality has a significant effect of fatigue lives.

Figures 11 and 12(11) correlate the multiaxial LCF lives of cruciform specimens with the Mises strain, BMK strain, LE strain and COD strain for SUS304 stainless and 1Cr–1Mo–1/4V steels, respectively.

For SUS 304 stainless steel shown in Fig.11, the Mises strain correlates almost all the multiaxial fatigue data within a factor of two scatter band but does not the data at $\phi=-1$. The Mises strain estimates these data conservatively. This trend of the data correlation agrees with that shown in Figs.1 and 2 using the hollow cylinder specimens in tension/torsion tests. The BMK strain gives a satisfactory correlation for the data in the range of $-1 \leq \phi \leq -0.5$ but yields a large scatter of the data in the range of $0 \leq \phi \leq 1$. The data in that range locate far below a factor of two scatter band. Such an unconservative estimate may be resulted from the discontinuity of the BMK strain at $\phi=-0.5$, which will be discussed later. The LE strain gives a satisfactory data correlation for SUS304 stainless steel in all the principal strain ratio range of $-1 \leq \phi \leq 1$

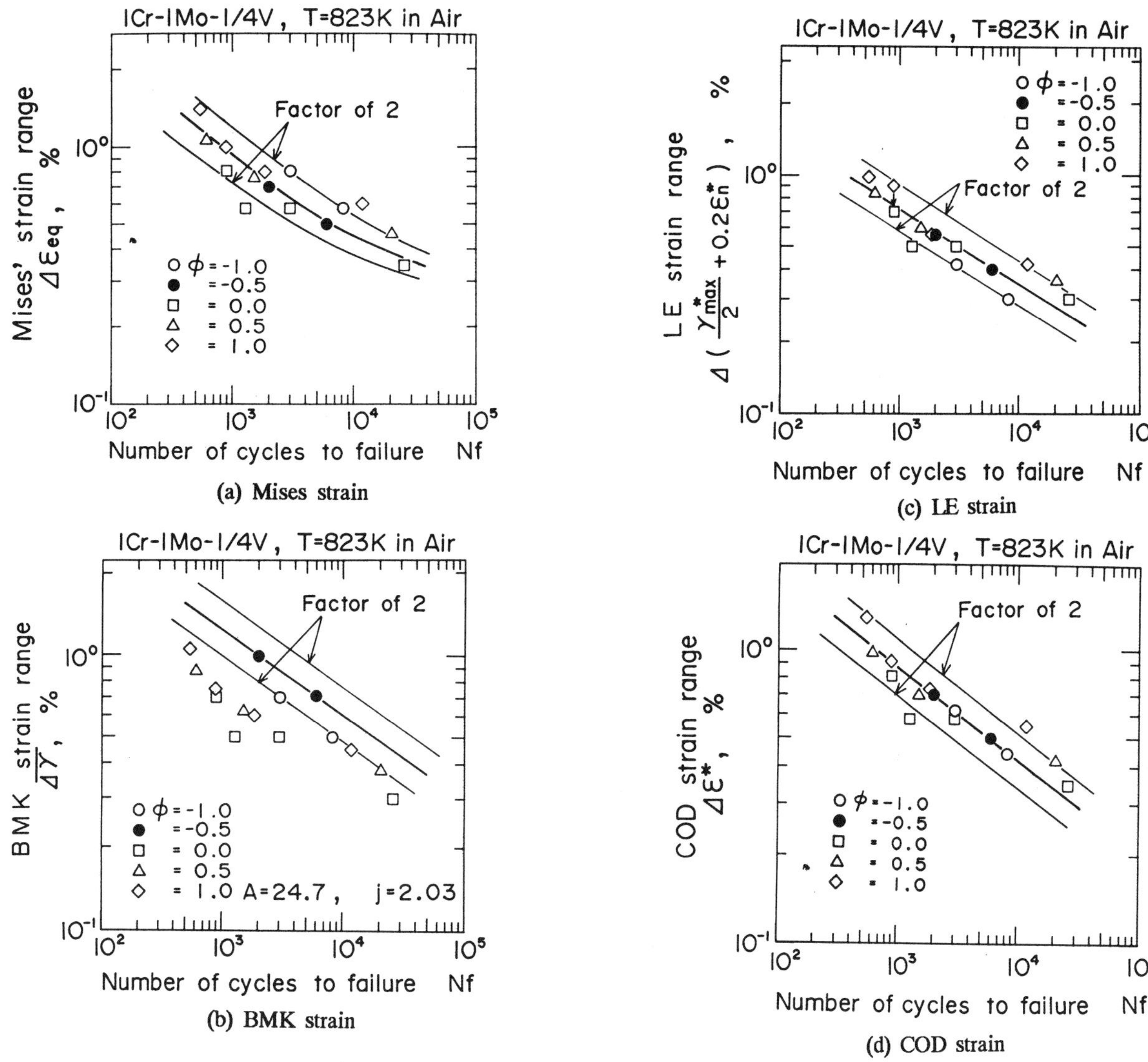


Fig.12 Correlation of multiaxial fatigue lives of 1Cr–1Mo–1/4V cruciform specimens with (a) Mises strain, (b) BMK strain, (c) LE strain and (d) COD strain.

although some data are located in the lower side. The COD strain range correlates the fatigue data within a factor of two scatter band and the scatter of the data is smaller in comparison with the LE strain.

For the correlation of 1Cr–1Mo–1/4V steel shown in Fig.12, the general trend of the correlation is similar to that for SUS304 stainless steel. The Mises strain correlates the fatigue data within a factor of two scatter band except the two data points. The BMK strain yields a large scatter of the correlation as well as the correlation for SUS304 stainless steel. The LE strain gives a satisfactory correlation in all the range of $-1\leq\phi\leq1$. In the correlation with the COD strain, the scatter of the data appears to be slightly large in comparison with SUS304 stainless steel, but the COD strain correlates the fatigue data within a factor of two scatter band.

In conclusion, the maximum principal strain, with which the data correlation is not presented here, the LE and COD strains give a satisfactory data correlation for SUS304 stainless steel and 1Cr–1Mo–1/4V steel. Among them, the LE and COD strains yield somewhat smaller scatter of the data. The Mises strain gives a satisfactory data correlation in the range of $-0.5\leq\phi\leq1$ but does not in the range of $-1\leq\phi\leq-0.5$ for SUS304 stainless steel. For 1Cr–1Mo–1/4V steel, the Mises strain correlates all the data within a factor of scatter band. The BMK strain yields a poor data correlation because of the discontinuity at $\phi=-0.5$. It correlates the fatigue data into two groups, depending on the principal strain ratio range.

**Comparison of Multiaxial Fatigue Strain Parameters.** Figure 13(11) compares the multiaxial fatigue strain parameters used for the correlation. The ordinate is normalized by the principal strain in y-direction($\varepsilon_y$). The LE and COD strains monotonously increase with increasing $\phi$. The maximum principal strain ($\varepsilon_1$) takes the value of 1.0 in the range of $-1\leq\phi\leq0$, but it increases with increasing $\phi$ in the range of $0<\phi<1$. The Mises strain decreases with increasing $\phi$ in the range of $-1\leq\phi\leq-0.5$, but conversely increases with increasing

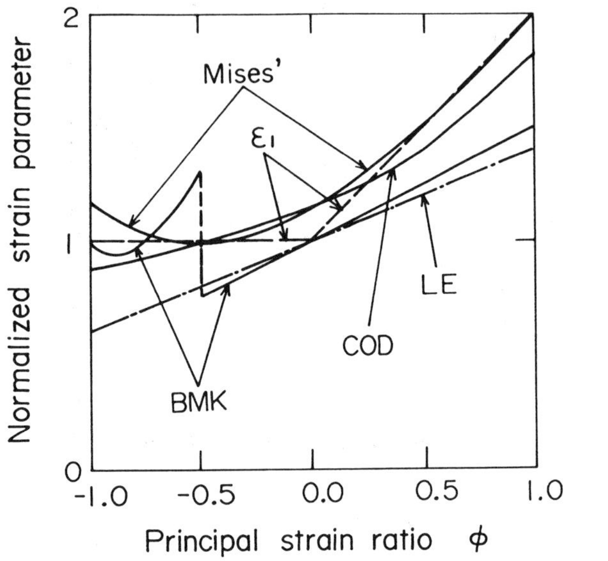


Fig.13 Variation of various multiaxial fatigue parameters with principal strain ratio.

$\phi$ in the range of $-0.5 \le \phi \le 1$. The BMK strain has a parabolic shape in the range of $-1 \le \phi \le -0.5$ but monotonously increases with increasing $\phi$ in the range of $-0.5 < \phi < 1$. The BMK stain has a discontinuity at $\phi = -0.5$.

As shown in Figs.9 and 10, the experimental fatigue lives decreased with increasing $\phi$. Thus, an appropriate parameter which correlates fatigue data with a small scatter in a whole principal strain ratio range must increase with increasing $\phi$. The parameter which meets this condition is the LE and COD strains. The Mises strain exhibits opposite behavior to the experimental results in the range of $-1 \le \phi \le -0.5$, whereas it expresses well the trend of the experimental data in the range of $-0.5 \le \phi \le 1$. The BMK strain correlates the experimental data into two groups; one group is the data in the range of $-1 \le \phi \le -0.5$ and the other group the data in the range of $-0.5 \le \phi \le 1$. Since the BMK strain has a discontinuity at $\phi = -0.5$, it gives a large scatter of the data.

**Multiaxial Low Cycle Fatigue Data of Superalloys.** The data correlation discussed above is for ductile heat resistent materials. The multiaxial low cycle fatigue parameter developed for ductile materials is not necessarily applicable to superalloys because the deformation and fracture mechanisms of superalloys are sometimes different from those of ductile materials. Superalloys generally have a large yield stress and deform elastically.

Figure 14(16) correlates the tension/torsion multiaxial low cycle fatigue lives of Mar-M247 directionally solidified (DS) and Inconel 738 conventional cast superalloys. The shape of the specimen was a hollow cylinder (ID. 5mm, O.D. 8mm, G.L. 5mm). The specimen axis agreed to the (001) crystallographic direction and the number of columnar grains were about 15 in the gage length. Fatigue lives of DS specimens can be correlated with the Mises strain within a factor of two

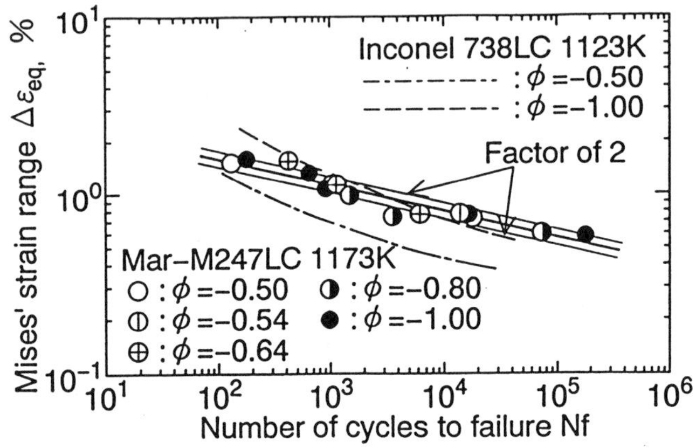


Fig.14 Correlation of tension/torsion fatigue lives of Mar-M247 directionally solidified and Inconel 738 conventional cast superalloys with the Mises strain.

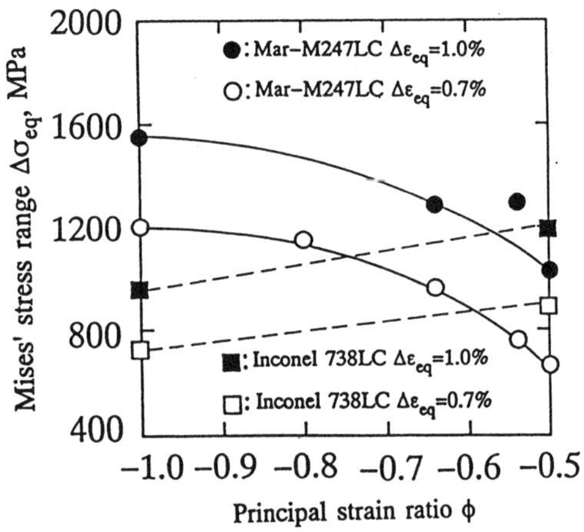


Fig.15 Variation of the Mises stress of the superalloys with principal strain ratio.

scatter band, whereas those of Inconel 738 conventional cast specimens cannot be correlated as shown by dotted lines in the figure. Thus, the effective strain parameter of DS specimens differs from that of the conventional cast superalloy.

The reason for the difference of the effective parameter between the DS and conventional case specimens is resulted from the difference in stress response. For the DS specimens, there is strong anisotropic deformation caused by crystallographic texture. Figure 15(16) compares the Mises stress of Mar-M247 DS and Inconel 738 conventional cast superalloys as a function of the principal strain ratio. The Mises stress of Inconel 738LC exhibits constant amplitude against the principal strain ratio, whereas that of Mar-M247 DS superalloy increases with decreasing the principal strain ratio ; the larger stress was applied in the test at the smaller principal strain ratios. Thus, in the test at the smaller principal strain ratio for Mar-M247 DS superalloy, the larger stress was applied in comparison with the test at the larger principal strain ratio, so the fatigue lives in the former test approached to those in the latter test.

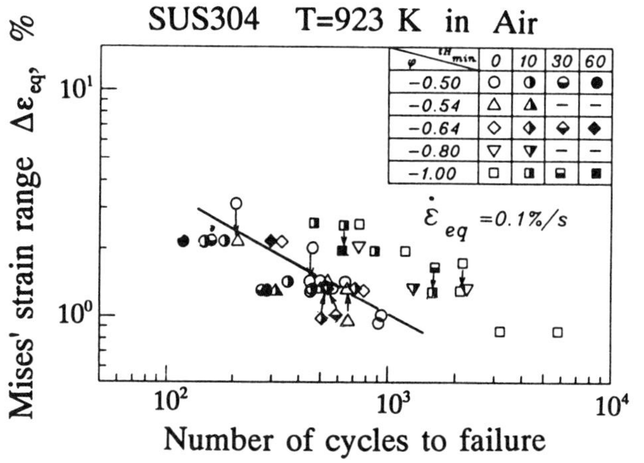


Fig.16 Tension/torsion creep–fatigue lives for SUS304 specimens correlated with the Mises strain.

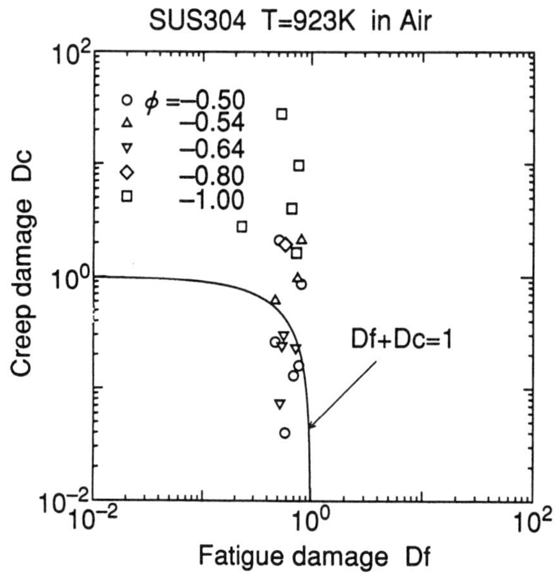


Fig.17 Creep and fatigue damage diagram in tension/torsion tests for SUS304 specimens calculated by linear damage rule.

**Multiaxial Creep–Fatigue.** Life prediction in multiaxial creep–fatigue is another important topic in practical design. However, extensive study of multiaxial creep–fatigue is scarce. The reason for the small number of studies is resulted from the difficulty of experiments but systematic accumulation of experimental results is essential to this research field.

Figure 16(17) is an example of creep–fatigue tests under combined tension and torsion. The data is correlated with the Mises strain. The figure shows that the failure life decreases with increasing the hold–time at each principal strain ratio.

Figure 17(17) is the creep–fatigue damage diagram, calculated by linear damage rule. In calculating the multiaxial creep damage, the uniaxial creep rupture data obtained in NRIM(18) were used. The multiaxial creep damage was evaluated on the Mises stress base ; the multiaxial creep rupture data was estimated by simply replacing the uniaxial stress in the creep rupture diagram by the Mises stress. The figure indicates that the creep damage increases as the principal strain ratio decreases. In the $\phi$=−1 test, the creep–fatigue test in pure torsion, the creep damage takes significantly large values more than 10. This large creep damage in torsion tests is attributed to the fact that the evaluation of the creep damage on the Mises equivalent stress does not accurately estimate the creep damage in multiaxial stress states. Development of the method for the accurate multiaxial creep damage estimation based on the experimental data is necessary.

## Nonproportional Multiaxial Low Cycle Fatigue

**Nonproportional Low Cycle Fatigue.** Nonproportional loading sometimes occurs in practical components under the combined loading of mechanical and thermal stresses. ASME Code Case(1) has been used as a design criterion for the nonproportional fatigue, but recent studies have shown that the Code Case estimates the unconservative lives for the nonproportional fatigue. In order to examine the effect of nonproportional loading on fatigue lives, fatigue tests under 13 nonproportional strain paths were carried out for SUS304 stainless steel hollow cylinder specimens (O.D.11mm, I.D.9mm, G.L. 4.2mm) at room temperature. The 13 strain paths are shown in Fig.18(19). The 13 strain paths were programmed to help study several effects of nonproportional loading shown in Table 1(19).

Table 2(19) and Table 3(19) list the fatigue lives Case 0 and 13 nonproportional tests. Nonproportional tests were carried out at two strain ranges of 0.5% and 0.8% to confirm the characteristics of fatigue lives in elastic and plastic strain dominant regions. In all the 13 strain paths, the normal strain had the same amplitude as the shear strain. A complete loading along the strain paths shown in Fig.18 was counted as one cycle for all the Cases except Cases 3, 4 and 13. In Cases 3, 4 and 13 a complete loading was counted as two cycles.

Figure 19(19) correlates the nonproportional fatigue lives under the 13 strain paths with the equivalent strain defined in ASME Code Case(1). In the figure, a factor of two scatter band is shown by the lines based on the push–pull data, i.e., Case 0 data. The equivalent strain in the Code Case is shown as

$$\Delta\bar{\varepsilon}_{ASME}=Max[\{(\varepsilon_A-\varepsilon_B)^2+(\gamma_A-\gamma_B)^2/3\}^{1/2}] \qquad (4)$$

Fatigue lives of Cases 5–7 and Case 11 locate within a factor of scatter band. The fatigue data in all the other strain paths are arranged unconservatively. Especially, fatigue lives of Cases 3, 4, 12 and 13 are 8–30 % of those in Case 0. Thus, the life prediction based on the ASME Code Case for nonproportional loading becomes unconservative for most of strain paths.

Instead of the ASME Code Case, the equivalent strain

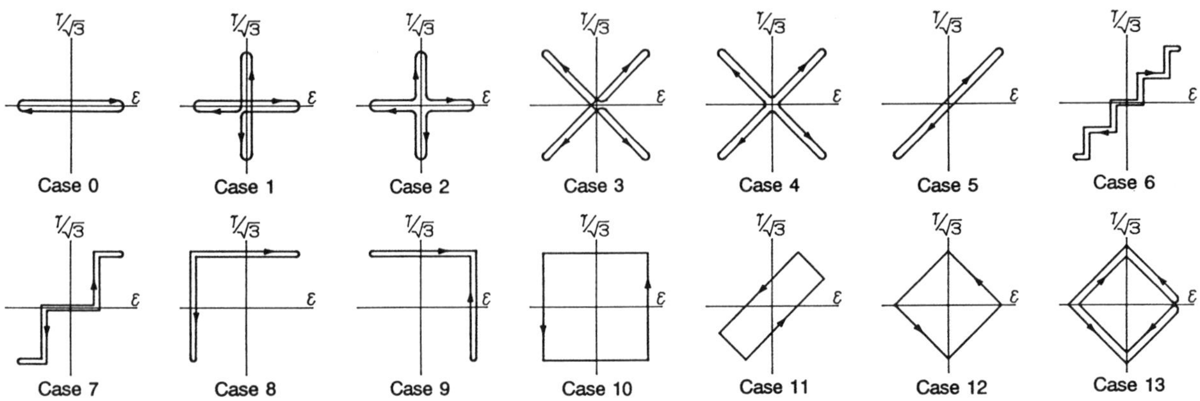


Fig.18 Nonproportional 13 strain paths.

Table 1 Nonproportional factors examined in the 13 strain paths.

| | Aspect | Strain path number |
|---|---|---|
| ① | Effect of two principal strain directions | 0↔1, 2 5↔3, 4 |
| ② | Effect of out-of-phase loadings(0°, 45°, 90°) | 5↔11↔12 |
| ③ | Full reversed loading or zero-to-maximum loading | 1↔2 3↔4 |
| ④ | Mean strain effect | 1↔8, 9 8↔9 8↔9↔10 |
| ⑤ | Step length effect | 5↔6↔7↔8, 9 |
| ⑥ | Rotating or turn arround loading | 12↔13 |

based on the strain path length has been proposed(20). This equivalent strain is the strain defined as the length of strain path shown as follows.

$$\Delta\bar{\varepsilon}_{path}=\int_{1/2cycle}d\bar{\varepsilon},\quad \bar{\varepsilon}=(\varepsilon^2+\gamma^2/3)^{\frac{1}{2}} \qquad (5)$$

The correlation of the fatigue data with the path equivalent strain is shown in Fig.20(19). In contrast to the correlation using $\Delta\varepsilon_{ASME}$, the equivalent path strain correlates the nonproportional fatigue data rather conservatively except the data of Cases 12 and 13. Some of the data are correlated too conservatively. Case 6 showed six times larger fatigue lives than those in Case 0.

Itoh *et. al.* (19) have proposed the energy parameter for correlating the nonproportional fatigue lives. The energy parameter is represented as follows.

$$\bar{E}=\sqrt{(\Delta\sigma_{max}\,\Delta\varepsilon_{max})^2+(\Delta\tau_{max}\,\Delta\gamma_{max})^2} \qquad (6)$$

In this equation, $\Delta\sigma_{max}$ and $\Delta\varepsilon_{max}$ are the maximum normal stress and strain ranges, and $\Delta\tau_{max}$ and $\Delta\gamma_{max}$ are the maximum shear stress and strain ranges. Figure 21 indicates the data correlation with the energy parameter expressed by equation (6). All of the nonproportional fatigue data can be correlated within a factor of two scatter band, whereas a few data locate

Table 2 Fatigue lives in Case 0 tests.

| Strain range Δε, % | | 0.5 | 0.65 | 0.8 | 1.0 | 1.2 | 1.5 |
|---|---|---|---|---|---|---|---|
| Path number | Case 0 | 49000 | 23400 | 7100 | 1500 | 690. | 540 |

Table 3 Summary of failure lives under 13 nonproportional loading paths.

| Path number (Case) | | 1 | 2 | 3 | 4 | 5 | 6 | 7 | 8 | 9 | 10 | 11 | 12 | 13 |
|---|---|---|---|---|---|---|---|---|---|---|---|---|---|---|
| Strain range Δε, % | 0.5 | 9500 | 20000 | 2400 | 3400 | 17500 | 9700 | 18000 | 2050 | 2950 | 2600 | 14400 | 4750 | 3200 |
| | 0.8 | 1260 | 2100 | 820 | 900 | 3200 | 2700 | 1700 | 470 | 660 | 320 | 1930 | 710 | 1000 |

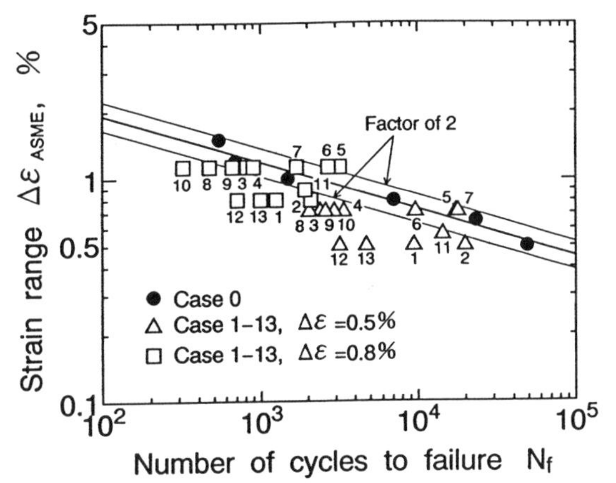


Fig.19 Correlation of nonproportional fatigue lives with the equivalent strain used in ASME Code Case.

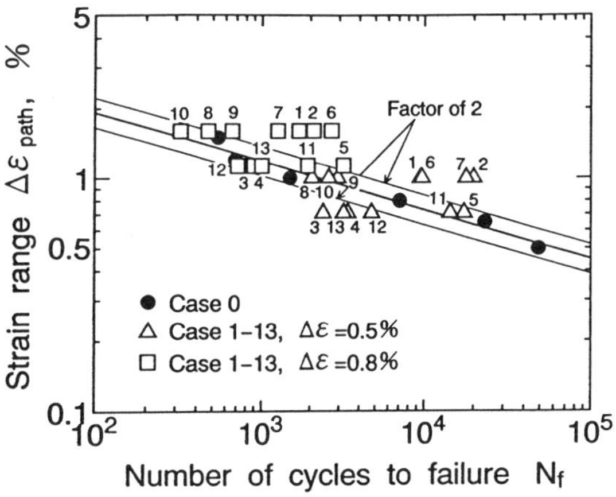


Fig.20 Correlation of nonproportional fatigue lives with the equivalent strain based on strain path length.

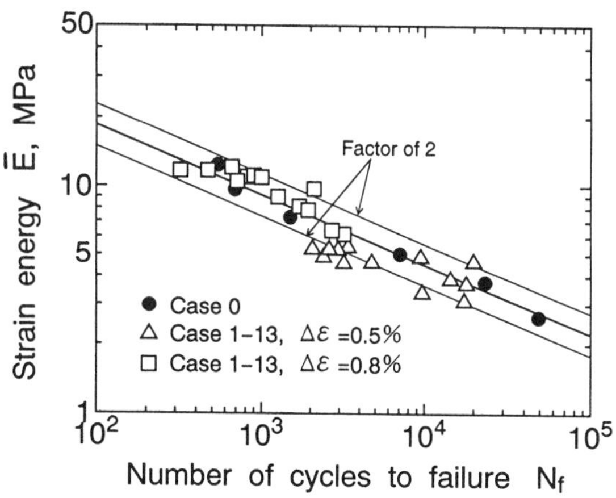


Fig.21 Correlation of nonproportional fatigue data with the equivalent energy parameter.

just outside the scatter band. For the life prediction of nonproportional fatigue lives using equation (6), the maximum normal/shear stress range should be known as well as the strain range. The maximum stress range can be calculated by the inelastic constitutive equation but it is not a simple procedure in general. The multiaxial nonproportional low cycle fatigue parameter that is described only by the strain components should be developed in future.

Figure 22 represents the correlation of nonproportional fatigue lives for the cruciform specimens with the COD strain. In nonproportional tests using the cruciform specimens, severe nonproportional straining was impossible and only the nonproportional straining with the fixed principal strain direction(21). The reduction in fatigue lives under this kind of nonproportional loading was not large, so large safety factor margin is not necessary for this type of nonproportional loading.

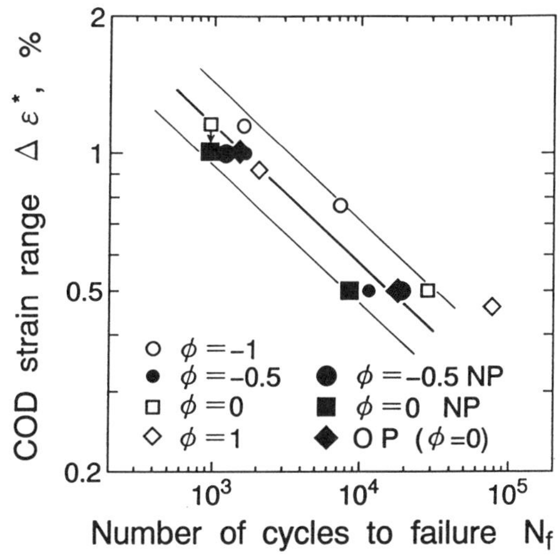


Fig.22 Nonproportional fatigue lives for SUS304 cruciform specimens correlated with COD strain.

## Conclusions

This paper stated the design criteria for multiaxial creep–fatigue, especially focused on the life prediction of the proportional and nonproportional fatigue lives. The main conclusions obtained in this paper is as follows.

(1) The data correlation of proportional low cycle fatigue lives in a full range of multiaxial stress states was extensively discussed in relation with the applicability of the multiaxial fatigue parameters. Among the parameters, the Mises equivalent strain could not correlate the proportional fatigue lives within a factor of two scatter band. The Mises strain range gave a discrepancy of data correlation for the data at the principal strain ratio between −0.5 and −1.0. The COD strain as well as the LE strain could correlate the proportional fatigue data within a factor of two scatter band.

(2) The data correlation of proportional multiaxial creep–fatigue data was discussed. The linear damage rule was effective for the life prediction of tension creep–fatigue data but ineffective for the tension/torsion creep–fatigue lives. The COD strain was effective in estimating the multiaxial fatigue damage but the Mises equivalent stress was not effective for creep damage evaluation in multiaxial stress states. The creep damage drastically increased as the shear stress component increased under tension/torsion creep–fatigue tests. This results indicate that the Mises equivalent stress is not an appropriate parameter to correlate multiaxial creep rupture data.

(3) Nonproportional fatigue lives were extensively discussed. The ASME Code Case gave the unconservative estimate to the life prediction for nonproportional fatigue, whereas the path strain gave too conservative estimate. The energy parameter appropriately correlated nonproportional low cycle fatigue lives.

## References


1 *Cases of ASME Boiler and Pressure Vessel Code, Case N-47, Class 1, Section 3, Division 1*, ASME, 1978.

2 Sakane, M., M. Ohnami, and M. Sawada, *Trans. ASME J. Eng. Mater. Technol.*, Vol.109, 236–243(1987).

3 Sakane, M., M.Ohnami, and M.Sawada, *Trans. ASME, J. Eng. Mater. Tech.*, Vol.113, 244–253(1991).

4 Garud, Y. S., *Trans. ASME J. Eng. Mater. Technol.*, Vol.103, No.2, 118–125 (1981).

5 Brown, M. W., and Miller, K. J., *Fatigue Eng. Mater. Struct.*, Vol.1, No.2, 217–229 (1979).

6 Kandil, F. A., M. W. Brown, and K. J. Miller, *Mech. Behave. Nucl. Appl. Stainless Steel Elevated Temperature*, 203–209 (1982).

7 Lohr, R. D., and E. G. Ellison, *Fatigue Eng. Mater. Struct.*, Vol.3, No.1, 1–17 (1980).

8 Hamada, N., M. Sakane, and M. Ohnami, "Fatigue Under Biaxial and Multiaxial Loading," *Mech. Eng. Publications Limited London*, pp.239–247 (1991).

9 Ohnami, M., M. Sakane, and N. Hamada, *Journal of the Society of Material Science, Japan*, Vol.35, No.390, 230–240 (1986).

10 Hamada, N., M. Sakane, and M. Ohnami, *Proc. of 30th Japan Conger. on Mater. Res.*, 69–75 (1987).

11 Itoh, T., M. Sakane, and M. Ohnami, *Trans ASME, J Eng. Mater. Technol.*, (submitted).

12 Sakane, M., M. Ohnami, T. Kuno, and T. Itsumura, *Journal of the Society of Material Science, Japan*, Vol.37, No.414, 340–346 (1988).

13 Itoh, T., M. Sakane, and M. Ohnami, *Transactions of the Japan Society of Mechanical Engineers*, Vol.56, No.527, 45–52 (1990).

14 Sakane, M., and M. Ohnami, *"Fatigue Under Biaxial and Multiaxial Loading," Mech. Eng. Publications Limited London*, pp.265–278 (1991).

15 Itoh, T., M. Sakane, and M. Ohnami, *Proc. 6th International Conference on Mechanical Behavior of materials*, Vol.2 367–272 (1991).

16 Shimomizuki, K., N. Shirafuji, M. Sakane and M. Ohnami, *Proc. 29th Symp. Strength of Mater. at High Temp.*, 21–25(1991).

17 Hamada, N., *Proc. 29th Symp. Strength of Mater. at High Temp.*, 26–31(1991).

18 *Creep Data Sheet NO4B, National Research Institute for Materials*, Japan(1986).

19 Itoh, T., M. Sakane and M. Ohnami, *Proc. of Conference on Materials and Mechanics*, No.917–71, Vol.B, JSME, 425–427(1991).

20 Sakon, T., H. Wada and Y. Asada, *Trans. Smirt–9, L*, 267(1987).

21 Itoh, T., M. Sakane, M. Ohnami, Y.Takahashi and T. Ogata, *Proc. This Conference.*

*Proceedings of the Fifth International Conference on Creep of Materials, Lake Buena Vista, Florida, USA, 18-21 May 1992*

# Creep Strength and Creep Damage Considerations for Life Cycle Engineering

**D.A. Woodford**
Materials Engineering Department
Rensselaer Polytechnic Institute
Troy, New York

ABSTRACT

Traditional methods for design of high temperature stressed components require extensive long time creep rupture testing. For mature technologies this approach has worked well for setting initial design allowables but has major limitations as a basis for estimating remaining life of operating components. Moreover, for advanced technologies using new materials there is in many cases no experimental basis for the initial design analysis. This paper describes a unified approach to both initial life analysis and remaining life assessment based on establishing a set of appropriate minimum performance criteria. In principle, the approach can be calibrated with traditional methods where the data are available. The approach specifies separate tests for creep strength performance and fracture resistance. For the former, the stress relaxation test is proposed as a primary basis for performance assessment, and examples of the applicability are described. For the latter, several tests may need evaluation. In particular, the importance of microstructural damage and embrittlement phenomena, including those associated with environmental attack, are emphasized.

TWO MAJOR CHALLENGES in high temperature structural applications are the need to predict long term service creep behavior based on short term laboratory tests, and the assessment of the remaining life of currently operating components. For mature technologies, e.g. boiler pressure vessel tubes, design stresses have been established based on extrapolation of creep rupture data, with appropriate safety factors, for a variety of commonly used ferritic and austenitic steels[1]. Consequently, for these materials, extensive data have been generated over many years, and a substantial experience base in operation has been accumulated. Implicit in this approach is that stress rupture lives in laboratory tests of solid cylindrical specimens can be directly translated to design lives of pressurized tubes. As a consequence, an enormous effort has been directed towards optimizing time-temperature parametric approaches for life extrapolation of test data. Moreover, this same approach to life prediction has been favored as a basis for predicting remaining life of components in operation, i.e. stress rupture tests of samples taken from such components[2,3].

There are a number of reasons why this approach to remaining life prediction may be unsound[4]:

- Component failure is often localized with little or no material degradation or damage remote from the failure.

- Cracks frequently initiate from the surface so that any post-exposure property measurement on material taken from the interior has limited value.

- Interactions with the operating gaseous environments, which may have profound effects on crack initiation and propagation, are generally ignored.

- The changing stress in a constant load test is not normally accounted for in summing life fractions to predict remaining life[5,6].

- Time to rupture in an unnotched ductile alloy is principally a measure of deformation behavior rather than fracture resistance.

• The sources of statistical scatter in laboratory tests (alignment, temperature control, specimen geometry) are quite different from those in service.

Thus, for these mature technologies, although the primary design analysis is well-founded on extensive data and experience, estimating remaining life is still hampered by both technical and conceptual problems.

For less mature technologies such as high temperature design of critical gas turbine components made from superalloys, the situation is less well defined. Basic stress rupture data for currently used alloys (such as IN 738 for industrial gas turbine blades) are available, and extrapolated lives for rupture, or for some limiting creep strain specified by design, can be made[7]. However, much of the data are not in the open literature.

As more advanced alloys (e.g. GTD 111 for industrial gas turbines) and processes (e.g. directional solidification and monocrystal growth) are introduced, the data become progressively scarcer, and manufacturers become more reluctant to share the results of costly testing. Thus, equipment users, in particular, seek short time tests which will allow them to participate in the alloy selection process.

For remaining life assessment of gas turbine blades, the manufacturers depend on their proprietary data bases, extensive experience, and largely empirical rules to recommend part replacements. The users again are interested in reliable methods of remaining life assessment as a decision support basis for making the best technical and economic decisions regarding part repair or replacement.

Finally, the emerging technologies involving the use of new materials such as ceramics, intermetallics and composites have virtually no basis for initial design or remaining life assessment. Only recently has equipment become available with sufficiently good alignment capability, temperature control and strain measuring sensitivity. Very little data exist on long term creep behavior, and there is a clear need to develop methods which can, at reasonable costs and in short times, provide a basis for setting allowable stresses for component design and durability evaluation.

The purpose of this paper is to describe a unified approach to life prediction involving accelerated testing. The approach is termed "Design for Performance"[4], and in principal can be applied both to the initial design and to remaining life prediction after service exposure. Hence, it is considered to be a contribution to life cycle engineering. Moreover, it should be applicable to all materials classes and for technologies from the mature to the undeveloped.

## DESIGN FOR PERFORMANCE

The fundamental problem in using the same criterion (i.e. estimated stress-rupture life) for initial part design life and remaining life analysis is what has been called "The Remaining Life Paradox"[4]. This may be stated: when a component fails, the material of the component has a finite life, sometimes approaching the original part design life, at the operating conditions of the component. A corollary is that a remaining life estimated from a sample taken from an operating component may bear no relationship to the actual component remaining life.

This paradox may be avoided by establishing separate criteria for strength, fracture, oxidation resistance, etc. We seek criteria that establish the minimum performance requirements when new, and which may be monitored periodically throughout the life of the part. The concept is shown schematically for three limiting performance criteria in figure 1. Each criterion should be measurable in a short time inexpensive test. To validate the methodology it should be calibrated with the traditional approach for mature technologies involving alloys (e.g. creep resistant low alloy steels). It may then be applied to develop new materials, where accelerated testing is necessary, with some confidence. The selected tests may be component-specific, dependent on the likely progress of damage and probable failure cause. However, some tests are emerging as being broadly applicable for many components within this framework. The progress to date for developing a short term strength criterion and short term embrittlement criterion are now reviewed.

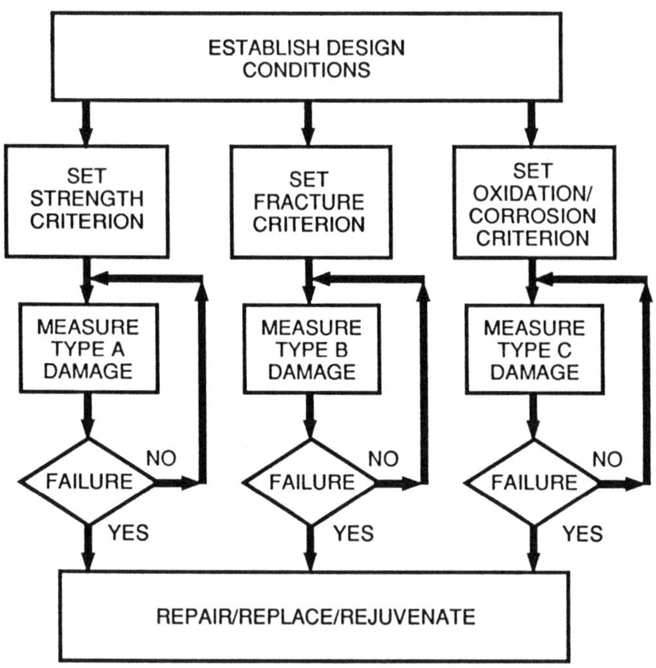


Fig. 1. Design for Performance

## CREEP STRENGTH EVALUATION

For low alloy steels the simplest short term test is indentation hardness, measured either at the temperature of interest or at room temperature[8,9]. However, even though good correlations have been reported (see figure 2) for ferritic steels, the approach is not applicable to austenitic steels or nickel-base superalloys.

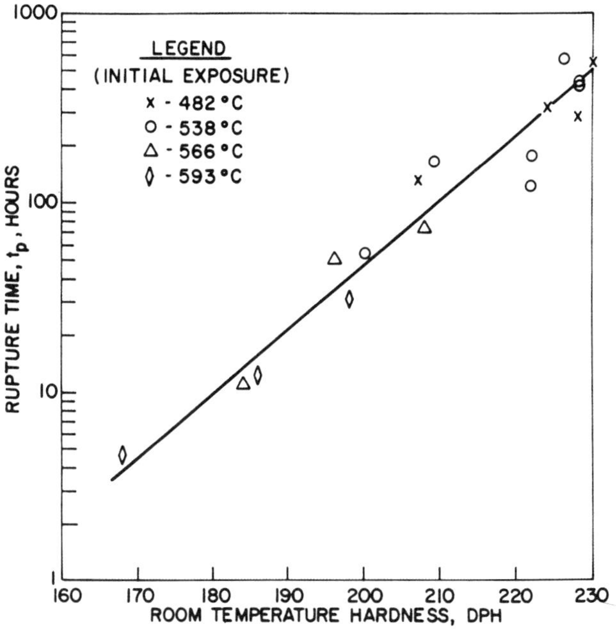


Fig. 2. Correlation between post-exposure rupture time at 240 MPa and 538°C and room temperature hardness for Cr-Mo-V steel[7].

An alternative approach, which could be generally applicable, is to generate a stress vs. strain-rate response at high temperature using a carefully controlled stress relaxation test. Initially proposed by Hart[10] as a basis for his plastic equation of state, and later supported by extensive experimental measurements[11,12] the test can be conducted several ways. The simplest is to lock the test machine actuator or crosshead at a prescribed stress and monitor the stress relaxation with time. This approach requires a precise value for the machine compliance[11]. To overcome this, one can monitor both stress vs. time and strain on the specimen vs. time, but the analysis is quite time consuming[12]. The method now proposed requires sophisticated and sensitive test equipment but offers considerable advantages in analysis simplicity. Moreover, it produces the least amount of creep strain during the test so several runs on a single specimen can be conducted even on materials, such as ceramics, with limited ductility.

The principle of the method involves loading rapidly at temperature to the prescribed stress in a closed loop testing machine, switching to strain control on the specimen and monitoring the relaxation of stress. The inelastic (principally creep) strain-rate is calculated from the following equations:

$$\epsilon_e + \epsilon_I = \epsilon_t = \text{constant}$$

$$\dot{\epsilon}_I = -\dot{\epsilon}_e = \frac{1}{E}\frac{d\sigma}{dt}$$

where $\epsilon_e$ is the elastic strain, $\epsilon_I$ is the inelastic strain (principally creep strain), $\epsilon_t$ is the total strain, $\sigma$ is the stress, and E is the elastic modulus measured during loading. Using this procedure, stress vs. strain-rate curves may be generated covering up to five orders of magnitude in strain-rate in a test lasting less than five hours.

Figure 3 shows the results of a single test lasting a few hours on a Cr-Mo-V steam turbine rotor steel. Four test runs on this same specimen define a curve over 12 orders of magnitude in strain-rate (see figure 4). Rather than present the data on a temperature compensated strain-rate scale a separate strain-rate scale is preserved for each test temperature. Note that test 7 at 811°K is a repeat of test 4 after the two intervening runs at 755°K and 700°K. The small displacement of the two curves reflects a small quantifiable change in the mechanical state.

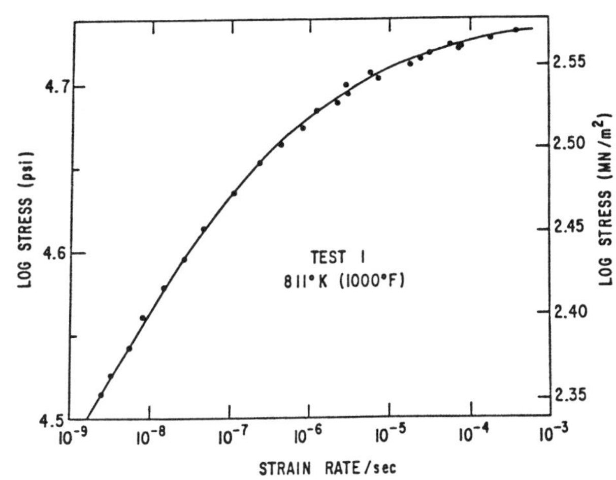


Fig. 3. Stress vs. strain-rate response at 538°C for a Cr-Mo-V steel[12].

To establish a preliminary calibration with conventional life analysis, the stress-rupture data for this alloy are shown in figure 5. Data to 70,000 hours at four temperatures are shown plotted against the Larson-Miller parameter with a constant of 20 in figure 5(a), and figure 5(b) shows a generalized plot to incorporate the stress dependence of the "constant" 20 using the graphical optimization procedure (GOP)[13].

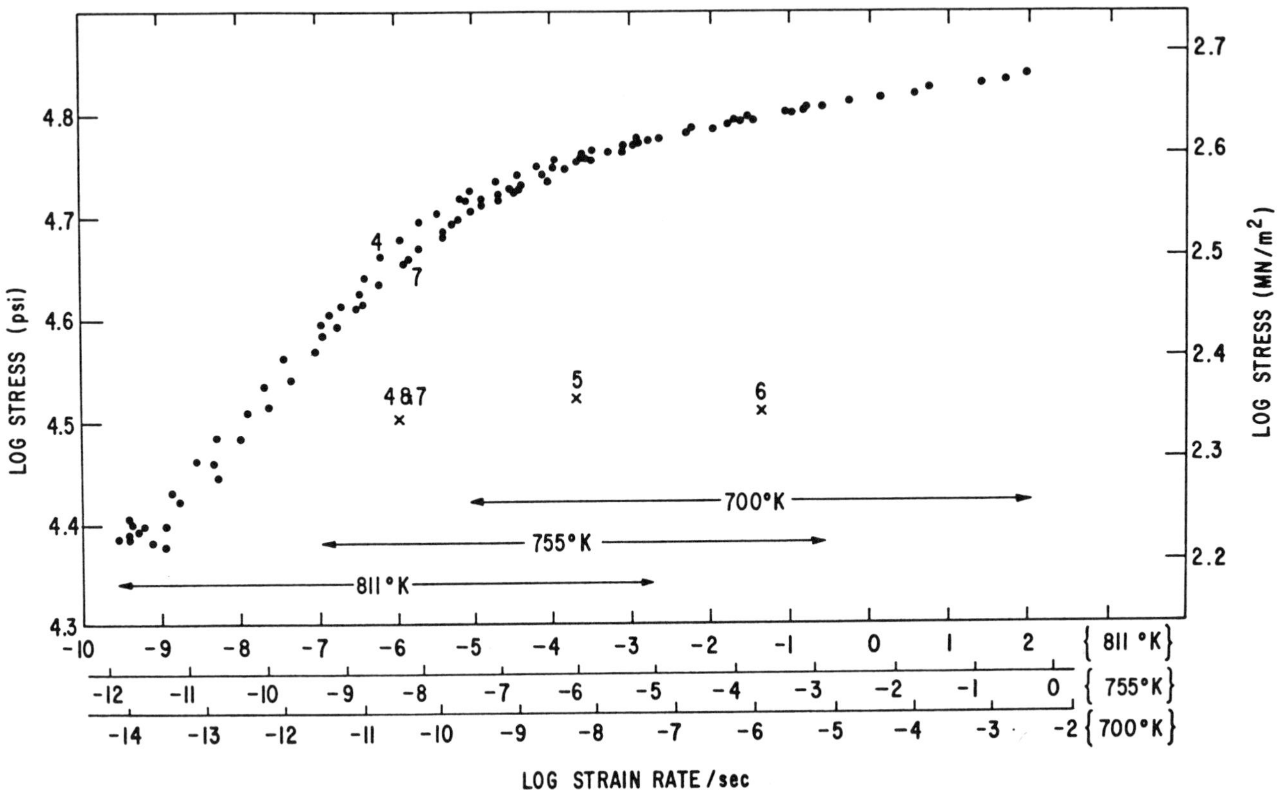


Fig. 4. Superimposition of stress vs. strain-rate data at three temperatures for a Cr-Mo-V steel achieved by horizontal displacement indicated by crosses[12].

To compare the data of figures 4 and 5 the ASME boiler and pressure vessel code[1] is used to identify a possible calibration procedure. For example, the two criteria for design based on stress for a creep rate of 0.01% in 1000 hours and 67% of the stress for rupture in 100,000 hours may be extracted from these data. Table 1 summarizes the results and shows good potential for the stress relaxation approach to accelerated life analysis in terms of creep strength. The ratios of the stress values taken from the extrapolated stress rupture data and the relaxation data range from 1.1 to 1.27.

To explore the use of this approach, for both initial life and remaining life, some recent data generated on serviced 2¼Cr-1Mo steel boiler tubes are examined[14]. Three conditions were studied for material taken from a thick section boiler T-piece after 167,000 hours service:

1). Material taken from the least damaged thick section was re-normalized from 970°C for 2 hours and tempered at 720°C for 9 hours, and taken to be undamaged reference material.

2). Some of this material was machined into large creep specimens which were creep tested at 630°C and 78.5 MPa for various strains and times to establish levels of damage. Small specimens were cut from these tested specimens for relaxation measurements.

3). Some material was taken from the T-piece near the inner surface where the damage was considered to be the greatest. This was evaluated as material damaged in service.

Table I. Design comparison for creep life analysis for Cr-Mo-V steel.

| Temperature | | ASME Code | | Relaxation Test |
|---|---|---|---|---|
| °K | °F | MPa | (2/3 MPa) | MPa |
| 700 | 800 | 331 | 222 | 300 |
| 755 | 900 | 220 | 148 | 194 |
| 811 | 1000 | 138 | 92 | 109 |

A series of stress relaxation tests at 600°C for all the exposure conditions was completed. Since a small amount of strain is accumulated during a relaxation run, a second test was conducted on each specimen under the same conditions. This gave an indication of the extent to which the mechanical state (i.e. the measure of damage) changed during each run.

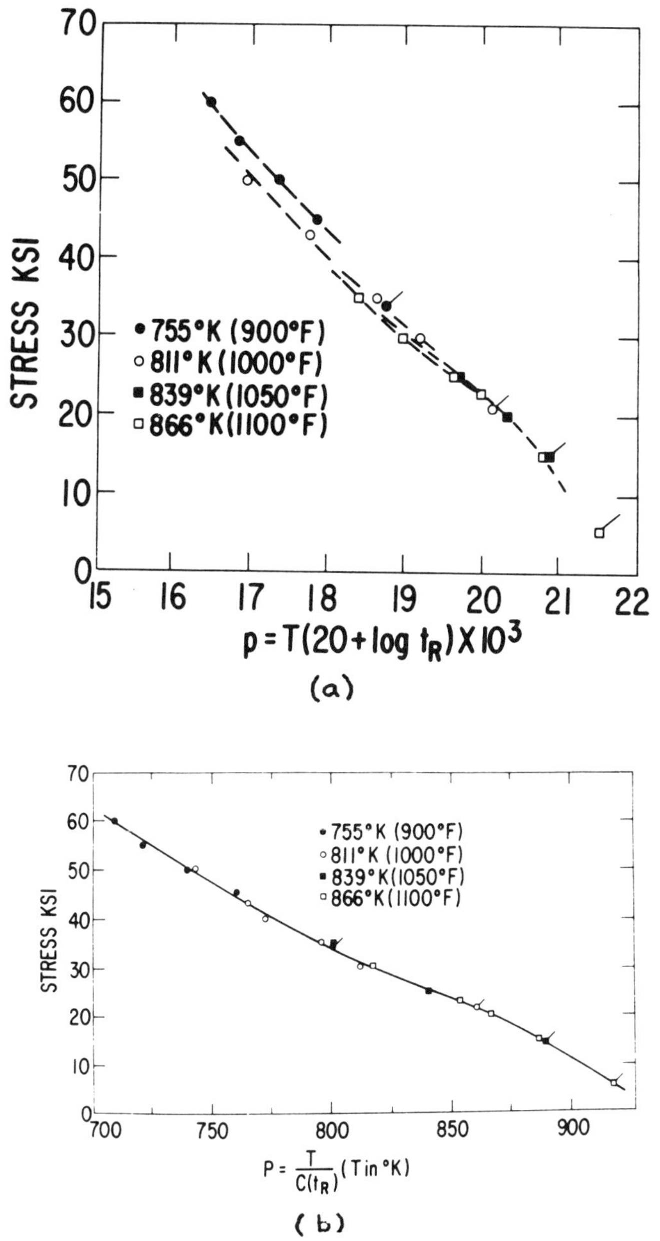


Figure 5. Stress rupture data for Cr-Mo-V steel plotted a) vs. the Larsen-Miller parameter and b) vs. a generalized parameter using the GOP method.

Table II summarizes the prior exposure conditions and figure 6 shows the results for the undamaged and service exposed conditions showing a substantial reduction in creep strength.

Unfortunately, no long term creep rupture data were available for this heat, but extensive data for the alloy have been published[15]. At 600°C the extrapolated average stress for rupture in 100,000 hours is in the range 30 to 45 MPa, dependent on the extrapolation procedure used. The design stress according to the ASME code would thus be on the order 25 MPa (taking two-thirds of the average). At 570°C it may be estimated to be about 37.5 MPa which is close to the calculated hoop stress of 34 MPa for the T-section operating at this estimated temperature[14]. As with the Cr-Mo-V data a calibration may be tried using a creep rate of .01% in 1000 hours; i.e. $2.77 \times 10^{-11}$ $sec^{-1}$ from the relaxation test. This requires some extrapolation of fitted equations but the results are given in Table III. Recalling that these tests were run at 600°C and the design stress at this temperature has been estimated to be on the order 25MPa (based on an average stress for 100,000 rupture of 37.5 MPa), the stresses in table III are of the correct order. The service exposed condition had degraded considerably and may effectively have exhausted the safety factor for creep strength. This stress value could clearly be identified with a limiting performance criterion for safe operation. Information which would aid the selection of an appropriate limiting value would include some indication of the kinetics of degradation. For example, was most of the damage developed in the early stages of service or is it a progressive weakening? As with the Cr-Mo-V data, there is a clear promise of accelerated testing to evaluate both long term performance and current state of damage for life assessment.

Further studies are now planned for nickel-base superalloys to cover the intermediate maturity technology where some long term data exist but may not be generally available to gas turbine users. There is also a need to evaluate new alloys and processes.

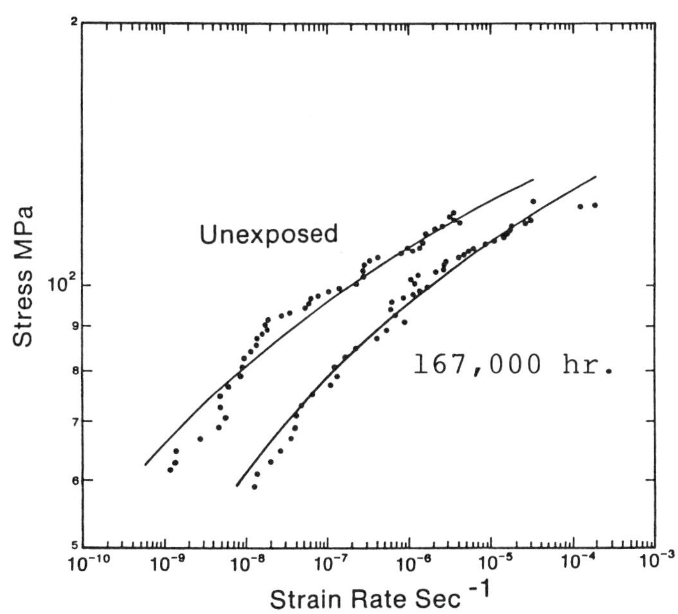


Fig. 6. Stress vs. strain-rate curves for the undamaged and service exposed condition for 2¼Cr-1Mo steel at 600°C.

Table II. Creep Test Exposure Conditions

| Specimen Type | | Creep Stress (MPa) | Temperature (°C) | Time (h.) | Creep Strain % |
|---|---|---|---|---|---|
| 1. | Undamaged | --- | --- | 0 | 0 |
| 2. | Damaged | 78.5 | 630 | 70 | 0.248 |
| 2. | Damaged | 78.5 | 630 | 156 | 0.44 |
| 2. | Damaged | 78.5 | 630 | 279 | 6.8 |
| 3. | Service Exposed | 34* | 570 | 167,000 | ? |

*Estimated hoop stress based on internal pressure of 16.9 MPa

Table III. Calculated Stress for a Creep Rate of $2.77x10^{-11}sec^{-1}$

| Creep Exposure Time, h | Run #1 (MPa) | Run #2 (MPa) | Average |
|---|---|---|---|
| 0 | 42 | 51 | 46.5 |
| 70 | 62 | 67 | 64.5 |
| 70 | 68 | 71 | 69.5 |
| 156 | 57 | 63 | 60 |
| 279 | 40 | 50 | 45 |
| 167,000 | 19 | 18 | 18.5 |

For the emerging technologies involving the use of ceramics or other advanced materials, no long term data exist. Very recently, preliminary relaxation tests on alumina have been reported[16]. Figure 7 shows results for a sequence of runs from increasing stresses at 1150°C. All data were generated on a single specimen. The lower stress runs show a considerable amount of recoverable anelastic strain on unloading. However, at the two highest stresses the relaxation provides a unique curve and the strain was principally non-recoverable. Without any extrapolation the data reach strain rates of $10^{-9}sec^{-1}$. This is equivalent to 3%/year obtained in a few hours with a maximum inelastic strain accumulation of about .05% during a run.

The potential for this approach to performance evaluation of silicon nitride is currently being evaluated.

## CREEP EMBRITTLEMENT EVALUATION

Setting performance limits for initial evaluation and in-service monitoring of the resistance to fracture is likely to require component-specific testing. The objective, as with creep strength evaluation, is to identify short time inexpensive tests. Tensile ductility, impact resistance, sustained load crack growth measurements, and fatigue resistance may all be useful in specific applications. One test which may have quite general applicability for metallic materials is the constant displacement rate (CDR) notched bar test[17].

The CDR test involves closed loop control on displacement rate across a notch so that crack growth usually occurs in a stable manner. The time to fail in such a test, or the notch displacement at failure, has been shown to distinguish among heats that may show varying tendencies to embrittle in long-term stress rupture tests

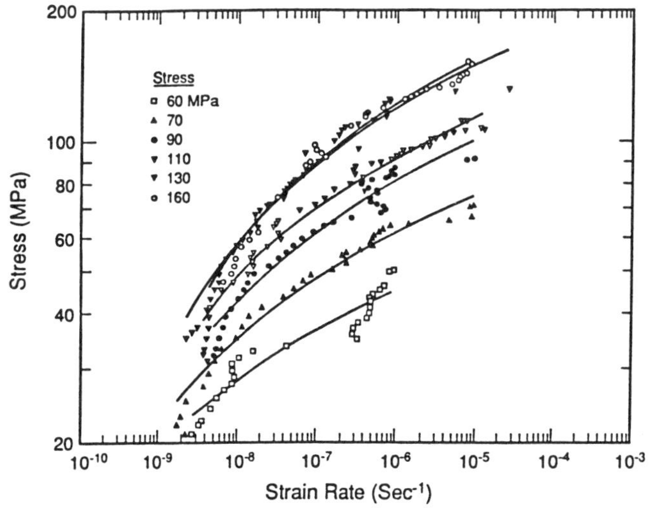


Fig. 7. Stress vs. strain-rate behavior at 1150°C for a series of relaxation tests for alumina from increasing stresses[16].

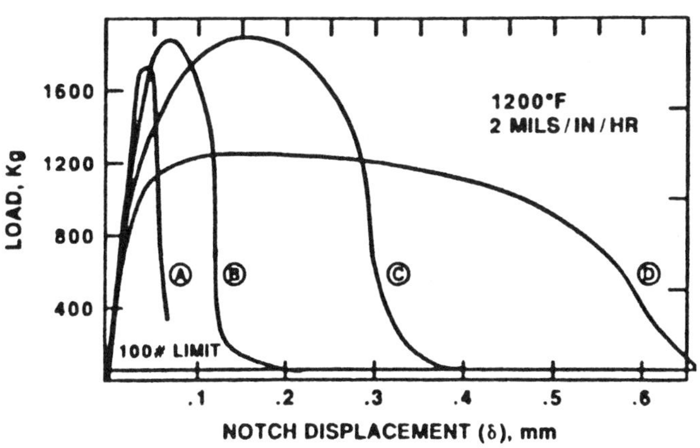


Fig. 8. Load vs. notch displacement for different Cr-Mo-V steels at 649°C and a notch displacement rate of $1.4 \times 10^{-5}$ mm/sec.[17]

(see figure 8). This test remains to be convincingly demonstrated as a useful tool for service damage evaluation.

Some phenomena which lead to embrittlement in service, e.g. segregation of embrittling species to grain boundaries, or intergranular penetration of embrittling species from the environment, may be subtle and difficult to monitor. However, considerable use has been made of monitoring creep cavitation which is often related to grain boundary chemistry. Near-surface cavitation can be studied in operating power plant components using replication methods[18,19], and criteria for repair or replacement established in terms of the extent of this damage. This approach, in fact, fits in well with the concept of Design for Performance depicted in figure 1.

In recent years it has become clear that, in nickel base alloys, this form of damage is linked to environmental influence. In particular, intergranular oxygen penetration in nickel reacts with carbon or carbides at grain boundaries and, if the temperature is sufficiently high, forms carbon dioxide gas bubbles which act as creep cavity nuclei[20,21,22].

The link between oxygen attack and intergranular cavitation was first demonstrated in commercially pure nickel. Figure 9 shows two populations for stress rupture. The shorter lives are for tests in air where extensive cavitation leads to intergranular failure. The longer lives are for specimens which are coated to prevent oxygen penetration, or decarburized by exposure in dry hydrogen, before testing. Both approaches eliminate the oxygen-carbon reaction, and failure is by necking to a point with no intergranular cavitation.

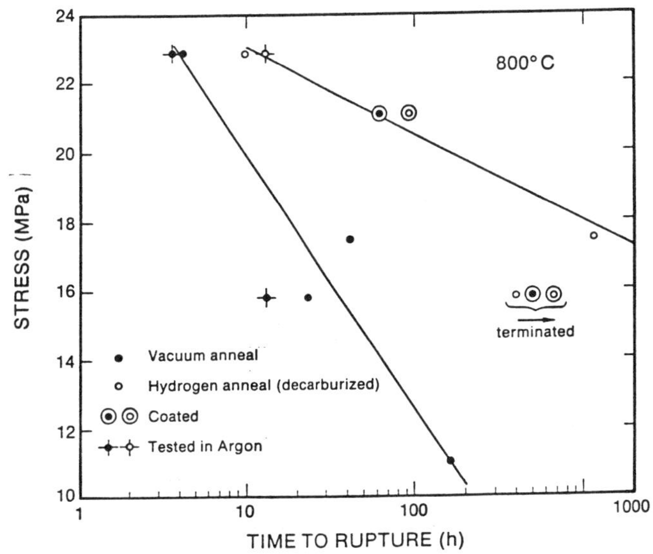


Fig. 9. Stress-rupture results on Ni270 at 800°C showing effect of stress enhanced nucleation and growth of gas bubbles. Testing in air except where indicated[22].

The significance of environmental damage for superalloys is now well-recognized for both oxygen[23] and sulfur attack[24]. This can be clearly demonstrated by pre-exposure in air which allows oxygen penetration and embrittlement. Figure 10 shows the effect of various environments and a protective coating on stress rupture life of the nickel base alloy IN 738. The loss in stress rupture life or associated tensile ductility could be used as a measure of performance degradation in situations where the intergranular embrittlement may not be easily revealed by standard microstructural observations.

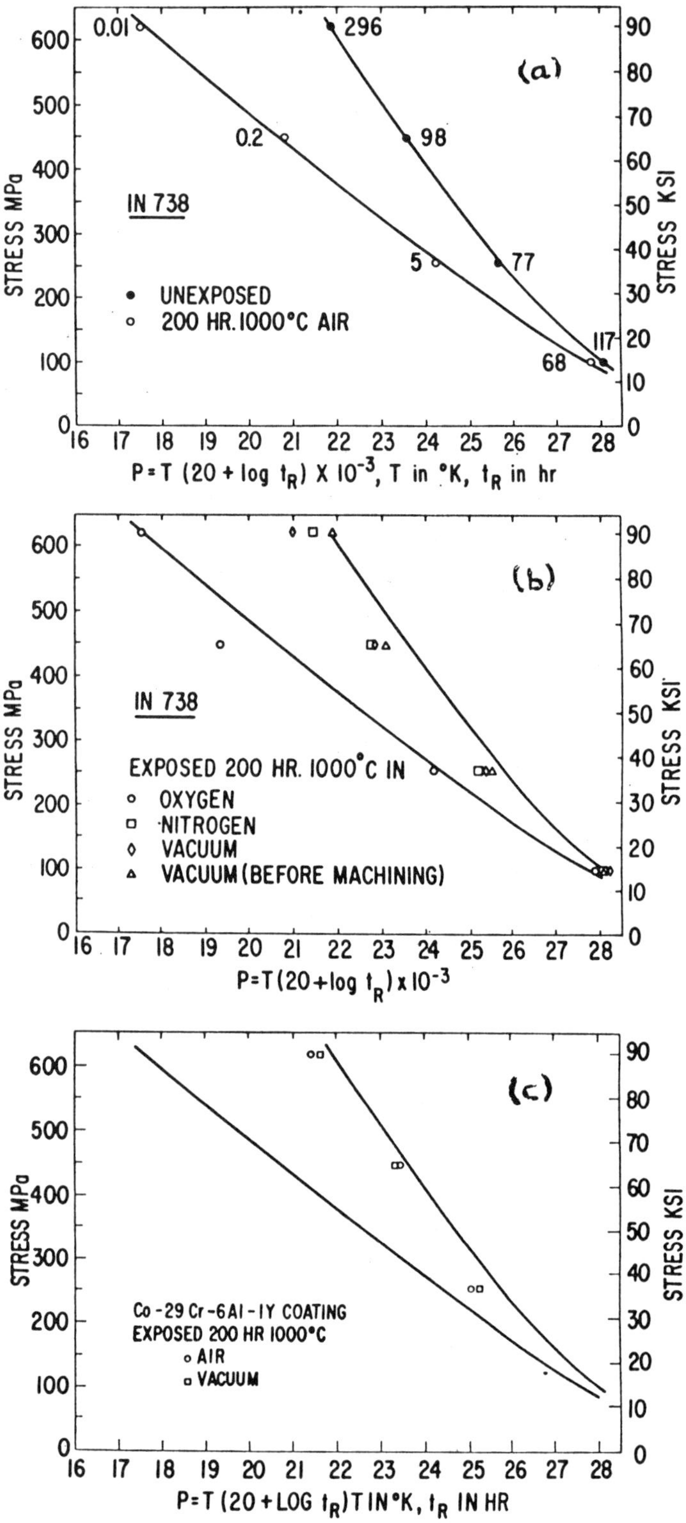


Fig. 10. a) Effect of air exposure for 200 hrs. at 1000°C on rupture life of IN738 plotted in terms of the Larsen Miller parameter. Individual lives are indicated in hours for tests at 700, 800, 900 and 1000°C; b) Exposures in oxygen, nitrogen and vacuum; c) Exposures in vacuum and air of specimens plasma sprayed with a Co-29Cr-6Al-14 coating.

Environmental damage may develop in joining operations. For example, in heat affected zones of welds, residual stress relief can lead to cavitation nucleated as gas bubbles[25]. It has recently been studied in the nickel alloy IN 625 which is very resistant to oxygen attack because of its high chromium content. However, in a heat exchanger application requiring the fabrication of thin sheets (~1mm) using diffusion bonding, the processing can lead to oxygen penetration[26]. This is true even in an industrial vacuum. Figure 11 shows a taper section through the surface after exposure in a low partial pressure of oxygen at high temperature. The voids were subsequently opened by tensile testing at room temperature. Figure 12 shows that the same features can be developed by simultaneous creep deformation during the high temperature exposure. These cavities are clearly nucleated as a result of the environmental interaction. The extent of damage can be monitored in this case by microstructural evaluation or by mechanical testing. Figure 13 shows the effect on fatigue life of various partial pressures of oxygen during a simulated diffusion bonding heat treatment.

A more striking effect of a combination of deformation and environmental interaction at high temperature leading to loss in fatigue life is illustrated in figure 14. Three orders of magnitude reduction in cycles to failure resulted from 1% creep extension at 982°C. The same deformation in a vacuum (of unspecified quality) resulted in less degradation. This observation led to a limiting strain being specified in the initial design of gas turbine buckets using this alloy[7].

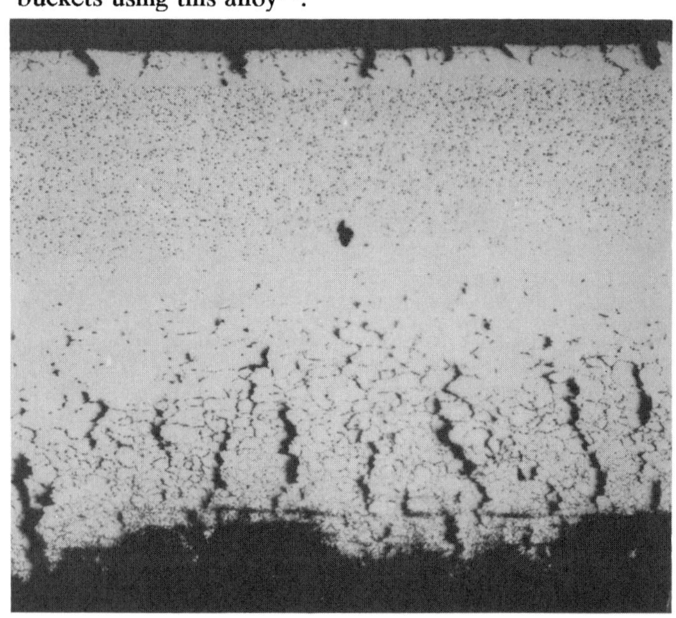

Fig. 11. Surface intergranular damage in thin section IN625 after exposure at 1100°C for 24 hours at $10^{-9}$ atm. oxygen and RT tensile testing. A normal surface section and a tapered section (6x) are shown in the same micrograph at 50x optical magnification.

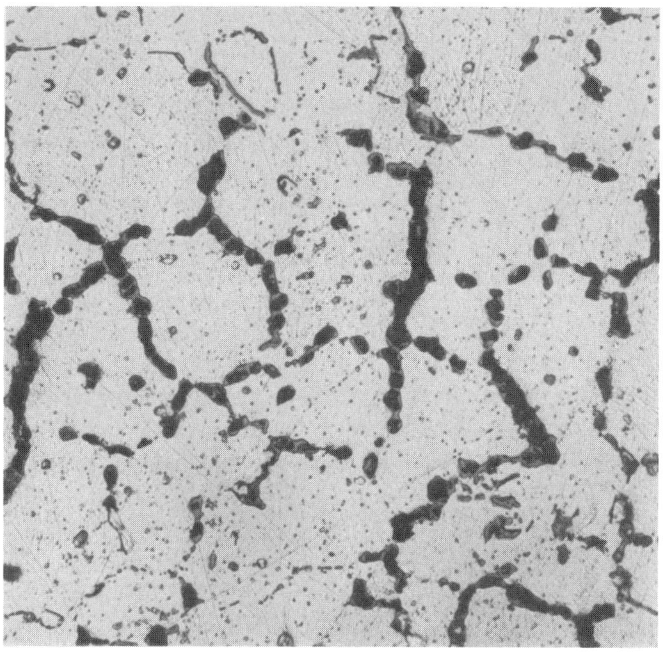

Fig. 12. Intergranular voids near the surface of IN625 produced by simultaneous deformation and exposure in air at 1100°C for 24 hours (10x taper and 500x optical magnification).

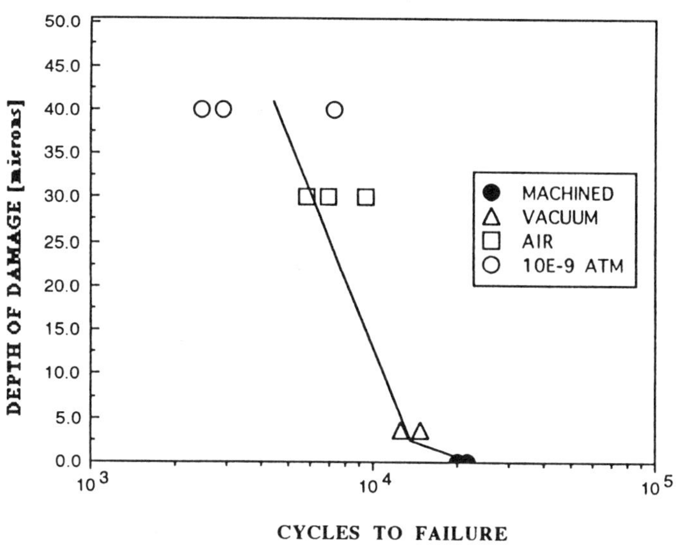


Fig. 13. Fatigue lives at RT for IN605 loaded at 735/140 MPa after annealing at 1100°C in various environments.

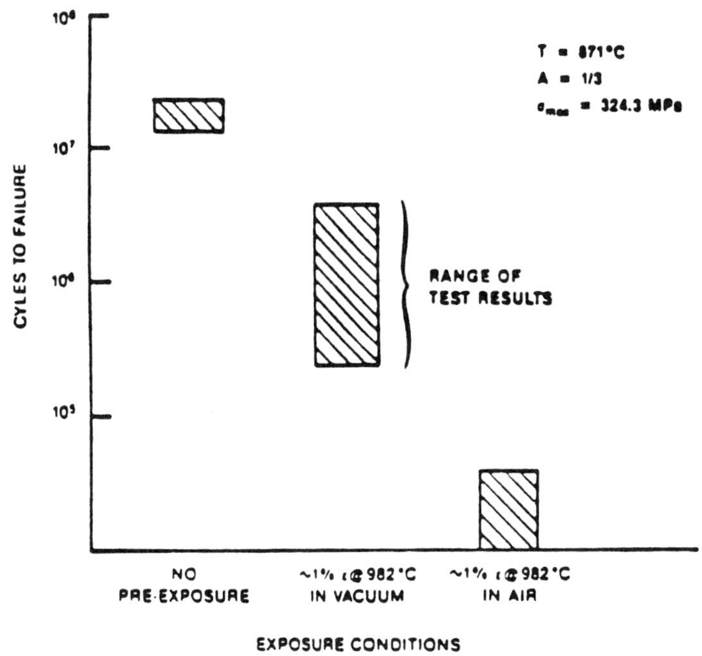


Fig. 14. Effect of creep exposure on HCF life of IN-738 at 870°C[7].

Damage and embrittlement associated with environmental interactions are very dependent on alloy chemistry. High chromium levels and small additions of boron, in particular, are effective in reducing oxygen attack[27]. Overlay coatings which produce a stable oxide scale may prevent embrittlement (see figure 10(c)). However, once the coating is breached, embrittlement may occur. Moreover, if cyclic stresses are present, then environmental acceleration of high temperature fatigue crack propagation (time-dependent fatigue) must be factored into the life assessment[24,28].

## SUMMARY

An attempt has been made to provide a common framework for both initial design life analysis and remaining life assessment in terms of separate design limiting performance criteria for creep strength and creep fracture. In principle, this approach can cover both mature and advanced technologies. It may also be applicable to established engineering alloys and new materials.

The approach uses accelerated testing to establish the initial state of the material used in high temperature components, and also to monitor changes in the state with service. This is in contrast to traditional creep rupture analyses which attempt to provide time-dependent properties which can quantify the initial or remaining part lives. In situations where extensive long term creep rupture data are available, calibration between the two approaches may be possible.

Stress relaxation testing, and the generation of stress vs. strain-rate curves, are proposed for creep strength evaluation. Several examples are drawn from low alloy steels and a ceramic (alumina) to show how this test could be applied. For embrittlement evaluation, several short time mechanical tests are proposed. It is also believed that direct microstructural evaluation, particularly that associated with environmental interactions, could be an important aspect of the framework.

The new approach has distinct advantages for remaining life analysis in operating energy conversion systems. It may also be an attractive alternative as a basis for initial design where new materials and/or new processes are involved.

## REFERENCES


1- ASME Boiler and Pressure Vessel Code, Section 1, American Society of Mechanical Engineers, N.Y.

2- Toft, L.H. and Marsden, R.A., "Structural Processes in Creep", ISI Report No. 70 (1961).

3- Melton, K.N., Mat. Sci. Eng., 59, 143-149 (1983).

4- Woodford, D.A., "Int. Conf. on Fossil Power Plant Rehabilitation", ASM Inter., p. 149, (1989).

5- Woodford, D.A., "Int. Conf. on Creep and Fatigue in Elevated Temperature Applications", I. Mech. E. paper 180.1 (1973-74).

6- Hart, R.V., Met. Tech., 4, p. 442, (1978).

7- Embly, G.T. and Kallianpir, V.V., Minnowbrook Conf. on Life Prediction for High-Temperature Gas Turbine Materials (1985).

8- Goldhoff, R.M. and Woodford, D.A., ASTM, STP 515, p. 89, (1972).

9- Cane, B.J., High Temp. Techn., May, p. 217, (1983).

10- Hart, E.W., Acta Met., vol. 18, p. 599 (1970).

11- Hart, E.W., and Soloman, H.D., Acta Met., vol. 21, p. 295 (1973).

12- Woodford, D.A., Met. Trans. A., vol. 6A, p. 1693 (1975).

13- Woodford, D.A., "Int. Conf. on Creep", ASME, ASTM, I. Mech. E., JSME, 11-20 (1986).

14- Woodford, D.A., Van Steele, D., Hyder, M.J., and M. Kitagawa, Mat. Sci. & Techn., to be published.

15- Smith, G.V., "Symposium on 2¼ Chrome 1 Moly Steel in Pressure Vessels and Piping", The Metals Properties Council, Edt. Schaefer, A.O., 1-26 (1971).

16- Woodford, D.A., Van Steele, D., and Hyder, M.J., J. Amer. Cer. Soc., vol. 74, p. 3142-44 (1991).

17- Pepe, J.J., "Int. Conf. on Fossil Power Plant Rehabilitation", ASM Int., 39-46 (1989).

18- Neubauer, B., and Wedel, U., "Advances in Life Prediction Methods", Edt. Woodford, D.A., and Whitehead, J.R., ASME, 307-314 (1983).

19- Gooch, D.J., Shammas, M.S., Brett, S.J., Coleman, M.C., and Stevens, R.A., "Int. Conf. on Fossil Power Plant Rehabilitation, ASM Int. 1-6 (1989).

20- Bricknell, R.H., and Woodford, D.A., Acta Met., 30, 257-264 (1982).

21- Dyson, B.F., Acta Met., 30, p. 1639 (1982).

22- Woodford, D.A., Bricknell, R.H., Scripta Met., 23, 599-601 (1989).

23- Woodford, D.A., Met. Trans., 12A, 299-308 (1981).

24- Whitlow, G.A., Beck, C.G., Viswanathan, R., and Crombie, E.A., Met. Trans. A, vol. 15A, p. 23 (1984).

25- Iacocca, I., and Woodford, D.A., Met. Trans. A., 19A, 2305-2313 (1988).

26- Rimlinger, C., and Woodford, D.A., "Superalloys 718, 625 and Various Derivatives", TMS, 653-662 (1991).

27- Woodford, D.A., and Bricknell, R.H., "Embrittlement of Engineering Alloys", Academic Press, NY, 157-196 (1983).

28- Chang, K-M., "Effects of Load and Thermal Histories on Mechanical Behavior of Materials", edt. by Liaw, P.K., and Nicholas, T., TMS, p. 13 (1987).

*Proceedings of the Fifth International Conference on Creep of Materials, Lake Buena Vista, Florida, USA, 18-21 May 1992*

# The Mechanical and Microstructural Stability of Austenitic Stainless Steels Strengthened by MC-Forming Elements

**R.W. Swindeman and P.J. Maziasz**
Metals and Ceramics Division
Oak Ridge National Laboratory
Oak Ridge, Tennessee

ABSTRACT

A high-strength, ultrafine precipitate strengthened (HT-UPS) steel, basically a 14%Cr-16%Ni-2%Mo-Fe alloy with V, Ti, and Nb additions, was found to have higher creep strength than the other MC-forming steels of the same class due to the stability of the precipitate distribution on dislocations produced by warm or cold work prior to service. Under constant-load creep conditions, the microstructure of the HT-UPS steel was found to be stable for long periods of time, and creep rates changed very little over thousands of hours. Studies were made to determine if the strength of the HT-UPS alloy would degrade by variable temperatures and stresses conditions that could disturb the precipitate-stabilized dislocation microstructure. Experiments included variable stress creep testing, repetitious and combined creep-relaxation testing, and creep-fatigue testing. Temperatures were in the range 600 to 800°C and times extended to beyond 20,000 h in some cases. Stress reductions and increases indicated that strain hardening, rather than softening, prevailed over much of the creep life. Relaxation data indicated some hardening from one run to the next early in the testing, but eventual stabilization of relaxation response. Similarly, creep-fatigue cycling produced a small increase in the stress range and relaxation strength early in life and then a gradual degradation of strength afterward.

INTEREST IN HIGH PERFORMANCE AUSTENITIC STAINLESS STEELS for superheater tubing has been stimulated by the need for higher steam temperatures to improve fossil power plant thermodynamic efficiencies. A number of new and improved alloys have been investigated in Europe and Japan, and many of the alloys contain MC-forming elements such as titanium, niobium, and vanadium. Examples, include old alloys, such as type 17-14CuMo stainless steel (1), and Esshete 1250 (2), Sandvik 12R72HV® (3), and new alloys such as ST3Cu® stainless steel (4), NF709® stainless steel (5), and HR3C® stainless steel (6), and fine-grained type 347 stainless steel (7). In the United States, a lean stainless steel (less than 20% chromium) developed for nuclear applications was proposed for fossil energy applications by Maziasz (8). This steel, as well as other niobium-bearing steels, requires a high solution treating temperature (>1150°C) to dissolve carbides that may form during fabrication. The MC carbides then precipitate on dislocations during service exposure, although in some instances the MC is found together with other carbides, like $M_{23}C_6$ or $M_6C$, particularly along grain boundaries. Since the creep properties tend to be controlled by the particle spacing, size, and stability, a high dislocation density and a stable fine precipitate size tends to maximize creep strength. Various fabrication routes have been developed to optimize thermal-mechanical treatment with respect to creep strength and ductility of MC-forming steels, and these routes have been guided by metallurgical considerations. Aging and post-creep test microstructures have been examined in many of the alloys to establish time-temperature-precipitation diagrams that identify the kinetics of the precipitation process (3-12), while the rate of particle growth has also been established in some instances (5,9,12). With the exception of Esshete

1250, surprisingly little is known about the shape of the creep curves for the MC-forming alloys. Further, the influence that variable stress and temperature may have on the stability has not been examined in any detail. This paper is an overview of studies at the Oak Ridge National Laboratory that address these topics for the HT-UPS steel.

## MATERIAL

Fourteen heats of the steel were produced whose compositions in weight percent ranged as follows: 0.07-0.11C, 0.10-0.65 Si, 1.6-2 Mn, 15-16 Ni, 13-16 Cr, 0.1-0.3 Ti, 0.1-0.2 Nb, 0.1-1.0 V, 2.0- 2.5 Mo, 0.01-0.08 P, 0.001-0.007 B, 0.002-0.02 S, 0.01-0.03 N, and 0-2 Cu, balance Fe. Twelve heats were cast as laboratory heats by argon induction or vacuum induction melting practices. These laboratory heats weighed 20 to 50 kg and were hot rolled or cold rolled to 13-mm thick plates. Two larger heats were fabricated as tubing. The tubes were either 50-mm or 65-mm diameter with 7-mm or 13-mm wall thickness. The plates were evaluated in several conditions: hot-rolled, cold-rolled, and annealed at temperatures in the range 1110 to 1200°C. The hot rolling was performed by annealing the plates at 1200°C between passes and producing enough reduction in thickness (approximately 10%) to raise the room temperature yield strength to around 400 MPa. The cold rolling was performed on plates that had been solution treated at 1200°C prior to cold rolling with 10% reduction in thickness. Typical room temperature yield strengths were in the range of 450 to 700 MPa, with the optimum strength being around 500 MPa. The tubing was evaluated in several conditions: as-extruded, cold-forged (5% reduction of wall thickness), cold-sunk (5 and 10% reduction of wall thickness), and cold-pilgered (5% reduction of wall thickness). The extrusion was performed at 1200°C and produced a microstructure and strength typical of annealed material. Cold worked tubing exhibited room temperature yield strengths in the range of 480 to 670 MPa.

## SPECIMENS AND TESTING METHODS

Specimens were machined parallel to the rolling direction of plates and parallel to the axial direction of tubing. Tensile specimens and most creep and relaxation specimens were threaded-end bars with a reduced section length of 32 mm and a diameter of 6.3 mm. High strain resolution creep specimens, used for some variable stress and temperature testing, were threaded-end bars with a gage section of 51 mm and a diameter of 6.3-mm. Fatigue specimens were button-end bars with a testing section of 19 mm and a diameter of 6.3 mm. Tensile and creep tests were performed in resistance heated furnaces. Fatigue specimens were inductively heated. Extensometers were attached to the gage length of tensile specimens, high resolution creep specimens, and fatigue specimens. Averaging extensometers were attached to the shoulders of creep specimens and relaxation specimens.

Altogether, more than 200 high-temperature tests were performed that covered temperatures from 600 to 800°C and times to more than 50,000 hours. However, variable stress creep tests were limited to a few specimens, and the longest test time was beyond 30,000 hours. Relaxation tests generally lasted 100 hour or less, and the longest fatigue testing time was near 2000 hours.

## TENSILE BEHAVIOR

Typical tensile curves at 700°C are shown in Fig. 1. These indicated the response of the yield curve to various thermal-mechanical processing variables. For example, the annealed materials and the hot-extruded pipe exhibited yield strengths in the range of 90 to 120 MPa. Hot-rolled steels exhibited yield strengths in the range of 240 to 310 MPa, while the cold-worked steels exhibited yield strengths that ranged from 310 to 490, depending on the level of cold work. The uniform strain diminished with the increasing yield strength and was reduced to less than 2% for the highest levels of cold work. However, the reduction of area exceeded 30% for all processing conditions.

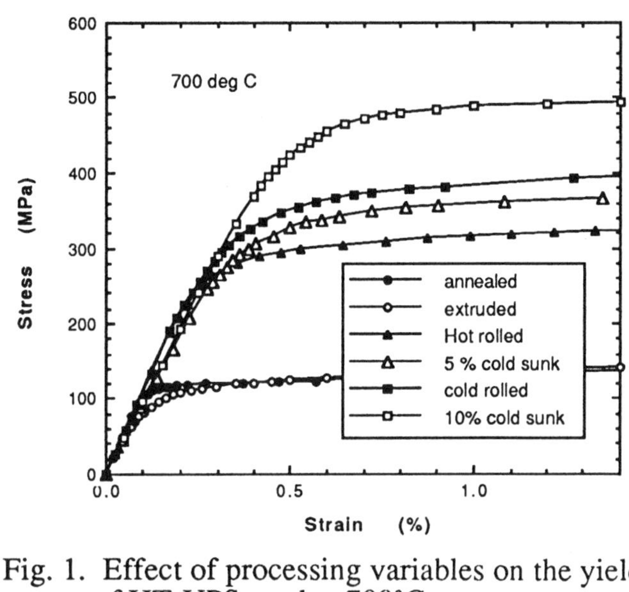


Fig. 1. Effect of processing variables on the yield curves of HT-UPS steel at 700°C.

CONSTANT LOAD CREEP BEHAVIOR

Although it was the original intent to evaluate the materials in the annealed condition, it was observed early in the testing program that the worked materials exhibited better creep properties, and that benefits extended for long times at high temperatures. A comparison is shown in Fig. 2 between creep curves at 700°C and 170 MPa obtained from annealed and cold-rolled plates. Specimens from the same heat were tested in the annealed (1200°C) and 10% cold-rolled conditions. The annealed specimen exhibited more primary creep, but the minimum creep was nearly the same as for the cold-rolled material. The period over which the minimum creep rate persisted was much greater for the cold-rolled material, and this phenomenon extended the rupture life by more than a factor of four. Tertiary creep in both specimens started at less than 1% creep strain. The creep rate increased rapidly after the start of tertiary creep, which occurred within 75 to 90 percent of the rupture life for the cold-rolled condition. In contrast, tertiary creep started at 50 or 60 percent of the rupture life for the annealed material.

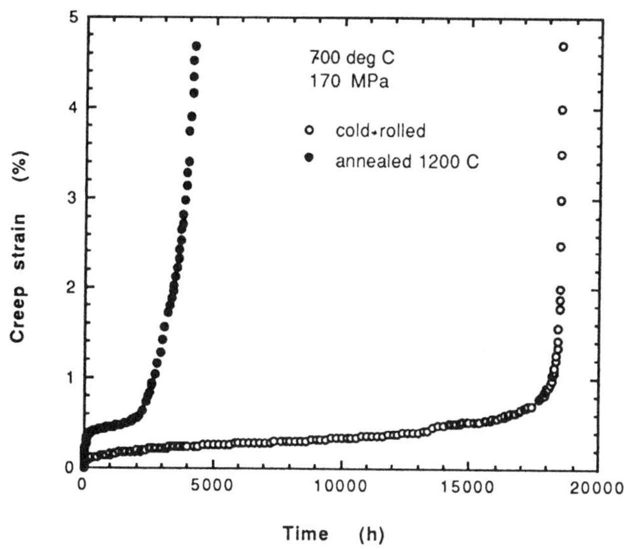


Fig. 2. Effect of cold work on creep at 700°C and 170 MPa.

In the absence of cyclic stresses and temperatures, the benefits of hot and cold work were retained for long times at temperatures to at least 700°C. Fig. 3 compares stress-rupture data for worked materials with data for annealed materials. Although both conditions produced large variations in life due to a wide range of processing histories in each category, the worked materials exhibited considerably greater strength than annealed materials over the entire stress range that was examined. This advantage extended for times to 50,000 hours. Fig. 4 compares minimum creep rate data at 700°C. Again, the hot-and cold-work materials generally exhibited greater strength than the annealed materials. As shown in Fig. 2, however, the minimum creep rate for the annealed material was often similar to that for worked material. The advantage of the prior deformation persisted over a broad range of testing conditions, and this advantage is illustrated by the rupture life data in Fig. 5. Here, the log stress versus the Larson Miller parameter for rupture has been plotted for material in the annealed or worked conditions. A parametric constant of 20 was assumed. For high values of the parameter, which correspond to long times and high temperatures, the worked materials were approximately 50% stronger.

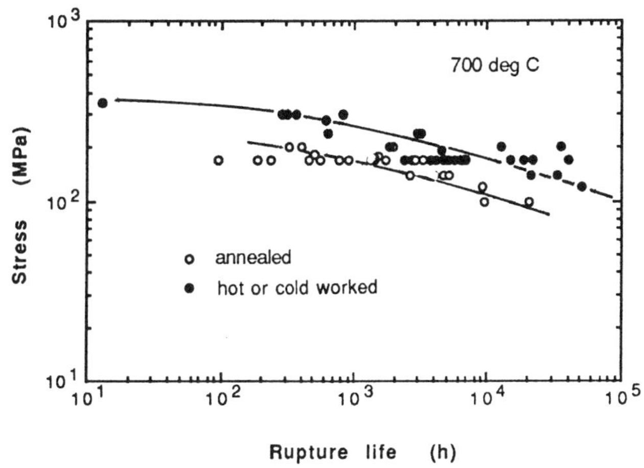


Fig. 3. Effect of work on the stress rupture life at 700°C.

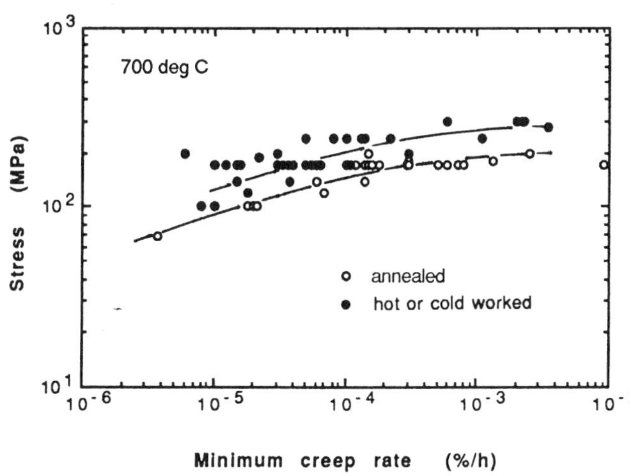


Fig. 4. Effect of work on the minimum creep rate at 700°C.

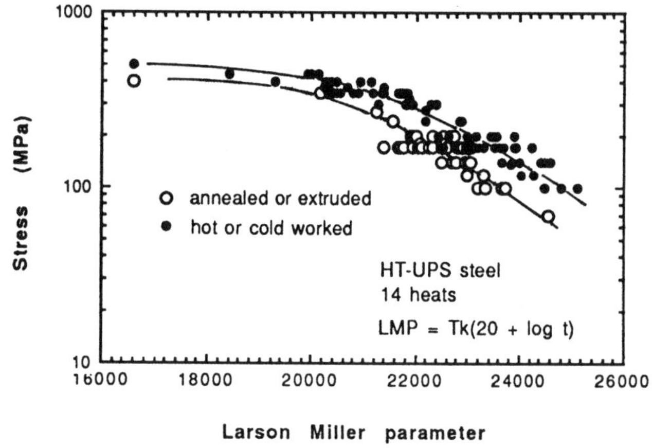


Fig. 5. Log stress versus Larson Miller parameter for 14 heats of HT-UPS steel.

Benefits from cold work have been seen in other MC-forming steel (9,11), but the HT-UPS steel was very much improved. A comparison strengths is shown in Fig. 6 for several of steels mentioned earlier. Again, the comparison is on the basis of log strength versus the Larson Miller parameter, assuming a parametric constant of 20. The optimized constant may differ from 20 for some steels, but an effort was made to minimize the influence of the selection by using the rupture strength data for 10,000 hours in developing the trends shown in the Fig. 6. The strength advantage of hot or cold work is clearly revealed at high values of the Larson Miller parameter. In making use of the greater rupture strength of worked materials, an issue to be resolved is whether or not stress-cycling or thermal-cycling would degrade such deformation-induced strength and produce rupture strengths more like annealed materials.

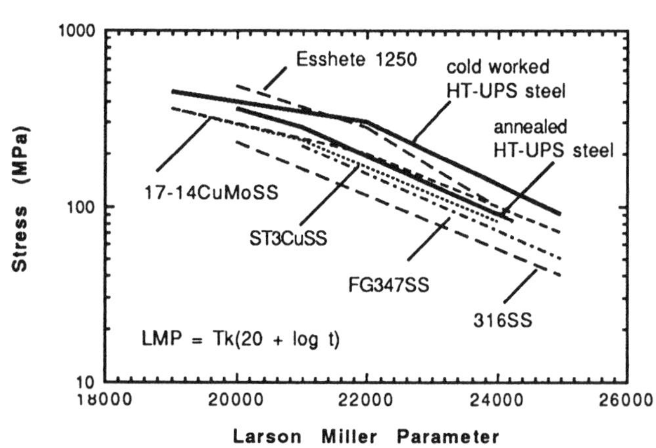


Fig. 6. Comparison of Log stress versus Larson Miller parameter for several steels.

## EFFECT OF STRESS AND TEMPERATURE CHANGES

Change-of-stress creep tests performed on HT-UPS steel involved simple creep recovery periods, step-stress tests with upward and downward transients, multiple reload relaxation tests, and interspersed tensile, creep, and relaxation tests. An example of a long-time test which was interrupted by periods of recovery creep at near zero stress is shown in Fig. 7 for a test at 700°C and 100 MPa on a material annealed at 1200°C. The figure shows the creep strain and log creep rate versus time. The spikes in the creep rate correspond to the higher creep rates after each reloading following a recovery period. Early in the test, the creep rate after recovery period diminished to values lower than before recovery. However, the minimum creep rate was attained after a few thousand hours and the creep rate gradually increased thereafter. Creep recovery periods did not have much of an effect on the overall trend of the creep curve, and rupture occurred in approximately 21,000 hours. A similar test was undertaken on hot-rolled material, but with fewer recovery periods. The minimum creep rate ($10^{-5}$ %/h) was achieved in approximately 5,000 hours.

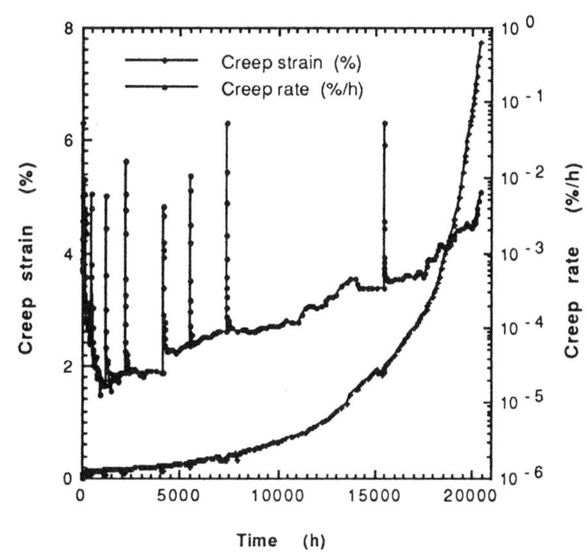


Fig. 7. Effect of recovery periods on the creep of annealed material at 700°C and 100 MPa.

Little or no increase in creep rate was observed over the next 50,000 hours. At that time the specimen had accumulated 0.7% creep strain. Another specimen from hot-rolled material was tested at 700°C under more variable loading conditions, and the results have been plotted in Fig. 8. Here, the stress and log creep rate have been plotted against time. Nine recovery creep periods were interspersed over the 20,000 hours, and ten periods of creep at lower and higher stresses in the range of 50 to 150 MPa were introduced. Some of the high-creep-rate data for upward stress transients and negative creep rates for downward stress transients have been deleted for clarity in the figure. Again, the creep rate at 100 MPa diminished toward a minimum in around 5,000 hours. After both upward and downward stress transients, the creep rate returned to the value typical of the constant load creep rate for hot-rolled material at 100 MPa ($10^{-5}$%/h). The test was discontinued after 21,000 hours with approximately 0.7% accumulated creep strain. Similar change in stress tests were performed to beyond 30,000 hours at 650°C on annealed and hot-rolled materials. Trends indicated no significant softening in either material as a result of creep recovery, and stress increases generally produced hardening after the stress was reduced. Similar observations were made for other heats of HT-UPS steel that had been cold-rolled plate or cold-sunk tubing.

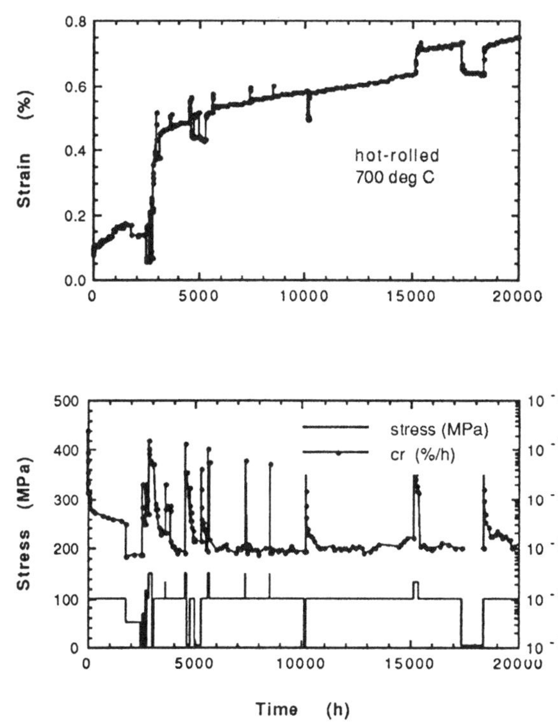


Fig. 8. Effect of downward and upward stress transients on creep of hot-rolled material at 700°C and 100 MPa.

The recovery creep periods obtained from tests on the annealed and hot-rolled specimens produced some interesting results which are shown in Fig. 9. Here, the recovery creep strains have been plotted against the log of the recovery creep time for various times during the tests. Early in the test on the annealed material, the recovery creep was small over a period of 24 hours. With increasing time and creep strain, the recovery creep strain increased, and by 15,000 hours the recovery strain reached 20% of the elastic strain component. There was no indication of the exhaustion of recovery creep over a 100 hour period. The recovery creep characteristics of the hot-rolled material were similar to those for annealed material. In fact, the recovery creep curves for hot-rolled material at 17,000 hours superimposed on the recovery curve for annealed material at 15,000 hours. Since the creep stress prior to recovery was the same (100 MPa) for the two materials and the creep rates differed by more than a factor of fifty, the recovery creep rate appeared to be insensitive to the basic creep strength of the material.

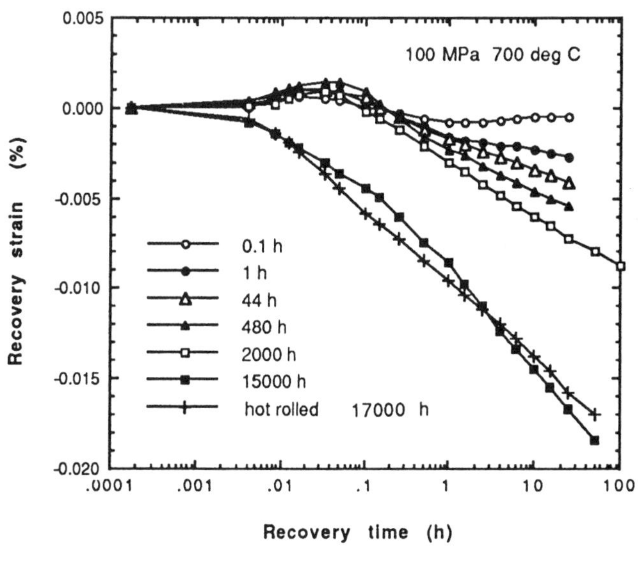


Fig. 9. Evolution of recovery creep with time or strain in annealed material at 700°C and 100 MPa.

A number of tests were performed in which the temperature was increased or decreased. When these changes were made at low strains and during long-time tests, periods of creep at higher temperatures did not degrade of creep strength at lower temperatures. An example of creep behavior of cold-rolled material is shown in Fig. 10. This figure includes plots of both creep strain and creep rate versus time for a relatively high stress of 300 MPa. After 11,300 hours at 600°C, the temperature was increased to 650°C. Creep rate increased at first, then diminished with time. Around 11,900 hours, the temperatures was increased to 700°C for a short period. This produced about 0.05% creep strain. After returning to 650°C, the creep rate continued to diminish. The upward transient was repeated around 13,200 hours with similar results. In a few tests, the influence of large upward temperatures excursions were examined. These large excursions produced large increases in creep strains, and tertiary creep resulted.

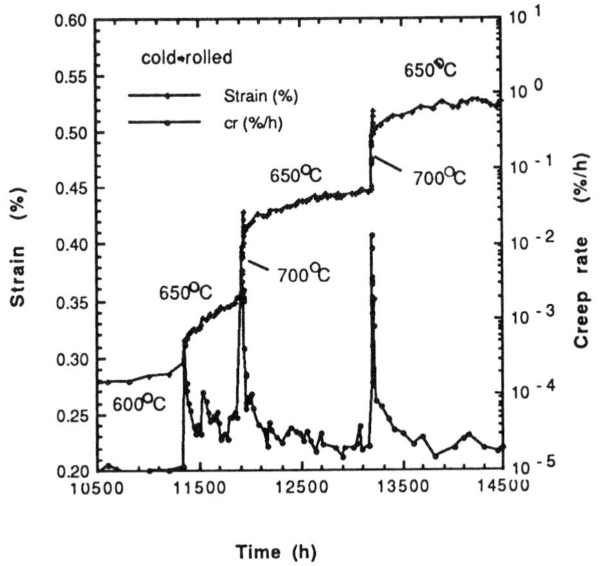


Fig. 10. Effect of upward temperature transients on creep of cold-rolled material at 650°C and 300 MPa.

## RELAXATION

Relaxation tests were performed on several heats of HT-UPS steel. These were multiple run tests in which the specimens were strained to beyond the yield strength in each run in order to examined the effect of accumulated time and strain on the flow stress and relaxation curve. The annealed materials generally exhibited significant hardening in the first few runs while the cold-rolled materials were relatively stable. Fig. 11 compares stress relaxation curves at 700°C for annealed and cold-rolled material from the same heat. Consistent with the yield curves in Fig. 1, the starting relaxation stress for annealed material was near 100 MPa. Only 10% relaxation of stress occurred over 24 hours, and the final stress was near 90 MPa. The accumulation of relaxation strains and some aging effects increased the yield stress of the annealed material to 200 MPa, and by the fourth run the relaxation increased to about 30%. The stress was around 130 MPa in 24 hours. Included in Fig. 11 is another curve for annealed material that was obtained from a specimen that was exposed to aging and small periods of creep for 1270 hours. Again, the yield stress was around 200 MPa. The relaxation curve was similar to that obtained from the specimen that underwent repetitive relaxation cycles. Consistent with the yield curves in Fig. 1, the cold-rolled material exhibited a higher yield strength than the annealed material. The material relaxed from near 300 MPa to near 200 MPa in 24 hours. Subsequent runs produced a slight increase in the yield strength and an increase in the relaxed strength at 24 hours. Similar trends were observed at 650 and 600°C.

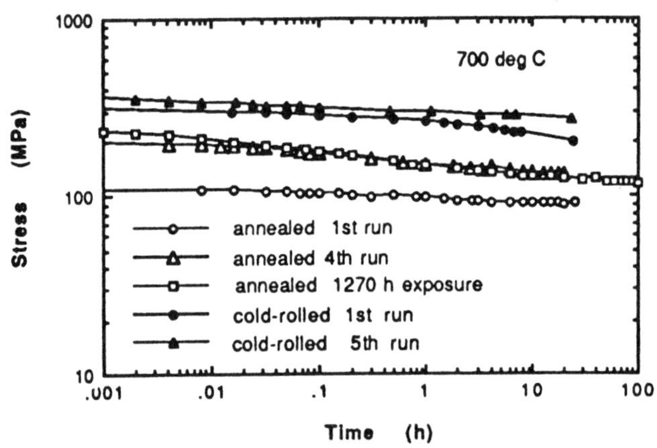


Fig. 11. Relaxation response of annealed and cold-rolled material for multiloading at 700°C.

Relaxation data for another heat of more highly cold-rolled material are shown in Fig. 12. Here, the temperature was changed from one run to the next, and test times extended to 100 hours. Temperatures ranged from 600 to 800°C and two specimens were tested at 600 and 650°C. The relaxation curves revealed a consistent pattern and showed no evidence of instability. When the relaxation data were converted to stress versus creep rate data and compared to data from constant load creep tests, relaxation data corresponding to times beyond 10 hours merged with minimum creep rate data. This trend was consistent with the findings of Black, et al. (13).

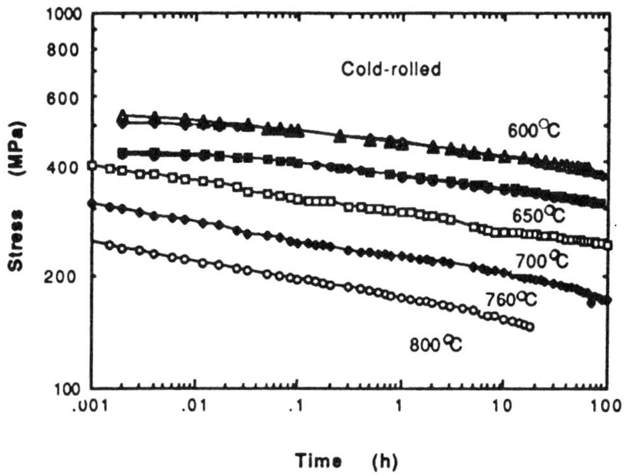


Fig. 12. Relaxation response of cold-rolled material tested at changing temperatures.

## REVERSED LOADINGS

Reversed plastic strains produced by fatigue loadings have the potential of accelerating metallurgical processes and may promote coarsening of precipitate-stabilized dislocation networks. A few creep-fatigue tests were performed on a cold-rolled HT-UPS steel to investigate stability at 600 and 650°C. Fig. 13 summarizes the results of two such tests at 650°C. Here cycling was performed at 1 and 0.55% strain range with a 0.5 hour hold time introduced at the tensile strain limit. The figure shows the stress range and the relaxed stress amplitude versus cycles. The specimen tested at 1% strain range exhibited hardening for the first twenty cycles. A gradual softening ensued, but the stress range remained above the first cycle range for most of the

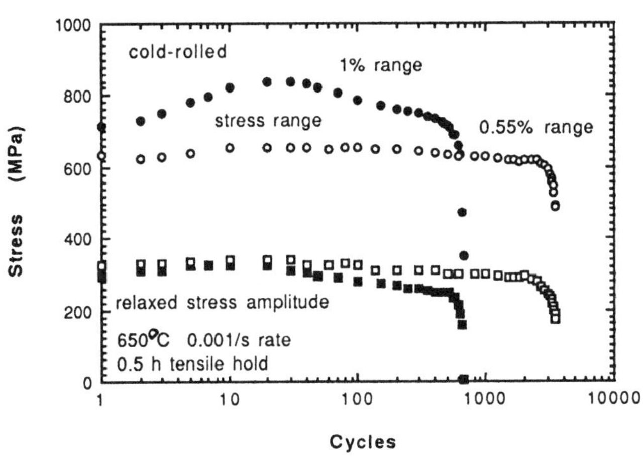


Fig. 13. Stability of the stress versus cycles behavior of cold-rolled material tested in creep-fatigue at 650°C.

life. The relaxed stress amplitude exhibited less hardening and fell below the first cycle relaxed stress amplitude around 100 cycles. The specimen tested at 0.55% strain range experienced less hardening and softening. Stable behavior was observed for over half of the life, and the relaxed stress amplitude was higher than that observed for the tests in the 1% strain range. Tests at 600°C exhibited hardening for most of the fatigue lives.

## MICROSTRUCTURE

The microstructure of cold or hot worked HT-UPS steel, observed by optical microscopy, consisted medium-to-coarse grains containing annealing twins and a few large TiN particles. After long-time exposure at high temperature, particles appeared within the matrix and on grain boundaries. This microstructure is shown in Fig. 14, which represents a region near the fracture of specimen from a hot-worked heat that ruptured after more than 50,000 hours at 700°C and 117 MPa. No grain boundary cavitation was observed, although a few transgranular fissures were observed in the necked region. Most of the metallurgical changes were on a scale too fine to be observed in the optical microscope.

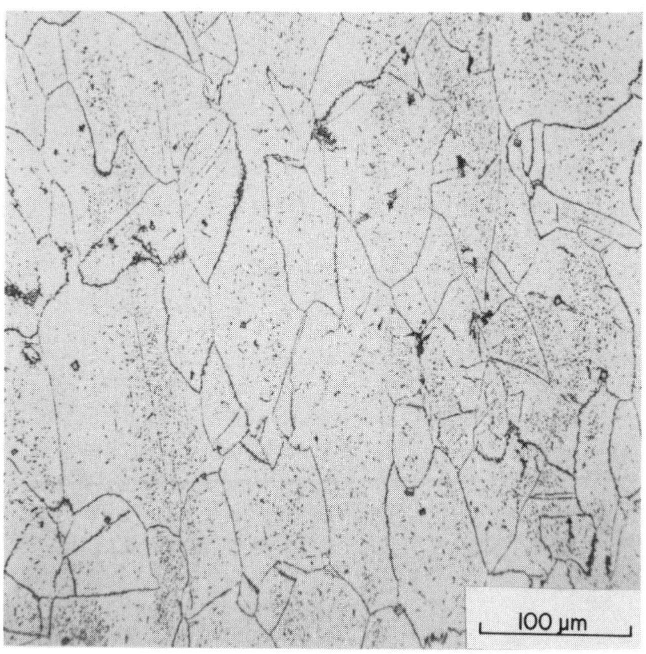


Fig. 14. Photomicrograph of hot-rolled material that ruptured after 50,000 h at 700°C and 117 MPa

Fig. 15 is a transmission electron micrograph (TEM) representing microstructure of grains in the gage section (away from the necked fracture region) of a specimen of cold-rolled material that failed after more than 18,000 hours at 700°C and

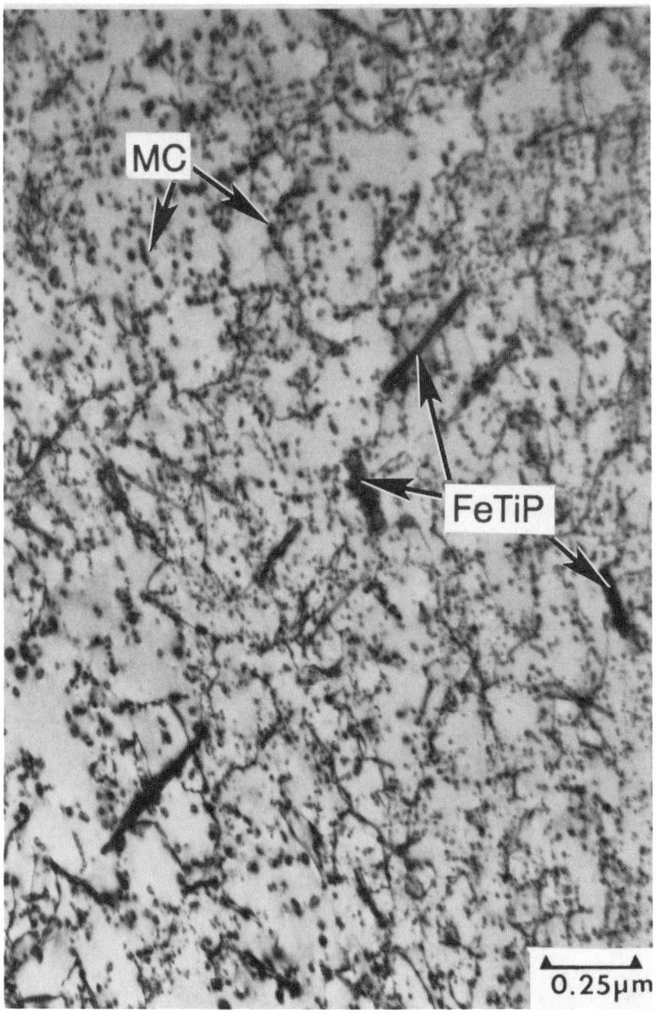


Fig. 15. TEM of the matrix region of cold-rolled material that ruptured after 18,000 h at 700°C and 170 MPa.

170 MPa. The needles were identified as phosphides (FeTiP) while the fine precipitates were MC which contained titanium (~30%), molybdenum (~27%), chromium (~18%), and niobium (~12%). Vanadium, iron, and nickel made up the remainder of the metal content of the MC phase. The size of the MC particles (8-24 nm in diam.) after 18,000 hours was essentially the same as the size observed after a 1000 hours at 700°C and 170 MPa (15). The grain boundary structure of this 18,000 h specimen is shown in Fig. 16. Here, $M_6C$ particles were observed along the grain boundary which had MC at the core. Large Laves phase particles (up to 3 μm in length) extended into the matrix from the grain boundaries. Quite likely, the precipitates observed in light microscopy (Fig. 14) were Laves phase particles. The structure of some boundaries provided evidence of a time-dependent degradation process, and this is shown in Fig. 17. Recrystallized

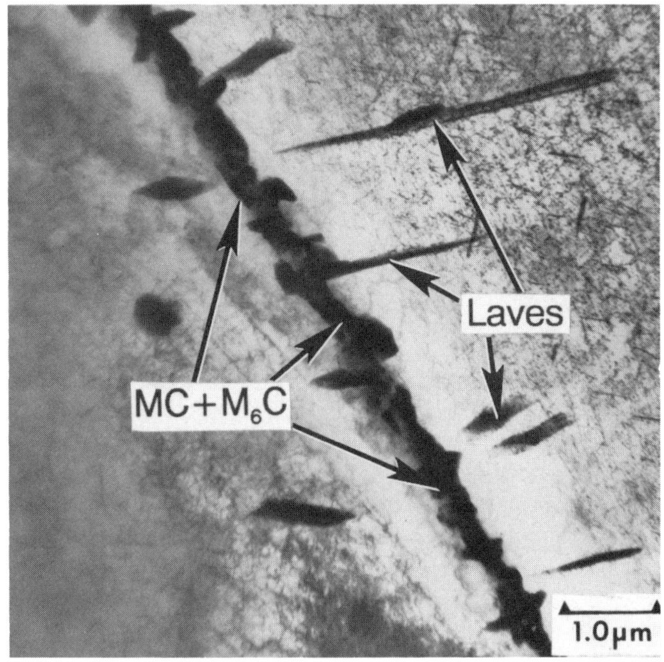


Fig. 16. TEM of a typical grain boundary region in cold-rolled material that ruptured after 18,000 h at 700°C and 170 MPa.

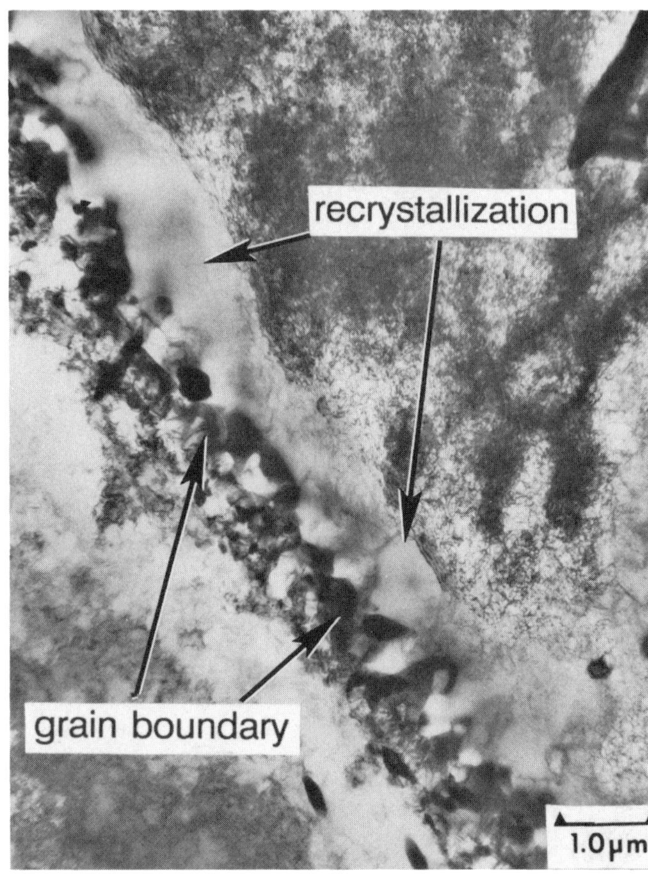


Fig. 17. TEM showing recrystallization near the grain boundary of a cold-rolled specimen that ruptured after 18,000 h at 700°C and 170 MPa.

regions appeared along some boundaries that extended 1 to 2 µm into one of the grains. The dislocation network density was reduced in the recrystallized regions, and MC particles were dissolved. In the boundary small cavities were observed whose diameters were typically 200 nm. These showed no evidence of linking together in any of the specimens that were examined.

## DISCUSSION

The complex evolution of microstructure in the HT-UPS steel has been discussed by Maziasz and coworkers in some detail (8,14,15). Several types of carbides were observed, and the elemental constituents changed with time and temperature. In spite of these changes, the size and distribution of the MC in the worked HT-UPS alloys was found to be relatively stable. Todd and Ren (12) measured the growth rate of MC at 700 and 900°C in a cold-rolled HT-UPS steel. Most of the growth occurred within the first 100 hours. After 3000 hours at 700°C, the MC particles were typically around 9 nm in size. They attributed the decrease in growth rate to a solute exhaustion mechanism and achievement of a stable, complex MC phase composition involving Ti, V, Nb, and Mo.

The variable stress and temperature testing outlined in this paper indicated that the microstructures of cold- or hot-worked HT-UPS steel remained stable for long periods of time. Hardening was observed upon stress and temperature increases during creep, most likely due to dislocation pinning interactions with the stable, fine precipitate dispersions. Relaxation testing indicated increasing strength with repeated loadings and short periods of creep or plasticity did not degrade the strength. It would seem, however, that the ability of MC to pin new dislocations produced by straining would have limits, especially if the solute exhaustion mechanism proposed by Todd and Ren holds true. The exhaustion of a hardening mechanism may account for the initiation of tertiary creep at low strain. Also, the recrystallization along grain boundaries should lead to more rapid creep and may be the degradation mechanism that leads to failure in materials with prior work. Although rupture occurred shortly after tertiary creep in the cold-worked alloys, the absence of creep cavitation led to a fairly high reduction of area in the necked region of specimens. The good local ductility should produce good creep-fatigue resistance. In the annealed material, which lacked the starting dislocation microstructure, the period of minimum creep rate was shorter than for cold-work material, and tertiary creep produced a lower creep life.

A number of challenging issues remain to be resolved before warm or cold-worked materials can confidently be brought into use for high-temperature applications. The ability to consistently produce homogenous work in components must be demonstrated. The ability to produce weldments that have matching strength must also be demonstrated. Material specifications that require high yield strengths must be developed. It is important to note, however, that new advanced steels can consistently show benefits from cold work and develop microstructures not manifested in more conventional steels.

## CONCLUSIONS

Microstructures in cold-worked HT-UPS stainless steel are stable for long times at temperatures to 700°C.

The rupture strength of cold-worked HT-UPS alloy is approximately 25% greater than that for annealed material.

Stress and temperature transients do not significantly degrade the strength of HT-UPS steel.

## ACKNOWLEDGEMENTS

This research was sponsored by the U.S. Department of Energy, Fossil Energy Advanced Research and Technology Development (AR&TD) Materials Program, DOE/FE AA 15 10 10 0, Work Breakdown Structure Element ORNL-2(B) at the Oak Ridge National Laboratory operated by MARTIN MARIETTA ENERGY SYSTEMS, INC. for the U.S. DEPARTMENT OF ENERGY under contract DE-AC05-84OR21400.

## REFERENCES


1- Tamura, M., et al., "Tempaloy 17-14CuMo/NAC CR35A Composite Tubing for the Superheaters of Advanced Coal Fired Boilers," Nippon Kokan Technical Report No. 49 (1987)
2- Townsend, R. D., "Advances in Material Technology for Fossil Power Plants," p. 11, ASM International, Metals Park, OH (1987)
3- Padilha, A. F., G. Scantz and K. Anderko, J. Nuclear Materials 105, 77-92 (1982)
4- Sawaragi, Y. and S. Hirano, "New Alloys for Pressure Vessels and Piping", p. 141, PVP-Vol. 210, Am. Soc. Mech. Engrs., New York (1990)
5- Takahashi, T., et al., in Proceedings on Second Int. Conf. on Improved Coal Fired Power Plants, EPRI publication GS-6422, Palo Alto, CA, 1989, p. 41-1.
6- Sawaragi, Y., H. Teranishi and K. Yoshikawa, in Proceedings Int. Conf. on Creep, Tokyo, Japan, 1986, p. 238-44.
7- Teranishi, H., et al., in Proceedings on Second Int. Conf. on Improved Coal Fired Power Plants, EPRI publication GS-6422, Palo Alto, CA, 1989, p. 33-1.
8- Maziasz, P. J., J. Metals 41, 14-22 (1989)
9- Kesternich, W. and D. Meertens, Acta Met. 34, 1071-82 (1986)
10- Kimura, H. and Y. Minami, in Proceedings Int. Conf. on Creep, Tokyo, Japan, 1986, p. 221-28.
11- Murray, J. D., J. Hacon and P. H. Wannell, "High Temperature Properties of Steels", p. 403, Pub. 97, Iron and Steel Inst., London (1967)
12- Todd, J. A., and J-C Ren, "New Alloys for Pressure Vessels and Piping", p. 189, PVP-Vol. 210, Am. Soc. Mech. Engrs., New York (1990)
13- Black, et al., "Collection and Uses for Relaxation Data in Design", p. 1, PVP-Vol. 172, Am. Soc. Mech. Engrs., New York (1989)
14- Maziasz, P. J., "Micon 86 Optimization of Processing, Properties, and Service Performance Through Microstructural Control", p. 116, STP 979, Am. Soc. Test. Matls, Philadelphia, Penn. (1988)
15- Maziasz, P. J., R. W. Swindeman, and J. A. Todd, "Properties of Stainless Steels in Elevated Temperature Service", p. 189, PVP-Vol. 132, Am. Soc. Mech. Engrs., New York (1987`)

*Proceedings of the Fifth International Conference on Creep of Materials, Lake Buena Vista, Florida, USA, 18-21 May 1992*

# Dependence of Irradiation Creep of Zircaloy-2 at 340 K on Microstructure

**R.J. Klassen, A.R. Causey, and N. Christodoulou**
Atomic Energy of Canada, Limited
Chalk River Laboratories
Chalk River, Ontario, Canada

## ABSTRACT

Irradiation induced creep and growth in non-cubic metals causes anisotropic dimensional changes that can limit the service life of nuclear reactor components made from these alloys. This paper deals with the effect of texture and microstructure on the creep properties of Zircaloy-2 at around 340 K. In-reactor creep tests were performed using a bent beam stress relaxation technique on specimens from extruded and cold-drawn tubes as well as annealled and cold-rolled strip, manufactured by 17 different procedures. The creep rate was found to increase with dislocation density. The analysis of the data was done using a self-consistent model that relates the behaviour of single crystals to that of polycrystals. The approach accounts for the intergranular stresses that are generated by anisotropic deformation of the individual grains in the polycrystal. The single crystal creep properties were derived using the measurements from one polycrystalline material. Then the single crystal values were used to predict the creep behaviour of the remaining 16 materials. The predicted creep compliances were similar to the measured values. We conclude that, at temperature of approximately 340 K, irradiation creep can be attributed to climb-assisted dislocation glide on prism, basal and pyramidal slip systems. The experimental findings are also consistent with a single crystal creep rate proportional to the square root of the dislocation density.

**IRRADIATION INDUCED CREEP OF ZIRCONIUM ALLOYS** has been observed during the operation of nuclear power reactors [1-7]. This phenomenon is of particlular interest in the CANada Deuterium Uranium (CANDU) reactor [6]. The CANDU reactor core contains a lattice of horizontal fuel channels in a calandria vessel containing a heavy water moderator. Each fuel channel has an outer tube, the calandria tube, which separates the hot inner pressure tube (containing the fuel and coolant) from the cold moderator. The calandria tube is fabricated from 1.5 mm thick Zircaloy-2 sheet. During reactor service the calandria tube operates at approximately 340 K, under low axial end loads as well as bending stresses. Under these conditions, deformation due to thermal creep is negligible, however, the calandria tube undergoes irradiation-induced creep as a result of exposure to neutron flux. The resulting dimensional changes (particularly sag) of the tube can be life limiting. It is, therefore, important from a design perspective to determine the effect of microstructure on the irradiation-induced creep in Zircaloy-2 at low temperatures.

In-reactor uniaxial creep tests of zirconium alloys [7-11] have shown that, over the stress range from 0 to about 150 MPa, the irradiation creep rate, in direction "d", is related to the applied stress, $\sigma_d$, and the fast neutron flux, $\phi$ by

$$\dot{\varepsilon}_d = C_d \, \phi \, \sigma_d \, . \tag{1}$$

Here $C_d$ is the creep compliance and is related to the material properties as well as the test temperature. The following expression for stress relaxation of a beam constrained to a constant radius of curvature can be obtained by integrating Hooke's law and applying Eq. (1) [10]

$$\sigma_d/\sigma_{od} = D \exp(-C_d E_d \phi t). \tag{2}$$

In this equation $\sigma_d$ and $\sigma_{od}$ are the stresses at time t and at time t = 0, D is an integration constant, $E_d$ is the Young's modulus and $\phi t$ is the fast neutron fluence at constant flux $\phi$. The creep compliance, $C_d$, is the parameter of interest from the stress relaxation tests since it reflects the steady state creep rate and is a function of the material properties, i.e. crystallographic texture and dislocation density.

In-reactor experiments indicate that the creep behaviour of zirconium alloys is anisotropic due to the crystal structure and the pronounced crystallographic texture in the wrought materials. In a polycrystal the deformation of each grain is constrained by the collective deformation of its neighbouring grains. Hence, inhomogeneous elastic and inelastic deformation must occur to maintain the continuity of the polycrystal. The experimentally measured irradiation creep behaviour of the material is the result of the collective deformation of the individual grains.

This paper describes an investigation of the effect of microstructure on the irradiation creep of Zircaloy-2 at 340K. In-reactor stress relaxation tests were conducted on materials produced by various fabrication routes including five materials previously reported in reference [9], which were given more irradiation time. These results, along with dislocation density and crystallographic texture measurements, were analyzed to derive single crystal creep parameters in a self-consistent model which were then used to predict the variation in creep properties in polycrystalline materials resulting from different microstructure.

## EXPERIMENTAL PROCEDURE

**Materials.** The 17 materials tested, all Zircaloy-2 (Table 1), are listed in Table 2. They are sub-divided into two catagories (i) rolled strip and (ii) extruded tube.

**Table 1** Typical chemical composition of Zircaloy-2.

| Alloying Element | wt% |
|---|---|
| Sn | 1.2-1.7 |
| Fe | 0.07-0.20 |
| Cr | 0.05-0.15 |
| Ni | 0.03-0.08 |
| Zr | Balance |

The rolled strip was fabricated by forging ingots at 1300 K to rectangular billets 100 mm thickness. The CW 50 and CJ strips were cold-rolled by 50 and 40 percent respectively from the billet stage. The remaining rolled strip materials were reduced to 3.5 mm thickness by hot rolling at 1060 K and further reduced to 1.4 mm thickness by cold-rolling. Periodically during and after cold-rolling, the strip was annealed in vacuum at 1005 K.

Variations in the fabrication route which resulted in differences in dislocation densities, crystallographic textures, or grain sizes, are listed in Table 2. Specimens cut from fabricated CANDU calandria tubes are included as rolled materials. The calandria tubes were fabricated from Zircaloy-2 sheet by brake forming, welding and annealing. A final sizing introduced small amounts of cold-work. The microstructure of the rolled material consisted of large, equiaxed, hcp α-Zr grains.

The extruded tubes were manufactured from forged billets and extruded at approximately 1073 K. Several extruded materials in Table 2 were given various amounts of cold-drawing while the remainder were in the as-extruded condition. The extruded material had grains elongated in the direction of extrusion.

**Table 2** List of materials used in this investigation.

| Specimen Number | Comments |
|---|---|
| Rolled Strip | |
| CW | 1% Cold rolled |
| CJ | 40% cold rolled |
| CW 50 | 50% cold rolled |
| 20 SR[†] | Reject calandria tube, ≈ 0.5 % cold work |
| 20 | |
| 1939 SR | Reject calandria tube, ≈ 0.5 % cold work |
| 1939 | |
| 758 SR | Reject calandria tube, ≈ 0.5 % cold work. |
| 758 | |
| 767 | Reject calandria tube, annealed |
| 55 | Reject calandria tube, ≈ 0.5 % cold work |
| 55 CW[‡] | Reject calandria tube, ≈ 2.0 % cold work |
| Extruded Tubing | |
| 390 | Extruded and cold drawn ≈ 20% |
| 485 | Extruded and cold drawn ≈ 15% |
| 501 | Extruded and cold drawn ≈ 15% |
| FT | As extruded |
| RT | As extruded |

[†] SR 24 hr anneal in vacuum at 400°C
[‡] CW Specimens cold-worked about 2% by rolling

**Microstructural Characterization.** The crystallographic texture and the dislocation density were measured by x-ray diffraction techniques [11]. The textures are expressed in terms of the resolved fraction of basal pole normals in the longitudinal, transverse, and radial (or normal) working directions $F_L$, $F_T$, and $F_R$ [5].

Typical basal pole figures are shown in Figure 1. Figure 1(a) is the pole figure of calandria tube material 767 and is a typical texture of the rolled sheet. Figure 1(b) is the pole figure of material 390 and is typical of the extruded materials.

The dislocation density was estimated by measuring the amount of line broadening of the $(10\bar{1}0)$ and the (0002) reflections which gives the <a> and the <c> components of the total dislocation density respectively. Table 3 lists the measured <a> and <c> component dislocation densities, $\rho_{<a>}$ and $\rho_{<c>}$, as well as the resolved fractions of basal pole normals.

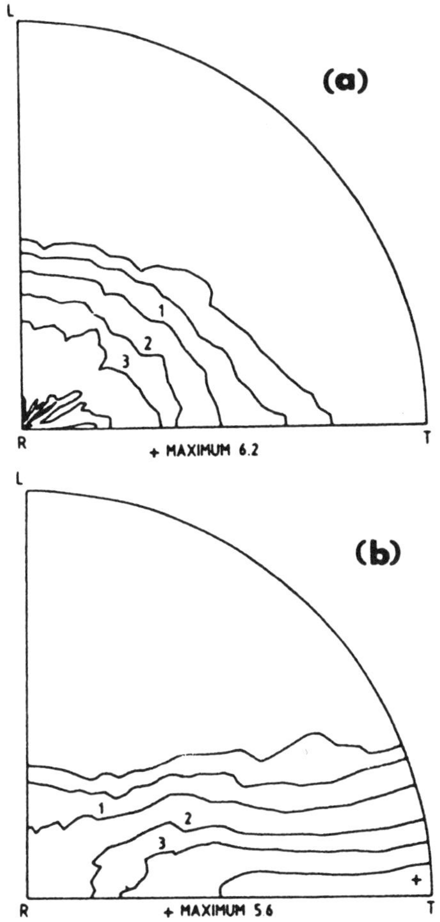


Fig. 1 Basal pole figures of specimens: a) rolled strip 767, and b) extruded tube 390.

**Stress Relaxation Testing.** Rectangular beams, 75 x 3.75 x 0.75 mm, were prepared with the long dimension either parallel or normal to the working direction. The CW 50 strip was of sufficient thickness to cut beams from the direction normal to the rolling plane in addition to the longitudinal and transverse directions. Additional specimens were also cut from the CJ strip with their length 45° from the normal direction in the normal-longitudinal plane.

Tests were performed in the National Reactor Universal (NRU) at CRL in flowing water at approximately 340 K and a fast neutron flux between 1.2 and 1.8 x $10^{17}$ n/m$^2$/s (E > 1 MeV) [9]. The specimens were bent in jigs to fixed curvatures which gave initial outer fibre stresses up to 200 MPa. At intervals of approximately 3000 hrs of irradiation, the specimens were removed from the reactor and from the jigs and the unconstrained radius of curvature, R, was measured. The amount of stress relaxation, as given by the unrelaxed stress ratio, was determined by

$$\sigma/\sigma_o = (1/R - 1/R_h) \,/\, (1/R_o - 1/R_h) \qquad (3)$$

where $R_o$ is the initial unconstrained specimen radius and $R_h$ is the radius of the specimen holder. Typically 3 to 6 stress relaxation specimens from each material were tested in each direction.

**Table 3** Resolved basal pole fractions and dislocation densities of the Zircaloy-2 materials.

| Specimen Number | Texture | | | Dislocation Density ($x10^{-14}$ m$^{-2}$) | |
|---|---|---|---|---|---|
| | $F_T$ | $F_R$ | $F_L$ | $\rho_{<a>}$ | $\rho_{<c>}$ |
| Rolled Strip | | | | | |
| CW | 0.24 | 0.63 | 0.13 | 0.04 | 0.01 |
| CJ | 0.27 | 0.64 | 0.09 | 1.96 | 0.69 |
| CW 50 | 0.32 | 0.59 | 0.09 | 1.92 | 0.80 |
| 20 SR | 0.33 | 0.59 | 0.08 | 0.12 | 0.01 |
| 20 | | | | 0.13 | 0.01 |
| 1939 SR | 0.16 | 0.72 | 0.12 | 0.43 | 0.06 |
| 1939 | | | | 0.70 | 0.12 |
| 758 SR | 0.24 | 0.66 | 0.10 | 0.28 | 0.03 |
| 758 | | | | 0.77 | 0.10 |
| 767 | 0.17 | 0.72 | 0.11 | 0.14 | 0.02 |
| 55 | 0.20 | 0.69 | 0.11 | 0.12 | 0.02 |
| 55 CW | 0.18 | 0.69 | 0.12 | 1.17 | 0.16 |
| Extruded Tubing | | | | | |
| 390 | 0.59 | 0.32 | 0.09 | 2.64 | 0.89 |
| 485 | 0.44 | 0.49 | 0.08 | 2.61 | 0.52 |
| 501 | 0.54 | 0.40 | 0.06 | 2.29 | 0.45 |
| FT | 0.45 | 0.47 | 0.08 | 0.24 | 0.02 |
| RT | 0.42 | 0.48 | 0.11 | 0.18 | 0.13 |

## EXPERIMENTAL RESULTS

Figures 2 and 3 show plots of $\ln(\sigma/\sigma_o)$ versus $\phi$t for the calandria tube 767 and the extruded tube 390. These plots are typical of the observed stress relaxation behaviour.

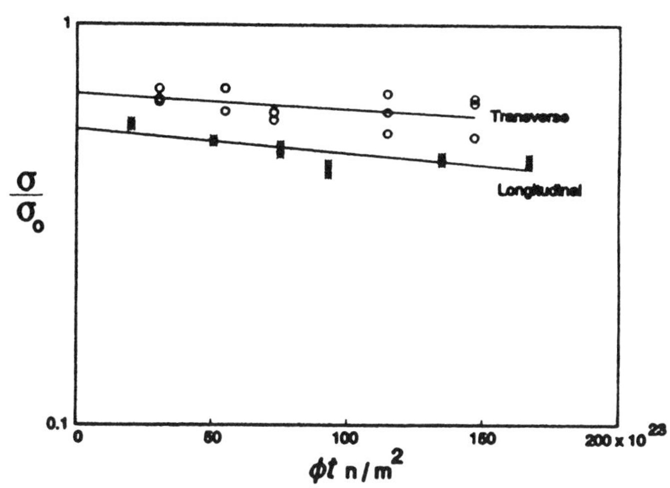


Fig. 2 Variation in stress ratio with fluence for calandria tube material 767.

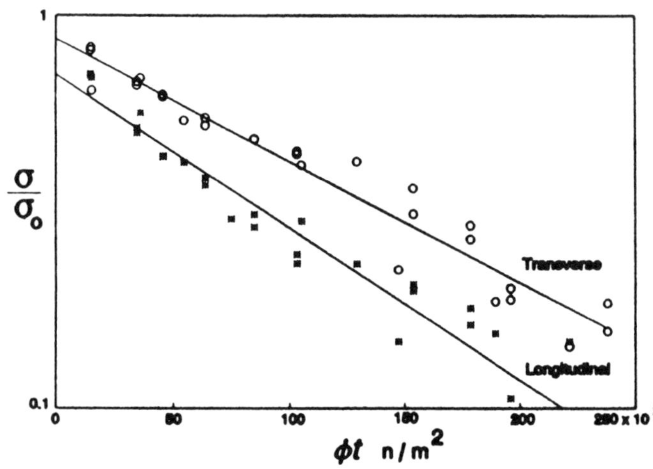


Fig. 3 Variation in stress ratio with fluence for extruded tube 390.

The plots illustrate the linear dependence of $\ln(\sigma/\sigma_o)$ with $\phi t$ over most of the irradiation. The slope, $G_d = C_d E_d$, of the linear region of the stress relaxation plot represents the steady state relaxation rate.

Figure 4 shows plots of $\ln(\sigma/\sigma_o)$ versus $\phi t$ for specimens from the CW 50 strip in the longitudinal, transverse, and normal orientations. The creep compliances calculated from these data were used to derive the single crystal creep compliances using the self-consistent creep model as discussed below.

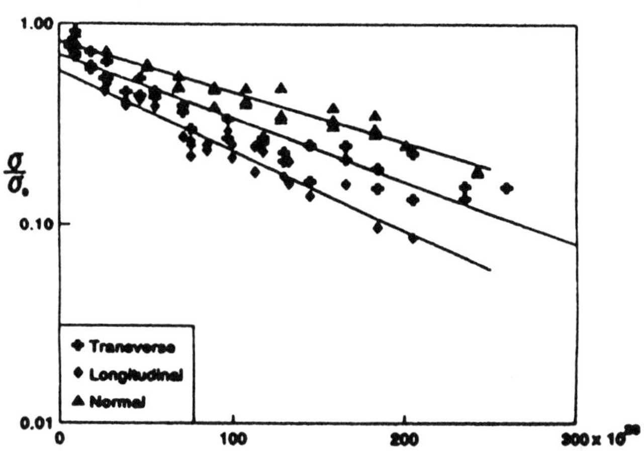


Fig.4 Variation in stress ratio with fluence for CW 50 specimens in the longitudinal, transverse and normal directions.

The results in Table 4 were determined from experiments over a longer irradiation time than the data in [9]. This improves the accuracy of the results since more measurements were made in the steady state creep stage of the tests. Nevertheless, the relaxation rates of the materials tested are comparable to values reported previously for Zircaloy-2 materials [9]. The materials showed a trend towards increased relaxation rates with increasing dislocation density and higher rates in directions parallel with the working direction than normal to it. Table 4 gives Young's Moduli, $E_d$, the measured relaxation rates, $G_d$, and the creep compliances, $C_d$.

**Table 4** Young's modulus, Relaxation Rate and Creep Compliance for the Zircaloy-2 specimens.

| Specimen Number | Young's Modulus (GPa) | | Rel'n Rate $10^{26}$ $(n/m^2)$-1 ± 95% C. L | | Creep Compliance $x10^{31}(n/m^2)^{-1}MPa^{-1}$ | |
|---|---|---|---|---|---|---|
| | $E_L$ | $E_T$ | $G_L$ | $G_T$ | $C_L$ | $C_T$ |
| Rolled Strip | | | | | | |
| CW | 94.4 | | 1.6±0.3 | | 1.7 | |
| CJ | 93.7 | 93.5 | 7.6±0.5 | 5.8±0.6 | 8.1 | 6.2 |
| | | 93.6† | | 5.0†±0.3 | | 5.3† |
| CW 50 | 94.4 | 94.8 | 9.1±1.1 | 7.2±1.2 | 9.6 | 7.6 |
| | | 94.6‡ | | 6.0‡±0.8 | | 6.3‡ |
| 20 SR | 94.1 | 94.8 | 1.5±0.7 | 1.7±0.3 | 1.6 | 1.8 |
| 20 | 94.1 | 94.8 | 3.1±0.5 | 1.8±0.7 | 3.3 | 1.9 |
| 1939 SR | 92.9 | 92.9 | 2.3±0.7 | 2.3±0.3 | 2.5 | 2.5 |
| 1939 | 92.9 | 92.9 | 5.2±1.5 | 4.3±1.2 | 5.6 | 4.6 |
| 758 SR | 93.2 | 94.3 | 2.5±0.5 | 2.7±0.5 | 2.7 | 2.8 |
| 758 | 93.2 | 94.3 | 4.6±1.1 | 4.6±1.7 | 4.9 | 4.9 |
| 767 | 93.1 | 92.9 | 1.4±0.6 | 0.9±1.0 | 1.5 | 1.0 |
| 55 | 93.2 | 93.1 | 1.2±0.3 | 0.7±0.7 | 1.3 | 0.8 |
| 55 CW | 93.4 | 93.1 | 1.9±1.4 | 3.7±0.5 | 2.0 | 4.0 |
| Extruded Tubing | | | | | | |
| 390 | 94.6 | 98.0 | 9.0±0.9 | 7.2±0.7 | 9.5 | 7.3 |
| 485 | 94.4 | 98.3 | 10.6±1.2 | 8.0±0.3 | 11.2 | 8.1 |
| 501 | 94.5 | 98.1 | 9.4±1.2 | 6.7±0.6 | 9.9 | 6.8 |
| FT | 94.5 | 96.3 | 3.3±0.7 | 3.9±0.7 | 3.5 | 4.1 |
| RT | 94.0 | | 4.1±0.9 | | 4.4 | |

† Normal orientation
‡ ST 45° to the Normal direction in the Transverse plane

Figure 5 shows a logarithmic plot of creep compliance derived from the relaxation rates versus the total dislocation density $\rho = \rho_{<a>} + \rho_{<c>}$ for the longitudinal and transverse specimens. A general increase in creep rate with dislocation density is shown in this figure. A weighted linear curve fit through the data was performed. The relaxation rates were weighted according to their relative measurement error, $\Delta C/C$. Therefore the material with low relaxation rates, where increased scatter in the data is observed, had less weight in the regression than the high relaxation rate data.

The following relationships for the longitudinal and transverse creep compliances as a function of ρ were calculated from the linear regression.

$$C_L = 1.45 \times 10^{-37}\ \rho^{0.47\pm0.05} \quad (4a)$$

$$C_T = 3.82 \times 10^{-37}\ \rho^{0.43\pm0.05} \quad (4b)$$

Typically the ratio $C_L/C_T$ is greater than 1 and increases with increasing dislocation density. This is evident in Figure 5 particularly at the high dislocation densities where the precision in the stress relaxation measurements is improved. In this region the data were mostly from the extruded tubing, which had a more transverse texture than the cold rolled sheet material (Figure 1 and Table 4). The trend in Figure 5 therefore illustrates the influence of both dislocation density and crystallographic texture on the irradiation creep compliance of Zircaloy-2.

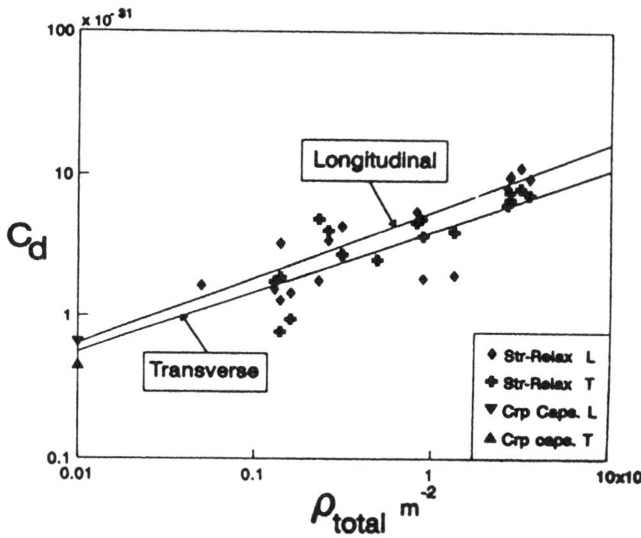


Fig. 5 Variation in creep compliance with total dislocation density.

## ANALYSIS AND DISCUSSION

The relationship between the grain-to-grain interaction and the resulting polycrystalline deformation has been recently treated by Woo and co-workers using a self-consistent model [12-16]. In this approach, each grain is treated as an inhomogeneous ellipsoidal inclusion embedded in a continium with the average elastic and irradiation deformation properties of the polycrystal. The intergranular compatibility condition is then described by the interaction between the inclusion and the continuum, taking into account the elastic and creep properties of the matrix and the inclusion.

The computer code SELF-POLY was developed [15] to predict the creep compliance tensor of the textured polycrystal when the single crystal creep properties are given. The steady-state strain rate under irradiation is given by:

$$\dot{E}_{ij} = K_{ijkl}\,\Sigma_{kl} + \dot{\Gamma}_{ij} \quad (5)$$

where **K** is the creep compliance tensor of the polycrystal, in units of $MPa^{-1}.h^{-1}$, and Σ and $\dot{\Gamma}$ are the external stress and the irradiation growth rate tensors respectively. The tensors **K** and $\dot{\Gamma}$ are given by suitable weighted averages of the corresponding properties of the grains and as a consequence they reflect the symmetry of the polycrystal. It has been shown [12-14] that **K** is completely defined if its six diagonal elements are specified. Eq. (5) implies that there is a linear relationship between the creep rate and stress, an assumption which is reasonable for the range of stresses in this study. An equation similar to Eq. (5) describes the creep and growth behaviour of the single crystal, namely:

$$\dot{\varepsilon}_{ij} = k_{ijkl}\,\sigma_{kl} + \dot{\gamma}_{ij}\ . \quad (6)$$

Due to the hexagonal symmetry of the lattice the single crystal creep compliance tensor **k** can be completely defined if its three eigenvalues $k_1$, $k_2$ and $k_3$ are determined. It has been shown earlier [14,16] that the $k_i$'s are functions of the dislocation density. The explicit form of the function **k**(ρ) depends on the creep mechanism. Woo et al [14] have shown that for a climb-assisted glide mechanism the single crystal creep compliances can be correlated with the slip systems that operate during irradiation. In particular, pyramidal glide contributes exclusively to $k_1$ and to a much lesser extent to $k_2$ and $k_3$. Prismatic glide contributes mainly to $k_2$ while basal slip contributes mainly to $k_3$. The code predicts **K** if the single crystal creep compliance tensor **k** and the texture of the polycrystal are given.

SELF-POLY has been modified [16] so that **k** is calculated provided that any three of the six compliances defining **K** and the texture are given. This technique was employed here. The single crystal creep compliance tensor $\mathbf{k}^{Ref}$ was determined using the CW 50 strip (Table 2). The behaviour of the rest of the materials was predicted using $\mathbf{k}^{Ref}$. In order to quantitatively predict the behaviour of the rest of the materials, the dependence of **k** on the dislocation density must also be known. If a climb-assisted glide creep mechanism (i.e. the I-Creep mechanism [14,16]) is assumed, then $\mathbf{k} \propto \sqrt{\rho}$. It follows that **K** will exhibit a similar dependence on dislocation density, as shown approximately in Figure 5. Based upon this dependence, the eigenvalues for the rest of the materials were calculated by means of:

$$k_i = k_i^{Ref}\,(\rho/\rho_{Ref})^{1/2} \quad (7)$$

where $\rho_{Ref}$ is the dislocation density of the CW 50. The single crystal compliances derived for the reference material are $k_1^{Ref} = 0.62$, $k_2^{Ref} = 1.11$ and $k_3^{Ref} = 0.32 \times 10^{-9}\ MPa^{-1}h^{-1}$. From the magnitude of the single crystal creep compliances and assuming that the strain producing mechanism is climb-assisted glide the following observations can be made. The relative contributions of pyramidal, basal and prism glide to the total creep strain can be estimated to be about 11% pyramidal, 12% basal and 77% prism slip in the case of Zircaloy-2 materials tested at 340 K, whereas in the case of Zr-2.5wt%Nb, tested at 570 K [16], the corresponding contributions are 4%, 0% and 96% respectively.

A ratio of $C_L/C_T$ higher than 1 is also predicted by the self-consistent model employed here with $C_L/C_T$ in the range $0.01 \times 10^{14} < \rho < 10 \times 10^{14}\ m^{-2}$ is between 1.15 and 1.51.

In order to check these observations based on the experimental measurements, the creep compliances of all the materials listed in Table 3 were calculated using the SELF-POLY model. Figure 6 shows a plot of predicted versus measured creep compliances confirming that the self-consistent approach combined with the assumption of I-creep does predict satisfactorily the creep behaviour of Zircaloy-2 materials irradiated at calandria tube temperatures.

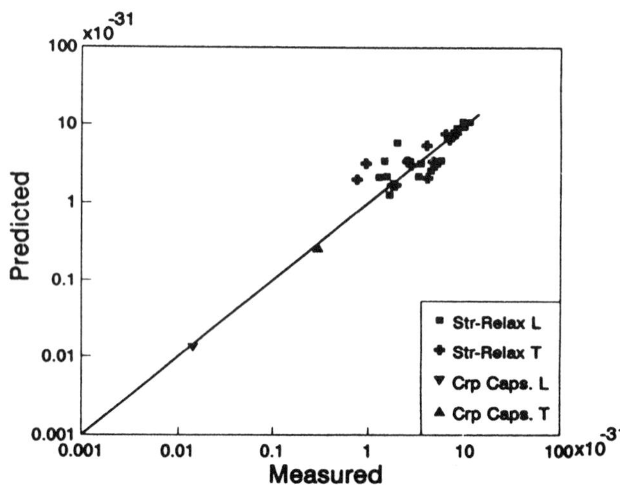


Fig. 6 Comparison of the measured creep compliance with the predicted compliance.

Results from an in-reactor biaxial creep experiment [17] using internally-pressurized capsules of annealed Zircaloy-4 tubing are of interest for testing the model since both the microstructure and stress system are quite different from those of the stress relaxation tests. Longitudinal and transverse creep compliances were derived from the measured diametral and axial strain rates of tubes with irradiated at 330 K in a fast neutron flux of $1.8 \times 10^{18}\ n/m^2/s$ and hoop stresses ranging from 0 to 200 MPa. The texture ($F_R$, $F_T$ and $F_L$ of 0.40, 0.49 and 0.11, respectively) and total dislocation density ($\rho \approx 1 \times 10^{12}\ m^{-2}$) were used in the SELF-POLY model to predict the creep compliance of the capsules. The correlation between the measured and predicted compliances for the capsules is included in Figure 6 and indicates that the model predicts well the creep behaviour of the material subjected to biaxial stress. In Figure 5 the predicted uniaxial compliances of the Zircaloy-4 tubing (calculated by the SELF-POLY model from the measured biaxial compliances and the texture and dislocation density) fall on the trend predicted by the stress relaxation data. This provides good support both for the use of stress relaxation data to study creep behaviour and for the creep-dislocation density correlations and the self-consistent model used here.

## SUMMARY

The microstructural dependence of irradiation creep of Zircaloy-2 at 340 K was investigated through in-reactor stress relaxation tests of materials manufactured by various thermo-mechanical routes. The creep rates were found to be approximately proportional to the square root of the dislocation density. The rates were also higher in the working direction than normal to it.

The irradiation creep behaviour of Zircaloy-2 was modelled with a self-consistent analysis that describes the polycrystal behaviour from single crystal properties. Single crystal creep compliance tensors were derived by using the polycrystalline creep compliances of a reference material and a linear relationship between the compliance and the square root of the dislocation density. We conclude that the creep strain is accommodated by climb-assisted glide mainly on prism slip systems with smaller contributions from basal and pyramidal slip systems.

The measured creep compliances for all the materials confirm that the self-consistent approach combined with the assumption of I-creep does predict satisfactorily the creep behaviour of Zircaloy-2 materials irradiated at calandria tube temperatures.

## ACKNOWLEDGEMENTS

The authors acknowledge F.J. Butcher, S.A. Donohue, and A.K. MacCormack for carrying out the stress relaxation tests, J.E. Winegar for providing the X-Ray diffraction measurements, R.A. Holt for helpful comments on the manuscript and the CANDU Owners Group for finacial support.

## REFERENCES

1) Zirconium in the Nuclear Industry, 7th Int. Symp., ASTM STP 939 (1987).

2) Dimensional Stability and Mechanical Behaviour of Irradiated Metals and Alloy, Proc. BNES Conf., London, 1983.

3) D.G. Franklin, G.E. Lucas, and A.L. Bement, ASTM STP 955 (1983).

4) V. Fidleris, At. Energy Rev., 13 (1975) 51.

5) W.K. Alexander, V. Fidleris, and R.A. Holt, ASTM STP 633 (1977) 344.

6) A.R. Causey, S.R. MacEwen, H.C. Jamieson, and A.B. Mitchell, Res. Mech. 2 (1981) 211.

7) P.A. Ross-Ross and C.E.L. Hunt, J. Nucl. Mater. 26 (1968) 2.

8) C.E. Coleman, A.R. Causey, and V. Fidleris, J. Nucl. Mater. 60 (1976) 185.

9) A.R. Causey, J. Nucl.Mater., 98 (1981) 313.

10) D.E. Fraser, P.A. Ross-Ross, and A.R. Causey, J. Nucl. Mater., 46 (1973) 281.

11) M. Griffiths and J.E. Winegar, Mat. Res. Symp. Proc., 209 (1991) 83.

12) C.H. Woo, J. Nuc. Mater., 131 (1985) 105.

13) C.H. Woo, "Materials for Nuclear Reactor Core Applications", British Nuclear Energy Society, (1987) 65.

14) C.H. Woo, A.R. Causey and R.A. Holt, invited paper in "Diffusion and Defect Data - Solid State Data", 1990, in press.

15) C.N. Tome, C.B. So and C.H. Woo, "Self-Consistent Calculation of Steady-State Creep and Growth in Textured Zirconium", to be published.

16) N. Christodoulou, C.H. Woo, A.R. Causey, C.N. Tome and R.A. Holt, "Derivation of Single Crystal Creep Properties from the Creep Behaviour of Polycrystalline Zirconium Alloys", to be presented at the 13$^{th}$ Riso Int. Symp. on Materials, September 1992, Roskilde, Denmark.

17) R.J. Klassen and A.R. Causey, to be submitted for publication.

*Proceedings of the Fifth International Conference on Creep of Materials, Lake Buena Vista, Florida, USA, 18-21 May 1992*

# Effect of Microstructural Parameters on the Creep Rupture Properties of Nickel-Base Superalloys

**M. Yoshiba**
Department of Precision Engineering
Tokyo Metropolitan University
Tokyo, Japan

ABSTRACT

In order to clarify an effect of the microstructural factors associated with a grain boundary such as a grain size and a grain boundary nature on the creep and creep rupture properties of precipitation-strengthening nickel-base superalloy, the creep rupture tests were carried out at 800 ℃ using two charges of Inconel 751 alloy specimens subjected to the various processes of heat treatment. The creep strength associated with the steady-state creep rate was found to be uniquely determined by a grain size dominating over the other microstructural factors such as a grain boundary nature. On the contrary, the creep rupture strength consisting of both the creep deformation and fracture processes was found to be affected by both a grain size and a grain boundary nature. In particular, the effect of grain boundary nature becomes more significant in the coarse grain size regime, in which a detrimental and a beneficial effects on the rupture properties were observed for a straight and a zigzag grain boundaries, respectively. The major microstructural factors affecting the creep and creep rupture properties were discussed in connection with the effects of the heat treatment and a carbon content in the alloy.

IT is well known that the microstructures associated with a grain boundary are very important factor determining the creep rupture properties of many kinds of heat-resisting alloys since a creep rupture takes place in many case at the grain boundary. Both a grain boundary nature and a grain size are of representative for such dominant factors. For the former, in particular, it was revealed by the authors research group that a zigzag grain boundary with the platelike coarse carbide precipitates is much effective for improving the creep rupture strength of both heat-resisting steels and nickel-base superalloys precipitation-strengthened together with an increased rupture ductility, mainly by virtue of the suppression of a grain boundary sliding (1-3).

On the other hand, it is very difficult to determine sololy a grain size dependence of the creep rupture properties for the precipitation-strengthening alloys, because the heat treatments adopted for controlling a grain size are capable of altering easily the other microstructures. Thus, for the precipitation-strengthened nickel-base superalloys no grain size dependence of the creep rupture properties is fully clarified as compared with that for the other metals and alloys with relatively lower strength by only the solid-solution strengthening such as stainless steels (4,5).

In the present study, a grain size was controlled for two charges of the nickel-base superalloys over a wide range in connection with a grain boundary nature by applying various processes of the sophisticated heat treatments, and the effects of such the microstructural parameters associated with a grain boundary on the creep rupture properties consisting of both the creep deformation and creep fracture processes were comprehensively investigated.

## EXPERIMENTAL

MATERIALS AND HEAT TREATMENTS. Two charges of the $\gamma'$ precipitation-strengthening nickel-base superalloys Inconel 751 with different

Table I. Chemical compositions of two charges of Inconel 751 alloys (mass %).

| Symbol | C | Mn | Si | S | Ni | Cr | Ti | Al | Fe | Cu | Nb+Ta |
|---|---|---|---|---|---|---|---|---|---|---|---|
| HC | 0.09 | 0.20 | 0.53 | 0.007 | Bal. | 16.07 | 2.05 | 1.10 | 5.74 | 0.05 | 1.18 |
| LC | 0.04 | 0.17 | 0.17 | 0.001 | Bal. | 15.42 | 2.14 | 1.03 | 7.87 | 0.11 | 0.99 |

carbon contents were used in this study. These alloys chemistry is listed in Table I. Hereafter, different charges of these two alloys will be termed as high carbon (HC) and low carbon (LC) alloys, corresponding to their carbon contents.

In order to control systematically both a grain size and a grain boundary nature, various processes of heat treatments were adopted on the round alloy rods with 14 mm dia.. Heat treatment diagrams are shown in Figure 1. Here, each notation denotes as follows:

O-S: one-stage solution and simple aging treatments
T-S: two-stages solution and simple aging treatments
O-D: one-stage solution and direct aging treatments
T-D: two-stages solution and direct aging treatments

A grain size was determined by the linear intercept method, and it varied in the range 57-285 $\mu$m in average diameter, depending mainly on the solid-solution temperatures.

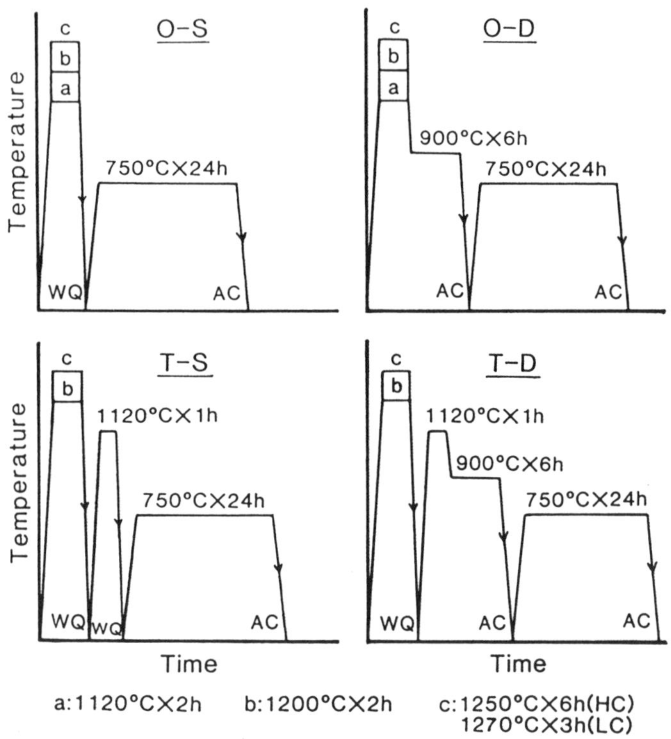


Fig.1 - Schematic diagrams of various heat treatments adopted.

Furthermore, the simple aging treatment for O-S and T-S resulted in the straight grain boundary with a great number of fine $M_{23}C_6$ carbide precipitates, while the direct aging treatment for O-D and T-D developed the zigzag grain boundary with the coarse Ti- and Nb-rich MC carbide precipitates (1,6). The second stage of solution treatment in the two-stages solution for T-S and T-D was additionally conducted at 1120 ℃; which appears the lowest temperature almost soluble both the $M_{23}C_6$ grain boundary carbides and the $\gamma'$ particles in a matrix, for the purpose of ensuring the same solid-solution condition prior to the subsequent aging treatments for all the specimens subjected to the first stage of solution treatment at different temperatures.

EXPERIMENTAL PROCEDURES. Creep rupture specimens were machined from the heat-treated rods to the smooth bar specimens with 5 mm in dia. and 30 mm in gage length. Constant load creep rupture tests were carried out at 800 ℃ in air under stresses in the range 150-300 N/mm$^2$, using the single lever creep testing machines. Temperature was controlled within ±2 ℃ over the gage length. Creep strain was measured by using mainly a linear variable differential transformer (LVDT) with a sensitivity of 1 $\mu$m and a stroke length of 4 mm; to obtain mainly the steady-state creep rate, together with a conventional dial gage with a minimum scale of 0.01 mm; to obtain an entire creep curve.

Metallographic examinations were made on the longitudinal sections of the specimens after heat treatments and creep rupture tests by means of the optical and scanning electron microscopies.

## RESULTS

CREEP CURVES. Typical creep curves under 250 N/mm$^2$ at 800℃ for different grain sizes of HC alloy are shown in Fig. 2. The pronounced primary, secondary and tertiary stages are distinguishable in all the creep curves, although the tertiary stage always occupies the major part of a creep strain. A grain size dependence of the creep rupture life is found to be considerably different between simple aging and direct aging; for simple aging it becomes maximum in the medium grain size of approximately 150 $\mu$m, while for direct aging it increases simply as a grain size increases. The creep rupture ductility also shows an opposed grain size dependence between different aging treatments; a grain size coarsening results in a rupture ductility decreased for simple aging but increased for direct aging. Furthermore, an introduction of the two-stages solution treatment is found to bring about a different effect on both the rupture life and ductility between simple aging and direct aging.

CREEP AND CREEP RUPTURE PROPERTIES. Figure 3 shows a relationship between the steady-state creep rate $\dot{\varepsilon}_s$ and the applied stress $\sigma$ for different grain sizes of HC alloy. For both simple aging and direct aging, the stress dependence of $\dot{\varepsilon}_s$ is found to be represented as the following power relation:

$$\dot{\varepsilon}_s = A\sigma^n$$

where A is a constant and n is the stress index. The values of n for simple aging and direct aging are in the range 4 ~7 and 5 ~8, respectively. These results are in agreement with the result by Sidey and Wilshire (7), who have reported the n value 7.4 for the creep

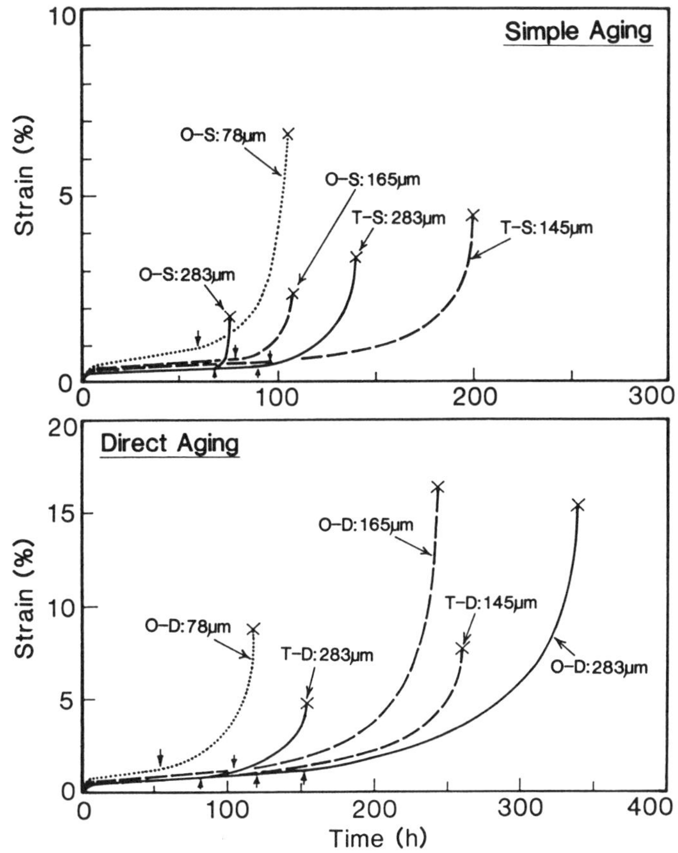


Fig.2 - Typical creep curves at 800 ℃ of the variously heat-treated HC alloy specimens.

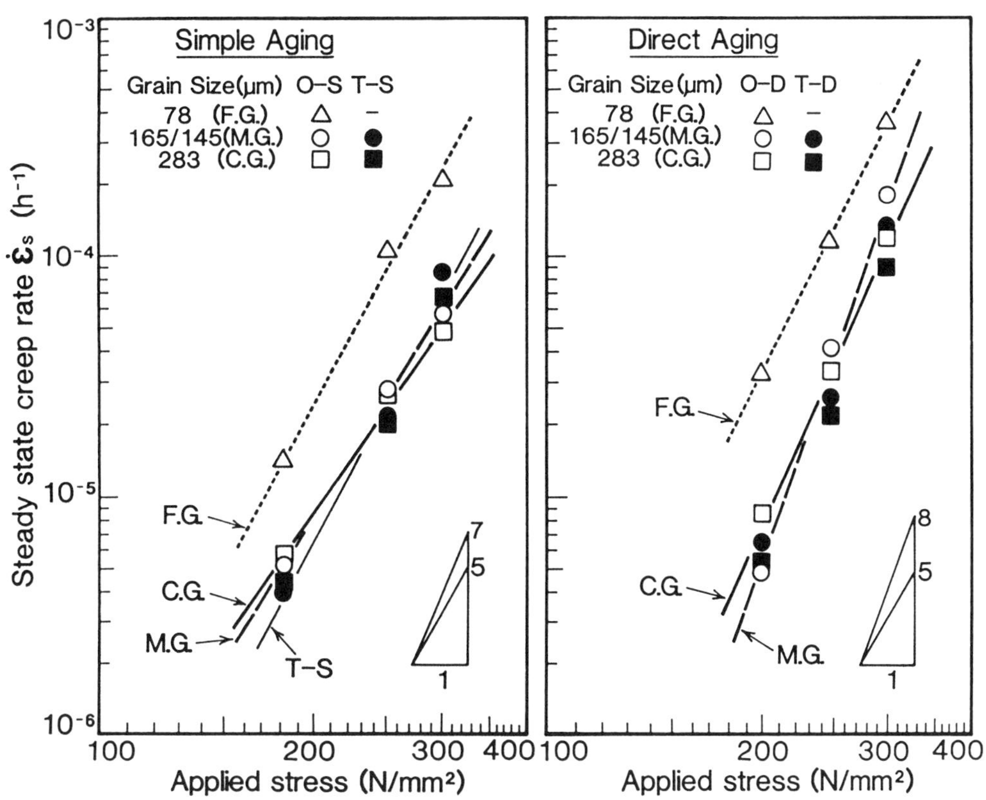


Fig.3 - Relationship between the steady-state creep rate and applied stress for the variously heat-treated HC alloy specimens.

test of Nimonic 80A at 750℃, and suggest for the creep deformation to be controlled by a recovery process such as a dislocation climb, regardless of a grain size. Furthermore, there is little difference in the stress dependence of $\dot{\varepsilon}_s$ between the one-stage and two-stages solution treatments.

The grain size dependence of both the creep rupture life and the steady-state creep rate for all the alloy specimens tested are

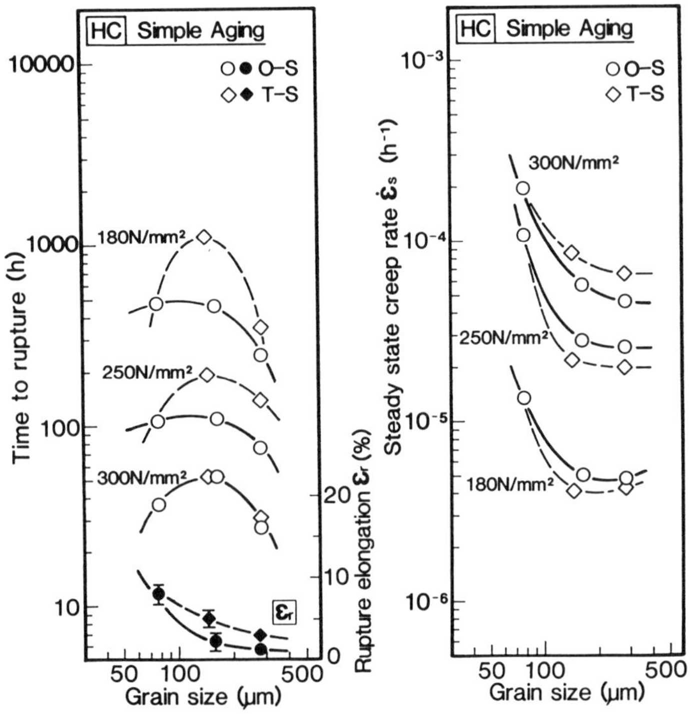


Fig.4 - Grain size dependence of rupture life and creep rate for the simply aged HC alloy specimens.

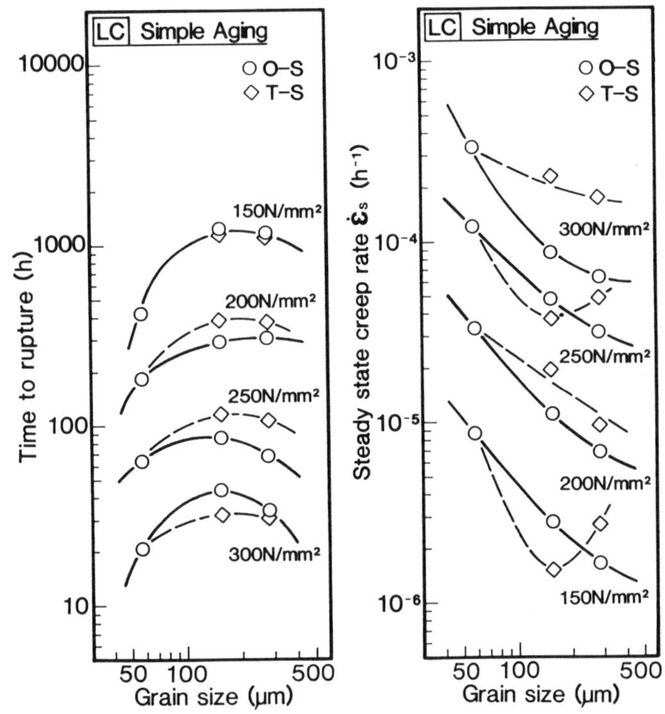


Fig.6 - Grain size dependence of rupture life and creep rate for the simply aged LC alloy specimens.

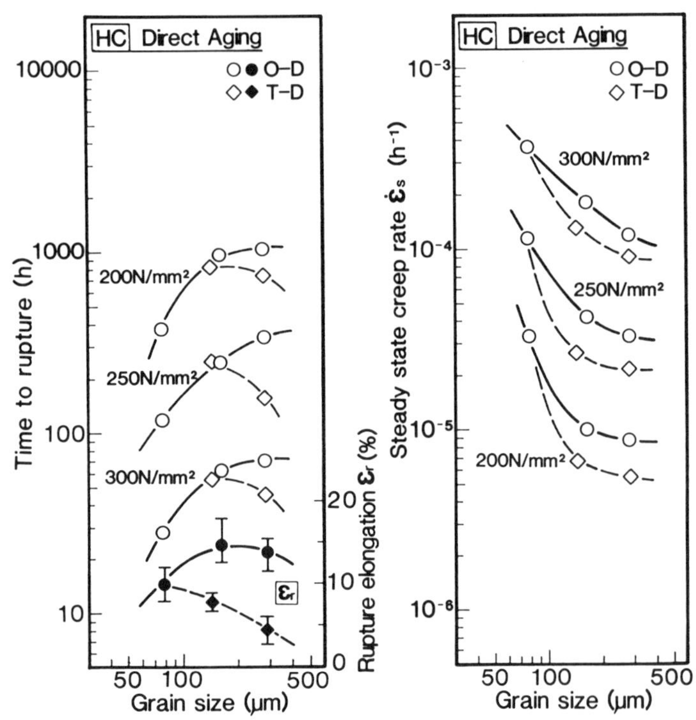


Fig.5 - Grain size dependence of rupture life and creep rate for the directly aged HC alloy specimens.

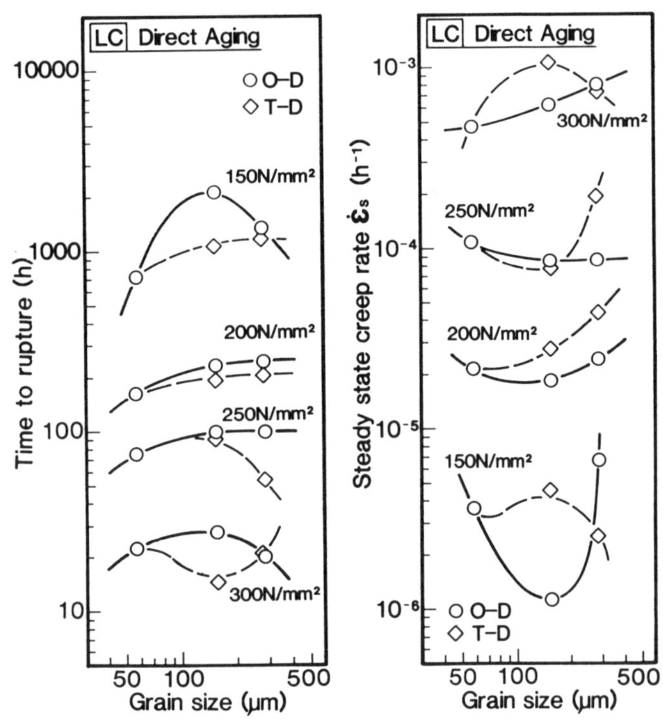


Fig.7 - Grain size dependence of rupture life and creep rate for the directly aged LC alloy specimens.

summarized in Figs. 4 ~7. For HC alloy, the rupture elongation also was additionally shown in Figs. 4 and 5. It can be seen that a grain size dependence of the rupture life differs considerably between simple aging and direct aging. Then, there is an optimum grain size to give the maximum rupture time for simple aging; in the medium grain size of about 150 μm. For direct aging, on the contrary, the rupture life tends to increase simply as a grain size increases in the case of O-D treatment, while for T-D treatment it has the same grain size dependence as simple aging. The creep rupture ductility tends to decrease simply as a grain size increases except for O-D treatment. On the other hand, the steady-state creep rate is found to have an unique grain size dependence regardless of different aging teatments; it decreases and tends to become constant as a grain size increases, although there is some exception for LC alloy. Such a grain size dependence of the steady-state creep rate is in agreement with the results early presented by Venkiteswaran and Taplin for Inconel X-750 (8), and by Tanaka et al. for the precipitation-strengthening 21Cr-4Ni-9Mn steel (9), although they have been obtained by adopting the non-consistent heat treatments.

CREEP RUPTURE MORPHOLOGIES. Typical creep rupture morphologies for both the fine-grained and coarse-grained specimens of HC alloy are shown in Figs. 8 and 9 for simple aging and direct aging, respectively. For the coarse grain size of 283 μm, the microstructures of the specimens subjected to the two-stages solution treatment are also shown for a comparison with those for the one-stage solution treatment. It was found that for the fine grain size of 78 μm the creep rupture morphology is essentially similar between simple aging and direct aging, in which a number of small cracks nucleate at the grain boundaries, in many case at their triple points, so as to result in the growth and coalescence to induce the final creep rupture in rather ductile manner. This is a representative creep rupture behavior usually observed in many systems of the ductile alloys. For the coarse grain size of 283 μm, on the contrary, the creep rupture morphology is found to differ considerably between simple aging and direct aging. Then, for simple aging only a small number of large but narrow cracks tend to develop so easily as to connect with the neighboring triple points. This suggests that the most predominant one of such the extended cracks can cause directly a premature fracture

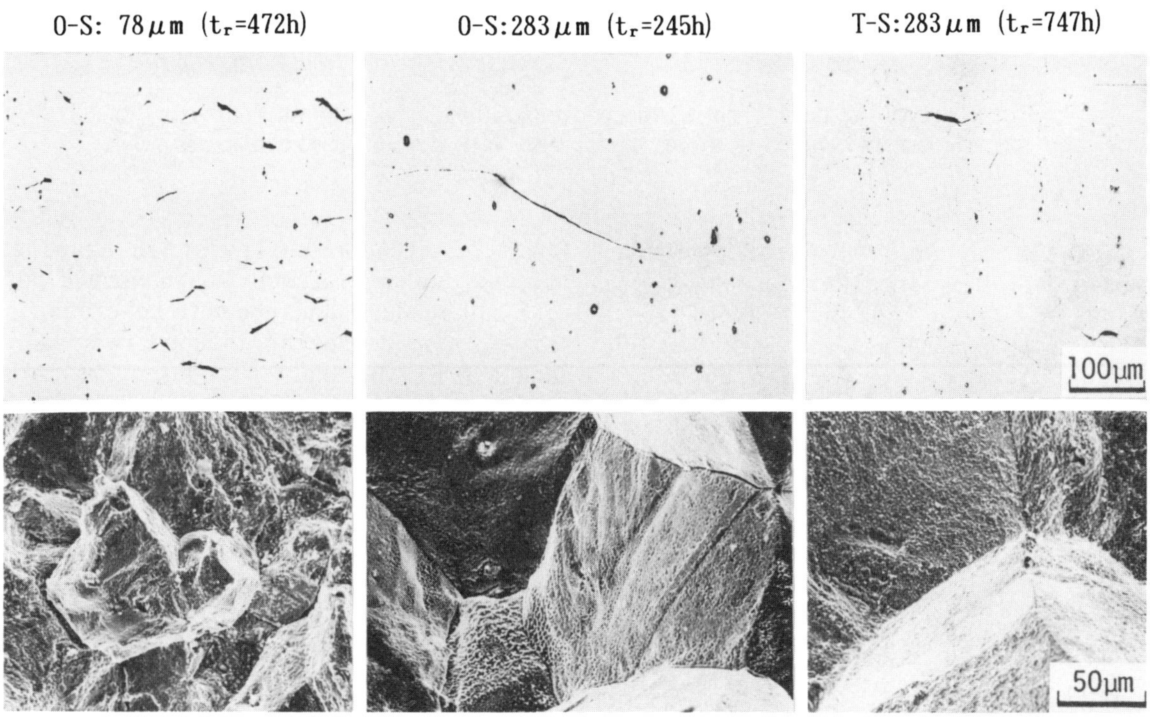


Fig.8 - Typical creep rupture morphologies under the stress 180 N/mm$^2$ for the simply aged HC alloy specimens with different grain sizes.

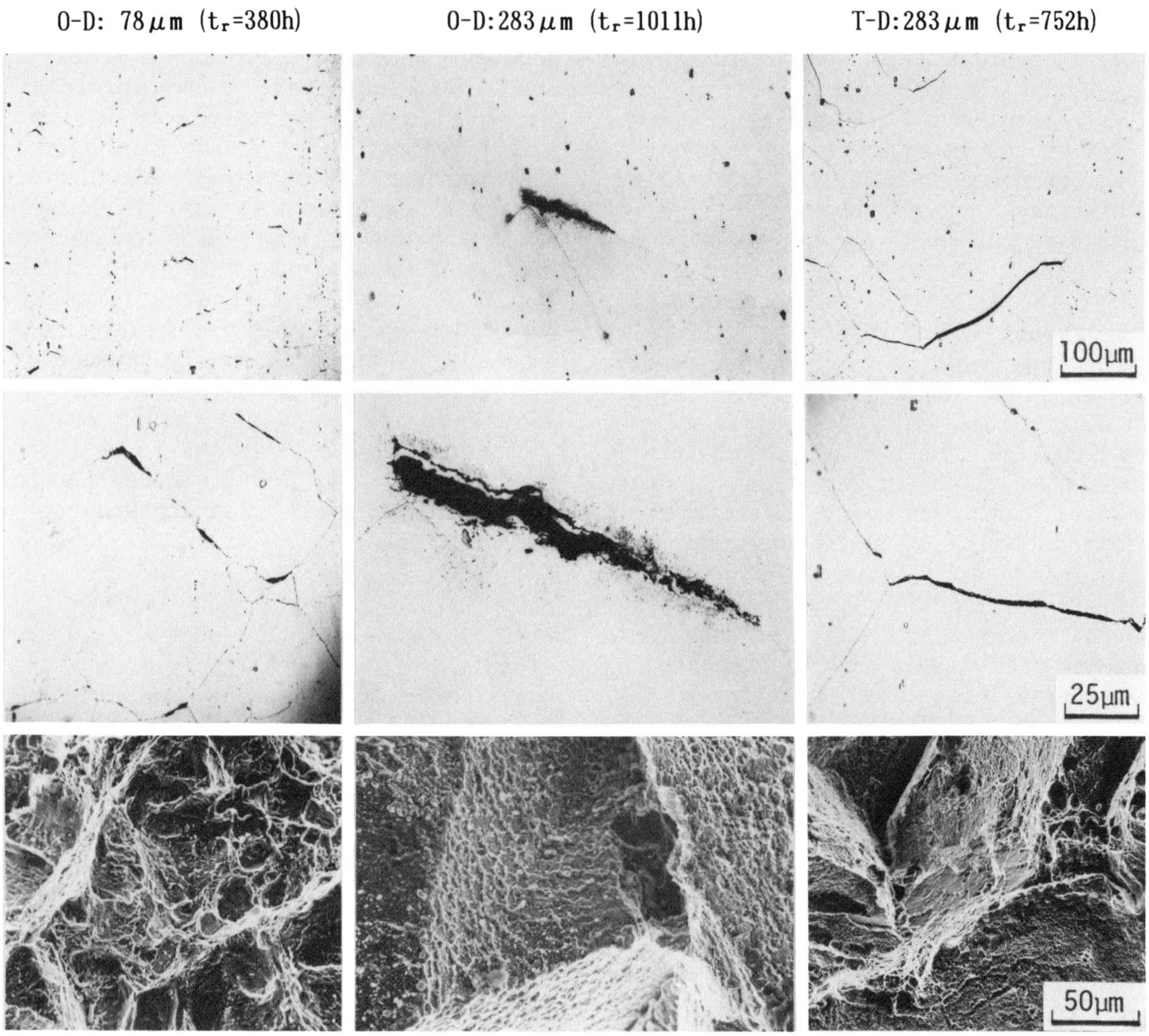


Fig.9 - Typical creep rupture morphologies under the stress 200 N/mm²
for the directly aged HC alloy specimens with different grain sizes.

in rather brittle manner. Fracture surface also indicates that a grain size coarsening for the one-stage solution treatment results in more smoothed grain boundary facets. For direct aging, on the other hand, a grain boundary crack is apt to grow more steadily with opening in the stress direction, which should bring about a fairly ductile fracture. Furthermore, fracture surface for direct aging suggests that a creep fracture takes place mainly by the grain boundary cavitation (1,2,6). However, an application of the two-stages solution treatment reduces such a cavity formation so as to cause a poor ductile fracture. These creep rupture morphologies observed for simple aging and direct aging are generally consistent with their creep rupture ductility data, as shown in Figs. 4 and 5, respectively.

From the metellographic examination, no microstructural change was observed both at the grain boundary and around the cracks such as a stress concentration-induced recrystallization and a grain boundary migration, which appear to bring about a localized alloy softening, for all the specimens investigated.

## DISCUSSION

### MICROSTRUCTURAL FACTOR DETERMINING CREEP STRENGTH.

It was found from Figs. 4~7 that the creep deformation process is uniquely controlled by a grain size dominating over the grain boundary nature, since a grain size dependence of $\dot{\varepsilon}_s$ was similar between simple aging and direct aging, regardless of different alloy carbon contents and solution processes adopted. Then, the creep strength; a resistance to the creep deformation, tends to increase and become constant as a grain size increases. It should be also reasonable to consider for the creep rate to be determined by the rate of recovery, because the stress index n of $\dot{\varepsilon}_s$ was in the range 4 ~8 regardless of different grain sizes and grain boundary natures, as shown in Fig. 3. The creep strength reduction in the fine grain size should be attributed to an enhancement of the diffusion-controlled recovery process, which is usually observed in many systems of the alloys (9,10).

On the other hand, a reason for the relatively constant creep strength in the medium and coarse grain sizes of more than approximately 150 μm is difficult to explain at present. So far as the present study is concerned, however, the creep strength reduction which can be induced by the micro-structural changes such as recrystallization, a grain boundary migration and a ledge formation at the stress-concentrated grain boundary is no need for considering even in the coarse grain size. Then it may be reasonable to consider that as a grain size increases the creep strength becomes saturated to a given level because of an effective stress for the creep deformation to be constant independently of a grain size (9).

MICROSTRUCTURAL FACTORS AFFECTING CREEP RUPTURE PROPERTIES. Creep rupture properties such as the rupture life and the rupture ductility were found to have a different grain size dependence between simple aging and direct aging, depending on different solid-solution treatment conditions, as already shown in Figs. 4 ~ 7. For the one-stage solution treatment, in particular, a grain size coarsening results in the rupture life reduction for simple aging, but in the life increasing for direct aging. However an application of the two-stages solution treatment leads to rather intermediate nature of result common to both simple aging and direct aging; an optimum grain size is obtainable for the maximum rupture life.

A reduction of the creep rupture life with decreasing in a grain size should be attributed to the increased steady-state creep rate due to an enhanced recovery process. On the contrary, a different grain size dependence of the rupture life in the coarse grain size regime between two kinds of aging treatments together with different solid-solution treatments is hardly explainable only by considering the creep rate.

Figure 10 shows the relationship between the steady-state creep rate and rupture life ($t_r$) for simple aging and direct aging with different grain sizes of HC alloy. The relation between $\dot{\varepsilon}_s$ and $t_r$ is found to be shown by a given Monkman-Grant equation as follows (11),

$$t_r = C\ \dot{\varepsilon}_s^{\,k}$$

where C and k are material constants, and $k \fallingdotseq -1$. For simple aging, a gap in the $\dot{\varepsilon}_s$-$t_r$ relation becomes larger with increasing in a grain size, although for T-S treatment a rather mitigated

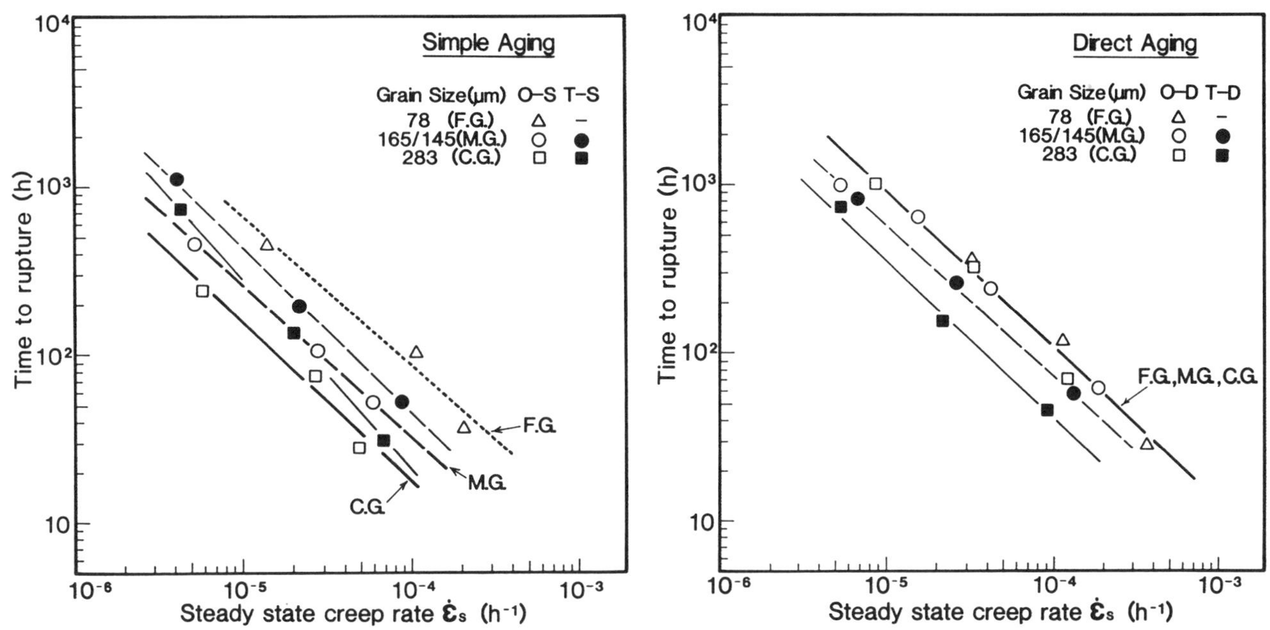


Fig.10 - Relationship between the steady-state creep rate and rupture life for the simply aged or directly aged HC alloy specimens with different grain sizes.

gap is obtained. This is consistent with a grain size dependence of the creep rupture ductility, and suggests for a coarse grain size to be more easily subjected a premature fracture in a brittle manner. This should be attributed to the fact that for the straight grain boundary produced by simple aging a grain size coarsening brings about very continuous $M_{23}C_6$ carbide precipitation with high density at the grain boundary because of a reduction of grain boundary area, so that a much enhanced grain boundary sliding can take place along there. Thus, for the straight grain boundary a resistance to the creep fracture becomes a critical factor determining the creep rupture strength in particular in the coarse grain size regime, as schematically shown in Fig. 11.

For direct aging, on the contrary, so far as O-D treatment is concerned there is no gap in the $\dot{\varepsilon}_s$-$t_r$ relation between different grain sizes, suggesting the same creep rupture behavior.

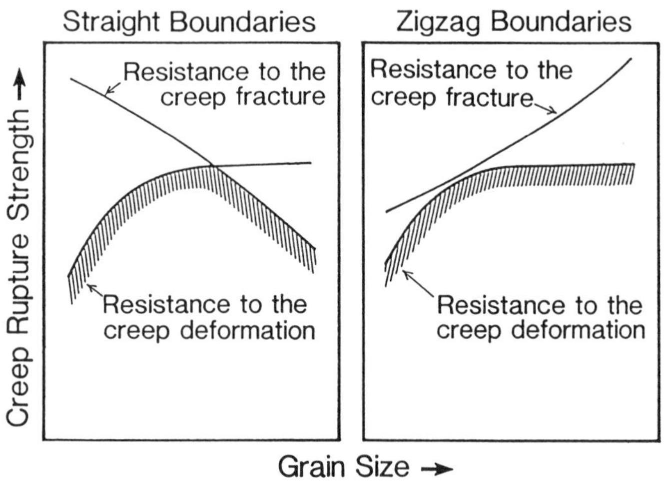


Fig.11 - Schematic representation showing a grain size dependence of the creep rupture strength for different grain boundary configuration.

However, T-D treatment brings about a gap similar to the case of T-S treatment, which is also related to the lowered rupture ductility with a grain size increasing.

Thus, direct aging (O-D) should be more effective for improving the creep rupture properties by combining with an increased grain size. Figure 12 shows a difference in the development of zigzag grain boundary by adopting different solid-solution temperatures prior to the direct aging. It is clear that the higher solution temperature is applied, the more fully the zigzag configuration develops. Such a difference in the zigzag configuration results from a difference both in a temperature gap between the solution and aging treatments which leads to the supersaturation of alloying elements such as carbon, and in a density of carbide precipitates associated with the grain boundary area. Provided for the zigzag grain boundary developed well enough, the creep cracking associated with a grain boundary sliding must be effectively suppressed (1,2). In this case, a criterion for determining the creep rupture strength should be always a resistance to the creep deformation over a wide range of grain size, as shown in Fig. 11.

From the metallographic examination, the two-stages solution treatment was revealed to modify the grain boundary nature. Figure 13 shows the morphology of grain boundary carbide precipitates after the second stage of solution treatment. As can be seen, rather coarse and spherical carbide precipitation takes place at this stage, which can alter the morphology of grain boundary carbide precipitates. Then a straight grain boundary becomes more roughened even for simple aging, while a zigzag grain boundary becomes more smoothed even for direct aging. Consequently, an application of the two-stages solution treatment should result in an intermediate nature of rupture properties for both simple aging and direct aging.

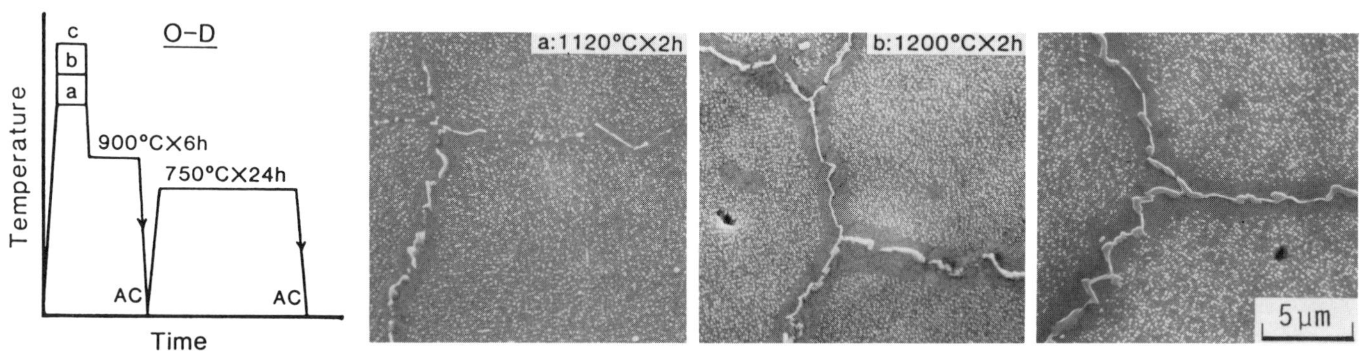


Fig.12 - Scanning electron micrographs of the zigzag grain boundaries produced by the direct aging (O-D) combined with different solution temperatures for HC alloy specimens.

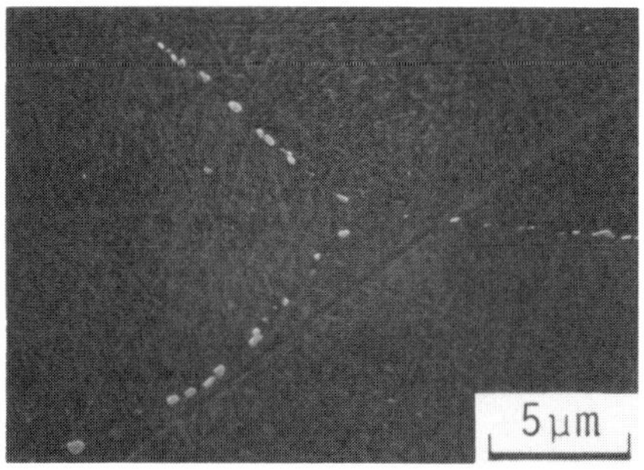


Fig.13 - Morphology of the grain boundary precipitates (carbides) after the second stage of solution treatment for HC alloy specimen. (The first stage solution temperature: 1250 ℃)

Furthermore, a carbon content in the alloy should affect mainly the creep rupture behavior by controlling the grain boundary carbide precipitation. Then increasing in a carbon content emphasizes the effects of grain boundary natures by enhancing the more dense carbide presipitates; for straight boundary it brings about the more stimulated brittle creep fracture, while for zigzag boundary it leads to the more effectively suppressed creep rupture.

## CONCLUDING REMARKS

The effects of the microstructural parameters associated with a grain boundary such as a grain size and a grain boundary nature on the creep and creep rupture properties were investigated for the $\gamma'$ precipitation-strengthening nickel-base superalloy Inconel 751 by adopting the various processes of heat treatments.

(1) The creep strength associated with the steady-state creep rate is uniquely determined by a grain size regardless of different natures of grain boundaries. The creep strength increases and tends to become constant with increasing in a grain size. A reduction of the creep strength in the fine grain size is closely related to an enhanced recovery process.

(2) The creep rupture strength; the rupture life, appears to be determined both by the creep strength; a resistance to the creep deformation, in the fine grain size regime and by the creep rupture ductility; a resistance to the creep fracture, in particular in the fine grain size regime. For the latter, the grain boundary nature such as its geometry is a significant microstructural factor dominating over a grain size. Then increasing in a grain size results in the different creep rupture strength dependingly on the grain boundary nature. For the much improved creep rupture strength, in particular, the zigzag grain boundary developed by direct aging is most effective in the case combined with the coarse grain size by the higher temperature solution treatment.

## ACKNOWLEDGMENTS

The author wish to acknowledge Prof. Emeritus O. Miyagawa of Tokyo Metropolitan University for valuable suggestion in performing the present study. The experimental work in this study was competently carried out by M.Eng. H. Taimura, Messrs. S.Suzuki, M.Shibata and W. Miki, who were formerly undergraduates at Tokyo Metropolitan University.

## REFERENCES

1. Miyagawa, O., M.Yamamoto and M. Kobayashi, in Proc. 3rd Int. Symp. on Superalloys, Seven Springs, PA, Sept. 1976, p.235-54.
2. Yamamoto, M, O.Miyagawa, M.Kobayashi and D. Fujishiro, Trans. Iron Steel Inst. Japan 19, 102-11 (1979)
3. Yoshiba, M. and O.Miyagawa, in Proc. 9th Int. Congress on Metallic Corrosion (ICMC), Toronto, Canada, June 1984, Vol.3, p.340-7.
4. Garofalo, F., W.F.Domis and F. von Gemmingen, Trans. Met. Soc. AIME 230, 1460-7 (1964)
5. Mannan, S.L. and P.Rodriguez, Metal Sci. 17, 63-9 (1983)
6. Weaver, C.W., J. Inst. Metals 88, 296-300 and 462-7 (1959-60)
7. Sidey, D. and B.Wilshire, Metal Sci. J. 3, 56-60 (1969)
8. Venkiteswaran, P.K. and D.M.R.Taplin, Metal Sci. 8, 97-104 (1974)
9. Tanaka, M., H.Fujita and D.Fujishiro, J. Iron Steel Inst. Japan (Tetsu-to-Hagane, in Japanese) 69, 1320-7 (1983)
10. Kondo, Y., T.Matsuo, T.Shinoda and R.Tanaka, J. Iron Steel Inst. Japan (Tetsu-to-Hagane, in Japanese) 65, 896-905 (1979)
11. Monkman, F.C. and N.J.Grant, Proc. ASTM 56, 593 (1956)

*Proceedings of the Fifth International Conference on Creep of Materials, Lake Buena Vista, Florida, USA, 18-21 May 1992*

# Effects of Grain Size on the Transition Behavior of an Al-Mg Solid Solution Alloy in the Power Law Creep Regime

**S-W. Nam**
Department of Materials
Science Engineering
KAIST
Taejan, Korea

**D-H. Lee**
Korean Institute of
Aeronautical Technology
Seoul, Korea

**K.T. Hong**
Division of Metals
KAIST
Seoul, Korea

ABSTRACT

*A mechanism of creep transition in power-law regime as a function of grain boundary deformation has been proposed for an Al-3at.%Mg solid solution alloy. The transition creep behavior was investigated at 573K ($0.62T_m$) under the stress range of 10 to 55 MPa, and a mechanism of grain boundary deformation under the dislocation creep has been developed. From these examinations, it has been concluded that in a polycrystalline aggregate the development of subgrains near grain boundaries may govern the transition behaviors, even though at higher stresses dislocations locally can break-away from their retarding solute clouds.*

**GENERALLY**, it has been accepted that Al-Mg substitutional solid solution alloy shows a transition from viscous glide ($n = 3$, where n is the stress exponent) to climb controlled creep ($n = 5$) at higher stresses in the power-law creep regime. The transition in the stress exponent has been interpreted in terms of a break-away of dislocations from their solute atmospheres [1-6].

In this study, however, it is observed that the stress for the transition of the stress exponent does not agree with the predicted break-away stress which is based on a model by Cottrell [7] or Friedel [8]. Besides, the present investigation obtains the significant results that the transition stress tends to decrease with decreasing grain size. Since the break-away concept is only concerned with dislocation and solute interactions within the interior of grain, it can hardly explain the dependence of the transition behavior on grain size. Therefore, it is necessary to suggest a mechanism governing the transition behavior which considers the grain boundary deformation (i.e. grain size). The purpose of this study is to verify the effect of grain size on the transition behavior of an Al-3at.%Mg alloy in the power-law creep regime. Particular attempts have been made to develope a mechanism of grain boundary deformation under the dislocation creep.

## DEVELOPMENT OF A MODEL FOR THE GRAIN BOUNDARY DEFORMATION

It is well known that during plastic deformation of polycrystalline materials the grain boundaries act as obstacles to dislocation movement, and the lattice dislocations interact with the grain boundaries creating new line defects which were called by Balluffi *et al.* [9] extrinsic grain boundary dislocations (EGBD). Since the grain boundaries containing EGBD possess higher energy, at high temperatures (above 0.5 $T_m$) they can be transformed to the low-energy equilibrium state by the disappearence of the grain boundary defects

During the recovery of grain boundary, motion of the EGBD with may also be linked to GBM and/or GBS. This process leads to the possibility of grain boundaries acting as a sink for the lattice dislocations [10]. During high temperature dislocation creep, in which the lattice dislocations can continually be brought into the grain boundaries, the dynamic recovery of the grain boundary structure may be occurred. Therefore, the changes of the EGBD density and the resulting modes of grain boundary deformation can be explained by a competition between the processes of both entering of lattice dislocations and their annihilation with the grain boundary structure.

Figure 1 shows the proposed model for this present work for the formation of EGBD by the interaction of lattice dislocation with a grain boundary, and a grain boundary structure with $b_n$-EGBD and $b_p$-EGBD which are annihilated by the climb and glide process in a grain boundary of length, 2L, during creep.

***Generation Rate of EGBD's,*** $d\rho_g^+/dt$. During creep, at a grain boundary, the entering rate of lattice dislocations,

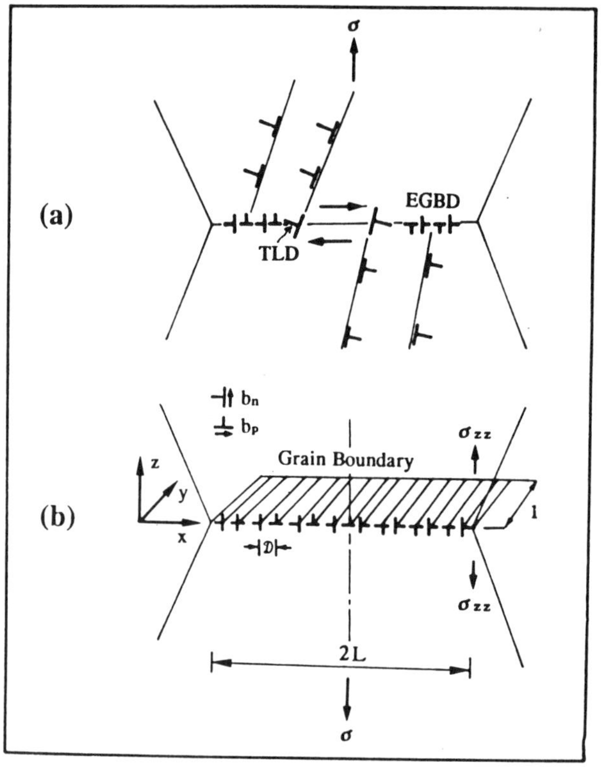


**Fig. 1.** A schematic diagram for the dynamic recovery of grain boundary structure.

$d\rho_L^+/dt$, may be proportional to the lattice dislocation density, $\rho_L^i$ and dislocation velocity with a solute atmosphere, $v_g$, so $d\rho_L^+/dt$ may be expressed as

$$\frac{d\rho_L^+}{dt} \propto \rho_L^i \cdot v_g \approx 2.5 \times 10^{-5} \sigma^3 \qquad (1)$$

where $\rho_L^i$ is experimentally given by a 2nd power function of stress [11] and $v_g$ is known to increase linearly with stress [3,7]. The constant value $2.5 \times 10^{-5}$ is calculated from the product of the two proportional constants ($C_1$ and $C_2$) for the relations $v_g = C_1 \cdot \sigma$ and $\rho_L^i = C_2 \cdot \sigma^2$. The trapped lattice dislocations are dissociated into a number of dislocations with small Burgers vectors [12,13], and the whole dislocations entered may not be transformed into EGBD on the grain boundary. Since these two processes may not be balanced equally, the density of the lattice dislocations entering grain boundary, $\rho_L^+$, may not be same but correspond to the density of EGBD generated on the grain boundary, $\rho_g^+$, so

$$\rho_g^+ = \zeta \cdot \rho_L^+ \qquad (2)$$

where the dimensionless constant, $\zeta$, is defined as the fraction of the generated EGBG with respect to the moving lattice dislocation in a grain. If $\zeta$ is approximately 0.1, from equation (1) $d\rho_g^+/dt$ is expressed as,

$$\frac{d\rho_g^+}{dt} = \xi \cdot \sigma^3 \qquad (3)$$

where, dislocation absorption coefficient, $\xi$, can be estimated as a constant value, $2.5 \times 10^{-6}$, provided $\zeta$ is not sensitive to the solute concentration and temperature.

***Annihilation Rate of EGBD***, $d\rho_g^-/dt$. The annihilation process of EGBD at a grain boundary can be taken for the recovery process of the grain boundary structure [14,15]. It has been postulated that the dynamic recovery of grain boundary involves the annihilatoin of EGBD by climb *via* lattice diffusion of vacancies at the triple points [16]. In this paper, the theoretical development of the annihilation rate of EGBD is based on the model proposed by Sangl and Tangri [16].

Assuming that a lattice dislocation is dissociated into $b_n$-EGBD and $b_p$-EGBD whose Burgers vector are normal and parallel to the grain boundary plane, respectively [17], as shown in Fig.1, the velocity of EGBD going towards a triple point may be controlled by climb of $b_n$-EGBD's rather than glide of $b_p$-EGBD along the grain boundary. In order to discribe the annihilation rate of EGBD, the vacancy concentration gradient around the triple point is given by the following relationship [18],

$$C(r) = C_o[1 - \frac{Gb_n\Omega\rho_n}{4\pi(1-\nu)kT} \cdot [1 - 2\ln(\frac{r_o}{L})] \cdot \frac{\ln(R_o/r)}{\ln(R_o/r_o)}] \qquad (4)$$

where, G is the shear modulus, $\nu$ is the Poisson's ratio, r is the distance from a triple point, $\rho_n$ is the density of $b_n$-EGBD, $\Omega$ is the atomic volume, $C_o$ is the vacancy concentration away from the triple point (at $r = R_o$, in the grain interior).

It is obvious that the climb of dislocations will be occurred by vancy annihilation at the triple point. With the knowledge of the vancy concentration profile, the climb rate, $v^L$, can be derived as follows:

$$v^L = -2\pi r \cdot \frac{D_v\Omega}{b_n} \cdot \frac{\partial C}{\partial r} \qquad (5)$$

where $D_v$ is the diffusion coefficient for vacancies. In time, $\Delta t$, the number of dislocations, $\rho_n \cdot 2\Delta t$, are annihilated, where $\Delta x$ is the climb distance at each triple point. If $\Delta \rho_n$ is the density of dislocations annihilated, then one can relate the climb rate to the annihilation rate as :

$$\frac{\Delta \rho_n}{\Delta t} = \frac{\rho_n}{L} \cdot \frac{2\Delta x}{\Delta t} \approx \frac{d\rho_g^-}{dt} \qquad (6)$$

$$\frac{d\rho_g^-}{dt} = \frac{2\rho_n}{L} \cdot v^L \qquad (7)$$

where $\rho_n$ is the density of $b_n$-EGBD generated on the grain boundary of length, L. Taking $\rho_{tot}$ is the total density of EGBD on the boundary of length, 2L, $\rho_n = \rho_p = \rho_{tot}/4$. Therefore, from the equations of (4), (5) and (7), the annihilation rate is given by

$$\frac{d\rho_g^-}{dt} = -\frac{G\Omega D_L[1 - 2\ln(r_o/L)]}{16L(1-\nu)kT\ln(R_o/r_o)} \cdot \rho_{tot}^2 \qquad (8)$$

where, $D_L$ is the atomic self-diffusivity ( $= \Omega C_o D_v$). It is recognized that the annihilation rate of EGBD's is dependent on the grain boundary length (which is related to the grain size); a larger grain size results in a decreased annihilation rate. Therefore it is expected that as the grain size increases, there is a growing tendency of irregular deformation of grain boundaries.

***Critical Stress for the Mode of Grain Boundary Deformation.*** If the annihilation rate on grain boundaries is fast enough, the equilibrium boundary deformation may be occurred continually in the form of massive migration and sliding of grain boundaries. If, however, the process of lattice dislocation absorption requires the incubation time [17], and the density of the dislocations entering grain boundaries exceeds the absorption capacity of grain boundaries, the lattice dislocations may pile up, and they will set up local stress concentration at the grain boundary. In this case, the grain boundary may be deformed irregularly, resulting in a wiggling of the grain boundary.

Therefore, for a grain size, the massive shape of grain boundary deformation is transformed into the serrated type with increasing stress at a certain stress level. The critical stress can be estimated from the following condition:

$$\left|\frac{d\rho_g^+}{dt}\right| = \left|\frac{d\rho_g^-}{dt}\right| \qquad (9)$$

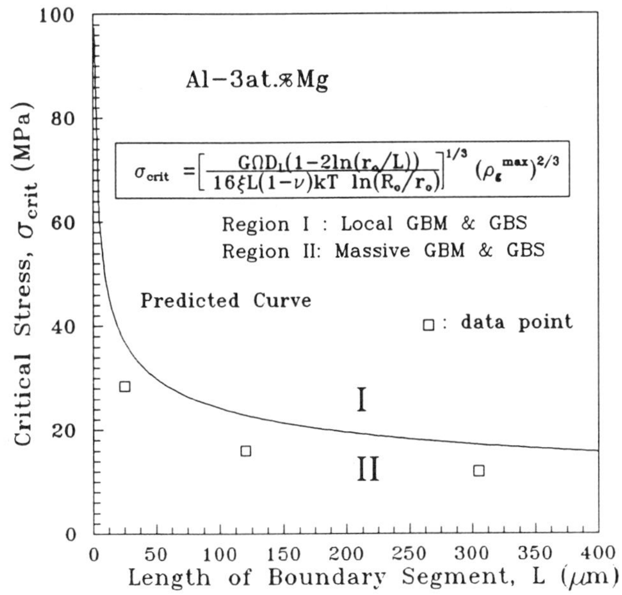


**Fig. 2.** Critical stress for grain boundary deformation computed from equation (9), plotted against grain boundary segement.

From the equations of (3) and (8), the critical stress, $\sigma_{crit}$, is given by

$$\sigma_{crit} = \left[\frac{G\Omega D_L[1 - 2\ln(r_o/L)]}{16\,\xi\,L(1-\nu)kT\ln(R_o/r_o)}\right]^{1/3} \cdot (\rho_g^{max})^{2/3} \qquad (10)$$

where $\rho_{tot} = \rho_g^{max}$; $\rho_g^{max}$ is the maximum density of EGBD during the recovery of the grain boundary structure ($5 \times 10^{13}\ m^{-2}$ [18]). Equation (10) can be

employed to predicted the mode of grain boundary deformation at a given grain size during dislocation creep.

For Al-3at.%Mg alloy, putting $\Omega = 12.5 \times 10^{-30}\ m^3$, $\nu = 0.34$, $G = 2.09 \times 10^4\ N/m^2$ (at 573K), and taking $r_o$ and $R_o$ as the Burgers vector b ($=2.86 \times 10^{-10}$ m) and the distance from the triple point where the vacancy concentration $C \simeq C_o$ ($R_o \simeq 10^4 r_o$ [18]), respectively, the critical stress, $\sigma_{crit}$, vs. the grain boundary length, L can be plotted as shown in Fig. 2. It can be seen from Fig. 2 that the $\sigma_{crit}$ is decreased with increasing grain size, and the experimental results are almonst consistent with the predicted values.

## EXPERIMENTAL PROCEDURE

The material used were the same as those reported in the previous papers [19,20]. Prior to the creep test the specimens were annealed at the various temperatures and

**Table I.** Annealing treatments and the resulting grain sizes

| Annealing temperature (K) | Time at temperature (min) | Average grain size ($\mu$m) |
|---|---|---|
| 640 | 120 | 80 |
| 700 | 240 | 400 |
| 760 | 360 | 1000 |

times to obtain a various grain sizes, and the results are summarized in Table I.

Optical microscopy was used to examine macroscopically the grain boundary and substructures developed during creep. After a specimen was deformed to the stage within the steady state region (up to about 30% creep strain), the specimen surface was mechanically polished, and electropolished. After this, the specimen was deformed an additioned 5% creep strain. The crept specimen was observed optically using a oblique illumination method to see the deformation mode along the grain boundary and within the grain.

For the investigation of the dislocation structure by TEM, a specimen was quenched by rapid air blowing under load to preserve the substructures formed during the creep deformation. The thin foils were examined by JEOL 200 CX transmission electron microscope operated at 160 kV.

At a given stress, if the subgrains are formed, they are beginning to be developed near the grain boundaries. In order to measure the volume fraction of the subgrain over the testing stress range for each grain sizes, an image analyzer was used. The subgrain portion was measured at more than 200 grains.

**Table II.** Experimental values of the stress exponent, n and the transition stress for the various grain sizes

| Average grain size ($\mu$m) | Stress exponent, n | | Transition stress (MPa) |
|---|---|---|---|
| | glide | climb | |
| 80 | 3.24 | 4.94 | 33 |
| 400 | 3.17 | 4.73 | 37 |
| 1000 | 3.20 | 4.69 | 42 |

## RESULTS AND DISCUSSION

Figure 3 shows the influence of the grain size on the steady state creep rate over the applied stress range. The plot in Fig. 3 indicates that there is the transition in the value of stress exponent from about 3 to around 5 as the applied stress increases and that the smaller the grain size the lower the transition stress. The transition stress and the stress exponent for the three grain sizes are listed in Table II. It is intuitively recognized that the transition

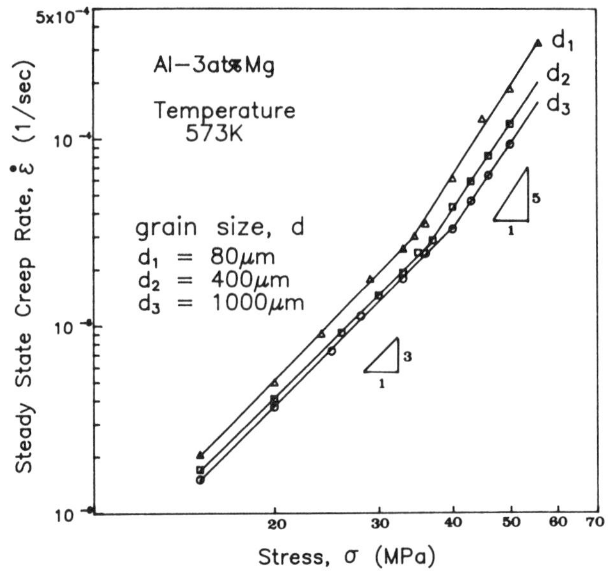


**Fig. 3.** The steady state creep rate vs the applied stress for an Al-3 at.%Mg alloy with three different grain sizes.

behavior of the exponent is dependent on the grain size, and closely related to the grain boundary deformation.

According to the model porposed in this paper, at a given grain size, if the applied stress, $\sigma$, is smaller than the critical stress, $\sigma_{crit}$ (Region II in Fig. 2), it is suggested that the subgrains are not formed but grain boundaries are migrated massively, however, if $\sigma$ is larger than $\sigma_{crit}$, most grain boundaries has to be serrated irregularly (Region I in Fig. 2). In the latter case, the piled-up dislocations will be rearranged to form groups of small angle boundaries, such as tilt boundaries, to relieve the stress concentration, which leads to the subgrain formation in the area adjacent to the grain boundary [21,22] indicating climb controlled creep. These are confirmed by the following experimental results.

For 80 $\mu$m grain size under the applied stress of 15 MPa, most grain boundaries have undergone massive migration without irregularities, and there is a uniform distribution of dislocations with no subgrains formed near the migrated boundary (Fig. 4). While for the specimens whose grain size is larger than 400 $\mu$m, it can be seen from Fig. 5 that under the same stress (15 MPa), the corrugated grain boundaries and the subgrains near boundaries are developed.

It has been accepted that if the creep deformation is controlled by a dislocation climb process, the characteristic substructure is a heterogeneous distribution of dislocations in the form of subgrains or cells [23,24]. Therefore, under the applied stress, in which most dislocations glide viscously due to the dragging of solute clouds, if there is a region of subgrains near the boundary, the creep deformation may be occurred in the parallel

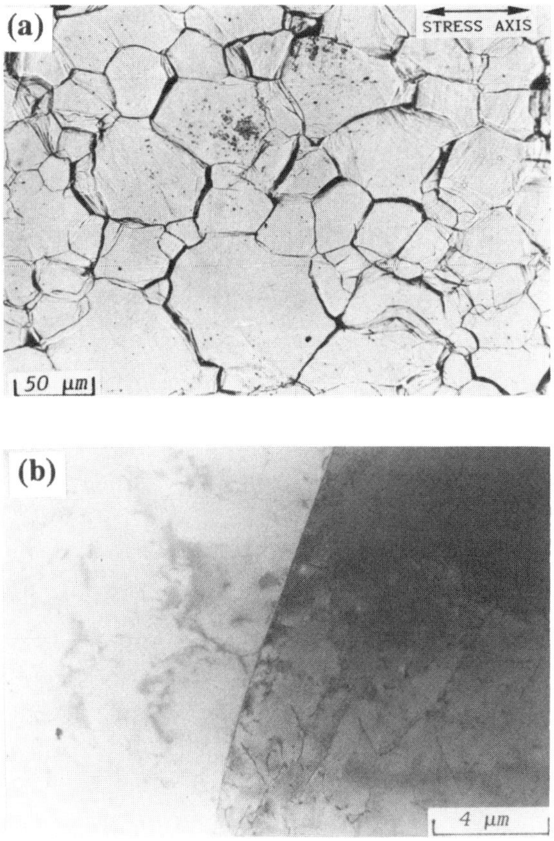


**Fig. 4.** Metallographic observations of (a) optical micrographs showing massive grain boundary migrations and (b) TEM micrographs of random distribution of dislocations for 80 $\mu$m grain size specimens. (creep condition: 15 MPa and 573 K and creep strain about 30 %)

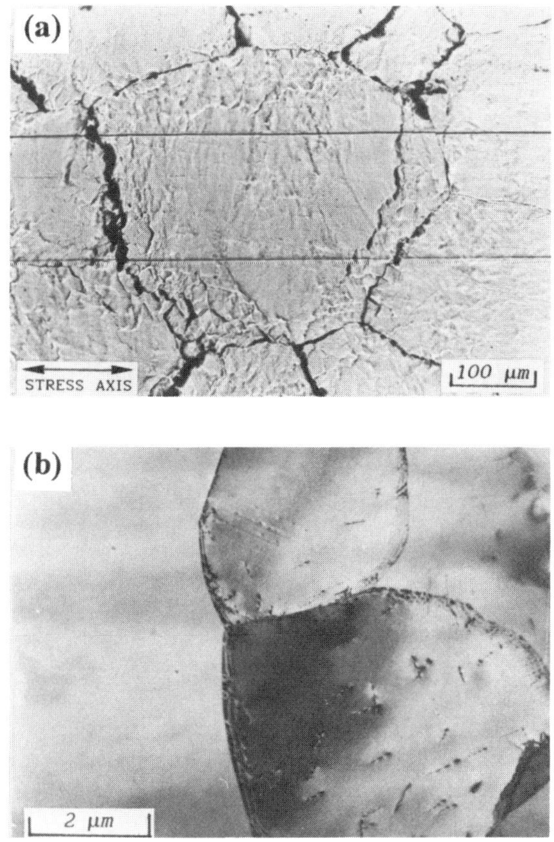


**Fig. 5.** Metallographic observations of (a) optical micrographs showing the irregularities in the shape of subgrains within and near the corrugated grain boundaries and (b) TEM micrographs of well-developed subgrains near the serrated grain boundary for 400 $\mu$m grain size specimens. (creep condition: 15 MPa and 573 K and creep strain about 30%)

processes of viscous glide of dislocations inside the grains and dislocation climb near the grain boundary where the subgrains are formed. At the higher stress level in power-law creep regime, the apparent creep mechanism may be determined by the relative volume fraction of climb and viscous glide controlled regions

Detailed measurements of the average volume fraction of the subgrains, $f_c$ , were undertaken to determine the dependence of $f_c$ on the applied stress for each grain size over the testing stress range, using an image analyzer. As shown in Fig. 6, a limiting $f_c$ value of zero at low stresses and unity at high stresses imply a sigmoidal type of transition; as the grain size decreases, $f_c(\sigma)$ becomes nearly a step function. Also, it should be noted that the observed transition stresses of the stress exponent exist within each transition stress range of $f_c$ for different grain sizes, and the plot of $f_c$ vs. stress shifts toward higher stress level as the grain size increases. Assuming that $f_c$ is a function of the applied stress and grain size, one may consider that the plot of strain rate vs. stress for different grain sizes (Fig. 3) should not reveal parallel lines. Since, however, $f_c$ does not increase linearly with increasing stress within the stress range occurring the transition in the stress exponent, and in each creep region $f_c$ is almost constant, as shown in Fig. 6, the apparent creep rate may be approximately exclude the dependence of $f_c$ on the applied stress and grain size. This result is very significant in that the apparent creep rate is considered to be controlled by an unique creep mechanism (i.e. viscous glide creep of n = 3 at low stresses or climb controlled creep of n = 5 at high stresses).

For the 80 $\mu$m grain size, the transition stress of $f_c$ nearly corresponds to the critical stress, $\sigma_{crit}$, in grain boundary deformation; below the tranistion stress of $f_c$,

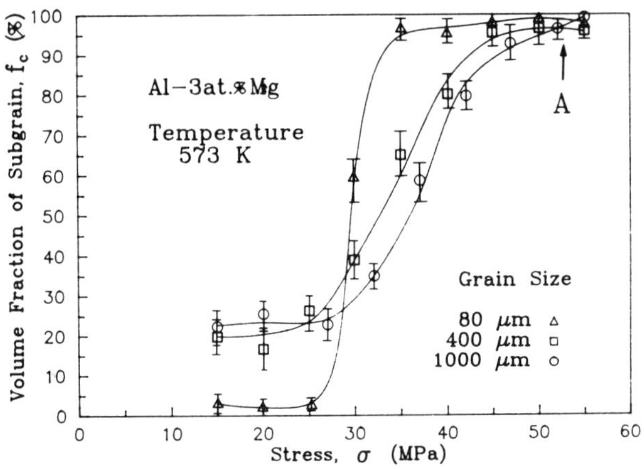


**Fig. 6.** The dependence of the volume fraction of subgrains near grain boundaries, $f_c$ on applied stress for different grain sizes [25].

about 32 MPa, the creep deformation is considered to be controlled by the viscous glide of dislocations, since $f_c \simeq 0$. For this very small grain size, even if only the narrow area along the grain boundaris is corrugated, $f_c$ may become larger than $(1 - f_c)$, shown in Fig. 6. This implies that some dislocations inside the grain still may glide viscously under this stress, the apparent creep deformation can be controlled by the dislocation climb process which is occurring in the subgrain region.

For the specimens whose grain size is larger than 400 $\mu$m, well-developed subgrains exist already near grain boundaries ($f_c \simeq 0.2$ in Fig. 6), because the critical stress required to deform a grain boundary massively is very low stress level ($\sigma_{crit} \simeq 10$ MPa in Fig. 2). Since, however, $f_c$ is still smaller than $(1 - f_c)$, the apparent creep deformation may be controlled by the viscous glide of dislocations. For 400 $\mu$m or 1000 $\mu$m grain size, when the creep stress becomes higher than a certain stress, the subgrain regions tends to also abruptly increase in a narrow stress region, as that of the 80 $\mu$m grain size. Therefore it is recognized that the transition in stress exponent may be occurred by the sudden expansion of the subgrain region adjacent to the grain boundaries. Finally at high enough stresses, subgrains are formed throughout the whole grain with serrated grain boundaries ($f_c \simeq 1$ in Fig. 6 and Fig. 7), and the apparent creep deformation mechanism seems to be controlled by the dislocation climb process.
From the surface observations of the crept specimens, at very high stresses (marked **A** region in Fig. 6), it is observed that subgrain formation may begin to occur simultaneously not only near grain boundaries but also

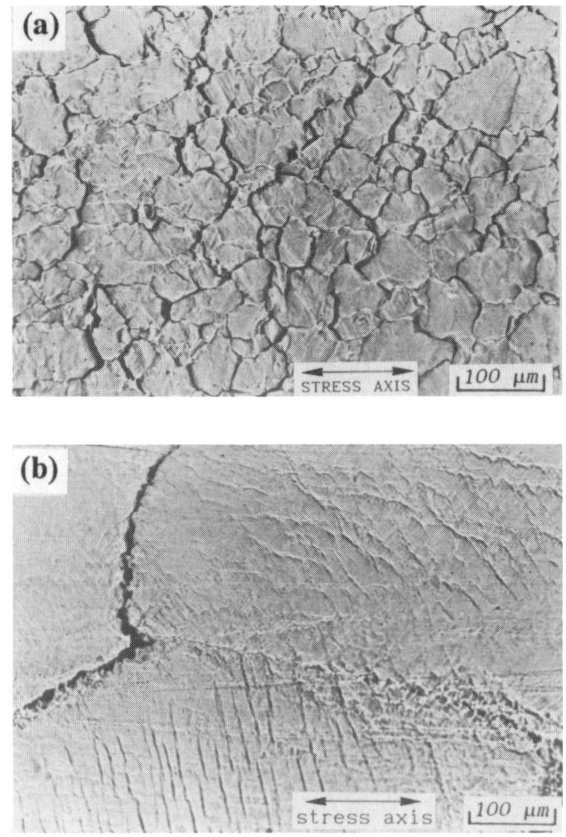


**Fig. 7** Optical micrographs showing the homogeneous distribution of subgrain like irregularities under the creep condition of 50 MPa and 573 K for (a) 80 lm and (b) 1000 lm grain size, respectively (creep strain 30%).

inside the grains. Accordingly it is inferred that under this stress level, dislocations can break away from solute atmospheres and then glide freely just like as in pure metals. However, as it is observed that the apparent applied stress inducing transition in $f_c$ and stress exponent is lower than the break-away stress predicted by the Cottrell's [9] or Friedel's model [10], the break-away of dislocations from their retarding solute clouds can occur only at region of the subgrain formation near grain boundaries under the high local stress.

## CONCLUSIONS

(1) Under the power-law creep condition, the mode of grain boundary deformation can be determined by a competition between the processes of entering of lattice dislocations and their annihilation with the grain boundary structure.
(2) For a given grain size, the massive migration of grain boundary deformation is suggested to be changing to

the serrated type with increasing stress at a critical stress as follows:

$$\sigma_{crit} = \left[\frac{G\Omega D_L[1 - 2\ln(r_o/L)]}{16\,\xi\, L(1-\nu)kT\ln(R_o/r_o)}\right]^{1/3} \cdot (\rho_g^{max})^{2/3}$$

(3) At a given applied stress level higher than $\sigma_{crit}$, the smaller the grain size, the larger the relative volume fraction of the subgrains near the grain boundaries. Thus, for the smaller grain size, the transition from n = 3 to n = 5 may occur at the lower stress.

(4) For the creep deformation of Al-Mg solid solution alloy, the transition stress of the stress exponent can be a function of grain size, and the grain size dependence of the transition stress may be well interpreted in terms of the development of subgrains near the grain boundaries.

## ACKNOWLEDGEMENT

This work is a part of the research project supported by the Korea Science and Engineering Foundation.

## REFERENCES

1. K.L. Murty, *Scripta Metall.*, 7 (1973) 899
2. H. Oikawa, K. Sugawara and S. Karashima, *Scripta Metall.*, 10 (1976) 885
3. T. Endo, T. Shimada and T.G. Langdon, *Acta Metall.*, 32 (1984)1991
4. P. Yavari and T.G. Langdon, *Acta Metall.*, 30 (1982) 2181
5. H.Sato and H.Oikawa, *Scripta Metall.*, 22 (1988) 87
6. H. Oikawa, H. Sato and K. Maruyama, *Mater. Sci. Eng.*, 75 (1985) 21
7. A.H. Cottrell, *Dislocations and Plastic Flow in Crystals*, Oxford Univ. Press, London, 1953, pp. 139
8. J. Friedel, *Dislocations*, Pergamon Press, Oxford, 1964, pp. 351-414
9. R.W. Balluffi, Y. Komen, and T. Schober, *Surface Sci.*, 31 (1972) 69
10. R.Z. Valiev, V.Yu. Gertsman, and O.A. Kaibyshev, *Phys. Status Solidi (a)* 97 (1986) 11
11. R. Horiuchi and M. Otsuka, *Trans. Japan Inst. Metals*, 13 (1972) 284
12. P.H. Pumphrey and H. Gleiter, *Phil. Mag. A*, 30 (1974) 593
13. J. Johannesson and A. Tholen, *Metal Sci.*, 6 (1972) 189
14. H. Gleiter and P.H. Pumphrey, *Mater. Sci. Eng.*, 25 (1976) 159
15. R.A. Varin, *Phys. Status Solidi (a)*, 52 (1979) 347
16. S. Sangl and K. Tangri, *Metall. Trans. A*, 20A (1989) 479
17. R.Z. Valiev, O.A. Kaibyshev, and Sh.Kh. Khannanov, *Phys. Status Solidi (a)* 52 (1979) 447
18. J.P. Hirth and J. Lothe, *Theory of Dislocations, 2nd*, ed. Wiley, New York, NY, 1982, p.76
19. D.-H. Lee, D.H. Shin, K.T. Hong and S.W. Nam, *Metall. Trans. A*, 22A (1991) 935
20. D.-H. Lee, K.T. Hong, and S.W. Nam, *Scripta Metall. Mater.*, 25 (1991) 823
21. J.L. Lytton, C.R. Barrett and O.D. Sherby, *Trans. AIME*, 233 (1965) 1399
22. S. Hashimoto and B. Baudelet, *Scr. Metall.*, 23 (1989) 1855
23. S. Takeuchi and A.S. Argon: *J. Mater. Sci.*, 11 (1976) 1542
24. W.D. Nix and B.Ilschner ; in *Proc. 5th Int. Conf. on the Strength of Metals and Alloys*, (edited by P. Hassen, V. Gerold and G. Kostorz) Pergamon press, 1979, vol. 3, pp. 1503-30
25. D. -H. Lee, K.T. Hong, D.H. Shin, and S.W. Nam, *Mater. Sci. Eng. A*, in press (1992)

*Proceedings of the Fifth International Conference on Creep of Materials, Lake Buena Vista, Florida, USA, 18-21 May 1992*

# Creep Rupture Properties and Microstructures of a New Ferritic W Containing Steel

**M. Ohgami, H. Mimura, and H. Naoi**
Nippon Steel Corporation
Chiba, Japan

**T. Fujita**
University of Tokyo
Tokyo, Japan

## ABSTRACT

The properties of a new ferritic W containing heat resistant steel for boiler tubes and pipes are presented. The replacement of a part of Mo in 9Cr-1Mo steels by W is very effective for increasing creep rupture strength (1-6). The creep rupture strength increases with increasing W content in the steel, and there is a obvious difference in the creep rupture strength of the steel which contained W from 0 up to 3%. The high W containing steel acheived high creep rupture strength. The high creep rupture strength of the high W containing steel is mainly due to the solid solution hardening by W. The tensile strength increases with increasing W content as same as the creep rupture strength. But the impact value decreases with increasing W content in the steel. So according to the combination of creep rupture strength, tensile strength and toughness, the optimum W content estimates to be about 2%. The amount of these precipitates increases during creep test. The precipitates are mainly $M_{23}C_6$ and Laves phase. These precipitates contain W. $M_{23}C_6$ mostly precipitates in the grain, and Laves phase precipitates in the grain and on the grain boundary.

THE OPERATING PRESSURE and temperature of fossil power plant have been increased continuously to achieve higher efficiency in this decade (7-9). Ferritic steels have an advantage of lower thermal stresses under cyclic operation owing to their smaller thermal expansion coefficients over austenitic stainless steels, so high strength ferritic steel is preferred in the ultra super critical power plant instead of austenitic stainless steels. For this purpose, 9% chromium creep resistant steels with high strength and corrosion resistance at high temperature, such as modified 9% chromium - 1% molybdenum (Mod.9Cr-1Mo) steel standardized in ASTM-A213, have been developed (10-11). The authors have developed a new ferritic W containing 9Cr-0.5Mo-1.8W-0.05Nb-0.2V steel for boiler tubes and pipes applications (12-14). This steel has higher creep rupture strength at 600° C than other ferritic steels including Mod.9Cr-1Mo steel. This steel is named NF616, and the main chemical composition of this steel is determined according to the optimum combination of creep rupture strength, toughness and tensile strength.

This study is concerned with the influence of W content on mechanical properties and microstructures.

## MATERIALS AND EXPERIMENTAL PROCEDURES

The chemical composition of ferritic W containing steels are shown in Table 1. These are characterized by small amount of Nb, V, N and B in addition to 9%Cr, 0.5%Mo, and 0.01～2.95%W.

Six heats of steels were melted by vacuum

Table 1 Chemical composition (mass%)

| Steel | C | Si | Mn | Cr | Mo | W | V | Nb | N |
|---|---|---|---|---|---|---|---|---|---|
| A | 0.10 | 0.08 | 0.49 | 9.00 | 0.50 | 0.01 | 0.21 | 0.05 | 0.0760 |
| B | 0.10 | 0.08 | 0.48 | 9.00 | 0.48 | 0.95 | 0.21 | 0.04 | 0.0675 |
| C | 0.10 | 0.08 | 0.47 | 9.02 | 0.49 | 1.24 | 0.20 | 0.04 | 0.0693 |
| D | 0.10 | 0.10 | 0.49 | 9.24 | 0.49 | 2.00 | 0.21 | 0.03 | 0.0675 |
| E | 0.09 | 0.09 | 0.48 | 9.32 | 0.49 | 2.45 | 0.21 | 0.03 | 0.0693 |
| F | 0.09 | 0.09 | 0.48 | 9.28 | 0.49 | 2.95 | 0.20 | 0.03 | 0.0774 |

induction furnace and casted into 20kg ingots. The ingots were reduced to 15mm thickness plates by rolling. The plates were normalized 1060° C for 60 minutes followed by air-cooling and then tempered at 780° C for 60 minutes. Tensile properties were tested at room and elevated temperatures. Creep rupture tests were carried out at 600 and 700° C. The specimens for tensile tests and creep rupture tests were machined from the center of thickness (1/2t) in the longitudinal (L) direction. Toughness in the as-tempered condition and after aging at 600° C for 3000 hours was examined by charpy impact testing. The specimens for impact tests were sampled at 1/2t in L direction. Microstructure was observed by an optical microscope and a transmission electron microscope The residues were extracted from as-tempered and aged specimens. The residues extracted from as-tempered and aged specimens were examined by chemical and X-ray diffraction analysis.

## RESULTS AND DISCUSSION

*Mechanical Properties.* Figure 1 shows the results of creep rupture tests of 9Cr-0.5Mo-W-Nb-V steels at 600 and 700° C. The creep rupture strength increased with increasing W content in the steel, and there was a obvious difference in the creep rupture strength from 0 up to 3%W containing steel. The creep rupture strength of these steels was extrapolated linearly at 600° C for $10^5$ hours. The creep rupture strength of 3%W containing steel at 600° C for $10^5$ hours was about 1.5 times as strong as that of W-free steel. Same tendency was recognized at 700° C. We investigated this difference of creep rupture strength owes to W content, especially almost of the difference is due to amount of solid solution of W and precipitates.

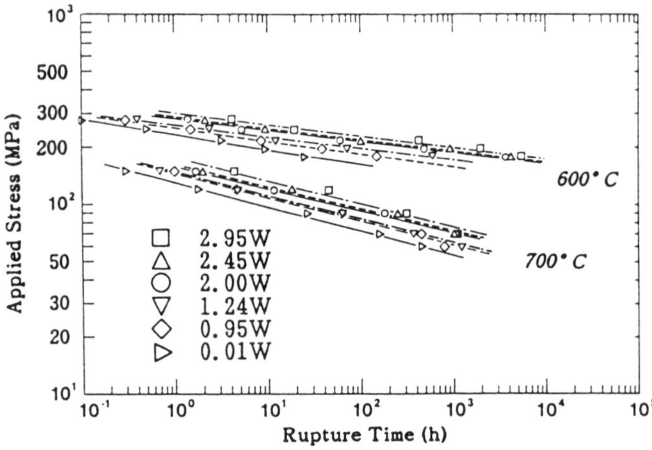


Figure 1 Creep rupture strength of 9Cr-0.5Mo-W steels tested at 600 and 700° C.

Figure 2 shows the results of tensile tests of W containing steels. There were a difference in the tensile properties from 0 up to 3%W containing steels. The tensile strength and 0.2% proof strength increased slightly with increasing W content in the steels. We considered this difference of the tensile strength of these steels at room temperature is the effect of solid solution hardening of W. But the elongation and reduction of area of these steels indicated almost same value at room temperature.

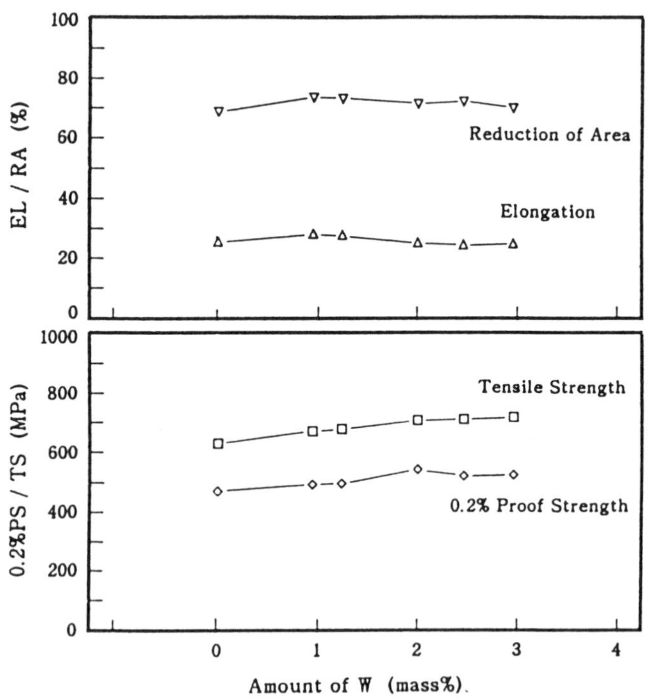


Figure 2 Tensile properties of 9Cr-0.5Mo-W steels tested at room temperature.

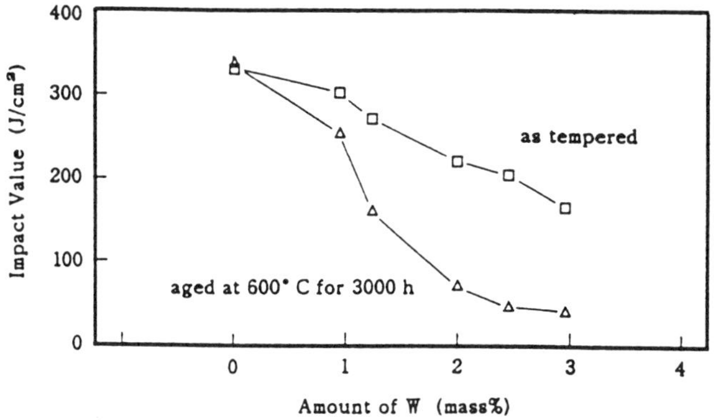


Figure 3 Toughness of 9Cr-0.5Mo-W steels tested at 0° C after tempering and aging at 600℃ for 3000 hours.

The results of impact tests is shown in Figure 3. The impact values of these steels at 0° C in as-tempered condition decreased with increasing W content. The impact value of W-free steel was about twice as high as that of 3%W containing steel. The tendency of impact properties was same after aging at 600° C for 3000 hours. After aging, the values of W-free steel was almost same as that value before aging. But the impact values of 3%W containing steel decreased with aging time. The deterioration of impact values after aging is obvious fact when the steel contains high W.

*Microstructers.* The microstructures of these steels are exhibited tempered martensite as shown in Figure 4. The grain size of these steels is almost same notwithstanding the difference of W content. The aspect of microstructures were a little different between 0.01%W containing steel and 2.95%W containing steel.

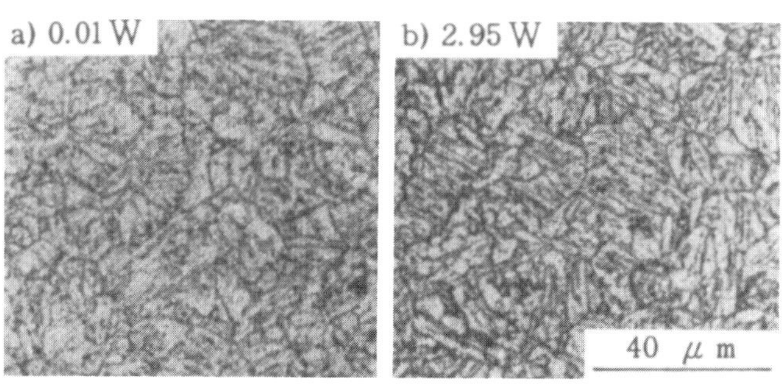


Figure 4 Optical microstructure of 9Cr-0.5Mo-W steels after tempering. a) 0.01W, b) 2.95W

Figure 5 shows the change in amount of W in precipitates and the amount of solved W with W content. The amount of solved W was increased with W content. Almost all of the W added in these steels existed in solution after normalizing and tempering. The amount of precipitated W is less than that of solved W. In the case of as-tempered condition, the solid solution hardening by W is very effective to be increasing creep rupture strength and tensile strength. Figure 6 shows the change in amount of W in precipitates and the amount of solved W with W content after aging at 600° C for 3000 hours. The amount of these precipitates increased during aging. But the amount of solved W was about twice to compare with that of precipitated W in aged these steels. The effect of solid solution hardening by W is maintained after aging. When the W content was up to 1%, the amount of precipitates was a little in comparison with when W content was to be over 1%.

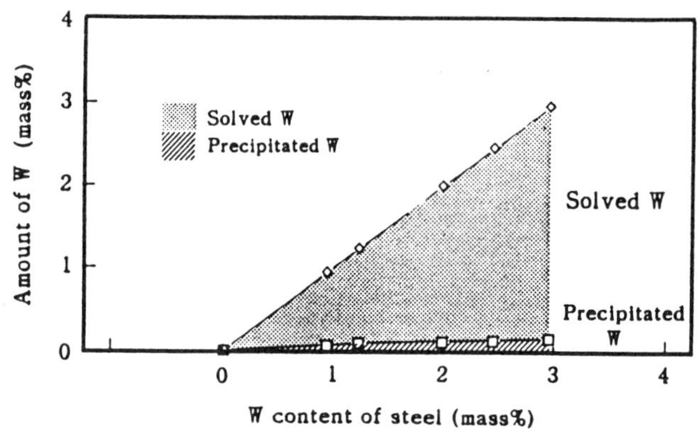


Figure 5 Change in amount of W in precipitates and the amount of solved W with W content after tempering.

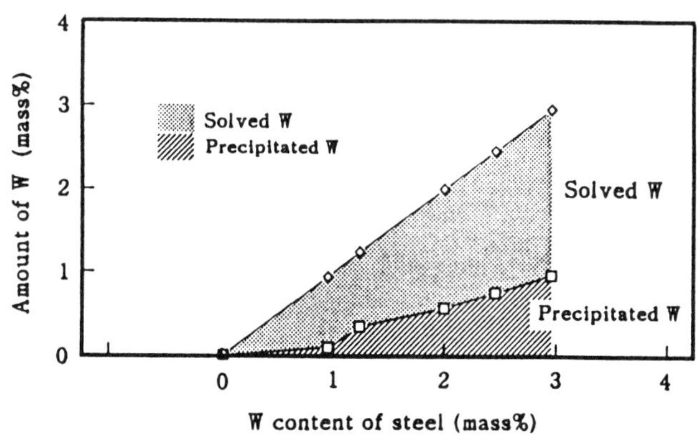


Figure 6 Change in amount of W in precipitates and the amount of solved W with W content after aging at 600° C for 3000 hours.

These precipitates was mainly $M_{23}C_6$ and Laves phase. These precipitates also included W.

Figure 7 shows the amount of the residue of 2%W containing steel with creep tests or aging time. The contents of precipitated Nb and V which form MX type carbonitride were hardly changed during creep test. The precipitates were determined by X-ray diffraction analysis of extraction. The size of MX type carbonitride was fine and hardly changed during creep tests. These fine and stable precipitates contribute to the high creep resistance. The contents of precipitated Fe, Cr, W and Mo were increased in the amount of residue of 2%W containing steel with creep tests. The amount of residue of crept specimens is shown in Figure 8.

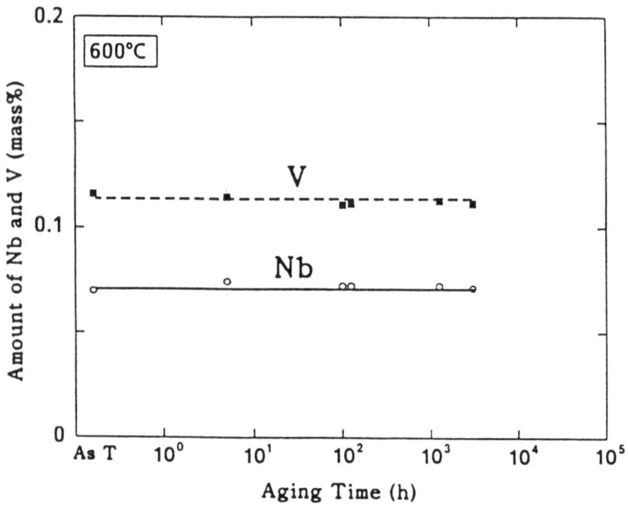


Figure 7 Change in precipitate alloying elements Nb, V with aging time at 600° C.

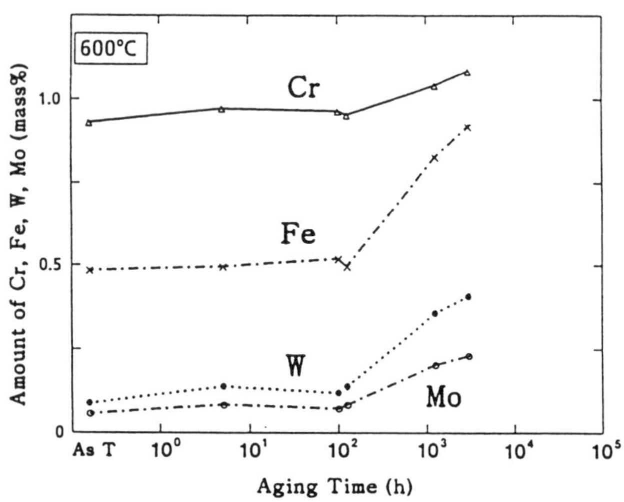


Figure 8 Change in precipitate alloying elements Cr, Fe, W, Mo with aging time at 600° C.

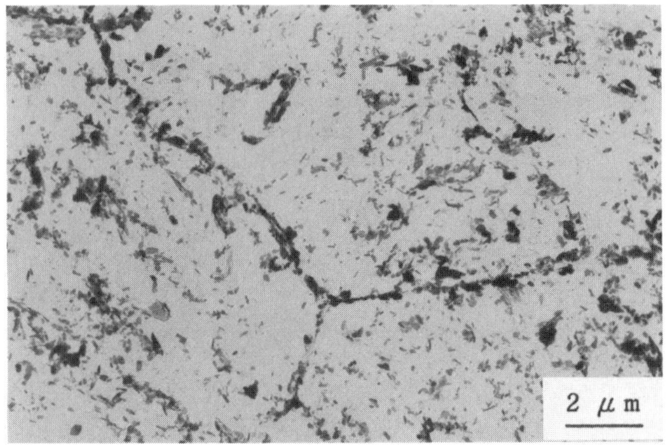


Figure 9 Transmission electron micrograph of extracted replica of 2%W containing steel after aging at 600° C for 3000 hours.

These alloying elements existed in $M_{23}C_6$ and Laves phase. These precipitates gradually condensed along prior austenite grain and lath/subgrain boundaries with creep tests or increasing the aging time. But the average size of the precipitates was almost same during creep tests or aging. Transmission electron microscope observation of precipitates of 2%W containing steel aged at 600° C for 3000 hours is indicated in Figure 9. In these precipitates, the fine and stable $M_{23}C_6$ precipitates in the grain is effective in increasing precipitation strengthening. This precipitation strengthening is one reason of increasing in the creep rupture strength in the long period test. The impact values decreased by aging, this reason is that these precipitates gradually condensed along prior austenite grain boundaries with aging time.

## CONCLUSIONS

(1) Creep rupture strength increases with increasing W content in the steel, and there is a obvious difference in the creep rupture strength between W-free steel and 3%W containing steel.
(2) The impact value decreases with increasing W content. The deterioration of impact value after aging is obvious when the steel contains high W.
(3) The optimum W content is estimated about 2% according to the combination of creep rupture strength, toughness and tensile strength.
(4) Almost all of the W added in these steels exist in solution after normalizing and tempering. The

amount of $M_{23}C_6$ and Laves phase precipitates which contents W increases during aging.

(5) The high creep rupture strength is achieved solid solution hardening by W and precipitation strengthening.

REFERENCES

(1) T. Fujita, T. Sato, and N. Takahashi : Trans. ISIJ, 18(1978), p115

(2) F. Abe, H. Araki : CAMP-ISIJ, 1(1988), p.1759

(3) H. Tanaka, K. Liu, K. Asakura, K. Shibata : CAMP-ISIJ 1(1988), p.1766

(4) K. Yoshikawa, A. Iseda, H. Teranishi, F. Masuyama, T. Daikoku, H. Haneda : High Temp Alloy (1987), p.247

(5) Y. Hosoi, N. Wada, S. Kunimitsu and T. Urita : Journal of Nuclear Materials, 141-143(1986), p.461

(6) T. Fujita : Metal Progress, 130, 2(1986), p.33

(7) H. Toda, H. Haneda, A. Hizume : Mitsubishi Power System Bulletin, MBB-82113E (1982)

(8) A. F. Armor, etal. : Proc. Americam Power Conference, Chicago, Illinois (April 27-29, 1981)

(9) A. F. Armor, R. I. Jaffee, R. D. Hottenstine : American Power Conference, Chicago (April 24-25, 1984)

(10) G. C. Bodine, JR, B. Chakravski, S. D. Harkness, C. M. Owens, B. Roberts, D. Vandergrift and C. T. Ward : Proc. of Inter. Conf. Ferritic Steels of Fast Reactor Steam Generators, BNES (1977), p.160

(11) P. Patriaka : Modified 9Cr-1Mo Steel Technical Program Data Package for Use in ASME Section I and II Design Analysis, ORNL Technology Transfer Meeting (1982)

(12) M. Sakakibara, T. Takahashi, H. Sakurai, T. Fujita : 1st. International Conference on Improved Coal-Fired Power Plants, Palo Alto (1986), 5-203

(13) H. Masumoto, H. Naoi, T. Takahashi, S. Araki, T. Ogawa, T. Fujita : 2nd. International Conference on Improved Coal-Fired Power Plants, Palo Alto (1989), 40-3

(14) H. Mimura, M. Ohgami, H. Naoi, T. Ogawa, Y. Sogo, T. Fujita : 3rd. International Conference on Improved Coal-Fired Power Plants, San Francisco (1991)

*Proceedings of the Fifth International Conference on Creep of Materials, Lake Buena Vista, Florida, USA, 18-21 May 1992*

# The Use of Miniature Specimen for Assessment of Residual Life by Temperature-Accelerated Creep Test

**K. Sonoya and M. Kitagawa**
Ishikawajima-Harima Heavy Industries Company, Ltd.
Tokyo, Japan

**M. Murata, H. Tanaka, and K. Yagi**
National Research Institute for Metals
Tokyo, Japan

## Introduction

To ensure high enough degree of precision in assessing the residual service life for high temperature equipment of thermal power stations, such as, for example, the utility boilers —a task which is urgently needed today—, it is generally held that creep rupture tests to be conducted for samples taken directly from the part concerned are necessary. The trouble is that provision of large regular-size test pieces is often impossible even when such a sample is available.

For this problem, we have devised a method of performing creep rupture tests with miniature test pieces that can be prepared from small samples. In this paper, we intend to discuss the efficacy of such a miniature test piece as a specimen of acceleration test.

## Experimental

**Testing Material and Specimen Design.** For experiments, two common boiler materials were selected: a 2.25Cr-1Mo steel to represent the creep-plasticized steels, and an 18Cr-8Ni steel to represent the creep-embrittled steels. Their chemical compositions are shown in Table 1.

The test piece was a round bar "miniature" specimen of 27mm in the overall length and 2mm in diameter at the gauge portion. Its design is presented in Fig. 1. Besides these, 6mm diameter regular-size (6mm diameter) specimens were prepared for comparison's sake.

**Testing Method.** The creep rupture tests were conducted in a creep tester of our own design(1), once

Table 1 Chemical composition of materials used

| Material | Chemical composition (wt.%) | | | | | | | | | | | | | |
|---|---|---|---|---|---|---|---|---|---|---|---|---|---|---|
| | C | Si | Mn | P | S | Ni | Cr | Mo | Cu | Ti | Aℓ | B | N | Nb+Ta |
| 2 ¼Cr−1Mo Steel (STBA24 (*1)) | 0.10 | 0.23 | 0.43 | 0.011 | 0.009 | 0.043 | 2.46 | 0.94 | 0.07 | — | 0.005 | — | .008 | — |
| 18Cr−8Ni Steel (SUS 304H TB (*2)) | 0.07 | 0.55 | 1.46 | 0.023 | 0.006 | 9.57 | 18.95 | 0.04 | 0.07 | 0.062 | 0.014 | 0.0018 | 0.0278 | <0.01 |

(*1) JIS G 3462−1978"Alloy Steel Boiler and Heat Excharger Tubes"

(*2) JIS G 3463−1978"Stainless Steel Boiler and Heat Exchanger Tubes"

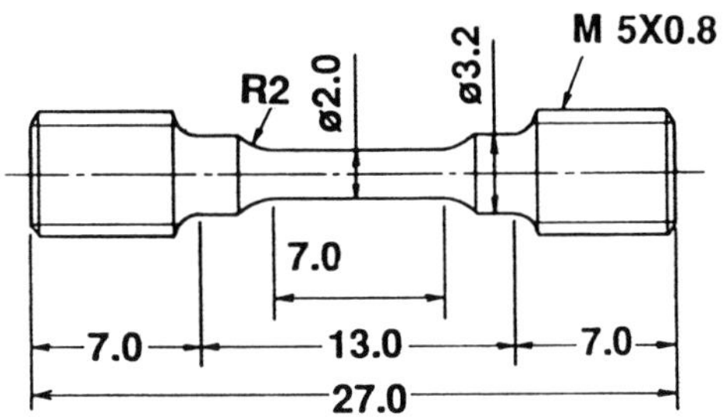


Fig. 1 Sampling location and dimension of miniature specimen

in an Ar atmosphere, then in ambient air to evaluate the effects of oxidation. Subsequently, the temperature-accelerated creep tests were conducted according to the ISO stress method (so as to estimate the creep life for the actual service temperature by running short time, constant load tests at higher temperatures) both with miniature test pieces (2mm diameter) and with regular test pieces (6mm diameter). The efficacy of the miniature specimen as a material for the temperature-accelerated test was examined on the data thus obtained by extrapolating the results to the range NRIM's long-term creep rupture data sheets (6mm diameter specimens)(2,3) cover.

## Results and Discussion

**Effect of Oxidation.** The results of the 2.25Cr-1Mo steel miniature specimen creep-rupture tests are presented in Figs. 2 and 3 in terms of the Larson-Miller parameter (P) respectively for the case of Ar-atmosphere and for the case of ambient air. Comparison of the two figures will show that the effect of oxidation becomes apparent as a decrease in the rupture time for P's larger than ca. $8\times10^4$.

On correcting for oxidation by taking the metal loss into account and by means of the linear damage accumulation law, we have obtained Fig. 4: the results of the ambinet air tests agree, when corrected appropriately as above, quite well with those of the Ar atmosphere tests. We conclude therefore that the decrease in the creep life observed in the ambient air test is due to increase in the true stress arising from loss of the load-carrying cross section. We find also that the miniature specimen in Ar atmosphere creep tests will give approximately the same creep rupture lives the regular specimens give in ambient air tests.

On the other hand, no such attrition of life due to metal loss on oxidation needs to be considered for 18Cr-8Ni, evidently owing to its ample resistance to air oxidation in the temperature range concerned.

**Temperature-Accelerated Creep Test.** *The case of creep-plasticized steel.* The 2.25Cr-1Mo steel, a typical

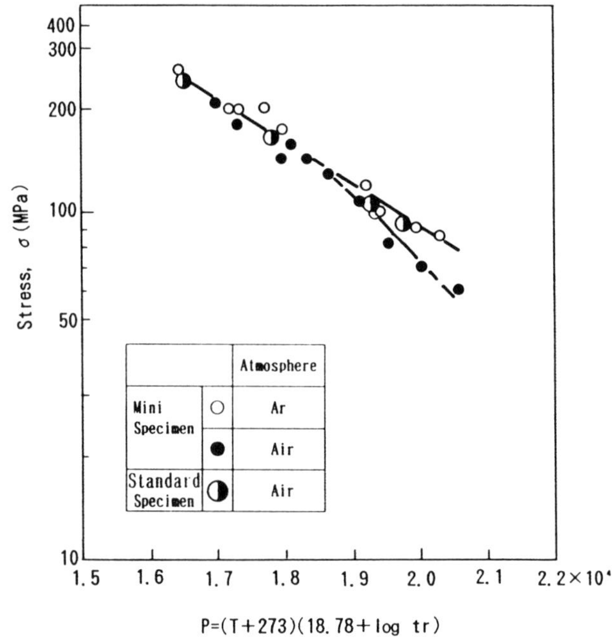


Fig. 2 Creep rupture life of $2\frac{1}{4}$Cr-1Mo steel

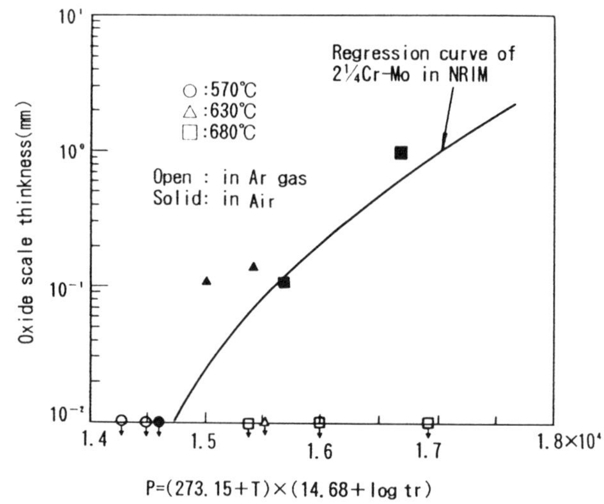


Fig. 3 Oxide scale growth kinetics for $2\frac{1}{4}$Cr-1Mo steel

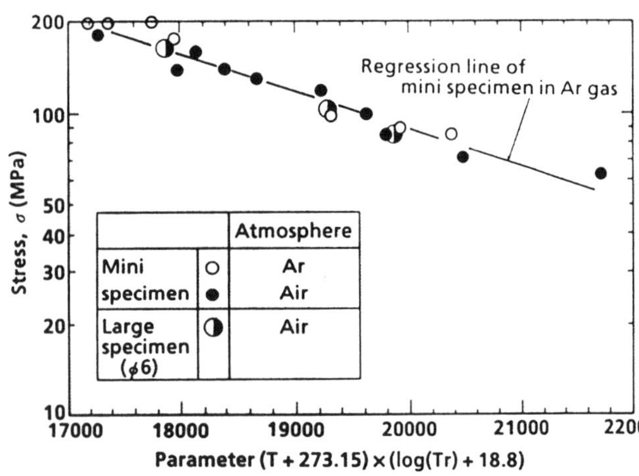


Fig. 4 Stress vs L.M. parameter for $2\frac{1}{4}$Cr-1Mo steel corrected for oxidation effect

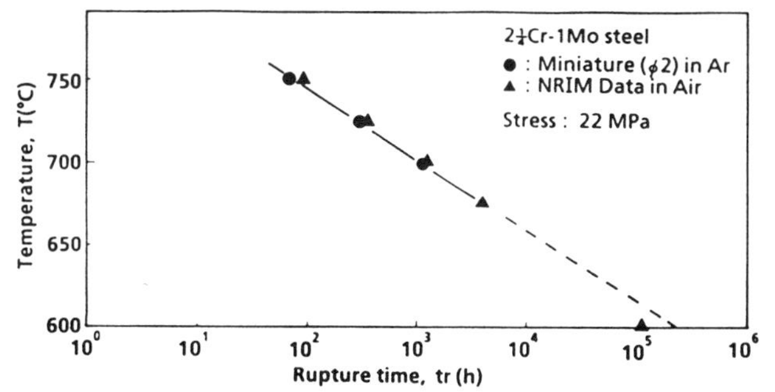


Fig. 5 Temperature vs. log tr in Iso-stress testing for 2¼Cr-1Mo steel

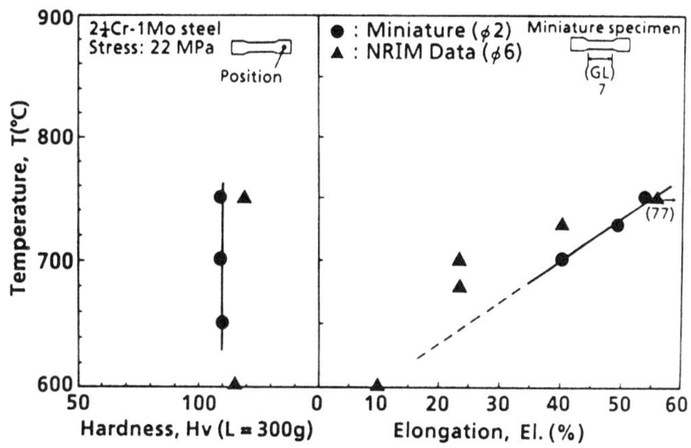


Fig. 6 Temperature vs. hardness and elongation for 2¼Cr-1Mo steel

creep-plasticized steel, gave results shown in Fig. 5 in the ISO stress method temperature-accelerated creep tests. Even though the miniature specimen agrees with the regular specimen satisfactorily enough for a stress of these short time data into the regime of the NRIM data sheet tends to overestimate the rupture lives for the miniature and for the regular specimens alike.

As a background for this observation, we offer changes in the hardness and in the rupture elongation in Fig. 6: even while hardness changes very little with temperature, elongation becomes the larger, the higher the temperature. Here, no conspicuous changes were observed in the microstructure up to 750°C, and the fracture mode was mainly intra-granular fracture, always ensuing a large cross sectional reduction.

Now, based on an observation that the form of creep curve changes with rising temperature in such a way as to give rise to larger creep rate, a commendable notion of correcting the rupture lifetime on the steady state creep rate has been proposed by Brear and McCarthy(4). Since the steady state creep rate appears to be proportionally relatable to the rupture elongation(4), moreover, we have corrected the observed creep data with the value of fracture elongation obtained for the given service temperature

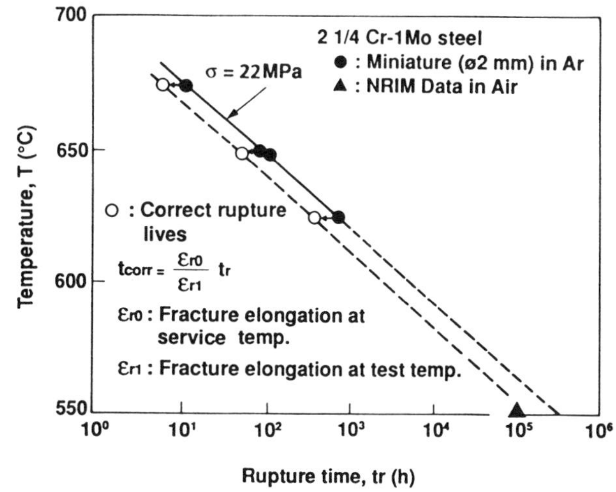


Fig. 7 Temperature vs. log tr for 2¼Cr-1Mo steel Iso-stress testing

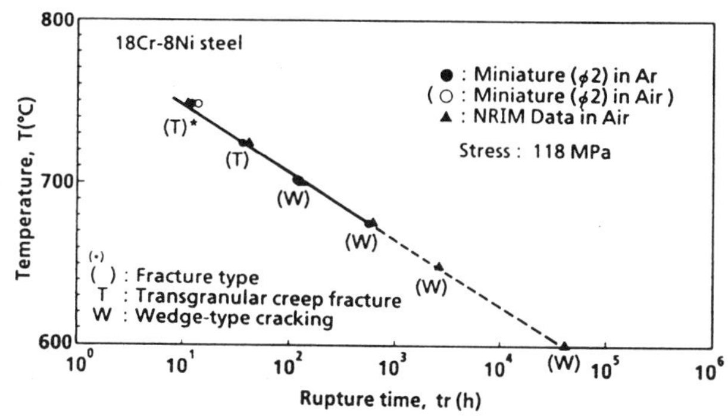


Fig. 8 Temperature vs. log tr in Iso-stress testing for 18Cr-8Ni steel

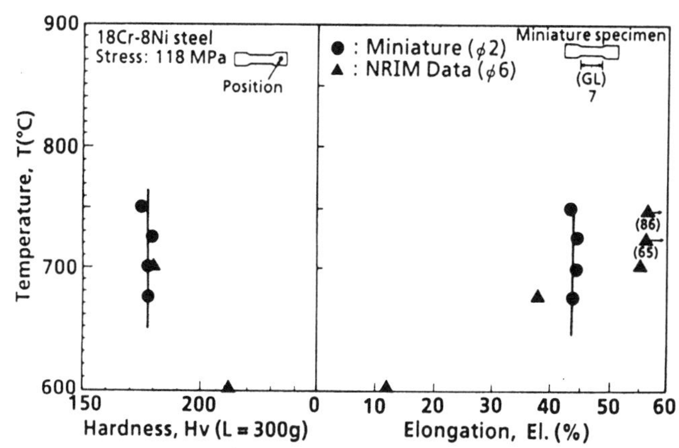


Fig. 9 Temperature vs. hardness and elongation for 18Cr-8Ni steel

as follows:

$$t_{corr} = (\varepsilon_{r0}/\varepsilon_{r1})\cdot t_r , \qquad (1)$$

where $\varepsilon_{r0}$ is the "standard" fracture elongation for the actual service temperature, i.e., the one that the NRIM data sheets(2,3) give, $\varepsilon_{r1}$, that obtained for the testing temperature, and $t_r$, the observed rupture life.

The results are shown in Fig. 7: considerable improvement in the accuracy of lifetime estimation

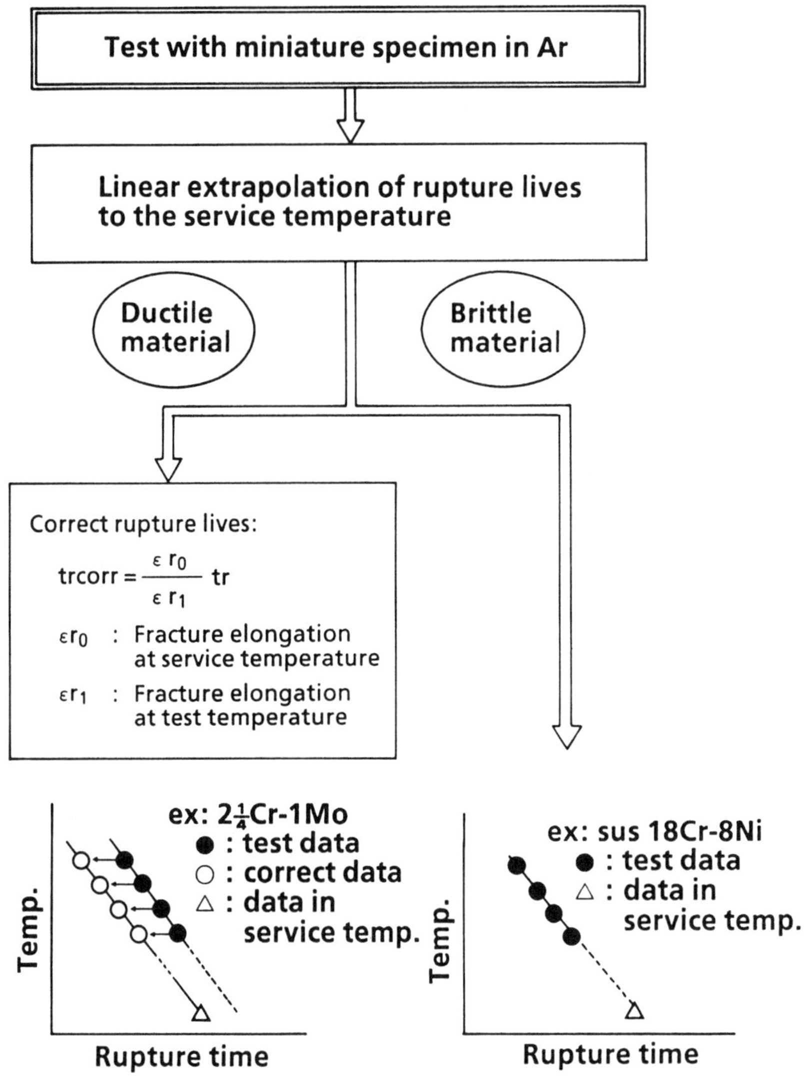


Fig. 10 Remaining lives assessment by creep rupture testing with miniature test piece

has been attained.

*The Case of Creep-Embrittled Steel.* The data for 18Cr-8Ni steel, a creep-embrittled material, are shown in Fig. 8. The results of the miniature specimens are in good accord with those of the regular specimens, and extrapolation from short time, high temperature side (for a stress of 118MPa) gives high degree of agreement with the NRIM data.

The change in hardness and rupture elongation are presented in Fig. 9. The hardness remains unvarying at Hv=170 level both for the miniature and for the regular specimens, except in the case of long-term (40,000 h) creep rupture at 600°C, where it is about 220 due undoubtedly to precipitation of $\sigma$ phse(3). Nevertheless, some disagreement is seen in the rupture elongation between these two sorts of specimen: where it remains substantially the same in the case of miniature specimen, the elongation tends to become the larger, the higher the temperature in the case of regular specimen. On the other hand, no apparent changes were observed both in the microstructure and in the fracture behavior for either specimen in the temperature range tested: the fracture mode was intragranular for temperatures above 700°C, but below it fracture took place by propagation of wedge-form cracks initiated at grain boundary triple junctions. Since precipitation of $\sigma$ phase occurs in the final stage of creep life, however, its influence on the rupture time should be considered practically nil.

*Residual Life Assessment with Miniature Specimen.* As presented above, the use of miniature specimen in conjunction with Ar atmosphere in the temperature-accelerated creep rupture test should afford a simple and effective means to assess the service life remaining in a damaged steel. The procedure of practicing this method is presented in Fig. 10: first, the object material is classified into two groups of creep-plasticized and creep-embrittled; perform the short time, temperature-accelerated tests according to the ISO stress method; then, if the material is of the former group, correct the observed rupture times on the rupture elongation as described in the previous section (if it is of the latter group, use the observed rupture lives uncorrected); and estimate the lifetime by extrapolating to the actual service temperature.

Although we believe this method to be reliable enough, we feel the need of proving its authenticity for wider range of stress.

## Conclusions

(1) Even when the steel is prone to adverse effects of oxidation, the miniature creep rupture test specimens (2mm diameter) can be protected from oxidation by conducting the test in Ar atmosphere, then they will give substantially the same creep strengths as the regular (6mm diameter) specimens do.

(2) The ISO stress method of short time, temperature-accelerated test conducted using the miniature specimen can be an apt means of estimating the long-term creep rupture life, hence the residual service life, with high enough degree of accuracy, provided:
that, for creep-plasticized material, such as 2.25Cr-1Mo steel, the observed rupture times be corrected by the ratio of the standard rupture elongation for the service temperature to the observed rupture elongation;
while that, for creep-embrittled material, such as 18Cr-8Ni steel, the observed rupture times be used without correction;
and that, for either kind of steel, the microstructure remain not affecting the creep strength.

## References

1) K. Sonoya and M. Kitagawa: *Zairyo*, 41-460 (1992) [in Japanese].
2) NRIM Creep Data Sheet No. 3B (1986).
3) NRIM Creep Data Sheet No. 4B (1986).
4) J.M. Brear and P. McCarthy: *Residual Life Assessment Methods, ERA Final Report* 89-0230.

*Proceedings of the Fifth International Conference on Creep of Materials, Lake Buena Vista, Florida, USA, 18-21 May 1992*

# Empirical Evaluation of Tensile Creep and Creep Rupture in a HIPed Silicon Nitride

**M.K. Ferber and M.G. Jenkins**
High Temperature Materials Laboratory
Oak Ridge National Laboratory
Oak Ridge, Tennessee

## ABSTRACT

Empirical evaluations of the long-term mechanical reliability of a hot-isostatically-pressed silicon nitride were conducted by measuring the tensile creep and creep rupture behaviours at 1150, 1260, and 1370° C. The stress and temperature sensitivities of the secondary (or minimum) strain rates were used to estimate the stress exponent and activation energy associated with the dominant creep mechanisms. Creep rupture behaviour was examined by allowing each creep test to continue until specimen failure.

The tensile creep/creep rupture data revealed two distinct failure mechanisms. At 1150° C, failure was controlled by a slow crack growth mechanism. At 1260 and 1370° C, the accumulation of creep damage as grain boundary cavities and cracks dominated the deformation behaviour whereby the creep rupture lives were controlled by the minimum strain rates in accordance with a Monkman-Grant-type relation.

MODERN SILICON NITRIDE (SN) CERAMICS are prime candidates for structural applications in advanced gas turbine engines (1,2). Hot-isostatically-pressed (HIPed) silicon nitrides utilizing yttria as the densification aid (Y-SN) are among the most promising candidates for these applications because of their excellent short-term strengths and fracture toughnesses. However, the successful utilization of these materials in such applications requires extensive characterization of their long-term mechanical reliability. For conditions involving high mechanical and thermal stresses at elevated temperatures, creep resistance is of particular concern. Excessive creep deformation can adversely affect the dimensional stability of the component, ultimately leading to loss of function. Furthermore, if damage evolution mechanisms such as cavitation are associated with the creep process, the resultant microstructural changes may lower the strength and lead to a time-dependent failure (i.e., creep rupture).

The analysis of creep data is typically based upon the assumption that the steady-state strain rate ($d\varepsilon_{ss}/dt = \dot{\varepsilon}_{ss}$) dominates the overall behaviour. For this case, the stress and temperature dependencies of $\dot{\varepsilon}_{ss}$ are given by the expression (3,4),

$$\dot{\varepsilon}_{ss} = A_o\,(\sigma_a/\sigma'_o)^n \exp(-Q_c/RT), \qquad (1)$$

where $A_o$ is a pre-exponential factor, $\sigma_a$ is the applied stress, $\sigma'_o$ is a normalizing parameter (i.e. $\sigma'_o$=1 MPa), n is the creep (or stress) exponent, $Q_c$ is the activation energy for creep, R is the universal gas constant, and T is the absolute temperature. The creep rupture life or time to failure, $t_f$, can be described by a similar equation (5),

$$t_f = B_o\,(\sigma_a/\sigma'_o)^{-N} \exp(-Q_f/RT), \qquad (2)$$

where N, $Q_f$, and $B_o$ now determine the stress and temperature dependencies of the dominant creep rupture mechanism. When failure is controlled by the accumulation of creep damage, $t_f$ is often found to be unique function of $\dot{\varepsilon}_{ss}$ independent of applied stress and temperature (6,7). The relationship between $t_f$ and $\dot{\varepsilon}_{ss}$ is adequately described by the Monkman-Grant relation (7),

$$t_f = C\,(\dot{\varepsilon}_{ss})^{-m}, \qquad (3)$$

where C and m are constants. Combining Eqs. 2 and 3 shows that $N = n \cdot m$.

Although the creep and creep rupture behaviours of silicon nitride have been the subjects of numerous studies (7-14), the application of these results to recently developed HIPed Y-SN ceramics is limited by several factors. First, much of the previous creep data were generated from flexure tests. The data analysis was generally based upon the assumption that the creep strain rates in compression and tension were equal (15), which as shown in Refs. 10, 11, and 16 is not valid for these materials. A second complicating

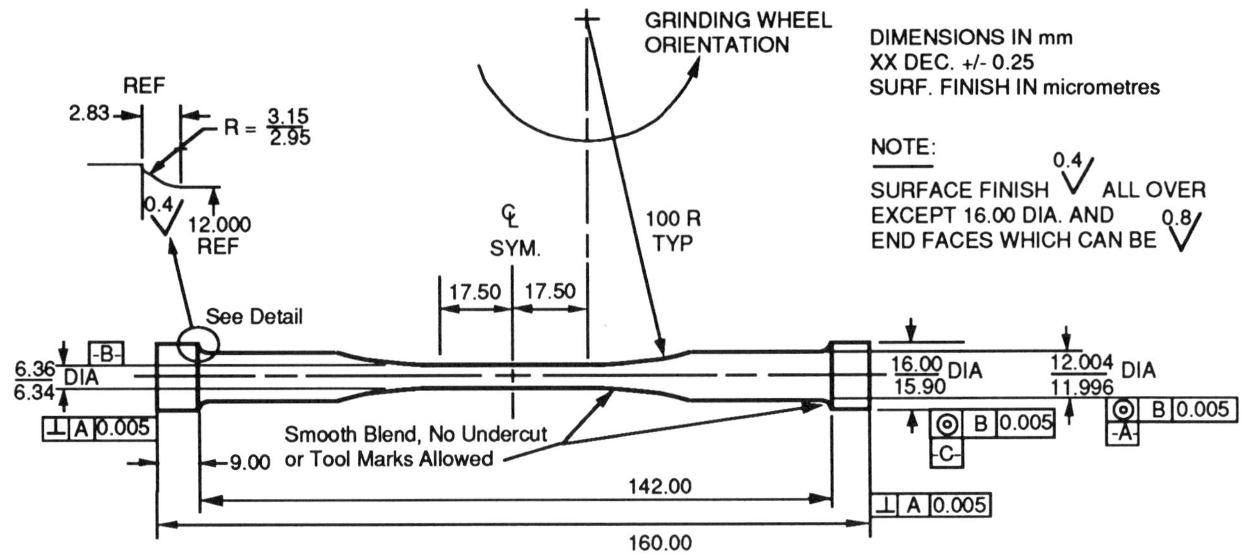


Fig. 1 - Dimensional specifications for the button-head tensile specimen.

factor is that most of the previous studies considered either creep or creep rupture, but not both. Consequently, a comprehensive understanding of the relationship between the creep and creep rupture behaviours (Eq. 3) was not possible. Finally, much of the earlier research involved the evaluation of hot-pressed or gas phase-sintered SN with relatively high second phase contents (MgO, $Y_2O_3$, $Al_2O_3$,etc.). Since the method of fabrication (17), type of densification aid (12), and amount of densification aid (18,19) influence creep and creep rupture resistances, the results of these studies would not necessarily reflect the behaviour expected for a HIPed Y-SN.*

In the present study the long-term mechanical reliability of a commercially-available HIPed Y-SN was evaluated by measuring the tensile creep and creep rupture behaviours at 1150, 1260, and 1370° C. The material is detailed followed by a description of the test set-ups and the empirical procedure. Creep/creep rupture behaviour is presented in which analyses of the stress and temperature sensitivities of the secondary (or minimum) strain rate are used to estimate the stress exponent and the activation energy associated with the dominant creep mechanism. Microstructural results are detailed in which key features of the fracture surfaces are described. Finally, possible creep mechanisms and creep rupture mechanisms and the applicability of several existing creep/creep rupture models are discussed for describing the stress and temperature sensitivity of this HIPed silicon nitride under static loading.

## EMPIRICAL PROCEDURE

**PY6 Silicon Nitride.** The material examined in this study was a commercial, HIPed silicon nitride designated PY6.* The PY6 was fabricated using ~6 wt% yttria as the densification aid. Typically, the microstructure consisted of 1 to 6 μm long acicular grains surrounded by equi-axed grains 0.1 to 1.0 μm in diameter. This grain morphology produced a relatively dense microstructure. The silicon nitride grains were generally in the form of $\beta$-$Si_3N_4$ and were separated by relatively thin layers of an amorphous yttrium silicate. The intergranular phase was also present in the triple points as a crystalline yttrium silicate. The exact phase composition of these intergranular compounds depends on the HIP conditions. Button-head tensile specimens (6.35 mm gage section diameter and 35-mm gage length) were fabricated to the specifications given in Fig. 1 from isopressed and HIPed rods 160-mm long and 20-mm in diameter.

**Tensile Test Set-up.** All tests were conducted on commercial electro-mechanical load machines having both load- and strain-control capabilities.** The desired load (strain)-time profiles were controlled by trapezoidal function generators. The specimen grips, which were configured for button-head specimens, were located outside the compact two-zone resistance-heated furnaces capable of generating maximum temperatures of 1600°C. The grips were attached to the load frame using hydraulic couplers to minimize bending moments. Specimen displacements were measured over 25-mm gage lengths with direct-contact extensometers employing remote capacitance sensors (see Fig. 2). By carefully controlling the temperatures of both the measurement hardware and grip cooling water (20) resolutions of ~0.5 μm could be achieved with these arrangements. Personal computers and data acquisition systems were used

* The chemistry and degree of crystallinity of the intergranular phase can also influence the creep and creep rupture behaviours (Ref. 17). In previous studies of silicon nitride, these two characteristics were not always determined.

* PY6 silicon nitride, GTE Laboratories, Inc., Waltham, Massachusetts, (1989-1990 vintage).

** Instron Model 1380 Low-Cycle Creep Fatigue Machine, Instron Corporation, Canton, Massachusetts.

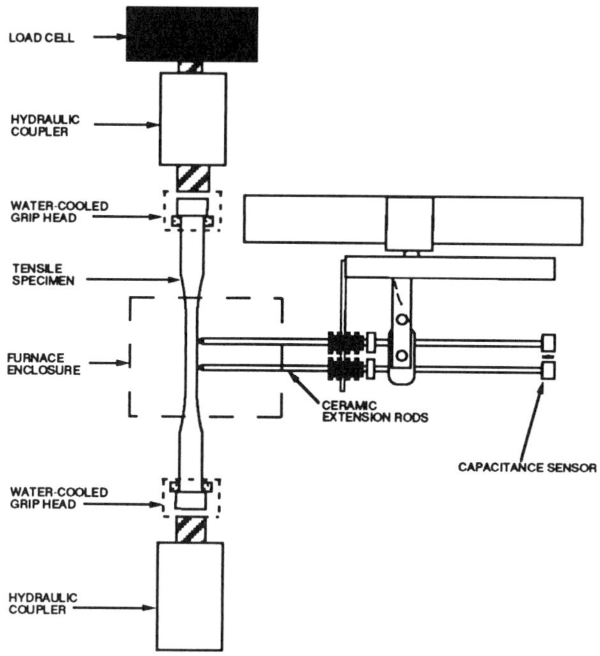


Fig. 2 - Schematic representation of loading configuration for tensile creep/creep rupture measurements.

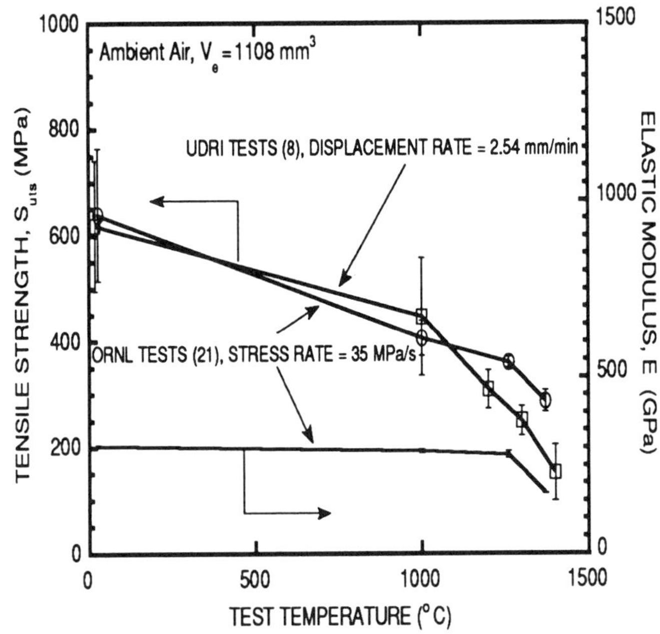


Fig. 3 - Temperature dependencies of tensile fracture strengths for PY6 silicon nitride. Data compared for the same effective tensile volumes (1108 mm$^3$) from Refs. 8 and 21.

to monitor and store various output signals including displacement, load, load error, and testing temperature. All tensile creep/creep rupture tests were conducted in ambient air under electronic load control. Specimens were strain gaged and the percent bending measured at room temperature to ensure that minimum bending was imposed at the lower loads and stresses applied during the creep tests. In all cases bending at the anticipated creep stresses was on the order of 3% of the uniaxial tensile stress.

Initial applied stress values for the creep/creep rupture tests were established using previously reported (8,21) fast fracture tensile strength versus temperature relations for PY6 with this same specimen geometry as shown in Fig. 3. Approximately six tensile creep/creep rupture tests of the PY6 were conducted at each temperature of 1150, 1260, and 1370°C. The lowest temperature examined, 1150° C, is believed to be near the critical temperature for softening of the secondary intergranular phase. The creep rupture behaviour at 1150° C would be expected to be dominated by slow crack growth since creep damage mechanisms are limited by kinetic considerations. The remaining two test temperatures examined, 1260 and 1370° C, are the design temperatures for the turbine rotor and stator inlet, respectively, for advanced gas turbine engine applications (1,2).

**Basic Material Characterization.** Several characterization techniques were used to examine both the as-received and tested specimens. For example, in order to observe the microstructure of the as-received material, several specimens were polished to a 0.25 μm finish and subsequently etched in molten potassium hydroxide for 1 to 4 minutes. The grain size and morphology were then determined with scanning electron microscopy (SEM).

Fracture surfaces resulting from creep/creep rupture tests were routinely examined with a low-power optical microscope. These observations were used to determine the nature of the creep rupture process (slow crack growth versus creep damage). Unfortunately for specimens tested at elevated temperatures, observations of the fracture surfaces made at higher magnification with SEM were obscured due to surface oxidation. However, it was possible to examine the cavitation process with SEM by generating secondary fracture surfaces in the gage section at room temperature. Because creep damage lowers the effective fracture resistance, this re-fracturing process is very effective at exposing such damage.

Variations in the phase composition (including both the silicon nitride and intergranular compounds) of the as-received PY6 were measured by x-ray diffraction. This technique was also used to examine changes in the phase composition resulting from the long-term creep exposures. For the tensile creep tests, these changes could be detected directly from the tested specimen by comparing the diffraction profile obtained from the gage section with that from the button-head region. Because the button head remained relatively cool (~200° C) during the test, the phases determined by x-ray diffraction should be representative of the as-received material. This approach minimized uncertainties arising from specimen to specimen variations in the phase compositions.

## RESULTS

**Creep/Creep Rupture Behaviour.** The time and stress dependencies of the tensile creep strain rates at 1150, 1260, and 1370° C are shown in Figs. 4a through 4c, respectively. There was a strong primary creep strain component in the 1260 and 1370° C curves. Previous studies (22,23) on silicon nitride have attributed the primary creep regime to viscoelastic deformation arising from grain boundary sliding accommodated by elastic deformation of grain boundary asperities. In the case of the 1150° C data, the large increase in strain at the onset of the test was due to the elastic strain component (900 - 980 με). A dominant steady-state region was clearly observed in the data generated at both 1150 and 1260° C. Specimen failure (vertical arrows) marked the end of this steady-state creep. However, at 1370°C a tertiary creep regime preceded failure. This tertiary creep could be attributed to the observed development of multiple cracks in the gage section.

The data generated at 1370° C exhibited a clear trend of increasing strain rate with increasing stress. However, as shown in Fig. 4b, the creep curves for the 1260° C tests were somewhat erratic in nature. In particular, the data generated at applied stresses of 125 and 150 MPa did not follow the trend of increasing creep strain rate with increasing stress. This behaviour, which was also apparent in the strain rate versus stress curve (Fig. 5), was thought to be a result of specimen-to-specimen variations in the crystallinity of the intergranular phase. For a given stress level, samples containing an amorphous intergranular phase would be expected to creep at faster rates than samples containing a highly crystallized grain boundary phase.

In order to measure the crystallinity of the intergranular phase, x-ray diffraction studies of the button-head and gage regions were conducted on samples tested at both 125 and 175 MPa (specimen designators P13 and P16, respectively). Because the tensile button head experienced relatively low temperatures (~200° C) during creep testing, the resulting x-ray diffraction data were expected to be representative of the starting composition. The major phase in all samples analyzed was β-$Si_3N_4$. However, the intergranular phase in sample P13 prior to testing was amorphous in character. X-ray data obtained from the gage section indicated that during the creep test, this amorphous phase crystallized into a δ-yttrium silicate (δ-$Y_2Si_2O_7$). The intergranular phase in specimen P16 prior to testing was already in the form of a crystallized $Y_2Si_2O_7$ phase. The creep exposure had no measurable effect upon the nature of this phase. These results suggest that the presence of the amorphous grain boundary phase in specimen P13 was responsible for the high strain rate. The specimen-to-specimen variations in the grain boundary phase crystallinity may have resulted from variations in the HIPing cycle.

Figure 5 shows log-log plots of the steady-state (or minimum) strain rate, $\dot{\varepsilon}_{ss}$ versus applied tensile stress, $\sigma_a$. Note that the stress dependency of the strain rate associated with the 1150° C tests exhibited more variability than that measured at 1260° C. As discussed below, the increased scatter in the 1150° C data was not due to variations in intergranular phase crystallinity but rather a change in failure mechanism. Additional support for this conclusion was provided by the creep rupture-time to failure characteristics.

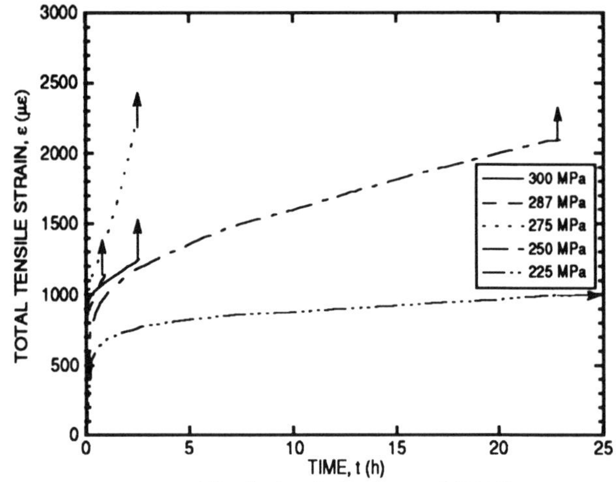


a) Strain-time behaviour at 1150° C

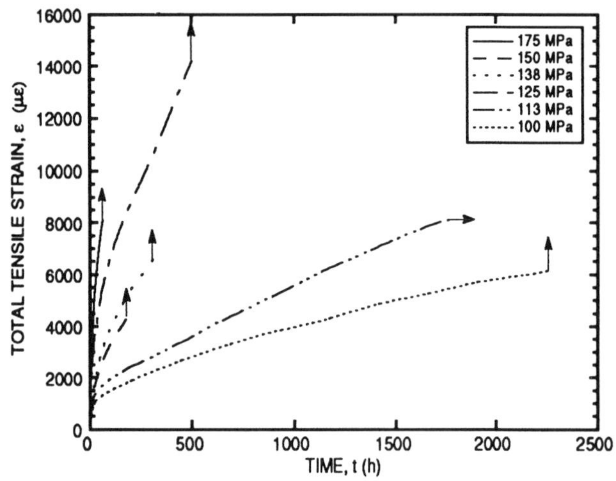


b) Strain-time behaviour at 1260° C

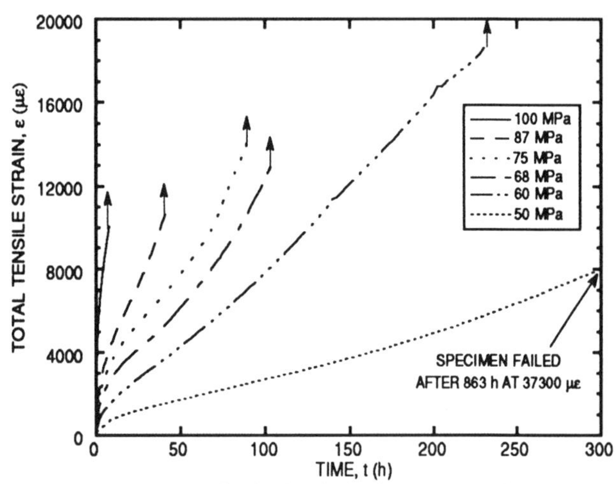


c) Strain-time behaviour at 1370° C

Fig. 4 - Stress and time dependencies of tensile creep strains measured at (a) 1150° C, (b) 1260° C, and (c) 1370° C for PY6 silicon nitride. Specimens tested at 1150° C ($\sigma_a$ = 225 MPa) and 1260° C ($\sigma_a$ = 113 MPa) did not experience creep rupture due to equipment problems.

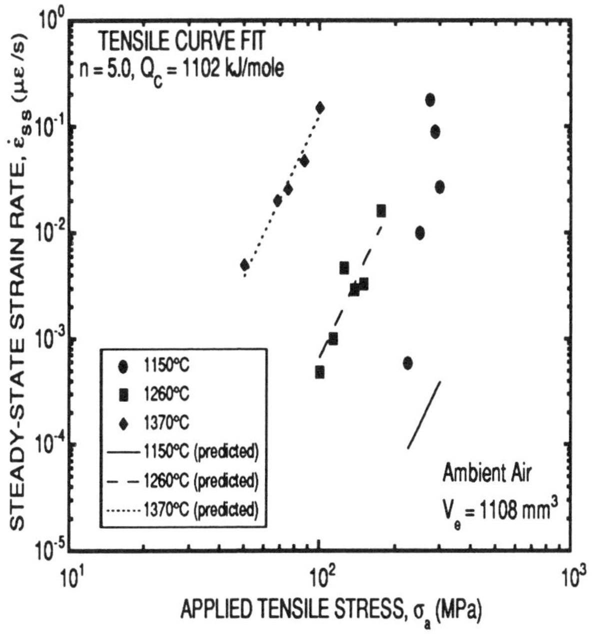


Fig. 5 - Stress and temperature dependencies of the tensile secondary (minimum) strain rate for PY6 silicon nitride. Solid and dashed lines represent the "best fit" of the empirical data (excluding the 1150° C results) to Eq. 1.

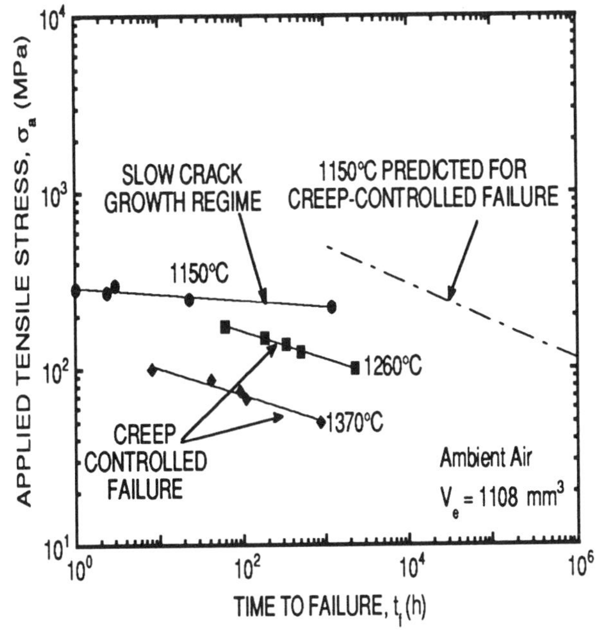


Fig. 6 - Tensile creep rupture behaviour of PY6 silicon nitride Tested at 1150, 1260, and 1370° C. The dashed line represents the creep rupture curve for the 1150° C tests predicted on the basis of creep damage accumulation and coalescence as estimated from Eqs. 1 and 3 using n = 5.0, $Q_c$ = 1102 kJ/mole, and m ≈ 1.

The n values (Eq. 1) calculated for each curve from separate regression analyses were 16.1 at 1150° C, 5.6 at 1260° C, and 4.7 at 1370° C. Based upon microstructural observations as well as results from previous creep studies of silicon nitride (7,9,10-13) one would conclude that the deformation at 1260 and 1370° C was due to diffusional creep (n = 1) accompanied by cavitation, grain separation by viscous flow, and possibly grain boundary sliding. As noted in the discussion section, these mechanisms are expected to give a nonlinear stress dependence ($2 < n < 4$). The extremely high value of n measured at 1150° C suggests a change in failure mechanism with decreasing temperature. Based upon results for other HIPed silicon nitrides (8,10) slow crack growth of pre-existing flaws or defects would be expected to control failure in the temperature range of 1000 to 1200° C. In this sense the strain rate values obtained at 1150° C may actually reflect compliance changes resulting from slow crack growth (24).

Values of $Q_c$ and n (Eq. 1) were estimated from a multiple-variable regression analysis of the tensile creep data in Fig. 5. The 1150° C results were excluded from this analysis since they were associated with a different failure mechanism. The resulting estimates of n and $Q_c$ were 5.0 and 1102 kJ/mole which provide a good description of the 1260 and 1370° C creep results as shown by the solid lines in Fig. 5. Activation energies measured for other SN ceramics (9,25) typically range from 550 to 700 kJ/mole. Possible reasons for this discrepancy are presented in a following section.

To examine the creep rupture behaviour, all tensile creep tests were continued until the specimens failed. The resulting tensile creep rupture relations are shown in Fig. 6. The tensile creep rupture exponents (N in Eq. 2) estimated from linear regression analyses were 18.0 at 1150° C, 6.4 at 1260° C, and 6.0 at 1370° C. The similarity of the creep and creep rupture exponents indicates that the failure process is controlled by the creep mechanism. This relationship is conveniently illustrated by expressing $t_f$ as a function of $\dot{\varepsilon}_{ss}$ (i.e., the Monkman-Grant type relation in Eq. 3). An application of the tensile experimental data to this relationship is shown in Fig. 7. Note that the data for the 1260 and 1370° C tests fall within a narrow band indicating that the creep rupture mechanisms were identical. Based upon SEM observations, the failure process probably involved the time-dependent accumulation of creep damage in the form of cavities and facet size cracks. Although the slope of the Monkman-Grant curve at 1150° C was similar to that of the higher temperature data, the curve was displaced to lower creep rupture-life values. This result implies that the creep rupture mechanism, while still controlled by the strain rate, was altered as the temperature was decreased. The fact that the fracture surfaces at 1150° C gave evidence of slow crack growth suggests that a localized creep crack extension process was responsible for the creep rupture behaviour.

Values of m estimated from the data in Fig. 7 were relatively independent of temperature and in the range 0.98 to 1.15. Taking m ≈ 1, which is consistent with values for other structural ceramics (10), leads to the prediction that N = n • m ≈ n in agreement with experimental observations.

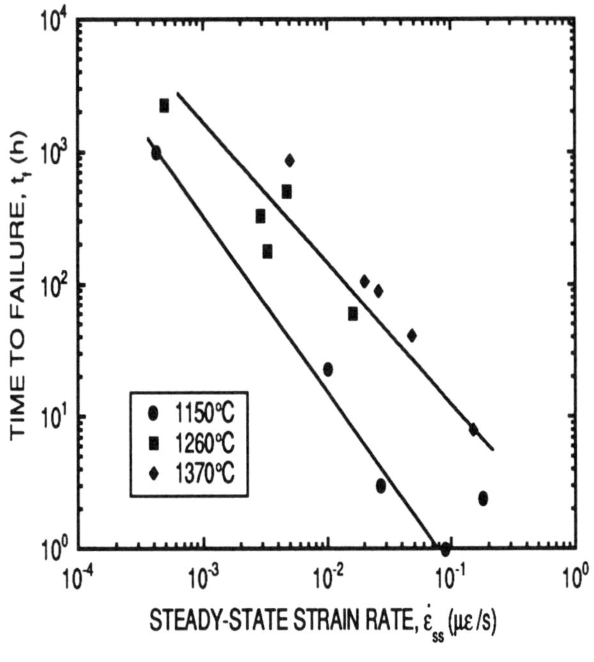


Fig. 7 - Dependence of tensile creep rupture life upon creep strain rate for PY6 silicon nitride. Monkman-Grant type relation provides a good description of the strain rate dependence of the creep rupture life for the PY6 at 1260 and 1370° C.

**Microstructural Results.** Low-power optical microscopy of the tensile fracture surfaces revealed a transition in the fracture markings as the test temperature increased from 1150 to 1370° C. At 1150° C the fracture surfaces contained well-defined mirror, mist, and hackle regions (Fig. 8). A region of apparent slow crack growth was also identified within the mirror region. SEM of this area (Fig. 8) showed the presence of microcracks, which were presumably involved in the slow crack growth process. As the applied stress decreased both the mirror and damage zone sizes increased. At 1260° C, the fracture surface was characterized by a relatively small creep damage zone (region of high surface roughness) originating from the gage section surface. This zone contained numerous microcracks extending over many grain facets and lenticular-shaped cavities located predominantly at two-grain junctions. Finally, as indicated in Fig. 8, the mirror size at 1260° C was significantly larger than that at 1150° C while the hackle region was smaller.

At 1370° C, the creep damage zone, which also appeared to originate from the surface, covered as much as 50% of the fracture surface with the mirror region comprising the remainder (Fig. 8). The absence of hackle markings suggests that the true mirror size exceeded the specimen dimensions. The creep damage zone was similar to that observed at 1260°C in that both cavitation and microcracking were predominate features. In addition, the sizes of the creep damage zone increased slightly with decreasing stress thus indicating a slight stress dependency of the damage accumulation process. These results are in agreement with those reported in other studies (11,26) of the creep deformation behaviour of silicon nitride.

An important difference in the failure behaviour at 1260 and 1370° C concerned the presence of multiple cracks in the gage sections of the 1370° C specimens. Therefore, as the temperature increased, the damage coalescence process became active throughout the gage section. This behaviour may have resulted from the increased oxidation at 1370° C. As indicated in Fig. 8, a 0.5 to 1 mm thick oxide scale was present along the outer surface of the 1370° C specimens. As discussed in the next section, surface oxidation may provide preferred sites for damage coalescence. This would also explain why the creep damage zone at 1260 and 1370° C appeared to originate from the gage section surface.

Detailed SEM examination of the cavities observed along the fracture surface of specimens tested at 1260 and 1370° C was prohibited by the oxide scale. However, these cavities could be readily observed by generating at room temperature a secondary fracture surface transverse to the test fracture surface (21). Extensive cavitation was found along two-grain junctions. These cavities, which were lenticular in shape, were present throughout the gage section volume. Furthermore, the extension of these cavities into the silicon nitride grains suggests that the growth mechanism involved solution/precipitation rather than viscous flow. No crack-like cavities were found, indicating that the rate of surface diffusion was relatively high (27). Finally, the number of cavities observed in the 1370°C specimens tended to decrease dramatically as the the applied stress decreased to 50 MPa which is consistent with a critical threshold for cavity nucleation (28).

A second relatively common feature observed along the 1260 and 1370° C secondary fracture surfaces was a skeletal pattern of the intergranular phase outlining the interfaces

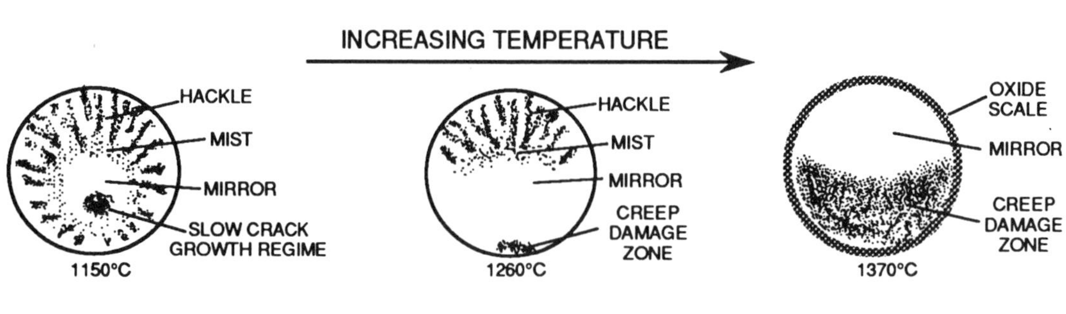


Fig. 8 - Summary of fracture surface characteristics for tensile creep rupture tests at 1150, 1260, and 1370° C.

between small grain clusters and a single large grain (21). The formation of this pattern may have been a consequence of grain separation by viscous flow of the boundary phase. Such deformation would lower the local fracture resistance of the intergranular phase thereby providing a preferred failure path during secondary fracture. This viscous flow process, as discussed in detail in Refs. 27 and 28, could ultimately lead to the formation of facet sized microcracks. The size of these microcracks was significantly greater than the average cavity diameter. However, the total damage generated by this viscous flow process may been been limited since potential damage sites were restricted to interfaces between large and small grains. Finally, the absence of extensive creep damage at 1150° C was a consequence of the increased viscosity of the intergranular phase.

## DISCUSSION

The stress and temperature sensitivities of the creep and creep rupture data as well as the microstructural observations provide critical insights into the nature of the underlying mechanisms controlling high temperature deformation and failure of PY6. The following sections provide limited discussion of these mechanisms.

**Creep Mechanisms.** Based upon results from previous creep studies on silicon nitride (7,9, 10-13) the accommodated creep deformation in the PY6 at 1260 and 1370° C was most likely controlled by grain boundary diffusion.* This mechanism would entail transport of species from the boundaries under compression to those under tension with no net change in volume Microstructural observations also revealed the presence of unaccommodated creep strain (i.e., creep damage), which resulted from both cavitation and viscous flow between small and large grain interfaces. Although grain boundary sliding is often cited as a deformation mechanism in the creep of silicon nitride, it was not expected to be the dominant or rate-controlling step during the creep of the PY6. This conclusion is based on a study by Lange (31) showing that sliding in the presence of a viscous intergranular phase is faster than the rate of grain deformation required to accommodate the sliding. During the first stage of creep, this accommodation is by elastic deformation while for long-term creep, grain deformation will be controlled by either diffusion or viscous flow. Grain boundary sliding may, however, play a secondary role by generating large transient stresses during abrupt sliding events, thus increasing the cavity nucleation rate (28).

One key indicator of the dominant creep mechanism is the creep stress exponent, n. In the present study n at 1260 and 1370° C was determined to be 5.0, which is greater than values (1 to 3) reported for similar silicon nitrides in previous flexural creep studies (9,12). This discrepancy is due to stress relaxation and redistribution in the flexure specimens (32). The fact that n is greater than that expected for pure diffusional creep ($n = 1$) is due to the stress dependency of the creep damage mechanisms. This damage gives rise to an additional (unaccommodated) strain rate component equal to the product of the steady-state strain rate in the absence of cavities and a parameter, $\rho$, which is proportional to the number of cavities (microcracks). It may be assumed that at a given stress, $\rho$ will increase from 0 (at time, $t = 0$) to some critical value, $\rho_c$, at failure. During this time if $\rho$ becomes significantly greater than 1, the strain rate may exceed the steady-state strain rate prior to failure. Such behaviour is consistent with tertiary creep deformation observed at 1370°C.

The stress dependence of $\rho$ can be estimated for two limiting cases. In the first case, if damage evolution is nucleation controlled (i.e., growth rate is relatively slow), then the number of similar sized cavities will be proportional to the product of the nucleation rate and the elapsed time. However, existing expressions (28) for the cavity nucleation rate reveal a very strong stress dependence [$\propto \exp(1/\sigma_a^2)$]. Estimates of the creep exponent based upon this stress dependence are unreasonably high ($n > 25$). For the second case , the number of nucleated cavities is assumed to be constant such that the growth rate controls damage accumulation. Time and stress-dependent changes in $\rho$ will be determined by the product of the time and the cavity generation rate. A previous analysis (3) of the cavity growth problem has shown that the cavity generation rate is proportional to $\sigma_a^c$ where c may be as great as 3. If $\rho >> 1$, then the creep stress exponent should be equal to 4, in reasonable agreement with the empirical results.

At 1150° C the creep exponent increased to 16.1 suggesting a change in the deformation mechanism. This hypothesis is confirmed by the results in Fig. 5, which illustrate poor agreement between the experimental strain rate-stress data at 1150° C and the curve predicted from the application of the 1370 and 1260° C data to Eq. 1. Microstructural observations of the 1150° C fracture surfaces revealed the presence of extensive microcracking within an apparent slow crack growth zone. Therefore, it is possible that measured deformations were due to compliance changes resulting from extension of these microcracks. This problem has been treated by Suresh and Brockenbrough (24) for the case of aligned slit cracks and random penny cracks. By assuming that the growth rate is a power-law function of the applied stress ($da/dt \propto \sigma_a^p$), the authors showed that the strain rate should be proportional to $\sigma_a^{p+1}$ for short-term creep and $\sigma_a^{2p+1}$ for long-term creep. Considering the relatively short times involved in the 1150°C tests, one would

---

* Although not directly examined in this investigation, creep by dislocation motion was not expected to be important at these temperatures. This conclusion based upon extensive TEM studies of a similar SN tested under similar conditions (Ref. 30).

expect the first relationship to apply. Further, by assuming that $p \approx N$,* the creep rupture exponent in Eq. 2, the predicted creep exponent is equal to 19 (i.e. 19 = N +1 where $N \approx 18$) which is in fair agreement with the measured valued of 16.1.

The activation energy, $Q_c$, describing the temperature dependence of the creep strain rate is a second key indicator of the rate-controlling mechanism. In the present study, possible mechanisms (for the 1260 and 1370° C data) include cavity nucleation, solution/precipitation of the silicon nitride in the intergranular phase, grain boundary diffusion, and viscous flow of the intergranular phase. Typical $Q_c$ values for silicon nitride ceramics (9,25) are in the range of 550 to 700 kJ/mole. Assuming that solution/precipitation is the underlying deformation mechanism, Raj and Morgan (25) showed that a major component to the activation energy is the heat of solution of silicon nitride (333 - 460 kJ/mole). However, a more recent analysis (9) of possible creep mechanisms has suggested that nitrogen is the rate-controlling step. This would explain the similarities in activation energies reported for creep and oxidation (33).

The experimentally determined $Q_c$ for PY6 was 1102 kJ/mole, which is considerably greater than values reported above. There are several possible explanations for this discrepancy. First, it is possible that the temperature dependence of the viscosity of the intergranular phase was the dominant factor in determining $Q_c$. Raj and Morgan (25) showed that for viscosity controlled creep, the activation energy can be quite high particularly near the glass transition temperature. A second possibility is that the strain rate at 1370°C was increased by the activation of additional volume generating mechanisms. In the present study, the two most likely sources for this incremental increase in strain rate were the generation of multiple gage section cracks and recrystallization of the intergranular phase.

**Creep Rupture Mechanisms.** The mechanisms responsible for specimen failure were more difficult to identify. In the case of the 1260 and 1370° C tests, failure was a result of the coalescence of creep damage followed by the generation of creep cracks. Although the predominant damage mechanisms were operative throughout the gage section, the accumulation process appeared to originate from the specimen surface (Fig. 8). This result suggests that the oxygen in the surrounding environment played a critical role in the coalescence stage. The increased oxidation rate at 1370°C would explain the fact that multiple surface cracks resulting from the coalescence process were observed at this temperature.

At 1150°C, damage in the form of microcracks was restricted to a well-defined region or zone (Fig. 8). Zone extension could be attributed to a slow crack growth mechanism similar to that observed for hot-pressed silicon nitrides (34).

Competition between the two failure mechanisms discussed above may have been responsible for the duality in the creep rupture-life characteristics measured for the PY6 (Fig. 6). Microstructural observations (Fig. 8) indicate that the size of the associated damage zone was a critical factor in determining the dominant mechanism. For example, slow crack growth, which controlled the creep rupture behaviour at lower temperatures and/or higher stresses (Fig. 6), was a result of subcritical extension of a localized microcrack zone. Fracture occurred when the product of the damage zone size, $a_Z$, and applied stress, $\sigma_a$, reached a critical value, possibly in accordance with a Griffith-type expression, $\sigma_a (a_Z)^{1/2} \propto K_{Ic}$ (i.e., stress-controlled fracture). In this sense, the localized microcrack zone was analogous to a large defect. As the temperature increased, the strain rate increased at relatively low stresses thereby promoting a rapid extension of the creep damage zone throughout the specimen. The creep-induced relaxation of crack-tip stresses precluded failure via the stress-controlled mechanism. In this case, fracture resulted when the strain associated with the creep damage reached a critical value dictated by the Monkman-Grant relationship (i.e., strain-controlled fracture).

The applicability of Monkman-Grant relationship to the experimental data has a number of interesting implications. First, it indicates that the rate of damage generation (cavitation and viscous sliding) was controlled by the bulk creep response of the material. In the case of cavitation, for example, such a situation arises when local diffusion processes associated with cavity growth are much faster that the bulk creep deformation of the surrounding material. The cavity growth thus becomes constrained by the ability of the surrounding material to accommodate the volume generation (3). The creep rupture life, which is determined by damage generation rate, will be controlled by the strain rate as well. The Monkman-Grant relationship also provides a powerful tool for predicting creep rupture lifetimes from creep deformation rates. As shown in Fig. 7, the creep rupture life is uniquely determined by strain rate independent of applied stress and temperature (for temperatures >1200° C).

## CONCLUSIONS

For PY6 silicon nitride the stress dependencies of the creep strain rates measured at 1260 and 1370° C were consistent with a diffusional creep mechanism accompanied by cavitation, grain separation by viscous flow, and possibly grain boundary sliding. Microstructural observations provided evidence of both cavitation and viscous deformation. This latter process was particularly evident along the interfaces between small grain clusters and a single large grain. The creep strain rates at 1260 and 1370° C also exhibited extremely strong temperature sensitivity as reflected by the activation energy of 1102 kJ/mole. This temperature sensitivity could be attributed to either the strong temperature

---

* Integration of the expression $da/dt = \sigma_a^{\,p}$ yields $t = \sigma_a^{\,-p} (a_f - a_i)$ where $a_f$ and $a_i$ are the final and initial flaw sizes, respectively. If one neglects the stress dependency of $a_f$, then comparison of this expression with Eq. 2 gives $p \approx N$.

dependency of the viscosity or the activation of additional volume generating mechanisms as the temperature increased from 1260 to 1370° C.

The creep rupture data revealed two distinct failure mechanisms. At 1150° C, failure was controlled by a slow crack growth mechanism. At 1260 and 1370° C, the accumulation of creep damage in the form of grain boundary cavities dominated the creep rupture behaviour. Oxidation also appeared to play a significant role in the accumulation process. In this temperature regime, the creep rupture life was controlled by the secondary (or minimum) strain rate in accordance with a Monkman-Grant-type relation.

## ACKNOWLEDGMENTS

Research sponsored by the U.S. Department of Energy, Assistant Secretary for Conservation and Renewable Energy, Office of Transportation Technologies, as part of the Ceramic Technology Project of the Materials Development Program, under contract DE-AC05-84OR21400 managed by Martin Marietta Energy Systems, Inc.

In addition, empirical work conducted at the High Temperature Materials Laboratory sponsored by the U.S. Department of Energy, Assistant Secretary for Conservation and Renewable Energy, Office of Transportation Technologies, High Temperature Materials Laboratory User Program, under contract DE-AC05-84OR21400 managed by Martin Marietta Energy Systems, Inc.

## REFERENCES

1 - H. E. Helms, R. A. Johnson, and L. E. Groseclose, AGT 100-Advanced Gas Turbine Technology Development Project, pp. 137-55 in "Proceedings of the Twenty-Third Automotive Technology Development Contractors' Coordination Meeting," P-165, Society of Automotive Engineers,Warrendale, Penn. (1986).

2 - D. Carruthers and L. Lindberg, Critical Issues for Ceramics for Gas Turbines, pp 1258-72 in "Third International Symposium on Ceramic Components and Materials for Engines," V. J. Tennery, ed., American Ceramic Society, Westerville, Ohio (1988).

3 - H. Riedel, "Fracture at High Temperatures," Springer-Verlag, Berlin, Heidelburg, Germany (1987).

4 - W. R. Cannon and P. E. D. Morgan, REVIEW: Creep of Ceramics, Part 1 Mechanical Characteristics, *J. Mater. Sci.*, **18**, 1-50 (1983).

5 - D. I. G. Jones, Phenomenological Analysis of Time-Dependent Mechanical Behavior of Some Ceramic Materials, *Ceram. Eng. Sci. Proc.*, **7** [7-8] 802-16 (1986).

6 - M. K. Ferber and M. G. Jenkins, Rotor Data Base Generation, pp 340-61 in "Ceramic Technology for Advanced Heat Engines Project Semiannual Progress Report for October 1990 Through March 1991," ORNL/TM-11859, Oak Ridge National Laboratory, Oak Ridge, Tenn., August 1991.

7 - D. C. Cranmer, B. J. Hockey, S. M. Wiederhorn, and R. Yeckley, Creep and Creep-Rupture of HIP-ed $Si_3N_4$, *Ceram. Eng. Sci. Proc.*, **12** [9-10] 1862-71 (1991).

8 - N.L. Hecht, Environmental Effects in Toughened Ceramics-Final Report, UDR-TR-91-15, University of Dayton, Dayton, Ohio, February 1991.

9 - J. A. Todd and Zhi-Yue Xu, The High-Temperature Creep Deformation of $Si_3N_4$-$6Y_2O_3$-$2Al_2O_3$, *J. Mater. Sci.*, **27**, 4443-52 (1989).

10 - M. K. Ferber, M. G. Jenkins, T. A. Nolan, and R. Yeckley, Creep-Fatigue of Structural Ceramics: I, Comparison of Flexure, Tension, and Compression Testing, pp. 343-51 in "The Thirty-Seventh Sagamore Army Materials Conference Proceedings," D.J. Viechnicki, ed., Materials Technology Laboratory, Watertown, Mass. (1991)

11 - M. K. Ferber, M. G. Jenkins and V. J. Tennery, Comparison of Tension, Compression, and Flexure Creep for Alumina and Silicon Nitride Ceramics, *Ceram. Eng. Sci. Proc.*, **11** [7-8] 1028-37 (1990).

12 - A. E. Pasto, W. C. Van Schailkwyk, and F. M. Mahoney, Creep Behavior of Yttria- and Alumina-Doped Silicon Nitride, pp 776-85 in "Third International Symposium on Ceramic Components and Materials for Engines," V. J. Tennery, ed., American Ceramic Society, Westerville, Ohio (1988).

13 - R. L. Yeckley and K. N. Siebein, High Temperature Cavitation of HIP Silicon Nitride, pp 751-65 in "Third International Symposium on Ceramic Components and Materials for Engines," V. J. Tennery, ed., American Ceramic Society, Westerville, Ohio (1988).

14 - N. J. Tighe, S. M. Wiederhorn, T. J. Chuang, and C. L. McDaniel, Creep Cavitation and Crack Growth in Silicon Nitride, pp 587-604 in "Deformation of Ceramic Materials II, Materials Science Research", Vol. 18, R. E. Tressler and R. C. Bradt, eds. Plenum Press, New York, New York (1984).

15 - G. W. Hollenberg, G. R. Terwilliger, and R. S. Gordon, Calculation of Stresses and Strains in Four-Point Bending Creep Tests, *J. Am. Ceram. Soc.*, **54** [4] 196-99 (1971).

16 - R. F. Krause, Jr., Comparison of Flexural Creep Rate with that Predicted from the Power-Law Parameters for the Compressive and Tensile Creep, submitted to the American Ceramic Society (1991).

17 - M. K. Ferber, T. N. Tiegs, and M. G. Jenkins, Effect of Post-Sintering Microwave Treatments Upon the Mechanical Performance of Silicon Nitride, *Ceram. Eng. Sci. Proc.*, **12** [9-10] 1993-2004 (1991).

18 - I. P. Tuersley, G. Leng-Ward, and M. H. Lewis, High-Temperature $Si_3N_4$-Based Ceramics, *Brit. Ceram. Proc.*, **26**, 231-46 (1990).

19 - Y. R. Xu, T. S. Yen, and X. R. Fu, Grain Boundary Tailoring of High Performance Nitride Ceramics and Their Creep Property Studies, pp 739-50 in "Third International Symposium on Ceramic Components and Materials for Engines," V. J. Tennery, ed., American Ceramic Society, Westerville, Ohio (1988).

20 - M. K. Ferber and M. G. Jenkins, Rotor Data Base Generation, pp 361 to 371 in "Ceramic Technology for Advanced Heat Engines Project Semiannual Progress Report for October 1988 Through March 1989," ORNL/TM-11239, Oak Ridge National Laboratory, Oak Ridge, Tenn., August 1989.

21 - M. K. Ferber and M. G. Jenkins, Evaluation of the High-Temperature Mechanical Reliability of a HIPed Silicon Nitride, submitted to the American Ceramic Society (1992).

22 - F. F. Lange, D. R. Clark, and B. I. Davis, Compressive Creep of $Si_3N_4$/MgO Alloys. Part 2: Source of Viscoelastic Effect, *J. Mater. Sci.*, **15**, 611-5 (1980).

23 - R. M. Arons and J. K. Tien, Creep and Strain Recovery in Hot-Pressed Silicon Nitride, *J. Mater. Sci.*, **15**, 2046-58 (1980).

24 - S. S. Suresh and J. R. Brockenbrough, A Theory for Creep by Interfacial Flaw Growth in Ceramics and Ceramic Composites, *Acta Metall. Mater.*, **38** [1] 55-68 (1990).

25 - R. Raj and P. E. D. Morgan, Activation Energies for Densification, Creep, and Grain-Boundary Sliding in Nitrogen Ceramics, *Comm. J. Am. Ceram. Soc*, **64** [10] C143-C145 (1981).

26 - T. Ohji, S. Sakai, M. Ito, Y. Yamauchi, W. Kanematsu, and S. Ito, "Yielding Phenomena of Hot-Pressed $Si_3N_4$, *High Temp. Tech.*, **5** [3] 139-44 (1987).

27 - A. G. Evans and A. Rana, High Temperature Failure Mechanisms in Ceramics," *Acta Metall*, **28**, 129-41 (1980).

28 - J. E. Marion, A. G. Evans, and M. D. Drory, High Temperature Failure Initiation in Liquid Phase Sintered Materials, *Acta Metall*, **31**, 1445-57 (1983).

29 - C. H. Hsueh and A. G. Evans, Creep Fracture in Ceramic Polycrystals II. Effects of Inhomogeneities on Creep Rupture, *Acta Metall*, **29**, 1907-12 (1981).

30 - T. A. Nolan, M.K. Ferber, and D. W. Coffey, Microstructural Characterization of Tensile and Flexural Creep Deformation and Fatigue in a $Si_3N_4$ Ceramic, presented at 15th Annual Conference on Composites and Advanced Ceramic Materials, Cocoa Beach, Florida, January 13-16, 1991.

31 - F. F. Lange, Non-Elastic Deformation of Polycrystals with a Liquid Phase, pp. 361-81 in "Deformation of Ceramic Materials," R. C. Bradt and R. E. Tressler, eds., Plenum Press, New York, New York (1975).

32 - D. F. Carroll, T-J Chuang, and S. M. Wiederhorn, A Comparison of Creep Rupture Behavior in Tension and Bending, *Ceram. Eng. Sci. Proc.*, **9** [7-8] 635-41 (1988).

33 - K. L. Luthra, Some Perspectives on Oxidation of Silicon Carbide and Silicon Nitride, *J. Am. Ceram. Soc.*, **74** [5] 1095-103 (1991).

34 - R. K. Govila, Uniaxial Tensile and Flexural Stress Rupture Strength of Hot-Pressed $Si_3N_4$, *J. Am. Ceram. Soc.*, **65** [1] 15-21 (1982).

*Proceedings of the Fifth International Conference on Creep of Materials, Lake Buena Vista, Florida, USA, 18-21 May 1992*

# Microstructural Effects on the Tensile Creep of Silicon Nitride

**S. Haig and W.R. Cannon**
Department of Ceramic Engineering
Rutgers University
Piscataway, New Jersey

**P.J. Whalen**
Allied-Signal, Inc.
Research and Technology Center
Morristown, New Jersey

**R.G. Rateick**
Bendix Engine Controls Division
South Bend, Indiana

## Abstract

The tensile creep behavior of two silicon nitride materials with similar compositions but different grain sizes was studied. A computer-controlled, pneumatically-loaded creep apparatus was used to assess the relative deformation responses between 69 and 173 MPa at 1100 °C and 1150 °C. The coarse-grained silicon nitride exhibited significantly greater creep resistance than the fine-grained material over the investigated range of stress and temperature. Since the grain boundary phases were found to be nearly identical in the two materials, the enhanced creep resistance for the coarse-grained material was attributed largely to the effects of microstructure coarsening on potential creep-controlling mechanisms (i.e. grain boundary sliding, viscous flow, cavitation, etc.).

ONE OF THE LEADING MATERIAL CANDIDATES for use in high performance structural applications is silicon nitride. Due largely to its covalent nature, silicon nitride possesses excellent potential as a strong, hard, creep-resistant high temperature material. However, the covalency of silicon nitride also necessitates the use of comparatively low-melting additives to achieve full densification. The resulting high temperature behavior of the sintered material is then limited to a great extent by the refractoriness of this second phase. For instance, the creep resistance of silicon nitride above 1000 °C has been shown to depend primarily on the nature of the residual grain boundary phase. [1-7] The rate-determining mechanism for creep in these materials has generally been found to be grain boundary sliding which is accommodated by diffusion of the viscous second phase or by cavitation.

Recently, a new class of high-toughness in-situ reinforced (ISR) silicon nitride materials has been developed which exhibits much coarser microstructures than those found in typical fine-grained high-strength silicon nitride. The higher toughness of the ISR silicon nitride material is due to the crack shielding effects (i.e. crack wake bridging) of the large high-aspect-ratio ( ~ 10) acicular grains which can be developed in these materials through composition control and high-temperature heat treatments. The coarse ISR silicon nitride microstructure has been previously shown to enhance the resultant compressive creep behavior.[8] The present study used a computer-controlled pneumatically-loaded creep apparatus to compare the tensile creep behaviors of two silicon nitride materials with identical compositions but different microstructures. All tests were conducted in the stress range of 69 - 173 MPa at 1100 °C and 1150 °C. The relative tensile creep responses of these two silicon nitrides will be discussed in this paper.

## Materials

The two silicon nitrides used in this study were developmental materials provided by Allied-Signal, Inc. in Morristown, New Jersey. Both materials contained precisely the same sintering additives (4 wt. % $Y_2O_3$ and 4 wt. % $MgAl_2O_4$), but they were processed differently in order to create distinct microstructures. The fine-grained material, designated AS44, was pressureless sintered to near (> 99 %) theoretical density. The coarse-grained material, designated AS440, was prepared by subsequently annealing the fine-grained material at high temperature in order to coarsen the microstructure. This high-temperature heat treatment was found to grow the grains uniformly by approximately an order of magnitude, but it did not significantly affect the resulting aspect ratio. The fine- and coarse-grained materials were thus assumed to have aspect ratios which were not statistically different. A more thorough description of the relative microstructures of these two materials has been presented elsewhere.[8] SEM micrographs of the two silicon nitrides used in this study can be seen in Figure 1. The change in grain size which results from the post-sintering annealing step is apparent.

X-ray analysis of the "as prepared" fine- and coarse-grained materials showed that they were composed mainly of ß-silicon nitride grains bonded together by an amorphous grain boundary phase. This residual grain boundary glass phase crystallized rather quickly in both materials at elevated temperatures. For instance, after only a two-hour anneal at 1100 °C , both silicon nitrides exhibited similar amounts of a second phase (possessing > 98% crystallinity) that was distinct from ß-silicon nitride. Although the XRD pattern of this second phase did not perfectly match any known indexed pattern, it did approximate that for the clinopyroxene family of silicates. Its failure to match a known pattern exactly is most likely due to the wide range of anion and cation solid solution which may occur at the grain boundaries in these materials to shift the resulting X-ray peaks. Subsequent annealing of both materials at 1100 °C did not increase the apparent amount of crystalline phase any further.

TEM research, however, suggests that not all of the residual grain boundary amorphous phase is crystalline after long-term exposure at 1100 °C. Dark-field TEM analysis of a crept fine-grained specimen indicated that some residual

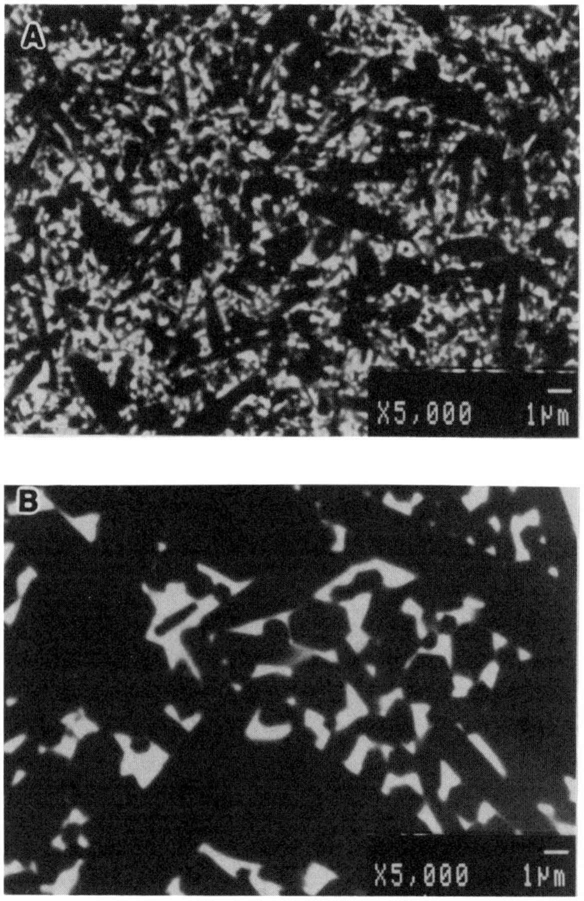


**Figure 1** Microstructures of fine (A) and coarse (B) grained $Si_3N_4$ showing relative grain sizes.

amorphous phase remained at the edges of triple-point pockets and along grain boundaries after more than 200 hours at 1100 °C and 69 MPa stress. Such a finding, which could not have been discovered solely through the use of XRD, corroborates previous reports that suggest complete crystallization of the grain boundaries may never occur as a result of grain-to-grain interactive forces.[9-12]

## Experimental Procedure

The tensile creep test apparatus closely resembled a unit developed several years ago at N.I.S.T. by Carroll et al.[13] It consisted mainly of a clamshell high-temperature furnace, a self-aligning tensile load train, a laser extensometer, and a sophisticated data acquisition and control system. A schematic of the load train can be seen in Figure 2.

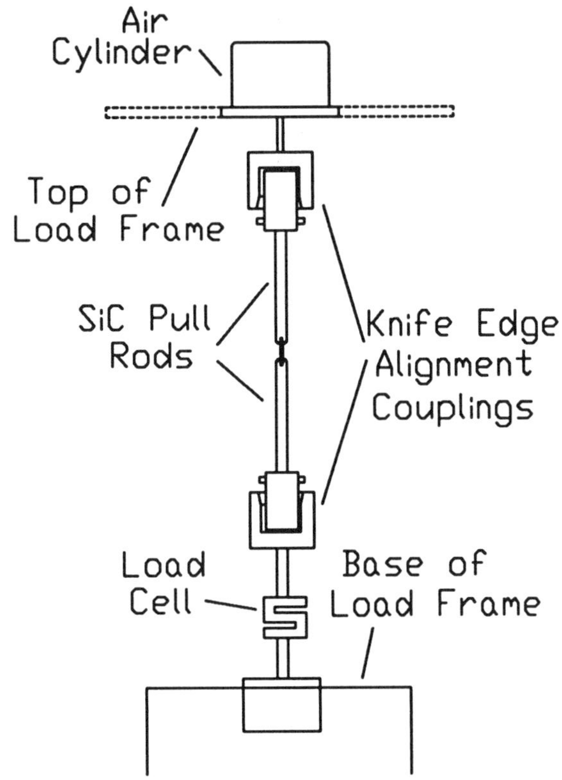


**Figure 2** Tensile creep apparatus load train.

Load application was achieved through use of a pneumatic air cylinder located at the top of the load train. An electric-to-pneumatic transducer served as a crucial interface between the data acquistion and control system and the load train itself. This transducer used analog input from the control computer to regulate pressure output to the air cylinder. Since the entire load train was fixed to the test frame at the bottom, a retraction of the air cylinder's pull rod served to apply a tensile stress to the test specimen. The actual applied load was then measured by means of a shear beam load cell. The silicon carbide pull rods extended into the hot zone of the clamshell high-temperature furnace, and the tensile specimen was affixed to these rods using silicon nitride pins in a pin-and-clevis configuration. Alignment of the load train was further aided at the top and bottom by the use of universal knife-edge alignment couplings.

A laser extensometer was used to monitor creep strain by means of a vertical scan of the test specimen's gauge section. Each specimen had silicon carbide marker flags glued on at room temperature, and these flags were ultimately held in place at test temperature by means of friction arising from the geometry of the flag's slots. This procedure for marking a creep specimen's gauge section has been previously described elsewhere.[13] The laser extensometer scanned the distance between the marker flags at a rate of approximately 100 scans/second and output an averaged value to the control computer every 2.5 seconds. The computer then further averaged 100 of these readings to yield a single flag-to-flag distance value with a sensitivity of ± 0.75 microns.

The tensile creep specimen itself was flat dogbone in shape and was also patterned after one originally developed at N.I.S.T.. It measured 50 mm from top to bottom and its gauge section had a cross-sectional area of 6.25mm$^2$. A schematic displaying precise specimen dimensions is shown in Figure 3. The attachment holes of the specimen were double-tapered in order to provide for improved alignment within the load train.

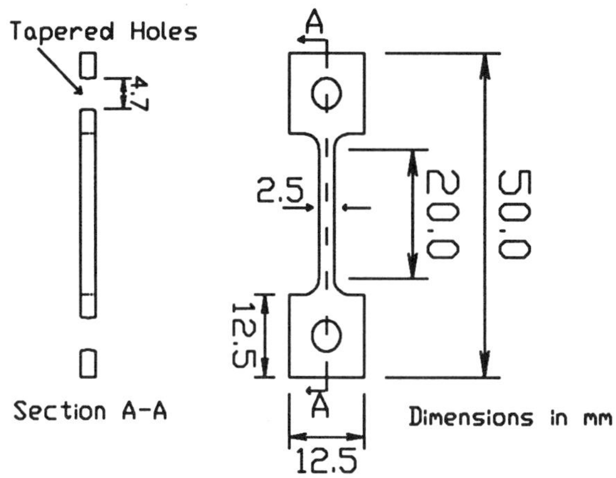


**Figure 3** Flat dogbone tensile creep specimen design.

All tensile creep samples were held at test temperature for 12 hours prior to application of the test stress. This was done to equilibrate the system and to assure a standard crystalline state at the time when the creep test stress was first applied. The creep specimen was also held under a 14 MPa tensile stress during this equilibration period so that the subsequent increase to test stress would not unseat the SiC marker flags from their positions. A stress of 14 MPa represented the minimum stress which could be reproducibly applied to a creep specimen due to the weight of the lower half of the load train.

Creep tests were conducted on the fine-grained and the coarse-grained silicon nitrides in the 69 - 173 MPa stress range at 1100 °C and 1150 °C in air.

## Results and Discussion

A comparison of the relative tensile creep behaviors of the fine-grained and coarse-grained silicon nitrides revealed both similarities and differences shared by the two materials. Representative creep curves for both materials at 1100 °C and 138 MPa are shown in Figure 4. It is clear from this graph that the general shape of the "strain vs. time" plots were similar for both silicon nitrides. No significant primary region was observed for either material, although a short (< 2 hrs) primary region was noticed in a few instances. In addition, no pronounced tertiary region of creep was evident in any of the tests conducted for this study. The majority of the time-dependent deformation was, therefore, observed to occur at a strain rate which varied little over the entirety of the test period. However, a slight decrease in strain rate was recorded at long times (> 100 hrs) for a few test specimens. Thus, in order to compare relative creep responses of the two materials, a " best-fit " slope was obtained through all data in the "strain vs. time" plots to yield a characteristic strain rate for a given set of test conditions.

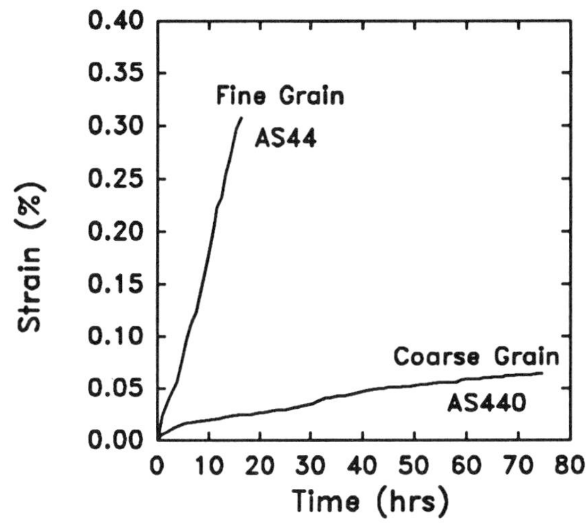


**Figure 4** Comparison of tensile creep behaviors for fine and coarse-grained $Si_3N_4$ at 1100 °C and 138 MPa.

As demonstrated by the plot in Figure 4, the coarsegrained silicon nitride exhibited greater creep resistance than the fine-grained material at 1100 °C and 138 MPa. The strain rates obtained from these plots for the fine- and coarse-grained materials were $5.8 \times 10^{-8}$/sec and $1.1 \times 10^{-9}$/sec, respectively. This enhanced creep resistance with increasing grain size is further illustrated by the strain rate data presented in Table 1. Under similar conditions of temperature and stress, the two silicon nitride materials exhibited strain rates which differed by approximately an order of magnitude (Samples E,H and Samples F,I). This type of difference in creep response had also been observed when these materials were previously tested in flexure and compression.[8]

The steady-state creep rates for both materials were assumed to follow the same empirical relation originally obtained for metals and other ceramic materials:

$$\dot{\varepsilon} = A \sigma^n e^{-\frac{Q}{RT}} \quad (1)$$

where $\dot{\varepsilon}$ = steady-state strain rate

$\sigma$ = applied stress

n = stress exponent

Q = apparent activ. energy for creep

**Table 1.** Tensile creep results for fine- and coarse-grained silicon nitride

| Material | Sample | Creep Stress (MPa) | Test Temperature (°C) | Strain Rate (x $10^{-9}$/sec) |
|---|---|---|---|---|
| Fine-grained $Si_3N_4$ | A | 69 | 1100 | 8.1 |
| | B | 69 | 1100 | 8.7 |
| | C | 104 | 1100 | 16.3 |
| | D | 138 | 1100 | 58.0 |
| | E | 69 | 1150 | 33.0 |
| | F | 104 | 1150 | 262.0 |
| Coarse-grained $Si_3N_4$ | G | 138 | 1100 | 1.1 |
| | H | 69 | 1150 | 1.8 |
| | I | 104 | 1150 | 13.4 |
| | J | 138 | 1150 | 77.0 |
| | K | 173 | 1150 | 222.0 |

and A is a constant which is dependent on the pre-exponential term of the diffusion coefficient, shear modulus, Burgers vector, and grain size of the deforming material. In order to obtain a stress exponent to describe creep in each material, log (strain rate) was plotted versus log (stress) at a given temperature to yield a slope which was equal to the stress exponent, n. The stress exponent plots for the fine- and coarse-grained materials are shown in Figures 5 and 6, respectively.

Based on preliminary results, it appears that the stress exponent for the fine-grained material underwent a marked increase (from $n \approx 0.78$ to $n \approx 4.52$) at a stress between 86 and 121 MPa at 1100 °C. A similar increase in stress exponent was previously reported for silicon nitride by Kossowsky et. al.[1] at a stress approximately equal to 104 MPa at 1150 °C. They attributed this sharp jump in stress exponent to a change in the controlling creep mechanism. At low stresses, cavities which formed as a result of grain boundary sliding were quickly filled by near-Newtonian flow of the viscous grain boundary phase ($n \sim 1$). However, they asserted that 104 MPa was most likely a threshold stress above which cavities were created at a much faster rate than they could be filled by the flow of the second phase. This would, therefore, change the controlling creep mechanism from diffusion-accommodated grain boundary sliding to cavitation-controlled creep at these higher stresses, and a different stress exponent would be observed.

Preliminary results indicate that a similar type of change in creep mechanism may be occurring in the fine-grained material used in this study between 86 and 121 MPa at 1100 °C. The activation energy obtained for the fine-grained material below the threshold stress was approximately 640 kJ/mole, which agrees well with values previously reported for diffusion-controlled creep in silicon nitride.[1-3] This value of 640 kJ/mole also corresponds to the approximate activation energy reported for the viscous flow of silicate glass. The coarse-grained material, however, has exhibited an activation energy on the order of 1100 kJ/mole, which is substantially higher than that expected for diffusionally-accommodated creep processes. Instead, this higher value is consistent with previous reports [5,14] for non-equiaxed silicon nitride which cited primarily cavitational creep processes.

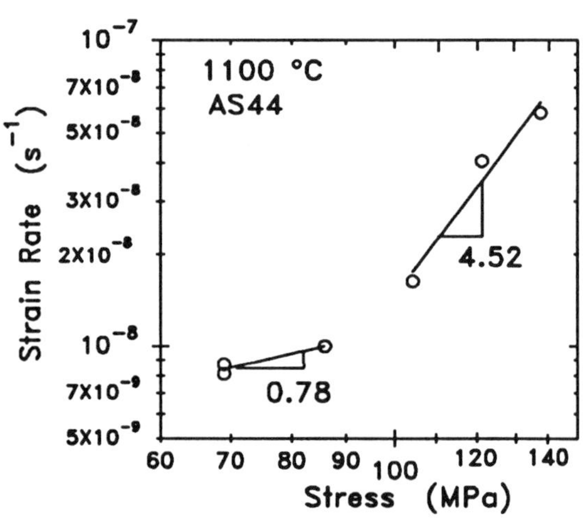


**Figure 5** Tensile strain rate dependence on stress for the fine-grained $Si_3N_4$ at 1100 °C.

The coarse-grained material did not exhibit any noticeable change in stress exponent between 69 and 173 MPa at 1150 °C (Fig. 6). The observed stress exponent (5.21) is similar to that obtained for the fine-grained material above its threshold stress, perhaps indicating similar creep mechanisms at work in both. If cavitation-accommodated grain boundary sliding were taken to be the operative creep mechanism at high stresses in these materials, then the observation of a

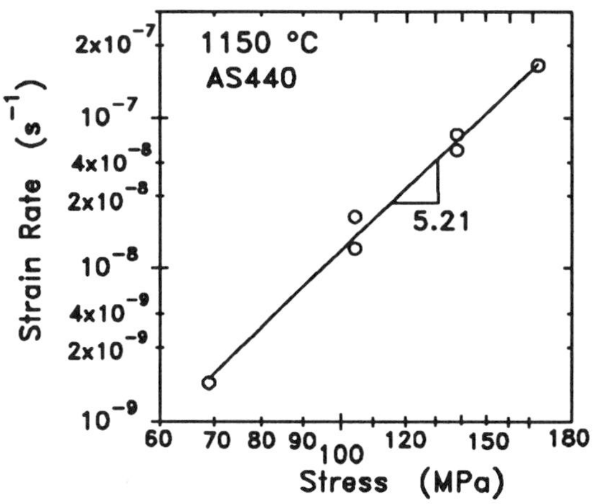


**Figure 6** Tensile strain rate dependence on stress for the coarse-grained $Si_3N_4$ at 1150 °C.

The enhanced creep resistance observed for the coarse-grained material used in this study can be rationalized in a number of ways. First, in terms of the idealized models to describe diffusional creep processes, an increase in grain size is accompanied by an increase in the distance between localized compressive and tensile stress regions. This would provide for a macroscopic strain rate which would be inversely related to grain size for a given aspect ratio. Second, at higher stresses, where cavitation is thought to be the rate-controlling creep mechanism for both materials, the strain rate would also be less for the coarse-grained material due to more effective crack wake bridging serving to limit boundary separation. For a given aspect ratio and grain boundary thickness in ß-silicon nitride, one would expect the effectiveness of crack wake bridging to scale with the grain size. Finally, if there exists a rotational component of creep associated with the re-arrangement of individual grains, each grain in the coarse-grained material would have to rotate through a greater distance to produce the same macroscopic strain as in the fine-grained material. Strain rate should once again be a reciprocal function of grain size.

higher threshold stress for cavitation in the fine-grained material would not be surprising. Lange [15] has previously addressed the effect of grain size on non-elastic deformation in polycrystals with a liquid boundary phase. He believed that the threshold stress for deformation via grain boundary separation should increase with decreasing grain size, since the size of the largest vapor bubble pre-existent in these materials should be proportional to the grain size. Therefore, it would be expected that, at a given stress, the fine-grained material would possess fewer cavities larger than a critical size for growth than the coarse-grained material.

The primary reason for enhanced creep resistance with increasing grain size in these acicular materials has not been clearly defined. There have been numerous models proposed to explain the creep of materials containing various amounts of a glassy grain boundary phase [15-19], but their applicability to these materials is limited due to the idealization of a complex microstructure to a simple periodic array of grains with uniform grain boundary thicknesses. The strain rates which arise from these models for primarily equiaxed grains are inversely related to grain size. This is presumably due to the increased diffusional distances and decreased number of grains along the length of the specimen associated with coarser microstructures.

However, the actual microstructures of the silicon nitrides used in this study more nearly approximate cylinders of differing lengths and aspect ratios randomly oriented within a viscous bonding phase. Thus, a shared boundary between two grains is a far less common feature in these materials than is a singular point of contact. Grain boundaries cannot, therefore, be considered to have uniform thicknesses throughout these materials. Rather, a wide range of grain boundary thicknesses will exist. Where strain energy considerations allow, the second phase will crystallize at elevated temperature to change the "effective" grain boundary thickness that can contribute to diffusional and cavitational creep processes at these localized spots.

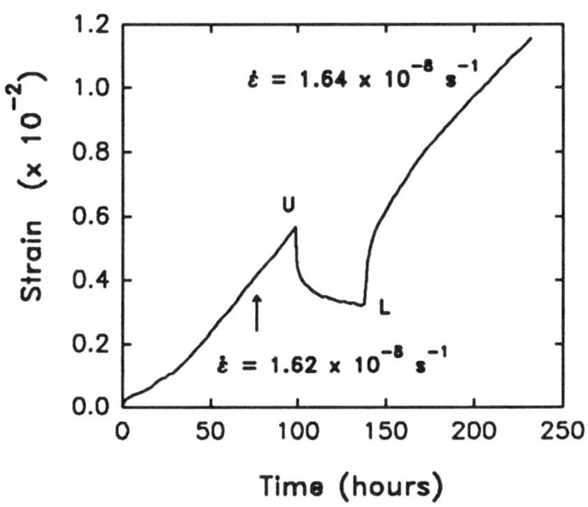


**Figure 7** A characteristic creep (95 MPa) and recovery (14 MPa) curve for fine-grained $Si_3N_4$ at 1100 °C. Creep stress was unloaded at Point **U** and re-loaded at Point **L**.

The present study has also determined that only a portion of the creep strain observed in these materials resulted from non-recoverable processes. The creep behavior of both materials at 1100 and 1150 °C also displayed an anelastic component which was recoverable when the creep stress was unloaded at test temperature. A typical creep/recovery curve for the fine-grained silicon nitride is shown in Figure 7. The creep specimen was unloaded at 1100 °C from 95 MPa to 14 MPa at Point U on the plot. An instantaneous elastic response by the material was subsequently followed by a time-dependent period of strain recovery. After the recovery rate dropped below $10^{-9}$/sec, the 95 MPa creep stress was then re-applied (Point L). Another elastic response was followed by an enhanced region of creep, which most likely resulted from the combined effects of the inverse of the recovery process and persistent creep. The post-recovery strain rate was then observed to decrease to a constant rate which was similar to its pre-recovery strain rate.

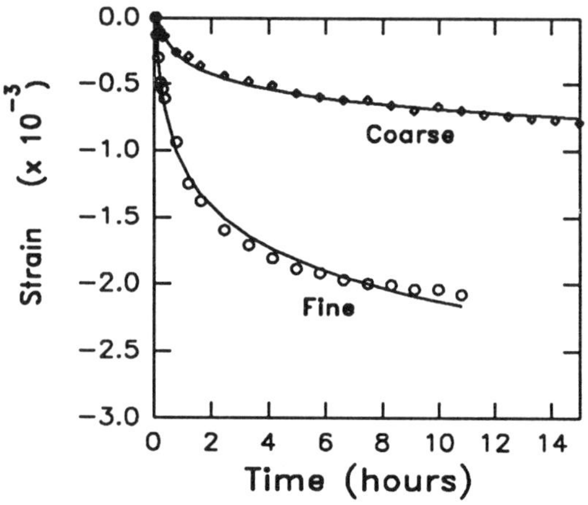


**Figure 8** Comparison of recovery behaviors for fine- and coarse-grained $Si_3N_4$ at 14 MPa and 1150 °C. Pre-strains (0.56%) and pre-stresses (121 MPa) prior to load removal were identical.

The percentage of recoverable deformation, after similar pre-strains in both materials, was also found to be more significant for the fine-grained material and approached 40 % under certain conditions. This is further illustrated by the plot in Figure 8, which compares the recovery behaviors of the fine- and coarse-grained silicon nitrides after similar pre-strains and pre-stresses at 1150 °C. This increase in recoverable deformation with decreasing grain size was thought to be attributable to an increased number of viscoelastic elements recovering in the material with the greater grain boundary area. A more detailed description of the recovery behaviors for these two materials has been presented elsewhere.[20]

## Conclusions

A comparison of the tensile creep behaviors of a fine-grained and a coarse-grained silicon nitride with identical compositions yielded enhanced creep resistance for the coarser microstructure. For two different sets of test conditions, the strain rates obtained for the coarse-grained material were less than those obtained for the fine-grained material by approximately an order of magnitude. Since the grain boundary phases were similar for both materials, the observed differences in creep behavior were most likely due to the effects of grain coarsening on diffusional and/or cavitational creep processes.

## References


1. Kossowsky, R., D.G. Miller and E.S. Diaz, *J. Mat. Sci.* **10**, 983-97 (1975).
2. Din, S.U. and P.S. Nicholson, *J. Mat. Sci.* **10**, 1375-80 (1975).
3. Birch, J.M. and B. Wilshire, *J. Mat. Sci.* **13**, 2627-36 (1978).
4. Dixon-Stubbs, P.J. and B. Wilshire, *J. Mat. Sci.* **14**, 2773-4 (1979).
5. Wiederhorn, S.M., D. Cranmer, D. Kauffman and D.E. Roberts, *N.I.S.T. Progress Report*, 447-52 (1990).
6. Lange, F.F., B.I. Davis and D.R. Clarke, *J. Mat. Sci.* **15**, 601-10 (1980).
7. Liu, K.C. and C.R. Brinkman, *ORNL Progress Report*, 359-72 (1990).
8. Whalen, P.J., C.J. Gasdaska and R.D. Silvers, *Ceram. Eng. Sci. Proc.* **11**, 633-48 (1990).
9. Raj, R. and F.F. Lange, *Acta Met.* **29**, 1993-2000 (1981).
10. Clarke, D.R., *J. Am. Ceram. Soc.* **70**, 15-22 (1987).
11. Cinibulk, M.K., G. Thomas and S.M. Johnson, *J. Am. Ceram. Soc.* **73**, 1606-12 (1990).
12. Kleebe, H. and M. Ruhle, Presented at the *93rd Annual Meeting of the Am. Ceram. Soc.*, April, 1991.
13. Carroll, D.F., S.M. Wiederhorn and D.E. Roberts, *J. Am. Ceram. Soc.* **72**, 1610-14 (1989).
14. Evans, R.W., T. Murakami and B. Wilshire, *J. Brit. Ceram. Trans.* **87**, 54 (1988).
15. Lange, F.F., "*Deformation of Ceramic Materials*", pp. 361-81, Plenum Press, New York (1975).
16. Raj., R. and M.F. Ashby, *Met. Trans.* **2**, 1113-27 (1971).
17. Tsai, R.L. and R. Raj, *Acta Met.* **30**, 1043-58 (1982).
18. Marion, J.E., A.G. Evans and M.D. Drory, *Acta Met.* **31**, 1445 (1983).
19. Thouless, M.D., *Res. Mechanica* **22**, 213-42 (1987).
20. Haig, S., W.R. Cannon and P.J. Whalen, To be published in *Ceram. Eng. Sci. Proc.* **13** (1992).

*Proceedings of the Fifth International Conference on Creep of Materials, Lake Buena Vista, Florida, USA, 18-21 May 1992*

# Study of Creep Crack Propagation Behavior of a Ni-Base Superalloy

C. Zhao
Central Iron and Steel Research Institute
Beijing, China

## ABSTRACT

Creep crack growth rates have been measured in single eade-notched tension specimen of a Ni-base superalloy at test temperature of 700-750°C. The results show that creep crack growth rate increases with increase of temperature for a giver stress intensity factor. Crack growth rate during the secondary creep crack growth region can be expressed by

$$\dot{a}_s \propto K_I^n e^{-\frac{Q}{RT}}$$

The addition of small amounts of Mg can not only notably increase the creep crack growth life,but also increase creep rupture life. Fractography by SEM showed that the path of creep crack growth was along the grain boundaries. However, the addition of small amounts of Mg did not changed the path of creep crack growth.

FAILURE OF HIGH TEMPERATURE COMPONENTS may occur by the propagation of preexisting cracks or flaws produced during processing. Therefore the study of creep crack growth behaviour of high temperature alloy is very important to aviod any catastrophic failure in service. Reported results (1,2) indicate that creep crack growth behaviour is very sensitive to chemical and test temperature variables of alloy. The purpose of this paper is to study the effect of small amounts of Mg and test temperature on creep crack growth behaviour of a Ni-base superalloy.

## SPECIMENS AND EXPERIMENTAL METHOD

The alloy tested in this study was vacuum melted and hot rolled into bars. The nominal chemical compositions for both heats of the alloy are shown in table 1. Obviously,the difference between both heats is only at small amounts of Mg.

Table 1. Chemical composition of the alloy (wt % )

| Materials | Without Mg | Containing Mg |
|---|---|---|
| C | 0.02 | 0.022 |
| Cr | 19.88 | 19.78 |
| Al | 0.65 | 0.66 |
| Ti | 2.49 | 2.47 |
| B | 0.001 | 0.005 |
| Ce | <0.001 | 0.0012 |
| Mg | / | 0.0015 |
| Ni | Bal. | Bal. |

The bars was heat treated at 1080°C for 8h,air cooled, then at 700°C for 16h, air cooled.The bars were machined into single edge-notched creep specimens as shown in Figure 1.

The specimens were tested in air in a lever-type constant-load tensile creep testing machine.The accuracy of the test temperature was controlled to within ±2 C during the test. The creep crack lengths at various test time were measured by a D.C. electrical potential drop technique. The direct current (normally 6 amperes) was supplied by a constant current source. The electrical

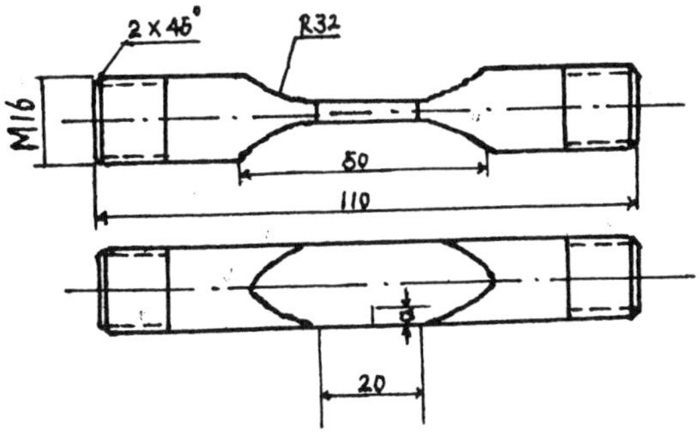


Fig.1. Geometry of single edge-notched specimen

potential was amplified and recorded automatically by the computer. The stress intensity factor $K_I$ under applied load P was calculated by (3)

$$K_I=\frac{YP}{B\sqrt{W}} \qquad (1)$$

where $Y=1.99(\frac{a}{W})^{\frac{1}{2}}-0.41(\frac{a}{W})^{\frac{3}{2}}+18.7(\frac{a}{W})^{\frac{5}{2}}-38.48(\frac{a}{W})^{\frac{7}{2}}+53.85(\frac{a}{W})^{\frac{9}{2}}$ and P is the applied load and B is the specimen thickness and W is the specimen width. The initial crack lengths were a little different,but the initial load were near 19.4KN.The test of the effect of test temperature on creep crack growth of the alloys were conducted with containing Mg specimen at 700,750°C. The test of the effect of small amounts of Mg on creep crack growth of alloy were conducted at 700°C. The creep crack growth rate of alloy was calculated by 'seven point incremental method'.

## RESULTS AND DISCUSSION

Creep Crack Growth Curve. Creep crack growth curve of alloy at constant initial load is shown in Figure 2. Obviously, the time to crack initiation and rupture of the alloy containing Mg are much longer than that of the alloy without Mg. In addition, the times to crack initiation and rupture of the alloy containing Mg decreases with increase of test temperature.

Sensitivity Of Creep Crack Growth To $K_I$ . The crack growth rates at constant temperature are plotted against stress intensity factor in Figure 3. It can be seen that process of creep crack growth for alloy could be divided into four stage: Initial stage and steady crack growth stage and accelerated crack growth stage and creep rupture stage. In addition, the addition of small amounts of Mg may increase the $K_I$ of the transient point from initial stage to steady stage. Small amounts of Mg can notably decrease creep crack growth rate of

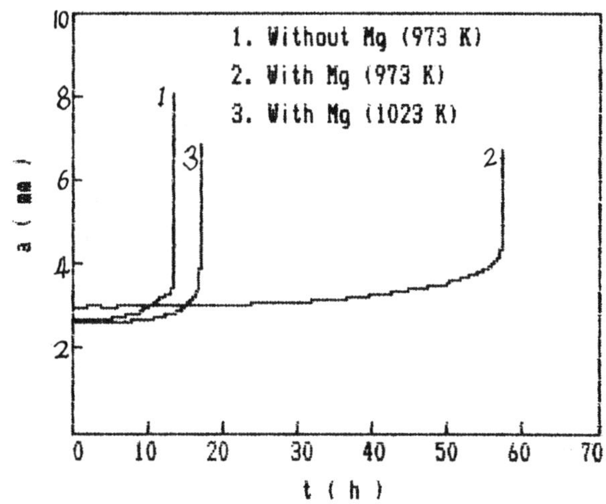


Fig.2. Creep crack growth curve of alloy

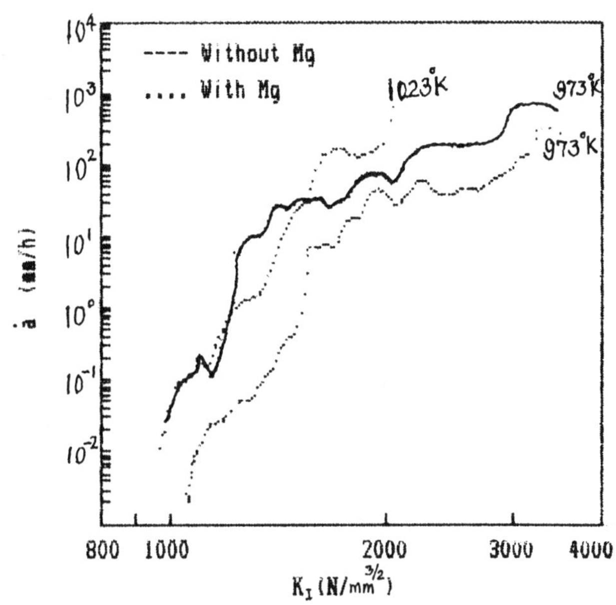


Fig.3. Variations of creep crack growth rates ȧ with stress intensity factor $K_I$

alloy. The rate of the creep crack growth increases as stress intensity factor is increased. It is shown that rate of the secondary crack growth can be represented by the Nortom equation:

$$\dot{a}_s = AK_I^n \qquad (2)$$

where n is the stress index for creep crack growth, A is the material constant. This is the same form of expression often used to describe the secondary creep strain rate. The value of n at each temperature was determined by linear regression analysis. Thus,the value of n for the alloy containing Mg is within 10.4-13.9 between 700-750°C. The calculation results shows that stress index of creep crack growth increases with the increase of test temperature. However,the results of S.F.Wu and Q.P.Kong (4) were that stress index of creep crack growth decreases as test temperature increase. These results show that the crack tip stress intensity factor distribution during creep crack growth is complicated.

Temperature Sensitivity Of Crack Growth. It can be seen from Figure 3 that the rate of secondary crack growth for the alloy containing Mg increases with increase of test temperature at constant $K_I$. Assuming an Arrhenius-type equation to apply,i.e.

$$\dot{a}_s = Be^{-\frac{Q}{RT}} \qquad (3)$$

where Q is the apparent activation energy for creep crack growth, B is the material constant, R is the gas constant,T is absolut temperature. This is the same form of expression often used to describe the secondary creep strain rate. From Eq.3 the apparent activation energy for creep crack growth may then be calculated by

$$Q = \frac{R \ln \dot{a}_{s1}/\dot{a}_{s2}}{\frac{1}{T_2} - \frac{1}{T_1}} \qquad (4)$$

where $\dot{a}_{s1}$ and $\dot{a}_{s2}$ is the secondary creep crack growth rates at $T_1$ and $T_2$ ,respectively.Thus,between 700 and 750°C, the value of Q for the alloy containing Mg is within 111.8-123.3 kcal/mol between 1150-1250 $N/mm^{3/2}$.

The calculation results shows that apparent activation energy of creep crack growth for the alloy containing Mg increases with the increase of K in the process of creep crack growth. However, Floeen (5) pointed out that the temperature dependence of crack growth rate was independent of stress intensity factor for Inconel 718 and results of Yokobori (6) were that apparent activation energy of creep crack growth decreased as K increased during creep crack growth. These results show that the thermal activation history is complicated.

By combining Eq.2 and Eq.3 a relationship of the form:

$$\dot{a}_s = CK_I^n e^{-\frac{Q}{RT}} \qquad (5)$$

is obtained for the rate of the secondary crack growth in the alloy containing Mg , where n=10.4-13.9, Q=111.8-123.3 kcal/mal in the temperature range 700 to 750°C.This expression is of the form used for describing the stress and temperature dependence of secondary creep deformationrate.

Metallography Analysis.Observations of the fracture surface by eye have indicated that the fracture surface of specimen was the plane surface. It shows that the crack was to propagate along the main crack. Fractography by SEM showed that the path of creep crack growth was along the grain boundaries. It also show that micro-cracknucleated ahead of the maincrack and that these joined to enable the crack to propagate. However, the addition of small amounts of Mg did not changed the path of creep crack growth.

## CONCLUSIONS

1. The addition of small amounts of Mg can not noly notably increase the creep crack initiation life and creep crack growth life, but also decrease creep crack growth rate.
2. The crack growth rate during the secondary creep crack growth region can be expressed by

$$\dot{a}_s \propto K_I^n e^{-\frac{Q}{RT}}$$

3. The path of creep crack growth of the alloy was along the grain boundaries. The addition of small amounts of Mg did not change the path of creep crack growth.

## REFERENCES


1. Encal Guo and Fengqin Xu, Proc. Int. Conf. on Mech. Behavior of Materals (ICM5), Vol.2,p957 (1987).
2. R.D.Nicholson, Mat. Sci. Engng., 22(1976)1-6.

3. Brown,W.F. and J.E. Strawley (1969),ASTM, Spec. Publ. No.41.

4. S.F.Wu and Q.P.Kong, Proc.Int.Conf. on Mech.Behavior of Materials (ICM5),Vol.2,p977 (1987).

5. Floreen ,S.Metall. Trans. A. 6A. 1741-1749 (1975).

6. Yokobri,T.Eng. Fract. Mech.,13,533-539 (1977).

*Proceedings of the Fifth International Conference on Creep of Materials, Lake Buena Vista, Florida, USA, 18-21 May 1992*

# Creep Behavior of Hastelloy XR2 at Very High Temperature

**Y. Kaji, Y. Muto, and K. Tachibana**
Japan Atomic Energy Research Institute
Ibaraki-ken, Japan

ABSTRACT

Creep constitutive equation was made for Hastelloy-XR2 at 900°C over the stress range of 45MPa to 100MPa. Hastelloy-XR2 contains 40ppm boron and is a strengthened version of Hastelloy-XR, which is a nickel base solution annealed alloy. The equation was envisaged to be applied to components like those in a very high temperature reactor where the material strength is low and an inelastic stress analysis is inevitable. The equation has a form of Norton type incorporating an internal back stress which is evolved by the equation based on a creep-recovery model. The equation was made to resolve into Garofalo-type equation for the constant stress condition by devising the parameters. The internal back stress and some of the parameters were determined by stress-dip tests. The others were determined by utilizing a data point in the primary creep curve. In order to examine its validity, the equation was applied to the results of creep tests both under the stress-increasing and the decreasing loads. The predicted creep behaviors were in fairly good agreement with those of the experimental results.

THE TEMPERATURE OF COMPONENTS in the Very High Temperature Gas-Cooled Reactor for the process heat or the gas-turbine applications becomes 900°C to 1000°C. Structural design in such high temperature region has to be based on an inelastic, that is, elastic-creep stress analysis because a small safety margin is allowed due to the low material strength. This requires a highly reliable creep constitutive equation.

In the High Temperature Engineering Test Reactor(HTTR) constructed by the Japan Atomic Energy Research Institute(JAERI), the highest temperature is 950°C and a nickel-base heat resistant alloy "Hastelloy XR" which is a modified version of the commercial alloy "Hastelloy X" is used. A creep constitutive equation of Garofalo type was prepared[1] whose parameters were determined by the experimental data obtained mainly in the National Research Institute for Metals[2]. In a commercial HTGR, however, the longer structural life of component will be required. Therefore, the research on creep constitutive equation has been continued in parallel with the design effort.

Constant stress-creep tests were conducted at 900°C for Hastelloy XR2 which was a modified version of HastelloyXR and contained around 40ppm boron additionally. A uniaxial creep constitutive equation was made which was also Garofalo type and applied to predict a creep behavior under a variable stress condition using a strain hardening law. The result was, however, insufficient. Then, a new constitutive equation was devised where a creep strain rate equation was the well-known Norton type[3] and an evolution equation of the internal back stress(hereinafter written as back stress) was based on a creep-recovery model[4]. Internal back stress parameter was determined by stress dip tests and then the other parameters were determined to conform to the above-mentioned Garofalo-type equation. The results of stress dip tests and the procedure determining the parameters are explained. The creep behaviors are calculated by both the new equation and the Garofalo equation for the variable stress loads. It is shown that the new constitutive equation predicts the creep behavior under the variable stress loads more exactly than the Garofalo equation.

CREEP CONSTITUTIVE EQUATION(GAROFALO TYPE)

The constant stress-creep tests were carried out at 900°C in air for Hastelloy-XR2 whose chemical composition is shown in Table 1. Lever-type test machines with dead loads were used. The load was imposed by many small weights( the minimum piece is 50g corresponding to a stress of 0.17MPa) which were decreased through manual operation so as to keep the stress constant in accordance with the value calculated from an elongation based on a constant volume assumption for the creep deformation. The Garofalo-type equation[5] is described as follows.

$$\varepsilon = \varepsilon_t\{1 - \exp(-rt)\} + \dot{\varepsilon}_{min}t \qquad (1)$$

where

$\varepsilon_t$ = asymptotic primary strain

$r$ = reciprocal of time constant

$\dot{\varepsilon}_{min}$ = minimum creep strain rate

The experimental values of $\varepsilon_t$, $r$ and $\dot{\varepsilon}_{min}$ are shown in Fig.1, Fig. 2 and Fig.3, respectively. Though a significant data-scattering is observed in the parameter $\varepsilon_t$, the parameters $r$ and $\dot{\varepsilon}_{min}$ can be fitted very well. Therefore, it can be said that this type must be the most adequate one for the equation of Hastelloy-XR2.

Substituting these numerical values, the Eq.(1) is given as follows.

$$\varepsilon = \varepsilon_t\{1 - \exp(-rt)\} + \dot{\varepsilon}_{min}t \qquad (2)$$

where $\varepsilon_t = 5.703\times10^{-2}\exp(-0.023\sigma)$

$r = 1.214\times10^{-3}\exp(0.0530\sigma)$

$\dot{\varepsilon}_{min} = 1.976\times10^{-14}\sigma^{6.083}$

Creep tests were conducted both for the constant load, that is, the stress increasing

Table 1 Chemical composition of Hastelloy-XR2 (Check analysis)

| Element | Wt% | Element | Wt% |
|---|---|---|---|
| Carbon | 0.07 | Chromium | 21.98 |
| Manganese | 0.86 | Cobalt | 0.05 |
| Silicon | 0.32 | Molybdenum | 8.83 |
| Phosphorus | <0.001 | Tungsten | 0.51 |
| Sulfur | 0.001 | Iron | 17.70 |
| | | Nickel | Balance |
| Aluminum | 0.02 | | |
| Titanium | 0.01 | | |
| Boron | 0.005 | | |
| Nitrogen | 0.006 | | |

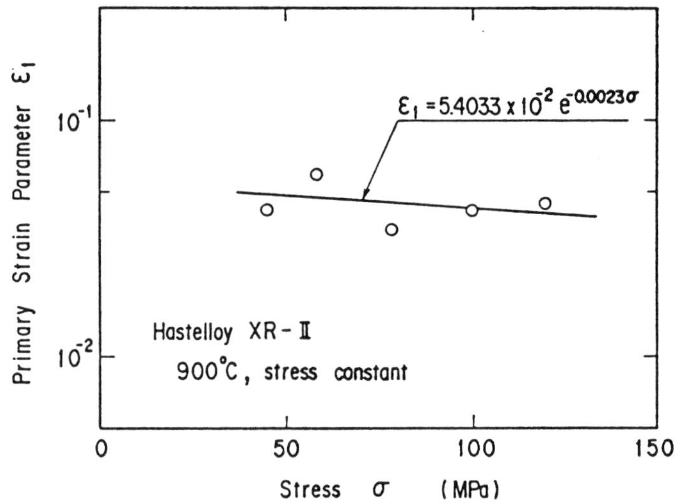


Fig. 1 Primary strain parameter $\varepsilon_t$ versus stress $\sigma$

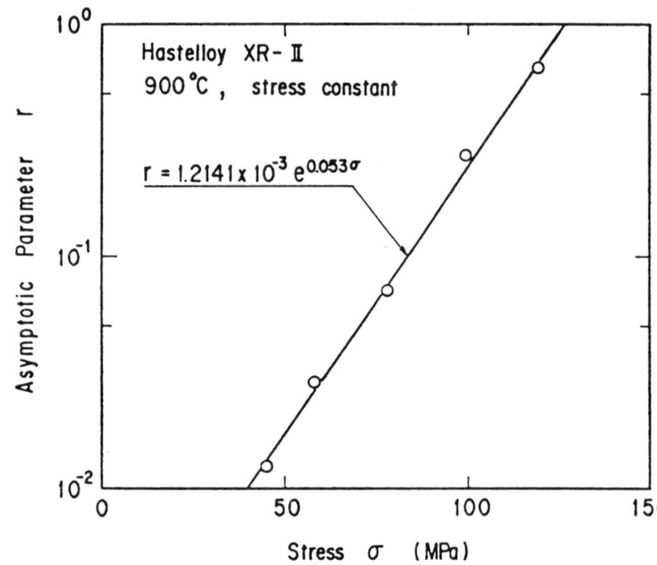


Fig. 2 Reciprocal time constant r versus stress $\sigma$

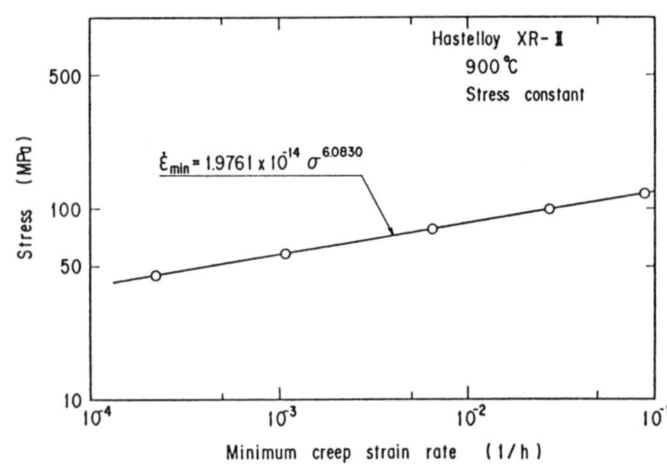


Fig. 3 Minimum creep strain rate $\dot{\varepsilon}_{min}$ versus stress $\sigma$

condition and for the stress decreasing condition. The results are shown in Fig. 4 and Fig. 5. The calculated curves by the Eq.(2) coupled with the strain hardening rule are also shown. Some differences are observed between the experimental curves and the calculated ones. In the case of stress increase, the value of experimental curve is larger than that of calculated one. On the other hand, the relation is reversed in the case of stress decrease. This tendency can be explained by considering of the back stress.

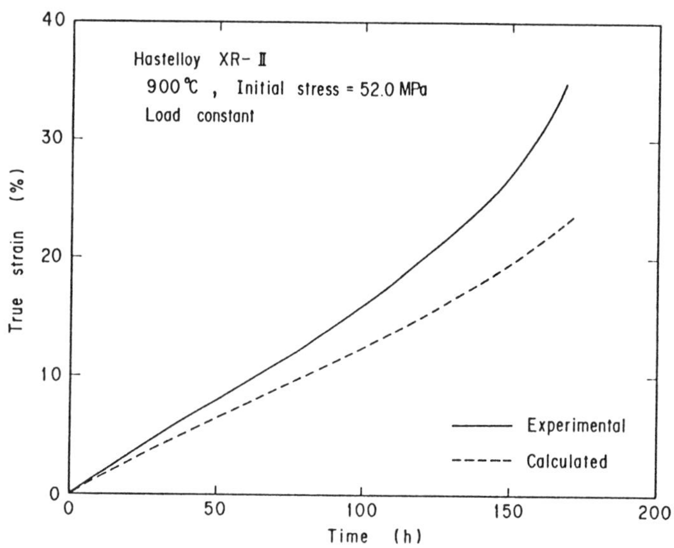


Fig. 4 Comparison between an experimental and a calculated creep curve for the stress increasing case

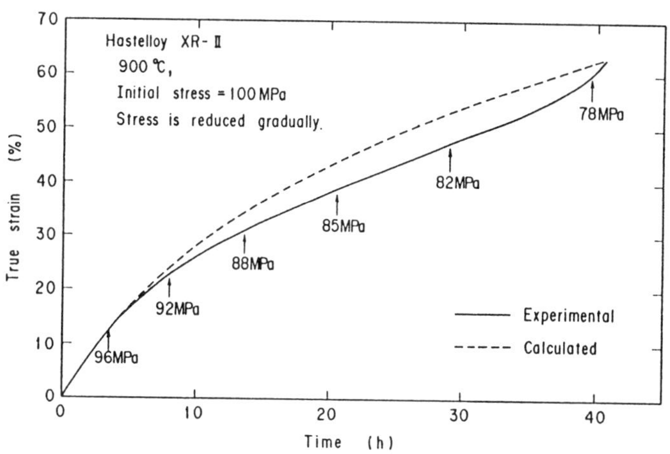


Fig. 5 Comparison between an experimental and a calculated creep curve for the stress decreasing case

## MEASUREMENT OF BACK STRESS

A back stress is usually measured by a stress-dip test[6]. In that test, an instantaneous zero strain rate is measured after the large stress reduction or estimated by an extrapolation of data for the smaller stress reduction. The same kinds of tests were carried out. The results are shown in Table 2 and some examples are illustrated in Fig.6. Negative strain rates were observed in the all cases. This fact is thought to verify the existence of back stress in the case of Hastelloy XR2 at 900°C.

Hitherto the back stress is said to increase with the applied stress, but the test results were different from the knowledge. The stress dependence of back stress seems to have no consistent characteristic as shown in Fig. 7, where the data of Inconnel-617 re-plotted from the work of Osthoff[7] are also shown. That alloy is also nickel base and gives almost the same creep strength as Hastelloy-XR2. Moreover, the back stresses in the secondary creep region are not necessarily greater than those in the secondary creep region.

In our tests, conventional dead load-creep test machines were used where the creep elongation was transmitted to the outside of electric furnace by extension rods and measured by differential

Table 2 Back stress measured by the stress-dip tests
(Hastelloy XR2, 900°C)

| Applied stress MPa | Primary creep region MPa | Secondary creep region MPa | Secondary creep region Average MPa | Tertiary creep region MPa |
|---|---|---|---|---|
| 100 | 30 | 27<br>25<br>36 | 29.3 | 33 |
| 78 | 38.5 | 38<br>40<br>33 | 37.0 | 33 |
| 58 | 34.3 | 37.4<br>44<br>46 | 42.5 | 50 |
| 45 | 36 | 30<br>30<br>33<br>25<br>35 | 30.6 | 28 |

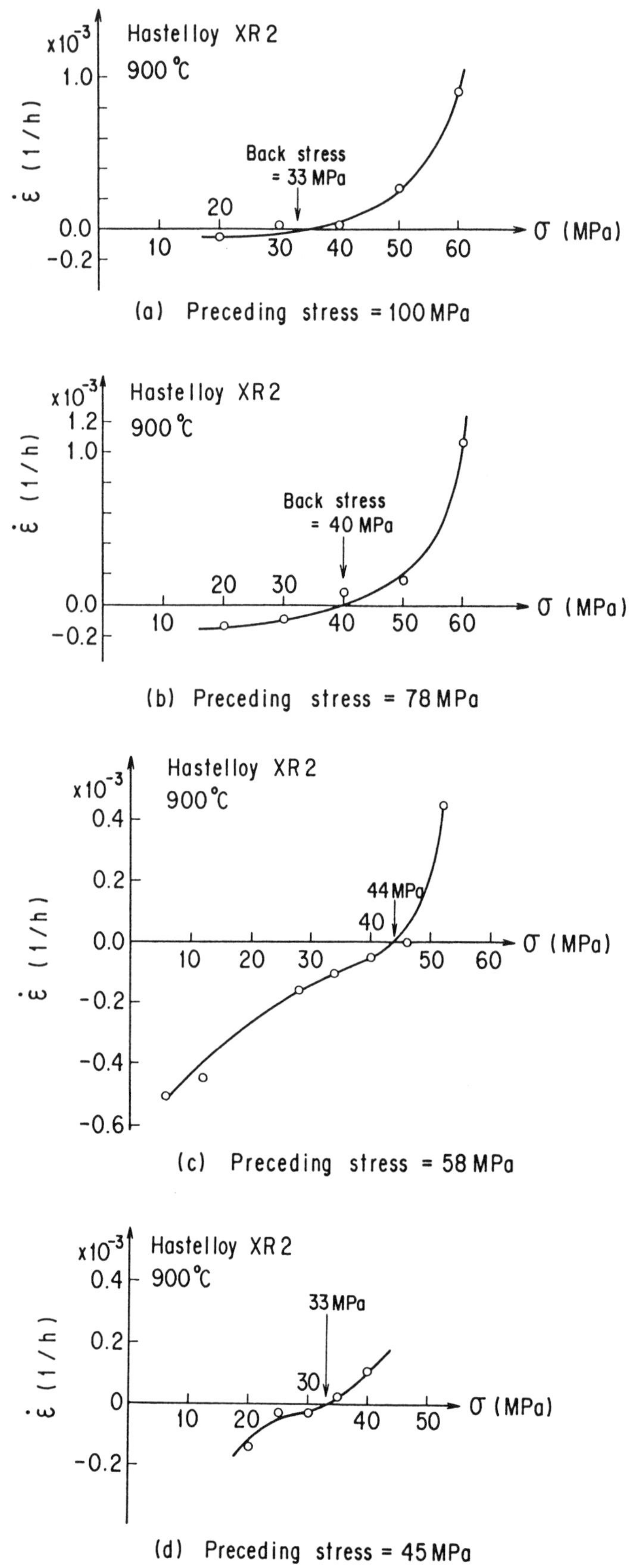


Fig. 6 Strain rate versus applied stress for the stress-dip tests

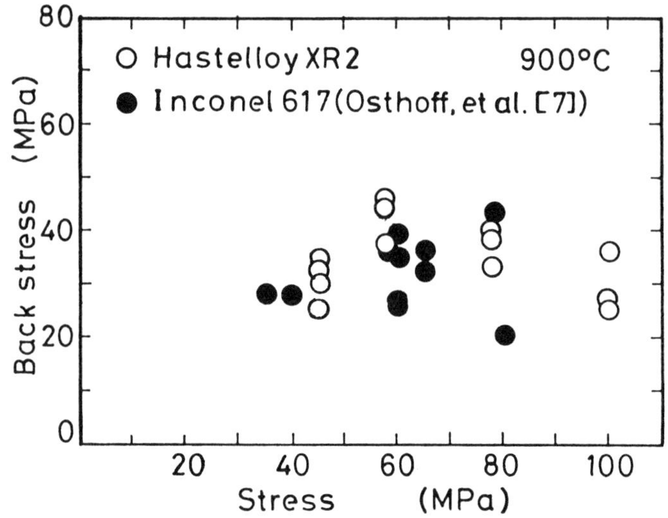


Fig. 7 Stress dependence of measured back stresses for Hastelloy-XR2 and Inconnel-617[7]

transformers. The elongation-data were recorded by digital printers at intervals of 1.0 min where the minimum indication of printers was 0.01mV which corresponded to the strain increment of $0.7\times10^{-5}$. After each stress reduction, a reduced stress was held during 10 min where the minimum change of strain became around $1\times10^{-4}$. Then it was raised to the original value and the next test was repeated after a suitable recovery time. The strain measurement was not sufficiently reliable if we consider that the strain could change eventually by $1\times10^{-4}$ owing to the change of ambient temperature. Therefore it could lead to an error of the elongation measurement. In addition, the variation of material strength property is inclined to be enlarged at very high temperature. Therefore, the strange result is thought to be attributed to an error in the measurement or to a variation of material property itself though the data-scattering is not necessarily larger than that of Osthoff.

Then, we decided to try another method. We measured creep strain rates for a little bit smaller stress reductions and for a longer time period. This method is similar to that of Cuddy[8]. The stress holding time was determined so as to get the value of strain-change of $1\times10^{-3}$. We can obtain the larger strain increment, which leads to the more reliable data of the creep-strain rate by this method. Though we cannot still eliminate the above-mentioned errors even by this method, a more plausible value can be estimated by modifying the experimental strain rate as follows.

$$\dot{\varepsilon} = (\dot{\varepsilon}_{exp})\times(\dot{\varepsilon}_{min,n})/(\dot{\varepsilon}_{min,exp}) \quad (3)$$

where $\dot{\varepsilon}$ = modified strain rate

$\dot{\varepsilon}_{exp}$ = experimental strain rate

$\dot{\varepsilon}_{min,n}$ = nominal minimum strain rate

$\dot{\varepsilon}_{min,exp}$ = experimental minimum strain rate

Table 3 and Fig.8 show the results of tests. The modified creep-strain rates after the stress reduction seem more converged and systematic and then more reliable. Using those data, we can calculate parameters including the back stress if we assume a form of equation. The assumption is inevitable which is one drawback of this method.

Table 3 Creep strain rates after the stress reductions

| Preceding stress MPa | Applied stress MPa | Strain rate | |
|---|---|---|---|
| | | Measured $10^{-2}$ 1/h | Modified $10^{-2}$ 1/h |
| 100 | 92 | 1.50 | 1.79 |
| | 84 | 0.729 | 0.886 |
| | 76 | 0.348 | 0.432 |
| | 69 | 0.141 | 0.178 |
| | 63 | 0.0620 | 0.0784 |
| | 58 | 0.0312 | 0.0384 |
| 78 | 70 | 0.264 | 0.301 |
| | 63 | 0.103 | 0.122 |
| | 57 | 0.0386 | 0.0450 |
| | 51 | 0.0165 | 0.0188 |
| | 46 | 0.00656 | 0.00712 |
| 58 | 54 | 0.0397 | 0.0540 |
| | 50 | 0.0206 | 0.0258 |
| | 46 | 0.0135 | 0.0164 |
| | 42 | 0.00444 | 0.00552 |
| | 38 | 0.00188 | 0.00234 |
| 45 | 42 | 0.00797 | 0.0117 |
| | 39 | 0.00339 | 0.00498 |
| | 36 | 0.00196 | 0.00288 |

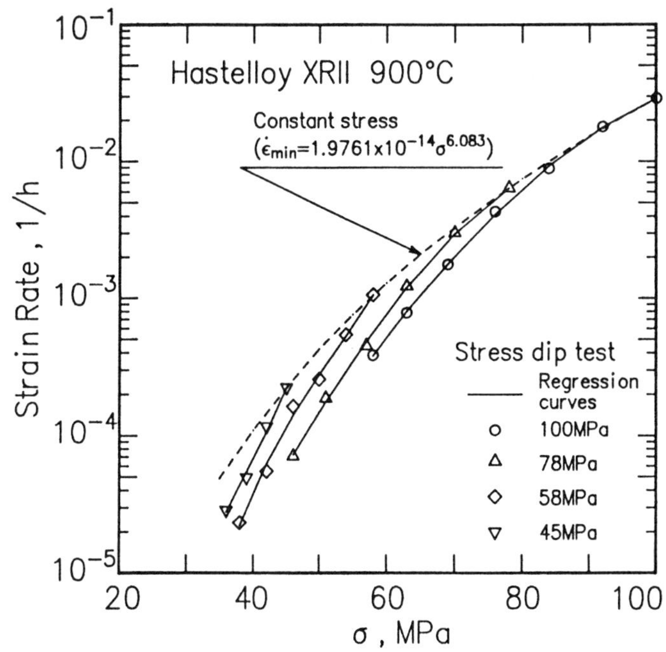


Fig. 8 Measured strain rates in stress-dip tests

## CREEP CONSTITUTIVE EQUATION(POWER LAW TYPE)

Hitherto, many types of creep constitutive equations have been proposed. Jonas[9] examined the validity of constitutive equation involving the back stress for the first time. Malinin and Khadjinsky[10] developed a multiaxial form and also an evolutional equation of the back stress. Miller developed a form incorporating both the back stress and a drug stress[11] and applied it to type 304 stainless steel. Krieg[12] improved the evolution equation and determined the value of parameters for aluminum at room temperature.

We selected the same form of creep constitutive equation as those in the above researches. In addition, that is devised so as to be resolved into the Garofalo equation for the constant stress load.

Firstly we assume that the creep strain rate and an evolution of the back stress are represented by the Norton-type equation and the recovery-creep model, respectively.

$$\dot{\varepsilon} = A(\sigma - \sigma_b)^n \quad (4)$$

$$\dot{\sigma}_b = B\dot{\varepsilon} - D\sigma_b{}^n \quad (5)$$

where $\varepsilon$ : creep strain
$\sigma$ : applied stress
$\sigma_b$: back stress
A, B, D and n : material constants
Dot denotes a derivative by time.

Secondly, the following were assumed.

a) $\sigma_b = \sigma_{b,0}$ at t=0,

b) $\sigma_b$ and $\dot{\sigma}_b$ becomes $\sigma_{b,max}$ and zero respectively at the secondary creep rate region,

c) $\dot{\varepsilon} = \dot{\varepsilon}_0$ at t=0.

From the Eqs.(4), (5),

$$\dot{\varepsilon}_{min} = A(\sigma - \sigma_{b,max})^n \quad (6)$$

$$\sigma_{b,max} = \{(B/D)\dot{\varepsilon}_{min}\}^{1/n} \quad (7)$$

Substitute the Eq.(7) to the Eq.(6),

$$\dot{\varepsilon}_{min} = A[\sigma - \{(B/D)\dot{\varepsilon}_{min}\}^{1/n}]^n \quad (8)$$

The minimum creep rate calculated by the Eq.(4) must coincide with that of the Eq.(2), which is re-written as follows.

$$\dot{\varepsilon}_{min} = A_s\sigma^n \quad (9)$$

Substitute the expression, the Eq.(8) becomes

$$(A_s/A)^{1/n}\sigma = \sigma - (BA_s/D)^{1/n}\sigma \quad (10)$$

The following equation is assumed to hold in order that the Eq.(10) may always be satisfied.

$$(A_s/A)^{1/n} = 1 - (BA_s/D)^{1/n} \quad (11)$$

Using the relationship, the Eq.(7) becomes

$$\sigma_{b,max} = \{1 - (A_s/A)^{1/n}\}\sigma \quad (12)$$

where A, B and D can be allowed to be functions of the applied stress $\sigma$.

Consequently, a new creep constitutive equation is described as follows.

$$\dot{\varepsilon} = A(\sigma)[\sigma - \sigma_b(\sigma)]^n \quad (13)$$

$$\dot{\sigma}_b = B(\sigma)\dot{\varepsilon} - D(\sigma)\sigma_b{}^n \quad (14)$$

$$\sigma_{b,max} = \{1 - [(A_s/A(\sigma)]^{1/n}\}\sigma \quad (15)$$

## DETERMINATION OF PARAMETERS

Changing a stress instantaneously in the secondary creep region, the Eqs.(13) and (15) must be

$$\dot{\varepsilon} = A(\sigma_i)[\sigma - \sigma_{b,max}(\sigma_i)]^n \quad (16)$$

$$\sigma_{b,max} = \{1 - [(A_s/A(\sigma_i)]^{1/n}\}\sigma_i \quad (17)$$

where $\sigma_i$ : initial applied stress preceding the stress change
$\sigma$ : applied stress after the stress change

From Eqs.(16) and (17),

$$A(\sigma_i) = [\{(\dot{\varepsilon})^{1/n} - (A_s)^{1/n}\sigma_i\}/(\sigma - \sigma_i)]^n \quad (18)$$

$$\sigma_{b,max} = \frac{\{\sigma - (\dot{\varepsilon}/A_s)^{-1/n}\}\sigma_i}{\sigma_i - (\dot{\varepsilon}/A_s)^{-1/n}} \quad (19)$$

In addition, the initial value of back stress must be related to the initial strain rate by the following equation.

$$\sigma_{b,0} = \sigma_i - \{\dot{\varepsilon}_0/A(\sigma_i)\}^{1/n} \quad (20)$$

From Eqs.(18), (19) and (20), the values of $A(\sigma_i)$, $\sigma_{b,max}$ and $\sigma_{b,0}$ are determined for each stress($\sigma_i$) by applying one set of stress dip tests shown in Fig.8. As there is some data scattering, a regression curve was calculated for each stress($\sigma_i$) and one set of stress and strain rate data was used.

Figures 9 and 10 show the results. The parameter A is found to be a function of stress and decreases with the stress. It is also found that the value of the back stress at the secondary creep

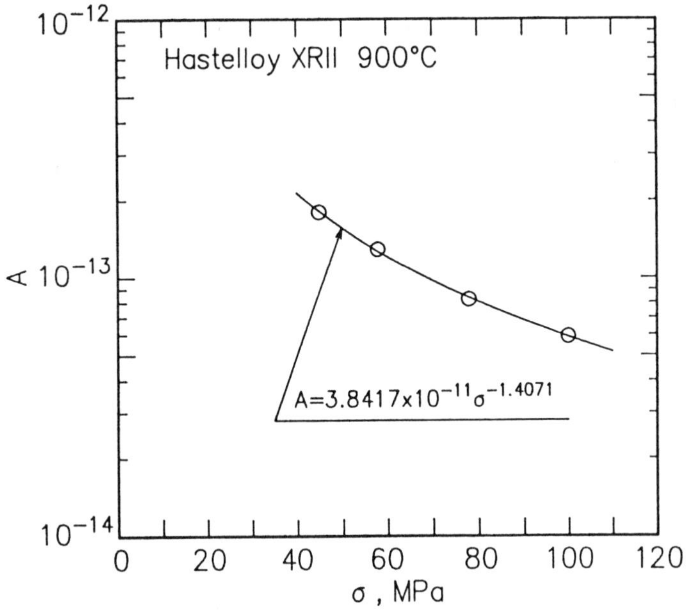


Fig. 9 Parameter A versus stress

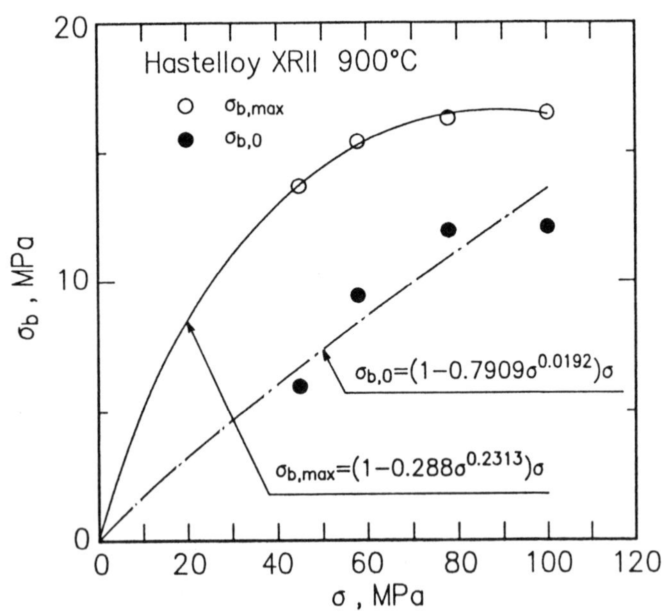


Fig. 10 Back stress versus stress

region($\sigma_{b,max}$) increases and saturates with the applied stress while the initial value of the back stress($\sigma_{b,0}$) increases proportionally with the applied stress. It should be pointed out that the back stress has a weak stress dependence above 60MPa and the values themselves are significantly lower than those shown in Fig.7. Observing the strain behavior in the case of large stress reductions, the strain decreases first and is kept constant during some period which is sometimes referred as incubation time. Then, it increases again. So, even in the case that a back stress is reached during short time which corresponds to the case of Fig.7, zero-strain rate cannot be observed if we notice the mean strain rate for the longer period which corresponds to the case of Fig.10. This fact can be explained if we consider two kinds of back stresses where the one has a short time constant and the other has a longer one.

Rearranging Eq.(11),

$$B/D = \{(1/A_s)^{1/n} - (1/A)^{1/n}\}^n \quad (21)$$

The time derivative of strain rate is represented as follows,

$$d\dot{\varepsilon}/dt = -An(\dot{\varepsilon}/A)^{(n-1)/n}[B\dot{\varepsilon} - D\{\sigma - (\dot{\varepsilon}/A)^{1/n}\}^n] \quad (22)$$

Using Eq.(21), Eq.(22) and each data set of strain rate and time derivative of strain rate, the values of B and D were determined as shown in Figures 11 and 12 respectively. The value of strain hardening constant B decreases with that of stress. On the other hand, the value of recovery constant D increases with that of stress.

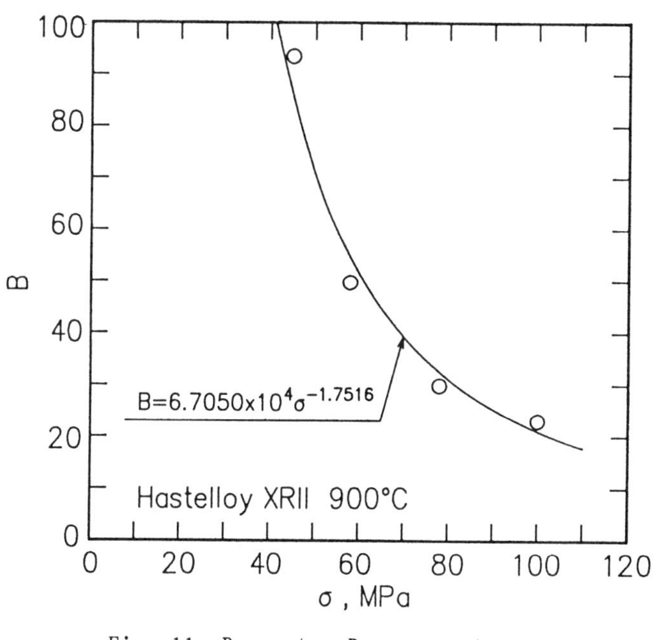


Fig. 11 Parameter B versus stress

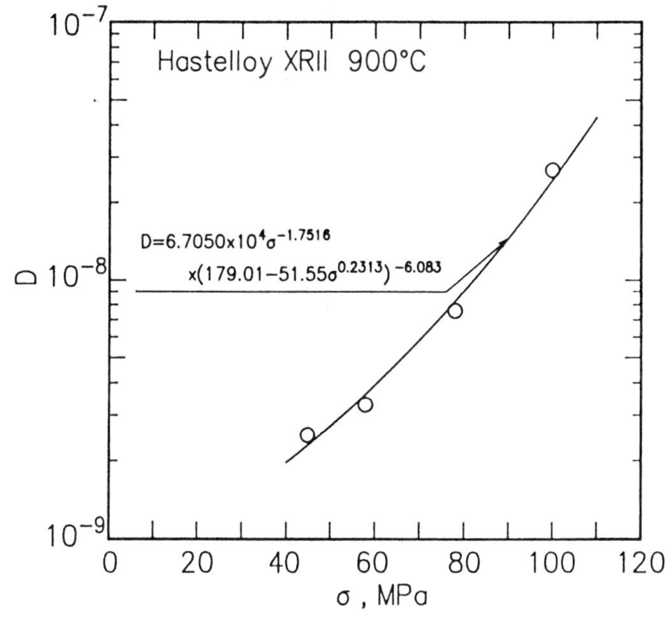


Fig. 12 Parameter D versus stress

The new constitutive equation is given as follows.

$$\dot{\varepsilon} = A(\sigma)[\sigma - \sigma_b(\sigma)]^n \quad (23)$$

$$\dot{\sigma}_b = B(\sigma)\dot{\varepsilon} - D(\sigma)\sigma_b{}^n \quad (24)$$

where $A(\sigma) = 3.8417\times10^{-11}\sigma^{-1.4071}$ (25)

$$n = 6.083 \quad (26)$$

$$B(\sigma) = 6.7050\times10^4\sigma^{-1.7516} \quad (27)$$

$$D(\sigma) = 6.7050\times10^4\sigma^{-1.7516} \times(179.01 - 51.55\sigma^{0.2313})^{-6.083} \quad (28)$$

$$\sigma_{b,0} = (1 - 0.7909\sigma^{0.0192})\sigma \quad (29)$$

VALIDITY OF THE CONSTITUTIVE EQUATION

In order to examine the validity of the constitutive equation, calculated creep strain behaviors were compared with the experimental data for the variable stress-loadings. The calculations were conducted by both the Garofalo-type equation and the new one. The strain hardening rule were used in the former case, while that was not necessary in the latter case because the mechanism of strain hardening was already incorporated as the evolutional function. In the calculations, a stress-value of the latest time mesh was assigned to that of parameters $A(\sigma)$

and $\sigma_b(\sigma)$ while the other parameters $B(\sigma)$ and $D(\sigma)$ adopted those of the time mesh. Figures 13 to 16 show the typical results. Figures 13 and 14 are the same ones with Figs.4 and 5 respectively. It is apparent the new constitutive equation represents the creep strain behavior more exactly.

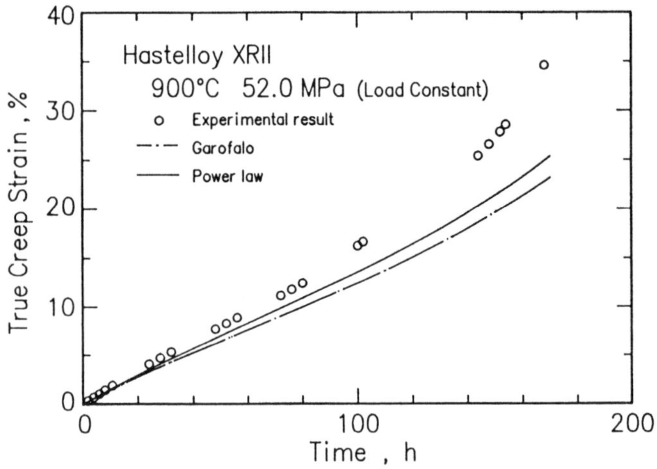


Fig. 13 Comparison between experimental data and calculated creep curves (Stress increase, Initial stress = 52MPa)

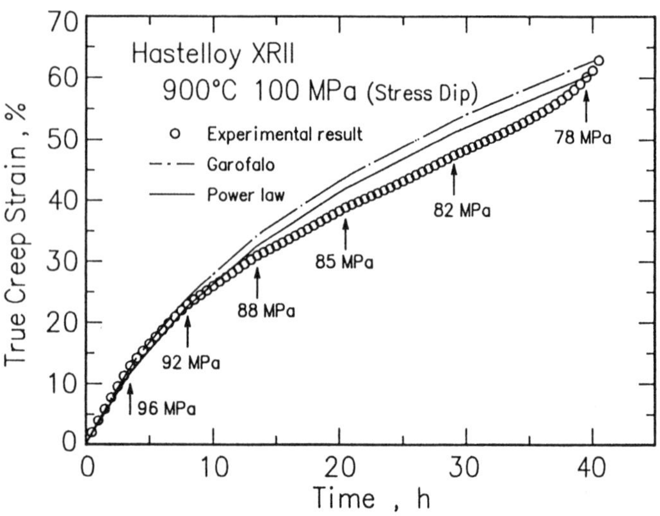


Fig. 14 Comparison between experimental data and calculated creep curves (Stress decrease, Initial stress = 100MPa)

In addition, the calculated creep curves by both the equations were compared in Figs. 17 and 18, for the cases of load constant tension tests and compression tests, respectively. Systematic differences between the equations are known from these figures where the curves of 35MPa are also shown though the reliability of those is low because

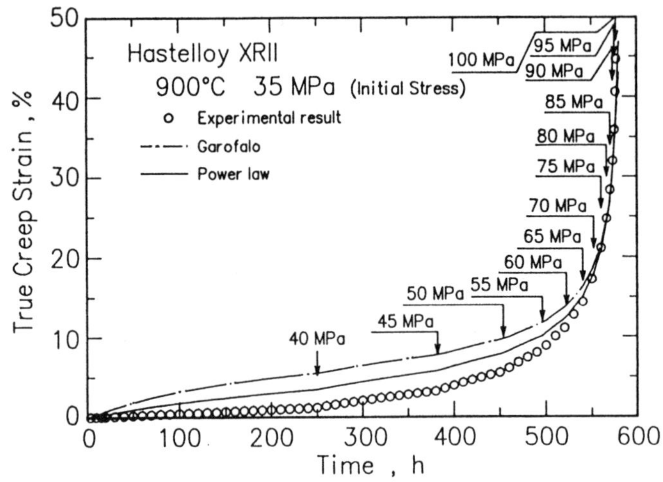


Fig. 15 Comparison between experimental data and calculated creep curves (Stress increase, Initial stress = 35MPa)

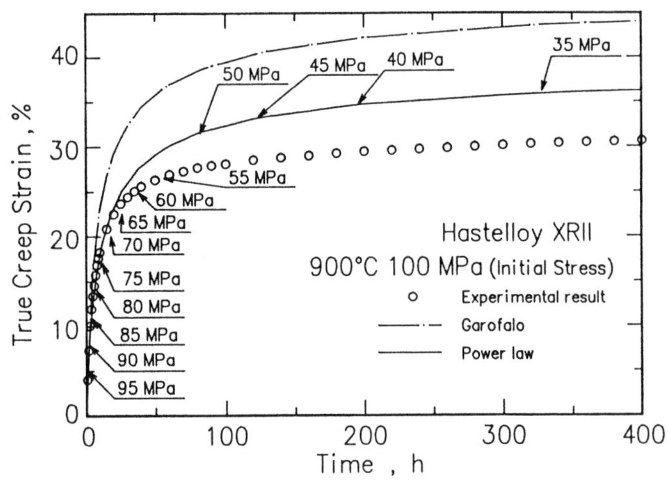


Fig. 16 Comparison between experimental data and calculated creep curves (Stress decrease, Initial stress = 100MPa)

35MPa exceeds the range where the experimental data exist.

## CONCLUSION

Creep constitutive equations were made for nickel base solution annealed alloy "Hastelloy-XR" at 900°C in air. Two types of equations were made. The first was Garofalo type and the second was Norton type incorporating an internal back stress which was evolved by creep-recovery model. The parameters were determined by constant stress-creep tests and stress-dip tests at the secondary creep region. Variable stress-creep tests were conducted to examine both the equations. The following were

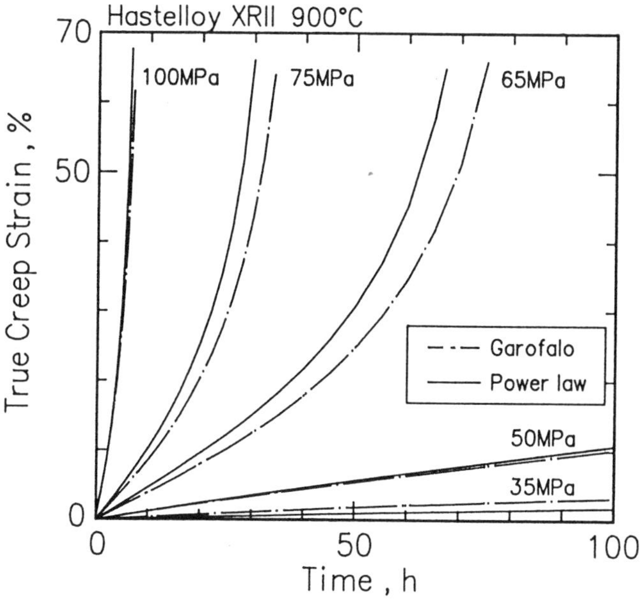


Fig. 17 Comparison between creep curves for tension tests by the two equations

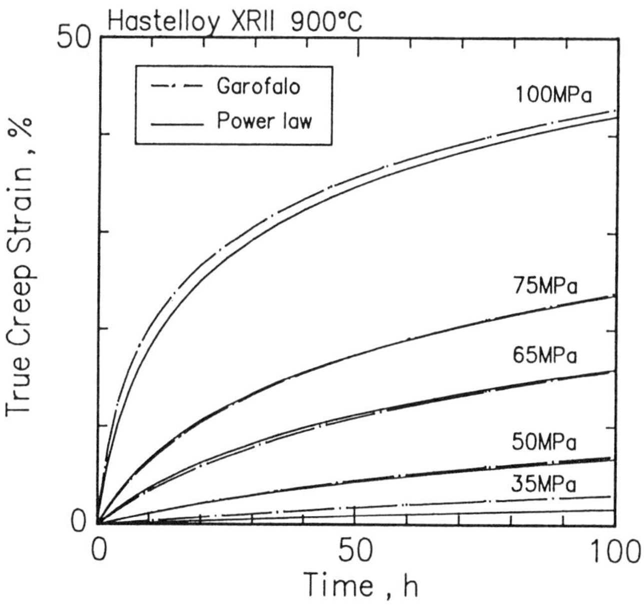


Fig. 18 Comparison between creep curves for compression tests by the tqo equations

revealed.
(1) Garofalo-type equation cannot represent the variable stress-creep behavior exactly.
(2) Internal back stress exists for Hastelloy-XR at 900°C.
(3) Initial value of the back stress at time= 0 must be considered in order to reduce the second equation to Garofalo equation for the constant stress-creep behavior.
(4) It is suggested that two kinds of back stresses exist where the one has a short time constant and the other has a longer one.
(5) The second equation can represent the variable stress behavior more exactly than the Garofalo equation combined with strain hardening rule.

## REFERENCES


1. Hada,K., Nishiguchi,I., Muto,Y. and Tsuji,H, Nucl. Eng. Des., Vol.132, 1-11 (1991)
2. Yokoi,S., Monma,Y., Kondo,T., Ogawa,Y. and Kurata,Y., JAERI-M 83-138 (1983)
3. Norton,F.H.,"Creep of steel at high temperatures", McGraw Hill, New York (1929)
4. Lagneborg,R., Metal Science J., Vol.3, 161-168 (1969)
5. Garofalo,F.,Fundermentals of creep and creep-rupture in metals, Macmillan, New York (1965)
6. Nix,W.D. and Ilschner,B.,"Strength of metals and alloys", Vol.3, Pergamon Press, 1503-1530 (1980)
7. Osthoff,W.,Schuster,H.,Ennis,P.J. and Nickel,H., Nucl. Technol., Vol.66, 296-307 (1984)
8. Cuddy,L.J., Metallurgical Transactions, Vol.1, 395-401 (1970)
9. Jonas,J.J., Acta Metallurgica, Vol.17, 397-405 (1969)
10. Malinin,N.N. and Khadjinsky,G.M., Int. J. Mech. Sci., Vol.14, 235-246 (1972)
11. Miller,A., ASME J. Eng. Mater. Technol., Vol.98, 97-105 (1976)
12. Miller,A., ASME J. Eng. Mater. Technol., Vol.98, 106-113 (1976)
13. Krieg,R.D., Inelastic behavior of pressure vessel and piping components, A.S.M.E., 15-28 (1978)

*Proceedings of the Fifth International Conference on Creep of Materials, Lake Buena Vista, Florida, USA, 18-21 May 1992*

# Transient Creep Deformation of a Single Crystal Nickel Base Superalloy After a Step Decrease in Load

**M. Maldini**
CNR-ITM
Balsamo, Italy

IN SPITE OF MANY efforts, an adequate understanding of the complicated mechanisms operating during the creep process is still lacking. For exemple, there are disagreements in the literature over the validities of rival theories to explain some important aspects of the creep behaviour of nickel-base superalloys, such as the creep curves dominated by an extensive tertiary creep rather than by a steady state, and the much greater dependence of characteristic creep rate on stress compared to that found in pure metals.

In order to improve this situation, the transient creep behaviour after stress changes has been used extensively to probe the detailed processes of creep. In addition, there is a more practical requirement to characterize the magnitude of perturbations following stress changes in order to describe creep deformation in a varying stress/temperature service cycle.

In this work, the results of an experimental investigation of transient creep following stress reduction over a range of conditions for the single crystal nickel-base superalloy SRR 99 are presented and analyzed.

A new model able to qualitatively describe the creep behaviour to a sudden change in stress has been analyzed in order to check if it is able to describe the experimental results.

## Material and experimental techniques

The experiments were carried out on single crystal superalloy SRR 99, which is designed for single crystal blades in advanced aeronautic gas turbines. It has the following nominal composition (wt%): 0.015 C, 8.5 Cr, 5.5 Al, 2.2 Ti, 5.0 Co, 9.5 W, 0.25 Mo, 2.8 Ta, balance Ni.

The creep specimens, supplied by FIAT Aviazione Spa, Turin, were cast to shape, partially homogenized at temperature slightly lower than solidus temperature and finally aged. Cooling following homogenization treatment and aging at 870°C produced cuboidal $\gamma'$ particles of 0.3 to 0.5 $\mu$m size. The hardening phases occupied about two thirds of the volume of the alloy. The specimens had cylindrical symmetry of 5.6 mm gauge diameter and 28 mm gauge length. Deviations of <001> crystalline direction from the specimen axis were within 12°.

Constant load creep tests and creep tests subjected to step changes in stress or temperature were carried out over a range of stresses at 850 and 950°C. Strain was measured from gauge length, using extensometers and capacitance transducers. All data were centrally recorded using a computerized data logging system.

Three thermocouples were placed in the gauge length allowing control of temperature gradient during creep. The furnaces were controlled by the central thermocouple.

## Results

The experimental results have shown that in a load reduction experiment, two types of transient stages are operative, depending on the amount of the stress reduction.

The observed transient behaviour after small reduction is shown in Figure 1a. After the instantaneous contraction a zero creep rate period appears. This is the expected behaviour for material when the rate controlling mechanism is the recovery process of dislocations network.

For larger stress reductions an anomalous creep transient has been detected (Figure 1b); such a behaviour has not been precedently reported for single crystal Ni-base superalloys. This anomalous creep transient can be separated in several stages.

a) A stage of time independent contraction $\epsilon_e$.

b) A stage of small time dependent contraction $\in_a$ depending on the stress reduction.
c) A period $\Delta t$, of no detectable creep whose duration depends on the stress reduction.
d) A local sigmoidal primary creep stage where strain rate decreases to a local minimum value.
e) Normal tertiary creep.

For the performed tests (Table 1), the items b and c occur over a time ranging from a minute to about 40 minutes, where the longest times are obtained for the largest stress reduction after small value of the total strain accumulated before the partial unloading. These times are invariably shorter than the cumulative periods of no detectable creep that are observed following equivalent sequential small stress reductions.

The transient occuring after the stress reduction, takes place over a strain range that can be larger than the elastic contraction on unloading. Figure 2 shows the local primary at 180 MPa/950 °C following stress reductions after different creep strain values at 400 MPa/950°C: the initial creep strain rate following the stress reduction increases with increasing creep strain at higher stress. Figure 3 shows the reductions in creep rate during the anomalous primary stage for the same creep tests of Figure 2.

During a complete unloading, two different types of transient behaviour have been observed depending on the stress level: unloading the specimen from a low stress level, after the elastic contraction, only a small backward strain occurred, while after a complete unloading from a high level of stress a backward strain rate is followed by a period of zero strain rate and then by a forward strain (Figure 4).

The experimental results have confirmed the results recently obtained at NPL , Teddington, on other superalloys (1), i.e. IN738ds and Nimonic 90. The latter is a wrought nickel base alloy and exhibits the anomalous primary behaviour when tested both in tension or in torsion indicating that the phenomenon is not an artefact of a single stress state.

A parallel set of experiments involving changes in temperature at a constant stress, rather than isothermal stress changes, showed no anomalous primary during transient creep.

## Discussion

The type of transient described in the present study has been reported previously both in pure metals and solid solutions (2-4).

Blum et al. (5) have modelled this transient behaviour in terms of heterogeneities in the substructure due to the develpment of subgrain boundaries. Essentially the material is considered to be a composite in which the creep behaviour is controlled by the deformation of the

Table 1 Summary of the principal features of transients following stress reductions from 400 MPa for single crystal nickel base superalloy SRR 99 (Temperature 950°C)

| Final Stress (MPa) | $\in_1$ (%) | Initial Strain Rate ($h^{-1}$) | Minimum Strain Rate ($h^{-1}$) |
|---|---|---|---|
| 180 | 1.0 | $1.3 \times 10^{-4}$ | $8.6 \times 10^{-6}$ |
| 180 | 2.3 | $6.5 \times 10^{-4}$ | $2.0 \times 10^{-5}$ |
| 180 | 2.5 | $2.3 \times 10^{-3}$ | $1.2 \times 10^{-4}$ |
| 180 | 8.0 | $4.4 \times 10^{-3}$ | $4.8 \times 10^{-4}$ |
| 220 | 0.8 | $1.9 \times 10^{-4}$ | $2.1 \times 10^{-5}$ |
| 220 | 5.2 | $6.0 \times 10^{-3}$ | $1.0 \times 10^{-3}$ |

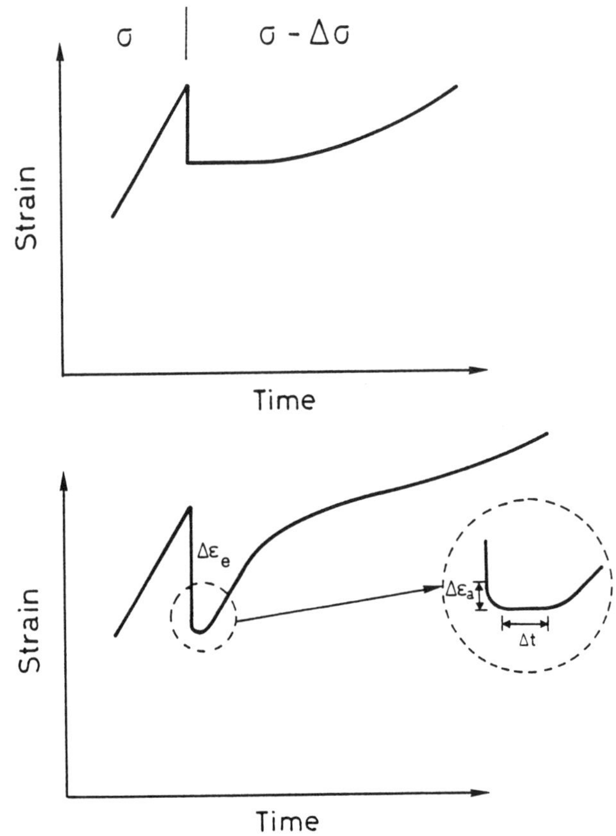


Fig. 1 - Schematic illustration of strain transient following a) small and b) large stress reductions.

"hard" regions in the neighbourhood of these subgrain boundaries. As the regions of low dislocation density, "soft" regions, are completely surrounded by the "hard" regions, they must deform simultaneously, in parallel, as a composite material consisting of hard and soft components. According to the composite model, initially introduced by Nix and Ilschner (6) for the interpretations of high temperature creep, during creep the stress is off-loaded from the "soft" regions, while the

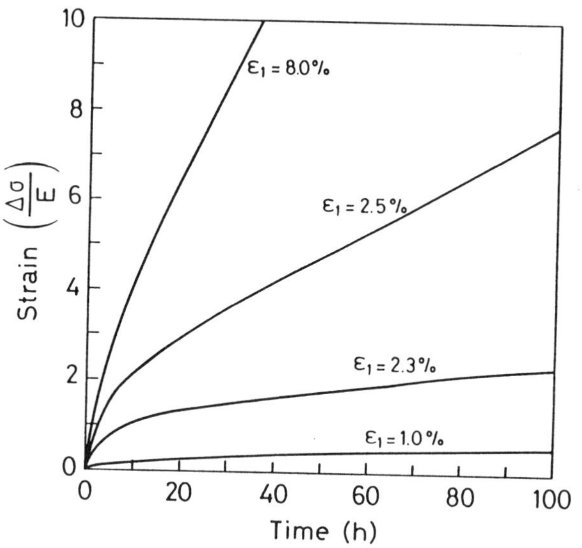


Fig. 2 - Local primary at 180 MPa following stress reductions after various creep strain, $\in_1$, at 400 MPa/950°C.
The strain is normalized by the elastic strain.

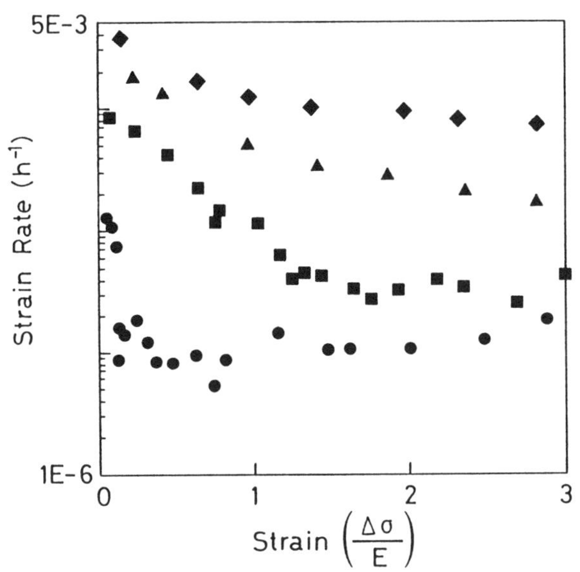


Fig. 3 - Strain rate behaviour during the anomalous primary for the creep tests of Figure 2.

stress in the "hard" zones is amplified up to reach an equilibrium that depends on the different creep strength of the two "phases". The strain transient after a partial or total decrease in load, can be explained by the redistribution of the stress between the two "phases" to reach the new equilibrium at the new applied stress. In particular, since after a complete unloading the forward straining occurs at zero applied stress, it can be explained only as the result of internal forward stresses acting in the hard regions. In order to have a quantitative description of the phenomenon, recently Vogler and Blum (7) have improved their model taking into explicit account the subgrain boundaries migrations. Of course the details of this model are inappropriate to nickel base superalloys, with high volume fractions of reinforcing $\gamma'$ particles, which do not develope an identifiable substructure. However it is possible to identify other microstructural inhomogeneities; it is possible, for example, to envisage a "composite" model where the "soft" and the "hard" phases are the matrix and the $\gamma'$ particles, respectively.

In this composite material, the relative volume fraction of the "hard" and "soft" phases must be constant, indipendent of the applied stress.
Recently Dobeš (8) has shown that such a composite model, with constant volume fraction of the two "phases", is able to provide qualitatively successful simulation of the observed anomalous transient. But also this model is inappropriate to describe the results presented in this paper, because the model cannot simulate an anomalous primary larger than the elastic contraction obtained during the partial unloading.

To overcome this problem a different mechanism must be envisaged, in particular it is known that deformation in nickel base superalloys and also in other particle strengthned materials, can happen by a variety of mechanisms, e.g.:

a) bowing of dislocations between particles
b) cutting of coherent particles by dislocations
c) climb of dislocations around particles.

When the local stress exceeds the threshold for the mechanisms a) or b) then the deformation will be

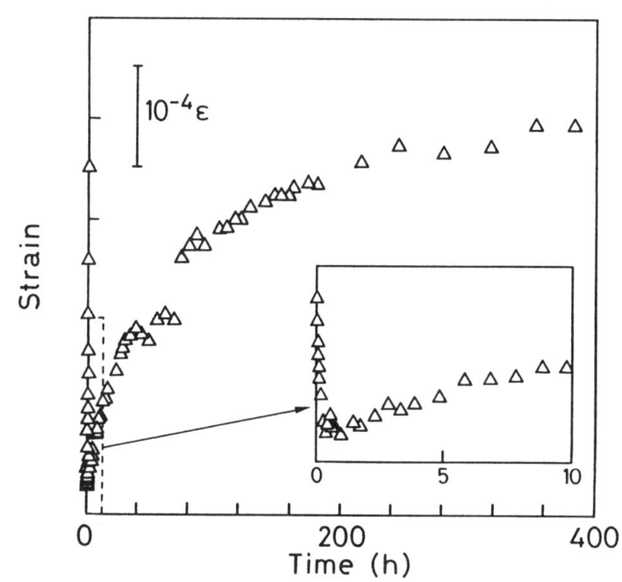


Fig. 4 - Strain behaviour after a complete unloading from 660 MPa at 850°C ($\in_1$= 9%).

much more rapid than by mechanism c) since they are glide rather than climb controlled processes. For lower stresses, when particle cutting or Orowan looping mechanims cannot occur, deformation will take place by mechanism c) alone. In a practical material there exist ranges of particle sizes and spacings between them, that means that at a single value of the applied nominal stress, different deformation mechanisms can coexist in the specimen.

Of course, the volume fraction where the fast mechanism is occurring, as suggested by Dyson and McLean (9), must be an increasing function of the stress: at the lowest stresses the slow mechanism is dominant in practically all the specimen, increasing the external applied stress, larger regions of the specimen must deform by fast mechanisms.

In the following discussion we consider the consequences of macroscopic creep being a combination of fast and slow mechanisms, where the relative fractions of material subject to the two mechanisms is dependent on the applied stress.

## The model

Consider a specimen with an applied stress $\sigma$ where $\phi$ is the volume fraction deforming by fast mechanisms (particle cutting or Orowan bowing) and $1-\phi$ deforming by slow mechanisms. Supposing that both the regions are deforming in parallel: that means that the total (elastic plus plastic) strain $\in$ must be the same in both regions and creep behaviour of the specimen can be simulated as in Figure 5a, where the bars F and S represent the fraction of the specimen straining by the fast and the slow mechanisms, respectively.

Representing by a power-law expression the stress dependence of the strain rate of both the regions, although alternative formulations could easily be considered, and supposing for sake of simplicity that the Young modulus, E, is equal in both regions, the strain behaviour of the specimen can be represented by the following couple of equations:

$$\dot{\in} = \dot{\sigma}_F/E + A\sigma_F^{n1} \qquad (1a)$$

$$\dot{\in} = \dot{\sigma}_S/E + B\sigma_S^{n2} \qquad (1b)$$

where $\sigma_F$ and $\sigma_S$ are the local stresses as operating on the bars F and S.

To ensure load equilibrium, the axial forces in the two regions must balance any external applied force. Hence the local stress on the two bars must be coupled through the applied stress, $\sigma$, by:

$$\sigma = \phi.\sigma_F + (1 - \phi).\sigma_S \qquad (2)$$

Derivating the relation (2) and supposing $\phi$ constant under conditions of creep ($\dot{\sigma}$=0) we obtain:

$$\phi\dot{\sigma}_F = -(1 - \phi)\dot{\sigma}_S \qquad (3)$$

then the stresses in the two regions are constant (at the equilibrium) or vary in opposite directions.
Combination of Equation 1 to Equation 3 yields:

$$\dot{\in} = \phi A\sigma_F^{n1} + (1 - \phi)B\sigma_S^{n2} \qquad (4)$$

The creep rate is simple the weigthed sum of the creep rate of the two bars with volume fractions as weight factors. Just after applying the load, the initial strain rate is given by Equation 4 with $\sigma_F = \sigma_S = \sigma$, then during the deformation process the stresses in the two bars change until the steady state creep rate, $\dot{\in}_{SS}$, is reached, when the stress in the two bars must satisfy both the Equation 2 and the relation:

$$\dot{\in} = A\sigma_F^{n1} = B\sigma_S^{n2}$$

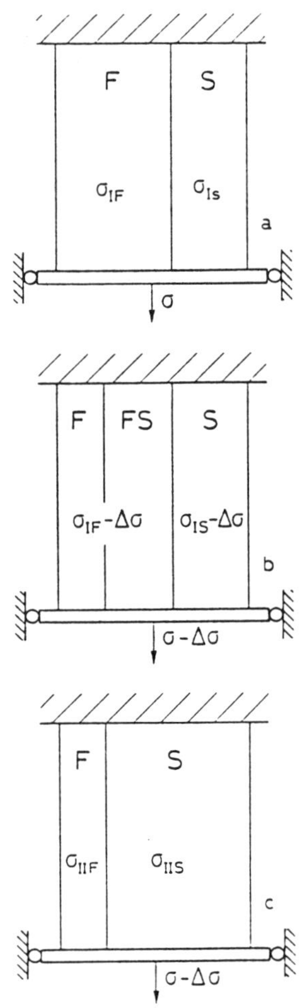


Fig. 5 - Rheological representation of the composite model (see text)

Reducing the load during the test, the change of the stress in the two regions will be only due to the elastic contraction:

$$\Delta\sigma_F = \Delta\sigma_S = \Delta\sigma$$

Dobeš (9) supposed that the volume fraction $\phi$ does not change after reducing the external load.

This simplification is not acceptable for the alloys studied in this work, at least for large stress reductions, but considering competitive fast and slow mechanisms, $\phi$ must be regarded as a variable that increases (decreases) with increasing (decreasing) the external stress.

We can now analyse the consequence of this model on the stress dependence of the steady strain rate, $\dot{\epsilon}_{SS}(\sigma)$, in conventionally constant stress tests, and the strain behaviour in a creep test where the applied initial stress is suddenly totally or partially reduced.

1) Although Equations 1 can both be represented by straight lines in a plot of log $\dot{\epsilon}$ versus log $\sigma$, the composite behaviour (Equation 4) takes the form shown schematically in Figure 6 for variable $\phi$. Real composite materials have fixed volume fractions of the constituent phases so that the creep behaviour follows one of the indicated curves a, b or c. However, in considering competitive deformation by fast and slow mechanisms, $\phi$ increases with increasing stress, consequently, in this model the variation of creep rate with stress will traverse the family of curves derived from Equation 4 as indicated in Figure 6. Depending on the relation of $\phi$ on $\sigma$, this can lead to an increasing slope of the log $\dot{\epsilon}$ vs log $\sigma$ plot as found typically in nickel-base superalloys.

2) After the initial loading, at $\sigma$, the specimen is simulated by the two bars in Figure 5a, where the bars F and S represent the volume fractions, $\phi_I = \phi(\sigma)$ and $1-\phi_I$, of material subjected respectively to fast and slow mechanisms.

When the stress is suddenly reduced to $\sigma - \Delta\sigma$ the local stress in the bars F and S becomes respectively $\sigma_{IF} - \Delta\sigma$ and $\sigma_{IS} - \Delta\sigma$. Meantime the volume fraction where fast mechanisms are active must achieve the lower characteristic value $\phi_{II} = \phi(\sigma - \Delta\sigma)$.

Practically, in a portion $\phi_{FS} = \phi(\sigma) - \phi(\sigma - \Delta\sigma)$ of the region initially deforming by fast mechanisms, the stress is reduced below the threshold stress for cutting or Orowan mechanisms and then the strain, in $\phi_{FS}$, at the reduced stress, is accumulated by slow mechanisms, i.e. following the Equation 1b. Then the two bar model, after the partial unloading, turns into a three bar model as shown in Figure 5b, with the initial condition $\sigma_{IIF} = \sigma_{IF} - \Delta\sigma$; $\sigma_{IIS} = \sigma_{IS} - \Delta\sigma$ and $\sigma_{FS} = \sigma_{IF} - \Delta\sigma$.

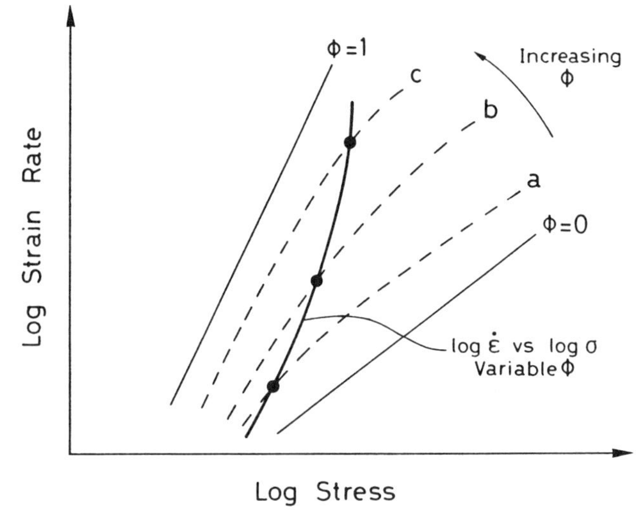


Fig. 6 - Strain rate vs. stress for a composite material with $\phi$ constant or increasing with the stress.

For a partial unloading the Equations 2 and 4 become respectively:

$$\sigma - \Delta\sigma = \phi_{II}\sigma_F + (1-\phi_I)\sigma_S + (\phi_I - \phi_{II})\sigma_{FS} \quad (5)$$

and

$$\dot{\epsilon} = \phi_{II}A\sigma_F^{n1} + (1-\phi_I)B\sigma_S^{n2} + (\phi_I - \phi_{II})B\sigma_{FS}^{n2} \quad (6)$$

Where $\sigma_F$, $\sigma_S$ and $\sigma_{FS}$ are the local stresses in $\phi_{II}$, $(1-\phi_I)$ and $(\phi_I - \phi_{II})$ respectively.

Accumulating strain, the stress in the region $\phi_{FS}$ tends to achieve the same value of the local stress of the region $1-\phi(\sigma)$ and the strain rate tends to reach the steady state at the reduced load, when the model becomes again a two bar model. All the sequences are shown in Figure 5.

Solving with numerical methods the Equations 5 and 6 it is possible to find the strain responce, in function of time, after a stress drop.

Without any effort to fit the calculation results, to the observed creep curve, in Figure 7 are represented a simulation of the strain, and the strain rate behaviour, after a partial unloading.

It is important to note that the model can predict anomalous primary larger than the elastic contraction during the unloadig.

Equations 5 and 6 can also predict the experimental behaviour after a complete unloading: in that case, the portion of the specimen initially deforming by fast mechanisms, will be in compression.

It is possible to suppose that in a portion, $\phi_{II}$, after the complete unloading, the stress in compression is high enough to activate fast mechanisms of deformation.

This situation can be again analyzed by means of Equation 6. Figure 8 shows a computed creep curve after a complete unloading.

The result documents that the model can predict the backward followed by the forward strain obtained experimentaly.

## Conclusion

a) The single crystal nickel-base superalloy SRR 99 exhibits an anomalous primary creep behaviour of decreasing creep rate following a wide range of stress reductions.

b) The anomalous primary transient occurs over a strain range that is dependent on the magnitude of the stress change and on the level of creep strain prior the stress reduction.

c) After a complete unloading, the strain behaviour depends on the value of the initial applied stress: when the initial stress is low (<400 MPa) only a backward strain occurs, while after a stress removal from high value, the backward strain is followed by a forward strain.

d) The observations are consistent with a model of creep where the macroscopic strain is due to the competition between fast and slow micromechanisms of dislocation/particle interaction.

## References


1- Maldini, M., A. Barbosa, B. F. Dyson and M. McLean, NPL Report DMA A126 January 1987

2- Blum, W., J. Hausselt and G. König, Acta Metall. 24, 293 (1976)

3- Ferreira, I. and R. G. Stang, Mater. Sci. Eng. 38, 169 (1979)

4- Langdon, T.G., R. B. Vastava and P. Yavary, in Proceeding 6th Int. Conf. on Strength of Metals and Alloys,p. 271, Pergamon Press, Oxford (1980)

5- Blum, W., A. Rosen., A. Cegielska and J. Martin, Acta Metall. 37, 2439-2453 (1989)

6- Nix, W. D. and B. Ilschner in Proceedings 5th Int. Conf. on Strength of Metals and Alloys, p. 1503, Pergamon Press, Oxford (1980)

7- Vogler, S. and W. Blum, in Proceedings 4th Int. Conf. on Creep and Fracture of Metals and Aloys, p. 65, The Institute of Metals, London (1990)

8- Dobeš, F., Scripta Metall. and Mat. 25, 2303-2306 (1991)

9- Dyson B. F. and M. McLean, NPL Teddington, Private communication


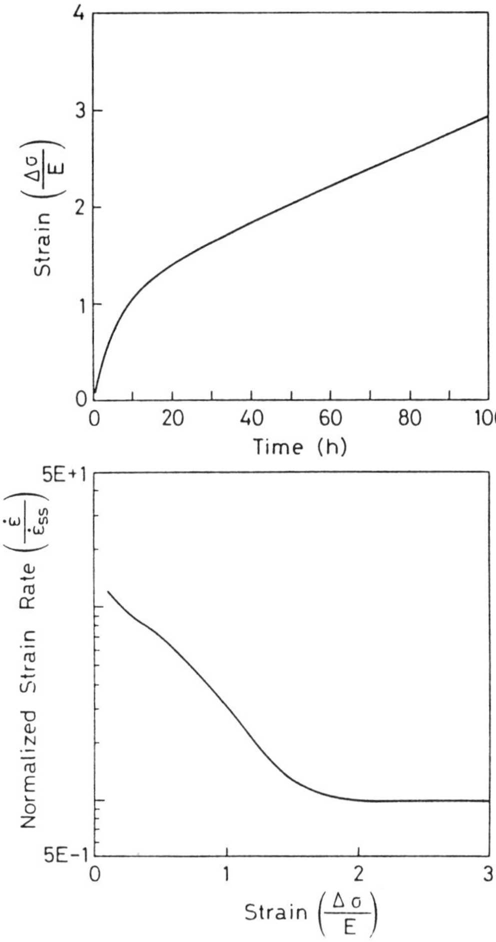


Fig. 7 - Strain and strain rate predictions of Equation 6 for a partial unloading after the steady state has been established at the highest stress. $A = 10^{-4}(h^{-1})$, $\sigma/E = 10^{-3}$ MPa, $A/B = 10^{+4}$ $n_1 = n_2$, $\Delta\sigma/\sigma = .5$, $\phi_I = .5$ and $\phi_{II} = .1$.

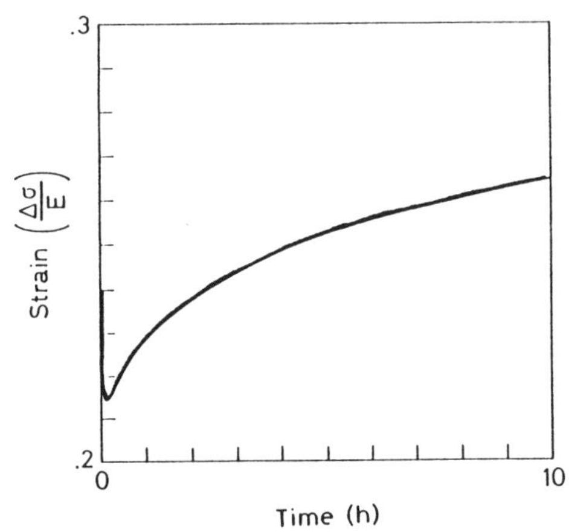


Fig. 8 - Prediction of Equation 6 for a complete unloading. $\phi_I$=.7, $\phi_{II}$ = .2, other values as Figure 7.

*Proceedings of the Fifth International Conference on Creep of Materials, Lake Buena Vista, Florida, USA, 18-21 May 1992*

# Effect of Fabrication Processes on the Creep Behavior of 9-12% Chromium Steels

**F. Arav and H.J.M. Lentferink**
Stork Boilers
Hengelo, The Netherlands

**C.F. Etienne and J.C. van Wortel**
TNO Research Institute
Apeldoorn, The Netherlands

ABSTRACT

This paper deals with the investigations carried out into the effects of welding and cold bending, particularly on long-term properties of the modified 9Cr steel T91/P91. Base metal from different suppliers and GTAW, SMAW and SAW weldments as well as short radius bends were subjected to microstructural, mechanical and (isostress) creep tests. Comparison is made with the properties of a 12Cr steel. Although the creep strength derived from the ASME Code could not be achieved, T91/P91 proved to be superior to the 12Cr steel at ≈600°C. The high creep strength of T91/P91 is caused by the presence of large quantities of very small incoherent Nb-V carbonitrides, which are very stable under creep conditions. A fall in creep strength was observed in weldments in the soft region of the HAZ and in cold-formed short radius bends. Beside its high creep strength, T91/P91 has an excellent weldability and cold formability. These properties make this steel very suitable for application in power generation components operating in the creep range.

CREEP IS A FAILURE MODE which plays an essential role in designing components operating at elevated temperatures. In common engineering practice the effect of creep is simply taken into account by using the allowable stresses which are also based on the creep properties of the material. These allowable stresses are, however, derived only from the creep properties of the base material, obtained by using uniaxial specimens. It is surprising to observe how little is known on the effect of the fabrication processes, such as welding and forming, on the creep behaviour of steels compared with the creep behaviour of the unprocessed material. For recent publications see [1], [2], [3], [4].

On the other hand, experience gained over many decades with components manufactured from common creep resistant steels gives no reason to expect serious premature creep damage, provided the design and fabrication requirements are met. The safety factor regarded in the derivation of the allowable stress appears to be sufficient to ensure safe and reliable operation.

Recently, in the power generation world, much attention is being paid to the modified 9Cr steel T91/P91, which has a considerably higher creep strength than its unmodified predecessor. The specified design stresses of T91/P91 are at temperatures above 550 °C even much higher than those of the 12Cr steel. Since the modified 9Cr steels owe their excellent creep properties to the precipitation of fine (Nb,V) carbonitrides, the question is justified to what extent these creep properties are affected by component fabrication methods such as welding, cold forming and subsequent heat treatment.

This paper deals with the investigations carried out by Stork Boilers, being a supplier of power plants, in close cooperation with the TNO Research Institute, into the effect of fabrication processes on the creep behaviour of T91/P91. In order to compare the outcome of the tests with the properties of 12Cr steel, a limited number of tests on 12Cr steel is carried out as well.

## PROPERTIES OF MODIFIED 9CR STEEL (T91/P91)

Like the German 12Cr steel X20CrMoV12 1 (abbreviated as X20), the T91/P91 steel is a fully martensitic air-hardening material with a hardness of HV10=200-250 and relatively small grains (≈ASTM 10). The main differences in chemical composition between the modified 9Cr steel (T91/P91) and 12Cr steel (X20) are the lower carbon content of modified 9Cr steel (0.1% instead of 0.2%) and the presence of niobium (0.06-0.1%) and nitrogen (0.03-0.07%).

In theory the creep properties and the good weldability are achieved by the well balanced chemical composition. The relatively low carbon content reduces the risk of cracking after welding (maximum hardness T91/P91 ≈HV450 versus maximum hardness X20 ≈HV600).

The large number of very small and stable Nb- and V-rich carbonitrides prevents excessive grain coarsening during welding. At high temperatures these carbonitrides are believed to stabilize the dislocation networks more effectively than carbides. This is because at temperatures in the creep range these carbonitrides are more stable than carbides.

According to the data derived from ASME Code [5] and DIN 17175 the $10^5$ h creep rupture strength of T91/P91 at 600 °C is approximately 60% higher than that of X20 (98 and 59 MPa respectively).

## OUTLINE OF THE INVESTIGATIONS

This project was focussed on the application of 9-12Cr steels in components for power generation, e.g. superheater tubing, superheater headers and steam piping.

For superheater tubing, GTAW girth welds were made between T91-T91, T91-X20, T91-2¼Cr1Mo and T91-TP347H steels. Welds between T91 tubes were made using modified as well as unmodified welding wire. All welds in these tubes were cooled down to ambient temperature prior to heat treatment. Tubes were also cold bent, using different bending radii. Some of the bends were heat treated after cold bending.

For application in headers and steam piping, SMAW and SAW weldments were made in thick-walled, large-diameter pipes.

In order to assess the degree of change in the properties due to welding and cold forming, the program also includes tests on the unprocessed base material. To get some indication of the dispersion of the properties of T91/P91, limited number of tests was performed on base materials from 4 other suppliers.

Table 1. Outline of the investigation program

| | Material | Supplier | Size |
|---|---|---|---|
| Base metal | T91 | A | 44.5x7.1 |
| | T91 | B | 25.4x2.8 |
| | T91 | G | 45x6.4 |
| | P91 | A | 380x50 |
| | P91 | C | 368x73 |
| | P91 | D | 343x74 |
| | P91 | E | 219x18.25 |
| | X20 | F | 44.5x7.1 |
| | X20 | F | 230x22 |
| | **Process** | **Materials** | **Consumables** |
| Weldments | GTAW 44.5x7.1 | T91/T91 | 9Cr unmod. |
| | | T91/T91 | 9Cr mod. |
| | | T91/X20 | 12Cr |
| | | T91/2¼Cr1Mo | 9Cr mod. |
| | | T91/TP347H | Nickel base |
| | SMAW | P91/P91, 380x50 | 9Cr mod. |
| | SAW | P91/P91, 380x50 | 9Cr mod. |
| | SMAW | X20/X20, 230x22 | 12Cr |
| Bends (cold bent) | T91 44.5x7.1 | R=60, no PBHT | |
| | | R=60, 740 °C/1h | |
| | | R=70, no PBHT | |
| | | R=80, no PBHT | |
| | | R=80, 740 °C/1h | |
| | X20 44.5x7.1 | R=60, no PBHT | |
| | | R=60, 740 °C/1h | |

For the prediction of the long-term behaviour of actual components fabricated from T91/P91 on the basis of the tests results, a comparison with the properties of a longer existing high-temperature steel was considered useful. For this purpose, the German 12Cr steel X20 was chosen, which is also a martensitic steel and has been satisfactorily applied in continental Europe for more than 25 years.

Beside the creep tests, which formed the major part of this research project, all base and processed materials were also subjected to extensive mechanical and microstructural investigations.

Table 2. Chemical composition of the investigated materials (weight %)

| Material | Dimensions | Supplier | C | Si | Mn | P | S | Al | Cr | Ni | Mo | V | Nb | N |
|---|---|---|---|---|---|---|---|---|---|---|---|---|---|---|
| T91 | 44.5 x 7.1 | A | 0.096 | 0.35 | 0.36 | 0.020 | 0.003 | 0.0102 | 8.36 | 0.11 | 0.928 | 0.20 | 0.07 | 0.0584 |
| T91 | 25.4 x 2.8 | B | 0.094 | | 0.45 | 0.014 | 0.004 | 0.024 | 8.11 | 0.09 | 1.03 | 0.20 | 0.072 | 0.053 |
| T91 | 45 x 6.4 | G | 0.100 | 0.41 | 0.37 | 0.002 | 0.004 | 0.01 | 8.52 | 0.18 | 0.93 | 0.22 | 0.08 | 0.034 |
| P91 | 380 x 50 | A | 0.097 | 0.33 | 0.44 | 0.016 | 0.003 | < 0.02 | 8.33 | 0.13 | 0.92 | 0.24 | 0.08 | 0.039 |
| P91 | 368 x 73 | C | 0.110 | 0.40 | 0.35 | 0.010 | 0.001 | <.005 | 8.68 | 0.12 | 0.93 | 0.21 | 0.09 | 0.068 |
| P91 | 343 x 74 | D | 0.090 | 0.46 | 0.47 | 0.020 | 0.002 | 0.008 | 8.95 | 0.12 | 0.95 | 0.18 | 0.074 | 0.043 |
| P91 | 219 x 18.3 | E | 0.090 | 0.46 | 0.35 | 0.014 | 0.005 | 0.015 | 8.39 | 0.13 | 0.95 | 0.22 | 0.10 | 0.030 |
| T91/P91 | Required (ASME) | | 0.08–0.12 | 0.20–0.50 | 0.30–0.60 | 0.02 max. | 0.01 max. | 0.040 max. | 8.00–9.50 | 0.4 max. | 0.85–1.05 | 0.18–0.25 | 0.06–0.10 | 0.030–0.070 |
| 12CrMoV12 1 | 230 x 22 | F | 0.18 | 0.23 | 0.52 | 0.016 | <0.005 | | 11.9 | 0.66 | 0.92 | 0.26 | | |
| | Required (DIN 17175) | | 0.17–0.23 | 0.23 max | 1.0 max | 0.03 max. | 0.03 max. | | 10–12.5 | 0.3–0.8 | 0.8–1.2 | 0.25–0.35 | | |

Table 3. Chemical composition of welding consumables (weight %)

| Weld Process | Type | Diameter (mm) | C | Si | Mn | P | S | Cr | Ni | Mo | V | Nb | Cu | W | Nb/Ta |
|---|---|---|---|---|---|---|---|---|---|---|---|---|---|---|---|
| GTAW | 9Cr unmod. | 2.0 | 0.063 | 0.62 | 0.54 | 0.008 | 0.011 | 8.85 | – | 0.99 | – | – | – | – | – |
| GTAW | 9Cr mod. | 1.6 | 0.08 | 0.20 | 0.95 | 0.006 | 0.006 | 9.02 | 0.69 | 0.90 | 0.18 | 0.04 | 0.15 | – | – |
| SMAW | 9Cr mod. | 3.2 | 0.06 | 0.33 | 1.52 | 0.005 | 0.003 | 9.25 | 0.92 | 1.07 | 0.17 | 0.03 | – | – | – |
| SMAW | 9Cr mod. | 4.0 | 0.06 | 0.36 | 1.56 | 0.005 | 0.003 | 9.40 | 0.90 | 1.07 | 0.17 | 0.03 | – | – | – |
| SAW | 9Cr mod. | 3.2 | 0.08 | 0.16 | 1.67 | 0.007 | 0.005 | 8.93 | 0.60 | 0.89 | 0.27 | 0.06 | 0.02 | – | – |
| GTAW | 12Cr | 2.0 | 0.21 | 0.40 | 0.60 | – | – | 11.3 | – | 1.0 | 0.30 | – | – | 0.45 | – |
| GTAW | Nickel base | 2.0 | <0.03 | <0.20 | 3.0 | <0.02 | <0.01 | 19–22 | >67 | – | – | – | <0.2 | – | 2–3 |
| SMAW | 12Cr | 2.5–4.0 | 0.18 | 0.3 | 0.8 | – | – | 11.0 | 0.5 | 0.9 | 0.3 | – | – | 0.5 | – |

The scope of the investigation program is summarized in table 1. The chemical composition of the investigated base metals is presented in table 2 and that of welding consumables in table 3.

## PROCEDURE FOR CREEP TESTING

In this project the assessment of the long-term creep behaviour was based on isostress creep tests, performed at 100 MPa and temperatures between 680 and 580 °C. The tests have been linearly extrapolated on a log time (h) versus temperature (°C) basis. According to the figures derived from the ASME Code an average rupture time of 100,000 h should be achieved at 98 MPa and 600 °C.

For tubes and pipes the creep specimens were taken in longitudinal direction. The specimens were uniaxially loaded at a constant load, corresponding with a nominal stress of 100 MPa.

The specimens for testing the weldments were taken in the same direction. Thus, the specimens taken from circumferential weldments were loaded perpendicular to the weld, so-called "crossweld-specimens".

The creep tests on bends were performed, using special ring-shaped specimens cut from the bend, which were loaded in circumferential direction, figure 1. This means also that the intrados and the extrados are loaded equally, since the cross sections were made equal. The advantage of this method is that the weakest zone of the bend can be detected (extrados or intrados). Furthermore, this facilitates a comparison between test results and operational loading conditions (highest principal stress is in circumferential direction). A disadvantage of this type of specimen is, however, the presence of an additional bending stress which leads to higher creep deformations on the inner surface, resulting in an overall lower creep performance with respect to the results from axial specimens. The tests on ring-type specimens were only performed to compare the results of bends with straight tubes and those results may not be interpreted as absolute values. Numerical creep analyses performed by Netherlands Energy Research Foundation (ECN) showed a maximum deviation in rupture times by a factor of 2.

The special test set-up for loading of the ring-shaped specimens is shown in figure 1.

## CREEP BEHAVIOUR

**Base metal.** The results of the creep tests on base materials are presented in figure 2. The data of most materials lie in a rather narrow scatter band. Rupture times at 600°C and 100 MPa are about one fourth of the expected time derived from the ASME Code. Only the material delivered by supplier C approaches the value derived from ASME (one test is still running). The rupture time of X20 base metal appears to be one fourth of the rupture time of T91/P91.

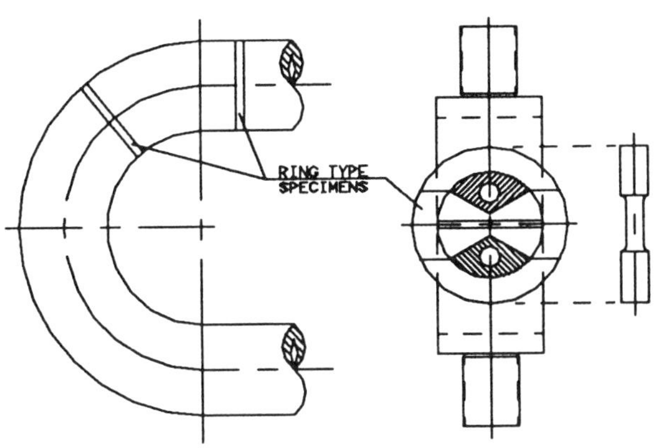


Fig. 1. Location and loading rig for creep testing of ring-type specimens.

In figure 3 the minimum creep rates for the base metals T91/P91 and X20 are presented. The creep rate of material C appears to be three times lower than that of the other suppliers, and the creep rate of X20 is approximately six times higher. Since the slope of X20 is steeper, the difference between X20 and T91/P91 decreases at lower temperatures and dissappears at about 540°C.

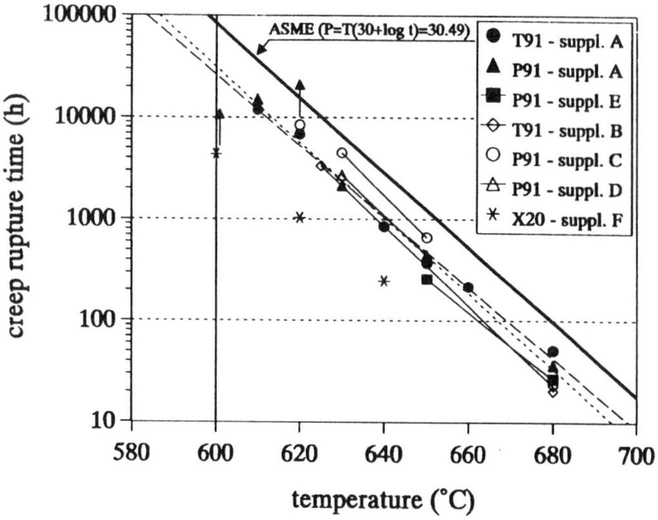


Fig. 2. Results of isostress creep tests on base materials (100 MPa, axial specimens).

The extrapolated times to 1% creep strain (t$_1$) at 600°C in base metals T91/P91 and X20 are presented in figure 4. At that temperature the time to 1% strain in material C is 2 times longer than that of the other materials, and that of X20 is 6 times shorter.

**Weldments.** The isostress creep tests on T91/P91 weldments have been restricted to the materials of supplier A. All welds were circumferential. The results of the creep tests are presented in figure 5. In all specimens rupture developed in the fine-grained part of the HAZ (near base metal). This location corresponds completely with the weakest zone of the weldment, as demonstrated by the hardness course in figure 6.

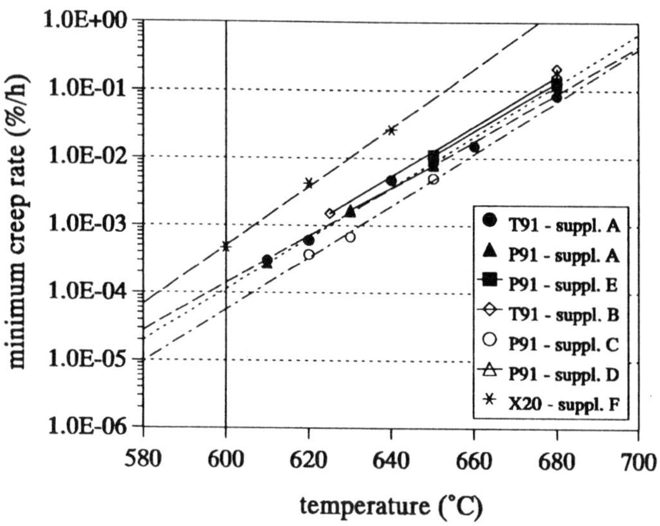


Fig. 3. Minimum creep rates for base metal T91/P91 and X20 (100 MPa).

The rupture location in the creep specimen of a T91 weldment made with unmodified wire is shown in figure 7. The fact that, even in this case, rupture takes place in the HAZ demonstrates that the shorter rupture time of the weldment in relation to the base metal is not caused by the lower creep strength of the weld metal, but by the properties of the base metal due to the heat input during welding. This phenomenon of rupture in the fine-grained HAZ is often referred to as Type-IV cracking.

The T91-TP347H dissimilar joint, welded with nickel base consumable, fractured in the HAZ of T91 in approximately the same time as the T91-T91 weldments. The dissimilar joint T91-X20 fractured in the fine-grained HAZ on the X20 side, the rupture time being half the rupture time of T91-T91 weldments. The dissimilar joint T91-2¼Cr1Mo, welded with modified wire, fractured in the soft decarbonized zone near the fusion line on the 2¼Cr1Mo side and not in the Type-IV zone near the base metal.

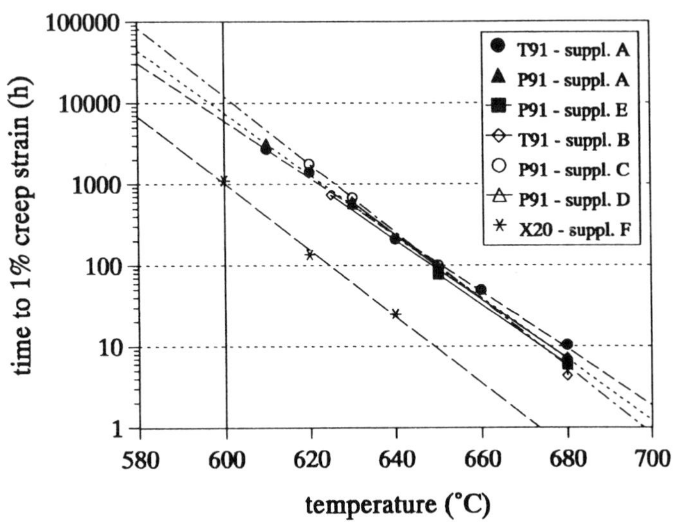


Fig. 4. Time to 1% strain for base metals T91/P91 and X20 (100 MPa).

**Bends.** The results of the creep tests on ring-shaped specimens of cold-formed bends are presented in figure 8. For comparison the results of the creep tests on the base metal of the same tubes, using axial specimens as well as ring-type specimens are also included.

The rupture times for the ring-type specimens of straight tube are somewhat shorter than those for the axial specimens. All bends fractured in the extrados and the creep rupture times were shorter than the rupture times of the axial and ring-type specimens of the straight tube. The creep rupture time of the bends decreases with decreasing bending radii (increasing plastic deformation), but this effect was not pronounced. A post bend heat

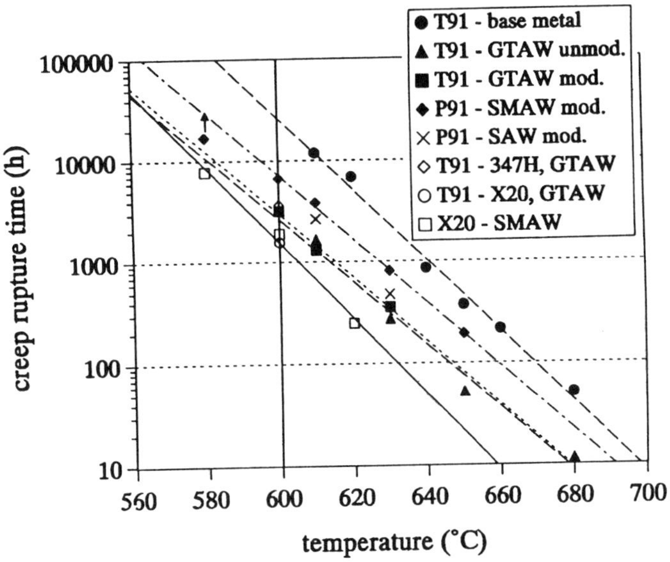


Fig. 5. Results of creep tests on weldments (100 MPa, cross-weld specimens, T91/P91 suppl. A, X20 suppl. F).

treatment (PBHT) of 740°C/1h hardly improves the creep rupture time at 600°C. The slope of the curve is, however, steeper compared with that of bends without PBHT. This suggests that a PBHT may be useful for service temperatures <580°C.

The results of the creep tests on cold-formed bends in X20 show the same tendency as in T91 material.

## DISCUSSION

The isostress creep tests enable the extrapolation of the results to longer rupture times or lower temperatures. The rupture time at 600°C and 100 MPa for base metal and weldments can be derived from the extrapolation for

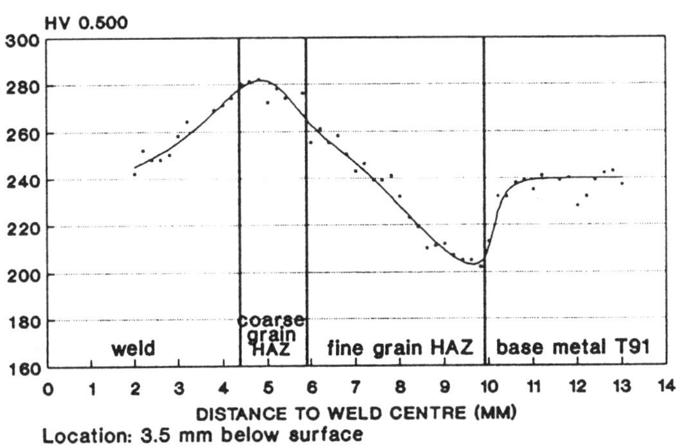


Fig. 6. Microhardness distribution in a T91 weldment (GTAW, 44.5x7.1).

temperatures not exceeding the results of the isostress creep tests by more than 10°C, as presented in figures 2, 5 and 8. In table 4 the results of the creep tests are presented according to this extrapolation. Moreover, the Larson-Miller parameters (with C=30, a value which could be proved to be realistic for the base metal) from the extrapolated results for 600°C at 100 MPa are presented in the second column. From these Larson-Miller parameters the creep rupture strengths for 100,000 h at 600°C are derived by extrapolation proportional to the ASME curve as indicated in figure 9 (there are now some indications that this method of extrapolation for the weldments leads to too optimistic results). These figures are presented in the third column of table 4. From these figures the ratio of the creep rupture strength between the processed material and base metal has been derived, as presented in the last column of table 4.

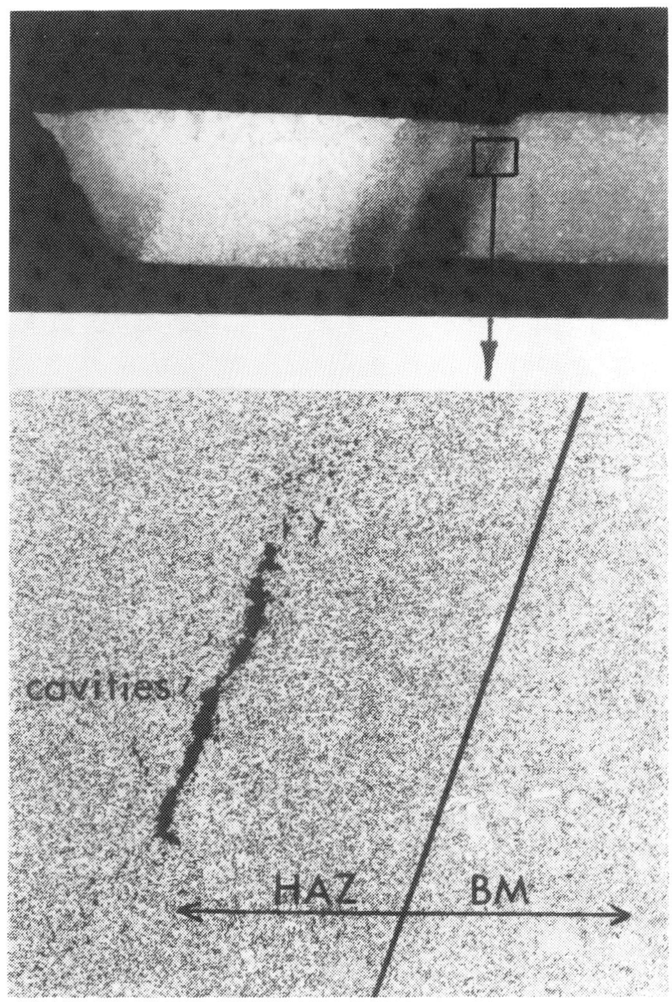


Fig. 7. Creep fracture of a T91 weldment in fine-grained HAZ near base metal (44.5x7.1, 100 MPa, 610°C).

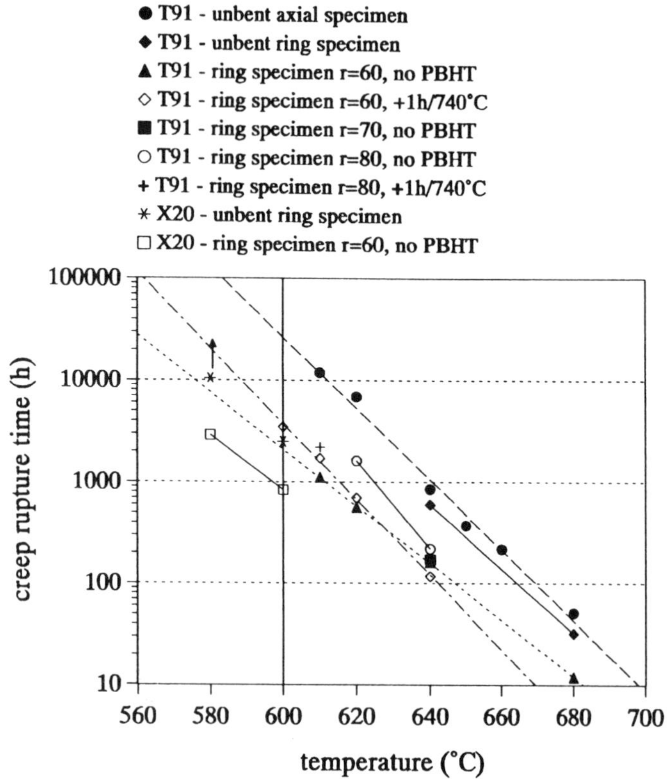


Fig. 8. Results of isostress creep tests on cold formed bends (100 MPa, supplier A, 44.5x7.1).

**Base metal.** The $10^5$ h creep strength at 600°C of most of the base materials investigated, is with ≈87 MPa ≈11% lower than that derived from the ASME Code (98 MPa). Only material C appears to give about 96 MPa, being almost equal to that derived from the ASME Code. This rather high creep strength of this material is probably caused by its higher carbon and nitrogen contents compared with the other base materials.

The extrapolated $10^5$ creep strength at 600°C of the tested X20 base material is approximately 62 MPa and 29% lower than that of the P91/T91 materials (≈87 MPa) and is 5% above the average value given in DIN 17175.

**Weldments.** The $10^5$ h creep strength of the T91/P91 cross-welds at 600°C is approximately 15% lower than the creep strength of the actual base materials. This decrease is caused by the soft fine-grained region in the HAZ, the so-called type IV region, and is more pronounced in GTAW weldments than in SAW and SMAW weldments. Because both the base metals A have the same creep rupture strength, it is expected that the difference in creep strength between these weldments is caused by differences in heat input during welding. The SMAW weldment was made with the lowest heat input and the GTAW weldment with the highest heat input ($10^5$ creep strength at 600°C, 77 and 73 MPa respectively).

Compared with the ASME values, the $10^5$ h creep rupture strength of the weldments at 600°C and 73-77 MPa is approximately 20-25% lower than the ASME values for base metal (98 MPa), but only 11-16% lower than the extrapolated value for the tested base metal.

For X20 the $10^5$ h creep strength at 600°C of weldments is also considerably lower than that of base metal. The limited number of tests suggests a decrease of 21% (from 62 to 49 MPa). This is consistent with the results described in [6].

**Bends.** The $10^5$ h creep rupture strength at 600°C for the cold-formed bend with a radius of 60 mm (R/D=1.3) is ≈13% lower than that for the base metal (73 and 84 MPa respectively). After a PBHT of 740°C/1h the $10^5$ creep rupture strength of the bend hardly increases (from 73 to 76 MPa).

The decrease of 12% (from 52 to 46 MPa) in $10^5$ h creep rupture strength for the X20 bend is comparable with that of the T91 bend. Such a drop is, therefore, not a typical T91 phenomenon. According to [1] and [7] the creep behaviour of cold-formed bends in T91 and X20 depends on the deformation grade and service temperature. A significant decrease in creep rupture strength can be expected at deformation grades >15-20%, which is in full agreement with the results of this investigation.

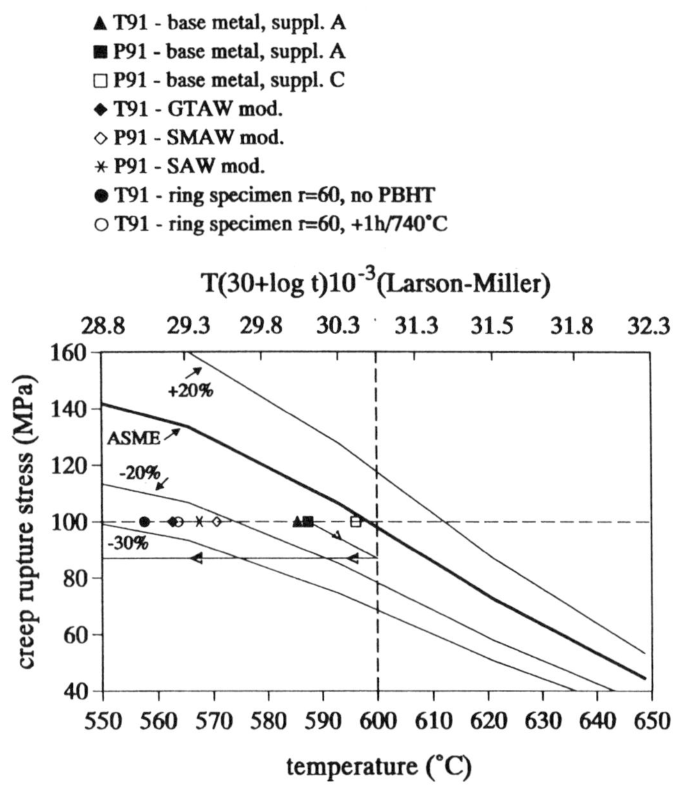


Fig. 9. Derivation of the $10^5$h creep strength at 600°C, using Larson-Miller parameter and values derived from ASME [5] (1.5 x allowable stress).

Table 4. Extrapolated results of isostress creep tests on base metal, weldments and bends

| | | | Rupture time $t_R$/600°C/100MPa (h) | $P_{LM}$(1) | Rupture strength $\sigma_R$/600°C/$10^5$h (MPa) | Ratio (%) of rupture strength processed metal/ base metal |
|---|---|---|---|---|---|---|
| T91/P91 | **ASME** | | 85,000 | 30.49 | 98 | |
| | **BASE METAL** | (axial specimen) | 30,000 | av. 30.11 | 87 | 100 |
| | **WELDMENTS** | (cross–weld specimen) | | | | |
| | GTAW | (44.5x7.1) | 3,000 | 29.22 | 73 | 84 |
| | SMAW | (380x50) | 6,700 | 29.50 | 77 | 89 |
| | SAW | (380x50) | 5,000 | 29.42 | 75 | 86 |
| | **BASE METAL** | (44.5x7.1, ring specimen) | 20,000 | 29.95 | 84 | 100 |
| | **BENDS** | (cold formed, 44.5x7.1, ring specimen) | | | | |
| | R=60, no PBHT | | 2,000 | 29.07 | 73 | 87 |
| | R=60, +740°C/1h | | 3,500 | 29.30 | 76 | 90 |
| X20 | DIN 17175 | | 5,000 | 29.40 | 59 | |
| | **BASE METAL** | (230x22, axial specimen) | 7,000 | 29.55 | 62 | 100 |
| | **WELDMENTS** | | | | | |
| | SMAW | (230x22, cross–weld specimens) | 1,910 | 29.05 | 49 | 79 |
| | **BASE METAL** | (44.5x7.1, ring specimen) | 2,650 | 29.18 | 52 | 100 |
| | **BENDS** | (cold formed, 44.5x7.1, ring specimen) | | | | |
| | R=60, no PBHT | | 850 | 28.75 | 46 | 88 |

(1) $P_{LM}=T(30+\log t)10^{-3}$ calculated from $t_R$/600°C/100MPa

## MICROSTRUCTURAL STABILITY

The presence of precipitates and dislocations in base metal of T91/P91 and the stability of these microstructural components after creep exposure, welding and bending have been investigated by means of transmission electron-microscopy.

The results are summarized in table 5. It can be concluded that the incoherent niobium-vanadium carbonitrides are responsible for the extra creep strength of T91/P91 and that these carbonitrides are very stable under creep conditions, figure 10. The coherent precipitates, on the other hand, are unstable and even not always present in the as-delivered condition.

The loss of creep strength in weldments could be explained by the rearrangement of lath martensite with a high dislocation density into equiaxed cells due to the welding process, figure 11. The loss of creep strength in cold-formed bends is a consequence of the deformation which causes accelerated destabilization of the dislocation structure during creep exposure.

## IMPLICATIONS FOR DESIGN

What are the consequences of the observed loss of creep strength in weldments and bends in practical applications, and how far should they be taken into account in the design stage? From the results of the investigations reported in this paper the following can be said.

In weldments the loss of creep strength has been observed in cross-weld specimens loaded perpendicularly to the circumferential seam, and all specimens broke in soft type-IV zone. This loading system is, however, not typical for cylinders loaded by internal pressure, where the hoop stress amounts to twice the longitudinal stress.

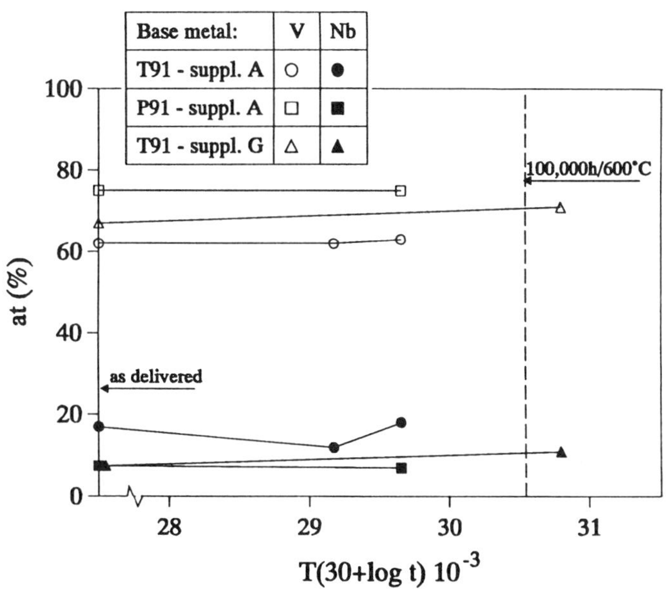


Fig. 10. Chemical composition of incoherent niobium-vanadium carbonitrides in three base metals in as-delivered condition and after creep exposure.

The effect of the soft zone on a pipe loaded by internal pressure can be compared with the effect of a very local decrease in wall thickness, which decrease has the same ratio as the loss of creep strength. If the part with smaller wall thickness has a relatively short length, the radial deformation and, as a result, the hoop stress in that part will be constrained by the adjoining parts. From the creep tests performed it appears that the length of the soft zone is rather short indeed, so it can be concluded that in the dimensioning of cylindrical parts loaded only by internal pressure there is no need to take the lower creep strength of the soft zone into account.

In case of additional axial loads, however, as in suspension tubes in steam boilers and steam piping subjected to thermal expansion, the longitudinal stresses should be limited. This means that, in case the longitudinal stress exceeds the value of the hoop stress multiplied by the ratio of the creep strength between the weldment and base metal, the allowable stress should be reduced with the same ratio.

Compared with X20 this approach seems to be somewhat conservative since X20 weldments show this drop as well, which is not being taken into account in design. Nevertheless, failures after a long period of application due to this drop are not known to the authors.

Concerning the cold-formed tube bends, the test results show a drop of creep strength in T91 comparable to that in X20. Since cold-formed X20 bends have been succesfully applied in boiler superheater tubing for many decades, T91 tube bends may be applied in the same way within the same field of applications.

Conservatism due to limited practical experience with T91/P91 and the possible higher service temperatures, however, requires a more careful approach, particularly in case of applications with larger diameters and if a higher degree of safety is required.

Table 5. Precipitates and dislocations in T91

| Precipitates or dislocations | Presence in base metal | Stability after creep exposure, welding and bending |
|---|---|---|
| M23C6 | • always present | not examined<br>(not responsible for extra creep strength) |
| Coherent precipitates | • MX–needles 15x3 nm<br>• Mostly present<br>(not in one supply) | • After creep exposure: no longer present<br>• In fine–grained HAZ: not present<br>• After bending: no change |
| Incoherent precipitates | • Many<br>• Niobium–vanadium carbonitrides<br>• 10–100 nm<br>• Two types<br>– Primary NbV–CN<br>* not dissolving during austenitizing<br>* Nb–rich: mainly 70Nb+20V+Fe+Cr (at%)<br>– Secondary NbV–CN<br>* appearing after tempering<br>* V–rich: mainly 70V+10Nb+Fe+Cr (at%) | • After creep exposure:<br>– more<br>– somewhat coarser (20–100 nm)<br>– composition remains nearly constant<br>• In fine–grained HAZ (as welded):<br>– many<br>– somewhat coarser (20–200 nm)<br>• In fine–grained HAZ after creep:<br>– somewhat less<br>– somewhat coarser (30–200 nm)<br>• After cold bending:<br>– no change<br>• After cold bending and creep exposure:<br>– many<br>– somewhat coarser |
| Dislocations | • Lath martensite<br>• High density | • After creep exposure<br>– lath –> equiaxed cells<br>• In fine–grained HAZ (as welded):<br>– no lath, but equiaxed cells<br>– relatively high density<br>• In fine–grained HAZ after creep:<br>– no lath –> equiaxed cells<br>– low density<br>• After cold bending:<br>– very high density<br>• After cold bending and creep exposure:<br>– lower density<br>– transition into equiaxed cells<br>– locally recrystallization |

a)

b)

Fig. 11. Comparison between the dislocation structure of P91 base metal and fine-grained HAZ (10,000x).
a) base metal
b) fine-grained HAZ

## CONCLUSIONS

The $10^5$ h average creep strength of T91/P91 base metal at 600°C is, according to the results of the isostress creep tests, about 87 MPa. This figure is 11% lower than the value derived from the ASME Code, but 40% higher than that of the 12Cr steel X20. The results of creep tests on base metals from different suppliers were for the greater part within a narrow scatter band.

The $10^5$ h creep strength at 600°C of weldments performed on cross-weld specimens appeared to be about 15% lower than that of the base metal. This drop is caused by the behaviour of the fine-grained HAZ, being a typical phenomenon for all ferritic and martensitic steels (type IV cracking).

The $10^5$ h creep strength at 600°C of cold-formed tube bends with low R/D ratio appeared to be about 10% lower than that of the base metal. The same holds for cold-formed X20 bends.

The high creep strength of T91/P91 is caused by the large quantity of very small incoherent niobium-vanadium carbonitrides, which proved to be very stable under creep conditions. The loss of creep strength in weldments is mainly caused by rearrangement of lath martensite into equiaxed cell structures during welding, and in case of cold-formed bends by an accelerated destabilization of the dislocation structure during creep exposure.

The workability of T91/P91 for welding (less critical than X20) and cold bending is very good. This aspect and the high creep strength make the use of this steel in power generation components operating in the creep range very attractive. It is, therefore, to be expected that T91/P91 will gradually take the place of X20 in those applications.

## ACKNOWLEDGMENTS

This project forms part of COST 501 II (Work Package 3, part F). The authors would like to thank the Netherlands Institute of Welding for incorporating this project in the national research project NIL-KACO, and the sponsors of that project and particularly the Dutch Ministry of Economic Affairs for their financial support. The electron-microscopical investigations were performed at Mannesmann Forschungsinstitut, Germany.

## REFERENCES

[1] W. Bendick, H. Weber, VDEh-Conference, Düsseldorf, November 1987.

[2] F. Brühl, K. Haarmann, G. Kalwa, H. Weber, M. Zschau, VGB Kraftwerkstechnik 12-1989, 1214-1231.

[3] F. Brühl, H. Cerjak, P. Schwaab, H. Weber, Steel Research 62 (1991), 2, 75-82.

[4] F. Brühl, H. Cerjak, H. Müsch, K. Niederhoff, Schweiβen und Schneiden 42 (1990) Heft 10, 515-520.

[5] ASME Boiler and Pressure Vessel Code, Section I, 1989.

[6] C. Coussement, M. de Witte, A. Dhooge, R. Dobbelaere, M. Steen, E. van der Donck and L. van Muysen, Revue de la Soudure 3, 1989, 18-32.

[7] A. Iseda, M. Kubata, Y. Haynase, S. Yamamoto, K. Yosikawa, The Sumitomo Search, 36, May 1988, 17-30.

*Proceedings of the Fifth International Conference on Creep of Materials, Lake Buena Vista, Florida, USA, 18-21 May 1992*

# Creep of Cold Deformed AISI 347 Stainless Steel

**D. D'Angelo**
ENEL-CRTN
Milano, Italy

**E. Evangelista, S. Spigarelli, and L. Kloc**
Department of Mechanics
University of Ancona
Ancona, Italy

**A. Rosen**
Department of Materials Engineering
Haifa, Isreal

## ABSTRACT

The creep and the rupture strength of pre-deformed AISI 347 stainless steel was investigated at 650 and 700°C using constant load creep tests. Initial applied stress was 127, 157 and 196 MPa. Considerable decrease in minimum creep rate and corresponding increase in creep life were found in specimens pre-strained to 5, 10 and 15%. For higher stresses, 10% pre-strained specimens showed the best creep characteristics, while, for lower stresses these characteristics improved monotonically with increasing pre-strain level. TEM observations revealed enhanced precipitation of fine stable NbC particles on dislocations in pre-strained specimens. The improved creep behaviour was attributed to the additional precipitation process occurring at the beginning of creep tests. The results indicates, that differences in creep behaviour between undeformed and deformed states becomes less important at lower stresses.

THE AISI 347 STABILIZED STAINLESS STEEL is one of the candidates for structural components in the superheater and reheater banks of power plant steam boilers due to its high stability at elevated temperatures. The construction of such components frequently imposes cold deformation, which can affect high temperature features of the material. Age hardening after cold work of AISI 347 was reported in the previous paper [1], and thus structural changes should be assumed. There are evidences in the literature, that cold deformation can cause dangerous loss of creep properties of high temperature resisting materials (see for instance ref. [2]). The aim of present research is to investigate the effect of different cold work level on creep behaviour of the AISI 347 stainless steel.

## MATERIAL AND EXPERIMENTAL PROCEDURE

Specimens for creep tests were prepared from the standard commercial tubes having 51mm outer diameter and 7.2mm wall thickness. The chemical composition of the alloy is summarized in Table I. The creep specimens were cylindrical with 5mm diameter, and 50mm gauge length. Some creep tests were performed on as received condition (stabilized) material, but most specimens were pre-deformed at room temperature by means of tensile machine to 5%, 10% or 15% of total strain.

The creep tests were carried out at 650°C and 700°C in air; temperature was stabilized within ±3°C. The constant load creep machines were used and all the tests were conducted to rupture under initial stress of 127 MPa, 157 MPa or 196 MPa. In all the tests, creep strain was measured by means of linear variable differential transducers and recorded continuously.

Thin foils for TEM observations were prepared by usual way. The ruptured specimens were cut at least 2mm from fracture surface. Dislocation densities were estimated by standard intercept method.

Table I. Chemical composition of AISI 347 stainless steel used for experiments. (in wt.%)

| Cr | Ni | Mn | Si | Nb | C |
|---|---|---|---|---|---|
| 17.3 | 10.42 | 1.52 | 0.42 | 0.52 | 0.048 |
| Ti | Ta | P | S | | |
| <0.005 | <0.01 | 0.025 | <0.005 | | |

## RESULTS

Creep curves obtained at different stresses and temperatures after various levels of pre-deformation are shown in Figs. 1a to 1d. All the creep curves are typical three-stage type with a relatively short primary stage and long tertiary stage. From the figures, strong effect of cold work on creep behaviour can be seen. Even 5% pre-straining leads to considerable reduction of creep rate, while creep life time is prolonged even with reduced rupture ductility.

The creep curves were fitted by the Theta-projection equation according to Evans, Parker and Wilshire [3] to obtain useful parameters as a minimum creep rate. Standard power-law equation was then employed and apparent activation

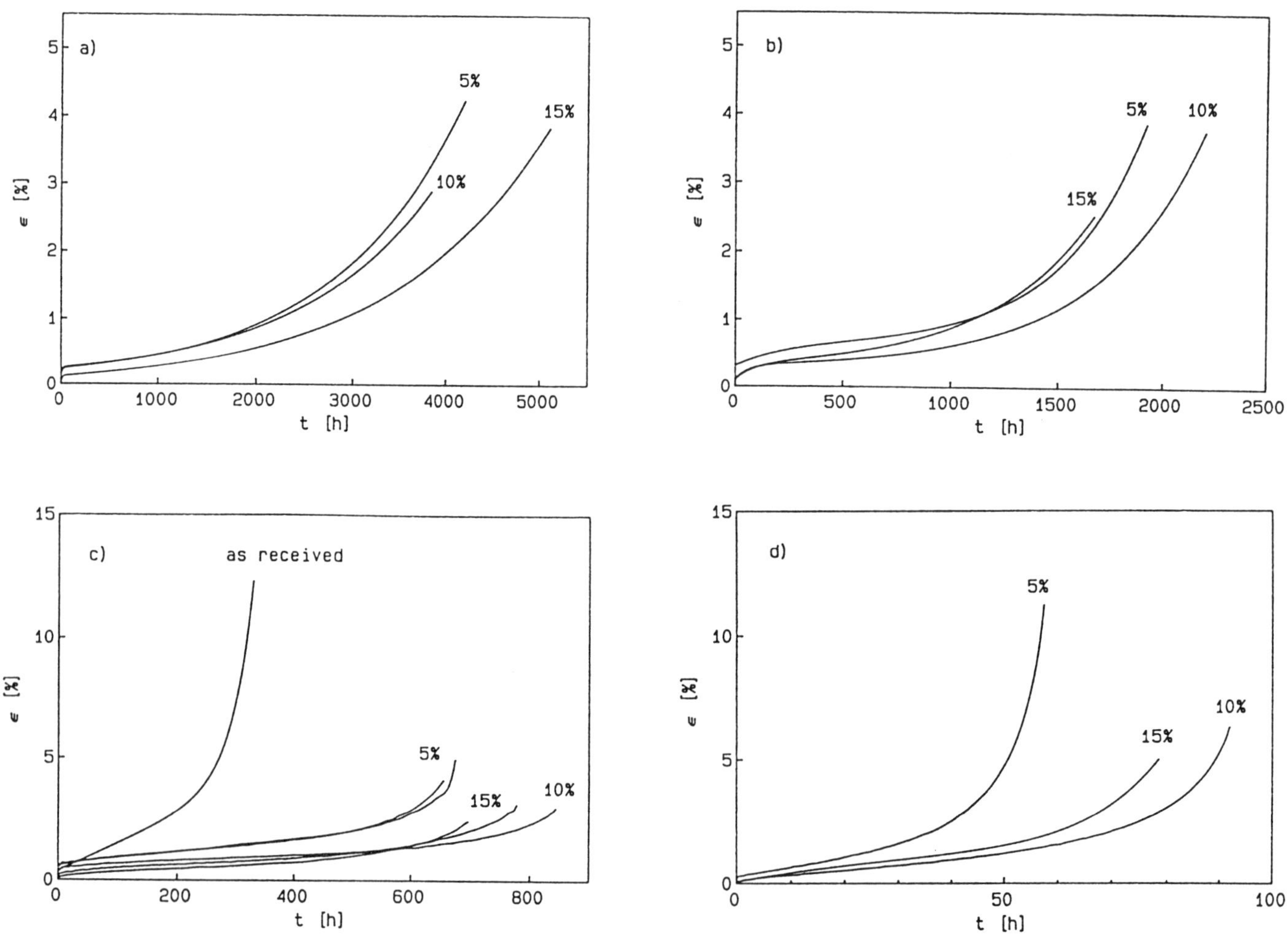


Fig.1 - Creep curves for AISI 347 pre-deformed to 5%, 10% or 15%. Test conditions: a) 650°C/127MPA, b) 650°C/157MPa, c) 650°C/196MPa, d) 700°C/196MPa

energy and stress exponent of minimum creep rate were determined. Results are summarized in Table II and depicted in Figs. 2 and 3. The stress exponent n is obtained from tests at 650°C and activation energy from tests at 196 MPa. It should be noted, that θ-parameters obtained here cannot be directly used in the θ-concept based databases (see for instance ref. [4]), because they are derived from constant load experiments. Austenitic stainless steels are classified by Loveday and Dyson [6] as materials, in which different behaviour under constant stress conditions can be expected in comparison with constant load creep tests.

Table II. The stress exponent of minimum creep rate n and apparent activation energy Q for different pre-straining.

| c.w. | n | Q [kJ/mol] |
|---|---|---|
| 5% | 5.7 | 450 |
| 10% | 4.7 | 496 |
| 15% | 4.0 | 483 |

The microstructure observations of crept specimens reveal regular three dimensional network of dislocations with weak tendency to form dislocation cell structure (Fig. 4) in some areas. A high number of fine NbC particles precipitated on dislocations was observed, as can be seen in Figs. 5 and 6. In specimens with creep life longer than about 500 hours, coarse intragranular particles are observed in some areas (Fig.7). Since they seem to be distributed very inhomogeneously, more detailed study is necessary to asses if they are produced by particle coarsening process. The results of measurements of dislocation density and fine NbC particles size and density are listed in tables III, IV and V. It should be emphasized that specimens crept at different stresses had different

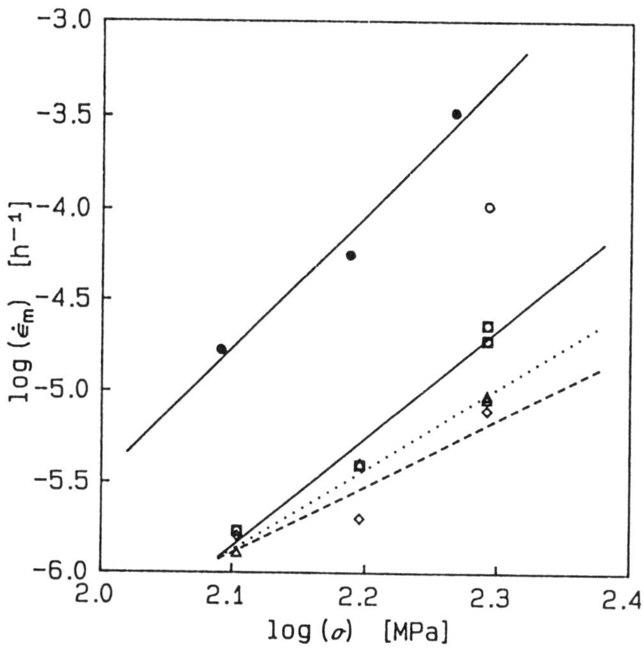


Fig.2 - Dependence of the minimum creep rate $\dot{\varepsilon}_m$ on applied stress σ at 650∘C.

—●— Needham and Gladman [5]
o without c.w. } this work
—□— 5% c.w.
---◇--- 10% c.w.
···Δ··· 15% c.w.

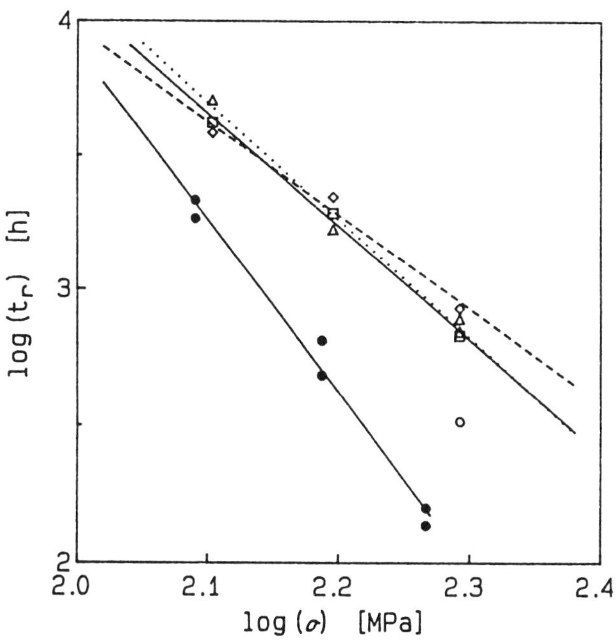


Fig.3 - Dependence of creep rupture time $t_r$ on applied stress σ at 650∘C.

—●— Needham and Gladman [5]
o without c.w. } this work
—□— 5% c.w.
---◇--- 10% c.w.
···Δ··· 15% c.w.

time to rupture. It is clear from the test results, that both dislocation and particle densities increase with increasing cold deformation before creep test. The homogeneity of particle dispersion also seems to increase with increasing cold work, though this parameter was not quantified in the present work. All the characteristics of the microstructure are very similar to those observed by Senior in aged AISI 347 weld metal [7].

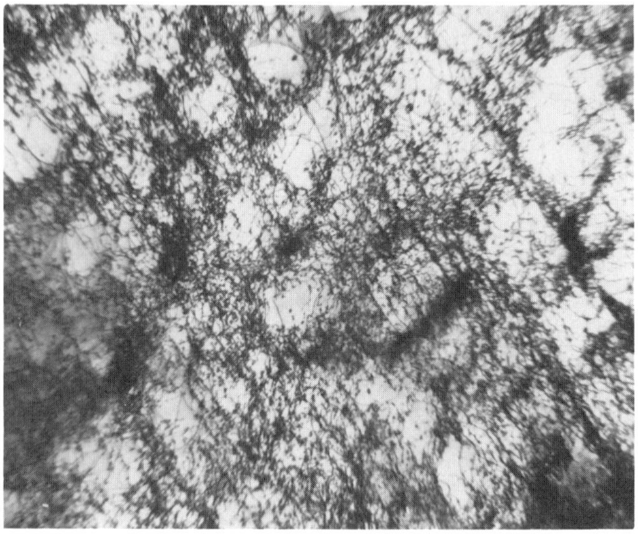


Fig.4 - Dislocation cell structure formation. TEM of specimen crept at 650∘C and 196 MPa, 10% prestrained.

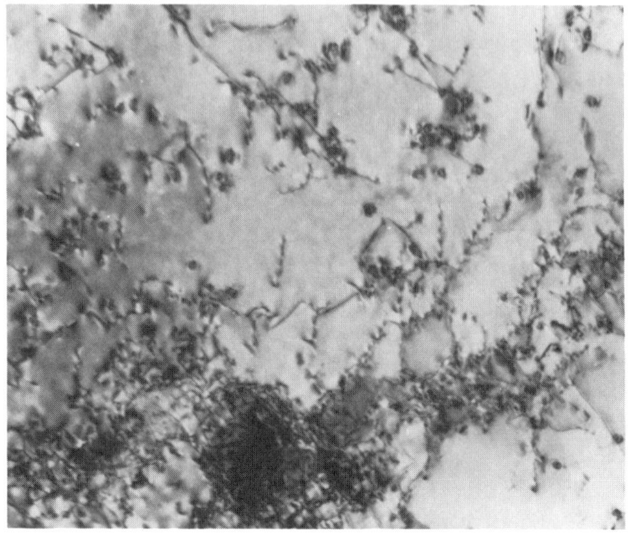


Fig.6 - Fine precipitates. TEM of specimen crept at 650∘C and 157 MPa, 5% prestrained.

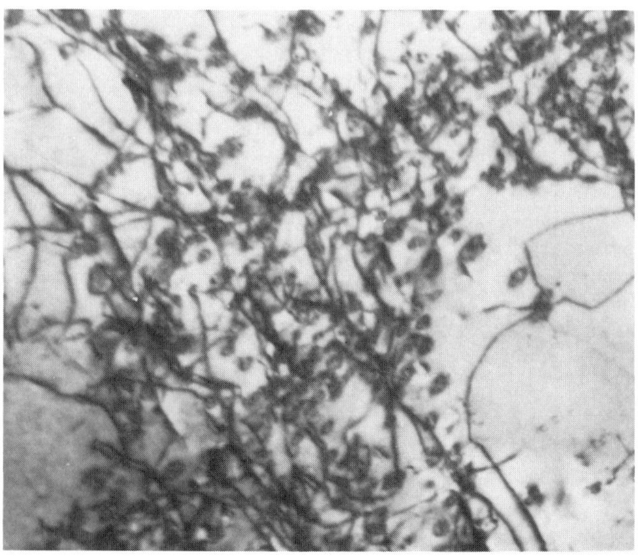


Fig.5 - Fine precipitates. TEM of specimen crept at 650∘C and 196 MPa, 10% prestrained.

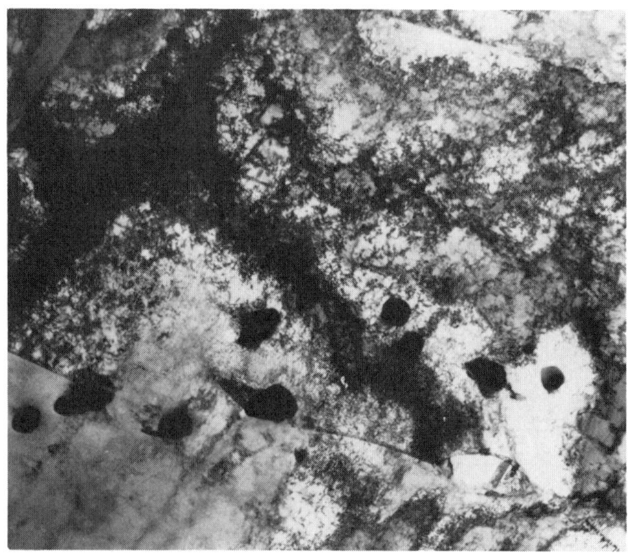


Fig.7 - Coarse intracrystalline precipitates. TEM of specimen crept at 650∘C and 127 MPa, 10% prestrained.

Table III. Dislocation densities in specimens crept at 650°C and at various stresses after various pre-strainig.

| | Creep stress [MPa] | | |
|---|---|---|---|
| | 127 | 157 | 196 |
| c.w. | Dislocation density $[10^{14} m^{-2}]$ | | |
| 0% | | | 3.2 |
| 5% | 3.0 | 2.5 | 3.5 |
| 10% | 4.8 | | 4.0 |
| 15% | 4.8 | 3.2 | 5.6 |

Table IV. NbC particle sizes in specimens crept at 650°C and at various stresses after various pre-strainig.

| | Creep stress [MPa] | | |
|---|---|---|---|
| | 127 | 157 | 196 |
| c.w. | Particle size [nm] | | |
| 0% | | | 9.0 |
| 5% | 8.5 | 6.0 | 8.7 |
| 10% | 8.0 | | 8.5 |
| 15% | 7.0 | 5.5 | |

Table V. NbC particles densities in specimens crept at 650°C and at various stresses after various pre-strainig.

| | Creep stress [MPa] | | |
|---|---|---|---|
| | 127 | 157 | 196 |
| c.w. | Particles density $[10^{21} m^{-3}]$ | | |
| 0% | | | 2.0 |
| 5% | 1.4 | 1.8 | 2.2 |
| 10% | 1.9 | | 2.5 |
| 15% | 2.2 | 2.5 | |

## DISCUSSION

The values of stress exponent and apparent activation energy for pre-strained specimens fit very well the values reported in the literature for stainless steels (see for instance ref. [8]). In figs. 8 and 9 the plots for Monkman-Grant [9] and Dobeš-Milička [10] relationships are shown. From the figures it is clear that cold deformation before creep have only small effect on the parameters of Monkman-Grant relation and negligible effect on Dobeš-Milička relation. All these results indicate that no qualitative change of creep strain and creep fracture mechanisms is caused by cold work before creep.

Nevertheless, considerable improvement of creep characteristics excluding ductility was found in pre-deformed specimens. The creep strengthening is due to fine NbC precipitates, which nucleate at dislocation network and stabilize it. Higher dislocation density in deformed specimens provide better conditions for particle nucleation, and consequently higher creep strength. The set of constitutive equations published recently by Senior [7,11] seems to provide good theoretical basis for quantitative analysis, but experimental result of decreasing stress exponent with increasing particle density contradicts strongly the Senior's basic idea of threshold stress caused by the irrecoverable dislocation network pinned by fine particles. In the model, threshold internal stress and consequently stress exponent should increase with increasing particle density, and that is just the opposite of what was observed experimentally. The fact that the stress exponent decreases with increasing pre-deformation level has one consequence which is important for practical applications. As can be seen from Figs. 2 and 3, where data of Needham and Gladman for the solution annealed AISI 347 are also included, differences

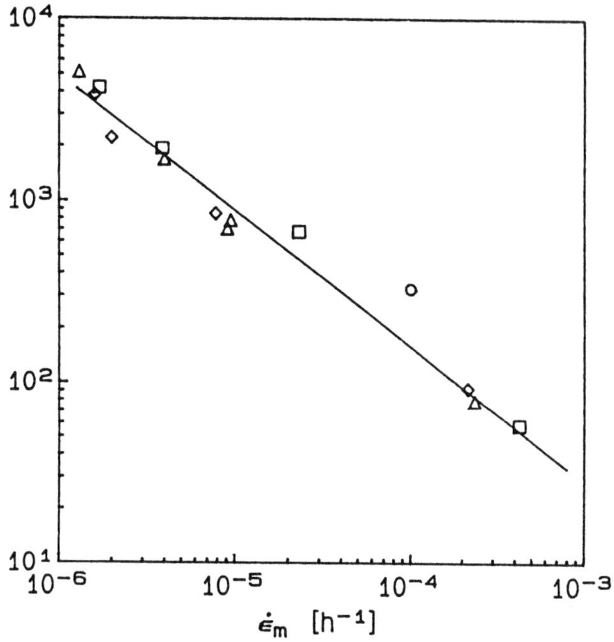


Fig.8 - Monkman-Grant relation. Different cold work levels:
o - 0%, □ - 5%, ◇ - 10%, Δ - 15%.

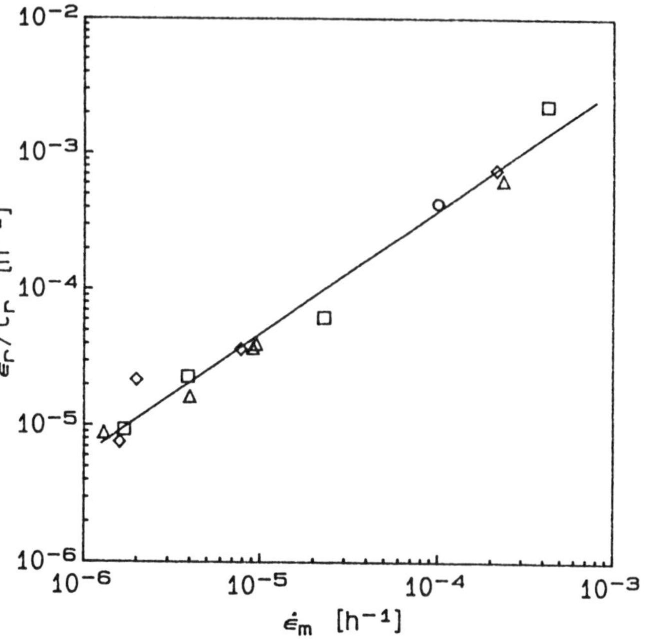


Fig.9 - Dobeš-Milička or modified Monkman -Grant relation. Different cold work levels:
o - 0%, □ - 5%, ◇ - 10%, Δ - 15%.

in creep behaviour tend to reduce at lower stresses.

It should be noted, that no explanation of slightly lower creep strength observed at 15% pre-strained specimens comparing with 10% ones yields from microstructural observations. More detailed study of precipitation kinetics and damage nucleation and growth will be necessary to shed light on this effect.

From the validity of Monkman-Grant and Dobeš-Milička relationships it follows, that no effective nuclei of cavities or other damage are introduced during cold pre-deformation up to 15%. But it is worthwhile to point out, that according to Ahn et al. [2] this conclusion may change drastically with the change in testing parameters.

## CONCLUSIONS

The results of creep and microstructural study of pre-deformed AISI 347 stainless steel can be summarized as follows:

i) Significant improvement of creep strength and life was found after cold work of 5 to 15% in comparison to original "stabilized" state. This improvement is due to enhanced precipitation of fine NbC particles on dislocations during primary creep period.

ii) The creep characteristics are similar to those typical for stainless steels. Hence, no qualitative change of operating creep mechanism is caused by the cold pre-deformation up to 15%.

iii) The influence of cold pre-straining on the creep strength seems to be non-linear at higher stresses. Current microstructural observations provide no explanation of this effect.

iv) The Monkman-Grant and Dobeš-Milička relationships are valid and they are not markedly influenced by the level of pre-deformation. Thus the creep damage development is governed by the creep strain process and no "a priori" damage occurs by pre-straining.

v) Constitutive equations developed by Senior [11] and used for AISI 347 weld metal are not applicable for the present results, because observed decrease of stress exponent with increasing particle density contradicts to the model.

vi) The results indicate, that differences in creep behaviour between "stabilized" and cold deformed states are less important at lower stresses, which are used in industrial applications.

## ACKNOWLEDGMENT

This work was partially funded by ENEL under a research contract.

## REFERENCES

1 - Y. Benharoe, D. D'Angelo, E. Evangelista, P. Mengucci and A. Rosen: in "High Temperature Materials for Power Engineering" (Proc. Conf.) Liège, Belgium. Kluwer Academic Publishers, Dordrecht 1990.

2 - Y. S. Ahn, K. Bothe, V. Gerold and B. Kempf in: B. Wilshire and R. W. Evans (eds.): "Creep and Fracture of Engineering Materials and Structures", Proc of 4thInt. Conf., Swansea, UK. The Institute of Metals, London 1990. p. 581

3 - R. W. Evans, J. D. Parker and B. Wilshire: in B. Wilshire and D. R. J. Owen (eds.): "Recent Advances in Creep and Fracture of Engineering Materials and Structures", Pineridge Press, Swansea 1982. p. 135

4 - Y. Z. Li, M. C. Nguyen and H. H. Over: in: B. Wilshire and R. W. Evans (eds.): "Creep and Fracture of Engineering Materials and Structures", Proc of 4th Int. Conf., Swansea, UK. The Institute of Metals, London 1990. p.961

5 - N. G. Needham and T Gladman: Metal Sci. 14 (1980), 64

6 - M. S. Loveday and B. F. Dyson: in: B. Wilshire and R. W. Evans (eds.): "Creep and Fracture of Engineering Materials and Structures", Proc of 4th Int. Conf., Swansea, UK. The Institute of Metals, London 1990. p.941

7 - B. Senior: Mater. Sci. Eng. A124 (1990), 159

8 - L. Bernard, E. Campo and S. Quaranta: "Ceep Behaviour of AISI 304 and 316 Stainless Steels and Influence of Cold Working", Proc. of Int. Conf. on "Physical Behaviour and Nuclear Application of Stainless Steels at Elevated Temperatures", Varese, Italy 1981.

9 - F. C. Monkman and N. J. Grant: Proc. ASTM 56 (1956), 593

10 - F. Dobeš and K. Milička: Metal Sci. 10 (1976), 382

11 - B. Senior: Mater. Sci. Eng. A130 (1990), 51

*Proceedings of the Fifth International Conference on Creep of Materials, Lake Buena Vista, Florida, USA, 18-21 May 1992*

# Tensile and Creep Properties of 3-D Braided PEEK/AS4 Graphite Composites

**J-N. Chu and F.K. Ko**
Fibrous Materials Research Center
Drexel University
Philadelphia, Pennsylvania

## ABSTRACT

In this study, 3-D braided poly(ether-ether-ketone)/AS4 graphite composites were preformed and processed. These composites were characterized using matrix digestion and wide angle x-ray diffraction to determine their fiber volume fractions and degrees of crystallinity. After physical characterization, these composites were evaluated mechanically through tensile tests at various temperatures.

Experimental results from the tensile measurements were compared to a fabric geometry model (FGM). This model predicts tensile strength based on knowledge of fiber and matrix properties, fiber volume fraction, and braiding angle. The model and experimental results are in good agreement.

In order to predict the creep properties of 3-D braided PEEK/AS4 composites, existing creep data on PEEK was incorporated into a quasi-linear viscoelastic model (QVM), and combined with the FGM to construct a viscoelastic model for 3-D braided composite materials.

## INTRODUCTION

Fiber reinforced thermoplastic matrix composites are increasing in popularity. Their popularity originates from reductions in processing temperatures and weight (relative to metals and ceramics), repairability, and high strength and toughness [1-5]. Among thermoplastic matrix materials, PEEK, or poly(ether-ether-ketone) is one of a newer generation of high performance engineering materials. PEEK is a semi-crystalline aromatic polymer with a glass transition temperature of 144°C and a melting point of 335°C. PEEK has very good chemical resistance and maintains much of its mechanical properties up to 300°C. Compared to most thermoplastic materials, PEEK can be used at higher temperatures (to 250°C) [6-9]; therefore, understanding its properties at higher temperatures is important. Like most polymeric materials, PEEK also exhibits time-dependent properties such as creep, stress relaxation, and energy dissipation. These time/temperature-dependent properties make studying the long-term performance of PEEK composites at elevated temperatures essential.

A wide variety of reinforcement fibers are used in PEEK composites, depending upon the application. For example, carbon fiber composites have been developed with short fibers [9], laminates [10], woven fabrics [11,12] and three-dimension braided structures [13]. Among these reinforcements, 3-D braided structures are known for their high damage tolerance and delamination resistance [14]. Previous studies [13] at room temperature show that 3-D braided PEEK/graphite composites have very good damage tolerance and tensile properties. In standard compression after impact tests [15], these materials showed higher compression after impact strength, and damage tolerance than laminated APC-2 composites. In this way, 3-D braided PEEK composites may be used to improve damage tolerance properties and delamination problems associated with APC-2 composites.

In order to increase the commercial applications of thermoplastic materials, their long-term properties must be well understood. However, thermoplastic composites have a relatively small database of information on their long-term behavior [4]. In the case of 3-D braided composites, long-term properties have not been examined till now. In this paper, the tensile and creep properties of 3-D braided PEEK/AS4 graphite composites are studied. First, a series of tensile tests at various temperatures are described to show their stress-strain behavior. Second, tensile properties of these materials are compared to the fabric geometry model (FGM) [14,16]. Third, available creep data for PEEK [7] is used to determine the time-dependent modulus at various temperatures using the quasi-linear viscoelastic model (QVM) [17-19]. Finally, a new model is developed which uses the above information to give creep properties of 3-D PEEK composites from a fabric geometry point of view.

## MATERIALS

In this study, Victrex® 150 G PEEK was used as the polymer matrix, and Magnamite® AS4 graphite was used as the reinforcing fiber. The physical and mechanical properties according to the manufacturer of these materials are presented in Table 1 [9,20].

Table 1. Materials Properties of PEEK and Graphite Fiber

| Properties | 150 G PEEK | AS4 Graphite Fiber |
|---|---|---|
| Density ($g/cm^3$) | 1.32* | 1.80 |
| Tensile Strength (MPa) | 95 | 3,795 |
| Tensile Modulus (GPa) | 4.5 | 235 |
| Poisson's Ratio | 0.37 | ---- |

*semi-crystalline

In order to ensure good matrix impregnation in the final composites, a commingled system consisting of 150G PEEK matrix in filament form and AS4 graphite fibers was used. The commingled yarns contain approximately 280 PEEK fibers and 6,000 graphite fibers. Ignoring voids, the initial graphite fiber volume fraction in each yarn is 62.7% [11].

The commingled yarns were braided into 3-D preforms at the Fibrous Materials Research Center at Drexel University. The 3-D braided preforms were 1.25 cm wide, 0.2 cm thick, with a surface braiding angle of ±20°. A detailed description of the preforming process can be found in ref. [14,21] .

Consolidation of the 3-D braided preforms was carried out in a Wabash® hydraulic hot-press using a carbon steel match die mold. Before consolidation, the 3-D braided preforms were carefully cut into 25 cm lengths. The time-temperature and time-pressure history of a typical sample during processing is shown in Figure 1.

After processing, the composites fiber volume fraction was examined using a matrix digestion method [21,22]. The fiber volume fraction in the processed composites was determined to be 63.60 (±1.20 )%.

The crystallinity index for the PEEK matrix in the processed composites was measured to be 32% using wide-angle X-ray diffraction. This method requires subtracting the diffraction from the fibers, and scaling the x-ray scattering from amphorous PEEK by a factor which is related to the crystallinity index. This method is detailed in ref [21].

## EXPERIMENTAL

Static tensile tests (according to ASTM standard D 3039-76) were conducted to obtain the ultimate tensile strength, modulus, and Poisson's ratio of the 3-D braided composites. These tests were conducted using an Instron 1127 Universal Tester. The dimensions of the specimens were 25.4 cm x 1.25 cm x 0.254 cm. A schematic of a specimen is shown in Figure 2. The strain induced from the applied stress was measured using strain gages. A 2100 system strain gage conditioner and amplifier system (manufactured by Measurements Group Inc.) was connected with a X-T plotter to record the strain data. The cross-head speed of the Instron was set at 0.05 cm/min.

Tensile tests were conducted at 25, 60, 120, and 180°C. Ten specimens were tested at each temperature. The results of these static tensile tests are summarized in Table 2.

Table 2. Summary of the mechanical properties of 3-D braided PEEK/AS4 graphite composites from the tensile testing results.

| Temp. (°C) | Stress (MPa) | Modulus (GPa) | Poisson's Ratio |
|---|---|---|---|
| 25 | 930.51 (±47.78) | 88.75 (±5.50) | 0.38 (±0.02) |
| 60 | 822.13 (±79.94) | 79.62 (±2.39) | 0.45 (±0.04) |
| 120 | 1027.59 (±58.38) | 83.67 (±1.31) | 1.14 (±0.02) |
| 180 | 1021.45 (±102.15) | 87.72( ±12.79) | 1.29 (±0.16) |

These results indicate that 3-D braided PEEK composites retain their tensile strength to at least 180 °C. Table 2 also shows that the Poisson's ratio increases with testing temperatures. After failure, the test specimens were cut using a diamond blade to examine the fracture surfaces by a scanning electron microscope (SEM). Figure 3 and 4 show the SEM micrographs of 25 °C and 180 °C tensile test fracture surfaces, respectively. The fracture surface of the 180 °C tensile test specimen shows more matrix flow then the 25 °C specimens.

## THEORETICAL ANALYSIS

### 1.TENSILE PROPERTIES

The FGM [14] was employed to determine the tensile properties of the 3-D braided PEEK/AS4 graphite composites. This model relates the mechanical properties of the 3-D braided composites to their fiber properties, matrix properties, fiber volume fraction and fiber orientation through a modified laminate theory. To use this model, a unit cell is identified. This forms the basis for the analysis. Figure 5 shows a typical unit cell structure for 3-D braided composites [16]. In general, the unit cell may be represented by diagonal rods representing the yarns. Their orientation depends upon the braiding angle.

The FGM assumes that each yarn in the unit cell is a unidirectional lamina and that the overall material property of the unit cell is a sum of the properties of all the yarns. The objective of FGM analysis is to determine the elasticity matrix for the unit cell. The elasticity matrix of each lamina in yarn coordinate system, C, is determined by the fiber properties, matrix properties and fiber volume fraction. The elasticity matrix of the lamina in the unit cell coordinate system is determined by a coordinate transformation. Therefore, the elastic matrix of the $i^{th}$ system of yarns, $[C_i]$, is expressed as:

$$[C_i] = [T_{\varepsilon,i}]\,[C_i]\,[T_{\varepsilon,i}]^{-1}, \quad (1)$$

where $[T_{\varepsilon,i}]$ is the Hamiltonian coordinate transformation tensor for the $i^{th}$ system of yarns

Finally, the elasticity matrix of the unit cell, $[C_S]$,can be obtained by summing up individual elasticity matrices of all the laminae

$$[C_S] = \Sigma\, k_i\, [C_i], \qquad (2)$$

where $k_i$ is the fractional volume of the $i^{th}$ system of yarns.

The FGM can be incorporated into general finite element analysis programs for solving the elastic matrix of the braided composite structures with complex shapes [16]. Figure 6 shows a comparison of the 25 °C experimental stress-strain relation for 3-D braided PEEK/AS4 composites with the FGM analysis. Figure 6 shows that the FGM analysis and experimental results are in good agreement. Therefore, FGM may be used to predict the mechanical properties of these 3-D braided PEEK composites.

## 2. CREEP PROPERTIES

In general, graphite fibers may be assumed to behave as a linearly elastic material, and creep is negligible at PEEK application temperatures [22]. Therefore, the creep behavior of 3-D braided PEEK/AS4 graphite composites is only attributed to the creep response of their PEEK matrix . The creep behavior of bulk PEEK has been studied under different stress levels [7,24,25]. In this study, a quasi-linear viscoelastic model (QVM) [17,18] was used to analyze the available PEEK creep data in order to determine the time-dependent modulus for various temperatures.

The QVM assumes the stress T(t) of a material at a given time is a sum of elastic and history dependent response. The history dependent response is reflected in creep, stress relaxation and dynamic mechanical behavior of the material. T(t) can be presents by the following equations:

$$T(t) = T^{e}[(\lambda)] + \int_{0}^{t} T^{e}[\lambda(t-\tau)]\dot{J}(\tau)^{-1}\, d\tau \qquad (3)$$

where, $T^{e}$ = elastic response of the applying stress
$\lambda$ = stretch ratio
J(t) =normalized creep function of time, J(0)=1
t = time

The normalized creep function can be expressed in terms of a continuous spectrum $S(\tau)$ as

$$J(t) = \frac{1 - \int_{0}^{\infty} \frac{S(\tau)}{1+S(\tau)} \exp\left[\frac{-1}{1+S(\tau)}\right] d\tau}{1 + \int_{0}^{\infty} \frac{S(\tau)}{1+S(\tau)}\, d\tau} \qquad (4)$$

$$\text{where, } S(\tau) = \frac{c}{\tau} \quad \text{for } \tau_1 < \tau < \tau_2 \qquad (5)$$
$$= 0 \quad \text{elsewhere}$$

Parameters c, $\tau_1$ and $\tau_2$ are obtained from experimental results. This spectrum has been used to represent the creep behavior of Kevlar, Nylon 6, spider silk and polypropylene fibers [18].

In this study, the creep data from Jones [7] was used to obtain the parameters needed in equation (5). The normalized creep function J(t) for the PEEK matrix at various temperatures can be obtained from equation (4). Therefore, the time-dependent modulus function of PEEK can be expressed as

$$E_m(t) = \frac{E_{mo}}{J(t)} \qquad (6)$$

where, $E_{mo}$ is the tensile modulus of PEEK.

Thus, the time-dependent modulus function, E(t), of 3-D braided PEEK/AS4 graphite composites can be obtained using FGM with the results from equation (6). The creep function of the 3-D composites can be expressed as

$$J(t)_{3D} = \frac{E_o}{E(t)}, \qquad (7)$$

where, $E_o$ is the tensile modulus of the 3-D PEEK composites.

Figure 8 shows creep behavior of the 3-D braided PEEK composites at various temperatures from the above theoretical analysis. Experimental stress relaxation test results of the PEEK composites were used to further examine the applicability of E(t). Figure 8 shows a comparison of the 120 °C experimental stress relaxation results with the theoretical analysis. The normalized theoretical stress relaxation function $G(t)_{3D}$ used in Figure 7 was obtained from E(t) divided by the experimental tensile modulus of the PEEK composites at 120 °C. Figure 8 shows that the experimental results and the theoretical analysis are in a good agreement. Therefore, the above analysis may be used to predict the creep and stress relaxation behavior of the 3-D braided PEEK/AS4 graphite composites.

## CONCLUSIONS

The tensile properties of 3-D braided PEEK/AS4 composites at various temperatures were studied. Experimental results indicated that the 3-D braided PEEK composites retain their tensile strength to 180 °C. Results from the tensile measurement were compared to the fabric geometry model (FGM) predictions. It was found that the FGM may be used to predict the tensile properties of these 3-D braided PEEK/AS4 composites.

Based on the combination of a quasi-linear viscoelastic model and the fabric geometry model, a new model was developed to predict the creep properties of 3-D braided PEEK/AS4 composites. This model also can be used to predict the stress relaxation behavior of the 3-D PEEK composites.

## ACKNOWLEDGEMENT

This study was partially supported by the Army Research Office and the Ben Franklin Center of Excellence for Textile Structural Composites. The help of Mr. Mitch Marmel in the preparation of this manuscript is also appreciated.

## REFERENCES

1. Barnes, J.A. and Cogswell, F.N. , *SAMPE Quarterly* , 20 ,22 (1989).
2. Belbin, G.R., Proceedings of the 12th John Player Lecture, Vol. 198, p.1,Institution of Mechanical Engineers, London (1984).
3. Rigby, R.B., Proceedings of the 27th National SAMPE Symposium p.747,SAMPE International, CA (1982).
4. Witzler, S., *Advanced Composites* , 5(5),50 (1990).
5. Halpin, J.C., "The role of the Polymeric Matrix in the Processing and Structural Properties of Composite Materials" p.3, Seferis,J. C. and Nicolais, L., Eds., Plenum Press, New York (1983).
6. Searle, O.B. and Pfeiffer, R.H , *Polym. Eng. Sci.,* 25(8), 474 (1985).
7. Jones, D.P., Leach, D.C. and Moore, D.R., *Polymer,* 26, 1385 (1985).
8. Hartness, J.T., Proceedings of the 14th International SAMPE Technical Conference, p. 26, SAMPE International, CA (1982).
9. Victrex High Temperature Engineering Thermoplastics Technical Information, ICI, Wilmington, DE.
10. APC-2 the Product of High Technology, ICI Fiberite, Laguna Hills, CA.
11. Commingled Yarns, BASF, Charlotte NC.
12. Olson, S.H,. *SAMPE J.* , 26(5) 31 (1990).
13. Ko, F.K., Chu, J.-N. and Hua, C. T., *J. Appl. Polym. Sci.* , 47, 501 (1991).
14. Ko, F.K., "Textile Structural Composites", p. 129, Chou, T.-W. and Ko, F. K., Eds., Elsevier Science Publishers, Netherlands (1989).
15. Boeing Specification Support Standard BSS 17260, Boeing, Seattle.
16. Liao, D., Ph.D. Thesis, Drexel University (1990).
17. Ko, F. F.-K., Ph.D. Thesis, Georgia Institute of Technology (1976).
18. Chen, H. Y.-L., Ph.D. Thesis, University of California, San Diego (1973).
19. Fung, Y.C., "Biomechanics", Springer-Verlag Inc., New York (1981).
20. Product Data, Magnamite® graphite fiber type AS4, Hercules Incorp., Wilmington, DE.
21. Chu, Jia-Ni, Ph.D. Thesis, in preparation.
22. ASTM Standards, D 3171-76, ASTM, Philadelphia, (1991).
23. Ha, S. K., Wang, Q. and Chang, F.-K., *J. Comp. Mater.* , 25, 334 (1991).
24. Ogale, A.A., Ph.D. Thesis, University of Delaware (1986).
25. Xinran, X., Ph.D. Thesis, Free University of Brussels (1987).

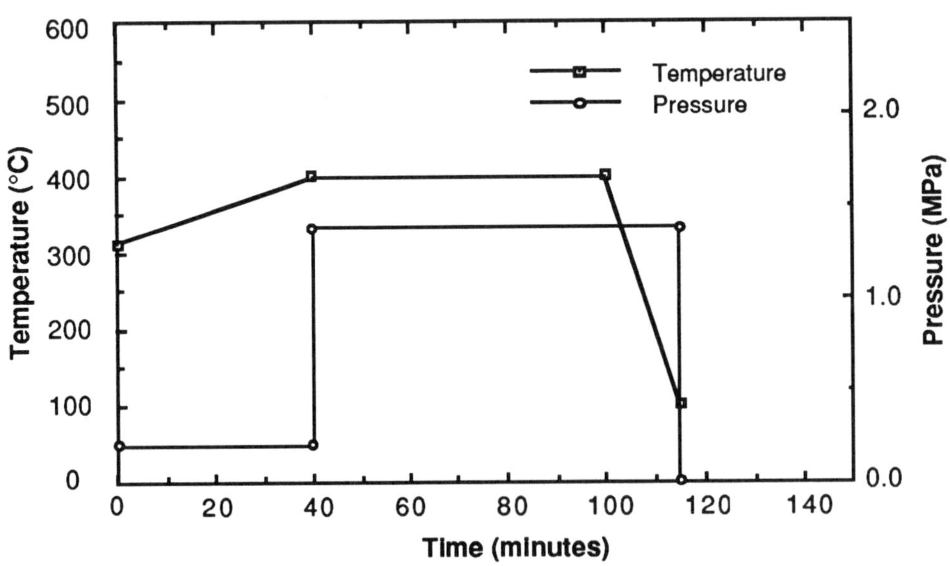


**Figure 1. Hot press processing history of 3-D braided PEEK/AS4 composites**

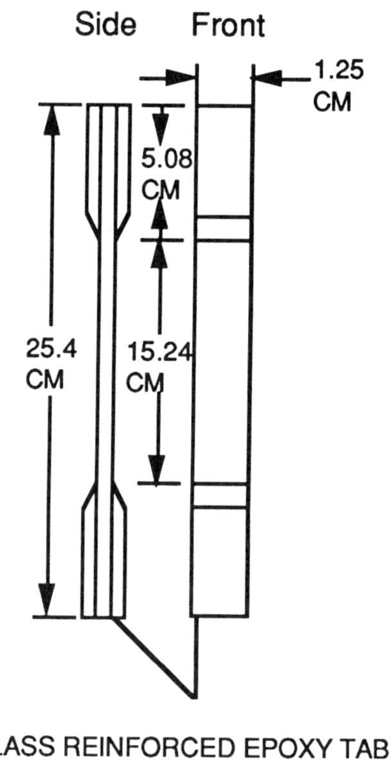


**Figure 2. A schematic of a tensile test specimen**

**Figure 3 SEM micrograph of 25 °C tensile test fracture surface**

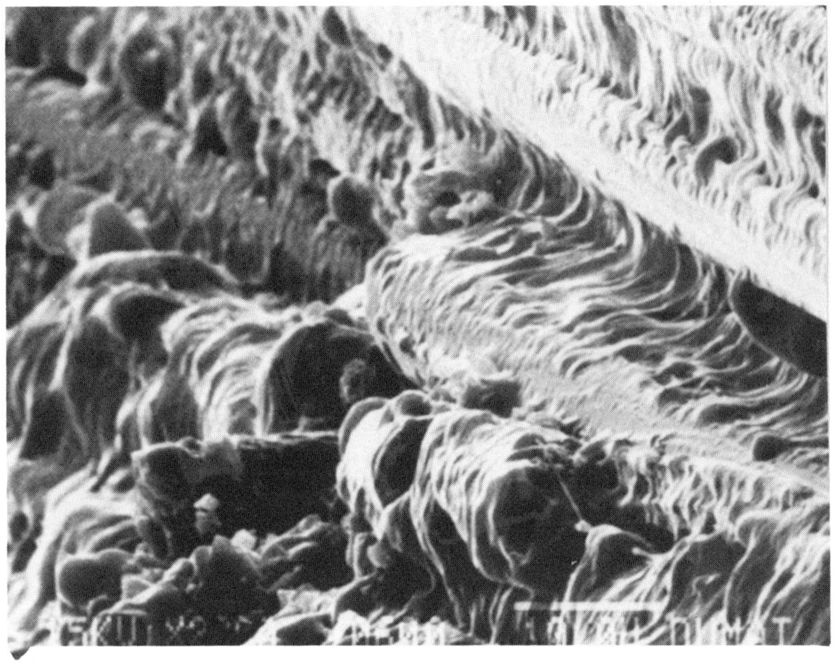

Figure 4 SEM micrograph of 180 °C tensile test fracturesurface

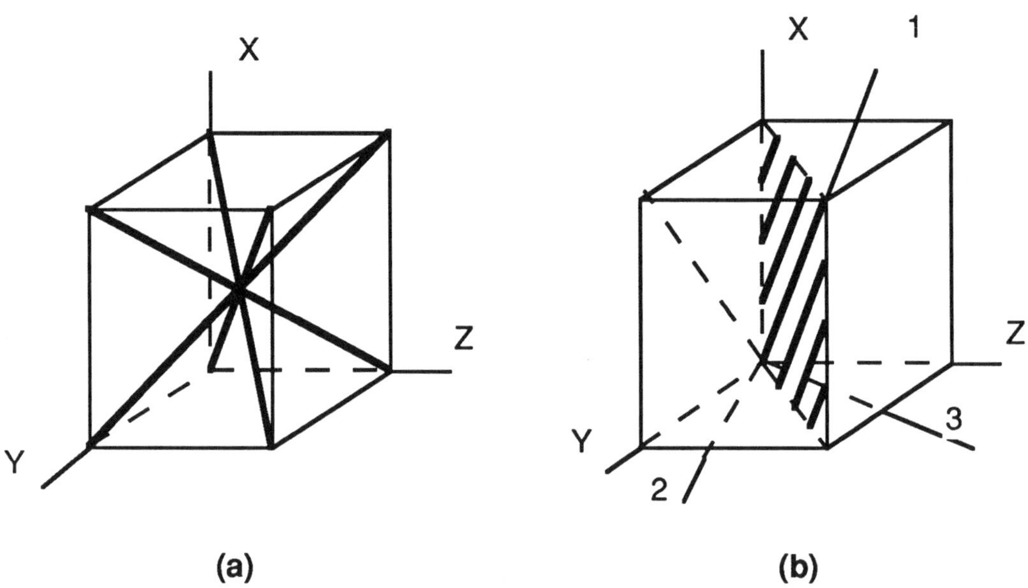


Figure 5. (a) The unit cell structure for 3-D braided composites
(b) A yarn system in the unit cell is treated as unidirectional lamina
(1,2,3) is the yarn coordinate system, (x,y,z) is the unit cell coordinate system

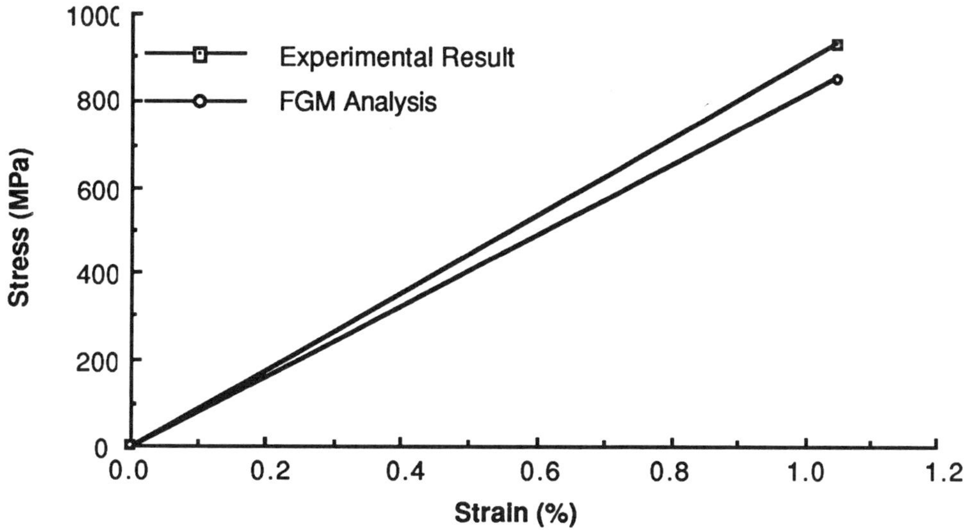


**Figure 6. Room temperature stress-strain relation of 3-D braided PEEK/AS4 composites**

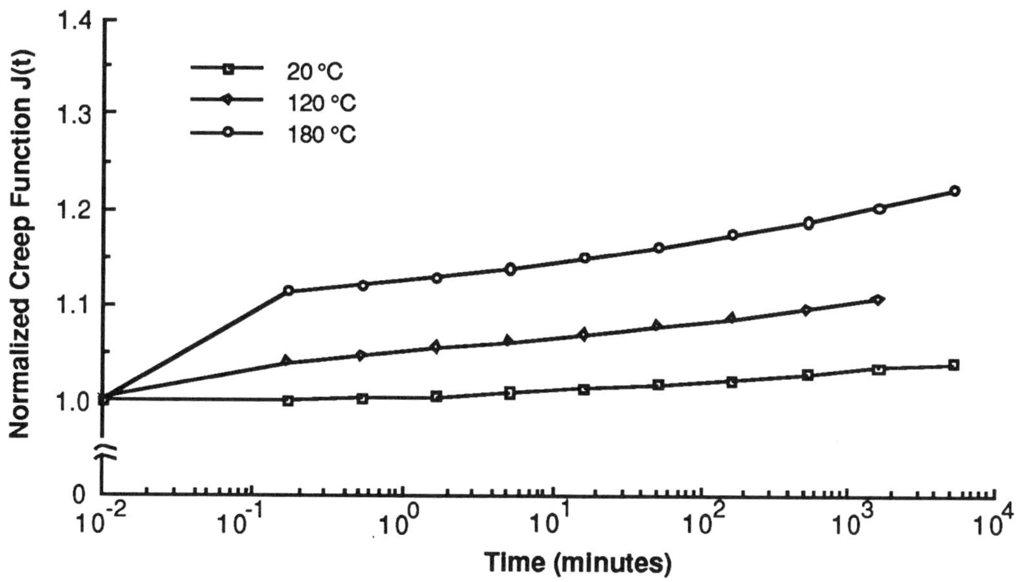


**Figure 7 Creep behavior of 3-D braided PEEK/AS4 composites**

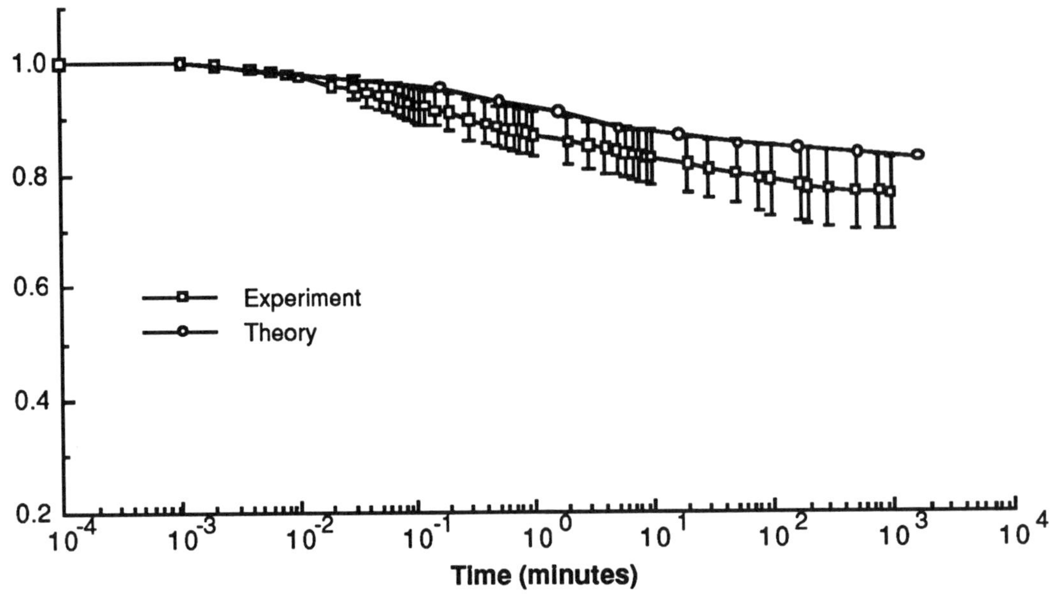


**Figure 8 Stress relaxation behavior of 3-D braided PEEK/AS4 composites at 120 °C**

*Proceedings of the Fifth International Conference on Creep of Materials, Lake Buena Vista, Florida, USA, 18-21 May 1992*

# Flexural Creep/Creep Rupture Testing of Thermoset/Thermoplastic Composites

**J. Raghavan and M. Meshii**
Department of Materials Science and Engineering
Northwestern University
Evanston, Illinois

## Abstract

Unidirectional laminates of a thermoset composite [AS4/3501-6] as well as a thermoplastic composite [AS4/J2] were loaded in a four point flexure fixture at room temperature. The creep deflection and creep rupture time were measured and compared. Thermoplastic composite exhibited appreciable creep as compared to thermoset composite. Also it exhibited an accelerated creep stage and failed by a tensile mode at all stresses. But thermoset composite failed abruptly by interlaminar shear. Both fracture modes were preceded by compressive buckling beneath the loading pins. It appears that four point shear test is not suitable for measuring the interlaminar shear strength of thermoplastic composites. The creep rupture data were fit to a logarithmic relation. Also both composites were subjected to a repeated stress relaxation test to study the influence of cyclic loading. Deceleration of relaxation rate with subsequent cycles was observed at low stresses while an acceleration of relaxation leading to rupture was observed at high stresses, in both composites.

THERMOPLASTIC COMPOSITES are actively pursued nowadays for defence and aerospace applications, thanks to their good thermal characteristics and high toughness of the thermoplastic matrix [1]. The desirable thermal properties of the these composites aid in manufacturing of thick composite components. The higher toughness increase the resistance to delamination failure which is plaguing the thermoset composites currently in use. Two such composites {[AS4/3501-6]& [AS4/J2]} currently under investigation were found to have a resin rich layer between the laminae. So a testing program was conducted to study the influence of such interlaminar layer on the creep/creep rupture of the composites. Browning et. al. [2] had developed a four point shear test and had shown that this yielded interlaminar shear fracture very well in thermoset composites. Flexure tests had been used by many other researchers to study the creep of composites [3,4,5].

However the flexural loading leads to a mixture of compressive, shear and tensile stresses in the flexural specimen. Vertical cracks beneath the loading pins due to compressive buckling, are always noticed to precede the horizontal interlaminar fracture of the specimens. Whitney et. al. [6] suggests that these vertical cracks induce interlaminar normal tensile stress near the crack tip which combines with the shear stress to initiate interlaminar failure. They are also of opinion that the compressive stress beneath the pins tend to suppress interlaminar shear failure mode.

Nevertheless four point shear test is one easy method of stressing the interlaminar boundaries and has been found to induce interlaminar fracture in thermoset composites [2]. Hence this test method was used in the current study on creep/creep rupture of polymer composites. Moreover a repeated flexural stress relaxation test method was used to study the influence of cyclic loading on the relaxation behavior of the composites.

## Materials and Testing Procedure

Two composite materials used in this investigation were carbon fiber [AS4] reinforced thermoset epoxy [3501-6] matrix composite [AS4/3501-6] and carbon fiber reinforced thermoplastic nylon [J2] matrix composite [AS4/J2]. Both were supplied as 24 ply [0] laminates by David Taylor Naval Research Center. The flexural specimen size was 3mm * 4mm * 50mm. A schematic diagram of the testing fixture is shown in Figure 1. A support to thickness ratio of 13.33 was used. The loading span was half of the supporting span. The fibers were oriented parallel to the longitudinal axis of the specimen and the laminate plane was parallel to the plane of loading i.e., perpendicular to the plane of the paper. For measuring time independent interlaminar shear strength and for relaxation testing, INSTRON 1125 was used with a cross head speed of 2 mm/min. An in-house fabricated dead weight loading frame was used for flexural creep tests. Since a hydraulic jack was used to elevate and lower weights, the loading rate could not be controlled closely.

Four point flexure fixture was fabricated in-house. Initial flexure tests were done using 4.5 mm diameter loading pins. But most of the specimens [AS4/3501-6] tested, developed vertical cracks below the pins. Those specimens which developed vertical cracks showed lower strength than those without vertical cracks and also a considerable scatter in the data was observed. So after some trial runs with larger pins, 10 mm diameter pins were finally chosen. The percentage of specimens that showed good interlaminar fracture was higher with 10 mm diameter loading pins. Also higher interlaminar shear strength values were obtained. The data reported here were obtained using 10 mm diameter loading pins.

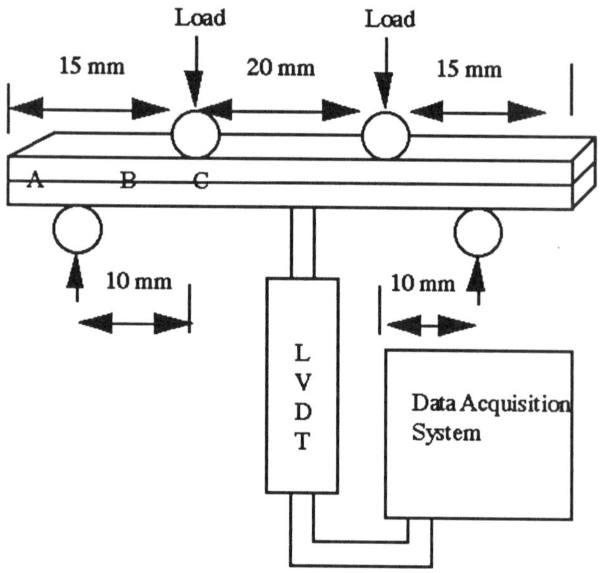


Fig. 1. A Schematic diagram of Four Point Flexture Test Fixture

Both creep and relaxation tests were carried out at room temperature. Stress levels used were 65-90% of the maximum flexural strength. The deflection at mid-span of the flexural specimens was measured by a Schaevitz HCA 250-Daytronic 3130 LVDT conditioner set up. The data was acquired and recorded by a Strawberry Tree data acquisition board built into 386 SX computer. The bending fixture was not instrumented for deflection measurements during relaxation tests. The repeated flexural stress relaxation test was very much similar to incremental strain test of Swain et. al. [7] or interrupted-ramp strain test of Fitzgerald et. al. [8], with minor differences. During this test, the INSTRON cross head was stopped after the stress on the specimen had reached the desired value. The relaxation in stress with time was recorded. After relaxation to a certain value, the specimen was reloaded at the same loading rate to the original stress level and allowed to relax again. Thus this testing procedure is different from interrupted strain test in that, the reloading stress level was maintained constant in subsequent cycles. But in the latter method, strain steps and relaxation time were kept constant during subsequent cycles. The advantage of the method used in the present study is that the loading pattern is very similar to the frequent loading-unloading encountered during tensile creep testing of composites. So it was envisaged that this study would lead to some understanding with respect to the effect of such loading patterns on the behavior of composites.

## Results and Discussion

The time independent flexural test data for thermoset and thermoplastic composites are tabulated in Table I. The flexural stress, strain and modulus values correspond to the outermost fiber on the tensile side of the flexural specimen. The interlaminar shear strength data for thermoplastic composite is not provided here for the reason that it did not fail by interlaminar shear as shown in Figure 2. The strength and modulus values were higher for thermoset composite as compared to thermoplastic composite.

Table I. Flexural test [at room temperature] data for thermoset [AS4/3501-6] and thermoplastic [AS4/J2] composites.

| Specimen | Flexural Strength MPa | Flexural Modulus GPa | Flexural Strain at Fracture | Interlaminar Shear strength MPa |
|---|---|---|---|---|
| *AS4/3501-6 :-* | | | | |
| As Received | 1656 | 133.5 | 0.0154 | 121.7 |
| *AS4/J2 :-* | | | | |
| As Received | 1198 | 88 | 0.0138 | ---- |

** All values are average of minimum 3 specimens

**Constant Load Creep Tests.** The creep and creep rupture results for both thermoset and thermoplastic composites are given in Figures 3-6 respectively. As can be inferred from figures 3 and 4, thermoplastic composites exhibited considerable ductility as compared to thermoset composite at a given stress level. An accelerated creep stage before fracture was also noticed with the former, whereas the latter fractured abruptly. The reason for this, as can be inferred from Figure 2 is the tensile failure with considerable yielding [Fig. 2 (b)]of the former as compared to the interlaminar fracture [Fig. 2 (a)[ of the latter.

Thermoplastic composite failed by tensile fracture of outermost fiber preceded by slow crack growth [buckled zone] down the thickness direction and significant compressive yielding beneath the loading pin. The buckled zone seems to have extended downwards till the tensile strength is reached at the outermost fiber on the tensile side, leading to the failure of the specimen. The time to fracture seems to be dependent on the progress of compressive yielding leading to initiation and growth of compressive buckling, as can be inferred from Figure 7. It can be observed from Figures 7 (a) & (b) that specimen at a stress of 765 MPa showed extensive yielding as compared to the specimen at a stress of 1082 MPa. This suggests that higher fracture time at lower stresses could be due to the delay in initiation of vertical crack and slower crack growth rate to a size needed for initiation of tensile fracture of the outer most fiber. As significant compressive yielding was noticed even at 64% of the maximum flexural strength, it appears that four point flexural shear test is not suitable for measuring interlaminar shear strength of composites with ductile thermoplastic matrices.

But thermoset composite failed by a sudden interlaminar shear fracture with no compressive yielding beneath the loading pin. A buckled zone [vertical crack] preceded the interlaminar shear failure and no meaningful correlation could be obtained between vertical crack depth values shown within brackets in Figure 3 and the time to failure. Since no compressive yielding was observed, it is thought that rupture time values at a given stress level might be increased if vertical crack is avoided. For example, another specimen [AS4/3501-6] tested at stress of 792 MPa, significantly lower than the stress values shown in Figure 3, had neither fractured nor shown any signs of vertical cracks, during a creep period of 10100 minutes.

**Creep Rupture Test.** Results of the test fit well to a logarithmic relation as shown in Figures 5 and 6. The scatter in data is larger in case of thermoset composite, probably due to the vertical cracks. Whether these creep rupture plots can be correlated to creep rupture under homogeneous tensile load is a problem currently under investigation.

The initial spread in the deflection/applied stress data of Figure 3 is thought at this time to be due to the uncontrolled

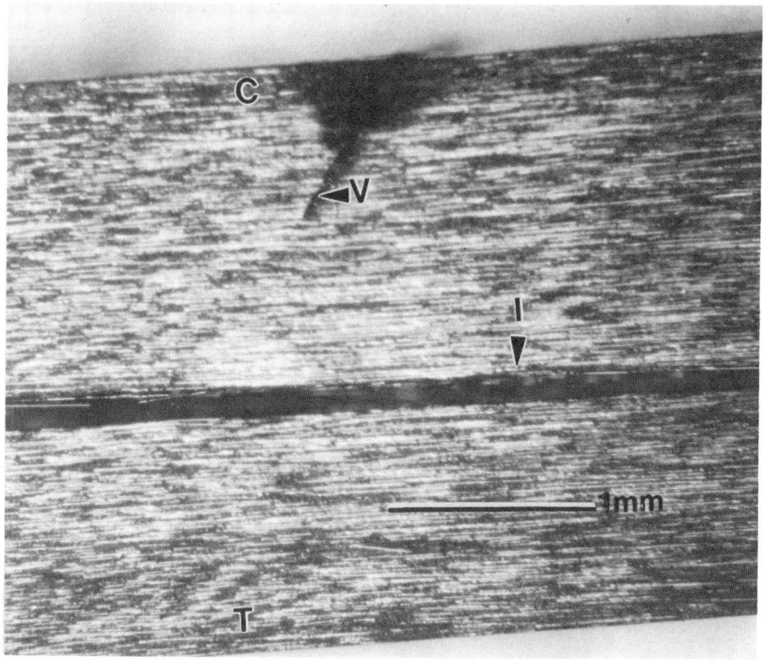


Fig. 2 (a). Fractured thermoset [AS4/3501-6] composite specimen viewed along thickness. T - Tensile side: C - Compressive side; V - Vertical crack; I - Interlaminar shear crack.

Fig. 2 (b). Fractured thermoplastic [AS4/J2] composite specimen. A - Viewed along thickness; B - Viewed along width; T - Tensile side; C - Compressive side; The arrow identifies the region of compressive yield beneath the loading pin.

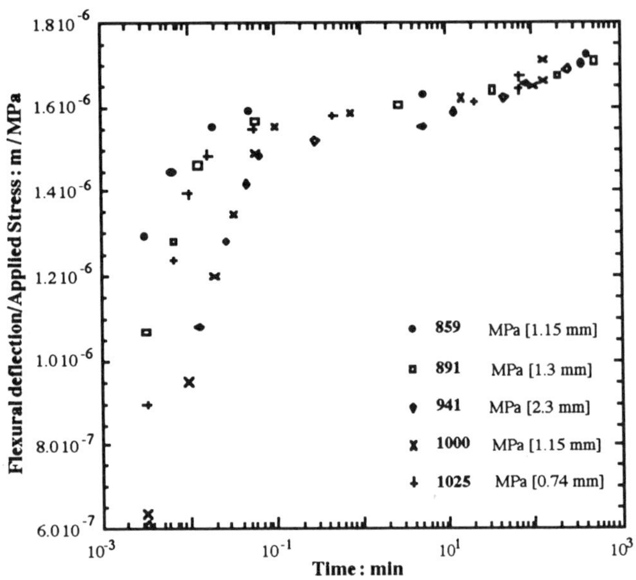


**Fig. 3 : Plot of flexural deflection / applied stress vs creep time of AS4/3501-6 composite at room temperature. The depth of vertical crack is given within brackets.**

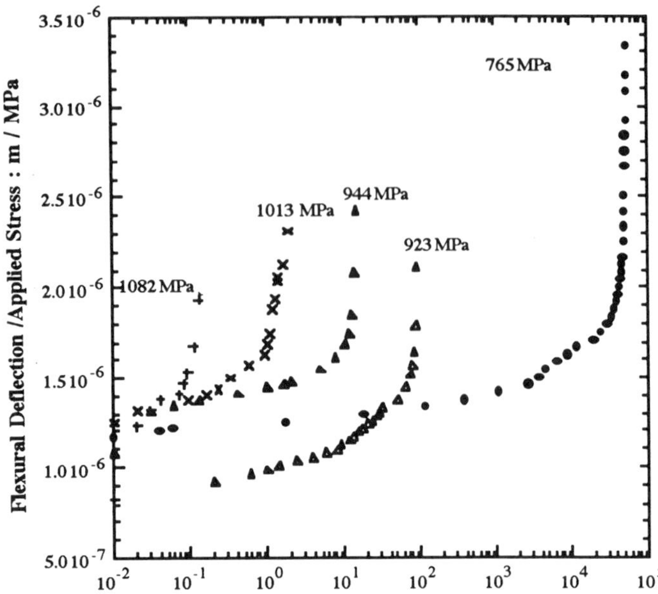


**Fig. 4 : Plot of flexural deflection /applied stress vs creep time of AS4/J2 composite at room temperature.**

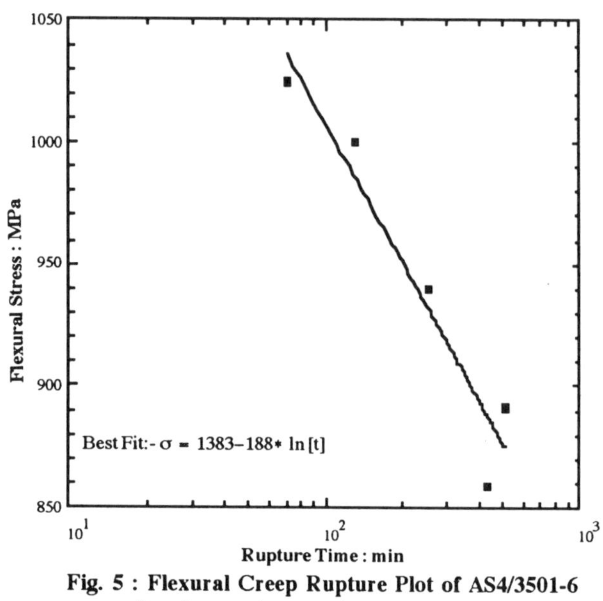


**Fig. 5 : Flexural Creep Rupture Plot of AS4/3501-6 composite at room temperature**

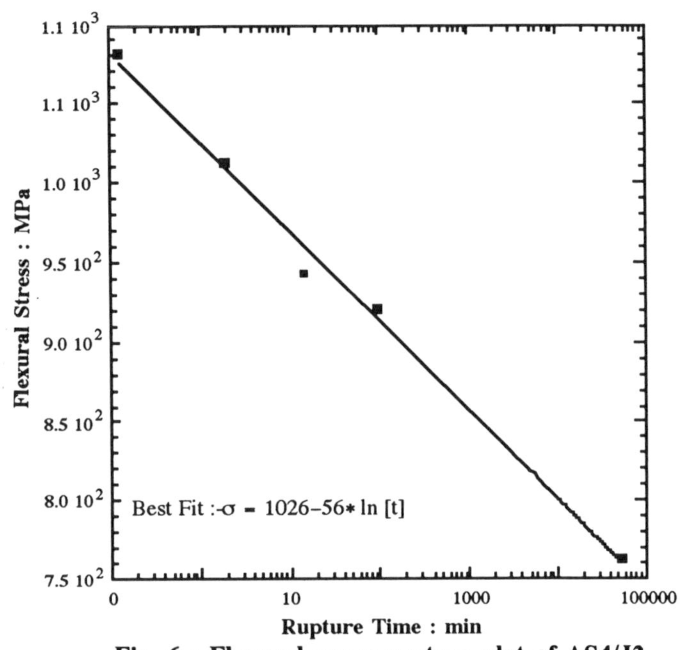


**Fig. 6 : Flexural creep rupture plot of AS4/J2 composite at room temperature**

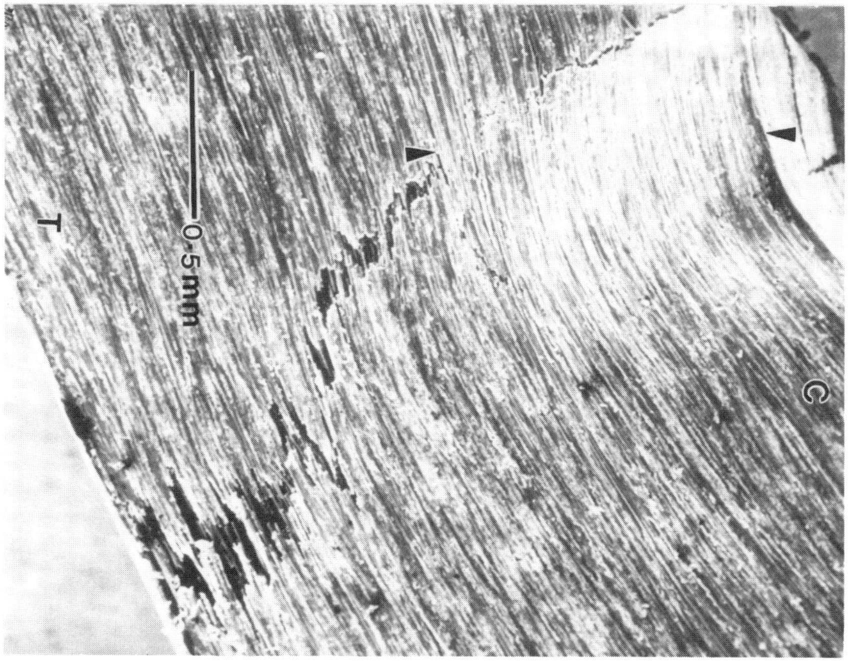


[a] Stress level :- 765 MPa; Rupture time :- 53544 minutes.

[b] Stress level :- 1082 MPa; Rupture time :- 0.13 minutes.

Fig. 7. SEM fractographs of thermoplastic [AS4/J2] composite specimens creep ruptured at room temperature [viewed along thickness]. T - Tensile side; C - Compressive side; The region between the arrows is the buckled zone.

loading rates at different stresses. **Scanning Electron Microscope Fractographs.** The SEM fractographs of thermoplastic composite as shown in Figure 7, have already been discussed in the previous section. The fractographs of AS4/3501-6 are given in Figures 8[a], 8[b] and 8[c]. They correspond to regions marked A, B and C in Figure 1 respectively. The smooth region as seen in Figure 8[a] is characteristic of tensile [mode I] fracture in composites. The source for the tensile normal stresses in this region could be the vertical cracks [buckled zone] as suggested by Whitney et. al. [6]. Also the hackles seen in Figure 8[b] are characteristics of interlaminar shear failure [9]. The buckled zone is a region of disintegrated resin as shown in figure 8[c].

SEM fractographs of specimens failed at different stresses are essentially same and no appreciable difference is noticed.

**Repeated Stress Relaxation Test.** The test results are given in Figures 9 and 10 for thermoset composite and in Figures 11 and 12 for thermoplastic composite. In both composites, at lower stress levels [figs. 9 & 11] the relaxation rate decreased with subsequent reloading cycles. Thus a deceleration effect was observed due to cyclic loading at low stress levels in both composites.

At higher stress levels as shown in Figures 10 & 12, the relaxation rate increased with subsequent reloading cycles ultimately resulting in fracture. Thus an acceleration effect was noticed at high stress levels due to cyclic loading. A systematic experimentation is underway to determine the stress ranges that exhibit these two responses.

## Conclusion

Four point flexure test was used to study and compare the creep/creep rupture behavior of thermoset [AS4/3501-6] and thermoplastic [AS4/J2] composites. The ductile thermoplastic matrix [J2] led to higher creep deflection and compressive yielding which made it difficult for the four point flexural shear test to induce interlaminar fracture in AS4/J2 composite as compared to brittle epoxy of AS4/3501-6. The creep rupture data fitted into a logarithmic relation with lesser scatter in data for thermoplastic composite than thermoset composite. The repeated flexural stress relaxation test revealed in both composites that repeated loading resulted in a deceleration of relaxation at lower stresses and an acceleration of relaxation at higher stresses.

## References

1. "Interlaminar Fracture of Composites", Edited by E.A. Armanios, Published by Trans Tech Publications Ltd., Switzerland, 1989.
2. Browning, C.E., Abrams, F.L. and Whitney, J.M., ASTM STP 797, 1983, pp. 54-74.
3. Greenwood, J. H., Composites, No.5, 6 (1975) pp. 203-206.
4. Nguyen, D.H., Wang, S.F. and Ogale, A.A., 34th Int. SAMPE Symp., May 1989, pp. 1275-1282.
5. Evans, J.T., Ningyun, W. and Chandler, A.W., Acta metall. mater., No.8, 38 (1990) pp. 1565-1572.
6. Whitney, J.M. and Browning, C.E., Fifth Int. Conf. on Exp. Mech., June 1984, pp. 97-101.
7. Swain, R. E., Reifsnider, K.L. and Vittoser, J. , ASTM STP 1059, 1990, pp. 390-403.
8. Fitzgerald, J. E. and Vakili, J., Exp. Mech., No.12., 13 (1973) pp. 504-510.
9. Morris, G.E., ASTM STP 696, 1979, pp. 274-297.

## Acknowledgement

This work is being sponsored by Office of Naval Research under contract N00014-89-J-3048. The authors would like to acknowledge the kind support of Dr. Vasudevan of ONR and Dr. Tom Juska of David Taylor Naval Research center.

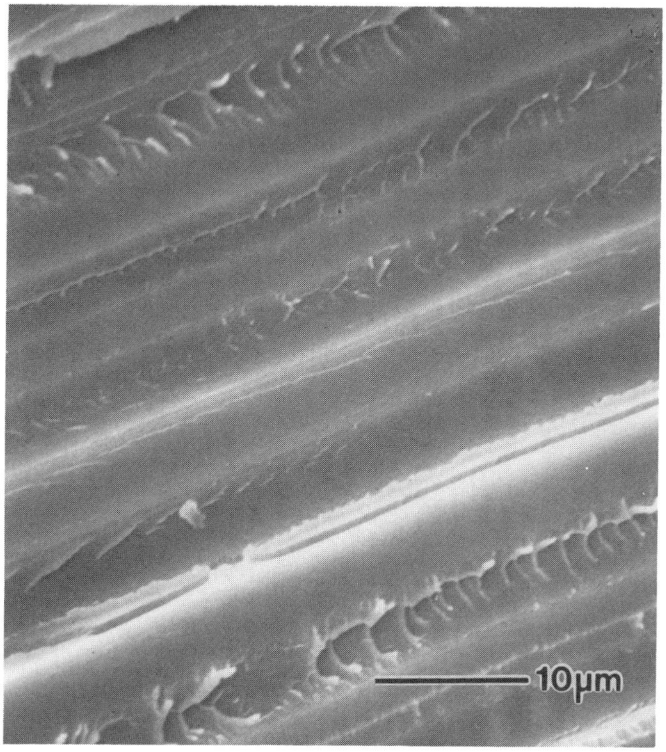


[a] Region A

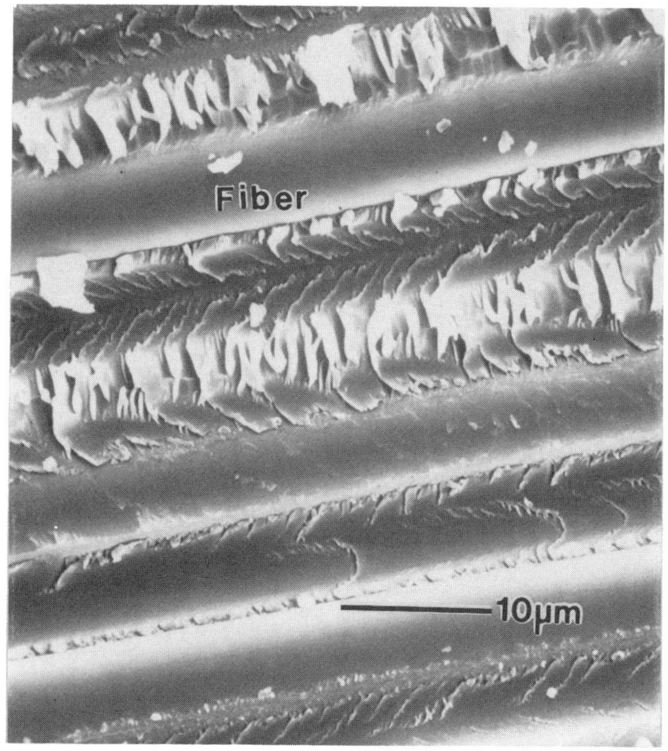


[b] Region B

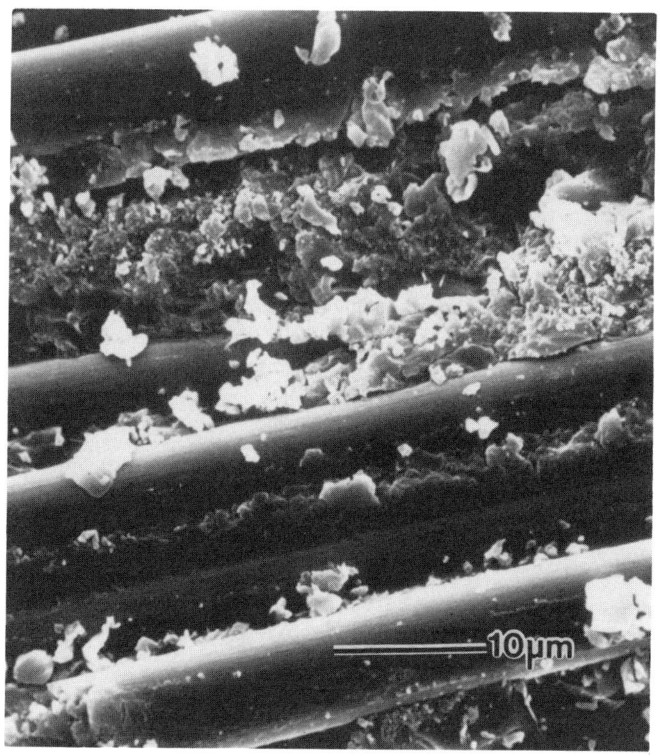


[c] Region C

Fig. 8. SEM fractographs of thermoset [AS4/3501-6] composite specimens creep ruptured at room temperature, corresponding to the regions marked in Figure 1.

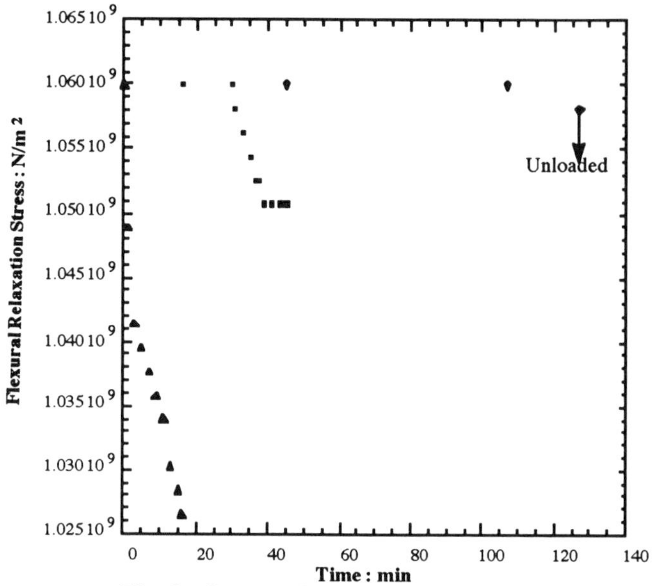


**Fig. 9 : Stress relaxation behavior of AS4/3501-6 composite at room temperature during repeated relaxation test. The loading rate is 2905 N/min. Initial stress level is 1060 MPa.**

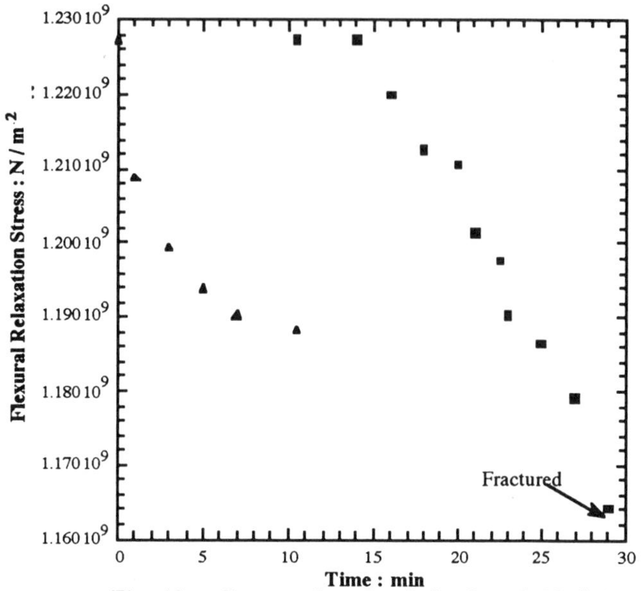


**Fig. 10 : Stress relaxation behavior of AS4/3501-6 composite at room temperature during repeated relaxation test. Loading rate is 2905 N/min. Initial stress level is 1227 MPa.**

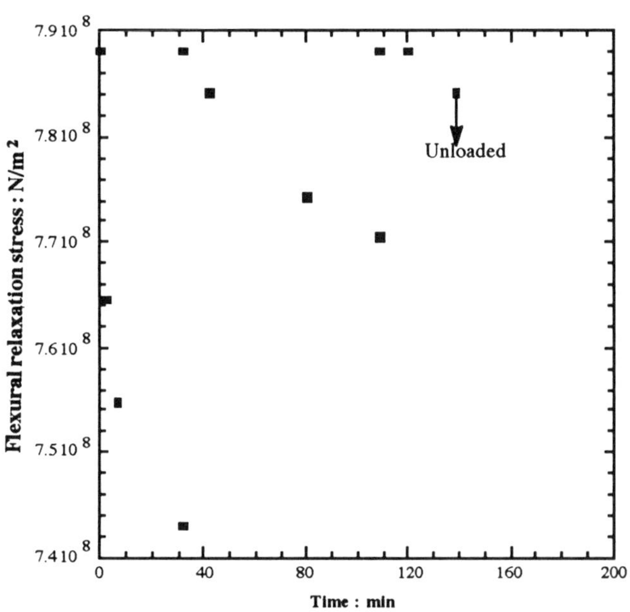


**Fig. 11 : Stress relaxation behavior of AS4/J2 composite at room temperature during repeated relaxation test. The loading rate is 2287 N/min. Initial stress level is 788 MPa.**

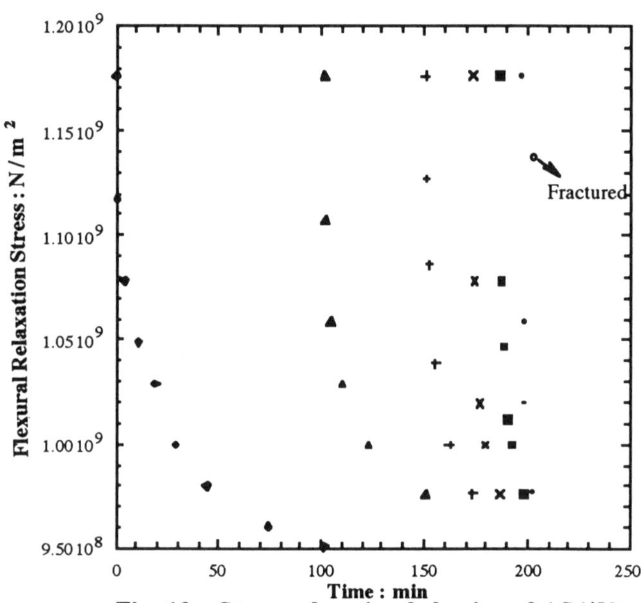


**Fig. 12 : Stress relaxation behavior of AS4/J2 composite at room temperature during repeated relaxation test. The loading rate is 2287 N/min. Initial stress level is 1176 MPa.**

*Proceedings of the Fifth International Conference on Creep of Materials, Lake Buena Vista, Florida, USA, 18-21 May 1992*

# High Temperature Deformation and Fracture Mechanisms in Monolithic and Particulate Reinforced Nickel Aluminide Processed by Spray Atomization and Co-Deposition

**H.K. Kim, X. Liang, E.J. Lavernia, and J.C. Earthman**
University of California
Irvine, California

ABSTRACT

Intermetallic-matrix composites (IMCs) have the potential of combining matrix properties of oxidation resistance and high temperature stability with reinforcement properties of high specific strength and modulus. The purpose of the present investigation is to develop a better understanding of the nature of creep fracture mechanisms in a $Ni_3Al$ composite reinforced with both $TiB_2$ and SiC particulates. In the present study, creep rupture specimens were tested under constant stresses ranging from 180 to 350 MPa in vacuum at 760°C. Examinations of tested specimen cross-sections receal that the high temperature damage in the unreinforced material is comprised of the plastic growth of pre-existing microcracks while damage in the composite is manifested by the nucleation and growth of cavities. The microstructural observations reveal that most of the cavities lie on the grain boundaries of the $Ni_3Al$ matrix as opposed to the large $TiB_2/Ni_3Al$ interfaces, suggesting that cavities nucleate at fine carbides that lie on the $Ni_3Al$ grain boundaries as a result of the decomposition of the $SiC_p$. This observation accounts for the longer rupture times for the monolithic $Ni_3Al$ as compared to those for the $Ni_3Al/SiC_p/TiB_2$ IMC.

## 1. INTRODUCTION

Intermetallic compounds have recently been the subject of considerable interest for high temperature applications. In particular, $Ni_3Al$ is one of the more promising intermetallic compounds as a result of its anomalous dependence of strength on temperature, low density and good oxidation resistance [1-5]. Also, the tensile ductility is improved with trace additions of boron in the material. More recently, investigators have sought to achieve further improvements in the properties of intermetallics, by utilizing them in intermetallic matrix composites (IMCs) containing continuous or discontinuous ceramic reinforcements [6-8].

Among the various processing methodologies that are currently being used to fabricate discontinuously reinforced IMCs, spray atomization and co-deposition offers several advantages [9, 10]. First, deleterious interfacial reactions between matrix and ceramic reinforcement can be controlled due to relatively low processing temperatures with highly efficient heat convection during atomization. Second, the deposited materials exhibit some of the characteristics associated with rapid solidification such as fine grained microstructures, increased solid solubilities, non-equilibrium phases and the absence of macro-segregation.

It has been shown that rupture of $Ni_3Al$ is facilitated by the nucleation, growth and coalescence of grain boundary cavities [11]. Thus, it follows that creep cavitation is also a controlling mechanism in determining the life of particulate reinforced $Ni_3Al$ alloys. One of the major limiting factors for the successful application of these composites at high temperatures is the formation of interfacial reactions between matrix and ceramic reinforcement during composite processing, as well as during service. In general, the development of an interaction zone is detrimental to the mechanical properties of the composite. However, a good understanding of the effect of the interfacial reactions on high temperature rupture mechanisms in $Ni_3Al$ matrix composites is still lacking.

The objective of the present study was to develop a better understanding of the nature of creep rupture mechanisms in monolithic $Ni_3Al$ and a $Ni_3Al$ matrix composite produced by spray deposition. Emphasis is placed in the present investigation on the roles of the products of the reactions between the reinforcements and the matrix in determining the life of specimens.

## 2. EXPERIMENTAL PROCEDURE

### 2.1. Experimental Materials

The nominal composition of the $Ni_3Al$ used in the present study (IC-396 designation) is 8.0 Al, 7.7 Cr, 0.85 Zr, 3.0 Mo, 0.005 B, bal Ni (in wt. pct.); the reinforcement SiC particulates (HCP α−phase) had an average size of 3 μm, whereas the $TiB_2$ particulates (C32 structure) had approximate average size of 10 μm prior to processing. During spray atomization and co-deposition, the matrix material ($Ni_3Al$) was disintegrated into a dispersion of droplets using high energy $N_2$. Simultaneously, $TiB_2$ and SiC particulates were co-injected into the atomized spray at a previously determined spatial location. The primary experimental variables used in the processing procedure are given in [12].

Both the unreinforced $Ni_3Al$ and IMC were hot extruded at 1200 °C with a extrusion ratio of 9:1 in order to eliminate micrometer sized pores that are normally present in spray atomized and deposited materials. The volume fraction of ceramic particulates was quantitatively characterized using two techniques: (a) image analysis and (b) acid dissolution. The results from both techniques indicate a total ceramic particulate volume fraction of 12.7% in the composite. Compositional studies of the $Ni_3Al$/ceramic interfaces were accomplished using a HITACHI S-500 microscope equipped with Energy Dispersive X-ray Spectroscopy (EDS). An x-ray diffraction pattern produced by a rotating specimen method was also used to determine the phase constitution in the materials.

### 2.2. High Temperature Testing

Tensile creep specimens were ground to size with a gage diameter and length of 3.18 mm and 15.88 mm, respectively. The hot-extruded unreinforced $Ni_3Al$ and composite specimens were tested under constant tensile stress conditions ranging from 180 to 350 MPa at 1033 K in a vacuum environment of $10^{-3}$ Pa, using a custom-built high temperature testing system. Cross-sections of the specimens were examined using SEM to determine the extent and location of creep damage within the microstructure of both the unreinforced and composite materials.

## 3. EXPERIMENTAL RESULTS AND DISCUSSION

### 3.1. Microstructural Observations

An optical micrograph of the unreinforced $Ni_3Al$ alloy is shown in Figure 1, revealing an elongated grain morphology, with an approximate mean grain width of 10 μm and an average grain aspect ratio of about 5. As shown in this figure, some microcracks were found along transverse boundaries in the unreinforced $Ni_3Al$ after hot extrusion. An x-ray diffraction pattern of the this alloy indicates that this material is not fully ordered, suggesting that a homogeneous distribution of both $Ni_3Al$ ($\gamma'$) and Ni-rich ($\gamma$) phases exist in the alloy.

Figure 2 shows the microstructure of the hot-extruded $Ni_3Al/SiC/TiB_2$ composite revealing a relatively homogeneous distribution of reinforcing particles and Ni-rich interfacial phases. These phases appear to have undergone extensive deformation and fracture during hot extrusion, but are still noted to be in intimate contact with $TiB_2$ particles. The ordered $\gamma'$ matrix grains recrystalized during hot extrusion, as indicated by their equiaxed morphology. The observation that $TiB_2$ particulates and Ni-rich phases are generally present at the grain boundaries of the matrix suggests that these particulates and interfacial phases may have effectively hindered grain growth in the matrix during hot extrusion.

SEM and EDS analyses showed that the matrix primarily consists of the $\gamma'$ phase, with interdispersed regions of $\gamma$. The $\gamma$ is highly irregular in shape, and was generally observed to be in intimate contact with $TiB_2$ particulates, suggesting that this phase precipitated heterogeneously around the $TiB_2$ particulates. The chemical reactions that occurred at the $Ni_3Al/TiB_2$ interface are evident from Figure 2b, where the $TiB_2$ particulates are surrounded by a well defined reaction zone layer. This reaction zone appears to have increased in thickness by consuming both the matrix and the $TiB_2$ particulates, leading to the formation of two distinct interfacial reaction layers (labeled I and II). Interfacial reaction layer I formed inside the $TiB_2$ particulate (i.e., at the $TiB_2$/reaction zone interface), while interfacial reaction layer II formed at the reaction zone/$Ni_3Al$ matrix interface. EDS analysis has shown that interfacial reaction layer I primarily consists of Ti, Ni, and Cr, and selected area diffraction using TEM showed that this layer is $Ni_3Ti$ which has a hexagonal crystal structure ($DO_{24}$-type). Interfacial reaction layer II is characterized by a very high Ni content with a relatively low Al content [12]. In addition, it is noted that microvoids, 0.4 - 0.9 μm long and 0.1 - 0.4 μm wide, were observed within the interface region.

X-ray diffraction studies conducted on the composite revealed the presence of SiC. However, these particulates were not evident from SEM observations combined with and EDS analysis. It appears that the SiC particulates were substantially reduced in size or eliminated during processing. In related studies, Chou and Nieh [13] noted that SiC reacts extensively with pure $Ni_3Al$ during annealing at 1000 °C, leading to the formation of multiple reaction layers in the interdiffusion zone at the interface. It appears that the temperatures during deposition (900-1300 °C) provide sufficient thermal energy for the alloying elements Cr, Mo, Zr to segregate to the $Ni_3Al$/SiC interface. These alloying elements are then able to replace the Si in the SiC lattice leading to the formation of other carbides. Concurrently, diffusion of Si into the $Ni_3Al$ matrix may also lead to the formation of compounds with Ni and Al, as reported by Chou and Nieh [13]. In addition, the high surface to volume ratio associated with the fine (3 μm) SiC particulates used in the present experiment may also have contributed towards their high reactivity.

### 3.2. Creep Behavior

Typical creep curves for both monolithic and composite specimens tested under 300 MPa are illustrated in Figure 3. The creep curves are not similar to those typically observed for other $Ni_3Al$ alloys in that both materials exhibited constant strain rate behavior with little indication of accelerating or "inverse" creep [14]. Normally, inverse creep is observed at small strains for $Ni_3Al$. $Ni_3Al$ generally exhibits a normal primary creep response indicating material hardening followed by an accelerating or "inverse" creep rate which has a minimum at relatively small strain. Inverse creep is then followed by steady state behavior [15]. In the current investigation, it is not clear why there is an absence of the inverse creep transition at low strain. It has been proposed that inverse creep is caused by the bowing out and subsequent glide of dislocations on the cube cross-slip plane at Kear-Wilsdorf locks [16, 17]. If this is the case then it is likely that enough dislocations are generated on the cube slip planes during hot extrusion to effectively bypass this process during creep.

Steady state strain rates are plotted against the applied stress in Figure 4. Least square fit analyses of the data in this figure give power law creep constants of

$\dot{\varepsilon} = 7.7 \times 10^{-37} \sigma^{3.44}$ for monolithic $Ni_3Al$

$= 8.9 \times 10^{-37} \sigma^{3.50}$ for $Ni_3Al/SiC/TiB_2$

where $\dot{\varepsilon}$ and $\sigma$ are in units of $s^{-1}$ and MPa, respectively. The experimental data reveal that the stress exponent for power law creep for both materials is about 3.5, suggesting that the creep deformation for both materials is predominantly controlled by dislocation glide processes. However, the composite specimens crept about 6 times faster than the monolithic $Ni_3Al$ specimens. It is not immediately clear why the composite was weaker than the unreinforced $Ni_3Al$ in terms of creep strength. Some possible explanations are offered below that are based on microstructural observations.

A general understanding of composite creep is obtained by considering creep strength from a continuum mechanics point of view. Composite hardening is known to be due to a difference in plastic strain of the components compensated by elastic strains which give rise to back stresses. It is often assumed that there is no difficulty in transferring load between matrix and reinforcement in order to prevent the development of local incompatibility stresses. However, weak interfaces can have a substantial effect on this load transfer and, as a result, on the creep behavior of composites. In the limit of zero interface strength, where no load can be transferred to the reinforcement, the composite is weaker than the matrix alone [18]. As mentioned before, the processing of the composite apparently results in the decomposition of the SiC particulates and the formation of other carbide phases, such as an $Cr_3C_2$-type carbide, and the formation of the Ni-rich interfacial phase at the $Ni_3Al/TiB_2$ interface. This could give rise to a weakening of the interfaces present in the composite.

The creep strain rate is determined by the number of mobile dislocations and their average velocity, when creep is controlled by dislocation glide processes, The initial creep rate should then increase with increasing dislocation density. However, even when glide processes govern the creep behavior, recovery processes will again attempt to reduce the dislocation density towards equilibrium levels for the applied stress imposed. For the composite, heterogeneous deformation at the interface between the particulates and matrix during hot extrusion apparently results in recrystalization and grain growth which in turn leads to a relatively low dislocation density. For the monolithic material, the same process seems to produce more homogeneous deformation in the matrix which is not sufficient to induce the recrystalization. These findings are based on the equiaxed grain morphology for the composite and the elongated grain structure for the monolithic material. Thus, it is reasonable to assume that monolithic $Ni_3Al$ has a higher dislocation density than the composite prior to testing. When creep is controlled by dislocation glide processes, the material with the higher initial dislocation density should creep faster. Therefore, it is unlikely that the difference in creep behavior, shown in Figure 4, is due to the difference in dislocation density.

Flinn [19] has proposed that the steady-state creep of $Ni_3(Al, Fe)$ polycrystals for the temperature range used in the present investigation is controlled by the thermal diffusion of neighboring atoms to annihilate the antiphase domain boundaries (APBs). Since the rate-controlling process is the diffusion of atoms to and from the APBs, the alloying elements affecting lattice diffusion could play a role in the creep behavior of $Ni_3Al$. In this case, the alloying elements may increase lattice diffusion and therefore increase the creep rate. However, it is not clear for the present composite whether Si, resulting from the decomposition of SiC, increases or decreases the lattice diffusion. It is possible that a reduction in the volume fraction of $Ni_3Al$, due to the formation of another phase with Si, might attribute to the weaker creep strength of the composite relative to that for the monolithic material.

Figure 1 shows that the monolithic material contains elongated grains with occasional microcracks on the transverse grain boundaries, whereas the composite has an equiaxed grain structure, as shown in Figure 2a. As mentioned before, the grain size of the composite is about 10 μm, whereas the grain width of the monolithic $Ni_3Al$ is about 25 μm with grain aspect ratio of 5. The finer and more equiaxed grain size present in the composite relative to that of the monolithic material may partially account for the faster creep rate of the composite, if grain boundary sliding makes a significant contribution to the overall deformation.

### 3.3. Creep Rupture Behavior

Figure 5 shows creep rupture lifetime for monolithic $Ni_3Al$ and the composite as a function of applied stress. The results indicate that the monolithic material exhibits a longer lifetime than the composite for a given stress. Therefore, it appears from this data that the mechanism which causes rupture is assisted by the presence of the reinforcing particulates. For the monolithic material examined in the present investigation, it appears that matrix creep deformation facilitates the growth of pre-existing microcracks. In this case it is reasonable to assume that the creep and fracture processes are closely linked. This is supported by the fact that the rate equations for the rupture time and the steady state strain rate have a similar dependence on stress. That is, at constant temperature and for a fixed microstructure:

$$t_r = A'\sigma^{-n'} \qquad (1)$$

and

$$\dot{\varepsilon} = A\,\sigma^{n} \qquad (2)$$

where $t_r$ is the rupture time, n = 3.5 and n' = 3.2 for the monolithic material. Constrained cavity growth is further supported by the fact that the Monkman-Grant product, representing the accumulated matrix strain, was found to be constant (=0.043). Thus, failure occurs when the specimens reach a critical matrix strain.

The composite material appears to undergo a somewhat different creep fracture process than is observed for the monolithic spray deposited material. It appears from the data that n = n' for this material. This finding will be discussed further in the following sections.

### 3.4. Cavitation Damage Morphology

In the case of the present composite material, the reinforcement/matrix interfaces as well as the grain boundaries are possible preferred cavity nucleation sites. Therefore, it is necessary to identify at which location cavities nucleate preferentially. Cross-sections of the ruptured specimens have been examined using SEM to determined the extent and location of creep damage within the microstructure of the composite. It is difficult to identify the preferred cavity nucleation site from the completely damaged microstructure after rupture due to the extensive microcracking in the gage section. Accordingly, SEM observations were made on tested material from the grip sections of specimens which experienced a much lower stress. A typical SEM micrograph of tested grip section material is shown in Figure 6. As shown in this figure, the SEM observations indicate that cavity nucleation and growth primarily occur in the composite at the $Ni_3Al$ grain boundaries as opposed to the $TiB_2/Ni_3Al$ interfaces of comparable size and orientation. It appears that the cavities nucleated on the grain boundaries at the location of fine carbides which resulted from the decomposition of SiC particulates. It has been shown in an earlier work that a large number of carbides exist on the grain boundaries in the composite [12]. It appears that the nucleation of grain boundary cavities on carbides occurs easily due to a relatively weak interfacial cohesive strength. On the other hand, the rare presence of cavities at the $Ni_3Al/TiB_2$ interface reaction zones indicates that higher cohesive forces exist at these interfaces. This finding is consistent with the microstructural observation of coherent interfacial bonding at the $Ni_3Al/TiB_2$ interface reported earlier [12].

Compared to the damage state at the high stress range (e.g.300 MPa, see Figure 7), the creep damage at the low stress range (e.g. 180 MPa, see Figure 8) consists of a smaller of cavities and microcracks. This could be due to the continuous cavity nucleation at lower stress state and the instantaneous cavity nucleation at high stress state. In the case of 180 MPa, it is more likely that cavities nucleated continuously as a result of a stochastic process which occurs throughout the creep life and triggers cavity nucleation. Therefore, there should be a large fraction of very small cavities and microcracks, compared to the case where most of the cavities are nucleated initially upon loading. This suggests that the cohesive strength of the weaker interface (i.e. SiC particulate) is greater than 180 MPa.

Plots of relative cavity frequency versus cavity size for both grain boundaries and matrix/$TiB_2$ interfaces are shown in Figures 9 and 10 for two different stress levels.. As shown in Figure 9, corresponding to 300 MPa, the relative frequency for the nucleation sites at matrix/$TiB_2$ interfaces was much less than that for the matrix grain boundaries. Figure 9 also indicates that the frequency of smallest cavities make up a relatively small portion of the entire cavity population. This suggests that much of the cavity nucleation occurred upon loading for the specimen tested to rupture at 300 MPa, causing a greater fraction of large cavities and microcracks. Therefore, it is reasonable to assume that the cohesive strength of the matrix/carbide interface is less than 300 MPa. However, the large fraction of small cavities in Figure 10 indicates that cavity nucleation occurs continuously over substantial fractions of the creep life for the specimen tested under 180 MP. From both Figures 9 and 10, it could be assumed that the interfacial cohesive strength between the carbides and matrix falls between 180 MPa and 300 MPa. This assumption will be discussed in the next section in terms of a rupture time analysis which is based on instantaneous cavity nucleation. Overall, the results indicate that a much greater lifetime would be achieved for the composite without the addition of the SiC during processing.

### 3.5. Rupture Time Analysis

If cavity nucleation occurs early in the creep life, the time to cavity coalescence can be obtained by integrating the cavity growth rate. Following Riedel [20], the time to cavity coalescence on an isolated boundary facet is given by

$$t_r = 0.006\frac{h(\psi)kT\lambda^3}{\Omega\delta D_b\sigma_I^{\infty}} + \frac{0.4h(\psi)(1+3/n)^{0.5}\lambda}{\dot{\varepsilon}^{\infty}(\sigma_I^{\infty}/\sigma_e^{\infty})d} \qquad (3)$$

where $h(\psi) = 0.61$, k is Boltzmann's constant, T is the temperature, $\lambda$ is the cavity spacing, $\Omega$ is the atomic volume, $\delta D_b$ is the product of grain boundary thickness and diffusivity, $\sigma_I^{\infty}$ is the applied normal stress, n is the stress exponent for power law creep, $\dot{\varepsilon}_e^{\infty}$ is the equivalent strain rate, $\sigma_e^{\infty}$ is the applied equivalent stress for axisymmetric loading, and d is the grain size. The first term of Eq. 3 represents the time to coalescence in the absence of the constraint, i.e. when diffusion is rate controlling. The second term takes into account the constraint on cavity growth by uncavitated regions. The average value of $\lambda$ (=1.8 μm) was determined from the SEM micrograph of tested grip section material.

Rupture time as a function of stress determined from the above equation for the composite material is indicated by the dashed line in Figure 5. It can be seen that the constrained cavity growth model predicts $t_r$ within the high stress range while it underestimated real $t_r$ by a factor of 2 in the low stress range. It was not clear why the observed rupture times were in good agreement with those predicted by the model only in the high stress range. But, as mentioned in the previous section, it seems that in the high stress range, there was instantaneous cavity nucleation which is consistent with the assumptions of the model. In the low stress range, however, there was significant continuous cavity nucleation which would result in longer rupture times than these predicted by that model.

Creep deformation in the matrix appears to also dominate the rupture process for monolithic $Ni_3Al$. This is supported by the relatively large crack opening displacements produced at the pre-existing microcracks (see Figure 11). In this case failure should occur when the creep strain within the grains reaches a critical value (i.e. M.G. constant = 0.043). Therefore, creep deformation should determine the lifetime. Rupture times for the monolithic material estimated in this manner are indicated by the solid line in Figure 5. Good agreement between measured and predicted rupture times are obtained using this constant creep strain to failure criterion.

## 4. CONCLUSIONS

1. The microstructural observations reveal that SiC reacts extensively with $Ni_3Al$ during spray deposition of a particulate reinforced $Ni_3Al$ matrix composite. This leads to the decomposition of the SiC particulates and to the formation of other carbide phases on the matrix grain boundaries. Also, a well defined reaction zone was observed at the $Ni_3Al/TiB_2$ interface which is composed of two layers, a Cr containing $Ni_3Ti$ phase layer and a Ni-rich phase.

2. An optical micrograph of the unreinforced $Ni_3Al$ alloy reveals an elongated grain morphology, with an average grain aspect ratio of about 5. Some microcracks were found along transverse boundaries in the unreinforced $Ni_3Al$ after hot extrusion. An x-ray diffraction pattern of the unreinforced IC-396 indicates that this material is not fully ordered, indicating that a homogeneous distribution of both $Ni_3Al$ ( $\gamma'$) and Ni-rich ($\gamma$ ) phases exist in the alloy.

3. The stress exponent for power law creep for both materials is about 3.5, suggesting that the creep mechanisms for both materials are primarily controlled by dislocation glide processes. However, the composite specimens crept about 6 times faster than the unreinforced $Ni_3Al$.

4. The hot extruded monolithic $Ni_3Al$ exhibited longer creep rupture lifetimes than the hot extruded $Ni_3Al/SiC/TiB_2$ composite. Microstructural observations indicate that small carbide particles located on the $Ni_3Al$ matrix grain boundaries, resulting from the decomposition of SiC particulates, provide cavity nucleation sites in the composite. This results in extensive grain boundary cavitation which, in turn, leads to rupture lifetimes that are shorter than those observed for the unreinforced $Ni_3Al$ material alloy

5. Creep deformation in the matrix appears to dominate the rupture process for both the unreinforced $Ni_3Al$ alloy and the composite by constraining cavity growth and microcrack coalescence.


## ACKNOWLEDGMENTS

The authors would like to acknowledge the Air Force Office of Scientific Research (grant no. AFOSR-90-0366) for their support of this work. Support for the image analysis system used in the present investigation by the National Science Foundation under grant no. MSS 8957449 is also gratefully acknowledged.


## REFERENCES


1. N.S. Stoloff, *Inter. Mat. Rev.*, 34, 153 (1989).
2. K. Aoki and O. Izumi, *J. Jpn. Inst. Met.*, 43, 1190 (1979).
3. V.K. Sikka, *in Casting of Near Net Shape Products*, edited by Y. Sahai, J.E. Battles, R.S. Carbonara and C.E. Mobley, AIME, Warrendale, PA, pp. 315 (1988).
4. C.T. Liu and V.K. Sikka, *J. Met.*, 5, 19 (1986).
5. K. Aoki, Mater. Trans. *JIM*, 31, 443 (1990).
6 J.M. Yang, W.H. Kao and C.T. Liu, *Mater. Sci. Eng.*, 107A, 81 (1989).
7. S. Nourbakhsh, F.L. Liang and H. Margolin, *Adv. Mat. & Manuf. Pro.*, 34, 57 (1988).
8. G.L. Povirk, J.A. Horton, C.G. Mckamey and T.N. Tiegs, *J. Met. Sci.*, 23, 3945 (1988).
9. M. Gupta, F. Mohamed and E.J. Lavernia, *J. Mat. & Manuf. Process.*, 2, 165 (1990).
10. M. Gupta, F. A. Mohamed and E.J. Lavernia, *in Advances in Processing and Characterization of ceramic Metal Matrix Composites*, edited by H. Mostaghaci, CIM/ICM, Pergamon Press, p. 236 (1989).
11. J.H. Schneibel and L. Martinez, *Acta Metall.*, 37, 2237 (1989).
12. X. Liang, H.K. Kim, J.C. Earthman and E.J. Lavernia, *J. Mater. Sci. & Eng.*, in press.
13. T.C. Chou and T.G. Nieh, *J. Mater. Res.*, 5, 1985 (1990).
14. J.H. Schneibel and J.A. Horton, *J. Mater. Res.*, 3, 651 (1988).
15. R.K. Ham, R.H. Cook, G.R. Purdy, and G. Willoughby, *Metal Sci.* 6, 205 (1972).
16. P.M. Hazzledine and J.H. Schneibel, *Script Metall.*, 23, 1887 (1989).
17. K.J. Hemker, M.J. Mills, and W.D. Nix, *Acta Metall.*, 39, 1901 (1991).
18. S. Goto and M. McLean, *Acta Metall.*, 39, 165 (1991).
19. P.A. Flinn, *Trans. Met. Soc. AIME*, 218, 145 (1960).
20. H. Riedel, *Fracture at High Temperatures*, Springer-Verlag, Berlin (1987).

Figure 1 - Microstructure of the unreinforced $Ni_3Al$ following hot extrusion at 1200 °C. Pre-existing microcracks are indicated by arrows.

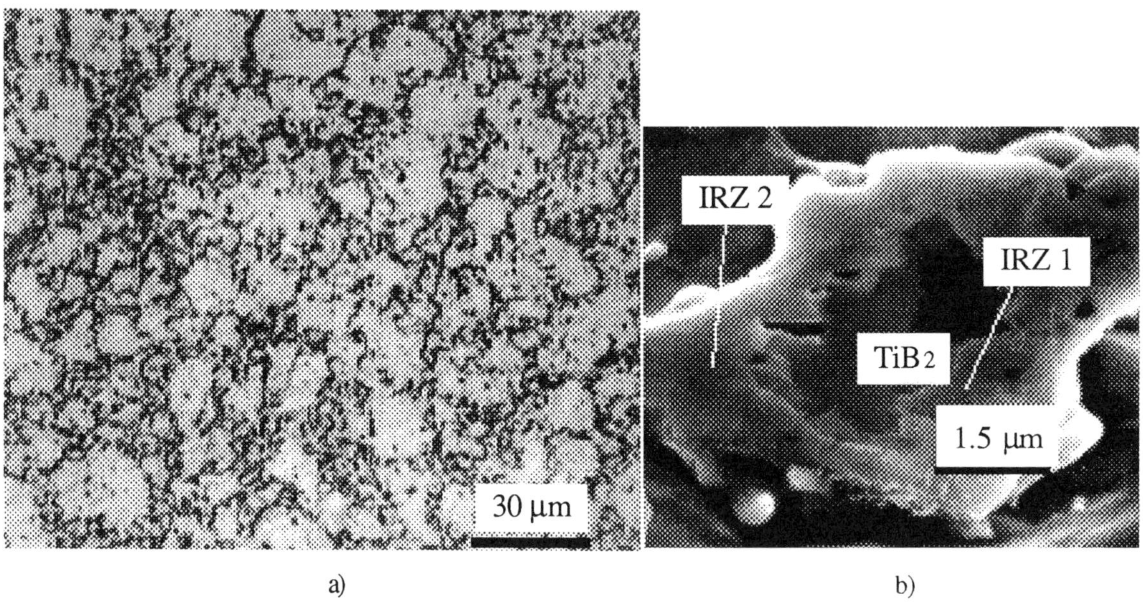


Figure 2 - SEM micrograph showing: a) the microstructure of the spray atomized and co-deposited $Ni_3Al/SiC_p/TiB_2$ IMC following hot extrusion at 1200 °C; and b) the interfacial reaction zones that form around the $TiB_2$ particulates.

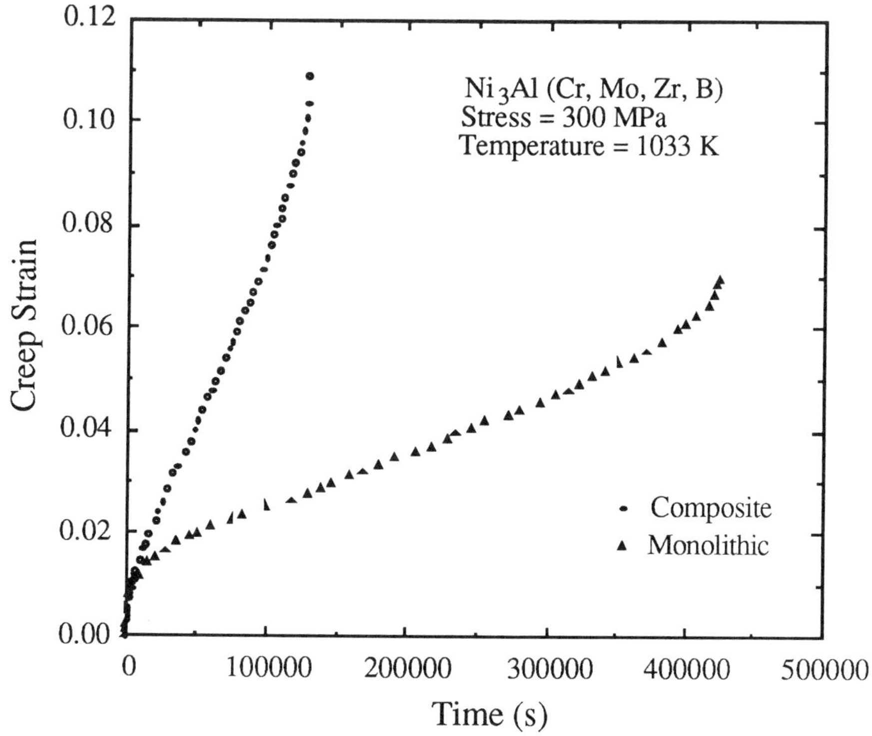


Figure 3 - Typical creep strain vs. time curves for the monolithic and composite materials.

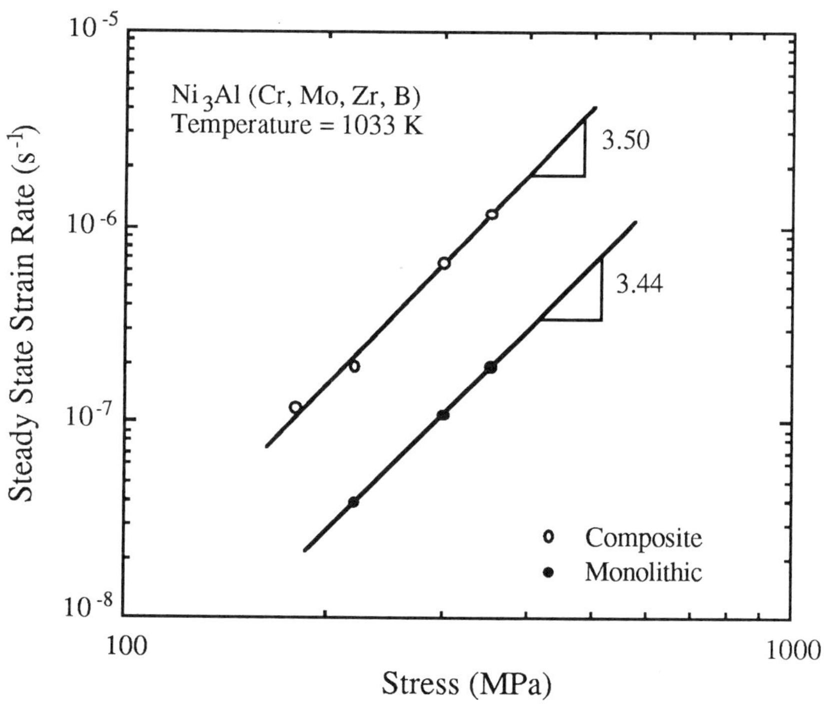


Figure 4 - Strain rate vs applied stress for the monolithic and composite materials.

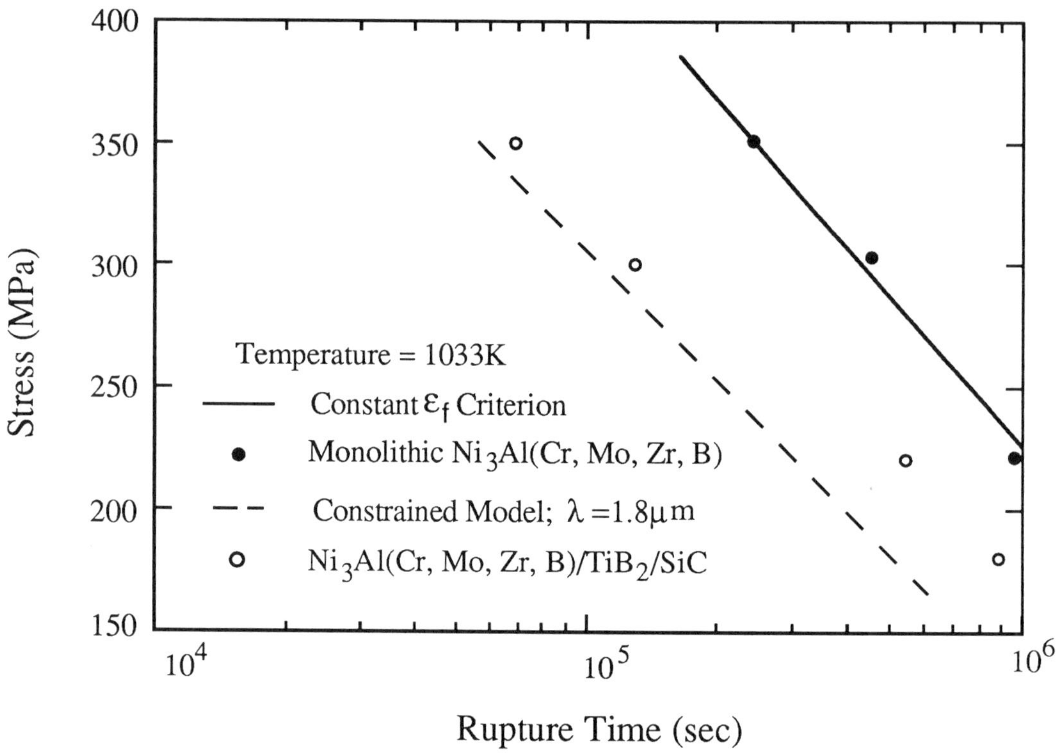


Figure 5 - Rupture time vs. applied stress for the monolithic and composite materials at 1033 K. The solid and dashed lines represent the values predicted by a constant creep rupture strain criterion and a constrained cavity growth model, respectively.

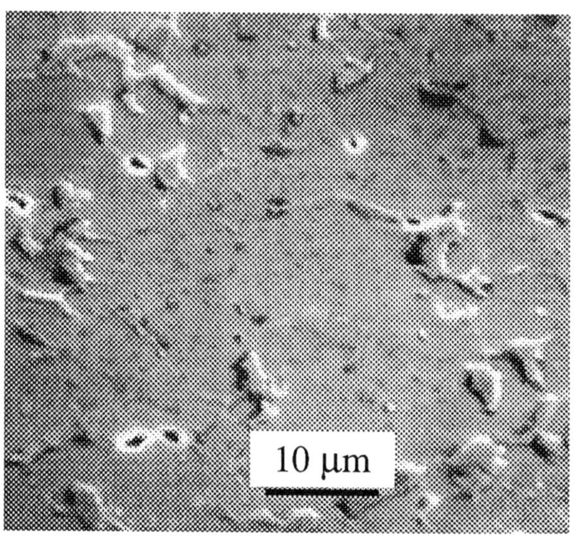


Figure 6 - SEM micrograph of a region in the grip section of a composite specimen tested under 300 MPa at 760°C. The tensile stress axis is in the vertical direction for this micrograph.

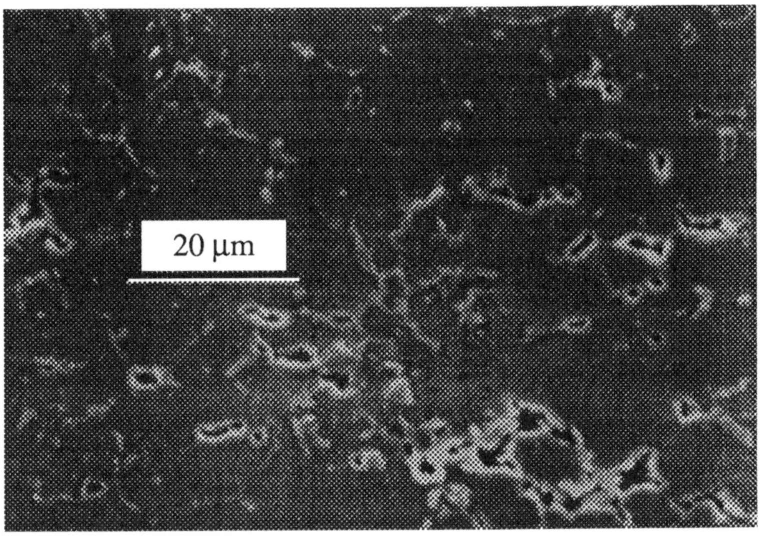


Figure 7 - SEM micrograph of a region in the gage section of a composite specimen tested under 300 MPa at 760°C. The tensile stress axis is in the vertical direction for this micrograph.

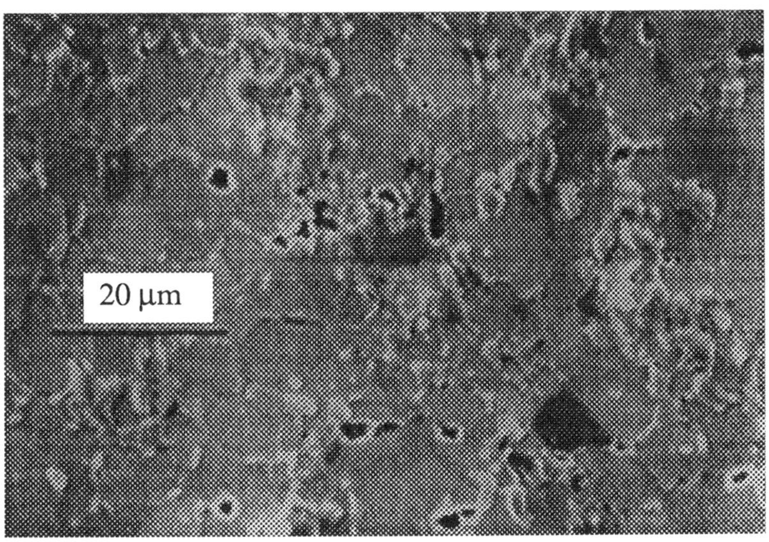


Figure 8 - SEM micrograph of a region in the gage section of a composite specimen tested under 180 MPa at 760°C. The tensile stress axis is in the vertical direction for this micrograph.

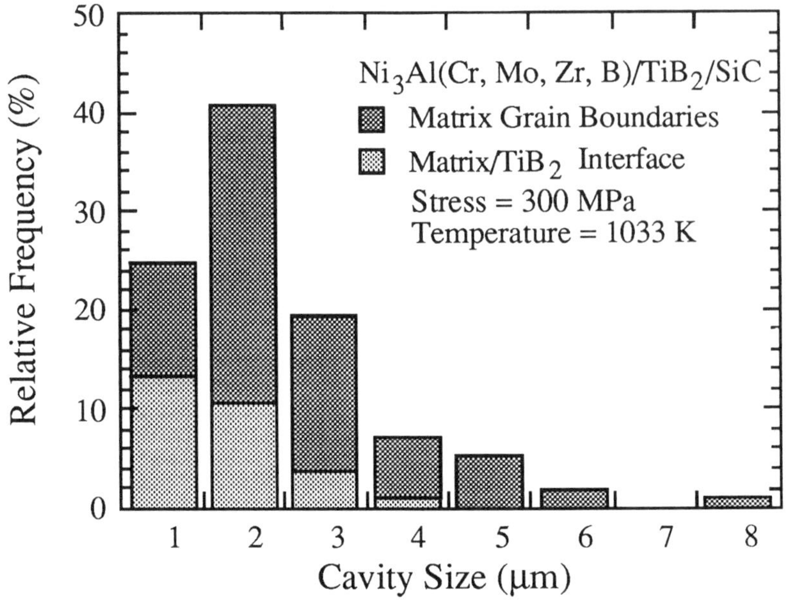


Figure 9 - Distribution of cavities in a composite specimen tested to rupture under 300 MPa.

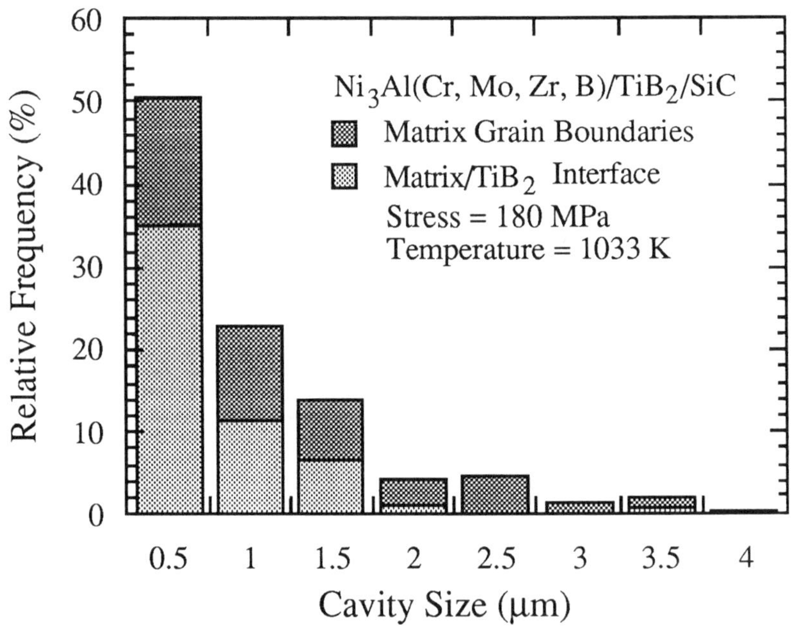


Figure 10 - Distribution of cavities in a composite specimen tested to rupture under 180 MPa.

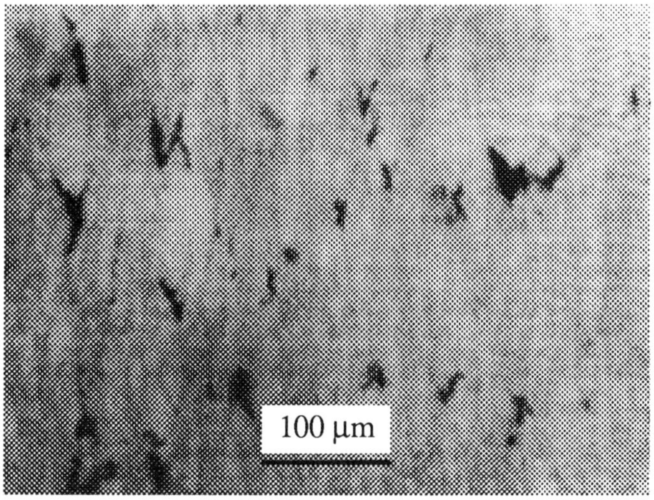


Figure 11 - The micrograph showing a polished cross-section of unreinforced $Ni_3Al$ specimen tested under 300 MPa at 760°C. The tensile stress axis is in the vertical direction for this micrograph.

*Proceedings of the Fifth International Conference on Creep of Materials, Lake Buena Vista, Florida, USA, 18-21 May 1992*

# A Model for Primary, Secondary and Tertiary Creep Rates

**M.N. Menon**
Garrett Auxiliary Power Division
Allied-Signal Aerospace Company
Phoenix, Arizona

## ABSTRACT

A comprehensive model has been developed for predicting creep rates and strains in primary, secondary and tertiary regimes. This method has been applied to modeling the creep behavior of cast, wrought and powder metallurgy (PM) nickel base superalloys, and in estimating stress relaxation and creep deformation in turbine engine components. This model can be used with finite element codes and requires only a nominal amount of data that are routinely generated for evaluating creep rupture of high temperature alloys. A uniqueness of this model is its incorporation of statistics for estimating all six parameters needed to completely describe a creep curve.

The materials modeled to date include CMSX-3 (single crystal), Mar-M 247 (directionally solidified and equiaxed), IN713LC, IN792, wrought Waspaloy and low carbon PM Astroloy.

EMPHASIS IN RECENT YEARS on development of accurate life prediction techniques has resulted in a need for models for calculation of time-dependent deformation in turbine engine components. This, in turn, has led to a need for models for predicting creep behavior of engine materials. This paper presents such a model developed for describing, as completely as would be necessary for the above goal, the strains and strain rates in creep of high temperature alloys.

## STAGES OF CREEP

Creep is the progressive deformation of a material under constant stress. A typical creep curve, depicting deformation under constant stress, is traditionally looked upon as having only three stages; primary, secondary and tertiary. However, in this comprehensive model, the traditional tertiary stage is split into a tertiary and a quaternary stage (as shown in Figure 1) to obtain a better fit. It seems appropriate to distinguish a fourth stage since it represents the final stage of damage accumulation in a creep test, and could be different mechanistically from the earlier stages of the traditional tertiary regime.

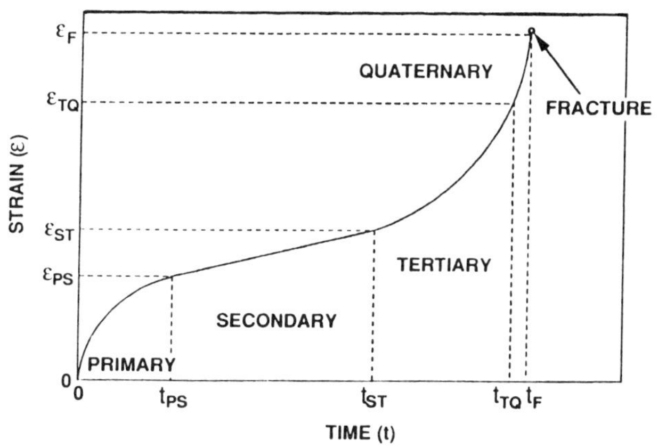


*Figure 1. Comprehensive Model Shows the Traditional Tertiary Stage Split Into a Tertiary and a Quaternary Stage.*

Briefly, the primary creep begins at loading and has a high initial creep rate. In this regime, creep rate gradually decreases with time until the secondary (steady state) regime is reached at the transition time, $t_{PS}$ and transition strain, $\varepsilon_{PS}$. The creep rate, $\dot{\varepsilon}_S$, in the second stage is the minimum reached in a creep test. The second stage continues until the creep rate starts to accelerate at time, $t_{ST}$ and strain, $\varepsilon_{ST}$, which marks the beginning of third or tertiary stage. Tertiary creep is presumed to occur until time, $t_{TQ}$ and strain, $\varepsilon_{TQ}$, are achieved, which marks the onset of the quaternary stage, the continuation of which leads to specimen fracture at time, $t_F$ and strain, $\varepsilon_F$.

## TWO-PART MODELING

Modeling is done in two parts. The first part involves modeling data from each specimen (test). The data consist of a series of monotonically increasing strain ($\varepsilon$) versus time (t) points at a specific load (stress) and temperature. This is accomplished by means of a program provided by Robert Geist[1].

This program fits a model with the aforementioned stages to the empirical creep data. The program parameters obtained for each specimen are then used in the second part for modeling four rate-constants and two transition times, which will completely describe the primary, secondary and tertiary regimes of the creep curve. The quaternary stage is not modeled in the second part of the program.

## PART I - MODELING OF SPECIMEN CREEP DATA

In modeling empirical creep data from individual specimens using the four-stage model, the primary, secondary, tertiary, and quaternary stages are fitted respectively with logarithmic, linear, exponential, and hyperbolic expressions, thus[1]:

Primary creep strain,

$$\varepsilon_P = B_L \text{Log}_e (A_L t + 1) \qquad o \le t \le t_{PS} \quad (1)$$

Secondary creep strain,

$$\varepsilon_S = \dot{\varepsilon}_S (t - t_{PS}) + \varepsilon_{PS} \qquad t_{PS} \le t \le t_{ST} \quad (2)$$

Tertiary creep strain,

$$\varepsilon_t = B_e \left\{ e^{A_E (t - t_{ST})} - 1 \right\} + \varepsilon_{ST} \quad t_{ST} \le t \le t_{TQ} \quad (3)$$

Quaternary creep strain,

$$\varepsilon_Q = \frac{B_H}{A_H - t} \qquad t_{TQ} \le t \le t_F \quad (4)$$

The fitting procedure developed by Geist[1] determines the constants $A_L$, $B_L$, $A_E$, $B_E$, $A_H$, $B_H$ and $\dot{\varepsilon}_S$, along with the transition points ($t_{PS}$, $\varepsilon_{PS}$), ($t_{ST}$, $\varepsilon_{ST}$) and ($t_{TQ}$, $\varepsilon_{TQ}$).

The input data file for this part of modeling consists of (for each specimen) individual specimen identification, the temperature and stress at which each specimen was tested, and strain-versus-time data. The output file from this portion of modeling has all of the aforementioned constants and transition times for each specimen, along with specimen identification, and the temperature and stress at which each specimen was tested. This output file is then used as input for the second part of modeling.

Creep data and the predicted line modeled from this first part for a Mar-M 247 specimen tested at 1144K, and an initial stress of 234 MPa, are shown in Figure 2. This figure also shows the creep rates calculated in the four regimes.

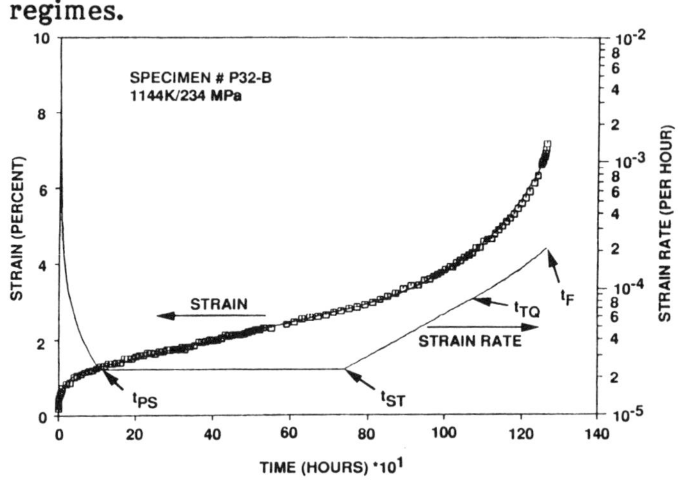


| | |
|---|---|
| $A_L = 1.4471$ | $t_{TQ} = 1037.2405$ |
| $B_L = 0.2562$ | $t_F = 1266.6000$ |
| $A_E = 0.0038$ | $A_H = 1592.7541$ |
| $B_E = 0.6121$ | $\varepsilon_{PS} = 1.2977$ |
| $\dot{\varepsilon}_S = 0.0023$ | $\varepsilon_{ST} = 2.7736$ |
| $t_{PS} = 108.8835$ | $\varepsilon_{TQ} = 4.0388$ |
| $t_{ST} = 740.1760$ | $\varepsilon_F = 7.1439$ |
| | $B_H = 2243.6272$ |

*Figure 2. Creep Data and The Predicted Line Modeled for a Mar-M 247 Specimen Tested at 1144K and 234 MPa.*

## PART II - MODELING OF RATE CONSTANTS AND TRANSITION TIMES

The product $A_L B_L$, and $\dot{\varepsilon}_S$, $A_L$, and $A_E$ are referred to as rate constants and characterize respectively the initial primary creep rate, secondary or steady-state creep rate, the rate at which primary creep rate approaches secondary creep rate, and the rate at which tertiary creep rate accelerates from secondary creep rate. This becomes evident by examining the equations for creep rates derived from equations 1 through 3.

Primary creep rate,

$$\dot{\varepsilon}_P = \frac{A_L B_L}{A_L t + 1} \quad (5)$$

Secondary creep rate,

$$\dot{\varepsilon}_S = \dot{\varepsilon}_S \quad (6)$$

Tertiary creep rate,

$$\dot{\varepsilon}_T = A_E B_E e^{A_E (t-t_{ST})} \quad (7)$$

Please note that at $t = t_{PS}$,

$$\dot{\varepsilon}_P = \dot{\varepsilon}_S = \frac{A_L B_L}{A_L t_{PS} + 1} \quad (8)$$

and at $t = t_{ST}$,

$$\dot{\varepsilon}_T = \dot{\varepsilon}_S = A_E B_E \quad (9)$$

Once $\dot{\varepsilon}_S$, $A_L$, and the product $A_L B_L$ are modeled, value of $t_{PS}$ can be obtained from equation 8 thus:

$$t_{PS} = \left[\frac{(A_L B_L)}{\dot{\varepsilon}_S} - 1\right]\left[\frac{1}{A_L}\right] \quad (10)$$

In order to completely describe a creep curve up to the end of the tertiary regime, models for the aforementioned four rate constants, along with those for the two transition times; times at the secondary to tertiary (S/T) and at the tertiary to quaternary (T/Q) transitions, are needed. The transition time from primary to secondary (P/S) is automatically provided by Equation 10, once rate constants are modeled.

**Modeling of rate constants.** $\dot{\varepsilon}_S$, $A_L B_L$, $A_L$ and $A_E$ are modeled based on an equation containing activation energy and a stress function as follows:

$$\dot{\varepsilon}_S,\ A_L B_L,\ A_L \text{ or } A_E = A_o\, e^{\frac{-Q}{RT}} \left[F\left(A_1 \cdot \frac{S}{E}\right)\right]^n \quad (11)$$

where

$A_o$ = a constant
Q = Activation energy (KJ/mole)
R = Universal gas constant (KJ/mol/deg K)
T = Temperature (deg K)
F = a function of modulus normalized stress
S = Stress in MPa
E = Young's modulus in MPa
$A_1$ = A constant
n = Stress exponent

Although the present model uses a Sinh function to account for stress dependence, the model also can incorporate other functions. Equation 11 can be rewritten as follows:

$$\text{eg: } \mathrm{Log}_e(\dot{\varepsilon}_S) = a_o + \frac{a_1}{T} + a_2\, \mathrm{Log}_e\left[\mathrm{Sinh}\left(A_1 \cdot \frac{S}{E}\right)\right] \quad (12)$$

where the correspondence with parameters in equation 11 are given by the following: $a_o = \mathrm{Log}_e(A_o)$, $a_1 = -Q/R$, and $a_2 = n$, and $A_1$ is the same constant referred to earlier. Multiple linear regression is used to obtain the best estimates for the parameters $a_o$, $a_1$, and $a_2$ · $A_1$ is arrived at by optimizing the standard deviation using an iterative procedure.

**Modeling of transition times.** Transition times are modeled by means of equations based on the Monkman-Grant expression[2], thus:

$$\mathrm{Log}_e(t_{ST}, \text{ or } t_{TQ}) = b_o + b_1\, \mathrm{Log}_e(\dot{\varepsilon}_S) \quad (13)$$

Linear regression is used to obtain the best estimates for the parameters $b_o$ and $b_1$.

## MODEL FOR CAST EQUIAXED Mar-M 247

**Test specimens and conditions.** Miniround creep specimens, 1.8 mm in diameter, were machined from cast turbine airfoils of Mar-M 247 [in the equiaxed (EA) condition] and were tested at eight temperatures ranging from 1033 to 1311K, and at stresses between 103 and 690 MPa. Data from 94 specimens were used to develop the model described below.

**Model.** Models for $\dot{\varepsilon}_S$, $A_L B_L$, $A_L$, and $A_E$ for EA Mar-M 247 are shown in Figures 3, 4, 5 and 6 respectively. The data encompass a temperature range of 1033K (1400F) to 1311K (1900F), and a stress range of 103 MPa (15 KSI) to 689 MPa (100 KSI). The ordinates in these figures represent the natural log of temperature-modified rate constants, arrived at by dividing the actual value of the rate

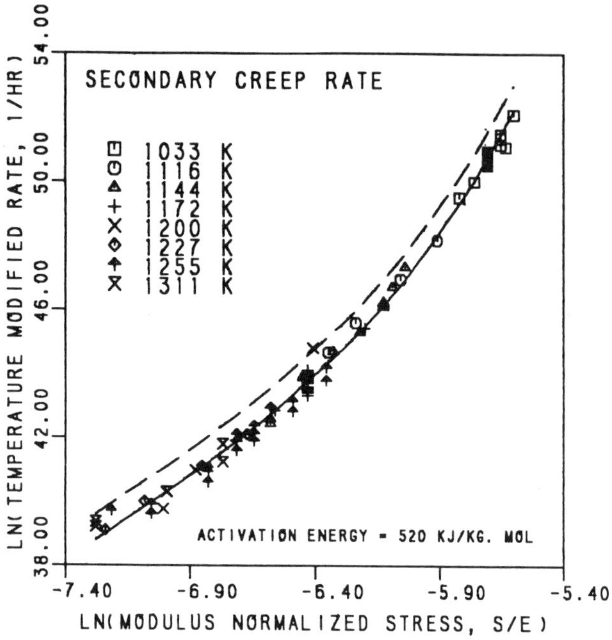


*Figure 3. Model for Secondary Creep Rate, $\dot{\varepsilon}_S$, for EA Mar-M 247.*

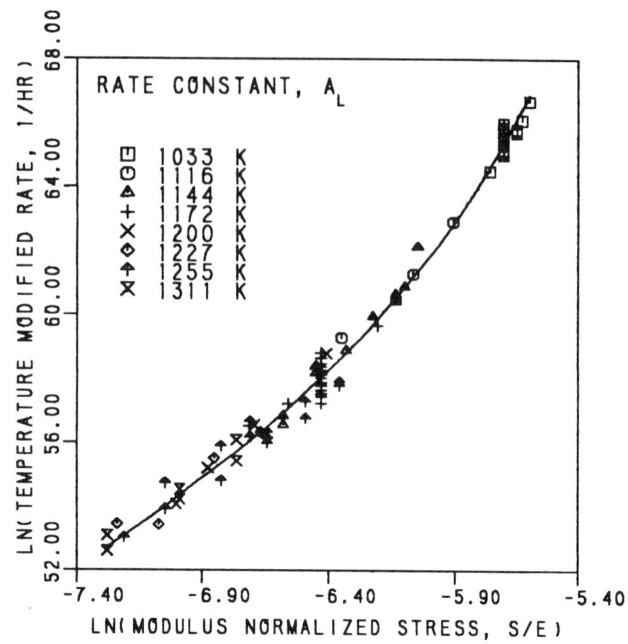


*Figure 5. Model for Rate Constant, $A_L$, for EA Mar-M 247.*

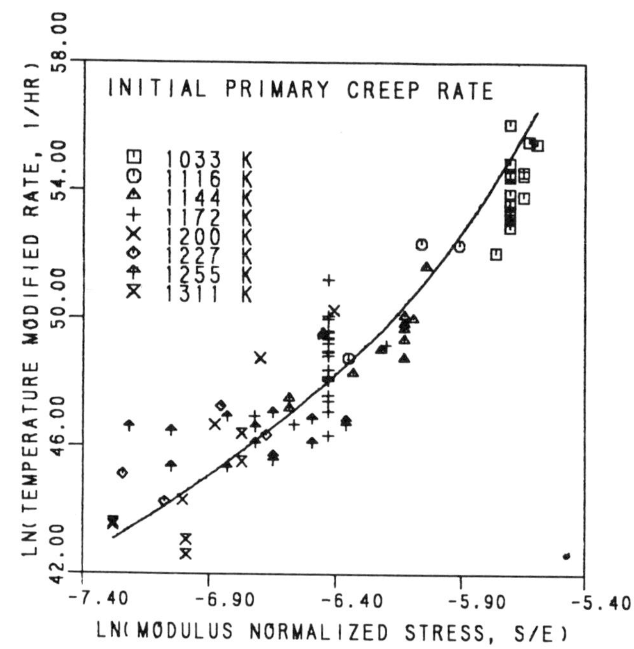


*Figure 4. Model for Initial Primary Creep Rate, $A_L B_L$, for EA Mar-M 247.*

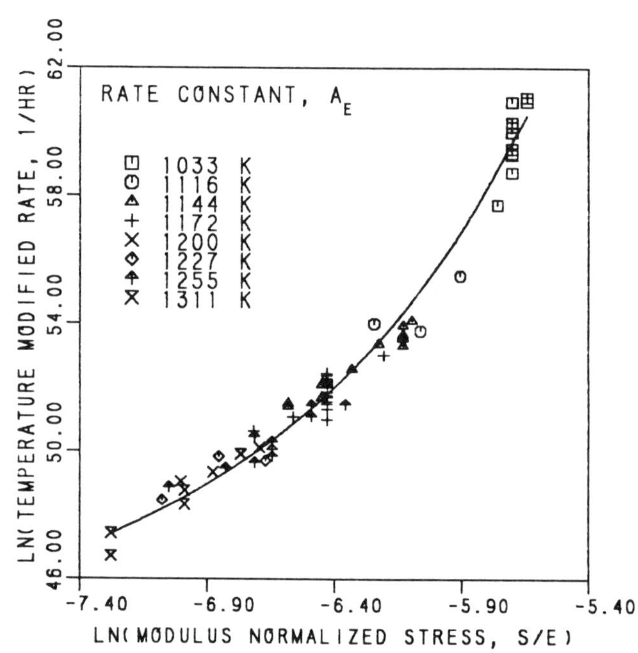


*Figure 6. Model for Rate Constant, $A_E$, for EA Mar-M 247.*

***TABLE I. PARAMETERS FOR THE FOUR RATE CONSTANTS FOR EA MAR-M 247***

| Rate Constant (1/hr.) | $A_o$ | -Q/R | $A_1$ | n |
|---|---|---|---|---|
| $\dot{\varepsilon}_S$ | $6.459 \times 10^{17}$ | 62,481 | 845 | 4.582 |
| $A_L B_L$ | $4.606 \times 10^{19}$ | 62,481 | 845 | 4.582 |
| $A_L$ | $1.105 \times 10^{24}$ | 66,751 | 810 | 4.982 |
| $A_E$ | $2.968 \times 10^{20}$ | 67,978 | 1900 | 2.446 |

constant by exp(-Q/RT). Table 1 provides the parameters $A_o$, -Q/R, $A_1$ and n for the four rate constants for EA Mar-M 247.

The average and maximum rate curves (dashed line - based on three times the standard deviation) are shown for the $\dot{\varepsilon}_S$ model only. Figures 7 and 8 show the models for transition times, $t_{ST}$ and $t_{TQ}$. The

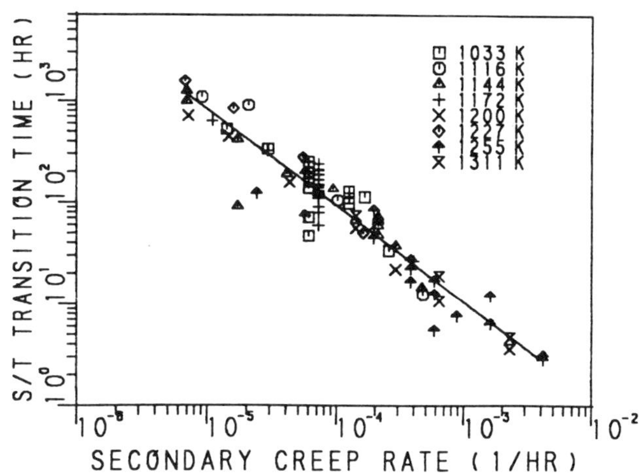


*Figure 7. Model for Secondary to Tertiary Transition Time for EA Mar-M 247.*

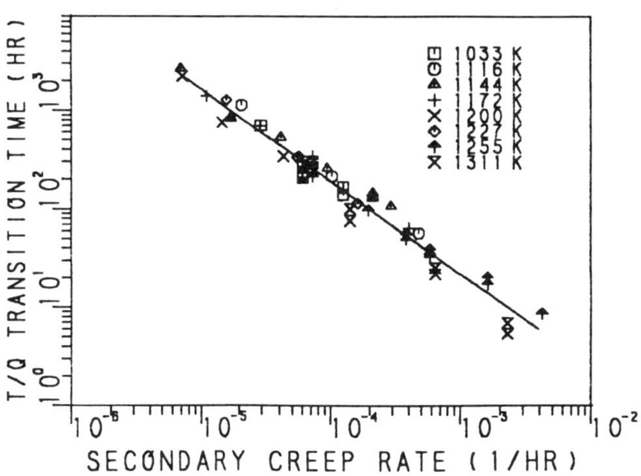


*Figure 8. Model for Tertiary to Quaternary Transition Time for EA Mar-M 247.*

respective values for $b_0$ and $b_1$ are -4.200 and -0.949 for $t_{ST}$ and -3.407 and -0.939 for $t_{TQ}$.

## PREDICTIONS

This comprehensive model may be examined on the basis of its ability to predict creep rates in the three regimes, and the three transition times. Figures 3, 7, and 8 show good agreement between the data and the average line for prediction of secondary creep rate, $\dot{\varepsilon}_S$, and the transition times, $t_{ST}$ and $t_{TQ}$. It is worth noting that the data represent 94 specimens from different components/heats, tested at an extensive range of temperatures and stresses.

Figures 9, 10, and 11 show the ratio of predicted-to-actual creep rates for each specimen in the primary, secondary and tertiary creep regimes respectively. In figures for both the primary and tertiary (Figures 9 and 11) regimes, each specimen is represented by a vertical line with symbols, and each line (specimen) is staggered on the abscissa for clarity. In the case of primary creep (Figure 9), the five symbols represent ratios of predicted-to-actual creep rates calculated at five

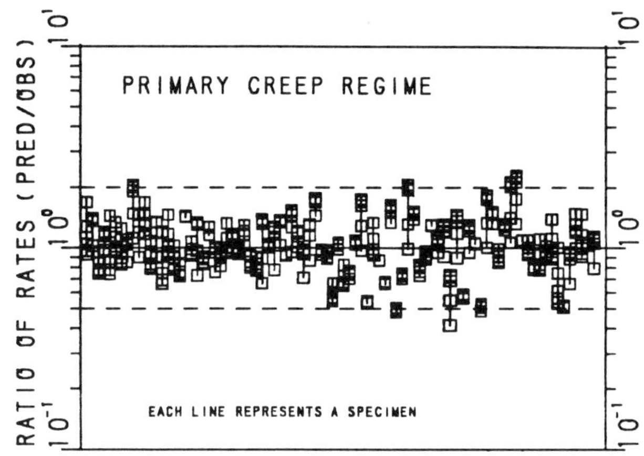


*Figure 9. Model Predicts Creep Rates in the Primary Creep Regime Within a Factor of Two.*

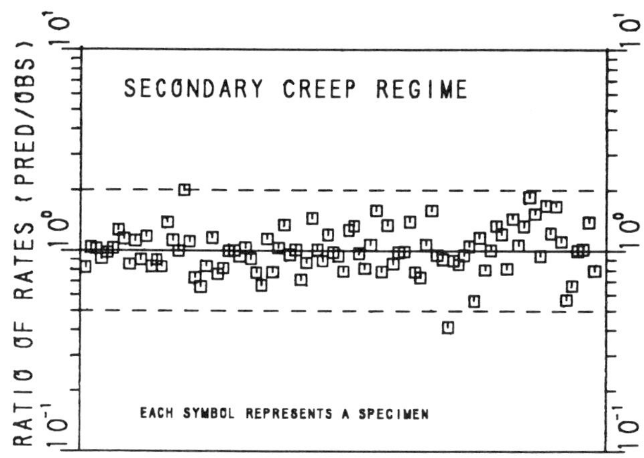


*Figure 10. Model Predicts Creep Rates in the Secondary Creep Regime Within a Factor of Two.*

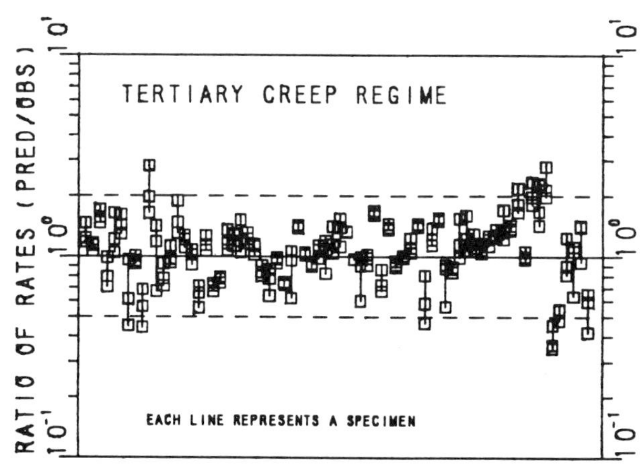


*Figure 11. Model Predicts Creep Rates in the Tertiary Creep Regime Within a Factor of Two.*

locations in the regime, (0.125, 0.25, 0.5, 0.75 and 1.0) of the predicted $t_{PS}$ value. The three symbols used in the case of tertiary creep (see Figure 11) represent ratios of predicted-to-actual rates calculated at locations 0.125, 0.25 and 0.5 of the tertiary regime. It can be observed that predictability of creep rates in the primary creep regime is within a factor of two (dashed lines). This also holds true in tertiary creep for most of the specimens. In secondary creep (Figure 10), predictability is well within a factor of two.

In addition, a comparison can be made between the predicted value (predicted by the final model through Equation 10), of transition time, $t_{PS}$, and its actual value obtained from the first part of the modeling in which data from each specimen were fitted with the four-stage model. Figure 12 shows this plot. Here again the model predictability is within a factor of two (dashed lines).

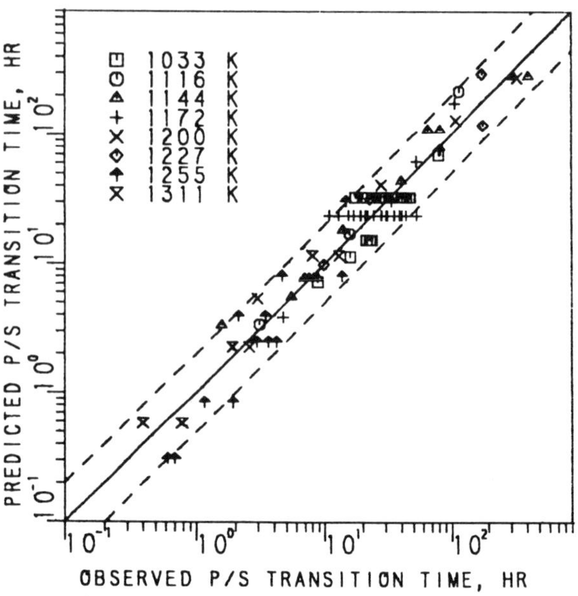


*Figure 12. Model Predicts Primary-to-Secondary Transition Within a Factor of Two.*

A comparison of strain-versus-time data for 17 specimens of EA Mar-M 247 tested at 1172K and 276 MPa, and the model prediction is shown in Figure 13. The curve showing model predictions is drawn only to the end of the tertiary stage as defined above. The model seems to adequately define the mean trend.

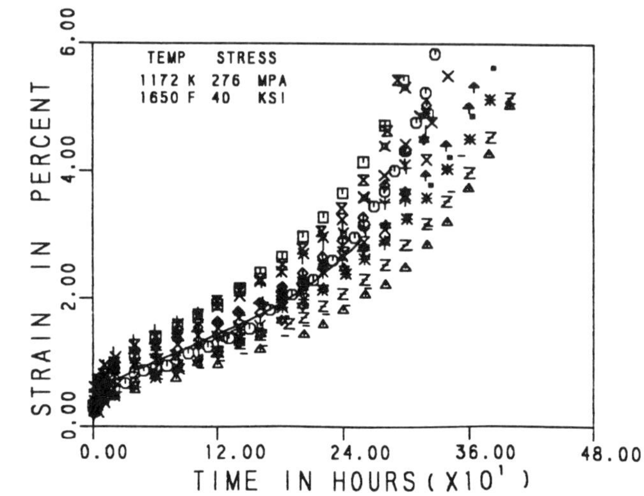


*Figure 13. Comparison of Mean Behavior Predicted by the Model With Specimen Data.*

## OTHER MATERIALS

Other materials that have been modeled include single-crystal CMSX-3 in the (100) orientation, Mar-M 247 in the directionally solidified form, cast IN713LC, cast IN792, wrought Waspaloy and low carbon PM Astroloy.

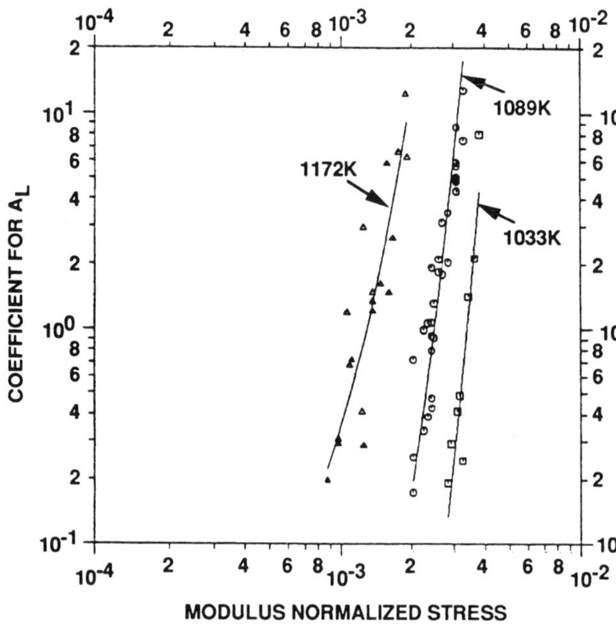


*Figure 14. Model for Rate Constant, $A_L$, for IN792.*

Figures 14 and 15 show the rate constant $A_L$ developed for IN792 and the rate constant $A_E$ developed for LC PM Astroloy.

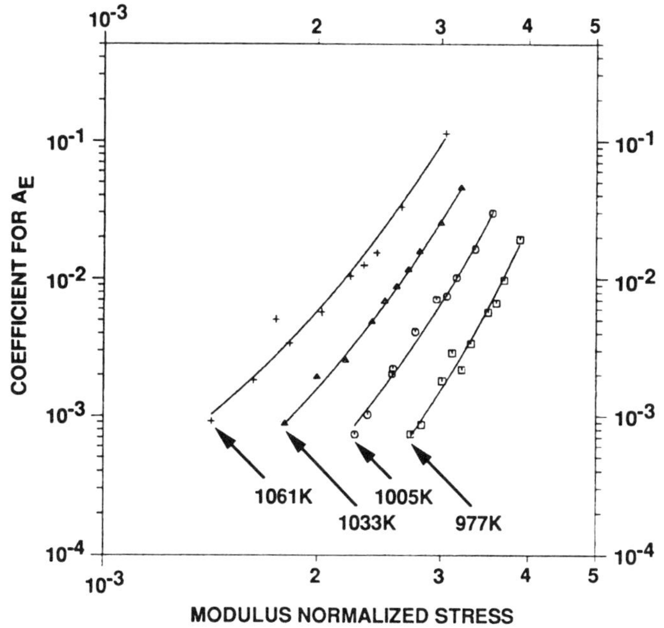


*Figure 15. Model for Rate Constant, $A_E$, for LC PM Astroloy.*

## APPLICATIONS

The creep rate model can predict the mean creep behavior of the material of interest. It has been successfully used in predicting the untwist and growth of turbine blades in an engine test for one of the alloys. It also has provided insight into the stress redistribution phenomena that occurs near bond joints in dual alloy combinations, and is currently being used in estimating stress relaxation and redistribution at stress concentrations in turbine components.

## ACKNOWLEDGMENT

The author is grateful to Professor Robert Geist of Clemson University for providing the computer program used in the first part for modeling specimen creep data.

## REFERENCES

1. Geist, R., Private Communication, Department of Computer Science, Clemson University, Clemson, South Carolina, (1986-1987).

2. Monkman, F. C., and N. J. Grant, Proc. ASTM, 56, 593-620, (1956).

Proceedings of the Fifth International Conference on Creep of Materials, Lake Buena Vista, Florida, USA, 18-21 May 1992

# An Assessment of a Ductility Exhaustion Model for Predicting Mixed-Mode Creep Crack Growth Rates and Directions

**T.H. Hyde**
The University of Nottingham
Nottingham, England

**A.C. Chambers**
Rolls-Royce
Derby, England

## Abstract

Under fatigue conditions, cracks tend to grow in directions which result in pure mode-I crack tip conditions. However, under creep conditions cracks subjected to initially mode-I conditions may bifurcate to produce mixed-mode crack tip stress states. In such materials, combined creep-fatigue crack growth is complex and in order to predict the resultant crack growth rate and direction, an understanding of the mixed-mode behaviour of cracks under creep and fatigue separately is required before the interaction effects can be understood.

In this paper, the C* approach used for correlating creep crack growth and which has been shown to correlated pure mode-I creep crack growth data is modified and applied to mixed-mode situations. The modification to the C* parameter is based on steady-state, mixed-mode crack tip stress fields, a multi-axial stress ductility exhaustion model and the inclusion of appropriate process zone sizes.

The accuracy of the modified C* approach is assessed by comparing predictions with experimental data obtained from pure-mode-I, pure mode-II and mixed mode ($K_I/K_{II} \approx 1.6$) tests carried out on Jethete M152, an aeroengine casing material, at 550° C. It is shown that with the appropriate multi-axial ductility exhaustion model, reasonably accurate predictions for the directions of creep crack growth are possible. By allowing for the differences in process zone sizes and for the thinning which occurs in the plane-stress, mode-I tests, the mode-I, mode-II and mixed-mode creep crack growth rates can be correlated using the C* parameter.

The small process zone associated with the mode-II tests, the elimination of specimen thinning, which can result in artificially high crack growth rates in mode-I tests and the additional information which can be obtained from the observed direction of crack growth in mode-II tests, suggest that mode-II creep crack growth testing may have significant advantages over mode-I testing.

THE COMPACT-MIXED-MODE (CMM) SPECIMEN, shown in Fig. 1, is capable of producing any combination of mode-I and mode- II crack tip conditions (1), including pure mode-I and pure mode-II conditions. However, for pure mode-I conditions, it would be easier to use a standard compact-tension (CT) specimen. Under elastic conditions, the mode-I and mode-II stress intensity factors are directly proportional to the load per unit thickness, P/B, inversely proportional to $W^{0.5}$, and are functions of loading angle, $\alpha$, and crack length ratio, a/W. The elastic calibration of the specimen and the method of producing initially straight fatigue pre-cracks have been described in detail (1).

The specimen was designed so that it will fit into standard furnaces, commonly used in elevated temperature static load creep testing machines, and in servo-hydraulic or servo-electronic machines which can be used for elevated temperature fatigue or combined creep and fatigue testing. Also, it has been shown (2, 3) that the DC potential drop technique can be used to monitor crack growth, provided the crack surfaces do not come into contact during the test: contact could occur, for example, when cracks are growing under pure mode-II conditions Self-calibration of each test is recommended to minimise experimental scatter.

Mixed-mode, pure mode-I and pure mode-II tests have been performed on Jethete M152 (typical composition 0.12% C, 0.8% Mn, 11.5% Cr, 2.5% Ni, 1.75% Mo, 0.3% V, Fe making up the balanced) at 550° C, under fatigue (2), creep (3) and combined creep and fatigue conditions (4). Under fatigue conditions, the cracks bend to a direction which results in pure mode-I crack tip conditions, after which, the crack growth rate per cycle is related to the stress intensity factor range. During the transition from mixed-mode to mode-I conditions the fatigue cracks appear to follow a path which is controlled by the maximum plastic tangential strain at the crack tip. Also, a suitably modified stress intensity factor range predicts the enhanced fatigue crack growth rates observed during these transitional stages of fatigue crack growth.

Mode-I creep crack growth rates have been shown to be related to the C* parameter for a wide range of materials (eg. 5-10) and test conditions. Further, C* variations throughout a test can be easily related to the load point displacement rates (5).

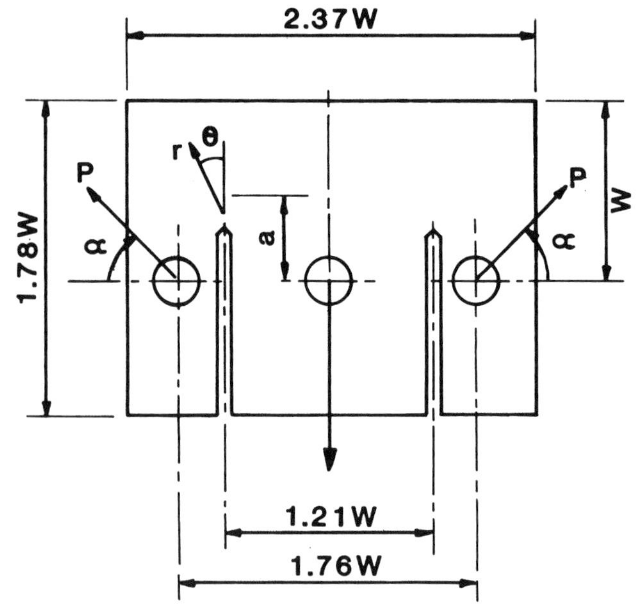


Fig. 1 - A compact-mixed-mode (CMM) specimen (Reported tests: W = 33 mm and thickness = 3.18 mm)

Hence obtaining experimental data which relates creep crack growth rates to C* is relatively simple. By using a similar approach to determine C* values from load-point displacement rates in pure mode-II and mixed-mode conditions, it was found (3) that for the same C* value, the pure mode-I crack growth rates were significantly higher (between 1.5 x and 6 x) than the pure mode-II and mixed-mode results for the Jethete at 550° C. Also, the direction of creep crack growth under mixed-mode conditions cannot be directly predicted using the C* approach. It was found that under pure mode-II conditions the creep cracks grew in a self-similar manner, ie. they continued to grow under pure mode-II conditions. Under mixed-mode conditions, with $K_I/K_{II} \approx 1.6$, the cracks bifurcated, one crack being at approximately 90° to the original crack and the other continuing in the original direction. The 90° crack was found to be the most dominant.

Pure mode-I and pure mode-II creep crack growth tests have also been carried out for 316 stainless steel at 600° C (11). In these tests, mode-I creep crack growth was found to continue in the same direction as the initial crack. However, the creep crack growth from initially mode-II conditions occurred at about 90° to the initial crack.

Under combined creep and fatigue conditions, in Jethete M152 at 550° C, the crack growth rates are generally greater than would be predicted by either linear or vector (taking into account the different directions obtained for pure fatigue or creep crack growth) summation of the corresponding fatigue and creep data (4). Also, the directions of crack growth were found to fall between the directions obtained under pure fatigue and pure creep conditions (4). The creep behaviour seems to have a more significant effect on the rate and direction of crack growth, under mixed-mode and pure mode-II combined creep and fatigue conditions, than the fatigue loading has.

From the above, it is evident that an understanding of the processes which control the rate and direction of creep crack growth, under pure mode-II and mixed-mode conditions, is an essential step towards the understanding of mixed-mode creep and mixed-mode combined creep and fatigue conditions. In this paper, a ductility exhaustion mode, which was used by Nikbin et al (12) to explain the differences in creep crack growth rates observed in plane strain and plane stress conditions, is extended to predict the directions of creep crack growth rates observed under pure mode-II and mixed-mode conditions. The model, which is based on an analysis of the ductile enlargement of voids (13), is also used to explain the differences in creep crack growth rates observed between pure mode-II, mixed-mode and pure mode-I tests having the same C* value. In addition, the differences in process zone size (12), necessary to explain the differences in the creep crack growth rates is determined.

## Experimental Creep and Creep Crack Growth Results for Jethete M152 at 550° C

Uniaxial creep and creep rupture tests were carried out, at 550° C, with stresses in the range 175 MPa to 400 MPa. After relatively short primary creep periods, typically less than 10% of the failure time, steady-state creep, for up to 50% of life, occurred, followed by tertiary creep until final failure. Primary creep strain, $\epsilon_{pc}$, and steady-state creep strain rates, $\dot{\epsilon}_{ss}$, are represented reasonably well by

$$e_{pc} = 6.9 \times 10^{-16} \sigma^{5.2} t^{0.37} \quad (1)$$

and

$$\dot{e}_{ss} = 2.93 \times 10^{-27} \sigma^{9.5} \quad (2)$$

where the stress, $\sigma$, and time, t, have units of MPa and h, respectively.

At the higher stress levels, when failure occurred in less than about 10 h, failure strains of about 20% were recorded. However, for lower stress levels, resulting in failure times between about 50 h and 1200 h, failure strains of about 40% were obtained.

A generalisation of the Harper and Ellison (14) method of determining C* for mode-I situations has been derived (15). This allows C* to be determined from the load line displacement rate, $\dot{\Delta}$, using the formula

$$C^* = \frac{n}{n+1} \cdot \frac{P\dot{\Delta}}{BW} \cdot \frac{1}{\sigma_R} \cdot \frac{d\sigma_R}{d(a/W)} \quad (3)$$

when n is the exponent of stress in the creep law (Eqs. 1 and 2) and $\sigma_R$ is the reference stress associated with the load-point displacement.

The use of approximate reference stresses, based on limit loads, has a negligible effect on the accuracy of C* values determined (3). With the usual notation for a compact tension specimen, Eq. 3 reduces to

$$C^* = \frac{n}{n+1} \cdot \frac{P\dot{\Delta}}{BW} \cdot \left[\frac{1}{(1+a/W)} + \frac{2}{(1-a/W)}\right] \quad (4)$$

For a CMM specimen, with a/W = 0.5, under pure mode-II conditions, ie. $\alpha$ = 106° in Fig. 1, C* is given by

$$C* = \frac{n}{n+1} \cdot \frac{P\dot{\Delta}}{BW} \cdot \frac{1}{(1-a/W)} \quad (5)$$

and under the mixed-mode conditions which would result is $K_I/K_{II} \approx 1.6$, ie. $\alpha$ = 90° and a/W = 0.5 in Fig. 1, C* is given by

$$C^* = \frac{n}{n+1} \cdot \frac{P\dot{\Delta}}{BW}\left[\frac{2}{(1-a/W)} - \frac{(1-a/W)}{(2a/W)^2+(1+a/W)^2}\right] \quad (6)$$

Taking n = 5.2 (see Eq. 1), Eqs. 4, 5 and 6 were used to obtain C* values from compact-tension and compact-mixed-mode test data; the results are shown in Fig. 2. Three mixed-mode ($K_I/K_{II} \approx 1.6$) and three pure mode-II tests, each having two cracks, were performed. The average crack growth rates and average C* values are shown in Fig. 2. It can be seen from Fig. 2 that the mode-II and mixed-mode results are significantly lower than the mode-I results. For the same C* value, the mode-I creep crack growth rates are about 4 times greater than the corresponding mode-II or mixed-mode creep crack growth rates.

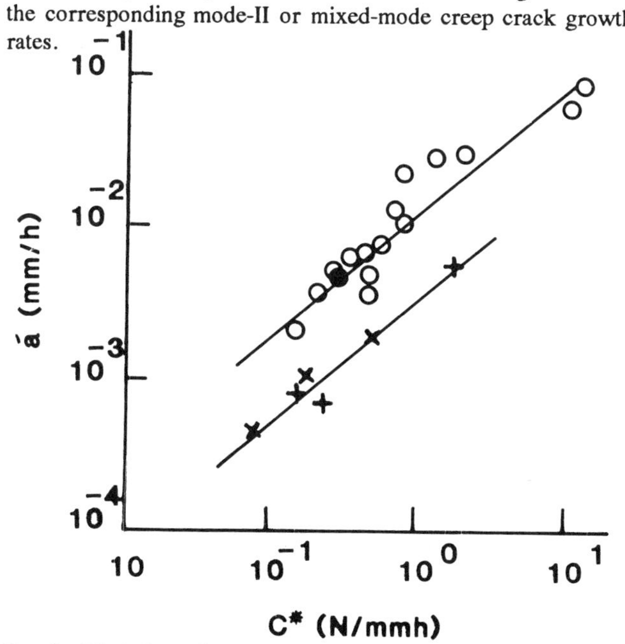


Fig. 2 - Variation of creep crack growth rate, a, with C* (0 = mode-I, + = mode-II, x = mixed-mode)

## Use of a Ductility Exhaustion Model to Predict the Directions of Crack Growth

The ductility corresponding to a multiaxial stress state, $\epsilon^*_{fo}$, has been related to the uniaxial stress state ductility, $\epsilon_{fo}$, using a model based on void enlargement (12, 13); the resulting expression is

$$\epsilon^*_{fo} = \frac{1.64\,\epsilon_{fo}}{\exp(3\sigma_m/2\sigma_e)} \quad (7)$$

where $\sigma_m$ and $\sigma_e$ are the hydrostatic and effective stresses, respectively.

The stress state and hence the limiting ductility in the vicinity of a crack tip varies with angular position. The variations of stresses, with angular positions, for various mixed-mode conditions have been obtained (eg. 16, 17). It is assumed that the effective creep strain is related to the effective stress according to Eq. 1. Therefore, the effective creep strain and the multiaxial ductility, $\epsilon^*_{fo}$, both depend on angular position at the crack tip. It is postulated that the direction of crack growth will occur in the direction at which the ductility is first exhausted, ie. in the direction for which $\sigma_e^{5.2} \exp(3\sigma_m/2\sigma_e)$ is the greatest. The variation of this quantity with angular position for the pure mode-I, the pure mode-II and the mixed-mode test conditions are shown in Fig. 3. From Fig. 3, it can be seen that the peak value for pure mode-I is at $\theta = 0$, ie. self similar crack growth. In the centre of the mode-I specimens, the creep cracks grew in the predicted direction. Under pure mode-II conditions, the peak value of the quantity, $\sigma_e^n \exp(3\sigma_m/2\sigma_e)$ occurs at $\theta = 10°$. In the experiments the cracks grew in a practically self-similar manner, ie. at $\theta \approx 0°$. Therefore, there is good agreement between the experiments and predictions for pure mode-II. For the mixed mode case, the peak value occurs at $\theta \approx -60°$. However, in the corresponding experiments, the dominant crack grew at $\theta \approx -90°$. Although the agreement in this case is not as good as that for the mode-I and mode-II cases, considering the variability in the crack growth directions for this loading condition, the agreement is encouraging.

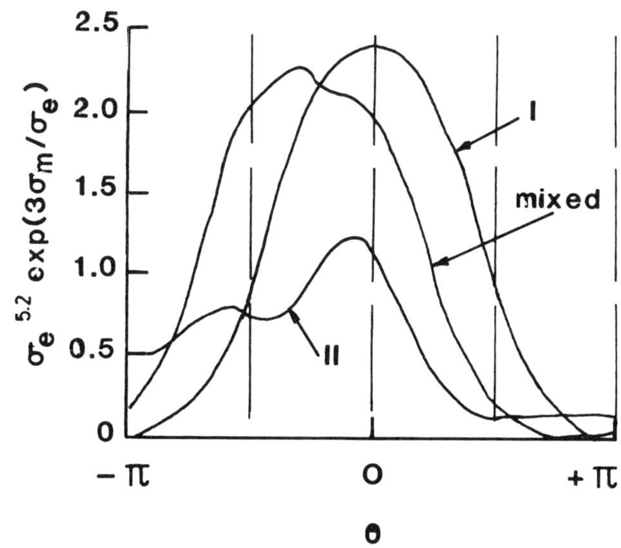


Fig. 3 - Variation of $\sigma_e^n \exp(3\sigma_m/\sigma_e)$ with $\theta$

## Use of the Ductility Exhaustion Model and Process Zone Size to Predict the Rates of Crack Growth

Substituting the expression for multiaxial ductility (Eq. 7) into the ductility exhaustion model used by Nikbin et al (12), which is used to related creep crack growth rates to C*, results in the expression

$$\dot{a} = \frac{n+1}{\epsilon_{f0}} \exp(3\sigma_m/2\sigma_e)\, r_c^{1/n+1}\, \dot{\epsilon}_o \left[\frac{C*}{I_n\, \sigma_o\, \dot{\epsilon}_o}\right]^{n/n+1} \quad (8)$$

where $r_c$ is the 'radius' of the so-called process zone, $\dot{\epsilon}_o$ and $\sigma_o$ are constants in the power law creep formulation, ie. $\dot{\epsilon}/\dot{\epsilon}_o = (\sigma/\sigma_o)^n$, and the $I_n$ values, which are numerically determined quantities that depend on the stress exponent (18) and the degree of mode mixity (16), are available in tabular form.

Therefore, a prediction of the ratios of the crack growth rates under mixed-mode and mode-I conditions can be obtained using Eq. 8, ie.

$$\frac{\dot{a}_{mm}}{\dot{a}_I} = \frac{(\exp((3\sigma_m/\sigma_e))_{mm}}{(\exp(3\sigma_m/\sigma_e))_I} \cdot \left[\frac{(r_c)_{mm}}{(r_c)_I}\right]^{1/n+1} \cdot \left[\frac{(I_n)_{mm}}{(I_n)_I}\right]^{n/n+1} \quad (9)$$

where the suffices mm and I refer to mixed-mode and mode-I conditions, respectively. Initially, the differences is the sizes of the process zones, $r_c$, for cases with the same C* value, but having mode-I, mode-II or mixed mode conditions will be neglected. On substitution of appropriate $I_n$ values and exp $(3\sigma_m/\sigma_e)$ quantities, under equal $r_c$ assumptions, Eq. 9 predicts that the mode-II and mixed-mode creep crack growth rates should both be less than 25% different from the corresponding mode-I values. Indeed, the predicted mode-II and mixed-mode crack growth rates are slightly greater than, rather than about 4 times lower than, the corresponding mode-I crack growth rates.

The ratio of the process zone sizes between the mode-I case and the mode-II case (or the mixed-mode case) would need to be between 15 and $10^5$ in order to explain the differences in creep crack growth rates (ie. between about 1.5 x and 6 x). Observations of the surface deformations for the pure mode-II tests (3) shows that very intensive shear zones are restricted to an extremely narrow strip of material which is practically co-linear with the initial crack. The width of this intense shear zone is estimated to be of order 0.1 mm. Under pure mode-I conditions, much larger, almost circular zones, of intense creep deformation were observed in compact tension specimens which were unloaded before large displacement occurred. The diameter of these large strain zones is estimated to be in the range 1 to 3 mm. These differences in process zone sizes could account for mode-I crack growth rates being between 1.5 and 2 times the mode-II rate. Similar, but small differences between the mode-I and mixed-mode creep crack growth rates could be attributed to differences in process zone sizes. Therefore, in order to account for the differences in the creep crack growth rates of the mode-I and mode-II (or mixed-mode) test, another factor of about 2 is required.

For the thin, practically plane stress, specimens (CT and CMM) used in the test (see Fig. 1), the in-plane stretching of material in the vicinity of the crack tips, under mode-I conditions, is accompanied by through-thickness thinning. Under pure mode-II conditions, the in-plane deformations are dominantly shear. Therefore, the in-plane deformations are not accompanied by through-thickness deformations. In the compact-tension (mode-I) tests, it is estimated that material thinning of up to 15%, in the process zones near the crack tips, occurred. The increase in stress due to this reduced thickness could result in a 2 to 2.5 times increase in creep strain rates, for $n \approx 5.2$. This phenomenon does not occur in the pure mode-II creep crack growth tests. The effect of this difference in behaviour, together with the effect of the differences in process zone sizes can completely account for the differences in creep crack growth rates in the pure mode-I and pure mode-II tests. In the mixed-mode ($K_I/K_{II} \approx 1.6$) tests, specimen thinning was found to be negligible. Therefore, similar differences between the mode-I and mixed-mode results, to those predicted between the mode-I and mode-II results, could be expected on the basis of specimen thinning.

## Conclusions

Apparent differences between creep crack growth data obtained from compact-tension (mode-I) tests and compact-mixed-mode (mode-II and mixed-mode) tests have been explained. By taking into account differences in process zone sizes and allowing for increased stresses due to specimen thinning under mode-I conditions, crack growth data from all tests can be correlated using the C* parameter.

The directions of creep crack growth were reasonably well predicted by using a ductility exhaustion model.

Mode-II testing appears to have significant advantages over mode-I testing. The thinning which occurs in plane stress, mode-I testing can result in artificially high creep crack growth rates. However, no thinning occurs in mode-II tests. Also, the process zone width in mode-II testing is significantly smaller than that in mode-I testing and additional information about stress-state dependence of creep crack growth can be obtained from the observed directions of crack growth obtained in mode-II testing.

## References


1 - Hyde, T.H. and A.C. Chambers, J. Strain Analysis, 23(2), 61-66 (1988).

2 - Chambers, A.C., T.H. Hyde, and J.J. Webster, Engineering Fracture Mechanics, 39(3), 603-19 (1991).

3 - Hyde, T.H. and A.C. Chambers, Materials at High Temperatures, 9(3) 127-38 (1991).

4 - Hyde, T.H. and A.C. Chambers, J. Strain Analysis, 27(1), 49-57 (1992).

5 - Ellison, E.G. and M.P. Harper, J. Strain Analysis 13, 35-51 (1978).

6 - Fu, L.S., Engineering Fracture Mechanics 13, 307-30 (1980).

7 - Sadananda, K. and P. Shahinian, Engineering Fracture Mechanics, 15, 327-42 (1981).

8 - Sadananda, K. and P. Shahinian, Metal Sci., 15, 425-32 (1981).

9 - Malik, S.N., Nuclear Eng. Design, 72, 359-71 (1982).

10 - Hyde, T.H., High Temperature Technology, 6(2), 51-61 (1988).

11 - Duan, Y., J.J. Webster and T.H. Hyde, 2nd Conf. on Creep of Brittle Materials, Leicester, England, 2-4 Sept. (1991).

12 - Nickbin, K.M., D.J. Smith, and G.A. Webster, Proc. R. Soc. Lond., A396, 183-97 (1984).

13 - Rice, J.R. and D.M. Tracey, J. Mech. Phys. Solids, 17, 201-17 (1969).

14 - Harper, M.P. and E.G. Ellison, J. STrain Analysis, 12, 167-79 (1977).
15 - Hyde, T.H., J. Strain Analysis 21(2), 117-19 (1986).
16 - Shih, C.F., Fracture Analysis, ASTM STP560, 187-210 (1974).
17 - Chambers, A.C., J.J. Webster and T.H. Hyde, Engineering Fracture Mechanics, to appear.
18 - Hutchinson, J.W., J. Mech. Phys. Solid, 16, 167-79 (1968).

*Proceedings of the Fifth International Conference on Creep of Materials, Lake Buena Vista, Florida, USA, 18-21 May 1992*

# A Creep Constitutive Model for Anisotropic Hardening Metals

**M. Kawai**
Institute of Engineering Mechanics
University of Tsukuba
Tsukuba, Japan

## Abstract

A creep constitutive model is proposed which is capable of describing creep deformation induced anisotropy. The creep model is characterized in terms of a unified constitutive structure, an additive effective stress and an auxiliary creep hardening rule. In the first place, a basic structure of unified type is developed in the general framework of irreversible thermodynamics; two thermodynamic potentials, a free energy function and a dissipation energy function, are introduced. A new additive representation of the internal state variables is assumed for isotropic and anisotropic hardening, and a quadratic form of Helmholtz free energy function is defined so that the associated thermodynamic forces are derived also in additive form. The evolution equations of the assumed internal state variables are of the Bailey-Orowan type and they are derived from a dissipation energy function. In the second place, to apply the unified model to multiaxial creep, an auxiliary creep hardening rule is formulated; it modifies the creep hardening processes described by the prototypal evolution equations. The auxiliary creep hardening rule is based on the concept of material memory surface in the total internal stress space, inside which the isotropic hardening is suppressed and the kinematic one is promoted. It is shown that the present approach to developing an anisotropic creep hardening model is flexible enough to be applicable to any model of pure kinematic-hardening type, and that it incorporates a feature of the ORNL creep model in a particular case. Finally, we show that the proposed model can predict fairly well the anisotropic creep behavior observed under repeated multiaxial nonproportional loading conditions.

ANISOTROPIC STRAIN HARDENING of metals is intrinsic to inelastic deformation. Once the anisotropy has developed with creep and plastic deformation, it has a significant influence on subsequent inelastic behavior. A typical example is the creep softening at uniaxial or pure-torsional stress reversals for type 304 or 316 stainless steel and $2\frac{1}{4}$Cr-1Mo steel [1-3]. Another example of creep anisotropy is observed after nonproportional stress changes for the above materials and aluminum alloys [4-10].

On the other hand, there is an increasing demand for analysis-based estimation of strength and/or life for high-temperature components. The local stress state of high-temperature structural components is generally characterized by multiaxial and unsteady conditions. Furthermore, the corresponding stress trajectory is not necessarily proportional. Hence, the inelastic analysis requires the adequate constitutive description of the inelastic behavior of engineering metallic materials; and, taking into account the experimental observations reported, modeling of anisotropic strain hardening becomes one of the most important issues. The modeling should include the discussion of the ease of material identification for practical application. It is also important to pursue the subject in parallel with the proper estimation of the existing constitutive models for general multiaxial inelastic deformation.

The predictive capability of creep constitutive equations of representative forms has been studied by Ohashi et al. [4], based on the creep tests performed for type 304 stainless steel at 650°C under several multiaxial nonproportional loading conditions. Their experimental results showed clearly the anisotropic nonstationary creep behavior, and it becomes more significant for larger rotations of the principal stress axes. The comparison with the isotropic hardening creep theory, such as the strain hardening theory [11], has shown its marked underestimation in the case of large stress changes which cause large creep softening. The pure kinematic creep theory [12], on the other hand, overestimated the creep softening. The experimental observation and the comparison with the calculated results of classical creep theories imply a significance of combined description of isotropic hardening and kinematic hardening. A simple combination between them [4], however, is not enough to predict accurately the creep anisotropy in the case of repeated nonproportional stress changes. It is because the isotropic hardening variable continuously grows. Hence, the issue for modeling an anisotropic creep model is how to combine them to obtain the most appropriate description.

For accurate description of creep anisotropy, there are three well-known models: the ORNL model [1] developed by Oak Ridge National Laboratory; the Murakami-Ohno model [13]; and the Mroz-Trampczynski model [14]. These creep models propose particular approaches for controlling isotropic and kinematic hardening. The ORNL creep model is a modified strain hardening theory which predicts the creep anisotropy with a memory of maximum creep strain; i.e., the patching of creep strain is controlled by the memory of prior maximum creep strain. This model is implemented into many major codes for inelastic analysis [15]. Murakami and Ohno have developed the ORNL model into a more sophisticated form which facilitates multiaxial creep analyses. They introduced a material memory surface in the creep strain space which is called the creep hardening surface; inside the surface the isotropic creep strain hardening vanishes. In the early version of the Murakami-Ohno model, the creep strain recovery could not be described due to the equation form of Bailey-Norton type. This shortcoming was removed in the sophisticated version [7] by using a superposition of recoverable and non-recoverable creep strains. Both the ORNL model and the Murakami-Ohno model employ the strain-rate function of the Bailey-Norton type and describe the creep anisotropy by discontinuous changes of a single scalar hardening variable. On the other hand, Mroz-Trampczynski [14] proposed a technique to describe the creep anisotropy by prescribing the evolution of back stress. They assumed a decomposition of the total back stress into two parts which are associated with a creep loading process and a creep reorientation process. The evolution equations of these back stresses are controlled by the memory surface in the back stress space which memorizes the maximum magnitude of back stress. The creep model was also sophisticated later by Mroz, Trampczynski and Hayhurst [16]. The validity of these creep models are shown based on the multiaxial non-steady creep experiments. Other possible approaches to the combined isotropic and kinematic hardening and the restriction of kinematic hardening are suggested in the studies by Ding and Findley [17], Robinson [18] and Farque et al. [19]. Some relevant discussions based on anisotropic creep potentials are also given by Ohnami et al. [5], Rees [20] and Hamanaka et al. [21].

Together with the effect of creep strain induced anisotropy, the effect of the prior plastic deformation is also of great significance. Such a history effect is called the creep-plasticity interaction. Experimental results and studies to incorporate the effect into creep constitutive models are reported by ORNL [1], Ohashi et al. [22-25], Rees [26] and Ikegami-Niitsu [27]. To avoid an ad hoc modification, however, an increasing interest is taken in a unified constitutive modeling for creep and plasticity. There are many unified models proposed so far, e.g. by Miller [28], Hart [29], Bodner [30], Krempl [31], Krieg et al. [32], Chaboche [33] and Freed-Walker [34,35]. The constitutive structures of these unified models have some similarities, which are discussed in detail by Krieg [36] and Inoue [37]. For the low stress level at which the effect of creep-plasticity interaction is small and the effect of creep strain induced anisotropy is dominant, these unified models are expected to predict well the creep strain induced anisotropy. According to the study of Ohashi et al. [4], however, the models which have the representative structure of these unified models fail to predict accurately the creep behavior under repeated multiaxial nonproportional loading conditions. This may be attributed to the combined isotropic-anisotropic hardening; in a simple combination of isotropic and anisotropic hardening, the continuous growth of isotropic hardening is an obstacle. Although the creep models mentioned above succeed in predicting creep anisotropy, their constitutive structure is not necessarily of typical unified type. Consequently, for more accurate description of anisotropic creep hardening in the framework of unified models, the Mroz-Trampczynski (-Hayhurst) creep model [14,16] the pioneer study on the subject, an incorporation of the technique developed in the creep modelings [7,13,14,16] will be necessary, and this is expected to promote further development of unified models.

The aim of the present study is to formulate a new hardening rule for a description of anisotropic creep behavior in the framework of unified models. The ease of material identification is carefully taken into account. The basic constitutive model of unified type is constructed using a combination of a new description of internal state variables and the formulation technique based on the irreversible thermodynamics. A free energy function is first defined in terms of the assumed internal state variables, from which the associated thermodynamic force variables and their state relations are derived. Another scalar function called a dissipation energy function is defined as an isotropic function of the thermodynamic forces, which leads to a description of prototypal evolution equations for the assumed internal state variables. Then, in order to enhance the predictive accuracy of the model for creep, an auxiliary hardening rule is developed which modifies the prototypal evolution equations, preserving the proposed unified constitutive structure. The auxiliary creep hardening rule introduces a material memory surface in the internal stress space; inside the surface the isotropic hardening is assumed to be suppressed and the kinematic hardening be promoted. Finally, the model characteristics and the predictive validity are discussed on the basis of the experimental results [4].

## Thermodynamic Constitutive Modeling

A fundamental constitutive model of unified type is first constructed which is applied to creep analyses in combination with an auxiliary hardening rule.

The approach employed is based on irreversible thermodynamics [38-42]. The present paper limits our attention to quasi-isothermal inelastic processes where temperature and temperature gradient terms are negligible. Also assumed is the infinitesimal deformation, so that the total strain is expressed by the sum of elastic and inelastic strain components. For the description of deviatoric tensors, the five-dimensional deviatoric vector representation [43] is employed; e.g., the vector components of the deviatoric stress $\mathbf{s}$ and the inelastic strain $\boldsymbol{\varepsilon}^p$ have the following relations with the tensor components: $\mathbf{s} = \{s_i\} = \{(3/2)s_{11}, \sqrt{3}(s_{11}/2+s_{22}), \sqrt{3}s_{12}, \sqrt{3}s_{23}, \sqrt{3}s_{31}\}$; $\boldsymbol{\varepsilon}^p = \{\varepsilon^p_i\} = \{\varepsilon^p_{11}, 2(\varepsilon^p_{11}/2+\varepsilon^p_{22})/\sqrt{3}, 2\varepsilon^p_{12}/\sqrt{3}, 2\varepsilon^p_{23}/\sqrt{3}, 2\varepsilon^p_{31}/\sqrt{3}\}$. The magnitudes of these vectors coincide with the invariants of von Mises type, which are denoted by $\|\mathbf{s}\| = [\mathbf{s}\cdot\mathbf{s}]^{1/2}$ and $\|\boldsymbol{\varepsilon}^p\| = [\boldsymbol{\varepsilon}^p\cdot\boldsymbol{\varepsilon}^p]^{1/2}$; the operation $(\cdot)$ represents the inner product of the vectors.

**Internal State Variables and Free Energy Function.** Start with an expression to describe the internal state which corresponds to a macroscopic inelastic deformation. At any moment of monotonic inelastic deformation, a simultaneous occurrence of isotropic and anisotropic hardening is postulated.

The current total internal state variable $\boldsymbol{\xi}$ is expressed in terms of two internal state variables $\boldsymbol{\alpha}$ and $\rho$ as

$$\boldsymbol{\xi} = \boldsymbol{\alpha} + \rho\mathbf{n} \tag{1}$$

where $\boldsymbol{\xi}$ and $\boldsymbol{\alpha}$ are deviatoric tensors of second rank and $\rho$ is a scalar [44,45]. The deviatoric tensor $\mathbf{n}$ is defined by

$$\mathbf{n} = \frac{\mathbf{s}}{\|\mathbf{s}\|} \tag{2}$$

which is a normalized tensor of the deviatoric part $\mathbf{s}$ of Cauchy stress $\boldsymbol{\sigma}$. It is assumed that $\mathbf{n} = \mathbf{0}$ if $\mathbf{s} = \mathbf{0}$.

The internal state described by Eq. (1) is identified with a point on the surface:

$$\begin{aligned}\bar{g}(\boldsymbol{\xi}) &= \bar{g}(\boldsymbol{\xi}; \boldsymbol{\alpha}, \rho)\\ &= \|\boldsymbol{\xi} - \boldsymbol{\alpha}\| - \rho = 0\end{aligned} \tag{3}$$

In order to define a thermodynamic local state law, i.e., the state relations among the assumed internal state variables and their associated thermodynamic forces, a Helmholtz free energy function $\psi$ is introduced. The thermodynamic forces are conjugate to the time rates of the internal state variables.

If uncoupled physical processes take place at the same time, the total free energy function is expressed by the sum of free energy functions for respective processes. Without considering damage effects, it may be assumed that the elastic and inelastic deformation processes are uncoupled. Then, the total free energy $\psi$ is expressed as

$$\psi = \psi_e(\boldsymbol{\varepsilon}^e) + \psi_p(\boldsymbol{\xi}) \tag{4}$$

where $\psi_e$ is the reversible elastic free energy and $\psi_p$ the irreversible inelastic free energy.

For linear elastic materials, the recoverable free energy is given by

$$\rho^* \psi_e = \frac{1}{2}\boldsymbol{\varepsilon}^e \cdot \mathbf{C}\boldsymbol{\varepsilon}^e \tag{5}$$

where $\rho^*$ denotes the mass density and $\mathbf{C}$ is the elastic modulus tensor of fourth rank. The Cauchy stress $\boldsymbol{\sigma}$, which is the thermodynamic force associated with the elastic strain, is defined by

$$\boldsymbol{\sigma} = \rho^* \frac{\partial \psi}{\partial \boldsymbol{\varepsilon}^e} = \mathbf{C}\boldsymbol{\varepsilon}^e \tag{6}$$

i.e., Hooke's law.

Regarding the inelastic free energy function, it is assumed that the isotropic and the anisotropic hardening are caused by different physical mechanisms and so they are uncoupled processes. The inelastic free energy function is then expressed by the sum:

$$\psi_p(\boldsymbol{\xi}) = \psi_{p1}(\boldsymbol{\alpha}) + \psi_{p2}(\rho) \tag{7}$$

where

$$\rho^* \psi_{p1}(\boldsymbol{\alpha}) = \frac{1}{2} H \boldsymbol{\alpha} \cdot \boldsymbol{\alpha} \tag{8}$$

$$\rho^* \psi_{p2}(\rho) = \frac{1}{2} H \rho^2 \tag{9}$$

The coefficient H is assumed to be constant in the present study. From this free energy function, the thermodynamic forces $\mathbf{a}$ and r, respectively associated with the internal state variables $\boldsymbol{\alpha}$ and $\rho$, are obtained so that

$$\mathbf{a} = \rho^* \frac{\partial \psi_{p1}}{\partial \boldsymbol{\alpha}} = H\boldsymbol{\alpha} \tag{10}$$

$$r = \rho^* \frac{\partial \psi_{p2}}{\partial \rho} = H\rho \tag{11}$$

Using Eq. (1), it follows that:

$$\rho^* \psi_p(\boldsymbol{\xi}) = \rho^* \psi_{p1}(\boldsymbol{\alpha}) + \rho^* \psi_{p2}(\rho)$$

$$= \frac{1}{2} H \boldsymbol{\xi} \cdot \boldsymbol{\xi} - \rho^* \psi_{p3}(\boldsymbol{\alpha}, \rho) \tag{12}$$

$$\rho^* \psi_{p3}(\boldsymbol{\alpha}, \rho) = H\boldsymbol{\alpha} \cdot (\rho \mathbf{n}) \tag{13}$$

Equation (13) expresses a coupling between the isotropic and anisotropic hardening; it is neglected in the present study. From Eqs. (12) and (13), the thermodynamic force $\mathbf{p}$ associated with $\boldsymbol{\xi}$ is expressed as

$$\mathbf{p} = \rho^* \frac{\partial \psi_p}{\partial \boldsymbol{\xi}}$$

$$= H\boldsymbol{\xi}$$

$$= H[\boldsymbol{\alpha} + \rho \mathbf{n}] \tag{14}$$

Therefore, an additive state relation

$$\mathbf{p} = \mathbf{a} + r\mathbf{n} \tag{15}$$

is satisfied among these thermodynamic forces. This additive relation in the space of thermodynamic force variable is equivalent to Eq. (1) in the space of internal state variable.

The thermodynamic force $\mathbf{p}$ given by Eq. (15) corresponds to a point on the surface

$$g(\mathbf{p}) = g(\mathbf{p}; \mathbf{a}, r)$$

$$= \|\mathbf{p} - \mathbf{a}\| - r = 0 \tag{16}$$

in the space of thermodynamic force.

It is noted that the independent variables are $\{\boldsymbol{\alpha}, \rho\}$ in the space of internal state variable, and equivalently $\{\mathbf{a}, r\}$ in the space of thermodynamic force variable.

Consider here the effective stress $\mathbf{s}^*$ defined by

$$\mathbf{s}^* = \mathbf{s} - \mathbf{p} \tag{17}$$

Substitution of Eq. (15) into Eq. (17) leads to the expression:

$$\mathbf{s}^* = (\|\mathbf{s}\| - r)\mathbf{n} - \mathbf{a} \tag{18}$$

From Eq. (18), we can see that the applied stress deviator $\mathbf{s}$ is modified isotropically and anisotropically in terms of r and $\mathbf{a}$, respectively. The isotropic hardening variable represents the pure material hardening and the anisotropic one represents the pure residual internal stress. When $\mathbf{s} = \mathbf{0}$, the driving force on inelasticity is given by $-\mathbf{a}$; i.e., $\mathbf{s}^* = -\mathbf{a}$.

It is important that the modification of the applied stress is made in the additive manner; this facilitates the understanding of flow stress decomposition, material hardening process and material identification. In the combined hardening models proposed so far, only the Chaboche model [33] employs such an additive expression.

**Dissipation Energy Function and Evolution Equations.** In order to prescribe evolution equations for the assumed internal state variables, a dissipation energy function $\Omega^*$ [46], which is convex, nested and positive, is introduced. In the present study, we assume the following form [47] to obtain the well-known hardening/softening format of Bailey-Orowan type [48,49]:

$$\Omega^*(\mathbf{s}^*, \mathbf{p}) = \Omega[U(\mathbf{s}^*)] + \Lambda[V(\mathbf{p})] \tag{19}$$

where each scalar function involved is isotropic. The effective stress $\mathbf{s}^*$ is defined in terms of Eq.(17) or Eq. (18).

From Eq. (19), the inelastic strain rate is expressed as

$$\dot{\boldsymbol{\varepsilon}}^p = \frac{\partial \Omega^*}{\partial \mathbf{s}} = \frac{\partial \Omega}{\partial \mathbf{s}^*}\frac{\partial \mathbf{s}^*}{\partial \mathbf{s}} + \frac{\partial \Lambda}{\partial \mathbf{p}}\frac{\partial \mathbf{p}}{\partial \mathbf{s}}$$

$$= \mathbf{FM} + \mathbf{R\Pi} \tag{20}$$

where

$$\mathbf{F} = \frac{\partial \Omega}{\partial \mathbf{s}^*} \tag{21}$$

$$\mathbf{R} = \frac{\partial \Lambda}{\partial \mathbf{p}} \tag{22}$$

$$\boldsymbol{\Pi} = \frac{\partial \mathbf{p}}{\partial \mathbf{s}} = r\frac{\partial \mathbf{n}}{\partial \mathbf{s}}$$

$$= \frac{r}{\|\mathbf{s}\|}[\mathbf{I} - \mathbf{n} \otimes \mathbf{n}] \tag{23}$$

$$\mathbf{M} = \frac{\partial \mathbf{s}^*}{\partial \mathbf{s}} = \mathbf{I} - \frac{\partial \mathbf{p}}{\partial \mathbf{s}}$$

$$= \mathbf{I} - \boldsymbol{\Pi}$$

$$= \left(1 - \frac{r}{\|\mathbf{s}\|}\right)\mathbf{I} + \frac{r}{\|\mathbf{s}\|}\mathbf{n} \otimes \mathbf{n} \tag{24}$$

In Eqs. (23) and (24), $\mathbf{M}$ and $\boldsymbol{\Pi}$ denote the fourth rank anisotropic tensors and $\mathbf{I}$ the fourth rank unit tensor. They lead to the following relations:

$$\mathbf{FM} = \mathbf{F}(\mathbf{I} - \boldsymbol{\Pi})$$

$$= \mathbf{F} - \mathbf{F\Pi}$$

$$= \left(1 - \frac{r}{\|\mathbf{s}\|}\right)\mathbf{F} + \frac{r}{\|\mathbf{s}\|}(\mathbf{F} \cdot \mathbf{n})\mathbf{n} \tag{25}$$

$$\mathbf{R\Pi} = \mathbf{R}\frac{r}{\|\mathbf{s}\|}(\mathbf{I} - \mathbf{n} \otimes \mathbf{n})$$

$$= \frac{r}{\|\mathbf{s}\|}[\mathbf{R} - (\mathbf{R} \cdot \mathbf{n})\mathbf{n}] \tag{26}$$

It is assumed that $\mathbf{n} = \mathbf{0}$ and $\boldsymbol{\Pi} = \mathbf{O}$, if $\mathbf{s} = \mathbf{0}$. When the inelastic deformation is strictly proportional, $\mathbf{x}\boldsymbol{\Pi} = \mathbf{0}$ for an arbitrary second rank tensor $\mathbf{x}$; in this case $\mathbf{M} = \mathbf{I}$, and so $\mathbf{FM} = \mathbf{F}$. Note that $\mathbf{M} = \mathbf{I}$, and $\dot{\boldsymbol{\varepsilon}}^p = \mathbf{F}$, if the hardening is purely anisotropic (i.e., r = 0).

On the other hand, the time rates of internal state variables $\boldsymbol{\xi}$, $\boldsymbol{\alpha}$ and $\rho$ are given by

$$\dot{\boldsymbol{\xi}} = -\frac{\partial \Omega^*}{\partial \mathbf{p}}$$

$$= -[\frac{\partial\Omega}{\partial \mathbf{p}} + \frac{\partial\Lambda}{\partial \mathbf{p}}]$$

$$= \mathbf{G} \quad (27)$$

$$\dot{\boldsymbol{\alpha}} = -\mu\frac{\partial\Omega^*}{\partial \mathbf{a}}$$

$$= -\mu[\frac{\partial\Omega}{\partial \mathbf{a}} + \frac{\partial\Lambda}{\partial \mathbf{a}}]$$

$$= \mu\mathbf{G} \quad (28)$$

$$\dot{\rho} = -(1-\mu)\frac{\partial\Omega^*}{\partial r}$$

$$= -(1-\mu)[\frac{\partial\Omega}{\partial r} + \frac{\partial\Lambda}{\partial r}]\cdot\mathbf{n}$$

$$= (1-\mu)\mathbf{G}\cdot\mathbf{n} \quad (29)$$

where

$$\mathbf{G} = \mathbf{F} - \frac{\partial\Lambda}{\partial \mathbf{p}}$$

$$= \mathbf{F} - \mathbf{R} \quad (30)$$

In the above expressions, the coefficients $\mu$ and $(1 - \mu)$ are introduced so that Eq. (1) is satisfied under uniaxial state of stress. The parameter $\mu$ gives the ratio of combined isotropic and anisotropic hardening; so it may be called the combined (or mixed) hardening parameter.

It is noted that the back stress $\mathbf{p}$, which prescribes the effective stress $\mathbf{s}^*$ in Eq. (17), depends on the current direction of the applied deviatoric stress $\mathbf{s}$. For this, the direction of inelastic strain rate does not coincide with the outward normal to the loading surface $U(\mathbf{s}^*)$ = const.; the inelastic flow takes place in the direction of $\mathbf{FM} + \mathbf{R\Pi}$. This is different from the conventional modelings where the flow direction is parallel to the unit outward normal $\mathbf{n}^*$ to the current loading surface. Only when $\mathbf{M} = \mathbf{I}$, however, the flow direction coincides with that of the conventional models, and the scalar function $\|\mathbf{F}\|$ is in agreement with the equivalent inelastic strain rate of von Mises type ($\|\mathbf{F}\| = F = \|\dot{\boldsymbol{\varepsilon}}^p\|$).

The time rates of the corresponding thermodynamic forces can be obtained by multiplying Eqs. (27) to (29) by a material constant H. In the space of thermodynamic force, the constitutive equations developed above can be summarized as follows:

$$\dot{\boldsymbol{\varepsilon}}^p = \mathbf{FM} + \mathbf{R\Pi} \quad (31)$$

$$\dot{\mathbf{p}} = H\dot{\boldsymbol{\xi}} = \dot{\mathbf{a}} + \overline{\dot{r}\mathbf{n}} \quad (32)$$

$$\dot{\mathbf{a}} = H\dot{\boldsymbol{\alpha}} = \mu H\mathbf{G} \quad (33)$$

$$\dot{r} = H\dot{\rho} = (1-\mu)H\mathbf{G}\cdot\mathbf{n} \quad (34)$$

These constitutive equations reduce to a simple form:

$$\dot{\boldsymbol{\varepsilon}}^p = \mathbf{F} \quad (35)$$

$$\dot{\mathbf{p}} = H\mathbf{G} \quad (36)$$

under monotonic proportional deformation.

From Eqs. (27) to (30), it can be seen that $\boldsymbol{\xi}$ is in agreement with the inelastic strain under monotonic proportional deformation if the recovery term $\mathbf{R}$ is neglected. Therefore, Eq. (1) may be interpreted in this case as a decomposition of the total inelastic strain into two parts which contribute to the purely isotropic hardening and to the purely anisotropic hardening. A similar interpretation can be given to the general case, if the quantity obtained from the integration of $\mathbf{G}$ in Eq.(30) is identified with an effective inelastic strain, which is the difference between the actual inelastic strain ($\int\mathbf{F}dt$) and the recovery strain ($\int\mathbf{G}dt$).

**Entropy Inequality.** The irreversible thermodynamics imposes the constraint of non-negative dissipation rate on inelastic deformation: i.e., the second law of thermodynamics.

In quasi-isothermal processes, the condition is approximated by the non-negative rate of the mechanical dissipation:

$$D_1 = \mathbf{s}\cdot\dot{\boldsymbol{\varepsilon}}^p - \mathbf{a}\cdot\dot{\boldsymbol{\alpha}} - r\dot{\rho} \quad (37)$$

The evolution of internal state, therefore, should occur only when $D_1 \geq 0$. Another interesting subject for study is to search for the domain in which the thermodynamic constraint is satisfied for the present model. Such a discussion is given by Malmberg [50] to a thermodynamic extension of the mechanical Robinson model [18].

## Auxiliary Hardening Rule for Creep

The constitutive model described above belongs to the category of the rate-dependent inelastic constitutive models or the so-called unified constitutive models which take into account the combined isotropic and anisotropic hardening. The proposed model, however, contains some essential difference from other existing formulations with respect to the description of the internal state variables.

As mentioned earlier, a simple combination of isotropic and anisotropic hardening employed in most unified models is not sufficient enough to describe accurately the observed non-steady creep deformation under multiaxial nonproportional repeated loading conditions [4]. Exactly the same problem is included by the proposed model. Therefore, to enhance the predictive accuracy of the proposed model in the application to creep analyses, we modify the evolution of internal state variables.

**Perfect Memory for Internal State.** The effects of creep anisotropy for the materials considered can be easily estimated in the creep tests under symmetric stress reversals. The typical behavior of type 304 stainless steel [4] is shown in Fig. 1. Almost the same results are reported by ORNL [1] and Ohno et al. [7].

As far as the experimental results shown in Fig. 1 are concerned, the creep strain rate increases at each stress reversal, and the creep strain amplitude shows no decrease with the cyclic loadings but even slight increase. Since the increase of the creep strain amplitude is not so large, we consider here as a first order approximation that the creep strain amplitude is constant with cycles. This implies two interpretations: (1) no internal change which results in the isotropic strain hardening; (2) cancellation of the isotropic hardening by the effect of anisotropic hardening.

The constitutive modeling depends on the interpretation of this phenomenon. The present study takes the standpoint of the first interpretation, and develops an auxiliary creep hardening rule which enables the prototypal model developed above to reproduce in a particular case the same amount of creep strain at each stress reversal.

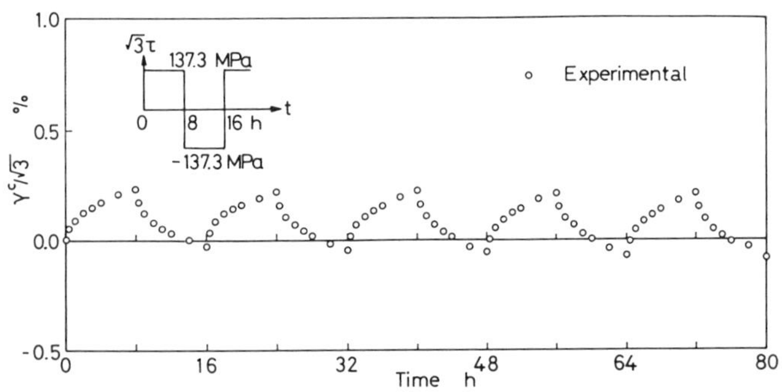


Fig. 1 Creep behavior under pure-torsional stress reversals

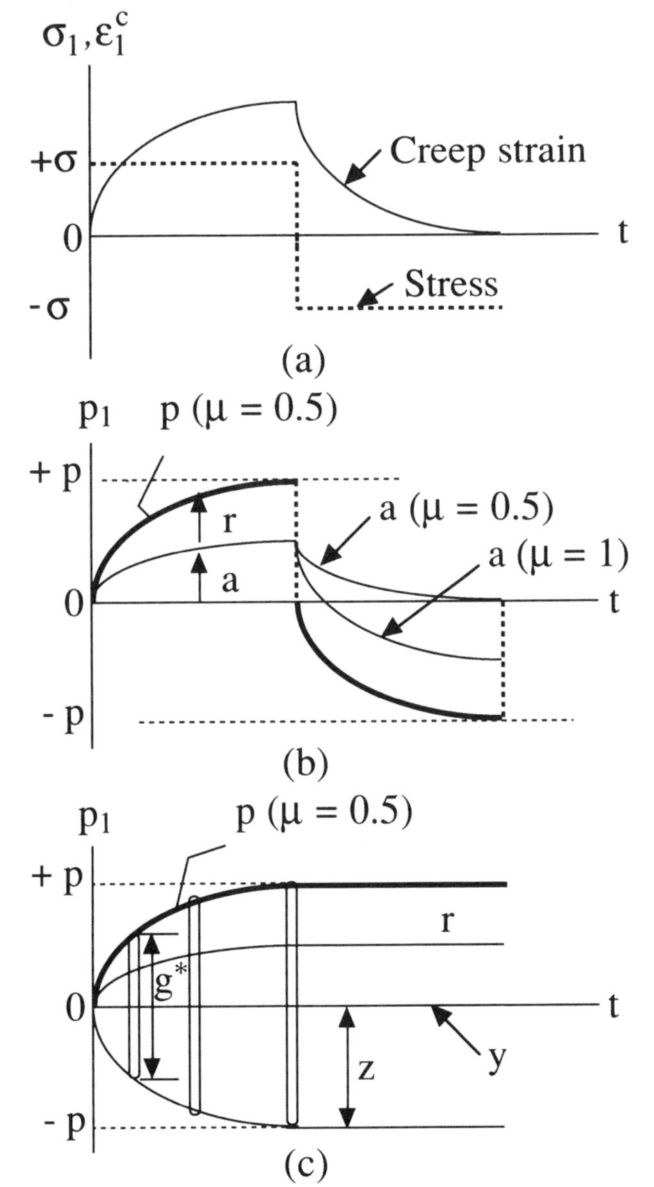


Fig. 2 Mechanism of the auxiliary hardening rule

For this purpose, a simplified case ($\mu = 0.5$) under uniaxial stress reversals is first considered (see Fig. 2). During the first creep loading ($+\sigma$), the hardening variables $p$, $a$ and $r$ change along the curves shown in Fig. 2(b); the magnitude of $a$ and $r$ is half of $p$ For the same amount of creep strain as in the first loading ($+\sigma$) to be predicted under the second loading ($-\sigma$), the total internal stress $p$ has to change along the bold curve in Fig. 2(b); it is noted that at the instant of the stress reversal the variable $p$ changes from +p to 0 due to the definition: $p = a + r \cdot \mathrm{sgn}(\sigma)$. According to the above interpretation, the isotropic hardening variable $r$ is assumed to be suppressed in the range $g^* = [-p, +p]$. To put simply $\dot{r} = 0$ in the range $g^*$, however, does not enable $p$ to follow the bold curve desired, because inside $g^*$ the variable $a$ changes along the fine curve labeled $a(\mu = 0.5)$ and $\dot{p} = \dot{a} = 0.5G$. So, for the total internal stress $p$ to change along the bold curve, the rate of $a$ in the range of $g^*$ should be increased by the amount of isotropic hardening rate disappearance so that $\dot{a} = G$.

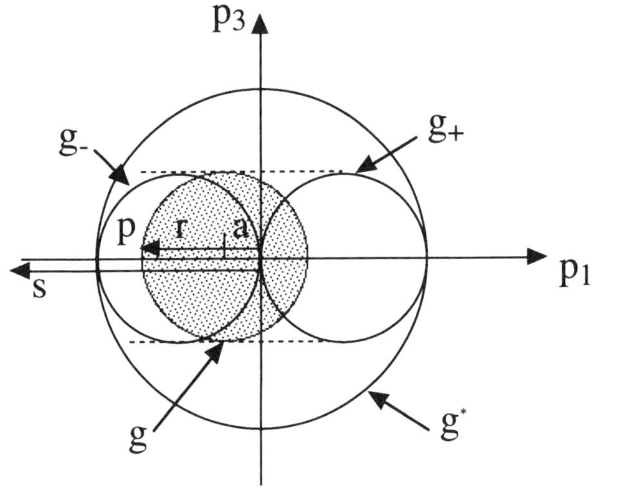


Fig. 3 Illustration of g-surface and g*-surface

Consequently, the above discussion can be generalized as follows. The formulation process consists of two parts which prescribe:

(A) the region of $g^*$ in the $\mathbf{p}$ space, in which the isotropic hardening does not take place ( i.e., $\dot{r} = 0$); and

(B) the rate of anisotropic hardening $\dot{\mathbf{a}} = \mathbf{0}$ inside the region $g^*$.

Regarding the part (A), the concept of material memory is applied, an idea originally developed by Chaboche et al. [51] and successfully applied to a creep modeling by Murakami-Ohno [13]. In the present study, the memory region in which the isotropic hardening does not take place is described by the hypersphere:

$$g^*(\mathbf{p}) = \|\mathbf{p} - \mathbf{y}\| - z = 0 \qquad (38)$$

where $\mathbf{y}$ expresses the center of the surface and z the radius of the surface. The condition $g^*(\mathbf{p}) \le 0$ is always satisfied for arbitrary $\mathbf{p}$; the point $\mathbf{p}$ is located exactly on the surface $g^*(\mathbf{p}) = 0$, or inside the surface $g^*(\mathbf{p}) < 0$.

For the location of the center $\mathbf{y}$ and the size z, the following evolution equations are assumed:

$$\dot{\mathbf{y}} = (2\mu - 1)H[g^*(\mathbf{p})]\mathbf{HG} = \begin{cases} (2\mu - 1)\mathbf{HG}, & g^*(\mathbf{p}) = 0 \\ 0, & g^*(\mathbf{p}) < 0 \end{cases} \qquad (39)$$

$$\dot{z} = 2\dot{r}H[g^*(\mathbf{p})] = \begin{cases} 2\dot{r}, & g^*(\mathbf{p}) = 0 \\ 0, & g^*(\mathbf{p}) < 0 \end{cases} \qquad (40)$$

where $H[\bullet]$ represents the Heaviside step function:

$$H[x] = \begin{cases} 1, & x \ge 0 \\ 0, & x < 0 \end{cases} \qquad (41)$$

For the part (B), the combined hardening parameter $\mu$ is controlled such that

$$\mu = 1 + (\mu_0 - 1)H[g^*(\mathbf{p})] = \begin{cases} \mu_0, & g^*(\mathbf{p}) = 0 \\ 0, & g^*(\mathbf{p}) < 0 \end{cases} \qquad (42)$$

This enhances the rate of the anisotropic hardening variable $\mathbf{a}$ inside the memory surface.

Fig. 3 gives the geometrical demonstration of the memory surface in the case $\mu = 0.5$. The g-surface and $g^*$-surface are described by the circles in the $\{p_1, p_3\} = \{\frac{3}{2}p_{11}, \sqrt{3}p_{12}\}$ plane. The first loading ($+\sigma$) corresponds to the enlargement of g- and $g^*$-

circles. After the first stress reversal, the circle $g_+$, which is the g-circle just before the reversal, moves toward $g_-$ inside the $g^*$-circle; during this process, $g_+$ and $g^*$ retain their sizes.

If $\mu = 0.5$, it follows that

$$\mathbf{y} = \mathbf{0} \tag{43}$$

$$z = 2r \tag{44}$$

for monotonic proportional deformation of initially isotropic materials. In this particular case, the center of the memory surface remains at the origin, and the prescribed region $g^*$ memorizes the maximum value of $\mathbf{p}$.

The auxiliary creep hardening rule, Eqs. (38) to (40) and (42), is formulated in the space of thermodynamic force. It can be also considered in the space of internal state variable.

**Recoverable Memory for Internal State.** In the auxiliary creep hardening rule developed above, the memory surface changes only when $\mathbf{p}$ moves out beyond the surface. Conversely, as long as $\mathbf{p}$ stays inside, the location and the size of the region once memorized do not change. In other words, the growing process of $\mathbf{p}$ continuously alters the memory; this corresponds to the erasure of memory of the past history. In the decreasing process of $\mathbf{p}$, by which $\mathbf{p}$ moves inside the region, on the other hand, the memory stays constant.

For cyclic inelasticity, the stepwise amplitude increase shows almost no influence of the prior deformation of lower amplitude on the subsequent steady inelastic behavior [52]. The above modeling will be appropriate in this case. For stepwise decrease in amplitude, the material response gradually approaches the steady state of virgin material attained at subsequent lower amplitude. It corresponds to the physical fact that the dislocation density caused by the prior deformation at higher stress level gradually decreases to the value associated with the subsequent lower stress level [53]. This evanescence of the memory effect is not taken into account in the above modeling, though an instantaneous effect of stress change is included for both stress increase and stress decrease.

To incorporate such a gradual memory erasure effect into the present model, we would have to modify it by including some recovery terms into the evolution equations of the memory surface, for example, as follows:

$$\dot{\mathbf{y}} = (2\mu - 1)H[g^*(\mathbf{p})]H\mathbf{G} - \mathbf{R}_y \tag{45}$$

$$\dot{z} = 2\dot{r}H[g^*(\mathbf{p})] - R_z \tag{46}$$

## Particular Forms of Creep Constitutive Equations

**Power Function Model.** Consider the following form of the dissipation energy function:

$$\Omega^* = \frac{K}{m+1}U^{m+1} + \frac{M}{\ell+1}V^{\ell+1} \tag{47}$$

$$U = \|\mathbf{s} - \mathbf{p}\| \tag{48}$$

$$V = \|\mathbf{p}\| \tag{49}$$

where K, M, m and $\ell$ are material constants.

For simple creep analyses which neglect the effects of transient stress changes, it may be assumed that $\dot{\mathbf{n}} = \mathbf{0}$.

Based on the unified constitutive framework developed in the present study, the following particular constitutive model for creep can be derived:

$$\dot{\boldsymbol{\varepsilon}}^p = F\mathbf{n}^* \tag{50}$$

$$\mathbf{p} = \mathbf{a} + r\mathbf{n} \tag{51}$$

$$\dot{\mathbf{a}} = \mu H\mathbf{G} \tag{52}$$

$$\dot{r} = (1 - \mu)H\mathbf{G}\cdot\mathbf{n} \tag{53}$$

where

$$F = K\|\mathbf{s} - \mathbf{p}\|^m \tag{54}$$

$$\mathbf{n}^* = \frac{\mathbf{s} - \mathbf{p}}{\|\mathbf{s} - \mathbf{p}\|} \tag{55}$$

$$\mathbf{G} = \mathbf{F} - \mathbf{R} = \dot{\boldsymbol{\varepsilon}}^p - M\|\mathbf{p}\|^\ell \frac{\mathbf{p}}{\|\mathbf{p}\|} \tag{56}$$

In deriving these relations, it is assumed that $r/\|\mathbf{s}\|$ is very small and hence $\boldsymbol{\Pi} \cong \mathbf{O}$, $\mathbf{M} \cong \mathbf{I}$.

Equation (56) represents the Bailey-Orowan type evolution equation. In place of this equation, the following evolution equation of Chaboche type [33,54] may be employed:

$$\mathbf{G} = \dot{\boldsymbol{\varepsilon}}^p - \frac{L}{H}\|\dot{\boldsymbol{\varepsilon}}^p\|\mathbf{p} - M\|\mathbf{p}\|^\ell \frac{\mathbf{p}}{\|\mathbf{p}\|} \tag{57}$$

where L is a material constant. In this equation, the strain hardening term is modified to be non-linear; so this format is referred to as the non-linear kinematic hardening [55].

The first creep model proposed in the present study consists of the constitutive equations, Eqs. (50) to (53), and the auxiliary creep hardening rule, Eqs. (38) to (40) and (42).

The creep model reduces to the form:

$$\dot{\boldsymbol{\varepsilon}}^p = K\|\mathbf{s} - \mathbf{p}\|^m \mathbf{n}^* \tag{58}$$

$$\dot{\mathbf{p}} = H\mathbf{G} \tag{59}$$

for a monotonic proportional deformation of initially isotropic materials, regardless of any value of $\mu$; the assumed decomposition of $\mathbf{p}$ into isotropic and anisotropic parts has no meaning in this case. This simple form facilitates material identification.

It is noted that without using the proposed auxiliary creep hardening rule we can simulate a pure kinematic hardening behavior by assuming $\mu = 1$, and a pure isotropic hardening behavior by using $\mu = 0$. The auxiliary hardening rule has no influence when $\mu = 1$, whereas it does when $\mu = 0$, so that the rule should be inactive to analyse the isotropic hardening behavior with the present model.

To explicitly express the thermodynamic constraint, it may be assumed

$$\hat{K} = KH[D_1] \tag{60}$$

$$\hat{M} = MH[D_1] \tag{61}$$

**Hyperbolic Function Model.** Applying the additive representation of the thermodynamic forces given by Eq. (15) and the auxiliary creep hardening rule, we can construct new creep models by modifying any creep model of pure kinematic-hardening type. For example, a pure kinematic-hardening model given by the following equations:

$$\dot{\boldsymbol{\varepsilon}}^p = f_1(\|\mathbf{s} - \mathbf{p}\|)\mathbf{n}^* \tag{62}$$

$$\dot{\mathbf{p}} = A\dot{\boldsymbol{\varepsilon}}^p - B\|\dot{\boldsymbol{\varepsilon}}^p\|\mathbf{p} - Cf_2(\|\mathbf{p}\|)\frac{\mathbf{p}}{\|\mathbf{p}\|} \tag{63}$$

can be transformed to a new combined hardening model just by substituting

$$F = f_1 \tag{64}$$

$$\mathbf{G} = \dot{\boldsymbol{\varepsilon}}^p - \frac{B}{A}\|\dot{\boldsymbol{\varepsilon}}^p\|\mathbf{p} - \frac{C}{A}f_2(\|\mathbf{p}\|)\frac{\mathbf{p}}{\|\mathbf{p}\|} \tag{65}$$

in place of Eq. (54) and Eq.(56), respectively.

In the first creep model, a power function was employed to represent the creep strain rate. However, an alternative expression for creep strain rate is given by the use of hyperbolic sine function [56]. In this case, the evolution equations of the assumed internal state variables cannot be derived from a potential function as gradients. From the empirical point of view, however, it is useful to consider a formulation of a hyperbolic function model which has the same feature as the power function model proposed.

The hyperbolic sine function model can be easily constructed. For example, consider a kinematic hardening model with the Bailey-Orowan type evolution equation [4]:

$$\dot{\boldsymbol{\varepsilon}}^p = K[\sinh\{\frac{\|\mathbf{s} - \mathbf{p}\|}{N_0}\}]^m \mathbf{n}^* \quad (66)$$

$$\dot{\mathbf{p}} = A\dot{\boldsymbol{\varepsilon}}^p - B[\sinh\{\frac{\|\mathbf{p}\|}{N_0}\}]^\ell \frac{\mathbf{p}}{\|\mathbf{p}\|} \quad (67)$$

where $N_0$ is not an isotropic hardening variable but a temperature dependent material constant. Following the same approach mentioned above, we modify the model to the combined isotropic and kinematic hardening model as:

$$\dot{\boldsymbol{\varepsilon}}^p = K[\sinh\{\frac{\|\mathbf{s} - \mathbf{p}\|}{N_0}\}]^m \mathbf{n}^* \quad (68)$$

$$\mathbf{p} = \mathbf{a} + \mathbf{r} \quad (69)$$

$$\dot{\mathbf{a}} = \mu A \mathbf{G} \quad (70)$$

$$\dot{r} = (1 - \mu) A \mathbf{G} \cdot \mathbf{n} \quad (71)$$

where

$$\mathbf{G} = \dot{\boldsymbol{\varepsilon}}^p - \frac{B}{A}[\sinh\{\frac{\|\mathbf{p}\|}{N_0}\}]^\ell \frac{\mathbf{p}}{\|\mathbf{p}\|} \quad (72)$$

By a combination of these constitutive equations with the auxiliary creep hardening rule, Eq. (38) to (40) and (42), a hyperbolic function version of the present modeling is completed. The same technique may be applied to obtain an exponential function model.

## Comparisons With Experimental Results

Finally, based on the experimental results reported by Ohashi et al. [4], the validity of the proposed creep models is discussed; although both the power function creep model and the hyperbolic function creep model were estimated, only the former is shown for the limited space.

Material constants involved with the power function model are determined using Eqs. (58) and (59) so as to fit them to uniaxial or pure torsional creep curves; in the steady state the assumption $m = \ell$ leads to the uniaxial relations:

$$\dot{\varepsilon}^p_s = K(1-\beta)^m \sigma^m \quad (73)$$

$$M = K(\frac{1-\beta}{\beta})^m \quad (74)$$

where $\beta = p_s/\sigma$.

The resulting values of material constants for the power function model are as follows:

$$m = \ell = 8.2,\ K = 1.41 \times 10^{-21}\ h^{-1}MPa^{-m},$$
$$H = 5393.6\ MPa,\ M = 5.1 \times 10^{-17}\ h^{-1}MPa^{-m} \quad (75)$$

**Creep Under Repeated Reversed Loadings.** In order to discuss the effect of the parameter $\mu$, the creep behavior under symmetric stress reversals is calculated for the cases of $\mu$ = 0.0, 0.25, 0.50, 0.75 and 1.0. Fig. 4 shows the calculated results for the power function creep model and the experimental results . From these calculations (solid line), we can see that the model behavior significantly depends on the value of the combined hardening parameter $\mu$. In the particular case that the combined hardening parameter $\mu$ is equal to 0.5, the same creep curve appears at each stress reversal. This behavior coincides with the prediction by the modified strain hardening theory of ORNL [1].

Also included in Fig. 4 in dashed line is the calculated results of the prototypal power function model, Eqs. (50) to (53), which does not accompany the auxiliary creep hardening rule. It is obvious that the prototype fails to predict accurately the creep strain change under stress reversals because of the development of isotropic hardening; almost the same trend will be found in the predictions by most unified models of combined hardening type. Consequently, the proposed auxiliary hardening rule succeeds in enhancing the predictive accuracy of the present model with unified structure under stress reversals.

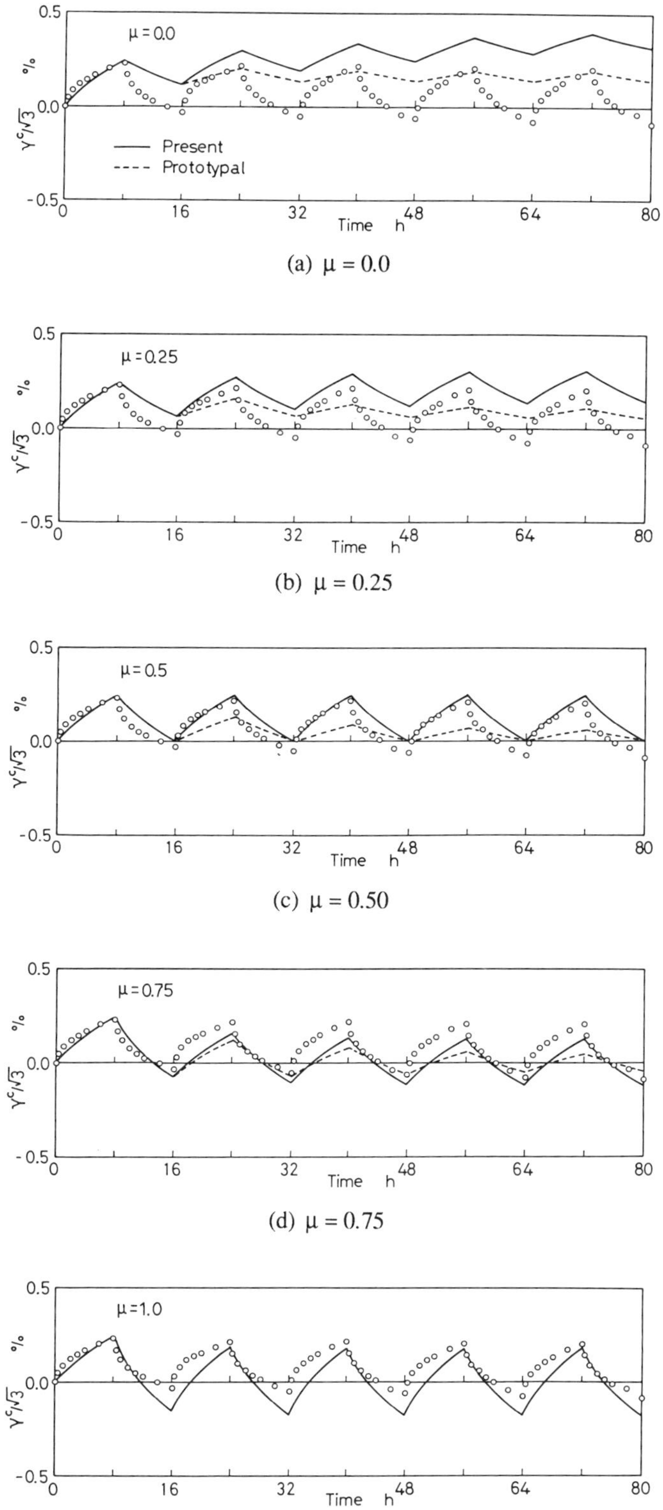


(a) $\mu$ = 0.0

(b) $\mu$ = 0.25

(c) $\mu$ = 0.50

(d) $\mu$ = 0.75

(e) $\mu$ = 1.0

Fig. 4 The $\mu$-dependence of the model behavior

As mentioned before, the prototypal model predicts completely isotropic hardening behavior when $\mu$ = 0 and completely kinematic when $\mu$ = 1. It is a feature of the proposed creep model that can estimate a pure isotropic and pure kinematic hardening behavior as special cases with the same material constants except $\mu$.

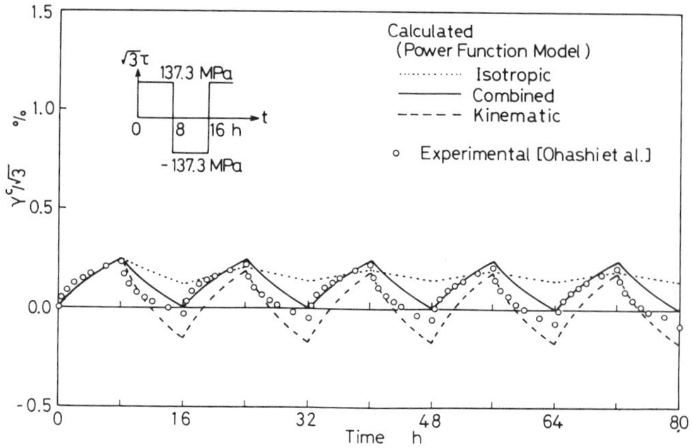


Fig. 5 Predicted results of the power function model for pure torsional stress reversals (θ = 180°)

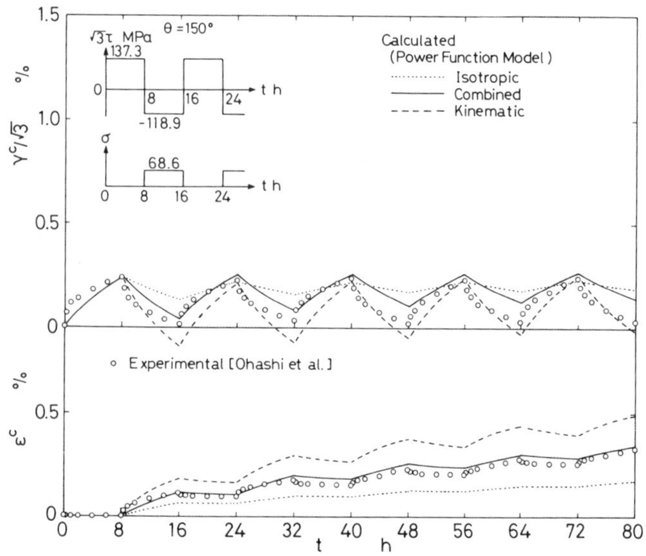


Fig. 6 Predicted results for repeated multiaxial nonproportional stress changes (θ = 150°)

Figure 5 shows in one figure the comparison among the combined hardening model (μ = 0.5, the auxiliary rule), the isotropic hardening model (μ = 0) and the kinematic hardening model (μ = 1) of the proposed power function model. From this figure, it is found that the predictions due to the combined and kinematic models gives a better prediction as compared with the isotropic hardening model (and also with the combined prototypal model without the auxiliary hardening rule shown in Fig. 4).

The hyperbolic sine function model, for the same analyses as in Fig. 5, results in a little larger increase in transient creep strain rate than the power function model. In this comparison, the combined model shows a better agreement with the experimental results than the other two models.

**Creep Under Repeated Multiaxial Nonproportional Loadings.** In what follows, the predictive capability of the proposed creep model for multiaxial nonproportional repeated loadings is discussed. Figs. 6 through 10 show the comparison between the calculated results of the power function model (μ = 0.5) and the experimental results [4]. These figures also include predictions of the pure isotropic and

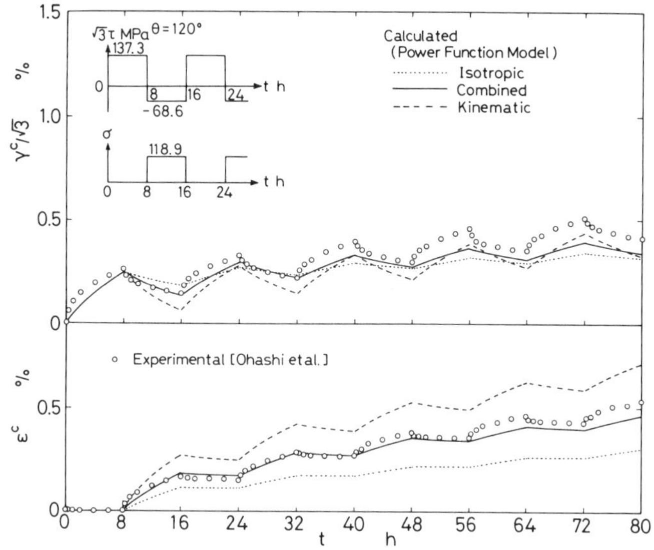


Fig. 7 Predicted results for repeated multiaxial nonproportional stress changes (θ = 120°)

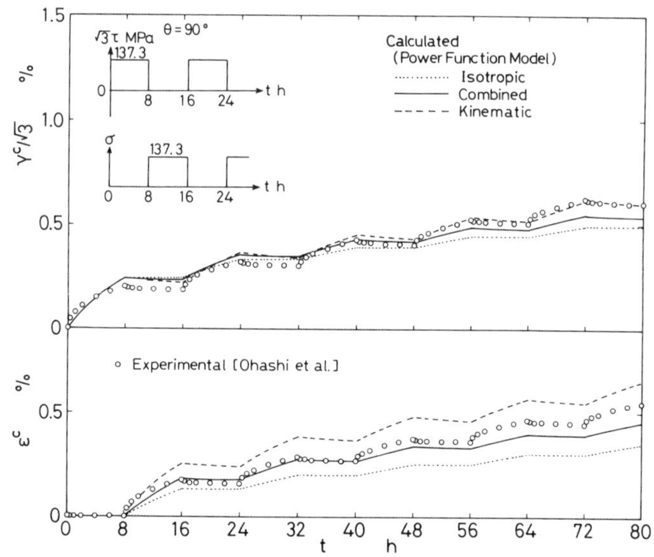


Fig. 8 Predicted results for repeated multiaxial nonproportional stress changes (θ = 90°)

pure kinematic hardening models, which are particular cases of the power function model. For larger rotations of the principal stress axes (θ = 150°, 120°, 90°), we can see that the proposed creep model reproduces fairly well the observed multiaxial creep behavior, as compared with the pure isotropic and pure kinematic hardening models. For small rotations, these models predict the creep behavior quite well and their difference is not so significant.

Examining the experimental results in detail, we can see the creep strain rate transiently increases at the moment just after each stress change. Such a temporal creep softening effect is not completely predicted by the present model. The gradual enlargement of difference between the calculated and experimental results with the cyclic loadings is attributed to this shortcoming. If we compare the calculated and experimental creep strain trajectories in the creep strain space, their discrepancy is much more exaggerated. These imply the significance of the transient

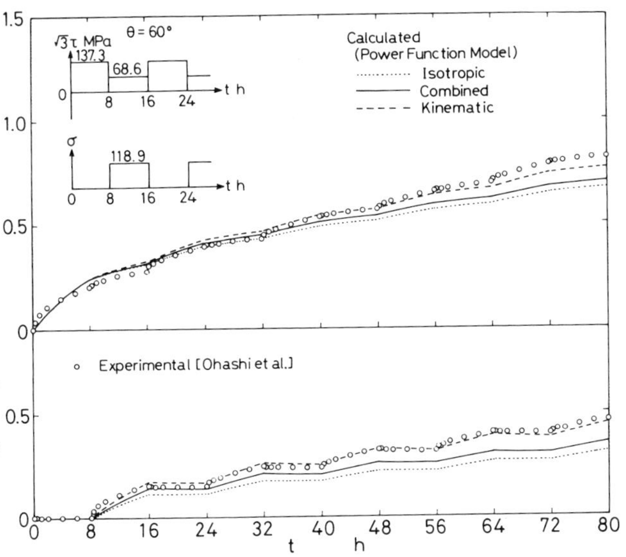


Fig. 9 Predicted results for repeated multiaxial nonproportional stress changes (θ = 60°)

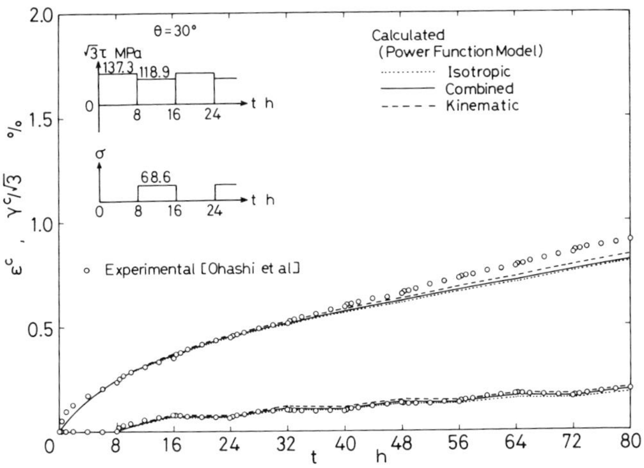


Fig. 10 Predicted results for repeated multiaxial nonproportional stress changes (θ = 30°)

softening and the necessity of further theoretical refinements of the present model.

**Mechanical Energy Dissipation.** The present creep model is formulated in the framework of irreversible thermodynamics, so that the energy conjugate relations are clearly defined. The mechanical dissipation energy can be calculated by Eq. (37); by using this equation we can determine whether the inelastic process considered satisfies the Clausius-Duhem inequality.

Neglecting the recovery terms, we can write the mechanical dissipation rate in uniaxial steady state as

$$D_1 = (\sigma - cp_s)\dot{\varepsilon}_s^p$$

$$= (1 - c\beta)\sigma\dot{\varepsilon}_s^p \qquad (76)$$

where $c = \mu^2 + (1 - \mu)^2$. In this relation, $c\beta$ represents the ratio of the stored energy to the total inelastic power; the stored energy is equivalent to the internal irreversible changes. The value $c\beta = 0.1$ is reported by Lee [57], and $c\beta = 0.16$ by Sluzalec [58]. In the present calculation, the material constants assumed lead to the value $\beta = 0.22$. The minimum value $c = 0.5$ is also obtained from $\mu = 0.5$. Therefore, the ratio of the stored energy is estimated to be $c\beta = 0.11$. Hence the structural changes in the present calculation are consistent with the usual estimation.

## Conclusion

In the present study, an anisotropic creep constitutive model was proposed using the formulation technique based on irreversible thermodynamics. The creep constitutive equations have a typical structure of unified models and include a new additive representation of combined effective stress. The present model can predict a purely isotropic hardening behavior and a purely kinematic hardening behavior in particular cases. The model developed reduces to a very simple form and the material constants involved can be easily identified.

For the application to multiaxial repeated creep, the auxiliary creep hardening rule was developed which limits the evolutions of the isotropic hardening and promotes the kinematic hardening in a memory surface. The material memory surface was defined in the internal stress space, inside which the growth rate of the isotropic hardening vanishes and that of the kinematic hardening increases by the amount of the isotropic hardening rate eliminated; i.e., the internal change occurs in a purely kinematic-hardening manner. Applying the auxiliary creep hardening rule with a particular value of the combined hardening parameter $\mu = 0.5$, we can simulate the same creep behavior as that due to the ORNL model for uniaxial or pure torsional stress reversals. Different behavior under stress reversals can also be predicted depending on the value of $\mu$. The auxiliary creep hardening rule can be applied to modify any constitutive model of pure kinematic hardening type. If the derivation of evolution equations from a potential function is not insisted on, then the present approach allows any form of the strain-rate and the recovery functions. Finally, the predictive accuracy of the proposed model was estimated by comparing with the complex creep behavior observed under multiaxial non-proportional repeated loadings. As the result, the anisotropic creep model proposed described fairly well the complex creep behavior and showed better agreement with experimental results as compared with the isotropic and kinematic hardening models.

The present model includes some similarities and differences with the promising creep models proposed by Murakami and Ohno [13] and Mroz, Trampczynski and Hayhurst [16]. In their models, the concept of material memory is introduced in the creep strain space for the Murakami-Ohno model and in the back stress space for the Mroz-Trampczynski-Hayhurst model. When $\mu = 0.5$, the present model memorizes the maximum value of the back stress. This is similar to the Mroz-Trampczynski-Hayhurst model. However, it is not so in the case of other values of $\mu$. The memory surface of the present model is identified with that in the space of the effective creep strain defined by the actual creep strain subtracted by the recovery strain. This is similar to the Murakami-Ohno model. The difference is that the present model prescribes back stress evolution, not a scalar hardening variable.

It is emphasized that the present model preserves a typical constitutive structure of unified type. This feature enables one to develop the present model to a fully unified version applicable to the combined creep and plastic deformation under monotonic and cyclic loading conditions.

## Acknowledgment


The author is grateful to Professor S. Murakami of Nagoya University for invaluable discussion. The financial support by the

Ministry of Education of Japan under a Grant-in-Aid for Encouragement of Young Scientist (No. 03750059) and from the project research of the University of Tsukuba for the fiscal year of 1991 are gratefully acknowledged.

## References


[1] Corum, J.M., W.L. Greenstreet, K.C. Liu, C.E. Pugh and R.W. Swindeman, ORNL-5014, Oak Ridge National Laboratory (1974).

[2] Fujino, M., F. Ishikawa and R. Ohtani, *Zairyo*, 29, (317), 113-119 (1980).

[3] Pugh, C.E. and D.N. Robinson, "Pressure Vessel and Piping -Design Analysis, Decade of Progress," p.171, S.Y. Zamrik and D. Diebich ed., ASME, New York (1982).

[4] Ohashi, Y., N. Ohno and M. Kawai, *Trans. ASME J. Eng. Mater. Technol.*, 104, (3), 159-164 (1982).

[5] Ohnami, M., K. Motoie and N. Yoshida, *Zairyo*, 18, (186), 226-232 (1968).

[6] Murakami, S., N. Ohno and H. Tagami, *Trans. ASME J. Eng. Mater. Technol.*, 108, (2), 119-126 (1986).

[7] Ohno, N., S. Murakami and T. Ueno, *Trans. ASME J. Eng. Mater. Technol.*, 107, (1), 1-6 (1985).

[8] Ohno, N., M. Kawabata and J. Naganuma, *Int. J. Plasticity*, 6, (3), 315-327 (1990).

[9] Ding, J.L. and W.N. Findley, *Trans. ASME J. Appl. Mech.*, 52, (3), 621-628 (1985).

[10] Findley, W.N., U.W. Cho and J.L. Ding, *Trans. ASME J. Eng. Mater. Technol.*, 101, (4), 365-368 (1979).

[11] Odqvist, F.K.G. and J. Hult, "Kriechfestigkeit Metallischer Werkstoffe," p. 159, Springer-Verlag, Berlin (1962).

[12] Malinin, N.N. and G. M. Khadjinski, *Int. J. Mech. Sci.*, 14, (4), 235-246 (1972).

[13] Murakami, S. and N. Ohno, *Int. J. Solids and Struct.*, 18, (7), 597-609 (1982).

[14] Mroz, Z. and W.A. Trampczynski, *Int. J. Solids and Struct.*, 20, (5), 467-486 (1984).

[15] Iwata, K., *Jour. Jpn. Soc. Mech. Eng.*, 90, (826), 1197-1201 (1987).

[16] Mroz, Z., W.A. Trampczynski and D.R. Hayhurst, *Int. J. Plasticity*, 4, (3), 279-299 (1988).

[17] Ding, J.L. and W.N. Findley, *Trans. ASME J. Appl. Mech.*, 51, (1), 133-140 (1984).

[18] Robinson, D.N., ORNL/TM-5969, Oak Ridge National Laboratory (1978).

[19] Farque, M.O. and M.M. Zaman, *Acta Mechanica*, 71, (1-4), 115-136 (1988).

[20] Rees, D.W.A., "Engineering Approaches to High Temperature Design," p. 57, B. Wilshire and D.R.J. Owen ed., Pineridge Press, Swansea (1983).

[21] Hamanaka, J., A. Suzuki and A. Tsuchiya, *Trans. Jpn. Soc. Mech. Eng.*, 47, (424A), 1354-1362 (1981).

[22] Ohashi, Y., M. Kawai and H. Shimizu, *Trans. ASME J. Eng. Mater. Technol.*, 105, (4), 257-263 (1983).

[23] Ohashi, Y., M. Kawai and T. Momose, *Trans. ASME J. Eng. Mater. Technol.*, 108, (1), 68-74 (1986).

[24] Kawai, M. and Y. Ohashi, *Proc. of the 4th Int. Conf. on Creep*, Tokyo, Japan, 459-464 April (1986).

[25] Kawai, M. and Y. Ohashi, *Proc. of the 2nd Int. Conf. on Constitutive Laws*, Arizona, U.S.A., 967-974 January (1987).

[26] Rees, D.W., "Creep and Fracture of Engineering Materials and Structures," p. 559, B. Wilshire and D.W.J. Owen, ed., Pineridge Press, Swansea (1981).

[27] Ikegami, K. and Y. Niitsu, *Int. J. Plasticity*, 1, (4), 331-345 (1985).

[28] Miller, A., *Trans. ASME J. Eng. Mater. Technol.*, 98, (2), 97-104 (1976).

[29] Hart, E.W., *Trans. ASME J. Eng. Mater. Technol.*, 98, (3), 193-202 (1976).

[30] Bodner, S.R. and Y. Partom, *Trans. ASME J. Appl. Mech.*, 42, (2), 385-389 (1975).

[31] Krempl, E., J.J. McMahon and D. Yao, *Mechanics of Materials*, 5, 35-48 (1986).

[32] Krieg, R.D., J.C. Swearengen and R.W. Rhode, "Inelastic Behavior of Pressure Vessel and Piping Components," p. 15, T.Y. Chang and E. Krempl ed., PVP-PB-028, (1978).

[33] Chaboche, J.L., *Bull. Acad. Polon. Sci., Sér. Sci. Tech.*, 25, (1), 33-42 (1977).

[34] Freed, A.D. and K.P. Walker, "Visco-plastic Behavior of New Materials," p. 1, D. Hui and T.J. Kozik ed., PVP-Vol. 184, (1989).

[35] Freed, A.D. and K.P. Walker, *Appl. Mech. Rev.*, 43, (5), S328-S337 (1990).

[36] Krieg, R.D., *Trans. of the 4th Int. Conf. on Struct. Mech. in Reactor Technol.*, San Francisco, U.S.A., M6/4 August (1977).

[37] Inoue, T., *JSME Int. Journal*, Ser. I, 31, (4), 653-663 (1988).

[38] Lemaitre, J. and J.L. Chaboche, "Mechanics of Solid Materials," Cambridge University Press, (1985).

[39] Freed, A.D. and J.L. Chaboche, NASA Tech. Mem. 102388, 1-19 (1989).

[40] Simo, J.C., "Cracking and Damage," p.440, J. Mazars and Z.P. Bazant ed., Elsevier Applied Science (1985).

[41] Germain, P., Q.S. Nguyen and P. Suquet, *Trans. ASME J. Appl. Mech.*, 50, (4b), 1010-1020 (1983).

[42] Kratochvil, J. and O.W. Dillon, *J. Appl. Phys.*, 40, 3207-3218 (1969).

[43] Il'yushin, A.A., *Prik. Mate. Mekh.*, 18, (6), 641-666 (1954)

[44] Kawai, M., "Creep in Structures," p.107-114, M. Zyczkowski ed., Springer-Verlag, Berlin, Heidelberg (1991).

[45] Kawai, M., *Trans. of the 11th Int. Conf. on Struct. Mech. in Reactor Technol.*, Tokyo, Japan, L23/3, 527-532 August (1991).

[46] Rice, J.R., *Trans. ASME J. Appl. Mech.*, 37, (3), 728-737 (1979).

[47] Ponter, A.R.S. and F.A. Leckie, *Trans. ASME J. Eng. Mater. Technol.*, 98, (1), 47-51 (1976).

[48] Bailey, R.W., *J. Inst. Metals*, 35, 27-43 (1926).

[49] Orowan, E., *J. West Scotland Iron & Steel Inst.*, 54, 45-53 (1946).

[50] Malmberg, T., *Trans. of the 11th Int. Conf. on Struct. Mech. in Reactor Technol.*, Tokyo, Japan, L24/1*, 539-550, August (1991).

[51] Chaboche, J.L., K. Dang Van and G. Cordier, *Trans. of the 5th Int. Conf. on Struct. Mech. in Reactor Technol.*, Berlin, Germany, L11/3 August (1979).

[52] Murakami, S., M. Kawai and Y. Ohmi, *Trans. ASME J. Eng. Mater. Technol.*, 111, (3), 278-285 (1989).

[53] Ochi, Y., S. Sasaki, A. Ishii and S. Saitoh, *Trans. Jpn. Soc. Mech. Eng.*, 53, (492A), 1508-1515 (1987).

[54] Chaboche, J.L., PhD Thèse, Université Pierre et Marie Curie, Paris (1978).

[55] Armstrong, P.J. and C.O. Frederick, CEGB Report RD/B/N731, Central Electricity Generating Board (1966).

[56] Garofalo, F., *Trans. AIME*, 227, 351-356 (1963).

[57] Lee, E.H., *Trans. ASME J. Appl. Mech.*, 36, (1), 1-6 (1969).

[58] Sluzalec, A., *Int. J. Non-Linear Mech.*, 25, (4), 395-403 (1990).

*Proceedings of the Fifth International Conference on Creep of Materials, Lake Buena Vista, Florida, USA, 18-21 May 1992*

# Evaluation of Shake Down Condition in Cyclic Elastic Follow-Up Tests with Simplified Inelastic Analyses

**K. Kobayashi**
Chiba University
Chiba, Japan

**T. Udoguchi**
The High Pressure Gas Safety Institute of Japan
Tokyo, Japan

## ABSTRACT

A method is presented whereby an inelastic strain range in cyclic elastic follow-up process is calculated with simplified inelastic analyses. These analyses consist of elastic plastic material with/without strain hardening and a relaxed stress after a certain period of time. Parameters needed for the calculation of an inelastic strain range are two stress ratios and an elastic follow-up strain amplification factor. When the simplified analytic model is used, a dimensionless value of the inelastic strain range produced in the cyclic elastic follow-up process can be easily expressed, and then the shake down condition in a structure can satisfy only the following requirement; $\xi \leq 1$, where $\xi$ is the stress ratio defined in this paper. In order to verify the condition of shake down, uni-axial cyclic elastic follow-up tests were carried out on 18Cr-8Ni steel at 700°C, and are compared with the proposed models. When the relaxed stress obtained from pure relaxation tests is used in the simplified analysis proposed, the dimensionless inelastic strain range can be estimated conservative.

A SIMPLISTIC DEFINITION OF ELASTIC FOLLOW-UP is given as follows: inelastic strain will be concentrated in a local section due to inelastic redistribution of stresses and strains in creep regime. If the elastic follow-up is not regarded in structural designs for elevated temperatures, the undesirable accumulation of inelastic strain may result in failure of structural components. ASME Boiler and Pressure Vessel Code Cases provide rules for materials and elevated temperature designs. In particular, the ASME Code, Case N-47, recognizes it to be a significant problem, and addresses the concern that a large amount of the elastic follow-up strain may be produced at higher stressed or more flexible components in the structure for high temperature use[1]. Although the Code requires that a large elastic follow-up strain should be taken into account in designs for elevated temperature, it provides no definite description on methods of its evaluation, nor lower limit of the amount of strain that should be accounted for.

Simplified inelastic analyses have been required for high temperature design to reduce the number of structural components where the detailed inelastic computation is needed. The authors studied the simple estimation of elastic follow-up behavior under the uni-axial condition. However, no consideration was given to the problem concerned with the stability of the components where the elastic follow-up occurred, and few experimental works were carried out on cyclic elastic follow-up tests yet.

Since many of machines and components employed at elevated temperatures are inspected at regular intervals, cyclic loading/deformation is induced in them. As a consequence, in order to ensure the safety of structural components within a certain life, it is desirable to maintain shake down in the structure, even if a local section in the structure experiences accumulation of inelastic deformation.

In this paper, the authors propose the condition of shake down and the estimation of the inelastic strain range under cyclic elastic follow-up tests using three kinds of simplified inelastic analyses. Cyclic uni-axial elastic follow-up tests are conducted on 18Cr-8Ni steel at 700°C, and the inelastic strain range produced are compared with results of the proposed analysis.

## ANALYTIC MODEL

The structure of uni-axial elastic follow-up model investigated is illustrated in Fig.1(a). The test specimen kept at an elevated temperature in creep regime is coupled in series to an end of a perfectly elastic element with spring constant K. A rigid tensile displacement $\delta_0$ is cyclically applied at an end of the assembly (Fig.1(b)). While the compulsive displacement is maintained constant for two hours, the elastic follow-up occurs: the stress in the specimen decreases, and the strain in the specimen increases with time (Fig.1(c) and (d)).

Figure 2 shows a typical stress-strain curve appeared at the first cycle and at the second re-loading process, where the path O→Q→R→T→U corresponds to the process 0→1→2→3→4 in Fig.1(b). If the behavior of the test material is elastic at the first loading, a virtual initial stress in the specimen is equal to the stress $\sigma_n$, representing the stress of the point P in Fig.2. However, the stress produced in the specimen at the starting point of elastic follow-up test (the point Q in Fig.2) was mostly beyond the elastic limit of the test material. Therefore, the initial strain $\varepsilon_0$ contained plastic strain. The path Q→R shows the elastic follow-up process.

The total strain range $\Delta\varepsilon_t$, the inelastic strain range $\Delta\varepsilon_{in}$, and stress amplitude $\Delta\sigma$ produced during the process of unloading and re-loading are defined as shown in Fig.2. Two dimensionless values p and $\xi$ which are introduced to normalize both the total strain range and the inelastic strain range are given.

$$p=E/k\ ,\quad \xi=\sigma_n/\overline{\sigma_y} \qquad \text{------------------------(1)}$$

where E is Young's modulus of the test material. The value of k means a tangent modulus at elastic follow-up in the stress-strain field. It depends on the stiffness of the elastic element employed. A triangle OPS is called as a triangle of elastic follow-up. The stress $\sigma_n$, elastically calculated at the initial loading on the triangle of elastic follow-up, represents the virtual stress of tests. Furthermore, the stress $\overline{\sigma_y}$ is defined as the sum of the yielding stress $\sigma_y$ and a relaxed stress $\sigma_r$ after a certain period of time. It was introduced to identify magnitude of the elastic follow-up test. Whether the inelastic strain is performed in cyclic elastic follow-up tests depends on magnitude of the virtual test stress $\sigma_n$.

**Model A [See Fig.3].** Figure 3 shows an analytic model for the evaluation of shake down, which consists of pairing elastic-perfectly plastic material and the relaxed stress $\sigma_r$. It was assumed that the stress $\sigma_r$ obtained from relaxation tests coincided with the stress in elastic follow-up tests, and also that the stress $\sigma_r$ was always constant despite the changes in magnitude of the virtual stress $\sigma_n$.

Shake down is indicated that when the cyclic loading/displacement is induced even in the structure where materials deform plastically, the response of the structure is elastic and then no inelastic strain will be detected during the cyclic

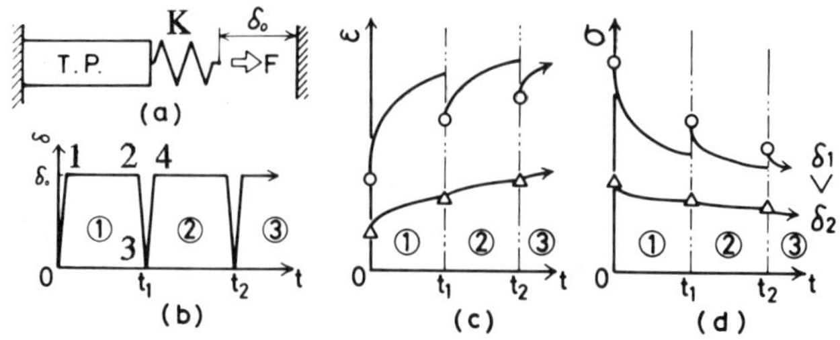


Fig.1 - Schematic view of the uni-axial cyclic elastic follow-up tests.
(a) Analytical model, (b)history of cyclic loading,
(c)strain response, (d)stress response of elastic follow-up.

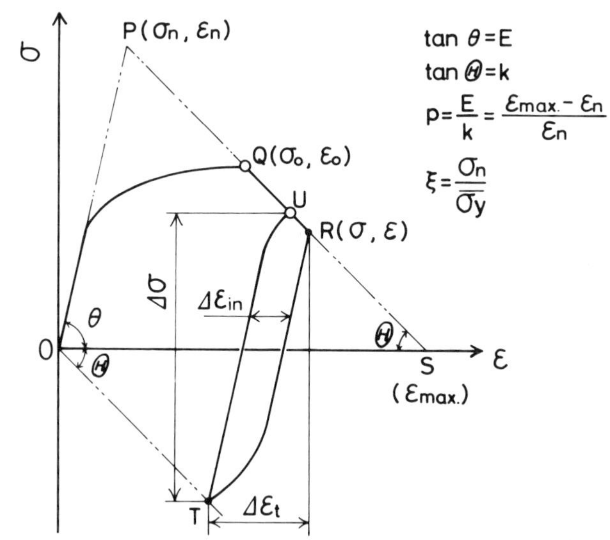


Fig.2 - Schematic illustration of cyclic elastic follow-up tests (Initial loading and the subsequent cycle).

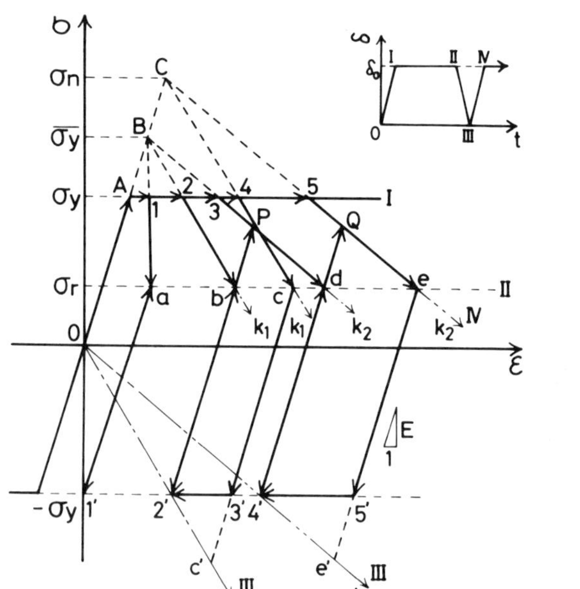


Fig.3 - Model A for the analysis.(Elastic-perfectly plastic material and a constant relaxed stress)

loading/displacement. When we complete the description of the condition of shake down using this model, only the following criterion will be required: $\xi \leq 1$. This is equivalent to the fact that when the virtual stress $\sigma_n$ in elastic follow-up tests is smaller than the value of stress $\overline{\sigma_y}$ in Fig.3, the stress-strain relationship between unloading and re-loading will behave in an elastic manner. In this paper hereinafter the authors referred to the stress $\overline{\sigma_y}$ as a critical stress of shake down.

Then, let us see results of some examples. In the case of pure relaxation which starts from the point B in Fig.3, the process at the initial loading is drawn as solid lines O→A→1, and that at relaxation 1→a. Responses to unloading and to re-loading correspond to paths a→1' and 1'→a, respectively. The stress-strain behavior in the elastic follow-up test with the tangent modulus $k_1$ is also represented as the path 2→b. Shake down is described in a similar way that adopted in the pure relaxation test: the process at unloading will be shown as the path b→2' and that at re-loading 2'→b. When the elastic follow-up test with the tangent modulus $k_1$ has the virtual stress $\sigma_n$ (the point C in Fig.3) which is beyond the critical stress $\overline{\sigma_y}$, the unloading process and the re-loading one are illustrated as paths c→3'→2' and 2'→P, respectively. As a consequence, inelastic strain $\Delta\varepsilon_{in}$ (=$\varepsilon_c$-$\varepsilon_b$ or $\varepsilon_{3'}$-$\varepsilon_{2'}$) is performed during unloading and re-loading. The total strain range $\Delta\varepsilon_t$ is calculated as an amount of strain( $\varepsilon_c$-$\varepsilon_{2'}$). It was also recognized that both magnitude of the strain ranges $\Delta\varepsilon_{in}$ and $\Delta\varepsilon_t$ depended on the stiffness of the elastic element k. The stress amplitude $\Delta\sigma$ performed in the cyclic elastic follow-up diagram is defined as a stress ($\sigma_P$-$\sigma_{2'}$). After normalization using the yielding strain $\varepsilon_y$, the strain ranges defined are listed in Table 1, where a parameter $\zeta$ is introduced as the stress ratio of the relaxed stress $\sigma_r$ to the flow stress $\sigma_y$:

$$\zeta = \sigma_r/\sigma_y \qquad \text{------------------------------} \quad (2)$$

In Fig.4, the dependence of dimensionless inelastic strain range $\Delta\varepsilon_{in}/\varepsilon_y$ on parameters p and $\zeta$ is examined. The dimensionless inelastic strain range $\Delta\varepsilon_{in}/\varepsilon_y$ is sensitive to the value of p, but it is a little affected by the growth of $\zeta$. If the stress $\sigma_n$ is the same in the two analyses, the values of $\xi$ will not coincide each other because of the fact that $\xi=1/(1+\zeta)$ exists. For example, when p was taken to be 2.0, the values of $\Delta\varepsilon_{in}/\varepsilon_y$ were calculated to be 2.1 at $\zeta$=0.75 and to be 3.6 at $\zeta$=0.25.

**Model B [See Fig.5].** Most of structural materials show the phenomenon of strain hardening. In order to describe the strain ranges mentioned in the former section more

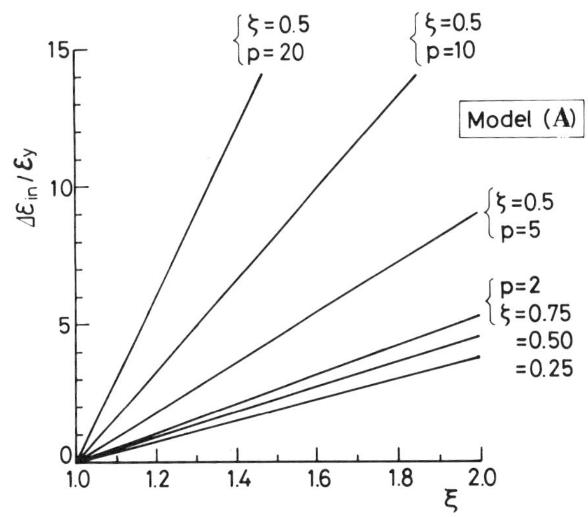


Fig.4 - Dependence of dimensionless inelastic strain range $\Delta\varepsilon_{in}/\varepsilon_y$ on $\zeta$, $\xi$, and p using the Model A.

Table 1. Dimensionless inelastic strain range $\Delta\varepsilon_{in}/\varepsilon_y$ and dimensionless total strain range $\Delta\varepsilon_t/\varepsilon_y$, using the simplified inelastic analyses proposed.

| Model | (A) | (B) | (C) |
|---|---|---|---|
| $\frac{\Delta\varepsilon_{in}}{\varepsilon_y}$ | $(1+\zeta)(1+p)(\xi-1)$ $(\xi\geq1)$ | $(1+\zeta)(1+p)(\xi-1)+\frac{\eta(1+p)^2}{1+\eta p}$ $(\xi\geq\xi_0)$ | $\frac{(1+\zeta)(1+p)(\xi-1)}{1+\eta p}$ $(\xi\geq1)$ |
| $\frac{\Delta\varepsilon_t}{\varepsilon_y}$ | $(1+\zeta)\{(1+p)\xi-p\}$ $(\xi\geq1)$ | $(1+\zeta)\{(1+p)\xi-p\}+\frac{\eta p(1+p)}{1+\eta p}$ $(\xi\geq\xi_0)$ | $\frac{1+\zeta}{1+\eta p}\{(1+p)\xi-(1-\eta)p\}$ $(\xi\geq1)$ |
| | $(1+\zeta)\xi$ $(\xi\leq1)$ | $(1+\zeta)\xi$ $(\xi\leq\xi_0)$ | $(1+\zeta)\xi$ $(\xi\leq1)$ |

$$\xi=\sigma_n/\overline{\sigma}_y, \quad \eta=h/E, \quad \zeta=\sigma_r/\sigma_y, \quad \xi_0=1-\frac{\eta(1+p)}{(1+\zeta)(1+\eta p)}, \quad \overline{\sigma}_y=\sigma_y+\sigma_r, \quad p=E/k$$

precisely, the model which consisted of linear hardening elastic-plastic material and a relaxed stress was introduced. Figure 5 shows a stress-strain diagram to calculate strain ranges produced in the cyclic elastic follow-up test. The aspect of strain hardening in compression is also considered as well as in tension.

Although the critical stress for the shake down was formulated, it varied with the stiffness of the elastic element $k_i$. For example, in the case of pure relaxation the critical stress which can attain shake down is represented as the point D in Fig.5; the initial loading process is drawn as the path $0 \rightarrow A \rightarrow 1$, and after the relaxation process of $1 \rightarrow a$ shake down is illustrated as the path a-1'-a. However, in the case that the elastic element with the stiffness $k_1$ (i.e. elastic follow-up amplification factor $p_1$) is employed in the elastic follow-up test, the critical stress is obtained at the point C in Fig.5. The critical stress $\overline{\sigma_{yi}}'$ is given as follows using a new parameter $\eta$:

$$\overline{\sigma_{yi}}' = \left(\zeta + \frac{1-\eta}{1+\eta p_i}\right) \sigma_y \quad \text{--------------(3)}$$

$$\eta = h/E \quad \text{------------------------------(4)}$$

where h showed tangent modulus after yielding of the material. Equation 3 has a maximum value at p=0 (pure relaxation).

The inelastic strain range $\Delta\varepsilon_{in}$ and the total strain range $\Delta\varepsilon_t$ produced during the cyclic elastic follow-up test was normalized using yielding strain $\varepsilon_y$. They are also listed in Table 1, where the critical stress $\overline{\sigma_{yi}}'$ used in this model is not equal to the critical stress $\overline{\sigma_y}$ introduced in the Model A. The lowest value of $\xi_0$ at $\Delta\varepsilon_{in}$=0 is obtained as

$$\xi_0 = 1 - \frac{n(1+p)}{(1+\zeta)(1+\eta p)} \quad \text{--------------(5)}$$

When the dimensionless strain ranges $\Delta\varepsilon_{in}/\varepsilon_y$ and $\Delta\varepsilon_t/\varepsilon_y$ were compared with those using the Model A, the results in this model had only an additional term to the results in the Model A. Moreover, it will be seen that both the second term of dimensionless strain range is much less than the first term of them.

**Model C [See Fig.6].** This model illustrated in Fig.6 is composed of linear hardening elastic-plastic material, and the relaxation stress $\sigma_r$ which depends on an amount of the strain. The modulus of the relaxed stress to the strain was regarded as the same as tangent modulus at strain hardening. The critical stress $\overline{\sigma_y}$ for shake down was equal to the critical stress $\overline{\sigma_y}$ defined in the Model A: $\overline{\sigma_y} = \sigma_y + \sigma_r$. The results of dimensionless strain ranges $\Delta\varepsilon_{in}/\varepsilon_y$ and $\Delta\varepsilon_t/\varepsilon_y$ obtained are listed in Table 1, too. The dimensionless strain range $\Delta\varepsilon_{in}/\varepsilon_y$ is smaller than the value using the Model A, and is equal to $1/(1+\eta p)$ times of the latter.

**Comparison of Dimensionless Inelastic Strain Range** In an attempt to change the parameters of p, $\zeta$, and $\eta$, the relationship between the dimensionless inelastic strain range $\Delta\varepsilon_{in}/\varepsilon_y$ and the stress ratio $\sigma_n/\overline{\sigma_y}$ (=$\xi$) are shown

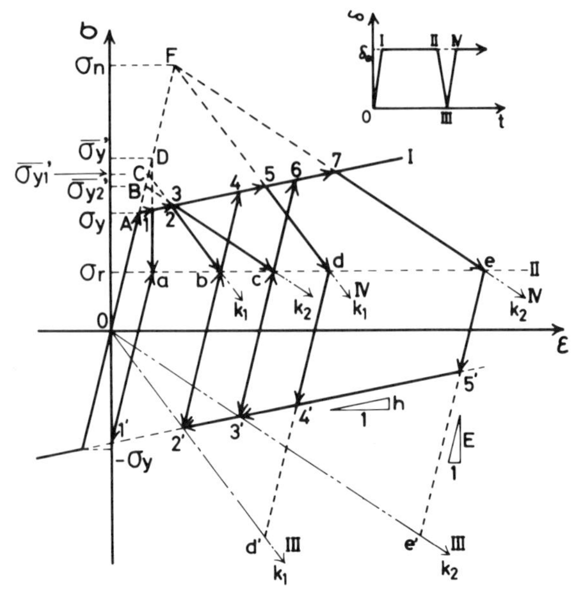

Fig.5 - Model B for the analysis. (Linear hardening elastic-plastic material and a constant relaxed stress).

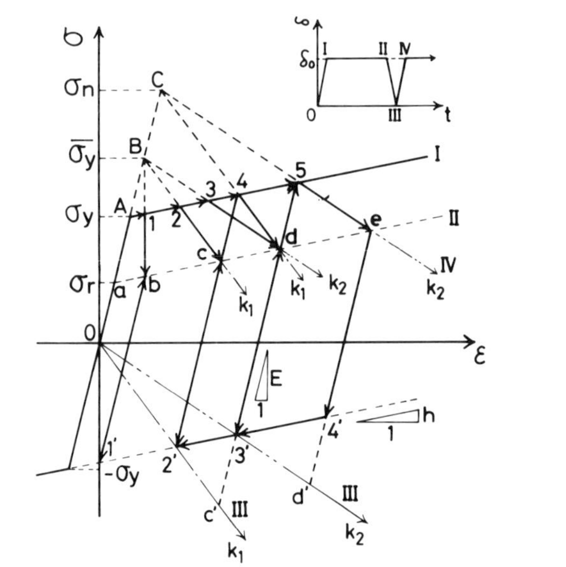

Fig.6 - Model C for the analysis. (linear hardening elastic-plastic material and a relaxed stress with a same hardening rate).

in Figs 7. Those indicate that there is little difference among the models proposed. When the strain ranges provided by the Model A were compared with the corresponding strain ranges by the Model B, the values obtained from the Model B were a little larger than the former ones. In this paper, the equation derived by the Model A was employed to evaluate the dimensionless strain ranges produced in the cyclic elastic follow-up tests, because the representation of both the critical stress for shake down and the equation to calculate the dimensionless strain ranges was much simpler and more convenient.

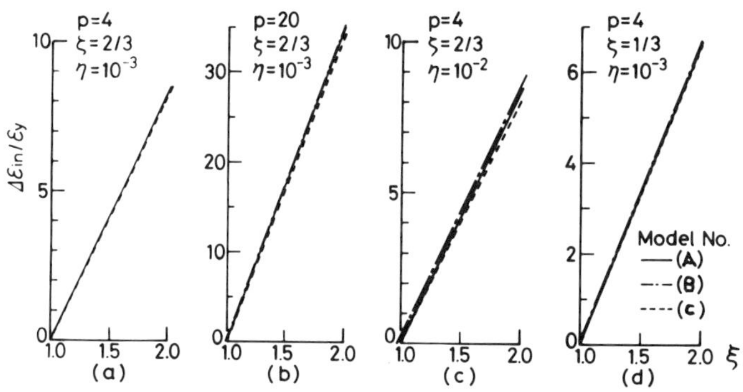


Fig.7 - Comparison of dimensionless inelastic strain range $\Delta\varepsilon_{in}/\varepsilon_y$ between the proposed analyses.

## EXPERIMENT

The material used was a standard 18Cr-8Ni steel (SUS 304) bar, and the experiment was carried out at 700°C. The diameter and gage length of the specimen machined were 40 mm and 8 mm, respectively.

Since detailed information on the testing apparatus has been reported in a previous paper[2], its outline was given here. Three elastic elements employed were a hard straight bar, a ring spring A, and a ring spring C. A stiffness of them in the stress-strain field was found to be 49, 30, and 6 GPa for the correspond-ing elastic elements. As a consequence, the elastic follow-up amplification factor p was expressed as about 2, 4, and 22, respectively.

In the tests, the compulsive tensile displacement was applied to the assembly. After keeping the displacement for two hours, it was completely removed. This process was repeated ten times in the cyclic elastic follow-up tests.

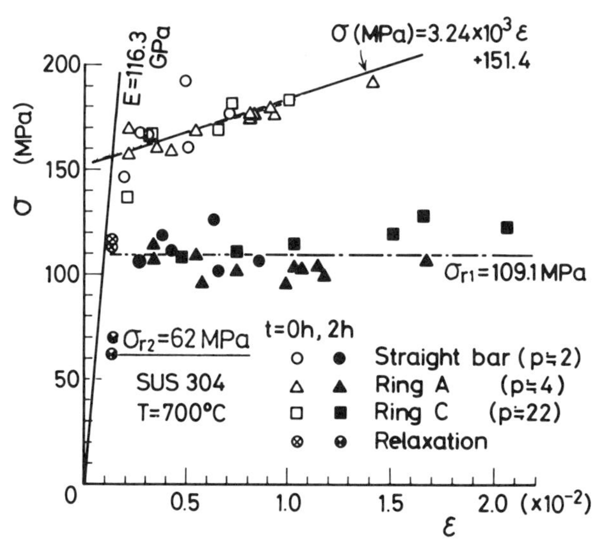


Fig.8 - Relationship between stress and strain at the start of elastic follow-up tests (open symbols) and at the time after the first test duration (closed symbols).

## RESULTS AND DISCUSSION

**Yield Stress $\sigma_y$ and Relaxed Stress $\sigma_r$.** First of all, both the yielding stress of testing material $\sigma_y$ and the relaxed stress $\sigma_r$ defined in the Model A are determined.

The open symbols in Fig.8 show the stress-strain relationship at the beginning of the first cycle of elastic follow-up tests and at that of the relaxation tests. On the other hand, the closed ones represent the states where the first test duration of 2 hour ended up. Although the relaxed stress obtained in the tests with the ring spring C seem to be a little dependent on the strain, the relaxed stress $\sigma_r$ is regarded as constant over the entire range of strain, and is found to be 109 MPa. We named herein it the relaxed stress $\sigma_{r1}$. Incidentally when we use the Model A described, the relaxed stresses obtained in pure relaxation tests are assumed to be equal to those in the cyclic elastic follow-up tests. However, the former values were found to be much smaller than the latter ones, and its minimum value was 62 MPa. We called it the relaxed stress $\sigma_{r2}$.

The yield stress of the material $\sigma_y$ is determined using the least square method, which is applied to the opening symbols except near the elastic line. The solid line in Fig.8 is thus obtained. As shown, the elastic curve is determined from the initial response of the test material. The yield stress $\sigma_y$ is introduced as an intersecting point of those solid lines. It is found to be 157 MPa, and then the yield strain $\varepsilon_y$ is equal to 0.134%.

As a consequence, when two relaxed stresses are used, critical stresses for shake down $\overline{\sigma_{yi}}$ are given as follows.

$$\overline{\sigma_{y1}} = \sigma_y + \sigma_{r1} = 157 + 109 = 266 \text{ MPa} \qquad \text{------------(6)}$$

$$\overline{\sigma_{y2}} = \sigma_y + \sigma_{r2} = 157 + 62 = 219 \text{ MPa} \qquad \text{------------(7)}$$

Then, parameters ξ and ζ in Table 1 are given as

$$\left.\begin{array}{ll}\xi_1 = \sigma_n/\overline{\sigma_{y1}} = \sigma_n/266, & \zeta_1 = \sigma_{r1}/\sigma_y = 0.694 \\ \xi_2 = \sigma_n/\overline{\sigma_{y2}} = \sigma_n/219, & \zeta_2 = \sigma_{r2}/\sigma_y = 0.398\end{array}\right\} \text{----(8)}$$

where the subscripts 1 and 2 were used to distinguish the use of the relaxed stress defined earlier.

**Properties of Cyclic Deformation** Figure 9 shows an example of cyclic stress-strain curves observed in cyclic elastic follow-up tests. The presented results indicate that the inelastic strain range produced in the hysteresis loop depends on the value of parameter $\xi$, which is the elastic ratio of the virtual stress $\sigma_n$ to the critical stress $\overline{\sigma_y}$. Although the parameter $\xi_2$ shows the condition of shake down more conservative than $\xi_1$ does, a small inelastic strain is detected even in the test where $\xi_2$ is equal to 1.03.

Figure 10 shows variations of $\Delta\varepsilon_t$, $\Delta\varepsilon_{in}$, and $\Delta\sigma$ related to Fig.9 during the cyclic elastic follow-up tests. The change of those ranges are found small in the cyclic elastic follow-up test, but a careful observation show us the cyclic softening of the test material. The results in the tests employed the ring springs A and C were similar to those tested with the use of the rigid straight bar.

**Evaluation of Inelastic Strain Range and Total Strain Range Using the Proposed Equation** Figure 11 illustrates the comparison of the test results with the calculation on the dimensionless inelastic strain range, using the parameter $\xi_1$. The bar in the figure shows the fluctuation of inelastic strain range during the tests. Straight lines indicate results of the representative calculation, where the parameter p is given 22, 4, and 2 when the rigid straight bar, the ring spring A and C are used as the elastic element, respectively. As shown, many test results are located on the condition of shake down: $\xi_1 \leq 1$, because the critical stress $\xi_1$ is large. In addition, the values of calculation (the open symbols) are usually smaller than those of experiments (the closed symbols). Then, an unconservative design could result if the designer chose the critical stress for shake down which was based on the mean value of relaxed stresses obtained from elastic follow-up tests. This implies that the good selection of the critical stress $\overline{\sigma_y}$ is very important for a reasonable design.

Figure 12 shows the evaluation of proposed equation for the same data of Fig.11, by using the critical stress $\overline{\sigma_{y2}}$ which are calculated with the help of the minimum stress obtained from pure relaxation tests. Since the critical stress $\overline{\sigma_{y2}}$ was larger than $\overline{\sigma_{y1}}$, the data smaller than $\xi_2 \leq 1$ decreased. The estimation was found to be conservative when it was applied for the data over $\xi_2 \geq 1$.

As the same as in Fig.12, the experiments of the dimensionless total strain range $\Delta\varepsilon_t/\varepsilon_y$ are compared with the calculation in Fig.13, by using the critical stress $\overline{\sigma_{y2}}$. Experimental values shown as the closed circles closely

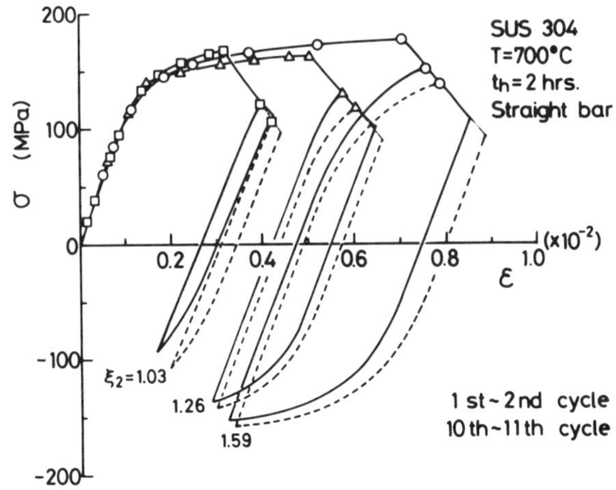


Fig.9 - Typical stress and strain curves obtained in cyclic elastic follow-up tests.

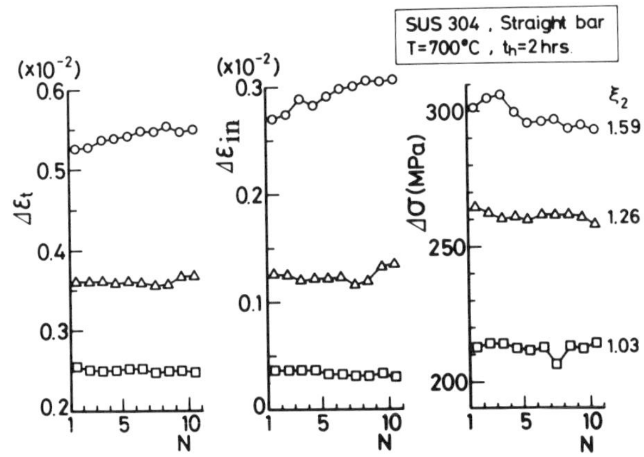


Fig.10 - Dependence of the total strain range $\Delta\varepsilon_t$ the inelastic strain range $\varepsilon_{in}$, and the stress range $\Delta\sigma$ on the number of cycles N for the cyclic elastic follow-up tests.

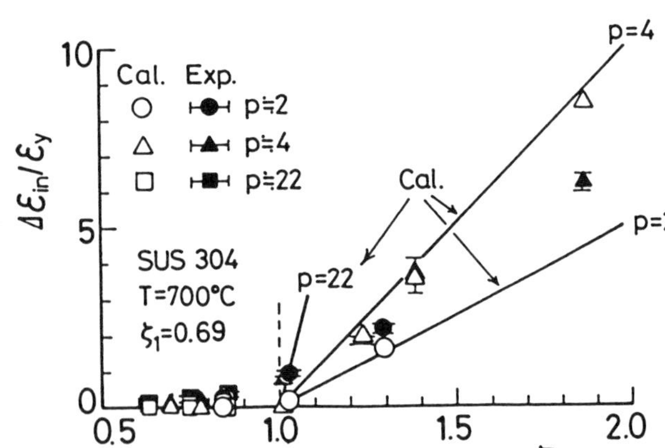


Fig.11 - Comparison between the dimensionless inelastic strain ratio $\Delta\varepsilon_{in}/\varepsilon_y$ and their calculations when the parameter $\xi_1$ is used.

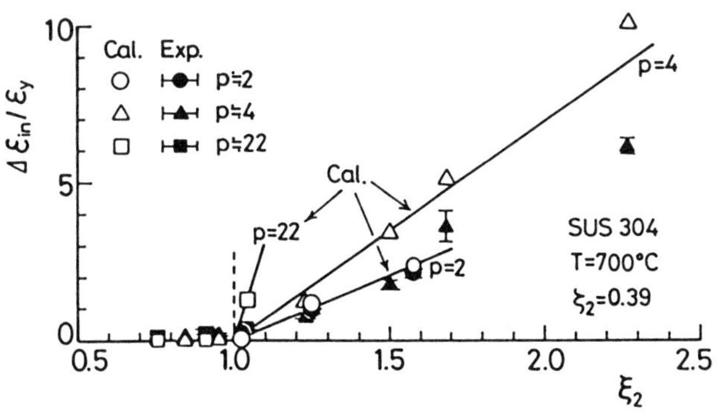


Fig.12 - Comparison between the dimensionless inelastic strain ratio $\Delta\varepsilon_{in}/\varepsilon_y$ and their calculations when the parameter $\xi_2$ is used.

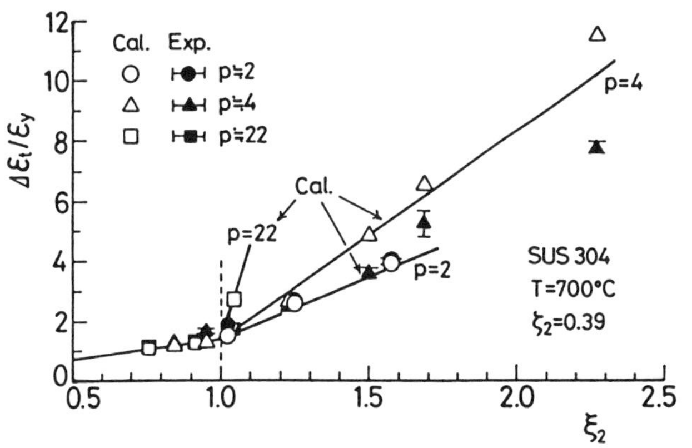


Fig.13 - Comparison between the dimensionless total strain ratio $\Delta\varepsilon_t/\varepsilon_y$ and their calculations when the parameter $\xi_2$ is used.

agree with the calculated ones shown as the open circles. The other calculations are somewhat larger than experiments. Thus, the critical stress $\overline{\sigma_y}$, expressed as the sum of the yield stress of the material $\sigma_y$ and the relaxed stress $\sigma_r$ based on the pure relaxation test, can be introduced to represent the condition of shake down, and will propose a rule for design at elevated temperatures.

## CONCLUSION

In order to estimate the inelastic strain range which is produced in the hysteresis loop during cyclic elastic follow-up tests, several simplified inelastic analyses are proposed, and are compared with some experiments. Thus, we conclude as follows:

(1) The model composed of elastic-perfectly plastic material and the relaxed stress is the simplest, and there is little difference between it and the others.

(2) Shake down is achieved when the virtual stress $\sigma_n$ is smaller than the critical stress $\overline{\sigma_y}$, which is defined as the sum of yield stress of the test material and the relaxed stress.

(3) When the relaxed stress obtained from pure relaxation tests after a certain time is used, both the inelastic strain range and the total strain range can be estimated conservatively.

## REFERENCES


1- ASME, *Boiler and Pressure Vessel Code, Section III, Case N-47-23*, 29 (1986)

2- Kobayashi,K.I., Y.Saitoh and T.Udoguchi, *Proc. of the 6th International Congress on Experimental Mechanics*, Portland OR., Vol.1, 95-100 (1985)

*Proceedings of the Fifth International Conference on Creep of Materials, Lake Buena Vista, Florida, USA, 18-21 May 1992*

# Creep Deflection Analysis of a Gas Turbine Combustion Transition Piece

**J.E. Barnes**
General Electric Company
Schenectady, New York

## ABSTRACT

The inelastic deflection of an industrial gas turbine combustion transition piece has been analytically calculated using the finite element method and correlated favorably to field measurements. This success required inclusion of startup and shutdown cycling of pressure and thermal loadings to obtain ratcheted deflections. Bilinear kinematic hardening and a three constant power law represented plasticity and creep behavior of the nickle base super alloy Hastelloy X. An incremental solution technique allowed proper distribution of plastic strain due to loading and creep strain due to increasing time. The commercial finite element computer code ANSYS was employed using a high performance workstation. Thus, the methodology can be used by design engineers without a high level of finite element expertise and extremely large computer resources. The methodology is a valuable design tool for new gas turbine development as well as consideration of the effects of numerous operating variables on creep life.

THE COMBUSTION TRANSITION PIECE is located at the midsection of the gas turbine between the compressor and turbine. A detailed view of the combustion system is shown in Figure 1. Fourteen of these combustors are located around the machine axis. The transition piece is about 64 cm. (25 inches) in length with an inlet diameter of 35.6 cm. (14 inches) and a thickness of 4.8 mm. (.188 inches).

Within the transition piece, gas at a temperature of 1140°C (2090°F) from the combustion liner flows to the turbine nozzles and buckets. The outside of this duct is exposed to .5 atm (7.5 psi) higher pressure than the inside and 350°C (655°F) temperature flow from the compressor. Thus, due to high metal temperatures, thermal gradients and pressure loading, creep is a critical design criteria.

In the early 1980's, GE transition pieces began exhibiting creep deformation as turbine firing temperatures had been increased. Excessive creep deflection causes a loss of sealing between the compressor air and combustion gas flows which degrades the turbine performance. Several design changes along with changing material from Hastelloy X to Nimonic 263 eliminated the creep deformation. Analytical representation of the creep improvement at that time was of limited success and ratcheting not believed to be occurring. This work corresponds to the older Hastelloy X design and correlates to the field data from that period. Nimonic 263 results are also given to further validate the method.

## ANALYTICAL PROCEDURE

The goal of this work was a production methodology for designing transition pieces that could be readily employed by design engineers. The commercial general purpose finite element analysis program ANSYS [1] was selected due to the widespread familiarity that engineers have with using this package.

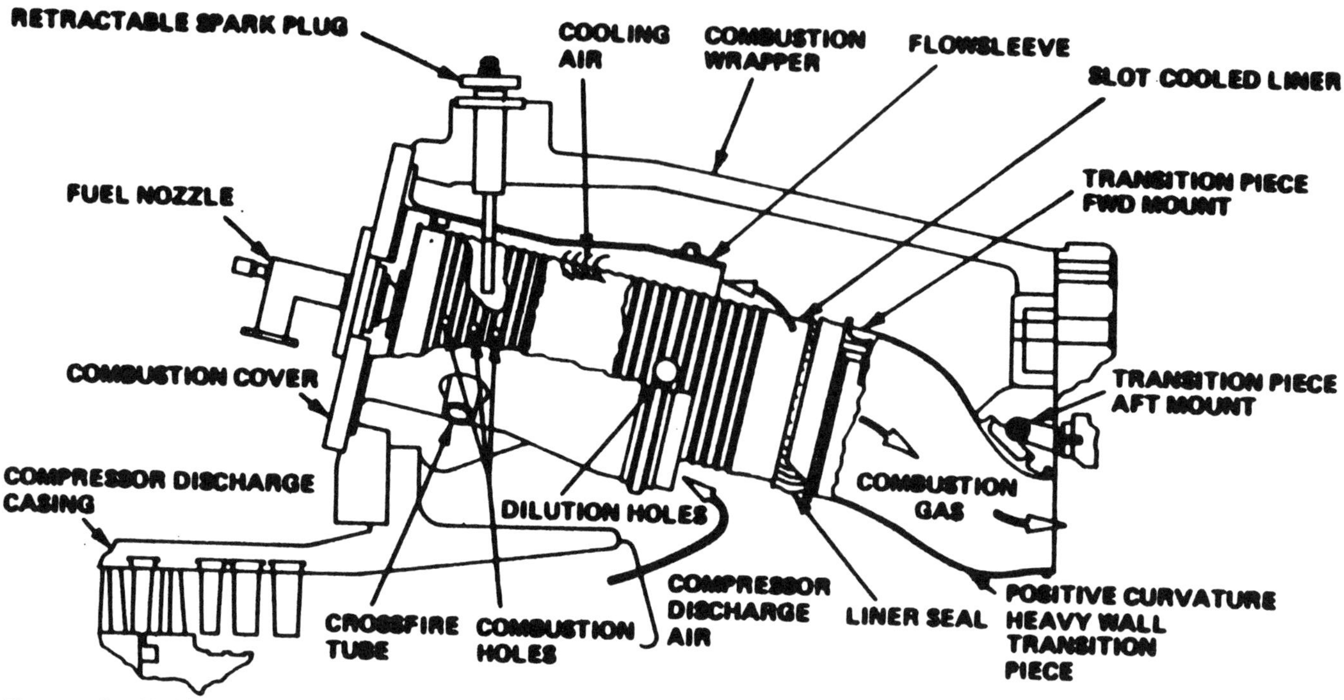


Figure 1. Gas Turbine Combustion System.

Advanced nonlinear finite element structural analysis was utilized for the deflection calculation. Representation of the complex geometry, boundary conditions, material constitutive behavior and turbine operating transients were all crucial to obtaining an accurate solution. Previous attempts at analytical creep computation for combustion hardware had not been very successful due to the over simplifications required by older limited computer resources.

The finite element model of the transition piece is shown in Figure 2. This figure shows a full transition piece for clarity but actually only a symmetric half was analyzed. The plastic quadrilateral shell element was used in the transition piece body section for modelling and solution efficiency. This shell allows nonlinear strain variation through the thickness direction to capture plastic and creep behavior. The eight noded brick element was used for the three dimensional attachments to the body such as the aft mount and frame. A total of 1392 elements and 1636 nodes were required to accurately represent the geometry. Restraints were applied at the symmetry boundary and at the actual support locations.

An iterative solution method is used for nonlinear material behavior. Plastic and creep strain are both computed based on the Von Mises equivalent stresses at each integration point at the start of the load step. The creep strains were determined from stresses computed after the plasticity solution had converged.

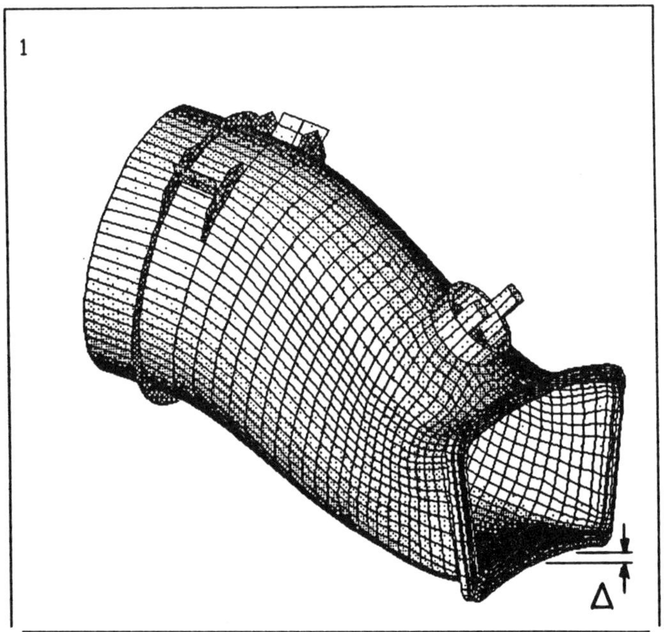


Figure 2. Transition Piece Finite Element Model.

The thermal and pressure boundary conditions were applied and removed gradually and creep time increments are suitably small because the solution is nonconservative with results being path dependent. The actual turbine startup and shutdown cycle is followed to allow load to accurately redistribute as higher stressed locations yield. Forty nine load steps were required to both load and unload the structure. Creep time steps were initially .0001 hours so that creep/ stress relaxation also accurately redistributes loads. These time steps increase based on the amount of creep strain in the previous iteration.

## MATERIAL CONSTITUTIVE RELATIONS

Hastelloy X material was analyzed because field data using this material showed significant creep deflections which allowed verification of the methodology. Temperature dependent material constitutive relations describing stress versus strain and creep strain versus time were input to the analysis.

**Plasticity.** Plasticity was modelled using bilinear kinematic hardening where the total stress range is equal to twice the yield stress; thus including the Bauschinger effect. Multilinear kinematic hardening behavior is available in ANSYS, but not for the plastic shell. Also, stress strain material data was not available. The tangent modulus was obtained as the slope through the .02 and .2% yield points using the numerical average from 1/2 inch plate data. This is a good approximation since a large amount of material strains to about .2%. Figure 3 is a plot of stress versus strain at six temperatures.

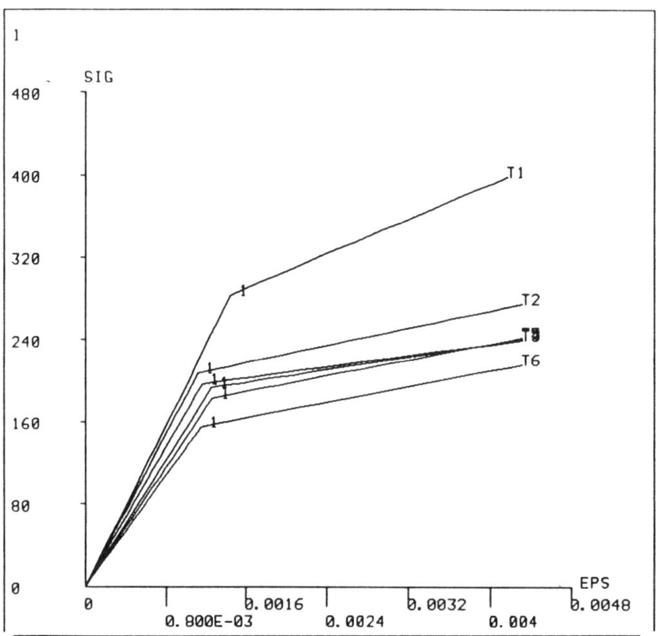


Figure 3. Hastelloy X Bilinear Stress (MPa) vs. Strain at Temperature (T1=10, T2=204, T3=427, T4=649, T5=760, T6=871°C).

**Creep.** Creep was represented by a 3 constant power law known as the Bailey Norton equation. This simple constitutive equation was tried initially because of the relative ease of the finite element solution and fitting of material property data. Only the primary and secondary creep regimes are typically represented with this equation.

The 3 constant power law creep function is commonly written:

$$\varepsilon = K\,(C_1\,\sigma)^N\,t^M$$

With the finite element method for small time increments the equation is differentiated with respect to time (considering $\sigma$ constant with time for the iteration):

$$\Delta\varepsilon = M\,K\,(C_1\,\sigma)^N\,t^{M-1}\,\Delta t$$

where: $\varepsilon$ = creep strain (in./in.)

$\sigma$ = elastic equivalent stress (psi)

$t$ = time (hrs.)

$C_1$ = .0001

M,K,N = material property constants determined experimentally.

Two sets of Hastelloy X creep properties were used in the analysis. Elements in the model experiencing high elastic plastic strain, ie. greater than about .2% were given "degraded" properties representing the tertiary creep regime. This degradation is believed to occur as strain in excess of .2% shortens the time in the primary and secondary stages [2]. Tertiary creep, increasing rate with time, is obtained by increasing the exponent M above 1.0. The high strain elements are located at the aft frame.

## BOUNDARY CONDITIONS

The base load steady state metal temperature distribution was computed using finite element analysis and correlated to measured data. Contours of metal temperature at base load are plotted in Figure 4. The steady state gas temperatures and heat transfer coefficients were then ratioed to determine the transient boundary conditions throughout the start−up and shut−down. Maximum metal temperature and thermal gradient through the aft frame versus time throughout the startup and shutdown is shown in Figure 5. Maximum metal temperature is 730°C (1350°F) at the aft frame end and a gradient of 175°C (350°F) exists through the aft frame.

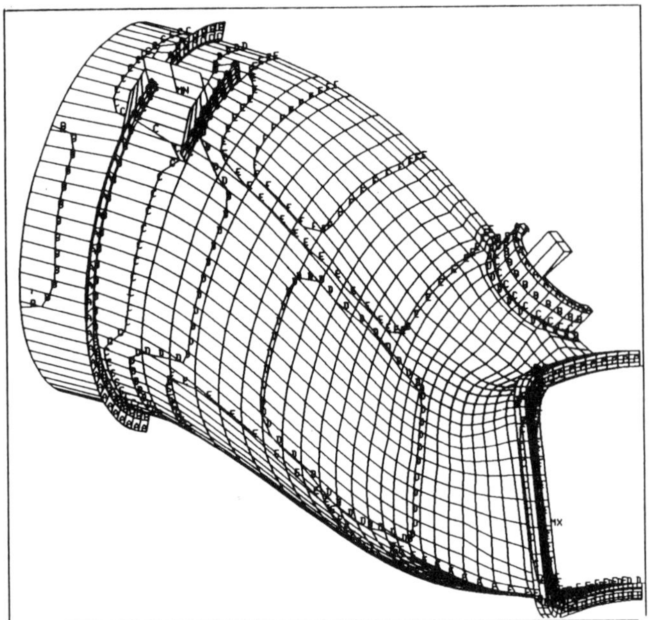

Figure 4. Metal Temperature Distribution ($T_{min}$=346, A=378, B=442, C=507, D=571, E=635, F=700, $T_{max}$=732°C).

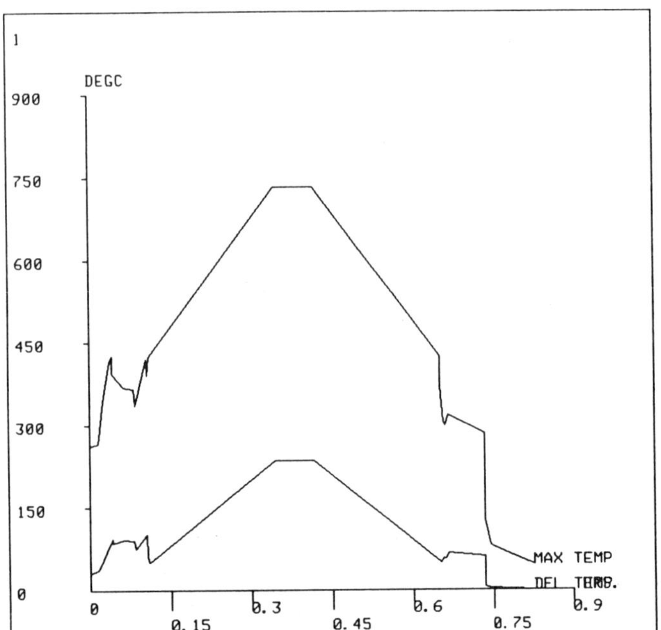


Figure 5. Maximum Metal Temperature and Thermal Gradient vs. Time throughout Startup and Shutdown.

An external pressure of .5 atm (7.5 psi) at full load operation was also applied. This load was also ratioed throughout the transients for the start–up and shut–down.

## RESULTS AND DISCUSSION

**Linear Elastic Results.** Linear elastic stress distributions for pressure only (primary) and pressure combined with thermal (secondary) loadings in the side of the aft frame are given in Figures 6 and 7. The combined loading has a maximum compressive stress of −522 MPa (−76 ksi) on the hotter internal surface of the transition piece side. Pressure only stress at this location is −41 MPa (−6 ksi). Basically, the thermal gradient through the aft frame causes the outer diameter, side and inner diameter sections of the aft frame to attempt to curve away from the hotter surface. Compatibility forces the three sections to have large compressive stress on the internal surfaces. The pressure only stress on the internal surface is compressive on the side panel but tensile on the inner and outer diameters.

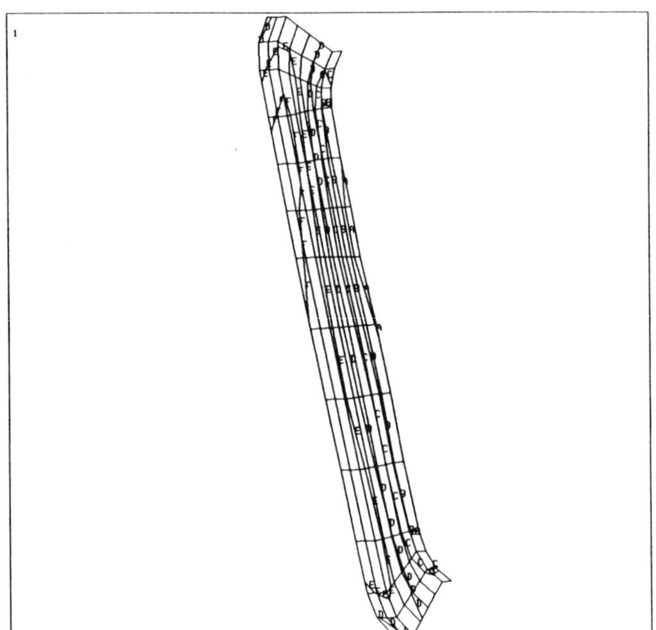

Figure 6.. Linear Elastic Stress on Side of Aft End of Aft Frame for Pressure Only Loading ($\sigma_{min}$=−41, A=−35, B=−22, C=−9, D=4, E=16, F=29, $\sigma_{max}$=36 MPa).

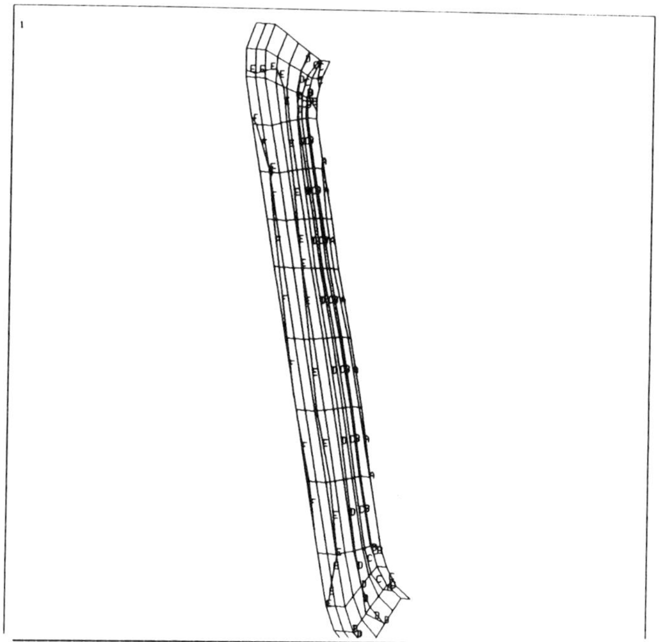

Figure 7. Linear Elastic Stress on Side of Aft Frame for Combined Pressure and Thermal Loading ($\sigma_{min}=-522$, A=446, B=295, C=144, D=8, E=159, F=310, $\sigma_{max}=386$ MPa).

**Hastelloy X Creep Results.** Deflection of the aft frame opening (defined on Figure 2) versus time from the elastic–plastic–creep analysis is shown on Figure 8 for five operating cycles of 200 hours each. The transition piece has closed 2.8 mm (.110 inches). Field measurements indicate about 3.0 mm (.120 inches) for the same time period.

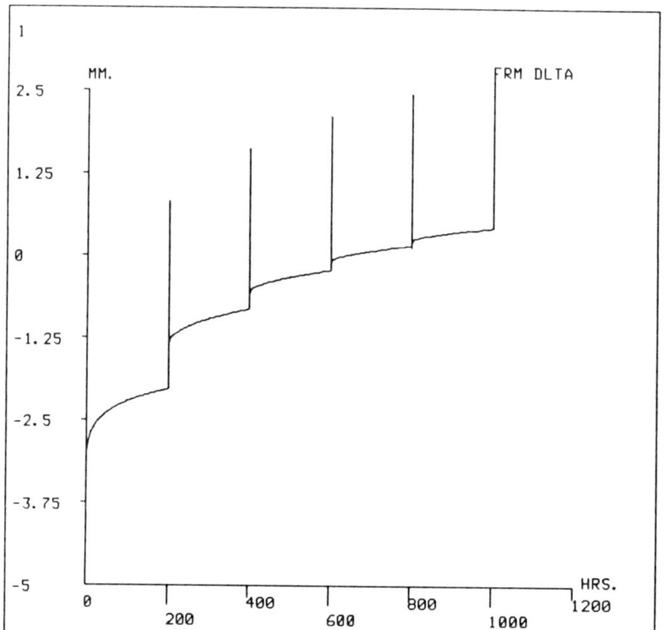


Figure 8. Deflection of Aft End Opening vs. Time; Positive is Closing.

Ratcheting has occurred because shutdown has caused a reversal in plastic strain. Figure 9 is a plot of the radial component of plastic strain versus time for an element at the aft frame internal surface; a location exhibiting the greatest inelastic deformation and causing the frame to close. Upon restarting, the elastic strain (stress) plotted on Figure 10 for the location described above which had relaxed prior to the shut–down, is now higher and a greater creep rate resumes. Creep strain versus time is plotted in Figure 11 for the same element and direction used above. Stress due to the loading caused by the thermal gradients continues to relax while the constant pressure loading induced creep progresses. Shakedown appears to be approaching as less plastic reversal occurs.

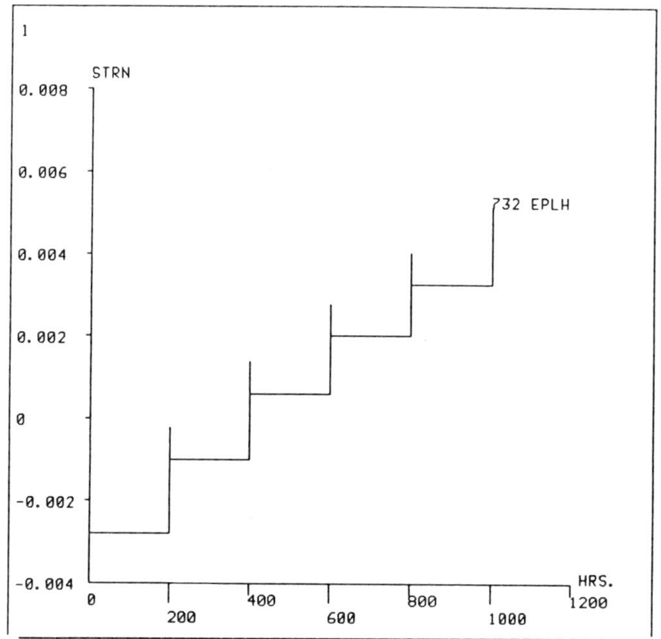


Figure 9. Radial Plastic Strain vs. Time.

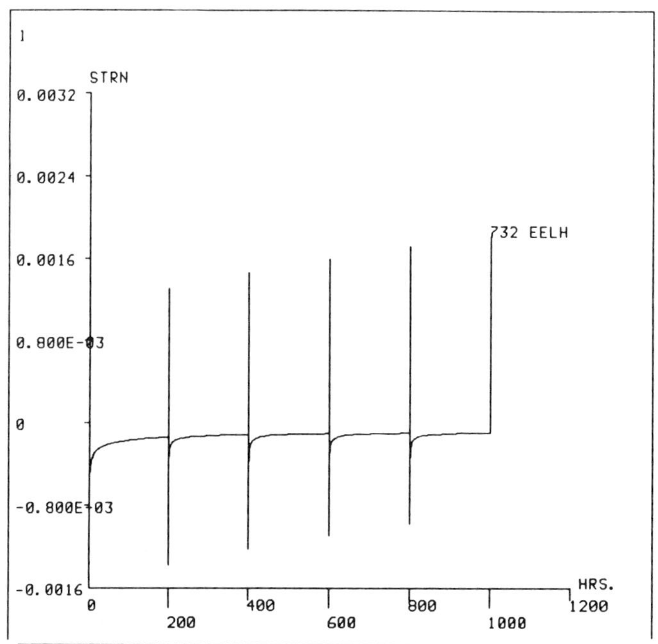


Figure 10.. Radial Elastic Strain vs. Time.

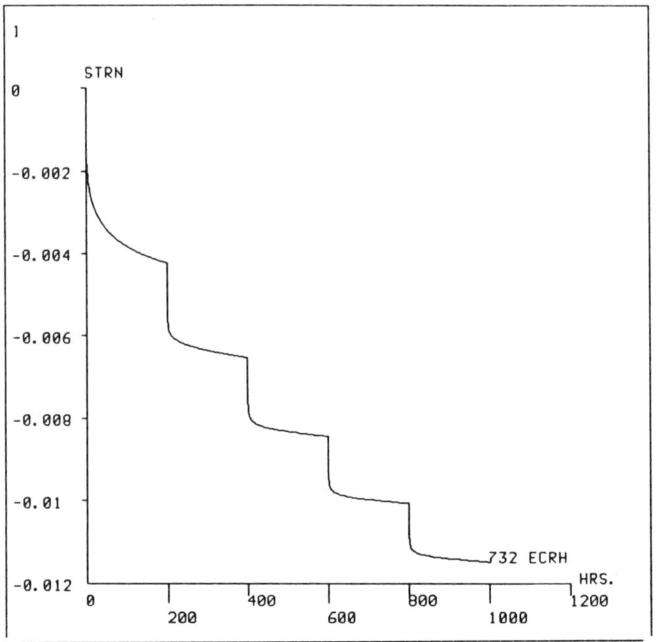


Figure 11. Radial Creep Strain vs. Time.

Figure 12 plots the same deflection of the aft frame opening for 1000 hours without load cycling. Closing deflection is only 1.2 mm (.047 inches). Therefore, ratcheting has over doubled the predicted permanent deflection.

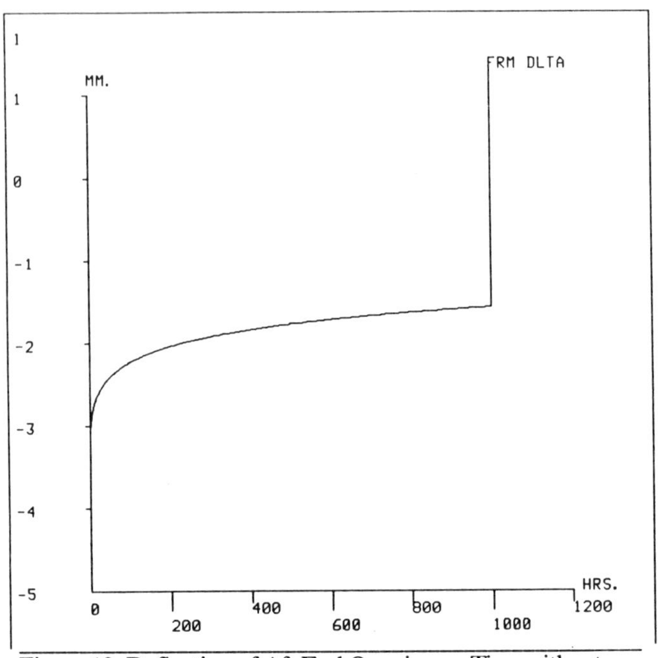


Figure 12. Deflection of Aft End Opening vs. Time without Startup/ Shutdown Cycling; Positive is Closing.

**Nimonic 263 Creep Results.** The same analysis was performed with Nimonic 263 material properties. This alloy has over twice the yield strength and is nearly 3 Larson Miller parameters at .5% creep superior to Hastelloy X. Field data indicates very little creep distortion for transition pieces constructed of this alloy. The analysis shows about .30 mm (.012 inches) of frame closure as shown on Figure 13. The difference in behavior can be seen in Figure 14, a plot of radial plastic strain versus time for the same location used earlier. Here, unloading reverses only .02% of the plastic strain compared to .26% for Hastelloy X. Consequently, elastic strain returns to the same level and creep resumes at about the same rate upon reloading as shown in Figures 15 and 16.

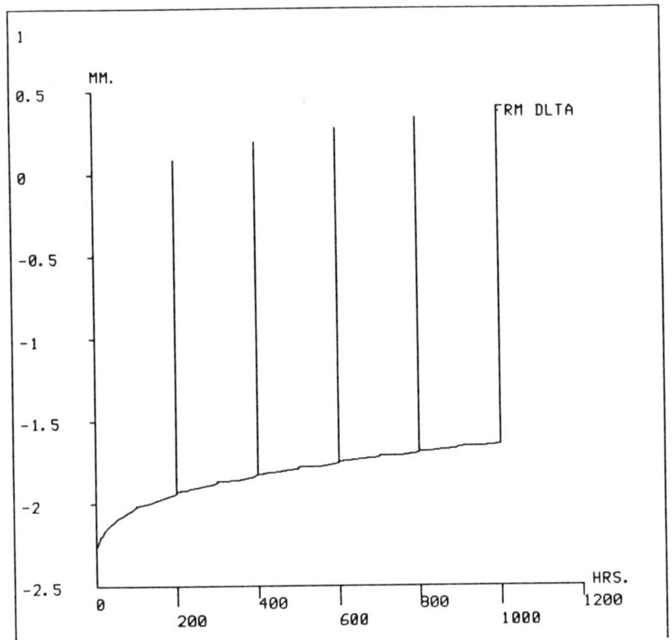


Figure 13. Deflection of Aft End Opening vs. for Nimonic 263; Positive is Closing.

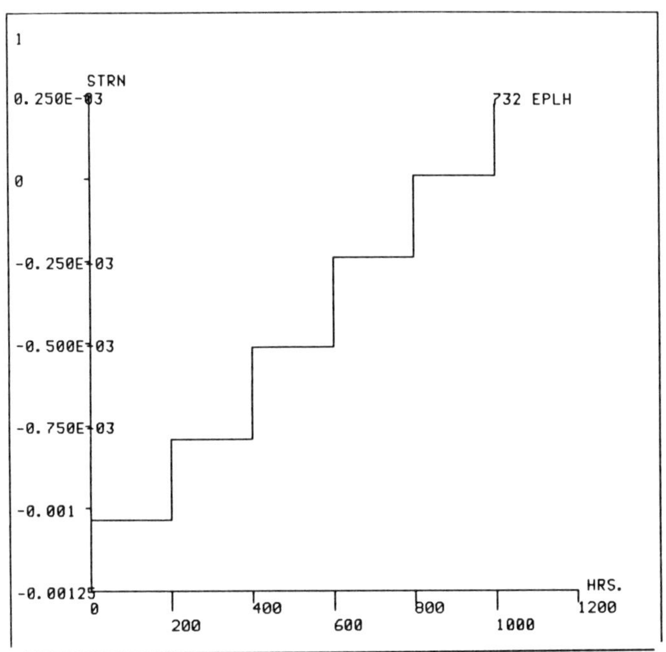


Figure 14. Radial Plastic Strain vs. Time.

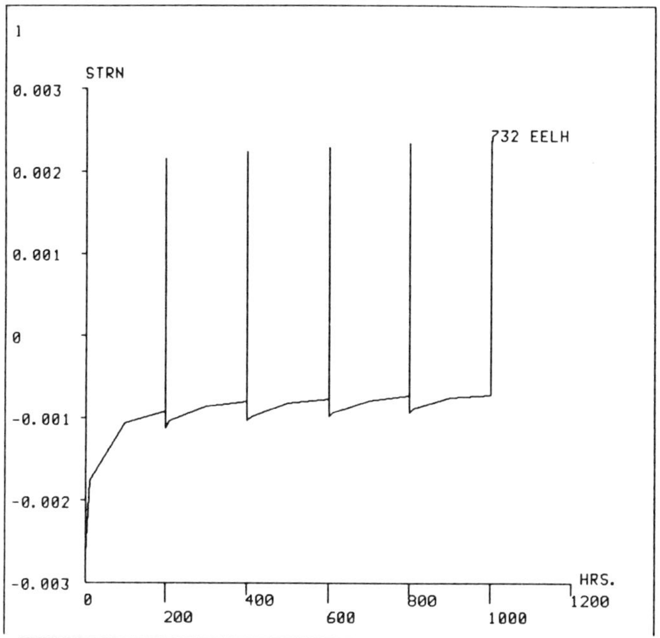


Figure 15. Radial Elastic Strain vs. Time.

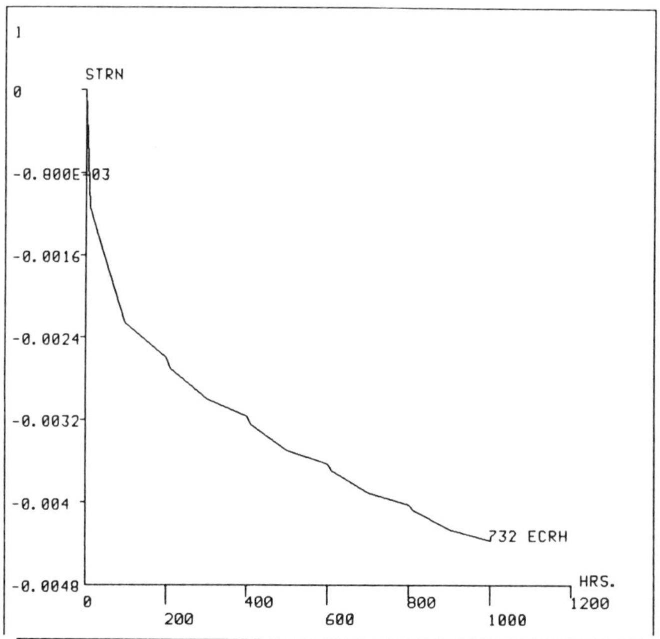


Figure 16. Radial Creep Strain vs Time.

**ASME Code Case 47.** Referring to the simplified method for bounding creep ratcheting strains in ASME Code Case 47 [3] using the maximum elastic stress location, Hastelloy X is operating in the ratcheting regime. Nimonic 263 is in the shakedown regime. However, this is a conservative approximation since the actual transition piece structure is redundant allowing lower stressed regions to carry more load as this high stressed area deforms.

## CONCLUSIONS

Inelastic finite element analysis of a gas turbine combustion transition piece which accounts for plastic and creep deformation has been completed. Comparison to field measurements show good correlation of the permanent deformation. Of particular importance was consideration of startup and shutdown cycling to obtain ratcheted strains.

This analysis methodology provides a valuable tool for designing next generation gas turbines as well as considering the effect of operating condition changes on transition piece creep life. In the past, engineers have been more dependent on component testing which is expensive and requires long times to obtain effects on creep behavior. This finite element analysis method is relatively straight forward to perform, was run on a high performance workstation and requires a minimum of materials data.

## REFERENCES

1. ANSYS Engineering Analysis System User's Manual, Revision 4.4, Volume 1, 2, Theoretical Manual, Swanson Analysis Systems,Inc., Houston Pennsylvania.
2. M. Ben−Amoz, "A Phenomenological Model of Cumulative Damage for Creep Rupture", General Electric Technical Memorandum, Schenectady, NY.
3. ASME Boiler and Pressure Vessel Code Case N−47.

*Proceedings of the Fifth International Conference on Creep of Materials, Lake Buena Vista, Florida, USA, 18-21 May 1992*

# Determination of Piping Acceptance Requirements Based on State-of-the-Art Creep Crack Growth Methodology

**J.M. Bloom and D.R. Lee**
Babcock and Wilcox Company
Alliance, Ohio

ABSTRACT

Design of fossil power plant equipment has been directed towards prevention of creep rupture based on components that were manufactured from materials that had met all quality and inspection requirements of the applicable ASME specification, and constructed in accordance with the requirements of the ASME Code. Experience has shown that some failures are by the growth of cracks from initial imperfections that are not rejectable by current ASME material specifications. This paper presents an alternate acceptance criteria based on state-of-the-art creep crack growth methodology developed for various pipe sizes under typical operating conditions above 510°C (950°F). Using the latest creep crack growth methodology, acceptance criteria were developed. The acceptance criteria was linked to the ratio of calculated hoop stress to allowable stress at the design temperature. Results are presented for a design life of twenty years for operating conditions of monthly, quarterly, and yearly cycling, representative of both peaking fossil power plant units and base loaded units. For peaking units, a defect of 5% of the wall is predicted to grow to failure in twenty years. For base loaded units current ASME material specifications appear reasonable. Two significant conclusions are made. One is that current ASME design approaches and material specifications can be nonconservative, when cyclic conditions are present and, two, the ASME Code allowable stresses at temperatures above 510°C (950°F) may be nonconservative, for certain operating conditions, if a small defect is initially present in the component.

THE TRADITIONAL APPROACH TO ELEVATED TEMPERATURE DESIGN of fossil power plant equipment has been directed to prevention of creep rupture in components that were manufactured from materials that had met all quality and inspection requirements of the applicable ASME specification, and constructed in accordance with the requirements of the ASME Code. In particular, Section I of the ASME Code does not provide rules for elevated temperature fossil power plant components containing imperfections. However, experience has shown that some failures are by the growth of cracks from imperfections that were not rejectable according to existing ASME material specifications.

When fossil power plant equipment manufacturers order piping for high temperature service, a standard inspection criteria for surface quality is used. ASME material specifications do not require a perfect product, but allow a certain level of imperfection due to seams, laps, scores, etc. Material specifications for SA-335 (P11) 1-1/4 Cr - 1/2 Mo-Si, and SA-335 (P22) 2-1/4 Cr - 1 Mo, currently allow a supplier of pipe to leave imperfections in the product up to 12-1/2% of the wall. Fossil power plant equipment manufacturers sometime receive piping which possess less than perfect surface quality, but which meets the ASME specifications. When the suitability of a finished piping component is challenged for its intended service, the equipment manufacturer has no firm technical basis to judge the product other than to say it met the specification as well as the design rules. This paper will present an alternate acceptance criteria based on state-of-the-art creep crack growth methodology. An acceptance criteria is developed for various pipe sizes under typical operating conditions above 510°C (950°F). At these temperatures, the growth mechanism has been shown to be creep crack growth.

## BACKGROUND AND ASSUMPTIONS

Previous work, funded by the Electric Power Research Institute (EPRI) (1), produced guidelines for the evaluation of longitudinal seam-welded steam pipes, based onassessments of creep damage in the welded regions of the pipe. This guideline, developed after the Mohave and Monroe steamline ruptures, was conservative in the use of end-of-life creep properties, crack modeling, and in the constraint assumptions for the cracked pipes. The creep crack growth prediction tool recommended by the EPRI guidelines was a computer code called SLIC (Steam LIne Cracks) (2). SLIC determines the time for a postulated initial flaw to propagate, due to creep crack growth, through the wall of a pipe operating at temperatures in the creep regime.

The version of SLIC recommended by EPRI at that time was limited to continuous operating (no cyclic loading) and secondary steady-state creep. However, it has been demonstrated (3) and will be further discussed in this paper, that cyclic operation, as well as primary creep can significantly shorten the life of elevated temperature components that are cycled as infrequently as four times a year. Initial tolerable flaw sizes for a fixed design life may be significantly reduced.

A creep crack growth based acceptance criteria was developed for attemperator bodies of size similar to steamlines. The results could be applied to steam pipes as well, provided the calculated hoop stress to allowable stress ratio and operating temperatures are the same. An attemperator is a piping component that is designed to inject cooling water spray into a steamline to cool, or temper, steam temperature on demand. An inner floating liner in the attemperator mitigates the effects of the numerous thermal shock cycles which the component is potentially subjected. For this study, therefore, only pressure cycles experienced by the outer pressure boundary of the component were considered. While flaws of specific interest are predominantly outside diameter (OD) flaws, the chosen flaw orientation (based on maximum stress and past steamline failures) for the development of the acceptance criteria was a longitudinal flaw on the inside diameter (ID) surface. This flaw produces a shorter life (4) and is more difficult to detect.

Material properties were based on pre-service (virgin) conditions for both the P11 and P22 materials, since the flaw acceptance criteria of interest in this study is for pipe manufacturing surface imperfections at the beginning of service. The material properties used in the EPRI guidelines (1) examples were based on severely degraded P11 base metal properties, taken from a 7 cm (2.8-inch) thick retired secondary superheater outlet header which had been in service for 24 years (5). The header was retired from service due to cracking in the ligaments between tubes. Since primary creep properties were deemed to be significant in the determination of initial tolerable defect depths, published data from Reidel and Detampel (6) were used for the P22 strain rate properties and material properties taken from the less degraded cold end of the retired outlet header were used for P11 strain rate properties. The creep crack growth rate properties of P22 for the virgin material were found to be almost identical to the base metal P11 virgin material (7). In an attempt to make the life predictions and the resulting acceptance criteria as useful as possible, the acceptance criteria was linked to the ratio of the calculated hoop stress to allowable stress at the design temperature. The calculated hoop stress was based on the equation in Section I of the ASME Boiler and Pressure Vessel Code and the allowable stress is taken from the Section I Rules for Construction of Power Boilers of the ASME Boiler and Pressure Vessel Code. These allowable stresses at temperatures above 510°C (950°F) are based on creep rupture data properties of these materials. In addition, actual production part dimensions and real operating boiler designs and conditions were used in this study as denoted by units 1 and 2 in this paper.

## METHODOLOGY OF CREEP CRACK GROWTH

A computer program, PCCREEP (8) developed at Babcock & Wilcox, was used to determine the alternate acceptance criteria based on state-of-the-art creep crack growth methodology. PCCREEP is based on the time-dependent fracture mechanics (TDFM) crack tip energy release rate (power) parameter, $C_t$, developed by Saxena (9). PCCREEP includes both cyclic operation and primary creep for axially oriented defects in pipes. Both ID and OD longitudinal flaws are modeled. For the ID flaws, pressure is assumed to act over the face of the defect. The $C_t$ formulations in PCCREEP include the latest published improvements in creep crack growth methods (4,10). Saxena and Liaw (5) have shown that the $C_t$ parameter gives the best correlations with creep crack growth rates, da/dt, from small-scale creep to steady-state creep through the relationship:

$$da/dt = bC_t^q \tag{1}$$

The remaining life of a component with a pre-existing defect under creep conditions can be obtained by integrating Eq.(1) numerically. Likewise, the determination of an initial flaw size is obtained by taking a pre-determined total life and back calculating the initial flaw size that will grow to failure in the pre-determined lifetime. Experience with life prediction of cracked fossil components has shown that 99% or greater of the total life is reached before the crack depth is 60% of the wall. Alternatively, the final crack size could be determined based on fracture toughness/instability considerations during a cold start. Fracture toughness data would be needed along with a crack instability analysis to determine the final flaw size. Based on the limited data in the literature on fracture toughness properties of P11 and P22, it is expected that the remaining life or initial tolerable flaw size would not be significantly affected by the additional crack instability analysis.

**Determination of $C_t$.** To integrate Eq. (1), a methodology for determining $C_t$ for the actual cracked fossil power plant component is needed. $C_t$ is made up of a transient term which is a function of the stress intensity factor, K, and a steady state term C* which is analogous to the J-integral used for elastic-plastic instability analyses. In fact, if strains and displacements in the definition of the J-integral are replaced by the strain rates and displacement rates, J and C* are identical. In addition, the elastic-plastic estimation scheme of Kumar, German, and Shih (11) can be used to obtain closed form expressions for C* for a pressurized cylinder with a longitudinal flaw of depth a. With $R_i$ as the inside radius and t the thickness, C* is given by:

$$C^* = (\sqrt{3}/2)^{n+1}\ AtH(a/t,R_i/t,n)p^{\,n+1} \tag{2}$$

where

$$H(a/t,n) = \frac{a/t\,(R_i/t+a/t)^{n+1}\; h_1(a/t,R_i/t,n)}{(1-a/t)^{n}} \tag{3}$$

where p is the internal pressure, $h_1$ is the non-linear calibration function obtained from deformation plasticity finite element solutions as functions of a/t, $R_i$/t, and n is the strain rate hardening exponent. Formulations for $C_t$ which include the transient behavior were proposed by Bassani, Hawk, and Saxena (12) and later shown by Bloom and Malito (3) as giving similar life predictions (to within 2%) to the earlier more complicated formulations of Saxena (5). For elastic-secondary creeping materials whose uniaxial material law can be represented by:

$$\dot{\varepsilon} = \dot{\sigma}/E + A\sigma^{n} \tag{4}$$

$C_t$ can be written as:

$$C_t = C^{*}\left[(t_T/t)^{\frac{n-3}{n-1}} + 1\right] \tag{5}$$

where the transition time $t_T$ is given by

$$t_T = \frac{(1-\nu^2)K^2}{(n+1)EC^{*}} \tag{6}$$

E is the Young's modulus and ν is Poisson's ratio. The stress intensity factors, K, can be found in handbooks (13) while the $h_1$ non-linear calibration functions can be found in (11) for limited geometries. The material properties A,n can be obtained from creep deformation tests.

*Primary Creep.* In materials that exhibit pronounced primary creep, the primary creep can have a significant impact on strain rates at elevated temperatures. This significance is especially important under cyclic loading conditions (3). Therefore, for the determination of tolerable flaws sizes for attemperator bodies or steamline piping, the inclusion of primary creep in the creep crack growth parameter $C_t$ is most important. If the component material at elevated temperature can be represented as an elastic-primary creep-secondary creeping material with the uniaxial material law:

$$\dot{\varepsilon} = \dot{\sigma}/E + A\sigma^{n} + \frac{[B(1+p)]^{1/(p+1)}\sigma^{m} t^{-p/(p+1)}}{(p+1)} \tag{7}$$

where B,p,m are the primary creep constants, then Eq. (5) is replaced by

$$C_t = \left[1 + (t_T/t)^{[1-2/(1+p)(m-1)]} + (t_2/t)^{p/(1+p)}\right] C^{*} \tag{8}$$

where $t_T$ is given by Eq. (6) and $t_2$ is given by:

$$t_2 = \left[\frac{(n+p+1)C_h^{*}}{(n+1)(1+p)C^{*}}\right]^{(1+p)/p} \tag{9}$$

$C_h$* is the value of $C_t$ when only primary creep dominates, and C* is the value of $C_t$ when power-law secondary creep dominates. The parameter $C_h$* is evaluated in the same manner as C* where (for a longitudinal ID flaw in a pipe):

$$C_h^{*} = [B(1+p)]^{1/(1+p)}\,(\sqrt{3}\,p/2)^{m+1}\,t\,H(a/t,R_i/t,m) \tag{10}$$

and H(a/t,$R_i$/t,m) is as given in Eq. (3) with "n" replaced by "m".

Bloom and Malito (3) have demonstrated that the above closed form expressions compare favorably to finite element results (14) using the material law given by Eq. (7) for a compact fracture specimen.

*Cyclic Operation.* Of equal importance to the formulation of $C_t$ is the loading history of the component of interest. Whether a fossil power plant is base loaded (continuous operation) or peak loaded (cyclic operation) will significantly effect the initial tolerable flaw size. Figure 1 illustrates a schematic plot of the pressure, temperature of a typical peak loaded power plant component during a startup and shutdown cycle. Note that for this situation, the value of $C_t$ decreased with time during the steady state operation and goes to zero at shutdown. At the next startup, the value of $C_t$ starts again at its original initial value. For startup and

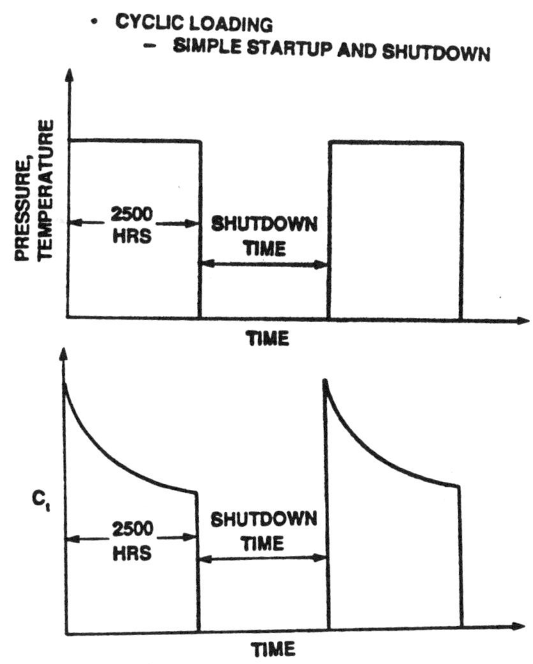


**Fig. 1 - Schematic illustration of the variation in $C_t$ with operating time following a start-up**

shutdown conditions, the crack growth rate is best estimated on a per operating cycle basis. The cyclic rate is da/dN and given by:

$$da/dN = \int_0^{OT} (da/dt)\,dt \tag{11}$$

where da/dt is obtained from Eq. (1). A numerical integration in time is performed in PCCREEP to obtain da/dN as a function of crack size. The values of da/dN are obtained through numerical integration between the limits of initial and final crack sizes to obtain life in cycles where "OT" is the operating time between shutdown in hours. Note that $C_t$ as formulated in either Eq. (5) or Eq. (8) cannot be defined for t=0 since the instantaneous response of the component does not involve any creep but is either elastic or elastic-plastic. Following the application of the load, there is a finite incubation time, $t_0$, during which crack growth is very slow. Yoon (15) has shown that this time, $t_0$, is related to the time required in generating the creep zone within the cyclic plastic zone. This time is of the order of 0.03 hour or smaller for cyclic loading conditions for primary creep. If instantaneous plasticity is allowed during cyclic loading, $C_t$ can be easily modified by replacing "t" by "$t+t_0$" thereby eliminating the problem of the singularity at t=0.

## RESULTS AND DISCUSSIONS

The acceptance criteria was developed based on an expected twenty year design life in terms of the ratio of calculated hoop stress to allowable stress at the design temperature. Both stresses are based on Section I of the ASME Code. Results are plotted in terms of the initial tolerable flaw size as a percent of wall thickness. The tolerable flaw size is the flaw depth that would grow to cause failure of the attemperator in twenty years. The flaw was assumed to be located on the ID, oriented in the axial direction, since previous work (4) showed the ID flaw to be more severe than the OD flaw. Acceptance criteria for both P11 at 538°C (1000°F) and P22 at 571°C (1060°F) are illustrated in Figures 2 and 3. Two attemperatures (as denoted by units 1 and 2 in Figures 2 and 3) with quite different thicknesses and inside radius to thickness ratios are presented for each material to illustrate the range of attemperator designs. Note that these differences in inside radius to thickness ratios become insignificant when plotted versus the stress ratio. One of the unexpected results seen from Figures 2 and 3, is that P11 has larger tolerable flaw sizes than P22. This is particularly noticeable for monthly cycling. However, for base loaded units (yearly cycling), the differences in initial flaw depths for P11 and P22 are small. This is because for continuous operation, the majority of time is spent in secondary creep where the P22 attemperator gives slightly larger tolerable flaw depths. This is a direct result of better secondary creep strain rate properties for P22, even at the higher 571°C (1060°F) temperature, than P11 at the 538°C (1000°F) temperature. However, when cyclic operation is considered (quarterly and monthly cycling) the majority of time is spent in the primary creep regime. In the primary creep regime, the creep strain rate properties of P11 at 538°C (1000°F) are equal or better than those of P22 at 571°C (1060°F). The primary creep properties used for P11 were obtained from testing of the cold end material from the RB-304 header (14) while the primary creep properties for P22 were obtained from published data (6). However, the P22 primary creep strain rates were obtained using deflection rates from a compact specimen and therefore may not be as reliable. Also contributing to these differences is the fact that the applied hoop stress is dictated by the allowable stress which is different for each material/temperature combination (P11/538°C (1000°F) , P22/571°C (1060°F)). If the temperature for P22 were decreased to 538°C (1000°F), while increasing the pressure to maintain the same stress ratio, it might be expected that the P22 attemperators would have larger tolerable flaw depths even for cyclic units, than for the P11 attemperators. However, just the opposite occurs. When the temperature is decreased and the stress ratio remains constant, the P22 attemperators have even smaller tolerable flaw sizes. The increase in stress allowed by the ASME Code more than offsets the decrease in creep crack growth rate due to temperature. These trends are supported by Becht and Buchheim (16). Both the findings of this paper and Becht and Buchheim (16) suggest that the design rules in the ASME Code may not be conservative for flawed components at elevated temperatures. Figure 2 shows that for P11 attemperators, acceptance standards depend strongly on whether the unit is base loaded or peak loaded giving initial tolerable flaw depths from a little greater than 5% of the wall to 15% of the wall for a stress ratio of one or less. For P22 attemperators, the acceptance standards shown in Figure 3 indicate that the stress ratio must be reduced to 0.9 or less to keep the tolerable flaw depth to 5% or more of the wall thickness for monthly cycling and a stress ratio of 0.95 or less for quarterly cycling. Acceptance standards for P22 base loaded units are consistent and even less stringent than the current ASME specifications of 12-1/2% of the wall thickness.

## CONCLUSIONS AND RECOMMENDATIONS

Two significant conclusions were drawn from the life prediction/flaw tolerance calculations:

- ASME pipe material specifications for surface inspection standards can be non-conservative when considering creep crack growth under cyclic conditions.

- The ASME Code allowable stresses and design rules, at temperatures above 510°C (950°F), may be nonconservative if a small defect is initially present in the component where cyclic service is anticipated.

These conclusions could lead to a recommendation that the acceptance criteria for SA-335 P11 and P22 piping, used in creep service, be tightened beyond what the ASME standards currently requires. An alternate approach would be to maintain the current material specification acceptance

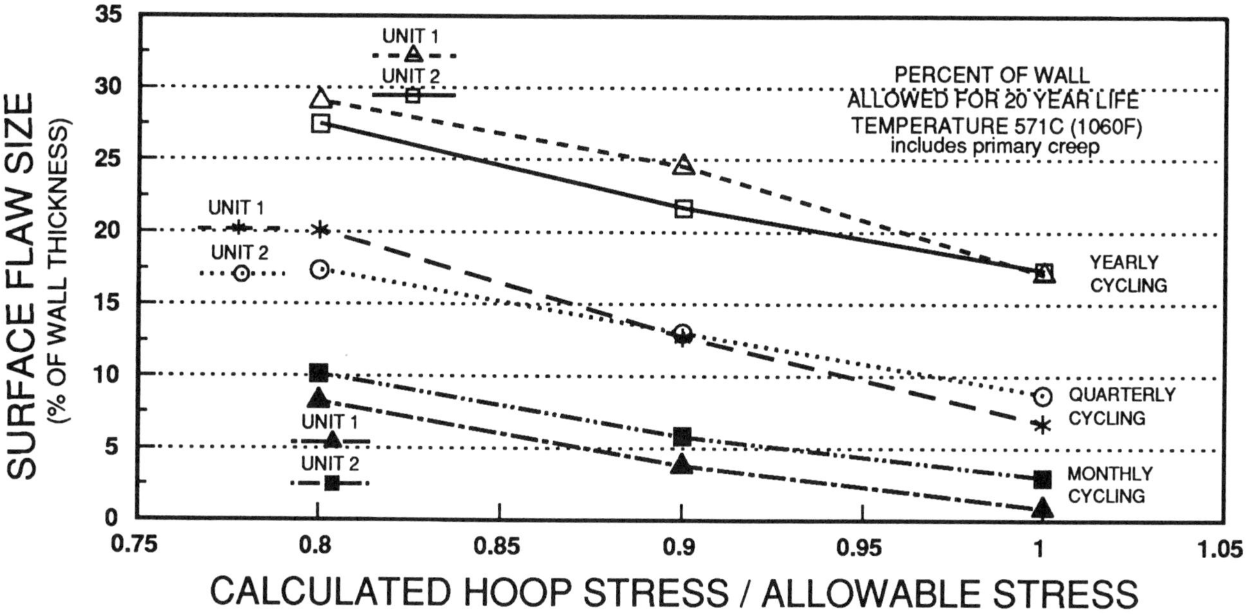


UNIT 1
Ri/t = 1.313
THICKNESS = 15.240 cm (6 in)

UNIT 2
Ri/t = 5.593
THICKNESS = 4.288 cm (1.688 in)

Fig. 2 %Wall VS Stress Ratio for P22

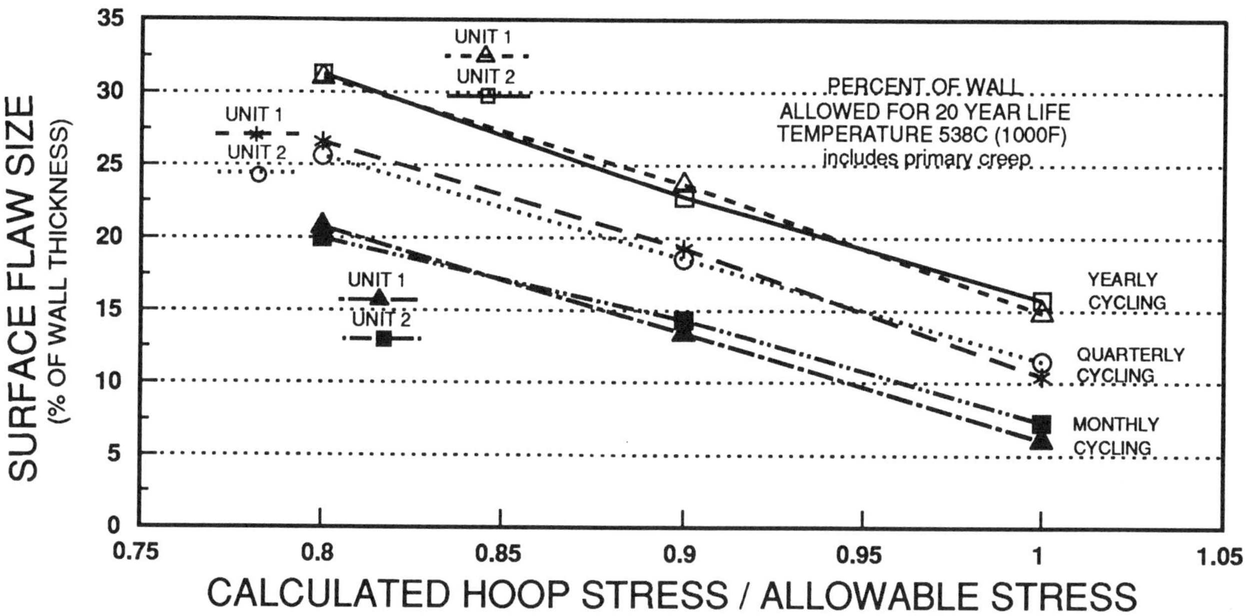


UNIT 1
Ri/t = 2.458
THICKNESS = 7.620 cm (3. in)

UNIT 2
Ri/t = 5.556
THICKNESS = 2.858 cm (1.125 in)

Fig. 3 %Wall VS Stress Ratio for P11

criteria, but instead impose more conservative design rules and allowable stresses on specific components designed for cyclic service in the creep regime.


ACKNOWLEDGEMENTS

This work was supported by the Fossil Power Division (FPD) of Babcock & Wilcox under the direction of J. M. Tanzosh.

The authors wish to express their gratitude to J. M. Tanzosh and M. Gold of FPD for their constructive suggestions in the preparation of this paper.


REFERENCES


1. "Guidelines for the Evaluation of Seam-Welded Steam Pipes" Electric Power Research Institute, Palo Alto, California, Final Report, November 1986, RP 2596-7.
2. "SLIC, A Computer Code for Analysis of Steam Line Cracks," Failure Analysis Associates, Palo Alto, California, September 1986.
3. Bloom, J.M. and Malito, M.L., "Using $C_t$ to Predict Component," Fracture Mechanics: Twenty-Second Symposium (Volume I), ASTM STP 1131, American Society for Testing and Materials, Philadelphia, PA., 1992.
4. Bloom, J.M. and Lee, D.R., "Applications of High-Temperature Assessment Methods to Operating Fossil Power Plant Components," Sixth Annual North American Welding Research Conference: Design and Fitness for Service of Welded Structures, Edison Welding Institute, Columbus, Ohio, October 1990.
5. Saxena, A. and Liaw, P.K., "Remaining Life Estimation of Boiler Parts - Crack Growth Studies," Electric Power Research Institute Final Report EPRI CS-4688 Project 2253-7, Final Report, July 1986.
6. Riedel, H. and Detampel, F.V., "Creep Crack Growth in Ductile Creep-Resistant Steels," International Journal of Fracture, Vol. 33, 1987, pp. 239-262.
7. Saxena, A., Han, J., and Banerji, K., "Creep Crack Growth Behavior in Power Plant Boiler and Steam Pipe Steels," Journal of Pressure Vessel Technology, Transactions of the ASME, May 1988, pp. 137-146.
8. Zien, H.M., and Bloom, J.M., "User's Guide for the Computer Program PCCREEP, Version 3," Babcock & Wilcox Research and Development Division, Alliance, Ohio, October 1987.
9. Saxena, A., "Creep Crack Growth Under Non-Steady-State Conditions," Fracture Mechanics, vol. 17, ASTM STP 905, American Society for Testing and Materials, Philadelphia, PA, 1986, pp. 185-201.
10. Leung, C.-P., McDowell, D.L., and Saxena, A., "Consideration of Primary Creep at a Stationary Crack Tip: Implications for the $C_t$ Parameter," Int. Journal of Fracture, Vol. 36, 1988, pp. 275-289.
11. Kumar, V., German, M.D., and Shih, C.F., "An Engineering Approach to Elastic-Plastic Fracture Analysis," EPRI NP-1931, Electric Power Research Institute, Palo Alto, California, July 1981.
12. Bassoni, J.L., Hawk, D.E., and Saxena, A., "Evaluation of the $C_t$ Parameter for Characterizing Creep Crack Growth Rate in the Transient Region," Third International Symposium on Nonlinear Fracture Mechanics, American Society of Testing and Material, ASTM STP 995, 1989, pp.7-26.
13. Tada, H., Paris, P.C., and Irwin, G.R., "The Stress Analysis of Cracks Handbook," Second Edition, Paris Productions Incorporated (and Del Research Corporation), St. Louis, Missouri, 1985.
14. Leung, C.-P., and McDowell, D.L., "Inclusion of Primary Creep in the Estimation on the $C_t$ Parameter," Int. Journal of Fracture, Vol. 46, 1990, pp. 81-104.
15. Yoon, K.B., "Characterization of Creep Fatigue Crack Growth Behavior Using the $C_t$ Parameter," Georgia Institute of Technology, Atlanta, GA., Ph.D. Thesis, June 1990.
16. Becht, C., and Buchheim, G.M., "Effect of Creep Crack Growth on Life of Elevated Temperature Stainless Steel Components," Journal of Pressure Vessel Technology, Transaction of the ASME, Vol. 11, August, 1989, pp. 266-274.

*Proceedings of the Fifth International Conference on Creep of Materials, Lake Buena Vista, Florida, USA, 18-21 May 1992*

# Crack Growth Assessments at High Temperature

**P.J. Budden**
Berkeley Nuclear Laboratories
Berkeley, Gloucestershire, England

## Abstract

Structures operating at high temperature may fail due to both creep and creep-fatigue mechanisms. Moreover defects can occur due to the manufacturing and welding processes or they can arise in service as a result of the operating conditions. Within the UK electricity generating industry, the requirement to perform assessments of the structural integrity of plant operating at creep temperatures has led to the development of the R5 Procedure (1). R5 is written in 7 volumes addressing various aspects of high temperature assessments. In particular, Volumes 4 and 5 address crack growth from pre-existing defects under conditions of steady creep and creep-fatigue loading, respectively. Volume 7 is a procedure for similar weldments and currently offers a simplified route for cracked and uncracked welds in CrMoV piping systems. This paper describes how improvements in the understanding of the high-temperature response of cracked structures have led to developments in the treatment of crack growth within R5 in four areas: treatment of transient effects; assessments for displacement-controlled loads; cracking in welds; and creep-fatigue crack growth under high-strain conditions.

STRUCTURAL INTEGRITY ASSESSMENTS of defective components operating at high temperatures require consideration of both creep and fatigue crack growth. Creep-fatigue interaction can occur by cyclic extension through creep-damaged material. Furthermore, component life may be limited by creep rupture (continuum damage failure) with little or no crack growth. Assessment procedures must take these different mechanisms into account.

The R5 Procedure (1) has been described elsewhere (2,3) and an earlier version of Volume 4 has been published in the literature (4). The scope of R5 has advanced with the issue of (1). Outlines of some of the developments in the treatment of crack growth within R5 are described in the following sections. The defect assessment procedures within R5 use reference stress methods and fracture mechanics parameters, together with experimentally-determined crack growth laws, to predict crack growth in structures. The time to creep rupture is estimated from the materials rupture curve at the appropriate reference stress. The effect of crack growth on creep rupture time is handled by updating the reference stress as the crack grows. Generalized creep laws can be used in R5, although much of the development is described here in terms of power-law creep for simplicity of presentation. In the case of power-law creep, creep strain rate, $\dot{\varepsilon}_c$, is related to stress, $\sigma$, by Norton's law

$$\dot{\varepsilon}_c = D\sigma^n \tag{1}$$

where $D$ and $n$ are material and temperature dependent constants.

The reference stress, $\sigma_{ref}$, is defined in R5, for a homogeneous medium as

$$\sigma_{ref} = P\sigma_y / P_L(a, \sigma_{ref}) \tag{2}$$

where $P$ is load and $P_L$ is the rigid plastic limit load corresponding to yield stress $\sigma_y$ in the component having a crack of length $a$. Then C* is estimated from

$$C^* = \sigma_{ref}\dot{\varepsilon}_c R' \tag{3}$$

where $\dot{\varepsilon}_c$ is evaluated at stress $\sigma_{ref}$ and creep strain $\varepsilon_c$, $R' = (K/\sigma_{ref})^2$ is a length scale and $K$ is the stress intensity factor. Eq.3 has been shown (5) to give a close approximation to C* for a range of geometries and stress exponents, $n$, in Eq.1. Modifications to Eqs.2,3 for the case of weldment cracking are described in this paper.

In Volume 5, creep-fatigue interaction is handled by partitioning creep and fatigue crack growth and incorporating the effect of creep damage during the dwell into the fatigue growth law. The cases where the structure is operating within strict shakedown limits and where a zone of contained cyclic plasticity is admitted, are considered separately in Volume 5. The latter case is discussed here.

## Transient Effects

It is well known that the singular stress field at the crack tip in an elastic power-law creeping material is of HRR type (6,7) and is characterized by the $C(t)$ integral following loading at time $t=0$, where $C(t) \to C^*$ as $t \to \infty$ and $C(t) \sim K^2/E'(n+1)t$ as $t \to 0$ (8). Here $E' = E, E/(1-\nu^2)$ in plane stress, plane strain, respectively, with $E, \nu$ the elastic constants. Ehlers and Riedel (9) and Ainsworth and Budden (10) have derived approximations to $C(t)$ for all $t$. In particular, (10) showed that

$$C(t)/C^* \simeq (1+\tau)^{n+1}/((1+\tau)^{n+1}-1) \quad (4)$$

where $\tau = E'C^*t/K^2 \equiv t/t_{red}$ is dimensionless time. From Eq.4, $C(t)$ is within 7% of $C^*$ at $\tau = 1$ when $n \geq 3$. Generalized creep deformation expressions can be handled by defining $\tau = \varepsilon_c(\sigma_{ref},t)/\varepsilon_e(\sigma_{ref})$ where $\varepsilon_e \equiv \sigma_{ref}/E$ is the reference elastic strain. This definition of $\tau$ reduces to $\tau = E'C^*t/K^2$ for the case of power law creep using Eq.3. Note that $\tau = 1$ corresponds to $\varepsilon_c = \varepsilon_e$ at the reference stress (which defines the redistribution time $t_{red}$).

**Validation of C(t) estimates.** Finite element calculations have been performed on a number of two-dimensional geometries. $C(t)$ and $C^*$ have been evaluated and compared with the reference stress formulation given by Eqs.3-4. Results for a compact tension (CT) specimen with crack length to specimen width ratio $a/w = 0.5$ and for a single-edge notched (SEN) plate with $a/w = 0.25$ under tension, bending and combined tension and bending are discussed in Volume 4 of (1). An example for the SEN under combined load controlled tension and bending is shown in Fig.1. The plane strain reference stress formulation agrees well with the computed values of $C(t)$. Note that the value of $C^*$ used for the normalization is the computed figure. Hence the discrepancy at $\tau \to \infty$ in Fig.1 is due to Eq.3 conservatively overestimating the steady state value of $C^*$. In the finite element calculations, elastic power law creep with $n = 5$ in Eq.1 was assumed. Similar conclusions hold for other geometries, the results of which will be published. The approximation of Eqs.3,4 tended to be more conservative for bend geometries.

**Crack Growth.** Models of creep crack growth suggest (11) that crack growth rate, $\dot{a}$, can be correlated with $C^*$ or $K$ depending on the magnitude of the dimensionless crack velocity, $\lambda = \dot{a}\sigma_{ref}/ER'\dot{\varepsilon}_c$. When $\lambda \leq 0.5$ and $n > 3$, $C^*$ correlations are expected. Then, for times $t > t_{red}$, crack growth rate $\dot{a}$ is often described by

$$\dot{a} = AC^{*q} \quad (5)$$

where $A$, $q < 1$ are material and temperature dependent constants. Materials data are generally given in this form. For times $t < t_{red}$ it is assumed here that $C^*$ in Eq.5 can be replaced by $C(t)$. Other models have been proposed (eg (12)) where $\dot{a} = \dot{a}(\Delta a, C(t))$ with $\Delta a$ the amount of crack advance, which predict the experimentally observed 'tails' in $\dot{a} - C^*$ data. It is considered that 'tails' are not yet sufficiently understood to be incorporated in the Procedure. Hence, within R5, they are currently neglected for conservatism.

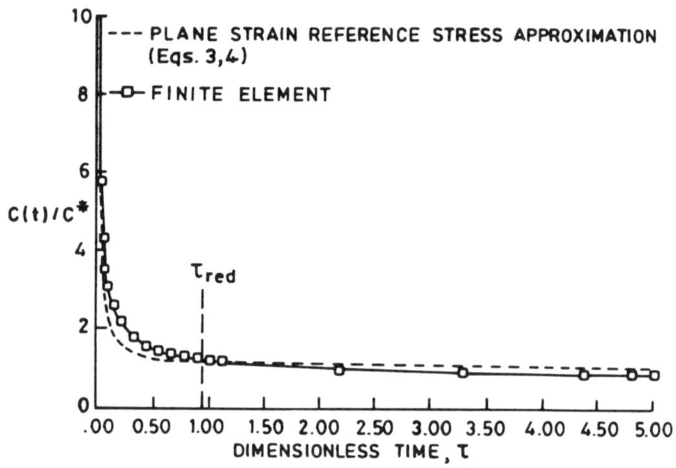


Fig.1 Comparison of estimates of $C(t)/C^*$ with finite - element results for SEN specimen with $a/w = 0.25$. Load - controlled tension, $P$, and bending, $M$ ($M/Pw = 0.125$).

In assessments, the enhanced growth rate during the transient period $t \leq t_{red}$, where $C(t) > C^*$, has to be accounted for. For structures which creep rapidly, practical assessment times exceed $t_{red}$ and then the transient effect appears as an additional increment in the total crack growth, $\Delta a$. This is covered by the recommendation within R5 that for total assessment times in excess of $t_{red}$, $\dot{a}$ is given by

$$\begin{aligned} \dot{a} &= 2\dot{a}(C^*),\ t < t_{red}\ (\tau < 1) \\ \dot{a} &= \dot{a}(C^*),\ t \geq t_{red}\ (\tau \geq 1) \end{aligned} \quad (6)$$

where $\dot{a}(C^*)$ is the experimentally determined growth law, such as Eq.5.

However at low temperatures within the creep range, $t_{red}$ exceeds practical assessment times. Similarly it may be conceded, due to lack of suitable materials data for the initiation event, that crack extension occurs on initial loading. In either case, creep crack growth assessments are required for times prior to redistribution. Then the averaged factor 2 on growth rate in Eq.6 is inappropriate since $\dot{a}/\dot{a}(C^*) \to \infty$ as $t, \tau \to 0$ and Eq.6 becomes

$$\dot{a} = f(\tau)\dot{a}(C^*),\ t < t_{red}\ (\tau < 1) \quad (7)$$

where $f(\tau) \to \infty$ as $\tau \to 0$. A suitable expression for $f$ is given below.

For small crack growth $\Delta a$, and for $q = n/(n+1)$, Ainsworth and Budden (13) showed that

$$\Delta a = AC^{*n/(n+1)} t_{red}((1+\tau)^{n+1} - 1)^{1/(n+1)}$$

that is,

$$\Delta a = AC^{*n/(n+1)} t(1 + 1/\tau),\ \tau = t/t_{red} \to \infty \qquad (8)$$

Then $n/(n+1)$ in Eq.8 can be generalized to q. Eq.8 expresses the total crack growth at times $\tau >> 1$ as the factor $1 + 1/\tau$ times the corresponding steady state growth (14). This led to Eq.6 for assessment times in excess of $t_{red}$. During the transient period $t \le t_{red}$ ($\tau \le 1$), Budden et al (15) showed that for the power-law material the normalized crack growth rate $\dot{a}/AC^{*q}$ could be estimated from the sum of short- and long-time limits:

$$\dot{a}/AC^{*q} \simeq 1 + (n+1)^{-q}\tau^{-q},\ 0 < \tau \le 1 \qquad (9)$$

When $q = n/(n+1)$ the estimate was an upper bound. For general creep crack growth and deformation expressions, $(n+1)^{-q}$ and $\tau$ were replaced by 1.445(1-q) and $\varepsilon_c/\varepsilon_e$, respectively, where the 1.445 came from bounding a particular function of n. Hence, for assessments at times less than $t_{red}$, $f$ in Eq.7 is given by the right-hand side of Eq.9. For small crack growth, $\Delta a << a$, Eq.9 can be readily integrated. For general crack growth, numerical integration is used with changes in crack size reflected in C* and $K$. When $\tau << 1$, the short-time limiting term alone is sufficiently accurate and then assuming there is no incubation and crack growth starts at $t = 0$:

$$\Delta a \simeq A(K^2/E'(n+1))^q (1-q)^{-1} t^{1-q}$$

or

$$\Delta a \simeq 1.445A(K^2/E')^q t^{1-q}$$

## Displacement Controlled Loading

Loading by a constant displacement remote from the crack tip is considered in Volume 4 of R5 (1). The corresponding load, that is the reaction at the fixture, then relaxes due to both creep and to compliance changes resulting from crack growth. Estimates of the load drop in a given time, the time to initiate creep crack growth and the amount of subsequent growth are derived here for an elastic-creeping material using reference stress methods. The effect of elastic follow-up is considered.

**Load Drop.** During a displacement-hold and prior to crack growth, the reference stress $\sigma_{ref}$ (Eq.2) is related to creep strain rate by

$$\dot{\sigma}_{ref} = -E\dot{\varepsilon}_c(\sigma_{ref})/Z \qquad (10)$$

where $Z$ is the elastic follow-up factor; $Z = 1$ and $Z \to \infty$ correspond to the limits of pure strain control and load control, respectively. We assume elastic power-law creep given by Eq.1 for the present. Then Eqs.1,10 can be integrated (13). Noting that $\sigma_{ref}/\sigma_{ref,0} = P/P_0$ prior to crack growth, where the subscript 0 denotes values at $t = 0$, the start of the dwell, the load drop $\Delta P = P_0 - P(t)$ satisfies

$$\Delta P/P_0 = 1 - (1 + (n-1)EC^*_0 t/K_0^2 Z)^{-1/(n-1)} \qquad (11)$$

$C^*_0$ is defined in terms of values at $t = 0$ by Eq.3. For generalized creep laws, $\Delta P$ follows by integration together with Eq.10. It will be shown that Eq.10 also holds for standard bend specimens when crack growth occurs. The reference stress estimate of C($t$) is given by Eq.3 using the instantaneous load from Eq.11 (13).

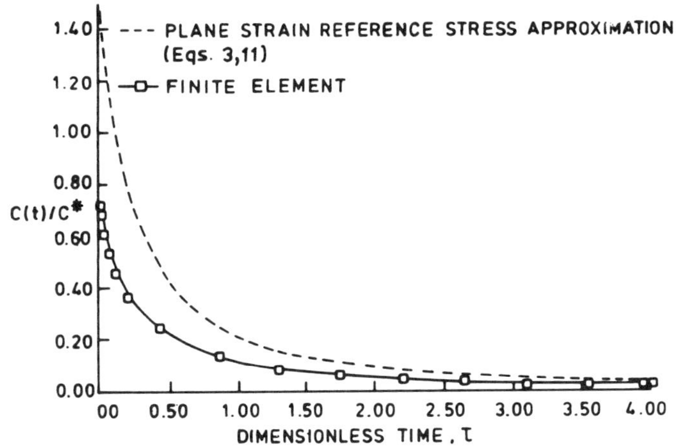


Fig.2 Comparison of estimates of C($t$)/C* with finite - element results for SEN specimen ($a/w = 0.25$). Displacement-controlled bending.

In Volume 4 of R5, validation of Eq.11 by finite element analysis is described. The geometries considered were a CT specimen with $a/w = 0.5$, and a SEN specimen with $a/w = 0.25$ under displacement controlled tension and bending. Each example supported the conclusion from Eq.11 that $(P_0/P)^{n-1} - 1$ is linear in time. Finite-element values of C($t$), computed in terms of a limiting value of a contour integral as the crack tip is approached, were also compared with the plane strain reference stress approximation. The latter conservatively over-predicted C($t$) in each case, supporting conservatism in practical assessments. Example results are shown in Fig.2 for the SEN specimen under bending.

**Initiation.** At initiation of crack growth, Ainsworth (16) showed that the reference creep strain $\varepsilon_{c,i} = \int_0^{t_i} \dot{\varepsilon}_c(\sigma_{ref}, \varepsilon_c)dt$ is approximately related to the initiation Crack Opening Displacement, $\delta_i$, by

$$\varepsilon_{c,i} = (\delta_i/R_0')^{n/(n+1)} - \sigma_{ref}(t_i)/E \qquad (12)$$

But, from integrating Eq.10,

$$\begin{aligned}\varepsilon_{c,i} &= (Z/E)(\sigma_{ref,0} - \sigma_{ref}(t_i)) \\ &\equiv (Z/E)\sigma_{ref,0}(\Delta P(t_i)/P_0)\end{aligned} \qquad (13)$$

where $\Delta P(t_i)$ is given by Eq.11. The time to initiate creep crack growth is then given by Eqs.12-13. The resulting expression can be simplified in two ways: (1) $\delta_i/R_0'$ is less than unity in general so that $n/(n+1)$ can be conservatively replaced by unity in Eq.12; (2) when $\varepsilon_{c,i} \geq \sigma_{ref}(t_i)/E$, that is $t_i \geq t_{red}$, $t_i$ follows conservatively from

$$(Z/E)\sigma_{ref,0}(\Delta P/P_0) = \frac{1}{2}(\delta_i/R_0')^{n/(n+1)}$$

**Growth.** As the crack grows, it follows from differentiating Eq.2 with respect to time that

$$\dot{\sigma}_{ref}/\sigma_{ref} = \dot{P}/P - (m'/m)\dot{a}/w \qquad (14)$$

where the prime denotes $d/d(a/w)$, $m(a/w) = P_L(a/w)/Bw\sigma_y$ is the ratio of the cracked- to uncracked-body yield loads and $w,B$ are the section width and thickness, respectively. Then since the sum of the elastic and creep displacements, $\Delta = \Delta_c + \Delta_e$, is fixed we have

$$\dot{\Delta}_e + \dot{\Delta}_c = 0 \qquad (15)$$

An expression for $\dot{\Delta}_e$ follows from the identity $\Delta_e = PC(a/w)/E'B$ with $C$ the compliance:

$$\dot{\Delta}_e/\Delta_e = \dot{\sigma}_{ref}/\sigma_{ref} + (C'/C + m'/m)\dot{a}/w \qquad (16)$$

Then using the relationship (10) $\dot{\Delta}_c/\Delta_e = E\dot{\varepsilon}_c/Z\sigma_{ref}$ it follows from Eqs.15,16 that

$$\dot{\sigma}_{ref}/\sigma_{ref} + (C'/C + m'/m)\dot{a}/w + E\dot{\varepsilon}_c/Z\sigma_{ref} = 0 \qquad (17)$$

For standard test (bend) specimens, Neale and Townley (17) have shown that $C'/C + m'/m \simeq 0$ and Eq.10 follows. However, in general an expression for $\dot{a}$ is required to integrate Eq.17, as is now shown.

From Eqs.3,5,17 with $q = n/(n+1)$ it follows that

$$\dot{\sigma}_{ref}/\sigma_{ref} + (C'/C + m'/m + \\ + (1/\phi)(\sigma_{ref,0}/\sigma_{ref})(R_0'/R')^q)\dot{a}/w = 0 \qquad (18)$$

with $\phi \equiv A(C^*_0)^q Z K_0^2/EC^*_0 w$. Then for small crack growth $\Delta a << a_0$, where $a_0$ is the initial crack size, $R_0'/R' \simeq 1$ and $C'/C + m'/m$ is evaluated at $a_0$. Hence using Eq.14 we have

$$\dot{P}/P + (P_0/P\phi + (C'/C)_0)\dot{a}/w = 0$$

which can be integrated to give

$$\Delta a = -(\phi w/\alpha)\ln(1 - \alpha\Delta P/(\alpha+1)P_0)$$

that is

$$\Delta a \simeq \phi w \Delta P/(\alpha+1)P_0, \quad \Delta P << P_0 \qquad (19)$$

Here $\alpha \equiv \phi(C'/C)_0$. Then Eq.19 can be conservatively approximated by

$$\Delta a = \phi w \Delta P/P_0. \qquad (20)$$

The result of Eq.20 is close to that of Eq.19 when $\alpha, \phi << 1$. Unless $Z$ is large, that is in the case of large elastic follow-up, $\phi << 1$. Eqs.19-20 show that crack growth is proportional to load drop for small growth.

In the limit $Z \to \infty$ of load controlled creep, Eq.20 becomes

$$\Delta a = AC_0^{*q}\Delta t$$

since $Z\Delta P \to const.\Delta t$

**Experimental Validation** Table 1 (from (1)) compares some experimental data from the literature for crack growth in ferritic steels under displacement control with the results of Eq.20. It can be seen that the agreement is generally reasonable. The non-conservatism in some cases is probably due to the large crack growth present. Full integration of Eq.18 accounting for changes in $K$ and $C^*$ is then required.

Table 1. Crack growth under displacement control.

| Material | Experimental Δa (mm) | Predicted Δa (mm) |
|---|---|---|
| ½CrMoV (22) Range of Initial Loads | 1.5-4.0 | 2.8 |
| | 7.0-8.0 | 5 |
| | 5.0-8.0 | 8 |
| | 12-17 | 13 |
| | 17-19 | 22 |
| 2¼Cr1Mo (23) | 0.4 | 0.22 |
| ½CrMoV (23) | 5 | 9 |

## Weldment Cracking

In high-temperature plant, defects are commonly associated with welds. Defects can occur as a result of the welding procedure itself or they can arise in service due to the operational conditions. Volumes 4 and 7 of R5 contain a route for assessing defective welds. An outline of the procedure is given here.

The procedure has been used (18) to compare experimentally-observed creep crack growth and failure in a cylindrical CrMoV pressure vessel, welded using 2CrMo electrodes, with the theoretical predictions. The vessel contained part-circumferential, part through wall outer surface defects pre-machined into the weld heat-affected zones (haz). Figure 3 shows some results for crack growth from one defect. Two different predictions of crack growth are given here: one which takes explicit account of the spatial distribution of the haz microstructure through which the pre-machined notch is assumed to grow; and secondly one which averages out the crack growth properties of the haz. The former comparison was given in (18). The average crack growth law adopted is Eq.5 with $A = A'$ where $A' = A_c(1 + \beta)/((A_c/A_R) + \beta)$ with $A_c, A_R$ corresponding to values of $A$ for the two assumed microstructures and $\beta$ the ratio of their spatial extents in the direction of crack growth. Note that $A' \rightarrow A_c, A_R$ as $\beta \rightarrow \infty, 0$, respectively, which correspond to the cases of single haz microstructures. In Fig.3 the times $t_{CD}$, $t_i$ are the predicted time to continuum damage failure at the cracked section and initiation time, respectively. The experimental points are from ultrasonic inspections.

The procedure accounts for two characteristics of welded joints: residual stresses and stress redistribution. Residual stresses occur in welds as a result of the welding thermal cycle. Residual stresses relax due to creep, so that long-term crack-tip stresses depend only on the mechanical load, defined by C*. However during the transient phase, $C(t)$ satisfies $C(t) \simeq K_{tot}^2/E'(n+1)t$ as $t \rightarrow 0$, where $K_{tot}$ is the total stress intensity factor at $t = 0$. Hence crack growth rates are higher when $K_{tot} > K$ with $K$ the stress intensity factor due to the mechanical load alone. Similarly the time to redistribute to the creep-dominated state is longer since $t_{red} \propto K_{tot}^2$. The relaxation of secondary stresses due to creep has been considered in (19) using the finite element method. Secondary stresses where $K_{tot} > K$ reduce the incubation time $t_i$. It is suggested that this can be accounted for by multiplying the elastic strain $\sigma_{ref}/E$ in Eq.12 by $(K_{tot}/K)^2$.

In a properly stress-relieved joint, residual stresses are small. In R5 Volume 4, residual stresses are ignored, pragmatically, if $\varepsilon_0 \equiv \sigma_0/E < 0.1\ (\delta_i/R')^{n/(n+1)}$, where $\sigma_0$ is the peak welding residual equivalent stress.

Stress redistribution from the initial elastic to the long-term creep states occurs at different rates in the differing metallurgical regions of the weldment. In the creep dominated limit, different stresses can then act on the various weld zones for the situation of the constrained pipe weld, and this needs to be accounted for in both creep rupture and crack growth calculations. Hence the homogeneous cracked-body reference stress is multiplied by a factor k, which differs between zones. For the pipe weld, when the hoop stress dominates, k values are estimated from the results of finite-element, uncracked-body creep analysis by comparing the computed equivalent stress in each zone with the plain pipe reference stress.

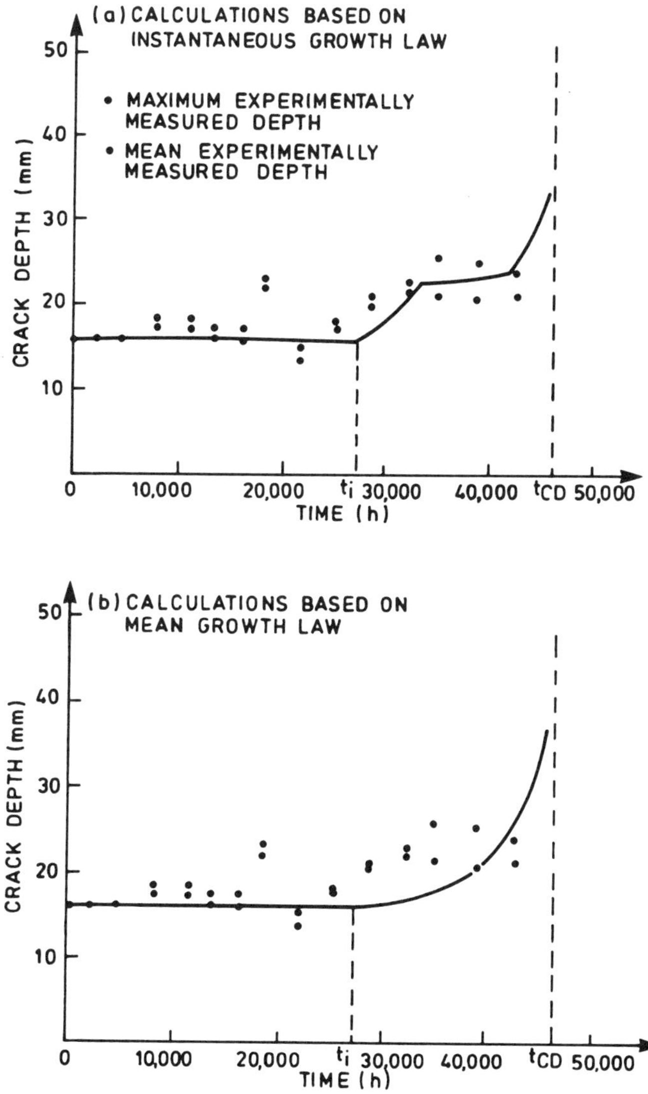


Fig.3 Comparison of calculated and measured crack growth for welded pressure vessel using instantaneous and mean crack growth laws.

Pragmatically, the outer surface zone centre line is chosen, as creep cracks generally grow from this surface. Alternatively, a collapse load can be determined for the inhomogeneous joint by assigning a 'yield stress' to each region which is proportional to the creep rupture stress for the desired service life (20). The individual reference stresses then follow directly from Eq.2 using the appropriate 'yield stresses'. The k factors are assumed to be constant as the crack grows. C* is then estimated from Eq.3 using the modified reference stress and the creep deformation properties of the cracked region (the factor $R'$ in Eq.3 is unaffected). Hoop strain compatability means that the strain rate in Eq.3 is approximately equal to the parent strain rate at the unfactored reference stress.

The heat-affected zone structure and corresponding values of k for a CrMoV pipe weld are given in (1). The k factor approach was used in (18), but further validation of the method is needed.

For cases of high axial stress or deep circumferential cracks, the net axial stress should be compared with the hoop reference stress and the higher of the two chosen for conservatism. The k factor is not appropriate in this case, and is set to unity.

## High Strain Fatigue

Creep fatigue crack growth under high cyclic strain range is addressed by Volume 5 of R5. It is assumed that the loading is such that the structure satisfies global shakedown, which means that the structure as a whole exhibits linear load-displacement behaviour, but that a region of contained cyclic plasticity can occur. A defect occurs within this cyclic zone. The approach is to use a continuous cycling fatigue crack growth law modified by a term dependent on creep damage accumulated during the dwell. The method is given in outline here.

For continuous cycling fatigue, the crack growth per cycle $da/dN_f$ is given by (21):

$$\begin{aligned}(da/dN_f) &= Ba_{min}^{Q}, \quad a \leq a_{min} \\ &= Ba^{Q}, \quad a_{min} \leq a \leq S\end{aligned}$$

where $a_{min}$, the depth below which the crack growth rate is taken to be constant, is set at 0.2mm. Here, S is the plastic zone size, and B,Q depend on material, strain range and environment and are experimentally determined. Often $Q \simeq 1$ and in the absence of data, Q = 1 is assumed. B then follows from continuous cycling fatigue data for the number of cycles to grow a crack to size $a$.

The number of creep-fatigue cycles, N, to grow a crack to length a is estimated from the corresponding continuous cycling endurance, $N_f$, by

$$N = (1/N_f + D_c)^{-1} \qquad (21)$$

where

$$D_c = \int_0^{t_h} \dot{\varepsilon}/\varepsilon_f(\dot{\varepsilon})dt.$$

is the creep damage per cycle evaluated at the surface of the uncracked body, $\varepsilon_f(\dot{\varepsilon})$ is the creep ductility at strain rate $\dot{\varepsilon}$ and $t_h$ is the hold period. Then differentiating Eq.21 with respect to $a$, the current growth rate follows from:

$$\begin{aligned}dN/da &= (dN_f/da)(1 + N_f D_c)^{-2} \\ &= (dN_f/da)(1 - ND_c)^2\end{aligned}$$

that is, substituting the accumulated creep damage

$$\Phi_c = \sum_{i=0}^{N} (D_c)_i$$

for $ND_c$,

$$da/dN = (da/dN_f)\ (1 - \Phi_c)^{-2.} \qquad (22)$$

It should be noted that Eq.(22) holds only while the growing crack lies within the cyclic plastic zone. An alternative approach should be adopted subsequently.

## Conclusions

Recent developments in R5 (1) for the assessment of crack growth at high temperature by both creep and creep-fatigue mechanisms have been described. In particular, methods have been described for the assessment of crack growth (i) at times prior to creep redistribution under steady creep load, where $C(t) > C^*$; (ii) under displacement control in creep where the amount of growth has been seen to be proportional to the load drop; (iii) for cracks in welds, where both stress redistribution and residual stresses are accounted for; and (iv) under creep-fatigue loading for a crack lying within a surface cyclic plastic zone.

## Acknowledgements

The contributions of colleagues within the electricity generating industry of the UK is gratefully acknowledged. This paper is published with the permission of Nuclear Electric plc.

## References

1- Goodall, I.W.(editor), "R5: Assessment Procedure for the High Temperature Response of Structures", Nuclear Electric plc (1990).
2- Ainsworth, R.A., pp79-90, Trans. SMiRT 10, Volume L, AASMiRT, Los Angeles (1989).
3- Goodall, I.W. and R.A. Ainsworth, pp303-311, "Creep in Structures", Proc. IV IUTAM Symposium, Springer Verlag, Berlin (1991)
4- Ainsworth, R.A., G.G. Chell, M.C. Coleman, I.W. Goodall, D.J. Gooch, J.R. Haigh, S.T. Kimmins, G.J. Neate, Fatigue Fract. Engng. Mater. Struct., 10, 115-127 (1987).
5- Miller, A.G. and R.A. Ainsworth, J. Engng. Fract. Mech., 32, 233-247 (1989).
6- Hutchinson, J.W., J. Mech. Phys. Solids, 16, 13-31 (1968).
7- Rice, J.R. and G.R. Rosengren, J Mech. Phys. Solids, 16, 1-12 (1968).
8- Riedel, H. and J.R. Rice, pp112-130, ASTM

STP 700, ASTM, Philadelphia (1980).
9- Ehlers, R. and H. Riedel, pp691-698, Proc Fifth Int. Conf. on Fracture, Vol.2, Pergamon Press, Oxford (1981).
10- Ainsworth, R.A. and P.J. Budden, Fatigue Fract. Engng. Mater. Struct. 13, 263-276 (1990).
11- Ainsworth, R.A., Intnl. J. Fracture, 20, 147-159 (1982).
12- Nikbin,K.M., D.J. Smith, and G.A. Webster, Proc. R. Soc., A396, 183-197 (1984).
13- Ainsworth, R.A. and P.J. Budden, Fatigue Fract. Engng. Mater. Struct., 13, 277-285 (1990).
14- Ainsworth, R.A., pp13-21, "Mechanics of Creep Brittle Materials-1", Elsevier Appld. Sci., London (1988).
15- Budden, P.J., D.W. Dean and G.E. Turner, pp90-99, "Mechanics of Creep Brittle Materials - 2", Elsevier Appld. Sci., London (1991).
16- Ainsworth, R.A., Intnl. J. Solids Structures, 18, 873-881 (1982)
17- Neale, B.K. and C.H.A. Townley, Intnl. J. Press. Vess. Piping, 5, 207-239, 1977.
18- Jones, M.R. and M.C. Coleman, 4th Intnl. Conf. Creep and Fracture of Engng. Matls. Struct., pp605-619, Pineridge Press, Swansea (1990).
19- Joch, J. and R.A. Ainsworth, Paper submitted for publication.
20- Ainsworth, R.A. and I.W. Goodall, Conf. Gas Cooled Reactors Today, BNES, Bristol (1982).
21- Skelton, R.P., pp337-380, ASTM STP 770, ASTM, Philadelphia (1983).
22- Neate, G.J., Mater. Sci. Engng., 82, 77-84 (1986).
23- Nikbin, K.M. and G.A. Webster, pp281-292, ASTM STP 942, ASTM, Philadelphia (1987).

*Proceedings of the Fifth International Conference on Creep of Materials, Lake Buena Vista, Florida, USA, 18-21 May 1992*

# Notch Geometry and Size Effects at Elevated Temperatures

**J.J. Pepe**
General Electric Company
Schenectady, New York

## Abstract

This paper reviews notch geometry and notch size effects at both elevated and ambient temperatures. By use of a constant displacement rate test procedure of both smooth and notched bars, it has been observed that the behavior trends are very similar in both temperature regimes. Constant displacement rate (CDR) creep tests were conducted by placing an extensometer on specimens at fixed locations. All CDR testing was conducted in a closed loop servo-hydralic testing machine. The same specimens were used for both constant displacement rate and constant strain rate tests, if the gauge length between the fixed locations of the extensometers were smooth, a strain could be defined and the test could be properly termed a constant strain rate test. If on the other hand the gauge length contained a notch, the effective gauge length could not be defined and thus the term constant displacement rate was introduced.

The major conclusion drawn from this review is the observed behavior trends are strongly dependent on the intrinsic material ductility.

ONE OF THE MOST IMPORTANT PARAMETERS for reliable performance of power generation equipment is adequate long time elevated temperature ductility. While the importance of long time ductility has been recognized for a long time, there is not a generally accepted ductility parameter used to characterize materials for long time service applications. Manjoine (1) has reviewed some of the more popular ductility indices used for materials characterization. Of the 21 indices discussed by Manjoine, the notch strength ratio (NSR) and the notch time ratio (NTR) are perhaps the most popular. The definition of these parameters are shown in Figure 1 by illustration. It should be clearly noted in Figure 1 that the notch strength ratio (NSR) is measured at a specific time, whereas the notch time ratio (NTR) is measured at a specific stress. By the nature of the notch bar and smooth bar rupture curves, the notch strength ratio is a function of time and the notch time ratio is a function of stress, i.e., neither of these ductility indices is unique to the material. Beyond this limitation, these ductility indices are also a function of temperature and perhaps more important, they are also a function of notch geometry. Accordingly, while they are the most popular elevated temperature ductility indices, the notch strength ratio and the notch time ratio have many limitations.

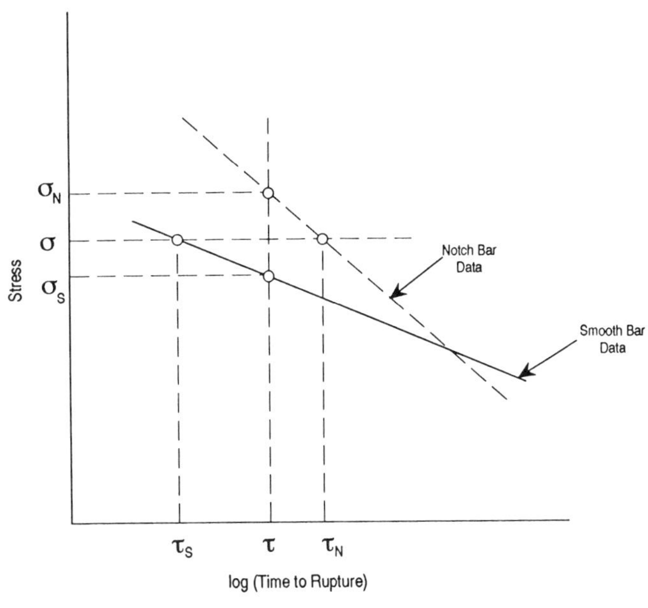


$$\text{Notch Strength Ratio (NSR)} = \frac{\sigma_N}{\sigma_S}$$

$$\text{Notch Time Ratio (NTR)} = \frac{\tau_N}{\tau_S}$$

Fig. 1 – Definition of Notch Ductility Parameters

One particularly important limitation of the notch strength ratio and notch time ratio for usage as a ductility index is the effect of notch geometry and size on the notch rupture behavior. There is a plethora of information in the literature (see references 2-17, for example) concerning the effect of notch geometry and size on notch rupture behavior. Close scrutiny of these data suggests a behavior trend between notch rupture behavior and intrinsic material ductility. It is the objective of this report to present an interpretation of selected literature data on notch geometry and size effects at elevated temperatures.

## Review of Literature Information

### High Temperature Effects

*1) Notch Sharpness.* The work of Davis and Manjoine (6) is perhaps the most complete documentation on the effects of notch geometry on high temperature rupture behavior and will be used here as representative literature data. Davis and Manjoine (6) conducted a systematic investigation on cylindrical notched bars with varying r/d and d/D ratios. Data from two alloys they studied are of particular interest because of their completeness. One alloy (Refractory 26, a NiCrMoCoTi material) was given three different heat treatments to produce various degrees of strength (ductility). The second alloy, a 12Cr3W steel was selected because of its good ductility and its stability at test temperature.

A representation of smooth and notch bar rupture data for one of these four variations (Alloy A) is displayed in Figure 2. Included on this figure are the rupture data for different ratios of r/d, unnotched (smooth) reduction in area at fracture, and details of the specimen geometry. It should be noted that for this alloy, the notch strength ratio (NSR), at a given time, initially increases with decreasing r/d. However, as r/d decreases further, the notch strength ratio experiences a peak and gradually decreases. The work of Davis and Manjoine shows that the NSR eventually decreases below one for all but the most ductile alloys as r/d decreases. This behavior is more evident in Figure 3, which is a plot of the NSR at 1000 hours versus r/d. Since the notch strength ratio at a give ratio of r/d is highest for Alloy D and lowest for Alloy B, it should be concluded that Alloy D is most ductile and Alloy B is least ductile.

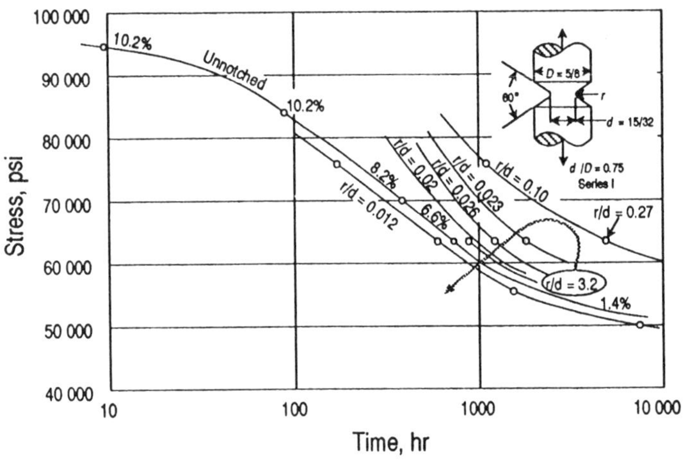


Fig. 2 – Rupture Curves for Notched and Unnotched Bars of Refractaloy 26 (Alloy A) at 1200°F. Davis & Manjoine, ASTM STP128

Davis and Manjoine note that the notch strength ratio developed from these rupture tests never exceeds 1.5, whereas the elastic stress concentration factor varies from a minimum of 1.5 up to a maximum of nearly 6 for the sharpest notch. This is to be expected because the elastic peak stresses at the notch root are reduced through stress relaxation very early in life due to plastic deformation by creep at elevated temperatures.

*2. Size Effect.* In contrast to the general agreement on the effects of notch geometry on notch rupture behavior, there are some divergent experimental data in the literature concerning size effects in rupture testing. It should be noted that "size effect" only is applicable to

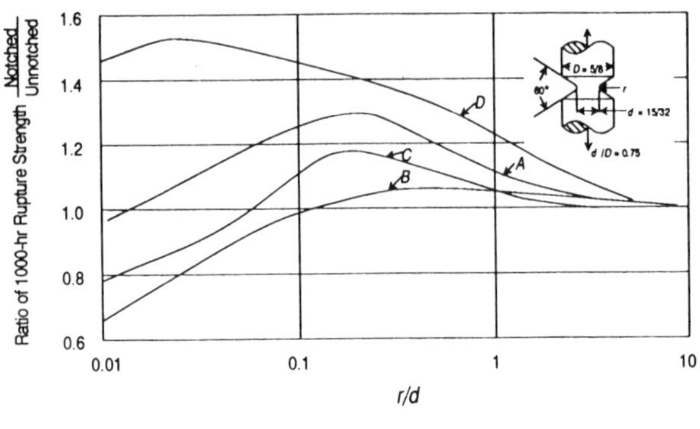


Fig. 3 – Influence of Radius of Curvature on Ratio of Rupture Strength of Notched to Unnotched Bars. Davis & Manjoine, ASTM STP128

notch bars, not smooth bars. For example, Table I is a listing of rupture data for smooth bars as reported by Schmieder (16) for a ferritic steel (ASTM-A-193-B16) heat treated to three different strength (ductility) levels. Careful examination of these data indicates that size has virtually no effect on rupture behavior beyond what could be taken as normal test scatter. In contrast, notch bar data for the same material(Table II), also reported by Schmieder (16), shows a rather dramatic effect of size on time to rupture. In general, Schmieder's notch bar data indicates that for the two most brittle materials, i.e., Condition A and Condition B, increasing size reduces the time to rupture. For Condition C (retempered & most ductile condition) no size effect is evident, albeit this conclusion is based on limited data.

There are two noteworthy but particularly conflicting size effect studies reported in the literature (13, 18). Manjoine (18) reported data suggesting that the time to rupture increased with size for a brittle CrMoV rotor material but decreased with size for a ductile CrMoV rotor material. Goldhoff, in contrast, on a similar CrMoV steel, reported that the 1000 hour notch strength ratio decreased with size for both ductile and brittle materials.

One of the difficulties faced by these previous investigators was the fact that they reported only those experimental observations made by themselves or by others previous to their own studies. None of these investigators proposed a unified interpretation of available results. Without a tentative model, they were at the mercy of experimental scatter and since their experimental data were limited, they could not easily justify ignoring any experimental data. Goldhoff did, however, extend his analysis to show analogies between his elevated temperature notch rupture data and similar behavior at room temperature.

### Ambient Temperature Effects

*1) Notch Sharpness.* The volume of literature data on notch bar behavior at ambient temperatures is at least as large as that at elevated temperatures. Fortunately, the interested reader can refer to two excellent reviews (12, 14) for a good summary of these studies. Of particular interest to this discussion is Figure 4 which was taken from the review article of Sachs, Sessler, and Brown (12). This figure shows the variation of notch strength ratio, as derived from room temperature tensile tests, versus notch sharpness for three materials of varying notch ductility. It should be noted that initially the notch strength ratio increases with notch sharpness regardless of notch ductility; however, eventually the notch strength ratio reaches a peak and then decreases with increasing sharpness. The maximum notch strength ratio, the rate of decrease, and the magnitude of the decrease from the peak all depend on the inherent ductility of the material. Also note, the maximum occurs at a lower value of notch sharpness for the most severely notch sensitive condition.

It is also interesting to note that Figure 4 has exactly the same shape as the summary plot of Manjoine (Figure 3) which was derived from creep rupture tests. The only difference being that in Figure 3 notch sharpness increases towards the left of the figure, whereas notch sharpness increases to the right in Figure 4. Accordingly, it should be concluded that the effect of notch geometry is identical regardless of whether testing is done at ambient temperatures or at elevated temperatures. The ultimate response will be dependent on the intrinsic ductility of the material at the test temperature.

It should be cautioned that there is no strong correlation between room temperature ductility and elevated temperature ductility. Therefore, while Figure 3 and Figure 4 show identical trends, they are only indicative of trends at the specific test temperature. There is no inference here that poor high temperature ductility infers poor low temperature ductility.

An interpretation of the notch bar behavior shown in Figure 4 was presented by Schwartzbart and Brown (19) in the form of Figure 5. They said, "Curve ABC is the function of triaxiality versus sharpness; it is also, therefore, the function of notch strength ratio versus sharpness for a material which does not crack before necking, and for which the necking strain is independent of sharpness and is always less than the notch ductility. Curve FBE is the nominal fracture strength function which is affected by notch ductility as well as triaxiality. The intersection of these two curves, at point B, is then the

Table I. Size Effect Study (from Ref 16)

Smooth Bar Data at 1000°F

Condition A (Most Brittle Condition) [1900°F, Oil Quench, 1200°F Temper]

| Size | Applied Stress Ksi | Time to Failure Hrs | Elong. % | Red. of Area % |
|---|---|---|---|---|
| 1 | 46 | 1336 | 2.5 | 4.7 |
| 2 | 46 | 1354 | 1.2 | 3.1 |
| 3 | 46 | 1087 | 1.1 | 2.4 |
| 1 | 33 | 7198 | 1.6 | 2. |
| 2 | 33 | 7820 | 1.1 | 2. |
| 3 | 33 | 7409 | 1.0 | 0. |

Condition B (Brittle Condition) [1750°F, Oil Quench, 1260°F Temper]

| Size | Applied Stress Ksi | Time to Failure Hrs | Elong. % | Red. of Area % |
|---|---|---|---|---|
| 1 | 50 | 1242 | 8.8 | 44.3 |
| 2 | 50 | 1140 | 19.1 | 43.4 |
| 3 | 50 | 1469 | 9.6 | 26.1 |
| 1 | 33 | 10930 | 7.8 | 19.0 |
| 2 | 33 | 12520 | 8.7 | 16.5 |
| 3 | 33 | 13074 | 8.7 | 13.3 |

Condition C (Most Ductile Condition) [1750°F, Oil Quench, 1300°F Temper]

| Size | Applied Stress Ksi | Time to Failure Hrs | Elong. % | Red. of Area % |
|---|---|---|---|---|
| 2 | 50.4 | 95 | 24.5 | 82 |
| 2 | 33 | 3510 | 19.2 | 76 |

| Size | 1 | 2 | 3 |
|---|---|---|---|
| Diam (Inch) | 0.197 | 0.472 | 1.14 |

Table II. Size Effect Study (from Ref 16)

Notch Bar Data at 1000°F (Geometrical Similar Series, Kt=5.9)

Condition A (Most Brittle Condition) [1900°F, Oil Quench, 1200°F Temper]

| Size | Applied Stress Ksi | Time to Failure Hrs |
|---|---|---|
| 2 | 34 | 418 |
| 3 | 34 | 480 |
| 2 | 25 | 2034 |
| 3 | 25 | 486 |
| 4 | 25 | 366 |
| 4T | 25 | 2117 |
| 2 | 17 | 24180 |
| 3 | 17 | 4328 |
| 4 | 17 | 2100 |

Condition B (Brittle Condition) [1750°F, Oil Quench, 1260°F Temper]

| Size | Applied Stress Ksi | Time to Failure Hrs |
|---|---|---|
| 2 | 50 | 665 |
| 2 | 50 | 1380 |
| 3 | 50 | 267 |
| 2 | 44 | 1155 |
| 3 | 44 | 1096 |
| 4 | 44 | 842 |
| 4T | 44 | 839 |
| 2 | 31 | 11560 |
| 3 | 31 | 5309 |
| 4T | 31 | 1018 |

Condition C (Most Ductile Condition) [1750°F, Oil Quench, 1300°F Temper]

| Size | Applied Stress Ksi | Time to Failure Hrs |
|---|---|---|
| 2 | 56 | 534 |
| 4T | 56 | 366 |
| 2 | 43 | 2376 |
| 4T | 43 | 2304 |

| Size | 1 | 2 | 3 | 4 |
|---|---|---|---|---|
| Diam (Inch) | 0.197 | 0.472 | 1.14 | 2.76 |

(4T indicates that these were triple notched specimens)

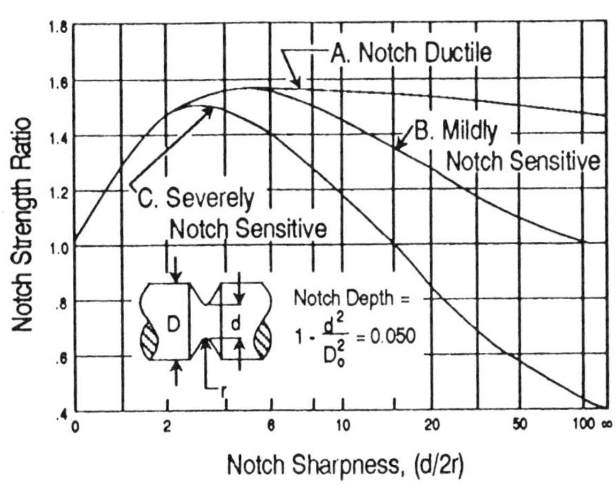


Fig. 4 – Notch Strength Ratio at a Constant Notch Depth (50 per cent) as a Function of the Notch Sharpness of a Notch Ductile, a Mildly Sensitive, and a Severely Notch Sensitive Material. (Sachs, Sessler & Brown, ASTM STP 260 [1959])

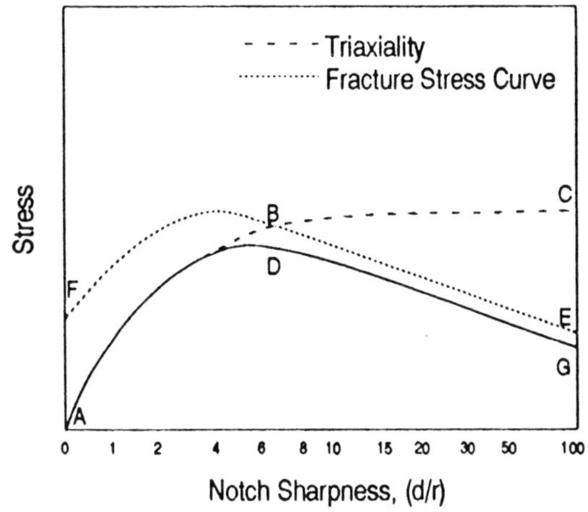


Fig. 5 – Schematic Representation of the Factors Determining the Shape of the Notch Strength Ratio Versus Sharpness Curve. (Schwartzbart & Brown, Trans. ASM [1954])

sharpness at which the notch ductility equals the necking strain and is the maximum of the notch strength ratio. Sometimes, however, the onset of cracking before final fracture can cause the observed stress strain curve to flatten at high strains. This causes the observed necking strain to change, giving an actually observed notch strength ratio curve of ADG with the maximum shifted from point B as indicated." Stated somewhat differently, the line ABC represents the effect of notch sharpness on the limit load of the notch bar. Presumably, this line could be calculated by finite element techniques using suitable constitutive relationships and an appropriate flow rule. The actual behavior, ADG, would deviate from predicted behavior when the combination of notch sharpness and the intrinsic material ductility exceeds a critical value giving rise to premature crack nucleation. The lower the strain for crack nucleation relative to the limit load strain, the greater the departure of actual versus predicted behavior.

A similar argument was discussed by McClintock (20) who points out for tensile loading, cracking may initiate before or after the point of maximum load which reflects the inherent strength of the material. He states that for a complete description of the structural behavior of a specimen, the entire load-deformation diagram as shown in Figure 6A is required. McClintock further states that since, for ductile materials, the maximum load occurs at a value of deformation which is small compared to the final deformation, the main features of the load displacement diagram are given by the maximum load and final deformation. As less ductile materials are considered, the final deformation is reduced without change in the maximum load (compare Figure 6C to 6B), while for still less ductile materials the maximum load will also be reduced because of early crack nucleation (compare Figure 6D with 6B and 6C).

While the above arguments relate to ambient temperature tests, Kloos and Kiehl (17) reported on the effect of notch geometry on the notch strength ratio derived from hot tensile tests at 550°C. Figure 7A summarizes these data for three different strength (ductility) levels.

It is interesting to note that the notch strength ratio increased with increasing notch sharpness in a manner similar to that shown in Figure 4 from room temperature data. Since these were short time tests, the Notch Strength Ratio (NSR) was always greater than one for these materials. In contrast, when the NSR was compiled at 3000 hours from creep data, different trends became apparent (see Figure 7B). For example, for Steel 1 (Zustand 1) the 3000 hour NSR curves looked similar to the short time tensile data, however, the 3000 hour NSR curves for Steel 3 (Zustand 3) indicated substantial notch weakening with increasing notch sharpness.

The different trends between the short time tensile results and the 3000 hour rupture results should not be too surprising. Figure 8 is

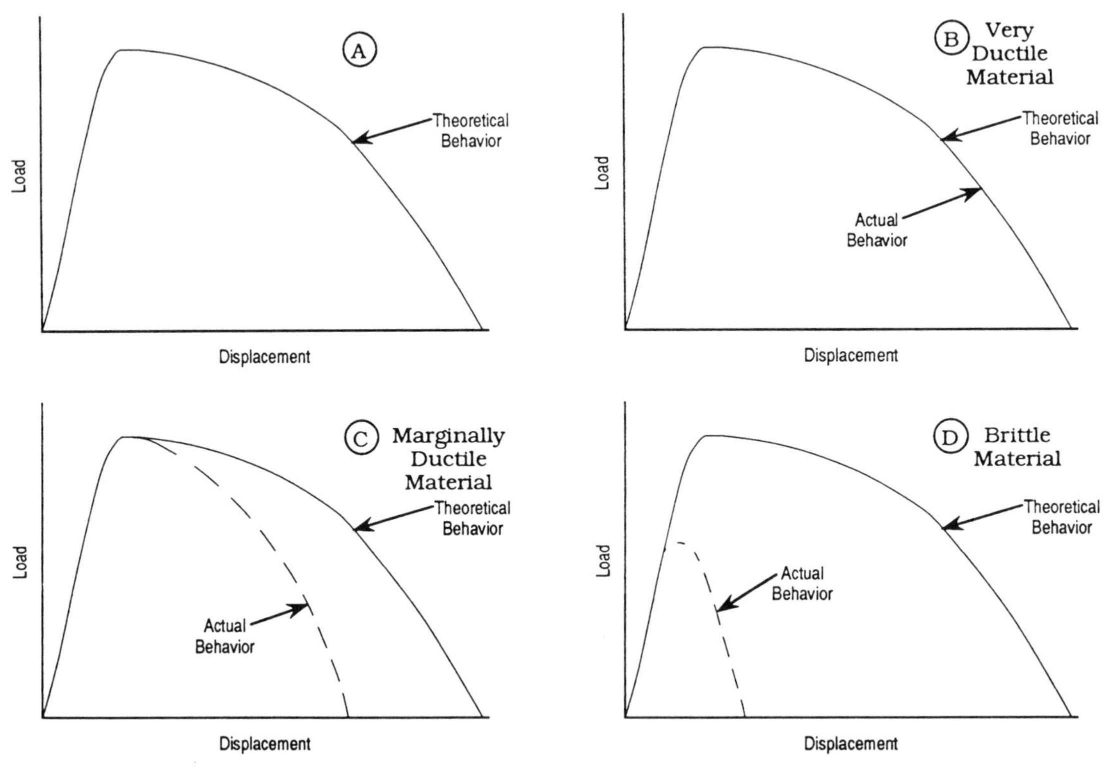


Fig. 6 – Hypothetical Tensile Behavior for Varying Ductility (After McClintock, 20).

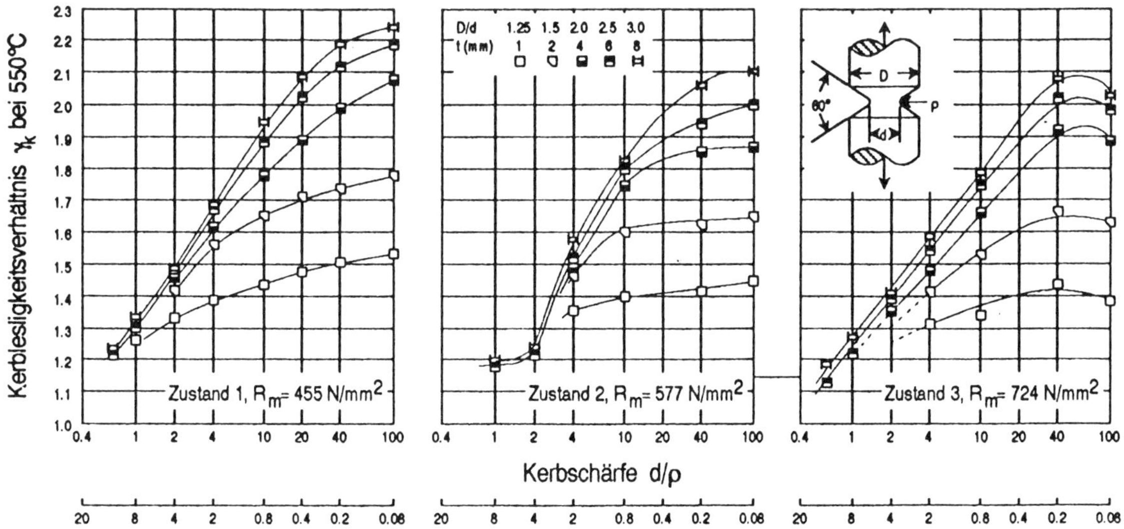


A. Effect of Notch Sharpness and Diameter Ratio on Notch Strength Ratio as Derived from Short Time Hot Tensile Tests

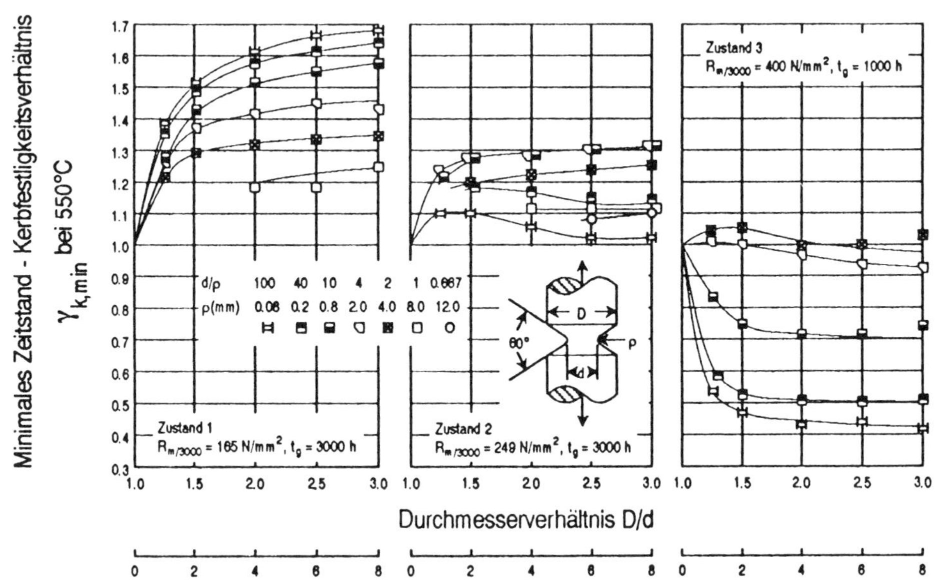


B. Effect of Notch Sharpness and Diameter Ratio on Notch Strength Ratio as Determined from Rupture Tests at 3000 hrs

Fig. 7 – Effect of Notch Sharpness and Diameter Ratio on Notch Strength Ratio (Ref. 17).

a plot of notch bar and smooth bar strength versus time to failure and shows both the hot tensile data (far left points) and the rupture data. Examination of this figure reveals why the tensile and rupture trends shown in Figure 7A and 7B differ. Presumably, if the hot tensile tests were conducted at a slow enough rate (i.e., long time tensile tests), or the rupture tests were conducted at a fast enough rate (i.e., short time rupture tests), the trends from both tests would be identical. This is an extremely important point emphasizing the importance of making comparisons over similar test time ranges. It also emphasizes the fact that notch ductility varies with time (strain rate).

*2) Size Effect.* Many past studies have been concerned with size effects on fracture at ambient temperatures. Of particular relevance to this discussion is Wundt's unified interpretation of room temperature strength of notched specimens as influenced by their size (21). Figure 9 is a schematic of Wundt's discussion presented in terms of the size effect on absolute notch strength. The curves shown encompass the sharply notched geometric size effect ranging from highly notch insensitive (Material A) to highly notch sensitive (Material C). No phenomenological explanation of the observed trends relative to fracture were presented by Wundt. However, it is speculated that the load displacement curves versus size could have looked similar to those shown in Figure 6. In Figure 6, the different load displacement curves resulted because of variations in ductility keeping the size and geometry constant. The analogy for size would be to keep the geometry and ductility constant but vary the size. Clearly, if this were a correct phenomenological interpretation of Wundt's diagram, the behavioral trends shown by Wundt could be rationalized by this reasoning.

Figure 10 schematically illustrates the above phenomenological interpretation of Wundt's diagram. In this figure Material A is sufficiently ductile that size cannot alter the limit behavior of the notched bars; Material B is sufficiently ductile to obtain limit behavior in bar size #1 but not sufficiently ductile to prevent premature crack nucleation and, therefore, deviation from limit behavior for bar sizes 2 and 3. Material C is sufficiently notch sensitive that limit behavior is not even possible for the smallest bar size (#1).

## A Unified Interpretation of Notch Bar Behavior

**Constant Displacement Rate Creep Testing Versus Constant Load Creep Testing.** While constant displacement rate creep testing is not common the author has been involved in this type of testing methodology for some time. This test methodology has resulted in the development of a short time testing procedure which can identify long time notch ductility tendencies in short times (23). Moreover there have been other limited studies relating the two testing methodologies (24-26). Why these studies are not specific and most of the constant displacement rate testing are limited to fast rates because of equipment limitations, they do imply that there are many interchangeable similarities between constant displacement rate creep testing and constant load creep testing. These are shown with the aid of Fig. 11. Figure 11A illustrates a series of typical isothermal constant load creep curves. Each curve represents a different initial load. It should be noted that total strain rather than plastic strain is plotted on the ordinate. Total strain will be used for the entire discussion of Figure 11. From the data of Figure 11A, it

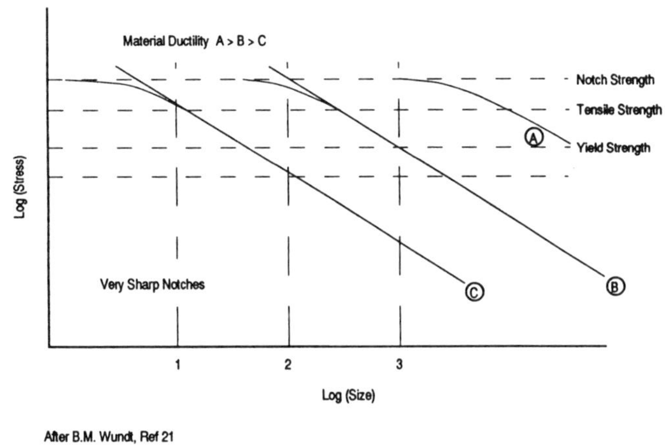


Fig. 9 – Effect of Size and Material Ductility on Notch Strength

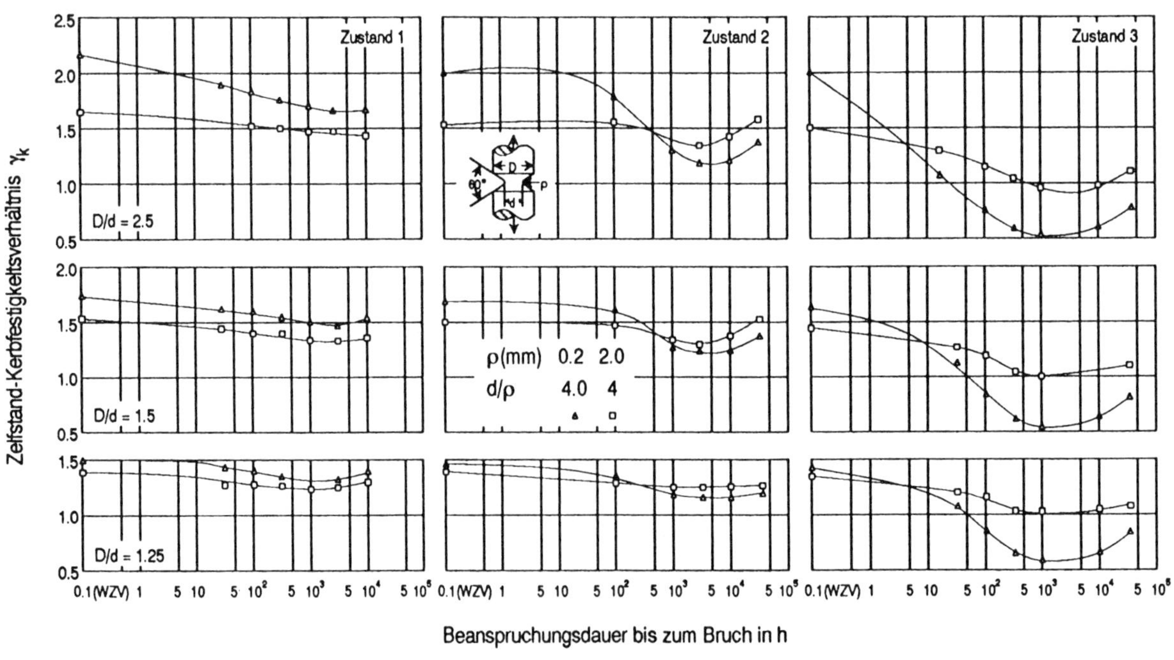


A. Rupture Behavior

Fig. 8 – The Effect of Notch Sharpness on Notch Strength Ratio Versus Rupture Time at 550°C (Ref. 17)

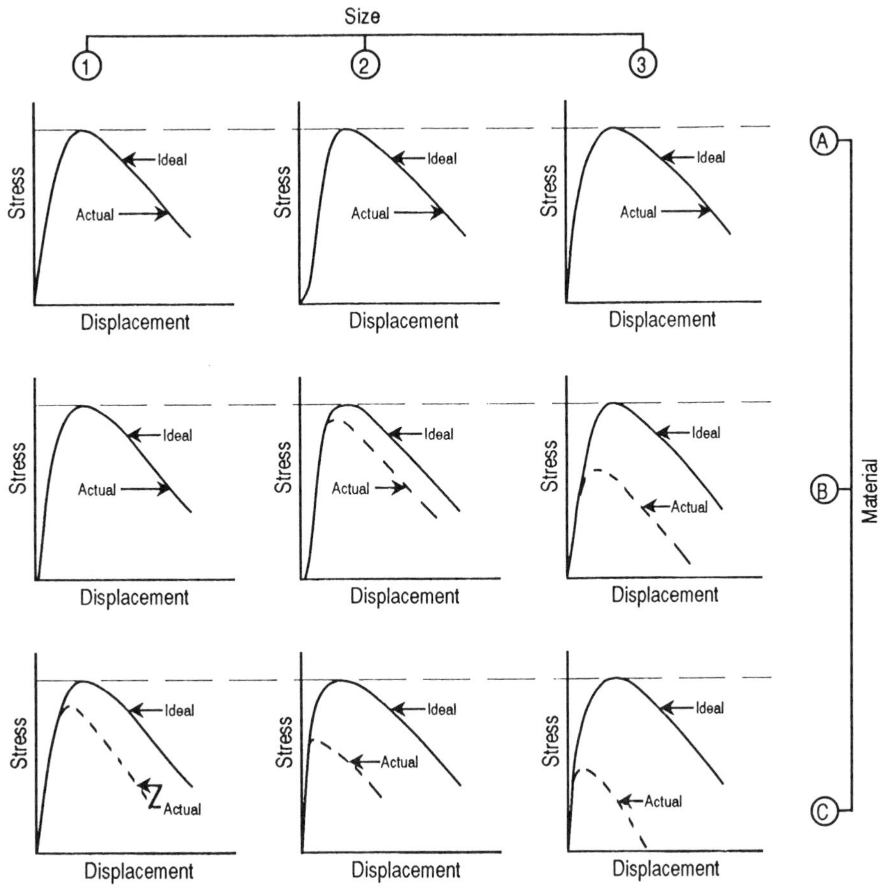


Fig. 10 – Phenomenological Interpretation on the Effect of Size and Ductility on Notch Strength as Described in Figure 9. (Solid Lines – Ideal Behavior, Dashed Lines – Actual Behavior)

is relatively straight forward to develop Figure 11B, which is a plot of the instantaneous strain rate versus total strain. From Figure 11B, the minimum strain rate can be identified for each level of initial applied stress (initial load/initial area). These data are replotted to develop Figure 11D. Typically, this plot yields a straight line relationship between log (stress) and log (minimum strain rate). This in turn results in the standard power creep law, i.e., $\dot{\varepsilon} = A\sigma^n$.

Figure 11D can also be developed directly from a properly conducted constant displacement rate tensile test. The only prerequisite would be identical placement of the extensometers on an identical specimen. The abscissa coordinate would be the applied displacement rate divided by the effective gage length (same as constant load creep procedure) to yield a strain rate to be used in Figure 11D. The appropriate stress would be the maximum load divided by the initial area (same as constant load creep procedure) to yield a stress to be used in Figure 11D. Figure 11C represents a series of typical stress-total strain plots developed at different strain rates. Having established a link between constant load creep data and constant displacement rate creep data, we are now in a position to examine salient ambient temperature notch bar data.

Using the relationships between constant load and constant displacement rate creep testing described above, an interpretation of notch geometry and size effects can be developed. This can be demonstrated schematically described with the aid of Figure 12 for geometry effects and Figure 13 for size effects. Figure 12A represents a hypothetical log (stress)-log (displacement rate) behavior over a wide range of displacement rates. The role of intrinsic material ductility simply places the material in a certain response regime. This is analogous to the excess temperature approach to brittle fracture discussed by Newhouse (22). For very ductile materials the log $\sigma$-log $\dot{\varepsilon}$ response of notch bars is shifted upward and parallel to the smooth bar curve. The magnitude of the upward shift is dependent on geometry (r/d and d/D). The various downward breaks in these curves at low displacement rates results because a critical ductility level was not met. The actual position of the breaks depend on a particular combination of notch geometry and intrinsic material ductility. Above these critical values, the notch bar can achieve its full strength capabilities as discussed in Figure 6. Below these critical values, the notch bar cannot reach its strength capabilities because of premature crack nucleation. An analogy similar to Figure 6 can be made for constant load creep rupture testing. For example, for a particular notch geometry, it should be anticipated that a limiting minimum displacement rate should be reached in the absence of cracking. However, if this limiting minimum displacement rate is not reached because of premature cracking, then an artificially high minimum displacement rate is measured leading to the downward break in the curves shown in Figure 12A. This point is more fully illustrated in Figure 14.

Figure 14A represents the ideal strain rate versus strain response of a ductile material. Figures 14B-14D show the expected deviation from ideal behavior as the intrinsic ductility decreases from very ductile to very brittle. This representation illustrates the effect of ductility on this behavior and is exactly analogous to the arguments of McClintock (20) described with the aid of Figure 6.

Continuing further, Figure 14E illustrates the expected response of conventional constant load creep tests for a series of different stress levels. In this figure it, is assumed that crack nucleation occurs after the minimum creep rate is reached for the stress levels $\sigma 1$, $\sigma 2$, $\sigma 3$. Consequently, the combination of stresses $\sigma 1$, $\sigma 2$, and $\sigma 3$ togeth-

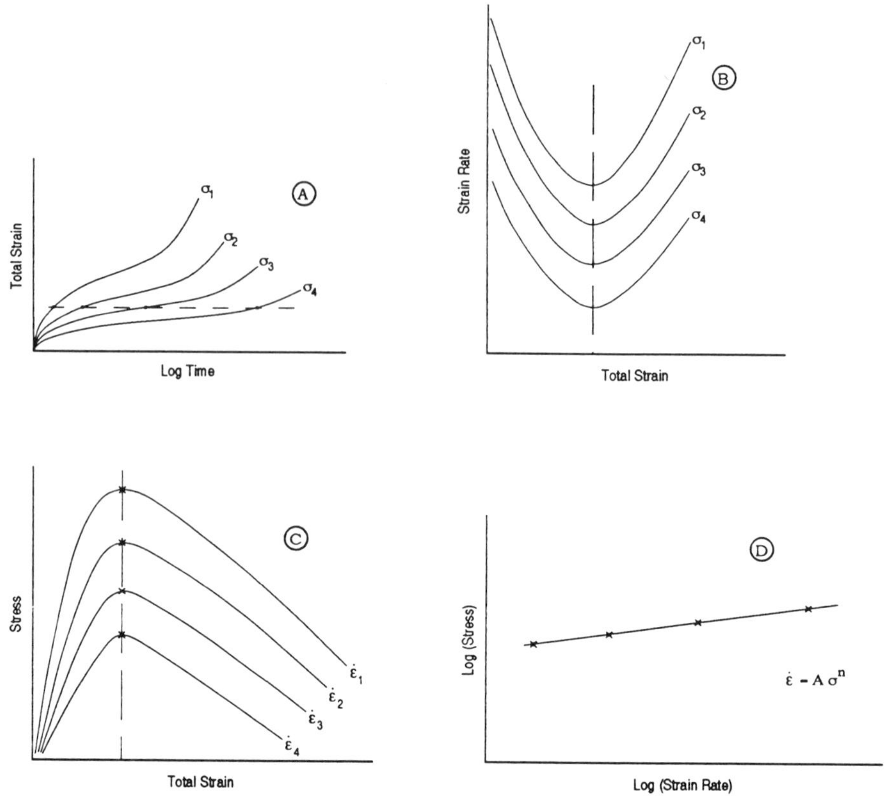


Fig. 11 – Interrelation Between Constant Load Creep and Constant Displacement Rate Creep Data Using Smooth Bars.

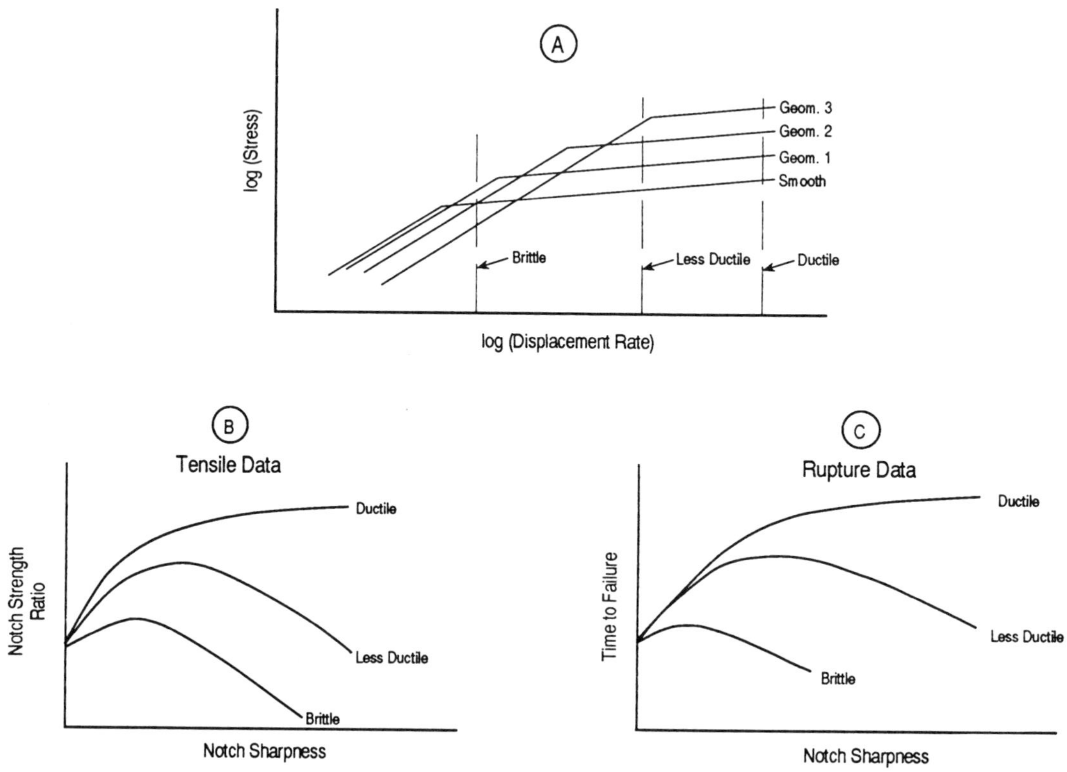


Fig. 12 – Geometry Effects (Size Constant)

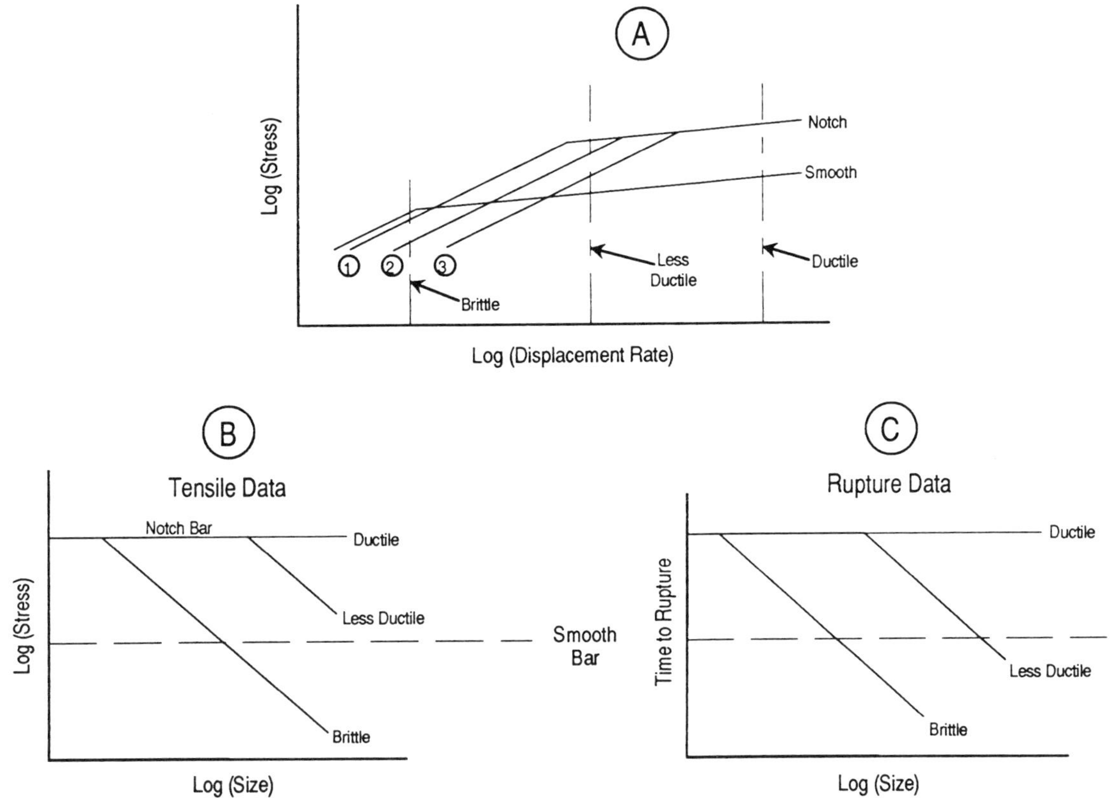


Fig. 13 – Size Effects (Geometry Constant)

er with the resulting minimum creep rates A, B, and C from a straight line on a Log $\sigma$ Log $\dot{\varepsilon}$ plot (Figure 14F). However, it is also assumed in this figure that cracks nucleate prior to the ideal minimum creep rate for the stress levels $\sigma 4$ and $\sigma 5$. For the stress level $\sigma 4$, the actual versus ideal minimum creep rate are D versus D' respectively; for the stress level $\sigma 4$, these are E versus E'. Consequently, the location of the actual minimum creep rates D and E on the Log $\sigma$ – Log $\dot{\varepsilon}$ plot result in a downward break as indicated. This phenomenological argument is applicable to both notch bars and smooth bars. The actual placement of the ideal curves on a Log $\sigma$-Log $\dot{\varepsilon}$ plot depends on the geometry (r/d and d/D). The downward breaks are dependent on geometry (r/d and d/D), size, and the intrinsic ductility of the material at the test temperature. Note that strain rate (displacement rate for notch bars) and ductility are used interchangeably for this discussion.

The qualitive explaination of these downward breaks in Log $\sigma$ - Log $\dot{\varepsilon}$ space offered above is probably the phenomonological explanation of the downward breaks often observed on stress rupture curves. These breaks have been attributed to an abrupt change in failure mode (27). It is suggested here that a more plausible, but similar, explaination is the loss of cross sectional area due to cracking before the strain limit typical of the minimum strain rate in test specimens tested at higher stresses, i.e., as discussed in Figure 14E.

If the curves shown in Figure 12A are entered from the strain rate axis (i.e., constant displacement rate (tensile) tests) the curves shown in Figure 12B should result. (Note these are similar to Figures 5 and 7A.) In contrast, if the curves shown in Figure 12A are entered from the stress axis (i.e., constant load creep tests, the curves shown in Figure 12C should result. (Note these curves are similar to those in Figure 3.) In Figure 12C, time to failure is assumed to be inversely proportional to minimum displacement rate. This is a reasonable assumption based on the extensive literature support of the Grant-Mockmann relationship (28), i.e., (time to rupture) multiplied by (minimum strain rate) = constant.

A similar schematic (Figure 13A) can be developed to explain size effects. In this figure, only one notch geometry is depicted. Consequently, for the most ductile condition, size should have no influence on either time to rupture or notch strength. The effect of increasing size results in a shift to the right (higher displacement rates) for the downward break in the curves. This suggests that increasing size makes crack nucleation easier and, therefore, prevents the attainment of the limit load potential if ductility is marginal at the time (strain rate) of interest.

If the curves shown in Figure 13A are entered from the strain rate axis (i.e., constant displacement rate (tensile) tests), the curves in Figure 13B should result. In contrast, if the curves shown in Figure 13A are entered from the stress axis (i.e., constant load creep tests), the curves shown in Figure 13C should result. Again, time to failure in Figure 13C is assumed to be inversely proportional to minimum displacement rate. At the present time, no reference is available to substantiate the proposed curves, Figure 13B and 13C. However, the data of Wundt for brittle fracture, together with the elevated temperature data of Schmieder (16), Goldhoff (13), and Manjoine (18), offer considerable indirect experimental evidence to give the aforementioned interpretation to size effects at elevated temperature substantial credence. Moreover, with the proposed model to work with, an experimental effort can be undertaken to obtain data to support or refute it.

## Summary

This report presents an interpretation of notch geometry and notch size effects on elevated temperature notch rupture behavior. A review of pertinent literature data derived at both ambient and elevated temperatures is presented. Analogies regarding behavior trends of notch bar results are made between both temperature regimes. The major conclusion drawn from this effort has been that the observed behavior trends

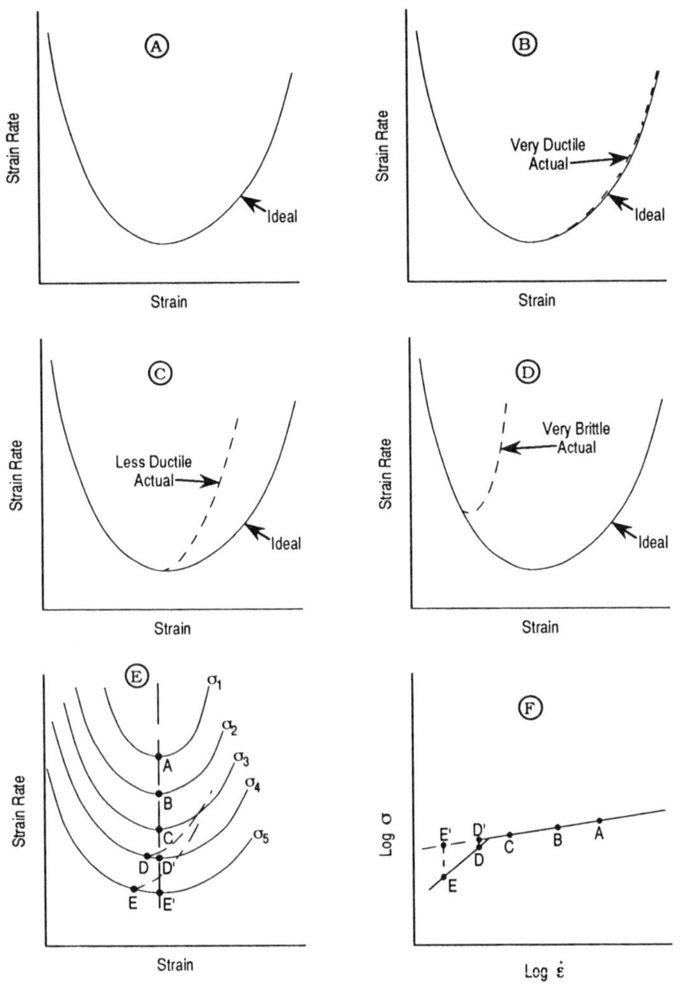


Fig. 14 – Effect of Intrinsic Ductility of $\dot{\varepsilon}$ - ε and Log σ - Log $\dot{\varepsilon}$ Behavior

are strongly dependent on the intrinsic material ductility. It was pointed out early in this report that notch bar behavior is often used to characterize material ductility. Accordingly, the use of notch bar behavior to characterize ductility should be used with caution. While an alternative ductility index has not been proposed, the unified interpretation developed should provide a framework toward a more complete understanding of notch bar behavior and the use of such information.

## References


1. Manjoine, M. J., "Ductility Indices at Elevated Temperature", Trans. ASME, April 1975, pp. 156-161.
2. Sachs, G. and Brown, W. F. Jr., "A Survey of Embrittlement and Notch Sensitivity of Heat Resistant Steels", Symposium on Strength and Ductility of Metals at Elevated Temperatures, ASTM STP, No. 128, 1952, pp. 6-20.
3. Brown, W. F. Jr., Jones, M. H., and Newman, D. P., "Influence of Charpy Notches on the Stress Rupture Characteristics of Several Heat Resistant Alloys", ibid., pp. 25-45.
4. Hull, F. C., Hann, E. K., and Scott, H., "Effect of Notch and of Hardness on the Rupture Strength of Discaloy", ibid., pp. 49-58.
5. Furman, D. E. and Talbot, A. M., "Notch Rupture Tests of Inconel X and Nimonic 80A", ibid., pp. 59-66.
6. Davis, E. A. and Manjoine, M. J., "Effect of Notch Geometry on Rupture Strength at Elevated Temperatures", ibid., pp. 67-92.
7. Siegfried, W., "Investigations into the Influence of Notches on Creep Strength at High Temperatures", ibid., pp. 93-130.
8. Frey, D. N., "Theory of Time Dependent Rupture and Interpretation of Some Stress Rupture Data", ibid, pp. 131-141.
9. Voorhees, H. R., Freeman, J. W., and Herzog, A. J., "Trends and Implications of Data on Notched Bar Creep Rupture", Trans. ASME, June 1962, pp. 207-213.
10. Couts, W. H. Jr. and Freeman, J. W., "Notched Rupture Behavior as Influenced by Specimen Size and Preparation", ibid., pp. 222-227.
11. Brown, W. F. Jr., "Summary of Session", ibid., pp. 243-245.
12. Sachs, G., Sessler, J. G., and Brown, W. F. Jr., "Influence of Stress Concentrations at Elevated Temperatures", ASTM STP No. 260, 1959, pp. 1-62.
13. Goldhoff, R. M., "Stress Concentration and Size Effects on a CrMoV Steel at Elevated Temperatures", Proceedings of Joint International Creep Conference, Vol. I, Institute of Mechanical Engineering, London, 1962, pp. 419 to 432.
14. Agogina, A. M., "Notch Effects, Stress State, and Ductility", Trans. ASME, Vol. 100, October 1978, pp. 348-355.
15. Schmieder, A. K., "Size Effect in Creep and Rupture Tests on Unnotched and Notched Specimens of CrMoV Steel", Current Work on Behavior of Materials at Elevated Temperatures, Publication G87, American Society of Mechanical Engineers, New York, 1974, pp. 125-155.
16. Schmieder, A. K., "Size and Effect During Rupture Tests of Unnotched and Notched Specimens of CrMoV and CrNi Steels", ASME-MPC 8, 1978, pp. 31-49.
17. Kloos, K. H. and Kiehl, H., "Influence of the Notch Geometry on the Creep Rupture Properties of Steel 21CrMoYS7 in Different Heat Treatment Conditions", Arch Eisenhuttenwes, Vol. 50, No. 6, June 1959, pp. 255-260.
18. Manjoine, M. J., "Size Effect on Notched Rupture Time", Trans. ASME, Journal of Basic Engineering, June 1962, pp. 220-221.
19. Schwartzbart, H. and Brown, W. F. Jr., "Notch Bar Tensile Properties of Various Materials and Their Relation to the Unnotched flow Curve and Notch Sharpness", Trans. ASM, Vol. 46, pp. 998-1020.
20. McClintock, F. A., "On Notch Sensitivity", Welding Journal, Welding Research Suppl., May 1961, pp. 202S-209S.
21. Wundt, B. M., "A Unified Interpretation of Room Temperature Strength of Notched Specimens as Influenced by Their Size", ASTM Publication 59-Met-9.
22. Newhouse, D. L., "Relationships Between Charpy Impact Energy, Fracture Appearance, and Test Temperature in Alloy Steels", Welding Journal Research Supplement, March 1963, pp. 105S-114S.
23. Pepe, J.J. and Gonyea, D.C., Constant Displacement Rate Testing at Elevated Temperatures, Proceedings: Fossil Power Plant Rehabilitation, Trends in Life Management and Reliability Technology, ASM International, Cincinnati, Ohio, 1989, pp 39-45.
24. Dotson, C.L., "The Effect of Slow Constant Strain Rates on the Tensile Properties of Selected Alloys at Room and Elevated Temperatures, Journ. of Engineering Materials & Technology, April 1974, pp 115-122.
25. Berkovits, A., Prediction of Inelastic High Temperature Materials Behavior by Strain Rate Approach, Transactions ASME, April 1974, pp 104-108.
26. Nechtelberger, E., et. al., Contributions to Rapid Determination by the Rajakovics Method of the Creep Behavior of Steel, Arch. Eisenhuttenwes, Vol. 44, No. 2, Feb 1973, pp 135-141.
27. Pavinich, W and Raj, R. "Fracture at Elevated Temperatures", Met. Transactions A, Vol 8A, Dec. 1977, pp 1917-1932.
28. Monkman, R. C. and Grant, N. J., "An Empirical Relationship Between Rupture Life and Minimum Creep Rate in Creep Rupture Tests", Proc. ASTM, 1956, Vol. 56, p. 593.

Proceedings of the Fifth International Conference on Creep of Materials, Lake Buena Vista, Florida, USA, 18-21 May 1992

# Practical Approach to J-Integral Estimation

**T. Sakon and H. Kaneko**
Mitsubishi Heavy Industries, Ltd.
Hyogo, Japan

## Abstract

The conventional estimation method of J-integral is investigated with the purpose of the practical use for the crack growth evaluation in high temperature structural components. A simple equation of J-integral for load-controlled type loading is derived based on Neuber's rule for notch problems. By using the equation J-integral can be estimated from the nominal stress and strain and the stress intensity factor. The application procedure for the deformation-controlled type loading is also proposed. The methods proposed are confirmed by several kinds of inelastic finite element analyses of two-dimensional cracked panels.

J-INTEGRAL ANALYSES are frequently required for the integrity evaluation of flawed structures in inelastic stress fields. Although J-integral can be calculated by the inelastic numerical analyses of flawed structure, a great deal of expense and work are often required for the inelastic analyses, and it is not practical thereby to perform such analyses for a lot of structures to be evaluated. For more practical and convenient application of J-integral, the fully-plastic solutions for various kinds of typical configurations have been obtained (1, 2), and they have been frequently used for the crack evaluation. However, it is not necessarily easy to apply them to the general subjects except the cases whose solutions have been already obtained. From more practical point of view, it is desirable that a J-integral estimation procedure has the following features;

- Applicability with any type of stress-strain relations (elastic-plastic constitutive equations),
- Using no correction factor obtained directly by inelastic numerical stress-strain analyses such as FEM.

Considering these points, this paper presents a conventional method for J-integral estimation which is derived from Neuber's rule used for estimating the inelastic stress/strain concentration at notch.

In this paper, the corrective of Neuber's rule will be discussed at first from the viewpoint on the improvement of the estimation accuracy, and the J-integral estimation method will be derived from the corrected Neuber's rule. The results of several confirmatory inelastic analyses will be successively shown for the load-controlled condition. Finally, the application procedure to the deformation-controlled loading problem will be also proposed.

## Inelastic Stress/Strain Concentration at Notch

Neuber's rule given by Eq.1 is widely used for the estimation of the stress and strain concentration at notch root in the elastic-plastic field.

$$\left.\begin{aligned} K_\sigma K_\varepsilon &= K_t^2 \\ K_t &= S_{max}/S_n \\ K_\sigma &= \sigma_{max}/\sigma_n \\ K_\varepsilon &= \varepsilon_{max}/\varepsilon_n . \end{aligned}\right\} \quad (1)$$

Here, $K_t$ is the elastic stress concentration factor and the suffix $n$ means the net section stress. Throughout this paper the elastically-calculated stress and strain are represented by $S$ and $e$.

Although Eq.1 has been widely used owing to its simple style, it has been pointed out that Eq.1 tends to overestimate the strain in general yield condition (3, 4). One of the authors has found by the analytical study that this shortcoming results from the multi-axial stress state in the net section. As shown in Fig.1, the stress component $\sigma_x$ perpendicular to the applied loading axis originates in the net section, so that the effective stress invariant $\sigma^*$ is generally lower than the axial stress component $\sigma_y$. This is more conspicuous in the plane strain or the axisymmetry conditions where $\sigma_z$ component exists. Assuming the Mises yield condition and flow rule, the general yielding will not start when the net section stress $\sigma_n$ comes

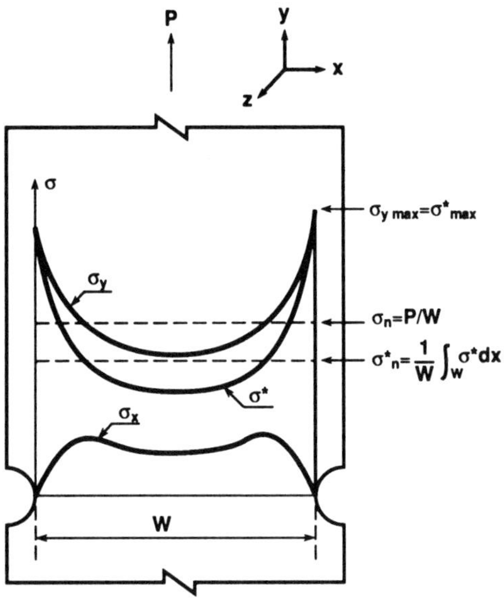


Fig.1 Schematic of stress distribution in the vicinity of notch.

up to the yield stress in the case such as Fig.1. One of the authors has therefore proposed to apply Neuber's rule based on the average effective stress in the net section in order to take such effect into account. In this procedure the following formulas are used instead of Eq.1.

$$\left.\begin{aligned} K_\sigma{}^* K_t{}^* &= (K_t{}^*)^2 \\ K_t{}^* &= S^*{}_{\max}/S_n{}^* \\ K_\sigma{}^* &= \sigma^*{}_{\max}/\sigma_n{}^* \\ K_\varepsilon{}^* &= \varepsilon^*{}_{\max}/\varepsilon_n{}^* \end{aligned}\right\} \tag{2}$$

where * means the Mises effective stress, and $\sigma_n{}^*$ can be estimated based on the elastic effective stress distribution. For instance the following formulas can be used for the case of Fig.1.

$$\sigma_n{}^* = \sigma_n \frac{S_n{}^*}{S_n} \tag{3}$$

$$S_n{}^* = \frac{1}{W}\int_W S^* dx \tag{4}$$

$$S_n = \sigma_n = P/W \tag{5}$$

And, $\varepsilon_n{}^*$ is calculated by substituting $\sigma_n{}^*$ to the uniaxial stress-strain equation.

Fig.2 shows the relationship between $\sigma_n$, $S_n$ and $\sigma_n{}^*$, $S_n{}^*$ obtained by the inelastic and elastic FE analyses for the double edge-notched plates, by which the validity of Eq.3 has been confirmed. Fig.3 shows the examples of the stress/strain concentration factors estimated from Eqs.1 and 2 and those obtained from the inelastic FE analyses of an axisymmetric notch where the stress multiaxiality is conspicuous. It is found that Eq.2 gives more accurate estimates

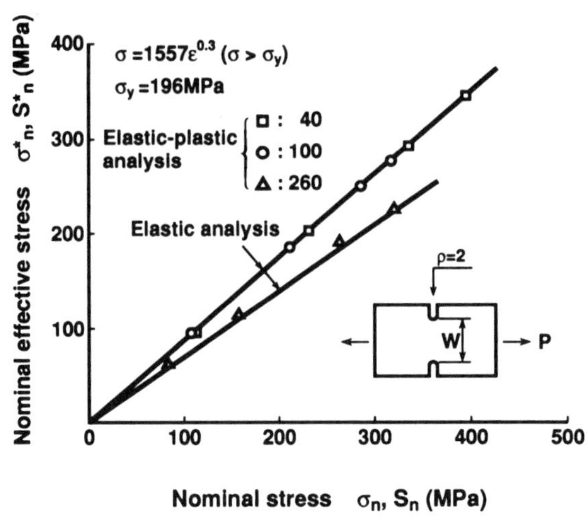


Fig.2 Relationship between $\sigma_n$ and $\sigma_n{}^*$ in elastic and inelastic conditions. (Double edge notched plate, Plane stress)

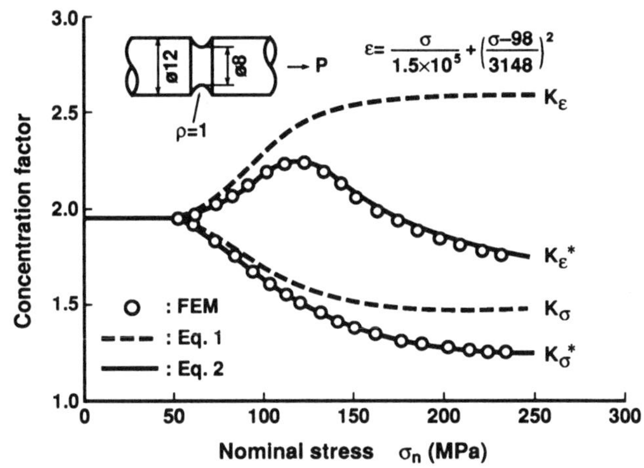


Fig.3 Comparison of stress/strain concentration factors estimated by Eqs.1 and 2.

in comparison with Eq.1. Fig.4 shows the examples of the distribution of $\sigma^*\varepsilon^*/\sigma_n{}^*\varepsilon_n{}^*$ in the net section. Little difference in the ratio of $\sigma^*\varepsilon^*/\sigma_n{}^*\varepsilon_n{}^*$ was observed for the various levels of nominal stress or between the elastic and the inelastic analyses, so that the following relationship can be assumed approximately.

$$\sigma^*\varepsilon^* = \sigma_n{}^*\varepsilon_n{}^* \frac{S^*e^*}{S_n{}^*e_n{}^*} \tag{6}$$

This equation indicates that the distribution of the product of stress and strain in the vicinity of notch root is approximately prescribed by that in the elastic stress field.

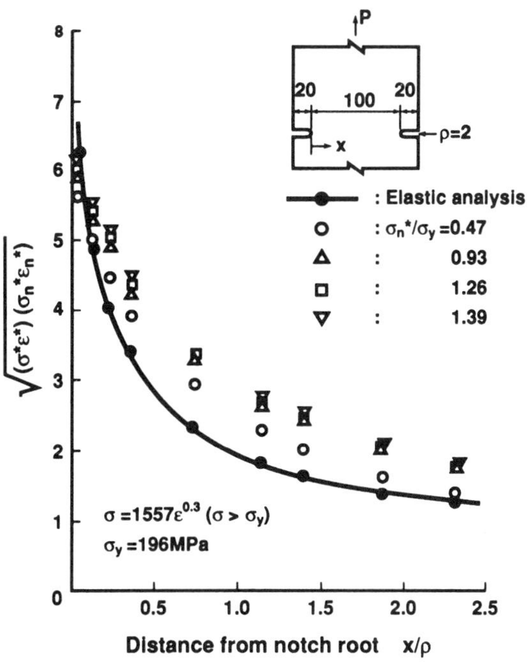


Fig.4 Distribution of the product of stress and strain in the vicinity of notch.

## Extension of Neuber's Rule to Crack Problem

The elastic J-integral $Je$ is connected with the stress intensity factor $K_I$ by the following equation. (Here, the plane stress condition is assumed.)

$$Je = K_I^2/E \tag{7}$$

Furthermore, $K_I$ is connected with the elastic stress concentration factor by

$$K_I = Sg \lim_{\rho \to 0}\left(K_t' \frac{\sqrt{\pi\rho}}{2}\right) \tag{8}$$

where, $Sg$ is the gross section stress and $\rho$ is the notch radius, and $K_t'$ is defined by

$$K_t' = S_{max}/Sg\,. \tag{9}$$

Assuming the plane stress condition, $S_{max}$ equals to $S^*_{max}$ at notch root. Then,

$$K_t' Sg = K_t^* S_n^*\,. \tag{10}$$

From Eqs.7, 8 and 10, $Je$ is given by

$$Je = S_n^* e_n^* \lim_{\rho \to 0}\left[(K_t^*)^2 \frac{\pi\rho}{4}\right]. \tag{11}$$

As mentioned in the previous paragraph, the inelastic stress and strain at notch root are approximately represented by the following formula.

$$\sigma^*_{max}\,\varepsilon^*_{max} = \sigma_n^*\,\varepsilon_n^*\,(K_t^*)^2 \tag{12}$$

This formula is effective not only for the notch root stress and strain but also for the distribution of stress and strain in the vicinity of notch as shown by Eq.6. By extending this behavior to the crack problem, it will be assumed that the inelastic distribution of the product of stress and strain at crack tip field can be specified by the elastic stress distribution. In such this case, the inelastic J-integral is given by the following formula as the first order approximation.

$$J = \sigma_n^*\,\varepsilon_n^* \lim_{\rho \to 0}\left[(K_t^*)^2 \frac{\pi\rho}{4}\right] \tag{13}$$

From Eqs.11 and 13, the following equation can be obtained.

$$J = Je \frac{\sigma_n^*\,\varepsilon_n^*}{S_n^*\,e_n^*} \tag{14}$$

Moreover, from Eqs.14 and 7,

$$J = \sigma_n^*\,\varepsilon_n^* \left(\frac{K_I}{S_n^*}\right)^2 \tag{15}$$

is obtained.

Using Eq.15 or Eq.14, the inelastic J-integral can be calculated based on the result of elastic analysis.

## Confirmatory Analyses for Load-Controlled Problem

In order to confirm the effectiveness of Eq.15, the inelastic FE analyses of the three kinds of cracked panels shown in Fig.5 were conducted. The general code MARC was

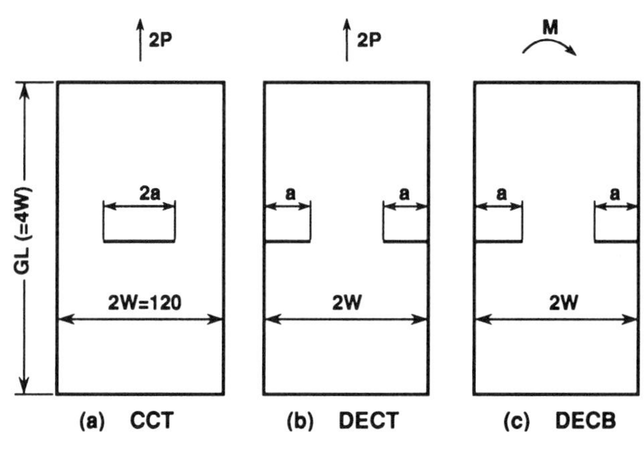


Fig.5 Cracked panels analyzed: (a) Center cracked panel under tension, (b) Double edge cracked panel under tension, (c) the same under bending load.

used for the analyses, and the plane stress condition was assumed. The stress-strain relation shown in Fig.6 was used, in the analyses, and it is given by

$$\varepsilon = \frac{\sigma}{196000} + \left(\frac{\sigma - 98.0}{294.0}\right)^{5.0}. \qquad (16)$$

The relationships between $J$ and the gross section stress $Sg$ obtained were shown in Fig.7, where $Sg$ was defined by

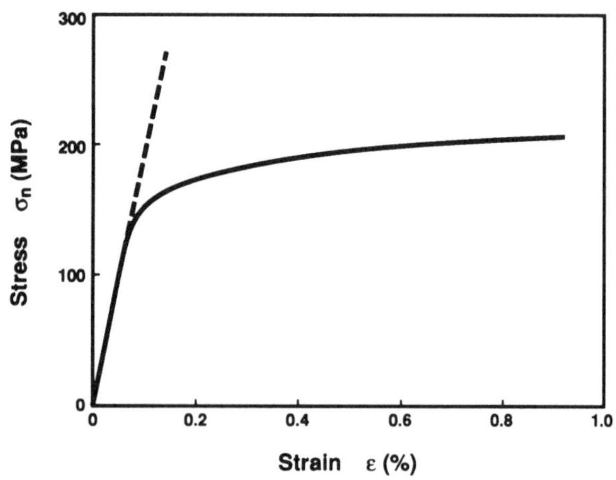


Fig.6 Stress-strain curve used for the bench mark analyses.

$$Sg = \begin{cases} \dfrac{P}{2W} & \text{(Tensile Load)} \\ \dfrac{3M}{2W^2} & \text{(Bending Load)} \end{cases} \qquad (17)$$

The solid lines in Fig.7 indicate the J-integral calculated by Eq.15, where $K_I$ and $S_n$* were determined from the first elastic increment of each FE analysis and $\sigma_n$* was calculated by Eq.4 for the tensile loading problems, finally $\varepsilon_n$* was obtained by substituting $\sigma_n$* to Eq.16.

For the bending load problem, on the other hand, the net section stresses were calculated by the following equations.

$$S_n{}^* = \frac{3}{2(W-a)^2}\int_{a-W}^{W-a} sgn(Sy)S^*\,x\,dx \qquad (18)$$

$$S_n = \frac{3M}{2(W-a)^2} \qquad (19)$$

$$\sigma_n = \frac{1}{K_n} S_n \qquad (20)$$

$$\sigma_n{}^* = \sigma_n \frac{S_n{}^*}{S_n} \qquad (21)$$

Here, $K_n$ in Eq.20 is the elasto-plastic section factor under bending load which can be obtained based on the elasto-plastic stress distribution in simple beam. The $K_n$ value calculated for the stress-strain relation of Eq.16 was shown

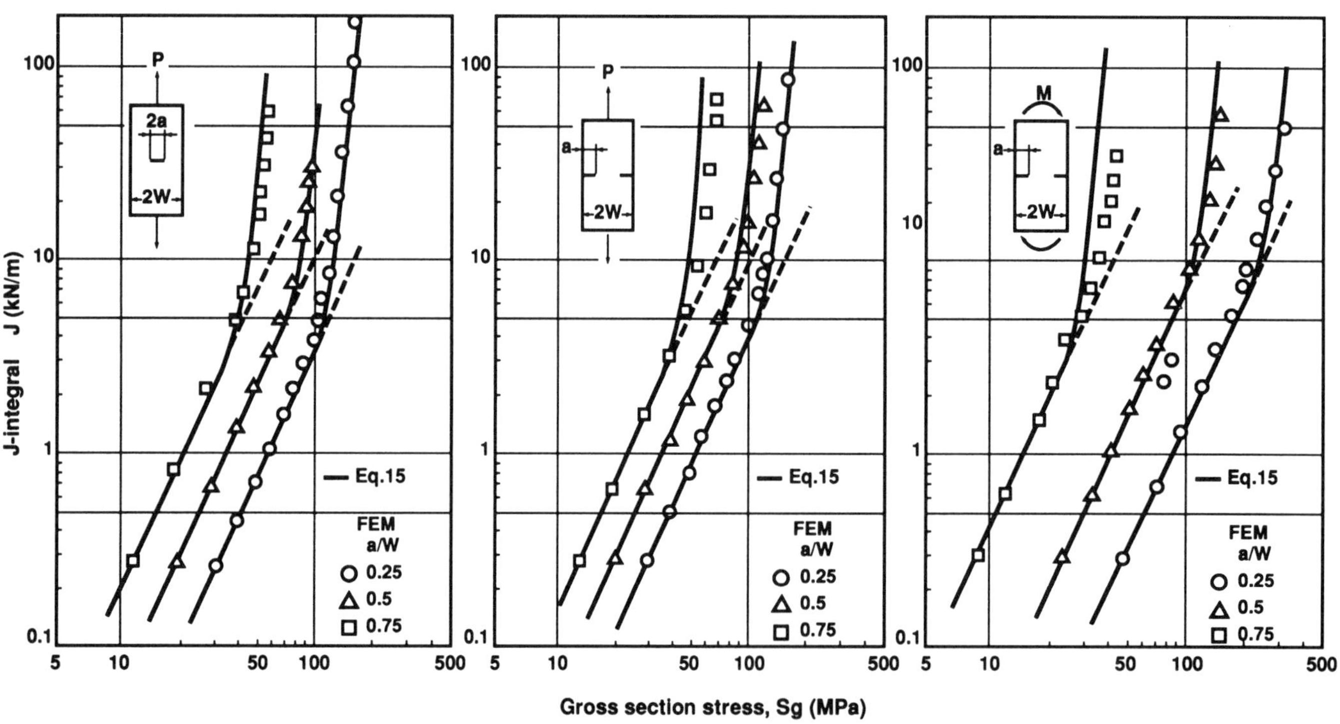


Fig.7 Relationships between J-integral and gross section stress obtained by inelastic FEM and Eq.15 for load-controlled condition.

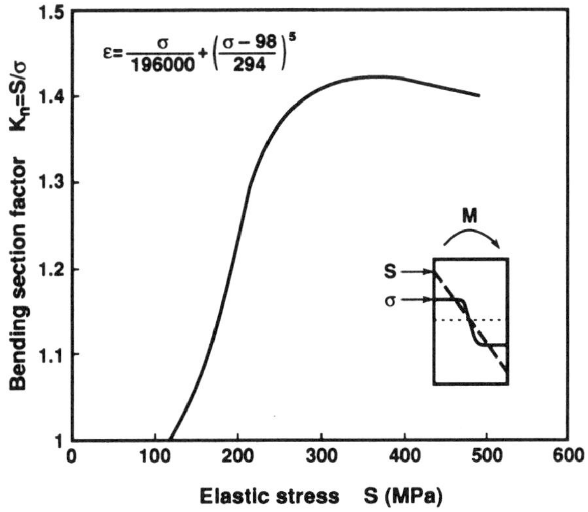


Fig.8 Bending section factor calculated for the stress-strain relation given by Eq.16.

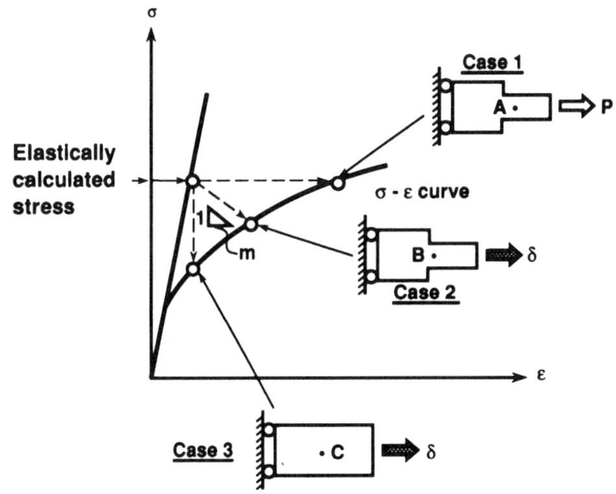


Fig.9 Schematic representation of the correlation between elastic stress and elasto-plastic stress and strain.

in Fig.8. If the stress-strain relation is given by power law ($\varepsilon = A\sigma^n$), $K_n$ of a rectangular sectional beam can be simply represented by

$$K_n = \frac{3n}{2n+1}.$$

As shown in Fig.7, the solid lines agreed nearly well with the results of the inelastic analyses, and it showed that Eq.15 proposed in this paper could be used as a practical J-integral estimation method. Although this investigation was conducted on the elasto-plastic problem, it is assumed through the analogy between plasticity and creep that this method is also applicable to the estimation of $C^*$ in general creep field by converting $\varepsilon_n^*$ to $\dot{\varepsilon}_n^*$.

## J-Integral Estimation for Deformation-Controlled Problem

The nominal stress in the inelastic condition should be known for estimating J-integral based on Eq.15. For the deformation-controlled loading condition such as thermal stress field, however, it is not necessarily easy to know the actual applied stress. In this paragraph, the $J$ estimation method for such cases will be discussed.

Fig.9 is the schematic figure of the relation between the elastically calculated stress and the actual inelastic stress and strain for three kinds of loading conditions. Case 1 is the beam subjected to an external tensile load, where the stress at point A can be exactly obtained by an elastic analysis. Case 3 is the beam uniform in section subjected to a forced displacement at its end. In this case, the true strain can be obtained by an elastic analysis. On the other hand, in Case 2 which is the non-uniform section beam subjected to a forced displacement, neither the true stress nor the true strain can be known by an elastic analysis. This kind of behavior is called "the elastic follow-up," and the cracked structures under the deformation-controlled loading generally correspond to this case.

If such elastic follow-up behavior, namely the slope $m$ in Fig.9 can be known for a given problem, the actual nominal stress $\sigma_n$ can be estimated. However, it is generally difficult to know the value of $m$ because it is changeable depending on structural configurations, loading conditions and material properties. In order to avoid such complexity the authors hypothesize that the elastic follow-up behavior can be estimated by Neuber's rule, and examine the effectiveness on J-integral estimation by bench mark analyses.

Provided that the elastic follow-up of the nominal stress and strain is specified by Neuber's rule, the relation between the elastically calculated stress and the inelastic stress and strain can be represented by

$$\sigma_n^* \varepsilon_n^* = \frac{(S_n^*)}{E}. \qquad (22)$$

Eq.14, therefore, results in the following simple form.

$$J = J_e. \qquad (23)$$

This equation means that the J-integral under the deformation-controlled loading condition can be estimated by the elastic $J$ analysis for a given displacement condition.

Fig.10 shows the cracked models for the bench mark analyses. The stress-strain relation was given by Eq.16.

Fig.11 shows the comparison between the results of the inelastic analyses and the estimates based on Eq.23. The horizontal axis shows the nominal strain in gage section obtained by dividing the displacement by the gage length of the models. The solid lines are the estimates based on Eq.23 and these are proportional to the square of $\varepsilon_{GL}$, namely the displacement applied. Although the J-integrals obtained

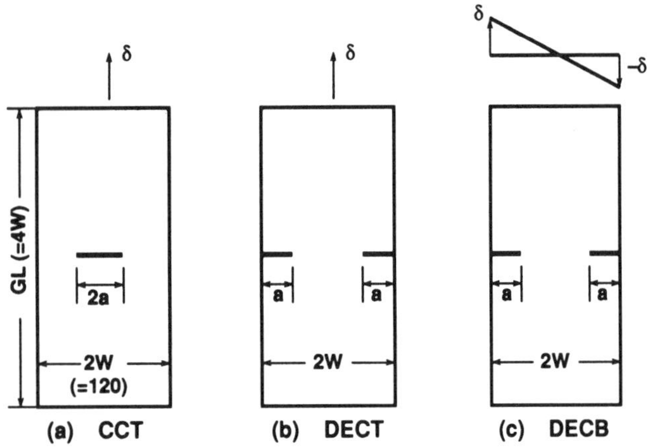


Fig.10 Cracked panels under deformation controlled condition for bench mark analyses.

by the inelastic analyses tended to be less than the lines with the increase of the nominal strain, the errors between them were not significant except the condition of $a/W$=0.75 subjected to the bending deflection.

From these analytical study, it is expected that the method proposed above is useful for the practical approximation of the J-integral under the deformation-controlled loading condition, though the assumption of Eq.22 for the elastic follow-up behavior is not well-founded theoretically.

On the other hand, the amount of the elastic follow-up ought to be affected with the ratio of $GL/2W$ in Fig.10 or Fig.11. The authors are studying on this point, and obtaining the result which shows the effect does not seem significant within the range of $a/W$=2~10.

## Discussion

The estimation errors of J-integral in Figs.7 and 11 were within a factor of about 1.5 on $J$, which was extremely rough comparing with the accuracy of the stress intensity factor formulas in the linear fracture mechanics. It is, however, considered that even such error is not intolerable for the application to the creep or low-cycle fatigue crack growth rate evaluation because of the following reasons.

The low-cycle fatigue and creep crack growth rates are expressed by the following equations.

$$\left.\begin{aligned}\frac{da}{dN} &= A_f\left(\Delta J_f\right)^{m_f}\\ \frac{da}{dt} &= A_c(C^*)^{m_c}\end{aligned}\right\}\qquad(24)$$

Where, it has been known from experimental data that the values of $m_f$ and $m_c$ are nearly equal to 1.5 and 1.0, respectively (5). Accordingly, it is expected that the J-integral estimation error of a factor of 1.5 corresponds to the estimation error within a factor of 1.5 to 2 on the crack growth rate or the crack propagation life, and it is compared to the predicability by which the life prediction method for creep, low-cycle fatigue or the interaction of them is generally judged reasonable.

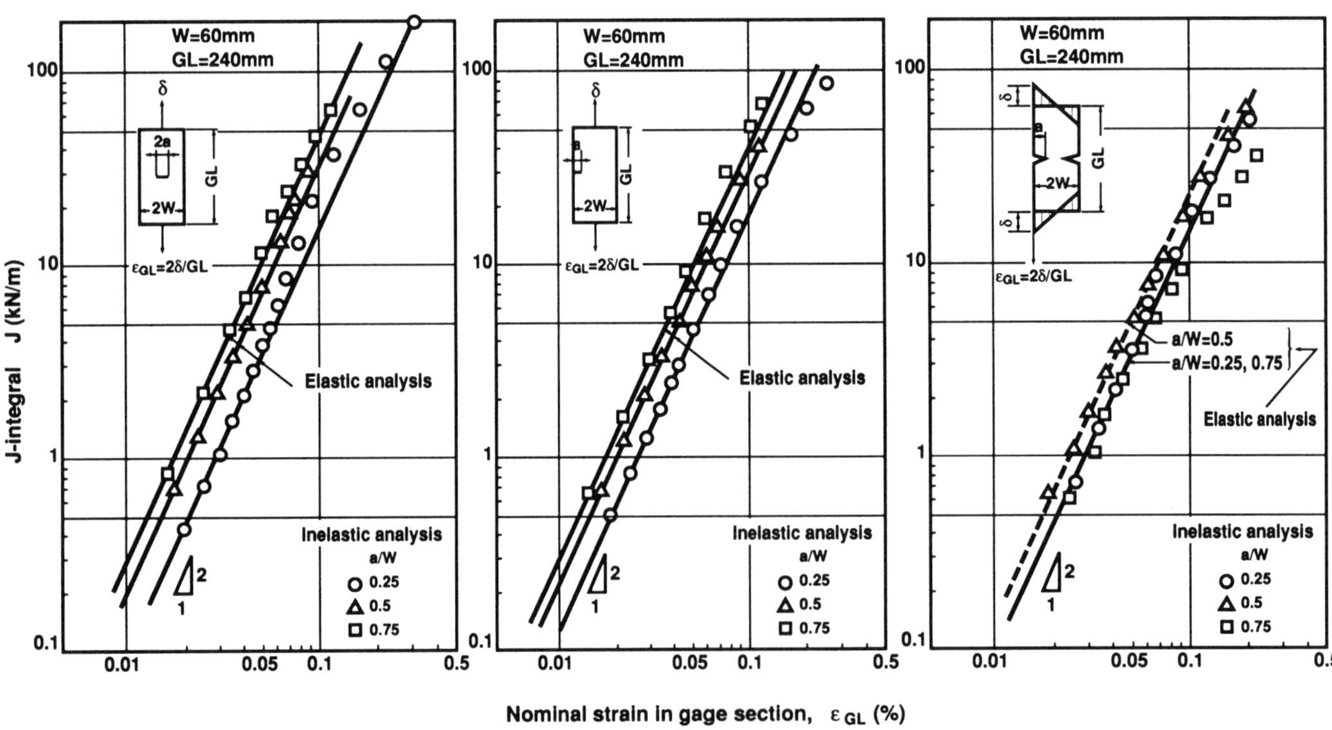


Fig.11 Relationships between J-integral and nominal strain in gage section obtained by inelastic FEM and Eq.23 for deformation-controlled condition.

Besides, it is the other merit of the proposed method that it can be used independently of the style of the stress-strain relation. It is quite possible that the approximation of the stress-strain relation by a specified equation such as Ramberg-Osgood type allows of the error about a factor of 2 on strain. In such case, the predicability of J-integral will be inevitably inferior.

From the consideration above the authors think the method proposed in this paper can be useful for the practical application.

## Conclusion

The practical J-integral estimation method was investigated focusing on the application to the creep and low-cycle fatigue crack growth evaluation in high temperature components. The results obtained are summarized as follows;

(1) The following equations were proposed based on the analogy with Neuber's rule for the inelastic stress/strain concentration at notch.

$$J = \sigma_n * \varepsilon_n * \left( \frac{Je}{S_n * e_n *} \right) = \sigma_n * \varepsilon_n * \left( \frac{K_I}{S_n *} \right)^2$$

— Load-controlled type loading.

$$J = J_e$$

— Deformation-controlled type loading.

(2) The equations above can be used independently of the style of the stress-strain relationship.

(3) In order to confirm the effectiveness of the above equations the inelastic analyses were conducted for several kinds of two-dimensional bench mark problems. It was shown in result that the predicability of the equations was about a factor of 1.5 on J-integral, which seemed to be equal to the practical application.

## References


1. Kumar, V., German, M. D. and C. F. Shih, Topical Report No. EPRI NP-1931, General Electric Company (1981).
2. Kumar, V., German, M. D., Wilkening, W. W., Andrews, W. R., deLorenzi, H. G. and D. F. Mowbray, Final Report No. EPRI NP-3607, General Electric Company (1984).
3. Ogura, K., Miki, N. and K. Ohji, Trans. JSME 47A (413), 55-62 (1981).
4. Ohnami, M. and M. Sakane, J. Soc. Mater. Sci., Jpn. 30 (336), 915-21 (1981).
5. Ohtani, R. and T. Kitamura, "High Temperature Creep-Fatigue – Current Japanese Materials Research; Vol.3," p.65, Elsevier Applied Science, London and New York (1988).

*Proceedings of the Fifth International Conference on Creep of Materials, Lake Buena Vista, Florida, USA, 18-21 May 1992*

# Damage Assessment and Integrity in High Temperature Plant

**T.L. da Silveira**
Tito Silveira Engenharia e Consultoria Limited
Rio de Janeiro, Brazil

**I. Le May**
Metallurgical Consulting Services Ltd.
Saskatoon, Saskatchewan, Canada

## Abstract

The process of damage accumulation in high temperature plant is discussed, and it is emphasized that creep is one of a number of processes that may be encountered. Cases from the steam generating, petrochemical and gas turbine sectors are described, and the operative damage mechanisms that led to failure are identified. It is emphasized that damage assessment and evaluation of integrity of high temperature plant and equipment depend on field inspections made during their operating life. These must involve a broad range of inspection techniques including field metallography and be carefully planned. It is pointed out that accumulated damage associated with failure is usually found in localized regions rather than being general, making its detection more difficult. Any cases of damage accumulation or of failure detected need to be analyzed fully so that the mechanisms responsible can be identified for further improvement of the damage detection procedures.

IN THE ASSESSMENT of the integrity of high temperature plant and equipment, including steam generators and petrochemical facilities, much emphasis is normally placed on the evaluation of creep damage. This rises as a natural consequence of the design procedure and codes followed, as these are based largely on estimations of creep strain and rupture life. Because creep damage develops over extended time at temperature, several procedures have been developed to characterize it on a quantitative or semi-quantitative basis (1-6). Depending on the classification of the severity of the damage observed, necessary repairs or replacements may be required and estimation made of the remaining safe life.

In a complex system operating at high temperature, accumulated damage is not confined to creep, although this is one of the processes leading to damage and must be considered. During operation at high temperature, diffusion processes involving interaction of the metallic components with their environment give rise to oxidation, hydrogen damage and microstructural changes in addition to creep. Also, thermal fatigue and thermal shock may occur, as a result of transient conditions.

As well as mechanisms of high temperature damage, other damage mechanisms may be operative during plant or equipment shutdown, and during the transient conditions when the system or equipment is cooling down or being brought back to service temperature. Such mechanisms include aqueous corrosion, stress corrosion cracking (SCC), hydrogen induced cracking, local cracking in brittle material owing to thermal or mechanical stress, and low cycle fatigue. Subsequent high temperature damage may be initiated from that produced at low temperature, after the system has been returned to its normal operating conditions.

Whether produced by a creep mechanism or by other means, damage is seldom uniform but is normally concentrated in localized areas, for example at local stress raisers, where changes in section occur, at nozzle attachments where there is local restraint, at bends in steam pipes, or at welds. There may be little relation between the nature and intensity of severe damage observed locally and that present in adjacent material, which is only mildly damaged and is typical of the overall condition of the material in the system. This emphasizes the need for careful consideration of points to be inspected and of the inspection techniques to be applied.

The assessment of damage in several high temperature steam generators and their component parts is discussed in this paper, as well as examples from high temperature equipment in the petrochemical and power generating sectors.

## Steam Generator Damage

Five specific examples of damage evaluation in steam generators are described. The first emphasizes the localized nature of creep damage; the following two cases describe damage in a single steam generator at a superheater header and in the steam drum, both resulting from mechanisms other than creep; the fourth example describes failure at the superheated steam outlet in a different steam generator, the mechanism being SCC; while the fifth relates to a failure in a superheater outlet, involving creep as one of the mechanisms occurring during progression of a crack to failure.

**Creep Rupture in a Superheated Steam Transfer Line.** Rupture occurred at a bend in a superheated steam transfer line between a header and a desuperheater of a boiler producing 230 t/h of steam at 540°C and 118 kPa. The boiler had operated for 77,000 h. The geometry of the failed region is shown in Figure 1. Rupture occurred along the outer bend radius of the 168 mm diameter tube, this being of 1 Cr, 0.5 Mo steel with a wall thickness of 14 mm. The design temperature of this tube was 490°C, but there is evidence that it was operating at a temperature much above 500°C.

Metallographic analysis of the material disclosed an advanced stage of creep damage accumulation in the form of local cracks, microcracks and aligned damage centers which show up as voids following upon repeated polish-etch cycles (7). The severity of the damage changed from level D at the surface cracks to level A within a very short distance, using the Neubauer criterion for classification of creep damage (2). Figure 2 shows an area where the damage is at level C. The overall distribution of damage around the crack centerline is shown in a cross-section through the bend in Figure 3, in which the extremely localized nature of the creep damage is apparent for this case.

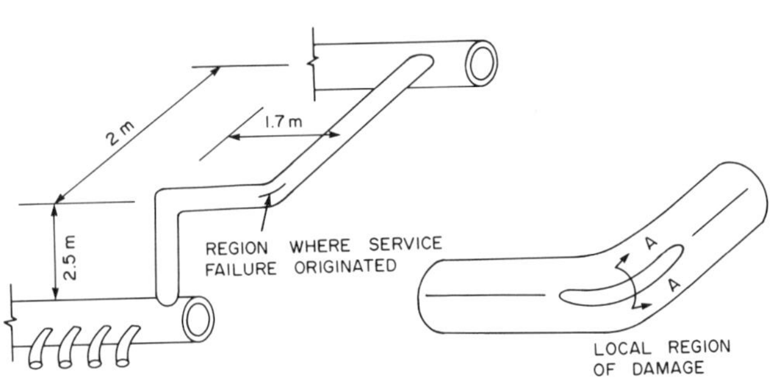


Fig. 1 - Schematic showing where failure occurred in superheated steam transfer line.

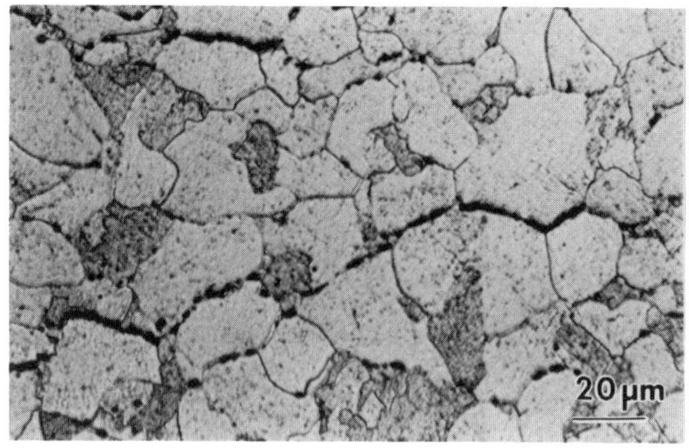


Fig. 2 - Specimen cut from the bend very close to the main crack, illustrating creep damage in the form of aligned damage center, shown as voids, with linking microcracks. Repeated polishing procedures. Nital etch.

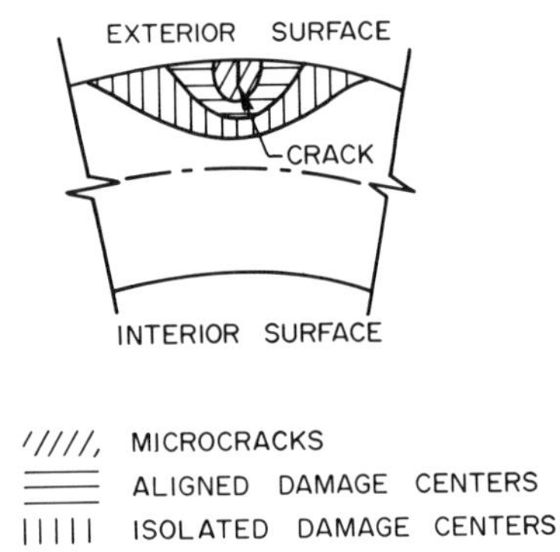


Fig. 3 - Schematic distribution of creep damage around main crack.

**Cracking at a Superheater Header Connection.** Examination of the header of the third superheater of a boiler producing 150 t/h of steam at 525°C and 118 kPa, disclosed extensive internal cracking at the connection to the tube joining this to a safety valve. Cracking was observed within the tube and in the thickness of the shell wall itself, the geometry being shown in Figure 4. The boiler had been in operation for approximately 160,000 h, and was shut down for inspection when the cracking was detected. The material involved was 2.25 Cr, 1 Mo steel, and the unit had been subjected to 115 shutdowns, although the number of load cycles was much greater because of hydrostatic tests and tests of the safety valves. The header was outside the radiation zone of the boiler.

Figure 5 shows the typical appearance of the surface cracks in the shell wall and entry to the safety valve connecting line: this shows a section parallel and adjacent to the surface in the polished and etched condition. Sections through the cracks were examined, and it was seen that the surface cracks penetrated to a depth of 10 mm in some cases (Figure 6). In Figure 6(a) the crack appears to be blunt and nonpropagating, while Figure 6(b) shows an adjacent sharp crack tip that is considered still to be growing. As may be observed in Figure 6, the cracks were filled with corrosion product at their tips.

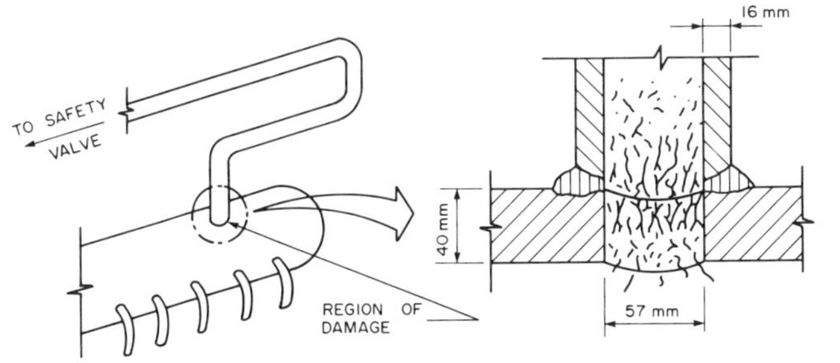


Fig. 4 - Arrangement of nozzle connecting safety valve line to superheater header, showing cracking.

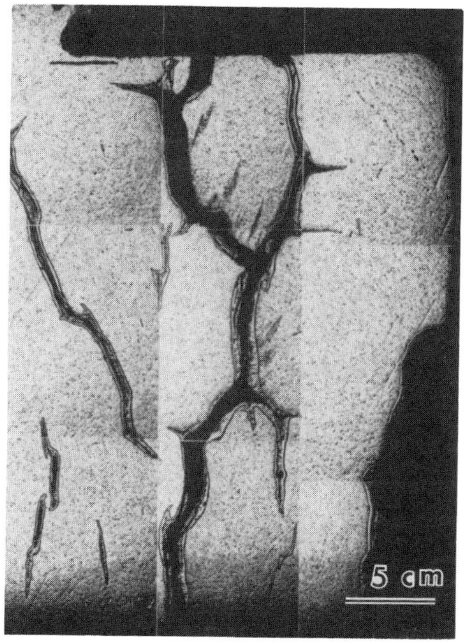


Fig. 5 - Section parallel and close to surface at nozzle connection seen in Fig. 4, showing cracking. Nital etch.

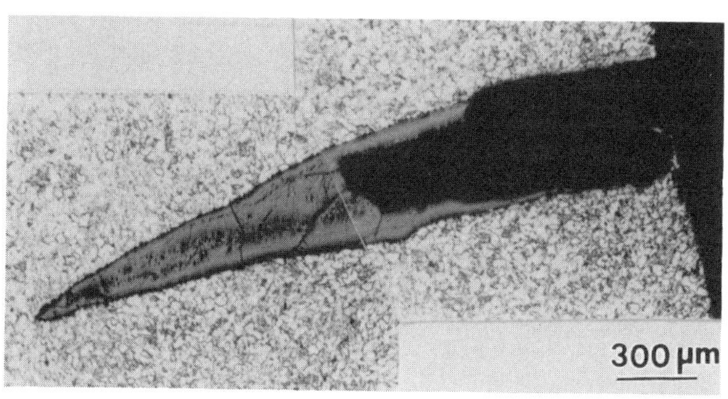


(a)

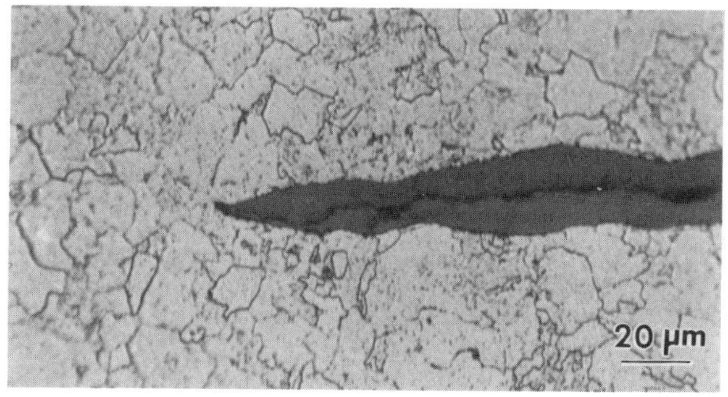


(b)

Fig. 6 - Sections through cracks at the nozzle: (a) blunted crack; (b) sharp tip of adjacent crack. Nital etch.

The initiation of the cracks was attributed to thermal shock, caused by the periodic return of condensate along the long connecting line (some 9 m long). This would "flash off" as steam when reaching the hot region at the entrance to the header, producing rapid surface cooling.

Propagation of the cracks is considered to be due to thermal cycling, together with periodic pressure cycles, producing growth by low cycle fatigue. This is likely to have been aided by corrosion within the cracks and by the wedging action caused by corrosion deposits at their tips.

**Cracking in a Boiler Steam Drum.** During the inspection of the boiler containing cracks at the superheater header connection (the previous case), cracking was also detected within the main steam drum. This was fabricated from a Mn-Mo-V low alloy steel. It operated with water and saturated steam at approximately 335°C.

Figure 7 shows the various areas in which cracking was detected. These were: at the nozzles connecting the tubes for the entry of steam and hot water to the drum; at the downcomers; and at the connection to the safety valve. All cracks had a similar morphology, running in a longitudinal direction along the drum from the cutouts in the shell.

From their orientation it was clear that all the cracks had developed under the influence of the hoop stress and were associated with the locally increased stress levels relating to the cutouts at nozzle and pipe connections. It was not possible to prepare metallographic sections through the cracks as the cracked areas were to be ground away and built up with weld metal if considered necessary. Thus, examination was made by replication at different stages during the grinding process conducted on the 68 mm thick shell. The maximum crack depth was determined to be approximately 2 mm.

At their ends the cracks were filled with corrosion products as shown in Figure 8, and their surfaces were seen to be very irregular. Following further grinding and polishing steps, the replicas showed the bottom of the cracks to be filled with corrosion products and to have rough edges (Figure 9).

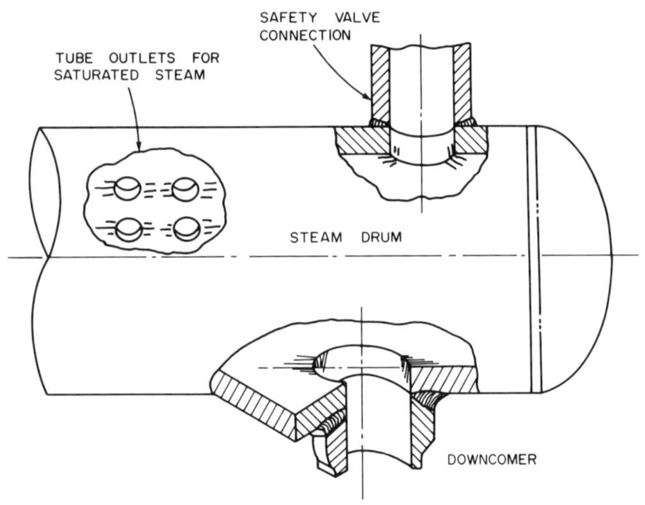


Fig. 7 - Showing cracking in steam drum.

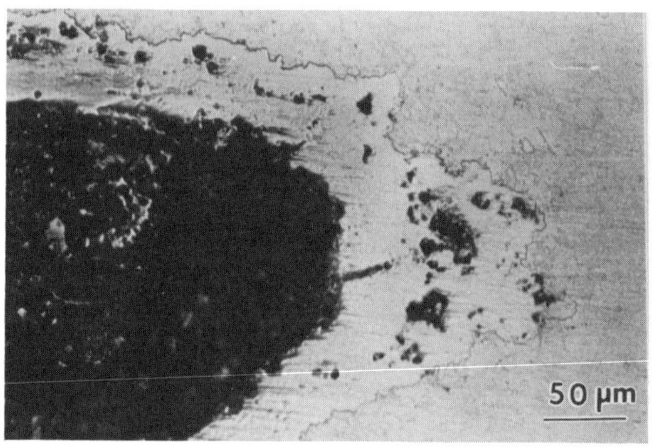


Fig. 8 - Replica of crack edge in steam drum after four grinding steps. Unetched.

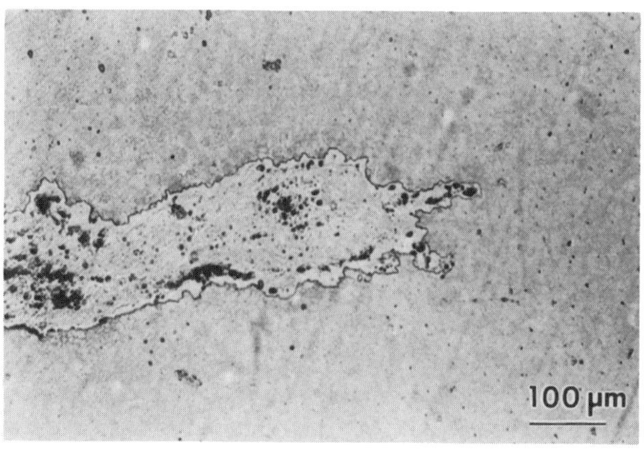


Fig. 9 - Replica of crack edge after eight grinding steps. Unetched.

The mechanisms of crack initiation and propagation cannot be fully defined, but low cycle fatigue is certainly involved together with corrosion. Crack initiation may be by pitting corrosion in association with high local stress, but this is speculative.

On the basis of the observed geometry of the crack tips, the process of crack growth is not a simple one due to fatigue only, but can most probably be attributed to corrosion fatigue.

**Failure at a Steam Outlet Collector.** Figure 10 shows the arrangement of the outlet from a superheater in a generator producing 80 t/h of steam at 400°C and 41 kPa. On two previous occasions there have been leakages, one at the connection to the soot blower pipe, the other at the connection to the vent. In the present case failure took place once more at the connection from the collector to the vent line used during start up. The material of construction was carbon steel, and the unit had 240,000 h of operation at the time of failure, with 99 shutdowns.

The two previous leakages had not been investigated in any depth, were attributed to defective welding, and repaired. Upon the later occurrence, a more complete failure investigation was undertaken. The interior view of the shaded area shown in Figure 10 is illustrated in Figure 11 after inspection with liquid penetrant. Widespread cracking on the inside is apparent, the most severe cracking being some distance from the nozzle connection in a downstream direction. Widespread cracking and pitting were also observed at the connections to the safety valve and soot blower. Pitting was most apparent on the downstream sides of the openings in the shell.

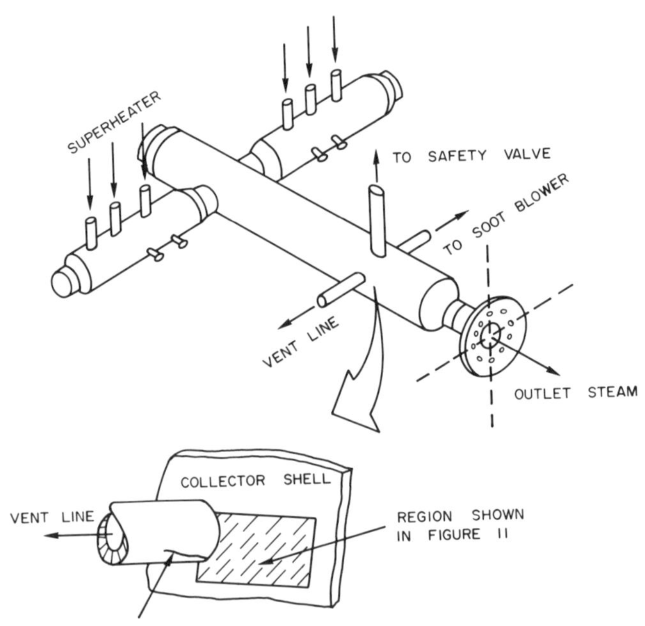


Fig. 10 - Arrangement of outlet steam collector showing failure region.

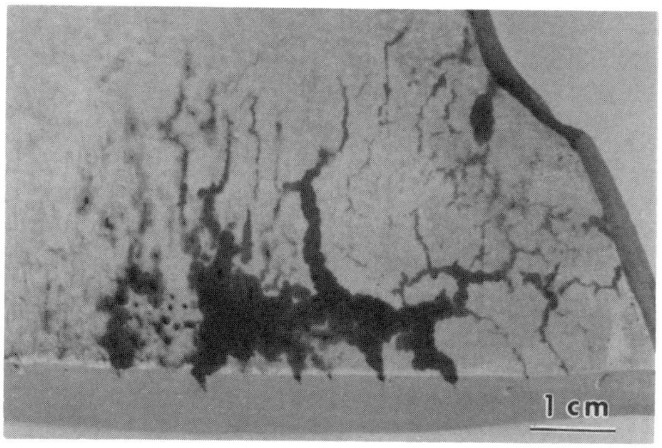


Fig. 11 - Inside of shaded area of Fig. 10 after removal and liquid penetrant inspection to show cracks.

Metallographic examination disclosed that the cracks shown in Figure 11 propagated deeply into the plate, were branching in nature, and contained products of corrosion. Figure 12 shows a view of the tip of one crack in a section through the plate shown in Figure 11. A section was cut from the top of the collector at the nozzle cutout for the safety valve connection. This showed cracking running into the plate from the cutout. Figure 13 shows this section viewed parallel to the surface of the collector shell: again there is extensive crack branching, typical of SCC.

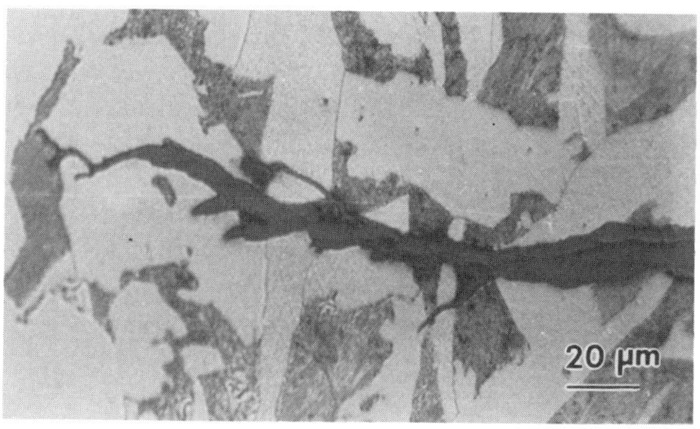


Fig. 12 - Tip of crack in section of plate shown in Fig. 11.

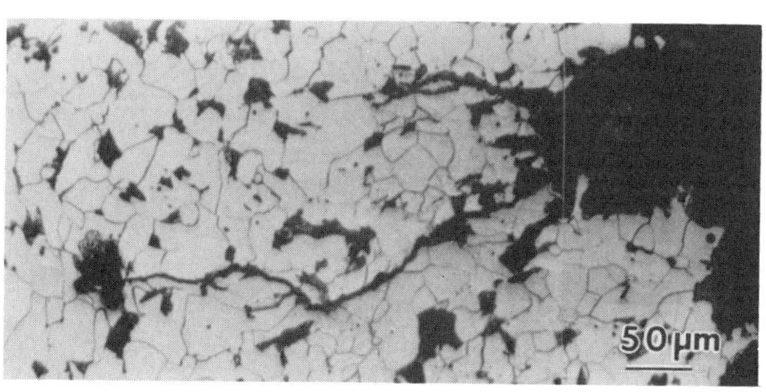


Fig. 13 - Section parallel to shell surface at safety valve nozzle connection. Nital etch.

In all the damaged areas the mechanism of failure involved surface pitting and subsequent SCC. This seems an unlikely situation in a component operating in an atmosphere of superheated steam, but when the long connections to the nozzles are considered, the matter becomes clearer. In all three lines condensation will take place and the condensation will return to the nozzles where it will be evaporated once more. It is probable that carryover in the form of boiler feedwater impurities will build up in these stagnant regions. Water droplets evaporating (particularly on the downstream sides of cutouts and nozzles on which they are likely to impinge) will be concentrated in salts, leading to pitting. SCC may start from these pits, progressing as seen to cause failure.

**Failure of a Superheater Outlet Tube.** After some 87,000 h of operation, failure took place in the bend of a steam pipe connecting a coil of the third superheater of a steam generator to the outlet steam collector. The unit operated at 538°C and 135 kPa, producing 400 t/h of steam. The 2.25 Cr, 1 Mo steel pipe in which failure took place was 50.8 mm in diameter with a nominal wall thickness of 8 mm. It connected to the AISI 321 superheater tube by means of a butt weld, and was one of 46 such parallel connecting tubes. The Cr-Mo tubing was situated outside the heat transfer zone of the superheater.

Figure 14 shows the outside of the bend, and the longitudinal split is clearly seen. Many parallel grooves are apparent, together with a heavy oxide deposit on the part of the tube adjacent to the failure. Approximately 10 of the outlet tubes were grooved in this way, being in the same region of superheater coils as the failed tube.

Fig. 14 - Outside of bend in superheater outlet tube, showing heavy oxide, parallel grooves and rupture.

The heavy layer of high temperature oxide suggests that the design metal temperature (543°C) was exceeded in a section of the superheater outlet tubes. Whether this was due to steam flow restriction or to differential heating of the superheater due to preferential hot gas flow is not known. Because steam temperature is recorded at the outlet steam collector, being an average from the mixing of steam from all superheater coils, no records of this local heating existed.

Metallographic sections showed the grooves to consist of cracks containing oxide, one having progressed through the wall to cause rupture. The initial mechanism appears to have been cracking of the thick oxide (probably during cooling-heating cycles), followed by thermal fatigue combined with oxidation during subsequent cycles. Approximately halfway through the wall thickness it was seen that the cracks had grown by creep, there being many wedge-type voids caused by grain boundary sliding: these were adjacent to the main crack leading to fracture and ahead of branch cracks. Figure 15 shows a branch crack and the wedge-type grain boundary voids: Figure 16 shows this latter feature in more detail.

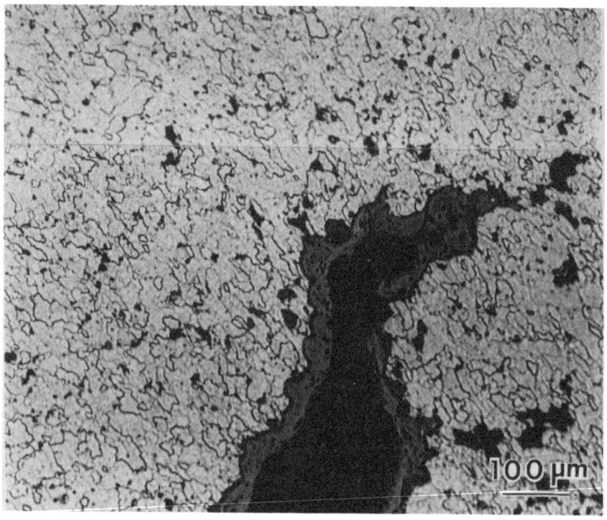


Fig. 15 - Branch crack from main fracture, approximately halfway through tube wall. Note grain boundary voids. Nital etch.

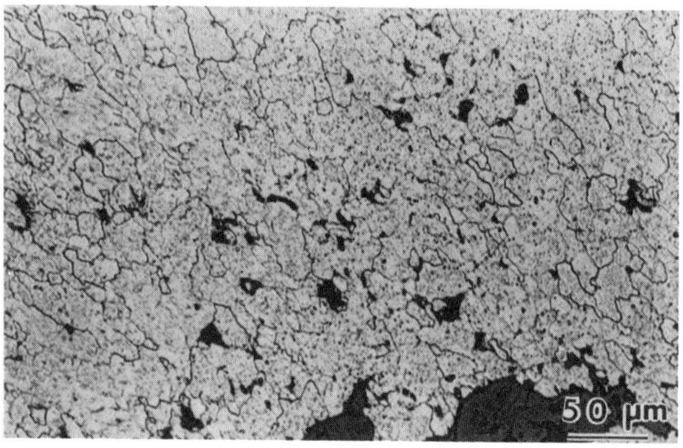


Fig. 16 - Wedge-type cracks seen in Fig. 15, shown in more detail. Nital etch.

At the inside surface of the tube the fracture showed extensive plastic deformation, and the overall sequence of failure may be given as follows:

> Overheating of the Cr-Mo outlet tubes → heavy oxidation → oxide cracking on thermal cycling → thermal fatigue cracking plus oxidation → creep controlled crack growth → rapid plastic deformation and rupture.

## Petrochemical Industry Examples

Two cases will be described. The first relates to the assessment of damage and remaining life in reformer furnace tubes in which creep is the major limitation on useful life; the second deals with local failure in a product transfer line, the mechanism being creep damage occurring on a local scale.

**Creep Damage in Welds of Reformer Furnace Tubing.** A sample tube was removed from a reformer furnace for life assessment after 69,000 h of service. The outlet hydrogen-rich gas resulting from the cracking of naphtha in the presence of steam had a temperature of approximately 850°C, the inlet temperature at the top of the tubes being 525°C.

Sections were cut from the tube, which was a spindle cast A297 Grade HK 40 (25 Cr, 20 Ni, 0.4 C) austenitic steel of 122.5 mm outside diameter and 10.5 mm nominal wall thickness. They were examined metallographically on transverse sections and on longitudinal sections through the butt welds joining the separate cast segments of the tube: examination was made to determine the extent of creep damage. This is known to be the main cause for life extinction in this class of equipment (8), whose useful lives may vary from 30,000 to 150,000 h.

Creep damage assessment is based on the detection of damage centers in the form of apparent voids distributed along grain boundaries perpendicular to the principal tensile stress and of microcracks resulting from their linkage. These features are revealed by repeated polishing and etching techniques, described elsewhere (7). The damage is mainly concentrated within the inner one third of the wall thickness.

Damage may be classified as of levels A to D, based on the distribution and quantity of damage centers and microcracks. Level A corresponds to the observation of some isolated damage centers, and is indicative of approximately one third of the life having been used up. Clearly aligned damage centers, classified as level B, correspond to one half of the life having been exhausted. Level C, representing two third of life consumed, is defined by the presence of microcracks linking the aligned damage centers. When the microcracks and aligned damage centers extend across 60 percent of the wall thickness, this is termed level D damage, and the useful tube life can be considered exhausted.

Figure 17 shows a plot of the level of accumulated creep damage along the tube length (9). It may be seen that the life of the tube is exhausted only in the two upper butt welds, whereas the spindle cast material shows a damage level B, corresponding to half the life, only, having been used up.

In Figure 18 a comparison is given between the damage observed in the two regions referred to above.

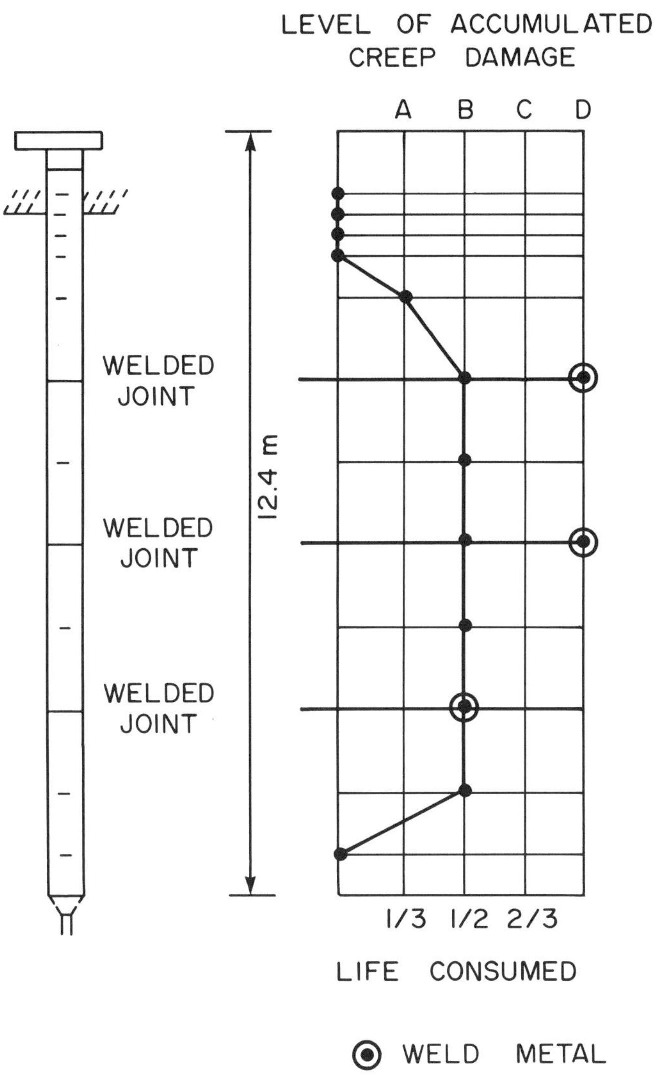


Fig. 17 - Plot of accumulated creep damage along reformer tube.

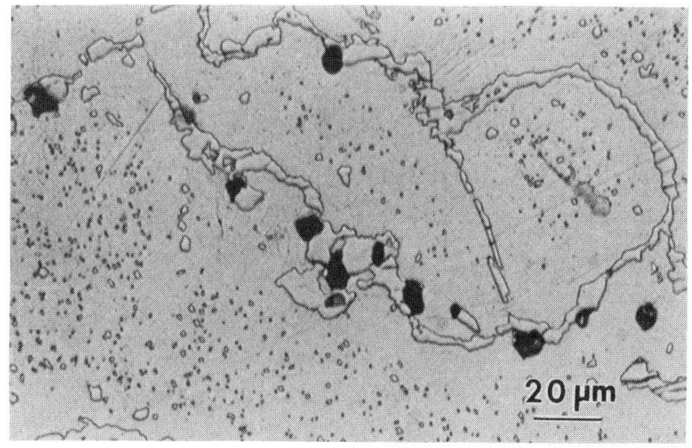


(a)

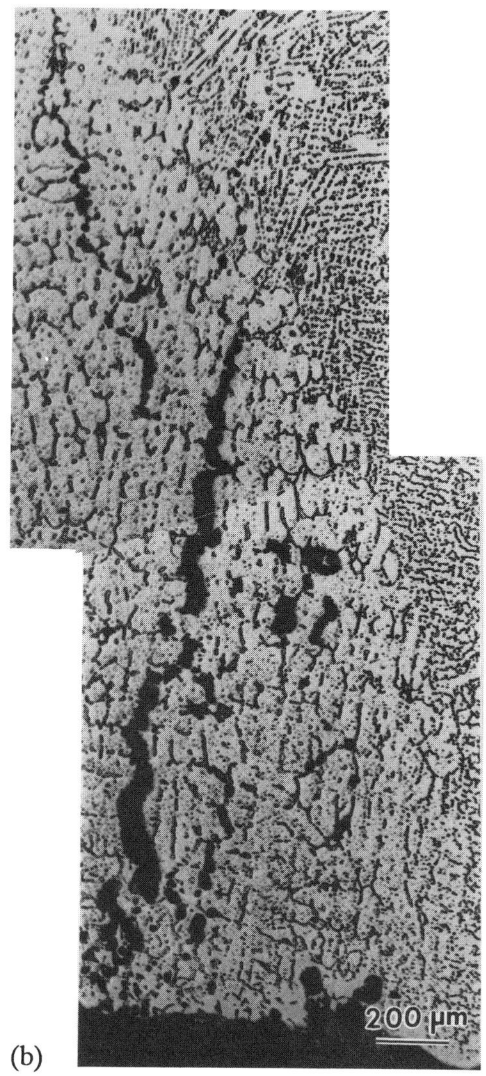


(b)

Fig. 18 - Showing (a) creep damage in base metal adjacent to upper welds and (b) creep damage in weld metal. Etched in Aqua Regia saturated with $CuCl_2$.

**Failure of Supports in a Product Transfer Line.** Leakage was detected at the welds between stiffening plates and the pipe in a transfer line carrying butane and related petrochemical compounds. The line and reinforcing rings were of AISI 316 stainless steel, the pipe being of 508 mm diameter and 6.25 mm wall thickness. The design temperature and pressure were 621°C and 2.75 kPa, respectively, while the operating conditions were 579°C and 1.03 kPa. The line was insulated. Failure occurred after approximately 90,000 h of operation, shutdowns being approximately two per annum. Figure 19 shows the arrangement of the pipe support regions and of the reinforcing plates. The cracking occurred at the toe of welds between these plates and the pipe.

There were many parallel cracks present and they were confined to a narrow area at and adjacent to the welds. There was no evidence of classical low stress creep damage showing up as damage centers or voids, and the microstructural damage observed was in the form of grain boundary sliding and resulting microcracks, ahead of and adjacent to the main cracks. Figure 20 shows a representative view of the pipe material outside of the heat affected zone.

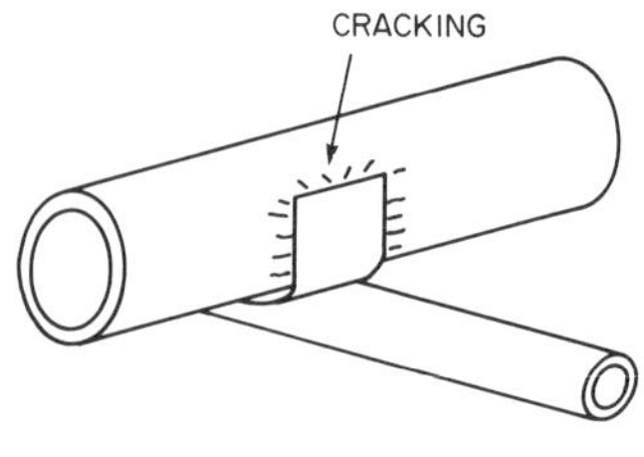


Fig. 19 - Schematic showing cracking at reinforcement at support for transfer pipe.

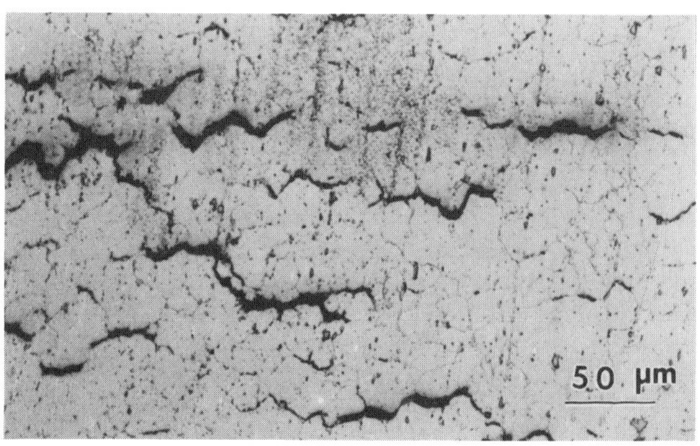


Fig. 20 - Showing creep damage in the pipe in the form of grain boundary sliding and cracking. Unetched.

The failure was attributed to repeated relaxation cycles of very high stresses of thermal origin resulting from the periodic shutdowns, temperature fluctuations during service, or both. These events would produce differential expansion and locally high stresses at the discontinuity in the section.

## Gas Turbine Damage

In the example presented, creep damage assessment had been conducted without due consideration of the likely mechanisms of damage being given.

**Damage in a Hot Gas Casing.** The hot gas casing of a gas turbine used for peak load power production had developed extensive cracking during operation. The operating time was 18,000 h, and it had been subjected to 1,600 operating cycles. The gas temperature on the hot side was 985°C, on the cold side 204°C, the material being AISI 321 stainless steel. Preliminary examination had been made previously to determine if creep voids were present, on the basis that this was one of the most critical questions concerning its possible repair and service. A view of one half of the casing is show in Figure 21. The authors were requested to determine optimum repair welding procedures on the premise that the material was basically sound and undamaged by creep.

Fig. 21 - General view of half of hot gas casing showing cracks indicated by dye penetrant.

First, it is apparent that overall damage by creep would not be expected as the deformation that had taken place was very limited. The cracking was the result of thermal fatigue, and such cracks can propagate at elevated temperature, with damage ahead of the crack tip occurring by means of very local processes of creep (10), as may be inferred from the immediately preceding case. Second, careful metallographic examination disclosed that there were heavy surface layers of carbides, such that the material was extremely brittle when subjected to bending. Figure 22 illustrates this microstructural feature.

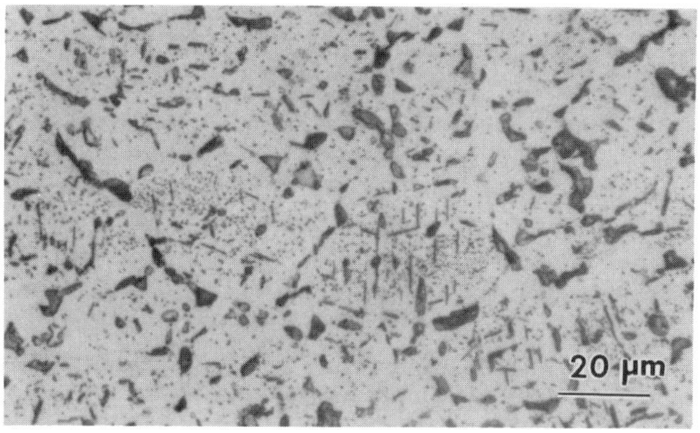


Fig. 22 - Microstructure of hot gas casing showing extensive carbide formation. Etched in Vilella's, methanolic aqua regia and Groesbeck's reagents.

Accordingly, although it was demonstrated that the casing could be welded successfully, it was suggested that the remaining useful life was effectively exhausted and that it should be replaced. Thermal stresses produced during operation would rapidly result in additional cracks.

## Discussion

The case studies provided as examples indicate clearly that many mechanisms of damage can lead to failure and shutdown in high temperature equipment. They include creep, but it is not the sole damage mechanism of practical concern.

Dealing first with steam generators, the case of creep rupture in the superheated steam transfer line emphasizes the local nature of creep damage that can occur. The outside of a bend in a superheater tube is a critical area at which creep damage may accumulate. Any inspection that involves in situ metallography must be conducted at exactly the right or critical position, or the presence of damage may not be detected. Also, if the preparation is not done in a correct manner to emphasize the presence of damage (7), little may be seen even though the material is close to failure. Thus, not only must the positions for in situ metallography be carefully chosen, but a sufficient number of areas must be examined to minimize the risk of missing a critical area.

The cracking at the safety valve connection in the superheater header, as well as the failure at the steam outlet collector, show the problems that can develop where there are long lines in which condensation may occur and return periodically to a superheater or other hot component. A design of this type is a poor one, and inspection of a steam generator having such features should include a careful evaluation of potential cracking in their vicinity. Although at first sight the idea of SCC or corrosion fatigue as damage mechanisms may seen inappropriate for components exposed to superheated steam, the analysis shows how they can come into play. The failures described suggest that, in this particular case, control of dissolved solids in the boiler feedwater may have been inadequate.

The failure in the boiler steam drum emphasizes the need for careful internal inspection of such components. In the event of cracking as described being detected, it may be appropriate to review the design carefully, with the object of reducing the local level of stress at the shell-nozzle connections.

The example of the failure at the superheater outlet tube of Cr-Mo steel is indicative of excess temperature of the steam coming from the heat transfer zone of the coil. This case is interesting in that it shows that many damage mechanisms may combine in the transition from fracture initiation to final failure. The presence of grain boundary sliding as an indication of creep damage is useful in the characterization of the stress level as high, and shows that the process of creep was not operative throughout the life of the equipment.

In the petrochemical industry examples, the use of damage assessment parameters in evaluating the reformer tube remaining life, shows the welds to be inadequate and to have a strength and creep resistance below those of the base metal. The product transfer line failure once again emphasizes the information available from an evaluation of the operative creep mechanism, namely grain boundary sliding, pointing to the periodic nature of the loading, with high residual stresses being present.

The evaluation of the gas turbine casing illustrates the need to review the state of the material of a component carefully, including the conducting of a sufficiently complete metallographic examination. Although thermal fatigue cracks growing at high temperature can progress by creep mechanisms at their tips, it is necessary to decide exactly what are the criteria by means of which evaluation of the material will be made. An examination in search of overall creep damage is likely to be unproductive if limited change of shape or dimensions of the component has occurred. Likewise, identification of local creep damage at the tips of thermal fatigue cracks is irrelevant, if they are, for example, to be ground out and repaired by welding.

In conducting an assessment of a structure operating at high temperature, careful planning at the outset is essential. A full range of inspection techniques must be considered and utilized as appropriate. Metallography is only one part of the process. In general it only provides information concerning external surfaces and at specific points, and the critical areas to be examined may be ones not envisaged as being likely ones at the design stage.

It is relevant to note that accelerated creep tests, or other mechanical tests on material removed from the structure, generally provide information concerning the overall state, but not the local state in areas of damage. Hence the important role of metallography.

Damage in local areas that may be detected by nondestructive inspection techniques is generally of a severe nature involving cracking or other macroscopic defects. The need to conduct such examinations in conjunction with visual and metallographic studies cannot be overemphasized.

In the examples described, design or operating deficiencies have been identified in several cases. These include the excessive temperature in the superheater steam transfer line, the problem of long stagnant unheated connecting lines, high local stresses at nozzle-shell connections, and excessive temperature in the unfired superheater outlet tube. As well as suggesting that the operating procedures be reviewed, these cases suggest that redesign (or reverse design) may be a worthwhile result of the study.

## Conclusions

1) The determination of fitness for purpose and the evaluation of the integrity of high temperature equipment depend on field inspections conducted during the operating life. The many possibilities of damage accumulation processes cannot be foreseen with sufficient confidence during the design state to preclude such inspection.

2) The obtaining of effective results from such field inspections depends on sound prior planning, involving a broad analysis of the potential damage mechanisms that may be operative, including creep. It depends also on utilization of an adequate range of inspection procedures and methods, one of which is field metallography.

3) It is essential to take into account that in the majority of cases, the state of accumulated damage associated with failure is concentrated in small regions when compared with the overall dimensions of the equipment.

4) In any evaluation it is of major importance to consider the actual operating conditions, and to examine in depth the causes of any individual failure or any specific case of damage accumulation that is detected during service life.

## References


1 - Robinson, E.L., *Trans. ASME* 60, 253-259 (1960).

2 - Neubauer, B., in Proceedings 2nd Int. Conf. on Creep and Fracture of Engineering Materials and Structures, pp. 1217-1268, Pineridge Press, Swansea (1984).

3 - Auerkari, P., in ASME Int. Conf. on Advances in Life Prediction Methods, pp. 359-356, ASME, New York (1988).

4 - Sidey, D., A.M. Abdel-Latif, H.J. Westwood and D.M.R. Taplin, *Can. Metall. Quarterly* 18, 49-56 (1979).

5 - Stubbe, J. and C. van Melsen, Restlevetids-Konference, Copenhagen, March 1980.

6 - Woodford, D.A., *J. Eng. Matls. and Technol.* 101, 311-316 (1979).

7 - da Silveira, T.L. and I. Le May, *Materials Characterization* 28, 75-85 (1992).

8 - Bezerra, A.F., A.C.C. Campos, D. Niculau, F.S. Moreira, J. dos S. P. Filho, M.C.G. Chavez and T.L. da Silveira, in Proc. 4th Braz. Symp. on Piping and Pressure Vessels, pp. 415-433, Salvador, Brazil (1986).

9 - Le May, I., T.L. da Silveira and N.A.M. Ferreira, in Proc. 11th Symp. on Eng. Applications of Mechanics, Regina, Sask., May 1992 (in press).

10 - da Silveira, T.L., F.S. Moreira, M.C.G. Chavez and I. Le May, in Microstructural Science, Vol. 16, pp. 579-591, ASM/IMS (1988).

*Proceedings of the Fifth International Conference on Creep of Materials, Lake Buena Vista, Florida, USA, 18-21 May 1992*

# An Engineering Evaluation of Service Degraded 316N Main Steam Piping Weldments

**S.R. Paterson**
Aptech Engineering Services, Inc.
Sunnyvale, California

## Abstract

A root cause and post-repair suitability-for-service evaluation of a main steam piping system was performed. The piping system of concern was fabricated from nitrogen strengthened austenitic stainless steel (ASTM A376 Type 316N). During a routine examination of the piping system (performed after 94,000 hours of 1000°F service), transverse cracks in some of the girth weldments were detected. The most severe cracks extended almost through the pipe wall thickness.

A root cause evaluation determined that none of the shop welded girth weldments contained indications, whereas, almost all of the field welded girth weldments contained significant cracks. It was found that the cracked field weldments contained high nitrogen and had a primary austenite solidification structure, whereas, the shop welds had low nitrogen levels and a primary delta ferrite solidification structure. Extensive metallurgical evaluations, destructive (including creep rate and creep crack growth rate) testing, and evaluation of nondestructive examination (NDE) data indicated that the cracks were characteristic of weld solidification cracks which extended during service by creep crack growth.

A detailed engineering evaluation of the repaired piping system indicated an inverse relationship between weld metal creep strength and weldment life (i.e., the weldments with the <u>highest</u> creep strength are expected to have the <u>shortest</u> lives). Probabilistic suitability-for-service models benchmarked against the field NDE data were used to prescribe an optimal reexamination interval.

DURING A ROUTINE CONDITION assessment inspection of a main steam piping system in a 600 MW supercritical boiler, large cracks were detected in many of the piping girth welds. All the cracks were located in field rather than shop fabricated girth welds. Twelve of nineteen field girth welds contained significant axial (transverse to weld) cracks. The largest cracks had depths of approximately 2 inches. This paper describes some of the most interesting aspects of the evaluations and corrective actions which followed from the discovery of these cracks.

## Preliminary Nondestructive Examination

The piping weldments were examined using visual, fluorescent penetrant, and ultrasonic test procedures that were in accordance with ASME Section V recommendations. These examination procedures missed the cracking that was present. The cracks were instead, first detected while performing replication metallography to characterize the extent of sigma phase and carbide evolution in the weld regions. By chance, the replicator detected surface connected cracks while performing his task. A modified ultrasonic procedure was developed which was capable of detecting and sizing the cracks.

The primary cracking of concern was outside surface connected, part through-wall, axial cracks which were centered at the weld centerline. Very shallow, intergranular attack was detected in some of the Type A316N spool pieces. Two girth weldments contained small weld root cracks. Three 5D bends contained some minor surface cracks which extended into the material by a few grain diameters. No rejectable indications were found in the dissimilar metal welds or the ferritic piping sections. Each type of rejectable indication was evaluated in some depth. This paper will focus on the part through-wall, axial weldment cracks found in the austenitic stainless steel piping field girth weldments.

## Description of the Piping System

The piping system of concern was designed to handle 1015°F, 4000 psig steam. Review of daily logs indicated a main steam piping operating temperature 1005°F ±5°F and pressure of 3520 ±20 psig. The unit had experienced only 140 startups; 73 were cold starts. Table 1 summarizes other salient operational data.

Figure 1 is a schematic of the piping system. The 18-inch diameter pipe branches from the boiler to the reducers which were fabricated from ASTM A335-P22 material. At this point there is a dissimilar metal weld downstream of which the piping is 17-inch OD x 1.849-inch minimum wall pipe fabricated from ASTM A376-A316N material. The stainless pipe branches combine at the wye block. The pipe downstream of the wye block was fabricated from 20-inch OD x 2.173-inch

Table 1-1

COMPILATION OF PLANT OPERATING DATA

- 94,000 total hours of operation
- 140 total starts, 73 cold and 67 warm or hot
- Normal operating pressure is 3520 ±20 psig
- Thermal transients of 10 to 15°F occur daily at 50% load
- Start-up ramp rates are typically less than 100°F/hour
- Hot restarts can subject the steam line to large and sudden thermal transients of 80°F/15 minutes on heating or 60°F/5 minutes on cooling
- Temperature increases of 50°F steam temperature on 5 minutes can occur; however, they are of short duration
- Up to 50°F temperature top-to-bottom variations can occur during cold starts boiling out the superheater
- Up to 100°F temperature variation from the two outlet legs can occur when balancing burners
- General cooling rates after a hot trip are approximately 280°F/4 hours

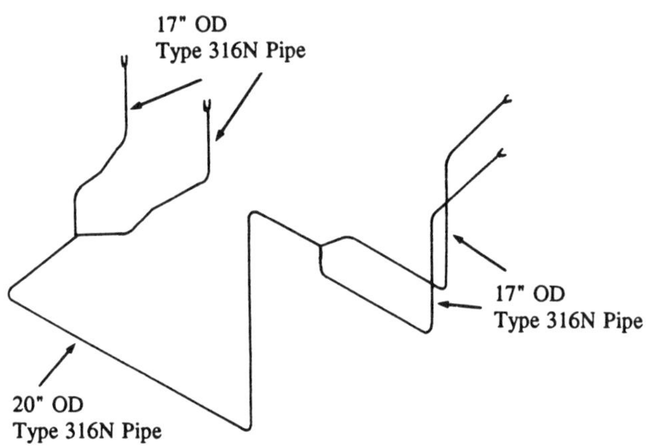


**Figure 1** - Schematic Showing the Sections of the Main Steam Piping Fabricated from Nitrogen Strengthened Austenitic Stainless Steel.

minimum wall ASTM A316N material. At the turbine throttle valve connections, there is a transition back to ASTM A335-P22 piping. The wye block and valve bodies were fabricated from ASTM A351, Grade CF8M, cast stainless steel.

## Events Which Followed the Discovery of Weld Cracks

The discovery of weld cracks, some of which penetrated almost through the pipe wall thickness launched extensive root cause and weld repair projects. Initial efforts included flaw sizing, piping support evaluations and stress analyses, boat sampling and metallurgical evaluations, operating history reviews, installation of trams and thermocouples (to supplement the existing piping displacement and thermal transient monitors), and measurement of the ferrite number of each weld (using a Severn gage). The evaluations indicated that the reliability of all the field girth welds were suspect. The utility made the decision to cut out each field girth weld and then rejoin the pipe section using a weld repair procedure which was capable of producing crack free, reliable field weldments. Short sections of new piping were added as required to make up for the lengths of piping lost in the process of removing the original field girth weldments. A total of 23 new girth welds (referred to in this paper as repair welds) were required to complete the piping system repairs.

It was at this juncture that the author and his colleagues became involved in this project. Our role was to determine the root cause of the cracking and to provide an estimate of the long-term integrity of the post-repair welded piping system. To achieve our objectives required bringing together all the prior information and supplementing it with additional field measurements, metallurgical evaluations and tests, characterization of the mechanical properties of the weldments, stress analyses, and suitability-for-service evaluations. These will each be discussed in-turn in the sections which follow.

## Field Measurements

**Ferrite Number Measurements.** It is well known (1) that the amount of retained delta-ferrite is a good indicator of the susceptibility of austenitic stainless steel weldments to solidification cracking. A delta ferrite level of at least 3FN will eliminate fissuring in weld metals deposited with austenitic filler metals E16-8-21 and E316 (1). It was clear from the metallurgical evaluations that the weld cracks initiated from preexisting solidification cracks. Severn gage measurements were made at four locations on each girth weld. Severn gage measurements were also used to verify that the repair welds had achieved an acceptable level of delta-ferrite. The original field girth welds contained no measurable delta-ferrite, whereas, the shop girth weld metal had ferrite number in the range of 3 to 5%. The repair welds produced ferrite numbers in the range of 2 to 12.5%.

**Residual Stress Measurements.** Blind hole residual stress measurements were made at nineteen locations on the main steam piping in the vicinity of old and repair weldments. A variation of ASTM E-837 was employed to make these measurements. An air-turbine driven carbide end mill was used for hole drilling. Incremental measurements to hole depths of 1/16 inch were utilized. The objective of these measurements was to determine if high secondary stresses, either still present from the fabrication or induced by service conditions, could be responsible for the cracking.

Extremely high tensile residual stress values were measured both near and away from weldments. In the service degraded pipe material well away from welds, the outside surface residual stress was approximately 40 ksi. Similar values were obtained for measurements made on one of the new sections of pipe. Measurements on both the inside and outside surfaces of an elbow which was removed from service had measured residual tensile stresses of approximately 30 ksi. Measurements made on the outside surface weld crown at the weld centerline of old shop and field girth welds produced values of approximately 30 ksi and 60 ksi, respectively. A value of 40 ksi was measured on a repair weld crown at the weld centerline.

The results from these measurements were somewhat baffling, especially with regard to the measured base metal values. Devarajan et. al. (2) reports similar results when turning 304L stainless steel at high feed and feeds, which might explain the source of residual stress in the new piece of pipe. Service exposure at temperatures above 1000°F for almost

100,000 hours would be expected to significantly relieve fabrication-induced residual stresses. Another possible explanation for the high residual stress levels in the service-degraded material was service-induced residual stresses, such as were reported to occur from severe thermal downshocks in the case of the Eddystone stainless steel main steam piping (3). If this were the case, then a compressive residual stress would be expected at the pipe inside surface.

To further evaluate the through-wall residual stress distribution in the old pipe material, two separate pieces of service-degraded pipes were measured. These measurements were made using the parting-out and splitting method described by Rosenthal and Norton (4). It was found (Figure 2) that the tensile residual stresses were present on both the internal and external pipe surfaces, thus ruling out thermal downshocks as their source. The results also indicated that the surface residual stresses in the piping die away very quickly below the surface.

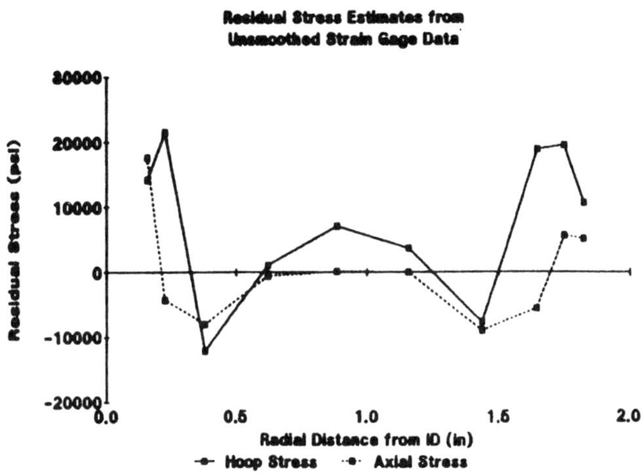


**Figure 2** - Through Thickness Residual Stress Distribution in Old Pipe Spool Piece.

## Metallurgical Evaluations

**Chemistry.** Chemical analyses were performed on the weld and adjacent base metals of the original shop and field girth welds, and some select repair welds. Table 2 provides a summary of the chemical analysis results. The base metal chemistry was within the anticipated range of ASTM A376 TYPE 316N material. The repair weld metal had a composition that was consistent with the use of ER 16-8-2 filler wire. The original shop and field weld metal fell within the expected range of E316 welding electrodes, except in the case of the field weldment; the amount of nitrogen, copper, and cobalt were high.

To evaluate the role that weld metal composition played in the cracking the nickel and chromium equivalents of each sample were determined and plotted on a Delong Constitution Diagram (Figure 3). The results were in excellent agreement with the Severn gage measurements discussed previously. The original field weld metal had significantly higher nickel equivalents and lower chromium equivalents than the shop or repair weld metals. The constitution of the field weld metal was such that these welds would be expected to solidify in a primary austenite mode and contain no retained delta-ferrite. Austenitic stainless steel weld metals which produce no retained delta-ferrite and solidify in a primary austenite mode are highly susceptible to solidification cracking (Figure 4).

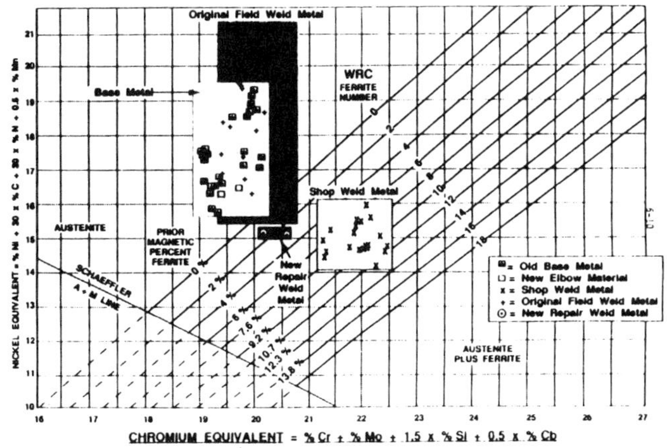


**Figure 3** - Delong Constitution Diagram Showing the Delta-Ferrite Potential of the Main Steam Pipe Base and Weld Metals.

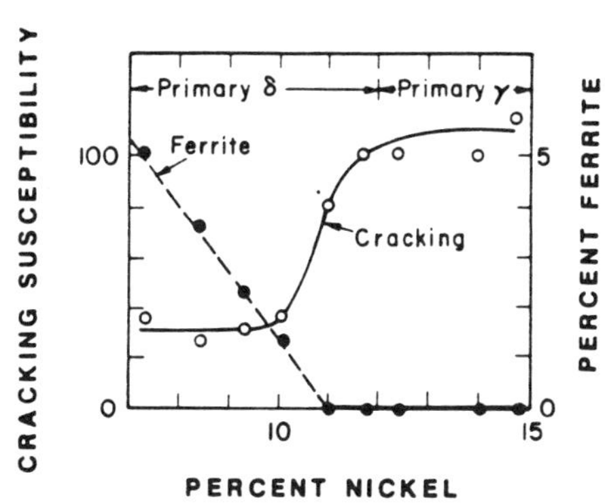


**Figure 4** - Solidification Cracking Susceptibility of 17.5% Cr Stainless Steel. Cracking Susceptibility is Higher When Austenite (γ) is the Primary Solidification Phase (11).

A more detailed evaluation of the weld metal chemistries revealed that nitrogen was the dominant element responsible for the nickel equivalent variations, which were in-turn responsible for the low delta-ferrite values in the field welds. A statistical study of the nitrogen concentrations (Figure 5) revealed an interesting trend. It was found that the welds with the highest nitrogen contents were field girth welds in the 20-inch diameter pipe which were welded in the 5G position. These were followed by 20-inch diameter pipe welds made in the 2G position, then the 17-inch diameter pipe welded in the 5G position, then the 17-inch diameter pipe welded in the 2G position. The shop welds and repair welds all had significantly lower nitrogen levels than the original field welds. It was also discovered that the 5G field welds were made using weave beads, whereas, the 2G welds contained stringer beads. It was surmised that the nitrogen contents in the field welds were to some extent influenced by the welding techniques used, with weave beads producing higher nitrogen contents than stringer beads. This might in-part be explained by the poorer arc voltage control inherent in producing welds with weave

Table 2

CHEMICAL ANALYSIS RESULTS

| Element | 17-Inch Outside Diameter Base Metal | 20-Inch Outside Diameter Old Base Metal | New Base Metal | 17-Inch Outside Diameter Shop Weld Metal | 20-Inch Outside Diameter Shop Weld Metal | 17-Inch Outside Diameter Field Weld Metal | 20-Inch Outside Diameter Field Weld Metal | New Repair Weld Metal | Specification For ASTM A376 Type 316N | Chemical Requirements For Weld Metal Deposited By E316 Covered Welding Electrodes | Chemical Requirements ER 16-8-2 Bare Electrode Materia |
|---|---|---|---|---|---|---|---|---|---|---|---|
| Carbon | 0.070 | 0.068 | 0.070 | 0.039 | 0.049 | 0.055 | 0.057 | 0.080 | 0.08 Maximum | 0.08 Maximum | 0.10 Maximum |
| Manganese | 1.69 | 1.64 | 1.60 | 2.02 | 1.83 | 1.92 | 2.06 | 1.81 | 2.00 Maximum | 0.5 to 2.5 | 1.0 to 2.5 |
| Phosphorus | 0.023 | 0.023 | 0.031 | 0.014 | 0.020 | 0.023 | 0.026 | 0.030 | 0.040 Maximum | 0.04 Maximum | 0.03 Maximum |
| Sulfur | 0.019 | 0.020 | 0.020 | 0.016 | 0.016 | 0.012 | 0.016 | 0.024 | 0.030 Maximum | 0.03 Maximum | 0.03 Maximum |
| Silicon | 0.35 | 0.27 | 0.62 | 0.77 | 0.72 | 0.32 | 0.42 | 0.38 | 0.75 Maximum | 0.90 Maximum | 0.30 to 0.65 |
| Nickel | 11.77 | 11.68 | 10.70 | 11.11 | 11.33 | 12.72 | 12.26 | 11.16 | 11.0 to 14.0 | 11.0 to 14.0 | 7.5 to 9.5 |
| Chromium | 16.86 | 16.63 | 16.33 | 18.36 | 18.18 | 17.14 | 17.05 | 17.22 | 16.0 to 18.0 | 17.0 to 20.0 | 14.5 to 16.5 |
| Molybdenum | 2.39 | 2.37 | 2.38 | 2.55 | 2.59 | [illegible].27 | 2.38 | 2.57 | 2.00 to 3.00 | 2.0 to 2.5 | 1.0 to 2.0 |
| Vanadium | 0.05 | 0.05 | -- | 0.03 | 0.03 | 0.06 | 0.06 | -- | -- | -- | -- |
| Columbium | 0.02 | 0.01 | -- | 0.01 | 0.01 | 0.01 | 0.01 | -- | -- | -- | -- |
| Titanium | 0.01 | 0.01 | -- | 0.01 | 0.01 | 0.02 | 0.03 | -- | -- | -- | -- |
| Cobalt | 0.09 | 0.19 | 0.18 | 0.03 | 0.055 | 0.118 | 0.136 | 0.070 | -- | -- | -- |
| Copper | 0.10 | 0.09 | 0.64 | 0.07 | 0.07 | 0.13 | 0.16 | 0.11 | -- | 0.5 Maximum | 0.5 Maximum |
| Aluminum | 0.005 | 0.005 | -- | 0.007 | 0.007 | 0.005 | 0.004 | -- | -- | -- | -- |
| Boron | 0.001 | 0.001 | -- | 0.001 | 0.001 | 0.001 | 0.001 | -- | -- | -- | -- |
| Nitrogen | 0.102 | 0.099 | 0.099 | 0.045 | 0.042 | 0.126 | 0.142 | 0.031 | 0.10 to 0.16 | -- | -- |
| Oxygen | 0.006 | 0.006 | -- | 0.113 | 0.107 | 0.081 | 0.056 | -- | -- | -- | -- |
| $Cr^{(1)}eq$: | 19.78 | 19.41 | 19.64 | 22.07 | 21.33 | 19.89 | 20.06 | 20.36 | -- | -- | -- |
| $Ni^{(2)}eq$: | 17.75 | 17.48 | 16.56 | 14.63 | 14.25 | 19.09 | 19.28 | 15.39 | -- | -- | -- |

Notes: (1) Creq = %Cr + %Mo + 1.5 x %Si + 0.5 x %Cb
(2) Nieq = %Ni + 30 x %C + 30 x %N + 0.5 x %Mn

beads. Previous studies of welds produced with a shielded metal arc welding process (5) indicate that high arc voltage (which corresponds to longer arc lengths) produce higher weld nitrogen contents.

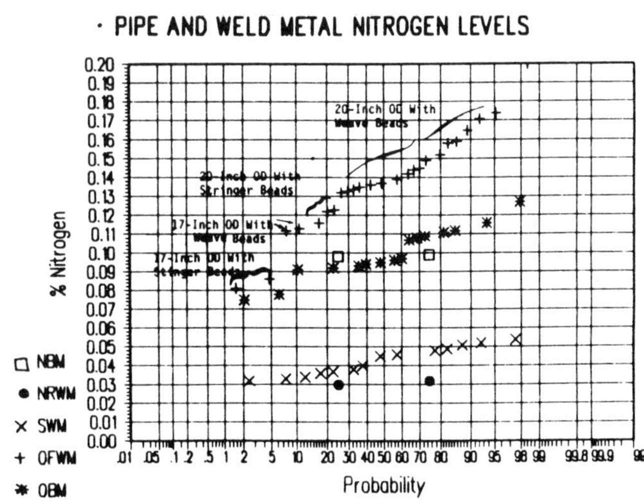


**Figure 5** - Probability Plot Showing the Weight Percent Nitrogen in Original Field Weld Metal, Base Metal, Shop Weld Metal, and Repair Weld Metal.

**Microstructure**. Detailed microstructural evaluations were performed on representative field, shop, and repair girth weldments. The results of these evaluations were consistent with the structures predicted from the Delong constitution diagrams. The weld metal of field weldments (Figure 6) consisted primarily of austenite. The grain boundaries were very pronounced. The microstructure was characteristic of an austenite solidification mode in which low melting point impurities are swept to the grain boundaries.

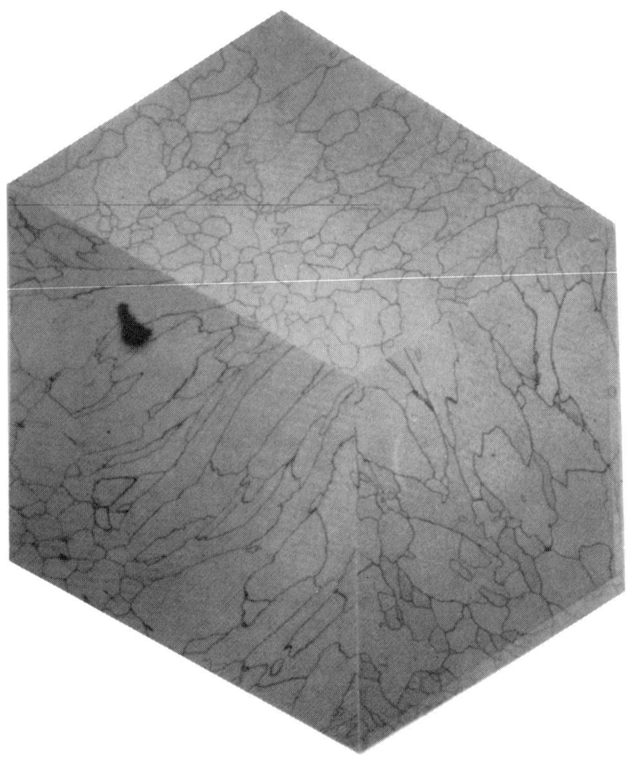

**Figure 6** - Microstructure of Weld Metal Deposit from a Field Girth Weld. The Microstructure Consists Primarily of Austenite. This Structure is Characteristic of a Weld Deposit with a Primary Austenite Solidification Mode.

The weld metal of the shop weldments was distinctly different. It had a structure which is characteristic of primary delta-ferrite solidification. In this case, the structure contains skeletal delta-ferrite within the grains which was surrounded by austenite (Figure 7). The grain boundaries were very indistinct.

This structure should be much more resistant to solidification or service-induced grain boundary cracking than the field weld metal structure.

**Figure 7** - Microstructure of the Weld Metal Deposit from a Shop Girth Weld. The Microstructure Consists of Skeletal Delta-Ferrite and Austenite. This Structure is Characteristic of a Weld with a Primary Delta Ferrite Solidification Mode.

**Flaw Characteristics**. The weld metal of the original field girth weldments contained an abundance of interconnected networks of solidification cracks. Cracking was interdendritic in the weld metal and intergranular in the base metal (Figure 8, see next page). Auger electron spectroscopy scans across primary cracks, secondary cracks, and uncracked grain boundaries in the weld metal revealed high concentrations of sulfur and phosphorus. Unfused regions with distinct secondary dendrite arm morphology were also observed (Figure 9). It was concluded that the cracking in these welds was for the most part solidification cracking. However, link-up and growth of small solidification cracks could not be ruled out. Both possibilities were considered in fitness-for service evaluations.

No cracking or grain boundary inhomogeneities were detected in the shop welds.

**Weld Characteristics versus Cracking Susceptibility.** There was a strong correlation between the cracking susceptibility (as inferred from both NDE results and destructive metallurgical testing) and the nitrogen content of the weld metals. It was found that the absence of delta-ferrite was a prerequisite but was not sufficient to delineate which welds cracked. For example, the NDE statistics (Table 3) indicate that none of the 17-inch diameter pipe field girth welds made in the 2G position contained cracks although these welds contained no delta-ferrite. The welds most prone to cracking were the 20-inch diameter pipe field weldments which contained the highest

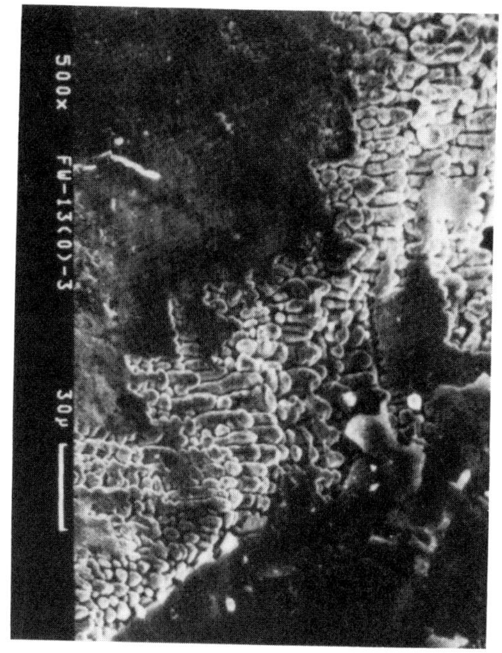


**Figure 9** - Scanning Electron Microscope Fractograph of a Section of the Cracking in the Original Field Weld Metal.

Table 3

NONDESTRUCTIVE EXAMINATION RESULTS STATISTICS

| Weld Type, Pipe Size, and Weld Position | Number in Piping System | Number Inspected | Number with Nondestructive Indications | Percentage with Weld Metal Indications |
|---|---|---|---|---|
| Shop Welds | | | | |
| 17-Inch OD and 20-Inch OD | 29 | 28 | 2 (Indication in Base Metal or Wye Block Material) | 0 |
| Field Welds | | | | |
| 17-Inch OD/2G | 5 | 5 | 0 | 0 |
| 17-Inch OD/5G | 6 | 6 | 5 | 83% |
| 20-Inch OD (2G and 5G | 8 | 6 | 6 | 100% |

Welding Positions:
2G (Horizontal - Pipe axis in vertical direction)
5G (All positions - Pipe axis in horizontal direction)

nitrogen contents. From a review of the metallurgical evaluations (Table 4), it appears that the amount of nitrogen present in the weld metal is a key indicator of cracking susceptibility. Welds with nitrogen levels above 0.1 wt% were prone to cracking. It was postulated that this trend results from both the austenite stabilizing and elevated temperature strengthening potential of nitrogen. In particular, the effect of nitrogen on the creep rate response of the weld metal was determined to be a key feature in the creep crack growth sensitivity of these weldments. It was found that excess nitrogen in the weld metal will retard the stress relaxation response of both the far-field (global) and near-field (near crack tip) stresses, thus, increasing the likelihood of crack growth and linkup. This is described in more detail in sections which follow.

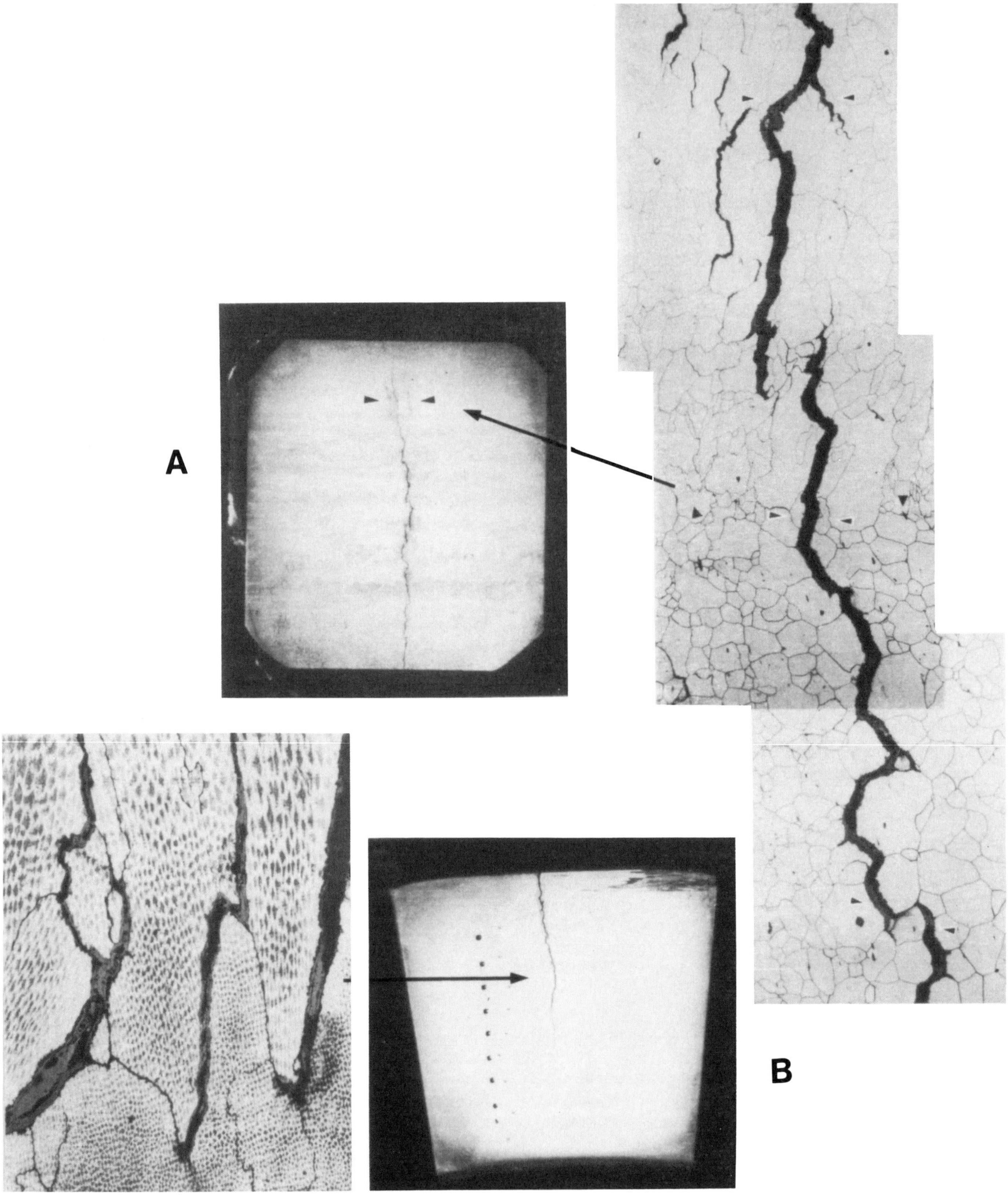


**Figure 8** - Photograph Showing the Propagation of Solidification Cracks (A) Viewed from the Weld Crown, (B) Viewed in Cross-Section Along the Pipe Axis.

Table 4

SUMMARY OF SALIENT METALLURGICAL CHARACTERIZATIONS

| Spool Piece/Weld Identification | Pipe Size (Inch) | Weld Position[(1)] and Bead Type | Cracking Characteristics | Percent Nitrogen in Weld Metal | Solidification Mode |
|---|---|---|---|---|---|
| FW1 or 2 | 17 OD | 2G/Stringer | None detected | 0.08 | Primary γ |
| FW7 | 20 OD | 2G/Stringer | Axial cracks | 0.12 | Primary γ |
| FW9 | 20 OD | 2G/Stringer | Axial cracks | 0.12 | Primary γ |
| F12 | 20 OD | 5G/Weave | Circumferential root crack | 0.14 | Primary γ |
| FW13 | 20 OD | 5G/Weave | Circumferential crack (opened during repair heat treatment) | 0.14 | Primary γ |
| FW18 | 17 OD | 5G/Weave | Axial cracks | 0.11 | Primary γ |
| SW12 | 20 OD | ?/Stringer | None detected | 0.04 | Primary δ ferrite (skeletal δ in dendrite cores) |

Notes: [(1)]Weld positions

Conclusions: 1. 5G Weld → Weave Beads → Highest Nitrogen Content
2G Weld → Stringer Beads → Low Nitrogen Content
2. 20-Inch OD pipe field welds have > N than 17-inch OD pipe welds

# Mechanical Properties

**Strength, Fracture Toughness.** Values of yield and ultimate strengths from TP316 and 316N materials were obtained from Refs. (5, 6, 7). Confirmatory tests of samples removed from the service-degraded weldments and tested at 70°F and 1015°F fell within the range of the literature values.

Fracture toughness values for service-degraded austenitic stainless steel base and weld metal were also extracted from the literature (8). A fracture toughness of approximately 55 ksi$\sqrt{\text{in}}$ for service temperatures was used as input in fitness-for-service evaluations. Confirmatory tests performed on the main steam piping material produced significantly higher toughness values.

**Creep Rate.** The inelastic, time dependent stress state including the stress relaxation response of the main steam piping weldments is controlled to a large extent by the time dependent (creep rate) response of the material. This is also true of the creep crack growth rate estimates. The well known "Norton Creep Rate Law" was used in both the inelastic stress and the creep crack growth rate analyses. This relationship is given as:

$$\dot{\epsilon} = A(T)\sigma^n \qquad (1)$$

where,

| | |
|---|---|
| $\dot{\epsilon}$ | = Minimum or steady state creep rate, $hr^{-1}$ |
| A(T) | = Norton Law pre-exponential constant |
| | = $A_o \exp(Q'/T)$ |
| $\sigma$ | = Stress, ksi |
| T | = Temperature, °R |
| n | = Norton Law exponent |
| $A_o$, $Q'$, n | = Empirical constants |

Creep rate data were extracted from Refs. (5), (6), and (7) and analyzed using nonlinear least squares regression methods to fit the data to the Norton creep rate law Eq. 1. These data comprised 135 data points which included Types TP316 and TP316N stainless steels with nitrogen contents that varied within the range of 0.04 to 0.20. Figure 10 documents the data used (white triangles) and the predicted median creep rate relationship (black squares). The best fit for the entire data set was obtained for the following values of $A_o$, $Q'$, and n:

$$A_o = \text{Exp}(36.765)_1\ hr^{-1}\ ksi^{-8.2}$$
$$Q' = -120{,}000°R$$
$$n = 8.2$$

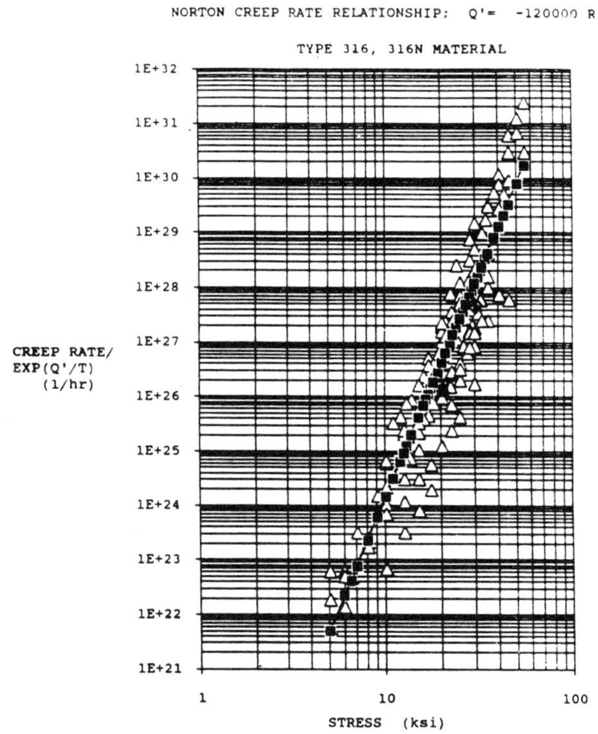


**Figure 10** - Creep Rate Data and Median Prediction for TP316 and TP316N Material.

With the term $\log_e [\dot{\epsilon}/\exp(Q'/T)]$ as the dependent variable the fitting parameters, $r^2$ for this relationship was 0.87. This suggests that the creep rate equation chosen and constants derived "explain" or model approximately 87% of the variability in the creep rate data analyzed. A study of the residuals, r (actual/predicted median value), from this data set was made to look for additional factors which might explain variability which was not already handled by this relationship. It was found that the nitrogen level of the specific test sample showed a negative correlation with regression equation residuals[1] (Figure 11). This supports the conclusion that Type 316SS material with high nitrogen contents (e.g., >0.09%) will have correspondingly slower creep rates. All other things being equal it would, thus, be expected that the field weld metal would have the slowest creep rate, followed by the pipe material, followed by the shop weld metal and repair weld metal.

To verify and refine the Norton creep rate relationship derived from literature, creep rupture specimens of material removed from the Williams Station main steam piping (including base metal, shop metal, and field weld metal) were tested. It was found (Figure 12) that the experimental data fell within the literature data scatter band except in the case of the field weld specimens.

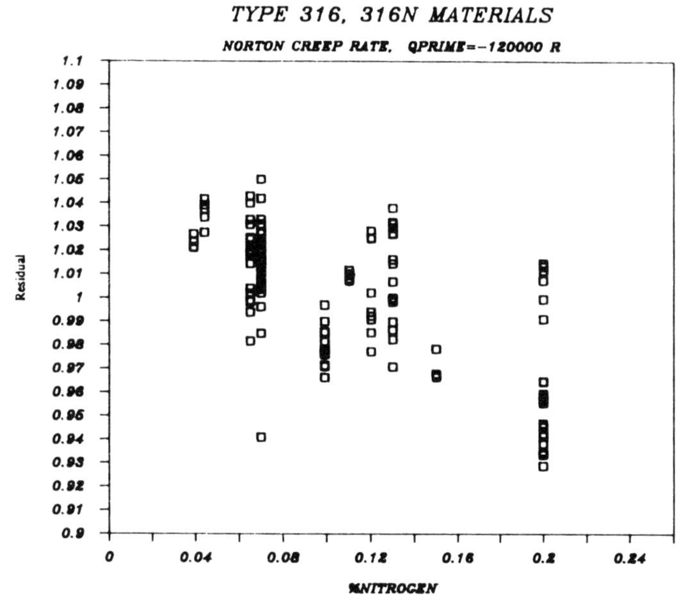


**Figure 11** - Percent Nitrogen Versus Measured Creep Rate/Predicted Median Creep Rate, r.

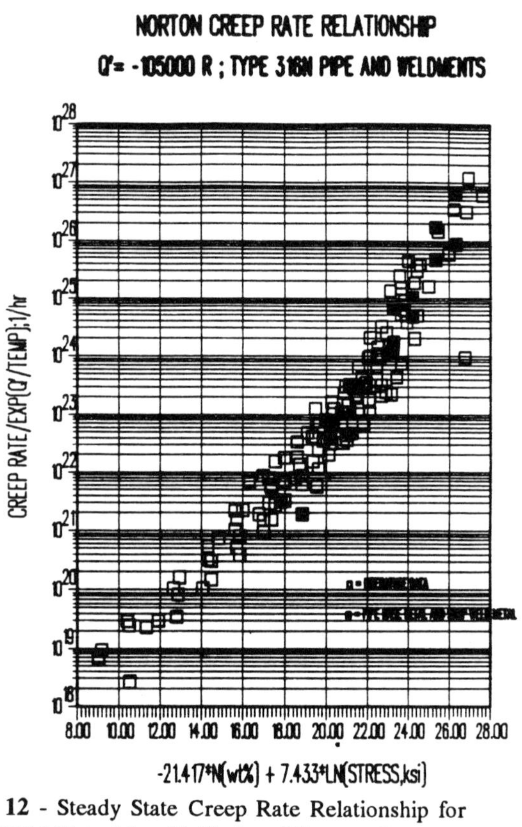


**Figure 12** - Steady State Creep Rate Relationship for Type TP316 Materials with Various Nitrogen Contents.

The measured creep rates of the field weld samples were significantly faster than expected (since the field welds have the highest nitrogen levels they should correspondingly have the slowest creep rates). Metallurgical evaluation of the field weld metal samples indicated that these samples contained significant preexisting cracks prior to testing. This was believed to be the cause of the high measured creep rates. The measured creep rate data derived from the field welds were, therefore, rejected. Instead, creep rates estimated from the literature for a material with approximately 0.15% nitrogen were used to model the behavior of the field weldments. The same approach was used to obtain creep relationships for base metal and shop welds (or repair welds).

By including a nitrogen term in Eq. 1, an improved creep constitutive equation was derived. It is given as:

$$\dot{\varepsilon} = \exp[32.646 - 21.417(\%N)] \times \exp\left(\frac{-105{,}000}{T}\right)\sigma^{7.433} \qquad (2)$$

Good agreement between these literature derived and experimental creep rates was found in the case of base metal and shop weld metal.

This relationship was used to estimate the stress relaxation behavior and creep crack growth rate response of the piping and weldments.

Proper selection of the nitrogen content of the weld electrodes is of obvious importance. This is true both from the standpoint of creep rate response and weldability. For thick section weldments for elevated temperature application, it would seem that slight undermatching of nitrogen contents would provide optimal weldability and inelastic stress response. The high nitrogen levels in the field weld metal produced a crack susceptible microstructure, an elevation in the stress in the weldment (due to an incompatibility between in the creep rate response of the weld and base metals), and slow stress relaxation response.

**Creep Crack Growth Rate.** Creep crack growth rate data for Type 316 material was measured from pipe base metal, shop weld metal, and field weld metal samples. First, estimates of the possible upper (plane strain) and lower (plane stress) bounds on creep crack growth rates were estimated using the theoretical ductility exhaustion model (9) (referred to as the Nikbin, Smith, Webster (NSW) model). The following theoretical relationships were derived:

Upper bound NSW curve (for a rupture ductility of 40%):

$$\frac{da}{dt} = 3.36(C^*)^{0.85} \qquad (3)$$

Lower bound NSW curve (for a rupture ductility of 40%):

$$\frac{da}{dt} = 6.72\ E - 2(C^*)^{0.85} \qquad (4)$$

These upper and lower theoretical bounds are shown in Figure 13.

where,

| | |
|---|---|
| da/dt | = Crack growth rate, inches/hour |
| C* | = Theoretical steady state creep crack growth parameter, in-kips/in$^2$-hr |
| Rupture Ductility | = Percent elongation in creep rupture tests |

To improve on our knowledge of the creep crack growth rate of the original field weld metal, the shop weldments, and the pipe base metal, six creep crack growth rate tests were performed. All tests were performed with side grooved compact tension samples, tested in air at 1015°F. It was found that the weld metals (field and shop had similar crack growth rates. The base metal on the other hand had a significantly slower crack rate (Figure 13).

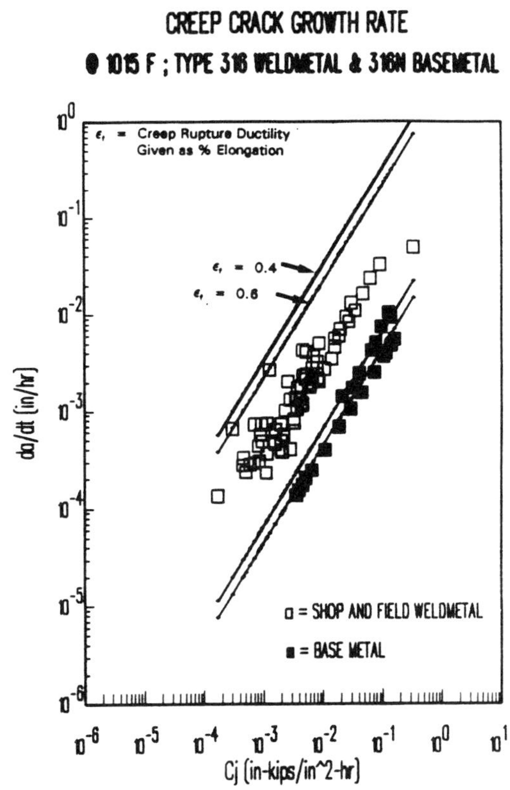


**Figure 13** - Experimental and Estimated Upper (Plane Strain) and Lower Bound (Plane Stress) Creep Crack Growth Rates for TP316 and TP316N Material.

The finding that the shop and original field weld have similar creep crack growth rates may seem to contradict our previous conclusion that the creep rate response of weld metal was erroneously high because the creep samples contained preexisting flaws. The field weld metal creep crack growth rate samples probably also contained preexisting flaws. Experimentally it was found that the creep crack growth rate of the field weld metal was "poorly behaved". One of two samples failed during fatigue precracking. The other had a very discontinuous crack growth rate (the white squares in Figure 13 which intersect the upper bound NSW predicted curves are field weld metal values). The occasional high creep crack growth rates probably occur when the crack tip of the test samples intersected preexistent flaws (thus, increasing the effective crack length and associated value of $C_j$). Since a relatively small region of material in the vicinity of the creep crack growth sample crack tip is controlling the crack growth rate, it is expected that some "valid" and some "invalid" crack growth rate data will be obtained from samples containing preexistent flaws. In this case, the lower bound crack growth data should accurately or conservatively represent the material in absence of flaws.

In contrast, the creep or creep rupture samples simultaneously expose a relatively large volume of material to the test environment (stress, temperature). The weakest zone within these samples gage length control the test result, thus, preexistent flaws produce creep rate results which are not at all representative of the material in absence of flaws.

The experimental median curve for weldments was determined to be:

$$\frac{da}{dt} = 1.58\ E - 1\ (C_j)^{0.82} \qquad (5)$$

The experimental median curve for pipe base metal was determined to be:

$$\frac{da}{dt} = 4.72\ E - 2\ (C_j)^{0.94} \qquad (6)$$

where, $C_j$ = Experimental creep crack growth rate parameter, in-kips/in$^2$-hr.

The experimental, $C_j$, relationships were input into our suitability-for-service evaluations. In addition, a sensitivity study to evaluate the affect of our assumption that $C_j \simeq C^*$ was made. For the interested reader, Ref. (10) provides a full description and comparison of $C_j$ and $C^*$.

## Stress Analyses

Estimates of the primary, secondary, and time dependent stresses in the vicinity of the main steam piping girth weldments indicated initial nominal stress magnitudes in the range of 40 ksi (ignoring near-surface residual stresses). In-service stress relaxation would be expected to reduce the stress field to a nominal value as low as 11 ksi. Time dependent finite element stress analyses indicated that in the steady state stresses in the original field welds would be 3 to 5 ksi higher than the shop or repair welds due to a detrimental creep rate mismatch. Due to uncertainties with regard to residual and time dependent stress, it was eventually decided that the fitness-for-service evaluations should be based on bounded rather than absolute stress magnitudes.

## Reliability Forecasts

**The Model.** A creep crack growth based computer model was developed to make conservative forecasts of the reliability of the shop and repair welded girth weld. The program was an enhanced version of the probabilistic, C* based creep crack growth code PCSTAR. The enhancements added for this project included a routine which allowed the far-field stresses to relax over time, a flaw shape correction factor, the ability to make estimates of "true flaw distributions" from the NDE data, and a routine to determine what family of initial and steady state stress values would predict the historical flaw distributions. A schematic of the program logic is shown in Figure 14 (see next page).

**Model Calibration.** To forecast crack growth in any weld, it is important to calibrate our models properly against the known behavior of the old welds. On this project, calibration was used mainly to reduce the effects of uncertainty of initial flaw size, applied stress, and initial residual stress.

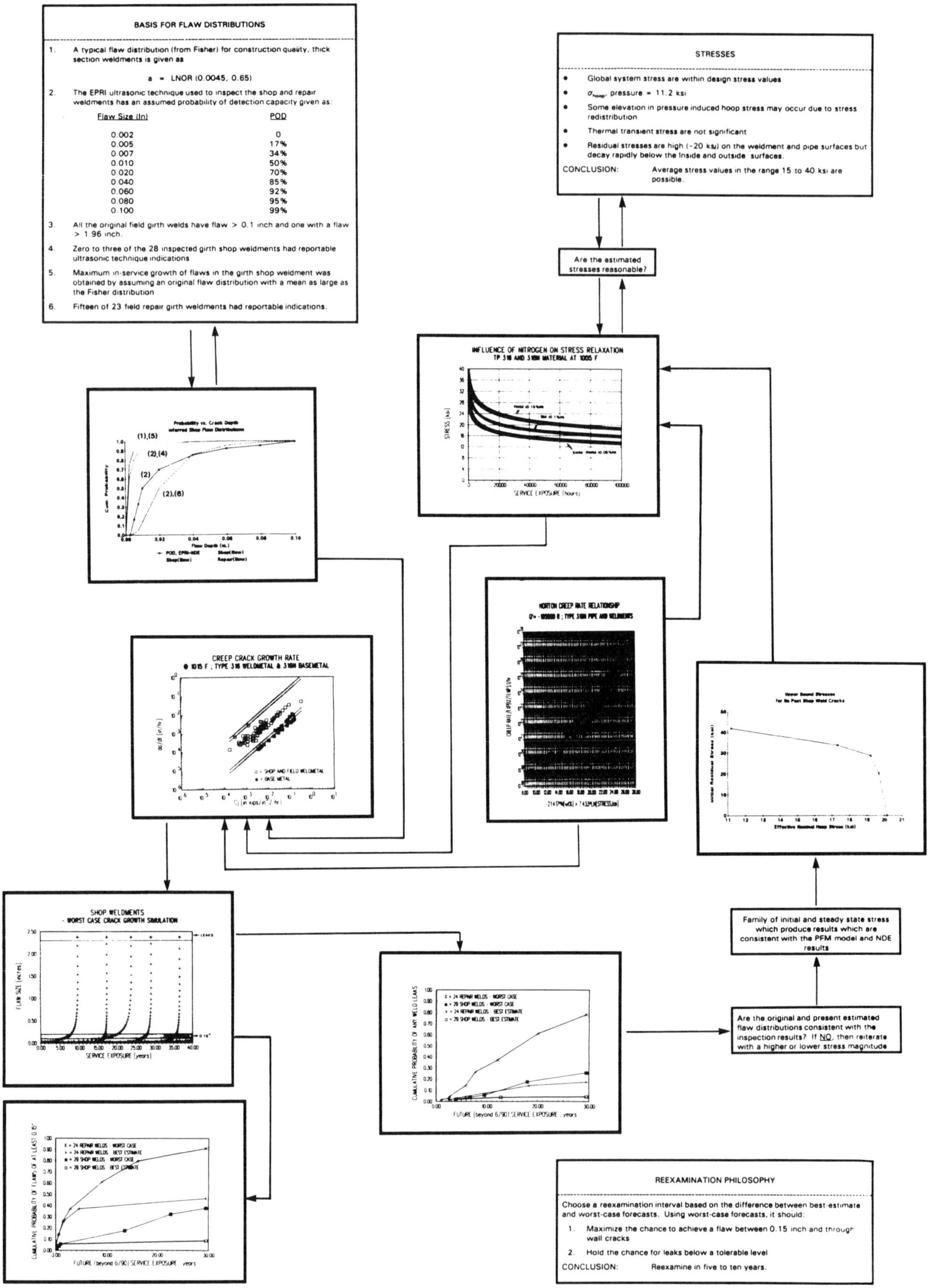


**Figure 14** - Schematic of the Logic Step Used to Forecast the Crack Growth and Leakage.

A key feature of this PCSTAR model was its ability to automatically benchmark itself against available NDE data. This capability along with it's Monte Carlo-based probabilistic forecasting capabilities provide an optimal approach for making crack growth forecast in those cases where input and crack growth modelling uncertainties are large.

**Past Performance of Field Welds.** In the case of the old field welds the model was considered to be valid only if it backcasts two observations:

1. The largest crack depth in 12 welds after 10.7 years of growth is the observed 1.96 inches.

2. Precisely 11 of 12 (ignoring the 2G 17-inch outside diameter pipe welds) inspected old field welds had a crack deeper than 0.1 inch after 10.7 years.

The analyses led to two major conclusions:

- Using reasonable stress levels (15 to 20 ksi nominal and 20 to 40 ksi with peak residuals), the median initial (pre-existing) flaw depth of old field welds was at least 0.1 inch. More likely the median depth was 0.25 inch.

- The maximum conceivable stress levels from our analyses and experiments (of tube steady state stress, $\sigma_{ss}$ = 25 ksi together with a peak stress, $\sigma_p$ = 70 ksi) never happened. At such stresses, even using flaw depth data for well-made welds, three leaks would have occurred in the first year of service. Also, the same run backcasting three sudden failures cannot produce cracks deeper than 0.1-inch cracks in four of the welds. Thus, the run is invalid because it matches neither observation and forecasts "all-or-nothing" behavior for over half the welds.

**Past Performance of Shop Welds.** A run simulating the 28 inspected shop welds was considered valid only if it backcasts one observation:

1. The flaw depth distribution in 28 welds after 10.7 years of growth is small enough such that zero weld metal indications were observed from the applied ultrasonic examination procedure.

Using estimates from Walker (12) and the Metals Handbook (13), the probability of detection (POD) of the ultrasonic examination procedures was modeled. Using the information from these references, a POD of 0.5 exists for 0.01-inch deep cracks and POD = 0.95 for 0.08-inch deep cracks, etc.

Using the POD model in a conservative way, a 95% upper confidence bound of the current flaw distribution was computed for the shop welds.

The stresses required to grow cracks from the estimated initial to estimated present depths in 10.7 years were computed. This lead to three major conclusions:

- The nominal steady state stress, $\sigma_{ss}$ required to match the history of the 28 inspected shop welds is less than 20.2 ksi.

- Given that $\sigma_{ss}$ is at least 11.2 ksi (the hoop stress for a pressure of 3500 psi) the initial shop weld residual stress $\sigma_r$ to match history is less than 42 ksi.

- Figure 14 shows calculated upper bound stresses for all other combinations of $\sigma_{ss}$ and $\sigma_r$.

**New Repair Welds.** Fifteen of the twenty-three new repair welds revealed reportable indications when inspected with the modified ultrasonic examination procedure. Figure 15 documents rough estimates of the flaw depths of these 15 "hits".

Since all 15 indications were considered acceptable, no attempt was made to remove them from the repair welds. This NDE data was used to estimate the initial flaw distribution on which reliability forecasts of the repair welds were made.

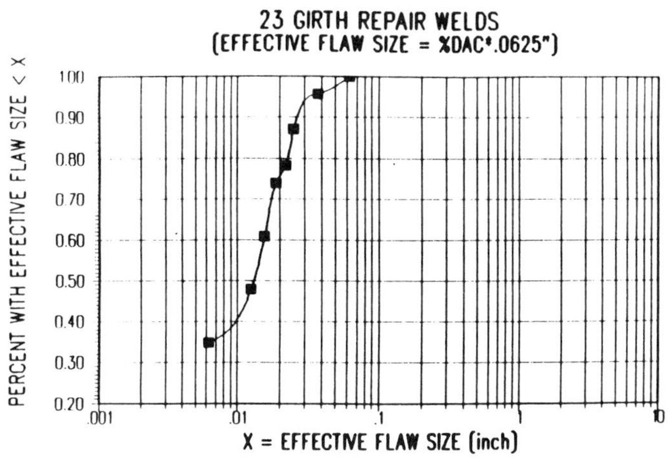


**Figure 15** - Distribution of Effective Flaw Sizes from the Ultrasonic Indications Reported from the Inspection of 23 Girth Repair Welds.

## Summary and Conclusions

Using input specified earlier in this section, PCSTAR was run to obtain two forecasts each for the shop and repair welds. These are labeled "worst case" and "best estimate" in Figure 14. Our broad parametric study shows that because of calibration, these forecasts are not sensitive to many errors and uncertainties.

For example, the forecasts are insensitive to the precise (plane stress versus plane strain) definition of C*. If we calibrate with a different definition, intermediate results (like upper bound stresses) do change. However, future failure rate estimates do not change very much when the altered C* definition is also used in the forecast.

Similarly, the forecasts are insensitive to assumed temperature if past and future temperatures are the same. If future temperatures increase, the forecasts must be changed.

A worst case forecast was made using the initial residual and uniform steady state stress combination which lead to the highest future failure (leak) rate predictions. For the shop welds, this was a nominal stress of 20.2 ksi with no residual stress. For the repair welds, this was a nominal stress of 19.7 ksi with an initial residual stress of 20.3 ksi (for a total initial stress of 40 ksi).

A best estimate forecast was made using through-thickness average stresses less than the upper bounds described above. For both shop and repair welds, we used an initial total stress of 40 ksi that drops off to the long-term value, $\sigma_m$ = 14.5 ksi. The stresses used, while still reasonable, are judged to be conservative for the shop and repair welds, and produce results which duplicate the weld NDE results.

Figure 14 documents each of these forecasts. The worst case model forecasts a 30% chance for one or more leaks in 23 repair welds occurring any time within the next ten years of service. For the same event, our best estimate forecasts a 7% chance of at least one leak.

For the 29 shop welds, the chances of a leak in the next ten years are 3% (best estimate) and 6% (worst case). The shop weld forecasts are more optimistic than those of the repair welds. This is mainly because of the difference in initial flaw distributions used.

Figure 14 also documents the chance of at least one 0.15-inch or deeper crack occurring. We choose this depth for the plot because it represents a detectable level of growth from flaws believed to exist today. Also, it discriminates clearly between our worst case and best estimate repair weld models. For example, the models start to deviate at five years of exposure and this deviation increases dramatically with age. This suggests that a five- to ten-year reinspection interval has a good chance of telling us which model is "right".

**Recommended Reinspection Interval.** The next reinspection interval should meet several criteria. Primarily, it should not pose an unacceptable safety risk. Since leakage was shown to precede rupture and signal the operator, we assume safety is not an issue.

Next, it should not pose an unacceptable economic risk. A complete economic analysis would weight the present value of the costs of each candidate inspection against its estimated benefits. The benefits are to reduce future cracking and leak probabilities. No economic evaluation was performed.

With safety not an issue and with no economic data, we can still limit the future failure rate intuitively. Our worst-case analysis is very conservative so any bad event for which it forecasts less than 50% chance probably will not happen. Even our so called best estimates for repair welds have conservatism built in. Our intuition is that best estimate leak probabilities of 10% for the entire weld population may be acceptable.

Using a 50% worst case chance of a leak and 10% best estimate as a guideline, any inspection interval up to 15 years would be acceptable. To err conservatively for this critical recommendation, we suggested an inspection interval of five to ten years.

As discussed above, an inspection ten years from now also would have a reasonable chance of discriminating between our worst case and best estimate crack growth forecasts.

## Acknowledgements

The work described in this paper required expertise in a number of disciplines. The efforts and expertise of the following individuals are acknowledged and appreciated: P. M. Besuner, M. J. Cohn, M. T. Cronin, S. Cunningham, G. R. Egan, K. Hayes, R. Smith, D. Vandergriff, R. Viswanathan, and J. J. Yavelak

## References

1 - American Welding Society, Welding Handbook, Volume 4, 7th Edition, "Metals and Their Weldability" Pp. 102-106 (1982).

2 - Devarajan, N., M. K. Asundi, and S. Somasundoram, "Experimental Method for Predicting Residual Stresses Due to Turning Stainless Steel," Experimental Techniques Pp. 22-26 (August 1984).

3 - Takashi, D., et. al., "Operating Experiences and Reliability Evaluation on Main Steam Line Pressure Parts of Ultra Super Critical (USC) Power Plant," Technical Review, Mitsubishi Heavy Industries, Ltd. (February 1985).

4 - Rosenthal, D. and J. T. Norton, "A Method of Measuring Tri-Axial Residual Stresses in Plates," Japanese American Welding Society Welding Supplement, Volume 24 (1945).

5 - Domis, W. F., "Creep and Creep-Rupture Properties of Types 304N and 316N Stainless Steels," American Society of Mechanical Engineers 71-Pet-34.

6 - Kadlecek, P., "Mechanical Properties Data on Hot Extruded 304N and 316N Stainless Steel Pipe," American Society for Testing and Materials STP 522, Pp. 3-34 (1969).

7 - Cullen, T. M. and M. W. Davis, "Influence of Nitrogen on the Creep-Rupture Properties of Type 316 Steel," American Society for Testing and Materials STP 525, Pp. 60-78 (1969).

8 - Mills, W. J., "Effect of Loading Rate and Thermal Aging on the Fracture Toughness of Stainless Steel Alloys," Department of Energy, DE 87-014377, KEDL-SA-3474-FP (May 1986).

9 - Nikbin, K. M., D. J. Smith, and G. A. Webster, "An Engineering Approach to the Prediction of Creep Crack Growth," Transactions of ASME, Volume 108, Pp. 186-191 (April 1986).

10 - Majumdar, B. S., C. E. Jaske, and M. P. Manahan, "Creep-Crack Growth Characterization of Type 316 Stainless Steel Using Miniature Specimens," Accepted for publication in the International Journal of Fracture Mechanics.

11 - Easterling, K., Introduction to the Physical Metallurgy of Welding, Butterworths (1983).

12 - Walker, S., Personal Communication with S. Paterson (November 1990).

13 - ASM Metals Handbook, 9th Edition, Volume 17 Nondestructive Evaluation and Quality Control, ASM International, P. 682 (1989).

*Proceedings of the Fifth International Conference on Creep of Materials, Lake Buena Vista, Florida, USA, 18-21 May 1992*

# Oxidation Cracking and Residual Creep Life of an Incoloy 800H Bottom Manifold in a Steam Reformer at 800 °C

**R.J. Gommans and K.F. Verheesen**
Materials and Corrosion Department
DSM Research
Geleen, The Netherlands

**J.H. Heerings**
TNO Industrial Research
Materials Technology
Apeldoorn, The Netherlands

ABSTRACT

Synthesis gas for ammonia production is made in a steam reformer furnace. The hydrogen-rich gas leaves the reformer through bottom manifolds, which operate at about 800°C and at an internal pressure of 3.1 MPa. The material of the bottom manifold is wrought Incoloy 800H. During a planned shut-down in 1990 it appeared that the bottom manifold parts had undergone diametrical expansion of up to 2% due to creep. Further, cracking at the outer diameter was found. It was decided to replace these parts.

Microscopical investigations showed that the cracking could not be caused by creep. It was found that the cracking was confined to a 4 mm deep coarse-grained zone (ASTM 0-1) at the outer diameter. The cracking appeared to be caused by strain-induced intergranular oxidation. When the cracks reached the fine-grained material the oxidation-cracks stopped, so that these cracks are -in fact- do not limit the residual life of the manifold.

In order to determine the residual creep life of the sound (non-cracked) bottom manifold material iso-stress creep tests were performed. It was found that tertiary creep started at 7% strain. The time-to-rupture was more than 100,000 hrs. The 7%-criterion will be used as end-of-life criterion.

It was concluded that the bottom manifold (and thus the furnace) could be used safely during the foreseen production period (2.5 years). In the next planned shut-down investigations will be performed to ensure further safe operation.

**Note** : "INCOLOY" is a tradename of the International Nickel Company

## INTRODUCTION

DSM Fertilizers operates several production units in The Netherlands. One of these units is a 1000 short ton/day ammonia plant designed by M.W.Kellogg and is situated in IJmuiden (NL).

Synthesis gas for ammonia manufacture is made in a top-fired furnace by reforming a steam/methane gas mixture over a nickel catalyst. The pre-heated gas mixture enters spun cast catalyst tubes (typically made of HK40 or HP40+Nb) and flows down through each tube and is collected in a bottom manifold. The partially reformed gases contain hydrogen, carbon monoxide, carbon dioxide, methane and steam. Then the gases flow towards the riser (typically made of HP40+Nb or Supertherm), and from there into a transfer line and on to the secondary reformer. Such a primary reformer is top-fired and contains 7 or 8 rows of 42 catalyst tubes each. A sketch of the furnace arrangement is shown in figure 1. Each catalyst tube and riser is joined to the manifold tubing with forged "weldolets". The manifold under discussion is made of Incoloy 800H; the metal temperature is about 800 °C (1475°F) at an internal gas pressure of 3.1 MPa (450 psi).

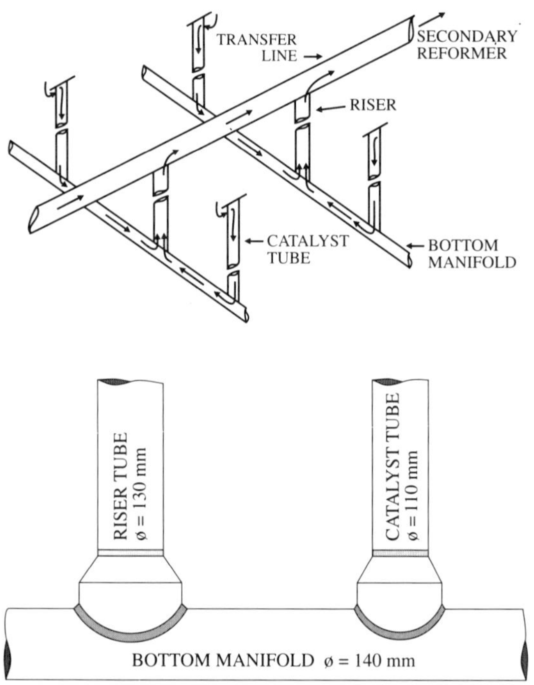


*Figure 1 – Schematic of primary reformer arrangement.*

The plant was taken into operation in 1969. In 1976 the bottom manifold showed severe bulging (up to 20% diametrical expansion) and creep damage. The cause of the damage was the fine grain-size (ASTM 6-7) of the Incoloy 800 Grade II used at the time. The manifolds were replaced in Incoloy 800H and taken into operation in 1977 [1]. From then on the manifolds were regularly inspected.

During a scheduled shut-down in 1987 a diametrical expansion of 1.1% was measured at one location. This could not be explained by the creep rate data. It was concluded that external stresses such as bending of the manifold caused this accelerated diametrical expansion. But finite element calculations [2] showed that external stresses could not occur if the spring hangers of the furnace were properly adjusted. Since this was the case, external stresses had to be ruled out.

During a planned shut-down in 1990 it appeared that the bottom manifold had undergone diametrical expansion of up to 2% due to creep. This expansion was uniform in diameter (no bulging), but locally extended over the full length of the manifold. Further, cracking at the outer diameter was found in axial direction at locations with a diametrical expansion of higher than 1%. Microstructural investigations using the replica technique indicated that creep was responsible for the cracking. It was decided to replace the parts with a diametrical expansion of more than 1%. After that, the reformer was taken into operation.

The bottom manifold parts that were taken out because of the cracking were used for extensive investigations on the following points :

- cause of the cracks in the manifold and weldolets,
- growth-rate of these cracks,
- maximum allowable deformation of aged Incoloy 800H,
- time to maximum allowable deformation.

## RESULTS OF THE INVESTIGATIONS

MATERIAL QUALITY The results of the chemical analysis of the material from the bottom manifold (see table 1) showed that it meets the requirements of Incoloy 800H. The hardness was 135 HV, indicating that no $\gamma$-fase had formed. The grain size in the bulk of the material was ASTM 4. However, in a zone of about 4.0 mm depth from the outside diameter a grain size of ASTM 0-1 was found. Thus, it was concluded that the material meets the requirements of Incoloy 800H.

Table 1. Chemical analysis of Incoloy 800H bottom manifold (in weight-%).
A. requirements – B. certificate – C. analysis [9]

| nr. | C | Ni | Cr | Al | Ti | Mn | Si | Cu |
|---|---|---|---|---|---|---|---|---|
| A | 0.05-0.1 | 30-35 | 19-23 | 0.15-0.6 | 0.15-0.6 | <1.5 | <1.0 | <0.75 |
| B | 0.07 | 31.9 | 20.0 | 0.40 | 0.40 | 0.93 | 0.12 | 0.69 |
| C | 0.062 | 31.8 | 20.1 | 0.41 | 0.35 | 0.93 | 0.15 | 0.70 |

DIMENSION MEASUREMENTS The following dimensions were measured: the outside diameter (OD), the circumference, and the wall thickness; see figure 2. These dimensions were taken every 50 mm (2") along the length of the removed parts.

The diameter measurements showed that the OD varied along the length of the manifold. The maximum diamtrical expansion measured was 2.1% at manifold F and 1.8% at manifold G.

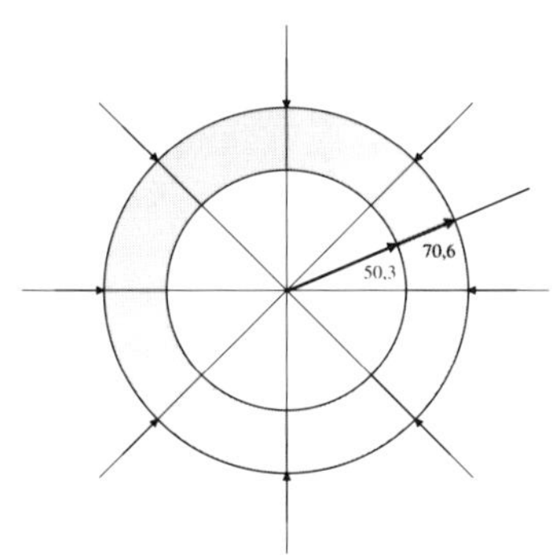


*Figure 2 – Sketch of the dimension measurements made.*
- *outer diameter: in 4 directions*
- *wall thickness: in 4 direction*
- *circumference*

From comparison of the measurements of the outside diameter and the wall thickness with the circumference, it was concluded that the diametrical expansion of the tube was uniform in circumference.

The wall thickness varied along the circumference; but it was always above the minimum required wall thickness of 20.3 mm (0.8").

LIQUID PENETRANT TESTING (dye-check) The cracks observed in the manifold were oriented in axial direction at locations with a diameter expansion of more than 1%. The length of these cracks was about 5-25 mm (0.2-1.0"); the cracks were not linked to each other.

MICROSCOPIC INVESTIGATIONS On the basis of the dimension measurements about 20 rings were taken out and prepared -as complete rings- for microscopic investigations. It was found that the crack depth was almost uniform (about 4.0 mm from the outside diameter) and that the cracks were confined to the coarse-grained zone (ASTM 0-1); see photograph 1. Penetration into the fine-grained zone (ASTM 4) was not observed. The difference between the coarse- and fine-grained zones is quite remarkable; see photograph 2. The cracks were oxidized and nitride formation had occurred along the sides of the cracks; see photograph 3. The oxide thickness along the crack diminished towards the crack tip. Nitride formation takes place as oxygen depletion in the crack occurs. Creep voids in the close vicinity of the crack and the crack tip were not observed.

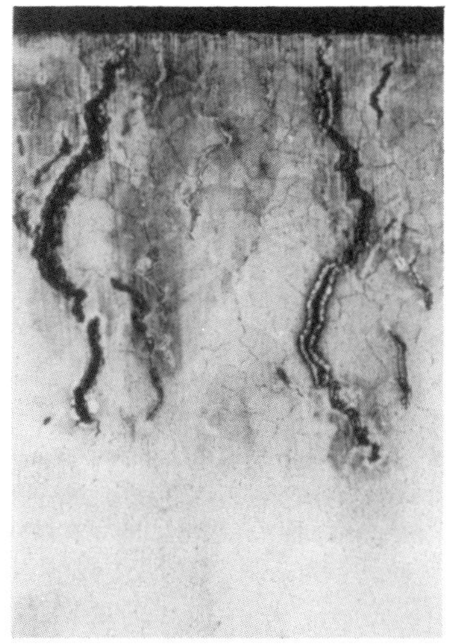

*Photograph 1 – Uniform crack depth of about 4 mm (1½") along the outer diameter of the manifold tube. Magnification = 8×*

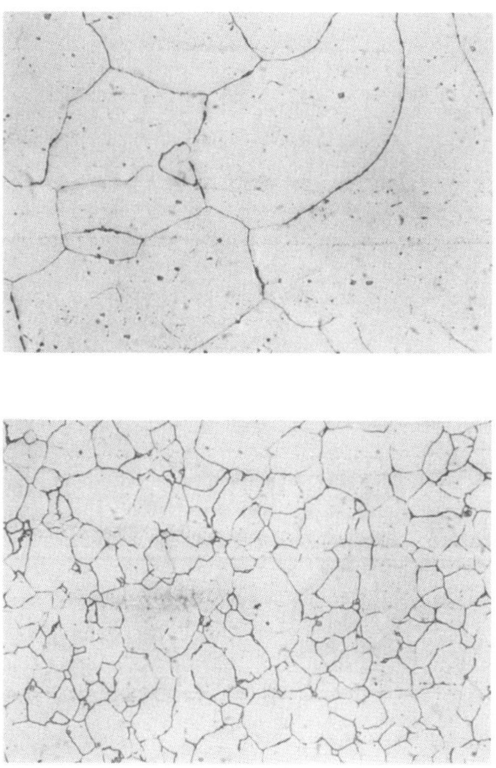

*Photograph 2 – Difference in grain size of the coarse- and fine-grained zones. Magnification = 50×*

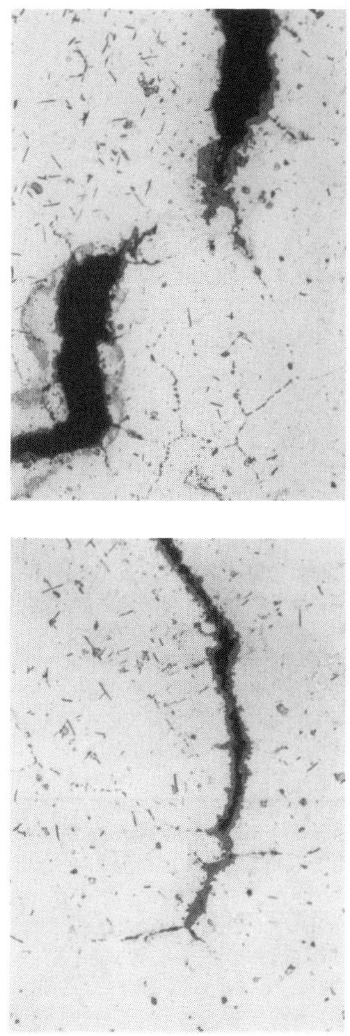

*Photograph 3 – Oxidation and nitride formation along the crack sides and the crack tip. Magnification = 100×*

ACCELERATED CREEP TESTING The iso-stress method is used for the accelerated creep testing. Samples from the manifold were taken in tangential direction, i.e. parallel to the hoop stress. This was done to create a test situation as close as possible to the real situation. The material taken from the manifold had undergone a diametrical expansion of 1.75% during service. A stress of 12 MPa was chosen, which is 20% higher than the hoop stress. Temperatures of 900 to 1015°C were used during the accelerated creep testing.

The test results [3] are given in figure 3. From extrapolation to the service temperature of 800 °C (solid line) it can be concluded that creep rupture occurs after more than 300,000 hrs.

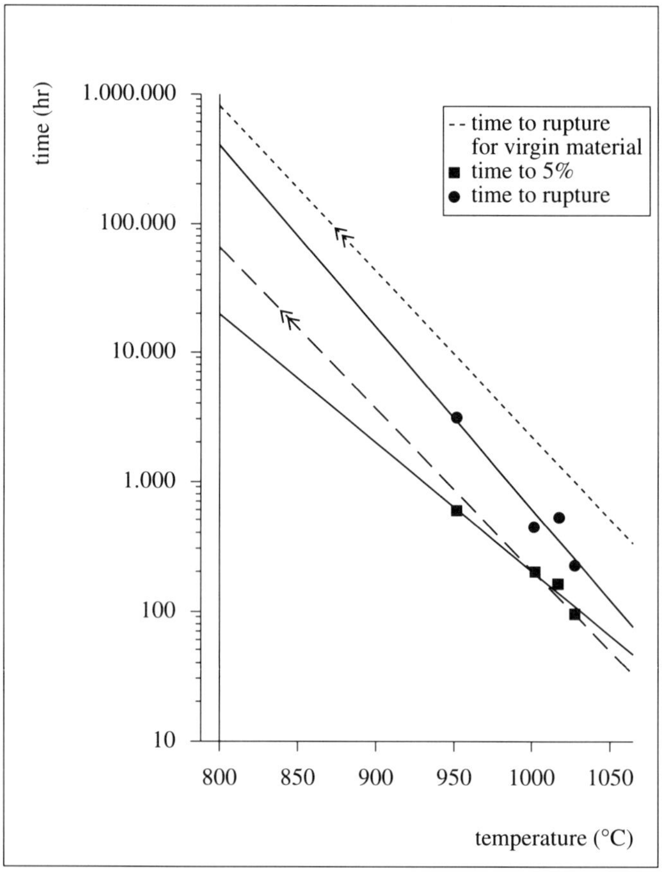


*Figure 3 – Results of the accelerated creep testing.*
*The itme to rupture and the time to 5% additional strain is plotted as a funciton of temperature.*

In the tests it was observed that tertiary creep starts at 5% elongation. Because the material had already undergone about 2% elongation during service, tertiary creep started at about 7% total elongation. The time to reach the additional 5% elongation (thus tertiary creep) can be estimated by extrapolation of the test results to the service temperature (solid line): 20,000 hrs. This is low compared to the time-to-rupture. Furthermore, it is observed that the line for the time to reach 5% is not parallel to the line for rupture. Normally, it is observed that these lines run parallel to each other. If a "parallel extrapolation" is made (dashed line), then the time to reach 5% elongation is 60,000 hrs. It was concluded that the time to reach tertiary creep is between 20,000 and 60,000 hrs. In our view, this will be on the conservative side because of the upperbound stress value chosen.

The secondary creep rate at the service temperature can be obtained by extrapolation from the iso-stress testing results at higher temperatures; see figure 4. At 800°C a creep rate of $2.5 \cdot 10^{-4}$ %/hr or $6.9 \cdot 10^{-10}$ 1/s is obtained.

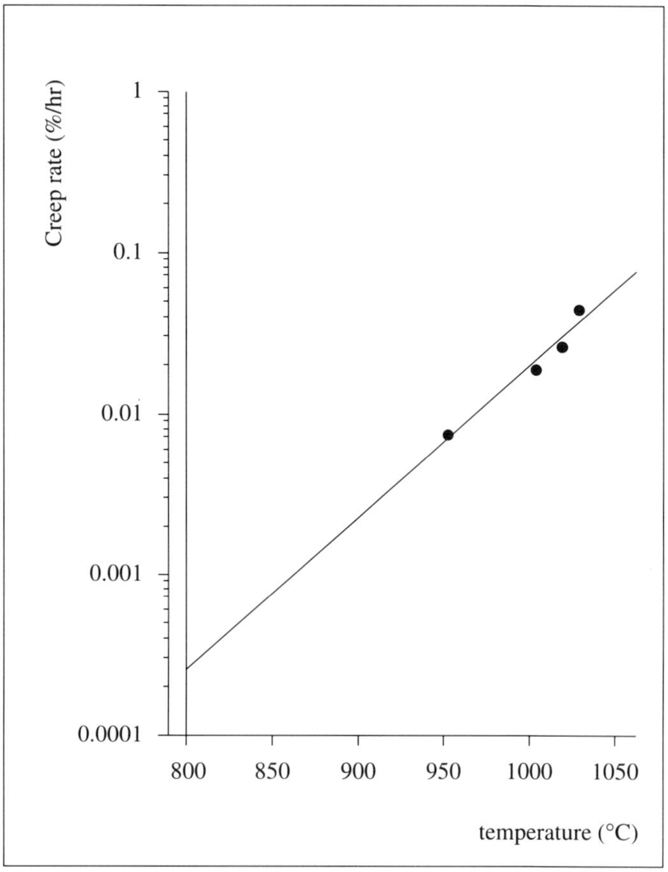


*Figure 4 – Results of the accelerated creep testing.*
*The secondary strain rate is plotted as a function of temperature.*

## DISCUSSIONS ON THE CAUSE OF THE CRACKS

COARSE-GRAINED ZONE The replaced parts of the manifold were made of materials from one heat number. The same heat number was available on stock, and was used for the replacement. From microscopic examinations it was observed that the same coarse-grained zone (without cracks) existed in the manifold which had been on stock for 13 years. Therefore, it must be concluded that the existence of the coarse-grained zone is due to the manufacturing method of the manifold (hot extrusion).

REPLICA TECHNIQUE The orientation of the oxidation cracks is in axial direction (thus similar to creep cracks). Using the replica technique in situ, the cracks were thought to be creep cracks. This confirms that information on the metal surface only is not sufficient to draw firm conclusions; information on the bulk metal is needed. This was obtained by taking boat samples.

CRACKS IN THE MANIFOLD From the microscopic investigations it was concluded that the cracking in the coarse-grained zone was caused by intergranular oxidation. It is postulated that the diffusion pathways for chromium diffusion in the coarse grains to the grain boundaries is too long. Therefore, there are low chromium contents at the grain boundaries of the coarse grains causing preferential attack. In the fine grains, the chromium level can be maintained at an acceptable level because of the shorter diffusion pathways.

Crack formation due to creep is not observed. Creep crack formation is not likely to occur, since the hoop stress at the inside diameter is much higher than at the outside diameter in such a thick-walled tube (9.5 compared to 6.5 MPa). Furthermore, the creep strength of the coarse-grained material at the outside diameter is higher than the creep strength of the fine-grained bulk material.

The oxidation cracks in the manifold occur only in the coarse-grained zone (ASTM 0) at the outer diameter of the manifold. They do not penetrate into the fine-grained zone (ASTM 4-5) of the bulk material.

As already mentioned, the cracks are oriented in axial direction. This is due to the hoop stress caused by the internal pressure. Because the oxidation cracks only occur at diametrical expansions of 1% or more, this type of oxidation could be called "strain-induced intergranular oxidation".

The cracks did not occur over the whole circumference. They only occurred on locations where the wall thickness (due to manufacturing) was the lowest.

## DISCUSSIONS ON THE RESIDUAL LIFE

OXIDATION The oxidation cracks do not penetrate the fine-grained zone of the manifold. Because of the observation that these cracks must have been present for a longer period, it was concluded that the cracks growth rate (if any) is very low. Thus, they will most probably not penetrate the fine-grained zone in the future. The oxidation cracks, therefore, will not affect the residual life of the fine-grained bulk material of the manifold.

NOTCH DUCTILITY An aspect which has not been discussed yet is the influence of notches (the oxidation cracks) on the residual creep life of the manifold. From literature, it is known that creep life is not impaired by notches: Incoloy 800H is a notch-ductile material [4].

DEFORMATION MECHANISM In the past it was thought that the acting deformation mechanism at 800°C was power-law creep. But, if the creep rate is plotted in relation to stress, it can be observed that the creep rate in service is a factor 350 higher than the creep rate obtained by extrapolation from a Larson-Miller data-set; see figure 5. This can be explained by the following. The creep rate of the data set is obtained from creep tests with relatively high stresses ($\sigma > 40$ MPa). The deformation mechanism at these relatively high stresses is power-law creep by climb. At low stresses however, a transition to diffusional creep must occur [5]. In [6] it is found that below 30 MPa diffusional creep is the acting deformation mechanism for Incoloy 800H at 800°C; see figure 6. By extrapolation from figure 6 it can be deducted that the creep rate is $5.0 \cdot 10^{-11}$ 1/s at the service stress of 10 MPa. This correlates very well with the actual creep rate during service: 2.1% in 13 years means a creep rate of $5.5 \cdot 10^{-11}$ 1/s.

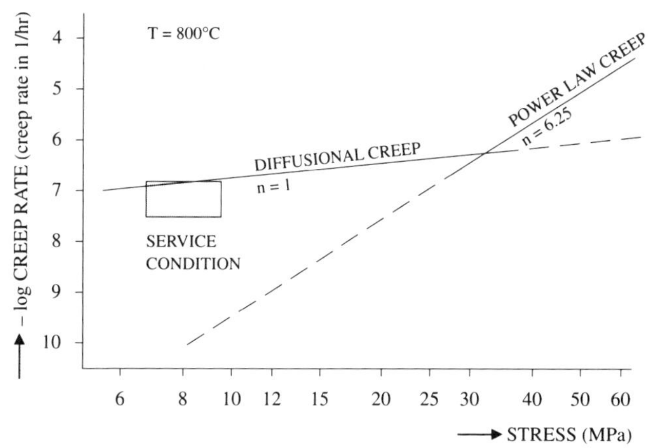


*Figure 5 – Creep rate of Incoloy 800H as a function of stress at a temperature of 800°C.*

*Data for power-law creep are taken from manufacturers (Inco, Sandvik) and from free literature. These data are put into a Larson-Miller equation by regression analysis.*
*Data for diffusional creep are taken from [6]. The creep rate in service is 2.1% in 13 years which equals $5.1 \cdot 10^{-11}$ 1/s or $1.8 \cdot 1^{-7}$ 1/hr.*

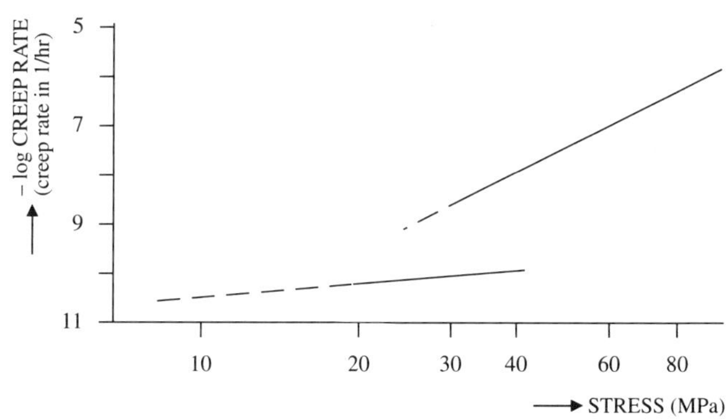


*Figure 6 – Creep rate as a function of stress [6].*

GRAIN SIZE With diffusional creep as deformation mechanism grain-size has a large influence on creep life. This already appeared in the damage case studied in 1977. In [1] a relation is given between the ASTM grain-size and the percentage of creep deformation in a given time (52,500 hrs). This can be re-written to a relation between creep rate and grain-size:

$$\log(\text{creep rate}) = 4.25 * \log(\text{ASTM grain-size number}) - 8.96 \quad (1)$$

The creep rate is in 1/s; the relation is valid for grain-sizes of ASTM 2-7. Because the ratio of service temperature to melting temperature is about 0.65, diffusion creep limited by grain boundary diffusion (Coble creep) is the most probable deformation mechanism for Incoloy 800H at 800°C at low stresses. This is in conformity with observations by Steen; see figure 7 [7].

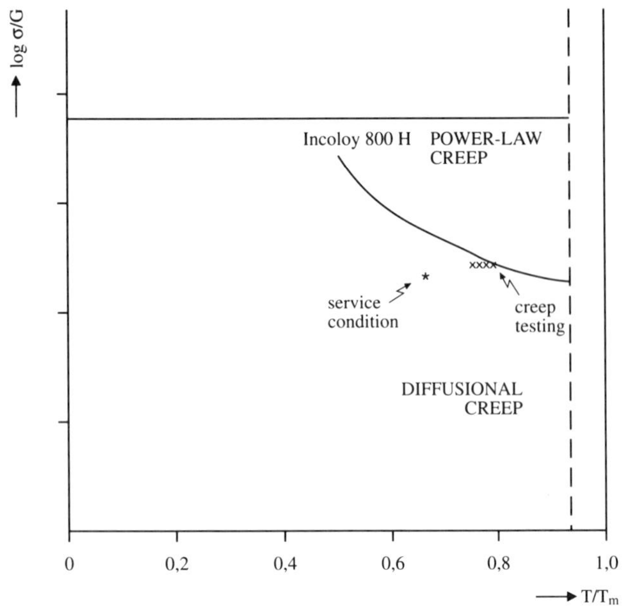


*Figure 7 – Deformation mechnaism map for Incoloy 800H [7]. The deformation mechanism is given as a function of homologue temperature (T/Tm) and homologue stress (σ/G).*

DEFORMATION CAPABILITY The deformation capability of aged materials is usually lower than for new materials. In the accelerated creep tests on the aged material it was observed that the tertiary creep regime starts at 7% elongation. This observation has been made earlier [8]. From the damage case in 1977 it is known that 20% diametrical expansion occurred without leakage or rupture. From microscopic investigations on the same damage case in 1977 it was found that void formation was only observed at 8% diametrical expansion or more. Therefore, the transition from secondary to tertiary creep was taken as criterion for rejection. In this case this means 7% diametrical expansion.

RESIDUAL LIFE In the above paragraphs it is stated that only creep deformation will determine the residual life of the manifold. The maximum allowable creep deformation is 7%.

From the accelerated creep tests it appears that the time to reach 7% diametrical expansion is between 20,000 and 60,000 hrs.

The extrapolated creep rate at 800°C is $2.5 \cdot 10^{-4}$ %/hr or $6.9 \cdot 10^{-10}$ 1/s; see figure 4. The creep rate obtained from the tests is therefore remarkably higher than the creep rate during service. This might be explained by the upperbound stress value chosen.

The data of the 1977 damage case [1] can also give an indication of the creep rate. Because ASTM grain-size 4 is the highest available grain size for the manifolds in the reformer, according to Eq.1 the maximum creep rate is $1.1 \cdot 10^{-10}$ 1/s.

In figure 8 the creep deformation of the manifold is given as a function of time. It can be seen that the creep rate given by the iso-stress creep testing is much higher than the creep rates during service and from the 1977 data. The actual creep rate will lie somewhere in between $9 \cdot 10^{-11}$ and $7 \cdot 10^{-10}$ 1/s.

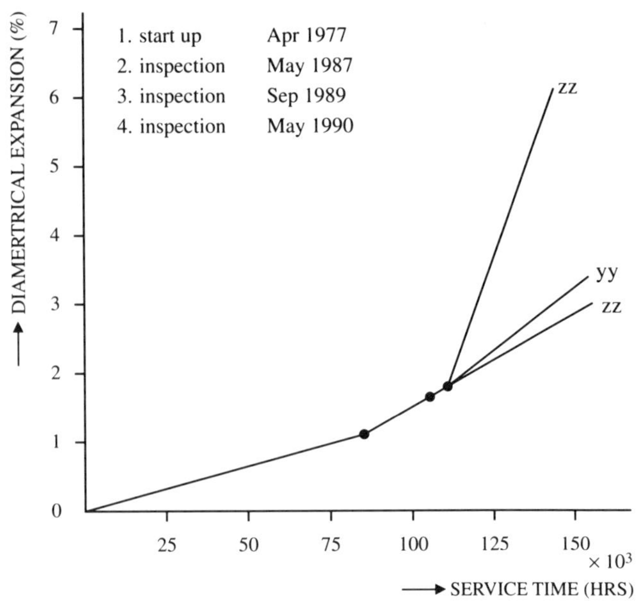


*Figure 8 – Creep deformation (diametrical expansion) of the bottom manifold as a function of service time.*

*line xx : momentaneous creep rate* $8.6 \cdot 10^{-11}$ *1/s*
*line yy : creep rate estimated from the 1977 damage case* $1.1 \cdot 10^{-10}$
*line zz : creep rate determined from the creep testing* $6.9 \cdot 10^{-10}$

The most conservative estimate for the time to reach 7% elongation will be 20,000 hrs (line zz). But, as can be seen in figure 8, this is quite unlikely to happen, because the extrapolation of the creep rate during service (line xx) yields much longer times to reach 7% elongation. Furthermore, the creep rate data of the 1977 damage case (line yy) yield similar estimates.

As indicated above, there is an uncertainty in the estimate for the creep rate, and thus also for the residual life. However, a better estimate could not be made at the moment. The diameter measurements during the next shut-down will give additional information for a more accurate determination of the actual creep rate. After that, a more reliable residual life estimate can be performed.

## INSPECTIONS DURING THE NEXT SHUT-DOWN

During the next scheduled shut-down of the plant in 1993 the manifolds will be examined thourougly. The following inspection points are planned :

a. measurements of diameter, wall thickness, and circumference,
b. boat samples of the replaced parts, which also had the coarse-grained zone at the outer diameter,
c. dye check (liquid penetrant test) of all heat-numbers to inspect for the occurrence of oxidation cracks,
d. if necessary, boat samples depending on the results of point c.

After completion of these investigations a better estimate can be given concerning the residual life of the manifold.

## REPLACEMENT OR NEW CONSTRUCTION

If the manifold has to be replaced (for whatever reason) Incoloy 800H will not be chosen for the new manifolds. Centrifugally cast alloys will be used with special welding techniques.

## CONCLUSIONS

* The cracks in the manifold are due to strain-assisted intergranular oxidation. This oxidation is strongly influenced by grain size.
* For this type of materials in furnaces, the replica technique can give misleading information. Therefore, taking boat samples is necessary.
* It had been concluded that tertiary creep starts at 7% elongation. This will be used as rejection criterion for the bottom manifolds in the future.
* Diffusional creep limited by grain-boundary diffusion (Coble creep) is the acting deformation mechanism for Incoloy 800H at 800°C and low stresses (typically $\sigma < 30$ MPa).
* The creep rate of the aged material is between $5 \cdot 10^{-11}$ and $7 \cdot 10^{-10}$ 1/s. The residual life, in this case defined as the time to reach 7% elongation, is between 20,000 and 300,000 hrs. From diameter measurements in the next shut-down, a better estimate of the creep rate and residual life can be made.
* During the next shut-down boat samples and liquid penetrant testing will be performed to search for the occurrence of oxidation cracking.
* For new construction or replacement, centrifugally cast materials will be used.


## ACKNOWLEDGEMENT

The authors would like to thank their colleagues both at DSM Research and at TNO for their technical assistance. The permission to publish the results of the investigations by DSM Fertilizers IJmuiden and DSM Research is gratefully acknowledged.


## REFERENCES


[1]. C.A. van Grieken, internal report, DSM Fertilizers IJmuiden (1977)

[2]. J. Prij, J. Broerse, ECN-report for DSM Fertilizers IJmuiden (1990)

[3]. J. Heerings, TNO-report for DSM Fertilizers IJmuiden (1991)

[4]. Battelle-Institute, "Materials for steam reformer furnaces" (1973)

[5]. H.J. Frost, M.F. Ashby, "Deformation-mechanisms maps", Pergamon Press (1982)

[6]. M.R. Diglio, H.Straube, K.Spiradek, H.P. Degischer, in: Proc. of COST-Conf. Liège on "High-temperature materials for power engineering (1990) p.545

[7]. M. Steen, Ghent State University, doctoral thesis (1986)

[8]. J. Heerings, TNO Industrial Research, unpublished research (1990).

[9]. K.D. Barr, INCO-report for DSM Research (1990)

*Proceedings of the Fifth International Conference on Creep of Materials, Lake Buena Vista, Florida, USA, 18-21 May 1992*

# Creep Analysis of Header Using Typical Load Cycles

**H. Koikegami, H. Kisamori, and S. Urabe**
Ishikawajima-Harima Heavy Industries Company, Ltd.
Yokohama, Japan

**K. Hada**
Japan Atomic Energy Research Institute
Ibaraki, Japan

## Abstract

This paper describes a procedure of inelastic creep analysis of manifold header using a three-dimensional FEM (Finite Element Method) model. Inelastic creep analyses at elevated temperature have been carried out on the manifold header of IHX (Intermediate Heat eXchanger) for HTTR (High Temperature engineering Test Reactor). The temperature of the manifold header is expected to be approximately 900°C during the steady state operation and the geometry of the header is complicated. It is difficult to meet the elevated temperature structural design criteria based on the simplified elastic analysis regarding creep damage evaluation. In this study, the reasonable creep damage was predicted using typical load cycles.

INELASTIC CREEP ANALYSES at elevated temperature have been carried out on the manifold header of IHX (Intermediate Heat eXchanger) for HTTR (High Temperature engineering Test Reactor), being constructed by Japan Atomic Energy Research Institute (JAERI). The temperature of the manifold header is expected to be approximately 900°C during the steady state operation. Furthermore, as the manifold header has a complicated geometry, it is one of the severest part of the IHX in terms of creep damage. Though primary stresses are limited within the lower level, it is difficult to meet the elevated temperature structural design criteria based on the simplified elastic analysis, especially the creep-fatigue damage criteria and limitation of the accumulated inelastic strain. To estimate the creep damage and the creep strain reasonably, more detailed creep analysis approach, which includes the stress relaxation and effects of load histories, is required. This approach has three analytical steps to calculate the creep damage at the evaluation points.

(A) In order to determine the load cycles for creep analysis, elastic thermal stress analyses are performed for several thermal transients. Then based on the effective transients concerning creep damage, the load cycles are obtained.

(B) Creep analyses are performed for load cycles obtained in previous step, using a practical three-dimensional FEM model. Three-dimensional model gives more detailed stresses and creep strains.

(C) Using element stresses and creep strains, the stresses and creep strains at evaluation points are extrapolated in space.

Moreover, creep damage and creep strains during service life are assumed by extrapolation method for time.

## Analytical Modelling

The IHX is a helium-helium heat exchanger operated at a very high temperature above 900°C. It transports the nuclear heat energy of primary helium gas to secondary helium gas, which is to be used in heat utilization system in future. The IHX has to meet the requirement on structural integrity as class 1 component of the pressure boundary during the service life, so it is one of the most important components in HTTR. The configuration of the IHX for HTTR is shown in Figure 1.

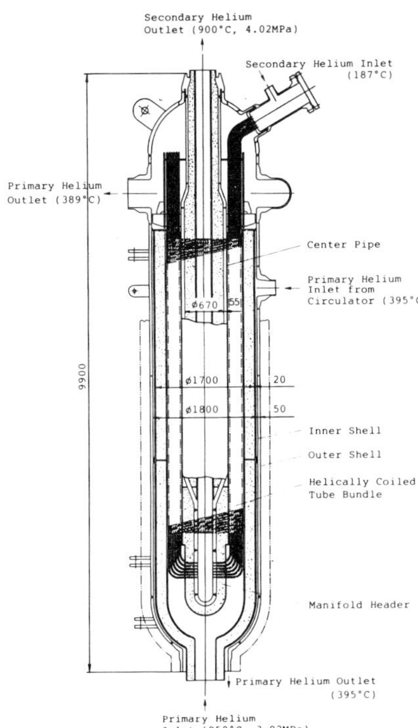


Fig. 1 Configuration of the IHX for HTTR

The geometry of manifold header is shown in Figure 2. Outer surface of manifold header is covered with thermal insulation to reduce the thermal stress and metal temperature during the steady state operation. Manifold header has the perforated region. The hole arrangement is uniform in the circumferential direction. The material of manifold header is nickel based Hastelloy alloy XR with high resistance for high temperature and corrosion in helium atmosphere, which is developed by JAERI.

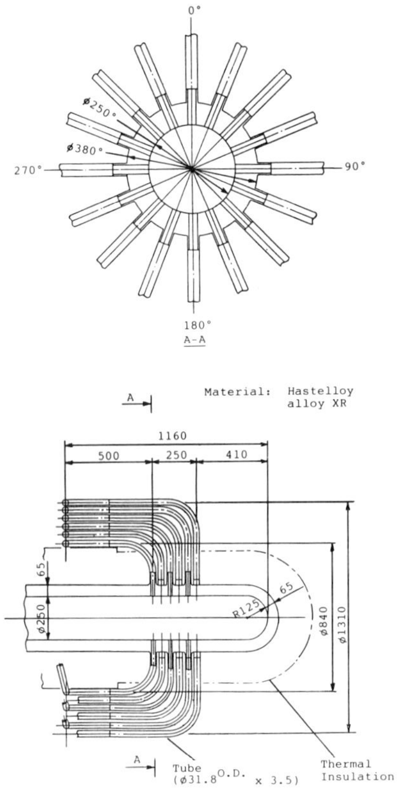


Fig. 2 Geometry of manifold header

**Finite Element Model.** Today, the elastic design and analysis of perforated structure such as manifold header (e.g. tubesheets) have progressed to a fair degree of refinement based on equivalent solid plate concept, which is described in the ASME Boiler & Pressure Vessel Code Section III A-8000. Furthermore, the concept has been extending to elevated temperature region using a two-dimensional finite element approach with inelastic analysis (1, 2).

On the other hand, in this paper a practical three-dimensional finite element model on manifold header, which requires the severest integrity, is applied in order to predict more detailed stresses and inelastic strains. The model image and evaluation points of manifold header are shown in Figure 3. Its model is a 1/32 section model considering symmetrical hole arrangement, and a hemispherical head connected with manifold header was treated as an equivalent cylinder in terms of heat capacity and weight. The finite element model consists of 800 elements (8-node three-dimensional linear isoprametric element) and 1275 nodes. The analysis was performed with MARC program.

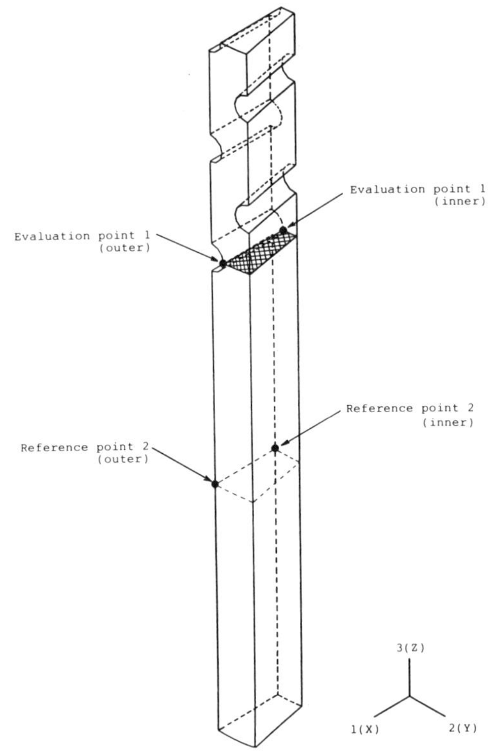


Fig. 3 Model image and evaluation points

**Constitutive Equation for Creep Analysis.** An equation-of-state approach for creep analysis was applied in these analyses. A uniaxial creep law of Hastelloy alloy XR based on Garofalo type equation, which was proposed by JAERI, is used.

$$\varepsilon_c = C \cdot \left(1 - e^{-rt}\right) + \dot{\varepsilon}_s \cdot t \tag{1}$$

where: $\varepsilon_c$ = Creep strain
$\dot{\varepsilon}_s$ = Steady creep strain rate
t = Time
C, r = Function of stress and temperature

The flow rule for multiaxial stresses is based on colinearity of creep strain rate and deviatoric stress tensors. For variable stress and temperature conditions, creep strains are based on the strain-hardening law. Furthermore, for load reversals the auxiliary procedures proposed by Oak Ridge National Laboratory (ORNL) in the U.S.A. were considered (3).

**Boundary Conditions.** Inner and outer surface of manifold header and inner surface of holes are considered as thermal boundary conditions. Concerning outer surface of manifold header, the thermal insulation which covered manifold was treated as equivalent heat transfer coefficient (i.e. equivalent thermal resistance). Outer surface of manifold header is exposed to 3.92 MPa external pressure, on the other hand inner surfaces of manifold header and holes are exposed to 4.02 MPa internal pressure. And dead weight is considered in the analyses.

## Results and Discussion

In case of structural integrity evaluation based on elastic analysis approach, it is relatively easy to calculate the stresses under loads combined complexly by combination of stresses under each load. On the other hand, the stresses are not able to be combined simply due to non-linearity of inelastic analysis, so it is necessary to consider whole load cycles. But it is impractical to perform the inelastic analyses for whole load cycles, so it is necessary to simplify the analyses considering conservatism.

**Elastic Analyses.** Figure 4 shows the temperature-time history during normal start-up, steady state and normal shutdown operation at evaluation point 1 and reference point 2 (see Figure 3). For normal operation (e.g. start-up, steady state and shutdown) the difference between the inner surface temperature response and the outer one is very small, and also the relative temperature difference between perforated region and solid region is small.

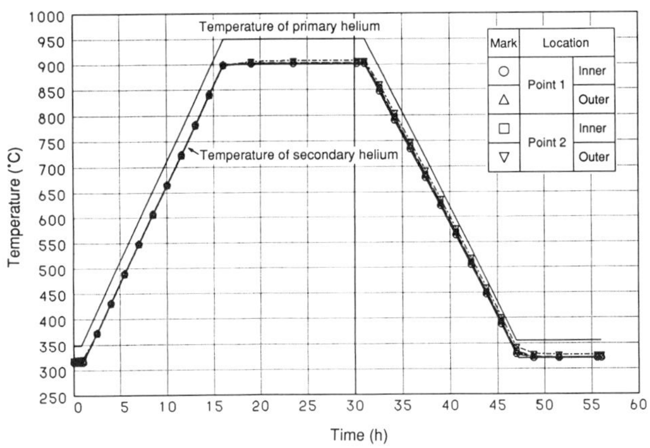


Fig. 4 Temperature-time history during normal operation

Though the temperature difference due to normal operation is small, the thermal stress is effective concerning creep damage. Figure 5 shows the relationship between the equivalent thermal stress and time at evaluation points, during the basic thermal cycle (i.e. start-up, steady state and shut-down). The stress due to basic thermal cycle is less than 15 MPa, but it is impossible to apply to the design guide with elastic analysis approach.

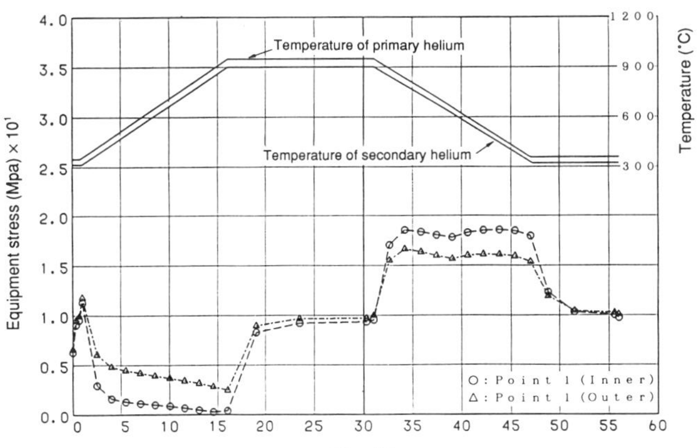


Fig. 5 Relationship between equivalent stress and time by elastic analysis

Figure 6 shows the equivalent stress distribution due to pressure loads of inner and outer surfaces. Combined stresses due to pressure loads and dead weight, i.e. primary stress, are about 0.4 MPa, which is equal to about 10 percent of maximum allowable stress intensity (So) in the structural design guide for class 1 component of HTTR proposed by JAERI (4, 5). Furthermore, it is found the creep damage due to only primary stresses is very small.

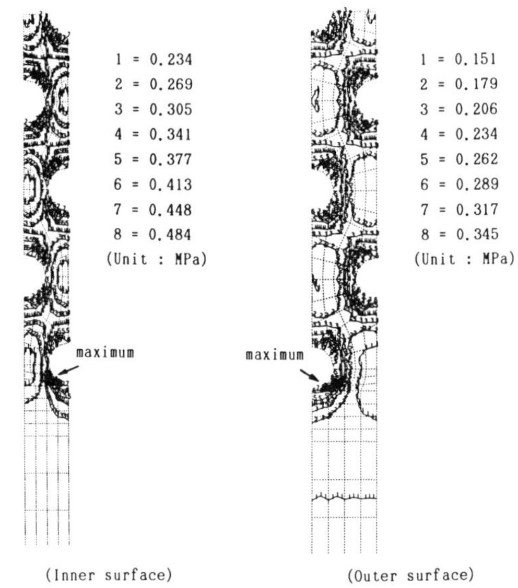


Fig. 6 Equivalent stress distribution due to pressure loads

An example of extrapolation and stress classification results for three-dimensional model is shown in Figure 7. Figure 7 shows the axial thermal stress distribution during steady-state condition. The extrapolation was performed using several elements near the evaluation point. The membrane stresses and linearized stresses were calculated using the forces and moments on the evaluation section shown in Figure 3.

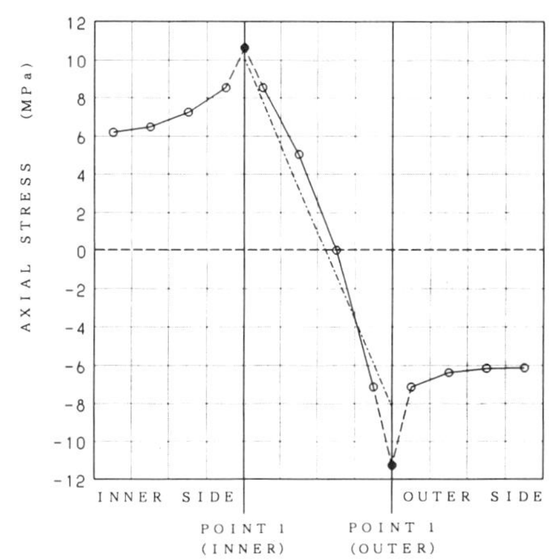


Fig. 7 Extrapolation and stress classification results at steady-state condition (axial component)

**Creep analyses.** Creep analyses of manifold header were carried out in order to evaluate the structural integrity in accordance with criteria limits such as accumulated inelastic strain and creep damage. The load histograms for creep analyses are shown in Figure 8. The histograms consist of three cycles. Each cycle consists of a start-up transient, a long period of steady state and shutdown transient. The holding time of steady state is 250 hours. It is operating time divided by the number of start-up cycles equally to be obtained initial stress conservatively. Moreover, creep analyses were carried out over 650°C so that significant creep strain was observed over this temperature.

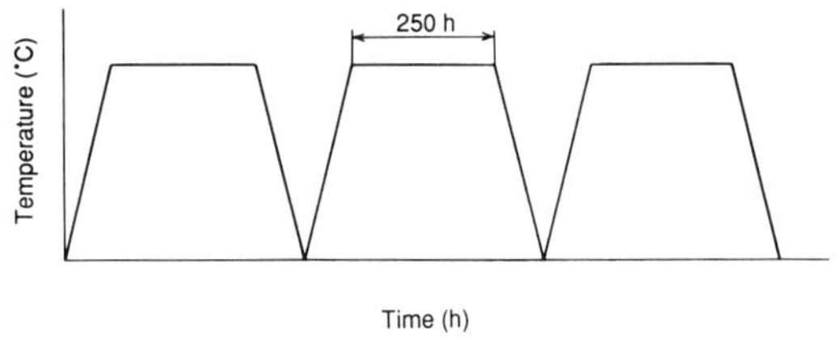


Fig. 8 Load histograms for creep analysis

Figure 9 shows the relationship between equivalent stress and time based on the histograms shown in Figure 8 at the evaluation points. The relationship between axial stress and time is shown in Figure 10. Axial stress is the most dominant component.

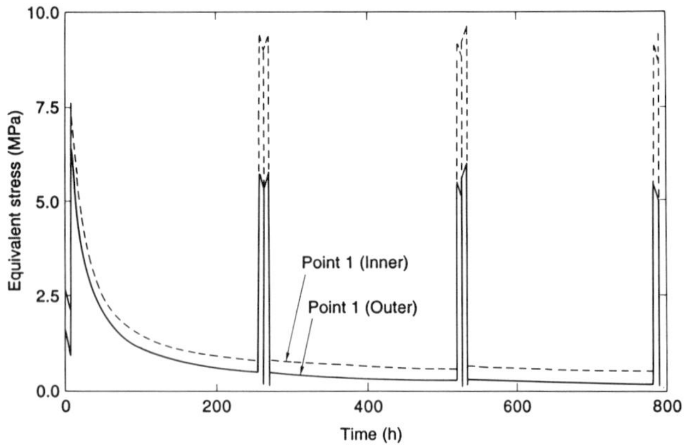


Fig. 9 Relationship between equivalent stress and time by creep analysis

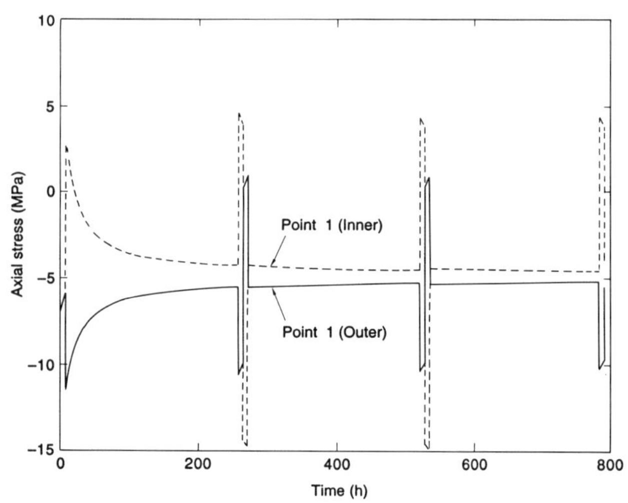


Fig. 10 Relationship between axial stress and time by creep analysis

It is found that the axial stress at evaluation point 1 (Inner) is negative value in the process of start-up operation, and then the stress changes into positive value at steady-state condition in Figure 10. The stress relaxes due to creep strain during steady state operation, and the axial stress grows to be larger than that of steady state in the process of shutdown operation. However, the stress obtained by creep analysis is smaller than that by elastic analysis shown in Figure 5. During restart-up operation, the large axial stress at evaluation point 1 (Inner) occurs due to stress reversal. But it is found the initial stress of second cycle grows to be smaller than that of first cycle, so it is assumed that the creep damage after second cycle is smaller and smaller.

Figure 11 and Figure 12 show the equivalent creep strain and the axial creep strain versus time respectively. Creep strains at evaluation points remarkably increase at the first cycle and the increment of creep strain at second or third cycle is smaller.

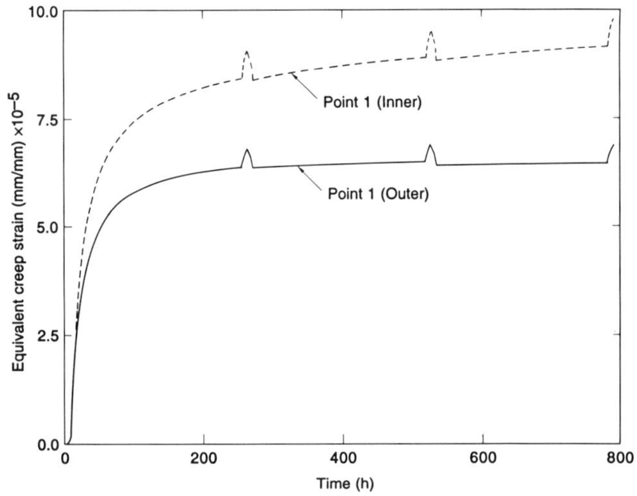


Fig. 11 Relationship between equivalent creep strain and time

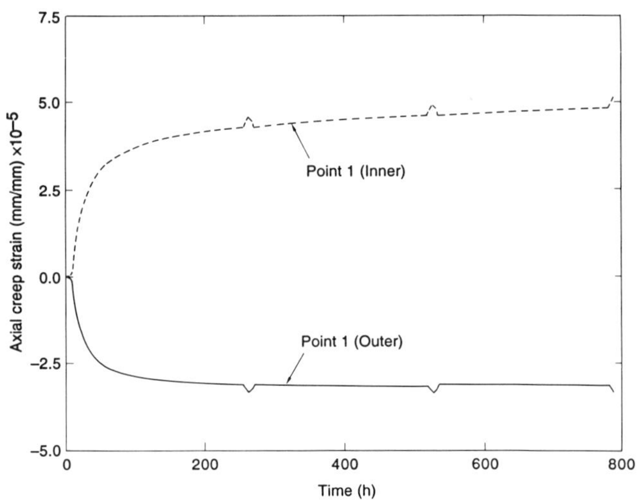


Fig. 12 Relationship between axial creep strain and time

**Creep Damage Evaluation.** The creep damage for histograms (see Figure 8) is shown in Table I. The creep damage is calculated by following equation (4).

$$D_c = 2 \cdot \sum \left( \Delta t / T_d \left( \sigma_{eq} \right) \right) \quad (2)$$

where: $D_c$ = Total creep damage
$\Delta t$ = Time mesh
$T_d$ = Allowable time from design creep rupture strength ($S_R$)
$\sigma_{eq}$ = Equivalent stress

In this equation, 2 is the safety factor of creep damage.

Table I Creep damage calculation results at evaluation point 1

| Thermal transient | Point | First cycle | Second cycle | Third cycle |
|---|---|---|---|---|
| Start-up | Inner | 0.0 | 0.0002 | 0.0002 |
| | Outer | 0.0 | 0.0001 | 0.0001 |
| Steady state | Inner | 0.0022 | 0.0002 | 0.0002 |
| | Outer | 0.0023 | 0.0002 | 0.0002 |
| Shutdown | Inner | 0.0001 | 0.0001 | 0.0001 |
| | Outer | 0.0001 | 0.0001 | 0.0001 |

(Remarks) Holding time is 250 hours each.

Since the creep damage at first cycle is the largest and the creep damage after second cycle is small enough, it is possible to predict the end-of-life creep damage using the results of first three cycles.

## Conclusions

In this study, a method of creep analysis of manifold header using a three-dimensional FEM model has been discussed. It was found the creep damage obtained by this approach was small enough and the approach was effective to predict the reasonable end-of-life creep damage. Furthermore, the extrapolation and classification methods of stresses or creep strains in regard to three-dimensional model were suggested and more detailed results were obtained.

## Acknowledgments

The results of the manifold header analysis were obtained by Ishikawajima-Harima Heavy Industries Company, Ltd. for Japan Atomic Energy Research Institute (JAERI) in connection with the design and development of Intermediate Heat Exchanger for High Temperature Engineering Test Reactor. Permission to publish this work is gratefully acknowledged. The authors would like to take this opportunity to thank the members of the technical staff of JAERI.

## References

(1) T. Igari, K. Setoguchi & S. Nomura, Simplified Creep Analysis of Perforated Plates under Steady Creep Condition, Nuclear Engineering and Design, Vol. 97 (1986) pp. 161-166

(2) D. H. Pai, M. B. Hsu, Inelastic Analysis of Tube sheets by the Finite Element Method, ASME 75-PVP-57 (1975)

(3) C. E. Pugh, Constitutive Equations for Creep Analysis of LMFBR Components, Advances in Design for Elevated Temperature Environment (ed. S. Y. Zamrik and R. I. Jetter), ASME (1975) pp. 1-13

(4) HTTR Designing Laboratory, Oarai Research Establishment Department of Fuels and Materials Research and Department of High-Temperature Engineering Tokai Research Establishment, High-Temperature Structural Design Code for the High-Temperature Engineering Test Reactor, JAERI-M 89-005 (1989.2)

(5) K. Hada, Y. Motoki & O. Baba, Development of Methods for Generating Design Allowable Limits for the HTTR High-Temperature Structural Design Code, JAERI-M 90-148 (1990.9)

*Proceedings of the Fifth International Conference on Creep of Materials, Lake Buena Vista, Florida, USA, 18-21 May 1992*

# Effect of Vanadium Addition on Degradation of High Temperature Service in 12Cr Ferritic Steels

**T. Matsuo and M. Kikuchi**
Tokyo Institute of Technology
Tokyo, Japan

**T. Watanabe and Y. Monma**
National Research Institute for Metals
Tokyo, Japan

## Abstract

The effect of vanadium addition to 12Cr type ferritic heat resisting steels on the degradation during high temperature creep deformation was studied by using two kinds of steels: a 12wt%Cr steel without vanadium and a 12wt%Cr-1wt%Mo-1wt%W steel with 0.3wt% vanadium. Creep-interrupted specimens were prepared by interrupting the first stage creep test under a stress of 18 Ms (5000 h) rupture life at several different times in the tertiary creep stage. The loss of creep resistance during the first stage creep test was measured through the enhanced stress creep test.

The loss of creep resistance during the first stage creep test was solely attributed to material degradation due to microstructural change other than mechanical damages. The creep rate of the specimen crept for 14.4 Ms (4000 h) in the first stage creep test increased by a factor of five in 12Cr steel, while ten times increase in the creep rate was observed in 12CrMoWV steel. In 12Cr steel with weaker creep resistance, the degradation proceeded due mainly to grain refinement through continuous evolution of subgrains and dynamic recrystallization. In 12CrMoWV steel with far stronger creep resistance, the loss of creep resistance proceeded due to local softening along the prior austenite grain boundary, which softening was caused by the formation of recovered zone with a very low density of precipitated $V_4C_3$ carbides and dislocations. Most of three orders of magnitude larger creep resistance in 12CrMoWV steel retained even after 14.4 Ms creep deformation.

FERRITIC HEAT RESISTING STEELS WITH 12% CHROMIUM have lately attracted world wide attention as potential rotor materials for advanced fossil fuel power generating plants[1),2)]. Modified 12Cr steels, which have recently been developed in Japan[3)~11)], are expected to increase the allowable temperature limit by at least 50 K, compared with the conventional 1Cr-1Mo-1/4V steel.

It is well established that the creep resistance of quenched and tempered ferritic heat resisting steels is greatly improved by precipitation strengthening due to carbides[12)~14)]. For this precipitation strengthening, vanadium carbide, $V_4C_3$, is most preferred because of its fine dispersion and its long term stability[14)]. A recent study on a 1Cr-1Mo-1/4V rotor steel by Kimura et al.[15)~17)] revealed that the creep resistance of the properly heat-treated 1Cr-1Mo-1/4V steel is more than four orders of magnitude larger than that of the steel in a well annealed state. Their study was based on an enhanced stress creep test for several creep-interrupted specimens at a few thousands hours crept times under a stress of ten thousands hours rupture life. Their study also revealed that the large creep resistance strengthened by the vanadium addition mostly retains even after high temperature creep exposure for 14.4 Ms at 873 K[16),17)]. The good stability of the creep resistance of the 1Cr-1Mo-1/4V steel was reflected in the observation that overall softening was greatly suppressed and that only local softening took place along prior austenite grain boundaries.

In modified 12Cr steels, vanadium is alloyed in addition to molybdenum, tungsten, niobium and nitrogen, for better creep strength. The creep resistance of the ferritic matrices in these steels is naturally larger than that in the 1Cr-1Mo-1/4V steel due to higher concentrations of alloying elements. The central question is, then, whether vanadium plays a similar role in modified 12Cr steels with higher matrix strengths to that in the 1Cr-1Mo-1/4V steel, through the precipitates of $V_4C_3$. Since main focus was placed on creep rupture strength in a stage of the development of these modified 12Cr steels, this study was made to clarify the effect of vanadium addition to 12Cr type ferritic heat resisting steels on degradation during high temperature creep by a similar technique used previously by Kimura et al.[15)~17)].

## Materials and Experimental Procedure

Two kinds of steels were used in this study[18),19)]. They were a 12wt%Cr steel without vanadium and a 12wt%Cr-1wt%Mo-1wt%W steel with 0.3wt% vanadium. The former is designated as 12Cr steel, while the latter as 12CrMoWV steel. Their compositions are shown in Table 1. Bars of the steels were quenched and tempered on the conditions shown in Table 1. These heat treatments gave prior austenitic grain sizes of 31 and 47 μm for 12Cr and 12CrMoWV steel, respectively. The heat-treated bars were machined to creep testing specimens with a size of 8 mm in diameter and of 50 mm in gage length.

Constant load creep tests were performed at 873 K in air under stresses of 47 and 137 MPa for 12Cr and 12CrMoWV steel, respectively. The rupture life on these conditions was 21.9 Ms ( 6084 h ) for 12Cr steel and 18.4 Ms ( 5124 h ) for 12CrMoWV steel. This test will be referred to as the first stage creep test.

Table 1. Chemical compositions(wt%), heat treatments and prior austenite grain diameters of 12Cr steel and 12CrMoWV steel.

| Steel | C | Si | Mn | P | S | Cr | Mo | W | V |
|---|---|---|---|---|---|---|---|---|---|
| 12Cr | 0.14 | 0.31 | 0.46 | 0.019 | 0.008 | 12.0 | 0.04 | – | – |
| 12CrMoWV | 0.23 | 0.32 | 0.69 | 0.016 | 0.009 | 11.2 | 1.00 | 1.13 | 0.30 |

| Steel | Heat treatments | | | Prior austenite grain diameter (μm) |
|---|---|---|---|---|
| 12Cr | Hot forged | | | |
| | 1223 K × 1h | → | Oil quenched | 31 |
| | 923 K × 2h | → | Air cooled | |
| 12CrMoWV | Hot rolled | | | |
| | 1123 K × 3h | → | Furnace cooled | |
| | 1223 K × 46min | → | Oil quenched | 47 |
| | 933 K × 2h | → | Air Cooled | |
| | 913 K × 2h | → | Air Cooled | |

The first stage creep tests were interrupted at several different times between 0.36 and 14.4 Ms to produce creep-interrupted specimens.

The creep-interrupted specimens were creep-tested at 873 K in air under a constant load with an enhanced stress of 84 MPa for 12Cr steel and 225 MPa for 12CrMoWV steel, to evaluate the loss of creep resistance during creep under a lower stress. This creep test under enhanced stresses will be called as the enhanced stress creep test. For the enhanced stress creep test, the creep-interrupted specimens were machined to a smaller size of 6 mm in diameter and of 30 mm in gage length, to avoid potential mechanical damages developed during the first stage creep test, such as the different reduction of cross sectional area from one specimen to the other and the voids and cracks at the surface of the gage portion of the specimens.

Microstructures of all the creep-interrupted specimens were examined with optical and transmission electron microscopy. The Vickers hardness of the creep-interrupted specimens was measured at room temperature under a load of 98 N. The half-value width of an X-ray diffraction peak at (211) of the creep-interrupted specimens was measured using a Cr-K$\alpha$ radiation.

## Results and Discussion

**Creep Rate - Time Curve in the First Stage Creep Test.** Figure 1 shows stress - time to rupture curves at 873 K for 12Cr and 12CrMoWV steels. The rupture life of 12CrMoWV steel is two orders of magnitude longer than that of 12Cr steel. A constant load creep test at a stress for 18 Ms (5000 h) rupture life, which is called the first stage creep test, was carried out to evaluate the loss of creep resistance during long term creep. Figure 2 shows creep rate - time curves of 12Cr and 12CrMoWV steels obtained by the first stage creep test under stresses of 47 and 137 MPa, respectively. Both creep rate - time curves consist of two stages: the transient and tertiary creep stages. No steady state creep stage was observed. In both cases, the tertiary creep stage occupied 90 per cent of the total rupture life, while the transient creep stage occupied only 10 per cent of the total rupture life.

There are two other points to be noted. Firstly, the minimum creep rate of 12CrMoWV steel was about half of that of 12Cr steel, although the rupture life of the former was slightly shorter than that of the latter. Secondly, the acceleration of the creep rate, defined as $d\dot{\varepsilon}/dt$, in the tertiary creep stage was larger in 12CrMoWV steel by a factor of two than in 12Cr steel.

The acceleration of creep rate is generally caused by both material degradation and reduction of load-bearing area. This study is concerned with the former. Indeed, the acceleration of the creep rate of the two steels in this study is mainly caused by material degradation, as explained in the following. Here, the material degradation means the loss of creep resistance due to microstructural changes other than mechanical damages. The reduction of load-bearing area is brought about by the formation of voids and cracks and by uniform elongation during creep under constant load. As to the former, neither void and nor crack was observed in the specimens creep-interrupted even at the later stage of the tertiary creep ( Figs. 6, 7, 9 and 10). As to the latter, the increase in the creep rate due to the reduction of cross sectional area brought about by creep deformation is negligibly small even at the later stage of the tertiary creep. The true strains at the time of 14.4 Ms (4000 h) were 0.016 and 0.010 in 12Cr and 12CrMoWV steel, respectively. Even in the case of the larger strain, the increase in the creep rate due to the reduction of cross sectional area is calculated to be 12%, since the stress exponent of the steel was 7 (Fig. 11). This increase

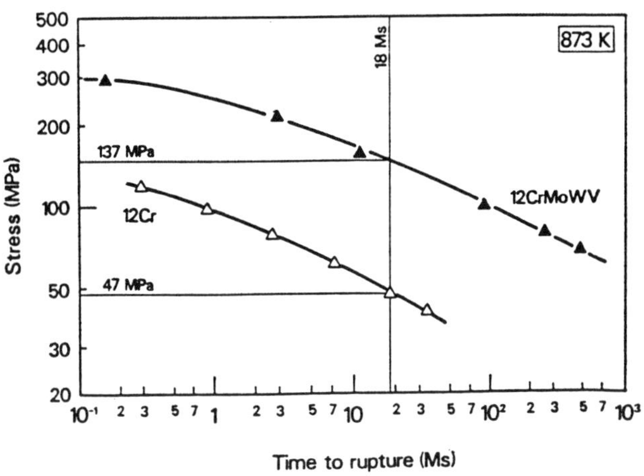


Fig. 1 Stress - time to rupture curves of 12Cr and 12CrMoWV steel at 873 K.

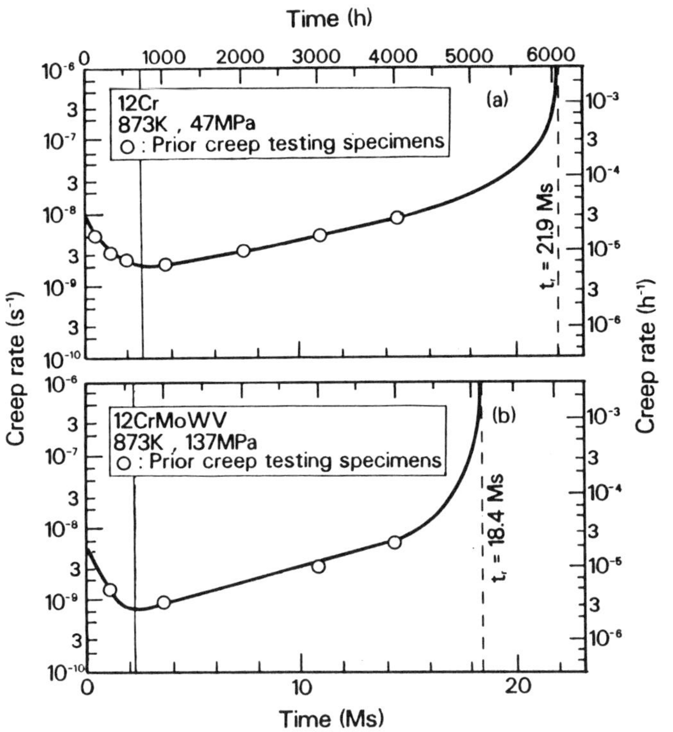


Fig. 2 Creep rate - time curves of 12Cr and 12CrMoWV steels at 873 K - 47 MPa and 873 K - 137 MPa, respectively.

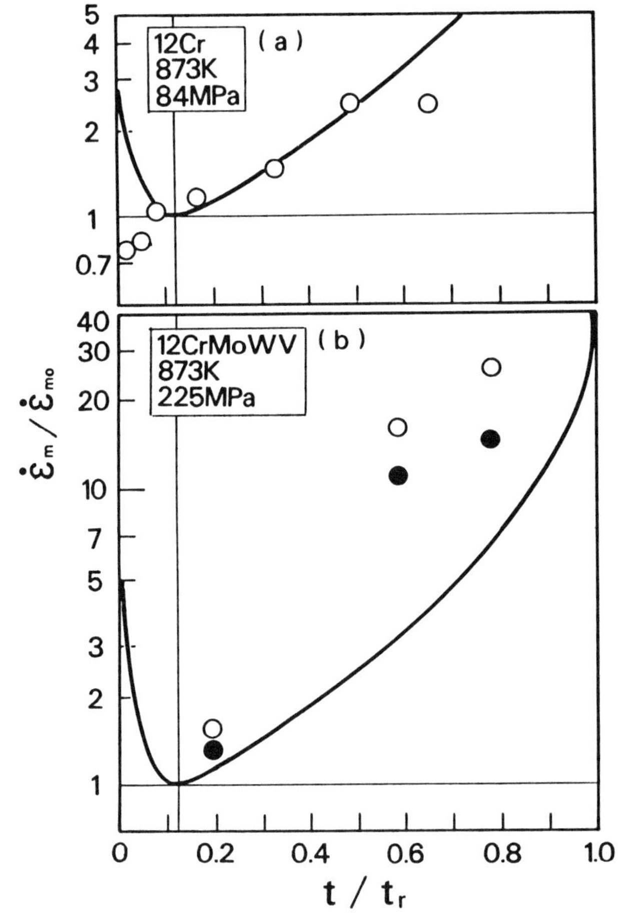


Fig. 3 Loss of creep resistance during the tertiary creep in the first stage creep test. Lines in the figure show normalized creep rate - normalized testing time curves, while circles indicate the degradation factors determined by the stress enhanced creep test.

should be compared with the observed increase in the creep rate by 400 per cent.

Thus, we regard the acceleration of the creep rate observed in the two steels as due solely to material degradation. In Fig. 3, the creep rate - time curves in Fig. 2 are replotted in a form of $\dot{\varepsilon}/\dot{\varepsilon}_m$ as a function of $t/t_r$. The former $\dot{\varepsilon}/\dot{\varepsilon}_m$ is the ratio of the creep rate against the minimum creep rate, while the latter $t/t_r$ is the normalized testing time against the rupture life. Here, is used the true strain rate calculated from the nominal strain rate. The $\dot{\varepsilon}/\dot{\varepsilon}_m$ was 5 and 10 in 12Cr and 12CrMoWV steel , respectively, at 0.8 of $t/t_r$. Again, it is to be noted that the minimum creep rate of 12CrMoWV steel was about half of that of 12 Cr steel.

**Evaluation of Creep Resistance by Enhanced Stress Creep Test.** The first stage creep tests at 873 K under the stress of 47 MPa for 12Cr steel and 137 MPa for 12CrMoWV steel were interrupted at several different times between 0.36 and 14.4 Ms, as shown by open circles in Fig. 2. By using these creep-interrupted specimens, the loss of creep resistance during the tertiary creep stage was measured with a creep test at the same temperature of 873 K under an enhanced stress. The enhanced stress was selected to be 84 MPa for 12Cr steel and 225 MPa for 12CrMoWV steel. It corresponds to a 1.8 Ms (500 h) rupture stress for the as-tempered specimen without being subjected to the first stage creep test, as shown in Fig. 1. The creep test for the creep resistance is called as the enhanced stress creep test.

The creep rate - time curves obtained by the enhanced stress creep test for the creep-interrupted specimens of 12Cr and 12CrMoWV steels are shown in Fig. 4. The minimum creep rate in the enhanced stress creep test became larger with increasing the creep-interrupted time in the first stage creep test in both 12Cr and 12CrMoWV steels. In the enhanced stress creep test, the minimum creep rate of the creep-interrupted specimen at the time of 14.4 Ms (4000 h) was five times larger in 12Cr steel than that of the as-tempered specimen without being subjectrd to the first stage creep test, while two orders of magnitude larger in 12CrMoWV steel.

The material degradation, which is equivalent to the loss of creep resistance in this study, was measured by the enhanced stress creep test. Degradation factor is defined as the ratio of the minimum creep rate of the creep-interrupted specimen in the tertiary creep stage, $\dot{\varepsilon}_m$, to the that of the creep-interrupted specimen at the time of the minimum creep rate, $\dot{\varepsilon}_{mo}$, in the first stage creep test. The measured degradation factor is plotted by open circles as a function of the normalized testing time in the first stage creep test for 12Cr and 12CrMoWV steels in Fig. 2. In 12Cr steel, the measured degradation factors lay well on the line, during the tertiary creep stage, which represents the acceleration of the creep rate in the first stage creep test. In 12CrMoWV steel, however, the measured degradation factors located far above the line.

It is to be noted that the minimum creep rate of the creep-interrupted specimens at the time of the minimum creep rate was larger than that of the as-tempered specimen in both steels in the

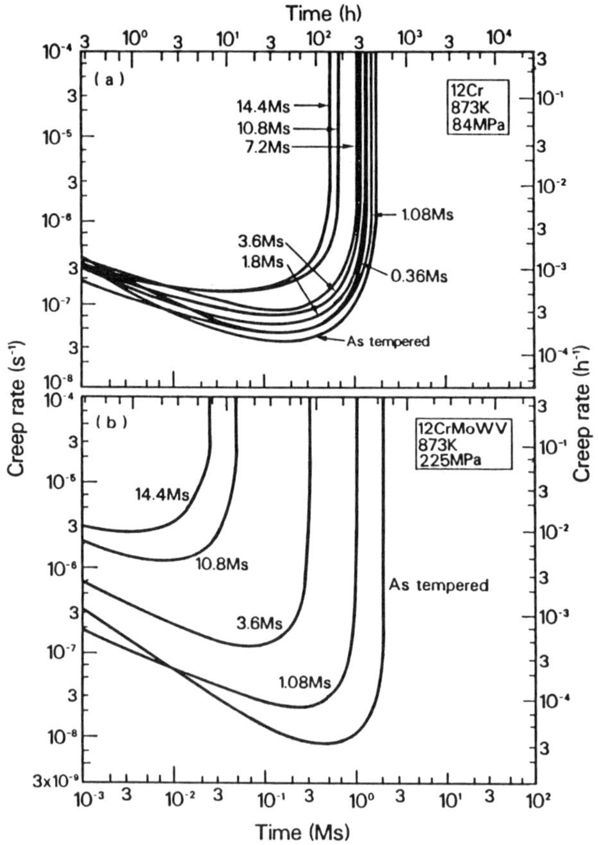


Fig. 4 Creep rate - time curves of the creep-interrupted specimens of 12Cr and 12CrMoWV steel at 873 K - 84 MPa and 873 K - 225 MPa.

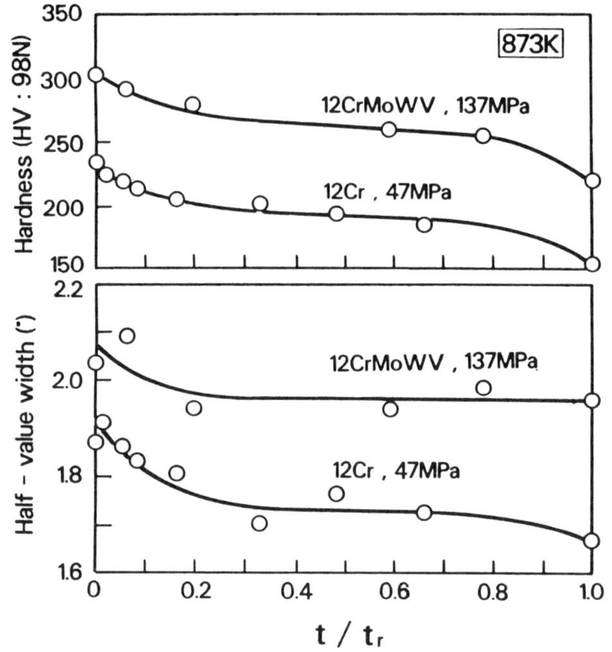


Fig. 5 Vickers hardness and half-value width of an X-ray diffraction peak as a function of the normalized testing time during the first stage creep test.

enhanced stress creep test. This is due to the progress of softening rather than hardening during the transient creep stage in the first stage creep test. One of a direct evidence for the progress of softening during the transient creep stage is given in Fig. 3. When the creep resistance was evaluated through the enhanced stress creep test, the creep-interrupted specimens during the transient creep stage exhibited the smaller minimum creep rate than the creep-interrupted specimen at the time of the minimum creep rate in the first stage creep test. The similar result was reported by Matsuo et al.[20] in the study of the transient creep stage of a quenched and tempered 1Cr-1Mo-1/4V rotor steel. An apparently unusual transient creep behavior, in which the creep rate decreased with the progress of softening, is explained by the gradual annihilation of a large number of pre-existing dislocations introduced during quenching.

Although the minimum creep rates in the enhanced stress creep test were used as a measure of degradation, the creep rate at a smaller strain than the strain at the minimum creep rate might be more suitable for the evaluation of the creep resistance. This is because additional softening due to further microstructural changes takes place during the transient creep stage in the enhanced stress creep test. Therefore, the degradation factor previously defined by the minimum creep rate is substituted by the similar ratio of the creep rate at a strain of 0.001 and the latter ratios are replotted by solid circles in Fig. 3. Agreement with the line became improved, but the solid circles still located above the line.

The Vickers hardness and the half-value width of an X-ray diffraction peak were measured to evaluate the overall softening during the first stage creep test. The overall softening is mainly caused by the coarsening of carbide precipitates and the annihilation of pre-existing dislocations. Measured values are plotted as a function of the creep-interrupted time in Fig. 5. They exhibited substantially larger values in 12CrMoWV steel than in 12Cr steel. They decreased with increasing the creep-interrupted time in the first stage creep test. The decrease in these values in the transient creep stage is in accord with the previous explanation that softening is accompanied by the transient creep stage. The large decrease in these values occurred in the transient creep stage rather than in the tertiary creep stage. The decrease in these values during the tertiary creep stage was relatively small so that the acceleration of the creep rate in this stage in the two steels cannot directly be explained by the overall softening measured by the hardness and the half-value width of the X-ray diffraction peak. It is to be noted that these values at the time of rupture in 12CrMoWV steel were still larger than those of 12Cr steel on an as-tempered condition.

**Loss of Creep Resistance Due to Overall Softening in the 12wt%Cr Steel without Vanadium.** Figure 6 shows optical micrographs of the creep-interrupted specimens of 12Cr steel. Laths of bainite (Fig. 6(a)) were further tempered during the first stage creep test through continuous evolution of subgrains, followed by dynamic recrystallization. The tempering process is better represented by transmission electron micrographs shown in Fig. 7. Intragranular carbides with a high density and a large number of dislocations within the bainitic laths introduced by bainite transformation shown in Fig. 7(a) substantially reduced their density at the time of the

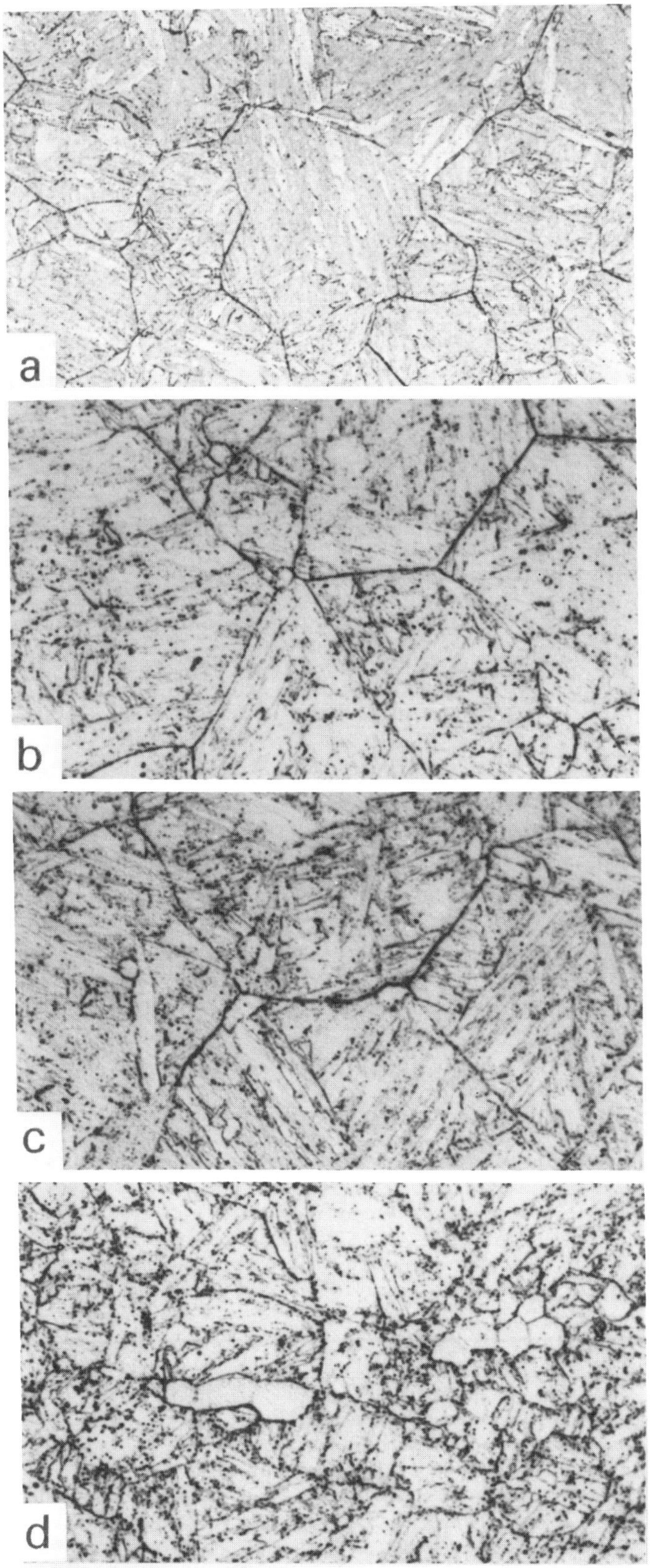


Fig. 6 Optical micrographs of 12Cr steel as-tempered(a), creep-interrupted during the first stage creep test at 3.6 Ms(b), 7.2 Ms(c) and 14.4 Ms(d).

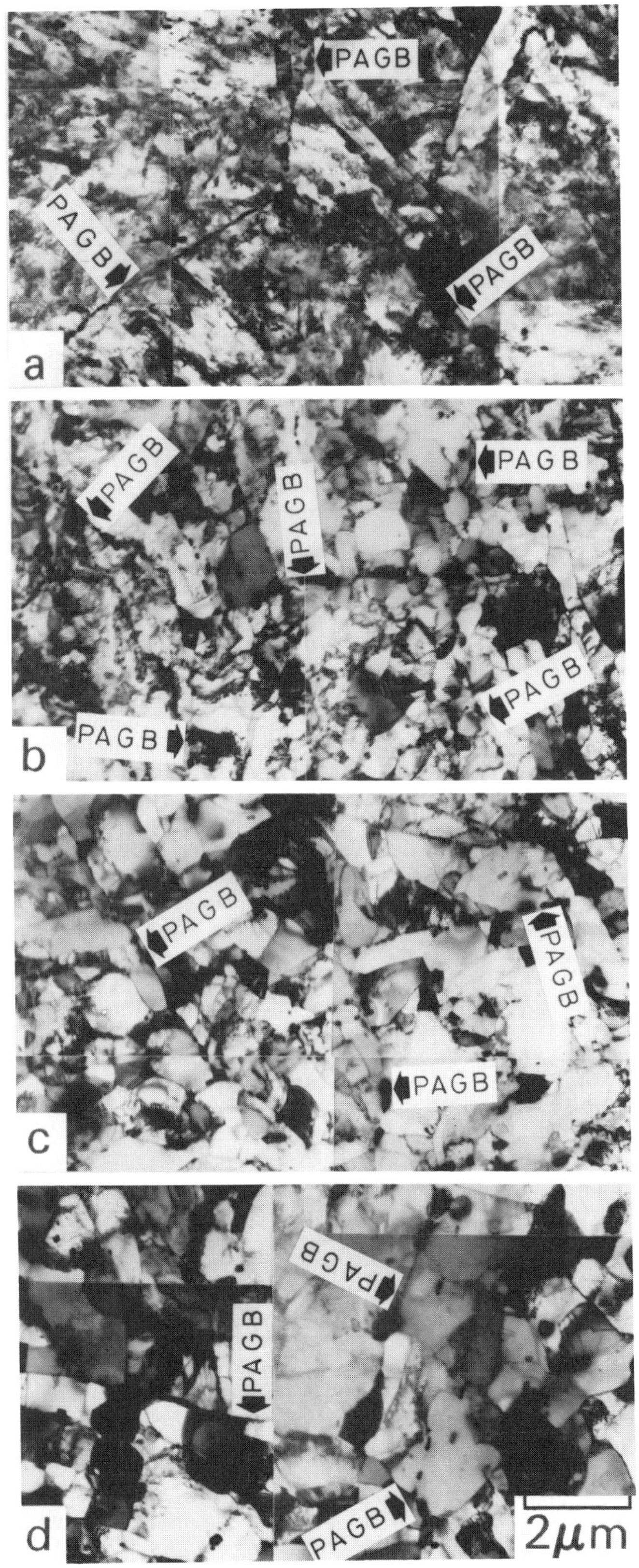


Fig. 7 Transmission electron micrographs of 12Cr steel as-tempered(a), creep-interrupted during the first stage creep test at 3.6 Ms(b), 7.2 Ms(c) and 14.4 Ms(d).

minimum creep rate shown in Fig. 7(b). The prior austenite grain boundary observed clearly in the as-tempered specimen became less clear. Subgrains developed uniformly within the laths. At the time of the minimum creep rate (Figs. 6(b) and 7(b)), the lath morphology clearly observed in the as-tempered specimen greatly changed by annealing and subgrains surrounded with subboundaries formed within laths. Further creep deformation led to the increase in the misorientation angle of the subboundaries and the size of the subgrains.

The loss of creep resistance during tertiary creep stage in quenched-and-tempered ferritic heat resting steels have been considered to be due to the decrease in the density of precipitated carbides through coarsening and/or due to the decrease in the density of dislocations, on the conditions that the formation and development of mechanical damages are somehow suppressed. In 12Cr steel, however, the overall softening was nearly complete during the transient stage due to the rapid coarsening of precipitated carbides. This observation together with the microscopic observation rules out a possibility that the coarsening of the carbides mainly resulted in the acceleration of the creep rate in the tertiary creep stage[21),22)].

It has empirically been accepted that the larger grain size leads to the longer rupture life. In a gamma single phase alloy, a few several investigators reported that the steady state creep rate is inversely proportional to the grain size in a diameter range less than 100 μm[23),24)]. Since the grain size is inversely proportional to the grain boundary area per unit volume, the steady state creep rate should be proportional to the grain boundary area fraction. The latter relationship is proved to be applicable to the case where the grain boundary becomes undulated through dynamic recrystallization. In the present case of 12Cr steel, the grain size became smaller with increasing the creep-interrupted time during the first stage creep test. This causes the acceleration of the creep rate: that is, the grain refinement due to subgrain evolution and dynamic recrystallization leads to the onset of the tertiary creep stage and further accelerates the creep rate in this stage. The grain boundary area fraction was measured optically by using the creep-interrupted specimens deeply etched to reveal the dynamically recrystallized grains. The ratio of the grain boundary area per fraction of the creep-interrupted specimens at the tertiary creep stage, S, to that of the specimen at the time of the minimum creep rate, $S_o$, $S/S_o$ is plotted as a function of the normalized testing time in Fig. 8. In this figure, the acceleration of the creep rate, $\dot{\varepsilon}/\dot{\varepsilon}_o$, is also drawn as a solid line. As shown in this figure, about half of the creep rate increase can be explained by the observed increase in the grain boundary area fraction.

One might wonder that the subboundary should act as a creep strengthener rather than a creep weakener. Recently, Terada et al.[25)] confirmed experimentally that the subboundary with large misorientation angles above a few degrees acts like the large angle grain boundary to reduce creep resistance. Based on this observation, we tried to reveal the grain boundary including the subgrain boundary with large misorientation angles by deep etching. A comparison between optical and transmission electron micrographs, however, made it clear that not all the grain boundaries were revealed in the optical micrographs. Therefore, the above discrepancy could be attributed to the inability to reveal all the grain boundaries in optical micrographs. If all the grain boundaries are taken into account, the better agreement is expected. Since the coarsening of the precipitated carbides is excluded as a mechanism of degradation, the onset of the tertiary creep stage and the acceleration of the creep rate in 12Cr steel is mainly due to the evolution of subgrain and dynamic recrystallization.

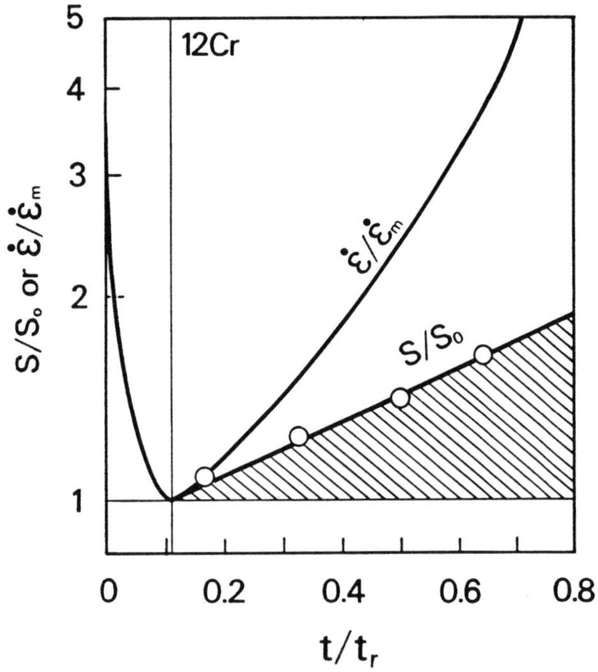


Fig. 8 A correlation between the normalized creep rate and grain boundary area fraction on 12Cr steel.

**Loss of Creep Resistance Due to Local softening in the 12wt%Cr-1wt%Mo-1wt%W Steel with 0.3wt% Vanadium.** Figure 9 shows the optical micrographs of the creep-interrupted specimens of 12CrMoWV steel. Unlike those shown in Fig. 6, not much change was observed during the tertiary creep in the first stage creep test in the optical micrographs. The lath morphology of bainite maintained up to 14.4 Ms (4000 h). Corresponding transmission electron micrographs are shown in Fig. 10. Contrary to the microstructures of 12Cr steel, the density of precipitated carbides and dislocations remained high within laths even after 14.4 Ms of the first stage creep test, as shown in Fig. 10(d). At the minimum creep rate stage of the first stage creep test, a recovered zone with a low density of precipitated carbides and dislocations started to form along the prior austenite grain boundary (PAGB). Average width of the recovered zone increased with increasing the testing time of the first stage creep test, as is well illustrated in Figs. 10(b)~(d). Like a 1Cr-1Mo-1/4V rotor steel[16),17)], the local softening due to the development of the recovered zone along the prior austenite grain boundary is mainly responsible to the acceleration of the creep rate.

**Resistance against Degradation -Comparison between 12Cr steel and 12CrMoWV Steel-.** Since the factors to determine the creep resistance of 12CrMoWV steel are different from those of 12Cr steel, the degradation mechanisms are naturally different in the two steels. Figure 11 shows stress - creep rate curves of 12Cr and 12CrMoWV steels. Bold lines are drawn connecting the following two creep rates; the minimum creep rate in the first stage creep test and the minimum creep rate in the enhanced stress creep test for the creep-interrupted specimen at the minimum creep rate stage. The latter was

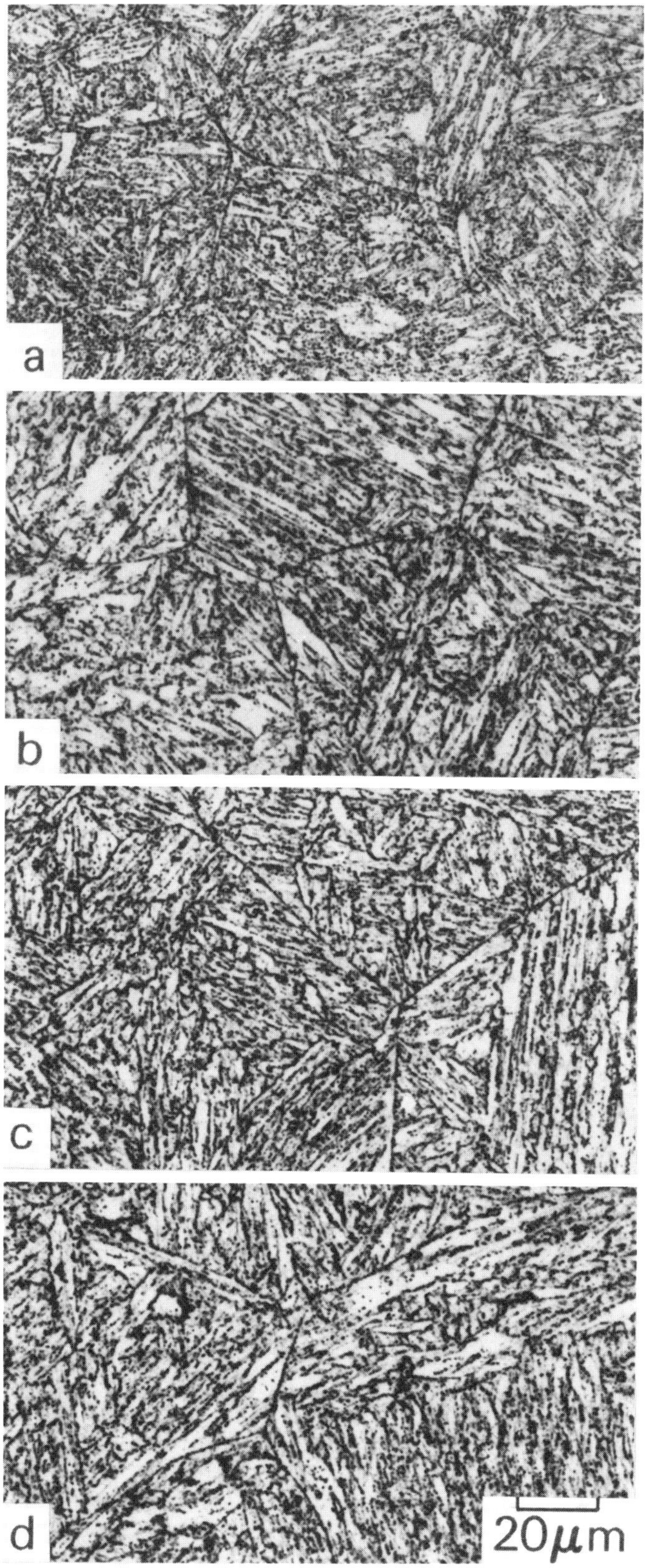


Fig. 9 Optical micrographs of 12CrMoWV steel as-tempered(a), and creep-interrupted during the first stage creep test at 3.6 Ms(b), 10.8 Ms(c) and 14.4 Ms(d).

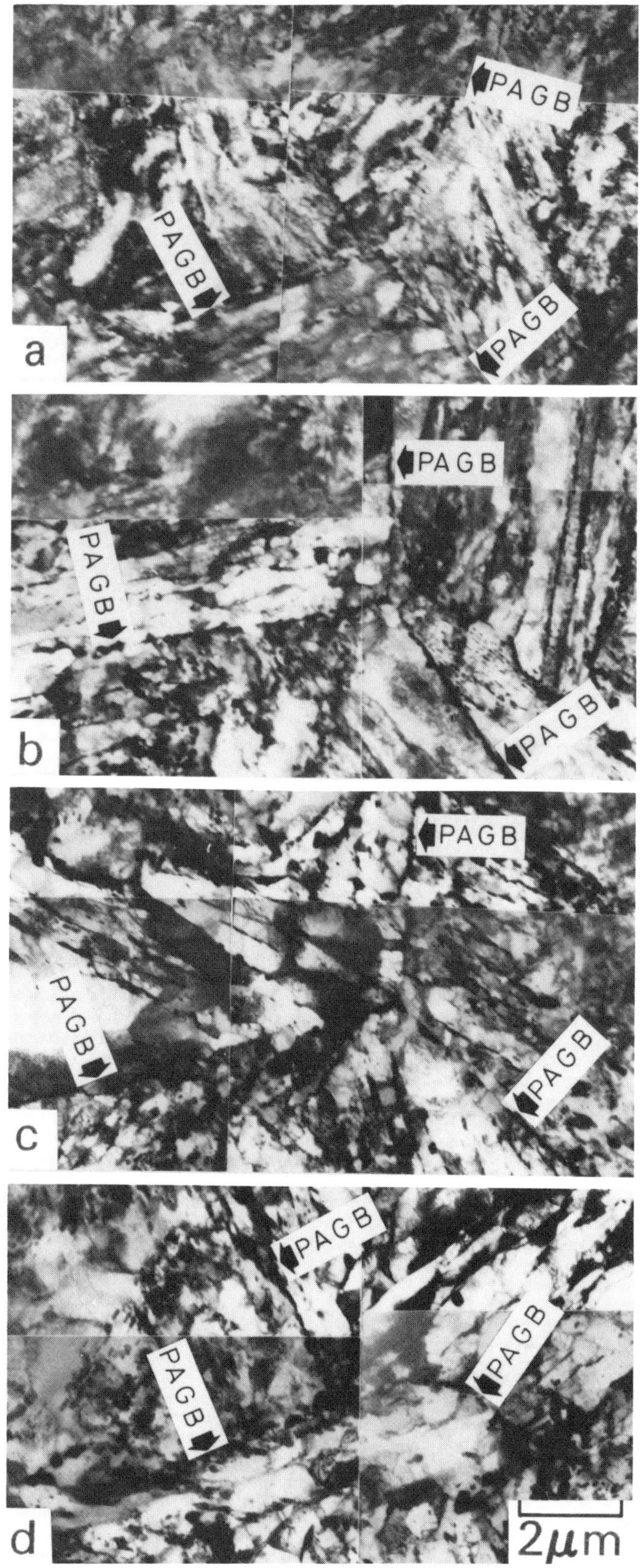


Fig. 10 Transmission electron micrographs of 12CrMoWV steel as-tempered(a), and creep-interrupted during the first stage creep test at 3.6 Ms(b), 10.8 Ms(c) and 14.4 Ms(d).

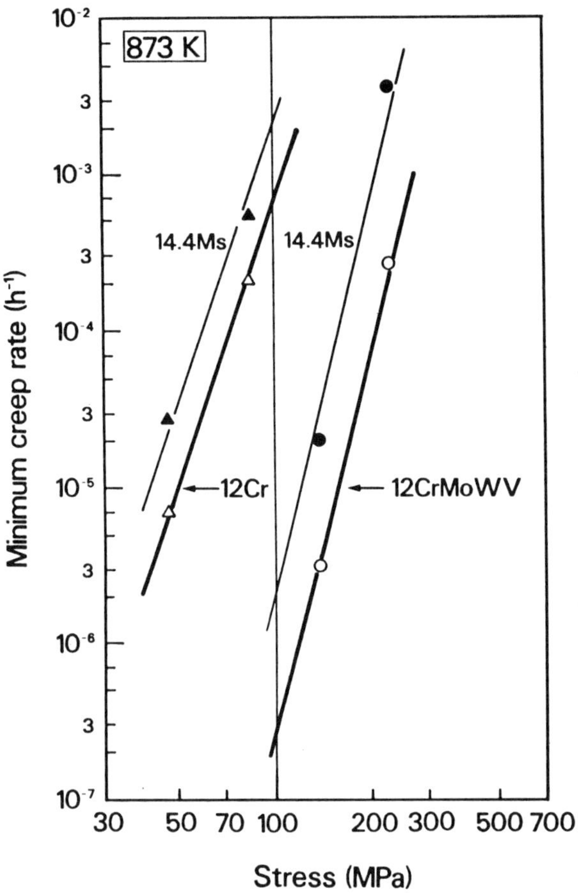


Fig. 11 Stress - creep rate curves of 12Cr and 12CrMoWV steel. Bold lines are drawn connecting the following two creep rates; the minimum creep rate in the first stage creep test and the minimum creep rate in the enhanced stress creep test for the creep-interrupted specimen at the minimum creep rate stage. The latter was estimated from the minimum creep rates of the creep-interrupted specimens at the time between 0.36 and 3.6 Ms. Creep rates at 14.4 Ms in the first stage creep test and the minimum creep rate of the creep-interrupted specimen at 14.4 Ms in the enhanced stress creep test are shown with thin lines.

estimated from the minimum creep rates of the creep-interrupted specimens at the time between 0.36 and 3.6 Ms. The creep rate at 14.4 Ms in the first stage creep test and the minimum creep rate of the creep-interrupted specimens at 14.4 Ms in the enhanced stress creep test are also plotted with thin lines in the figure. In the specimens with the microstructures at the minimum creep rate stage in the first stage creep test, the minimum creep rate of 12CrMoWV steel is more than three orders of magnitude smaller than that of 12Cr steel. The larger creep resistance of 12CrMoWV steel over 12Cr steel was attained by both solid solution strengthening due to molybdenum and tungsten and precipitation strengthening due to $V_4C_3$. Although the first stage creep test was made with different stress, the effect of stress on degradation did not produce order of magnitude difference in the creep resistance. As shown in Fig. 11 at 100 MPa, the creep resistance of 12CrMoWV steel after 14.4 Ms exposure on a creep condition is not only three orders of magnitude larger than that of 12Cr steel on a similar condition, but also more than two orders of magnitude larger than that of 12Cr steel on a creep condition up to the minimum creep rate.

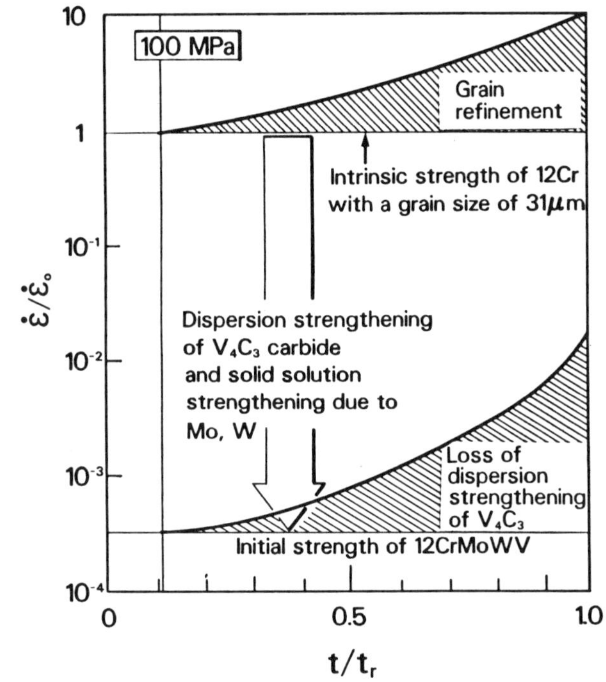


Fig. 12 Schematic illustration of degradation during the first stage creep test for 12Cr and 12CrMoWV steels.

The loss of creep resistance during the first stage creep test is schematically shown in Fig. 12. In this figure, $\dot{\varepsilon}/\dot{\varepsilon}_0$ is used as a measure of the loss of creep resistance. In 12Cr steel, on the one hand, the degradation proceeded due mainly to grain refinement through continuous evolution of subgrain and dynamic recrystallization. In 12CrMoWV steel, on the other hand, the loss of creep resistance proceeded due to local softening along the prior austenite grain boundary, accompanied by the decrease in the density of precipitated $V_4C_3$ carbides and dislocations in the recovered zone. If we compare the degradation factor, $\dot{\varepsilon}_m/\dot{\varepsilon}_{m0}$, between the two steels, the degradation factor of 12CrMoWV steel was twice as large as that of 12Cr steel. As shown in Figs. 11 and 12, the large creep resistance of 12CrMoWV steel is still maintained mostly by solid solution strengthening due to molybdenum and tungsten and partly by precipitation strengthening due to $V_4C_3$.

## Conclusions

The effect of vanadium addition to 12Cr type ferritic heat resisting steels on the degradation during high temperature creep deformation was studied by using two kinds of steels: a 12wt%Cr steel without vanadium and a 12wt%Cr-1wt%Mo-1wt%W steel with 0.3 wt% vanadium. Creep-interrupted specimens were prepared by interrupting the first stage creep test under a stress of 18 Ms(5000 h) rupture life at several different times in the tertiary creep stage. The loss of creep resistance during the first stage creep test was measured through the enhanced stress creep test.

The loss of creep resistance during the first stage creep test was solely attributed to material degradation due to microstructural change other than mechanical damages. The creep rate of the specimen crept for 14.4 Ms(4000 h) in the first stage creep test, which creep rate was measured through the enhanced stress creep test, increased by a factor of 5 in 12Cr steel, while ten times increase in the creep rate was observed in 12CrMoWV steel. The creep resistance of 12CrMoWV steel is, in an as-tempered condition, more than three orders of magnitude larger than that of 12Cr steel. The former steel is strengthened by solid solution strengthening due to molybdenum and tungsten and by precipitation strengthening due to $V_4C_3$. In 12Cr steel with weaker creep resistance, the degradation proceeded due mainly to grain refinement through continuous evolution of subgrains and dynamic recrystallization. In 12CrMoWV steel with far stronger creep resistance, the loss of creep resistance proceeded due to local softening along the prior austenite grain boundary, which softening was caused by the formation of recovered zone with a very low density of precipitated $V_4C_3$ carbides and dislocations. Most of three orders of magnitude larger creep resistance in 12CrMoWV steel retained even after 14.4 Ms creep deformation.

## References


1. Kougami, K. and H.Isaka, *Tetsu to Hagané, 76*, 1043-52 (1990).
2. Fujita, T., *Tetsu to Hagané, 76*, 1053-9(1990).
3. Yamada, M., O.Watanabe, Y.Yoshioka and M.Miyazaki, *Tetsu to Hagané, 76*, 1084-91(1990).
4. Shiga, M., Y.Fukui, S.Kirihara, R.Kaneko, F.Ito and S.Sugai, *Tetsu to Hagané, 76*, 1092-9(1990).
5. Takeda, Y., Y.Takano, H.Yokota, A.Hizume, T.Tsuchiyama, M.Kohno, S.Kinoshita and A.Suzuki, *Tetsu to Hagané, 76*, 1100-7(1990).
6. Fujita, T. and N.Takahashi, *Tetsu to Hagané, 60*, 1506-18(1974).
7. Takahashi, N. and T.Fujita, *Tetsu to Hagané, 61*, 2604-16(1975).
8. Gooch, D.J., *Met. Sci., 16*, 79-89(1982).
9. Iseda, A., H.Teranishi, K.Yoshikawa, F.Masuyama, T.Daikoku and H.Hada, *The Thermal and Nuclear Power, 39*, 897-907(1988).
10. Iwabuchi, Y., M.Murata, S.Yamakuro, M.Yamada and O.Watanabe, *Tetsu to Hagané, 76*, 1060-7(1990).
11. Takebayashi, K., Y.Tashiro, A.Fujita, M.Nakamura and Y.Takeda, *Tetsu to Hagané, 76*, 1068-75(1990).
12. Yokoi, S., N.Shinya and M.Koori, *J. Soci. Mater. Sci., 26*, 241-7(1977).
13. Abe, F., S.Nakazawa, H.Araki and T.Noda, *123rd Committee on Heat Resisting Metals and Alloys Rep., 30*, 17-27(1989).
14. Komatsu, S., S.Nakamura and M.Nakahashi, *J. Jpn. Inst. Met., 43*, 485-91(1979).
15. Kimura, K., T.Kisanuki, S.Komatsu, T.Matsuo and R.Tanaka, *Tetsu to Hagané, 71*, 1803-10(1985).
16. Kimura, K., T.Matsuo, M.Kikuchi and R.Tanaka, *Tetsu to Hagané, 72*, 474-81(1986).
17. Matsuo, T., K.Kimura, R.Tanaka and M.Kikuchi, *Proc. of the 4th Intern. Conf. on Creep and Fracture of Engineering Materials and Structures*, 477-86(1990).
18. *NRIM Creep Data Sheet, No.10A*, (1979).
19. *NRIM Creep Data Sheet, No.13A*, (1980).
20. Matsuo, T., K.Kimura, N.Ohi and M.Kikuchi, *Proc. of the 4th Intern. Conf. on Creep and Fracture of Engineering Materials and Structures*, 487-96(1990).
21. Bolton, C.J., B.F.Dyson and K.R.Williams, *Mater. Sci. Eng., 46*, 231-9(1980).
22. Williams, K.R. and B.Wilshire, *Mater. Sci. Eng., 47*, 151-60(1981).
23. Kondo, Y., T.Matsuo, T.Shinoda and R.Tanaka, *Tetsu to Hagané, 65*, 896-905(1979).
24. Garofalo, F., W.F.Domis and F.Gemmingen, *Trans. AIME, 230*, 1460-7(1964).
25. Terada, Y., T.Matsuo and M.Kikuchi, *Proc. of Intern. Conf. on Stainless Steels « Stainless Steels `91 »*, 618-26 (1991).

*Proceedings of the Fifth International Conference on Creep of Materials, Lake Buena Vista, Florida, USA, 18-21 May 1992*

# Creep Cavity Growth in Cr-Mo-V Forged Steel for Steam Turbine Rotor

**K. Iijima, M. Sukekawa, Y. Fukui, and R. Kaneko**
Hitachi Works of Hitachi, Ltd.
Ibaraki-ken, Japan

## ABSTRACT

Creep cavity growth for a commercial Cr-Mo-V steam turbine rotor material has been examined, to determine the creep fracture mechanism under plant service conditions. Creep cavity observations were made with a scanning electron microscope and cavity length, width and density were measured with an image analyzer. Creep cavities nucleated on grain boundaries nearly perpendicular to a stress axis, and this tendency was independent of creep damage. The mean cavity length was proportional to creep strain before cavities started to combined with each other. In the secondary creep stage, creep cavity growth depended mainly on grain boundary diffusion and cavities remained similar in shape. Assuming continuous cavity nucleation, mean cavity length was calculated by the Hull-Rimmer theoretical model of grain boundary diffusion growth and compared with experimental data. This comparison supported the observation that cavity growth depended mainly on grain boundary diffusion phenomena.

## NOMENCLATURE

a= center-to-center cavity spacing, $\mu$m
D=diffusion coefficient, $m^2s^{-1}$
f=fraction observed
k=Boltzmann's constant, $NmK^{-1}$
l=cavity length, $\mu$m
N=number of cavities per unit area, $mm^{-2}$
R=cavity radius, $\mu$m
S=constant
T=temperature, deg K
t=time, s
$\alpha$=constant
$\beta$=constant
$\delta Z$=grain boundary thickness, m
$\varepsilon$=creep strain
$\theta$=angle between cavited grain boundary and stress axis, deg.
$\sigma$=applied stress, MPa
$\Omega$=atomic volume, $m^3$

Subscripts

g=grain boundary
l=lattice
o=initial

Cr-Mo-V FERRITIC STEEL has been applied to steam turbine rotor shafts, because it has superior elevated temperature creep resistance under steam turbine service conditions. In Japanese thermal power plants, many steam turbine rotor shafts are kept in long-term service beyond the designed service life. Therefore correctly evaluating the remaining lifetimes of their components, especially turbine rotor shafts,is necessary. A study of their creep fracture mechanism should be important to understand creep damage in Cr-Mo-V turbine rotor shafts and develop a useful lifetime prediction method.

Based on the long-term creep test data, Shinya (1) found that creep fracture in Cr-Mo-V ferritic steel under steam turbine service conditions mainly occurred through formation of creep cavities on grain boundaries. Other studies have descrived integranular failure caused by cavitation associated with weld metal and heat affected zone structure in 21/4Cr-1/2Mo and 1Cr-1/2Mo ferritic steels (2,3). Our initial scanning electron microscopy (SEM) study also indicated that cavity nucleation occurred with less than 0.2 creep damage at 550°C under 216MPa creep test condition (4). Additionally creep cavities seemed to grow in the secondary creep stage and they started linking with each other at the beginning of the tertiary stage.

In the present work, we study the behavior of the creep cavity growth in Cr-Mo-V ferritic steel. In particular, the effect of a stress axis on cavity nucleation and the mechanism of grain boundary cavity growth are investigated through a theoretical method.

## EXPERIMENTAL

MATERIAL. The test material was taken from a Cr-Mo-V intermediate pressure rotor shaft which had been operated in a 600MW steam turbine power plant for 7889h. The chemical composition is shown in Table 1.

Table 1 Chemical composition (wt%)

| C | Si | Mn | P | S | Ni | Cr | Mo | V |
|---|---|---|---|---|---|---|---|---|
| 0.28 | 0.24 | 0.77 | 0.010 | 0.007 | 0.38 | 1.08 | 1.23 | 0.25 |

The shaft had undergone the following processes at the time of fabrication: heat treatment including quenching for 21h at 970°C then air-blast cooling and tempering for 54h at 655°C and finally furnace cooling. It contained bainiteand fine carbides, distributed homogeneously. Creep tests were carried out at 550°C under 216 and 245MPa, and at 600°C under 147MPa loads and they were interrupted for cavitation observation. Table 2 illustrates creep test conditions where creep damage was defined by the time fraction rule. Test samples were taken from the cooler journal part of the rotor shaft, where no creep failure had occurred, and machined into creep specimens with a 50mm gauge-length and 10mm guage-diameter as shown in Fig.1.

Table 2 Creep test conditions

| Temp. (°C) | Stress (MPa) | Creep damage | Time to rupture (h) |
|---|---|---|---|
| 550 | 245 | 0.16, 0.34, 0.50, 0.67 | 2786 |
| | 216 | 0.20, 0.42, 0.62, 0.81 | 7284 |
| 600 | 147 | 0.22, 0.44, 0.68, 0.88 | 2408 |

Creep damage=(Time)/(Time to rupture)

CREEP CAVITY OBSERVATIONS. A longitudinal section of the gauge-length was cut from each specimen after testing, and all surfaces of the section were carefully etched by 4% nital etchant after polishing. On each specimen at least 100 fields were observed on a 10mm square area parallel to the stress axis (Fig.2). Creep cavities were

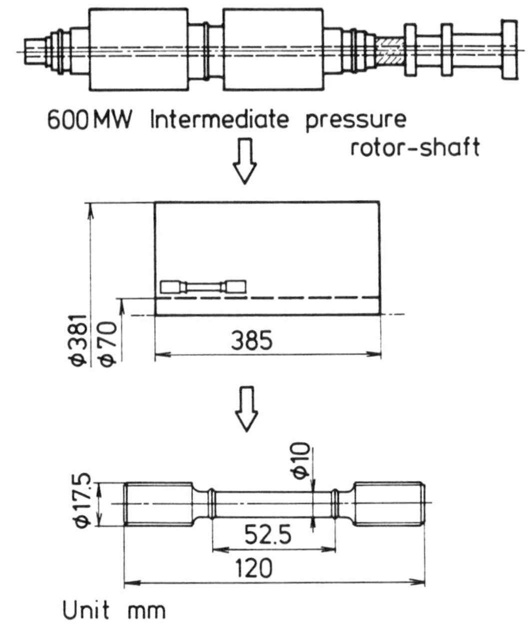


Fig. 1-Specimen design and orientation within rotor forging.

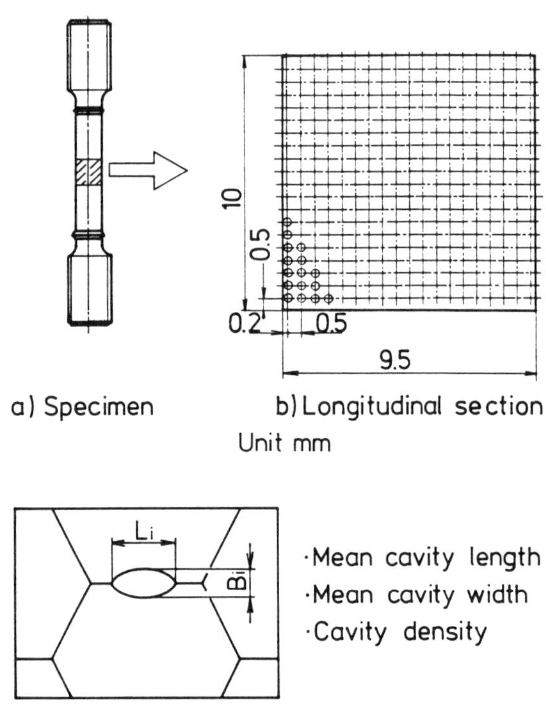


Fig. 2-Observation technique of intergranular creep cavitation.

observed at a magnificaton of 2000 with SEM, and mean cavity length, mean cavity width and the number of cavities per unit area were measured with an image analyzer, before creep cavities had linked with each other.

## EXPERIMENTAL RESULTS

CAVITY NUCLEATION BEHAVIOR. Cavity nucleation behavior, i. e. the number of cavities per unit area is presented in Fig.3 as a function of time. The cavity number densities increased with time under the three test conditions, where the time exponent was from 2.4 to 3.5. It seemed that creep cavities nucleated continuously under these creep conditions. In this work, grain boundary carbides were not always cavity-nucleation sites. Fig. 4 shows the relation between cavity number of densities and creep strain. It seemed that densities could be described by a power-law function of creep strain, with an exponent of approximately 2.5, and density was independent of the temperature and applied stress. This behavior differed from an earlier report (5) in which cavity density of a low carbon Astroloy linearly increased with creep strain. The stress

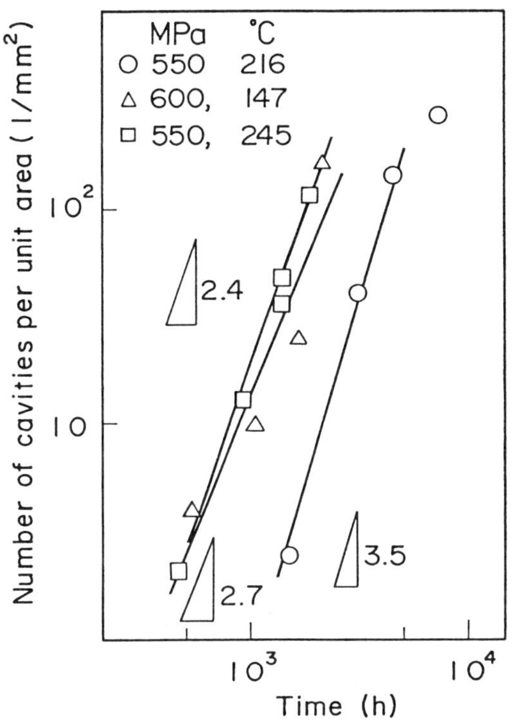


Fig. 3-Time dependance of creep cavity nucleation.

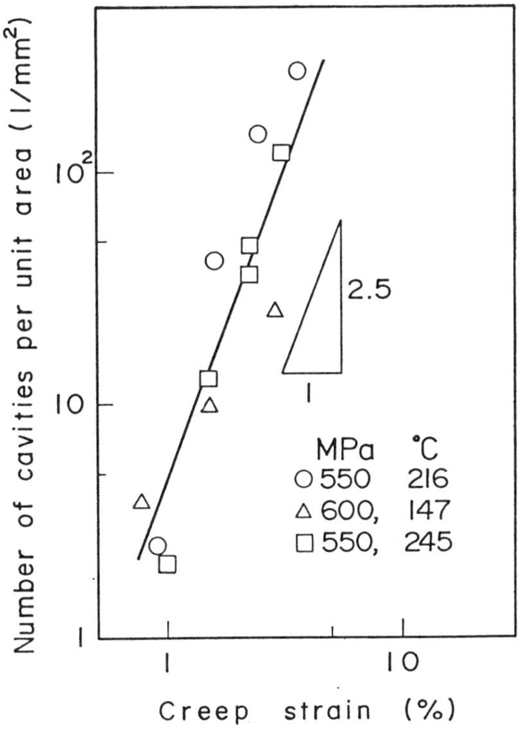


Fig. 4-variation of cavity density with creep strain.

axis dependent behavior of the cavited grain boundaries are illustrated in Figs. 5 and 6 under two stresses of 216 and 245MPa at 550°C . Cavited grain boundary fraction increased with the angle between the grain boundary and stress axis which indicated that creep cavities mainly nucleated on grain boundaries almost perpendicular to a stress axis, even through creep damage varied from 0.34 to

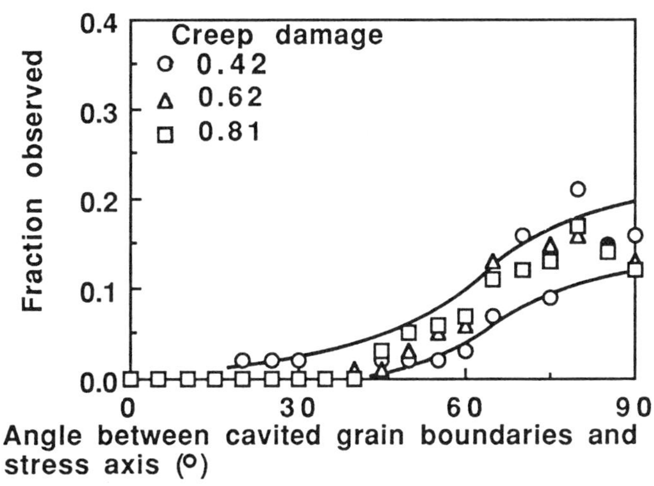


Fig. 5-Stress axis dependent behavior of cavited grain boundaries under 216MPa at 550°C.

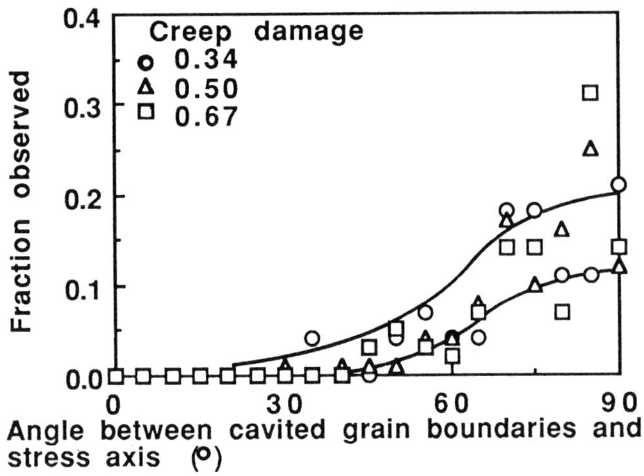


Fig. 6-Stress axis dependent behavior of cavited grain boundaries under 245MPa at 550°C.

0.81. It resulted in a mean angle between cavited grain boundaries and the stress axis of approximately 70 degree, defined by Eq. 1, where $f_i$ represents the fraction observed in the i th group having an angle $\theta_i$ between the cavited grain boundaries and stress axis.

$$\sin\theta = \Sigma f_i \cdot \sin\theta_i \qquad (1)$$

The results are summarized in Table 3. On the basis of these test data, cavity nucleation tended to have a stress axis dependance of the cavited grain boundaries, so that creep cavities did not distribute homogeneously on grain boundaries.

Table 3 Stress axis dependance of cavited grain boundaries

| Creep conditions | Creep damage | Mean angle between cavited grain boundary and stress-axis (degree) | sin $\theta$ |
|---|---|---|---|
| | 0.42 | 68.1 | 0.93 |
| 550°C/216MPa | 0.62 | 70.4 | 0.94 |
| | 0.81 | 69.0 | 0.93 |
| | 0.34 | 69.5 | 0.94 |
| 550°C/245MPa | 0.50 | 70.8 | 0.94 |
| | 0.67 | 71.4 | 0.95 |

CREEP CAVITY GROWTH BEHAVIOR. The time dependance of cavity growth in length is shown in Fig.7. Under these conditions, the mean cavity length on the grain boundaries was roughly a linear function of time and independent of the temperature and applied stress, because slopes of the three lines were 0.7 and 1.3 on a log-log scale. The same behaviorwas seen in cavity growth in length under 147MPa at 600°Cand 245MPa at 550°C. This suggested that a variation of mean cavity length might correlate to creep rupture life. Fig. 8 shows the relation between mean cavity length and creep strain during the secondary creep stage. Mean cavity length increased linearly with creep strain and this ment that creep

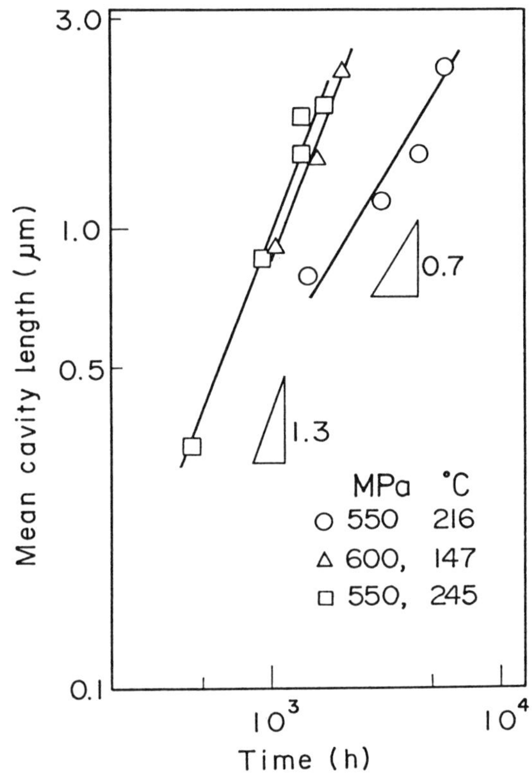


Fig. 7-Time dependence of cavity growth.

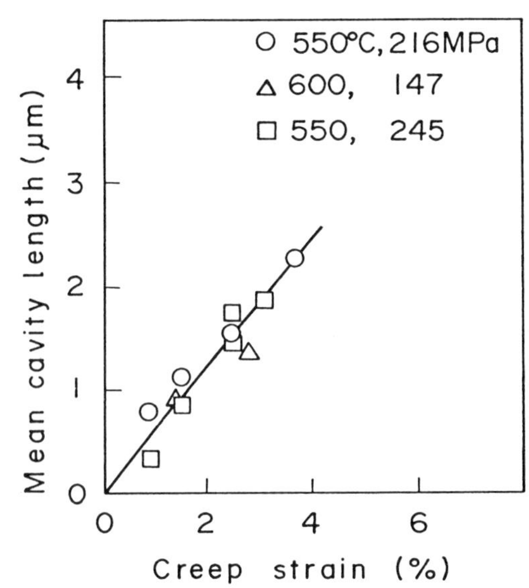


Fig. 8-Relation between creep strain and mean cavity length.

deformation was strongly dependent on creep cavity morphology and independent of temperature and applied stress. Therefore creep deformation seemed to have occurred by the same creep mechanism in this work.

## DISCUSSION

Several mechanisms for the growth of intergranular cavities during creep have been presented (6,7). For Cr-Mo-V steel, it is important to note that creep fracture is caused by diffusional cavity growth (1). In general, vacancy and atom transport in a crystalline solid state takes place by grain boundary diffusion and lattice diffusion, and the effects of the two diffusion mechanisms are dependent on temperature. The ratio of the number of atoms transferred from the cavity to the boundary by lattice diffusion to that by grain boundary diffusion, $\lambda$, is approximated by Eq. 2.

$$\lambda=(D_l \cdot R)/(D_g \cdot \delta Z) \qquad (2)$$

For bcc Fe at 550°C, $D_l$ and $D_g$ are $\sim 2\times10^{-20}\,m^2s^{-1}$ and $\sim 6\times10^{-15}\,m^2s^{-1}$ (8), respectively, while $\delta Z$ is between 2 and 3 nm. Then R must be 5 to $8\times10^2 m\mu$, which is an extreme high limit even if $\lambda=1$. The cavity length, which is similar to R, was observed as below 4 $\mu$m (cf. Fig.7), which means that grain boundary diffusion is more effective than lattice diffusion under the present conditions. Cavity length is a linear function of cavity width as shown in Fig. 9, and it suggests that cavities grow with a similar shape during the creep stage. On the basis of the results that (a) atom transport strongly depends on grain boundary diffusion and (b) cavities retain a similar shape during growth, the Hull-Rimmer Model (6) was applied to this work. A modification was made to account for the surface energy which was insignificant compared with the cavity size. The growth rate of an individual cavity in length, dl/dt, can be obtained from Eq. 3

$$dl/dt=(D_g \cdot \delta Z \cdot \Omega \cdot \sigma)/(k \cdot T \cdot a \cdot l) \qquad (3)$$

In Eq. 3 factors a and l are the only ones dependant on time. Thus, time dependance of center-to-center cavity spacing, a, was considered as follows, while taking into account continuous nucleation. Senior and et al. (9) have demonstrated hat the number of cavities per unit area, N, can be described by Eq.4 for 9Cr-1Mo steel.

$$dN/d\varepsilon=S(\varepsilon-\varepsilon_0) \qquad (4)$$

Then integration gives Eq.5 which shows that N vs $\varepsilon$ can be a parabolic relation

$$N=S(\varepsilon-\varepsilon_0)^2 \qquad (5)$$

For the test material, the creep strain exponent was approximately 2.5 as shown in Fig. 4. It is then assumed that N could also be described by Eq. 5. Additionally cavities were distributed on grain goundaries in the manner shown in Figs.5 and 6, so that N can be obtained by Eq. 6.

$$N=1/(a_g^2 \cdot \tan\theta) \qquad (6)$$

where $a_g$; Center-to-center cavity spacing on the grain boundary perpendicular to stress axis

From Eqs. 5 and 6 $a_g$ is given by Eq. 7 as a function of $\varepsilon$.

$$1/a_g= (S \cdot \tan\theta) \cdot (\varepsilon-\varepsilon_0) \qquad (7)$$

Substituting Eq. 7 into Eq.3 leads to the following expression for cavity growth rate.

$$dl/dt= (S \cdot \tan\theta) \cdot (D_g \cdot \delta Z \cdot \Omega \cdot \sigma) \cdot (\varepsilon-\varepsilon_0)/(k \cdot T \cdot l) \qquad (8)$$

where $(D_g \cdot \delta Z \cdot \Omega \cdot \sigma)/(k \cdot T)$=const.

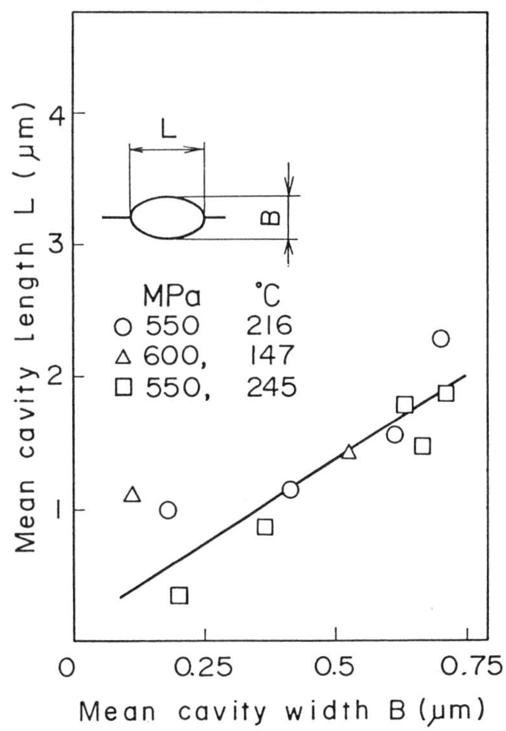


Fig. 9-Relation between cavity length and cavity width.

Creep strain, $\varepsilon$, is usually presented by a linear function of time, t, during the secondary creep stage. Thus, substituting this relation into Eq. 8 and subsequently intergrating cavity length leads to the conclusion that cavity length should be a linear function of time, where it is assumed that creep strain, $\varepsilon$, is much larger than the creep strain at the stage of cavity nucleation, $\varepsilon_0$. Then both cavity length and creep strain have a linear dependance on time during the secondary creep stage, and cavity length is also thought to be directly proportional to the creep strain. This prediction agrees well with experimental results as shown in Fig.8, which tend to support the explanation of creep in Cr-Mo-V steel by the above-mentioned grain boundary diffusion mechanism.

In this work variation in mean cavity morphology was observed and it was found that creep in Cr-Mo-V steel was mainly caused by a grain boundary diffusion. When estimating creep cavity growth theoritically it is necessary to evaluate stress and stress axis dependences of creep cavity growth. It is also thought that cavity nucleation behavior, such as nucleation mechanisms and time of nucleaton, should be considered.

## CONCLUSIONS

The stress-axis dependance of creep cavity nucleation and creep cavity growth behavior were experimentally studied for Cr-Mo-V steam turbine rotor material. The results are summarized as follows.

1. Intergranular creep cavities mainly nucleated on grain goundaries perpendicular to a stress axis, even if creep damage varied from 0.34 to 0.81 under 216 and 245MPa at 550°C.

2. Mean cavity length was roughly proportional to time and creep strain vs mean cavity length could also be approximated as a linear relationship. The number of cavities per unit area was described by a power law function of time, where the time exponent was between 2.4 and 3.5.

3. Creep cavity growth occurred by grain boundary diffusion, which could be explained by the Hull-Rimmer theoretical model.

## REFERENCES


1. N. Shinya, Report of the 123rd committee on Heat Resisting Metals, 22,69(1981)
2. Cane B. J. , Metal Science, 15, 295-301(1981)
3. Needham N. G. and Gladman T., Proceedings of the Second International Conference on Creep and Fracture of Engineering Materials and Structure, 1263-1276(1984)
4. K.Iijima, Y. Fukui and R. Kaneko, Proceedings of International Joint Power Generation Conference, PWR-10, 45-50(1990)
5. Capano M. A. , Argon A. S. and Chen I-W., Acta Metall., 37, 3195-3204(1989)
6. Hull D. and Rimmer D. E.,Philos. Mag. , 4, 673-687, (1959)
7. Dyson B. F., Metal Science, 349-353(1976)
8. Chang T-J., Kagawa K. I., Rice J. R. and Sills L. B., Acta Metall., 27, 265-284(1979)
9. Senior B. A., Noble F. W. and Eyre B. L., Acta Metall., 34, 1321-1327(1986)

*Proceedings of the Fifth International Conference on Creep of Materials, Lake Buena Vista, Florida, USA, 18-21 May 1992*

# Determination of Creep Damage in Steels

**U. von Estorff and H. Stamm**
Commission of the European Communities
Institute of Advanced Materials
Ispra, Italy

## Abstract

Remanent life assessment for pressurised components operating at elevated temperatures requires knowledge about the actual degradation of the material. In this paper studies are presented which address the problem of damage determination in steels exposed to creep. Ultrasonic (US-) methods backed up by quantitative metallographic studies have been applied to different materials in order to determine the damage state. For the case where creep damage manifests itself in the cracking of grain boundary facets, a damage variable related to the density and size of the microcracks can be identified. The relationship of the damage variable to ultrasonic measurements and metallographic findings is established. This allows the direct comparison between quantitative metallographic and ultrasonic results. The consistency of the results with predictions from the material models is demonstrated. The applicability of ultrasonic measurements to other types of microstructural damage, i.e. cavity formation on grain boundaries is discussed. Based on such laboratory tests, some conclusions concerning the advantages and limitations of metallographic and ultrasonic techniques for damage determination are drawn.

CREEP DEFORMATION in pressurised components operating in the high temperature regime involves a microstructural degradation of the material (in the following referred to as 'creep damage') which may lead to rupture, thus limiting the lifetime of components. Therefore, reliable methods for the detection of creep damage are indispensable both, to guarantee safe operation during the designed lifetime and to provide the basic information about the defect state of the service exposed material in order to estimate the possible lifetime extension of old plants.

In the case of creep damage which manifests itself as cavities on grain boundaries or intergranular microcracks various testing methods have been employed in the past such as metallography, measurements of density, hardness, electrical resistivity, ultrasonics, small angle neutron scattering etc. So far, only metallographic methods e.g. the replica technique are successfully used for the inspection of components. This technique is, however, restricted to applications at the surface of the testpieces and cannot be used to screen large areas. A principal problem of all the methods which inherently include a volume averaging is to discriminate between creep damage and other effects that influence the physical quantity used for damage characterisation. Therefore, it is important to develop each of the promising methods especially the non-destructive ones to a certain maturity. Only a well thought-out combination of different methods can lead to a reliable assessment of damage.

In recent years several efforts were undertaken to investigate the potential of non-destructive ultrasonic techniques for the detection of creep damage [1-17]. So far, such methods have mainly been carried out on a laboratory scale by examining test specimens and also damaged parts of components. Most of the current research activities aim at qualifying the US-methods by laboratory tests comparing the sensitivity of the techniques with other methods for damage detection. An overview on the current status has been given in [18].

In this paper the measurement of creep damage by non-destructive ultrasonic methods is discussed with regard to different materials. The materials and creep conditions are presented in the next section. The subsequent sections are devoted to a detailed description of the US-measurements and their theoretical interpretation. Finally, the results obtained are compared with findings from other techniques for damage determination such as metallographic examinations.

## Creep

US-measurements were performed on two austenitic steels which had been subjected to creep. The most complete testing series has been accomplished with the material described in the following.

**Mn-Cr austenitic steel.** A series of uniaxial creep tests has been performed on a Mn-Cr austenitic steel (trade-name AMCR 0033) which was supplied by Creusot-Loire as a 30 mm hot-rolled plate in the solution-annealed condition. This steel was considered as a candidate material for the first wall of a fusion reactor because of its low activity after neutron irradiation [19]. The chemical composition in weight-percent is given in Table 1. The grain size of the fully austenitic structure was 70 μm, determined by the lineal-intercept method. The creep tests were either continued until rupture or interrupted at various stages of the creep curve, in particular at the end of the secondary and during the tertiary creep regime. Flat specimens with two parallel faces and a gauge length of 50 mm were used. Initial stresses

$\sigma_0$ of 80 MPa, 100 MPa, 120 MPa and 150 MPa were applied at the test temperature of 923 K. Rupture times were obtained in the range from about one week for the highest load to about 1 year for the lowest load. The corresponding failure strains fell into the range from 1.5% to 7%. A total of 45 crept specimens was available.

Table I. Chemical composition of AMCR 0033 in weight-percent.

| C | Cr | Mo | Mn | P |
|---|---|---|---|---|
| 0.1 | 10.1 | 0.065 | 17.3 | 0.02 |
| **Cu** | **S** | **Si** | **Al** | **Pb** |
| <0.06 | 0.008 | 0.55 | <0.005 | <0.0001 |
| **Ni** | **B** | **V** | **N** | |
| <0.1 | 0.0025 | 0.015 | 0.2 | |

in weight-%

The secondary creep rates $\dot{\varepsilon}_{ss}$ could be represented by Norton's law

$$\dot{\varepsilon}_{ss} = B\sigma^n \quad \text{with} \quad B = 1.10\times10^{-19}(\text{MPa,h}), \quad n = 6.9 \qquad (1)$$

and the stress dependence of the rupture times $t_f$ by the relation

$$t_f = \frac{C}{\sigma^\nu} \quad \text{with} \quad C = 1.10\times10^{14}(\text{MPa,h}), \quad \nu = 5.4 \qquad (2)$$

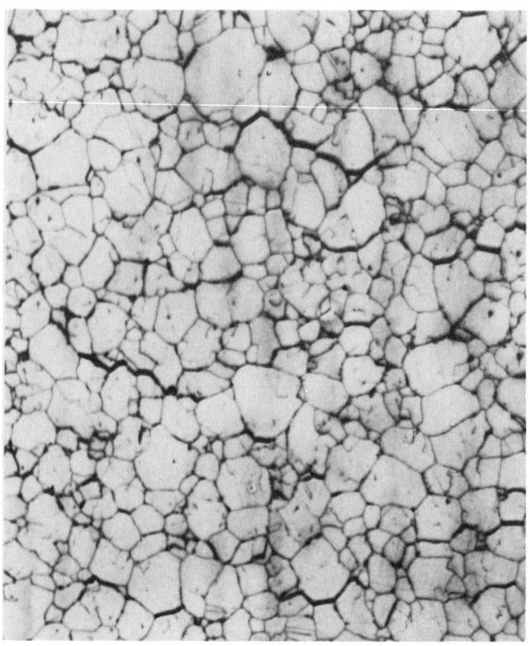

Fig. 1 Optical micrograph of crept AMCR-specimens close to the rupture surface showing intergranular microcracks

Metallographic examination of the specimens after creep indicated the appearance of intergranular microcracks as shown in Fig. 1. These microcracks in general consist of 1 to 2 grain boundary facets with a tendency to coalesce to longer cracks in the vicinity of the rupture surface. The crack faces are preferentially oriented perpendicular to the direction of stress as can be seen from Fig. 2 showing the empirical cumulative distribution function of the angle $\phi$ between the crack plane and the stress direction. This diagram has been obtained by evaluating more than 300 cracks in different specimens. The preferential orientation is indicated by the curved shape of F. The cracking of grain boundary facets has been found to be caused by the rapid formation of intergranular carbide precipitations at the test temperature [20]. This leads to a rather brittle behaviour of the material under creep which is also disclosed by the low average value of the strain ratio (creep damage tolerance)

$$\gamma = \frac{\varepsilon_f - \varepsilon_p}{\dot{\varepsilon}_{ss} t_f} \approx 1.6 \qquad (3)$$

where $\varepsilon_f$ is the failure strain and $\varepsilon_p$ the strain accumulated during primary creep.

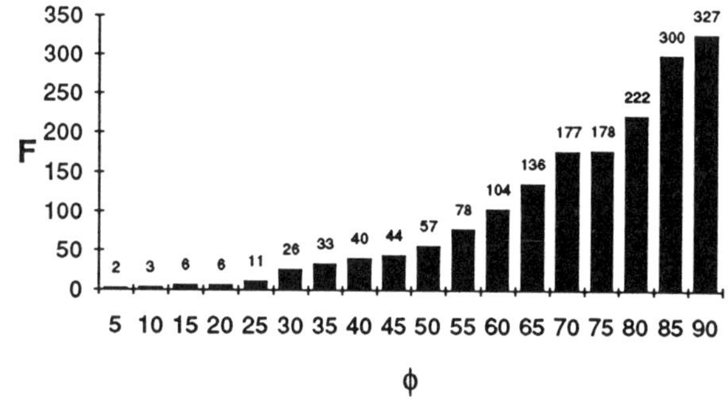


Fig. 2 Empirical cumulative distribution function of the angle $\phi$ between the crack faces and the stress direction in AMCR 0033

**Incoloy 800H.** This alloy is also known as Nicrofer 3220. Table 2 shows the chemical composition of the solution-annealed material. Uniaxial creep tests at 1073 K with initial stresses of 40 MPa and 50 MPa were performed at KFA Jülich [21], using flat specimens. The tests were interrupted after 1330 - 5122 h before rupture occurred. This resulted in creep strains of 0.16 - 4.39 % and in one case of 13.3%.

Table II. Chemical composition of Incoloy 800H in weight-percent.

| C | Ti | Mn | Al | Si |
|---|---|---|---|---|
| 0.07 | 0.42 | 0.87 | 0.47 | 0.06 |
| **Fe** | **Cr** | **S** | **Ni** | **Cu** |
| 43.88 | 19.72 | 0.003 | 33.83 | 0.68 |

in weight -%

## Ultrasonic measurements

Both ultrasonic velocity and attenuation measurements have been carried out in order to assess their potential for creep damage detection. Their application to the Mn-Cr austenitic steel will be detailed in the following. US-velocity measurements in Incoloy 800H will be discussed.

**US-velocity measurements.** The generation of flaws in the form of cavities or microcracks on grain boundaries influences the local compliance of a polycrystalline material. This causes, in general, a decrease in the US-velocity. Bulk waves are most often used for damage detection in order to obtain information on the damage state in the bulk material.

Due to the preferential orientation of the microcracks in AMCR 0033 (which is also observed in many other materials) a damage induced anisotropy results. Therefore after creep, even an initially isotropic material exhibits a transverse isotropic behaviour and the US-velocity depends on the propagation direction. Pure modes, i.e. modes where the particle velocity is either parallel or normal to the propagation direction exist only for special propagation directions. For waves propagating perpendicular to the axis of symmetry which is the direction of the applied stress, two shear modes with different propagation velocities polarised parallel ($v_{||}$) and perpendicular ($v_{\perp}$) to the axis of symmetry appear besides the longitudinal mode $v_l$. With $q_0$ characterising an initial anisotropy which may already be present in the undamaged material, the ratio

$$q = \frac{v_{||} - v_{\perp}}{v_{||}} - q_0 \quad , \qquad q_0 = \left( \frac{v_{||} - v_{\perp}}{v_{||}} \right)_{initial} \tag{4}$$

can be taken as a characteristic measure for the damage induced anisotropy [11, 14-16].

*Measurement technique.* The US-velocities have been determined by time-of-flight measurements using a pulse-echo method. In each specimen the measurements were executed in the specimen heads and along the gauge length before and after the creep test. Each echo was compared with the corresponding one obtained in a reference specimen of the same material and with the same geometry. The US-waves were generated with 0° highly damped transducers with a nominal frequency of 10 MHz for the longitudinal waves and 5 MHz for the polarised transverse waves. The wavelength of about 0.6 mm was much larger than the typical size of a broken grain boundary facet. The crystal diameter of the transducer was 6.35 mm (0.25 inch).

In order to improve the reproducibility of the velocity measurements micrometric monitoring was employed to enable an accurate positioning of the transducer and a compressed air system aided in the maintenance of a constant coupling layer thickness.

In principle, the attenuation measurements can be performed in the same way. However, since this method is very sensitive to the coupling conditions, the immersion technique must be applied and thus only longitudinal waves can be generated. A transducer with the same characteristics as for the velocity measurements has been used. Frequency spectra were recorded and the attenuation coefficients were calculated for each frequency from the decrease of the amplitudes of the two subsequent echoes.

*Results in AMCR 0033.* By means of the measures described above, a reproducibility of ±0.15% of the US-velocity values measured at a given point could be achieved. In the case of the anisotropy measure $q$ the reproducibility was even better. To separate the effects on the US-velocity which are not caused by creep damage, the influence of ageing during the different hold times at the temperature of the creep test and of a deformation induced texture have been assessed.

The changes in the US-velocities of the transverse waves polarised parallel to the stress axis as a function of the hold times at the test temperature are plotted in Fig. 3. The measurements were taken in the specimen heads which experienced no load during the test. As well as an increase, a decrease in the velocities with respect to the virgin state could be observed with a maximum change of about 0.5%. Similar results were obtained for the other wave modes $v_{\perp}$ and $v_l$ and the anisotropy measure $q$. Taking into account this fact, the detectability of creep damage is limited to velocity reductions greater than 0.5%.

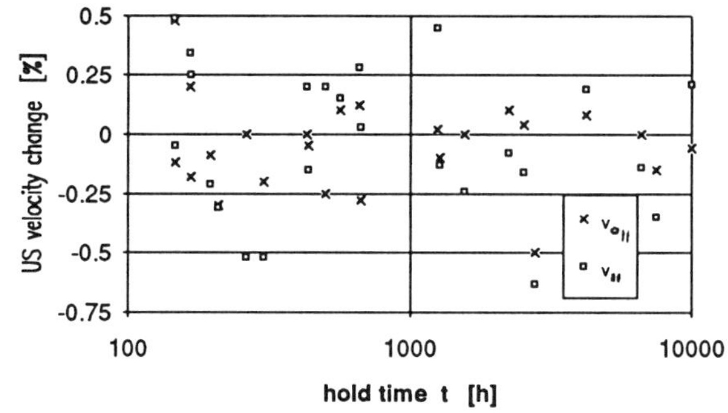


Fig. 3 Influence of the hold times at the creep temperature $T$ = 923 K on the US-velocities of transverse waves polarised parallel to the stress axis.

The effect of a texture induced by the creep deformation has been estimated by measuring the US-velocity in a specimen which ruptured after 0.25 h with a failure strain of 30% at a $T$ = 923 K with an initial applied stress $\sigma_0 = 300$ MPa . For propagation directions perpendicular to the stress axis no changes in the longitudinal wave velocity were observed. For the case of transverse waves polarised parallel and perpendicular to the stress axis a nearly uniform decrease of 1.7% and 0.7% respectively, was found within the gauge length leading to an anisotropy value of $q$ = 0.9%. It must be emphasised that these values were measured at a strain which was at least five times as high as the typical failure strains obtained in the creep tests at 150 MPa. In addition an elongation and a preferential orientation of the grains could be observed under the microscope which was not visible in the crept specimens. Therefore, contributions to the wave velocity changes due to pure creep deformation are negligible under the conditions of the creep tests described above.

A representative result for the decrease in the US-velocity as obtained after creep is shown in Fig. 4. The points represent the velocity changes with respect to the reference specimen. It can be seen that the US-velocity in the virgin specimen is rather uniform within the gauge length (differences less than 0.1%). After creep, the velocities of all wave modes display a decrease with a gradient towards the centre of the specimen. A similar result is found in ruptured specimens with the highest decrease at the measurement points next to the rupture surface. As expected from the theory, the velocity decrease for the longitudinal waves lies between the values for the shear waves polarised perpendicular (lowest decrease) and parallel to the stress axis.

The difference in the two shear wave modes is very marked. This gives rise to an anisotropy measure $q$ as shown in Fig. 4 which exhibits the same dependency on the position within the gauge length as the US-velocities. Another example is given in Fig. 5 where $q$-values obtained in a broken and an unbroken specimen are displayed. The creep test of the unbroken specimen had been interrupted after $t_i$ = 146 h whereas the other specimen ruptured after 167 h. The change in anisotropy is quite similar in both cases which indicates that the damage state in the unbroken specimen is rather advanced. As will be discussed more in detail in the following sections, $q$ is a characteristic measure of creep damage for the case of AMCR 0033, so therefore, it is convenient to present the US-velocity measurements in terms of $q$.

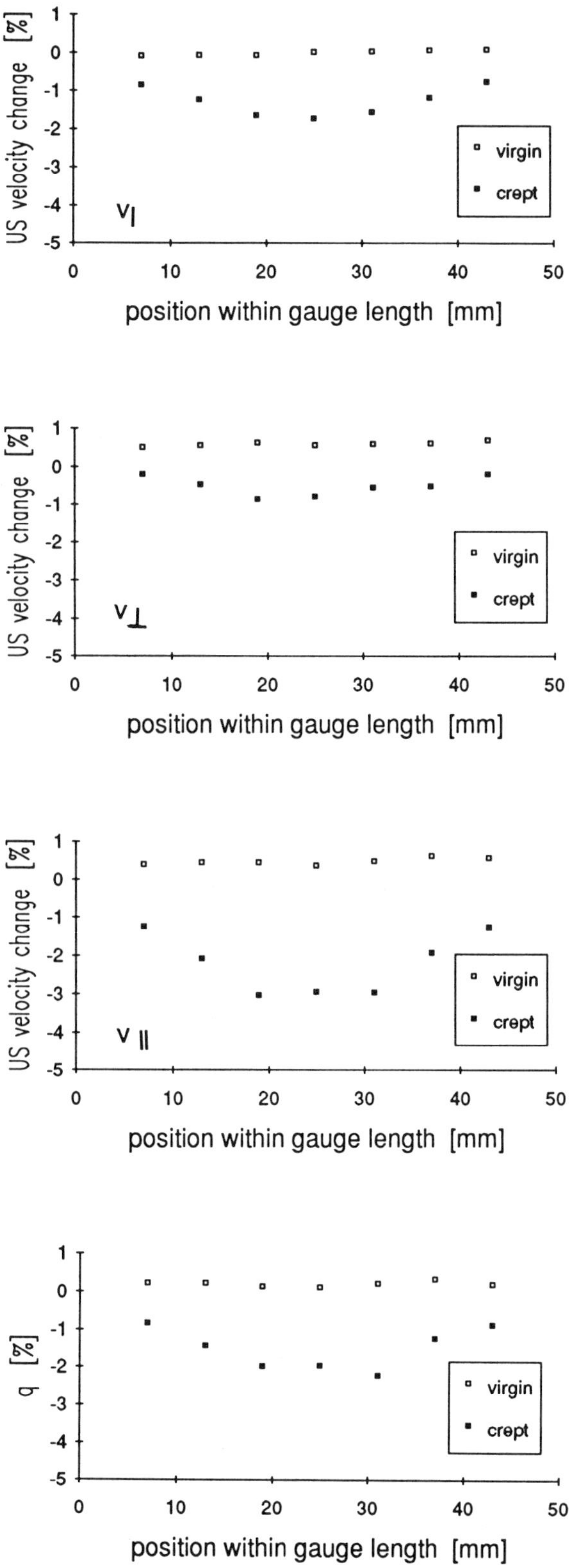


Fig. 4 US-velocity measurements as a function of the position within the gauge length of a non-ruptured specimen exposed to creep for 146h (tertiary creep) at $\sigma_0 = 150$MPa, $T = 923$K

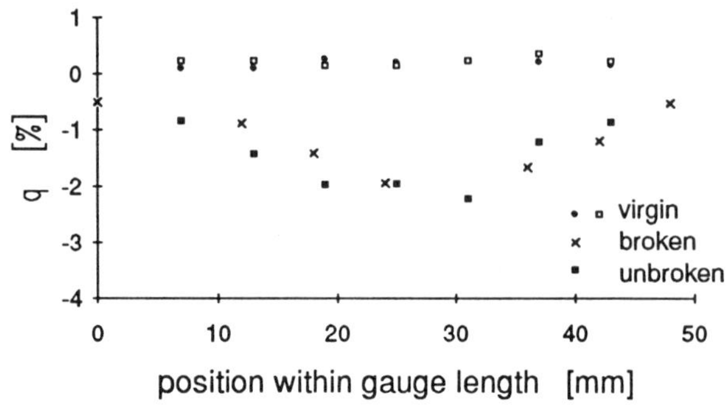


Fig. 5 Anisotropy measure $q$ in a broken and an unbroken specimen from a creep test with $\sigma_0 = 150$ MPa, $T = 923$ K, interruption time $t_i = 146$ h, failure time $t_f = 167$ h

*Results in Incoloy 800H.* Five unbroken specimens of creep tests were available for US-velocity measurements. A typical result is shown in Fig. 6. Here, the longitudinal wave velocity and the anisotropy measure $q$ as a function of the position in

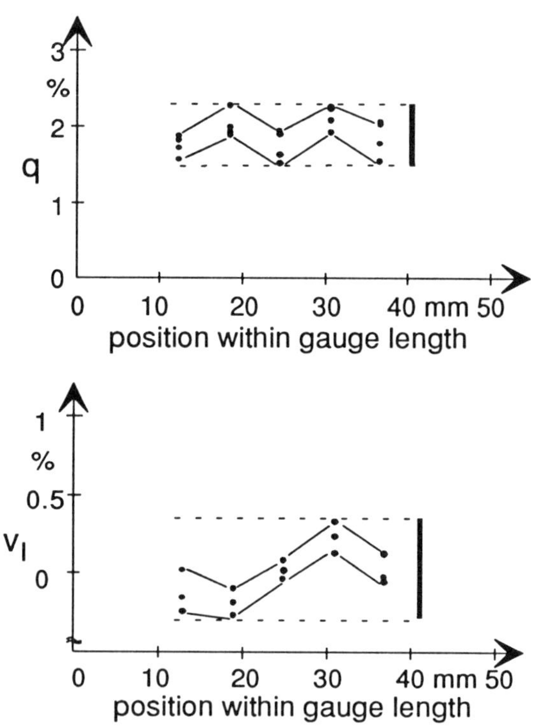


Fig. 6 Variations of longitudinal wave velocities and of the anisotropy measure $q$ in the gauge length of virgin 800H specimens

the gauge length of a virgin specimen is plotted. The relatively high $q$-value indicates a preferential orientation in the microstructure of the virgin material. Problems are encountered as far as the reproducibility of the measurements is concerned. These arise partly from the difficulties in obtaining the clear superposition of two successive echos which is necessary for the determination of the transition time of a signal. In addition, compared to other steels investigated, this alloy showed a significant inhomogeneity. In Fig. 6 the scattering range obtained by repeating several times the measurement at the same point is indicated. This lack of homogeneity and reproducibility can be observed in each specimen.

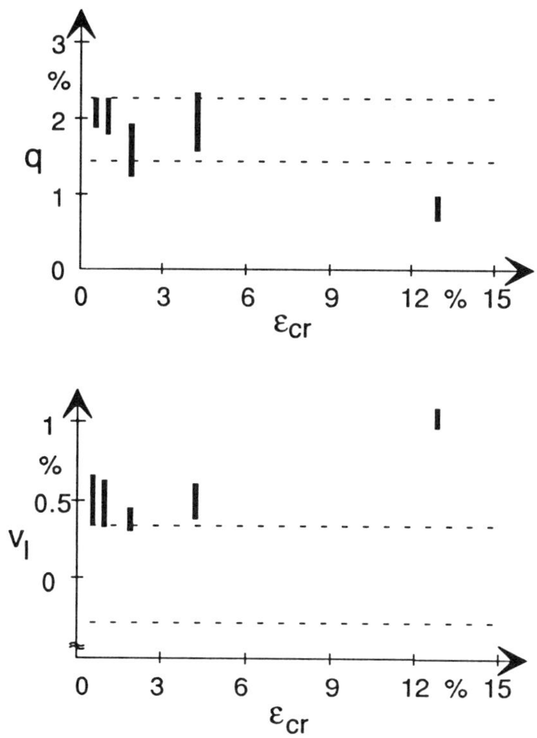


Fig. 7 Dependence of the longitudinal wave velocity and of the anisotropy measure $q$ on the creep strain $\varepsilon_{cr}$.

The scattering ranges of the US-velocity and $q$ measurements in the crept specimen are plotted in Fig. 7 as a function of the strain incurred during the creep test. As these specimens were not available before creep i.e. in the virgin state, it is not possible to say anything about the contribution of the initial inhomogeneity to the indicated scatter bands. By comparing the results with those shown in Fig. 6, only for the specimen with a deformation of 13.3 % are velocity and anisotropy changes found which fall clearly outside of the scatterband of the virgin specimen. As in the case of the AMCR-steel the anisotropy measure $q$ decreases, whereas the velocity of the longitudinal waves increases by about 1% when compared to the virgin material. This increase is contrary to the behaviour expected because of the formation of microcracks or voids. In [22-24] the microstructure of the alloy 800H was thoroughly investigated and it was shown that carbide precipitates formed as the temperature was increased which hardened the structure and lowered the creep rate. Whether the change of the longitudinal wave velocity is caused by the precipitates or by other phenomena, e.g. reorientation of grains due to the large strain, is not yet completely examined.

**US-attenuation measurements.** The attenuation of US-waves is very much influenced by the polycrystalline structure as has been discussed e.g. in [25,26]. Therefore, several attempts have been made to analyse attenuation and frequency spectra of reflected waves with respect to their sensitivity on the creep damage.

Disregarding the effects due to geometrical divergence of the sound field, the attenuation of the amplitude $A$ after propagation along the sound path $s$ can be described by

$$A = A_0 e^{-\alpha s} \qquad (5)$$

where $A_0$ is the initial amplitude. In polycrystalline alloys the attenuation coefficient

$$\alpha = \alpha_a + \alpha_s \qquad (6)$$

can be decomposed into two contributions $\alpha_a$ and $\alpha_s$ which are related to absorption e.g. by dislocation damping and to scattering from elastic discontinuities such as grain boundaries, precipitations, inclusions, voids etc., respectively. In a polycrystalline material, the main contribution to the attenuation coefficient is due to scattering.

All the experimental results reported here were obtained from work on AMCR 0033. Using the same type of signals as for the velocity measurements, the frequency dependence of $\alpha_s$ is determined by Rayleigh scattering with $\alpha_s \propto k^4$ where the wave number $k$ is related to the wavelength $\lambda$ by $k = 2\pi/\lambda$.

In Fig. 8 the frequency spectrum of the first and second backwall echoes in a virgin and a crept specimen are compared. Two features are apparent: (1) The intensity of the echo is significantly reduced in the crept specimen and (2) the maximum of the reflected intensity is shifted to a lower frequency.

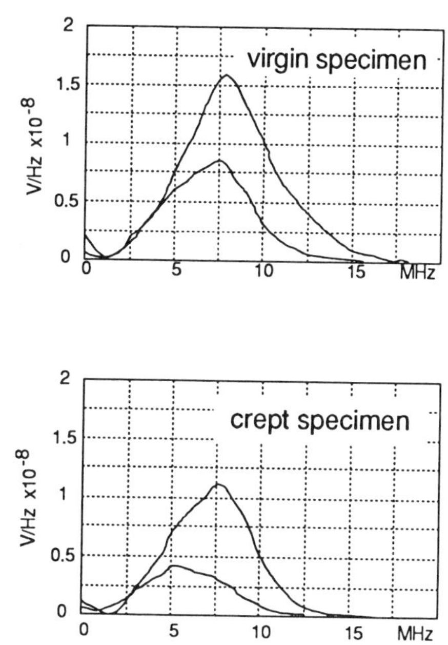


Fig. 8 Comparison between the frequency spectra of the first and second echo of a virgin and a crept specimen (creep at 923 K, $\sigma_0$ = 150 MPa, $t_f$ = 167h)

From measurements as shown in Fig. 8 the attenuation coefficient $\alpha$ can be determined as a function of the frequency. In order to see whether changes to the attenuation coefficients in crept specimens can be correlated with the US-velocity changes, $\alpha$ at the frequency of 10 MHz has been plotted in Fig. 9 as a function of the longitudinal wave velocity changes. An increase of $\alpha_s$ can be observed as $v_l$ diminishes although a considerable scatter has to be taken into account.

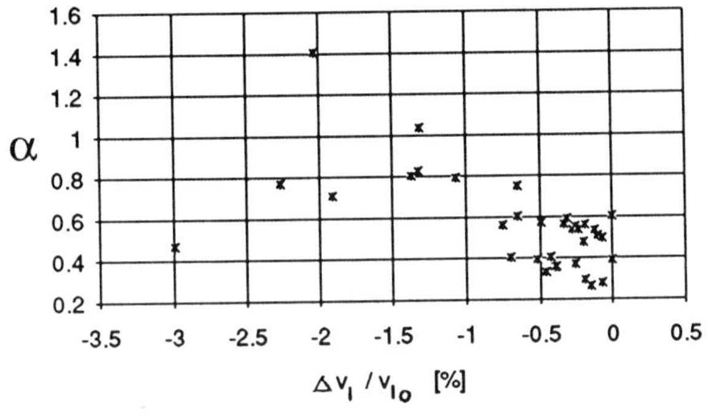


Fig. 9 Scattering coefficients as a function of the longitudinal wave velocity changes.

## Comparison with theory

The application of the US-measurements to the characterisation of creep damage requires some theoretical considerations in order to link the creep and US-results. The following discussion is based on the extensive data base obtained for AMCR 0033.

**Modelling of creep.** For the description of the creep behaviour in the secondary and tertiary stage, the phenomenological Kachanov-Rabotnov model [27] is well suited to many situations. In this model a macroscopic damage variable $D$ is introduced for the characterisation of the creep damage. $D$ equals zero in the undamaged material and failure is assumed to occur when $D = 1$. The constitutive model is given by the relations

$$\dot{\varepsilon}_{cr} = \frac{B\sigma^n}{(1-D)^n} \qquad \dot{D} = \frac{\sigma^\nu}{C(1+\mu)(1-D)^\mu} \tag{7}$$

for the creep strain rate $\dot{\varepsilon}_{cr}$ and the damage rate $\dot{D}$. $\sigma$ is the applied stress and $B$, $C$ as well as the exponents $n$, $\mu$, $\nu$ are the material constants. Their values are given in equations (1) and (2). The additional parameter $\mu = 17.4$ has been determined from the creep damage tolerance $\gamma$.

By integrating the second relation in equation (7), it can be shown that for a constant stress creep test, $D$ only depends on the normalised time $t/t_f$ and is independent of the applied stress. This has been examined by calculating the damage parameter from the creep curves recorded during the test by inversion of the flow law with

$$D = 1 - \left(\frac{\dot{\varepsilon}_{min}}{\dot{\varepsilon}_{cr}}\right)^{\frac{1}{n}}. \tag{8}$$

$\dot{\varepsilon}_{min}$ is the minimum creep rate measured in each test. Fig. 10 shows both the theoretical curve calculated with the material parameters given above and the damage parameter values calculated from the individual tests. Some scatter around this curve can be observed, however, there is no indication of a systematic separation of data points obtained at different stresses. The maximum strain was always less than 6.8%, so that the condition of constant stress during creep is approximately fulfilled.

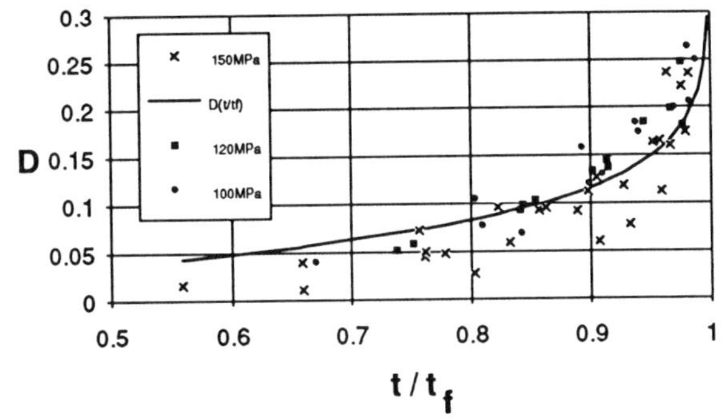


Fig. 10 Damage parameter $D$ determined from creep curves for different initial stresses $\sigma_0$. The solid line represents the theoretical curve calculated from eq. (7)

From Fig. 10 it can be seen that $D < 0.1$ for about 85% of the lifetime of a creep specimen. Under these conditions (see [3]), the flow law in eq. (7) can be compared with the creep rate as determined by Hutchinson [28] for a material damaged by a dilute concentration of circular microcracks. This allows a physically based interpretation of $D$ which is closely related to the microstructural damage mode observed in AMCR 0033. If the creep damage, in the form of grain boundary cracks, can be idealised by a dilute concentration of circular microcracks oriented perpendicular to the axis of the applied stress, then according to [28] the flow law can be given by

$$\dot{\varepsilon}_{cr} = B\sigma^n[1+\eta f(n)] \quad \text{with} \quad f(n) = \frac{3(n+1)}{\pi(1+3/n)^{1/2}} \tag{9}$$

where $B$ and $n$ are the known material constants characterising secondary creep. The damage is described by the crack density parameter

$$\eta = \frac{4\pi}{3} N a^3 \tag{10}$$

where $N$ is the number density of cracks with radius $a$. Developing the phenomenological flow law in eq. (7) into a Taylor series and comparing the term linear in $D$ with eq. (9) the relation

$$D \;=\; \frac{3}{\pi}\frac{n+1}{n}\frac{1}{(1+3/n)^{1/2}}\eta \;=\; 0.91\,\eta \tag{11}$$

follows ($n = 6.9$). This equation will prove to be useful when relating the Kachanov damage parameter to experimental measures of damage which can often be expressed in terms of $\eta$. For the case of varying crack sizes, $a^3$ in the $\eta$ expression has to be replaced by the average value $\langle a^3 \rangle$.

It should be noted that the analysis of the results presented here ignores the non-uniform deformation within the gauge length. In fact, strain gradients towards the centre of the gauge length are observed which are probably due to slight geometrical inhomogeneities in the virgin specimens (cross section reduction ≈ 0.5% near the centre). This is also the reason for the damage gradients observed by the ultrasonic measurements or by other detection techniques. The modelling as described here concentrates on the average effects within the gauge length and should be complemented by a more detailed analysis which requires numerical methods.

**Modelling of ultrasonic behaviour.** *Ultrasonic velocity.* With the frequencies chosen for the US-testing, the wavelength $\lambda$ is large compared to the average grain size and the characteristic dimension of the defects. Thus, dispersion effects or excessive scattering due to jumps of the wave impedance on grain boundaries are avoided. The wave velocity $v_i$ of the wave mode $i$ has the form

$$v_i = \sqrt{\frac{C_i}{\rho}} \tag{12}$$

where $\rho$ is the average mass density, $C_i$ is a combination of elastic constants depending on the material symmetry, on the wave mode $i$ and on the propagation direction [29].

For the calculation of the elastic constants of a material with defects such as microcracks or cavities (which are observed in many other materials as the typical form of creep damage), the Multiple Scattering Theory [7,30,31] or the Effective Medium Theory [32-34] can be used. The defects are then treated as a second phase with zero stiffness and density. For the case of a distribution of microcracks, the elastic moduli depend on the crack density parameter $\eta$. The density remains unchanged since an ideal crack has no volume and the opening of the crack can be neglected.

As discussed earlier, the microcracks in AMCR 0033 have a preferential orientation with respect to the applied stress. The solution for the elastic moduli of randomly dispersed, unidirectionally aligned circular cracks which can serve as an approximation for the elastic behaviour of damaged AMCR 0033 is given in [34]. From this solution, e.g., the velocities of transverse waves polarised parallel ($v_{||}$) or perpendicular ($v_\perp$) to the axis of the applied stress can be calculated according to

$$\frac{v_{||}}{v_{t0}} = \left[1+\frac{4(1-\nu_0)}{\pi(2-\nu_0)}\eta\right]^{-\frac{1}{2}} \tag{13}$$

and

$$\frac{v_\perp}{v_{t0}} \approx 1. \tag{14}$$

$v_{t0}$ is the transverse wave velocity and $\nu_0$ the Poisson's ratio in the undamaged, isotropic material. From this the anisotropy measure defined in eq. (4)

$$q = 1-\sqrt{1+\frac{4(1-\nu_0)}{\pi(2-\nu_0)}\eta} \tag{15}$$

can be deduced. As $q \ll 1$ was fulfilled in all the measurements, it is sufficient to consider in the following the linear approximation of (15),

$$q \approx -\frac{2(1-\nu_0)}{\pi(2-\nu_0)}\eta = -0.27\,\eta \tag{16}$$

The proportionality factor 0.27 follows if the value of $\nu_0 = 0.26$ for AMCR 0033 is inserted. This linear relationship coincides with the one that can be derived from [30].

Combining eqs. (11) and (16), the anisotropy measure $q$ can be directly related to the Kachanov damage parameter $D$ by

$$D = \frac{3(2-\nu_0)(n+1)}{2(1-\nu_0)n(1+3/n)^{1/2}}|q| = 3.38\,|q| \tag{17}$$

The maximum values of $q$ which were found in broken specimens close to the rupture surface approached about 2.5% from which the damage value $D \approx 0.085$ can be calculated using eq. (17). According to Fig. 10 this corresponds to a consumed lifetime fraction $t/t_f$ of more than 0.8 at the measurement point. The fact that the specimen is already broken is due to the inhomogeneous damage distribution within the gauge length. Fracture is caused by the progressively growing damage at the smallest cross sectional area. The damage growth in other parts of the gauge length lags behind because of the highly non-linear dependence of the damage growth rate on the actual damage value (see eq. (7)). The comparison of results in broken and unbroken specimens as shown in Fig. 5 supports this interpretation. In the unbroken specimen which has reached 87% of the lifetime of the broken one the same level of damage can be detected. An additional indication comes from the metallographical studies, showing that the appearance of damage concentrations (cracks comprising several broken grain boundary facets) is confined to distances of the rupture surface which are small compared to the beam width. The US-transducer cannot be placed closer to the rupture surface since the superposition of the reflected signals involves too many uncertainties due to the finite beam width. Therefore, the high damage values in the vicinity of the rupture surface cannot be resolved.

The theoretical relationships given in eq. (15) and (17) are based on the assumption that the microcracks are perfectly aligned with the crack planes oriented perpendicular to the applied load. The micrograph in Fig. 1 and the statistical evaluation in Fig. 2 demonstrate that this is only approximately fulfilled. Hence, a change in the wave velocity $v_\perp$ is observed, which is not predicted from eq. (14). The misalignment results in an underestimation of the crack density parameter $\eta$ obtained from eq. (15) as a consequence of inserting an experimental value measured for $q$. In the same manner the Kachanov parameter $D$ calculated from eq. (17) is underestimated.

Nevertheless, the theoretical interpretation yields a fairly good understanding of the observed phenomena. Although the models are rather simple, the order of magnitude of the effects is reproduced quite well. In addition, the US-measurements are consistent with the continuum mechanical approach for quantifying the damage state. In the next section, comparisons with quantitative metallographic results will corroborate the viewpoint adopted here.

*Ultrasonic attenuation.* The total attenuation coefficient in the Rayleigh range is almost completely determined by the contribution $\alpha_s$ due to scattering at elastic discontinuities such as grain boundaries and cracks. The scattering coefficient $\alpha_s^{(cr)}$ for randomly dispersed, unidirectionally aligned circular cracks is given in [30]. This result can be used for assessing the effect of grain boundary cracks in a polycrystalline material on the attenuation. For longitudinal waves incident in planes parallel to the crack faces with the wave number $k_l$ the relation

$$\alpha_s^{(cr)} = 0.06\,\eta\, a^3 k_l^4 \tag{18}$$

is obtained when material parameters for AMCR 0033 are inserted. The coefficient

$$\alpha_s^{(gb)} = 0.054\, d_g^3 k_l^4 \tag{19}$$

which accounts for the grain boundary scattering is assessed on the basis of results given in [35] with the wave velocities of AMCR 0033 and an anisotropy factor characterising the cubic symmetry of the austenitic grains. $d_g$ stands for the average grain size. From this the ratio

$$\frac{\alpha_s^{(cr)}}{\alpha_s^{(gb)}} = 1.11\,\eta\left(\frac{a}{d_g}\right)^3 \tag{20}$$

follows. Assuming that the two contributions to the wave scattering are additive, the change in the attenuation coefficient due to the presence of creep damage can be assessed. By taking an upper bound for the anisotropy measure $q = 5\%$ and an estimate of $a/d_g \approx 1$ (both values which are higher than the ones observed in AMCR 0033) it follows that $\alpha_s^{(cr)}/\alpha_s^{(gb)} \approx 21\%$. This estimate demonstrates that it is difficult to detect creep damage by attenuation measurements since the reproducibility is poor (about ± 20%). Only in the immediate vicinity of the rupture surface where the microcracks coalesce and high ratios $a/d_g > 1$ are observed can a distinct effect be expected. Such an observation fits well to the attenuation measurements discussed earlier, where for the high changes in the longitudinal US-velocity $v_l$ at points close to the rupture surface an effect can be noticed (see Fig. 9). However, considerable scatter of the results is envisaged which impedes a quantitative analysis.

## Comparison with metallographic results

Metallographic examination provides the most direct information on the defect state. However, for a meaningful comparison with other methods for damage detection, a quantitative evaluation of the damage visible on metallographic sections is necessary.

**'A'-parameter.** For the quantitative metallographic evaluation the so-called 'A'-parameter method [36] is applied. It is well suited for the description of a distribution of intergranular crack-like defects as observed in AMCR 0033. Taking a micrograph that contains the axis of the applied stress, a line parallel to this axis is drawn. The 'A'-parameter is calculated according to

$$A = \frac{\text{number of cracked facets intersecting the line}}{\text{total number of grain boundaries intersecting the line}}$$

'A'-parameter measurements have been performed in most of the crept AMCR 0033 specimens. After having accomplished the US-measurements, the specimens were cut into pieces along their lengths, parallel to the flat surfaces. 'A'-parameters were determined at the locations of the US-measurements by drawing several parallel lines in an area corresponding to the transducer section. In order to obtain a statistically meaningful value for $A$, about 400 grain boundaries have to be counted. The evaluation was done using a light microscope and a magnification of 100 times which proved to be sufficient. This has been verified by comparing optical images of the damaged zones with those of the same zone obtained in a scanning electron microscope. In both cases, the same facets were identified as cracked. The uncertainty of the $A$-values amounts to about ±0.03 due to uncertainties in the identification of cracked grain boundaries. This also explains the average value $A \approx 0.05$ in the undamaged specimen heads. Typical examples for the 'A'-parameter variation in the gauge length of broken specimens is given in Fig. 11.

**Relation to creep damage.** For a quantitative comparison with the creep and damage measurements reported so far, a relationship between the 'A'-parameter and the crack density parameter is desirable. By means of stereological considerations and using known results from stochastic geometry a linear relationship

$$A = \beta Q \frac{\sqrt{\langle s \rangle}}{\langle z_f \rangle} F(\sigma_{cr}, \sigma_{gbf}, \sigma_g)\, \eta \qquad (21)$$

between $A$ and $\eta$ can be derived [37]. $\beta$ and $Q$ are numerical factors depending on the stochastic model and on the orientation distribution, respectively. $\langle s \rangle$ is the average number of facets per grain and $\langle z_f \rangle$ the average number of connected facets forming a crack as seen in the 2D-metallographic section. The function $F$ contains information about the scatter of the sizes of the cracks, grain boundary facets and grains in terms of second order moments, i.e. variances $\sigma_{cr}, \sigma_{gbf}, \sigma_g$. $\langle z_f \rangle$ can be easily determined from the micrograph.

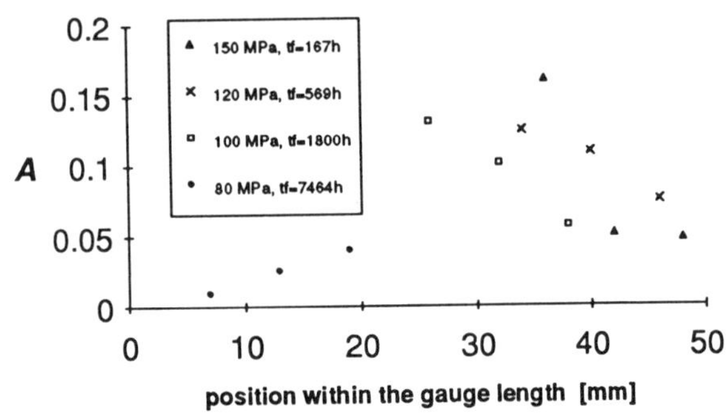


Fig. 11 'A'-parameter as a function of the position in the gauge length. $t_f$: failure time of the creep test

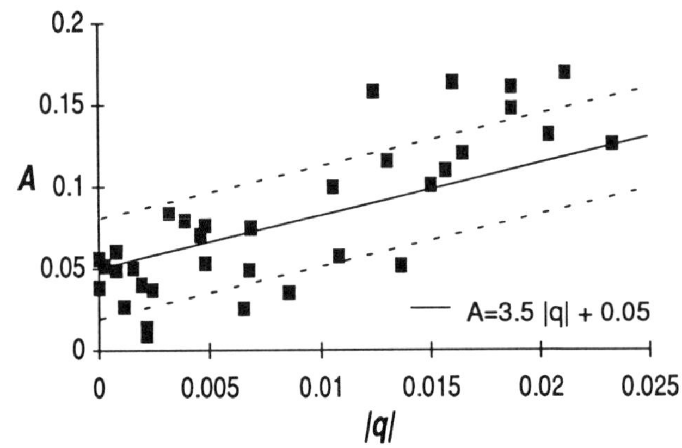


Fig. 12 'A'-parameter as a function of the experimentally determined anisotropy measure $q$. The solid and dashed lines show the theoretical result including the expected scatter range

**Comparison with experimental results.** Combining eq. (21) with either eq. (11) or (16) allows further comparisons between different methods for damage determination. In [37] the proportionality coefficient between $A$ and $\eta$ has been assessed on the basis of the AMCR 0033 results. It turns out that

$$A \approx 0.94\, \eta \qquad (22)$$

should be a reasonable approximation. From this the relations

$$A = 1.03\, D \quad , \quad A = 3.5\,|q| \qquad (23)$$

follow. The first equation in (23) states that the plausible assertion in [36] which identifies the 'A'-parameter with the Kachanov damage parameter is fulfilled for the type of creep damage observed in AMCR 0033. Similar to the case of the US-velocity measurements the $A$-values next to the rupture surface indicate a consumed lifetime fraction of about 90%, as can be seen from the examples in Fig. 11. This is slightly higher than in the case of the US-velocity measurements, however, the misalignment of the cracks has been taken into account in eq. (21). A more detailed comparison with the US-results is given in Fig. 12, where the experimentally determined relationship between $A$ and $q$ is plotted. The solid line represents the second equation (23) taking into account an offset value of 0.05 corresponding to the $A$-value measured in the undamaged material. The two dashed lines represent the uncertainty in the 'A'-parameter determination as obtained from repeated counting at the same measurement point. Most of the data are contained between the dashed lines indicating despite the scatter a fairly good correlation with the theoretical predictions.

## Discussion

The results obtained in AMCR 0033 clearly demonstrate that in this material non-destructive ultrasonic velocity measurements are sensitive enough to detect creep damage. Effects on the US-velocity such as the hold time at elevated temperature and texture formation incurred by inelastic straining have been assessed. The latter does not play a role at the creep strains considered here, and the former is lower than the typical US-velocity decrease observed after creep.

Metallographic examinations show that creep damage manifests itself in the form of intergranular microcracks. These cracks normally consist of one to two broken grain boundary facets and in broken specimens they coalesce to larger flaws only in the immediate vicinity of the rupture surface. The 'A'-parameter method was used to quantify the damage state at the places where the US-measurements had been performed.

The results of the creep tests, the ultrasonic damage measurements and the 'A'-parameter can be quantitatively interpreted by using appropriate models for the elastic and creep behaviour. Being based on an idealised crack distribution, i.e. randomly dispersed, unidirectionally aligned circular cracks, they depend on or can be related to the crack density parameter $\eta$. It has been shown that the results for $\eta$ which were obtained with the different experimental methods are consistent, thus indicating that the modelling reflects the principle characteristics of the damage mode. It also turns out that the reproducibility of the US-velocity measurement is superior to the quantitative metallographic evaluation.

Although the AMCR 0033 steel is not widely used as a structural material, the agreement between experiment and theory is instrumental in assessing the damage behaviour of other materials. Whereas in the present case only intergranular microcracks were observed, in many ferritic steels creep damage in the form of cavities on grain boundaries is found. While a cavitated grain boundary under constraint cavity growth conditions [38] has the same influence on the macroscopic inelastic strain rate as a microcrack, its influence on the elastic constants is much lower than the corresponding effect due to a completely broken facet. For this case the proportionality factors in the linear relationships between $q$, $A$ and $\eta$ will be quite different. Determining the 'A'-parameter as the number density of cavitated grain boundaries intersecting a testline parallel to the stress axis and relating it to the US-velocity changes will yield a much higher proportionality factor than in the $A \propto q$ relationship given in eq. (23). This could mean that for this case the resolution potential of the quantitative metallographic method is superior to the US-velocity measurement.

The 800H measurements show that there may be other phenomena which have to be taken into consideration. The lack of reproducibility of the US-measurements is a serious obstacle for the application of this method. In 800H, creep at the loads considered involves large strains before rupture occurs. The texture induced by the strains can influence the US-velocity to the same order of magnitude as the damage [6]. In addition, it is very likely under these conditions that the damage incurred is not typical of creep but, more characteristic of a ductile response.

This discussion underlines that the application of ultrasonic velocity measurements for creep damage determination may be more successful in one material than in another depending on the creep behaviour, the damage mode, scatter in the material behaviour etc. Much care has to be taken in the interpretation of the results which should be backed up by other methods for damage determination.

## Conclusions

Methods for damage detection have been applied to two different austenitic alloys, the Mn-Cr austenitic steel AMCR 0033 and the iron-based alloy INCOLOY 800H. A rather complete test programme of creep tests and damage measurements has been conducted on AMCR 0033. The results in 800H are more difficult to interpret because of the lack of reproducibility of the ultrasonic damage measurements and the possibly different type of damage induced during creep

For AMCR 0033 it has been demonstrated that the US-velocity results are quantitatively consistent with metallographic evaluations and can be interpreted quantitatively in terms of microstructural defect distributions. This is important for two reasons: firstly it allows a comparison with other methods for damage determination and secondly the results can be used to derive damage parameters which can be applied to lifetime prediction models. The ultrasonic attenuation measurements indicate that only a qualitative interpretation is possible.

The results of this study can be used together with the theoretical models available to assess the effect of creep damage, e.g. on the ultrasonic velocities, in other materials. It becomes obvious that depending on the principle damage mode and on the importance of the side effects on the damage measurement the application of a method for damage determination can be more successful in one material than in another. Therefore, a careful examination of the relevance of a method in a particular material is indispensable.


## Acknowledgements

The authors are grateful to Dr. P Schiller, Dr. F. Lakestani and Mr. R. Denis for helpful discussions and wish to express their appreciation for the help received in the laboratory to Mrs. M. Romor, Mr. F. Merli and Mr. N. Wächter.


## References


1- Brathe, L., *Scand. J. Metallurgy 7*, 199-203 (1978)

2- Stigh, U., Jansson, S., Brathe, L. and A. Dyne, *Revue d'acoustique 66*, 202-204 (1983)

3- Jansson, S. and U. Stigh, *J. Appl. Mech. 52*, 609 - 614 (1985).

4- Stigh, U. *J. Appl. Mech. 53*, 485-490 (1986).

5- Stigh, U. *Eng. Fract. Mech. 28*, 1-12 (1987)

6- Willems, H. and H. P. Degischer, "DGZfP-Tagung, Tagungsband, Teil I" p. 51-60 Lindau (1987).

7- Ledbetter, H. M., Fields, R. J. and S. K. Datta, *Acta metall. 35*, 2393-2398 (1987).

8- Willems, H., Bendick, W. and H. Weber, "Nondestructive Characterization of Materials II", p. 451-459 Plenum Publishing Corporation (1987).
9- Willems, H., Proc. of the Int. Conf. "Life Assessment and Extension", paper 2.4.3, The Hague/The Netherlands (1988)
10- Yoneyama, H., Nakashiro, M., Murakami, K., Shibata, S. and A. Ohtomo, *Ishikawajima-Harima Giho Engineering Review 28,* 306-311 (1988).
11- Lakestani, F. and P. Rimoldi, "Nondestructive Characterization of Materials", P Höller et al. Eds., p. 391-397, Springer-Verlag Berlin (1989).
12- Willems, H., *Mat. Res. Soc. Proc. 142,* 165-170 (1989).
13- Stigh, U. and P. Sotkovszki, *Advances in Fracture Research 5,* 3157-3164 (1989).
14- Stamm, H., Gariboldi, E., Lakestani, F., Rimoldi, P., Scholz, R. and U. von Estorff, "High Temperature Materials for Power Engineering, Part I", p. 737-748 Kluwer Academic Publishers, Dordrecht (1990)
15- Stamm, H. and U. von Estorff, "Fracture Behaviour and Design of Materials and Structures", Preprints of the 8th European Conference on Fracture, p. 1202 - 1208 (1990)
16- Hirao, M., Morishita, T. and H. Fukuoka, *Metallurgical Transactions A 21A,* 1725-1732 (1990).
17- Eggeler, G., Brear, J. M., Cane, B. J., Martinez-Oña, R., Bañes Fernandez, J., Borregan, J. M. and A. Navarro, Proc. 4th Int. Conf. on "Creep and Fracture of Engineering Materials and Structures", p. 931-940 University College Swansea (1990).
18- Stamm, H. submitted to *The European Journal of Research in Non-Destructive Testing and Evaluation'* (1991)
19- Piatti, G. and P. Schiller, *J. of Nuclear Materials 141-143,* 417-426 (1986)
20- Ruedl, E., Matera, R. and G. Valdré, *J. of Nuclear Materials 151,* 238-244 (1988)
21- Ennis, P. and H. Schuster, private communication, KfA-Jülich
22- Guttman V. and R. Bürgel, *Metal Science, 12,* 549-555 (1983)
23- Stickler, R. and G. Weidlich, *Prakt. Metallographie 20,* 174-192 (1987)
24- Degischer, H. P., Aigner, H., Danzer, R. and W. Mitter, "High Temperature Alloys for Gas Turbines and Other Applications 1986, Part II", p. 1371-1381, D. Reidel Publishing Company, Dordrecht 1986
25- Reynolds, W. N. and R. L Smith, *J. Phys. D: Appl. Phys. 17,* 109-116 (1984)
26- Smith, R. L. and T. E. Dixon, in: "Nondestructive Characterisation of Materials" P. Höller et al. eds., p. 368-375, Springer Verlag Berlin (1989)
27- Kachanov, L. M., *Izv. Akad. Nauk. S.S.R., Otd. Tekh. Nauk. 8,* 26-31 (1958)
Rabotnov, Y. N., "Proc. XII Congr. of Applied Mechanics" p. 342-349 Springer Verlag Berlin (1968)
28- Hutchinson, J. W., *Acta Metall. 31,* 1079-1088 (1983)
29- Auld, B. A. "Acoustic Fields and Waves in Solids", John Wiley & Sons, New York (1973)
30- Piau, M., *Int. J. Engng. Sci 18,* 549-568 (1980)
31- Sayers, C. M. and R. L Smith, *ULTRASONICS September 1982,* 201-205 (1982)
32- Budiansky, B. and R. J. O'Connell *Int. J. Solids Structures 12,* 81-97 (1976)
33- Benveniste, Y., *Mechanics of Materials 6,* 147-157 (1987)
34- Zhao, Y. H., Tandon, G. P. and G. J. Weng *Acta Mechanica 76,* 105-130 (1989)
35- Hirsekorn, S., *J. Acoust. Soc. Am 72(3),* 1021-1031 (1982)
36- Shammas, M. S. 'Proc. of the Conf. on Life Extension an Refurbishment of Steam Plants', Institution of Mechanical Engineers, London, 289-299 (1987)
37- Stamm, H. and U. von Estorff, to be presented at the International Conference 'Behaviour of Defects at High Temperatures', 30th March - 3rd April 1992, Sheffield U.K.
38- Dyson, B. F., *Material Science 10,* 349-353 (1976)

*Proceedings of the Fifth International Conference on Creep of Materials, Lake Buena Vista, Florida, USA, 18-21 May 1992*

# Creep Crack Growth Behavior of Ni-26Cr-17W-0.5Mo Alloy in Air and Helium Gas Environment at 1273 K

**M. Tabuchi and K. Yagi**
National Research Institute for Metals
Tokyo, Japan

**Y. Nakasone and T. Tanabe**
National Research Institute for Metals
Ibaraki, Japan

ABSTRACT

The present paper reports the effects of carbon content and environment on creep crack growth behavior in Ni-26%Cr-17%W-0.5%Mo alloys which have been developed as candidate structural materials for heat exchanger in Japanese high-temperature gas-cooled reactor. Creep crack growth tests are conducted on the alloys with different carbon contents in air and in helium gas environment at 1,273K using the CT specimen. Creep crack growth rate da/dt is evaluated by the C* parameter. The da/dt is faster in the higher carbon alloys than in the lower carbon alloys in each environment. The effect of carbon content on the da/dt vs. C* relations can be explained by the difference of creep ductility. The environmental effect on the da/dt vs. C* relations in 0.025%C and 0.085%C alloys is very small. For 0.003%C alloy, however, da/dt in He gas environment is much lower than in air. This difference is attributed to marked increase in creep ductility that this alloy showed in the He gas environment.

NI-26%CR-17%W-0.5%MO ALLOYS have been developed as candidate structural materials for heat exchanger in Japanese high-temperature gas-cooled reactors (HTGR)(1,2). Those alloys have good creep resistance because of precipitation strengthening by $M_{23}C_6$ and $\alpha$-W ; e.g., the 2,000h creep rupture strength is higher than 40MPa at 1,273K (2,3). Those alloys are fractured due to the growth of internal cavities and surface cracks along the grain boundaries at 1,273K (3). Therefore, it is necessary to understand the creep crack growth behavior of those alloys. Although the creep rupture properties of this alloy family have been studied, little has been known yet about the creep crack growth behavior at very high temperatures.

For improving safety and reliability of high temperature components, it is important to understand the creep crack growth behavior. There are many studies concerning the creep crack growth testing procedure and the method for evaluating creep crack growth rate using engineering fracture mechanics(4-6). The C* parameter (7-9) seems to be one of the best parameters to evaluate creep crack growth rate property in engineering materials. The creep crack growth tests at very high temperature, however, have scarcely been conducted. It is also interesting to investigate the applicability of the testing and evaluating methods of creep crack growth to the study of fracture at very high temperature.

The present paper reports the creep crack growth behavior in Ni-26%Cr-17%W-0.5%Mo alloys containing 0.003%, 0.025%, 0.06% and 0.085% carbon in air and in a simulated HTGR He gas environment at 1,273K using the CT specimen. The effect of carbon content and environment on the creep crack growth rate da/dt vs. C* correlation is discussed.

## MATERIALS AND EXPERIMENTAL PROCEDURE

**Materials and Specimens.** The chemical compositions of the alloys tested in this study are shown in Table I. These alloys have almost the same composition except carbon content. These alloys were water-quenched after solid solution treatment at 1,573K for 1h.

The specimen used for the present creep crack growth tests was the ASTM CT specimen of 50.8mm in width and 12.7mm in thickness. Fatigue pre-crack of about 2.5mm in length was introduced at room temperature, and side-grooves having a depth of 12.5% of specimen thickness were machined.

The creep and rupture properties were examined using the round bar specimen of 6mm in diameter and in 30mm in gauge length.

**Creep Crack Growth Test.** Creep crack growth tests were conducted using the CT specimens under constant load condition at 1,273K in air and in a simulated HTGR He gas environment. The composition of this impure He gas is shown in Table II.

The testing system for creep crack growth in the He gas environment is shown schematically in Fig.1. The He gas is supplied from a He gas recirculating loop. The He gas composition is kept the same during the tests.

Creep crack growth tests were conducted principally according to the guideline proposed by the VAMAS international collaborative research project (10). Creep crack length was measured using D.C. electrical potential technique. Pt wire was used for potential leads because of the high resistance to oxidization. Crack length was computed from the output voltage $V$ using Johnson's equation (11). Load line displacement between the upper and lower specimen-connectors $\delta$ was measured by a couple of linear gauges. The values of $V$ and $\delta$ were measured every 15sec, and transmitted into a micro-computer. Average values of all the data taken in 1h were recorded and used for the analysis of creep crack growth behavior.

**Creep Test.** Creep tests were conducted using the round bar specimens at 1,273K in air and in the simulated HTGR He gas environment.

Table I. Chemical compositions of the present Ni-26%Cr-17%W-0.5%Mo alloys(wt%).

| C | Cr | W | Mo | Ti | Al | B | Zr | Ni |
|---|---|---|---|---|---|---|---|---|
| 0.003* | 26.0 | 17.0 | 0.5 | 0.5 | 0.1 | 0.005 | 0.05 | Bal. |
| 0.025 | 26.0 | 17.0 | 0.5 | 0.5 | 0.1 | 0.005 | 0.05 | Bal. |
| 0.06 | 26.0 | 17.0 | 0.5 | 0.5 | 0.1 | 0.005 | 0.10 | Bal. |
| 0.085 | 26.3 | 17.0 | 0.5 | 0.5 | 0.1 | 0.005 | 0.05 | Bal. |

* No carbon was added; however, this amount of carbon is to be contained.

Table II. Composition of the helium gas used in the present study(ppm).

| $H_2$ | $CH_4$ | CO | $CO_2$ | $H_2O$ | O | N | He |
|---|---|---|---|---|---|---|---|
| 300 | 15 | 100 | 1 | 3 | N.D.* | N.D.* | Bal. |

* N.D.: Not Detected.

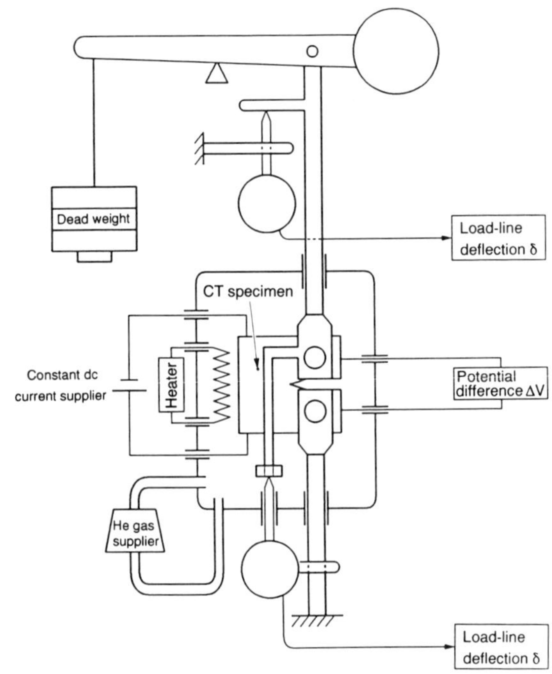


Fig.1- Creep crack growth testing system in the He gas environment.

**Evaluation of Creep Crack Growth Rate.** Creep crack growth rate is evaluated using the C* parameter that is defined by replacing the deflection and strain terms in the J-integral with deflection rate and strain rate. In the present paper, C* values were calculated according to the following equation (12);

$$C^* = \frac{n}{n+1}\frac{P\dot{\delta}}{B_N(W-a)}\left(\gamma - \frac{\beta}{n}\right) \quad \cdots\cdots (1)$$

where $n$ is the creep exponent, $P$ is the applied load, $W$ is the specimen width, $B_N$ is the specimen net thickness between sidegrooves, $a$ is the crack length and $\dot{\delta}$ is the load line displacement rate. $\gamma$ and $\beta$ are given by the following functions of $a$ and $W$ for the CT specimen used;

$$\gamma = \frac{2(1+\alpha)(1+a/W)/(1+\alpha^2)+\alpha(1-a/W)}{(1+a/W)+\alpha(1-a/W)} \quad \cdots\cdots (2)$$

$$\beta = \frac{\alpha}{\alpha + \frac{1+a/W}{1-a/W}} \quad \cdots\cdots (3)$$

$$\alpha = \sqrt{\{2a/(W-a)\}^2+2\{2a/(W-a)\}+2} - \{2a/(W-a)+1\} \quad \cdots\cdots (4)$$

RESULTS AND DISCUSSION

Creep and Creep Rupture Properties. Figure 2 shows the results of uniaxial creep tests conducted using the round bar specimens at 1,273K in air. As shown in Fig.2(a), the creep strength of the 0.06%C and 0.085%C alloys is higher than that of low carbon alloys. Grains in high carbon alloys are strengthened not only by α-W precipitates but also by fine $M_{23}C_6$ precipitates.

From the relationship between stress $\sigma$ and minimum creep rate $\dot{\varepsilon}$ shown in Fig.2(b), the creep coefficient $A$ and creep exponent $n$ in the following Norton's rule was obtained;

$$\dot{\varepsilon} = A\sigma^n \quad \cdots\cdots (5)$$

These values are necessary for the evaluation of C* values. The values of $A$ and $n$ are shown in Table III.

Figures 2(c) and (d) show the elongation and the reduction of area as a function of rupture time, respectively. Creep ductility is lower in higher carbon alloys.

Effect of Carbon Content on Creep Crack Growth Rate. Figure 3 shows the creep crack length vs. time relations obtained from the creep crack growth tests conducted in air at 1,273K for initial net section stress, σg = 83MPa. The time to fracture for the CT specimen of the 0.06%C and 0.085%C alloys is longer than that of the 0.025%C and 0.003%C alloys. The time to fracture for the CT specimens has the same tendency with the creep strength for the round bar specimens mentioned above. The creep crack growth rate at the primary and secondary stages is smaller in the higher carbon alloys. The higher carbon alloys

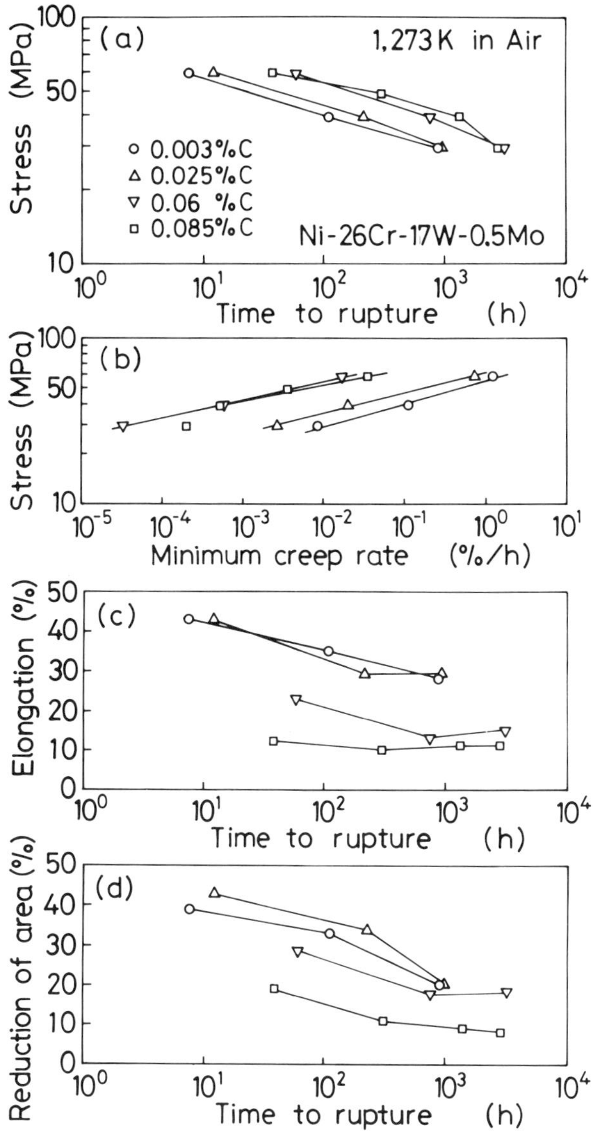


Fig.2- Results of creep tests using round bar specimens; (a)stress vs. time to rupture, (b)stress vs. minimum creep rate, (c)elongation vs. time to rupture and (d)reduction of area vs. time to ruprue

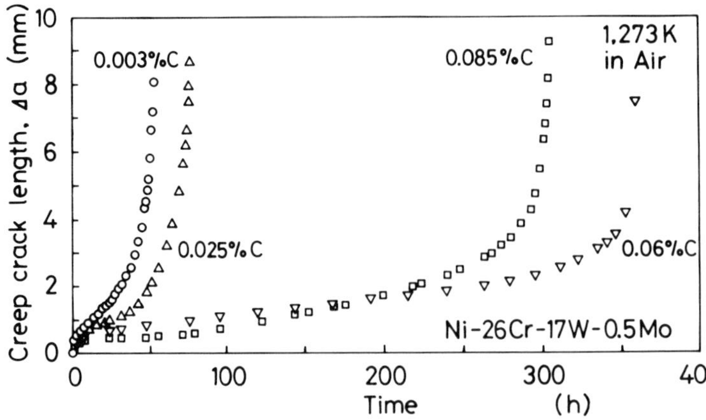


Fig.3- Creep crack length vs. time curves of the present alloys at 1,273K for 83MPa in air.

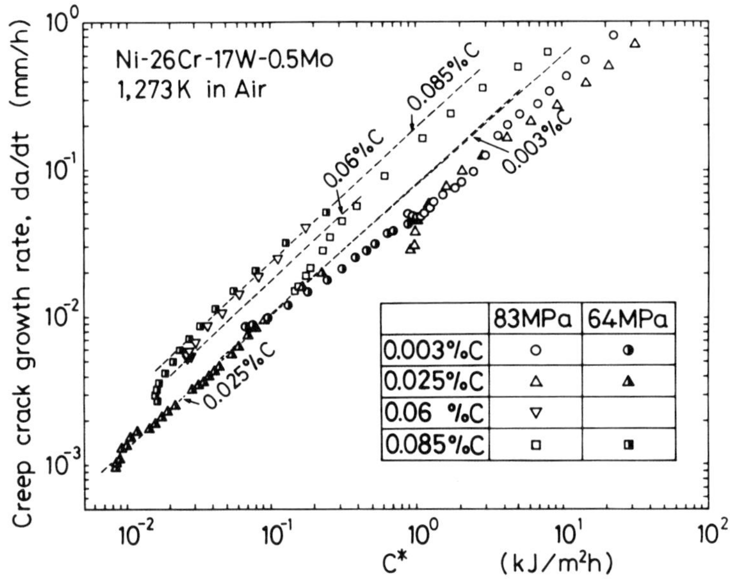


Fig.4- Relationship between creep crack growth rate vs. the C* parameter of the present alloys at 1,273K in air.

are hardened in the primary stage by the nucleation of the carbide $M_{23}C_6$ earlier than that of the α-W (13). Although the duration of secondary stage is longer for the higher carbon alloys, the creep crack growth rate at the accelerating stage is faster. The load line displacement vs. time relations have the same tendency as the curves in Fig.3.

Figure 4 shows the relationship between creep crack growth rate da/dt vs. the C* parameter obtained from the tests in air. The values of the C* parameter was calculated from eqs.(1) to (4). In the da/dt vs. C* relations obtained in this study, the complicated behavior that is called tail part (14) was observed at the early stage of creep crack growth. The data corresponding to this stage are omitted in Fig.4. Figure 4 also contains the results of the tests which were conducted at a different initial stress $\sigma_g$=64 MPa for the 0.003%C, 0.025%C and 0.085%C alloys. The effect of testing load on the evaluation concerning creep crack growth behavior can be neglected as shown in this figure.

The creep crack growth rate da/dt increases with increasing carbon content, and the da/dt in the 0.06%C and 0.085%C alloys is about three times higher than that in the 0.003%C and 0.025%C alloys. The addition of carbon markedly reduces the resistance to creep crack growth, whereas it improved the creep strength as shown in Fig.2.

Table III. Values of material constants and calculated da/dt by eq.(6) for the present alloys at 1,273K in air.

| Alloy | n | A ($MPa^{-n}h^{-1}$) | $I_n$ | $r_c$ (μm) | $\varepsilon_f^*$ | da/dt (mm/h) |
|---|---|---|---|---|---|---|
| 0.003%C | 7.1 | $4.17\times10^{-15}$ | 4.68 | 800 | 0.43 | $7.91\times10^{-2}\,C^{*\,0.88}$ |
| 0.025%C | 8.1 | $2.88\times10^{-17}$ | 4.59 | 800 | 0.43 | $8.09\times10^{-2}\,C^{*\,0.89}$ |
| 0.06 %C | 9.0 | $2.29\times10^{-20}$ | 4.52 | 200 | 0.20 | $1.46\times10^{-1}\,C^{*\,0.90}$ |
| 0.085%C | 10.4 | $1.45\times10^{-22}$ | 4.44 | 200 | 0.16 | $1.90\times10^{-1}\,C^{*\,0.91}$ |

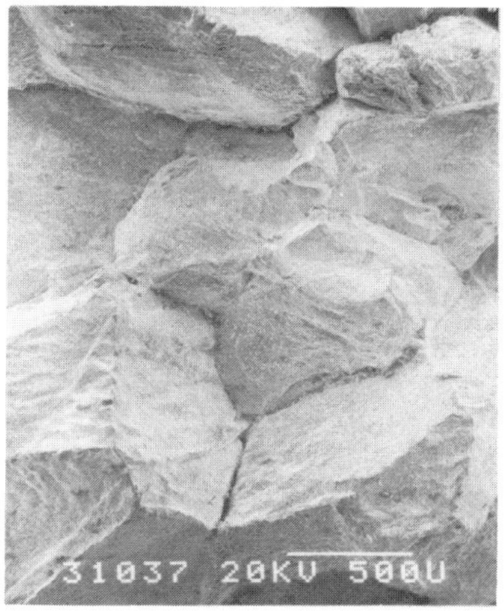


0.003%C

0.085%C

Fig.5- Fratured surfaces of the 0.003%C and 0.085%C alloys tested at 1,273K in air.

**Relationship Between Creep Crack Growth Behavior and Creep Ductility.** It is reported that the da/dt vs. C* relation depends on the ductility of a material (15,16). The high values of da/dt for the high carbon alloys is considered to be due to the low creep ductility shown in Figs.2(c) and (d).

According to Nikbin et al.(17), the creep crack growth rate is given as follows;

$$\frac{da}{dt} = \frac{n+1}{\varepsilon_f^*}(Ar_c)^{1/n+1}\left(\frac{C^*}{I_n}\right)^{n/n+1} \quad \cdots\cdots (6)$$

where $\varepsilon_f^*$ is the creep ductility, $r_c$ is the grain size and $I_n$ is a non-dimensional function of $n$. In this study, the values of $I_n$ was calculated by the following equation (15);

$$I_n = 10.3(0.13+1/n)^{1/2} - 4.8/n \quad \cdots\cdots (7)$$

Figure 5 shows the fractured surfaces of the 0.003%C and 0.085%C alloys tested in air. The specimens of all the alloys tested in this study showed the intergranular fracture. The grain size in the 0.003%C and 0.025%C alloys is larger than that in the 0.06%C and 0.085%C alloys as shown in Fig.5. In this study $\varepsilon_f^*$ was assumed to be the true fracture strain obtained from the uniaxial creep tests of the round bar specimens. The material constants in eq.(6) and the values of da/dt estimated from eq.(6) are tabulated in Table III. The estimated values of da/dt for each alloy were drawn in Fig.4 by the dashed lines. Equation(6) elucidates the experimental results that da/dt in the higher carbon alloys became faster. The values of $A$, $n$ and $r_c$ were different for each material, but the influence of these values on the value of da/dt was smaller than that of the creep ductility. The increase of creep crack growth rate with increasing carbon content was caused by the decrease of creep ductility with increasing carbon content.

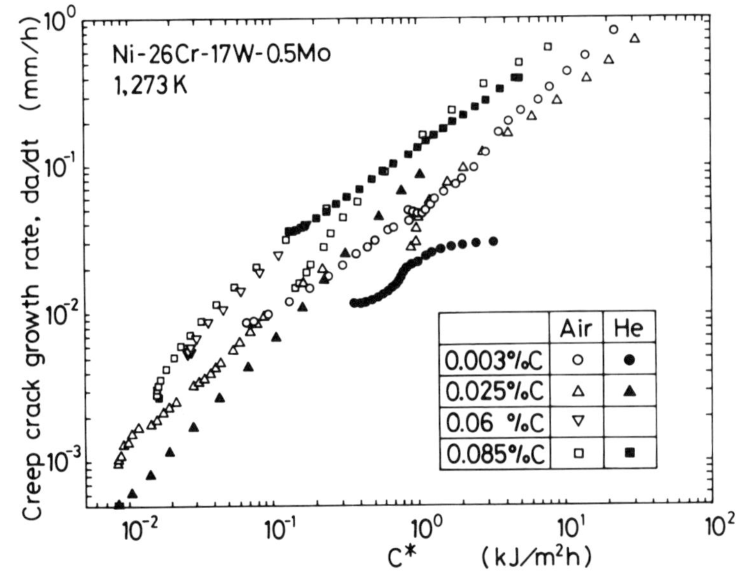


Fig.6- Relationship between creep crack growth rate and C* parameter in the present alloys at 1,273K in the He gas environment and in air.

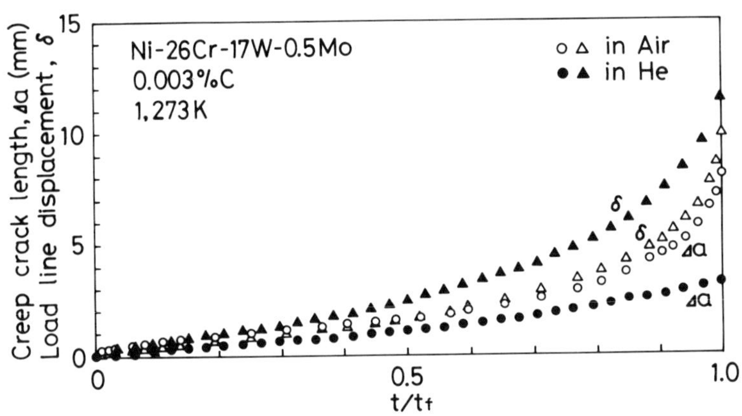


Fig.7- Creep crack length and load line displacement vs. normalized time $t/t_f$ curves of the 0.003%C alloy at 1,273K in the He gas environment and in air.

**Effect of Testing Environment on Creep Crack Growth Rate.** Solid symbols in Fig.6 indicate the relationship between da/dt vs. C* parameter obtained from the creep crack growth tests for $\sigma g$ = 72MPa at 1,273K in the He gas environment. The effect of testing environment on da/dt is small for the 0.085%C and 0.025%C alloys. For the 0.003%C alloy, however, da/dt in the He gas environment is about four-tenth that in air.

Figure 7 shows the changes in creep crack length and load line displacement with the non-dimensional time $t/t_f$ for the 0.003%C alloy in air and in the He gas environment. In order to compare the characteristics of the curves, time $t$ was normalized by the time to fracture $t_f$. In the He gas environment, the increase of crack length is much smaller than the increase of the load line displacement even in the accelerating stage. The deformation ahead of the crack tip in the He gas environment is larger than that in air. In the He gas environment, creep crack is blunted due to the large deformation ahead of the crack tip. Therefore, da/dt in the He gas environment becomes lower than that in air, when da/dt is evaluated by the C* parameter.

The creep tests were also carried out in the He gas environment using the round bar specimens of the 0.003%C and 0.06%C alloys. For the 0.06%C alloy, the creep ductility in the He gas environment

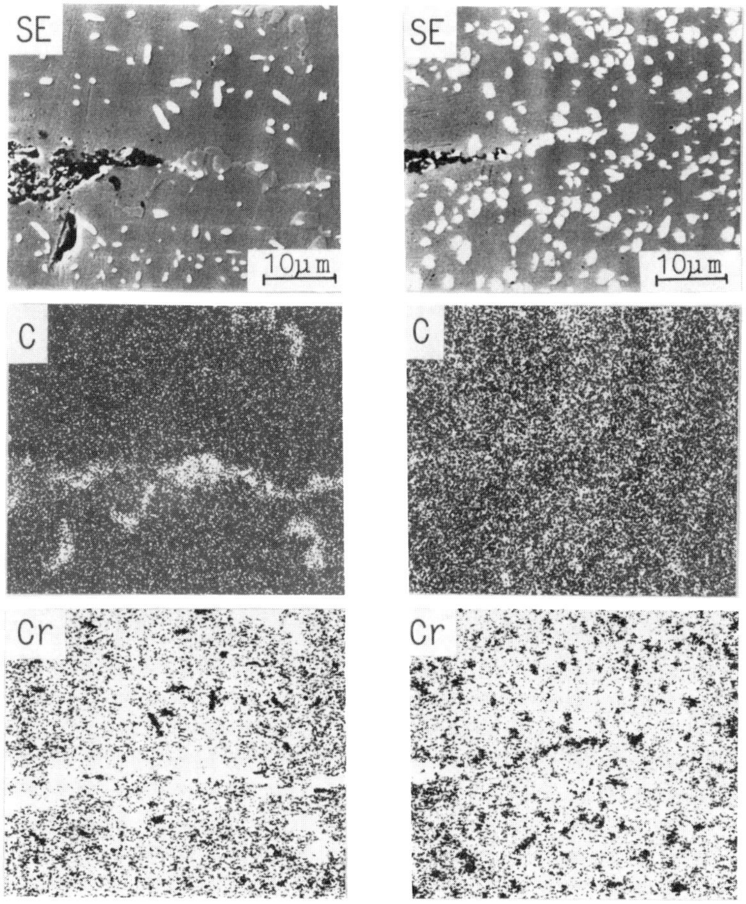


Fig.8- Results of the EPMA analysis of C and Cr near the crack tip of the 0.003%C alloy tested at 1,273K in air and in the He gas environment.

was nearly the same as that in air. For the 0.003%C alloy, however, the creep ductility in the He gas environment was higher than that in air; i.e., the reduction of area in the He gas was about 50% at 1,273K for 39MPa. It is reported that the elongation of a Ni-26%Cr-17%W alloy which does not contain Mo and C was as large as 140% in the same He gas at 1,273K, whereas its elongation in air is about 60%, less than half the ductility in the He gas (3). Assuming that the values of $A$ and $n$ be the same as those in air, and substituting R.A.= 50%, i.e., $\varepsilon_f^* = 0.69$ in eq.(6), da/dt is estimated as $4.9 \times 10^{-2} \cdot C^{*0.88}$ . This value is about six-tenth the estimated value in air shown in Table III. Although the estimated value does not coincide with the experimental value completely, equation (6) can predict the tendency that da/dt in the He gas becomes lower than that in air for the 0.003%C alloy. It is considered that the low value of da/dt in 0.003%C alloy in the He gas can be explained by its high creep ductility in the He gas.

Figure 8 shows the microstructure and the results of EPMA analysis of C and Cr near the crack tip for the 0.003%C alloy specimens fractured in the He gas and in air. The EPMA analysis detected carbides, probably in the form of $M_{23}C_6$ , along the grain boundaries in the crack wake only in the He gas environment for the 0.003% alloy (see Fig.8). This result may suggest that carburization took place selectively on the grain boundaries in the wake of the crack. Carbides initiated along grain boundaries have so good coherency with the matrix that it brings about stronger grain boundaries. This effect may be attributed to marked decrease in da/dt only in the 0.003%C alloy crept in the high-temperature He gas environment. Further investigation, however, is necessary about the mechanism of deceleration effect of the He gas environment on the da/dt vs. C* relation.

## CONCLUSIONS

The creep crack growth behavior in Ni-26%Cr-17%W-0.5%Mo alloys at 1,273K in air and in a simulated HTGR helium gas environment was investigated. The effect of carbon contents and testing environment on the creep crack growth rate da/dt vs. C* relation was discussed.

1. The value of da/dt was larger in the higher carbon alloys than that in the lower carbon alloys in each environment. The increase of da/dt with increasing carbon content was explained by the decrease of creep ductility with increasing carbon content.
2. The effect of testing environment on the value of da/dt was observed only for the 0.003%C alloy. For this alloy, the creep ductility in the He gas environment was markedly increased so that da/dt was decreased in the He gas environment.

## REFERENCES


1- Yoda,R. et al., "Final Report on the R & D of Heat-Resistant Superalloys for HTGR's Use" Jpn. Sci. and Tech. Agency (1981)

2- Tananabe,T., F. Abe, M. Okada, R. Hamano and K. Hiraga, Testu-to-Hagane, 75,167-174(1989)

3- Ohba,T., K. Yokokawa, E. Baba, K. Hiraga, K. Yagi and T. Tanabe, Report of 123rd Committee on Heat Resisting Metals and Alloys, 32,275-287(1991)

4- Koterazawa,R. and T. Mori, J. Soc. Mater. Sci.,Jpn., 26,948-954(1977)

5- Ohtani,R. and S. Taira, Trans. ASME J. Eng. Mater. Tech., 101,369-373 (1979)

6- Sadananda,K. and P. Shahinian, Eng. Fract. Mech., 15,327-342(1981)

7- Landes,J.D. and J.A. Begley, Mechanics of Crack Growth, ASTM STP, 590, 128-148(1976)

8- Harper,M.P. and E.G. Ellison, J. Strain Analysis, 12,167-179(1977)

9- Ohji,K., K. Ogura, Y. Katada, M. Takemoto and H. Nakajima, J. Soc. Mater. Sci., Jpn., 27,1089-1095(1978)

10- "VAMAS Creep Crack Growth: A State-of-the-art Report", Issue 1, ed.by Gibbons,T.B. (1989)

11- Johnson,H.H., Materials Reserach and Standards, 5,442-445(1965)

12- Ernst,H.A., Fracture Mechanics, 14th Symp. Vol.I, Theory and Analysis, ASTM STP 791, I-499-519(1983)

13- Nakasone,Y., K. Hiraga and T. Tanabe Proc. of FATIGUE 90, 1553-1558(1990)

14- Yokobori,A.T. and T. Yokobori, Eng. Fract. Mech., 31,931-945(1988)

15- Webster,G.A., "Engineering Approaches to High Temperature Design", ed.by Wilshire,B. and D.R.J. Owen, Pinerigde Press, Swansea, U.K., 1-55(1983)

16- Tabuchi,M., T. Ohba and K. Yagi, ISIJ International, 30,847-853(1990)

17- Nikbin,K.M., D.J. Smith and G.A. Webster, Trans. ASME, J. Eng. Mater. Tech., 108,186-191(1986)

*Proceedings of the Fifth International Conference on Creep of Materials, Lake Buena Vista, Florida, USA, 18-21 May 1992*

# Creep Crack Growth in Hot Corrosion

**V.M. Radhakrishnan**
Metallurgical Engineering Department
Indian Institute of Technology
Madras, India

ABSTRACT

Experimental investigations have been carried out to study the effect of hot corrosion on the creep crack growth behaviour of typoe 316 L stainless steel under plane stress condition. Crack growth takes place under steady state condition of deformation and the presence of corrodents increases the crack growth rate. The energy rate line integral appears to give a good description of crack growth rate both in air and in hot corrosion conditions at each temperature. The steady state energy rate, $C_s*$ , is found to be inversely related to the rupture time, $t_r$ , in the form

$$C_s^* t_r = R$$

where R is a material constant.

CREEP life is an important consideration in the design of high temperature components. In many power plant components like superheaters, heat exchangers etc., creep deformation takes place in the presence of aggressive environments. Corrosive elements such as sodium chloride, sulphides etc., find their way into the system through the combustion processes. In such conditions creep deformation and damage accumulation are accelerated in the components. In the cases where an initial defect is present, cracks under these creep and hot corrosion conditions will nucleate and grow with time, thus reducing the useful life of the component. Hence it becomes essential to characterize the creep crack growth (CCG) behaviour of the material under hot corrosion (HC) condition. Many parameters like stress intensity factor, net section stress energy rate line integral etc., have been tried to correlate the CCG rate [1-5]. Among them the energy rate line integral, C*, had been found to be more appropriate, specially when the crack nucleation and growth start in the steady state condition and all the transient effects are negligible.

In the present investigation, the creep crack growth behaviour has been studied in an austenitic stainless steel under hot corrosion condition in the temperature range of 700 °C - 800 °C. The CCG rate has been analysed through the energy rate line integral C*.

EXPERIMENTAL

The material used is Type 316 L stainless steel, given by Indira Gandhi Centre for Atomic Research, Kalpakkam, India, and the nominal composition ( by wt % ) is given below.

C : 0.012 %, Mn : 1.33 %,
Cr : 16.9 %, Ni : 11.8 %,
Mo : 2.6 %, S : 0.007 %,
Fe : balance

Creep crack growth experiments have been carried out on SEN type of specimens under constant load conditions. Since the project is to understand the CCG behaviour in plane stress condition, the specimens were in the form of plates of 150 mm long, 15 mm wide and 1.0 mm thick. A sharp starter notch of depth 3 mm and root radius 0.2 mm was introduced at the edge in the mid-section of the specimen.

The specimen was heated by radiation in a split tubular furnace, having provisions for thermo-couple and crack growth measurements. The CCG experiments were carried out at specimen temperatures of 700 °C, 750 °C, and 800 °C. The specimen temperature was measured and controlled by a chromel-alumel thermocouple. The load on the specimen was calibrated by means of a proving ring. After the application of the load, the load point deflection, Δ, and the crack length, a, were continuously monitored by means of a dial gauge and a travelling microscope, both with a least count of 0.01 mm. Details of the creep crack growth studies are given elsewhere [6].

The CCG studies under corrosive conditions were carried out with a coating of 75% $Na_2SO_4$ and 25 % NaCl mixture at the mid section of the specimen. The salt deposition on the surface of the specimen was achieved by spraying saturated salt solution over the sample kept at 150°C. The deposition obtained was uniform at 2 mg $cm^{-2}$. Preliminary tests carried out to ensure the adherence of the coating when the specimen was exposed to stress and temperature, indicated that the

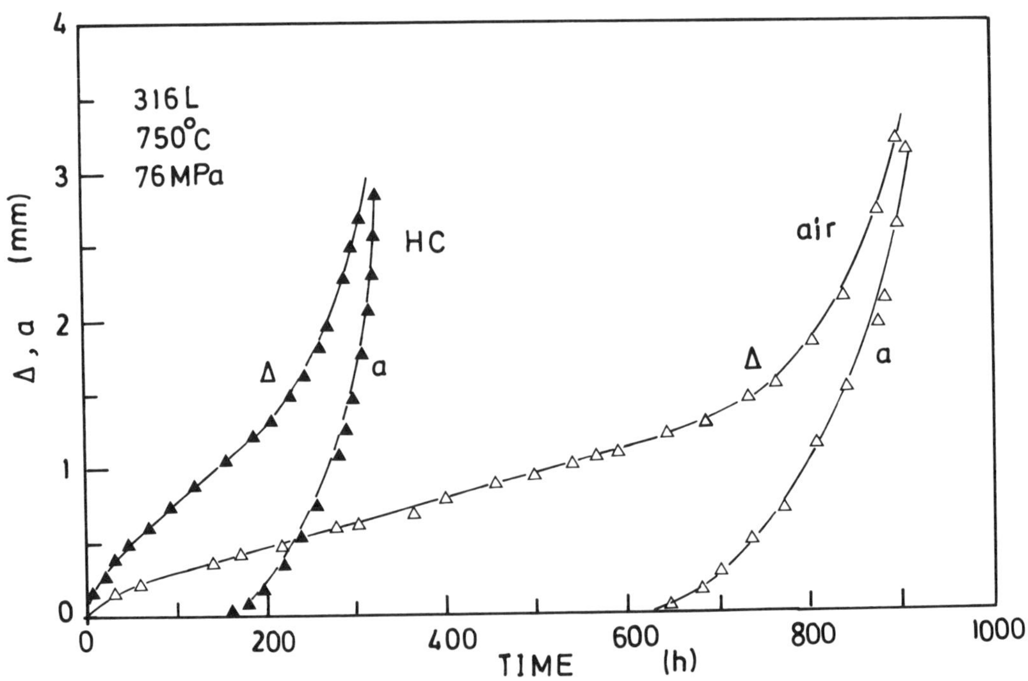


Fig.1. Variations of crack length 'a' and deflection 'Δ' with time.

coating thickness used gave the best adherence without any portion peeling of the surface.

After the experiments, the fracture surface was analysed by light microscope, and SEM. X-ray analyses were also carried out to determine the corrosive products formed. Details of these studies are given elsewhere [7].

## RESULTS AND DISCUSSION

CREEP CRACK GROWTH. Figure 1 shows the increase in load point deflection, $\Delta$, and the crack length with time at 750 °C. Both air and hot corrosion conditions are shown in the figure. It can be seen that in both the cases, the crack nucleation and propagation take place after the steady state in the deflection rate is well established. Further, the crack growth and the deflection are accelerated due to the presence of hot corrosion condition for the same applied load. This type of behaviour has been observed in all the cases.

The parameter C*, as defined by Webster[5], has been used in the present study. Since the crack initiation and growth took place only after the steady state condition has been established in the entire material and the transient effects were negligible, the parameter, $C_t$, suggested by Saxena [8] has not been tried.

The load point deflection rate due to creep, $\dot{\Delta}_c$, has been calculated taking into account the

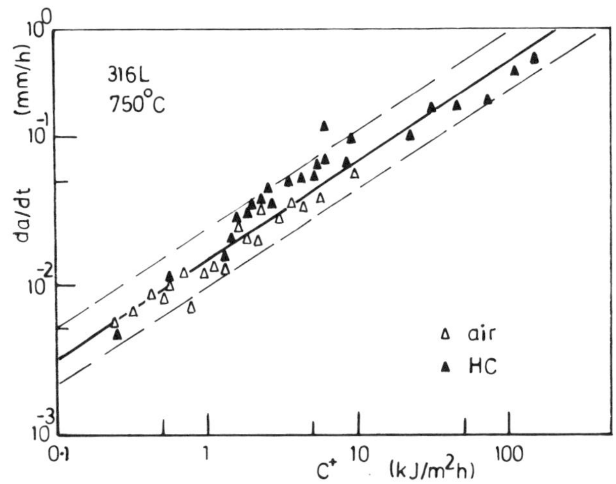


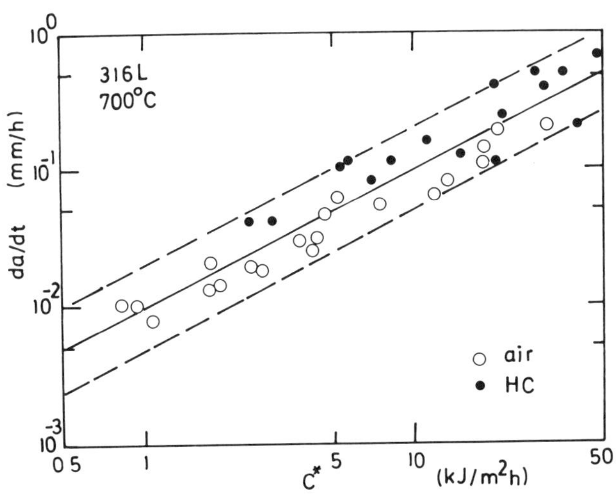


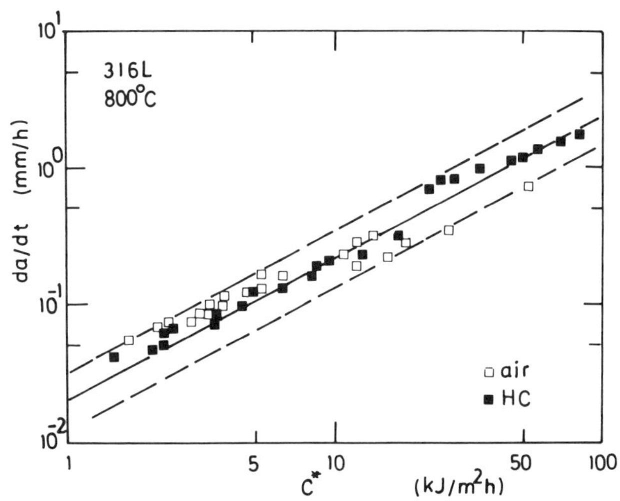


Figs. 2(a), (b) and (c).
Relation between crack growth rate, da/dt, and the energy rate line integral, C*, at 700, 750 and 800 °C.

components due to elastic and plastic compliances. The C* parameter is defined as

$$C^* = \frac{n}{(n+1)} \frac{P \dot{\Delta}_c}{B(W-a)} \qquad (1)$$

where W and B are the width and thickness of the specimen. The values of the creep exponent 'n' at 700 °C, 750 °C and 800 °C have been taken as 5.5, 5.0 and 4.3 [6,7]. The relation between the CCG rate, da/dt, and the energy rate line integral, C*, is shown in Figs. 2(a), (b) and (c) on log-log plot for 700 °C, 750 °C and 800 °C. The data points of both air and hot corrosion (HC) fall within the scatter band of +/-2. The relation between the CCG rate, da/dt and the energy rate line integral C* can be given in the form

$$da/dt = A\ (C^*)^r \qquad (2)$$

where A is a constant = 0.01, 0.015 and 0.02 and the exponent 'r' = 1.0, 0.6 and 1.0 for the temperatures 700, 750 and 800 °C. The crack growth rate is in mm/hour and the energy rate line integral is in $kJ/m^2.h$. There is scatter in the data and there is no systematic trend in the relation with respect to increase in temperature. As such it may be difficult to use the relation for the purpose of design and life prediction of a component having a crack growing under the creep condition.

It has been shown [9,10] that the steady state energy rate line integral $C_s$*, may have a relation with the rupture time, $t_r$. This may be possible specially when the crack growth takes place after a steady state condition is established, as it is in the prersent case. Thus, the relation between steady state value of the energy rate line integral, $C_s$* and the rupture time has computed and shown in Fig.3. The relation yields a straight line on the log-log plot with a negative slope equal to one. The relation can be written as

$$C^*_s \cdot t_r = R \qquad (3)$$

where $C^*_s$ is in $kJ/m^2$.hour and rupture time is in hours. R is a constant whose value is obtained as 300 $kJ/m^2$. This relation can be used to predict the life the material if the crack growth starts after the transient conditions have subsided. This relation has not been verified for cracks starting in the non-steady state conditions.

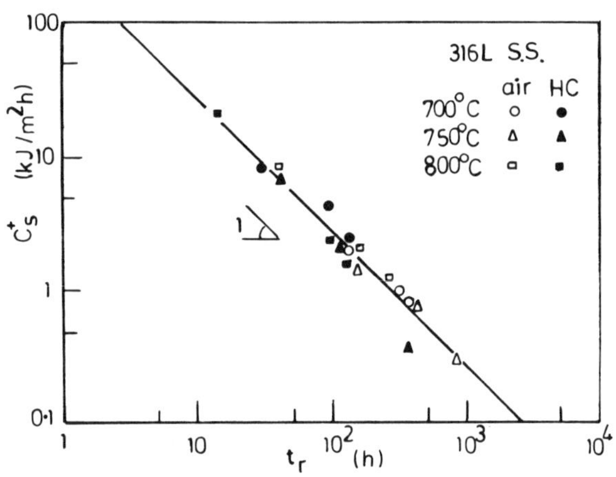


Fig.3. Relation between steady state line integral and rupture time.

Based on the above relation the stress and rupture time for a component having a flaw of non-dimension value a/W, can be given as [11]

$$\sigma\ [t_r]^{(1/n+1)} = [R/C_o]^{(1/n+1)}\ f(a/W) \qquad (4)$$

where f(a/W) is a function of the flaw size. $C_0$ is constant. The relations between the applied stress and the rupture time and the stress and the steady state line integral, $C^*_s$, are shown in Figs. 4 and 5. The slope of the $\sigma$ versus $t_r$ relation on the log-log plot plot has been found to be nearly equal to 1/(1+n), as indicated in the figure, thus conforming to the above relation. The above relation can be obtained for different flaw sizes, thereby giving the life of any component having a particular flaw size and designed for a given stress level.

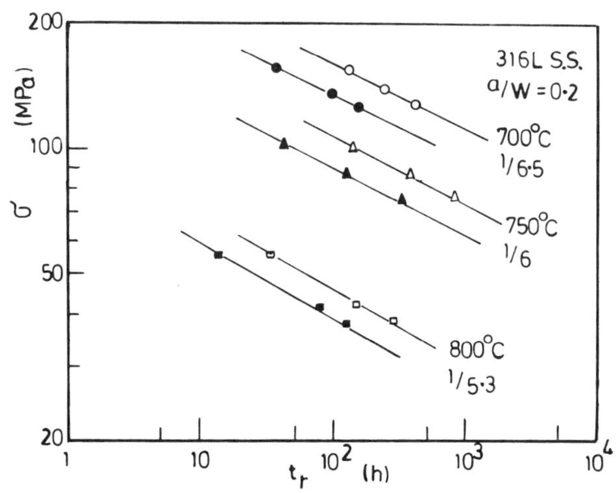


Fig.4. Relation between stress and rupture time of a material. a/W = 0.2

CORROSION ATTACK AND FRACTOGRAPHY. Figs.6(a) and (b) show the light and SEM micrographs of the exposed specimens tested in air and Figs. 7(a) and (b) that of the specimen exposed to HC conditions. The difference between the damaging mechanisms in specimens crept in air and in hot corrosion condition can be clearly seen in these fractographs. Under the hot corrosion condition, grain boundary attack has been clearly observed.

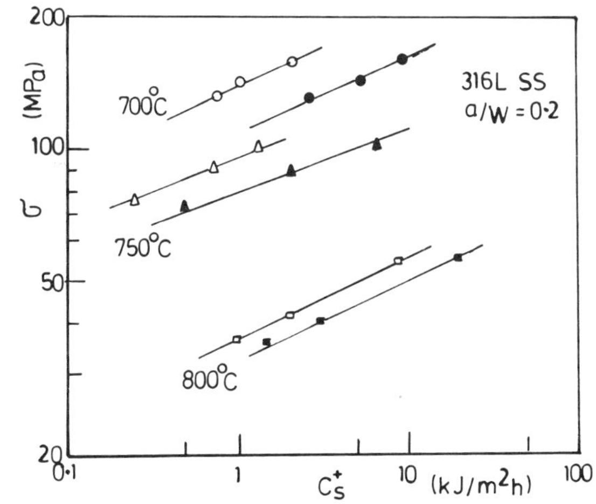


Fig.5. Relation between stress and steady state line integral for flaw size a/W = 0.2

A mixture of $Na_2SO_4$ and NaCl was found to have a more deleterious effect than $Na_2SO_4$ or NaCl acting separately in the case of creep [7,12]. The X-ray analysis of the fractured region showed the formation of corrosion compounds, typical among them are given below:

| | | |
|---|---|---|
| Strong | : | $Cr_2Fe_2O_3$ , $\gamma$-$Fe_2O_3$. |
| Medium | : | NiO, $Ni_3S_2$, NiS. $FeNiS_2$. |
| Weak | : | $Fe_2CrS_4$, $NiCr_2S_4$. $Fe_3O_4$, $NaNiO_2$ NiO. |

Chromium in the matrix gets depleted. The formation and penetration of the corrosion species along the grain boundaries weaken the material [7,12,14]. Visual inspection of the hot corroded samples coated with a mixture of sodium sulphate and sodium chloride, showed that the scales formed were porous, non-adherent, and flaky. The fractured area clearly showed intergranular attack, confirming the penetration of corrodents along the grain

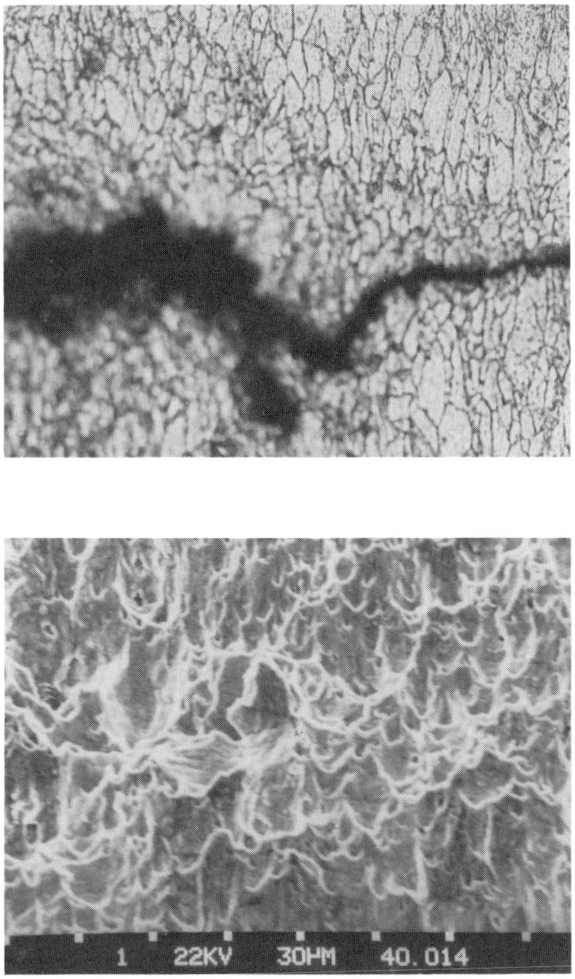


Fig. 6(a) and (b).
Light micrograph (X500) and SEM of specimen crept in air at 750 oC.

boundaries resulting in the formation of nickel sulphides around the penetrating crack. Sodium chloride with sodium sulphate has a marked effect on the creep properties with increased CCG rate and reduced rupture life. Chloride being a highly reactive species, readily reacts with the oxides formed on the alloy surface. It also reacts with the alloying elements and with the grain boundary carbides. Volatile species are produced during chloride attack and as they migrate out of the grain boundaries to the free surface, they generate

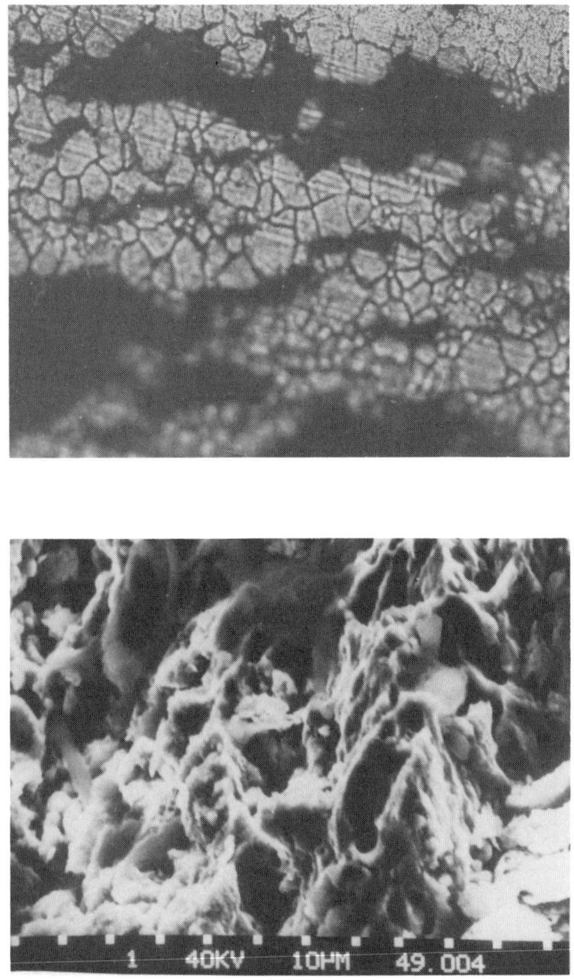


Fig. 7(a) and (b).
Light micrograph (X500) and SEM of specimen crept in air at 750 oC. Grain boundary attack by the corrodents and formation of pits can be observed in these specimens.

numerous pits at the grain boundary. Low melting eutectics like Ni-$Ni_3S_2$ (m.p = 635°C) form which penetrate along the grain boundary and weaken it. Based on these findings a model for creep crack growth in hot corrosion is proposed.

CRACK GROWTH MODEL. A model for creep rupture life under hot corrosion has been proposed [13] based on the grain boundary weakening due to the penetration of low melting corrosion products. This model has been extended in the present study, to the cracking of the material just ahead of the main crack thus making it propagate through a small distance $\Delta a$. The model of the crack growth during the time $\Delta t$ (HC) is shown in Fig.8.

At the tip of the main crack a cavity of size 'r' nucleates under the action of the stress $\sigma$ and the deflection rate, $\Delta$. Considering a miniature specimen of width, W, the damage $\Phi$ is defined as r/W. The damage $\Phi$ grows with time 't' and is given by

$$1 - (1-\Phi)^{((n/(n+1))+1} = A_1 (\sigma/\sigma_o) . t \qquad (5)$$

where $\sigma$ is the applied stress, $\sigma_o$ is the grain boundary strength and n is the creep exponent [13]. $A_1$ is a constant. $\Phi$ increases in dimension and reaches the critical size, $\Phi_c$, at the stress level $\sigma$, given by

$$\Phi_c = 1 - (\sigma/\sigma_o) \qquad (6)$$

Failure of the miniature specimen takes place, thereby increasing the size of the main crack by the width of the miniature specimen, namely, W. The growth of the damage $\Phi$ with time 't' is shown in the figure by the drooping curve. Now the strength of the grain boundary $\sigma_o$, may remain constant if the environmental condition does not affect the grain boundary strength. However, in hot corrosion the corrosive species penetrate the grain boundary and weaken it. Thus the GB strength $\sigma_o$ starts decreasing with respect to time, or in other words, the ratio $\sigma/\sigma_o$ will increase. This will reduce the size of the critical damage $\Phi_c$ required for failure of the miniature specimen. This is shown by the upward curve in the figure.

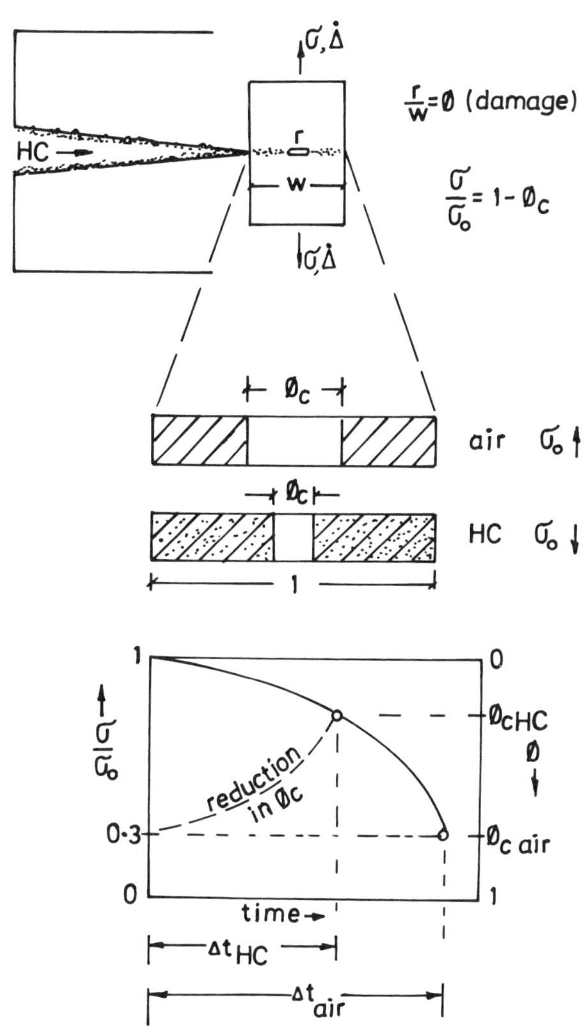


Fig.8. Creep crack growth Model in hot corrosion condition.

When $\sigma/\sigma_o$ -t curve meets the $\Phi$-t curve failure takes place and the corresponding time of failure is $\Delta t$(HC). If $\sigma_{oa}$ is the actual strength of the grain boundary at any time, its relation with the initial strength $\sigma_o$ is given by

$$\sigma_{oa} / \sigma_o = 1 - A_2 t^{1/2} \qquad (7)$$

The constant $A_2$ is dependent on the diffusion properties of the corrodents and hence depends on the corrosive medium, corrosive species formed and the temperature [13].

Identifying the width of the miniature specimen $W = \Delta a$ and its failure time as $\Delta t$, the crack growth rate, $\Delta a/\Delta t$ can be computed with eqns. (5) to (7) and related to the other parameters like diffusion properties of the corrosive medium, temperature, stress etc., and a quantitative relation can be developed. The above model can give us insight into the damage accumulation and cracking mechanism as the crack growth proceeds. However, creep propagation approach to life prediction requires the knowledge of terminal criterion when failure will take place. Hence, for the purpose of engineering design and life prediction, the relation developed in eqns (3) and (4) can be effectively used.

## CONCLUSIONS

From the investigations carried out to study the creep crack growth behaviour of Type 316L stainless steel under hot corrosion condition of a mixture of 75 % $Na_2SO_4$ + 25 % NaCl in the temperature range of 700 °C to 800 °C, the following conclusions are arrived.

1. The crack nucleation and growth take place in the steady state condition of deformation both in air and in hot corrosion.

2. Under hot corrosion condition the creep deformation and the crack growth are accelerated.

3. The energy rate line integral C* appears to give a good description of the creep crack growth rate and a relation of the type

$$da/dt = A\,(C^*)^r$$

appears to be valid. However, the values of the constant A and the exponent 'r' are dependent on the temperature. The constant 'A' increases with increasing temperature, but the variation in the exponent 'r' with temperature does not show any systematic trend.

4. There appears to be a relation between the steady state value of the energy rate line integral, $C^*_s$ and the rupture time, $t_r$ given in the form

$$C^*_s \cdot t_r = R$$

where R is constant and is equal to 300 with C* in $kJ/m^2h$ and time in hours. This relation could be used for the life prediction of components which have cracks starting in the steady state condition of creep deformation.

5. The above correlation between the steady state energy rate integral and the rupture time leads to a relation between the applied stress and the rupture time in the presence of a flaw of size (a/W) in the form

$$\sigma\,[\,t_r\,]^{1/(n+1)} = [R/C_o\,]^{1/(n+1)}\,f(a/W)$$

Thus the rupture life under different operating stress conditions can be obtained if the flaw size in the component is known.

6. Hot corrosion attacks the grain boundaries of the material. The compounds formed lead to a depletion of the strengthening elements in the matrix and weakening of the grain boundaries. This results in an accelerated crack growth and earlier failure of the material.

7. Based on the above findings, a model is proposed for creep crack growth in hot corrosion.

## ACKNOWLEDGMENT

The author is very much thankful to Dr. P. Rodriguez, Head, Metallurgy Programme, Indira Gandhi Centre for Atomic Research, Kalpakkam, India, for his kind help in providing the material and to Department of Atomic Energy, Govt. of India, for their financial help through their sponsored project.

## REFERENCES

1. Fu, L.S., Engg. Fract. Mech., 13(2), 307-330 (1980)
2. Sadananda, K. and P. Shahinian, Metal Science,15, 425-432(1981)
3. Ellison, E.G., and M.P. Harper, J.Strain Analysis, 13, 35-51 (1978)
4. Malik, S.N. Nucl Engng Design, 72, 359-371 (1982)
5. Webster, G.A., in Engineering Approaches to High Temperature Design. (eds) B. Wilshire and D.R.J. Owen. Pineridge Press, Swansea, U.K., 1-55 (1983)
6. Kamaraj, M. ' Studies on creep and creep crack growth in austenitic stainless steels and their weldments.' Ph.D. Thesis, I.I.T., Madras, India. (1990)
7. Suryanarayanan, V., 'Some Studies on the Effect of Hot Corrosion on Creep'. Ph.D. Thesis, I.I.T., Madras, India. (1991)
8. Saxena, A., Creep Crack Growth under Non-steady State Condition. 17th ASTM Symposium on Fracture Mechanics, Albany, N.Y. (Aug. 1984)
9. Radhakrishnan, V.M. and M. Kamaraj. Mater. Sci. Engng., A127, L15-L18 (1990)
10. Radhakrishnan, V.M. and M. Kamaraj. Proc Inter Conf. Mech. Behaviour of Materials ICM-6, (eds) M. Jono and T.Inoue, Pergamon Press, 2, 297-302 (1991)
11. Radhakrishnan, V.M., Kamaraj, M and V.V. Balasubramaniam. Trans. A.S.M.E., JEMT, 113(3), 303-306 (1991)
12. Suryanarayanan, V., Iyer, K.J.L., and V.M. Radhakrishnan, J. High Temperature Technology, 7(1), 33-40 (1989)
13. Suryanarayanan, V., Iyer,K.J.L, and V.M. Radhakrishnan, J. High Temperature Technology, 7(2), 91-96 (1989)
14. Hartnagel, W., Bauer, R and H.W. Grunling, Proc. Sym. on Corrosion and Mech. Stress at High Temperature. (eds). V. Guttman and M. Merze. Appl. Sci.Publ.,London. 257-273 (1980)

*Proceedings of the Fifth International Conference on Creep of Materials, Lake Buena Vista, Florida, USA, 18-21 May 1992*

# Assessment of the High Temperature Performance of Steel Tubing Through the Understanding and Application of Damage Mechanisms

**L.H. Bisbee**
Failure Analysis Associates, Inc.
Alexandria, Virginia

**J.D. Parker**
Department of Materials Engineering
University College
Swansea, United Kingdom

## Abstract

The design of boilers within fossil-fuelled electricity generating plant is such that operating metal temperature increases as steam passes through the superheater and reheater stages. Thus, alloys more resistant to high temperature operation must be utilized as service conditions become more arduous. Problems may then arise if the necessary changes in material occur too far through the cycle and alloys are operated beyond the correct range. Failures will then occur first in regions where the conditions are most severe in relation to a specific tubing material.

The present paper reviews damage mechanisms typically encountered in the high temperature sections of power boilers. This review indicates that the damage mechanisms of low alloy ferritic steels and transition welds between ferritic and austenitic steels are well established. Moreover, generally accepted approaches for assessment of these locations have been developed. However, while the failure of austenitic steel tubes is frequently related to the formation of sigma phase, no procedure for performance monitoring is currently available. A detailed methodology for assessment of austenitic steels is proposed based on observations made from in-service tubing and the results of a programme of laboratory experiments.

BOILER TUBING in fossil-fuelled power plants is manufactured from a wide range of steels. Typically the highest temperatures occur in the superheater and reheater sections where metal temperatures up to about 600°C are encountered. A single tube failure does not represent a major safety hazard. However, every failure constitutes a financial problem and if multiple or repeated failures occur then significant costs may be accumulated. These costs are partly the result of the cost of carrying out the necessary repair and partly because of costs associated with the loss of availability of a particular unit. In general it is important to establish a boiler maintenance strategy which will keep these losses within a defined limit. The appropriate limit should be developed for a specific plant within a particular utility. Thus, for a large, high efficiency unit with a relatively low generating cost, it is usually financial beneficial to prevent all forced outages. However, for a low efficiency unit in a system with spare capacity, occasional failures do not usually lead to significant financial penalties.

Boiler tube failures have been the leading cause of loss of availability in the 1980s (1). Furthermore, it is frequently the case that many failures occur in similar tubes over and over again. Thus, one approach to development of a maintenance strategy is to perform a statistical evaluation of the failures with the view to estimating future performance (2). Widespread repair or replacement of tubes is then planned when the statistics suggest that an unacceptable rate of failures will occur. Whilst in some situations this relatively simple approach may be adequate, in general accurate assessment of performance must be based on root cause analysis of failures. Performing a root cause analysis offers a number of specific benefits.

- Failures may occur for many different reasons. Simply performing statistical analysis on all failures will prevent an accurate pattern from being developed.
- Problems are usually the result of deficiencies in material selection or operation or both. If these deficiencies are not identified, simple replacement of like with like will lead to future failures at an unnecessarily high rate.
- Knowledge of damage mechanisms should allow a worthwhile inspection strategy to be developed. Remedial actions can then be based on information gathered during planned maintenance periods rather than from failures which result in forced outages.

The present paper describes the failure modes observed for the range of steels used in the superheater and reheater sections of fossil-fuelled power plant. In each case an appropriate methodology for life assessment based on the specific damage mechanisms identified is discussed.

## Low Alloy Steel Tubing

**Failure Mechanism in Service.** The vast majority of superheater and reheater tubing is manufactured from $1Cr\frac{1}{2}Mo$ and $2\frac{1}{4}Cr1Mo$ steels. Design approaches are such that operating lifetimes well in excess of 200,000 hours would be expected. However, operating experience indicates that failures well within this period frequently occur. Failures have been detected very early in operating life, up to about 10,000 hours. In general these failures occur by high temperature creep as a consequence of over-temperature operation. These problems are usually related to steam flow irregularities leading to the local metal temperatures being significantly in excess of recommended values. Long term failures are also related to creep, however damage is usually accelerated by corrosion or erosion. An example of a typical long-term failure in a superheater tube is shown in Figure 1.

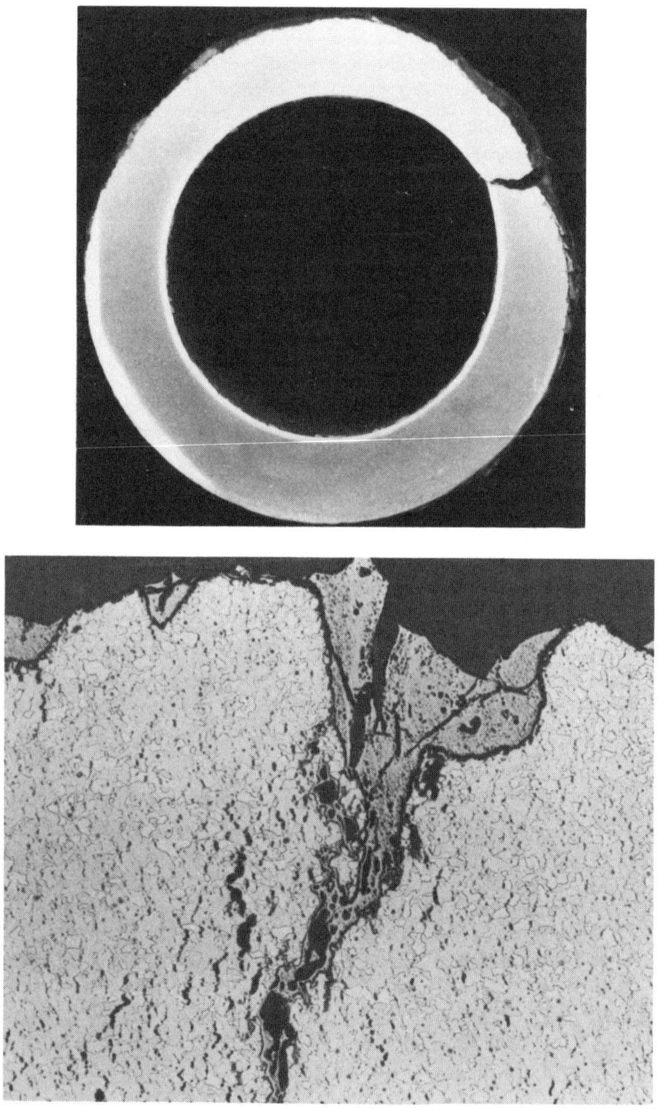

Fig. 1 - Cross section showing wall thinning and creep cracking typical of a long-term tube failure in low alloy steels, x1 (a) and x100 (b).

Consideration of Figure 1a shows that the faces of this tube which have been directly in the path of the hot gases have experienced metal loss through accelerated corrosion. This metal loss has increased the local hoop stress in the tube so that creep cavitation and cracking have developed, Figure 1b.

**Tube Life Assessment.** Assessment methodologies for low alloy steel tubing must therefore take into account damage due to metal loss and creep. A number of approaches which combine the elastic stress equation for hoop stress in a tube wall with oxidation rate and creep life have been suggested. The relationship developed by workers at Ontario Hydro is typical of these expressions. In this case, the service life of a tube, $t_s$, is given by the equation

$$t_s = \frac{1}{K}\left\{1 - (K(n-1)t_r + 1)^{\frac{1}{1-n}}\right\} \qquad ...(1)$$

where n is the stress sensitivity of the creep rupture life, $t_r$ is the rupture time of the material and K is the wall thinning rate. The required value of K is obtained from the expression

$$K = \frac{(W_{max} - W_t)}{W_{max}t}$$

where $W_{max}$ is the original wall thickness and $W_t$ is the wall thickness after time, t.

Equation 1 has been used to develop relationships between wall thinning rate and rupture life for specific tube geometries, materials and operating conditions. A schematic representation of a typical relationship is shown in Figure 2. Clearly at rapid rates of wall thinning, life is dominated by the corrosion/erosion processes, whereas at low values of K the service life is controlled by creep.

This type of model was developed to prevent repeated forced outages. In its most simple form it can be applied after a single tube failure to provide guidance as to future performance. Following a tube failure, $t_s$ and n will be known and K can be estimated following laboratory examination of the failed tube. Application of equation 1 will therefore allow $t_r$ to be calculated for a given material and thus the relationship between wall thinning rate and service life can be established for a given bank of tubes. This relationship then forms the basis for analysing wall thickness measurements made on similar tubing. The approach has two main advantages:

- it should provide conservative assessments since the model assumes that all tubes are operating at the same temperature. In fact monitoring of tube metal temperatures has shown that a distribution of temperatures will be present in any operating superheater or reheater. Since both creep and corrosion are temperature dependent processes, the first tube to fail in a specific region will be operating under the worst case conditions.
- the model is relatively insensitive to materials creep data so that post exposure testing of material is not required. In fact it is now generally accepted that creep testing of ex-service tubes does not significantly improve prediction of plant performance.

Refinements of this basic approach have attempted to combine measurement of the tube metal thickness with measurement of the steam-side scale present on the inside of the tube. The thickness of this scale can then be used to estimate the operating temperature of a given tube (4).

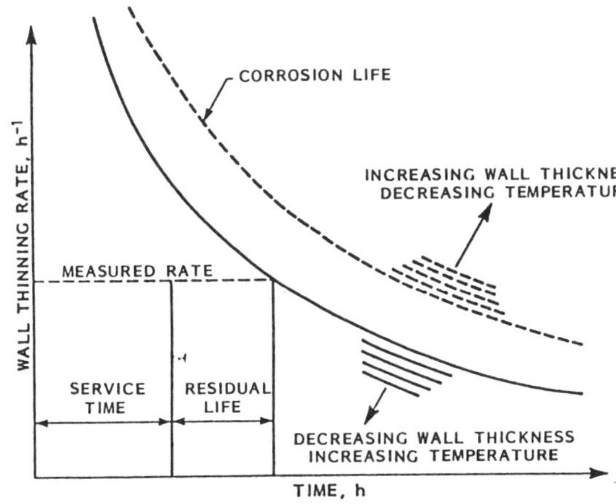


Fig. 2 - Relationship between wall thinning rate and rupture life calculated using equation 1 (3).

## Dissimilar Metal Welds

**Failure Mechanism in Service.** The metal temperature within a superheater or reheater section will increase with the steam cycle. Thus, alloys more resistant to high temperature operation must be utilized as the service conditions become more arduous. In the sections operating at the highest temperatures, tubing is frequently manufactured from high alloy stainless steels. The transition from low alloy steel to stainless steel therefore necessitates the fabrication of a dissimilar metal weld. This fabrication process is complicated since, in addition to their different creep and oxidation behaviour, low alloy and stainless steels also exhibit differing coefficients of thermal expansion and different carbon solubility. Furthermore, because of the different composition, care must be taken to avoid the formation of brittle microstructures at the weld interface. Plant welds have been successfully manufactured by a number of different methods, including

- the formation of a pressure weld without added filler metal
- application of fusion welding using a stainless steel filler metal
- application of fusion welding using a nickel based alloy filler.

Experience has shown that the high temperature life of weldments which involve an interface between low alloy and stainless steels is generally lower than that of the welds manufactured using nickel-based electrodes. Comparison of laboratory tests carried out under controlled conditions is given in Figure 3 (5). These data show that for joints between $2\frac{1}{4}$Cr1Mo and TP321H steels, welds manufactured using E309 stainless steel exhibit a life approximately half that observed with a nickel based filler. However in both

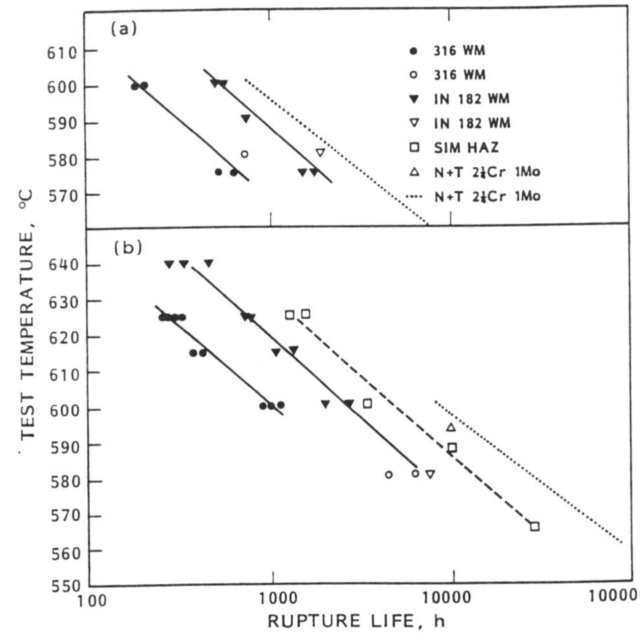


Fig. 3 - Uniaxial creep rupture data for austenitic steel and nickel based dissimilar metal welds at stresses of 100 MPa (a) and 62 MPa (b) (5).

cases the high temperature life is less than that of the $2\frac{1}{4}$ Cr1Mo steel. Furthermore, service experience with these welds is variable with significantly different lives obtained from nominally similar weldments operating under nominally similar conditions.

High temperature damage invariably involves the nucleation and growth of creep cavities with failure occurring with low overall ductility, Figure 4. However, the details of the creep cavities differ. Joints welded with austenitic filler exhibit creep voids and microcracks in the heat affected zone of the low alloy steel, Figure 5, whereas in the nickel-based welds damage occurs predominantly at the fusion boundary. The relatively short life detected with the stainless steel weld metal is attributed to the large different in coefficient of thermal expansion between the low alloy steel

Fig. 4 - Typical low ductility failure of a dissimilar metal weld.

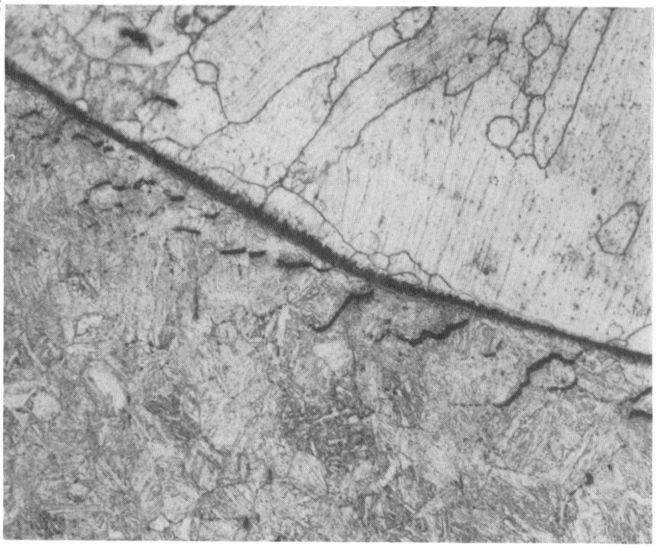

Fig. 5 - Creep cavitation and microcracking in the heat affected zone of a dissimilar metal weld produced with an austenitic steel filler (x100).

and the weld metal. With damage in the nickel based welds developed as the result of carbide formation at the weld interface.

**Tube Life Assessment.** Methods of assessment of dissimilar joints must consider the type of weld present and the operating stress and temperature. Analysis of laboratory tests and plant experience led to the following expression being developed (5)

$$\frac{10^3}{T} = Constant + 0.21\log\sigma + 0.058t_f \qquad ...(2)$$

where T is the temperature, $\sigma$ is the stress and $t_r$ is the life. The constant in equation 2 takes the value of 0.59 and 0.57 for austenitic and nickel based welds respectively. It should be emphasized that equation 2 has been used successfully to predict the performance of dissimilar metal welds taking the appropriate value of $\sigma$ as the elastic axial stress. Thus, no estimates of stresses due to the difference in thermal coefficient of expansion are required. Instead, damage introduced by the relaxation of these stresses is included in the test data which forms the basis of this equation.

Application of equation 2 to determine the lives of dissimilar metal welds in service requires knowledge of the materials of fabrication and joint design which should be available, estimates of operating temperature which can be obtained from plant thermocouples or measurement of steam-side scale in the low alloy steel tubing. The value of elastic axial stress is critical to the estimated life. In situations where tubing is properly supported with correct clearances available to prevent stresses being developed from restrained thermal expansion, axial loads may be simply calculated from geometry and internal pressure. If system loading is suspected then analytical procedures must be adopted to calculate these stresses. However, where such methods have been used, accurate assessment of performance has been obtained.

## Stainless Steel Tubing

**Failure Mechanism in Service.** The highest temperature regions of a superheater/reheater require the use of stainless steels with the necessary corrosion and creep resistance, eg TP321 or 347. Despite apparently exhibiting properties well in excess of those required to achieve operation of greater than 200,000 hours failures have been documented within this period. Typical long-term failures occur in a generally brittle manner leading to a thick-lipped rupture, Figure 6. The low ductility observed is attributed to the presence of sigma phase which forms during service. Since the sigma phase develops at grain boundaries, it can act as nucleation sites for creep void formation. Furthermore, this phase is itself brittle so that networks of sigma phase will significantly reduce fracture toughness.

Evaluation of failed tubes confirms that creep cavitation is associated with grain boundary sigma phase. However, failed tubing does not allow detailed information regarding crack initiation and growth to be established. Examination of an ex-service tube without macroscopic cracking has provided significant new information in this area. A number of ring samples were prepared from TP321 stainless steel tubing to allow documentation of the microstructures and damage present after approximately 200,000 hours in service.

Representative optical micrographs from the ex-service material are shown in Figure 7a, b and c for the outside surface midwall location and inside surface respectively. In general the microstructure observed is austenite with grain boundary precipitates. Scanning electron microscopy combined with energy dispersive X-ray analysis suggested that the larger, geometrically shaped particles were chromium rich whereas the rounder, semi-continuous precipitates contained significant levels of titanium. It therefore appears that the larger particles are sigma phase with the semi-continuous precipitates being carbides. The distribution of the structures present was different at each location. Thus, at the outside surface of the tube the grain size varied from approximately 10μm to 50μm, Figure 7a, with the vast majority of grains being towards the lower end of this range. Rounded creep cavities were observed with the majority of cavities detected on the relatively straight region of boundary between neighbouring grains. The majority of sigma phase detected was at grain boundary triple points. Structure of this type was observed for approximately 20% of the tube wall thickness. A range of grain sizes was also observed in the central zone of the tube wall. In this region, grains appeared to be either very large, approximately 100μm in diameter, or small, approximately 15μm in diameter. Sigma phase was again observed at triple point junctions. However, the majority of creep damage detected was in the form of triple point wedge cracks, Figure 7b. Approximately 60% of the tube wall thickness exhibited this form of structural distribution. The remaining 20% of the wall thickness towards the inside surface of the tube is shown in Figure 7c. The grain sizes in this zone appear to be even finer than observed at the outside. Sigma phase is again present at the grain boundary triple points, however, very few creep voids were observed in this region.

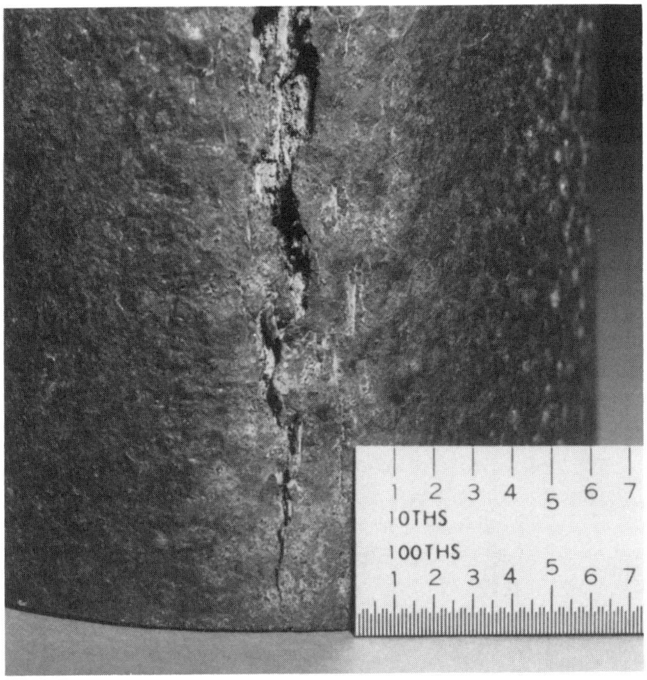


Fig. 6 - Typical thick-lipped failure of a stainless steel superheater tube.

**Tube Life Assessment.** The microstructural examination has confirmed that high temperature failure of tubes of this type is related to the nucleation and propagation of creep damage. Yet for the nominal operating hoop stress and temperature of the tubing lifetimes well in excess of $1 \times 10^8$ hours would be expected from standard stress rupture data. Furthermore even for operating temperature 100°C above nominal the lifetime is still greater than $10^6$ hours. For the present tubing extreme service conditions cannot explain the levels of damage observed. Moreover no corrosion/erosion of the tube material was detected so that the tube wall thickness at failure exceeded the minimum design value. An appropriate method of tube life assessment must therefore take into account the apparently anomalously short failure times and explain why the damage appears to develop in mid-wall regions rather than at the tube surface.

A rational assessment method has been developed. This approach considers the variation of grain size observed in ex-service tubing and the effect of grain size variations on creep deformation and fracture characteristics. Creep testing carried out on a number of 18Cr-8Ni type steels has shown that creep strength and ductility depend on grain size (6,7). Thus the deformation rate of fine grained material, average grain size ~10μm, has been observed to be an order of magnitude faster than the creep rate measured in relatively coarse grained samples, average grain size ~100μm (6). This effect of grain size on creep rate may be explained since it is frequently observed that for long term tests the activation energy for creep is similar to that for grain boundary diffusion (8). Thus, at the long times involved with the service tubing a faster rate would be expected in fine grained materials because of the relatively high levels of grain boundary area present.

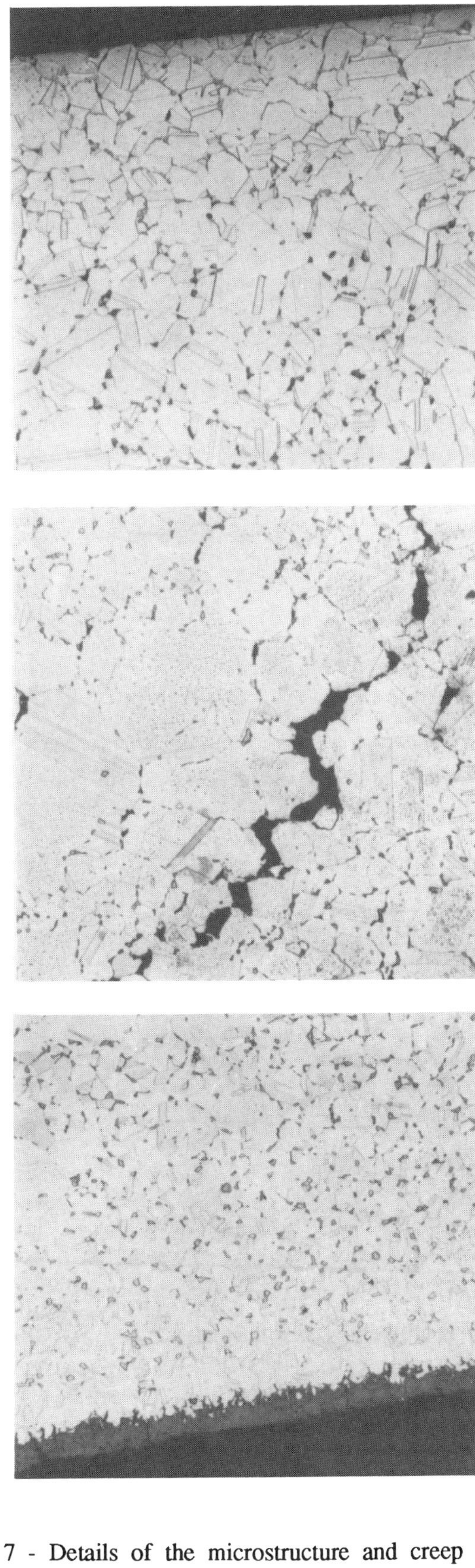

Fig. 7 - Details of the microstructure and creep damage observed in a TP321 superheater tube after long-term service, at the outside (a), at the midwall (b) and at the inside (c) (x150).

The difference in creep deformation rate in zones of different grain size then offers an explanation for the observed service behaviour. Previous analysis has examined the creep performance of thick-walled, cylindrical components with zones of different creep strength (9). This work considered the deformation behaviour of components where three zones of different creep strength were present. Even where the creep rate of these zones varied by up to an order of magnitude no significant changes in overall deformation characteristics were detected compared with that of homogeneous components with average properties. The compatibility of creep behaviour was achieved because the stress distribution established in the component wall varied such that the zones of high creep strength became loaded compared to regions of lower creep resistance. Thus, in the present tubes the coarse grained regions observed towards the centre of the wall would be expected to sustain significantly higher stresses than the fine grained regions at the surfaces. Since the microstructure contains large amounts of brittle sigma phase these high stresses would tend to promote the formation of wedge-type grain boundary micro-cracks. These cracks then propagate relatively rapidly since the presence of networks of sigma phase leads to a reduction in fracture toughness.

The different grain sizes present in TP321 tubes appear to be a consequence of the fabrication procedures used in manufacture. Clearly an accurate method of assessment must consider the effect of manufacturing methods on grain size and take account of these variations in grain size on deformation and fracture behaviour. Studies are in progress to quantify these effects and in combination with further plant data will lead to the validation of the present life prediction model.

## Concluding Remarks

Failures in high temperature tubing can lead to significant losses of availability in boilers of fossil-fuelled power plant. Cause analysis of failed tubes forms the basis for planning maintenance and repair/replacement schedules for specific units. Review of operating experience suggests that whilst long-term rupture of tubing is related to creep the form of damage developed varies. Thus for low alloy steels service life depends on both creep and wall thinning. Adequate models of assessment are available and in combination with surveys of tube wall thickness provide the information necessary to prevent overly conservative replacement or forced outages. The lifetime of dissimilar metal welds is known to be a function of joint type and operating stress and temperature. Preventing axial loads on these welds is important in prolonging service life. However, methods of predicting performance are available where axial loading is present. Failure of creep resistant stainless steel tubing occurs in part because of embrittlement due to the formation of sigma phase. However, the lifetime of these tubes cannot be simply based on application of operating hoop stress to minimum rupture data. Instead, it appears that accurate predictions of performance must take into account the variability of grain size and creep strength of different zones with the tube wall.

## References


1 - Dooley, R.B., "Summary of World Wide Approaches to Life Extension and Assessment", Proc. Int. Conf. on Fossil Power Plant Rehabilitation, ASM, 1989.

2 - Davidson, J., "The Use of BTF Statistics to Estimate Future Incidcnce of Failures", Proc. "Boiler Tube Failures in Fossil Power Plants", Electric Power Research Institute, (1988).

3 - Moles, M.D.C., D.V. Leamans and H.J. Westwood, "Residual Life Estimation of Boiler Tubing in Thermal Power Plants", Proc. Int. Conf. on Engineering Aspects of Creep, Vol. II Inst. of Mech. Engrs., (1980).

4 - Foulds, J.R., D.I. Roberts and R. Viswanathan, "Methods of Estimation of the Temperature of Reheater and Superheater Tubes in Fossil Boilers", Conf. on Boiler Tube Failures in Fossil Power Plants, R.B. Dooley and D. Broske eds., Electric Power Research Institute, Atlanta, (1987).

5 - Parker, J.D., T.P. Byrne and R.B. Dooley, "Ontario Hydro's Experience with Dissimilar Metal Welds in Boiler Tubing", Proceedings: Seminar on Dissimilar Metal Welds in Fossil-Fired Boilers.

6 - Garafalo, F., "Fundamentals of Creep and Creep-Rupture in metals", McMillan, New York, (1965).

7 - Kinoshita, K., T. Mimino and I. Minegishi, "Improved superheater tubing material - Ti and Nb bearing Austenitic Steel", Creep and Fatigue in Elevated Temperature Applications, Inst. of Mech. Engrs., C162/73, (1973).

8 - Beckitt, F.R., T.M. Banks and T. Gladman, "Secondary Creep Deformation of 18%Cr-10%Ni Steel", Creep Strength in Steel and High Temperature Alloys, The Metal Society, London, (1974).

9 - Coleman, M.C., J.D. Parker, D.J. Walters and J.A. Williams, "The deformation behaviour of thick-walled pipe at elevated temperatures", Third Int. Conf. on Mechanical Behaviour of Materials, K.J. Miller ed., Pergamon Press, (1979).

*Proceedings of the Fifth International Conference on Creep of Materials, Lake Buena Vista, Florida, USA, 18-21 May 1992*

# Failure Analysis of a Set of 1CrMoV Studs from a Steam Turbine and Techniques for Its Assessment

**J.J. Hickey and J.H. Bulloch**
ESB
Dublin, Ireland

ABSTRACT

*The present report describes the results of a failure analysis of a series of CrMoV steel turbine studs which had experienced a service lifetime of some 50,000 hours. It was observed that certain studs suffered complete fracture while others showed significant defects located atthe first stress bearing thread. Crack extension was the result of marked creep embrittlement and reverse temper embrittlement, RTE. Selected approaches were examined to assess the effects of RTE on the material toughness of selected studs. It was observed that Auger Electron Microscopy results which indicated the extent of grain boundary phosphorus segregation exhibited a good relationship with ambient temperature charpy data. The Electrochemical Polarisation Kinetic Reactivation, EPR, approach, however, proved disappointing in that the overlapping scatter in the minimum current density, $I_r$, for an embrittled and a non-embrittled material was such that no clear decision of the toughness properties was possible by this approach.*
*The initial results obtained from Small Punch testing showed good agreement with other reported data and could be related to the FATT. Indeed this Small Punch test, combined with a miniature sample sampling method represents an attractive approach to the toughness assessment of critical power plant components.*

*DURING A ROUTINE MAINTENANCE OVERHAUL of a 250MW steam turbine the removal of the intermediate pressure inner cylinder casing revealed the presence of a number of cracked and through section fractures flange studs. The studs, sixteen in total, were fabricated from a 1CrMoV steel approximately 140mm in diameter and 700mm in length and were subjected to an operational temperature of 490$^{o}$C. Prior to removal of the casing these particular studs had operated for 50,000 hours with around 600 and 300 hot and cold starts respectively. Also, ultrasonic inspection was carried out on the end face of the studs prior to removal in order to assess the location, extent and orientation of the cracking characteristics. It was observed that of the 16 studs, eight were completely fractured, four contained significant cracks and the remainder were defect free. The studs which had fractured or cracked in service did so at the first stress bearing thread in the lower half casing. The location and orientation of the cracking events are shown in Figure 1.*

*The present paper describes a*

*detailed failure analysis aimed at assessing the extent of creep embrittlement and the role played by in service toughness degradation through reverse temper embrittlement, RTE, on the cracking process. A number of techniques aimed at assessing RTE effects in a 1CrMoV steel are also investigated.*

## *CREEP-REVERSE TEMPER EMBRITTLEMENT INTERACTION EFFECTS*

*Reverse temper embrittlement is the result of the segregation of trace solute elements such as P, Sn, As and Sb to grain boundaries which cause a lowering of the grain boundary cohesive energy. RTE affects a materials fracture toughness (1) while other phenomena such as Hydrogen embrittlement (2), intergranular fatigue crack extension (3), 500°F embrittlement (4) and stress corrosion cracking (5), can be subjected to impurity induced or assisted grain boundary decohesion. However, the influence of grain boundary segregation, i.e., RTE, on creep is unclear. This probably is the result of the differences between the mechanisms of low and high temperature (creep) fracture, viz., in low temperature RTE fracture, initiation and propagation of microcracks are rate controlling while at high temperatures the nucleation and growth of creep cavitation, dislocation mobility and grain boundary sliding are important.*

*Wilkinson et al (6) observed that the ductility of P-doped 1 1/4 Cr1Mo steels were higher than those without P additions and suggested that the presence of P at grain boundaries enhance the nucleation of grain boundary cavities while retarding their subsequent growth by reducing boundary self diffusion. Takasugi & Pope also reported the beneficial effects that P had on hot ductility properties. Recently Yu and Grabke (7) observed that the minimum creep rate was increased with P additions. In addition to this effect of P on cavity and carbide formation it was evidenced that P increased the density of Manganese Sulphide non-metallic inclusions which served as prime nucleation sites for creep voids.*

*Finally Burke and Howe (8) have recently reported that lowering the Phosphorus content from 0.02 to 0.007% essentially decreased the creep rupture life, which they suggested was the result of loss of solid solution hardening, and marginally increased the rupture ductility. This increase in rupture ductility was thought to be due to the reduction of phosphorus at grain boundaries.*

## *EXPERIMENTAL TECHNIQUES*

*The studs were manufactured to the French specification 28CDV5-08 and this together with the bulk chemical composition of two studs are given in Table 1.*

*The studs were originally supplied in the oil quenched (1 hour at 1000°C) and tempered (700°C for 4 hours) condition.*

*A selection of the cracked and fractured studs were subjected to detailed metallographic and fractographic examinations. The fracture surfaces of the studs were covered by a high temperature oxide which was removed by immersing the fracture in a solution of inhibited hydrochloric acid for 10 minutes in an ultrasonic bath.*

*A series of charpy vee-notch impact test specimens were machined from selected studs and full fracture appearance transition temperature, FATT, curves were obtained.*

*Small bar samples, 25mm long, 2mm x 2mm, were machined from broken charpy specimens and fractured at liquid nitrogen temperatures in a scanning auger microprobe, SAM. Ten separate grain boundaries were analysed, utilising auger spectroscopy (9 ) in each sample. An incident electron beam current of $5 \times 10^{-7}$ A was used at a magnification of x200 and the point* analysis was always employed and checked routinely with area analysis.

Electrochemical polarisation kinetic reactivation, EPR, tests were conducted on a series of broken charpy specimens utilising a portable Toshiba EPR testing unit (10). The anodic polarisation tests were conducted, using picric acid and a surfactant as the electrolyte, by sweeping the electric potential from the natural electric potential to 400mV at 1mV/sec and holding at this potential for 2 mins to allow for a sufficient passive state film to form. The potential is then swept back to near the neutral potential at a similar rate and the corresponding current measured and plotted against the potential.

A series of small thin disc specimens, 6mm dia. and 0.5mm thick, were machined from the undamaged portion of broken charpy test specimens. The small punch test (11, 12) was performed on an Instron tensile testing machine. The load train contains the punch/die set which holds the small disc test sample, environmental or temperature chamber and loads which record the punching force. The Instron cross head speed was kept constant at 0.02mm/sec. The specimens were punched until a 10% or greater decrease occurred in the peak load. The area under the load/deflection curve was assessed to estimate the energy which is called the small punch energy which is used to estimate $T_{sp}$, the small punch transition energy. Foulds et al (13) has demonstrated that a good correlation existed between $T_{sp}$, and the FATT for CrMoV steels.

## EXPERIMENTAL RESULTS

FRACTOGRAPHY & METALLOGRAPHY. Examination of the studs revealed that they had either one or two discrete "in service" portions of crack extension. In the case of the two crack studs a discrete crack occurred at opposite ends of the diameter and were located on the inside and outside of the casing. From a macroscopic fracture examination it was observed that four fracture types were in evidence, see Figure 2; viz;

Type 1 - A double sided crack followed by complete fracture in service.

Type 2 - A single crack followed by complete failure during service.

Type 3 - A double sided crack which did not fracture in service.

Type 4 - A single fracture face in service.

The results of the oxide stripping procedure used in the present study are shown in Figure 3. As can be seen, this figure illustrates an oxidised portion of the fracture face of a stud before and after oxide stripping and it is clear that the oxide was successfully stripped from the fracture face without destroying or altering the underlying original fracture surface. Figure 4 shows the intergranular nature of the initiation fracture process at the thread root location while a detailed view illustrating extensive secondary intergranular cracking, is shown in Figure 5.

Evidence of grain boundary creep cavitation showing the coalescence or linking of individual creep cavities to form an incipient crack is portrayed in Figure 6. Indeed no transgranular failure was observed on any of the fracture surfaces.

The fractures were always in a plane normal to the axis of the stud and located at the root of the first stress bearing thread in the end which was screwed into the bottom half casing. The initiation sites of the sub-critical cracks contained numerous grain boundary cavities located on prior austenite grain boundaries which frequently contained intergranular microcracks, see Figure 7. Discrete cavity linkage to form incipient microcracks were also evident. The initial severe level of creep damage, i.e., cavitation, is followed by a lesser degree of cavitation continuing through to the second, third and fourth stress bearing threads. Extensive creep cavitation was also observed in the bulk section of the threaded studs which fractured while cavitation was also observed at the thread location of the uncracked studs.

The tempered microstructure of both the damaged and the uncracked studs exhibited no apparent differences while the hardness values varied marginally, i.e., from 260 to 300 H.V. Large variations in the average prior austenite grain size, i.e., 38 to 80μm, however, were observed.

IMPACT ENERGY TESTS. A selection of the studs were impact tested and it was clear that the RTE studs exhibited low room temperature charpy energies which varied from 5 to about 30 joules while a series of other non-embrittled CrMoV studs (14) showed energies in excess of 100 joules. Partially embrittled studs had charpy energies of around 70 joules at ambient temperatures.

A selected stud which had an FATT value of +80°C was subjected to a de-embrittlement. The FATT of the de-embrittled material was -65°C showing that the embrittlement due to phosphorus segregation to grain boundaries, was significantly reversible. It is known that embrittlement is greater when grain boundary creep cavitation is present in an already embrittled material (15). It is suggested that the applied deformation leads to irreversible grain boundary damage, but that such damage is not enough to embrittle the material. On the other hand this effect becomes more significant in a material that has already embrittled by grain boundary segregation. It is clear thus, that the embrittlement of the studs is both reversible, due to phosphorus segregation to grain boundaries, and irreversible due to grain boundary creep damage.

AUGER ELECTRON MICROSCOPY. This data only recorded phosphorus as the damaging solute element and the extent of grain boundary phosphorus segregation was portrayed as a % of a grain boundary monolayer of phosphorus. The embrittled studs exhibited phosphorus levels which varied between the range 0.15 to 0.38 while a partially embrittled stud from another investigation (14) had 0.05 of a monolayer. The influence of % phosphorus monolayer on the room temperature charpy energy, RTCE, for the present study and other reported data (14, 16, 17) are shown in Figure 8. From this figure it can be seen that there is a significant correlation between % monolayer of phosphorus and the degree of embrittlement. Although the data exhibited a fair degree of scatter at impact energies below about 20 joules the data can be described by the following expression :

$$\text{RTCE (Joules)} = \frac{4}{(\% P_{mon})}$$

An example of a phosphorus map, from which the % phosphorus monolayer was assessed, and its companion fracture surface is shown in Figure 9. Note that the regions where no phosphorus was evident was transgranular in nature.

EPR TESTS. In an effort to evaluate the validity of the EPR test in assessing the toughness of a material, a total of fourteen repeat tests were conducted on theservice embrittled material which had an FATT of +80°C and the de-embrittled material which exhibited an FATT of -65°C. It was found that the embrittled material exhibited $I_r$ values which ranged between -0.74 to -1.85μamps averaging -1.35μamps while the de-embrittled material showed $I_r$ data between -0.89 to -2.36μamps averaging -.173μamps. A typical EPR test curve is shown in Figure 10 while the scatter and overalpping nature of the embrittled and de-embrittled material EPR data is shown in Figure 11. Clearly, for the present material and FATT toughness range -65°C to +80°C, the EPR test does not represent a realistic non-destructive toughness assessment method.

SMALL PUNCH TEST. To date only one stud has been subjected to small punch testing. This embrittled stud,

FATT = +123°C, data is shown in Figure 12, together with other CrMoV steel data (12) and Ni-Cr steel results (13). From this figure it can be seen that the present data exhibited good commonality with the other reported data and that the small punch transition temperature can be related to the charpy FATT by the expression :

$$FATT\ (^{\circ}C) = 438 + 2.48\ (T_{sp})$$

## CONCLUDING REMARKS

It has been shown that with a series of CrMoV steel studs which had been in service for around 50,000 hours, four types of defects were identified at the first stress bearing thread in the end which was screwed into the bottom half casing. Much evidence of significant creep damage has been observed on the fracture surfaces, i.e., primary and secondary macro-cracking as well as isolated cavities and linked cavities (incipient micro-cracks). As well as irreversible creep embrittlement the material was temper embrittled through the diffusion of phosphorus solute atoms to prior austenite grain boundaries; such embrittlement is however reversible, viz., RTE.

A series of assessments of the influence of RTE on the toughness of the CrMoV studs were conducted using (i) charpy impact tests, (ii) auger electron microscopy, (iii) electro-chemical polarisation kinetic reactivation, EPR and (iv) the small punch test. It was observed that a significant relationship existed between ambient temperature charpy impact energies and the extent of grain boundary phosphorus segregation expressed as a % of a monolayer. At low energy values, the data scatter became a limiting factor in the assessment. However, at such values the material was severely embrittled and the data scatter was much less in the practically important regime, i.e., the transition region.

The EPR results, however, were disappointing in as much as that when extremely embrittled and de-embrittled materials were examined the data points significantly overlapped and theaverage $I_r$ value only changed by 28% for a 145°C change in the FATT value.

Although only limited small punch test data is available to date, this data exhibited excellent commonality with other reported data and as a result this technique represents a viable method of assessing the toughness properties of a recently developed miniature sample sampling system (18) it now makes direct assessment of critical components by the small punch test, both a feasible and attractive approach.

## REFERENCES


1. C.J. McMahon & V. Vitek, Acta Metall, 27, (1979) pp 509-517.
2. K. Yoshino & C.J. McMahon, Metall. Trans., 5A, (1974) pp 363-374.
3. R.O. Ritche & J.F. Knott, Metall Trans, 5A (1974) pp 182-190.
4. C.L. Briant & S.K. Banerji, Metall Trans., 12A, (1981) pp 309-317.
5. J. Kupper, H. Erhart & H.J. Grabke, Corr. Science, 21, (1981) pp 227-238
6. D.S. Wilkinson, K. Abiko, N. Thyagarajan & D.P. Pope, Metall Trans., 11A, (1980), pp 1928 - 1933.
7. Jin Yu & H.J. Grabke, "Effects of P-addition on creep of 1CrMoV steels", metal science, 17, (1983) pp 389-396.
8. P.R. Burke & A.A. Howe "Creep Rupture Properties of Welded 1/2 Cr 1/2 Mo 1/4 steels with additions of Titanium and two levels of phosphorus", Creep & Fracture of Engineering Materials & Structures, Procs, 4th Inter Conf, Swansea, April, 1990, pp 635-650.
9. R.K. Wilde, Vacuum, 36, (1976) pp 441-448.

10. Y. Murakami, A. Yaegashi, E. Nishida, T. Shoji, H. Takahashi, "Non-destructive evaluation of materials degradation after long term operation in chemical plants by means of electro-chemical method". Conf. "Life Assessment and Extension", The Hague, Netherlands, June 1988, pp 130-135.
11. M.P. Manahan, A.S. Argon & O.K. Harling. J. Nucl. Materials, 103, (1981) pp 1545-1552.
12. Jai-Man Baik, J. Kameda & O. Buck "Small Punch Test Evaluation of Intergranular Embrittlement of an Alloy Steel", Script. Metall., 17, (1983), pp 1443-1447.
13. J.R. Foulds, C.W. Jewett & R. Viswanathan, "Miniature Specimen Test Technique for FATT", Inter Power Generation Conf. San Diego, Oct, 1991, 91-JPGC-PWR-38.
14. J.H. Bulloch & J.J. Hickey, ESB, Ireland, unpublished work on CrMoV studs subjected to 120,000 hours service.
15. J.J. Hickey, "Failure Analysis of a set of 1CrMoV studs from a 250MW steam turbine after 50,000 hours service. Inter. Conf. Failure Analysis ASM Conf. Montreal, Canada, July 1991, to be published.
16. C.J. McMahon, D.H. Gentner & A.H. Vcisik, "Investigation of the Effects of Grain Size and Hardness in Temper-Embrittled 2 1/4 Cr1MoV steel" Trans ASME 106, (1984), pp 66-70.
17. R. Viswanathan & R.I. Jaffee, Toughness of CrMoV steels for steam turbine rotors, Trans ASME, 105, (1983), pp 286-294.
18. J.D. Parker & L.H. Bisbee, "Monitoring in-service changes in Material Properties. "Acceptance Testing and Monitoring of Steam Plant", I Mech., E. London, Oct. 1991, paper No. 2.

**TABLE 1 : CHEMICAL COMPOSITION OF THE VARIOUS BARS**

| | C | S | P | Cr | Mo | V |
|---|---|---|---|---|---|---|
| Specifi-cation | 0.25-0.31 | ≤0.030 | ≤0.030 | 1.00 - 1.50 | 0.60 - 0.90 | 0.20 - 0.40 |
| Bar A | 0.290 | 0.008 | 0.019 | 1.25 | 0.70 | 0.380 |
| Bar B | 0.288 | 0.012 | 0.010 | 1.40 | 0.68 | 0.325 |

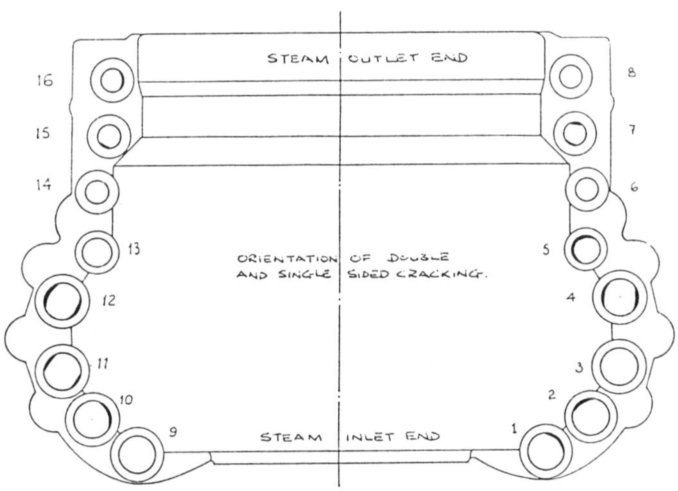


FIG. 1
I.P. INNER CASING FLANGE

FIG. 2

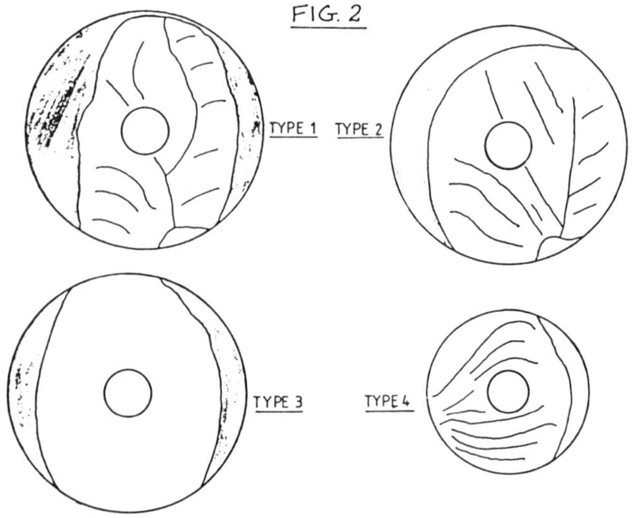

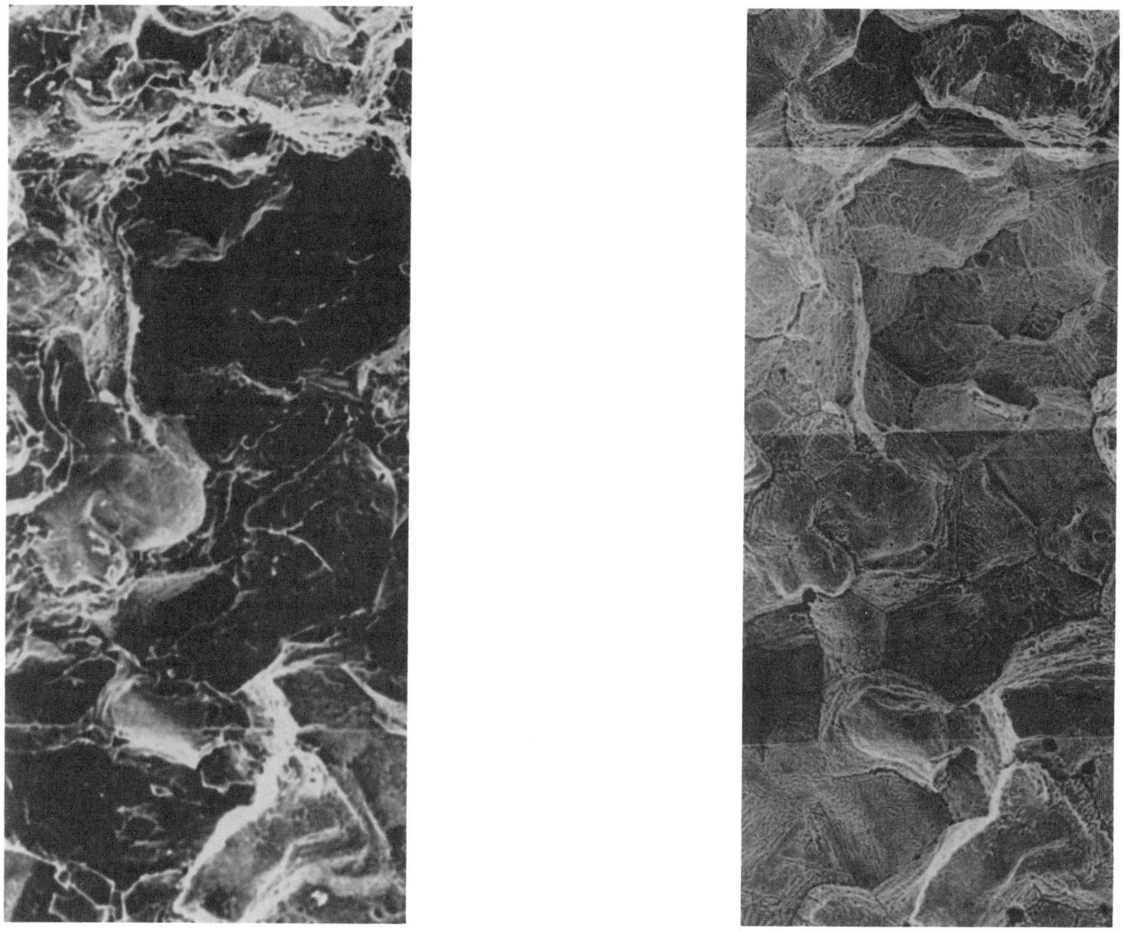

OXIDISED FRACTURE SURFACE

OXIDE STRIPPED FRACTURE SURFACE

FIGURE 3. FRACTURE SURFACE DETAILS

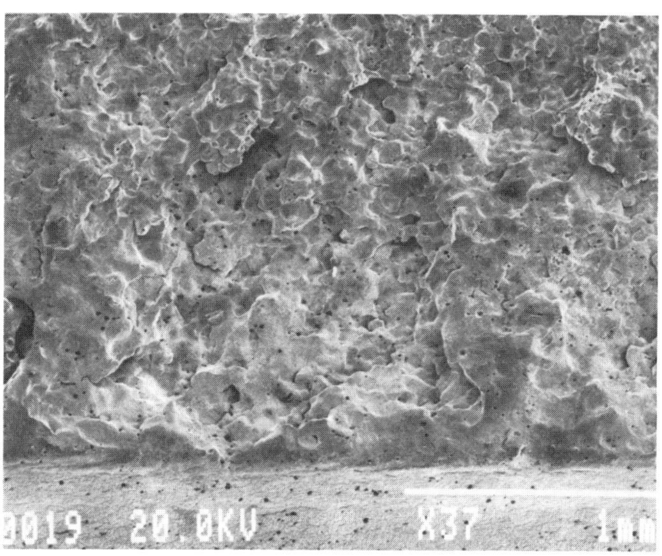


FIGURE 4. GENERAL VIEW OF FRACTURE INITIATION

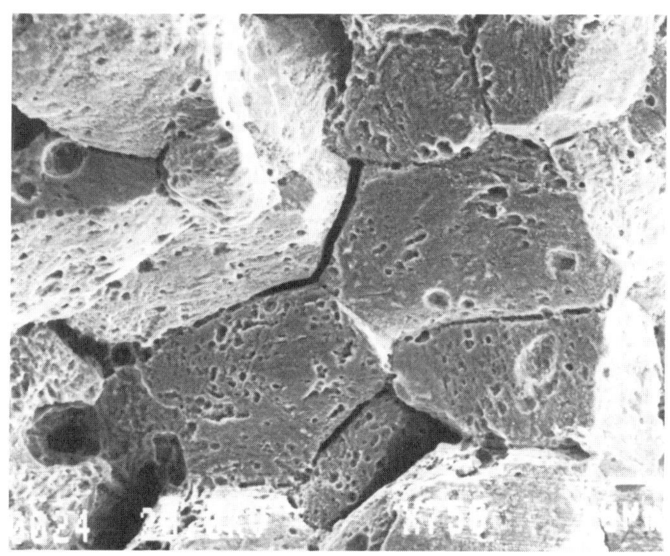

FIGURE 5. DETAILED VIEW OF SECONDARY CRACKING

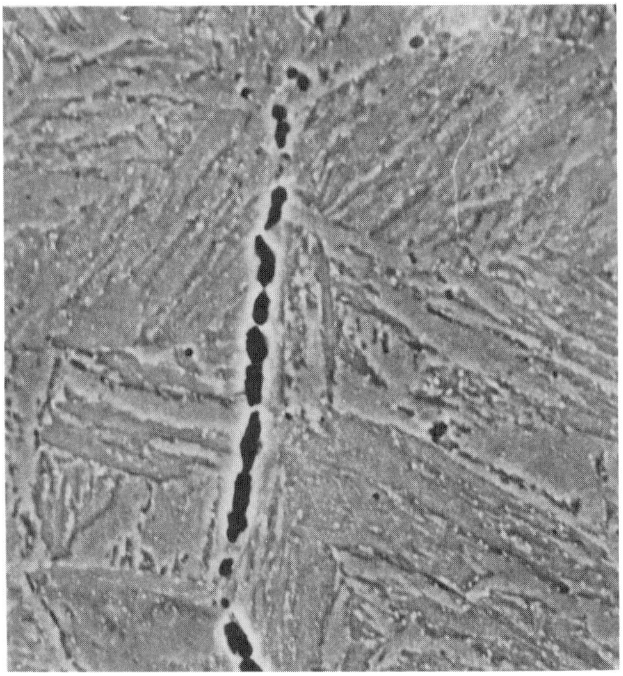

FIGURE 6. CAVITY COALESCENCE AT GRAIN BOUNDARIES.

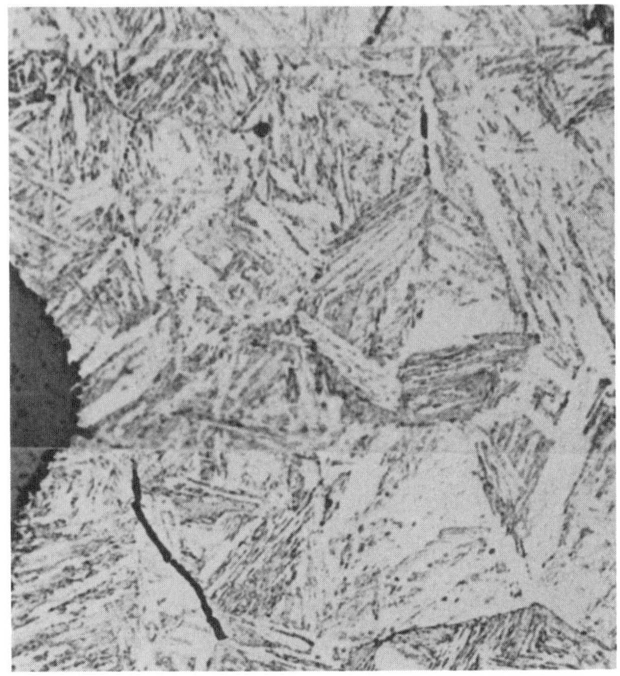

FIGURE 7. MICROCRACK FORMATION AT GRAIN BOUNDARIES.

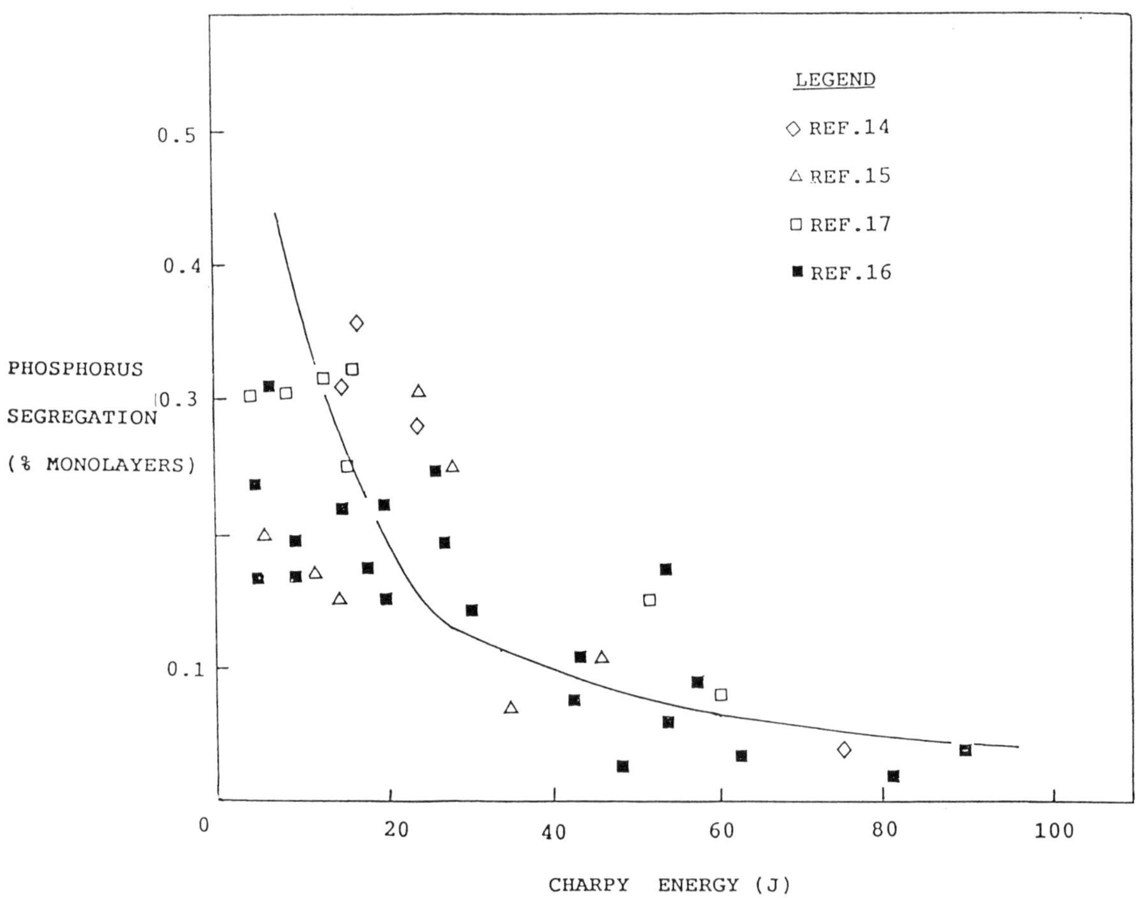


FIGURE 8. RELATIONSHIP BETWEEN PHOSPHORUS GRAIN BOUNDARY SEGREGATION AND R.T. CHARPY ENERGY.

(a) Phosphorus Dot Map X1000

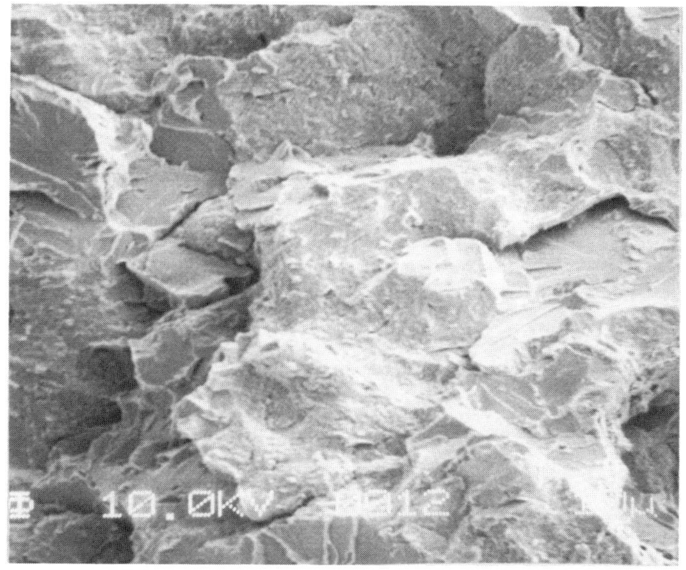


(b) Fracture Surface X1000

**FIGURE 9 : DETAILS OF INFORMATION ACQUIRED BY AUGER ELECTRON SPECTROSCOPY.**

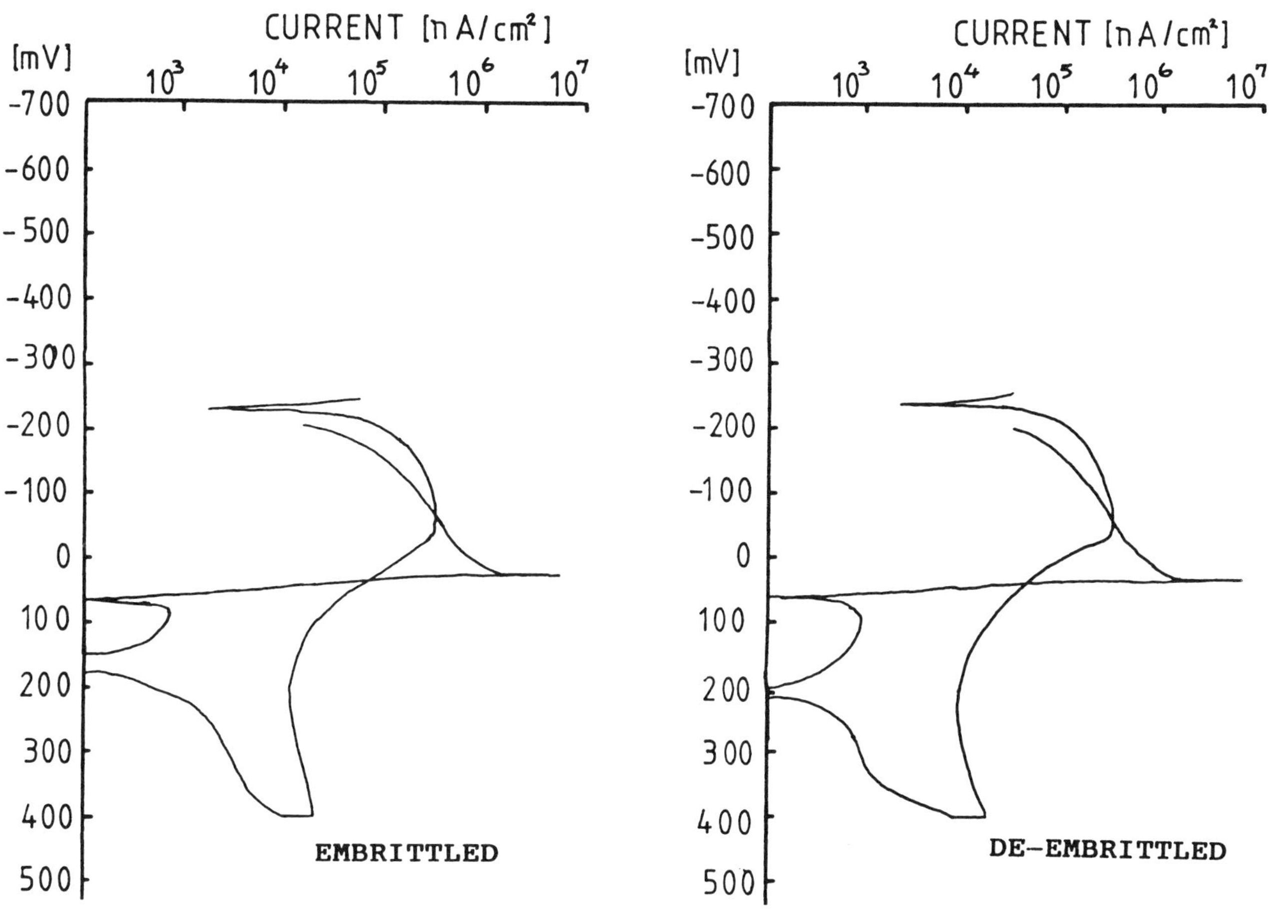


**FIGURE 10. TYPICAL EPR TEST CURVES FOR EMBRITTLED AND DE-EMBRITTLED STUDS.**

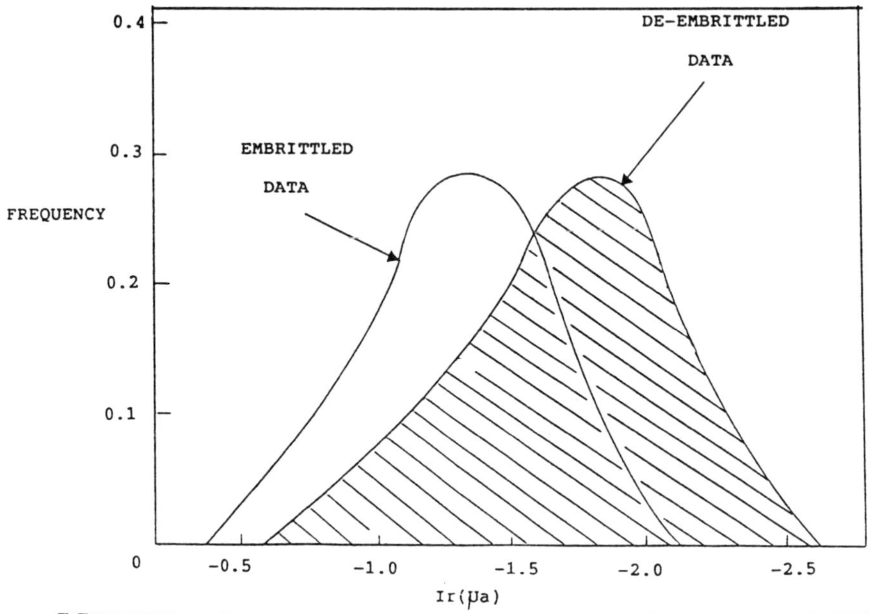


**FIGURE 11. SCATTER IN Ir, THE MINIMUM CURRENT DENSITY, FOR SERVICE EMBRITTLED AND DE-EMBRITTLED STUD.**

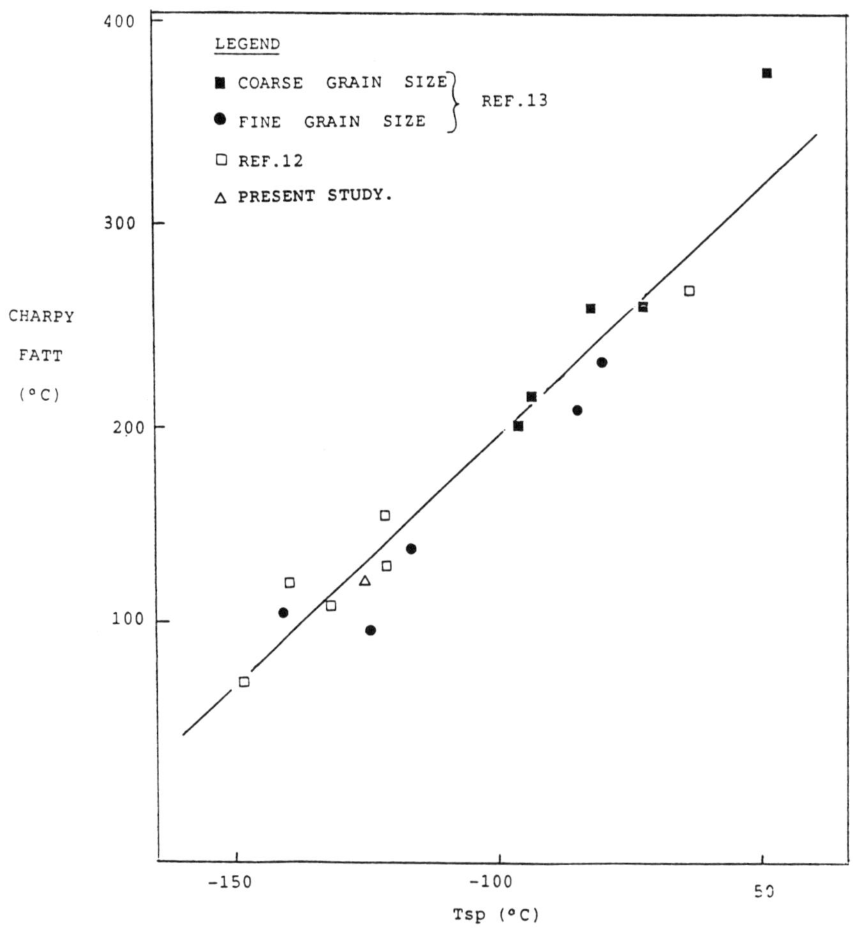


**FIGURE 12. RELATIONSHIP BETWEEN CHARPY FATT AND Tsp (SMALL PUNCH TEST TRANSITION ENERGY).**

*Proceedings of the Fifth International Conference on Creep of Materials, Lake Buena Vista, Florida, USA, 18-21 May 1992*

# Nonproportional Multiaxial Low Cycle Fatigue Using Cruciform Specimen at Elevated Temperature

**T. Itoh**
Graduate School
of Ritsumeikan University
Kyoto, Japan

**M. Sakane and M. Ohnami**
Department of
Mechanical Engineering
Ritsumeikan University
Kyoto, Japan

**Y. Takahashi and T. Ogata**
Central Research Institute of
Electric Power Industry
Tokyo, Japan

## Abstract

This paper assesses the multiaxial low cycle fatigue under proportional and nonproportional loadings. Proportional and nonproportional low cycle fatigue tests were carried out using SUS304 stainless and 1Cr–1Mo–1/4V steel cruciform specimens at high temperatures. Fatigue tests using the cruciform specimens can achieve a wide range of strain multiaxiality which cannot be performed in tension/torsion tests using hollow cylinder specimens. The Mises' equivalent strain was not effective for correlating the multiaxial low cycle fatigue lives while the equivalent strain range based on crack opening displacement was effective for the correlation. Nonproportional loading did not largely reduce fatigue lives for the cruciform specimens which is attributed to the un-change of the slip direction caused by the nonproportional loading. The crack morphology was extensively discussed from not only the surface crack direction but also the crack inclination into the specimen.

Multiaxial low cycle fatigue has been studied using hollow cylinder specimens subjected to tension/torsion loading (1-3). This type of tests, however, provides a limited strain biaxiality. The strain biaxiality achieved by tension/torsion tests are in the range of $-1 \le \phi \le -0.5$, where $\phi$ is the principal strain ratio and is defined as $\phi = \varepsilon_x/\varepsilon_y$. $\varepsilon_x$ and $\varepsilon_y$ are the minimum and maximum principal strains, respectively. Practical components, like pressure vessel and turbine blade, sometimes receive multiaxial loadings beyond that range and the multiaxial fatigue research which covers a widely range strain multiaxiality is required for the safety and reliability of components.

The authors have developed a multiaxial low cycle fatigue machine using the cruciform specimen (4-8). This machine can perform multiaxial low cycle fatigue tests in the principal strain ratio range of $-1 \le \phi \le 1$, which is a full range of multiaxial strain states. The authors have carried out the multiaxial low cycle fatigue tests using cruciform specimens for SUS304 stainless and 1Cr–1Mo–1/4V steels, and discussed the applicability of the multiaxial fatigue parameters (6-8).

The aim of this paper is to describe the proportional and nonproportional multiaxial low cycle fatigue lives of the cruciform specimens. Proportional and nonproportional low cycle fatigue tests were carried out using the SUS304 stainless steel cruciform specimens and the applicability of multiaxial low cycle fatigue parameter was discussed. The crack mode was also discussed from the detailed observation of crack direction on the specimen surface and into the specimen.

## Experimental Procedure

Figure 1 shows the shape and dimensions of the specimen together with the coordinates employed here. This paper defines two angles $\theta$ and $\alpha$ to express the angle between the crack plane and xy plane, and that between the crack propagation direction and z axis on the crack plane. Shape of the cruciform specimen was determined in FEM analyses so that the stress and strain are distributed uniformly along the gage length. The variation of stress and strain along the 8–mm gage length was less than 10 %. The detailed description of FEM analyses has been noted in the previous papers (4,6).

Experimental apparatus used was an electric–hydraulic servo machine for the cruciform specimen (4,6). Figures 2 and 3 show the general view and control flow of the apparatus. The apparatus has four actuators and four servo controllers to generate widely ranged multiaxial stress/strain states.

The cruciform specimen requires extensometry in two directions, x and y directions. Two extensometers were pressed to the specimen through the windows of the electric resistance furnace. Power of the furnace was 2.5 KW. Temperature variation along the gage length was within 5 K.

The material tested was SUS304 stainless and 1Cr–1Mo–1/4V steels. Strain controlled proportional and nonproportional multiaxial low cycle fatigue tests were carried for

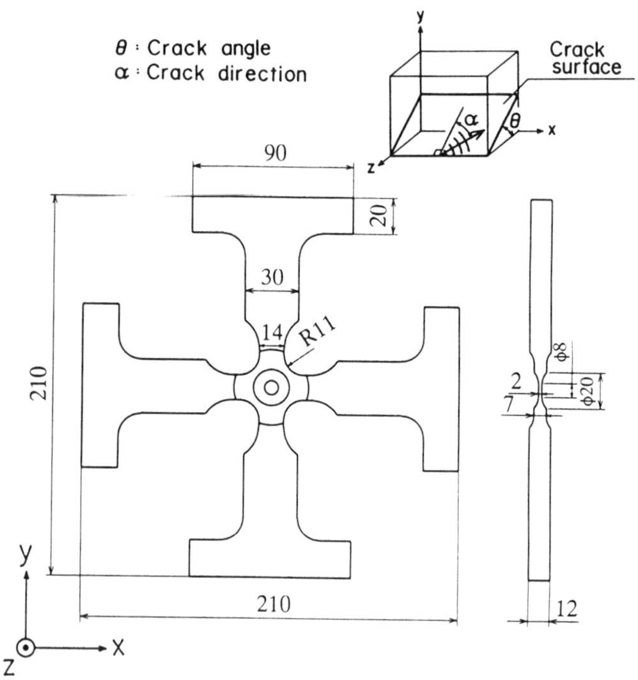


Fig. 1 Shape and dimensions of the specimen tested (mm).

SUS304 stainless steel at 823 K and 923 K and for 1Cr–1Mo–1/4V steel at 823 K. This paper employs the principal strain ratio ϕ to express the strain multiaxiality. ϕ is defined as $\varepsilon_x/\varepsilon_y$ ; $\varepsilon_x$ and $\varepsilon_y$ are the principal strains in x and y directions in the coordinates shown in Fig. 1.

Strain biaxiality at respective ϕ's is schematically shown in Fig. 4. In the figure, the ϕ=−1 test corresponds with the reversed torsion test using a hollow cylindrical specimen and ϕ=−0.5 test the uniaxial push-pull test, assuming that the Poisson's ratio is equal to 0.5. The ϕ=1 test is the equi-biaxial tension and compression test.

Combined tension/torsion multiaxial low cycle fatigue tests using hollow cylindrical specimens only cover the principal strain ratio range of $-1 \le \phi \le -0.5$, but the multiaxial fatigue tests using the cruciform specimens can cover all the strain biaxiality as shown in Fig. 4. The strain wave used in proportional tests was a fully reversed triangle at the strain rate of 0.1 %/s on the Mises' equivalent strain base.

Figure 5 shows strain paths used in the nonproportional tests on an $\varepsilon_x$–$\varepsilon_y$ diagram, where the dotted line indicates the Mises' strain constant ellipse. The ϕ=−0.5NP nonproportional test had loading sequence of $\Delta\varepsilon_y$=1 % and $\Delta\varepsilon_x$=−0.5 % followed by $\Delta\varepsilon_y$=−0.5 % and $\Delta\varepsilon_x$=1 % if the test was carried out at $\Delta\varepsilon_{eq}$=1 %, where $\Delta\varepsilon_{eq}$ is Mises' strain range. The ϕ=0NP also had similar loading sequence but had the loading sequence of $\Delta\varepsilon_y$=0.87 % and $\Delta\varepsilon_x$=0 % followed by $\Delta\varepsilon_y$=0 % and $\Delta\varepsilon_x$=0.87 %. The minus sign means the straining in the opposite direction. The ϕ=OP test had a 90 degrees phase difference between $\varepsilon_x$ and $\varepsilon_y$. In the former two tests, the direction of the principal strain axes was fixed but the

Fig. 2 General view of the test apparatus.

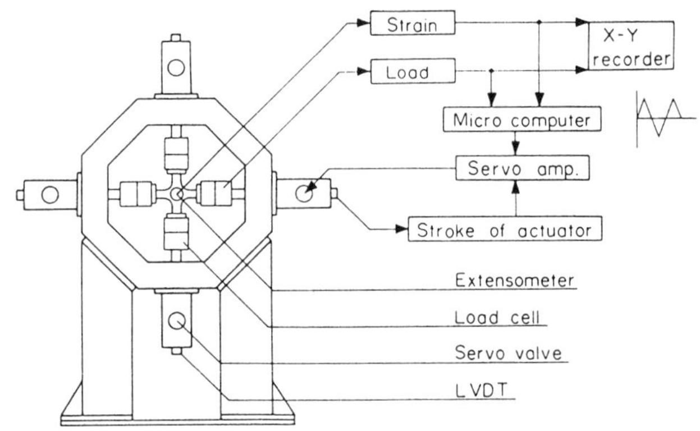


Fig. 3 Control flow of the test apparatus.

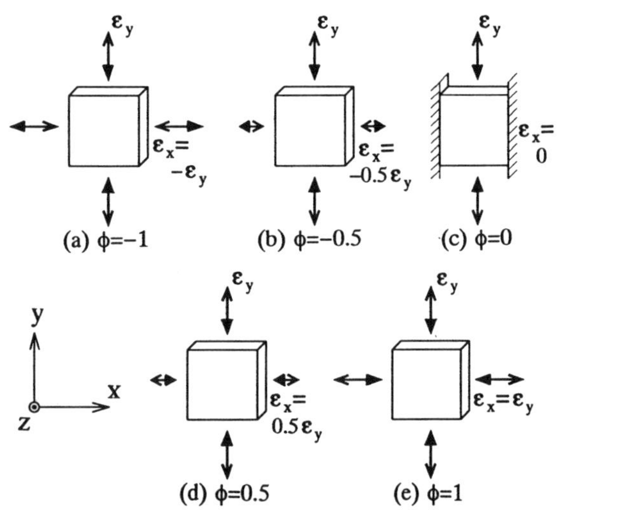


Fig. 4 Schematic graph showing the definition of the principal strain ratio. Open and solid arrows indicate the respective loading directions.

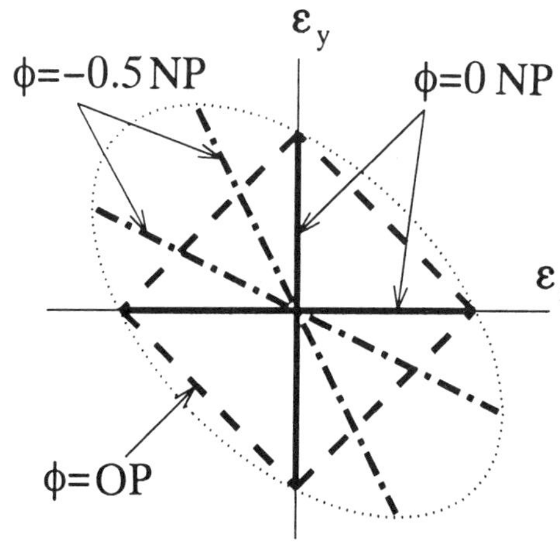


Fig. 5 Strain paths in nonproportional tests.

maximum principal direction changed at each cycle.

Failure cycle was defined as the cycle at which the tensile load normal to the crack decreased to 95 % from saturation for SUS 304 stainless steel, and as the cycle to start a rapid load drop for 1Cr–1Mo–1/4V steel.

## Experimental Results and Discussion

**Proportional Multiaxial Low Cycle Fatigue Life.** Figures 6 (a) and (b) (8) represent the experimental relationship between failure cycle and principal strain ratio in $\Delta\varepsilon_y$ constat tests for SUS304 stainless steel and 1Cr–1Mo–1/4V steel. The figures show that failure lives drastically decreased with increasing $\phi$. These results indicate that the strain biaxiality has a significant effect on failure lives.

Figures 7 (a) and (b) (8) correlate the fatigue lives with the Mises' strain. In these figures, a factor of two scatter band is shown based on the $\phi$=–0.5 test, i.e., uniaxial tension–compression test. For SUS304 stainless steel, most of the data can be correlated within a factor of two scatter band but the data in the $\phi$=–1 test are out of the scatter band. For 1Cr–1Mo–1/4V steel, on the other hand, fatigue lives can be correlated within a factor of two scatter band although the $\phi$=–1 data are located conservatively. The discrepancy of the data correlation with the Mises strain also has been reported in combined tension/torsion tests for SUS304 hollow cylinder specimens (9).

Figures 8 (a) and (b) (8) show the multiaxial low cycle fatigue lives for SUS304 stainless steel and 1Cr–1Mo–1/4V steel correlated with the equivalent strain based on crack opening displacement (COD strain) (8,10-14). The COD strain physically expresses the intensity of crack opening displacement in multiaxial stress states and can be equated as follows.

$$\Delta\varepsilon^* = \beta\,(2-\phi)^{m'}\,\Delta\varepsilon_y \qquad (1)$$

$$\beta = 1.83,\ m' = -0.66$$

The figures indicate that the COD strain is effective for the multiaxial fatigue data correlation in all the principal strain ratio ranges.

Many parameters have been proposed for the data correlation of multiaxial low cycle fatigue lives (15). Comparison of the typical parameters is shown in Fig. 9, where the ordinate is normalized by the principal strain in y direction. The experimental data shown in Fig. 6 decreased with increasing $\phi$, so the parameter which decreases with

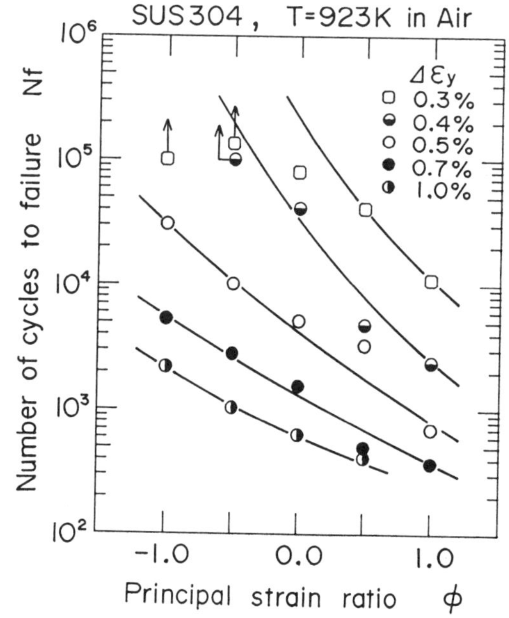


(a) SUS304 steel

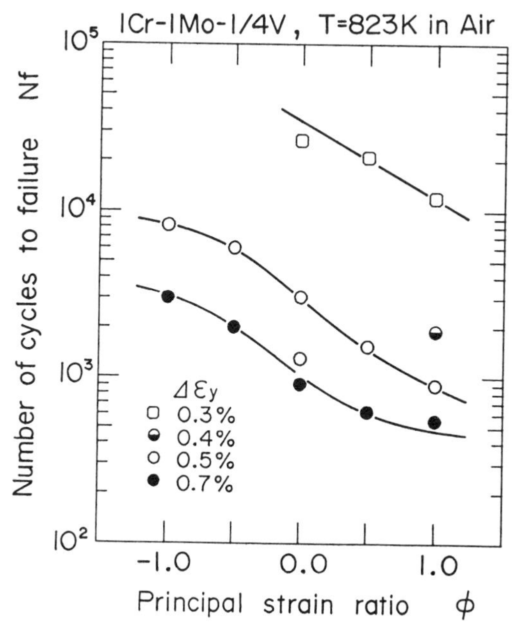


(b) 1Cr–1Mo–1/4V steel

Fig. 6 Relationship between multiaxial fatigue life and principal strain ratio in proportional tests. (a) SUS304 stainless steel at 923 K, (b) 1Cr–1Mo–1/4V steel (8).

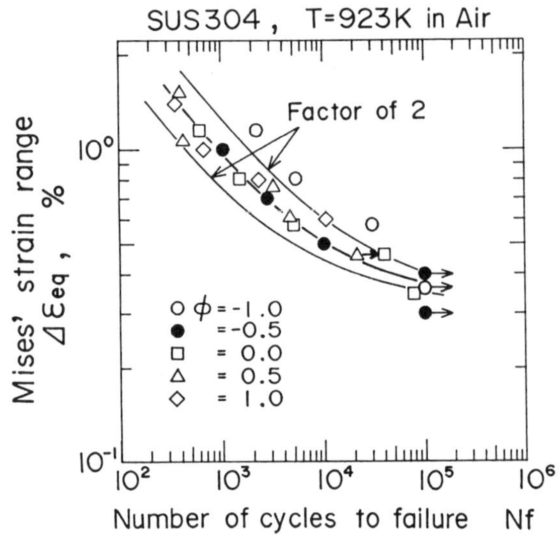


(a) SUS304 steel

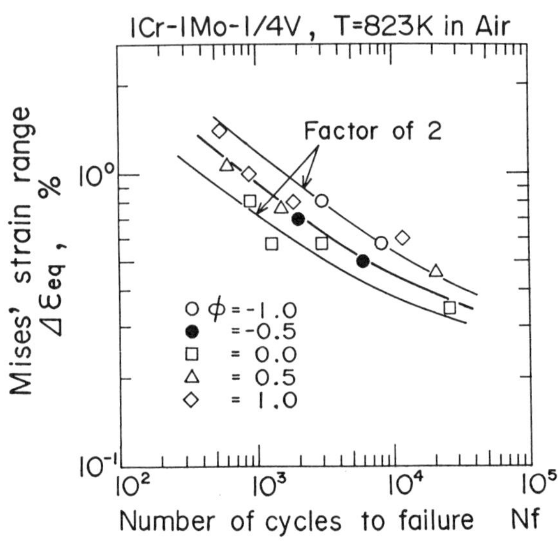


(b) 1Cr–1Mo–1/4V steel

Fig. 7 Correlation of multiaxial fatigue data with the Mises' strain range in proportional tests. (a) SUS304 stainless steel at 923 K, (b) 1Cr–1Mo–1/4V steel (8).

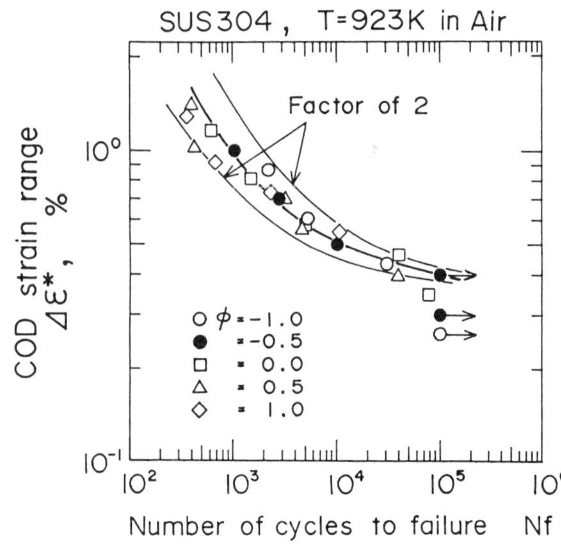


(a) SUS304 steel

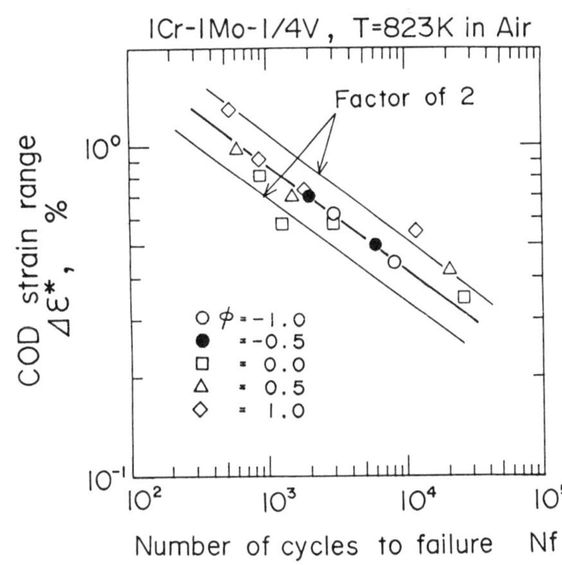


(b) 1Cr–1Mo–1/4V steel

Fig. 8 Correlation of multiaxial fatigue data with the COD strain range in proportional tests. (a) SUS304 stainless steel at 923 K, (b) 1Cr–1Mo–1/4V steel (8).

increasing $\phi$ is an appropriate parameter to correlate multiaxial low cycle fatigue lives. The parameter which meets this condition is only the $\Gamma^*$–plane parameter (LE strain) (16) and the COD strain. The Mises' strain also meets the condition in the principal strain ratio range of $-0.5 \le \phi \le 1$, but conversely decreases as $\phi$ increases in the principal strain ratio range of $-1 \le \phi \le -0.5$. Thus, for the data correlation with the Mises' strain range, the data at $\phi = -1$ result in out of a factor of two scatter band. The $\Gamma$–plane parameter (BMK strain) (17,18) has a discontinuity at $\phi = -0.5$, so the parameter could not correlate the multiaxial fatigue data in all the principal strain ratio range.

**Nonproportional Multiaxial Low Cycle Fatigue.** Figure 10 shows the data correlation in nonproportional tests together with the data in proportional tests for SUS304 stainless steel. The test material shown in Fig. 10 was SUS304 stainless steel but was the material in the different batch from that shown in Fig. 7. The data at $\phi = -1$ locate out of a factor of two scatter band as well as the data correlation as shown in Fig. 7.

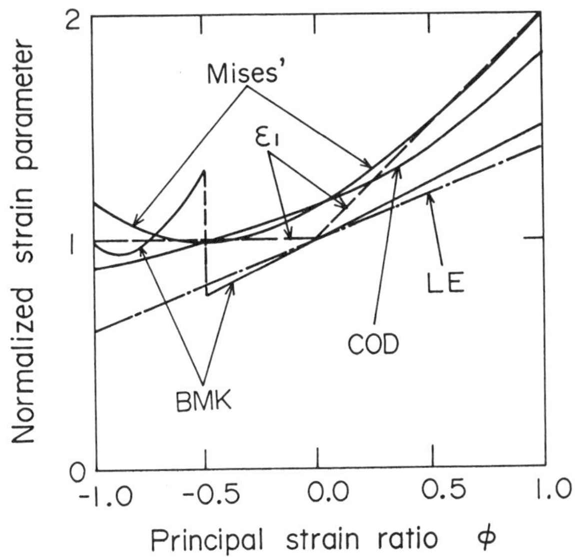


Fig. 9 Comparison of multiaxial fatigue strain parameters.

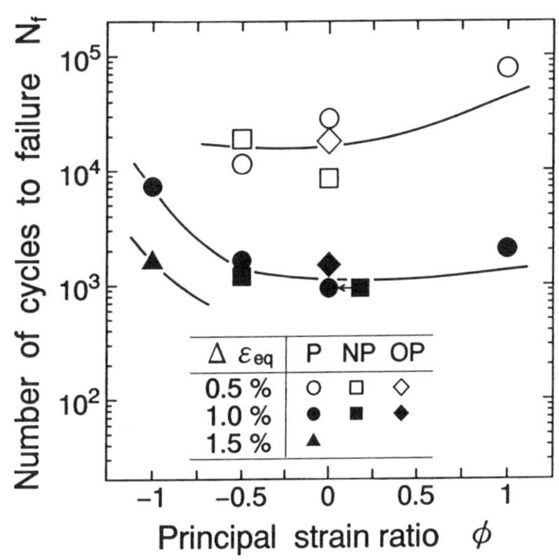


Fig. 11 Relationship between multiaxial fatigue life and principal strain ratio in proportional and nonproportional tests.

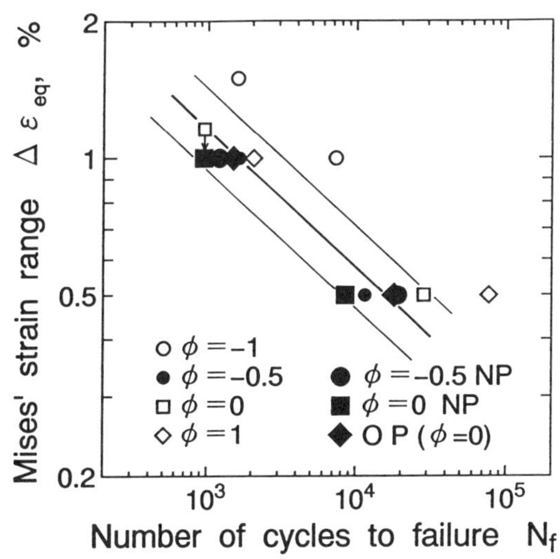


Fig. 10 Correlation of the multiaxial fatigue data with the Mises' strain in proportional and nonproportional tests.

Figure 11 is a replot of the data shown in Fig. 10 and shows the variation of the failure lives with the principal strain ratio. Experiments were performed under the Mises' strain constant condition in both the proportional and nonproportional tests. Fatigue lives in the three nonproportional tests were not smaller than those in the proportional tests, and they showed almost the compatible fatigue lives to the proportional ones. Thus, the nonproportional loading using the cruciform specimens does not give a large reduction in fatigue life.

These experimental results with respect to the nonproportional fatigue do not indicate that the nonproportional loading, in general, does not additionally reduce the fatigue life. In fact, a drastic reduction has been reported (19-24) under nonproportional loading in tension/torsion tests in comparison with the proportional loading. The small reduction in fatigue life using cruciform specimens in nonproportional tests is attributed to the following reason.

In nonproportional fatigue tests using cruciform specimens, the direction of the three principal strain is fixed and only the maximum principal strain direction is changed to the other principal strain direction with proceeding cycles. Nonproportional loading which has a continuous change of principal strain direction is impossible using the cruciform specimens. Thus, the maximum shear stress direction did not change largely and no significant interaction between the slip systems occurred. No large additional hardening occurred in nonproportional fatigue which results in small reduction in fatigue life.

**Crack Mode in Multiaxial Low Cycle Fatigue.** The crack propagation direction was examined on the specimen surface and into the specimen. Figure 12 shows the crack propagation direction observed on the specimen surface in proportional and nonproportional tests at $\Delta\varepsilon_{eq}$=1 %. The crack was classified into two groups with respect to crack length. One was the main crack which led to the fracture and was larger than 1 mm in length. The other was the subcrack which was smaller than 1 mm in length.

Main cracks in all the proportional and nonproportional tests except the $\phi$=−1 test propagated in x or y direction. The main crack in the $\phi$=−1 test propagated in the direction of 45 degrees to x or y axis. The crack in the $\phi$=−1 test propagated in the direction of the maximum shear strain, so this crack

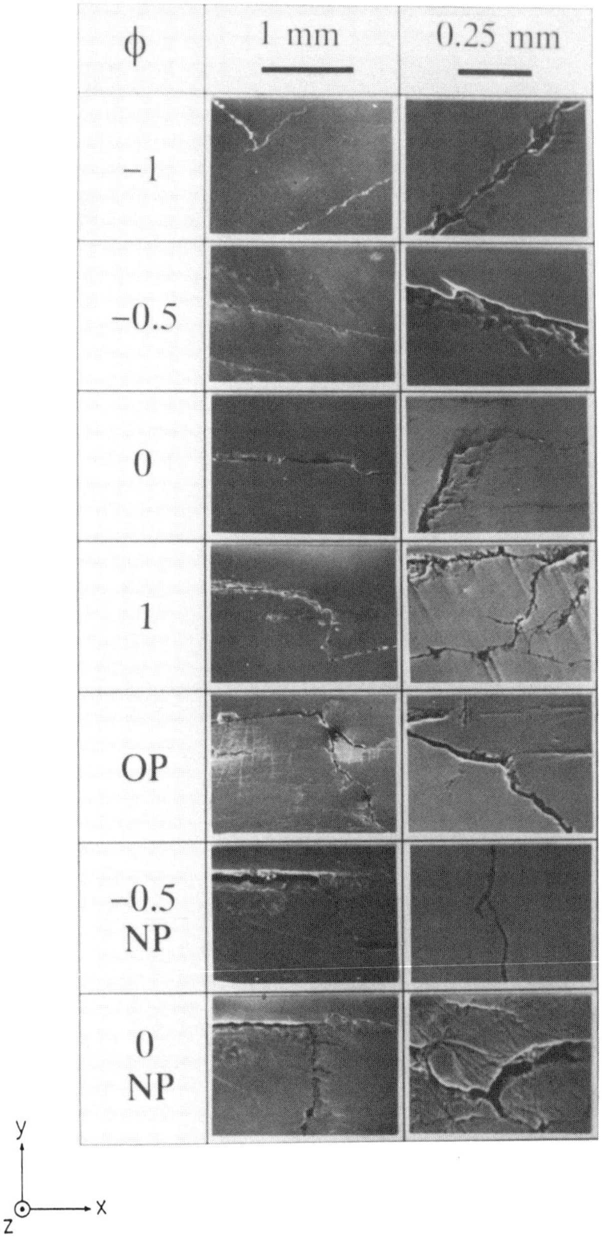


Fig. 12 Crack propagation direction on the specimen surface at failure at $\Delta\varepsilon_{eq}$=1 %.

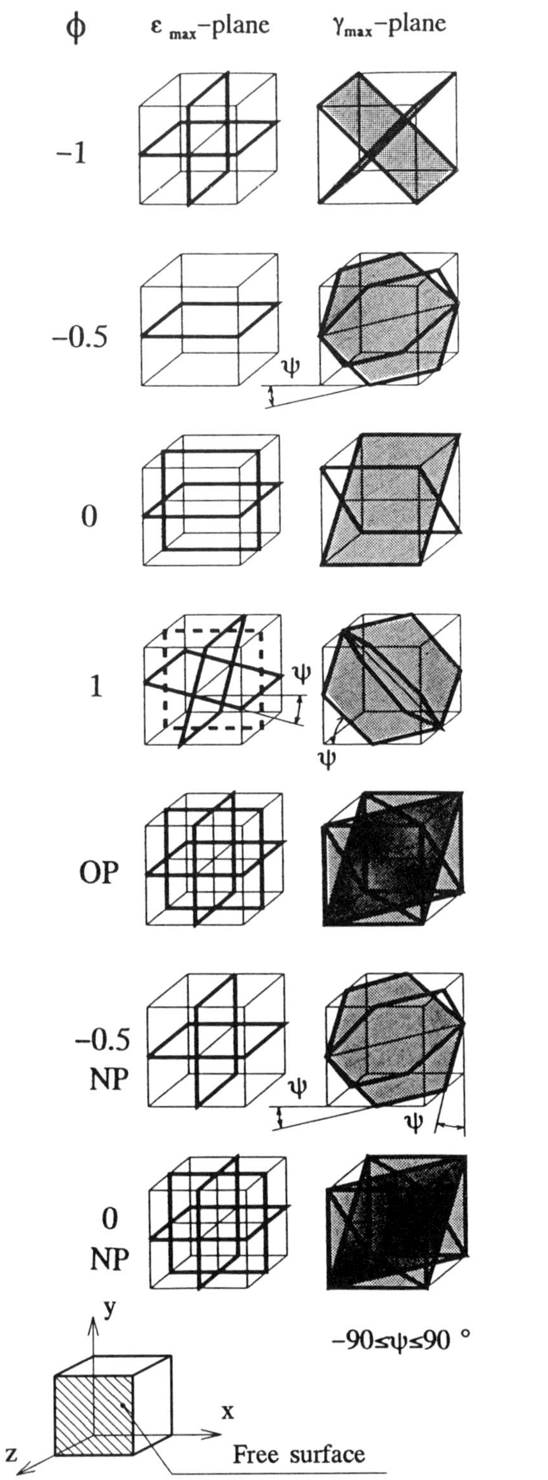


Fig. 13 Crack planes in multiaxial fatigue. Broad lines show the maximum principal strain or maximum shear strain plane. Shaded plane are the plane on which the crack was found in the final stage of propagation.

was mode II (in−plane shear mode). However, cracks in the range of $-0.5 \leq \phi \leq 1$ could not be determined from only the crack direction on the specimen surface. In addition to the crack direction on the specimen surface, the inclination angle into the specimen is necessary, which will be discussed later. The subcrack direction was different from the direction of the main crack in some tests. In $\phi$=1, OP and 0NP tests, subcracks propagated in many directions as shown in Fig. 12.

Figure 13 shows the fracture planes, following the similar classification of the crack plane suggested by Brown and Miller (25). Planes surrounded by broad lines indicate the maximum principal strain plane ($\varepsilon_{max}$−plane) or the maximum shear strain plane ($\gamma_{max}$−plane) except the $\phi$=1 test. In the $\phi$=1 test, the maximum principal strain plane is parallel to the

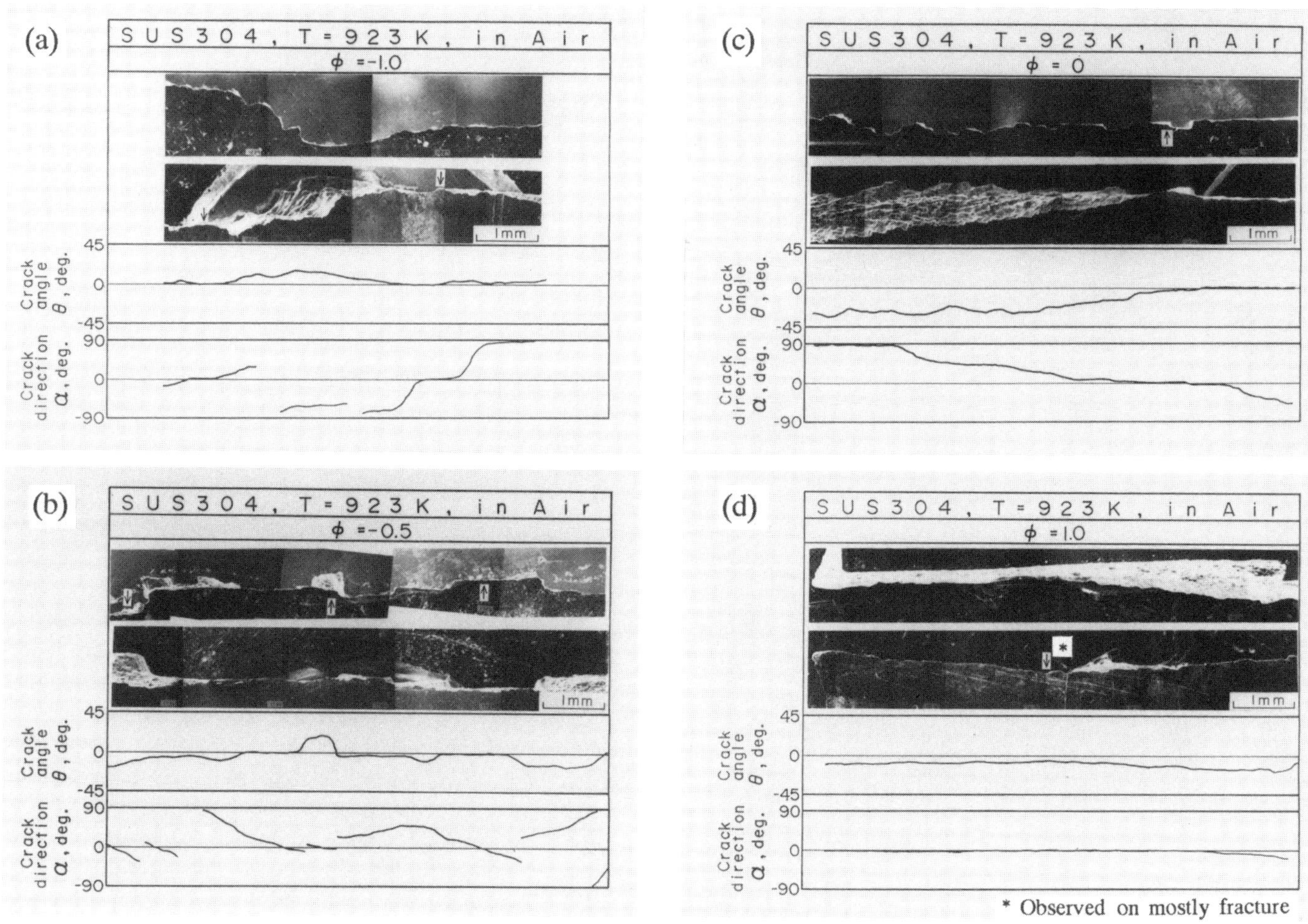


Fig. 14 Variation of the crack angle θ and crack propagation direction α for SUS304 stainless steel at 923 K, $\Delta\varepsilon$=0.7 % (5).

free surface (xy plane) indicated by dashed lines. Broad lines in this case indicate the second maximum principal strain plane. The shaded planes in the figure are the planes on which the main crack was observed in the final stage of propagation, and it will be mentioned later.

The crack direction of $\gamma_{max}$-plane on the specimen surface agrees with that of the $\varepsilon_{max}$-plane in all the tests except the $\phi$=−1 test, so it is impossible to determine the crack mode from the crack direction on the specimen surface. It should be determined considering the crack direction into the specimen. In the $\phi$=−1 test, on the other hand, the crack propagates in the different direction on the specimen surface depending on the crack mode, so the crack mode can be determined from only the crack direction on the specimen surface.

Figures 14 (a)–(d) (5) show the variation of the angles θ and α with crack extension in the principal strain ratio range of $-1\leq\phi\leq1$ at $\Delta\varepsilon_y$=0.7 % for SUS304 stainless steel at 923 K. θ is the angle between crack plane and z axis, and α the angle between crack propagation direction and z axis on the fracture plane as shown in Fig. 1. Arrows in the figures indicate crack initiation sites determined by observing striation direction.

Figure 14 (a) shows the result in the $\phi$=−1 test. In that test, two cracks were initiated at the points indicated by arrows. In the early stage of propagation, the cracks propagated in the α=0° direction on the θ=0° plane, which means that the crack were initiated on the specimen surface and propagated into the specimen. However, the crack changed the direction as the crack length increased. As the crack length increased, the crack changed the direction to increase or decrease the angle α. The final crack direction was α=±90°. Although the crack direction changed as the crack length increased, the crack remained on the θ=0° plane.

As indicated in Fig. 14 (b), cracks in the $\phi$=−0.5 test were initiated on the specimen surface in the α=0° direction on the θ=0° plane. The crack direction α increased or decreased with the crack extension. The increase in θ was observed partly, but most of the crack plane did not change and was the θ=0° plane. In this case, crack initiation occurred at the

Table 1 Summary of the crack plane.

| Angle $\alpha$ deg. | Principal strain ratio $\phi$ | | | | | | |
|---|---|---|---|---|---|---|---|
| | -1 | -0.5 | 0 | 1 | OP | -0.5NP | 0NP |
| 0 | $\gamma_{max}$ | $\varepsilon_{max}$ | $\varepsilon_{max}$ | $\varepsilon_{max}$ $\gamma_{max}$ | $\varepsilon_{max}$ $\gamma_{max}$ | $\varepsilon_{max}$ $\gamma_{max}$ | $\varepsilon_{max}$ $\gamma_{max}$ |
| 90 | $\gamma_{max}$ | $\gamma_{max}$ | $\gamma_{max}$ | — | — | — | — |

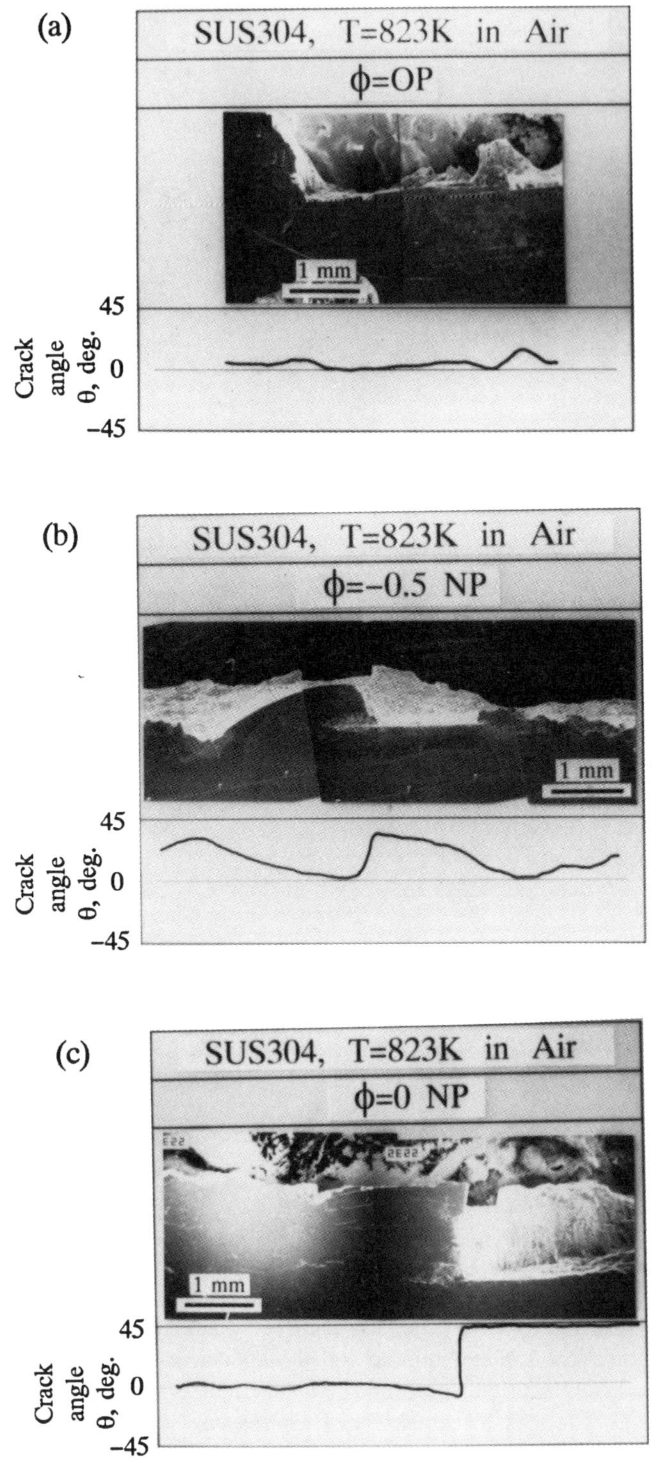


Fig. 15 Variation of the crack angle $\theta$ in nonproportional tests at $\Delta\varepsilon$=1 %.

several points.

Figure 14 (c) shows the result in the $\phi$=0 test. Cracks in this test initiated on the specimen surface propagated in the direction of increasing $\alpha$. As $\alpha$ increased, the angle $\theta$ decreased simultaneously. Finally the cracks propagated in the $\alpha$=90° direction on the $\theta$=−45° plane.

Figure 14 (d) shows the crack propagation in the $\phi$=1 test, where many cracks were initiated on the specimen surface. These cracks propagated in the $\alpha$=0° direction on the $\theta$=0° plane without changing the crack direction and plane. In this test, the crack initiated on the specimen surface propagated only into the specimen.

From the observations shown in Fig. 14, it is concluded that, in the test with the principal strain ratio between −0.5 and 0, the angle of the crack plane, $\theta$, well corresponded to the crack direction $\alpha$. When the crack propagated in the $\alpha$=0° direction, the crack was on the $\theta$=0° plane. However, as the crack changed the propagation direction, the crack plane also changed. In the final stage of propagation, when the crack propagated in the $\alpha$=±90° direction, the crack plane was the $\theta$=±45° plane. Furthermore, the cracks were always initiated in the $\alpha$=0° direction on the $\theta$=0° plane. After the crack initiation, the crack changed the direction $\alpha$ to ±90°. At the same time the crack plane also changed to incline to $\theta$=±45° plane.

Figures 15 (a)–(c) show the crack angle $\theta$ in nonproportional tests at $\Delta\varepsilon_{eq}$=1 %. Crack propagation behavior in these tests differed from that in proportional test. Many cracks initiated on the specimen surface propagated into the specimen in the direction normal to the specimen surface. Especially, many cracks on the different planes were observed. After the extension, the material between the two cracks which were propagated from the opposite surface was sheared to fracture. Such a result was also observed in the $\phi$=1 test at 823 K.

Table 1 summarizes the relation between the crack propagation direction $\alpha$ and crack plane. In the $\phi$=−1 tests, cracks propagated in the $\alpha$=0° or 90° direction on the maximum shear strain plane ($\gamma_{max}$–plane). In the principal strain ratio range of $-0.5 \leq \phi \leq 0$, the crack plane was the maximum principal strain plane ($\varepsilon_{max}$–plane), when crack propagated in the $\alpha$=0° direction, and it was the maximum shear strain plane when they propagated in the $\alpha$=90° direction. In the $\phi$=1 test and in all nonproportional tests, cracks propagated in the $\alpha$=90° direction on the $\varepsilon_{max}$ or $\gamma_{max}$–plane. In the final stage of propagation, the crack plane in all the tests was $\gamma_{max}$–plane as shown by the shaded planes in Fig. 13.

## Conclusions

(1) In $\Delta\varepsilon_y$ constant tests for SUS304 stainless and 1Cr-1Mo-1/4V steels, fatigue lives drastically decreased with increasing the principal strain ratio $\phi$. The strain biaxiality had a significant effect on low cycle fatigue lives.

(2) The equivalent strain range based on COD could correlate the multiaxial fatigue lives in the principal strain ratio range of $-1\leq\phi\leq1$. The COD strain range can be equated as

$$\Delta\varepsilon^*=\beta\,(2-\phi\,)^{m'}\Delta\varepsilon_y$$

where the constants, $\beta$ and m', take values of 1.83 and -0.66, respectively.

(3) The nonproportional loading did not give a large reduction in fatigue lives. The small reduction in fatigue lives in nonproportional tests was attributed to the fact that the three principal strain directions were fixed in the tests using the cruciform specimen.

(4) Cracks in the $\phi=-1$ test propagated on the maximum shear strain plane ($\gamma_{max}$-plane) when the crack propagated in the $\alpha=0°$ or 90° direction. In the principal strain ratio range of $-0.5\leq\phi\leq0$, when cracks propagated in the $\alpha=0°$ direction, the crack plane was the maximum principal strain plane ($\varepsilon_{max}$-plane), and when crack propagated in the $\alpha=90°$ direction, the crack plane was the $\gamma_{max}$-plane. In the $\phi=1$ and all nonproportional tests, cracks propagated in the $\alpha=0°$ on the $\varepsilon_{max}$ or $\gamma_{max}$-plane.

## References


1 Brown, M. W., and K. J. Miller, ASTM STP, No.770, 482-499 (1982).

2 Brown, M. W., and K. J. Miller, ed., "Biaxial Multiaxial Fatigue," *Mech. Eng. Publications. Limited London*, (1989).

3 Ohnami, M., M. Sakane, and N. Hamada, *Journal of the Society of Material Science, Japan*, Vol.35, No.390, 230-240 (1986).

4 Sakane, M., M. Ohnami, T. Kuno, and T. Itsumura, *Journal of the Society of Material Science, Japan*, Vol.37, No.414, 340-346 (1988).

5 Itoh, T., M. Sakane, and M. Ohnami, *Transactions of the Japan Society of Mechanical Engineers*, Vol.56, No.527, 45-52 (1990).

6 Sakane, M., and M. Ohnami, "Fatigue Under Biaxial and Multiaxial Loading," *Mech. Eng. Publications Limited London*, pp.265-278 (1991).

7 Itoh, T., M. Sakane, and M. Ohnami, *Proc. 6th International Conference on Mechanical Behavior of materials*, Vol.2 367-272 (1991).

8 Itoh, T., M. Sakane, and M. Ohnami, *Trans ASME, J Eng. Mater. Technol.*, (submitted).

9 Sakane, M., M. Ohnami, and M. Sawada, *Trans. ASME J. Eng. Mater. Technol.*, Vol 109, 236-243 (1987).

10 Hamada, N., M. Sakane, and M. Ohnami, *Fatigue Eng. Mater. Struct.*, Vol.7, No.2, 85-96 (1984).

11 Hamada, N., M. Sakane, and M. Ohnami, *Bull. Japan Soc. Mech. Eng.*, Vol.28, No.241, 1341-1347 (1985).

12 Hamada, N., M. Sakane, and M. Ohnami, *Journal of the Society of Material Science, Japan*, Vol.34, No.377, 214-220 (1985).

13 Hamada, N., M. Sakane, and M. Ohnami, *Proc. of 30th Japan Conger. on Mater. Res.*, 69-75 (1987).

14 Hamada, N., M. Sakane, and M. Ohnami, "Fatigue Under Biaxial and Multiaxial Loading," *Mech. Eng. Publications Limited London*, pp.239-247 (1991).

15 Garud, Y. S., *Trans. ASME J. Eng. Mater. Technol.*, Vol.103, No.2, 118-125 (1981).

16 Lohr, R. D., and E. G. Ellison, *Fatigue Eng. Mater. Struct.*, Vol.3, No.1, 1-17 (1980).

17 Brown, M. W., and Miller, K. J., *Fatigue Eng. Mater. Struct.*, Vol.1, No.2, 217-229 (1979).

18 Kandil, F. A., M. W. Brown, and K. J. Miller, *Mech. Behave. Nucl. Appl. Stainless Steel Elevated Temperature*, 203-209 (1982).

19 Ohnami, M., and N. Hamada, *Journal of the Society of Material Science, Japan*, Vol.30, No.335, 822-828 (1981).

20 Ohnami, M., M. Sakane, and N. Hamada, ASTM STP, No.853, 622-634 (1985).

21 Nitta, A., T. Ogata, and K. Kuwabara, *Journal of the Society of Material Science, Japan*, Vol.36, No.403, 376-382 (1987).

22 Nitta, A., T. Ogata, and K. Kuwabara, *Journal of the Society of Material Science, Japan*, Vol.37, No.414, 334-339 (1988).

23 Nitta, A., T. Ogata, and K. Kuwabara, *Journal of the Society of Material Science, Japan*, Vol.38, No.427, 416-422 (1989).

24 Doog, S. H., D. F. Socie, and I. M. Robertson, *Trans. ASME J. Eng. Mater. Technol.*, Vol.112, 456-464 (1990).

25 Brown, M. W., and K. J. Miller, *Fatigue. Eng. Mater. Struct*, Vol.1, 231-246 (1979).

*Proceedings of the Fifth International Conference on Creep of Materials, Lake Buena Vista, Florida, USA, 18-21 May 1992*

# Hold Time, Strain Rate and Environmental Effects on Near Eutectic Sn-Pb Solder Under Conditions of Thermomechanical Fatigue

**D.R. Frear, W.B. Jones, and N.R. Sorensen**
Sandia National Laboratories
Albuquerque, New Mexico

## Abstract

We present results that correlate microstructure and mechanical evolution to variations of deformation rate, hold time and environmental effects on the thermomechanical fatigue (TMF) behavior of 60Sn–40Pb solder. The results are used to define valid conditions for performing accelerated TMF tests. TMF tests at deformation rates of $5.6 \times 10^{-4} s^{-1}$, $2.8 \times 10^{-4} s^{-1}$ and $2.1 \times 10^{-4} s^{-1}$ were performed. Deformation rates greater than $2.8 \times 10^{-4} s^{-1}$ result in fewer cycles to failure. At low deformation rates, the microstructure heterogeneously coarsens at cell boundaries. At higher rates, the deformation mechanism changes and heterogeneous coarsening occurs at a strain concentration in the joint, independent of the microstructure. TMF tests with hold times of 0, 3 and 6 min. at the temperature extremes were performed. At hold times 3 min. or longer the recrystallization at cell boundaries is annealed, resulting in heterogeneous coarsening. With no hold times the TMF life was greatly enhanced as a result of limited coarsening. The effect of the oxygen environment was explored. The TMF life in the presence of oxygen was found to extend the fatigue life. Valid acceleration conditions for a TMF test of solder are: a deformation rate of $2.8 \times 10^{-4} s^{-1}$ or lower, with hold times of 3 min. or longer.

SOLDERS USED IN ELECTRONIC PACAKGES act as electrical and mechanical interconnections that often constrain materials with different thermal expansivities. In service, the electronic packages encounter cyclical thermal fluctuations caused by internal heating of the package, or by temperature fluctuations in the ambient environment. The changes in temperature, combined with the differing thermal expansivity of materials used in the electronic package, result in low cycle thermomechanical fatigue (TMF) of the solder joints. It is well established that the microstructure of near eutectic Sn–Pb solder joints heterogeneously coarsen under conditions of TMF in shear [1–7]. The cyclic strain localizes in the joint resulting in the recrystallization and growth that forms a heterogeneous coarsened band. Eventually, cracks form in the coarsened bands and solder joints fail. The same heterogeneous coarsening process has been observed in near eutectic Sn–Pb solder joints deformed under conditions of creep and isothermal fatigue in shear [8–11].

It is important to understand and fully characterize the mechanics of solder joints under conditions of themomechanical fatigue in order to predict the lifetime of the solder using fundamental principles. The characterization of solders under these conditions must be performed experimentally.

Thermomechanical fatigue tests are performed on solder joints to examine the effect of cyclic strain and temperature on solder. Under a controlled setting, it is possible to determine metallurgical mechanisms and make lifetime predictions for solder joints. Thermal cycles are typically on the order of a few cycles a day for up to 30 years of life. To perform these experiments in a timely fashion, the testing rates must be accelerated. There are three variables that can be changed to accelerate a fatigue test on solder:

a) increase the temperature extremes beyond what is typically encountered,

b) increase the rate at which the temperature and strain are imposed (increase the deformation rate), and

c) decrease the time spent at the temperature extremes (decrease the hold times).

Option a) is not a viable alternative; by changing the temperature extremes the deformation mechanisms in the solder change and the testing results can not be applied to predict real use conditions. Methods b) and c) are therefore commonly used. However, it has yet to be demonstrated whether these methods of test acceleration cause solder joints to behave in the same fashion as those in the field.

We present results that correlate the effect of hold time and deformation rate on solder joint behavior both from a microstructural and mechanical perspective. The results show what factors are reasonable for experimental acceleration to approximate real use conditions. We also will present preliminary results on the effect of environmental exposure (namely oxidation of the solder) on thermomechanical fatigue behavior. If environmental exposure is a major factor in the thermomechanical fatigue failure of solder joints, this mechanism must also be accounted for appropriately.

## Experimental Procedure

Thermomechanical fatigue tests were performed on 60Sn—40Pb (wt. %) solder joints using a procedure that is described in greater depth elsewhere [3]. A brief summary of the test method is given below.

The specimen used to test solder joints is shown in the schematic drawing in Figure 1. The specimen consists of 18 electrically isolated solder joints that, when the specimen is gripped on the ends, deforms in simple shear. Strain is imposed upon the solder joints by a servo—hydraulic loadframe operated under strain control. Thermal fluctuations are induced by a chamber that fits around the specimen in the loadframe. Air circulated around the specimen is heated with resistive heating coils, and cooled by vaporized liquid nitrogen. The strain and temperature are digitally controlled by a computer. The temperature extremes are –55° to 125°C at 10% shear strain. The thermal cycle consists of a

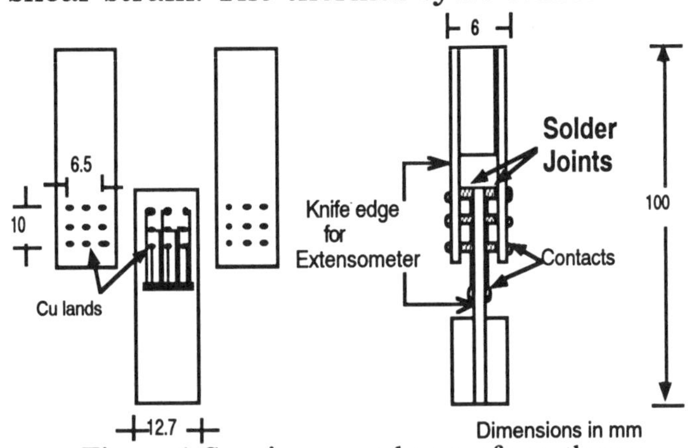


Figure 1 Specimen used to perform thermomechanical fatigue of solder joints.

ramp in strain and temperature to the elevated temperature extreme and hold period, a ramp down, and another hold at the low temperature extreme.

Failures are monitored electrically using an event detector. The event detector measures spikes in resistance of greater than 500Ω for periods of time less than 1 μs.

Event detector data along with load, temperature, and strain are collected and stored digitally.

To examine the effect of environment TMF tests were performed in an oxygen rich environment. Oxygen (95% pure) was introduced to the specimen at a flow rate of 40 x $10^4$ $mm^3s^{-1}$ through a piece of teflon tubing situated adjacent to the solder joints. Oxygen at the solder joints was 40% of the total gas content. A coupon of 60Sn—40Pb coated Cu was placed behind the specimen as a control that was weighed before and after the test to determine weight gain due to oxide formation.

Experiments were performed at three deformation rates and hold times. Two deformation rates and one hold time were selected to evaluate the effect of oxygen. These test conditions are summarized in Table I.

*Table 1: Thermomechanical Fatigue Test Conditions*

| Test | Strain Rate | Hold Times (sec.) | Gas Content |
|---|---|---|---|
| A | $5.6 \times 10^{-4}$ | 0, 3, 6 | --- |
| B | $2.8 \times 10^{-4}$ | 0, 3, 6 | --- |
| C | $2.1 \times 10^{-4}$ | 0, 3, 6 | --- |
| D | $5.6 \times 10^{-4}$ | 3 | $N_2$, $O_2$ |
| E | $2.8 \times 10^{-4}$ | 3 | $N_2$, $O_2$ |

Specimens were sectioned after testing to optically characterize the microstructure. Standard metallographic techniques were used to polish the specimen down to 1 μm. A final polishing/etching step was performed in a slightly acidic colloidal suspension of 0.1 μm alumina to highlight the solder microstructural details.

## Results and Discussion

**Effect of Deformation Rate.** The three deformation rates examined were 5.6 x $10^{-4}s^{-1}$, 2.8 x $10^{-4}s^{-1}$ and 2.1 x $10^{-4}s^{-1}$ with a hold time of 6 min. at each temperature extreme. Typical hysteresis loops of load *vs.* strain for deformation rates of 5.6 x $10^{-4}s^{-1}$ and 2.8 x $10^{-4}s^{-1}$ after 5 cycles are shown in Figure 2. Data from the hysteresis loops were used to plot the high temperature load maximum, low temperature load maximum, and the load after the high temperature hold as a function of the number of cycles. Plots of cyclic load *vs.* number of cycles are shown in Figure 3 for the three deformation rates.The solder joints deformed at 5.6 x $10^{-4}s^{-1}$ failed after 50 cycles while the two slower rates the solder joints failed after 200 cycles. This represents a four–fold increase in life by slowing the deformation rate by only a factor of two.

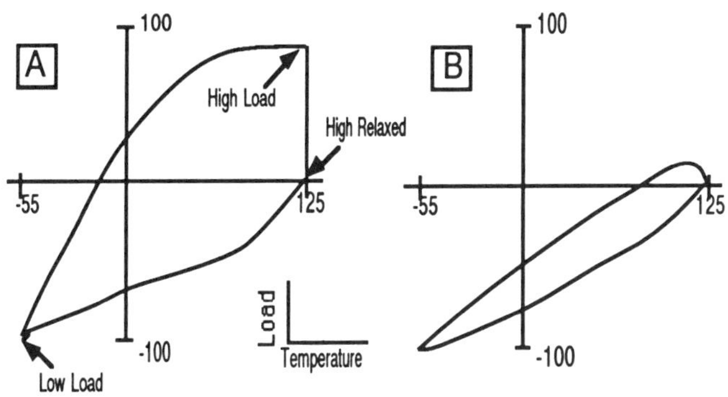


Figure 2 Hysteresis loops of load *vs.* temperature for: A) $5.6 \times 10^{-4}/s^{-1}$. and B) $2.8 \times 10^{-4}/s^{-1}$.

Figure 4 shows optical micrographs of solder joints after 100 cycles at 5.6 x $10^{-4}s^{-1}$, and after 300 cycles for the other two deformation rates. At the 5.6 x $10^{-4}s^{-1}$deformation rate, Figure 4A, the solder has heterogeneously coarsened parallel to the direction of imposed shear strain. The coarsening appears to be independent of the microstructural features of the solder; and cuts across the eutectic cells. After 100 cycles, the joint has failed with cracks running though the coarsened band.

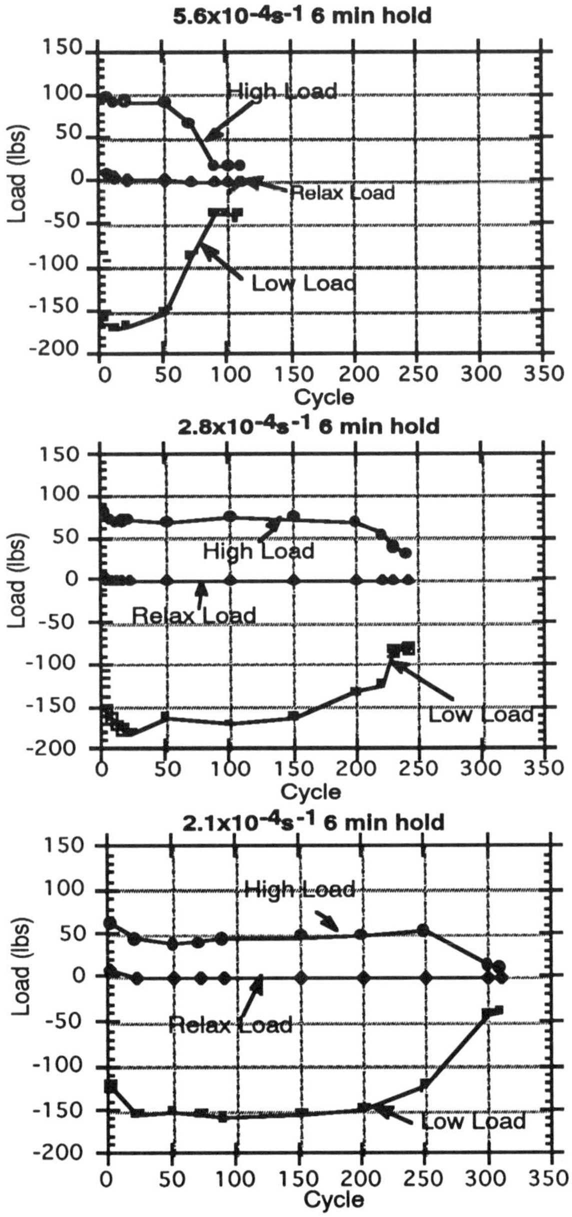


Figure 3. Plots of load *versus* number of cycles at a 6 minute hold time for A) $5.6x10^{-4}s^{-1}$, B) $2.8x10^{-4}s^{-1}$ and C) $2.1x10^{-4}s^{-1}$.

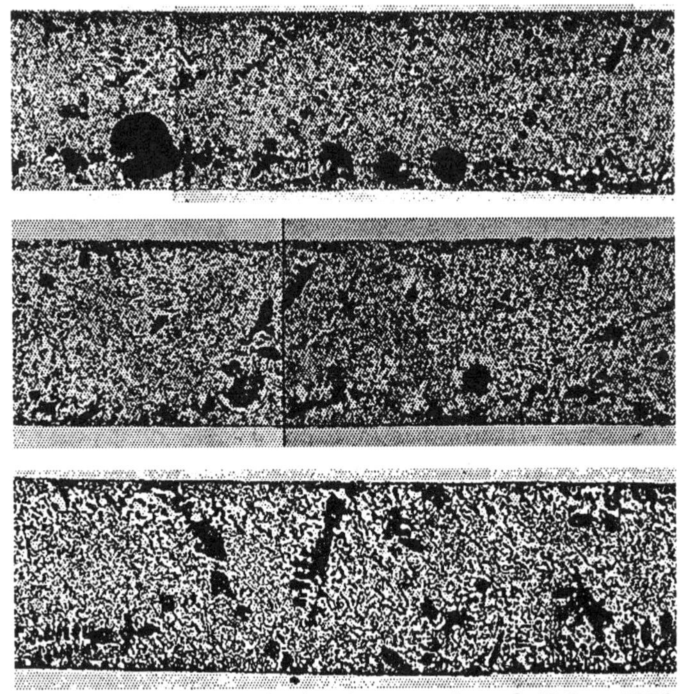

Figure 4. Optical micrographs of 60Sn–40Pb solder joints thermomechanically fatigued with 6 minute hold times at A) $5.6x10^{-4}s^{-1}$ after 100 cycles, B) $2.8x10^{-4}$ $s^{-1}$after 300 cycles and C) $2.1x10^{-4}s^{-1}$after 300 cycles.

At the two slower deformation rates, the microstructural evolution of the solder differs from that at 5.6 x $10^{-4}s^{-1}$. Heterogeneous coarsening occurs at the eutectic cell boundaries that are more or less parallel to the direction of imposed shear strain. The coarsening does not cut across the cells. Cracks form in these joints after 300 cycles.

Based upon these results, there is a difference in the damage mechanism between the fast deformation rate of 5.6 x $10^{-4}s^{-1}$ and rates slower than 2.8 x $10^{-4}s^{-1}$. The mechanism for each can be explained as follows.

*Slow Deformation Rate.* In the as-solidified state, the cell boundaries are slightly coarsened and are the weakest part of the microstructure. At the slower deformation rate, the eutectic cells slide and rotate relative to one another to accommodate the strain [6,7,12]. This method of strain accommodation imparts damage only to the cell boundaries. At the low temperature regime of the thermal cycle the damage, in the form of defects, causes the cell boundaries to recrystallize. No stress relaxation occurs during the hold time at the low temperature; annealing of the solder microstructure occurs upon heating. At the slow deformation rate, the damage anneals faster than additional damage can be imparted which results in a lower maximum load at the high temperature, Figure 2B. The load relaxes to zero in the high temperature portion of the cycle and the recrystallized regions at the cell boundaries coarsen. In subsequent cycles, deformation concentrates at cell boundaries because they remain the weakest part of the microstructure.

*Fast Deformation Rate*. At faster deformation rates (> $5.6 \times 10^{-4}s^{-1}$), the strain is imparted to the solder microstructure too quickly to be accommodated at cell boundaries. The cells do not have enough time to slide and rotate to take up the strain. In this case heterogeneous coarsening occurs at the stress concentration in the joint which is independent of the microstructure. The strain also concentrates in this region parallel to the direction of imposed shear. The recrystallization process occurs during the low temperature portion of the cycle at the region of high strain. At the high temperature portion of the thermal cycle, damage is imparted faster than it can be annealed. During the high temperature hold, the damage is annealed and the recrystallized structure coarsens. The damage of subsequent cycles accumulates in the band of coarsened material.

The fatigue life at a $5.6 \times 10^{-4}s^{-1}$ deformation rate is shorter than the two slower rates because there is a change where damage is concentrated. The strain is heterogeneously distributed in both cases. However, at the slower deformation rates, the damage is distributed over a greater volume. The volume of coarsened material at cell boundaries is greater than a planar band through the center of the joint. Therefore, the same amount of damage is concentrated into a smaller volume at the faster deformations rates and leads to more rapid failure.

The deformation rate effect has been studied extensively on the isothermal fatigue behavior of 60Sn—40Pb [13–16]. These studies typically used a sawtooth wave form and simply varied frequency. Solomon [13–15] and others [16] observed that as frequency decreases, the isothermal fatigue life also decreases. This is caused by imparting additional damage to the solder joint by creep and stress relaxation processes at lower frequencies. These phenomenon occur in isothermal fatigue because the damage mechanisms are the same throughout each fatigue cycle. During TMF, however, there are two operative mechanisms: low temperature damage accumulation (recrystallization), and high temperature annealing (coarsening). An isothermal fatigue cycle can experience only one of these mechanisms.

Fox et al. [17] have observed a decrease in life with lower frequencies during TMF of 60Sn—40Pb solder. The difference between this observation and our results is that Fox et al. used a sawtooth wave form with no hold times. As we discuss later, the incorporation of a hold time dramatically changes the response of the solder to TMF deformation.

TMF studies on stainless steels also show that slower deformation rates shorten the fatigue life [18]. These stainless steel results cannot be used in the interpretation of solder results for several reasons. Stainless steels, at slower deformation rates, are prone to creep cavity formation and growth. 60Sn—40Pb solder does not form creep cavities during TMF in shear. Furthermore, there is extensive microstructural evolution in solders that is not found in steels. Therefore, a comparison between the two materials is problematical, at best.

A significant conclusion that can be made from these tests is the deformation rate at which an accelerated fatigue test is valid. Deformation rates slower than 2.8 x $10^{-4}s^{-1}$ result in a microstructural and mechanical evolution that can be compared with that caused by service conditions.

**Effect of Hold Time.** The effect of hold time was explored at one deformation rate, 2.8 x $10^{-4}s^{-1}$, and three times: no hold time, 3 m, and 6 m. The hold times represent the time spent at each temperature extreme. The no hold test had, in effect, a sawtooth waveform. Figure 5 shows the plots of load *versus* number of cycles for the three hold times. Figure 6 shows the resultant microstructure of the three hold times.

Figure 5A shows that when there is no hold time at either temperature extreme, the solder joints have a very long fatigue life. This was found to be true at all deformation rates less than 2.8 x $10^{-4}s^{-1}$. (At deformation rates of 5.6 x $10^{-4}s^{-1}$ and faster the hold time had no effect on life; all samples failed after 50 cycles.) With a 3 min. hold time, or longer, the fatigue life decreased from over 600 cycles to 200 and remained constant for longer hold times.

The cell boundaries coarsen during the hold time leading to eventual failure (Figures 6B and 6C). Microstructural evolution is limited in solder joints that had no hold times, Figure 6A. When there is no hold time, the solder joint coarsens homogeneously. During a hold time at the high temperature portion of the thermal cycle the load relaxes to zero in 2 minutes. The load relaxation indicates that heterogeneous growth occurred at cell boundaries.

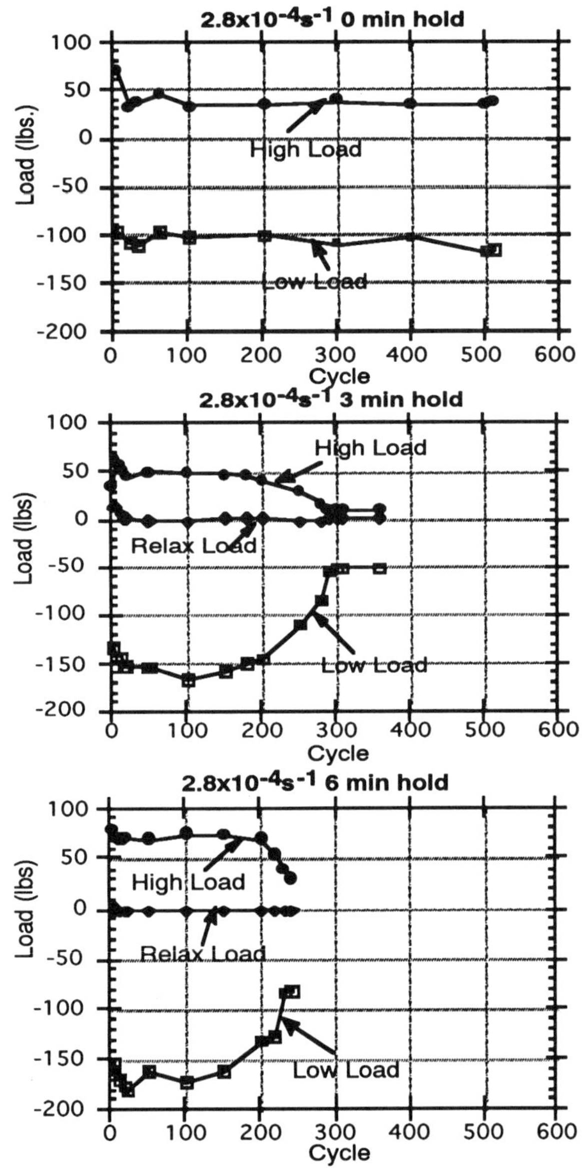


Figure 5. Plots of load *versus* number of cycles at a deformation rate of $2.8x10^{-4}s^{-1}$for A) no hold, B) 3 minute and C) 6 minute hold times.

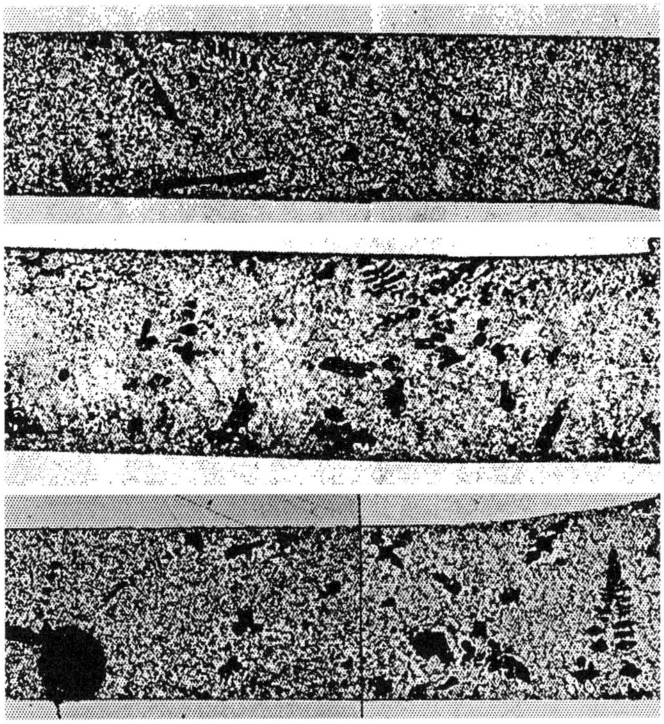

Figure 6. Optical micrographs of solder joints after TMF at a deformation rate of $2.8x10^{-4}s^{-1}$for A) no hold, B) 3 minute and C) 6 minute hold times.

The results of the hold time tests show that a hold time long enough to allow complete load relaxation is necessary to create a valid accelerated fatigue test. In tests over a temperature range of –55° to 125°C a hold of at least 3 min. at the high temperature is sufficient. Little or no load relaxation occurs at –55°C so a hold at this temperature is not critical. If TMF tests are performed at other temperature extremes, especially where extensive load relaxation can occur at the low temperature portion of the cycle, the hold time must be of sufficient length to allow the load to relax to zero.

**Effect of Oxygen on the TMF Behavior of 60Sn–40Pb Solder.** The presence of oxygen during isothermal fatigue has been found to decrease the lifetime of Pb–Sn solders. The majority of this evidence has been gathered on Pb and Pb–rich Sn–Pb alloys [19–21]. Snowden [20] found that as the oxygen partial pressure was reduced, there was a sharp increase in the room temperature life of pure Pb. Snowden also found that placing a protective coating of grease or In plating dramatically improved the fatigue lifetime of pure Pb. Berriche et al.[19] found that a moderate vacuum of 2 x $10^{-3}$ torr improved the fatigue lifetime of 96.5Pb–3.5Sn over that of samples fatigued in air. Later, Berriche [21] found that the oxygen environment only appeared to play a significant role in the isothermal fatigue behavior of 96.5Pb–3.5Sn in strain ranges below 0.8% and the presence of oxygen enhanced intergranular cracking of the Pb. Lacey and Woodford [22] explored the effect of oxygen on 60Sn–40Pb solder under room temperature, load control conditions. They studied wire samples in air and with a protective grease coating. The protected samples had an increased life of an order of magnitude at lower stress levels.

Despite the apparent strong effect of oxygen on fatigue life there have not been many studies on the mechanisms of oxidation of Sn–Pb solders. Particularly lacking is information on alloys at, or near, the Sn–Pb eutectic. What data do exist are mainly the work of Konetzki et al. [23–26] and Frankenthal and Siconolfi [26–27].

Konetzki et al. [23–26] studied both eutectic solder and a Pb–rich 99Pb–1Sn solder. The oxides formed were PbO on the Pb and SnO on Sn. Both oxides have a tetragonal structure with closely matching lattice parameters. The oxides were thin, only 1.5 to 2 nm thick. The oxide forms an inversion layer where the amount of preferentially oxidized Sn increases with depth into the sample, reaches a maximum, then decreases into the bulk. This phenomenon was explained as being due to the oxide of the solute (Sn) being more stable than the solvent (Pb) and oxygen ions diffusing more rapidly than cations [23]. This inversion layer was found for both eutectic and Pb—rich solders. The oxidation kinetics were found to be logarithmic up to 90°C then parabolic until the solder melted [23-26]. In the Pb—rich solder, Sn was found to preferentially segregate to the Pb grain boundaries. The Sn preferentially oxidizes both at the surface and down the Pb grain boundaries. The PbO forms in regions away from the grain boundaries.

Konetzki et al. [25] found that in the eutectic Sn—Pb alloy PbO forms on the Pb phase, away from the Sn precipitates and Pb grain boundaries, and pure SnO forms on the Sn phase [25]. Preliminary evidence shows that there is a small amount of internal oxidation along the Pb/Sn phase boundaries. Frankenthal and Siconolfi [27] found that during the oxidation of 60Sn—40Pb solder the Sn preferentially oxidizes. If the surface is depleted in Sn, then PbO forms. Later work by Frankenthal and Siconolfi [26] found that the Pb—rich regions that contain Sn do preferentially oxidize at Pb grain boundaries, similar to that found by Konetzki et al. [23–26].

Based upon their observations Konetzki et al. [23] propose that the fatigue life of solder is degraded in the presence of oxygen, not by the surface oxide (which is much too thin to have a significant effect), but by SnO grain boundary oxidation that limits grain boundary movement. However, no experimental evidence was available to confirm this theory.

The effect of oxygen on the TMF behavior of 60Sn—40Pb solder was explored at a deformation rate of 2.8 x $10^{-4}$ $s^{-1}$ and 5.6 x $10^{-4}$ $s^{-1}$with hold times of 3 minutes at each extreme. The oxygen accounted for 40% of the environment at the sample. Surface oxidation was minimal as characterized by weight gain measurements on test coupons. The results of the load *versus* number of cycles for the oxygen environment are compared with the same TMF parameters without oxygen in Figure 8. The fatigue life, in the presence of oxygen, exceed 250 cycles. An optical cross section micrograph of a 60Sn—40Pb solder joint after 300 TMF cycles is shown in Figure 9. The joint is cracked through the heterogeneous coarsened bands. However, the coarsening does not extend through the length of the joint. Experiments performed at the faster deformation rate of 5.6 x $10^{-4}$ $s^{-1}$ had the same mechanical and microstructural results and without oxygen (Figures 3A and 4A) and failed after 50 cycles.

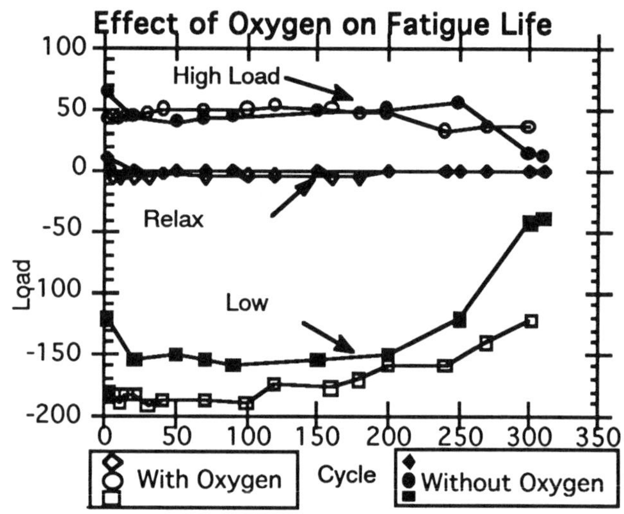


Figure 8. Plot of load *versus* number of cycles at a 3 minute hold time and deformation rate of $2.8x10^{-4}s^{-1}$ for tests with oxygen (open symbols) and without oxygen (closed symbols).

Our experiments indicate that, if anything, oxygen extends the fatigue life of the solder joint. There is evidence that the presence of oxygen suppressed the grain coarsening seen in the absence of oxygen. There is also evidence that the cracking behavior has been modified. The samples tested in nitrogen exhibited cracking that initiated at both edges of the sample. The cracks then progressed inward, and resulted in failure when the two met at the center of the sample. Associated with the cracking is a coarsening of the microstructure. When oxygen is present, less coarsening is observed. The sample still cracks, but the crack progresses either from the center or from one side of the sample. The fact that the side of the sample being subjected to the oxygen stream did not crack suggests that the oxygen may be playing a role in preventing crack formation.

One can speculate about the role of oxygen in this process. It is likely that enhanced diffusion along the grain boundaries occurs. It is also evident that preferential oxidation of the Sn would be expected. If the oxidation process, when coupled with TMF, results in small SnO particles, then the presence of these particles may alter the deformation mode. This is particularly true if the oxide is concentrated at grain boundaries. They may also act to limit the amount of grain growth possible. We intend to investigate these possibilities in the future.

The fact that oxygen was not detrimental to the fatigue life of the solder joint was unexpected. In our experiments, the sample was bombarded with nitrogen and oxygen. However, both gases were dry. If environmental exposure assists in the fatigue failure of solder joints, then other species that oxygen may be important. In particular, humidity may play a role. In most atmospheric corrosion processes, humidity can be very important. As the humidity level increases, the number of layers of adsorbed water increases. At some relative humidity (typically ~60% RH), the thin film of water begins to behave as a bulk electrolyte and the corrosion (oxidation) reactions can be enhanced considerably. In the future we plan to investigate the effect of water vapor on the oxidation and TFM of solder. At that time, we will be able to assess the contribution of environment to the TFM of solder joints, and determine an effective accelerated test configuration.

## Summary and Conclusions

Thermomechanical fatigue tests over a temperature range of —55° to 125°C were performed to examine the effect of deformation rate, hold time and oxidation on 60Sn—40Pb solder. At deformation rates slower than 2.8 x $10^{-4}s^{-1}$ heterogeneous coarsening occurs at eutectic cell boundaries. At faster deformation rates, a strain concentration develops in the solder joint because the eutectic cells cannot slide and rotate fast enough to accommodate the imposed strain. At faster deformation rates, heterogeneous coarsening occurs independent of microstructural features. Real use conditions for solder joints in electronic packages have slow thermal cycles, so a valid accelerated TMF test should have a deformation rate no faster than 2.8 x $10^{-4}s^{-1}$.

The microstructural and mechanical evolution of solder joints with hold times greater than 3 min. were similar. At 3 min. the load relaxes to zero at the high temperature portion of the thermal cycle. The deformation anneals and causes the microstructure to heterogeneously coarsen. With no hold times, the TMF life was found to be greatly enhanced. This enhanced fatigue life was attributed to insufficient time for annealing and coarsening to occur in the heterogeneous coarsened regions.

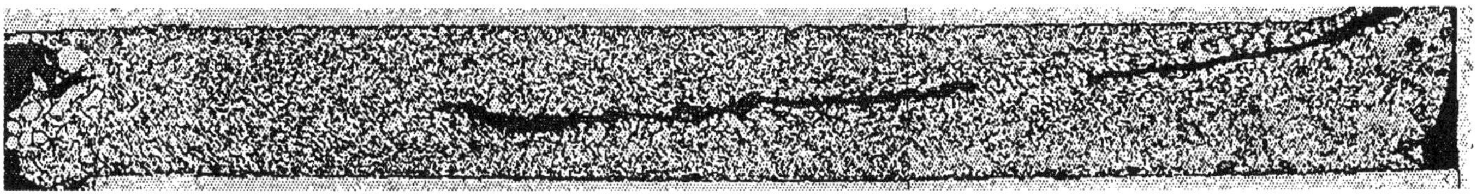

Figure 9 Optical micrograph of a 60Sn–40Pb solder joint after 300 cycles in an atmosphere of 40% oxygen at a deformation rate of 2.8 x $10^{-4}$ $s^{-1}$ and a hold time of 3 minutes at each extreme.

The effect of oxygen on the TMF behavior of solder is not totally understood. At worst there was no degradation in cycles to failure. At best, there may be a beneficial affect of oxygen on the fatigue process. However, this may not indicate the effect of environment on fatigue. Future experiments are designed to determine this.


### Acknowledgments

The authors would like to thank Jan Finch for performing the thermomechanical fatigue tests, Alice Kilgo for the metallography, and Mike Hosking for his insightful comments on the text. This work performed at Sandia National Laboratories supported by the U.S. Department of Energy under contract DE–AC04–76DP00789.


### References


1. Wild, R. N., Welding Res. Supp., 51, 521s (1972).
2. Bangs, E. R., Beal, R. E., Welding Res. Supp., 377s (1975).
3. Frear, D. R., IEEE Comp. Hybrids. Manufact.. Tech., CHMT–12, 492–501 (1989).
4. Frear, D. R., Jones, W. B., Proc. NEPCON West '90, p. 1340 (1990).
5. Frear, D. R., Proc. 40th ECTC Conf.,40, 510 (1990).
6. Frear, D. R., "Solder Mechanics–A State of the Art Assessment", edt.D. R. Frear, W. B. Jones, K. R. Kinsman, p. 191–237, TMS (1991).
7. Lee, S. M., Stone, D. S., Proc 40th IEEE ECTS, 40, 491 (1990).
8. Frear, D. R., Grivas, D., McCormack, M., Tribula, D., Morris, J. W. Jr., Effect of Load and Thermal History on Mech. Behavior, TMS, p. 113 (1987).
9. Frear, D. R., Grivas, D., Morris, J. W. Jr., Proc. 3rd Electron. Packag. and Corr. in Microelectron. Cong., P. 269 (1987).
10. Tribula, D., Grivas, D., Frear, D. R., Morris, J. W. Jr., ASME J. Electron Packag., 111, 83 (1989).
11. Tribula, D., Grivas, D., Frear, D. R., Morris, J. W. Jr., Weld. Res. Supp, Oct. p. 404s (1989).
12. Stone, D. S., Lee, S. M., SMART Conf. (1990).
13. Solomon, H. D., Electronic Packaging: Materials and Processes, edt. J. A. Sortell, ASM, 29–49 (1985).
14. Solomon, H. D. ASTM STP 942, 342–369 (1988).
15. Solomon, H. D. J. Electon. Packag. 111, 75–90 (1989).
16. Vaynman, S. Fine, M. E., Jeannotte, D. A., "Solder Mechanics–A State of the Art Assessment", edt.D. R. Frear, W. B. Jones, K. R. Kinsman, p. 155–189, TMS (1991).
17. Fox, L. R., Sofia, W., Shine, M. C., IEEE CHMT, 8, 275–282, (1985).
18. Majumdar, S., Jones, W. B., "Solder Mechanics–A State of the Art Assessment", edt.D. R. Frear, W. B. Jones, K. R. Kinsman, p. 273–360, TMS (1991).
19. Berriche, R., Vaynman, S., Fine, M. E., Jeannette, D. A., "Electronic Packaging and Corrosion in Microelectronics", edt. M. E. Nicholson, ASM, 169 (1987).
20. Snowden, K. U., Acta Met., 12, p. 295 (1964).
21. Berriche, R., "Environmental Effects of Time Dependent Fatigue of Low Tin–High Pb Solder",Ph.D. Thesis, Northwestern University (1990).
22. Lacey. T. G. Lacey and Woodford, D. A., Scripta Met., 22, 1543–1545 (1988).
23. Konetzki, R. A. and Chang, Y. A., J. Mater. Res., 3, 466–470 (1988).
24. Konetzki, R. A, Chang, Y. A. and Marcotte, V. C., J. Mater. Res., 4, 1421–1426 (1989).
25. Konetzki, R. A., Zhang, M. X., Sluzewski, D. A. and Chang, Y. A., J. Electron. Packag., 112, 175–178 (1990).
26. Frankenthal, R. P. and Siconolfi, D. J., Corrosion Science,21, 479–486 (1981).
27. E. Frankenthal, R. P. and Siconolfi, D. J., J. Vac. Sci. Technol., 17, 1315 (1980).

*Proceedings of the Fifth International Conference on Creep of Materials, Lake Buena Vista, Florida, USA, 18-21 May 1992*

# A New Prediction Method of Fatigue-Creep Interaction Behavior from a Tensile Total Strain Energy Viewpoint for High Strength Superalloy

**N. Matsuda and S. Umezawa**
Mechanical Engineering Research Laboratory
Hitachi, Ltd.
Ibaraki-ken, Japan

## ABSTRACT

A new life prediction method is proposed for high strength materials such as a cast Ni-based superalloy under fatigue-creep interaction (FCI) conditions based on tensile total strain energy behavior. This new life prediction method is termed "frequency modified total strain energy parameter." The total tensile strain energy, including large elastic strain energy and the time under tensile stress loading, are used to define this tensile strain energy parameter. It is found that strain controlled FCI and cyclic creep failure life prediction can be estimated using this parameter. Creep rupture strength data are plotted using the best fitted curve of strain controlled FCI failure life data.

Furthermore, not only failure life but also surface crack initiating and propagating behavior are found to be expressed by this parameter. The surface crack propagation rate is found to be a function of this parameter and surface crack half length. Thus, this new life prediction method may be used to determine surface crack propagation behavior and initiation life .

UNDERSTANDING OF FCI has been recognized for many years as a vital importance in the design and operating of high temperature equipment subjected to thermal and load transient cycles. Numerous prediction laws have been proposed for FCI life in metallic materials for example strain range partitioning method (1), frequency modified fatigue life approach (2) and frequency modified damage function approach (3). These prediction rules have been shown to provide a good method for correlating with high temperature fatigue tests. Various strain controlled cycles using triangular, trapezoidal and sinusoidal waveforms are employed in FCI testing. Prediction laws which can estimate FCI life employing all waveforms are not well developed. That is, there are certain prediction laws that cannot estimate the effects of only compression hold, or sinusoidal waveform on FCI life. Prediction laws that evaluate FCI life use only the inelastic strain component criterion. These damage laws do not allow accurate life assessment, because the elastic strain range is much larger than the inelastic strain range in FCI behavior of high strength materials, such as a Ni-based superalloys.

In this paper, strain controlled F-F, F-S and S-F waveforms and also tensile or compressive strain hold tests are performed on a cast Ni-based superalloy. A new life prediction method is proposed for high strength materials under FCI conditions based on tensile total strain energy behavior. The new damage rule's applicability to the prediction of cyclic creep life and to the evaluation of the correlation between FCI life and creep rupture strength is also examined.

In general, low cycle fatigue consists of two stages: micro-crack initiation and crack propagation to failure. It is important to examine both initiating and propagating phenomena of micro-cracks to estimate the life of a smooth specimen. Initiating and propagating phenomena of surface cracks in a smooth specimen are observed, and a new life prediction method based on these surface cracks is also proposed for this material under FCI conditions .

## MATERIAL AND TEST PROCEDURES

The material used in this investigation was a cast Ni-based superalloy, which was used in its cast form as a blade in a gas turbine. The Ni-based superalloy was obtained in the form of cast round bars. Smooth test specimens for low cycle fatigue tests were ground from each bar. Dimensions and geometry of the test specimen are shown Figure 1. The chemical composition and the mechanical properties of the test material are given in Tables I and II. The test program described in this paper may be summarized as follows;

-Strain controlled FCI test
-Load controlled cyclic creep test
-Static creep test

The FCI tests were conducted using strain controlled triangular, sinusoidal, and trapezoidal waveforms. The applied fast strain rate was 0.1%/s and slow strain rate was 0.001%/s in the triangular and sinusoidal waveforms. The tests using the trapezoidal waveform were performed with a hold time of 10 and 60 min. in tension and 10, 60 and 480 min. in compression. The load controlled cyclic creep tests were performed using a trapezoidal waveform. When the specimen was loaded in tension and held at a constant stress, tensile creep strain occurred. When the tensile creep strain reached a constant value ($\varepsilon_{max}$ = 0.25% or 0.5%), the specimen was then loaded in compression and held under a compressive stress. When the compressive creep strain reached a constant value ($\varepsilon_{max}$ = -0.25% or -0.5%), the specimen was again loaded in tension. The strain rate value corresponded to 0.1%/s.

The static creep tests were conducted at a high temperature. Deformation was measured in the gauge length by an extensometer. Specimen heating in all tests was performed by an induction heating system or electric furnace. Temperatures were monitored by spot-welding Pt-Rh thermocouples to the specimen. The test temperatures were 815 and 900 °C, and controlled to within ±5 °C along the gauge length.

Table I Chemical composition

(Wt %)

| C | Cr | Co | Mo | W | Ti | Aℓ | B | Ni |
|---|---|---|---|---|---|---|---|---|
| 0.15 | 13.9 | 9.2 | 4.0 | 4.0 | 4.9 | 3.02 | 0.014 | Bal. |

Table II Mechanical properties

| Temp. (°C) | Tensile strength (MPa) | 0.2% offset stress (MPa) | Elongation (%) | Reduction of area (%) |
|---|---|---|---|---|
| R.T. | 951 | 614 | 5.9 | 7.4 |
| 871 | 580 | 487 | 24.7 | 31.9 |

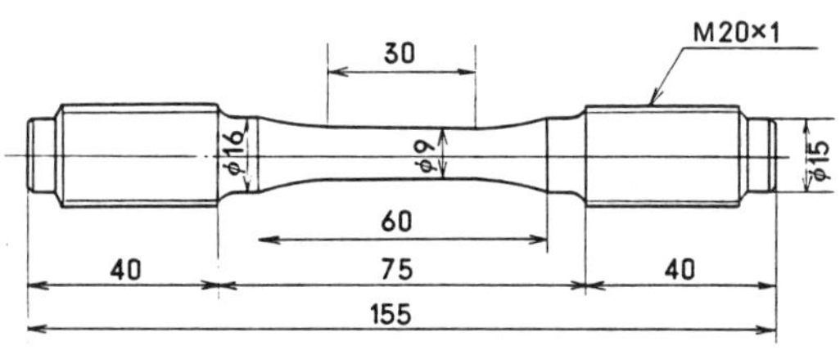


Fig.1 Specimen for fatigue-creep interaction tests

# TEST RESULTS

**Strain Controlled FCI Tests.** Figure 2 shows the relation between total strain range and cycles to failure obtained during strain controlled FCI tests at 815 °C. FCI life of F-F, F-S and S-F waveforms were about the same value at $\Delta\varepsilon$ = 1.0%, but FCI life of the specimen under an applied F-F waveform was greater than that of F-S, S-F, and the trapezoidal waveform specimens. In addition, specimen life under tensile hold times were greater than those under compressive hold times. Figure 3 shows the relationship between inelastic strain range, elastic strain range and cycles to failure under FCI conditions. The elastic strain component forms the bulk of the total strain range in spite of FCI tests under high total strain range, for example $\Delta\varepsilon$ = 0.5% or 1.0%.

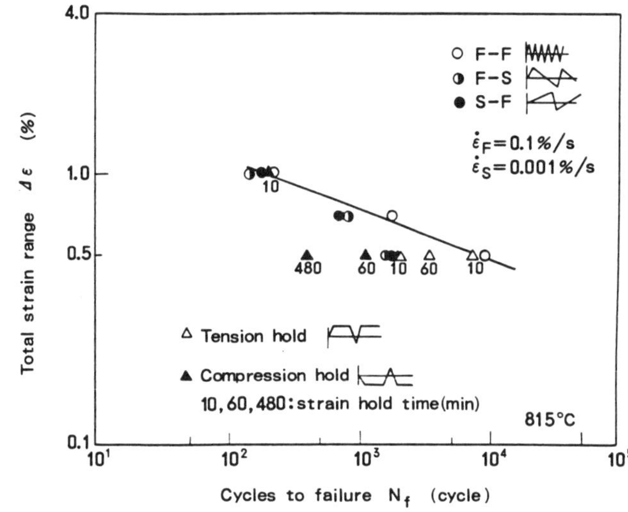


Fig.2 FCI failure life of Ni based superalloy at 815°C

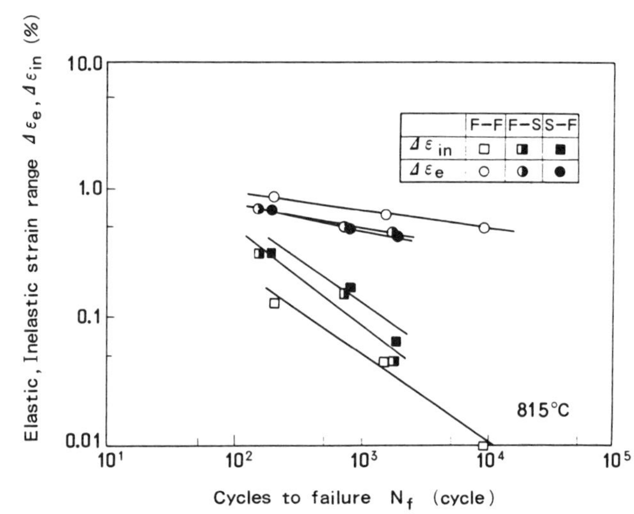


Fig.3 Relationship between elastic,inelastic strain range, and cycles to failure under sawtooth waveform

Figure 4 shows the change of maximum and minimum stress with strain cycling under FCI conditions at 815 °C. Cyclic stress ranges of all waveform (F-F, F-S, S-F, ten. hold and comp. hold) are about the same, but the tensile maximum stress values are different. The mean stress is in tension stress under F-S waveform and conversely in compressive stress under S-F waveform. This mean stress develops as a result of stress relaxation occurring during hold times. This same phenomenon was originally proposed by W. J. Ostergren (3).

Figure 5 shows an example of surface crack propagation curves of smooth bar specimens under FCI conditions at 815 and 900 °C. Correlating the log of surface crack length and number of cycles in a crack length that is longer than a constant value in each crack propagation curve gives a linear relationship. The constant value of crack length is about 1.0 mm at 815 °C. Figure 6 shows photographs of surface cracks under F-F, F-S and S-F waveforms at 815 °C. Transgranular failure occurred under the F-F waveform and intergranular failure occurred under F-S and S-F waveforms. Grain size can be measured in the photographs of Fig. 6 and is about 1.0 mm.

Low cycle fatigue consists of two stages crack initiation and crack propagation to failure of smooth specimens. Cycles to crack initiation was defined as the number of cycles needed for the surface crack length to grow from. crack to the grain size of this material. The grain size and the constant value of crack length that correlates the log of surface crack length and number of cycles in each crack propagation curve which give a similar linear relationship in each case.

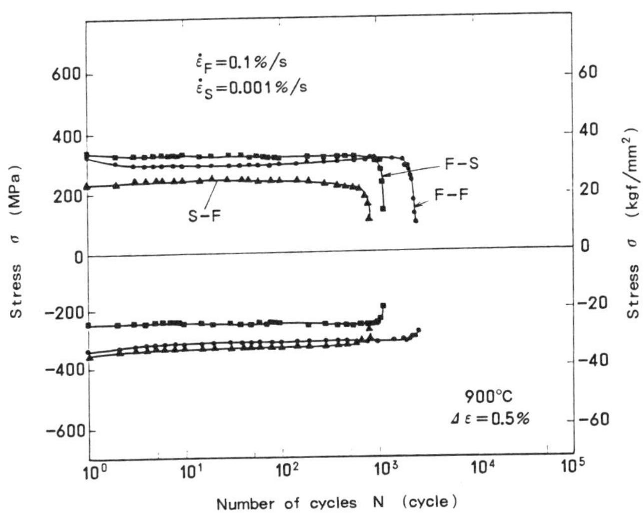


Fig.4 Change of stress with strain cycling under sawtooth waveform

**Load Controlled Cyclic Creep Test.** Figure 7 shows the relation between stress range and cycles to failure under cyclic creep test conditions at 815 °C. Stress range was directly proportional to cycles to failure. This tendency was different from the relation between stress (or strain) range and cycles to failure under usual load (or strain) controlled

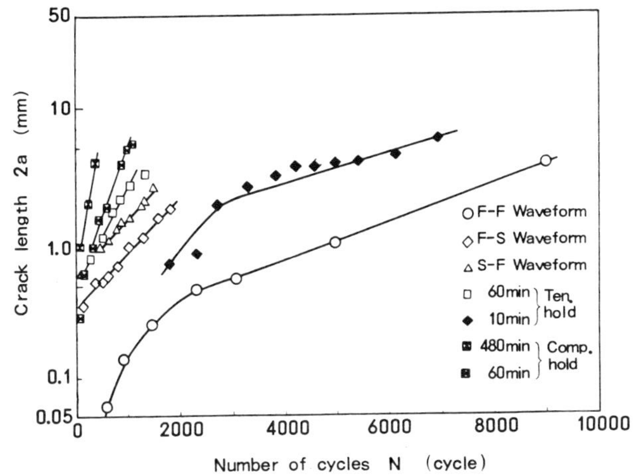


Fig.5 Surface crack propagating behavior on the smooth round bar specimens under FCI conditions at 815°C and total strain range of 0.5%

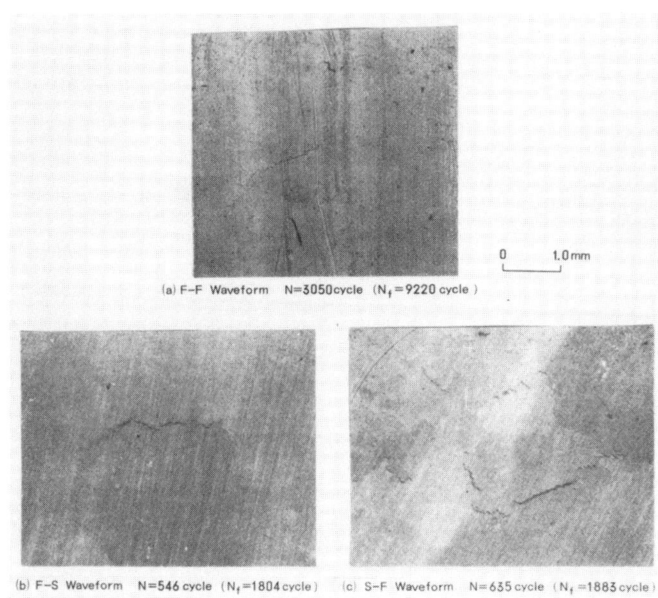


Fig.6 Typical surface cracks of smooth round bar specimens

fatigue tests. Figure 8 shows change of tensile hold time at each cycle with cyclic stress range. In this case, tensile hold time per unit cycle was inversely proportional to number of cycles. This was caused by the fact that the specimen was damaged cumulatively by each stress cycle.

Further, tensile hold time was inversely proportional to test stress. Figure 9 shows a hysteresis loops for these test conditions. Since this material has a high 0.2% proof stress, there is in elastic region within the stress conditions of this test. Therefore, this hysteresis loop is similar in form to a parallelogram, and all of the inelastic strain component is due to creep strain . Time that was needed to reach a constant creep strain value was longer if the applied stress was low.

**Static Creep Test.** Figure 10 shows the creep curve of this material at 800 and 815 °C. Four creep tests were performed between 343 and 441 MPa giving rupture times between 200 and 5000 hr, and rupture elongation between 8 and 15%.

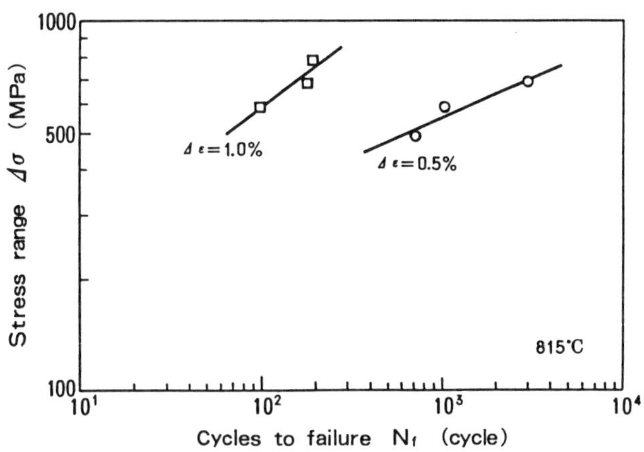


Fig.7 Stress range versus cycles to failure under cyclic creep test at 815 °C

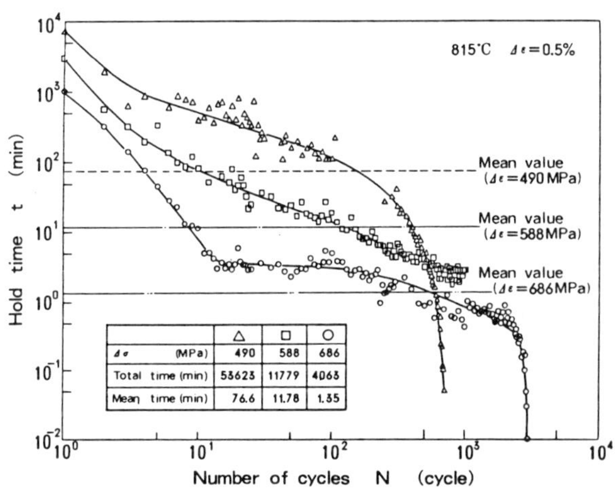


Fig.8 Hold time per one cycle versus number of cycles under cyclic creep test at 815°C

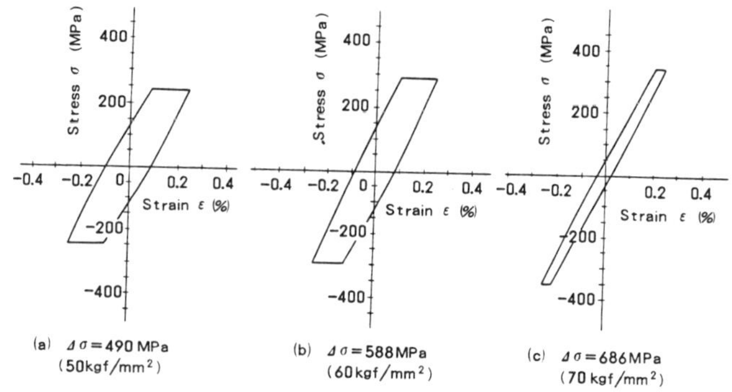


Fig.9 Hysteresis loops under cyclic creep test at 815°C and strain range of 0.5%

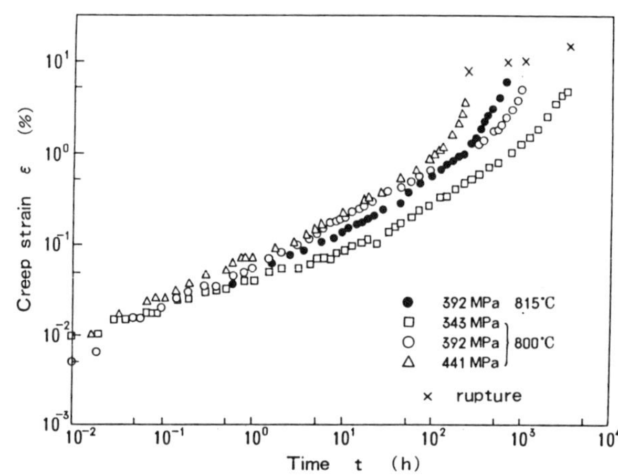


Fig.10 Creep curves at 800 and 815°C

## EVALUATION OF FCI FAILURE LIFE USING THE TOTAL STRAIN ENERGY CRITERION

**Strain Controlled FCI Failure Life.** As previously mentioned, strain controlled test results found compression hold times to be more damaging than the equivalent hold times in tension for this material, as shown Fig. 2. The net tensile hysteresis energy is now being used to evaluate FCI life using the frequency-modified damage function rule developed by W. J. Ostergren(3). However, for high strength materials, this damage rule seems to require some modifications in order to obtain appropriate life assessments, because the elastic strain range is much larger than the inelastic strain range in the FCI of these materials.

In the this paper, strain controlled FCI tests with trapezoidal and sinusoidal waveforms were performed on a Ni-based superalloy. The total strain energy including the large elastic strain energy is formulated, and evaluation of FCI life is obtained using this energy parameter. Figure 11 shows the relation between the tensile total strain energy and cycles to failure under FCI at 815 °C. At high tensile total strain energy values, cycles to failure have same value in spite of the testing waveform. However, at low tensile total strain energy values, test results show that life without a hold time is greater than that with hold times. Thus, one may question how the theory may be modified for frequency and hold times to become significant. The treatment of frequency under triangular and trapezoidal waveforms was explained in the frequency-modified damage function approach. However, treatment of frequency under a sinusoidal waveform was not explained in this approach.

In this paper, modifications for frequency, independent of test waveforms, are done using loading times under tensile stress. If we consider low cycle fatigue to be the primary mechanism of crack propagation, then a reasonable assumption would be that deformation occurring in the region of the tensile stress, with the crack open, contributes to the state of damage by propagating the crack. Thus, the tensile total strain energy ($\Delta W_{TT}$) and the loading time ($\tau_{TT}$) in tension are used to define this tensile strain energy parameter. A life prediction equation can be expressed as follows.

$$\Delta W_{TT}\cdot(\tau_{TT})^{n}\cdot N_f^{m} = C \qquad (1)$$

where, m, n, C are material constants. and $N_f$ is the number of cycles to failure. On the other hand, the relation between the tensile inelastic strain energy and the cycles to failure is defined as follows, since there are different effects causing dependence of cycles to failure on tensile inelastic strain energy.

$$\Delta W_{TP}\cdot(\tau_{TT})^{n'}\cdot N_f^{m'} = C' \qquad (2)$$

where, m′, n′, C′ are also material constants.

Multiple regression analysis was performed to assess the relationship between strain energy and cycles to failure and is expressed numerically to improve life estimation accuracy. The constants m, n, C, m′, n′, C′ are determined by this regression analysis. Figure 12 compares test and regression analysis results. Correlating between $\Delta W_{TT}\cdot(\tau_{TT})^{n}$ and cycles to failure under FCI conditions gave a linear relationship on a log-log graph. In addition, the relationship between $\Delta W_{TP}\cdot(\tau_{TT})^{n}$ and cycles to failure is also linear. These results show that the frequency-modified tensile total strain energy law can be used for correlating strain controlled triangular, sinusoidal, and trapezoidal waveform test results under FCI conditions for high strength materials such as Ni-based superalloys in which the elastic strain range is much larger than the inelastic strain range in the FCI. The material constant n obtained by multiple regression analysis is 0.077.

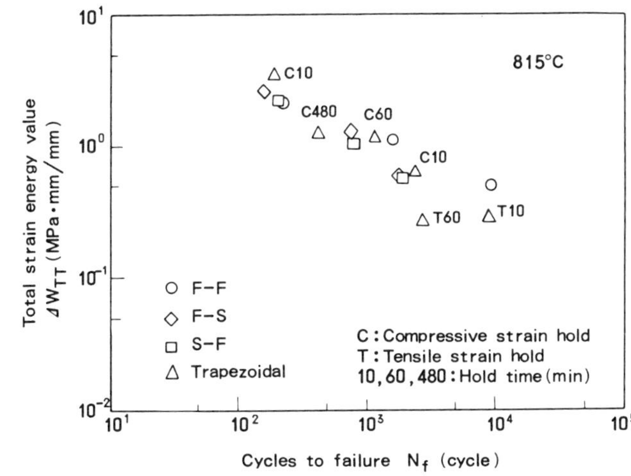


Fig.11 Total strain energy value versus FCI life at 815°C

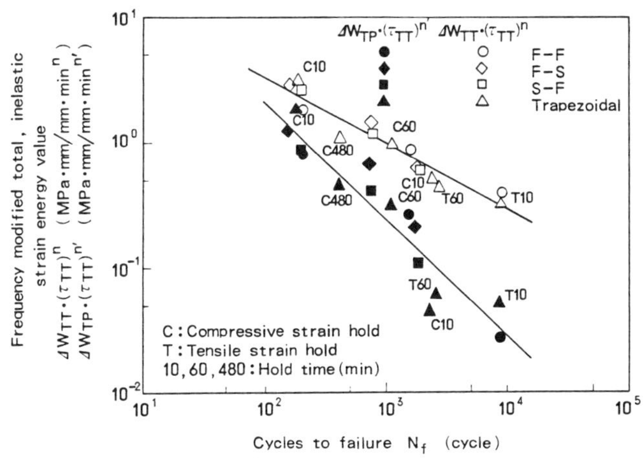


Fig.12 Frequency-modified inelastic and total strain-energy value versus FCI life at 815°C

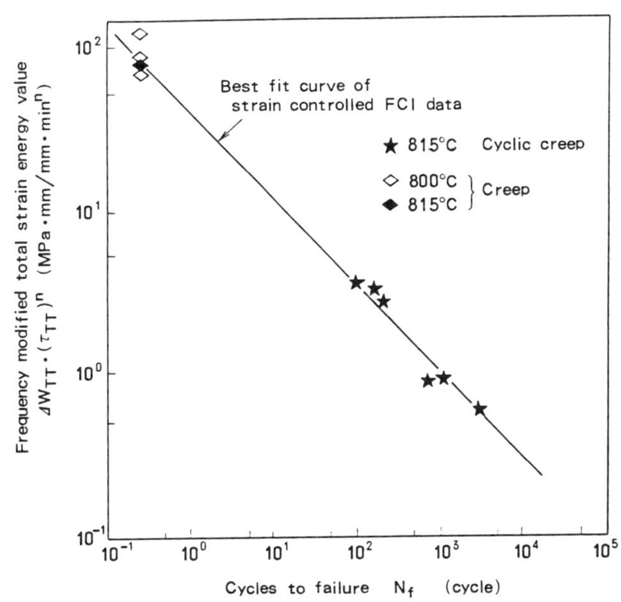


Fig.13 Comparison between creep rupture strength ,cyclic creep life and the best fit curve of strain controlled FCI life at 815°C

**Load Controlled Cyclic Creep Life And Static Creep Strength.** In this section, it will be shown that the frequency-modified total strain energy can be used for correlating load controlled cyclic creep and static creep test results.

First, the frequency-modified total strain energy is defined under a load controlled cyclic creep test. The tensile total strain energy at every cycle is held at the same value during total life in a cyclic creep test since tensile stress and strain were specified. However, the hold time per cycle is varied at every stress cycle such as shown Fig. 8. Therefore, the loading time value under tensile stress per cycle is an average value in which total tensile hold time during total life is divided by the number of cycles to failure. The material constant n is 0.077 which is found from the strain controlled FCI test.

Next, the frequency-modified total strain energy is determined under a static creep test. The applicability of this parameter to evaluate the correlation between creep rupture strength and strain controlled FCI life is then examined. The tensile total strain energy is defined as the product of creep stress ($\sigma_c$) and rupture elongation ($\varepsilon_{Cf}$). The loading time of tensile stress which is equal to creep rupture time ($\tau_{cr}$), Nf is 0.25 cycles and the material constant value is again 0.077. Figure 13 shows evaluated results with cyclic creep life and the best fit curve of results from strain controlled FCI life. It was found that the strain controlled FCI and cyclic creep life for this material could be estimated from the proposed parameter. Additionally, creep rupture strength data is plotted with the best fit curve of strain controlled FCI life data. As a result of this examination, the material constant C can be defined by

$$C = 0.25^{m} \cdot \sigma_C \cdot \varepsilon_{Cf} \cdot \tau^{n}_{cr} \quad (3)$$

## A FAILURE LIFE PREDICTION LAW BASED ON SURFACE CRACK INITIATION AND PROPAGATION BEHAVIOR USING THE TOTAL STRAIN ENERGY CRITERION

In previous section, it was shown that small cracks were initiated very early and that low cycle fatigue is the primary crack propagation phenomenon. That is, most of the cycles required to fail a smooth specimen are involved in propagating a micro-crack to the critical crack size. In this section, a new prediction rule is proposed based on the surface crack initiating and propagating behavior using the total strain energy criterion.

In Fig. 5, correlation between the log of the surface crack length (2a) and number of cycles (N) is linear at crack lengths longer than the grain size. The slope is defined by d(log2a)/dN. This material shows no softening or hardening with strain cycling because maximum and minimum stresses are held at a constant value during the entire life of a specimen in FCI tests as shown Figure 4, and the total strain energy value is constant during the total life of a specimen.

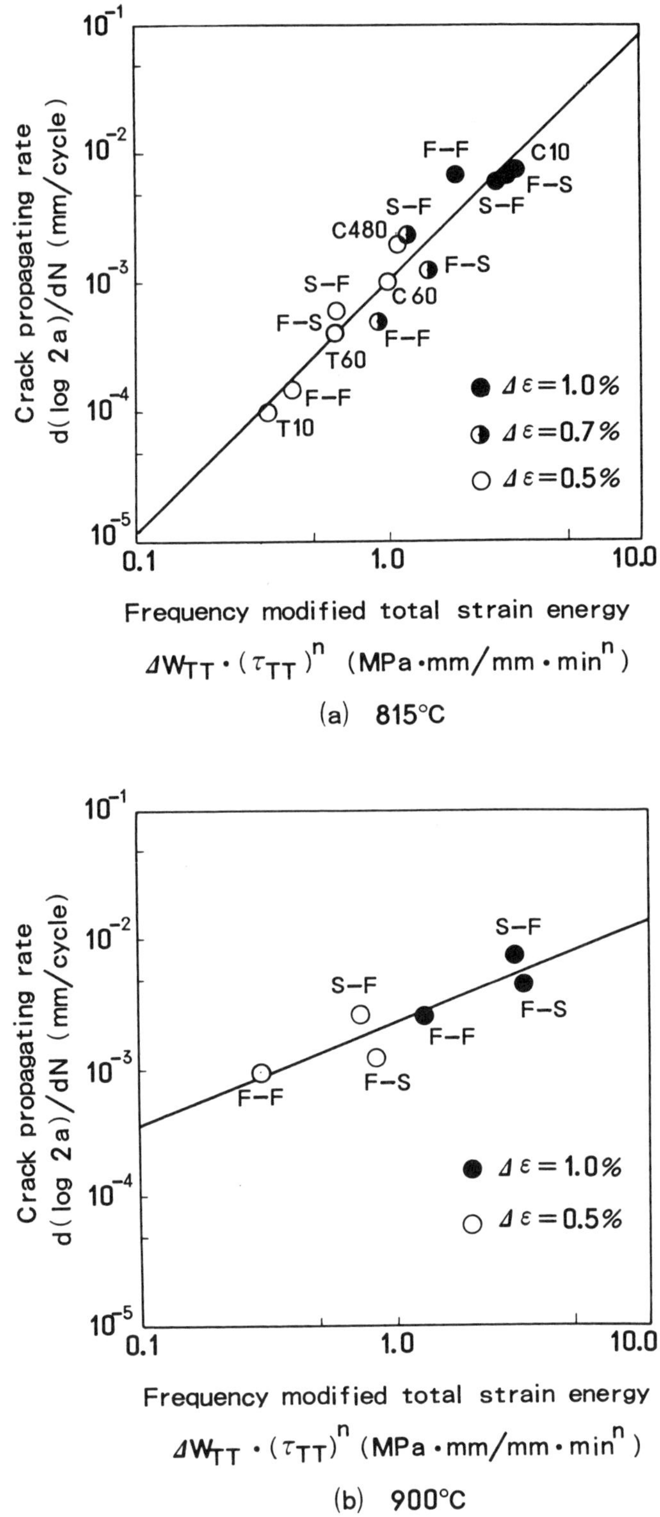


Fig.14 Relationship between surface crack propagating rate and frequency-modified total-strain energy value under FCI conditions at 815°C

Figure 14 shows plots of d(log2a)/dN and the frequency-modified strain-energy value. Correlating d(log2a)/dN and this parameter gave a linear relationship on a log-log graph. This relation may be written as

$$d(2a)/dN = B(\Delta W_{TT}\cdot(\tau_{TT})^n)^{\ell}\cdot(2a) \qquad (4)$$

where B and n are material constants.

On the other hand, Figure 15 shows the relationship between this parameter and cycles to surface crack initiation ($N_{C_{0.2}}$, $N_{C_{0.5}}$ and $N_{C_{1.0}}$) which are obtained from crack propagation curves, where 0.2, 0.5 and 1.0 are surface crack lengths. The relation between cycles to surface crack initiation and this parameter is also linear on a log-log graph and is expressed by the following equation.

$$Nc = A(\Delta W_{TT}\cdot(\tau_{TT})^n)^{m_1} \qquad (5)$$

Not only failure, but also surface crack initiation life and surface crack propagating behavior can be expressed using the total stain energy, and the prediction life law using surface crack initiating and propagating behavior is examined. The number of cycles to crack initiation ($Nc_{2a_0}$) is defined as the number of cycles when a surface crack length has propagated to a length approximately equal to the grain size. The number of cycles ($N_{2a}$) which propagate micro-cracks to an arbitrary surface crack length ($2a_N$) is given by

$$N_{2a} = Nc_{2a_0} + \int_{2a_0}^{2a}(dN/d(2a))d(2a) \qquad (6)$$

Using equations (4) and (5), the expression for $N_{2a}$ can be written as:

$$N_{2a} = A(\Delta W_{TT}\cdot(\tau_{TT})^n)^{m_1} + B'(\Delta W_{TT}\cdot(\tau_{TT})^n)^{-\ell}\cdot(\log(2a_N)-\log(2a_0)) \qquad (7)$$

The number of cycles to failure of a smooth specimen is defined as the number of cycles in which arbitrary surface crack length ($2a_N$) propagates to a critical size 2af. The number of cycles to failure can be calculated by substituting $2a_f$ for ($2a_N$) of Eq.(7). For example, when the critical size of a surface crack ($2a_f$) is 10 mm and the grain size is 1.0 mm, the number of cycles to failure under FCI conditions at 815 and 900 °C for the Ni-based superalloy can be predicted by Eq.(7). Figure 16 shows a comparison between predicted and observed life. Agreement between predicted and observed life was generally within a factor of two.

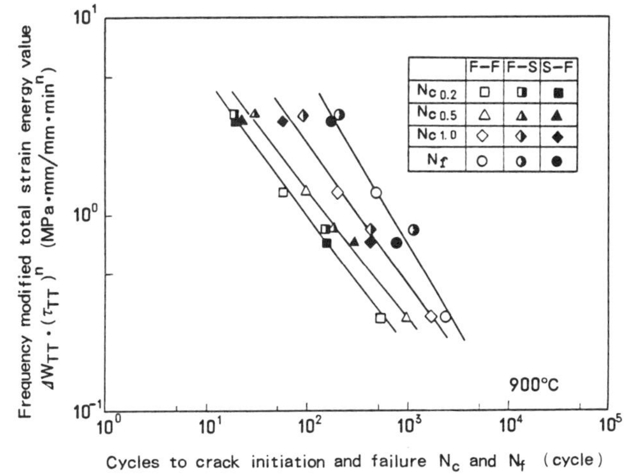


Fig.15 Dependence of frequency-modified total-strain energy on number of cycles to surface crack initiation and failure under FCI conditions at 900°C

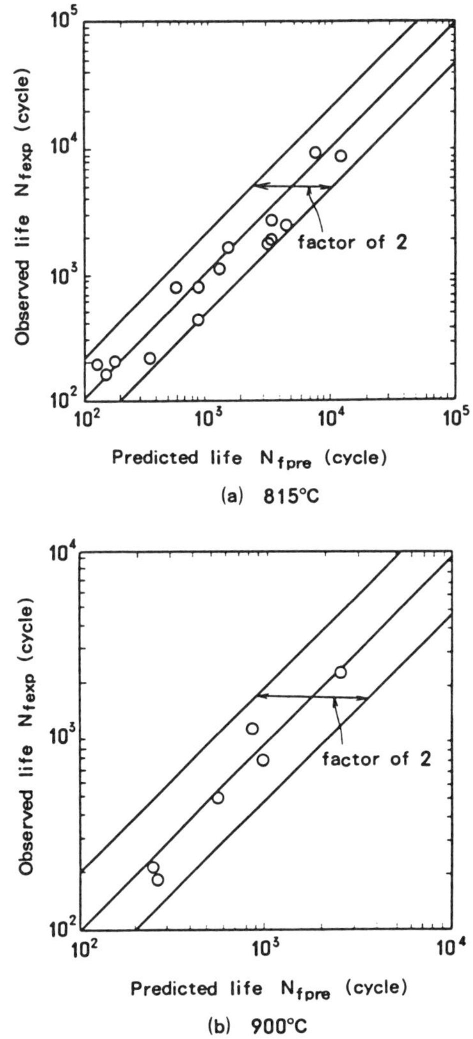


Fig.16 Comparison between observed and predicted life on frequency-modified total-strain energy parameter

## CONCLUSION

The results of this study lead to the following conclusions.

(1) A frequency-modified total-strain energy parameter is formulated to evaluate failure life of Ni-based superalloys under the FCI condition. The total tensile strain energy ($\Delta W_{TT}$) and the time ($\tau_{TT}$) under tensile stress loading are used to define this tensile total strain energy parameter. A failure life prediction equation can be expressed as follows:

$$\Delta W_{TT} \cdot (\tau_{TT})^{n} \cdot N_f^{m} = C$$

where, m, n, C are material constants. and $N_f$ is the number of cycles to failure. A similar relation between micro-crack initiation life and total strain energy parameter was also obtained.

(2) It is found that cyclic creep life can also be estimated by this parameter. In addition, creep rupture strength data is plotted with the best fit curve of strain controlled FCI life data ($\Delta W_{TT} \cdot (\tau_{TT})^n$ vs $N_f$). and the material constant C can be defined as :

$$C = 0.25^{m} \cdot \sigma_C \cdot \varepsilon_{Cf} \cdot \tau^{n}_{cr}$$

where, $\sigma_c$ is creep stress, $\varepsilon_{Cf}$ is rupture elongation and $\tau_{cr}$ is creep rupture time.

(3) Surface micro-crack propagation behavior was also found to be related to this parameter under FCI conditions. The surface crack propagation rate (d(2a)/dN) is a function of this parameter and surface crack length (2a) :

$$d(2a)/dN = B(\Delta W_{TT} \cdot (\tau_{TT})^{n})^{\ell} \cdot (2a)$$

where, B, n, $\ell$ are material constants.

(4) A new life prediction method taking into account of surface crack propagation behavior and initiation life can be defined as :

$$N_{2a} = A(\Delta W_{TT} \cdot (\tau_{TT})^{n})^{m_1} + B'(\Delta W_{TT} \cdot (\tau_{TT})^{n})^{-\ell} \cdot (\log(2a_N) - \log(2a_o))$$

If the grain size ($2a_o$) and number of cycles ($Nc_{2ao} = A(\Delta W_{TT} \cdot (\tau_{TT})^{n})^{m_1}$) when the surface crack length has reached $2a_o$ are known, the number of cycles ($N_{2a}$) of a certain crack length (2a) can be determined using the above equation.

## REFERENCES


1. Manson, S. S., Halfold, G. R., and Hirschberg, H. M., Symposium on design for elevated temperature environment. ASME, 1971

2. Coffin, L. F. Jr., Proc. Air force conf. on fatigue and fracture of aircraft structures and materials. AFFDL TR 70-144 1970

3. W. J. Ostergren, J. of Testing and Evaluation, JTEVA, Vol.4, No.5, 1976

*Proceedings of the Fifth International Conference on Creep of Materials, Lake Buena Vista, Florida, USA, 18-21 May 1992*

# The Influence of Creep on the Fatigue Life of Nickel Base Superalloys

**R. Danzer**
Institut für Metallkunde und Werkstoffprüfung
Montanuniversität Leoben
Leoben, Austria

**ABSTRACT**

In many applications the service temperature of metals and alloys is limited by the occurrence of serious creep deformation. In general, creep tests are performed at constant stress and temperature. Therefore creep is hard to recognize if stresses and temperatures change with time, as it is the case during fatigue loading. A parameter called r is defined which indicates the occurrence of significant steady state creep strain in general loading situations. If $r \approx 1$, the complete inelastic strain is steady state creep, and if $r >> 1$, only a small fraction of it is steady state creep.

By means of a large set of fatigue tests, performed on several superalloys, it is shown that fatigue life time is given by the material's creep properties, if cyclic inelastic strain is steady state creep ($r \approx 1$). In this case, the simple linear life fraction rule for creep damage (AC– rule), which needs no adjustable parameter, can be used to calculate fatigue life.

For $r >> 1$, the AC–rule overestimates fatigue life time. This systematic deviation, which increases with the increase of r, is used to define a new rule, which needs one adjustable (empirical) parameter. It is shown that the new rule describes all measured fatigue data fairly well. For obvious reasons, it is restricted to loading situations, where creep still has some influence. These conditions are characterized by $r < 10^6$.

IN ORDER TO INCREASE THE EFFICIENCY OF COMBUSTION ENGINES and gas turbines, the service temperature has to be increased. This implies that many high temperature material phenomena such as creep, corrosion and structural stability, become increasingly important (1–3). These phenomena have a strong influence on the performance of components and limit the service temperature of the materials. They may reduce fatigue life time (number of cycles of failure) by several orders of magnitude (4,5,7).

This paper is focused on fatigue properties at high temperatures, where creep has some influence on fatigue life time (the so–called creep fatigue interaction regime). In general, this happens at temperatures higher than 1/3 of the melting temperature. In this regime, fatigue life time strongly depends on some loading parameters, which, at low temperatures, have a minor influence or no influence at all, e.g. the test frequency, the cycle shape and the mean stress (5,6,7). There are three approaches to take the influence of these parameters into account:

The first approach is an attempt to generalize the methods established to predict low temperature fatigue life time. At low temperatures, fatigue life time is limited by the occurrence and the propagation of cracks. Using fracture mechanical models (8), one can show that the number of cycles of failure, $N_f$, depends on the inelastic strain range, $\Delta\epsilon_{in}$. This leads to the well known Manson Coffin (MC) rule (9,10):

$$\Delta\epsilon_{in} = c_1 N_f^{-c_2} \quad . \qquad (1)$$

Failure is defined by the occurrence of cracks of a length of several 100 $\mu$m. $c_1$ and $c_2$ are material parameters. For example, the influence of frequency (which is important at high temperatures) can be considered in a given frequency range by the introduction of an additional empirical parameter, leading to the frequency modified MC–rule (11,12):

$$\Delta\epsilon_{in} = c_3 N_f^{-c_4} \cdot (1/\tau)^{c_4(c_5 - 1)} \quad . \qquad (2)$$

$\tau$ is the cycle period and $1/\tau$ the frequency. $c_3$, $c_4$ and $c_5$ are material parameters.

The influence of mean stresses can be considered using the Ostergren rule (13) or its frequency modified version ($\Delta\epsilon_{in}$ in eqs. 1 and 2 is replaced by the tensile part of the area under the hysteresis loop). Other examples for rules which are generalizations of eq. 1 are the different versions of the strain range partitioning method (14,15,7).

Generalized MC–rules are sometimes claimed to model situations where corrosion has some influence on high temperature fatigue (16). Then, due to a localized corrosion attack, the formation of cracks is promoted. Riedel has shown that eq. 2 also describes the unconstrained growth of creep pores under cyclic loading conditions (17). Although the modified rules are not based on physical models, they describe, in a wide range of loading conditions, some key features of the material's behavior. Nevertheless, they are based an a limited set of data and, moreover, on a limited range of experience. Therefore, they should not be applied beyond this range.

The second group of methods to predict cyclic life time is concerned with loading situations where creep is the most serious damage mechanism. Creep damage can be the nucleation and growth of pores, a softening of the material (e.g. due to a coarsening of the microstructure, a dissolution of particles or the multiplication of dislocations) or internal or external corrosion, which can be enhanced due to deformation (e.g. in (4), a more than 100–fold increase of the corrosion rate during creep, compared with the stress free state, was observed). The later situation also promotes the nucleation and the growth of cracks.

More than 50 years ago, Robinson (18,19) has calculated the creep rupture life for changing temperatures and stresses, assuming that in any time interval the accumulated damage is t/L, independent of the damage in the other time intervals, t being the time spent at a stress and temperature, and L being the rupture life according to the same stress and temperature. This rule

$$\sum_i t_i/L_i = 1 \text{ at failure,} \qquad (3)$$

is known as the linear life fraction rule for creep damage (accumulation of creep damage; AC–rule). The index i refers to constant stress and temperature. This rule has been successfully applied to describe the influence of small load changes on the life time of tensile loaded structures, e.g. internal pressurized pipes. Taira applied this rule to low cycle fatigue (LCF) tests with high mean stress (20), and Spera to LCF–tests at high temperature and low test frequency (21,22), using the equation:

$$\int_0^{t_f} \frac{dt}{t_{f,c}(\sigma,T)} = N_f \cdot \int_0^{\tau} \frac{dt}{t_{f,c}(\sigma,T)} = 1 \quad \text{at failure,} \qquad (4)$$

where $t_f$ is the time of failure in the actual high temperature test (whereas $t_{f,c}$ is the time of failure in a creep test). $N_f$ is the number of cycles of failure. In (23) the conditions necessary in order to get the AC–rule are discussed in great detail. In some cases, eq. 4 describes the cyclic life time very accurately but in some cases it does not (because the necessary conditions do not apply). For low temperatures and high frequencies Spera proposed to use eq. 1, which indicates the break down of the AC–rule. This idea is correct but not very helpful, because it requires many complicated and expensive fatigue tests at different temperatures, strain rates and cycle types in order to define the validity limits of the AC–rule. A simple way to avoid these problems is the 10 % rule (21). Here it is assumed that 10 % of the cyclic life time yielded by eq. 4 is a lower limit for the real life time. In fact this observation may be correct for a special set of data but it cannot be generalized in such a simple way (5,7).

In summary, one can say that the AC–rule is useful in order to replace cost–intensive fatigue tests by cheaper creep tests. This is possible in the creep but not in the creep fatigue interaction regime. The proper definition of the border between these regimes is still unclear; to give such a definition is an important intention of this paper.

The methods of the third approach are a combination of the AC–rule and a linear life fraction rule for fatigue damage, which is based on a measured S–N curve (7). A prominent example of such a method of life prediction is the ASME Code (25):

$$\sum_j \left[\frac{N}{N_d}\right]_j + \sum_k \left[\frac{t_h}{t_d}\right]_k = D. \tag{5}$$

$N_d$ and $t_d$ are fatigue and creep test results, respectively, N is the number of cycles and $t_h$ is a hold period (part of the load cycle, where stresses and temperatures are constant). The indices j and k refer to a fatigue and creep loading state respectively. The parameter D characterizes damage. If the damage due to both types of loading can be superposed linearly, $D = 1$ at failure. If the creep damage and the fatigue damage interfere, D can be much smaller than one. For design applications, the value of D at failure can be selected in such a way that a safety factor is incorporated. Additional safety is reached due to the definition of $N_d$ and $t_d$: they are lower limits of scatterbands of large sets of data, the data set of $N_d$ contains fatigue tests with different cycle shapes, which are typical for the service loading. In principle, the contribution of $t_h/t_d$ is taken into consideration twice, because fatigue cycles with hold times have to be included within the data set $N_d$.

Nevertheless, a guarantee for a safe design based on the third group of rules can only be given, if all fatigue cycles occurring in service also occur in the data set $N_d$. Due to the enormous testing costs, this will hardly be the case in practice. So, as in the case of the second group of rules, there is a need for the proper definition of the limits of applicability.

The tasks of this paper are to define the share of creep deformation during a high temperature fatigue cycle and, using this result, to define the limits of applicability of the AC– rule of creep damage. Then a generalization of the AC–rule is developed. The new rule can be applied both in the creep dominated and in the creep fatigue interaction regime.

## EXPERIMENTAL DATA

In this paper, data of the Nickel–base superalloys Waspaloy (wrought), IN738LC (cast) and PM Astroloy are analyzed. The data were made available in the cooperate European research programs COST 50 and COST 501 (1–3,5,7). They are described in detail in (26). The data set which pertains to IN738LC is very large (it contains more than 150 fatigue (26) and about 70 creep data (27)). The fatigue data cover the whole range of loading conditions expected to occur in highly loaded gas turbine blades. Most of the data were measured at 850 °C. The fatigue data set contains experiments with all usual cycle shapes (e.g. stress and strain rate controlled cycles, symmetric and asymmetric cycles, cycles with several types of hold times, with and without a mean stress, with frequencies between $1 \cdot 10^{-4}$ Hz and $2 \cdot 10^{2}$ Hz as well as thermal and thermomechanical cycles). The number of cycles of failure ranges from $3 \cdot 10^{0}$ to $10^{8}$, the time of failure from $2 \cdot 10^{1}$ s to $5 \cdot 11^{6}$ s.

An important measure for the quality of a data set is its inherent scatter (scatter due to the material's inhomogeneities and experimental problems). It can be quantified using the logarithmic standard deviation, s. For a set of logarithmic normal distributed data, 68 % (96 %) of all data are within a range of $x \pm s$ around their mean value, x (or 96 % within a range of $x \pm 2s$). In the case of the IN738LC, the logarithmic standard deviation (due to the inherent scatter) of the creep data set is $s = 0.16$ (27) and that of the fatigue data set is $s \cong 0.20$ (26). The inherent scatter of the data measured on Waspaloy is similar, on PM Astroloy it is smaller ($s \sim 0.15$) (28).

## CREEP STRAIN ACCUMULATION DURING UNSTATIONARY LOADING CONDITIONS

In general, creep is measured at constant stress and temperature. Figure 1 shows a schematic sketch of a creep curve (a: strain over time and b: strain rate over time). The strain immediately after the loading ($\epsilon_0$) has an elastic and a plastic component. At sufficiently high temperatures, the specimen continues to deform when the load is applied. This deformation is called creep. In the primary creep region (I), the dislocations are rearranged to reach a steady state structure (there exists a steady state between strain hardening and recovery (29)), which stays constant during the secondary creep (II). In this region the dislocation structure is optimized to withstand the external loads; as a consequence the deformation rate is a minimum. The secondary creep is also called stationary, steady state or minimum creep. During creep deformation, creep damage (e.g. the occurrence and growth of pores) accumulates. At the beginning of the tertiary

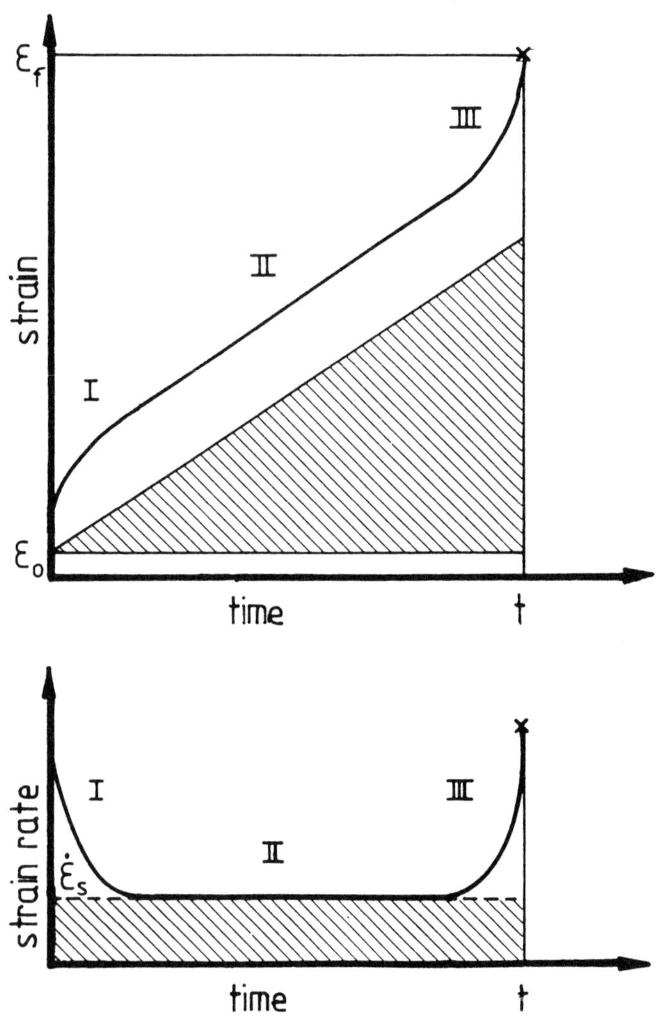


Fig. 1 – Schematic sketch of a creep curve. The symbols are explained in the text.

creep region, the damage has reached such a high level that it causes a measurable reduction of the load bearing cross section of the specimen. The effective stress in the undamaged material and – as a consequence – the creep rate increase.

The steady state creep rate ($\dot{\epsilon}_s$) only depends on the material's structure, the applied stress and the temperature, but not on the time after loading. For a general load situation, the ratio of the inelastic deformation rate, $\dot{\epsilon}_{in}(\sigma,T,t)$, to the corresponding steady state creep rate, $\dot{\epsilon}_s(\sigma,T)$, is used to quantify the deviation of the material from - the steady state conditions ($\sigma$: stress, T: temperature, t: time). The mean deviation from the steady state creep condition within a time interval ($\tau$) can be described using the parameter r (30):

$$r(\tau) = \frac{\epsilon_{in}}{\epsilon_s} = \frac{\int_0^{\tau} |\dot{\epsilon}_{in}| \cdot dt}{\int_0^{\tau} |\dot{\epsilon}_s| \cdot dt} = \frac{\int_0^{\tau} \left| \dot{\epsilon}_{in}\left[\sigma(t), T(t), t\right] \right| \cdot dt}{\int_0^{\tau} \left| \dot{\epsilon}_s \left[\sigma(t), T(t)\right] \right| \cdot dt} , \quad (6)$$

$\epsilon_{in}$ being the inelastic strain and $\epsilon_s$ the corresponding steady stata creep strain. In the following, the r–parameter is used to describe creep, stress rupture and fatigue test results.

**Creep tests.** In the time interval between loading and failure ($\tau = t_{f,c}$), the r–parameter yields: $r(t_{f,c}) = \epsilon_f/(\dot{\epsilon}_s \cdot t_{f,c}) \equiv r_0$. $\epsilon_f$ is the fracture strain. The steady state creep rate stays constant, because the external stress and the temperature are constant. $r_0$ equals the parameter used by Dobes and Milicka (31) to predict the rupture life of creep tests. In general, $r_0$ is between 1 and 5 (for the IN738LC creep data set, it is 2.7). It has been shown by Ashby and Dayson that $r_0$ depends on the working damage mechanism (32).

**Stress rupture tests.** These tests are performed at constant temperature and strain rate ($\dot{\epsilon}$). In a good approximation, $r(t_f) \cong \dot{\epsilon}/\dot{\epsilon}_s(\mathrm{UTS},T)$ , where $t_f$ is the time of failure and UTS is the ultimate tensile strength (30). Figure 2 shows the parameter $r(t_f)$ for a hot working steel and a cast superalloy depending on the test temperature. It can clearly be seen that, for increasing temperature, $r(t_f) \to 1$ . At very high temperatures (if $r(t_f) \sim 1$), almost all inelastic deformation is steady state creep.

**Fatigue tests.** Figure 3 shows some fatigue test results measured on IN738LC at 850° C (26). The fatigue cycles shown are of the symmetrical saw tooth type with a test frequency of (about) 1 Hz and with a mean stress of (nearly) zero. During this type of test, the stress changes at any time. The parameter r is calculated for the tensile part ($\tau_t$) of a typical cycle. If the stress amplitude increases, $r(\tau_t) \equiv r_t \to 1$. At very high amplitudes almost all inelastic deformation is steady state creep. The same holds true with increasing test temperature, mean stress or cycle period.

**Conclusions.** The parameter r can be used to describe the high temperature deformation behavior of materials. There exists a regime of loading conditions, where the inelastic deformation is steady state creep. This happens, if the mobility of the dislocations is sufficiently large to allow an immediate adjusting of the dislocation structure on the external load changes. Then, the dislocation structure is always near its steady state structure and the inelastic deformation rate is always near the steady state creep rate. This situation is promoted by loading situations which cause a high mobility of dislocations (high temperatures, high stresses), or which cause slow load changes (low test frequencies, low stress or strain rates, high cycle periods, hold times, etc.).

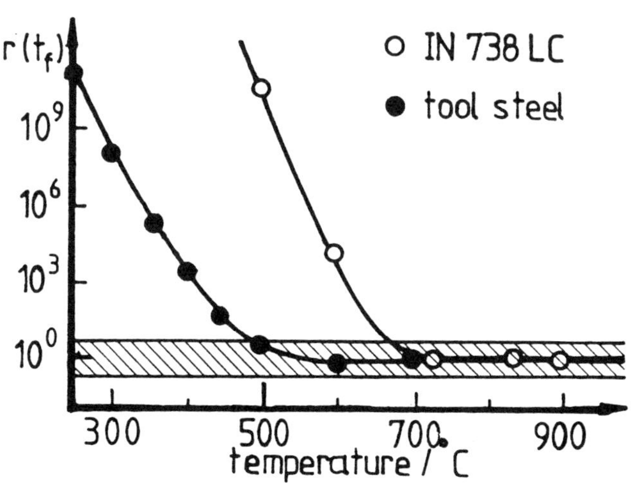


Fig. 2 – High temperature stress rupture test results.

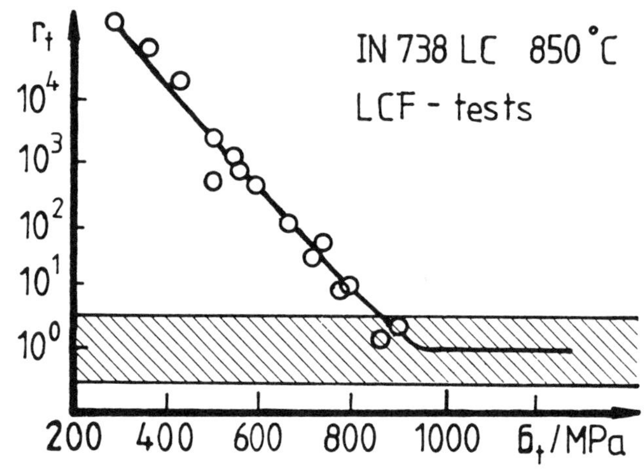


Fig. 3 – High temperature low cycle fatigue test results. At high stress amplitudes, almost all inelastic strain is steady state creep ($r \sim 1$).

## LINEAR LIFE FRACTION RULE FOR CREEP DAMAGE

It has been mentioned earlier that the linear life fraction rule for creep damage (AC) can only be used to describe fatigue tests with a sufficiently high mean stress (20), a sufficiently high temperature (21) or with a sufficiently low test frequency (21). On the one hand these arguments are reasonable in the light of the above arguments, because steady state creep is promoted under the proposed conditions. On the other hand the arguments are not helpful, because the limit of applicability of the rule is not quantified. Therefore, in an earlier paper, the author has proposed (23) to use the r–parameter in order to decide whether or not the AC–rule is applicable: the AC–rule is applicable if $r_t \sim r_0$ (in practice, the rule can be used for $1 < r_t < 10$). Then, the mean deformation state during the fatigue test is comparable to that occurring during the creep test, and it is assumed that the same kind of damage occurs in both types of test.

Figure 4 shows creep and fatigue test results of IN738LC measured at 850° C. In the case of the creep data (shaded scatter band), the applied stress is plotted over the time of failure. The fatigue data are marked by numbers which indicate the order of magnitude of the r–parameter: e.g. the number "3" indicates that $10^3 \leq r < 10^4$. The maximum tensile stress, $\sigma_t$, (measured in a mean fatigue

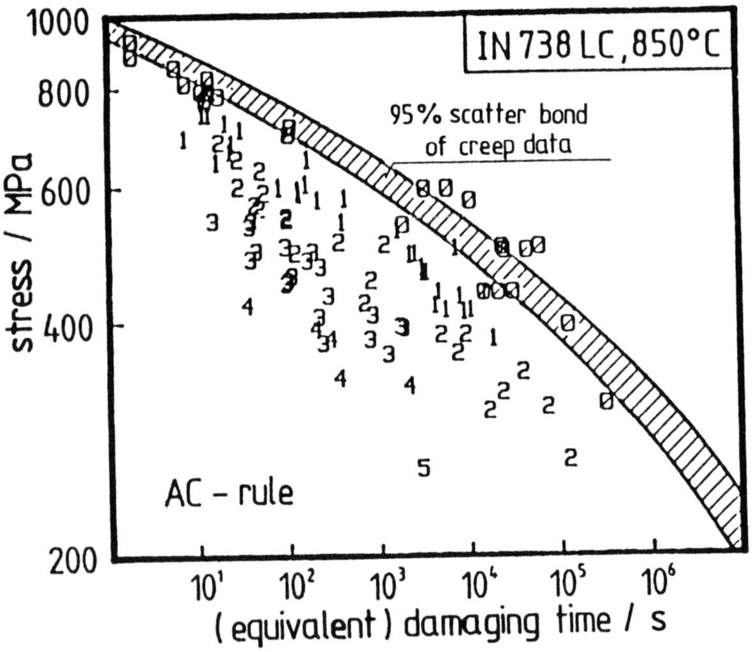


Fig. 4 – Tests on IN738LC at 850° C. Shaded band: Proof stress of creep data against time of failure. The numbers refer to fatigue tests (see text). The maximum tensile stress in the cycle is plotted over the equivalent time of failure, $\tau_{eq} \cdot N_{f,meas}$ (corresponding to the AC–rule).

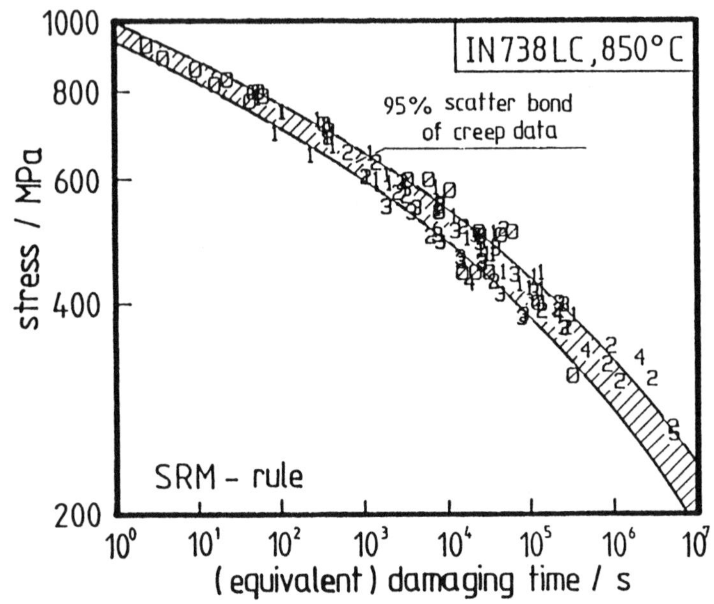


Fig. 6 – Same data as in Figure 4. The equivalent time of failure is defined according to the SRM–rule).

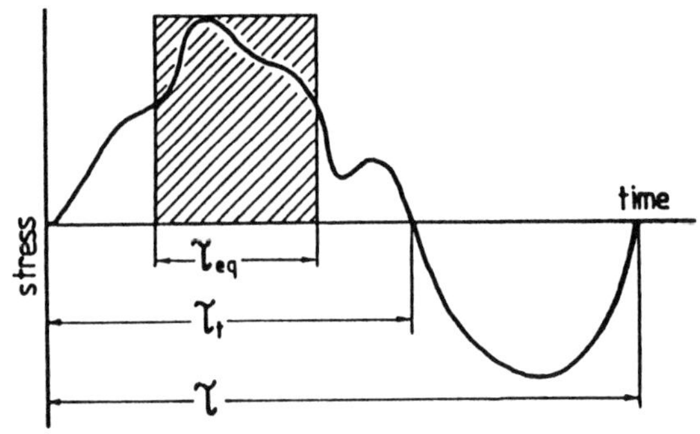


Fig. 5 – Schematic sketch of the equivalent damaging cycle period.

cycle) is plotted against the equivalent damaging time, $t_{f,eq} = \tau_{eq} \cdot N_f$, (according to the AC–rule). Within the equivalent damaging time, the same amount of creep damage occurs that would accumulate in a creept specimen (loaded at the stress $\sigma_t$ and at the temperature $T_m$; $T_m$: maximum temperature in the fatigue cycle) as it accumulates in the fatigue test:

$$t_{f,eq} = t_{f,c}(\sigma_t, T_m) \quad . \tag{7}$$

The equivalent damaging cycle period is: $\tau_{eq} = t_{f,eq}/N_f$ . Its definition is shown in Figure 5. Using eq. 4 for $N_f$ , we get:

$$\tau_{eq} = t_{f,c}(\sigma_t,T_m) \cdot \left[ \int_0^{\tau} \frac{dt}{t_{f,c}(\sigma,T)} \right]^{-1} \quad . \tag{8}$$

Figure 4 shows that all fatigue data symbolized by "0" coincide with the creep data. This indicates that, for these data, the AC– rule describes the cyclic life time of fatigue tests fairly well: It follows that the linear life fraction rule for creep damage can be used, if a reasonable part of the inelastic deformation is steady state creep. One may expect that the life time calculated with the AC–rule only coincides with the life time of creep data, if the same deformation state occurs in both types of tests. There–

fore, the condition for the validity of the AC–rule is:

$$r_t = r_0 \quad . \qquad (9)$$

In practice, the AC–rule can be applied, if $1 < r_t < 10$.

For the other fatigue data, it holds: $r_t >> r_0$. Then, the AC–rule systematically overestimates the time of failure of fatigue tests (see Figure 4). The same behavior is observed in the other materials.

Table 1 summarizes the results of fatigue tests with $r < 10$. $n_0$ is the number of data, s the logarithmic standard deviation and $d_{max}$ is the maximum of $N_{f,cal}/N_{f,meas}$ or its inverse ($N_{f,cal}$ calculated life time) for the data set under investigation.

Only a small part of all data can be described by means of the AC–rule (in the other cases, $r_t > 10$; i.e. the stress changes were to quick or the temperature was to low to allow a quick adjustment of the dislocation structure to the new stress state). No parameter adjustment is necessary. Life time can be calculated, if the stress and temperature history of a typical cycle is known. If the strain history is also known, the r–parameter, which is used to define the limits of applicability of the rule, can also be calculated (for special loading conditions, life time as well as r can be calculated from the stress and strain amplitudes, (7, 33)). The quality of prediction is good, but s and $d_{max}$ are clearly higher than the values expected due to the inherent scatter of the data sets, because in each experiment the parameter $r_t$ was (slightly) different from $r_0$. Nevertheless, prediction quality is sufficient.

**Summary.** If most of the inelastic deformation during a fatigue test is steady state creep, fatigue life time can be calculated by the simple linear life fraction rule for creep damage. This situation is promoted by high temperatures, stresses or low deformation rates and characterized by $r_t \sim r_0$. If the r–parameter is much higher ($r_t >> r_0$), the AC–rule overestimates fatigue life time (e.g. in Figure 4, the test marked by the symbol "5" has a measured life time, which is less then 1/100 of that calculated by the AC–rule). The parameter r can easily be calculated, if the stress, strain and temperature history of the fatigue cycle is known.

**Table 1:** AC–rule: statistical data to describe the quality of life prediction

| **Material** | **$n_0$** | **s** | **$d_{max}$** |
|---|---|---|---|
| Waspaloy | 2 | 0.41 | 2.3 |
| PM Astroloy | 7 | 0.07 | 1.3 |
| IN738LC | 30 | 0.39 | 4.9 |

## A LIFE TIME PREDICTION RULE FOR THE CREEP FATIGUE INTERACTION REGIME

As discussed in the introduction, there exists no well founded life time rule for the loading regime, where creep and fatigue damage may interact. The systematical overestimation of fatigue life time by the AC–rule with increasing r–parameter (see Figure 4; the same occurs in the other investigated materials) is used to fill this gap (7, 33): eq. 4 is modified in such a way that, for data with a high value of $r_t$, the calculated life time is reduced amd for data with $r_t = r_0$, the calculated life time is equal to the AC–life time:

$$N_{f,srm} = \left[ \int_0^{\mathcal{T}} \frac{1}{t_{f,c}(\sigma,T)} \cdot \left[ \frac{\dot{\epsilon}_{in}(\sigma,T,t)}{r_0 \cdot \dot{\epsilon}_s(\sigma,T)} \right]^{\nu} dt \right]^{-1} , \qquad (10)$$

$\nu$ is a material parameter.

The rule is called strain rate modified accumulation of creep damage, (SRM–rule; 7,33). Of course, stress and temperature may be time dependent. For many cycle shapes (e.g. saw tough cycles and for sinoidal cycles, see (7,33), eq. 10 can be approximated by:

$$N_f \approx N_{f,ac} \cdot \left[ \frac{r(\tau_t)}{r_0} \right]^{-\nu} \quad . \qquad (11)$$

Figure 6 shows the same data as Figure 4, but the equivalent damaging time is defined according to the SRM–rule. For most of the fatigue data, eq. 10 is approximated by eq. 11. Now, all creep and all fatigue data coincide, which indicates a good prediction quality of this rule.

In order to characterize the prediction quality of the SRM–rule, the fatigue test results of each material were divided into two subsets. The first set, which was selected as small as possible (it contains between 3 and 5 fatigue test results, depending on the material), is used to adjust the material parameter $\nu$. For all materials characterized so far, $\varkappa$ is between 0.5 and 0.8 (for IN738Lc, it is 0.71). Then the fatigue life time of the experiments, which are in the second subset, has been calculated using this parameter. It should be mentionied that the set for IN738LC also contains data of experiments performed at several different (but high) temperatures and under thermomechanical conditions. The results are summarized in Table 2.

**Table 2:** SRM–rule: statistical data to describe the quality of life prediction

| **Material** | $n_0$ | s | $d_{max}$ |
|---|---|---|---|
| Waspaloy | 29 | 0.28 | 4.3 |
| PM Astroloy | 9 | 0.18 | 2.0 |
| IN738LC | 140 | 0.28 | 5.2 |

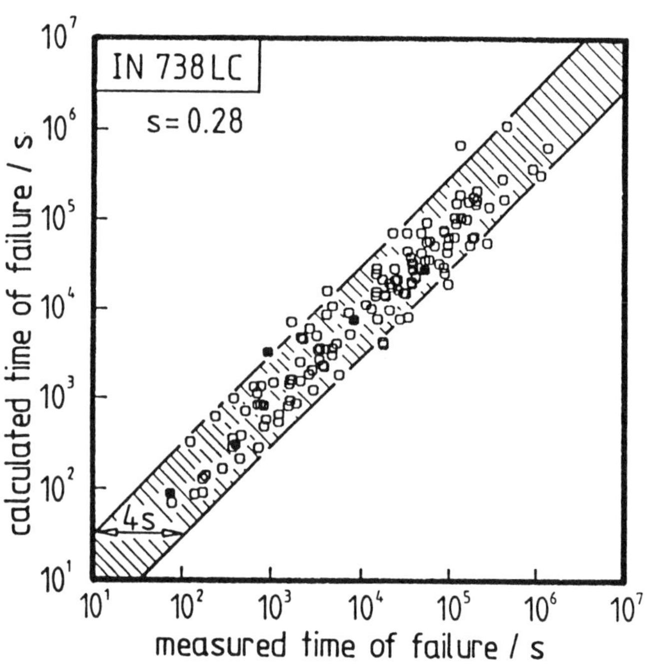


Fig. 7 – Calculated (according to the SRM–rule) against the measured life time for fatigue measurements on IN738LC. Full symbols: data used to adjust the parameter $\nu$; open symbols: these data are calculated without any further parameter adjustment.

For all three materials, the scatter of the data only slightly exceeds the inherent scatter of data sets. The same holds for the parameter $d_{max}$. Its values equal those expected for logarithmic normal distributed data due to the standard deviation and due to the number of experiments. Both show that the prediction quality of the SRM–rule is excellent. This is also shown in Figure 7, which compares,for IN738LC, the SRM–life time and the measured life time. In addition to the data shown in Figure 6, the data set contains tests performed at several different test temperatures, thermal fatigue tests and thermomechanical fatigue tests.

## LIFE TIME PREDICTION

An important reason for the development of life time rules is the intention to reduce the number of necessary experiments, to reduce the necessary testing time and, moreover, to reduce costs. In the case of the SRM–rule, this is done by replacing expensive fatigue tests by relatively cheep creep tests and a few short time fatigue experiments.

Figure 8 shows an evaluation of eq. 11 for some fatigue data measured on IN738LC. The parameter $r_t$ is plotted over $N_{f,ac}/N_{f,meas}$ in a double logarithmic scale. $N_{f,meas}$ is the measured number of cycles of failure. In this diagramm, eq. 11 is represented by a straight line, the parameter $\nu$ beeing the slope of the line. A point at the left hand side of the line is given by the condition: $N_{f,ac}/N_{f,meas} = 1$ at $r_t = r_0$. As mentioned earlier, $r_0$ is determined by creep tests. A second point of the line at the right hand side of the diagramm can be determined using a few fatigue data with a high $r_t$–parameter. Favourably, these tests can be performed at high frequencies (they should be far away from steady state conditions) and do not last for a very long time. So far, the experimental experience is restricted to the loading regime characterized by: $1 < r_t < 10^6$. Therefore, the application of the SRM–rule should be restricted to this regime.

The SRM–rule interpolates fatigue data with $r_t$–values between these two limits. Fortunately, the life time of fatigue data with a low or an intermediate $r_t$–value may be much higher than the life time of high frequency fatigue tests with a high $r_t$– value. Then, an interpolation in $r_t$ can lead to an extrapolation in the life time. This is shown in Figure 9 for the fatigue data of PM Astroloy. The full data points were used for parameter adjustment. Open data are calculated using

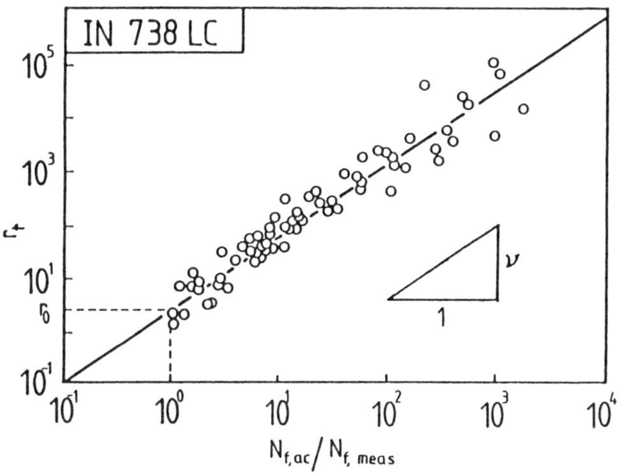


Fig. 8 – Fatigue tests on IN738LC at temperatures between 750°C and 850°C. The parameter $r_t$ is plotted against $N_{f,ac}/N_{f,meas}$ in a double logarithmic scale. The straight line corresponds to the SRM–rule, eq. 11.

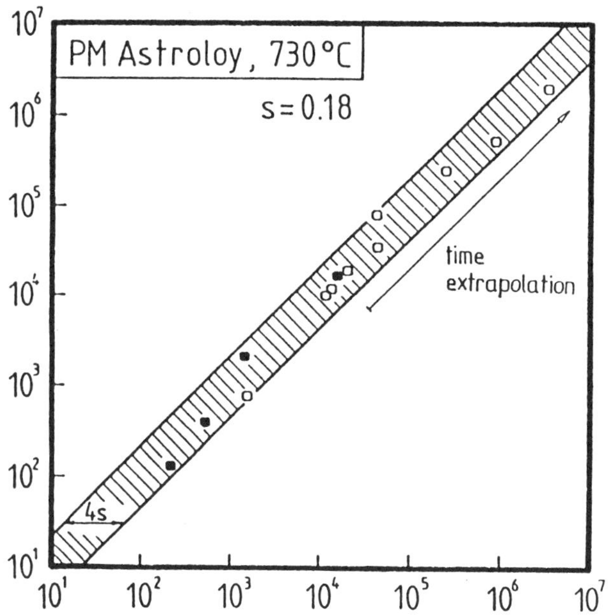


Fig. 9 – Fatigue data of PM Astroloy at 730°C. The plot shows the calculated over the measured life time. Data marked by full symbols are used for parameter adjustment, data plotted with open symbols are calculated without any further parameter adjustment.

this value of the parameter $\nu$. The life time of the longest test, which is predicted using this procedure, is 220–times longer than the life time of the longest fatigue test used for parameter adjustment. The scatter of the data used for parameter adjustmen2t is equal to the scatter of the extrapolated data.

## CONCLUDING REMARKS

The r–parameter can be used to quantify the share of steady state creep in the inelastic deformation of a material. For any general loading situation, the r–parameter can easily be calculated if the steady state creep rate (depending on stress and temperature) of the material and the histories of stress, strain and temperature of the fatigue test are known. For typical loading situations the input information necessary to calculate the value of r can be reduced to a large extent. For symmetrical LCF–tests for example, r can approximately be calculated, if only the stress and the strain amplitude of the LCF–cycle are known (7,33). There are regions of loading parameters, where all inelastic deformation is steady state creep. This material behavior is promoted by high temperatures, stresses and low stress rates. The situation is characterized by the condition: $r \sim 1$.

If $r_t \sim r_0$, the deformation state is similar to the deformation state in a creep test. Then the simple linear life fraction rule of creep damage (AC–rule) can be used to calculate fatigue life time. In the other loading regimes, the r– parameter is higher (or much higher) than $r_0$ and the AC–rule systematically overestimates the cyclic life time. This behavior is used to define a new life time prediction rule (called SRM–rule), which is applicable in the creep fatigue interaction regime characterized by the condition: $1 < r < 10^6$. The range of applicability is given by this condition.

To apply the SRM–rule, the creep data of the material, the parameter $\nu$ (determined by a few high–temperature high–frequency fatigue data) and the stress and strain histories of the fatigue cycle have to be known. Again, for typical loading cycles, an approximate calculation of life time can be performed, if only a few characteristic points of the cycle, e.g. stress and strain amplitudes, are known.

So far, experience with the application of this rule is restricted to a few superalloys and, up to the gain of further experimental experience with other materials, the application should be restricted to this material class.

## REFERENCES


1. Bachelet E. ed., "High Temperature Materials for Power Engineering 1990", Kluwer Academic Publ., Dordrecht, The Netherlands, 1990
2. Betz W. ed., "High Temperature Alloys for Gas Turbines and other Applications 1986", D. Reidel Publ. Comp., Dordrecht, The Netherlands, 1986
3. Coutsouradis D. ed., "High Temperature Alloys for Gas Turbines", Appl. Science Ltd., London, UK, 1978
4. Reuchet J. and L. Remy, Met. Trans. 14A, 141–149, 1983
5. Fischmeister H., R. Danzer and B. Buchmayr, in (2), 495–549, 1986
6. Francois M. and Remy L., Fatigue Fract. Engng. Mater. Struct., 14, 115–129, 1991
7. Danzer R., "Lebensdauerprognose hochfester metallischer Werkstoffe bei hoher Temperatur", Gebrüder Borntraeger, Berlin, Germany, 1988
8. Bressers J., W. Hoffelner and L. Remy, in (2), 441–468, 1986
9 Coffin Jun. L.F., Trans. ASME 76, 931–950, 1954
10. Manson S.S., NACA TN–2933, Cleveland, Ohio, USA, 1953
11. Coffin Jun. L.F., Proc. Air Force Conf. on Fatigue anf Fracture of Aircraft Structures and Materials, AFFDL TR 70–144, 301–311, 1970
12. Coffin Jun. L.F., MPC–3, ASME, New York, USA, 349, 1976
13. Ostergren W.J., Journal of Testing and Evaluation, 4, 327–339, 1976
14. Manson S.S., G.R. Halford and A.J. Nachtigall, NASA TM X–71737, San Francisco, USA, 1975
15. AGARD Conf. Proc., 243, "Characterisation of Low Cycle High Temperature Fatigue by Strain Range Partitioning Method", Aalborg, Denmark, ISBN 92–835–0220–5, 1978
16. Nat. Mater. Advis. Board, Report NMAB–347, Natl. Acad. Sci., Washington, USA, 1980
17. Riedel H., "Fracture at High Temperatures", Springer Verlag, Berlin, Germany, 1986
18. Robinson E.L., Trans. ASME, 60, 253–259, 1938
19. Robinson E.L., Trans. ASME, 74, 777–780, 1952
20. Taira S., in "Creep in Structures", N.J. Hoff ed., Springer Verlag, Berlin, Germany, 96–119, 1962
21. Spera D., Thesis, Univ. of Wisconsin, USA, 1968
22. Spera D.A., NASA TN D–5317, 1969
23. Danzer R., Res Mechanica, 13, 63–76, 1985
24. Nickel H., F. Schubert, H.J. Penkalla and G. Breitbach, Int. J. Pres. Ves. and Piping, 47, 167–192, 1991
25. ASME Boiler and Pressure Vessel Code, Case–47–15, 1979
26. Danzer R., "Life Time Prediction of the Gas Turbine Materials PM Astroloy, MAR M–509, Waspaloy and IN738LC in the Creep Fatigue Interaction Regime", Internal Report, Inst. f. Metallkunde, Montanuniversität Leoben, 1985
27. Buchmayr B., Thesis, Montanuniversität Leoben, Austria, 1983
28. Danzer R. and J. Bressers, Fatigue Fract. Engng. Mater. Struct., 9, 151–168, 1986
29. Ilschner B., "Hochtemperatur– Plastizität", Springer Verlag, Berlin, Germany, 1973
30. Danzer R., Res Mechanica, 12, 259–273, 1984
31. Dobes F. and K. Milicka, Metal Science, 10, 382–384, 1976
32. Ashby M.F. and B.S. Dayson, in "Advances in Fracture Research (Fracture 84)", Valluri S.R. ed., Pergamon Press, Oxford, UK, 3, 1984
33. Danzer R., Z. Metallkunde, 78, 19–31, 1987

*Proceedings of the Fifth International Conference on Creep of Materials, Lake Buena Vista, Florida, USA, 18-21 May 1992*

# Stochastic Nature on Initiation and Early Growth of Intergranular Creep-Fatigue Cracking

**R. Ohtani, T. Kitamura, and N. Tada**
Kyoto University
Kyoto, Japan

ABSTRACT

The spatial and temporal scatter of the initiation and early grwth of small cracks is attributed to the random distribution of fracture resistance and driving force f or each crack caused by the microstructural inhomogeneity of the material.An intergranular surface cracking is a typical mode of creep fatigue fracture in polycrystalline steels and allys at elevated temperatures. A small crack of the size of about one grain boundary facet initiates at a very early stage of fatigue, following the increase in its number and length. The experimental observation conducted mainly on a Type 304 stainless steel were summerized to give evidence of stochastic nature on the crack initiation and the early growth. Then, the numerical analysis, which followed a Monte Carlo method based on a simple statistical model of combined damage and fracture mechanics, was performed for characterizing the process of creep fatigue cracking. Emphasis was put on the parameter, Ko /Fo, which was the ratio of the driving force for crack growth to that for crack initiation, in order to represent the state of cracking.

AN ATTEMPT TO DERIVE LIFE LAWS of high-temperature low-cycle fatigue in smooth specimens has been made on the basis of the fracture mechanics equations on crack growth rates(1,2). The effect of time or cycle dependence, stress or strain waveforms, and stress and elastic strain contribution on the failure life of creep fatigue was well described qualitatively. But the failure laws derived were not always sufficient for the quantitative and accurate estimation. This was supposed to be due to the lack of consideration of both the to crack initiation and the spatial and temporal scatter of small crack initiation and growth.

The investigation on small cracks was started on this this account, yet it has been taking much time and trouble. The measurement of the number of cracks during creep fatigue tests gives us information about stochastic characteristics on crack initiation; the distribution of time to crack initiation and fracture resistance of grain boundaries,the effect of grain boundary inclination to the stress axis, the effect of strain rates and strain waveforms and so on. The measurement of each crack length and its increase with increasing the number of fatigue cycl es shows evidence of stochastic nature of each crack growth; the distribution of crack growth rate, the difference in the driving force between crack initiation and crack growth, the behavior of discrete crack growth, the frequency of crack arrest and coalescence and so on.

The results clarified by the experiments were analyzed as one of the inverse problems by using numerical simulation. Characterization of the process of creep fatigue failurer, classification of the fracture mode, and signification of the failure life are given in relation to the cracking.

## MAIN FEATURES OF EXPERIMENTAL RESULTS

Experimental results on some low alloy steels, austenitic stainless steels and Ni- and Ti-base allys are summerized as follows.

[1] Integranular surface cracks initiate at the unit of about one grain boundary facet as a result of the growth and coarescence of creep cavities in most polycrystalline steels and alloys with relatively fine grains (3) as shown in Figure 1.

[2] Surface cracks initiate at an early stage of fatigue; less than 10% of the failure life, and increase in the number in near proportion to the number of cycles (4,5), as shown in Figure 2.

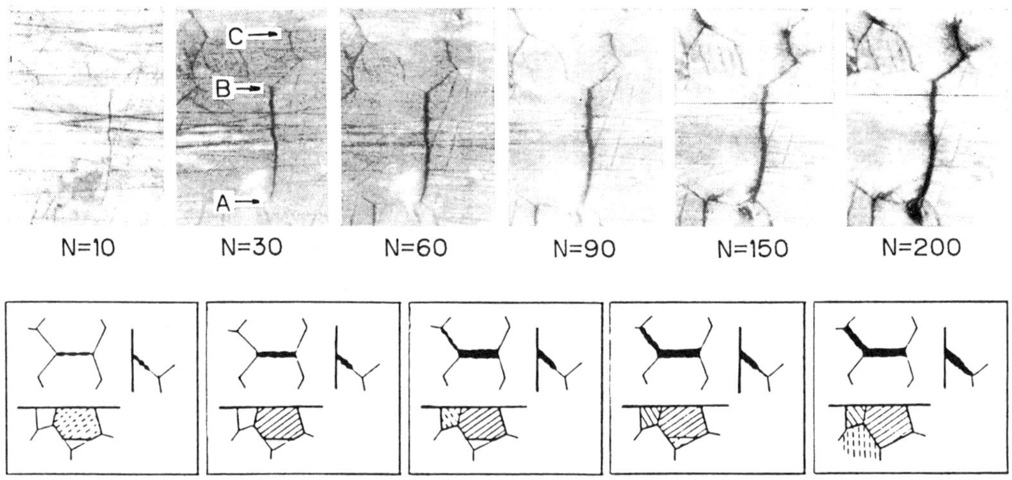


Fig. 1. *In situ* observation of a surface crack initiating at the grain boundary AB and growing to the adjoining ones BC in the high temperature microscope with servo-hydraulic fatigue testing equipment, Type 304 Stainless steel at 600℃ in a vacuum, and schematic representation of the cracking.

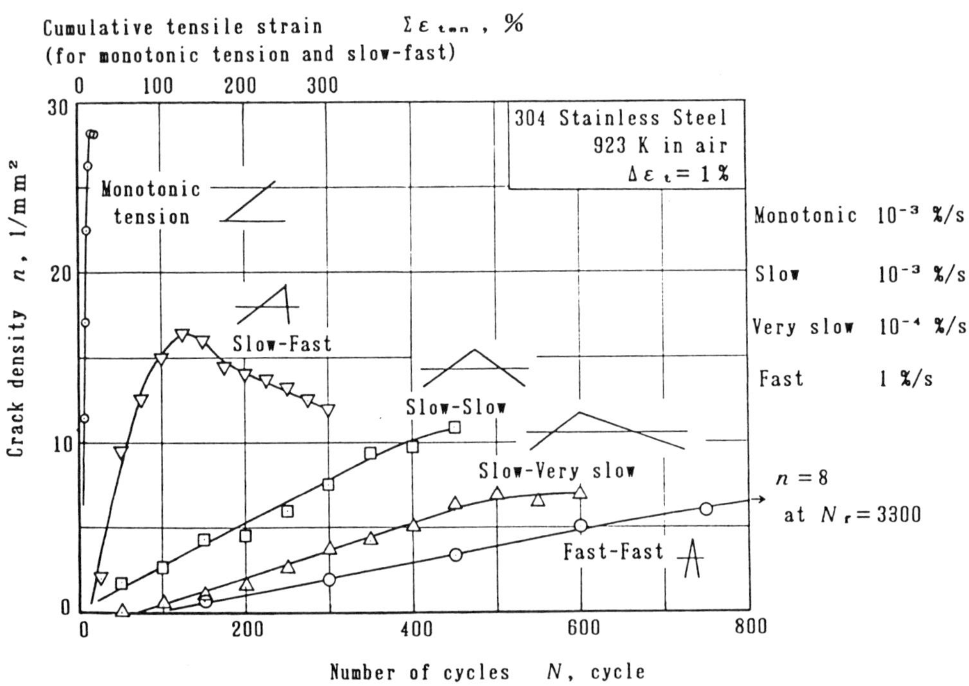


Fig. 2. Effect of strain waveform on the increase in the number of surface cracks.

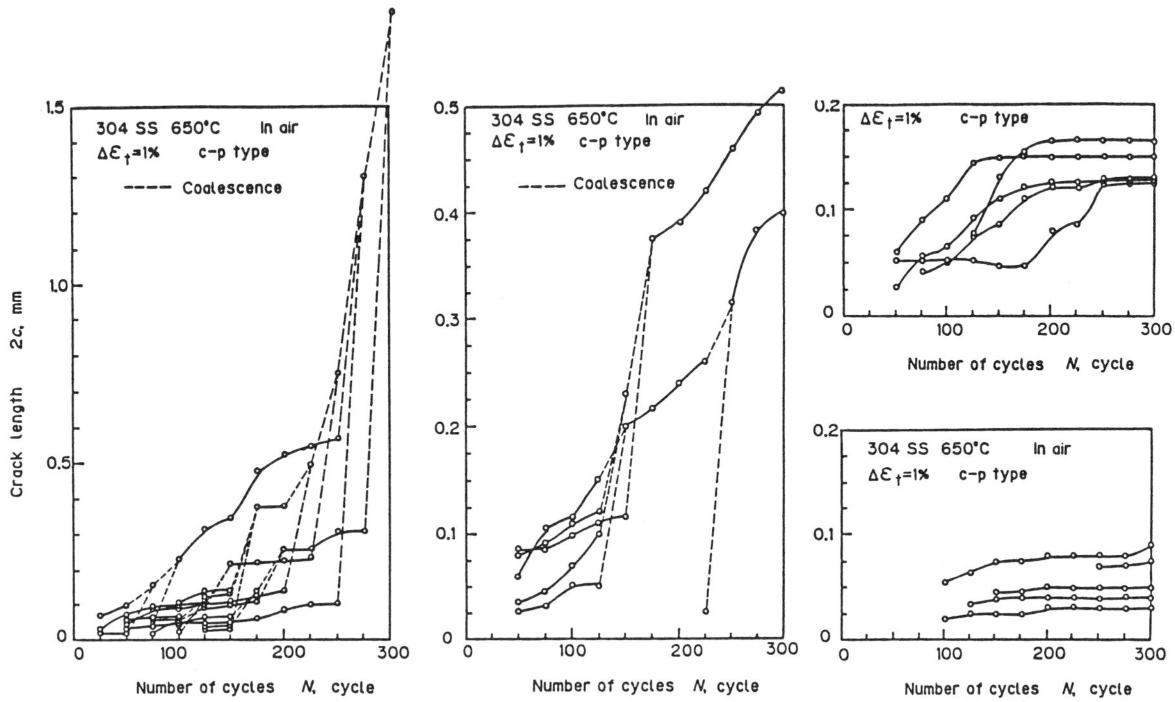


Fig.3. Growth curves and retardation curves of small surface cracks.

[3] The early growth is arrested at the grain boundary triple junction for most of small cracks, and the rest grows discretely as single cracks or with the coalescence of adjoining cracks (3-6) as shown in Figure 3.

[4] The growth rate of "microstructurally small crakcs" has a wide distribution. The upper bound is more than ten times as fast as the extraporated rate of large cracks (4-7) as shown in Figures 4 and 5.

[5] The reterdation of crack growth disappears and the rate distribution tends to converge when the half crack length reaches about 100$\mu$m or two or three grain diameters as shown in Figures 4 and 5.Cracks over 500$\mu$m half length ( ten grain diameters ) can be regarded as "mechanically large cracks" which act on the deterministic fracture law (8).

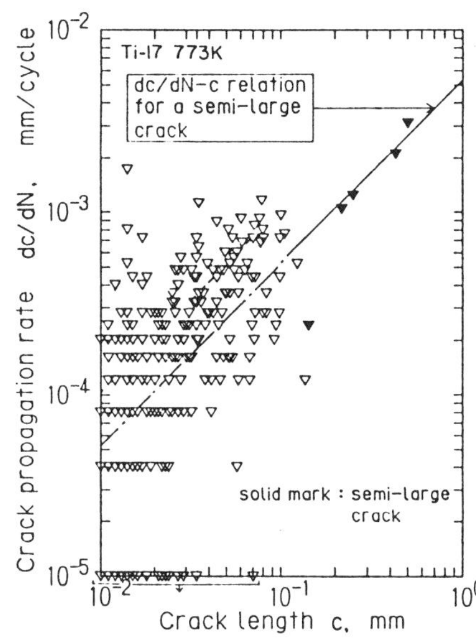


Fig.4. Scatter of crack growth rates.

[6] The environment or the surface oxidation has a strong effect on crack initiation, while it has little on crack growth (4,5), as shown in Figure 5. The inclination of grain boundary cracks is the same; the crack being perpendicular to the applied stress tends to be preferential as the crack length increases (9,10) as shown in Figure 6. Moreover, the compressive strain rate in the reverse cycle affects the crack initiation, although it hardly does the crack growth (11-13) as shown in Figures 2 and 5.

[7] In the case of very high test temperatures and/or very slow strain rates, cracks initiate uniformly inside the material. Such inner cavitation and cracking are apt to yield the acceleration of crack growth and the

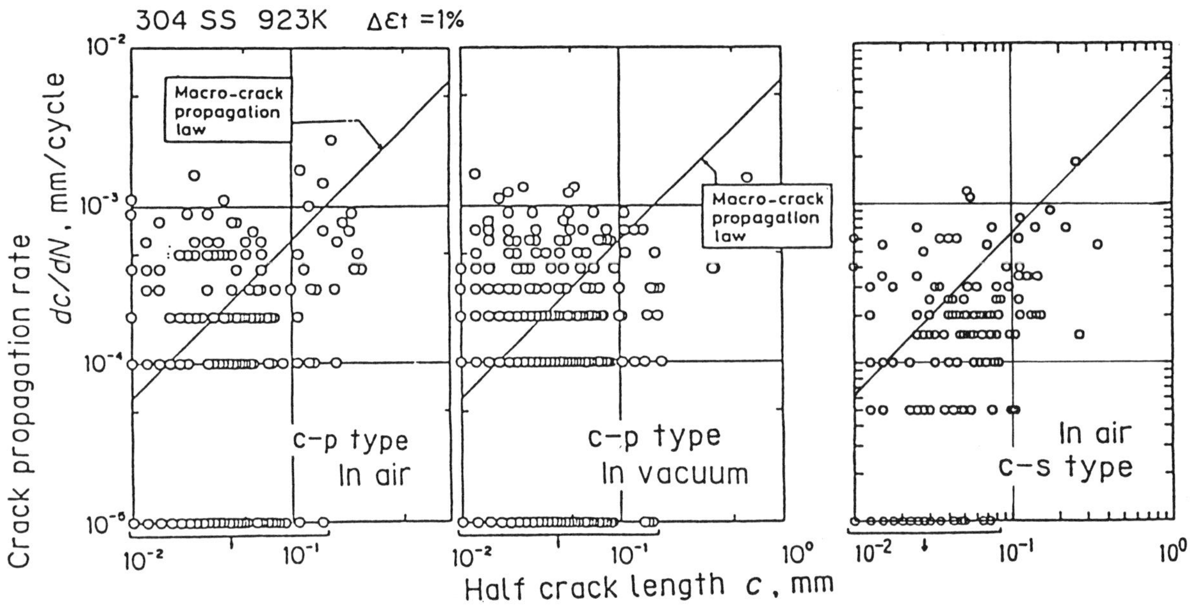


Fig.5. Effect of environment and strain waveform on the crack growth rate.

shortening of fatigue life (12,13). Discussion on this type of fracture will not be made in the present study.

Figure 7 shows schematic representation of the mode and process of cracking for three typical cases of fatigue loading.

MODELS AND METHODS OF NUMERICAL SIMULATION

Quasi-Three Dimensional Model. In reflection to the observation, a method of numerical simulation of surface cracks along grain boundaries which was shown in Figure 7(b), is proposed. General description is first made on a quasi-three dimensional model of grain boundaries which have their own fracture resistance and are subjected to the driving force corresponding to the external loading (10, 14-16).

First, grain boundaries at the surface of a polycrystalline material are numerically constructed by means of a computer on the basis of an isotropic grain growth model (9,15). As shown in Figure 8, nuclei of grains are generated at random points and time in a fixed surface area, and they are expanded as circles at a constant speed until the front of two adjacent nuclei meets, where a grain boundary is formed. Appropriate random variables are chosen for the numerical simulation so as to

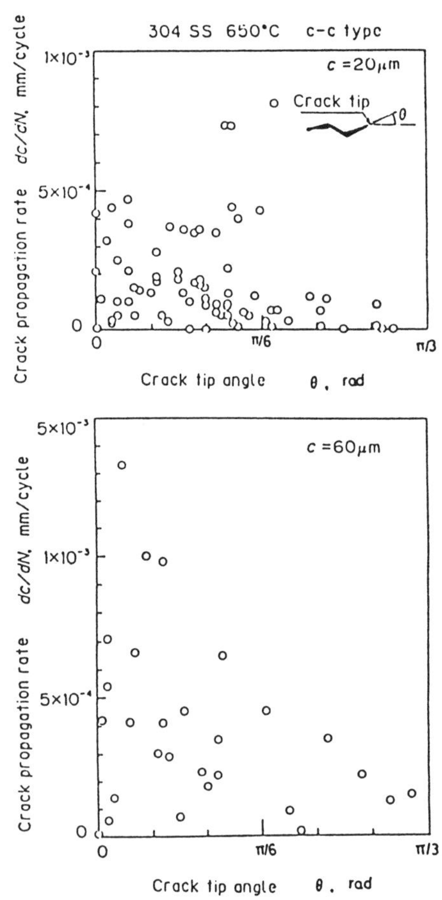


Fig.6. Effect of crack-tip inclined angle on the crack growth rate.

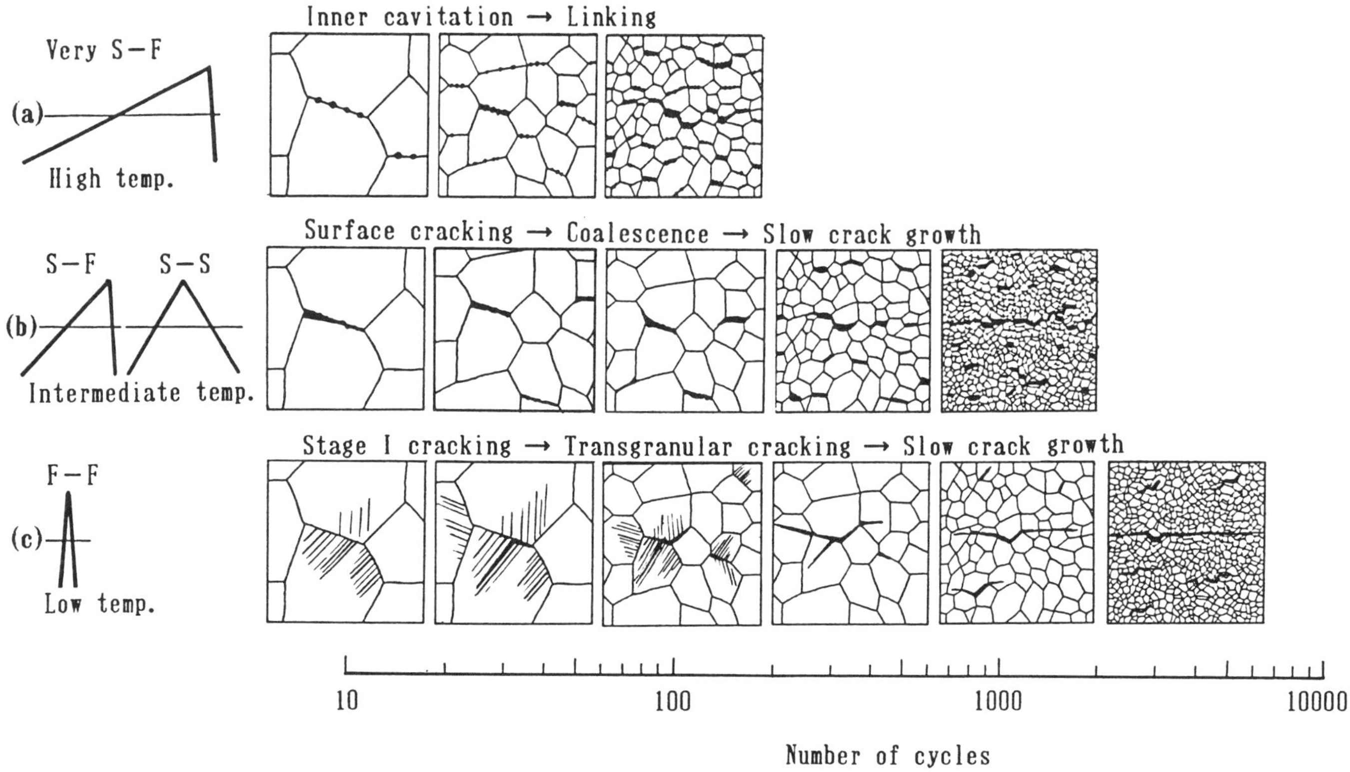


Fig.7. Classification of the mode and process of cracking in relation to the fatigue loading.

fit the the mean grain size and the inclined angle of grain boundaries, $\theta g$, statistically with the actual ones. The cumulative probability of the surface inclined angle, $F(\theta g)$, is expressed as

$$F(\theta g) = (2/\pi)\,\theta g, \tag{1}$$

which was confirmed to agree well with the numerical one(9). The grain boundary structure thus obtained is a two dimensional one on the material surface (x-z plane). Accordingly, the inward inclination (to y-direction) of the grain boundaries only appearing at the surface is given to each boundary. The angle, $\theta g'$, is expressed statistically as a cumulative probability function, $H(\theta g')$, as follows (9,15).

$$H(\theta g') = \sin\,\theta g'. \tag{2}$$

Here, $\theta g$ and $\theta g'$ are measured as angles to the plane perpendicular to the applied stress axis (z-plane).

Second, it is assumed that each grain boundary has its own fracture resistance, R,of individually different magnitude.The magnitude depends on many factors such that

R=R(grain boundary structures and properties,

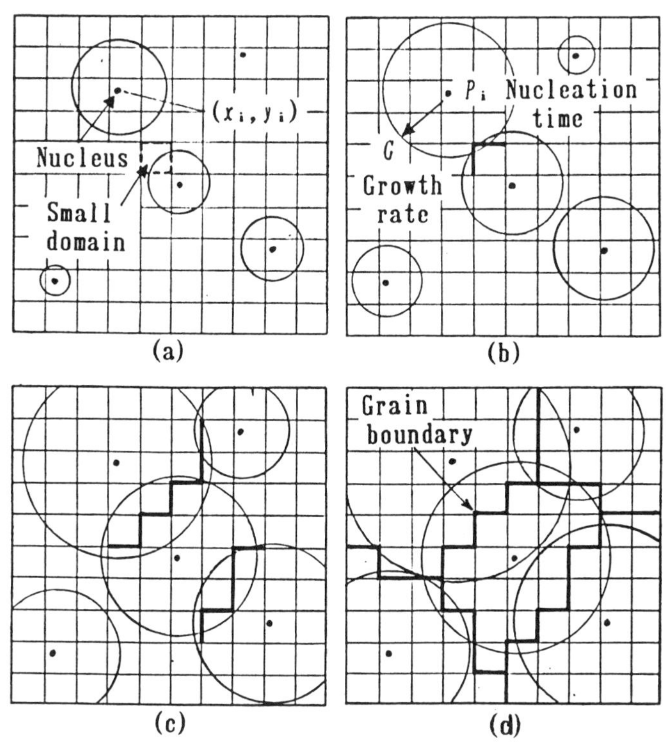


Fig.8. Computational procedure of two dimensional, isotropic grain boundary formation.

constraint of surrounding grains,
surface conditions and environments,
heat treatment and aging,
precipitation and degradation,
etc.). (3)

Thirdly, the driving force for fracture, F, is imposed on each grain boundary in every time-interval or every fatigue-cycle as shown in Figure 9. Although the driving force is essentially stochastic variables, it is handled as a deterministic variable by imposing the random effect only on the fracture resistance, R, because it has been clarified that this

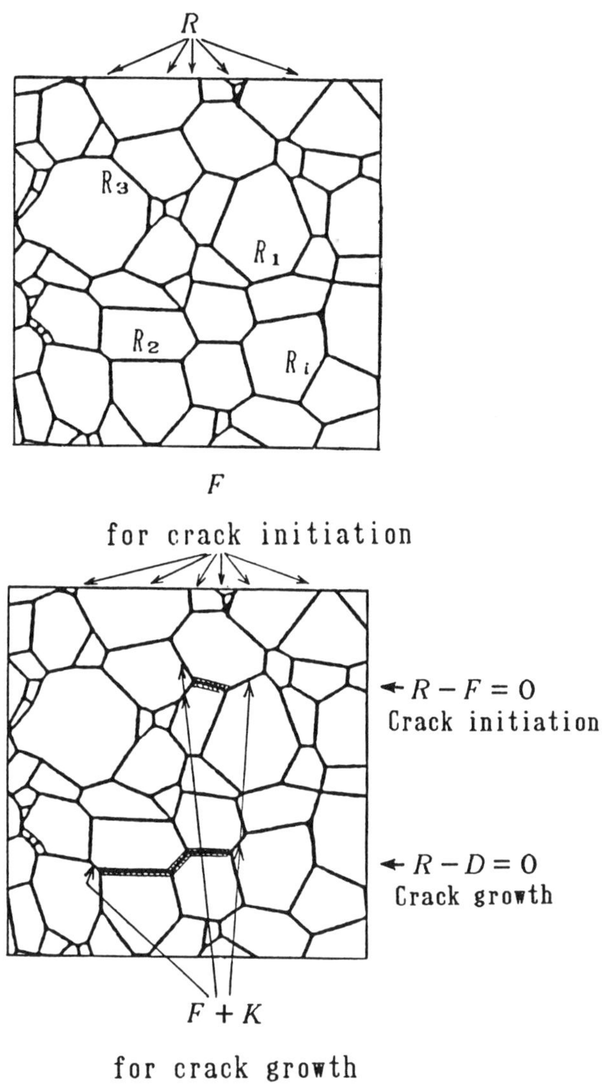


**Fig.9. A part of simulated grain structure and description of computational procedure for giving the fracture resistance, *R*, and the driving forces, *F*, to all grain boundaries and *F+K*, to crack-tip grain boundaries.**

treatment vertually gives the identical result with that of the simulation in which both R and F are stochastic (17). Then, the force, F, is assumed to be a function of the normal stress acting to each boundary, $\sigma n$, as

$$F = Fo \cdot (\sigma n / \sigma)^m \ . \quad (4)$$

Here,

$$\sigma n = \sigma / (\tan^2 \theta g + \tan^2 \theta g' + 1), \quad (5)$$

and

$$m = n, \quad (6)$$

where n is the stress exponent of creep rate (15). Fo in Eq.(4) is a main part of the driving force given as a function of the applied tensile stress, $\sigma$, and the tensile creep rate, $\dot{\varepsilon}_c$, as (7)

$$Fo = Fo(\sigma, \dot{\varepsilon}_c) = k_1 \cdot (\sigma \dot{\varepsilon}_c)^2, \quad (7)$$

where $k_1$ is a material constant but depends on the strain waveforms (or the compressive creep rate ).

The fracture resistance, R, decreases by a magnitude corresponding to the driving force, F, at every number of cycles of fatigue. Grain boundary fracture takes place at the residual resistance being equal to zero. Small cracks thus initiate here and there in order of the value of R.

Fourthly, it might be rational to assume two different driving forces, F and K, as the experimental results show that, once a crack initiates, it tends to grow easily due to its stress intensity, and that the effect of the grain boundary inclination, the magnitude of stress or strain range and the strain waveform on the crack initiation differ from those on the crack growth (14). Then, the fracture driving force, D, is given to each grain boundary in the form of

$$D = F + K, \quad (8)$$

where K is the additional driving forth for crack growth. In regard to the driving forth for large crack growth, the creep J-integral, $J^*$, or the creep J-integral range, $\Delta Jc$ (the accumulation of $J^*$ in one cycle of fatigue ), is considered to be a probable parameter for the nonlinear fracture mechanics (1,2,18). On this basis, K is determined to be in proportion to $J^*$ or $\Delta Jc$ on the assumption that the crack length is relatively small as compared with the material width or thickness (ligament).

$$K = Ko \cdot (c / lg), \quad (9)$$

where c is the half surface-crack-length, lg is the mean grain-boundary-length, and Ko is given as

$$Ko = Ko(\sigma, \dot{\varepsilon}_c) = k_2 \cdot (\sigma \dot{\varepsilon}_c), \quad (10)$$

in which $k_2$ is a material constant.

Until the grain boundary fractures, K=0 for c=0 in Eq.(9) and D=F from Eq.(8). Once it fractures and the crack initiates,the adjacent grain boundaries to the crack are subjected to the forth, D=F+K, as shown in Figure 9.

Numerical Simulation. Procedure of the numerical simulation is as follows (14,15):

[1] Random grain boundaries are numerically generated in the surface area containing about 100 grains.

[2] The value of R cannot be determined as many factors in Eq.(3) are unknown variables. Therefore,

R = random number of a uniform distribution from 0 to 1 (11)

are chosen and assigned to each grain boundary as the value of fracture resistance.

[3] As nondimensional expression is given by Eq.(11) for the value of R, the positive value of the driving forth, D, cannot be determined. Moreover, it is difficult to fix the magnitude of F and K through any simple test, although rough prediction of the value of K is possible using the value of $J^*$ estimated throgh the cyclic stress-strain response. Eventually, it could not help being estimated inversely by trial and error so as to fit the actual crack data in the present study.

[4] The residual fracture resistance, R-D, is calculated at every cycle,and the crack initiation and growth are judged by whether the residue is possitive or not.

[5] When two adjacent cracks happen to coalesce, the crack is assumed to be a large crack of semi-circular shape immediately.

[6] More precise simulation is done on the assumption that the driving force, D, is not given to grain boundaries within the stress relaxation zone,which is expressed by a circle of the diameter being equal to the distance between both tips of a pre-existing crack.

The results of the numerical simulation were shown in Refs.(14,15)for a Type 304 steel tested in the creep fatigue condition of 1% total strain range for the slow-fast strain waveform($10^{-3}$-1 %/s) at the temperature of 650 °C in a vacuum environment.The change in crack density, the angular distribution of initiated cracks, the scatter of crack growth rates and the distribution of crack lengths were in good agreement with the experimental observation, when the values of Fo, Ko and m were chosen as

$$Fo=8.0 \times 10^{-3},\ Ko=4.5 \times 10^{-3},\ m=10. \quad (12)$$

Computer aided visual representation of multiple surface cracks is shown in Figure 10 in comparison with the photos of actual cracks.

One Dimensional Model. Before proposing the quasi-three dimensional model, the authors have published a one dimensional mode for a basic method numerical analyses(19,20). Points of difference in the model are as follows.

[i] Grain boundaries between two adjacent triple points on the specimen surface are projected and linked on a straight line perpendicular to the applied stress axis,and its length, lg, are given as random variables of a normal distribution.

[ii] The driving force,F,in Eq.(4) is given as

$$F = Fo, \quad (13)$$

since grain boundary inclined angles, $\theta g = \theta g' = 0$, and so Eq.(8) becomes

$$D = Fo + Ko \cdot (c/lg). \quad (14)$$

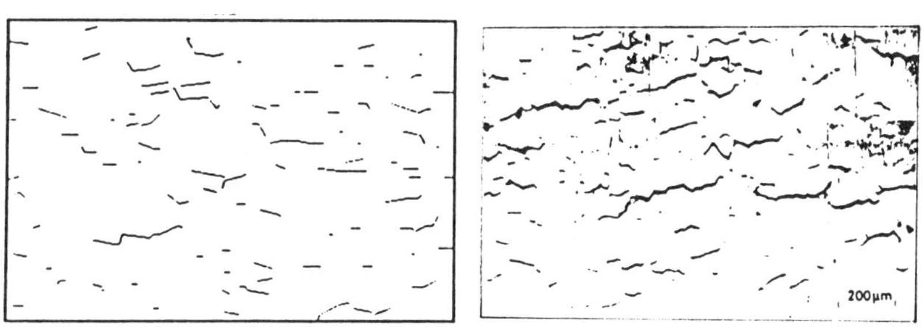


Fig. 10. Visualized figures of simulated cracks and observed ones, the number of cycles to failure: $N_f$=388.

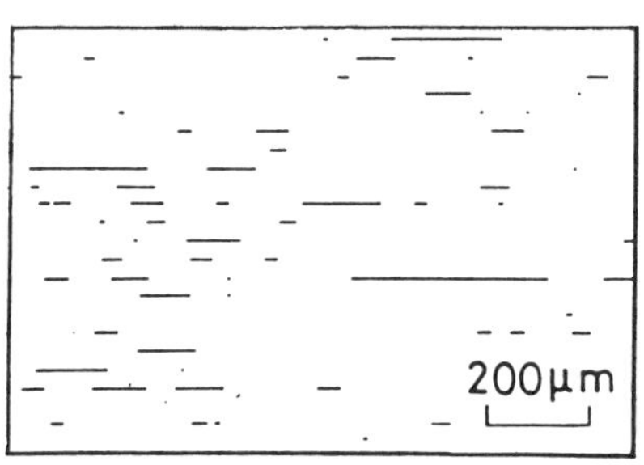


Fig. 11. Simulated cracks based on one dimensional model.

Figure 11 shows the surface cracks in the one dimensional simulation and the observation on the same cracking material shown in Figure 10. In this case, the values of Fo and Ko are as follows:

$$Fo = 6.4 \times 10^{-4}, \quad Ko = 4.3 \times 10^{-3}. \quad (15)$$

Comparing the values with those of the quasi-three dimensional simulation, Fo is 1/8 times as large as Fo in Eq.(12), whereas both Ko's are about the same magnitude. Looking at the main issue, other results of the simulation ( 19,20) are quite analogous to those of the quasi-three dimensional model. Therefore, by using the one dimensional model, a parametric analysis was carried out for characterizing the failure process, classifying the fracture mode and signifying the failure life of creep fatigue in relation to the cracking. The results are the following.

## PARAMETRIC ANALYSIS CREEP FATIGUE CRACKING

Difference Between Fo and Ko. The dependence of the driving forces, Fo and Ko, on the increase in crack density is shown in Figure 12, where the crack density, n, is defined as the number of cracks per unit surface area. This representation is similar to Figure 2. It is found that the value of Fo has a great influence on the increasing rate of crack density, dn/dN, whereas the value of Ko scarcely does. It was already confirmed that the crack growth was not controlled by Fo but by Ko to the contrary (14,15,19,20).

Supposing that the material fails when a longest crack has grown up to 1mm half length on the surface,the contribution of the driving forces, Fo and Ko, to the failure life, Nf, is calculated numerically and shown in Figure 13. It is found from Figure (b) that Nf becomes small with increasing Ko, which might be logical as Ko is the driving force for crack growth rate. On the other hand, Figure (a) indicates that Nf becomes small to the same extent when Fo increases by the same multiples as Ko. It is worth noticing that the increase in the number of initiated cracks shortens the failure life not only because the growth rate of a main crack is accerelated by the coalescence of many small cracks but also because the probability of the early initiation of growing cracks is increaed.

Parameter,Ko/Fo. It seems that the values of Fo and Ko independently bring about a change in the failure life. However, attention should be paid to the calculated results showing that the state of cracking can be characterized in terms of the value of the ratio of two driving forces, Ko/Fo. When the values of Fo and Ko in fatigue condition A are twice as large as those in fatigue condition B having equal magnitude of Ko/Fo, the cracking situation repre-

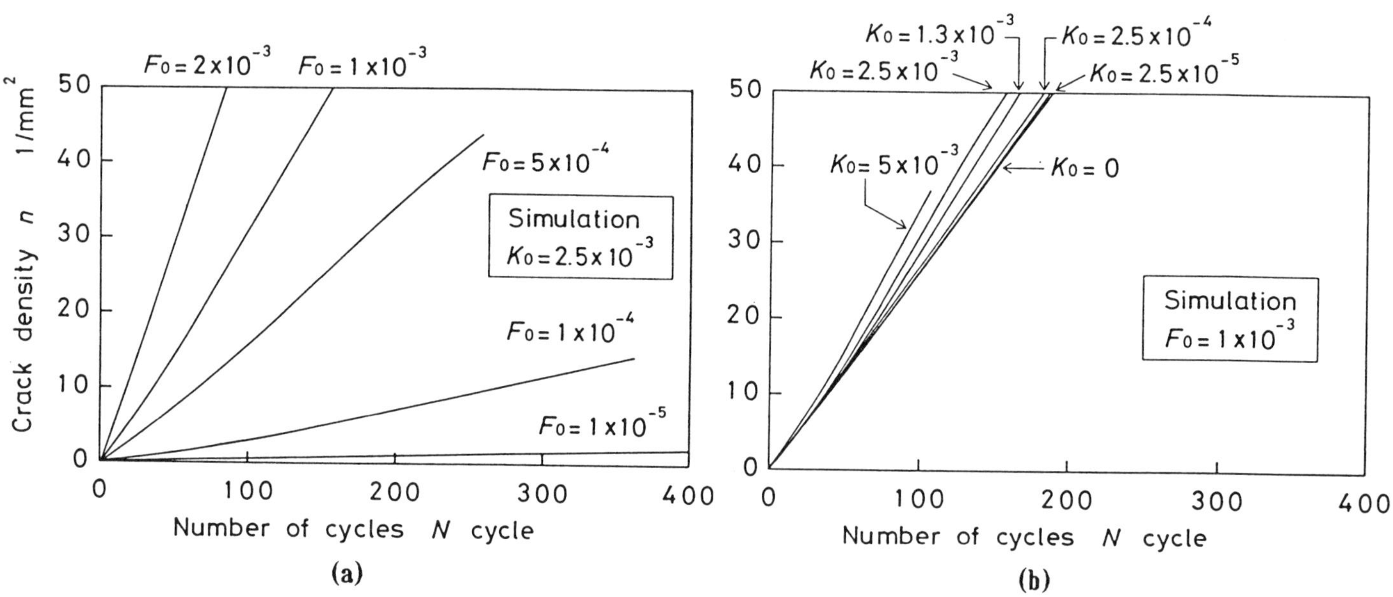


Fig. 12. Dependence of (a) $F_0$ and (b) $K_0$ on the increase in crack density.

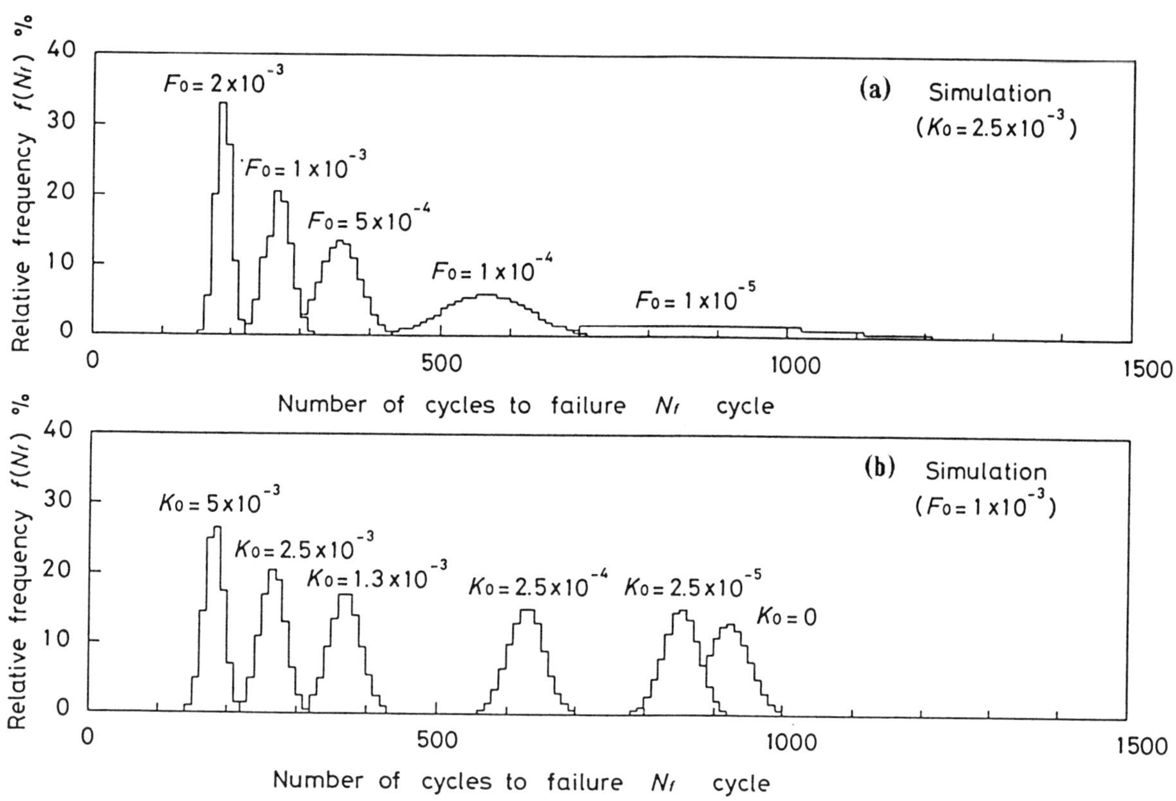


Fig. 13. Dependence of (a) $F_0$ and (b) $K_0$ on the failure life and its distribution.

sented by the crack density and the distributions of crack length and growth rate at the number of cyckes of $N_1$ in condition A, is identical to that of condition B at the number of cycles of $2N_1$. The failure life in condition A is consequently shorter by half than condition B. If only the normalized fatigue cycle is taken as the ratio of fatigue cycle to mean fatigue life, the ratio, Ko/Fo is significant of the parameter representing the state of cracking. Figure 14 shows the failure modes represented by the parameter. It can be known from the experimental results that the magnitude of Ko/Fo ranging from 1 to 100 is realistic. In the case of the value being close to 1, it is characteristic of the failure mode that the material surface is dense with small cracks. It can, therefore, be called "crack initiation type". When the value is large enough to be nearly 100, the failure mode changes to "crack growth type", in which the limited number of cracks initiates to grow as

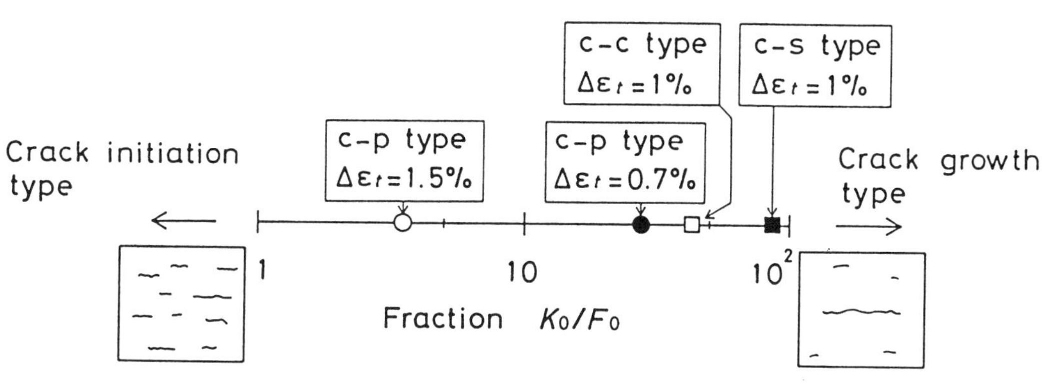


Fig. 14. State of creep fatigue cracking represented in terms of the parameter, $K_0/F_0$, and the experimental results of Type 304 stainless steel showing the change in the value in corespondence with the difference in strain ranges and strain waveforms.

Table I. Test conditions and the values of $F_0$ and $K_0$.

| Type | Strain waveform | $\Delta\varepsilon_t$(%) | $\dot{\varepsilon}_{ten}$(%/s) | $\dot{\varepsilon}_{com}$(%/s) | $N_f$ | $F_0$ | $K_0$ | $K_0/F_0$ |
|---|---|---|---|---|---|---|---|---|
| ① c-p |  | 1.5 | $1x10^{-4}$ | $2x10^{-2}$ | 100 | $3x10^{-3}$ | $1.1x10^{-2}$ | 3.6 |
| ② c-p |  | 0.7 | $1x10^{-4}$ | $2x10^{-2}$ | 480 | $8x10^{-5}$ | $2.3x10^{-3}$ | 28 |
| ③ c-c |  | 1.0 | $2x10^{-3}$ | $2x10^{-3}$ | 800 | $4x10^{-5}$ | $1.8x10^{-3}$ | 44 |
| ④ c-s |  | 1.0 | $2x10^{-3}$ | $3.2x10^{-4}$ | 600 | $2x10^{5}$ | $1.8x10^{-3}$ | 88 |

a single crack.

The experimental data in Figure 14 come from four kinds of the creep fatigue test of a bar specimen of 10mm-diameter of a Type 304 stainless steel conducted at the temperature of 923K (650 C) in air (7). Other test conditions and the values of Fo and Ko are listed in Tables I and II. Here, the numerical simulation for these tests were conducted on the basis of the same procedure in the parametric analysis shown in Figures 12 to 14, where the projections of the grain boundary length, lg, are prescribed as random variables of a normal distribution with the average of 25$\mu$m and the standard deviation of 20$\mu$m.

The creep fatigue in the waveform of c-p type(slow-fast) generally has the largest value of Fo under the same magnitude of the total strain range, $\Delta\varepsilon_t$. The larger the value of $\Delta\varepsilon_t$ is, the larger becomes the value of Fo in any type of strain waveform. The slower compressive strain rate in the c-s type waveform tends to diminish the value of Fo, whereas it does not change the value of Ko. Among the four types tested, the failure mode of the c-p type at $\Delta\varepsilon t$ = 1.5% corresponds to a typical crack initiation type, and the c-s type is close to a crack growth type.

It is also found from Table II that the value of Fo is more changeable than that of Ko depending on the strain range as well as the strain waveform. This suggests that the crack initiation is more sensitive to the loading condition than the crack growth.

REFERENCES


1- Ohtani,R.and T.Kitamura, J.Soc.of Materials Science,Japan, 34-382, 843-49 (1985).

2- Ohtani,R.,T.Kitamura,N.Nitta and K.Kuwabara, Low Cycle Fatigue,ASTM STP 942,1163-80(1988).

3- Ohtani,R.,T.Kitamura,H.Murayama and T.Watabe, J. Soc.of Materials Science, Japan, 36-408, 974-79 (1987).

4- Ohtani,R.,T.Kinami and H.Sakamoto, in Proc. Int.Conf. on the Role of Fracture Mechanics in Modern Technology,Fukuoka Japan,June1986, North-Holland,1987, pp.381-92.

5- Ohtani,R. and T.Kitamura, in Proc. 4th Int. Conf. on Creep and Fracture of Engineering Materials and Structures,Swansea,April 1990, The Institute of Metals,London, pp.791-802.

6- Ohtani,R. and T.Kitamura, in Proc. 7th Int. Conf. on Fracture(ICF7), Houston,March 1989, Pergamon, pp.213-21.

7- Ohtani,R.,T.Kitamura,M.Abe,Y.Kuriyama,and H. Miki, J.Soc.Materials Science,Japan, 39-440, 529-35 (1990).

8- Ohtani,R.,T.Kitamura,N.Tada,and S.Iio,to be presented at Int.Conf. on Low Cycle Fatigue, Berlin, Sept.1992.

9- Kitamura,T.,N.Tada,Y.Kuriyama, and R.Ohtani, Trans.JSME,Ser.A, 56-524, 702-7 (1990).

10-Kitamura,T., R.Ohtani and N.Tada, to be presented at Int.Conf.on Behavior of Defects at High Temperatures, Sheffield, March 1992.

11-Ohtani,R.,N.Tada and T.Hashimoto, J.Soc.of Materials Science,Japan,38-432, 89-94(1989).

12-Ohtani,R.,T.Kitamura and W.Zhow, ibid., 40-457, 1290-96 (1991).

13-Ohtani,R.,T.Kitamura and N.Tada, Materials Science and Engineering, A143, 213-22(1991).

14-Ohtani,R.,T.Kitamura and N.Tada, in Proc. 4th Int.Con.on Fatigue(Fatigue 90),Honolulu Hawaii,July 1990, MCEP Bermingham, Vol.IV, pp.2143-48.

15-Tada,N.,T.Kitamura and R.Ohtani,Trans.JSME, Ser.A, 56-524, 708-14 (1990).

16-Ohtani,R.,T.Kitamura and N.Tada, in Proc. 6th Int.Conf.on Mech. Behav. of Materials (ICM-6),Kyoto Japan,July 1991,Pergamon,Vol. 2, pp.205-12.

17-Kitamura,T. and R.Ohtani, Trans.JSME,Ser.A, 56-523, 196-202 (1990).

18-Ohtani,R.and T.Kitamura, "High Temperature Creep-Fatigue" , CJMR(Soc.Mat.Sci.,Japan), Vol.3,1988, Elsevier, pp.65-90.

19-Ohtani,R.,T.Kitamura,H.Murayama and N.Tada, Trans.JSME,Ser.A, 54-503, 1312-16 (1988).

20-Ohtani,R., T.Kitamura and N.Tada, "Structural Design for Elevated Temperature Environments" ,ASME PVP-Vol.163, 1989, p.123.

*Proceedings of the Fifth International Conference on Creep of Materials, Lake Buena Vista, Florida, USA, 18-21 May 1992*

# Near-Threshold Fatigue Crack Growth Behavior at High Temperatures

**K. Ohji and S. Kubo**
Department of Mechanical Engineering
Osaka University
Osaka, Japan

**Y. Nakai**
Department of Mechanical Engineering
Kobe University
Kobe, Japan

## ABSTRACT

Near-threshold fatigue crack growth behavior in type 304 stainless steel was investigated at 650°C and 550°C. An alternate current (A.C.) electric potential technique was applied to the measurement of crack length in center-cracked specimens made of the stainless steel at high temperatures.

The crack growth behavior under constant amplitude loadings was investigated. The threshold value for the cycle-dependent crack growth was determined by decreasing stress intensity range $\Delta K$ gradually. Temperature dependence was observed in the threshold J-integral range $(\Delta J_f)_{th}$ of the cycle-dependent crack growth; the threshold value at 650°C was lower than that at 550°C. Above the threshold values fatigue crack growth rates *da/dN* were expressed in terms of the J-integral range $\Delta J_f$ or stress intensity range $\Delta K$.

To investigate the character of the threshold for the cycle-dependent crack growth, crack growth experiments were conducted using two-step variable amplitude loadings, in which a high-range load was introduced periodically in sub-threshold low-range loads. To exclude the effect of crack closure, high stress ratios were employed. It was found that fatigue cracks grew even at J-integral ranges below the threshold values under constant amplitude loadings. The crack growth rates $(da/dN)_L^{e*}$ under low load-range were estimated by assuming that the crack growth rates under high load-range could be predicted by using the crack growth characteristic curve obtained under constant amplitude loadings. It was found that the threshold value decreased by the intermittent applications of the high-range loads.

FOR DESIGN AND SAFETY ASSESSMENT of structures and their components for high-temperature use, knowledge of fatigue crack growth behavior at high temperatures is indispensable. Fracture mechanics has been applied to the characterization of crack growth at high temperatures and many research reports have been accumulated (1-17).

The present authors examined the fatigue crack growth behavior in type 304 stainless steel under creep conditions (18-25). It was found that the fatigue crack growth rates under creep conditions were either purely time-dependent or purely cycle-dependent, depending on the test conditions employed. A threshold was identified in the cycle-dependent crack growth under constant amplitude loadings even in the creep regime (18,19,22). In spite of the importance of the fatigue threshold and near-threshold fatigue crack growth behavior at high temperatures for evaluating long term life of structures, they have not been fully investigated.

In the present study the near-threshold cycle-dependent crack growth

Table I. Chemical composition of the material tested

(wt.%)

| C | Si | Mn | P | S | Ni | Cr | Fe |
|---|---|---|---|---|---|---|---|
| 0.04 | 0.60 | 1.69 | 0.036 | 0.001 | 8.29 | 18.34 | Bal. |

Table II. Mechanical properties at ambient temperature of the material tested

| Tensile Strength (MPa) | Elongation (%) | Reduction in Area (%) | Hardness ($H_B$) |
|---|---|---|---|
| 618 | 50.2 | 62.0 | 164 |

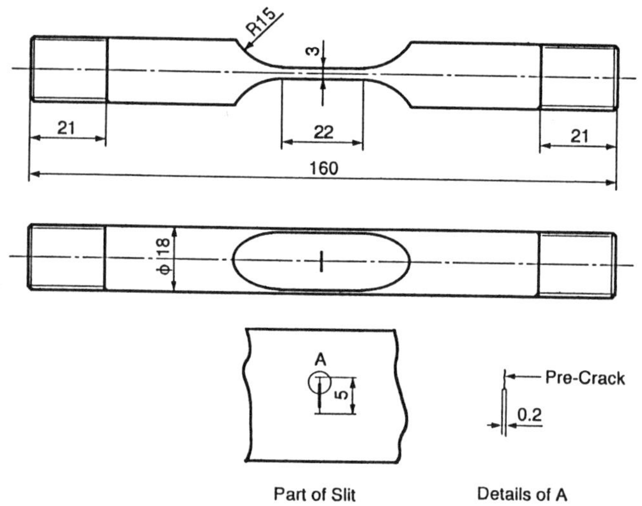


Fig. 1 - Geometry of a center-cracked specimen

behavior in type 304 stainless steel was investigated at 650°C and 550°C. Temperature dependence of the threshold under constant amplitude loadings was examined. Fatigue crack growth under two-step variable amplitude loadings was discussed to investigate the nature of fatigue threshold at high temperatures.

## EXPERIMENTAL PROCEDURES

Type 304 stainless steel was used, whose chemical composition and mechanical properties at ambient temperature are listed in Tables I and II, respectively. Tests were conducted at 650°C and 550°C in air with center-cracked specimens. The specimen geometry is shown in Fig. 1. As a crack starter, a slit was made by electrical-discharge machining, and specimens were precracked by fatigue loadings of stress ratio of 0.1 at ambient temperature, the total length of the precrack being about 5 mm. The specimens were heated by an induction heater, the temperature being controlled to be 650°C or 550°C by a controller with a microprocessor.

Fatigue crack growth experiments were conducted by using a computer-controlled servo-hydraulic machine, whose load capacity was 98 kN. Crack growth behavior under constant amplitude loadings was examined by constant load-amplitude tests and $\Delta K$-decreasing tests. Sinusoidal stress wave forms were used

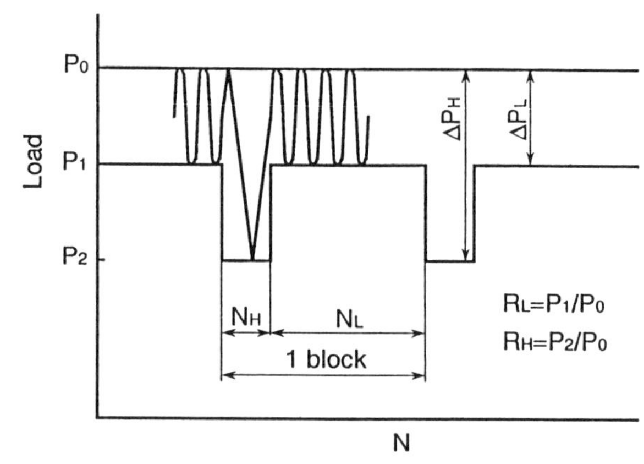


Fig. 2 - Two-step variable amplitude loadings

for these tests. The loading frequency was set at 30 Hz. To eliminate the effect of crack closure high value of stress ratio $R$ was employed: $R$ was set at 0.5. In the $\Delta K$-decreasing tests load was decreased for every incremental crack advance of 50 $\mu$m so that decrease rate $dK/da$ of stress intensity factor with crack advance satisfied $(1/K)(dK/da) = -0.08\ m^{-1}$ in accordance with ASTM

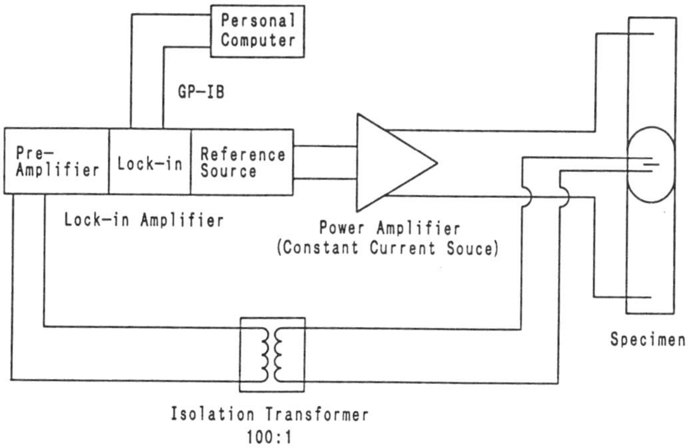


Fig. 3 - Schematic diagram of a system for measuring crack length using A.C. potential

standard E 647-89.

For the two-step variable amplitude loadings a wave form composed of sinusoidal and triangular waves shown in Fig. 2 was employed. Low load-range $\Delta P_L$ and high load-range $\Delta P_H$ were selected so that the corresponding stress intensity ranges $\Delta K_L$ and $\Delta K_H$ were lower and higher than threshold stress intensity range $\Delta K_{th}$ under constant amplitude loadings. Ratio of $\Delta P_H$ and $\Delta P_L$ was selected to be 2. To minimize the effect of crack closure, high stress ratios were also used in the two-step variable amplitude loadings. Stress ratio $R_H$ for high load-range $\Delta P_H$ was set at 0.4 and stress ratio $R_L$ for $\Delta P_L$ was set at 0.7. Number of cycles of the high-range and low-range loads were $N_H$ = 1 and $N_L$ = 16. Loading frequencies for the high-range and low-range loads were $\nu_H$ = 10 Hz and $\nu_L$ = 25 Hz, respectively.

For the study of the near-threshold crack growth behavior, automatic and high-accuracy measurement of crack length had to be established. An alternate current (A.C.) electric potential technique was developed in this study for measuring crack length in center-cracked specimens made of the stainless steel at high temperatures. The system of A.C. potential measurement is shown in Fig. 3. A reference signal generated by a function generator in a lock-in amplifier was amplified by a power amplifier. An A.C. current of 1.0 A was supplied to the specimen through shielded wires as is

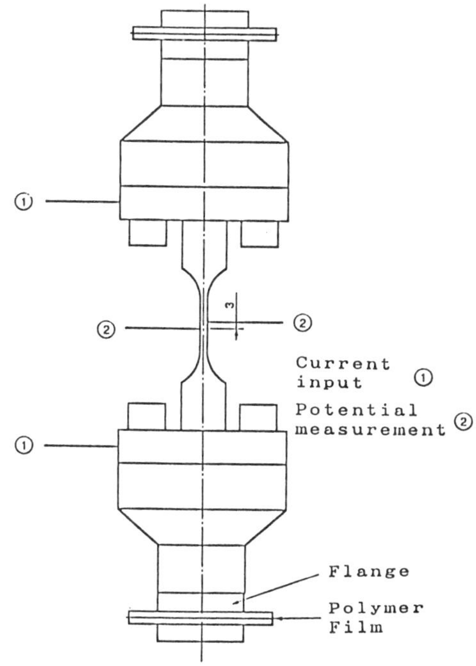


Fig. 4 - Positions of current input and A.C. potential measurement

shown in Fig. 4. Probes for the potential measurements were spot-welded to the specimen at two opposite positions with respect to the slit, the distance between the positions being 3 mm. To insulate the specimen from the servo-hydraulic testing machine, polymer films were inserted in two flanges as can be seen in Fig. 4.

Stress intensity factor was evaluated using Tada's formula for center-cracked specimens (26). The effect of crack closure was excluded by using high stress ratios for crack growth experiments, and therefore the effective stress intensity range $\Delta K_{eff}$ was equal to $\Delta K$. Since no significant nonlinearity in load $P$-displacement $\delta$ relationship was observed, the J-integral range $\Delta J_f$ was evaluated by using the following equation in accordance with Dowling's method (27).

$$\Delta J_f = \Delta K_{eff}^2/E = \Delta K^2/E \qquad [1]$$

Young's modulus $E$ of the material was given as follows.

$E$ = 150 MPa at 650°C

$E$ = 158 MPa at 550°C

The modified J-integral $J^*$ (28,29),

which is equivalent to the C*-integral (30,31), was evaluated experimentally using a simple estimation formula (32,33) which was expressed as

$$J^* = \sigma_{net}\dot{\delta} \qquad [2]$$

for center-cracked specimens, where $\sigma_{net}$ and $\dot{\delta}$ are net section stress and load-point displacement.

## CRACK LENGTH MEASUREMENT

Relationships of A.C. potential reading $e$ with crack length $a$ observed at ambient temperature and at 650°C are shown in Figs. 5(a) and 5(b), respectively. In the figures $e$ is normalized by its value $e_o$ at a reference crack length $a_o$. The dashed lines in the figures represent Johnson's theoretical relation of D.C. potential reading for a center-cracked plate (34), which is expressed as,

$$e/e_o = \cosh^{-1}\{\cosh(\pi L/2W)/\cos(\pi a/2W)\} / \cosh^{-1}\{\cosh(\pi L/2W)/\cos(\pi a_o/2W)\} \qquad [3]$$

where $W$ and $L$ denote half specimen width and half distance between potential measuring points, respectively.

As can be seen in Fig. 5(a) noticeable frequency effect was not observed in $e/e_o$ for frequencies ranging from 93 Hz to 940 Hz, and the relationship between $e/e_o$ and relative crack length $a/W$ agreed well with that predicted by the theoretical one for D.C. potential readings (Eq. [3]). This may be due to the fact that type 304 austenitic stainless steel used in this study was a paramagnetic material, and hence skin effect of A.C. current was very small. In Fig. 5(b), which shows the relationship between $e/e_o$ and $a/W$ at 650°C, frequency effect was observed. This frequency effect may be attributed to that the potential measuring wires could not be fully shielded at high temperatures. The frequency effect, however, was small and the behavior of $e/e_o$ could be approximated by Eq. [3] for frequencies lower than 200 Hz. In this study, therefore, frequency was set at 93 Hz and Eq. [3] was utilized as the calibration curve for determining the crack length from A.C. potential reading.

## FATIGUE CRACK GROWTH BEHAVIOR

### Constant Amplitude Loadings

The near-threshold crack growth behavior under constant amplitude loadings was examined in air at 650°C and 550°C by $\Delta K$-decreasing tests and $\Delta P$-constant tests. To obtain threshold values for fatigue crack growth, the $\Delta K$-

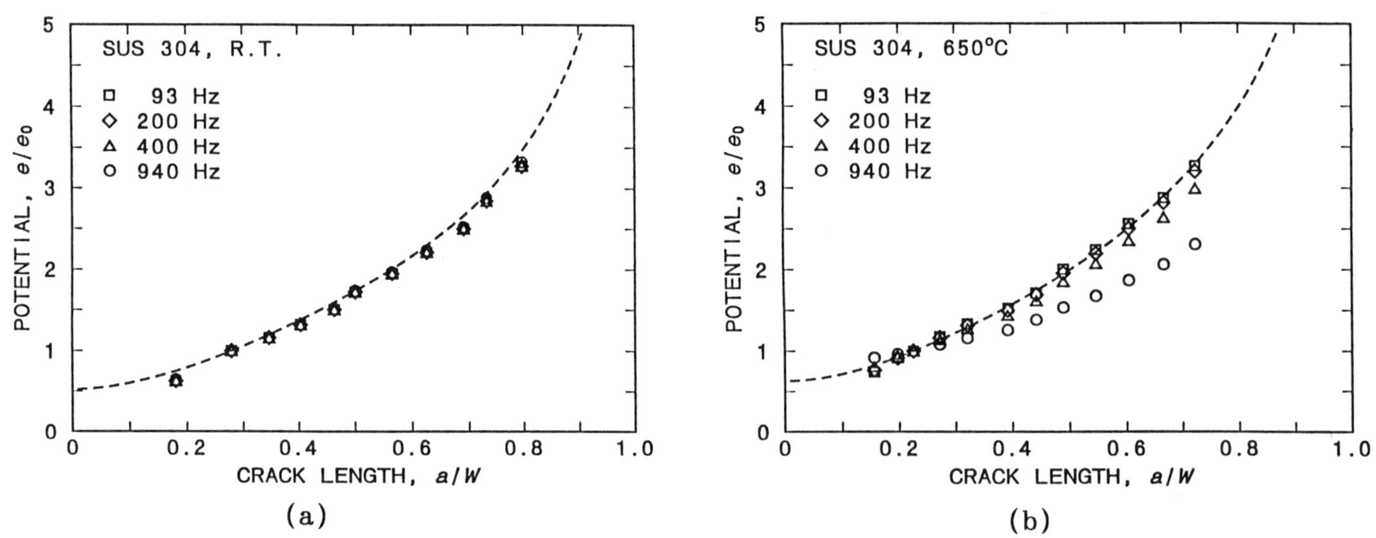


Fig. 5 - Effect of current frequency on the relationship between A.C. potential and crack length:
(a) At ambient temperature; (b) At 650°C

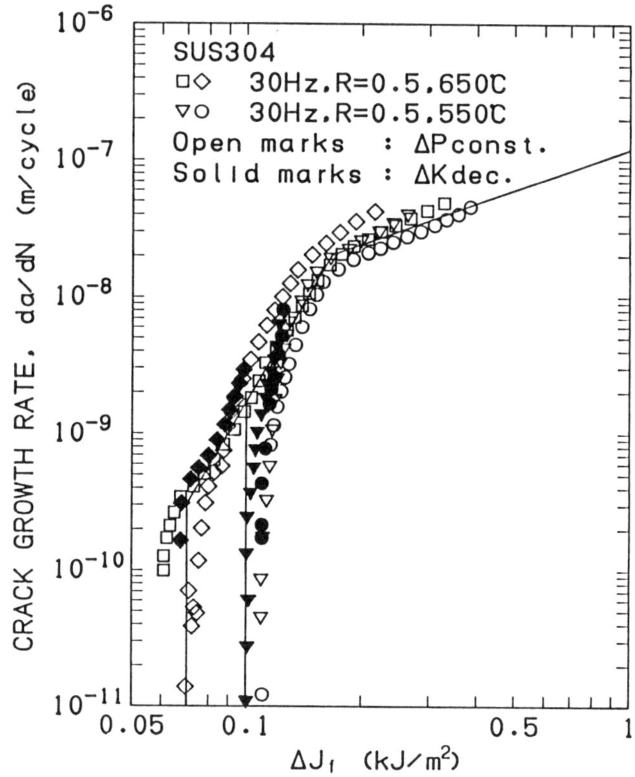


Fig. 6 - Crack growth rate per cycle *da/dN* under constant amplitude loadings plotted against J-integral range $\Delta J_f$

decreasing tests were conducted until crack growth rate *da/dN* decreased and attained a value lower than $10^{-10}$ m/cycle. Then, the applied load was increased by a small amount and constant load-amplitude tests was made.

Figure 6 shows the relationship between crack growth rate *da/dN* and the J-integral range $\Delta J_f$. The data for crack advances smaller than 1 mm from the starter fatigue crack, which were prone to show higher growth rates than those of steady crack growth, were excluded in the figure. Data obtained by the $\Delta K$-decreasing tests were shown by solid symbols, open symbols with the same shape being data obtained by the $\Delta P$-constant test following the corresponding $\Delta K$-decreasing tests.

Threshold value was dependent on test temperature. Threshold J-integral range $(\Delta J_f)_{th}$ is given by

$$(\Delta J_f)_{th} = 0.07 \text{ kJ/m}^2 \quad \text{at } 650°\text{C} \qquad [4]$$

and

$$(\Delta J_f)_{th} = 0.10 \text{ kJ/m}^2 \quad \text{at } 550°\text{C} \qquad [5]$$

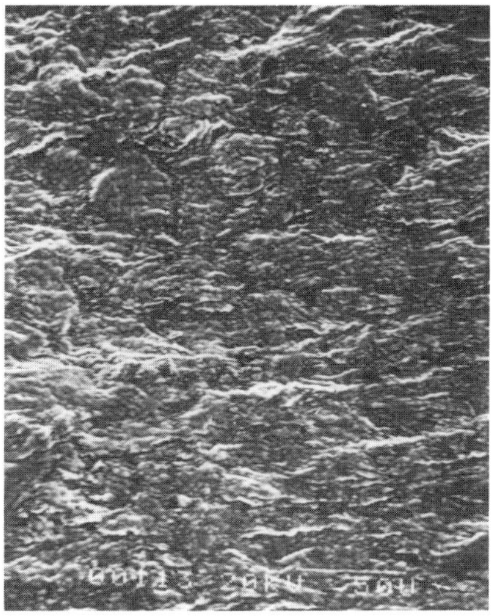

(a)

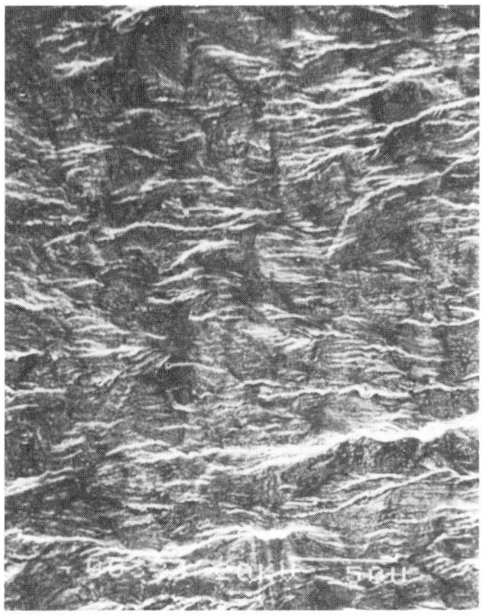

(b)

Fig. 7 - Fracture surfaces observed under constant amplitude loadings:
(a) At 650°C (*da/dN* = $4\times10^{-9}$ m/cycle);
(b) At 550°C (*da/dN* = $7\times10^{-9}$ m/cycle)

It was reported by Ogura et al. (35,36) that the threshold value of type 304 stainless steel increased with test temperature ranging from ambient temperature to 500°C. It is therefore considered that the threshold $(\Delta J_f)_{th}$ or $\Delta K_{th}$ attained its maximum at a temperature around 550°C.

In the range of $\Delta J_f$ > 0.167 kJ/m², *da/dN* was expressed in terms of $\Delta J_f$, or equivalently in terms of $\Delta K$. The *da/dN*-$\Delta J_f$ relation was almost independent of test temperature and coincident with that

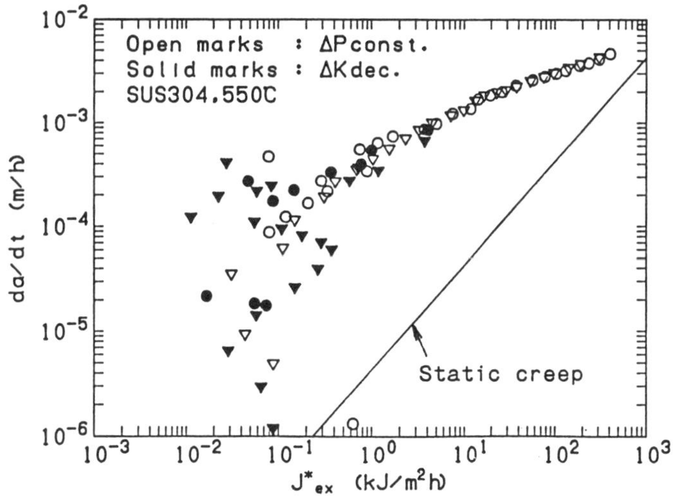


Fig. 8 - Crack growth rate per time *da/dt* at 550°C under constant amplitude loadings plotted against the modified J-integral *J**

obtained in the previous studies (19-21). In the range of $(\Delta J_f)_{th} < \Delta J_f < 0.167$ kJ/m² another power relation between *da/dN* and $\Delta J_f$ seemed to prevail. The *da/dN* was then expressed as,

$$da/dN = 1.17\times10^{-4}\Delta J_f^{4.84} \quad [6]$$
$$\text{for } 0.167 \text{ kJ/m}^2 > \Delta J_f > (\Delta J_f)_{th}$$

and

$$da/dN = 1.2\times10^{-7}\Delta J_f \quad [7]$$
$$\text{for } \Delta J_f > 0.167 \text{ kJ/m}^2$$

$(da/dN$ in m/cycle, $\Delta J_f$ in kJ/m²$)$

It was pointed out by the present authors (18,19,22) that pseudo thresholds induced by time-dependent crack growth were observed under certain conditions. To assure that the threshold and near-threshold crack growth in this study were cycle-dependent ones, fractographical examination and crack growth characterization in terms of the modified J-integral *J** were made.

Figures 7(a) and 7(b) show photographs of fracture surfaces observed at 650°C and 550°C using a scanning electron microscope. It is seen that the fracture was of transgranular type. The correlation of crack growth rate per time *da/dt* at 550°C with the modified J-integral $J^*_{ex}$ evaluated experimentally by using Eq. [2] is shown in Fig. 8. Static creep crack growth rate and time-dependent fatigue crack growth rate of the stainless steel can be expressed in terms of *J** as (18,19)

$$da/dt = 4.3\times10^{-6}J^*$$

$(da/dt$ in m/h, $J^*$ in kJ/m²h$)$

This relationship is shown by a solid line in the figure. Although there is a scatter in the range of small *J**, the values of *da/dt* are much higher than those of the results for static creep crack growth or time-dependent fatigue crack growth. These are typical behaviors for the cycle-dependent fatigue-type crack growth (18,19,22). Accordingly, it can be concluded that the pseudo thresholds induced by time-dependent creep-type crack growth did not appear under the loading conditions employed in this study.

**Two-Step Variable Amplitude Loadings**

To investigate the character of the fatigue threshold for the cycle-dependent crack growth at high temperatures, crack growth experiments were conducted using two-step variable amplitude loadings, in which a cycle of high load-range $\Delta P_H$ was introduced periodically in the sub-threshold low-range loads as shown in Fig. 2. In Figs. 9(a) and 9(b) crack growth rate $da/dN_H$ per high-range load cycle is plotted against $\Delta J_{f.H}$, which signifies the $\Delta J_f$ value evaluated for the high load-range $\Delta P_H$. Data plots corresponding to crack advances smaller than 1 mm were excluded again. Solid lines in the figure show crack growth rate predicted by Miner's linear cumulative law of crack growth rate, i.e. by assuming that crack growth rates for $\Delta P_H$ and $\Delta P_L$ were expressed by Eqs. [6] and [7], and that no crack growth occurred when $\Delta J_f$ went below the threshold value $(\Delta J_f)_{th}$ given by Eqs. [4] and [5]. The prediction can be written as follows.

$$da/dN_H = 0 \quad [8]$$
$$\text{for } (\Delta J_f)_{th} > \Delta J_{f.H}$$
$$(\text{for } (\Delta J_f)_{th}/4 > \Delta J_{f.L})$$

$$da/dN_H = 1.17\times10^{-4}\Delta J_{f.H}^{4.84} \quad [9]$$
$$\text{for } 0.167 \text{ kJ/m}^2 > \Delta J_{f.H} > (\Delta J_f)_{th}$$
$$(\text{for } (\Delta J_f)_{th} > 0.042 \text{ kJ/m}^2 > \Delta J_{f.L} > (\Delta J_f)_{th}/4)$$

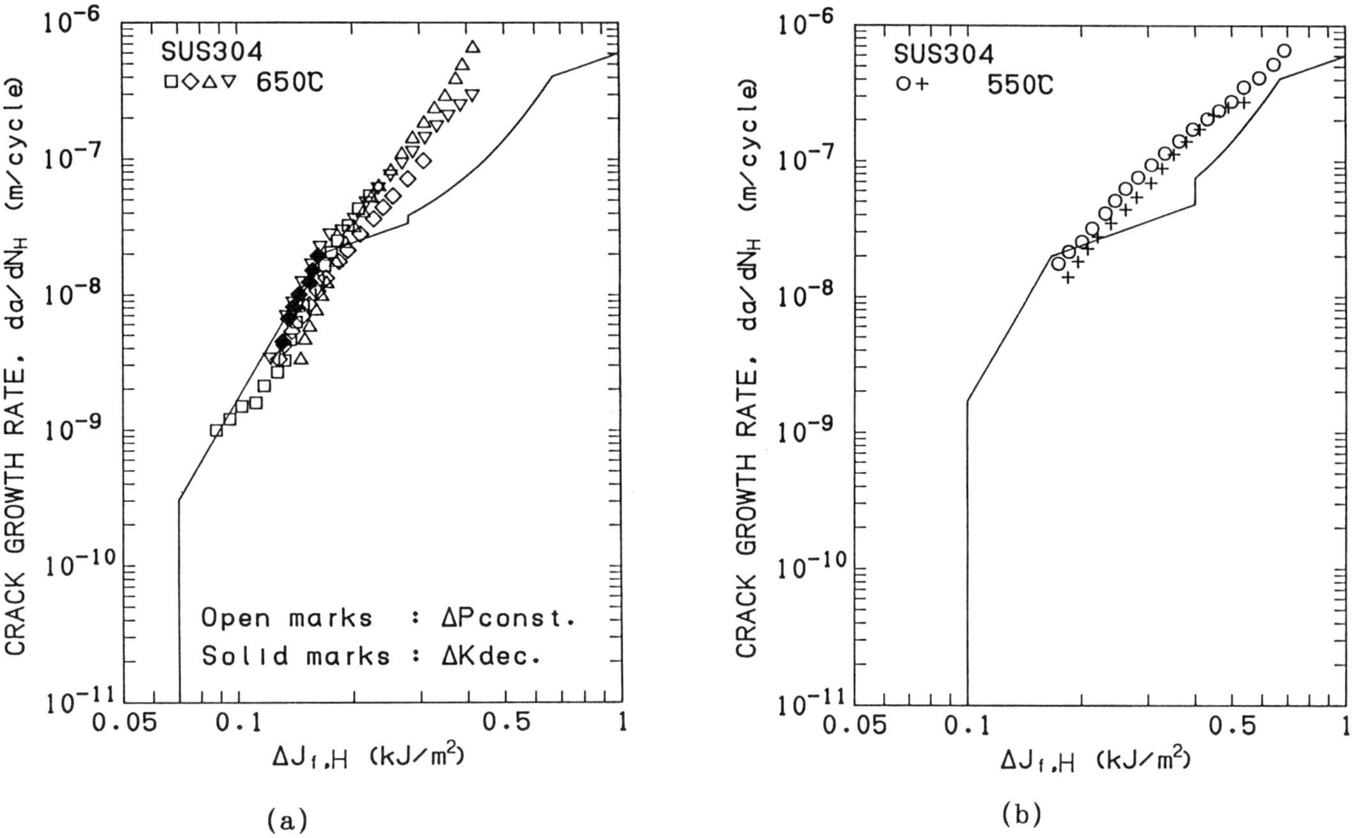


Fig. 9 - Crack growth rate per high-range load cycle $da/dN_H$ under two-step variable amplitude loadings as a function of $\Delta J_{f,H}$ evaluated for high load-range $\Delta P_H$:
(a) At 650°C; (b) At 550°C

$$da/dN_H = 1.2\times10^{-7}\Delta J_{f.H} \qquad [10]$$
for $4\times(\Delta J_f)_{th} > \Delta J_{f.H} > 0.167$ kJ/m²
(for $(\Delta J_f)_{th} > \Delta J_{f.L} > 0.042$ kJ/m²)

$$da/dN_H = 1.17\times10^{-4}\Delta J_{f.L}^{4.84}\times N_L/N_H + 1.2\times10^{-7}\Delta J_{f.H} \qquad [11]$$
for 0.67 kJ/m² $> \Delta J_{f.H} > 4\times(\Delta J_f)_{th}$
(for 0.167 kJ/m² $> \Delta J_{f.L} > (\Delta J_f)_{th}$)

and

$$da/dN_H = 1.2\times10^{-7}(\Delta J_{f.L}\times N_L/N_H + \Delta J_{f.H}) \qquad [12]$$
for $\Delta J_{f.H} > 0.67$ kJ/m²
(for $\Delta J_{f.L} > 0.167$ kJ/m²)

($da/dN_H$ in m/cycle, $\Delta J_{f.H}$ in kJ/m², $\Delta J_{f.L}$ in kJ/m²)

As can be seen from these figures, experimental crack growth rates exceeded predicted values based on Miner's rule in the region of 0.7 kJ/m² $> \Delta J_{f,H} >$ 0.2 kJ/m². This increase in crack growth rate means that the crack grew at J-integral ranges below its threshold $(\Delta J_f)_{th}$ for constant amplitude loadings. Similar results were already obtained in fatigue crack growth at ambient temperature (37,38). This phenomenon may be attributed to the decrease in the fatigue threshold caused by the intermittent introductions of high-range loads. On the other hand, in the range of 0.2 kJ/m² $> \Delta J_{f,H}$ of Fig. 9(a) experimental data lie on the estimated curve, suggesting that for very low load-ranges the corresponding J-integral ranges $\Delta J_{f.L}$ were too low for the crack to give a noticeable advance even in the presence of the decrease in threshold value.

Estimation of the crack growth rate under low load-range was made by subtracting crack growth due to high load-range from total crack growth. The estimated crack growth rates $(da/dN)_L^{e*}$ can be expressed as,

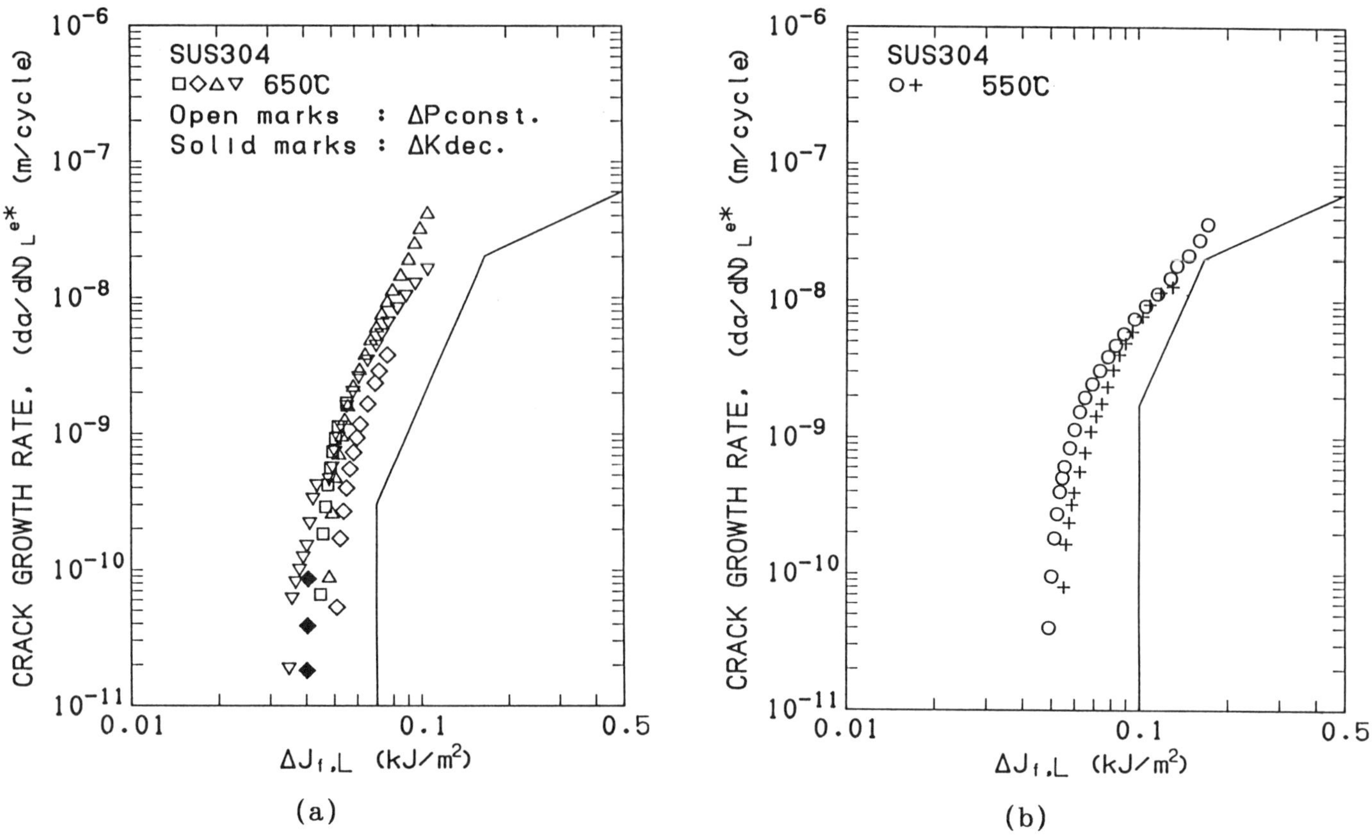


Fig. 10 - Estimated crack growth rate $(da/dN)_L^{e*}$ under low load-range $\Delta P_L$ plotted against $\Delta J_{f,L}$: (a) At 650°C; (b) At 550°C

$$(da/dN)_L^{e*} = \{(da/dN)_{HL}\times(N_H+N_L) - (da/dN)_{\Delta Jf.H}\times N_H\}/N_L \quad [13]$$

where $(da/dN)_{HL}$ denotes measured mean crack growth rate in one block of the two-step loadings, $(da/dN)_{\Delta Jf.H}$ being crack growth rate under high load-range estimated by substituting $\Delta J_{f.H}$ into Eqs. [6] and [7].

Figures 10(a) and 10(b) show the estimated crack growth rate $(da/dN)_L^{e*}$ under low load-range plotted against J-integral range $\Delta J_{f.L}$ evaluated for the low load-range $\Delta P_L$. Solid lines in the figures show the crack growth characteristics under constant amplitude loadings expressed by Eqs. [6] and [7]. It is seen that the threshold value decreased by a factor of 2 by the intermittent applications of the high-range loads. Since the value of $(da/dN)_L^{e*}$ is an average crack growth rate under low-range loads and the threshold value recovers during the application of low-range loads, the apparent fatigue threshold is expected to increase and to approach to the thresholds under constant amplitude loadings for large numbers of $N_L$.

## CONCLUSIONS

Fatigue threshold and near-threshold cycle-dependent crack growth behavior at high temperatures in air was investigated with type 304 stainless steel. An alternate current (A.C.) electric potential technique was utilized for measuring crack length. Crack growth experiments were made at 650°C and 550°C. The threshold J-integral range $(\Delta J_f)_{th}$ at 650°C was lower than that at 550°C. Above the thresholds crack growth rates were expressed in terms of the J-integral range $\Delta J$ or stress intensity range $\Delta K$. The character of the threshold for the cycle-dependent crack growth was examined by fatigue crack growth experiments under two-step variable amplitude loadings, in which a high-range load were introduced periodically in sub-threshold low-range loads. Fatigue crack was found to grow even at J-integral ranges below the threshold value under constant amplitude loadings. The estimated crack growth rates $(da/dN)_L^{e*}$ under low-range loads showed that the threshold value decreased by the intermittent introduction of the high-range loads.

## ACKNOWLEDGEMENTS

The present authors would like to acknowledge that this research was partly supported by the Grant-in-Aid for Co-operative Research (principal investigator: K. Ohji) from the Ministry of Education, Culture and Science.

The authors would like to express their appreciation to Mr. H. Enoki and Mr. S. Shiotani for their help with the experiments.

## REFERENCES

1- Siverns, M.J. and A.T. Price, *Nature*, **228**, 760-1 (1970)
2- Haigh, J.R., *Mater. Sci. Eng.*, **20**, 213-23 (1975)
3- Ohji, K., K. Ogura, S. Kubo and Y. Katada, *Proc. 21st Japan Congr. on Mater. Research*, Soc. Mater. Sci., Japan, 99-104 (1978)
4- Ellison, E.G. and M.P. Harper, *J. Strain Analysis*, **13**, 35-51 (1978)
5- Pilkington, R., *Met. Sci.*, **13**, 555-64 (1979)
6- Taira, S., R. Ohtani and T. Komatsu, *J. Eng. Mater. Technol.*, **101**, 162-7 (1979)
7- Koterazawa, R. and T. Mori, *Int. Conf. on Engineering Aspects of Creep, Sheffield,* IME, 219-224 (1980)
8- Sadananda, K. and P. Shahinian, *Creep-Fatigue-Environment Interaction* (Pelloux, R.M. and N.S. Stoloff, eds.), Metall. Soc. AIME, 86-111 (1980)
9- Larsen, J.M. and T. Nicholas, *ASTM STP* **791**, II-536-52 (1983)
10- Ohji, K., S. Kubo, N. Yamakawa and T. Kuri, *Trans. Japan Soc. Mech. Engrs.*, **50**, 1218-27 (1984)
11- Ellison, E.G., *"Subcritical Crack Growth due to Fatigue, Stress Corrosion and Creep"* (Larsson, L.H., ed.), Elsevier, 531-61 (1984)
12- Ellison, E.G. and G.G. Musicco, *ibid.*, 403-48
13- Okazaki, M., I. Hattori, T. Ikeda and T. Koizumi, *J. Eng. Mater. Technol.*, **107**, 346-55 (1985)
14- Hyde, T.H., B. Kubba, K.C. Low and J.J. Webster, *J. Strain Analysis*, **20**, 101-10 (1985)
15- Saxena, A., *ASTM STP* **905**, 185-201 (1986)
16- Yokobori, A.T., Jr., T. Yokobori, T. Kuriyama, T. Kako and Y. Kaji, *Int. Conf. on Creep, Tokyo*, JSME, 135-40 (1986)
17- Dimopulos, V., K.M. Nikbin and G.A. Webster, *Metallurg. Trans.*, **19A**, 873-80 (1988)
18- Ohji, K., S. Kubo, T. Kuri and W.S. Joo, *Int. Conf. on Creep, Tokyo*, JSME, 297-302 (1986)
19- Ohji, K., S. Kubo, T. Kuri and K. Nishizaki., *Trans. Japan Soc. Mech. Engrs.*, **51**, 2146-53 (1985)
20- Ohji, K., S. Kubo and Y. Manabe, *J. Soc. Mater. Sci., Japan*, **35**, 924-9 (1986)
21- Ohji, K., S. Kubo, H. Kinugasa and T. Gotoh, *J. Soc. Mater. Sci., Japan*, **37**, 951-6 (1988)
22- Ohji, K. and S. Kubo, *"High Temperature Creep-Fatigue"* (Ohtani, R., M. Ohnami and T. Inoue, eds.), Current Japanese Materials Research Vol. **3**, Elsevier Appl. Sci., 91-113 (1988)
23- Ohji, K. , S. Kubo, Y. Nakai and T. Gotoh, *Preprint of Japan Soc. Mech. Engrs.*, No. 890-50, 99-100 (1989)
24- Ohji, K. and S. Kubo, *"Fractography"* (Koterazawa, R., R. Ebara and S. Nishida, eds.), Current Japanese Materials Research Vol. **6**, Elsevier Appl. Sci., 105-23 (1990)
25- Ohji, K. and S. Kubo, *"Creep in Structures IUTAM Symp. Cracow/Poland"* (Zyczkowski, M., ed.), Springer, 253-68 (1991)
26- Tada, H., P. Paris and G. Irwin, *"The Stress Analysis of Cracks Handbook"*, Del Res. Corp., p.2.2 (1973)
27- Dowling, N.E., *ASTM STP* **601**, 19-32 (1976)
28- Ohji, K. , K. Ogura and S. Kubo, *Preprint of Japan Soc. Mech. Engrs.*,

No. 740-11, 207-9 (1974)
29- Ohji, K. , K. Ogura and S. Kubo, *Trans. Japan Soc. Mech. Engrs.*, **42**, 350-7 (1976)
30- Landes, J.D. and J.A. Begley, *ASTM STP* **590**, 128-48 (1976)
31- Nikbin, K.M., G.A. Webster and C.E. Turner, *ASTM STP* **601**, 47-61 (1976)
32- Kubo, S., *"Study on Mechanics of Crack Initiation and Growth"*, Doctoral Thesis, Osaka University, (1975)
33- Ohji, K. , K. Ogura and S. Kubo, *Trans. Japan Soc. Mech. Engrs.*, **44**, 1831-7 (1978)
34- Johnson, H.H., *Mater. Research Stands.*, **5**, 442-5 (1965)
35- Nishikawa, I., T. Gotoh, Y. Miyoshi and K. Ogura, *JSME Int. J.*, Ser. I, **31**, 92-9 (1988)
36- Ogura, K. and I. Nishikawa, *"Fatigue 90, Proc. 4th Int. Conf. Fatigue and Fatigue Threshold, Honolulu"* (Kitagawa, H. and T. Tanaka, eds.), Mater. Compon. Eng. Pub., III, 1413-8 (1990)
37- Kikukawa, M., M. Jono and Y. Kondo, *Trans. Japan Soc. Mech. Engrs.*, Ser. A, **47**, 468-75 (1981)
38- Kikukawa, M., M. Jono and Y. Kondo, *"Advances in Fracture Research"* (Francois, D., ed.), Pergamon Press, **4**, 1799-806 (1980).

*Proceedings of the Fifth International Conference on Creep of Materials, Lake Buena Vista, Florida, USA, 18-21 May 1992*

# Development of a New 0.1C-11Cr-2W-0.4Mo-1Cu Steel for Large Diameter and Thick Wall Pipe for Boilers

**A. Iseda, Y. Sawaragi, and S. Kato**
Sumitomo Metal Industries, Ltd.
Amagasaki, Japan

**F. Masuyama**
Mitsubishi Heavy Industries, Ltd.
Nagasaki, Japan

## ABSTRACT

A 0.1C-11Cr-2W-0.4Mo-1Cu large diameter and thick wall pipe (HCM12A) which has higher strength than P91 and superior weldability to X20CrMoV121 has been developed for use in conventional and ultra super critical (USC) boilers. In the developed steel, the addition of Cu less than 2mass%Cu is effective to suppress δ-ferrite formation and to increase the resistance to softening in HAZ of welded joints, without deterioration in creep rupture strength and ductility. Excess Ni addition with more than 0.4mass% deteriorates the long-term creep rupture strength because Ni accelerates the precipitation of Laves phase. In addition, substituting high W for a part of Mo and a slight amount of B addition show strongly effective in the long-term creep rupture strength. The manufactured pipe with 350mm dia. and 50mm thick shows superior creep strength to P91 and can be applicable for boilers.

HIGH CHROMIUM FERRITIC heat resistant steels with 9-12 mass%Cr have superior high temperature strength and hot corrosion resistance as compared with low alloy steels such as P22 (2 1/4-1Mo). These steels have high heat-transfer coefficient and low thermal expansion coefficient, which are powerful effect on improvement of the thermal fatigue resistance of thick wall pipe, as compared with austenitic stainless steels such as TP304H and TP316H. Many 9-12Cr steel tubes and pipe have been developed for use in fossil fuel power generation boilers, chemical plants and fast breeder reactors. Among them, Modified 9Cr-1Mo steel designated as ASME P91 and T91 ranks one of the highest strength and is widely used for heat exchanging tubes and large diameter and thick wall pipe. Recently, the 9Cr-0.5Mo-1.8W steel (NF616) has been developed. The strength of NF616 shows superior to P91 due to high W content[1].

X20CrMoV121 designated as DIN 17175 is one of typical 12Cr steels for tube and pipe materials. The strength of X20 is, however, lower than those of the above 9Cr steels. Furthermore, weldability and formability of X20 are not sufficient because of high carbon content with 0.2 mass%. To improve these disadvantages, a 0.1C-12Cr-1Mo-1W-V-Nb steel tube (HCM12) which contains δ-ferrite has been developed[2)-5)]. The creep rupture strengths of the base metal and the welded joint of HCM12 show higher than those of P91 and T91[6]. This steel tube has been used as superheater and reheater for a power plant for 9 years[2]. The superior strength and corrosion resistance of HCM12 have been proved by evaluation after service exposure[4)5)].

As for large diameter and thick wall pipe, however, the existence of δ-ferrite decreases toughness. In addition, recent power generating plants are required the higher strength ferritic steels for thick wall pipe at high temperatures than those of conventional 9-12Cr steels in order to maximize the heat efficiency and to reduce the thermal fatigue during the two shift operation.

To improve these disadvantages of conventional 9-12Cr steels, the new 0.1C-11Cr-2W-0.4Mo-1Cu steel (HCM12A) for large diameter and thick wall pipe has been developed by Sumitomo Metal Industries Ltd. and Mitsubishi Heavy Industries Ltd. This paper describes the effects of alloying elements on toughness, creep rupture strength and weldability of 12Cr steels. The developed steel pipe was trial manufactured and the properties were evaluated.

## DEVELOPMENT PHILOSOPHY OF 0.1C-11Cr-2W-0.4Mo-1Cu STEEL

Fig. 1 shows the philosophy for the development of the new 0.1C-11Cr-2W-0.4Mo-1Cu steel (HCM12A) applied to large diameter and thick wall pipe. To improve corrosion resistance as compared with 9Cr steels, 11mass%Cr was selected for this steel. Finely dispersed precipitates of VN and Nb(C,N) and solid-solution strengthening due to high W and low Mo addition were effective in creep strength[7]. These strengthening elements were optimized by various experiments[3)4)]. A slight amount of boron has powerful effect on the long-term creep rupture strength. Ni, which can reduce the formation of δ-ferrite, is lowered than 0.4mass%, because it is confirmed to decrease in the long-term creep strength as discussed later.

Improving poor weldability of conventional X20 steel, carbon content was reduced to 0.12mass% or less considering weldability and toughness. Although δ-ferrite is effective in SCC resistance and in improving welded HAZ strength, δ-ferrite significantly decreases toughness. Therefore, the thick wall pipe should be reduced δ-ferrite. In order to suppress δ-ferrite formation, it can be achieved by decrease in Cr equivalent[8] lower than 9mass% as shown in Fig.2. Cu was chosen for this purpose instead of Ni, Mn and carbon which decrease in the creep rupture strength. In addition, Cu addition less than 2mass% maintains high $Ac_1$ temperature as compared with Ni and nitrogen as shown in Fig.3. Lowering Si with less than 0.1 mass% can suppress the heating embrittlement which is caused by Laves phase precipitation during a long-term heating[9].

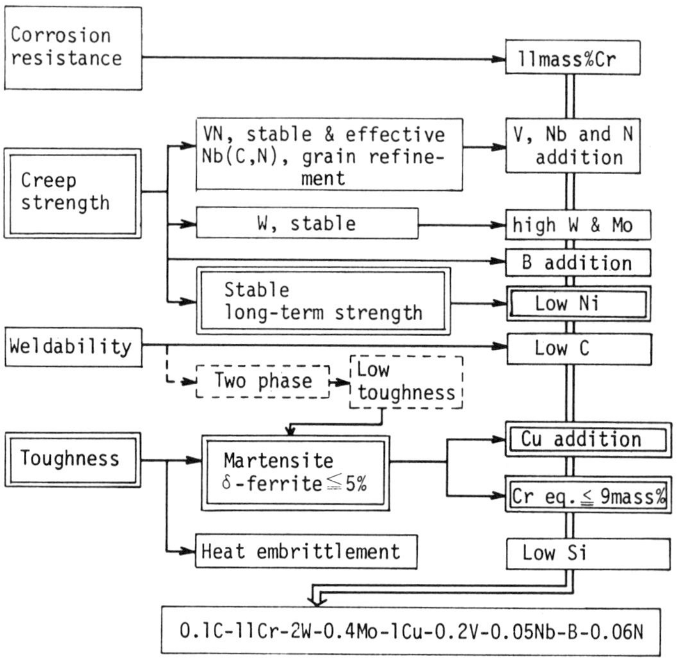


Fig. 1 Philosophy for development of 0.1C-11Cr-2W-0.4Mo-1Cu large diameter and thick wall pipe (HCM12A).

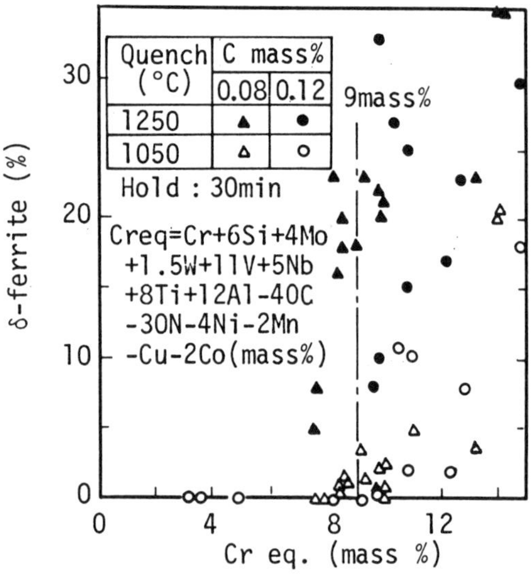


Fig. 2 Relation between Cr eq and δ-ferrite content of 11Cr-W-Mo-V-Nb-Cu steels.

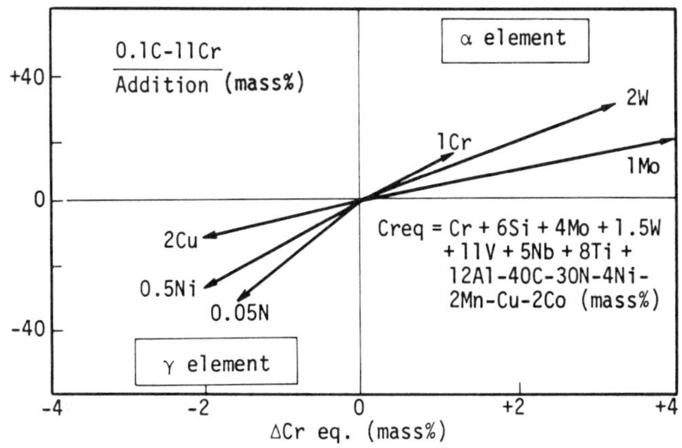


Fig. 3 Effects of alloying elements on $Ac_1$ transformation temperature and Cr equivalent for 0.1C-11Cr steels.

## EXPERIMENTAL PROCEDURE

Table 1 shows the chemical composition range and typical compositions of test materials. These steels were melted in a 150kg vacuum induction furnace, forged to 15mm thick plates, normalized at 1050°C for 1h and tempered at 770°C for 3h or 800°C for 1h. Charpy impact tests were conducted using 10 x 10 x 55mm, 2mm-V notched specimens. Creep rupture tests were carried out using 6mm dia. x 30mm gauge length tensile specimens at 550-700°C. Microstructural observation after creep rupture was conducted by using a carbon extracted replica and a thin foil by the transmission electron microscope. The extracted residues before and after creep tests were identified by X-ray diffraction and the chemical compositions of them were analyzed. The residues were electrolytically extracted at the constant current of $10mV/cm^2$ for 5h in the non-aqueous solution. In order to evaluate the softening behavior in the heat affected zone (HAZ) of welded joints, simulated HAZ specimens were prepared by heating at 830-950°C for 1min and air-cooling, and then were post-heat treated at 740°C for 5h.

Table 1. Chemical compositions of typical test materials (mass%)

| Steel | C | Si | Mn | P | S | Ni | Cr | Mo | W | V | Nb | Cu | B | N |
|---|---|---|---|---|---|---|---|---|---|---|---|---|---|---|
| A | 0.04 / 0.2 | <0.1 | 0.5 | <0.02 | <0.01 | 0 / 1.2 | 10 / 12 | 0 / 2.1 | 0 / 2.4 | 0.2 | 0.05 | 0 / 3 | 0 / 0.004 | 0.06 |
| B | 0.12 | 0.07 | 0.52 | 0.015 | 0.002 | 0.32 | 11.12 | 0.42 | 2.15 | 0.25 | 0.07 | 0.87 | 0.0018 | 0.067 |
| C | 0.11 | 0.07 | 0.57 | 0.014 | 0.001 | 0.84 | 10.50 | 0.38 | 1.89 | 0.22 | 0.05 | 0.94 | 0.0028 | 0.061 |
| D | 0.10 | 0.06 | 0.51 | 0.012 | 0.001 | 1.18 | 11.04 | 0.27 | 2.39 | 0.25 | 0.07 | 0.99 | 0.0021 | 0.069 |
| E | 0.12 | 0.07 | 0.59 | 0.012 | 0.002 | 0.12 | 11.05 | 0.40 | 2.01 | 0.20 | 0.06 | 1.96 | 0.0017 | 0.067 |

## EFFECT OF δ-FERRITE ON TOUGHNESS

Fig. 4 shows the relationship between δ-ferrite and Charpy absorbed energy at room temperature for 0.1C-11Cr-W-Mo-Cu steels. The increase in δ-ferrite tends to decrease in toughness even for longitudinal direction of thin plate. Considering thick wall pipe, δ-ferrite should be reduced to less than 5vol%. As shown in Fig.2, δ-ferrite can be suppressed less than 9mass%Cr equivalent for 1050°C normalizing.

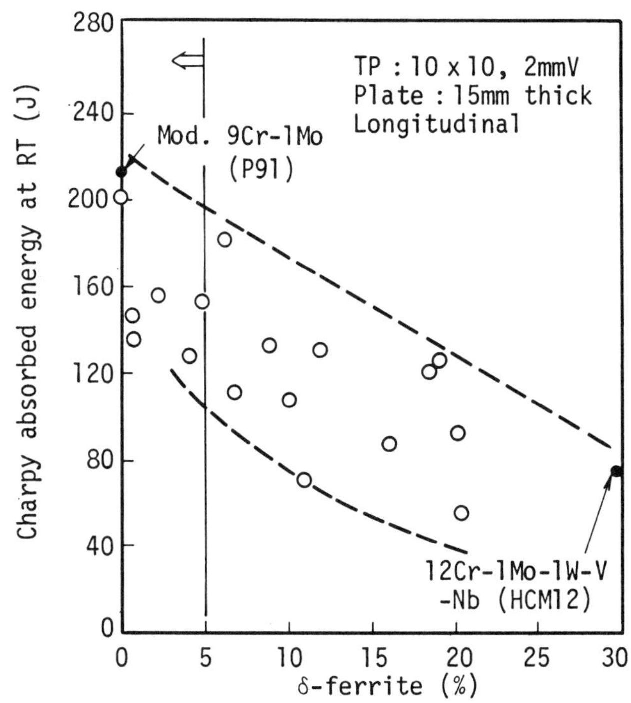


Fig. 4 Relationship between δ-ferrite content and Charpy absorbed energy at room temperature. (1050°C x 1h AC + 800°C x 1h AC)

## EFFECTS OF ALLOYING ELEMENTS ON CREEP RUPTURE STRENGTH

Optimum amounts of W and Mo were previously discussed and determined as Mo + 1/2W=1.5mass% and 1mass%Mo and 1mass%W at 600°C for 12Cr steels[5]. At high temperature over 600°C, substituting high W for a part of Mo is effective[10)11)].

Fig. 5 shows the relationship between stress and creep rupture time of 0.1C-11Cr-W-Mo-0.25V-0.05Nb steels. The optimum amounts of V and Nb were added based on the previous experiments [2)5)]. A slight increase in strength was observed by substituting W for a part of Mo. As considered with strength and Cr equivalent, 2mass%W and 0.4mass%Mo were selected for the developed steel.

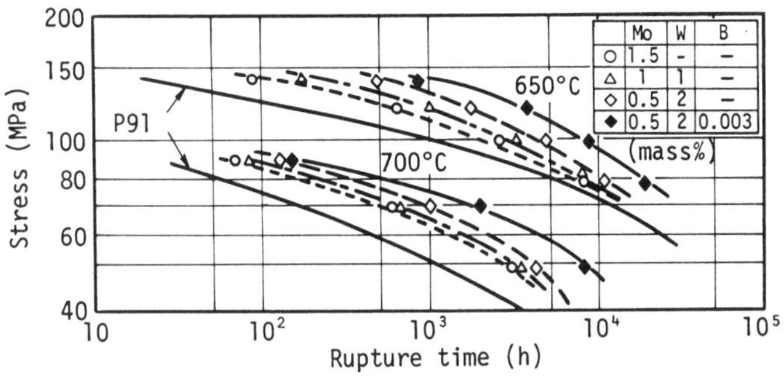


Fig. 5 Effects of Mo,W and B on creep rupture strength of 0.1C-11Cr-Mo -(W)-(B)-0.25V-0.05Nb steels.

In addition, it was confirmed that the creep rupture strength was significantly improved by a few amount of boron and that the strength of the boron containing steel kept stable for over $10^4$h at 650 and 700°C.

Fig. 6 shows the effect of Ni on creep rupture strength of 0.1C-11Cr-2W-0.4Mo-1Cu steels (steel B, C and D listed in Table 1). At the shorter period up to 2000h, these steels have almost the same strength. At the longer period over 2000h, the increase in Ni decreased the creep rupture strength significantly. Fig.7 shows the effects of Cu and Ni contents on the extrapolated creep rupture strength at 600°C for $10^5$ h, which was given by the Larson-Miller parametric method. Ni decreased the creep rupture strength over 0.4mass%. In contrast, Cu did not decrease the strength up to 2mass%. These differences are discussed later.

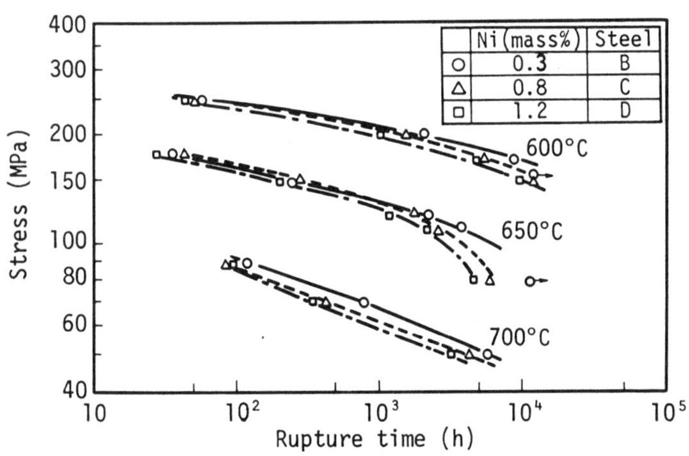


Fig. 6 Effect of Ni on creep rupture strength of 0.1C-11Cr-2W-0.4Mo-1Cu steels.

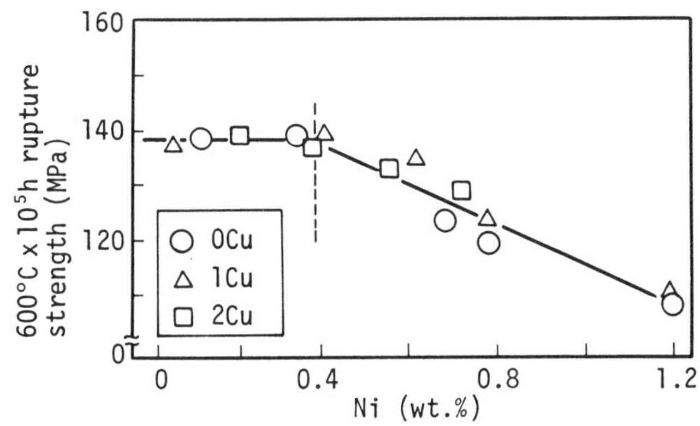


Fig. 7 Effects of Cu and Ni on the extrapolated creep rupture strength of 0.1C-11Cr-2W-0.4Mo-Cu steels.

Considering both high toughness and high creep strength, Cu is effective due to suppress δ-ferrite formation and to be not harmful for creep strength. Conclusively, the optimum contents of Cu and Ni were selected 1mass%Cu and 0.35mass%Ni. respectively.

## WELDABILITY AND TEMPER SOFTENING IN SIMULATED HAZ

As shown in Fig.8, the effect of carbon on the susceptibility to cold weld cracking are compared by the y-groove restraint weld cracking test (Japanese Code JIS Z3158) for 11Cr-2W-0.4Mo-1Cu-0.2V-0.05Nb-0.003B-0.06N steels. The susceptibility to cracking decreased by lowering carbon content. The cracking was completely suppressed when the preheating temperature was adjusted at over 150°C and carbon content was controlled less than 0.12mass%. These results confirmed that the developed steel which contained 0.12mass%C or less shows almost the same weldability as conventional low alloy steels such as P22 and that shows superior weldability to X20.

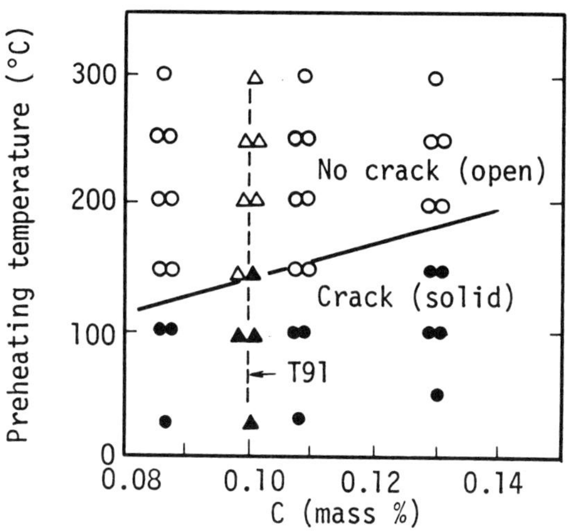


Fig. 8 Result of y-groove restraint weld cracking tests of 11Cr-2W-0.4Mo-1Cu-V-Nb-0.003B steels.

As shown in Fig.9, the temper softening behavior of various Cu contents was evaluated by simulated HAZ heating at above and below $Ac_1$ temperatures. The $Ac_1$ temperatures of test materials decreased from 840 to 820°C in the proportion to increase in Cu. The minimum hardness was observed in the heating just above $Ac_1$ temperature for each steel. The hardness of the simulated HAZ increased with increase in Cu content. The difference in hardness after normalizing and tempering was small for these steels, because Cu does not harden for the steel after high temperature tempering. The addition of excess Cu, however, deteriorates creep ductility.

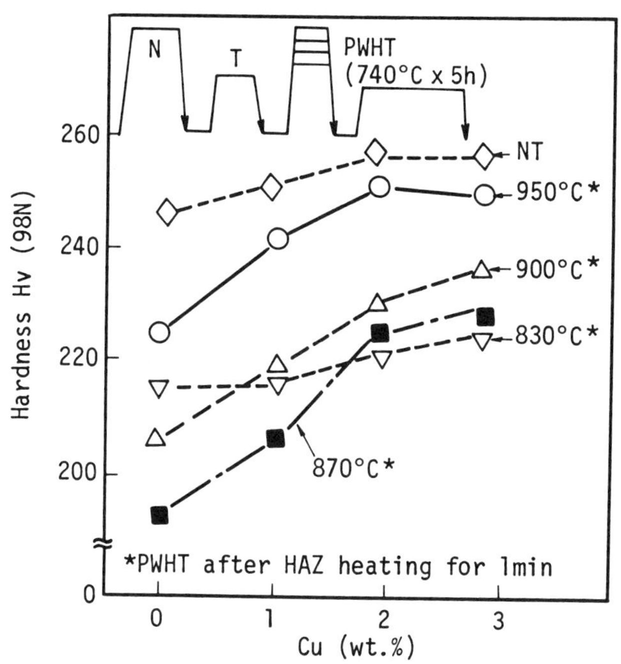


Fig. 9 Effect of Cu on hardness of 0.1C-11Cr-2W-0.4Mo-Cu steels after simulated HAZ heating.

**EFFECTS OF Ni AND Cu ON MICROSTRUCTURE AFTER CREEP**

Fig. 10 shows transmission electron micrographs of extracted replicas of the steels with 0.3mass%Ni (steel B) and 1.2mass%Ni (steel D) after creep ruptured at 650°C and 108MPa. Fine and coarse carbide, nitride and Laves phase which consists of $(Fe,Cr)_2(W,Mo)$ were observed as shown in Fig.11. These precipitates were almost the same size for both steels. However,

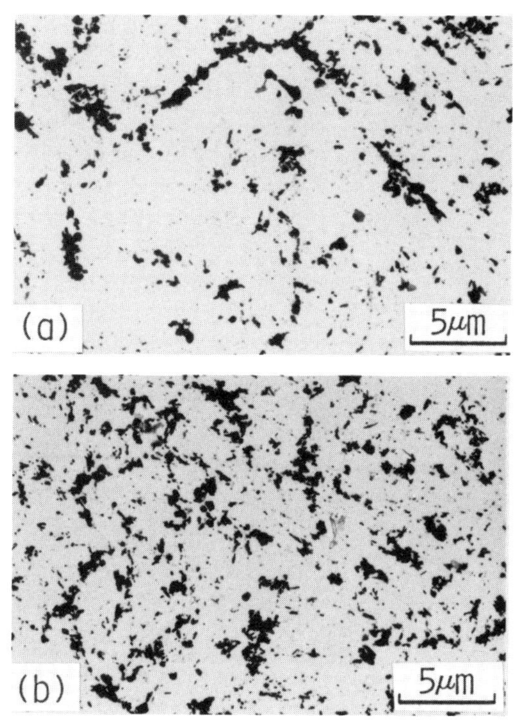


Fig. 10 Transmission electron micrographs of extracted replicas of 0.1C-11Cr-2W-0.4Mo-1Cu steels after creep ruptured at 650°C and 108MPa. (a) 0.3mass%Ni (b) 1.2mass%Ni

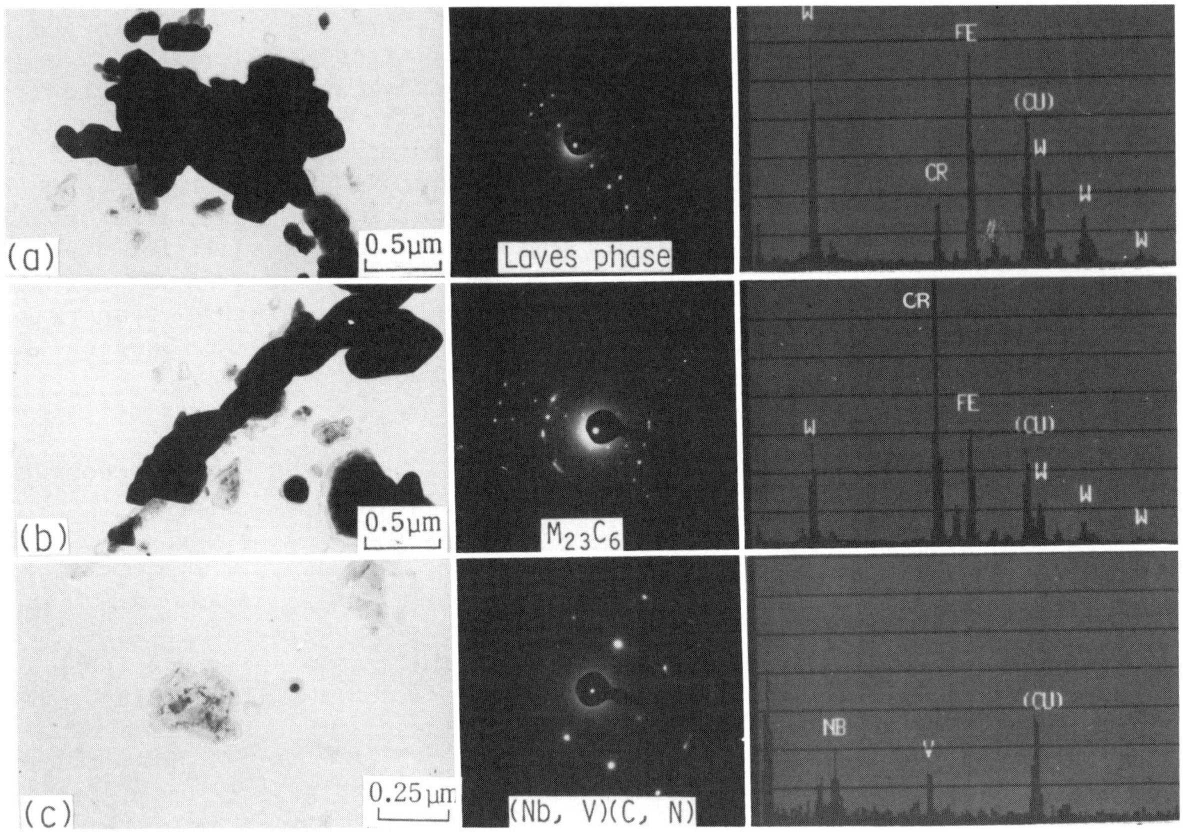


Fig. 11 EDX analysis and diffraction pattern of precipitates of the steel B after creep ruptured at 650°C for 3627h.
(a) grain boundary, (b) lath boundary, (c) within lath

the precipitation in the steel D was significantly large as compared with the steel B, even though the steel D creep-ruptured for 2053h which was significantly shorter than 3627h for steel B.

The amounts of extracted residues after creep rupture are shown in Fig.12. In the condition of as-normalizing and tempering, the amounts of residues and their compositions for the steel D were almost the same as those for the steel B. These residues were identified as $M_{23}C_6$, VN and NbC by X-ray diffraction. As for the creep-ruptured specimens, the amounts of Fe and W in the residues increased for both steels and especially the steel D had higher amounts of residues than the steel B.

It is suggested that precipitation during creep was accelerated by excess Ni addition. In Fig.12, **the** amounts of residues increased after creep. The atomic fraction of the increased residues after creep is almost the same for each steel. Therefore, the increase in residure during creep is mainly Laves phase, which is also identified as shown in Fig. 11.

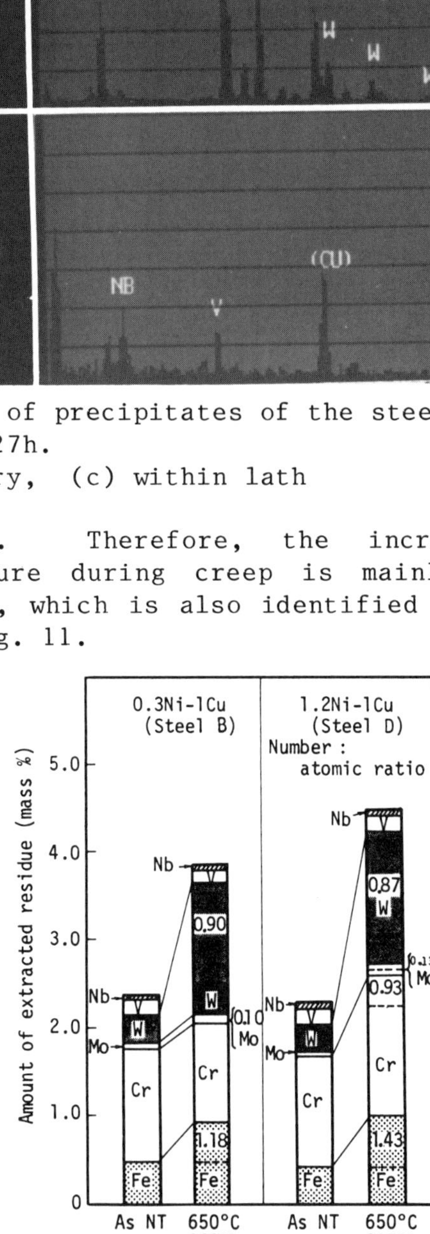


Fig. 12 Effect of Ni on the amounts of extracted residues after creep rupture at 650°C and 108MPa.

The steel D with 1.2mass%Ni shows larger amounts of residues than those of the steel B with 0.3mass%Ni. It is reported that carbide and nitride completely precipitated during tempering for 9-12Cr steels[9]. Therefore, Laves phase is considered to precipitate mainly during creep. because the composition in $M_{23}C_6$ may be not to change remarkably during creep[9]. It is concluded that the steels, which contain excess Ni with higher than 0.4mass%, decrease the creep rupture strength, because that the Laves phase precipitation is accelerated by excess Ni addition. In addition, the solution strengthening effect of W decreases and then the martensitic structure changes to polygonal ferrite which reduces the creep rupture strength at long-term creep stage. As for the similar effect of Ni, it is reported that the creep strength is reduced by Ni addition due to coarsen fine carbides during creep[12].

As for Cu addition up to 2mass%, any difference of the amounts and compositions in extracted residues was not confirmed. However, in the case of 2mass%Cu (steel E) after creep ruptured at 650°C and 98MPa for 5306h, fine Cu phase precipitated on the lath boundaries and within lath as shown in Fig.13. No difference of precipitation of Cu phase was observed for low Ni and high Ni steels. These precipitates are thought not to be effective in creep strength.

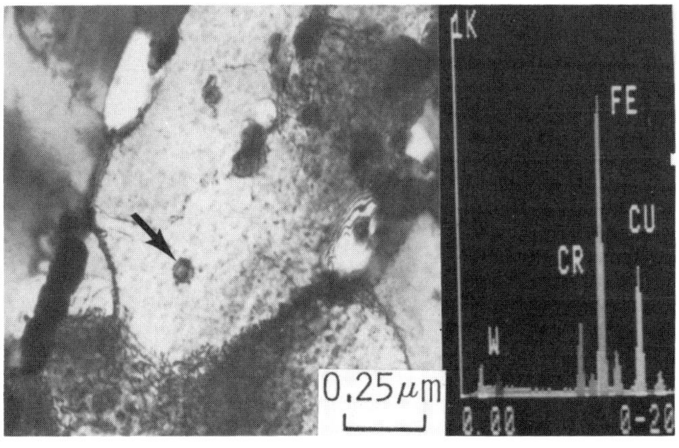


Fig. 13 Transmission electron micrographs of thin foil of the steel E (2mass%Cu) after creep-rupture at 650°C and 98MPa for 5306h. Fe, Cr and W peaks indicate the matrix composition.

Considering with creep ductility, Cu should be reduced less than 2mass% to suppress the precipitation of Cu phase and to improve ductility

## PROPERTIES OF HCM12A LARGE DIAMETER AND THICK WALL PIPE

Based on the above experiments, the optimum compositions were selected for the developed steel (HCM12A). Table 2 shows typical compositions and specification of composition range for HCM12A steel pipe. The test materials were melted in an electric furnace, refined by the VOD process and casted to ingots. Some ingots were remelted by the ESR process and the others were directly forged and rolled to 350mm dia. and 50mm thick pipe. These pipes were normalized at 1050°C and then tempered at 770°C.

Table 2. Typical compositions of HCM12A steel pipe

| Heat | C | Si | Mn | P | S | Ni | Cr | Mo | W | V | Nb | Cu | B | N |
|---|---|---|---|---|---|---|---|---|---|---|---|---|---|---|
| VOD | 0.11 | 0.02 | 0.56 | 0.013 | 0.001 | 0.33 | 10.80 | 0.42 | 1.94 | 0.19 | 0.05 | 0.87 | 0.0030 | 0.053 |
| VOD/ESR | 0.12 | 0.05 | 0.64 | 0.016 | 0.001 | 0.32 | 10.61 | 0.41 | 1.87 | 0.21 | 0.05 | 0.86 | 0.0022 | 0.064 |
| HCM12A (Spec.) | ≤0.15 | ≤0.70 | ≤0.70 | ≤0.030 | ≤0.020 | ≤0.70 | 10.00/12.60 | 0.20/0.60 | 1.50/2.50 | 0.15/0.30 | 0.02/0.10 | ≤2.5 | ≤0.0050 | 0.020/0.100 |

Fig.14 shows Charpy impact properties for the longitudinal direction. The ESR process is effective in purification of ingots to remove inclusions and macrosegregation. Both processes, however, show almost the same toughness and toughness is sufficiently high for practical use. Fig.15 shows tensile properties at temperatures from 20°C to 650°C. Both pipes have almost the same and sufficient strength and ductility. Fig. 16 shows creep rupture properties for both steels as compared with plate. Creep strengths for these pipes are stable up to the long-term creep stage and the same as those for plate. The average strength curves by means of Larson-Miller parametric analysis show remarkably higher strength than those of P91. The extrapolated strength at 600°C for $10^5$h is about 130MPa. Based on the properties of the manufactured pipe, it is concluded that the developed HCM12A steel has higher strength than conventional P91 steel and can be applicable for large diameter and thick wall pipe for use in conventional and USC power generation boilers. To confirm the manufacturing for pipe with various size and process and to collect the practical engineering data, further study is progressing.

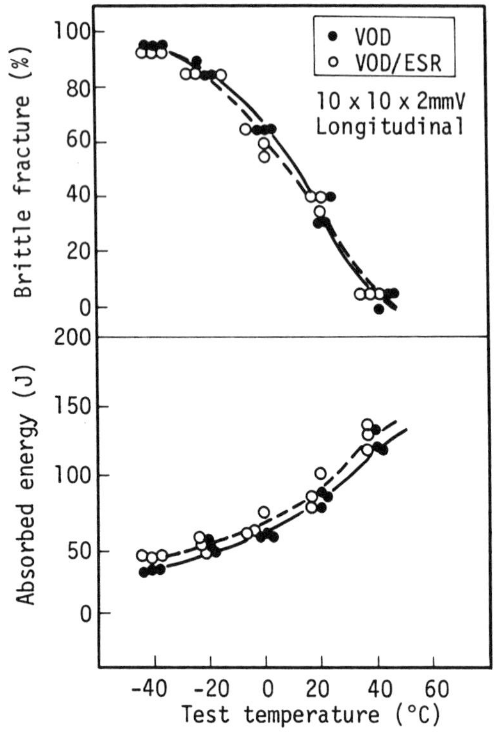


Fig. 14 Charpy impact properties of HCM12A steel pipe.

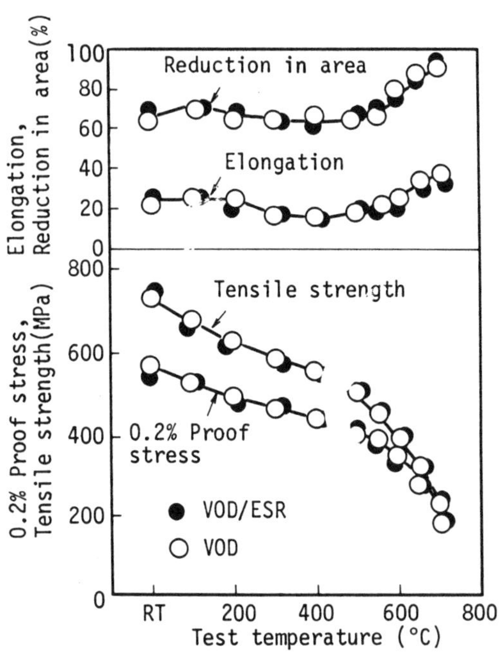


Fig. 15 High temperature tensile properties of H2M12A pipe.

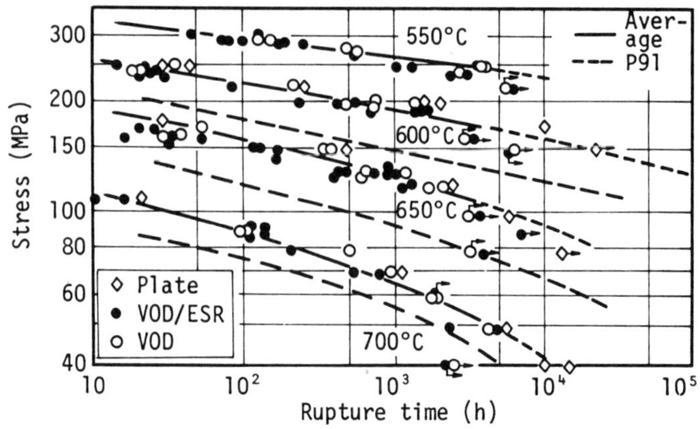


Fig.16 Creep rupture properties of HCM12A pipe and plate.

## CONCLUSIONS

1) The 0.1C-11Cr-2W-0.4Mo-1Cu steel pipe (HCM12A) has newly developed for use in power generation boilers. This steel contains 1mass%Cu to improve toughness due to suppress δ-ferrite less than **5vol%.** Cu is proved to increase the resistance to softening in HAZ and to keep stable the creep rupture strength. In the case of 2mass%Cu, fine Cu phase precipitates on grain boundaries and within lath martensite after long-term creep at 650°C.
2) Ni addition with more than 0.4mass% decreases the long-term creep strength at 650 and 700°C. It seems that the solid-solution strengthening effect of W is reduced due to the acceleration of Laves phase precipitation.
3) Substituting W for Mo, 2mass%W and 0.4mass%Mo addition is effective in the creep rupture strength at over 600°C. A few amount of boron can strongly improve the strength for the developed steel.
4) Cold weld cracking of weldment for the developed steel is prevented when the preheating is conducted at over 150°C, which shows the same weldability as low alloy steels and superior to X20CrMoV121 due to decrease in carbon less than 0.12mass%.

5) Thick wall pipe of the developed steel is proved to have higher creep rupture strength than that of P91 and sufficient toughness for practical use in boilers.

## REFERENCES


1) H.Mimura, M.Ohgami, H.Naoi, Y.Sogo, T.Ogawa and T.Fujita: Third Int.Conf. on Improved Coal-Fired Power Plants, San Francisco, Apr.2-4 (1991)
2) K.Yoshikawa, A.Iseda, H.Teranishi, F.Masuyama, T.Daikoku and H.Haneda: Int. Conf. on High Temp. Alloys, Petten, Oct.15-17 (1985), 247.
3) K.Yoshikawa, A.Iseda, M.Yano, F.Masuyama, T.Daikoku and H.Haneda: First Int. Conf. on Improved Coal Fired Power Plants, Palo Alto, Nov. 19-21 (1986), 5-185.
4) A.Iseda, Y.Sawaragi, H.Teranishi, M.Kubota and Y.Hayase: The Sumitomo Search, No.40, Nov. (1989), 41.
5) F.Masuyama, T.Daikoku, K.Yoshikawa and A.Iseda: Int. Conf. on Advances in Materials Tech. for Fossil Power Plants, EPRI, Illinois (1987), Session 4, Paper No.5.
6) A.Iseda, Y.Sawaragi and F.Masuyama: Heat-Resistant Materials, Proc. 1st Int. Conf., Fontana, USA, Sep.23-25 (1991), 577.
7) A.Iseda, H.Teranishi and F.Masuyama: Tetsu-to Hagané, 76 (1990), 1076.
8) P.Patriarca, S.D.Harkness, J.M.Duke and L.R.Cooper: Nuclear Tech., 28 (1976), 516.
9) A.Iseda, H.Teranishi and K.Yoshikawa: Tetsu-to-Hagané, 76 (1990), 2190.
10) T.Fujita: COST-EPRI Workshop, Schaffhausen, (1986)
11) H.Yasuda, M.Sakakibara, T.Takahashi, H.Naoi and H.Masumoto: The Thermal and Nuclear Power, 39 (1988), 517.
12) T.Marrison and A.Hogg: Creep Strength in Steel and High-temperature alloys, Sheffield, Sep.20-22 (1972), 242.

*Proceedings of the Fifth International Conference on Creep of Materials, Lake Buena Vista, Florida, USA, 18-21 May 1992*

# Effects of Post-Weld Heat-Treatment on Creep Behavior of 2.25Cr-1Mo Ferritic Steel Base, Weld Metal and Weldments

**K. Laha, K.B.S. Rao, and S.L. Mannan**
Materials Developments Division
Indira Gandhi Centre for Atomic Research
Tamilnadu, India

## ABSTRACT

In this paper, experimental results involving the effects of post-weld heat-treatment (PWHT) and test temperature on creep behaviour of 2.25Cr-1Mo ferritic steel are presented. The material conditions examined include base metal, weld metal and composites(comprising base,weld and heat affected zone (HAZ)). Creep rupture tests were conducted at 773 and 823 K over a stress range of 80-300 MPa. Specimens for creep testing were taken from single-V-weld pads, fabricated by shield metal arc welding process using basic coated 2.25Cr-1Mo electrodes. PWHTs of 973 K/1h and 1023 K/1h were given to the weld pads prior to machining the creep samples. Microstructure and hardness in the as-welded, post-weld heat-treated and creep-tested conditions were evaluated. In both the PWHT conditions, the creep rupture strength displayed by three material states is in the order: weld metal > base metal > composites. Base, all-weld and composites subjected to PWHT of 1023 K/1h yielded lower rupture strength values than the respective material conditions with a PWHT of 973 K/1h at lower applied stresses. Composite specimens deformed at higher creep rates while the weld metal deformed at lower rates than the base material. Higher PWHT generally caused larger steady state creep rates in all the three material conditions. Creep rupture ductility decreased with increasing rupture time. Higher PWHT temperature produced larger values of percentage elongation. The rupture ductility exhibited by the base material was significantly higher than that shown by weld metal and composite specimens. At all the test conditions, failure resulted by transgranular fracture. Failure in the composite specimens occurred in the intercritical region of the HAZ.The influence of microstructure on creep failure of composite specimens is critically examined.

## INTRODUCTION

2.25Cr-1Mo ferritic steel is used extensively as a structural material in the petroleum, chemical process and power generating industries. The creep properties of 2.25Cr-1Mo have been extensively studied in the annealed (1), quenched-plus-tempered(2) and normalised -plus-tempered conditions (1,3). Under creep conditions, the performance of 2.25Cr-1Mo weldments is considered to be a life limiting factor. Information on creep property data available in the literature on 2.25Cr- 1Mo weld metal has been reviewed recently by Lundin et al(4). However,the information on creep behaviour of weldments consisting of three interacting component parts, the weld metal, its attendant HAZs and the adjacent base material is lacking. In actual structures, fabricated by welding, a high percentage of the failures have been reported to occur in the microstructurally altered and inhomogenious region in the base metal,i.e, the HAZ (5-7). The detailed microstructure in the HAZ is extremely complex and is controlled by the interaction of thermal fields, produced by the heat input from the welding process, and the phase transformation and grain growth characteristics of the materials being welded (8-10 ).Further modifications in microstructure can occur as a result of tempering either during the later stages of welding and post-weld heat-treatment (PWHT) or during service (8).These microstructures which generally vary from wrought base material through transformed HAZs to cast weld metal, can have greatly different mechanical properties. As a consequence cracking can arise both in the weld metal and HAZ when the weldment is subjected to stresses and strains associated with service conditions. This study is concerned with the effects of PWHT and test temperature on the creep behaviour of 2.25Cr-1Mo ferritic steel base material, weld metal and weldments i.e composite specimens. Creep properties of weldments were compared with those of base and all-weld metals determined under identical stress conditions at 773 K and 823 K. In view of the scarcity of creep property data on 2.25Cr-1Mo welds and weldments, the present results are expected to serve as a valuable source of information in the design of 2.25Cr-1Mo components fabricated by weldinng.

## EXPERIMENTAL DETAILS

**Welding Procedure and Specimen Details.** 2.25Cr-1Mo steel in normalise plus-tempered condition (1223 K for 17 min,AC and 1003 K/1h, AC) was supplid by Thyssen Stahl Aktiengsell, FRG. Details of welding procedures adopted are discussed esle where (8). The chemical compositions of base metal and deposited weld metal are given in Table.1. Subsequent to welding, the weld pads were examined, non-destructively by X- radiography to ensure their soundness and was later subjected to the post-weld heat-treatments at 973 K for 1h and 1023 K for 1h followed by air cooling. Base metal, weld metal and composite creep specimens of 5 mm gauge diameter and 50 mm gauge length were machined from the weld pads subjected to PWHTs.

**Table.1 Chemical composition** (wt%)

| Element | Material | |
|---|---|---|
| | Base Metal | Weld Metal |
| C | 0.06 | 0.05 |
| Si | 0.18 | 0.40 |
| P | 0.008 | 0.020 |
| S | 0.008 | 0.012 |
| Cr | 2.18 | 2.30 |
| Mo | 0.93 | 1.10 |
| Fe | Balance | Balance |

**Metallography and Hardness Testing** Metalographic examination of the polished and etched sections of weldments was carried out in the as-welded, post-weld heat-treated and creep-tested conditions by using Richert MeF optical microscope. The etchant used was 2% nital. Fractography of creep tested samples was carried out employing a Philips PSEM 501 scanning electron microscope. Vickers hardnes measurements under a load of 50 gms were carried out on as-welded, post-weld heat-treated and creep-tested composite specimens.

**Creep Testing.** Uniaxial constant-load creep tests were conducted at 773 and 823 K in air on base, all-weld and weldment specimens in both the PWHT condition. The tests were performed employing stresses in the range 80-300 MPa. The temperature along the gauge portion of the specimens was controlled to better than ± 2K. The creep strain was monitored continuously during the creep tests.

## RESULTS

**Creep Behaviour.** Fig.1 describes the typical creep curves of the base metal, all-weld metal and composite . In both the PWHT conditions, the creep behaviour of base, weld as well as composites was characterised by a small instantaneous strain on loading, followed by well-

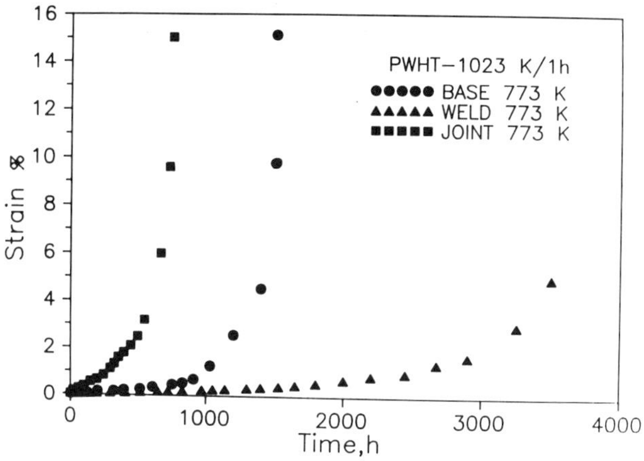


Fig.1 Creep curves for base, weld and composite specimens(PWHT 1023 K/1h) at 773 K and 200 MPa

defined primary, secondary and tertiary creep regions. The composite specimens exhibited shorter primary and secondary stages of creep compared with the base and weld metals. The varition of steady state creep rate with applied stress at 773 and 823 K is shown in Figs.2 and 3 respectively. Generally, composite specimens deformed at higher creep rates while weld metal showed lower creep rates than base material in both the PWHT conditions. The steady state creep rate generally increased with increase in PWHT temperature.The stress dependence of the steady state creep rate could be described by a power law relation $\dot{\varepsilon}_s = A\sigma^n$ where n is the stress index and A is material constant. Stress exponent n was found to have lower values for higher PWHT temperature.

The log-log varition of rupture life with applied stress at 773 and 823 K are shown in Figs. 4 and 5 respectively.

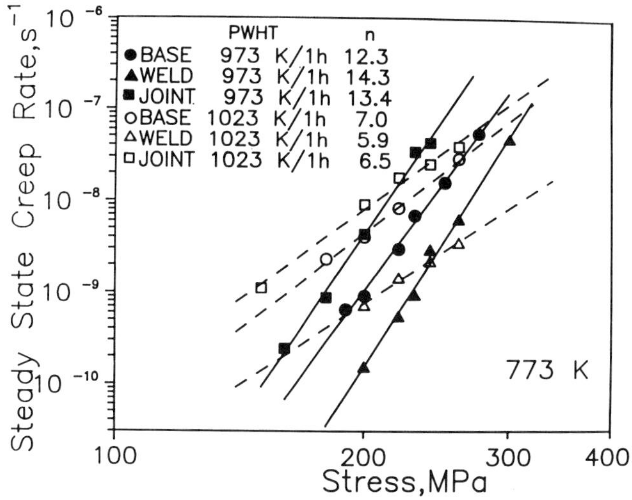


Fig.2 Steady state creep rate as a function of applied stress for base, weld and composite specimens at 773 K.

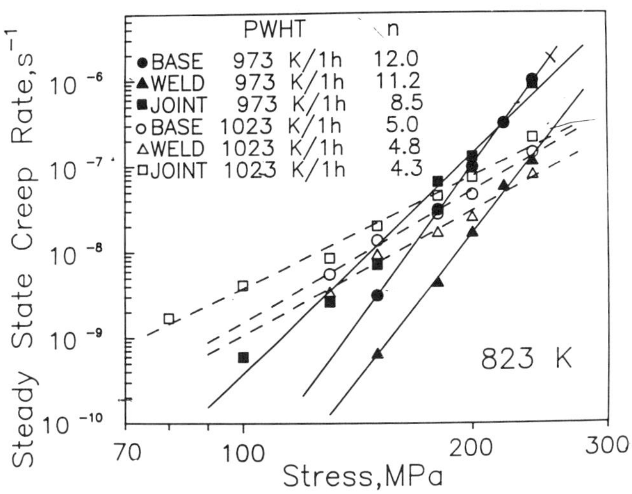


Fig.3. Steady state creep rate as a function of applied stress for, weld and composite specimens at 823 K.

In both the PWHT conditions at 773 K (Fig.4) the weld metal showed better creep rupture life compared to base material, while composite specimen displayed lower life. At 823 K too, composites were found to have lower rupture life than the base metal at all stress levels investigated (Fig.5). At 823 K the weld metal rupture life was found higher than the base material only above a transition stress(Fig.5). The transition stress decreased with increase in PWHT temperature (Fig.5). Higher PWHT temperature was found detrimental with respect to rupture life in all the material conditions specially at lower applied stresses.

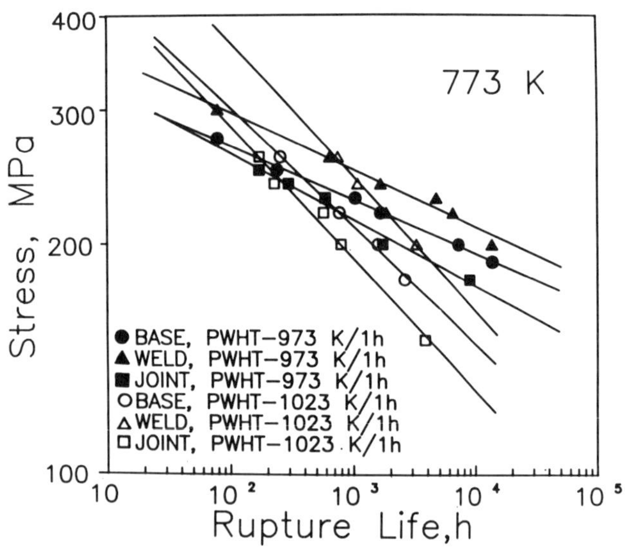


Fig.4.Applied stress vs. creep rupture life plots at 773 K.

Creep rupture ductility (%elongation) of all the material conditions decreased with increasing rupture time (Fig.6). Higher PWHT caused larger values of percentage elongation. Irrespective of PWHT and testing temperature, generally the rupture elongation exhibited by the three material conditions is in the order: base > weld > composite.

**Microstructure.** The microsturcture of 2.25Cr-1Mo base metal in the normalised and tempered condition was composed of 70% granular bainite and 30 %

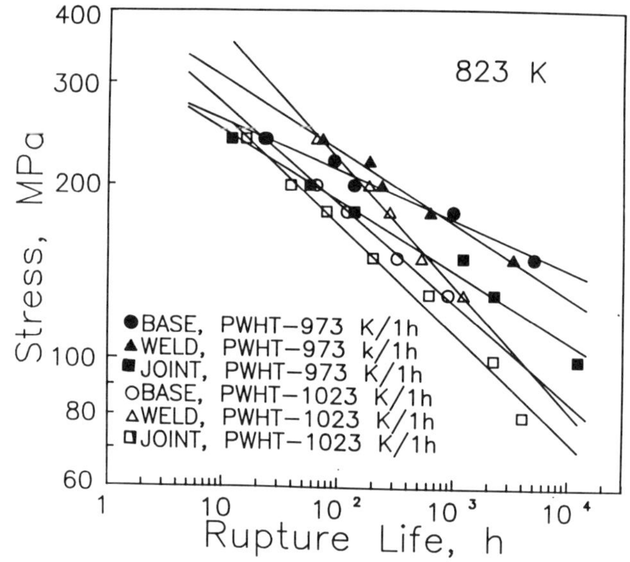


Fig.5. Applied stress vs. creep rupture life plots on log-log scale at 823 K.

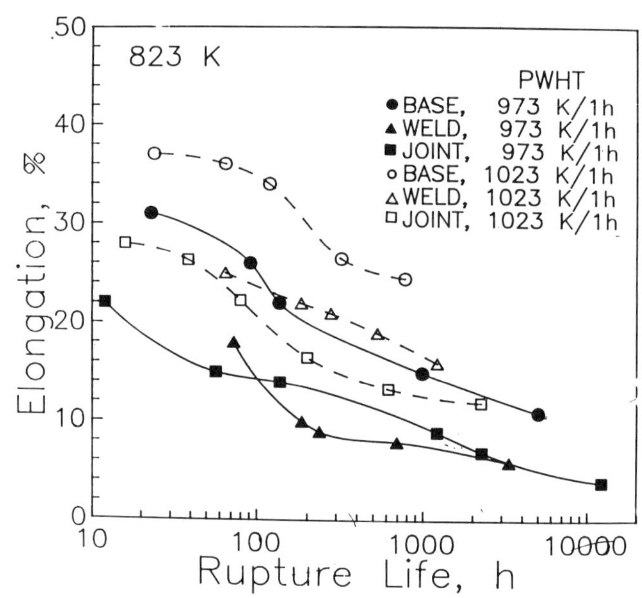


Fig.6. Variation of rupture elongation with rupture time at 823 K.

proeutectiod ferrite with carbides on prior austenite grain boundaries (Fig.7a). On subjecting to PWHTs, the grain boundary carbides coarsened and fine carbides precipitated both in ferrite and bainite regions (Figs.7b). The weld metal microstructure revealed upper bainite characterised by lath like arrangement of ferrite (Fig.8a). Lath boundaries were decorated with carbides. PWHT caused coarsening of ferrite laths and lath boundary carbides with fresh precipition of fine carbides within the laths (Fig.8b). The microstructure in the HAZ consisted of coarse-grained (100 µm)bainite near the fusion boundary followed by fine-grained (15 µm) bainite and intercritical region adjacent to the base metal (Fig.9a). In the intercritical region, the microstructure was made up of uniformely distributed carbides in a ferrite phase (tempered bainite) surrounded by austenite transformation products. After PWHT, the carbides present in the coarse-grain and fine-grain bainitic regions coarsened spherodised. Additional precipitation of

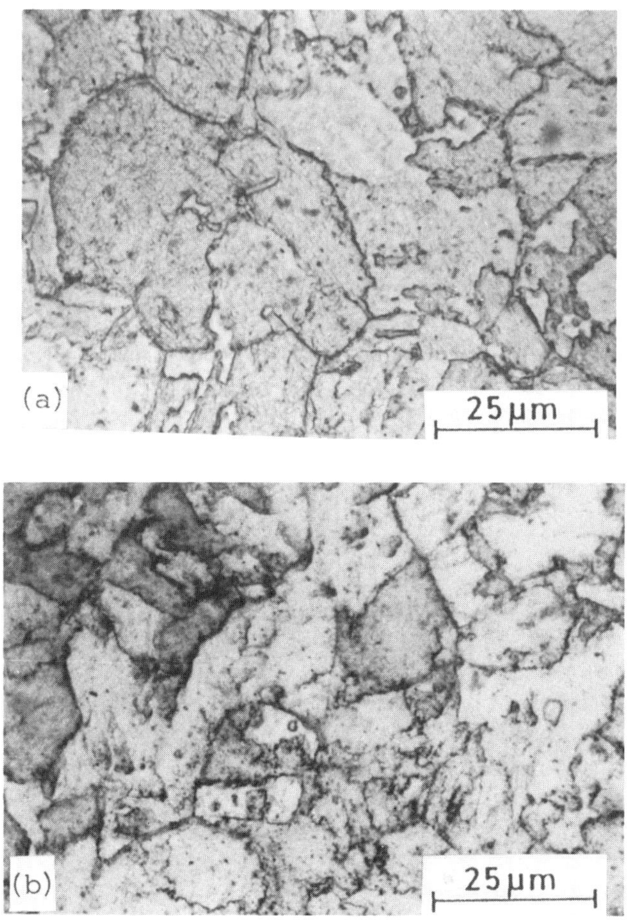


Fig.7. Microstructure of base metal: (a) normalised plus tempered (b) after PWHT (1023 K/1h ).

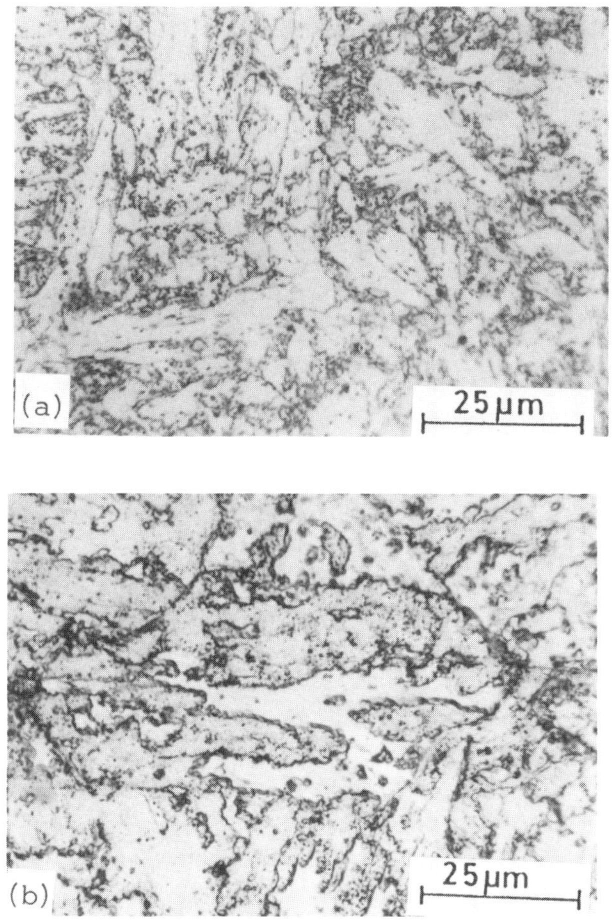


Fig.8. Microstructure of weld metal: (a) as-welded (b) after PWHT (1023 K/1h).

fine carbides was also observed. In the intercritical region the microstructure underwent a change into uniformly dispersed carbides in a ferrite matrix (Fig.9b). Genearlly, coarsening of microstructure was more pronounced at higer PWHT temperature in all the material conditions.

**Fractographic Observations.** Base, weld and composite specimens failed by predominantly transgranular fracture (Figs.10). In all the material conditions, the appearence of the fracture surface was typical of ductile failure, showing many intragranular dimples. Optical metallography of longitudinal section of creep tested composite specimens with lower PWHT temperature and at 823 K and 100 MPa revealed intergranular cavities in the coarse-grain bainitic region of HAZ (Fig.11a). Though the longitudinal

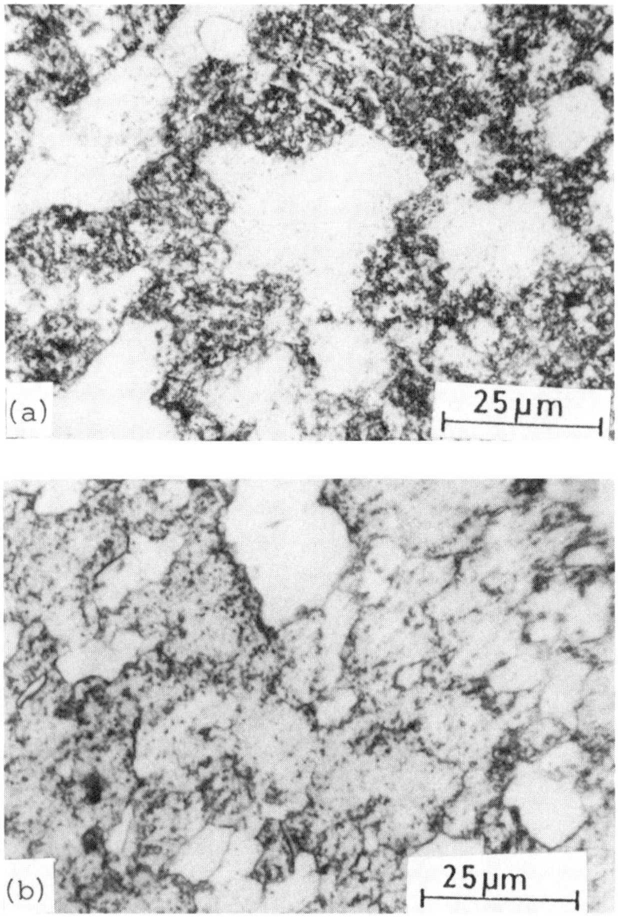


Fig.9.Microstructure of intercritical region of HAZ (a)as welded (b)after PWHT (1023K/1h).

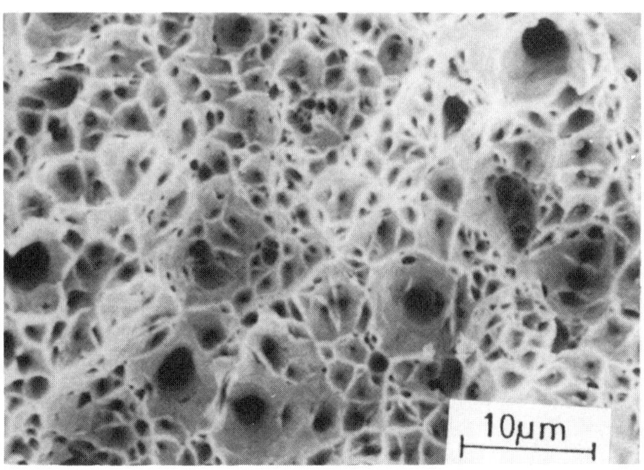


Fig.10.SEM fractographs of compocite specimen (PWHT 1023 K/1h)creep tested at 823 K and 150 MPa.

sections of the base and weld metal failed to reveal intergranular wedge cracks or cavities characteristic of creep damage by optical microscopy, scanning electron microscopy of the creep tested specimen surface exposed by fracturing at liquid nitrogen temperature revealed few intergranular cavities (Fig.11b).

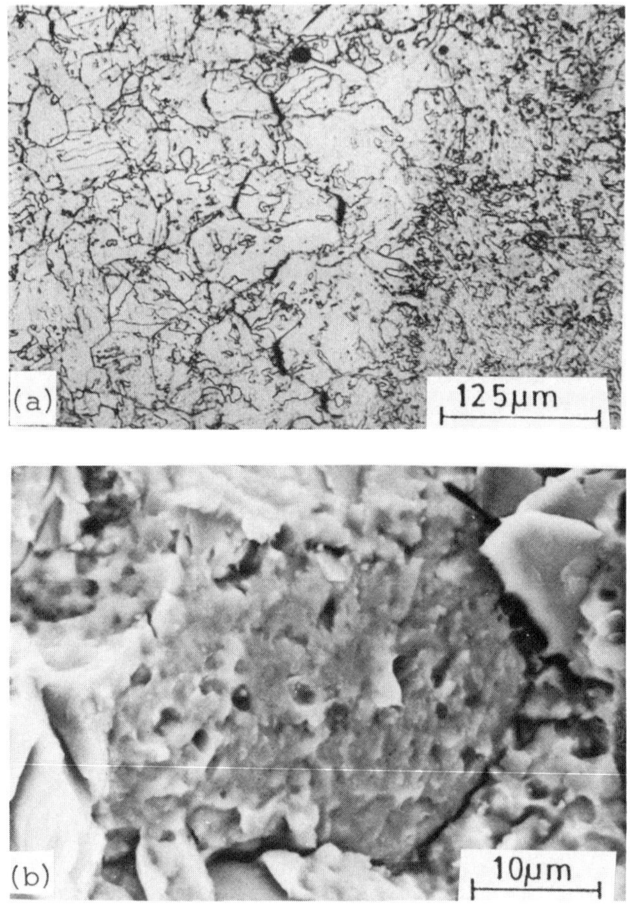


Fig.11. (a) Creep cavities in coarse grain region of HAZ ( PWHT-973 K/1h,100 MPa,823 K),(b)SEM micrograph showing creep cavities in base metal(PWHT-973 K/1h, 150 MPa, 823 K).

## DISCUSSION

**Microstructure Prior to Creep Testing.** Microstructure of the weldments has been discussed in details earlier(8). Granular bainitic structure in the normalised-plus-tempered base metal(Fig.8a) (6,11-15) was reported to occur when the cooling rates during continuous cooling are low (16). Becker and Nutting (13) constructed isothermal diagram showing sequence of carbide formation on tempering the normalised steel. According to this diagram, $Fe_3C$ and $Mo_2C$ carbides are expected to be present in the normalised and tempered (1003 K/1h) base metal. $Mo_2C$ has been noticed to precipitate uniformly in the matrix (13,14). PWHT of the base metal at 973 K and 1023 K for 1h caused thickening of carbides at grain boundaries and led to additional precipitation of finely distributed carbides probably $Mo_2C$ in the intragranular regions of the bainite-proeutectiod ferrite microstructure (Fig.7b). Microstructural changes following PWHT of 973 K/1h have been reported earlier (8). A higher PWHT led to greater coarsening of the constituents of microstructure compared to low PWHT.

The lath like ferrite morphology in weld metal (Fig.9a) is a characteristic feature of upper bainite. $Fe_3C$ and $Mo_2C$ are the only carbides reported to occur in the as-welded microstructure (4). PWHTs of weld metal caused the coarsening of ferrite laths, additional precipitation probably of $Mo_2C$ within the laths and coarsening of prior existing $Fe_3C$ at lath and grain boundaries (Fig.9b). The coarsening of lath structure in welds during PWHT has also been observed by several others (4,11).

The HAZ developed three principal microstructures. These microstructures have been shown to depend mainly on the peak temperatures attained in the different regions during multipass welding (6). The regions closest to the fusion boundary experience peak temperatures well above $Ac_3$. At these high temperatures the carbides which impede the austenite grain growth dissolve and hence coarse grained austenite would develop. This transforms on cooling to give a fully coarse-grained bainitic structure.The grain size decreases with increasing distanance from the fusion line as the peak temperature experienced decreases. When the thermal cycle peak temperatures experienced are in the range between $Ac_1$ and $Ac_3$, only partial transformation to austenite takes place. Consequently, on cooling, the resulting microstructure in this intercritically annealed regions would be a mixture of austenite transformation products, surrounding the volumes of untransformed ferrite which has been merely tempered during thermal cycle (Fig.9a).

The types of carbides that can exist in the individual zones of the HAZ have been examined to a limited extent. The virtual absence of $Mo_2C$ in the intercritical region has been noted by several researchers (5,17). Roy and Lauritzen (5) have reported $M_7C_3$, $M_{23}C_6$ and $M_6C$ in HAZ but did not clarify the specific zones in which they occur. After PWHTs the HAZ microstructures generally showed carbides of globular morphology which are more pronounced at higher PWHT. The carbides possessing globular morphology have been identified as $M_{23}C_6$ (5). Comparision of Fig.9a and Fig.9b clearly reveal that austenite transfromation products present in the intercritical region in the as-welded state had decomposed into ferrite and carbides during PWHT.

**Correlation between microstructures and hardness.** Microhardness profiles across the weldments (Fig.12) reflected the microstructural variations observed in different regions. A rapid fall in hardness was seen as the distance from the fusion boundary was increased. PWHTs reduced the hardness of weld metal zone and HAZ considerabely. Higher PWHT temperature led to more reduction in hardness due to the coarsening of various microstructural features. Within the HAZ the intercritical region exhibited the lowest hardness compared with coarse- and fine-grained bainite regions. This may be attributed to the virtual absence of fine accicular $Mo_2C$ and reduced bainite content in the intercritical regions. Further reduction in the hardness of HAZ in creep tested weldment samples can be associated with the coarsening of intra- and inter-granular carbides. The base metal owes its lower hardness to the higher amount of proeutectoid ferrite in the microstructure compared with the weld and HAZ microstructures.

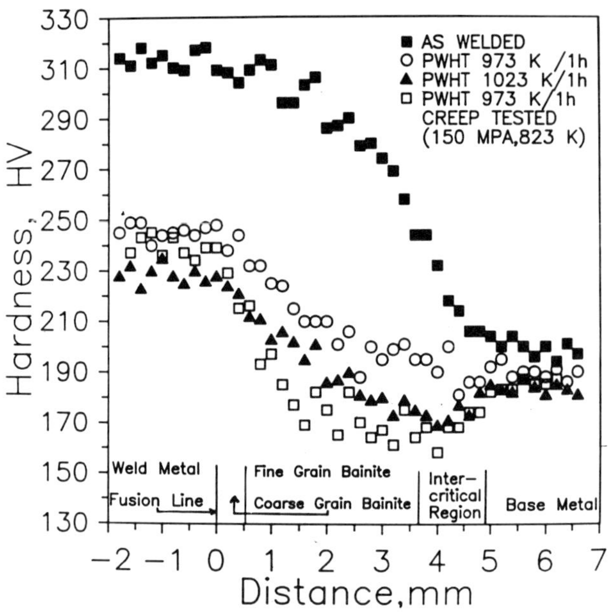


Fig.12. Microhardness profile across the weldments.

**Creep properties.** At 773 K, in both the PWHT conditions, when compared at the same stress level, the rupture lives exhibited by three material conditions are in the order: weld metal > base metal > composite (Fig. 4). All material conditions exhibited predominantly transgranular fracture (Fig.10) with a little evidence of intergranular creep damage. These observations clearly exclude the possibility that the variation in creep damage is responsible for the differences in rupture lives. In this context, the observed varition in rupture behaviour has been found to have good correlation with the initial microstructures and microstructural modifications that occur during creep testing of the base, weld and composite specimens. In the weld metal,the numerous boundaries of lath bainite along with the fine dispersion of precipitates in the matrix (Fig.8) contribute to the overall strength by serving as obstacles to the dislocation movement. Several others (12,18,19) have also indicated that 2.25Cr-1Mo derives its high temperature creep strength from the fine distribution of acicular $Mo_2C$ particles in the intragranular regions. However, it has been pointed out that $Mo_2C$ is more stable in ferrite than in bainite (1,13). In both the PWHT conditions, at 823 K, although the fully bainitic weld metal with the fine distribution of $Mo_2C$ in it displayed a high rupture strength compared to the base metal at shorter times, the strength advantage of weld metal at longer times could not be sustained (Fig.5). The lower creep rupture strength of fully bainitic weld metal at longer times can be associated with the rapid transformation of $Mo_2C$ to one or more of the complex carbides i.e ($M_{23}C_6$, $M_7C_3$, $M_6C$ ).The trainsition stress below which weld metal creep strength (Fig.5) is inferior than the base metal was found to decrease with increase in PWHT temperature. This may be associated with relatively more stable microstructure produced by higher PWHT temperature. However, at the lower temperature of 773 K the weld metal maintained its higher rupture strength over the entire stress range investigated, probably because the temperature was not high enough to induce the transformation of $Mo_2C$ to higher order carbides.

Creep tested composite specimens in both the PWHT conditions, revealed the localization of failure in the intercritical regions of HAZ (7,8). The reduction or total absence of $Mo_2C$ has been suggested to be the cause of the failures in the HAZs of composite specimens (5). Chilton et al.(6) have reported lower creep resistance and shortest rupture life for the simulated intercritical structure compared to base metal. Higher PWHT temperature generally led to poor creep rupture strength for all the material conditions. This can be attributed to coarsening of the microstructures. This is also reflected in the variation of microhardness in various conditions.

It must be pointed out that, in this study, composite specimens in both the PWHT conditions gave rupture lives approximately around 10000 hours. The conclusion that the intercritical region of the HAZ is the weakest link in weldments is applicable to these rupture lives. However, at the design stresses of components it is expected that precipitation reactions might even out differences in strength levels that arise due to the existance of different precipitates in different regions of the weldment(1).The tendency of the rupture lives of the weld metal and composite specimens to converge (Figs.4 and 5) as the stress is reduced can be attributed to microstructural changes. It is possible that the location of failure in the HAZ might shift to the less ductile coarse-grain region of the HAZ (20-22) at the intended service life in view of the succeptibility of this region to creep damage (Fig11.a).

## CONCLUSION

The investigations conducted on the creep rupture behaviour of 2.25Cr-1Mo base metal, weld metal and composite specimens at 773 and 823 K in the two post-weld heat-treatment conditions (973 K/1h and 1023 K/1h) have led to the following conclusions.

1. Creep behaviour of the base, weld and composite specimens was characterised by well defined primary, secondary and tertiary creep regions in both the PWHT conditions.
2. The stress ($\sigma$) dependence of the steady state creep rate ($\dot{\varepsilon}_s$) was found to obey a power law relationship, i.e $\dot{\varepsilon}_s = A\sigma^n$. Composite specimens showed higher creep rates while weld metal deformed at lower rates than base material. Higher PWHT temperature caused larger steady state creep rates in all the three material conditions.
3. In both the PWHT conditions,the weld metal displayed significantly higher creep rupture strength, while composite specimens exhibited inferior creep strength. Higher PWHT temperature yielded lower rupture strength values in the respective material conditions.
4. Creep rupture ductility (%elongation) decreased with increasing rupture time in all the three material conditions. PWHT at higher temperature yielded generally larger values of percentage elongation. The rupture ductility of the base metal was significantly higher than that of the weld metal and composite specimens.
5. All material conditions exhibited transgranular fracture characterised by dimples. Fracture in the composite specimens had occurred in the inter-critical regions of HAZ.The reduction or total absence of $Mo_2C$ has been suggested to be the cause of failure in HAZ of composite specimens.

## Acknowledgments

The authors wish to thank Dr. Placid Rodriguez, Head, Metallurgy and Materials Programme for his keen interest in this work and for many useful discussions. Experimental help provided by Mrs. K.S.Chandravathi in conducting the creep tests is gratefully acknowledged.

## REFERENCES

1. Murphy, M.C. and G.D.Branch, J. Iron Steel Inst., 209, 546-561 (1971)
2. NRIM Creep Data Sheet No.36, (1985), National Research Institute for Metals, Tokyo.
3. NMIR Creep Data Sheet No.11A, (1980), National Research Institute for Metal, Tokyo.
4. Lundin, C.D., S.C.Kelly, R.Menon and B.J.Kruse, Welding Research Council Bulletin No.315, 1-66 (1984).
5. Roy,p. and T. Lauritzen, Welding J.-Res. suppl., 65, 45s-47s (1986).
6. Chilton,I.J., A.T.Price and B.Wilshire, Met.Technol., 11, 383-391 (1984).
7. Arnswald,W.,R.Blum,B.Neubauer and K.E.Poulson, in Proc. Intl. Conf. on Creep, Tokyo, Japan, April 1986, pp 367-372.
8. Laha,K., K. Bhanu Sankara Rao and S.L.Mannan, Mater. sci. Engng., A129, 183-195 (1990).
9. Alberry, P.J. and W.K.C.Jones, Met. Technol.,4, 360-364 (1977).
10. Nippes, E.F., Welding.J-Res. Suppl., 38, 1s-18s (1959).
11. Serrano,O.V.,G.R.Edwards and R.H.Forst, ASTM Spec. Tech. Publ., 755, 275-296 (1982).

12. Klueh, R.L., J.Nucl. Mater., 54, 55-63 (1974).
13. Backer, R.G. and J. Natting., J. Iron Steel INst., 192, 257-268 (1959).
14. A.G.Hippsley, C.A., Met. Sci., 10, 137-147 (1981).
15. Woodhead, J.H. and A.G.Quarrel, J.Iron Steel Inst., 203, 605-620 (1965).
16. Habraken, L.T. and M. Economopoulos, Proc. Symp. on Transformation and Hardenability in Steels, Climax Molybdenum, Dordrecht, 1967.
17. Smith, E., B.E.Blanchard and R.L.Apps, in Proc. Conf.on Welding of Creep Resistant Steels, The Welding Institute, Cambridge, 1970, pp 79-89.
18. Abdel-Latif, A.M., J.M.Corbett and D.M.R.Taplin, Met.Sci., 16, 90- 96 (1982).
19. Orr, J., F.R.Beckitt and G.D.Fawkes, in The Physical Metallurgy of Chromium-Molybdenum Steels for Fast Reactor Boiler, BNES, London, 1978, pp 91-109.
20. Cane, B.J. and R.S.Fidler, in Proc. British Nuclear Energy Society Conference, London, 1977, pp 31.1-31.5.
21. Laha, K., K.Bhanu Sankara Rao and S.L.Mannan, Tensile Deformation and Fracture Behaviour of Base, Weldments and Heat Affected Zone Structures of 2.25Cr-1Mo steel, IGCAR-Kalpakkam. (1991)-unpblished results.
22. Cane, B.J., in Proc.Symp. Creep Resistance Steels for Power Plants. B.H.E.L, Corporate R and D divn, Hyderabad, 1983, pp 15-30.

*Proceedings of the Fifth International Conference on Creep of Materials, Lake Buena Vista, Florida, USA, 18-21 May 1992*

# An Approach to Prolonging the Creep Rupture Life of Dissimilar Metal Welds

**W. Zuqian and C. Xinzi**
Shanghai Power Equipment Research Institute
Shanghai, China

## ABSTRACT

In this paper the premature failure mechanism of dissimilar metal welds between austenitic and ferritic steels operating at elevated temperature are investigated, and an effective approach to prolonging the creep rupture life of dissimilar metal welds is provided.

DISSIMILAR METAL WELDS (DMWs) between ferritic and austenitic steel tubes are commonly used in superheaters and reheaters of utility boilers. It is clearly known from published literature and various internal reports that the premature failure of DMWs operating at elevated temperatures occurs frequently after only a few years service. The main factors causing the DMWs failure are generally considered to be the difference in expansion coefficients of the base metals and weld metal and the carbon migration across the weld interface. So, from the middle of 1950's, instead of anstenitic filler metals, the Ni base alloys were usually selected for welding the dissimilar metals. The life of DMWs joined with Ni base alloy was longer than that with austenitic filler metals. But unfortunately the DMWs welded with Ni base filler matals also fractured in power stations. According to the survey of US utilities, conducted by P. E. Hass in 1981[1], the mean time to initial failure was 74, 000 hrs for austenitic filler metal, and 100,000 hrs for Ni base filler metal, whereas the shortest time to initial failure was 29, 000 hrs for austenitics and 40, 000 hrs for Ni base filler metal. Therefore, the DMWs failure still remains a problem with no satisfactory solution so far.

In order to solve the problem of mismatch of thermal expansion coefficients, J. F. King and others[2] have suggested to insert a 800 Ni alloy transition piece having a medium thermal expansion coefficient between austenitic and ferritic steels into DMWs to form a transition joint, in which the coefficients of thermal expansion were changed step by step in a smoother way. For this purpose, the Cr16Ni8Mo austenitic filler metal was used to join 800 alloy with austenitic steel, and the Ni base filler metal to join 800 alloy with ferritic steel. But, same as the DMWs welded directly with Ni base filler metals, the fracture of the above mentioned welds also occurred on the fusion boundary between Ni base alloy weld and ferritic steel, the application of transition joint in this case was not of any significance. Another question was carbon migration, as a result of carbon diffusion from base metal to HAZ, after a long term operation, the decarbonized zone in HAZ disappears, therefore, R. H. Ryder and others[3] considered that the carbon migration was neither the main cause leading to the premature failure of DMWs. So the DMWs' failure was still an unsolved serious problem.

Based on the results of research and long time practical experience of SPERI and some our boiler manufacturers[4,5], a general conclusion can be made that the mismatch of the creep rupture strengths of the ferritic and austenitic steels is the main cause, leading to the premature failure of DMWs. Therefore, in order to reduce the difference of creep rupture strengths among the various zones of the welds, it is recommended to insert a transition piece of an intermediate creep rupture strength into DMWs and it is also required that the filler metals selected should have an intermediate creep rupture strength between those of the two joined base metals, so as to eliminate a big jump in creep rupture strengths among various zones.

The results of our investigation have showed that the creep rupture occurred on the fusion boundary can be prevented by the application of the transition joint and the creep rupture life of DMWs with transition joint may be increased to the same level as of ferritic weld between similar base metals.

The aim of this paper is to provide an approach to preventing the fusion boundary fracture and to prolonging the creep rupture life of DMWs. The major subjects covered in this paper are as follows:

1. Selection of transition piece material and comparison of creep rupture lifes of DMWs with and without transition joint.
2. Selection and evaluation of filler metals joining the transition piece to austenitic and to ferritic steels.
3. Investigation and analysis of fracture characteristic and failure mechanism of DMWs.

## THE INFLUENCE OF FILLER METALS ON THE CREEP RUPTURE CHARACTERISTIC OF DMWs

The tension creep rupture test was used to investigate the creep rupture characteristic of DMWs between the 304 austenitic steel and 12Cr1MoV ferritic steel. It is a test with more severe conditions as compared with the practical ones.

The chemical compositions of testing materials are shown in tab. 1. The thickness of the steel plates was 20 mm. Before welding the ferritic 12Cr1MoV steel was normalized at 1000℃ for 1hr. and tempered at 760 ℃ for 3hrs., and the austenitic 304 steel was in delivery solid solution condition.

The ferritic CrMoV and Ni base Inconel electrodes were selected for dissimilar metal welding, chemical compositions of these deposit metals are also listed in tab. 1. The V-groove angle of test plates was 60° with root gap of 2 mm, the preheat temperature before welding was 150℃, after completion of welding, the test plates were tempered at 730℃ for 2hrs. Then the creep rupture specimens with a diameter of 10mm were cut from test plates and machined.

The creep rupture test temperature was $580 \pm 3$℃, the test results of two series of specimens are shown in tab. 2. In this table, the creep rupture test results of 12Cr1MoV similar metal welds are also given. It can be seen from tab.2 that the creep rupture strengths of the DMWs extrapolated to $10^5$ hrs are much lower than that of 12Cr1MoV similar metal welds, inspite of whether CrMoV filler metal or Inconel filler metal was used in DMWs. It should be noted, that the fracture occurred along the fusion boundary between 304 austenitic steel and ferritic CrMoV weld, when the CrMoV filler metal was used; and the fracture occurred along the fusion boundary between 12Cr1MoV steel and Ni base Inconel weld in case of Inconel filler metal. But for both filler metals, the section reduction ratio at the fracture boundary was very low, nearly equal to zero.

It should be noted, that the reason why the fracture occurred along the different

Tab. 1 Chemical compositions of test plates and deposit metals

| Material | Typ | C | Si | Mn | S | P | Cr | Mo | V | Ni | Nb | |
|---|---|---|---|---|---|---|---|---|---|---|---|---|
| | | Element (wt%) | | | | | | | | | | |
| base metal | 12Cr1MoV | 0.10 | 0.23 | 0.52 | 0.011 | 0.021 | 1.18 | 0.30 | 0.25 | | | |
| | | | | | | | | | | W | Ti | B |
| | 12Cr2MoWVTiB | 0.12 | 0.65 | 0.55 | 0.005 | 0.015 | 1.91 | 0.58 | 0.37 | 0.47 | 0.15 | 0.003 |
| | 304 | 0.10 | 0.55 | 1.26 | 0.005 | 0.023 | 18.50 | | | 9.10 | | |
| | 347 | 0.08 | 0.60 | 1.55 | 0.007 | 0.019 | 18.70 | | | 10.80 | 0.70 | |
| | CrMoV | 0.08 | 0.43 | 0.75 | 0.015 | 0.015 | 1.05 | 0.62 | 0.20 | | | |
| Welding Material | Cr5MoVNb | 0.06 | 0.26 | 0.94 | 0.009 | 0.020 | 4.73 | 0.76 | 0.41 | | 0.19 | |
| | E309 | 0.11 | 0.46 | 1.36 | 0.010 | 0.014 | 23.50 | | | 12.30 | | |
| | | | | | | | | | | | | Fe |
| | Inconel | 0.08 | 0.34 | 6.90 | 0.008 | 0.012 | 16.50 | | | bal | 2.60 | 5.50 |

Tab.2 The creep rupture test results of DMWs between 12Cr1MoV and 304 steels

| Welding material | test temperature | creep rupture strength | fracture location |
|---|---|---|---|
| CrMoV | 580℃ | 42 MPa | fusion boundary between 304 steel and weld metal |
| Inconel | 580℃ | 45 MPa | fusion boundary between 12Cr1MoV steel and weld metal |
| CrMoV(for similar metal welds of 12Cr1MoV steel) | 580℃ | 72 MPa | 12Cr1MoV base metal and/or its HAZ |

fusion boundaries of the DMWs lies mainly in the great difference of the creep rupture strengths on both sides of the fractured fusion boundary, where the great difference in strength leads to the high stress

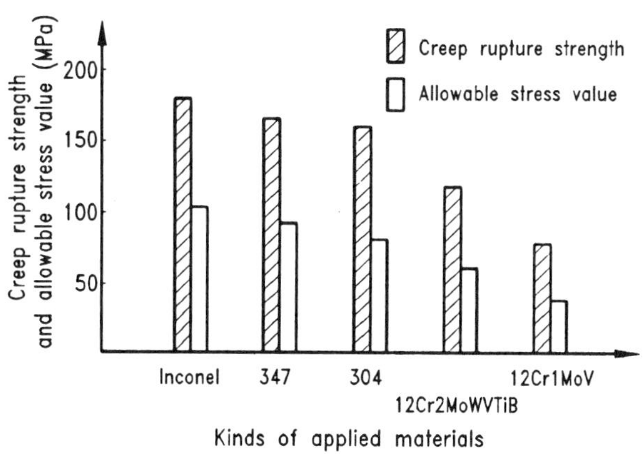


Fig. 1 The creep rupture strength and the allowable stress value of some commonly used materials at 580℃

concentration at the boundary. Fig.1 Shows the creep rupture strengths and the allowable stress values of some common used materials at 580℃.

Therefore, under the above mentioned matches, the failure naturally occurred on the fusion boundary between ferritic 12Cr1MoV steel and Inconel weld or between austenitic 304 (347) steel and CrMoV weld, where the difference of creep rupture strength was greater.

Consequently, if the weld metal having an intermediate creep rupture strength between those of the two base metals was selected to join the DMWs, what would be the effect? For this purpose, the electrode Cr5MoVNb was selected as filler metal. The chemical composition of Cr5MoVNb deposit metal is also given in table 1. It can be seen from the chemical composition, that the creep rupture strength of Cr5MoVNb deposit metal was obviously higher than that of

12Cr1MoV steel.

The creep rupture curve of the DMWs joined with Cr5MoVNb electrode and tested at 580℃ is shown in fig.2. The creep rupture curve of that with Inconel is also shown in this figure for comparison. Here the dark cycle indicates that the fracture occurred along the fusion boundary between 12Cr1MoV steel and weld, the bright cycle indicates that the fracture occurred along the fusion boundary between 304 steel and weld. As shown in fig.2, although the rupture time of the DMWs joined with Cr5MoVNb under different stresses was longer than that with Inconel, the fracture still occurred along the fusion boundary. Because the creep rupture strength of Cr5MoVNb deposit metal was approximately the mean value of two joined metals, so the fracture occurred sometimes along the fusion boundary between ferritic steel and Cr5MoVNb weld, sometimes along the fusion boundary between austenitic steel and Cr5MoVNb weld. Therefore, even though the difference of creep rupture strengths between base metal and weld has been some what reduced, but the fracture along the fusion boundary still exist. However, a clear idea may be got from this experience, that is, the failure problem can be solved through the gradual variation of creep rupture strengths.

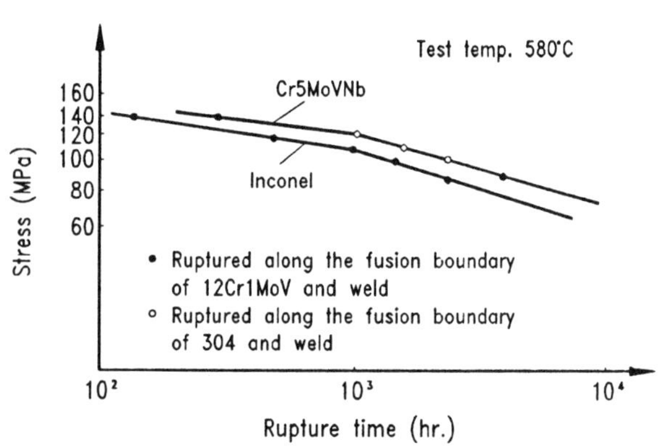


Fig. 2 Creep rupture curves of the DMWs joined with Cr5MoVNb and/or Inconel filler metals

## THE CREEP RUPTURE STRENGTHS OF THE DMWs WITH TRANSITION JOINT

In order to reduce the difference of creep rupture strengths among various parts of the welds, it was considered to insert a transition piece of an intermediate creep rupture strength into DMWs. Moreover, it was required that the filler metals selected should have an intermediate creep rupture strength between that of the two joined metals, so as to form a gradual variation of creep rupture strengths, as illustrated in fig.3.

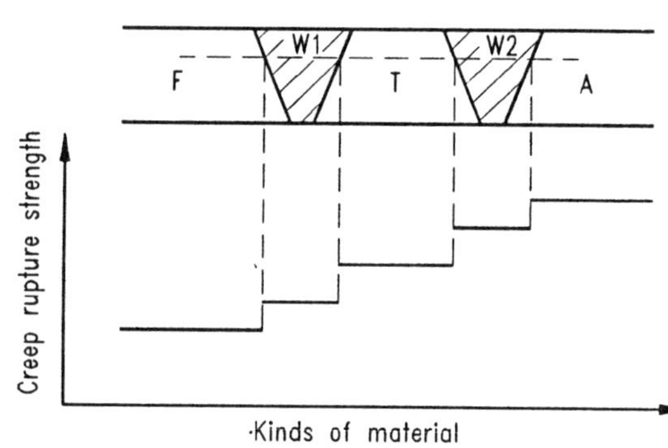


Fig. 3 A DMWs of gradually varied creep rupture strength

Therefore, the 12Cr2MoWVTiB steel which is widely used in our power plants was selected as the material for transition piece. The two joined metals were austenitic 347 steel and ferritic 12Cr1MoV steel respectively. The CrMoV electrode was selected to join the transition piece and 12Cr1MoV, and the Inconel and/or E309 were selected to join the transition piece and 347. The chemical compositions of above mentioned steels and deposit metals are also listed in tab.1.

The plates and transition piece were butt-welded together under a preheating temperature of 150℃. The post-weld heat treatment was at 730℃ for 2 hrs. Then, the welded plate was cut and machined to creep rupture specimens, as illustrated in fig.4

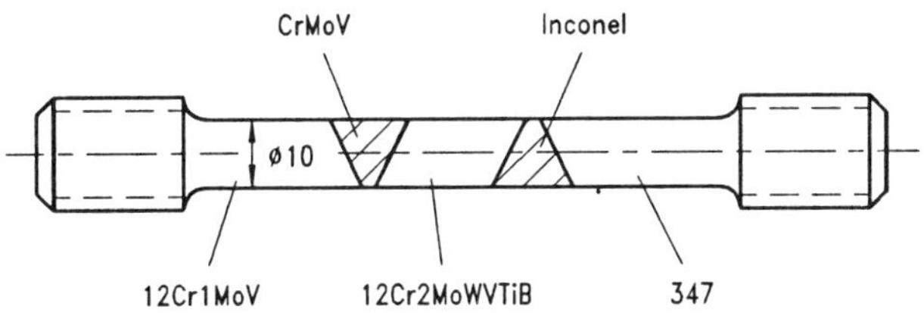


Fig. 4. Creep rupture specimen with transition joint

The creep rupture test results of the above mentioned specimens under 580℃ are shown in fig.5.

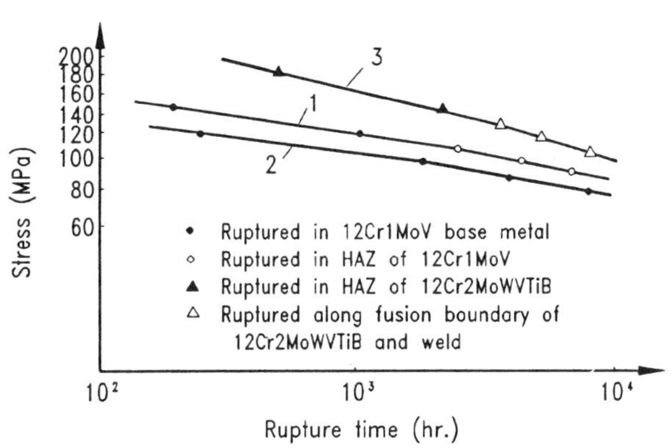


1. 12Cr1MoV + CrMoV + 12Cr2MoWVTiB + Inconel + 347
2. 12Cr1MoV+ CrMoV + 12Cr2MoWVTiB +E309 +347 (test temperature 580℃ for curves 1.2)
3. 12Cr2MoWVTiB +Inconel +347 (test temperature 600℃ for curve 3)

Fig. 5 Creep rupture curves of DMWs with transition joint

In order to make a comparison, the creep rupture curve of the single DMWs 347 + Inconel + 12Cr2MoWVTiB under 600℃ is also given. The experimental results indicated that for DMWs with transition piece, the fracture occurred without exception in 12Cr1MoV base metal and/or its HAZ, inspite of whether Ni base or austenitic filler metals were used in austenitic side.

Because the creep rupture strength of austenitic side of specimen at 600℃ was much higher than that of 12Cr1MoV steel and its HAZ at 580℃, as shown in fig.5 curve 3, so the fracture would not occur in the austenitic part of the specimen at 580℃.

In brief, after adopting the transition joint the fracture moved from weld fusion boundaries to 12Cr1MoV base metal and/or its HAZ. Moreover, the creep rupture strength of DMWs was equal to that of 12Cr1MoV similar metal welds. Therefore, the problem of premature failure of DMWs has actually got a good solution.

## DISCUSSION

The premature failure of austenitic/ferritic DMWs, operating at elevated temperature has a specific characteristic, i.e, the fracture occurs along the fusion boundary. So, in order to prevent the DMWs failure, the problem of creep rupture at fusion boundary should be solved. The regularity of the creep rupture of DMWs along fusion boundary is such that when the creep rupture strength of weld is equal to that of ferritic base metal, the welds will fracture along the fusion boundary of austenitic side, when the creep rupture strength of weld is equal to austenitic base metal, the welds will facture along the fusion boundary of ferritic side.

With regard to the thermal expansion, the thermal expansion coefficient of Inconel (14.9) is close to ferritic 12Cr1MoV steel (14.0), and there is a relatively great difference between that and austenitic 304 steel (18.9), but the fracture occurred along fusion boundary of ferritic steel. Therefore, the difference in thermal expansion coefficients is not the major cause. The mismatch of creep rupture strengths should be the main cause of DMWs failure.

The sketchs of two different matches of creep rupture strengths are shown in fig.6

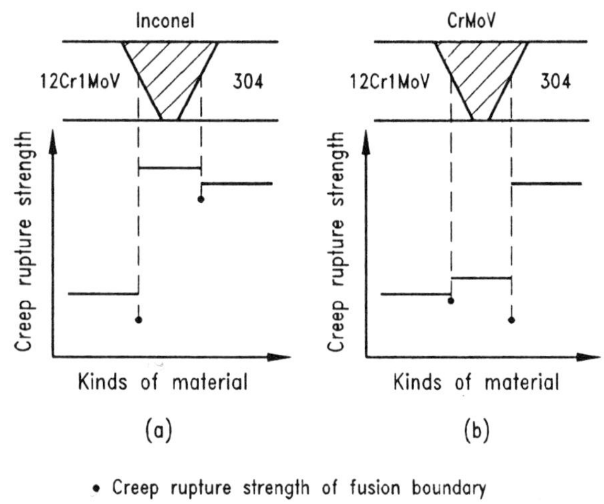


Fig.6 The sketch of creep rupture strengths match among the various parts of ferritic/austenitic DMWs

It can be seen from this figure, the weakest zone of the whole DMWs is the fusion boundary, where the difference of creep rupture strengths was the greatest.

To insert a transition piece having an intermediate creep rupture strength into DMWs has many adventages, as it reduces considerably the difference of creep rupture strengths between two sides of weld interface; moves the rupture location out of the fusion boundary to the ferritic base metal and/or its HAZ; which consequently makes the strength of DMWs equal to that of ferritic similar metal welds. Fig.7 shows the sketch of the match of the creep rupture strengths of the DMWs with transition joint. This figure indicates, though there is also a lower value creep rupture strength along the fusion boundary between Inconel weld and 12Cr2MoWVTiB transition piece, but its real strength is much higher than 12Cr1MoV itself and/or it HAZ. Therefore, the fusion boundary between transition piece and Inconel (or austenitic) weld would no longer be the weakest zone. For the whole DMWs, the weakest zone was the 12Cr1MoV base metal and its HAZ.Besides, as shown in the figure, even though the difference of creep rupture strengths between Inconel weld and 12Cr2MoWVTiB transition piece was greater than that between E309 weld and 12Cr2MoWVTiB transition piece, the creep rupture strengths of fusion boundaries for these two welds differed very slightly. It was attributed to the fact, that the thermal expansion coefficient and the carbon migration also had some effect on the failure. In general consideration, it is better to select Inconel as welding material of austenitic side. As a result, applying a transition piece of intermediate creep rupture strength can be considered an effective approach to prolonging the creep rupture life of DMWs.

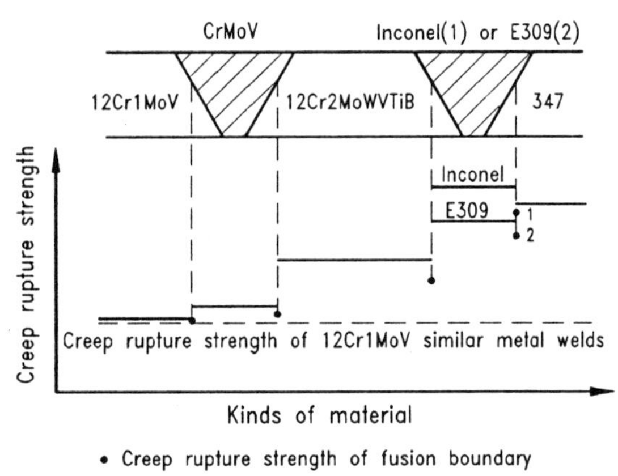


Fig.7 Sketch of the creep rupture strengths' match of DMWs with the transition joint

At present, the DMWs with transition joint of intermediate creep rupture strength are progressively used in our power station boilers and no failures were reported up to now.

## CONCLUSION

1. The mismatch of the creep rupture strengths between materials is the main cause leading to the premature failure of austenitic/ferritic DMWs operating at elevated temperature.
2. The characteristic of DMWs failure is such that the fracture occurs along the fusion boundary, where the difference of creep rupture strengths is greater.
3. To insert a transition piece having an intermediate creep rupture strength into DMWs is an effective approach to prolonging the creep rupture life.
4. After inserting a transition piece the fracture location of DMWs moved to ferritic base metal and/or its HAZ and the creep rupture strength reached the same level as similar metal welds.

## REFERENCES


1. Haas P.E., "Results of Industry Wide Survey on Dissimilar Metal Weld Performance", American Electric Power Service Corp., Canton, Ohio (1978)

2. King, J. F. , M. D. Sullivan and G. M. Slanghter, Welding Journal 56( 11 ), 354s-58s (1977)

3. Ryder, R.H., C.C.Li, R. Viswanathan and J.Dimmer, in Proceedings Conf. on Electrical Utility Research, Chicago, Illinos, Apr. 1984.

4. Cong Xinzi, The Transactions of the China Welding Institution 10(1), 19-28 (1989)

5. Wang Dongfen and Cao Rouli, Han Jie No1 1-6 (1988)

*Proceedings of the Fifth International Conference on Creep of Materials, Lake Buena Vista, Florida, USA, 18-21 May 1992*

# Behavior of 2.25Cr-1Mo Base Material and (Repair) Weldments at High Temperature

**M. De Witte**
Laborelec
Linkebeek, Belgium

**C. Coussement**
Belgian Welding Institute
Gent, Belgium

ABSTRACT

The paper gives an overview of the main results of a Belgian industrially sponsored research project on high temperature materials and their weldments.
Accelerated uniaxial creep tests have been performed on new and service exposed 2¼Cr1Mo (P22) and 1Cr½Mo (P11) base material. The results of constant strain rate (CSR) tests, (constant load) isostress and isothermal creep tests are compared. It is shown that the results of extrapolated minimum creep rate, time to 1 % strain, time to rupture are in very good agreement for the explored testing techniques on service exposed material, provided that no microstructural instability occurs. For new material, the continuously evolving microstructure results in different predictions by the CSR and isostress methods.
Accelerated uniaxial creep tests have been performed on crossweld specimens from new welds on service exposed 2¼Cr1Mo (P22) material. Based on the experimental results a failure location map has been drawn. At stresses and temperatures relevant for the service conditions failure is to be expected in the HAZ.
Finally a component creep test on a service exposed 2¼Cr1Mo steel vessel containing a new longitudinal weld (simulating a repair weld) and a nozzle has been carried out. Failure is occurring at the crotch corner of the non-reinforced nozzle, although the stresses there redistribute under creep conditions. The reference stress at the nozzle (crotch corner and saddle point) is considerably higher than across the longitudinal weld. An evaluation table of the various NDT methods is drawn.
Barkhausen noise and ultrasonic techniques do not seem to be very promising for early creep damage monitoring.

MOST OF THE OLDER FOSSIL FIRED POWER PLANTS exhibit a rather good thermal efficiency and a good availability. As they gradually reached operation times even beyond 200 000 hours, and at least some of them are planned to operate up to 300 000 hours, life assessment has become for more than a decade a topic of increasing concern. Reducing the risk of in service failure and hence costly unplanned outages and avoiding unnecessary replacements are amongst its goals. Major components should only be replaced when reaching their end of economic and safe operational life. Furthermore a replacement or repair should be performed during regularly planned overhauls.

From failure experience "damage-sensitive" locations on components can be identified to be heat affected zones (HAZ) of weldments, presenting potentially both a reduced creep strength and geometrical inhomogeneities. The material behaviour and its creep properties have a major influence on life assessment. In this context an industrially sponsored research project on the behaviour of high temperature materials and their weldments was designed and performed, aiming at :

- verification of the weldability of service exposed material
- comparison and evaluation of (short term) creep tests
- evaluation of the creep strength of weldments under uniaxial loading and in a component
- verification of the performance of NDT techniques for monitoring creep damage.

## Uniaxial Accelerated Creep Tests On New And Service Exposed Base Material

Two types of uniaxial accelerated creep tests have been performed within the research project. They are isostress creep tests performed at the anticipated design or operating stress but at temperatures above that of service and constant strainrate (CSR-)tests ([1], [2], [3], [4]) performed at the operating temperature but at higher strainrates than in service conditions. The relative merits and limitations of both short term test methods have been evaluated through a comparison with isothermal creep test results of longer test duration. In literature a large number of publications deals with these testing techniques ([5], [6], [7], [8]).

2¼Cr1Mo pipes (343 mm outer diameter x 24 mm thickness) in new and service aged conditions were used for the tests. The specified heat treatment for the new material was normalizing at 925–950°C followed by slow cooling (< 100°C/h) to 650°C, which resulted in a ferrite–pearlite structure. The aged material was removed from service after 123 000 hours at approximately 565°C and 41 bar.

**Service Exposed Material.**

*Isostress Creep Tests.* The isostress tests have been carried out at Laborelec at the service stress level of 27,5 MPa and temperatures ranging from 620 to 750°C. During the tests, the creep strain was continuously measured. In a constant load test, the true creeprate ($\dot{\epsilon}$) and the true creep strain ($\epsilon$) during the stationary stage are related by the following equation (assuming constant volume during inelastic deformation) :

$$\ln \dot{\epsilon} = \ln \dot{\epsilon}_0 + n\epsilon \quad (1)$$

where $\dot{\epsilon}_0$ represents the original creep rate and n is the Norton exponent. According to the above equation, the secondary or stationary creep stage is represented by a straight line in a ($\ln\dot{\epsilon},\epsilon$)–plot, with a slope equal to the Norton exponent n. This behaviour is indeed observed for the isostress tests in the temperature range from 680°C to 705°C. This indicates that for these tests, the basic assumptions are met and that specifically the microstructure has not changed from that obtained under service conditions, at least not for testing times up to reaching the stationary stage. On this basis, it is expected that valid extrapolations can be made for the service–exposed material for the parameters minimum (or in this case stationary) creeprate and time to one percent creep strain. The occurrence of microstructural instability is both time and temperature dependent.

*CSR Tests.* The CSR tests have been carried out by the Belgian Welding Institute at a temperature of 580°C over the strainrate range from $10^{-8}$ to $10^{-5}s^{-1}$. For extrapolation purposes the straight line relationship obtained at the lowest strainrates in a ($\sigma_i,\dot{\epsilon}_t$ 1/3)–plot (where $\sigma_i$ is the saturation stress) is used. The extrapolation of the CSR–data in the (stress,temperature)–field is based on the linear relationship between the applied stress ($\sigma_a$) and the internal stress (saturation stress) ($\sigma_i$) in the relevant range and the experimentally determined value of ΔH. Because of this, the extrapolation is carried out for the thermodynamically stable microstructural condition.

Consequently, it yields creeprate predictions at given stresses and temperatures corresponding to the equilibrium microstructure at temperature. The time to rupture predictions obtained from the CSR–tests using an indirect approach based on equation (1) over a range of temperatures and stresses are plotted in figure 1, where the isostress extrapolations based on the "stable" temperature range are also shown. A good agreement can be observed.

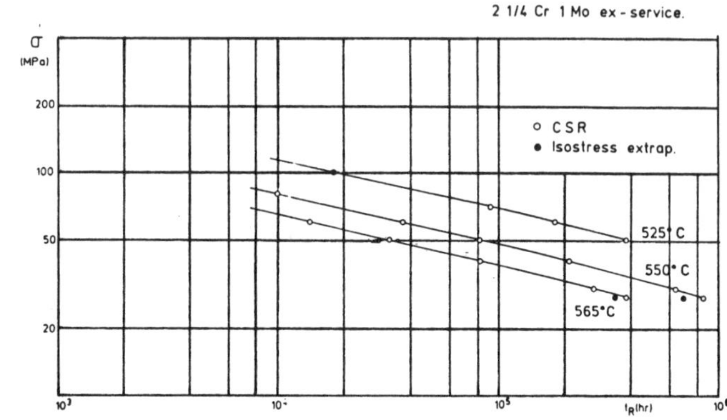


Fig. 1 – Comparison of CSR with "stable" isostress time to rupture predictions.

**New P22 Material.** In contrast to the service–exposed 2¼Cr1Mo material whose microstructure corresponds to the thermodynamic equilibrium structure at the service temperature, the microstructure of the new 2¼Cr1Mo material evolves continuously with time at temperature. This continuous evolution causes the appearance of a minimum rather than a stationary creeprate ($\dot{\epsilon}_m$ instead of $\dot{\epsilon}_s$). Due to this continously evolving microstructure, it cannot be expected that the microstructures in the isostress tests at different temperatures are similar. Consequently, extrapolations of any of the parameters $\dot{\epsilon}_m$, $t_1$ or $t_R$ for the new material should be treated with caution [8].

The CSR–method allows to predict the minimum creeprate corresponding to the thermodynamic equilibrium microstructure.

Within this project, medium term isothermal creep tests have been performed at 550°C, 575°C and 625°C on the new 2¼Cr1Mo material. The CSR method allows a direct comparison between the predictions and these experimental data, since both stress and temperature can be covered in the extrapolation.

The isostress data on the other hand only allow a comparison at the stress of 27,5 MPa. The predicted CSR and isostress rupture lives are compared to the experimental lifetimes in figure 2. As can be observed, the CSR predictions describe the experimental data very well. The anticipated conservatism of the isostress extrapolations for new material is confirmed.

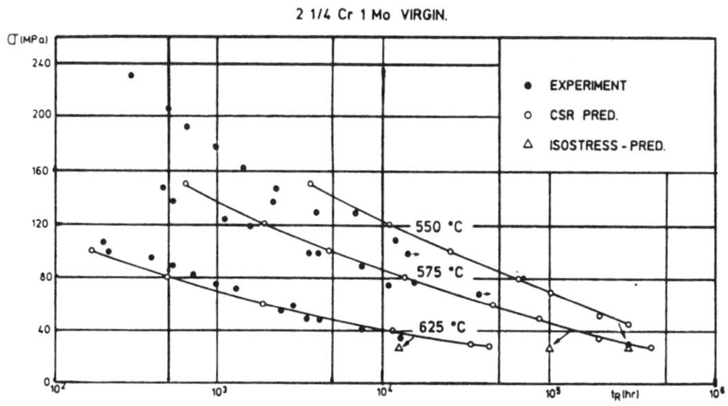


Fig. 2 – Comparison of CSR and isostress rupture life predictions with isothermal creep rupture test results.

## Reheat Cracking

In order to evaluate the reheat cracking susceptibility, isothermal tensile tests on weld simulated specimens, which had previously been subjected to a weld thermal cycle with different peak temperatures, were conducted at different test temperatures ranging from 400 to 800°C. 2¼Cr1Mo (P22) steel is slightly susceptible to reheat cracking, with a ductility dip around 650°C. At temperatures around and above 700°C, the material has an excellent ductility. No special precautions during welding are necessary.

## Uniaxial Accelerated Creep Tests On Weldments In Service Exposed 2¼Cr1Mo Steel

Building further on the positive experiences to make long term creep extrapolations for base materials from the results of short term accelerated creep tests, an attempt is made in this part of the research project to extend these short term approaches to weldments.

**CSR Tests.** The CSR tests have been carried out on the service-exposed 2¼Cr1Mo base material and on the stress-relieved crosswelds at 550 and 600°C. For the latter tests, two experimental set-ups have been used. The first is identical to that for testing of base materials. It consists of furnace heating of the specimen and testing with a 25 mm gauge length. In this case, the whole gauge length of the specimen experiences a uniform temperature and consequently, the imposed deformation over the 25 mm gauge length, containing base metal, deposited weld metal and heat affected zone, will be concentrated in the constituent with the weakest creep resistance. Consequently, using furnace heating and a 25 mm gauge length, the full weldment is tested. In the second experimental set-up, the specimen is heated through the Joule effect. This heating method creates an axial temperature gradient along the gauge length. A constant temperature prevails only in an approximately 10 mm wide zone centered on the control thermocouple. When this heating method is combined with an extensometer with a small gauge length of 12,5 mm, it is possible to test a particular constituent of the weldment, specifically the deposited weld metal.

The weld metal has a far greater creep strength than both the weldment and the base material. Since the only zone in the weldment which is not explicitly taken into account in the CSR-tests is the HAZ, the low weldment strength must necessarily be due to the HAZ. However, the strength of the HAZ is not that low that the weldment strength decreases below that of the base metal. The CSR rupture life predictions for the HAZ at 600°C are shown in figure 3. Conservative rupture predictions are obtained as they are based on the assumption that the failure time of the weldment equals that of the weakest constituent (HAZ) [9].

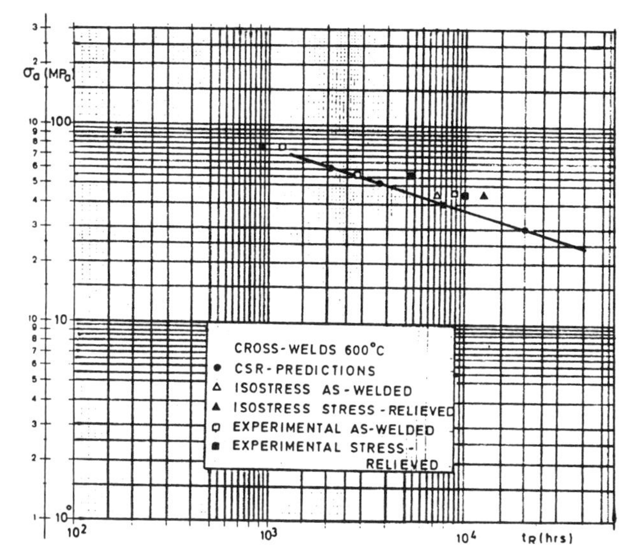


Fig. 3 – Comparison of CSR and isostress time to rupture predictions with isothermal test results on crossweld specimens.

**Isostress Tests.** The isostress creep tests have been carried out on the service-exposed 2¼Cr1Mo base material as well as on crossweld specimens. As for the CSR-tests, the crossweld specimens were taken perpendicular to a new circumferential weld in a service-exposed 2¼Cr1Mo pipe. Whereas the CSR-tests have only been performed on specimens in the stress-relieved condition, the isostress tests have been carried out on as-welded as well as stress-relieved specimens. The isostress results for the crosswelds in the as-welded and stress-relieved condition virtually coincide for all the parameters considered, viz. minimum creeprate, time to one percent strain and rupture time. This indicates a negligible influence of stress-relief on creep properties.

**Isothermal Creep Tests.** Isothermal creep tests with durations extended up to 10 000 hours have been performed at 600°C on the as-welded and stress-relieved new crosswelds in the service-exposed 2¼Cr1Mo pipe material. On the basis of the obtained results, no distinction can be made between the strength in the as-welded and in the stress-relieved condition. The experimental rupture times are superimposed on the CSR- and isostress predictions in figure 3. It can be observed that the predictions fit the experimental data reasonably well. The creep strength of the weldments lays within the DIN scatterband for 2¼Cr1Mo base material.

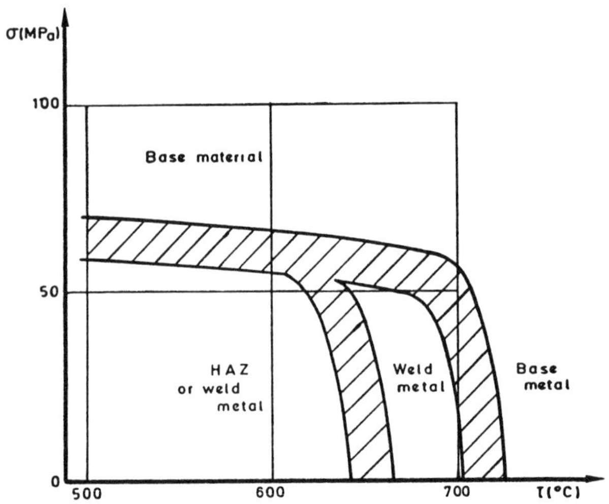


Fig. 4 – Rupture location map for service aged 2¼Cr1Mo crossweld creep tests.

A tentative failure location map can be drawn from the experimental results as shown in figure 4. This map is independent of the application of a PWHT. Above 700°C–720°C failure occurs in the base material at all stresses. At lower temperatures, failure occurs still in the base material for stresses above about 60–70 MPa. For intermediate temperatures (600/625°C < T < 700°C) and stresses below 60 MPa uniaxial specimens fail in the weld metal. At stresses below 60 MPa and for temperatures lower than 600°C (i.e. real service conditions), failure is to be expected in the HAZ, although failure in the weld metal is still encountered in the uniaxial cross weld creep tests.

## Component Creep Test

The aim of the component creep test is to check the reference stress controlling creep deformation and failure behaviour of components containing welds. Advantage is taken of the test to verify the performance of capacitance strain gauge measurements, Barkhausen noise, replication and ultrasonics to follow up creep behaviour.

The component creep test has been performed on a Ø 268 mm pipe with a wall thickness of 9 mm and with a length of 695 mm, containing a nozzle (fig. 5). The component has been in service as part of an intermediate pressure turbine piping for about 200 000 hours at 540°C and at an internal pressure of 30 bar. A longitudinal weld has been produced in the pipe on the opposite side of the nozzle for obtaining the maximum principle stress (the hoop stress) perpendicular to the weldment.

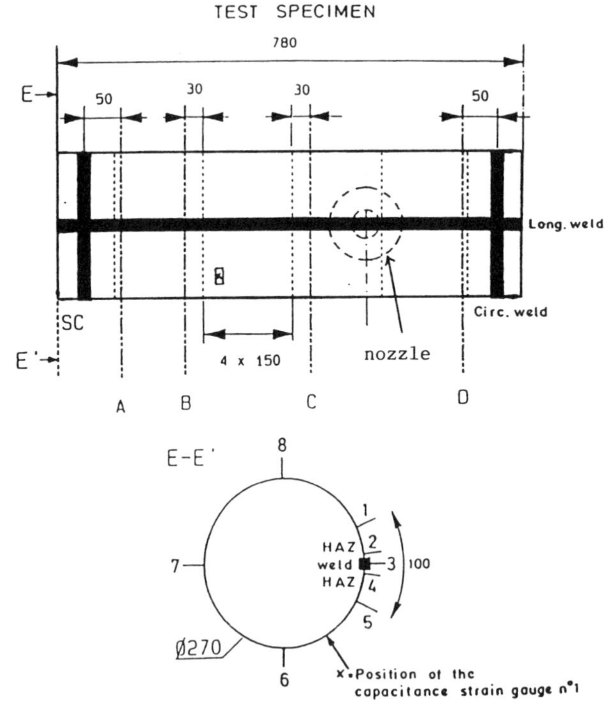


Fig. 5 – Schematic view of the test vessel.

In order to measure local hoop strain at the longitudinal weldment two capacitance strain gauges of the CERL-PLANER type were installed on the weld metal and in the heat affected zone. As a reference, a third gauge was installed on the base material away from the weldment. For measuring the macroscopic deformation of the component small shallow holes were punched at the outer surface as indentations for gauge measurements of the axial and diametral deformations and strains across the longitudinal weldment. A particular measurement grid was provided around the nozzle.

The internal pressure creep test has been performed at 600°C and 40 bar. This results in a mean diameter hoop stress of 56 MPa. The skeletal point stress (equivalent steady state von Mises stress) is about 49 MPa. At the nozzle the reference stress calculated according to TRD 301, 5.3.1 is 68 MPa.

The failure of the component occurred twice at the weld of the nozzle which was only monitored by macroscopic strain measurements.

A first fracture occurred after about 3400 equivalent hours of creep testing at the right angle position of the nozzle suggesting that the reference stress at the failure location reaches about 60 MPa. Repair welding was performed by inserting a new (larger, i.e. Ø 85 x 6,5 m) nozzle, increasing the stress level at the nozzle according to TRD301 to about 80 MPa. The occurrence of the second failure after only 1246 h extra creep testing (i.e. at about 37 % of the first time to failure) can be explained by this higher stress level. The apparent reference stress in this case is about 70 MPa. Qualitatively, it can be stated that the observed failure location at the non-reinforced nozzle could be expected. In fact the maximum stresses arise at the crotch corner connection. Although they redistribute there under creep conditions, the failure location is not shifted to the saddle point where the stress concentration is not significantly creeping away. In both positions the stress level is significantly higher than across the longitudinal weld.

Strain measurements using capacitance strain gauges are giving reliable results as far as they are correctly used (after ageing for limiting the drift, e.g.) with proper protections (against dust, oxidation, ...) and with adequate measuring equipment. For larger strain (and strain rates), the deformation can be followed using mechanical macroscopic deformation measurements (> 0.5 %). The skeletal point stress (von Mises stress) is a fairly good reference stress for the description of the creep deformation of a straight pipe with or without longitudinal weld. The creep strain and damage are essentially concentrated at the circumferential position of the vessel corresponding to the nozzle. The rupture location (in the nozzle weldment) is in agreement with what could be expected from the failure location map in figure 4.

Post failure ultrasonic examination did not reveal any damage indication outside the failure area. Post failure replica analysis in the vicinity of the main cracks at the vessel-nozzle connection indicates that very local creep damage occurred. Extensive destructive metallographic examinations at the failure location reveal that transverse weld metal cracking was the main damage. The creep cracks develop further into the HAZ and base metal. The damage distribution around the nozzle is in good agreement with the results of time dependent finite element caculations on unreinforced cylinder-nozzle connections. This longitudinal weld does not present any creep damage detectable by replicas, ultrasonic or Barkhausen noise measurements.

## Conclusions

Following conclusions can be put forward from the research project.

- No special precautions are necessary during welding of 2¼Cr1Mo steel.
- The occurrence of microstructural instability during isostress tests can easily be checked in the following two ways :
  - the shape of the $(\ln \dot{\epsilon}, \epsilon)$-plot,
  - a change in the temperature dependence of the Monkman-Grant product obtained by consecutively incorporating the data at lower temperatures.
- Provided that the temperature range in isostress tests on ex-service material is restricted to those tests where no microstructural instability occurs, the isostress [T (°C)] and CSR-predictions are very similar.
- For new material, the continuously evolving microstructure results in different predictions by the CSR and isostress methods. The isostress rupture life predictions tend to be conservative.
- On crossweld specimens from new welds on service exposed 2¼Cr1Mo (P22) material, both CSR and isostress creep tests yield rupture life predictions which agree relatively well with the isothermal data. No influence of the postweld heat treatment has been found on the creep strength of the weldments.
- A rupture location map has been drawn.
- The creep strength of repair weldments is not significantly reduced compared to original weldments, provided that all creep damage has been removed before repair welding and that the equivalent stress is not increased by repairing.
- Barkhausen noise and ultrasonic techniques do not seem to be very promising for creep damage monitoring. For strains below 0,5 %, capacitance strain gauges allow reliable measurements. For larger strains mechanical macroscopic deformation measurements can be used. Replication allows the detection of creep cavities. In P22 steel cavitation occurs in a rather late stage of creep life.

## Acknowledgements

The authors wish to thank Ir. A. Vinckier, Professor at the Gent University, Ir. A. Dhooge, Director of the Belgian Welding Institute, and the sponsoring companies AIB-Vinçotte, Laborelec, Mannesmann-Carnoy, Monsanto and Solvay for their permission for publication of this paper. Last but not least the financial support of the I.W.O.N.L. and the N.F.W.O. is acknowledged.

## References


[1] Steen M., Provost W., Dhooge A., "Fifth International Conference On Pressure Vessel Technology", Vol. II, ASME, New York, p 1121 (1984)

[2] Steen M., Ph.O., "Dissertation", Univ. of Gent (in Dutch) (1984)

[3] Steen M., Provost W., Dhooge A., "Res Mechanica, 18, 35 (1986)

[4] Hart R.V., "Metals Technology", 1, 1 (1976)

[5] Stubbe J., Van Melsen C., "Restlivetids Konference", Copenhagen, p 29–69 (March 1980)

[6] Cane B. "High Temperature Technologie", 1, 215 (1983)

[7] Melton R.N., Materials Science and Engineering, 143 (1983)

[8] De Witte M., Steen M., "Third International Conference on Creep and Fracture of Engineering Materials and Structures", Swansea, p 773–802 (April 1987)

[9] Steen M., De Witte M., Coussement C., "Fourth International Conference on Creep and Fracture of Engineering Materials and Structures", Swansea, p 663–690 (April 1990)

*Proceedings of the Fifth International Conference on Creep of Materials, Lake Buena Vista, Florida, USA, 18-21 May 1992*

# Creep-Fatigue Assessment of a Welded Steel Component

**R.H. Priest, P.M. Cairns, D.N. Gladwin, and D.A. Miller**
Nuclear Electric
Bristol, England

## ABSTRACT

Some power generating components are subject to combined creep and fatigue loading during their service lifetimes. To assess the integrity of such components, Nuclear Electric plc have developed the so-called 'R5' procedure (1). The application of R5 to predict the response of Type 316H steel superheater pipework is discussed. An example of service-induced cracking is shown. Finite element predictions of the service loading, commensurate with the observed cracking mechanism, are described. Ductility exhaustion, as specified in R5, is then used to rationalize the observed incidence of cracking. Features tests are also described. These were mounted to support the creep-fatigue initiation calculations and have revealed some of the likely conservatisms in the R5 method. Finally, the results of laboratory experiments to characterise crack growth response under creep-fatigue conditions are given. These data may be used with the R5 approach to demonstrate the integrity of the pipework in the presence of cracking.

THE BOILERS in some of the UKs advanced gas cooled reactors are positioned within the main pre-stressed concrete pressure vessels. The boilers are supported in such a way that movement due to the thermal expansion of the boiler units, particularly during reactor start-ups and shut-downs, are accommodated. To transfer superheated steam from the boilers to the generator turbines, steam is passed from a number of sub-headers, through so-called 'tailpipes' into fixed superheater penetrations which pass through the pressure vessel wall (Figure 1); there are four tailpipes per superheater penetration. Due to the differential movement between the boiler units and the penetrations, creep-fatigue loadings are experienced by the tailpipes during service.

A plan view of the tailpipe/penetration configuration is shown schematically in Figure 1; the two shortest tailpipes are shown shaded. The tailpipes are manufactured from Type 316H stainless steel welded with Type 316 weld metal to Type 316H stainless steel stubs at the superheater penetration end, Figure 2. Under normal

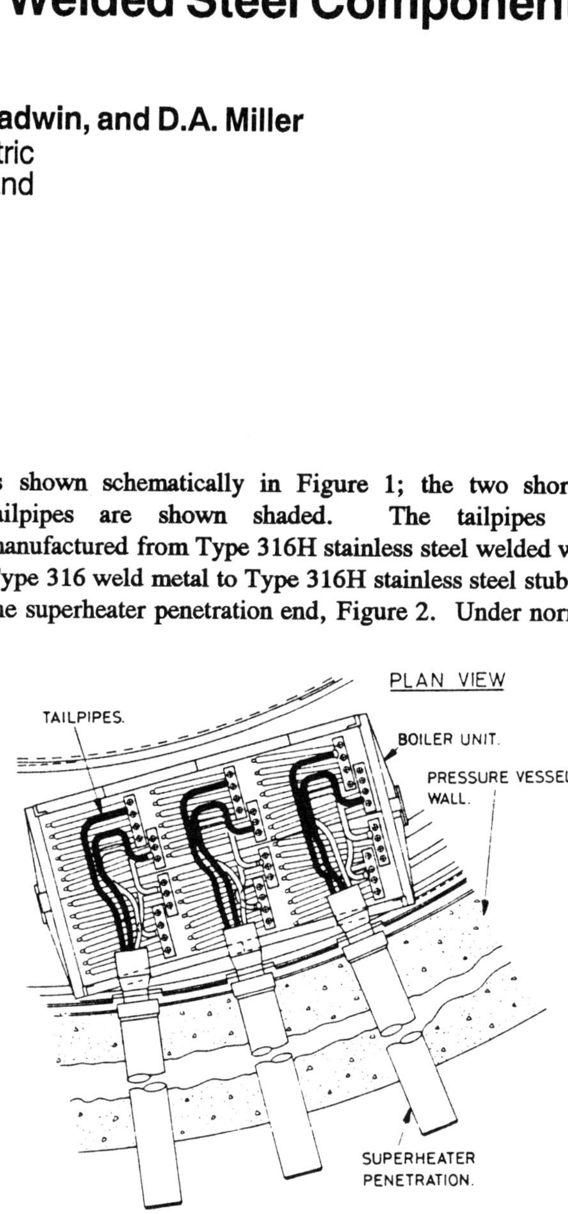


Fig. 1 - Schematic of tailpipe/penetration configuration

operating conditions, the pipework operates at 515 to 530°C. Limited cracking has been observed in a small number of tailpipes, in the region adjacent to the weld fusion line on the stub side of the tailpipe to stub weld. To understand the damage mechanisms, in order to predict the occurrence of cracking and to gauge the rate of crack growth, a number of avenues of work have been followed; these are outlined in the following sections.

First, the characteristics of the cracks found in-service were established by metallographic sampling and replication techniques. Then the likely loading history producing this cracking was modelled using finite element analyses. This information was used to predict the onset of cracking due to creep-fatigue using the ductility exhaustion route recommended in R5 (1). These predictive calculations were supported by accelerated features testing. Finally, crack growth rates have been characterised in order to demonstrate structural integrity in the presence of part-penetrating defects. To do this, creep-fatigue crack growth rate data were generated using standard laboratory specimens. These tests, along with the resulting crack growth data obtained, are also described.

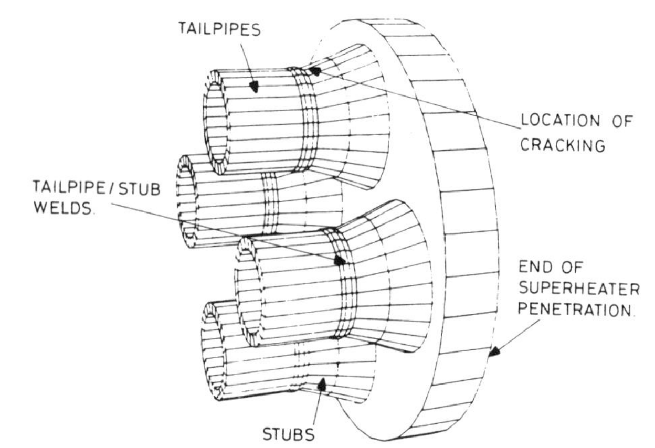


Fig. 2 - Detail of Tailpipe/Stub Geometry

### Characteristics of Cracking Found In-Service

The presence of tailpipe cracking has been revealed during routine eddy current inspection of the tailpipe/stub weld region during biennial reactor overhauls; the eddy current method reliably detects defects ≥5 mm long and 2 mm deep. More detailed information on the exact location and type of cracking was obtained by performing in-reactor acetate replication of the cracked regions of the tailpipes; this work was performed at an ambient temperature of 60-70°C, requiring a 'hot' replication technique to be devised (2). Subsequent examination of the replicas, using an optical microscope, typically revealed intergranular cracking adjacent to the weld fusion line within the heat affected zone (HAZ) of the stub material. The location of cracking was thus at a "metallurgical notch" at a geometric discontinuity (i.e. the tailpipe/stub weld).

Further information on the morphology and depth of cracking was obtained by extracting small 'boat' samples of the cracked area prior to a weld repair being effected. The boat samples were sectioned, polished and etched and examined using an optical microscope. This revealed the cracking to be typical of that caused by a creep-dominated creep-fatigue mechanism, Figure 3. In the main cracking was shallow and although discrete cavitation was observed ahead of the main crack tip, in the circumferential direction this was limited in extent and was shallow (< 100 $\mu$m).

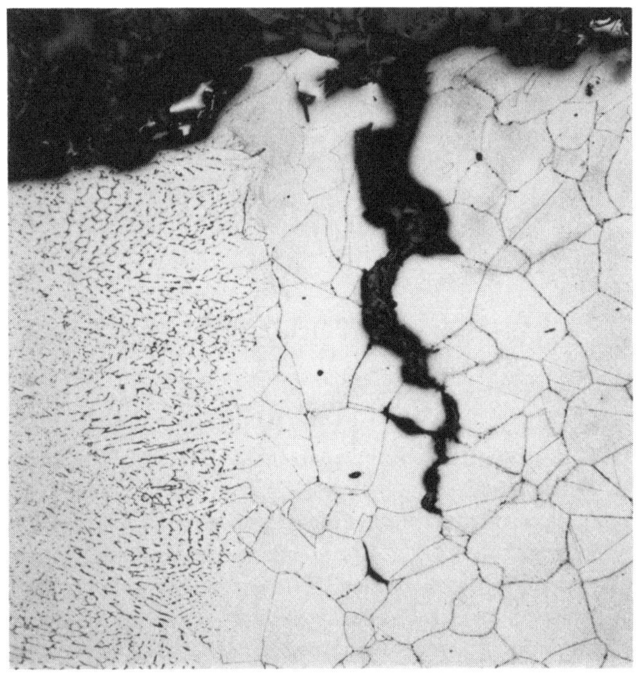

Fig. 3 - Intergranular Cracking Adjacent to Weld Fusion Line (x 128).

### Crack Initiation

**Loading History.** In order to assess the time to creep-fatigue crack initiation, the loading experienced by the tailpipe/stub welds in-service had to be calculated. Elastic analyses of the tailpipes were therefore carried out using the finite element codes ABAQUS (3) and BERSAFE (4). Analyses using ABAQUS 'piping' type elements produced details of the overall response of the tailpipe system and showed that the shortest and hence stiffest tailpipes experienced the highest loads, Figure 4. The loadings applied to the ABAQUS pipe idealisation were the rotations and displacements, corresponding to shut-down and normal operating conditions, produced by the relative thermal movements of the boiler unit and superheater penetrations, Figure 4. The ABAQUS analysis thus permitted the principal loading, primarily a global bending moment at the tailpipe/stub welds, to be established.

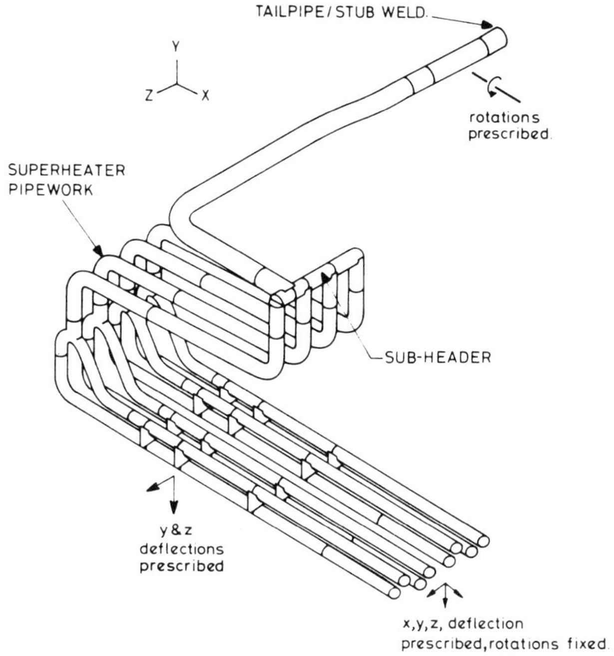


Fig. 4 - ABAQUS Piping Element Idealisation

To perform a creep-fatigue initiation calculation, the stress at the start of the dwell (hold) period was needed, as was a measure of any elastic follow-up during the dwell and the total fatigue strain range, see below.

For defect free structures subject to creep-fatigue loading, the R5 procedure (1) provides a simplified inelastic method of analysis based on shakedown. This approach was adopted and used the elastic stresses local to the tailpipe/stub weld. These were calculated using BERSAFE axisymmetric and fourier analyses, assuming the principal loadings to be those associated with the major service cycle, ie. start-up/shut-down. In this context it is noteworthy that during construction, tailpipes were welded to penetrations which had a permanent 'set', to offset thermally-induced loadings under normal operating conditions. During calculations it was assumed that the pre-sets used were such that the loading modelled was minimised; different assumptions about the level of pre-set would mean that the tailpipes would experience higher loadings than those modelled.

To demonstrate shakedown, a residual stress field was defined; it was taken to be that caused by a fixed end rotation at the tailpipe/stub weld. This was added to the elastic stress field to produce a stress at the start of the dwell, the shakedown reference stress, which was used in the creep damage assessment.

Subsequent behaviour in creep during the dwell depends upon the extent of elastic follow-up. The R5 procedure characterises elastic follow-up by a factor Z. The creep strain, $\epsilon_c$, accumulated during the creep dwell is related to Z by:

$$\epsilon_c = \frac{Z\Delta\sigma_{rD}}{E} \qquad (1)$$

where $\Delta\sigma_{rD}$ is the stress relaxation during the creep dwell and E is the Young's modulus. R5 provides methods for determining Z.

A strain range is also required for fatigue assessment. The strain range calculated elastically was increased to account for creep and local plastic strains due to yielding at the location of interest.

**Damage Calculation.** Using the R5 procedure, a linear damage summation law was assumed such that the number of cycles to the initiation of a creep-fatigue crack, $N_f$, was given by;

$$N_f = [1/N_o + \frac{Z\Delta\sigma_{rD}}{E\epsilon_l}]^{-1} = [D_F + D_C]^{-1} \qquad (2)$$

where $\epsilon_l$ is the creep ductility and $N_o$ is the mean continuous fatigue endurance relevant to the calculated total fatigue strain range. $D_F$, the fatigue damage, was thus defined as the reciprocal of $N_o$. The ductility exhaustion method was used to quantify the creep damage, $D_C$. This requires a knowledge of creep ductility, $\epsilon_l$, as a function of strain rate. However, the conservative adoption of $\epsilon_l$ as the lower shelf ductility operating throughout the dwell period was used in the current work.

The fatigue damage was found to provide a very small contribution to the total level of damage. Calculations of the creep damage term, however, did not predict initiation of cracking in-line with service experience. The reason for this may be the materials property data used in the analysis (for example, mean creep deformation response was assumed) or it may be due to the aforementioned assumptions regarding the level of pre-set. Sensitivity studies, as required by R5, are thus continuing to investigate the effect on total damage of changes in tailpipe loading, temperature and materials properties.

**Features Tests.** By way of support for the finite element work, a number of features tests were mounted. The features were full size replicas of the tailpipe to stub welds, fabricated from material of correct specification and welded according to the original weld procedure. In the main, constant load tests were performed with the tailpipe specimens being tested as one metre long cantilever beams: Tests were loaded such that the moment experienced at the tailpipe/stub weld was comparable with that expected in service. However, testing was performed with the tailpipe/stub weld at 620 ±5°C in order to reduce test times.

Displacement/time plots were produced during the tests to monitor the loading history. One result is shown in Figure 5. A comparison has been made between this load/displacement response and that predicted by finite element analysis using ABAQUS. In the analysis, the shell finite element idealisation was of the tailpipe as an encastre beam. The 150 mm length of tailpipe adjacent to the stub weld was analysed using elastic-plastic-creep; the remaining 850 mm length was modelled elastically. Figure 5 shows remarkably good agreement between the measured load/displacement response and that predicted for a temperature of 615°C. The tests were interrupted and the weld/stub region of the specimens inspected using dye penetrant and replication to reveal any cracking present. The interruption points at which inspections for initiation were carried out were every ~3000 hours.

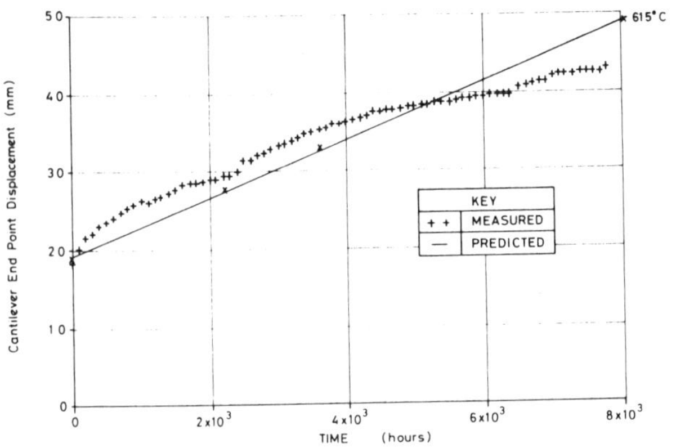


Fig. 5 - Load-Displacement Response of Features Tests

Calculations were performed based on ductility exhaustion, to estimate the time to crack initiation. Using a lower bound creep ductility for Type 316H parent material at 620°C of 6%, ductility exhaustion predicts crack initiation within 18500 hours. This is in line with the observation that after ~ 9000 hours only isolated grain boundary cracking was observed in the features specimens. Notwithstanding this observation, it is noted that equivalent load histories have caused cracking in-service. The discrepancy between the features tests results and the service experience may be due to either;

1) The fact that relatively high ductility material has been used in the features tests, whereas the ductility of some of the material used in service pipework may have been closer to the lower bound, or

2) The possibility that the loading on the service component is more severe than that modelled due to assumptions made about the set-up of the tailpipes.

These aspects are being considered as part of the sensitivity studies mentioned earlier.

## Crack Growth

Since cracks have been observed in a number of superheater tailpipes, estimates of crack growth rates are required in order to insure against the occurrence of fully penetrating defects being produced between inspection overhauls. The R5 assessment procedure (1) permits the calculation of remaining life using a creep-fatigue crack growth rate equation of the form;

$$da/dN = da/dN_f + da/dN_c \qquad (3)$$

where the total crack growth rate, $da/dN$, comprises the sum of a cyclic contribution, $da/dN_f$, characterised by a Paris-type law;

$$da/dN_f = C\Delta K_{eq}{}^m \qquad (4)$$

and a creep crack growth contribution, $da/dN_c$, characterised by C*;

$$da/dN_c = \int_0^{t_h} AC^{*q}\, dt \qquad (5)$$

In Eq. 4 $\Delta K_{eq}$ is the stress intensity factor range for which the cracks are open and includes any necessary plasticity correction (5). The coefficients C and m are found by experiments. In Eq. 5 $t_h$ is the dwell duration, during which creep crack growth is important, and C* is the energy dissipation rate function characterising crack tip stress and strain rate fields. Expressions for C* pertaining to constant displacement hold periods, typical of those induced during thermal cycling, have been given elsewhere (6). The coefficients A and q are once more found by experiment.

To obtain values for C, m, A and q of relevance to superheater tailpipe cracking, crack growth experiments were conducted using 12.5 mm thick x ¾T compact tension specimens at 530°C. Because the observed cracking in plant has been associated with the HAZ of the tailpipe/stub welds, Type 316H steel was used for the test work in both solution treated and simulated HAZ conditions; Table 1 gives details of the material used, heat treatment and test conditions. Constant load creep crack growth experiments were performed in addition to displacement controlled, fully reversed creep-fatigue tests which included constant displacement hold periods at maximum displacement of up to 190h. The test configuration and procedures used are described in full elsewhere (6). Following testing, specimens were sectioned longitudinally and examined metallographically. Cracking was found to be similar to that exhibited in service-cracked tailpipes.

Table 1 - Details of Material, Heat Treatments and Tests Performed

Composition:

| | Element | | | | | | | |
|---|---|---|---|---|---|---|---|---|
| | C | Si | Mn | S | P | Mo | Ni | Cr |
| Wt% | 0.034 | 0.38 | 1.66 | 0.012 | 0.013 | 2.33 | 13.48 | 17.54 |

Heat Treatments:

| | |
|---|---|
| Solution Treated | 1050°C for 1h, AC. |
| Simulated HAZ | 1050°C for 1h, AC + 3% strain + age at 600°C for 1270 h. |

Test Details:

| | |
|---|---|
| Solution Treated and Simulated HAZ | All testing at 530°C. Continuous cycling (2h per cycle) + 22h, 94h and 190h const, displ. hold period tests. Static load creep crack growth tests also performed. |

**Cyclic Data.** Data obtained from the cyclic contributions to crack growth are plotted as a function of $\Delta K_{eq}$ in Figures 6 and 7 for 316H parent and simulated HAZ material, respectively. The results from the HAZ material exhibited remarkably little scatter and cyclic crack growth rates conformed to an upper bound previously defined for fatigue crack growth in Type 316 steels (7);

$$da/dN_f = 1.4 \times 10^{-10} (\Delta K_{eq})^3 \quad m.cycle^{-1} \qquad (6)$$

where $\Delta K_{eq}$ is in MPa$\sqrt{m}$. Data from the parent material were broadly in line with the HAZ results but exhibited more scatter. Furthermore, a possible trend towards increasing cyclic rates with hold time is apparent, yielding an upper bound equation;

$$da/dN_f = 2 \times 10^{-9} (\Delta K_{eq})^3 \quad m.cycle^{-1} \qquad (7)$$

**Creep Crack Growth Data.** Creep crack growth data from both materials conformed to similar trends, Figures 8 and 9. Furthermore constant load creep crack growth data compared well with those data obtained from the hold periods in the creep-fatigue tests for both materials. Once more, a previously measured upper bound crack growth rate law was found to hold (8);

$$da/dt = 0.023\ C^{*0.81} \quad mh^{-1} \qquad (8)$$

Eq. 8 can be used with Eqs. 5, 7 and 3 to describe total crack growth rates and hence remaining life. For example, assuming that a 5 mm long semi-circular defect just escaped detection by eddy current inspection, the amount of crack growth predicted over a 2 year inspection interval period would be ~ 2 mm. This can be shown to be sufficiently small to have minimal structural integrity implications.

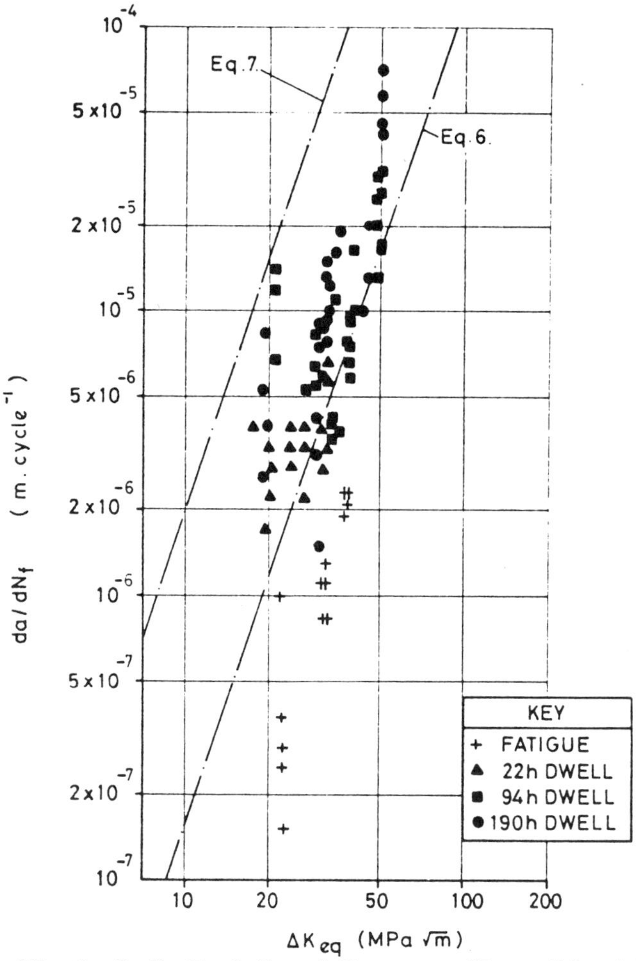

Fig. 6 - Cyclic Crack Growth Response of Parent Material

It should be noted that in Eq. 8, C* refers to experimentally measured values. In practice, reference stress approximations to C* have been found to accurately predict measured C* values (9), their use significantly simplifying the application of the creep crack growth rate laws to assessments.

## Summary and Conclusions

Limited cracking has been observed associated with a small number of Type 316H superheater tailpipe welds operating at 515-530°C. Examination of defective service components using a hot replication technique and metallographic sampling has confirmed the cracking to be due to creep dominated creep-fatigue.

Finite element analysis has been used to permit the prediction of the onset of cracking using the ductility exhaustion approach specified in R5 (1).

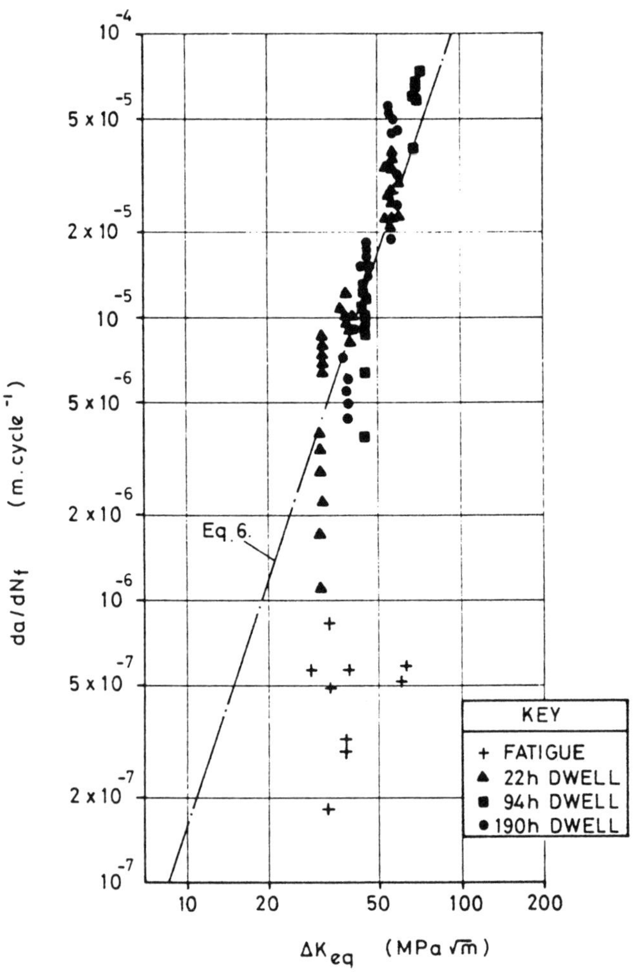


Fig. 7 - Cyclic Crack Growth Response of Simulated HAZ Material

A number of features tests were mounted in which replica tailpipes were exposed to accelerated temperature testing in an attempt to further support the numerical analysis. These tests have likely conservatisms in nature the R5 approach. Pessimisms are being investigated using sensitivity analysis.

Creep and creep-fatigue crack growth experiments were performed on 316H parent and simulated HAZ material at 530°C to obtain crack growth rate equations. The following equation was found to bound the data from both materials;

$$da/dN = 2 \times 10^{-9} (\Delta K_{eq})^3 + \int_0^{t_h} 0.023 C^{*0.85} dt \quad (9)$$

where da/dN is the total crack growth rate in $m.cycle^{-1}$, $\Delta K_{eq}$ is the plastically corrected stress intensity factor range for which the cracks are open (MPa √m), $t_h$ is the hold time in hours and C* is the crack tip stress and strain rate characterising parameter ($MPa\ mh^{-1}$). The equation can be used with the R5 procedure to demonstrate the integrity of the superheater tailpipes in the presence of cracking.

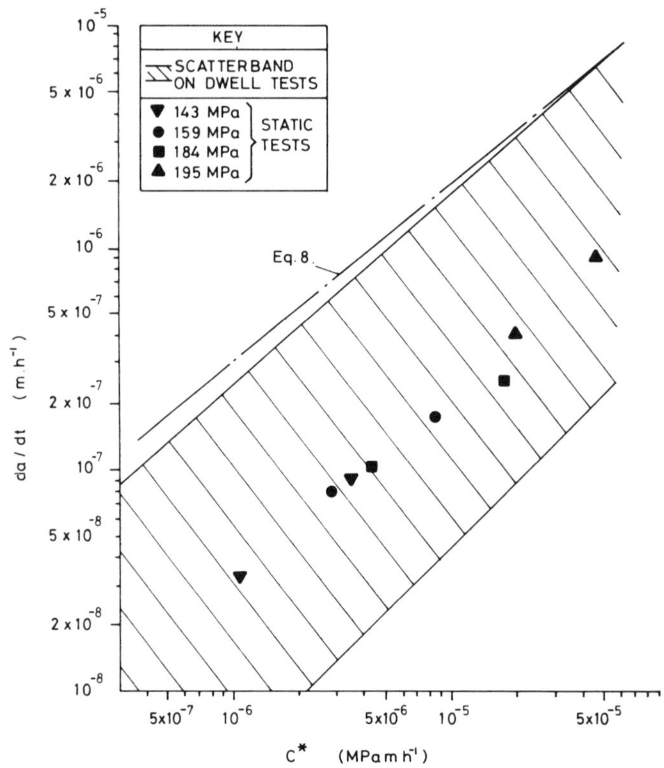


Fig. 8 - Creep Crack Growth Rates in Parent Material

## Acknowledgement

This paper is published by permission of Nuclear Electric plc.

## References

1. Goodall, I.W., Goodman, A.M., Chell, G.G., Ainsworth, R.A. and Williams, J.A., (Eds.), *Nuclear Electric Assessment Procedure R5, Issue 1* (1990).

2. Neale, D.P., *Nuclear Electric Memorandum* TD/SIP/MEM/1070/90 (1990).

3. ABAQUS, "Vol. 1 : User's Manual", Hibbitt, Karlsson and Sorensen., Inc., Providence R.I.

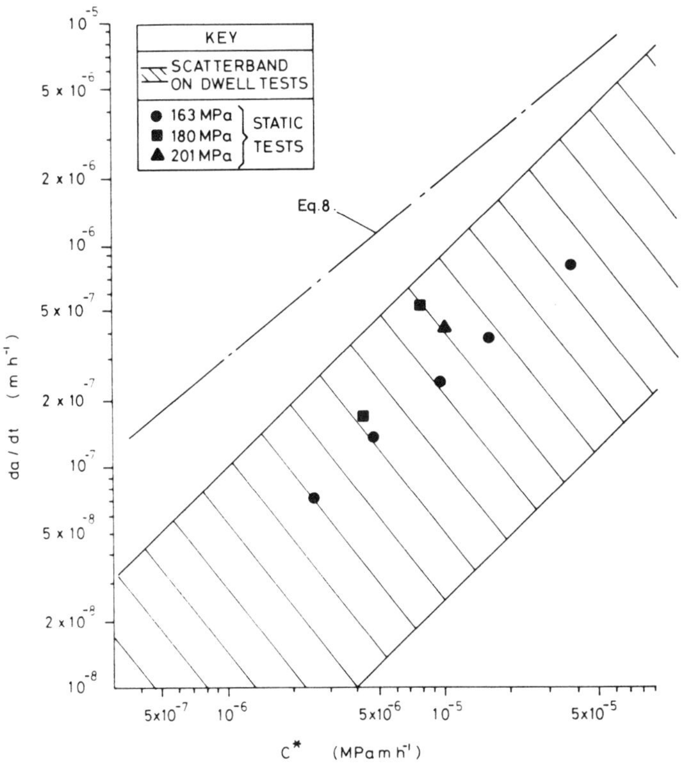


Fig. 9 - Creep Crack Growth Rates in Simulated HAZ Material

4. Hellen, T.K., "A User's Guide to BERSAFE ELASTIC", Berkeley Nuclear Laboratories, U.K., (1990).

5. Skelton, R.P., Beech, S.M., Holdsworth, S.R., Neate, G.J., Miller, D.A. and Priest, R.H., *Proc. Int. Conf. Behaviour of Defects at High Temps.* (1992).

6. Gladwin, D.N., Miller, D.A. and Priest, R.H., *Mat. Sci. and Tech.*, 5, 40-51 (1989).

7. Marshall, P., "Austenitic Stainless Steels - Microstructure and Mechanical Properties", Elsevier Applied Science, London (1984).

8. Curbishley, I., "Data Sheet on Creep Crack Growth in Type 316 and 316L(N) Steel", AEA Risley Internal Report (1990).

9. Priest, R.H., *Proc. Engng. Appl. of Modern Plasticity*, Nagoya, (1989).

*Proceedings of the Fifth International Conference on Creep of Materials, Lake Buena Vista, Florida, USA, 18-21 May 1992*

# Creep Crack Initiation and Growth at Natural Defects in Heavy Section Forgings

**K. Kussmaul, H. Theofel, and K. Maile**
Staatliche Materialprufungsanstalt
Stuttgart, Germany

**K-H. Mayer**
MAN Energie GmbH
Nurnberg, Germany

**C. Berger**
Siemens AG, Energieerzeugung KWU
Mulheim, Germany

**G. Gnirss**
ABB Power Generation AG
Mannheim, Germany

ABSTRACT

In order to improve the evaluation for the acceptance of ultrasonic indications in heavy forgings, the behaviour of defects under creep conditions was investigated. Large scale specimens with defects were removed from three rejected turbine shafts having unallowable ultrasonic indications. Extensive nondestructive examinations were carried out during sampling and production of the specimens as well as during the course of the creep tests.

Fractographical post-examinations on the exposed natural defects more or less affected by creep damage and creep crack growth were conducted. Defects which were not influenced by creep and findings on defects in form of transcrystalline ductile cracks, intercrystalline creep damage and creep cracks were found. The findings could be characterized according to the load level and creep test time as well as to the size and type of defect. It was proved that the nominal stress has to be considered as the decisive parameter for failure (ligament damage) for most specimens. Evaluations with the stress intensity factor K for creep crack initiation and creep crack growth showed for specimens with higher K-values a good agreement between the behaviour of specimens with defects and fracture mechanics specimens under creep loading conditions. The comparison of UT-indications with the true defect sizes confirms former results in relation to the width of the defect.

HEAVY FORGINGS CONTAIN NATURAL DEFECTS and imperfections such as segregations, nonmetallic inclusions and shrinkage cracks. These defects occur during manufacture due to physical and chemical effects (1). Influencing the material properties and local loading states, they may have an effect on the operational performence and safety of components. Forgings for thermal power plants have to be subjected to a complete ultrasonic testing (UT) in order to locate the internal defects, which can certainly be detected at a defect size of $\geq 1\ mm^2$. The allowable limits for UT-indications are amplitudes, corresponding to the indication of a flat bottom hole (FBH) with diameters of 5 mm for single flaws or 2 mm for clusters. About 25 % of the forgings manufactured exceed these limits and after a more sophisticated UT and a fracture mechanics analysis approximately one third of them have to be rejected.

In order to have an improved assessment of the UT-findings with taking into consideration the size, position and distribution of the flaws as well as the assessment of defects by means of linear elastic fracture mechanics methods (LEFM),

Table 1: Description of selected turbine shafts

| Sym. | Material | Dimensions [mm] | Melting process | Weight of ingot (melt) [tons] | $R_{p0,2}$ at RT [MPA] | $A_v$ at RT, ISO-V-Spec. [J] |
|---|---|---|---|---|---|---|
| K | 26 CrMoV 6 11 | ø 656/828<br>L 1500 | Electro-furnace, vacuum carbon deoxidation | 140<br>(150) | rim: 663<br>core: 642 | rim, tang.: 108<br>core, rad.: 24 |
| T | 30 CrMoNiV 5 11 | $ø_{max}$ 1035<br>L 6630 | Electro-furnace, vac. ingot casting | 50<br>(114) | rim: 689<br>core: 674 | rim, rad.: 12<br>core, rad.: 6 |
| U | 20 CrMoNiV 4 7 | $ø_{max}$ 891<br>L 4208 | Electro-furnace | 25<br>(26,4) | rim.: 648<br>core: 633 | rim, rad.: 49/150<br>core, rad.: 12/ 28 |

investigations were conducted within the framework of joint programmes (2).

PROCEDURE AND TEST MATERIAL

In order to examine the behaviour of natural internal defects under creep conditions large scale specimens were removed from large rejected forgings. The procedure is schematically shown in Figures 1 and 2. After the first UT (U I) specimen blanks were accurately removed followed by the machining of large scale specimens after the second UT (U II). Relatively large specimens with 100 mm diameter, cf. Figure 3, were chosen to attain component-like loading conditions in the relation of defect size to loaded cross-section.

Prior to creep testing the third UT (U III) was carried out. This was the basis for the interpretation of the additional UT carried out after different periods of creep time (U IV). The creep tests were carried out at 530 °C test temperature and should have been terminated before rupture of the specimens. For the fractographical and metallographical investigations of the defect surfaces and the related fracture faces fracture tests under liquid nitrogen were performed resulting in cleavage fracture.

Data about three 1%CrMoV rotor forgings investigated concering the

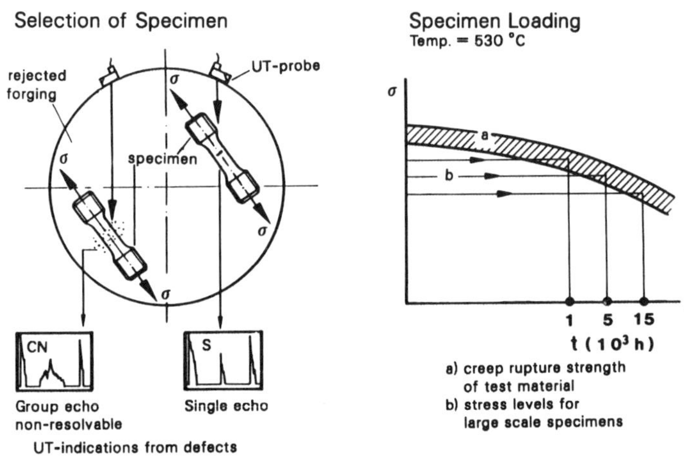


Fig. 1 - Schematic representation of specimen removal and loading.

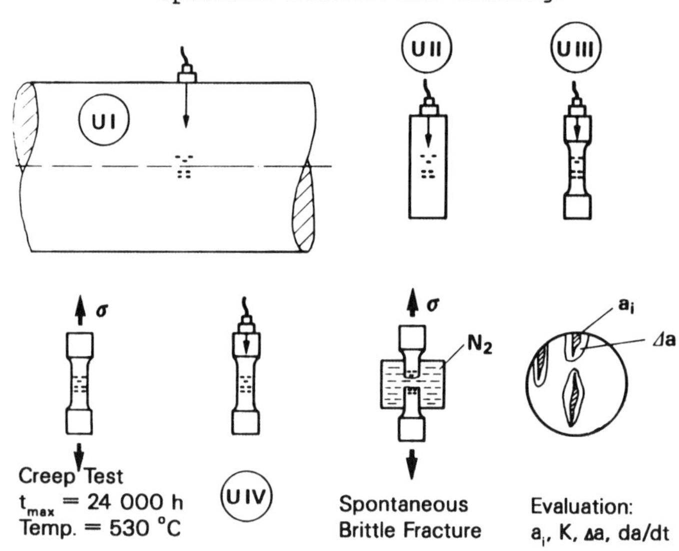


Fig. 2 - Schematic representation of accompanying ultrasonic testing U I to U IV.

manufacture, ingot weight and mechanical properties can be found in Table 1. Forging K is a part of a LP-turbine rotor. Forgings T and U are HP-turbine rotors.

In forging K clusters of resolvable indications and single indications up to 5 mm FBH were found in the ultrasonic test U I. They were concentrically located around the shaft axis in an area of approx. 120 mm diameter. Forgings T and U were rejected due to UT-indications which exceeded the above mentioned allowable limits with maximum echo amplitudes of 3.6 mm FBH (resolvable clusters) in the case of T and of 7 mm FBH (single echo) in the case of U.

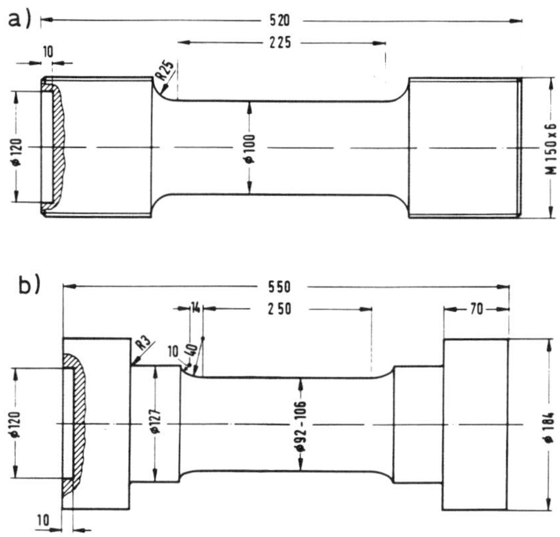


Fig. 3 - Shape of large scale specimens, a) screw head, b) shoulder head.

## CREEP TESTS

The results of the creep tests are compiled in Table 2. The maximum loading time was 24 000 h.

The creep curves of the specimens removed from shaft K are plotted in Figure 4. In addition small scale specimens were tested as well. A relative good coincidence in the course of the curves of large and small scale specimens can be observed. The reduced rupture elongation for longer rupture times of small scale specimens could have been enhanced by very small defects in the specimens cross section. These defects have not been recorded at UT (i.e. FBH is ≤ 1 mm).

Table 2: Creep tests at large scale specimens

| Specimen | σ (MPa) | time (h) | $\varepsilon_p$ (%) |
|---|---|---|---|
| K3 | 310 | 1115 | 3.7 |
| K1 | 262 | 4824 | 3.1 |
| K2 | 262 | 6497 | 2.5 |
| TB | 300 | 8990 | 3.1 |
| TC | 300 | 3441 | 4.9 |
| TD | 262 | 13335 | 7.3 |
| TA | 262 | 24000 | 1.7 |
| UD | 320 | 457 | 2.2 |
| UA | 280 | 4012 | 3.1 |
| UB | 280 | 5026 | 2.9 |
| UC | 240 | 16637*) | 3.4 |

*) Specimen broken in creep test

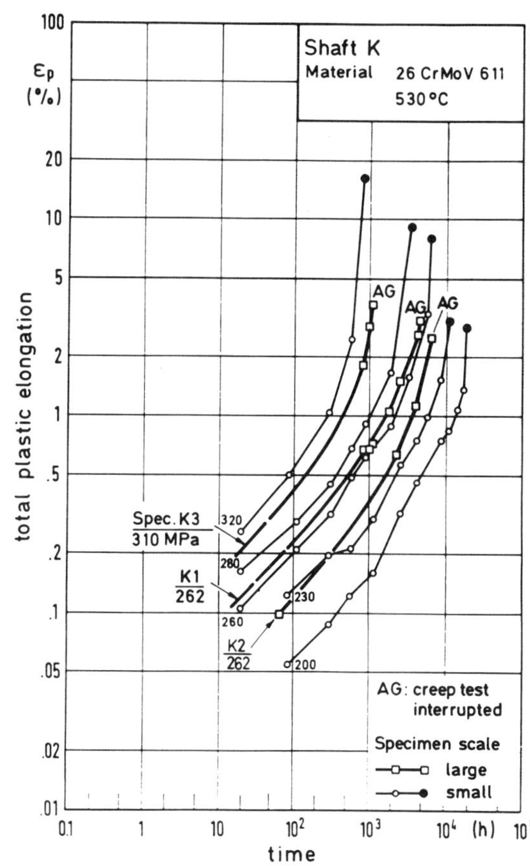


Fig. 4 - Creep curves for specimens from shaft K.

Figure 5 shows the creep curves for the specimens taken from shaft T. The creep curves of the specimens from shaft U are represented in Figure 6. The specimen UC shows lower creep strain than the small scale specimen with the same load.

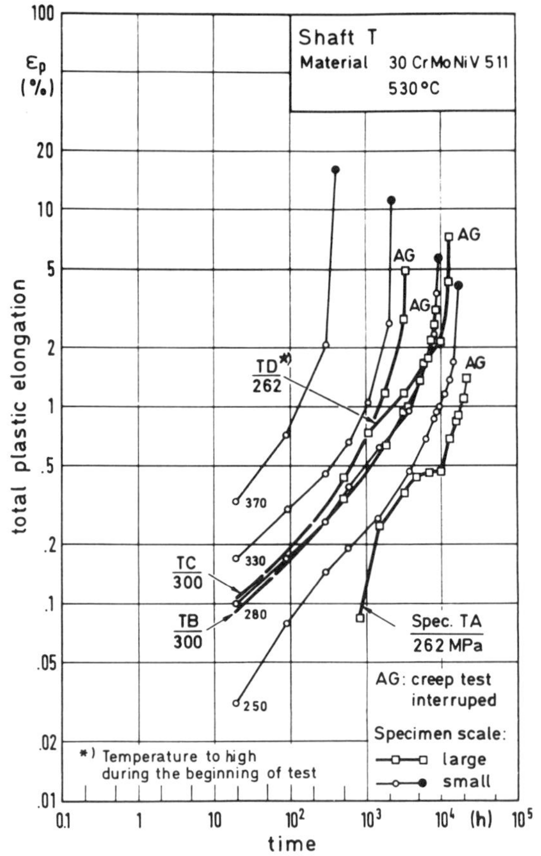


Fig. 5 - Creep curves for specimens from shaft T.

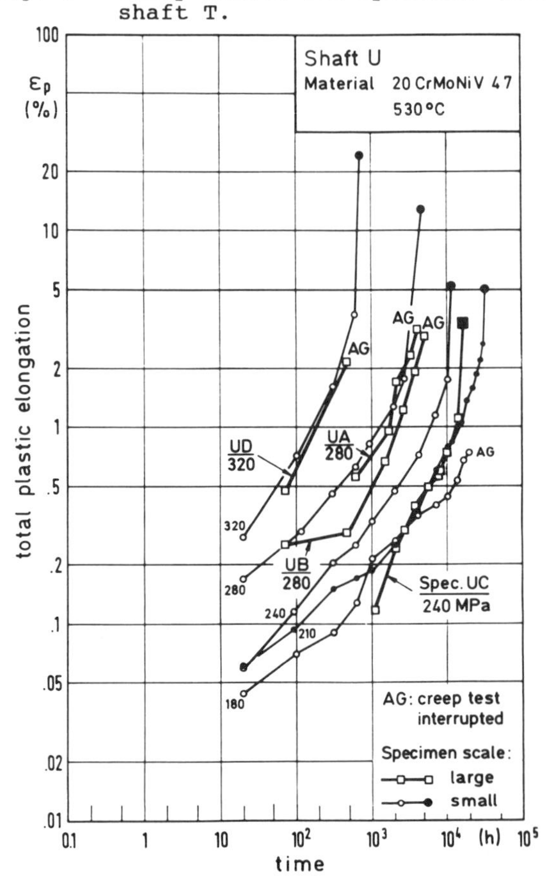


Fig. 6 - Creep curves for specimens from shaft U.

The large scale specimen broken after 16 637 h showed a reduced creep rupture ductility behaviour due to large areas with clusters in the centre of the shaft.

## RESULTS

DEFECTS AT HIGH STRESSES - Figure 7 shows a relatively large defect with a lot of small defects (inclusions) in the neighbourhood. The high stress applied (specimen UD, $\sigma$ = 320 MPa, t = 457 h) generates a distincitve gaping of the defect. At the edge of the defect a trans-crystalline ductile crack (TDC) was initiated. In the further crack growth several small neighboured defects were incorporated.

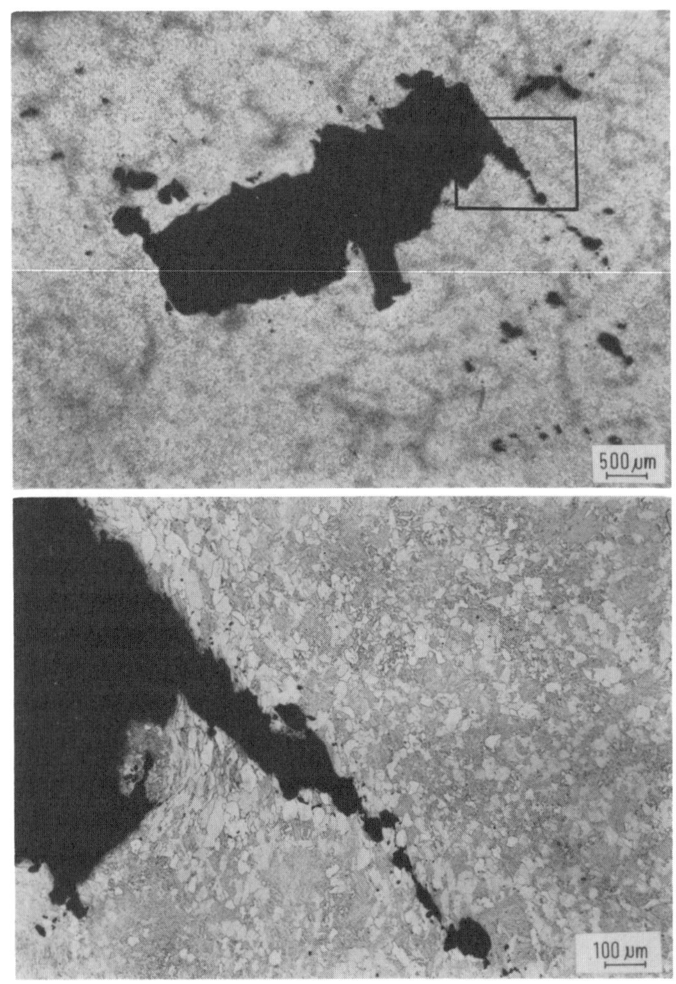


Fig. 7 - Longitudinal cut through gaping defect in specimen UD.

CLUSTERS AT MEDIUM AND LOWER STRESSES - INCLUSIONS AND FREE SOLIDIFIED SURFACES - In Figure 8 a detail of a defect area consisting several free solidified surfaces, covered with inclusions, is shown. The stress applied (specimen UA, $\sigma$ = 280 MPa, t = 4 012 h) leads to failures of the ligaments between the adjacent free solidified surfaces in the above mentioned TDC--mode. This creates a something larger overall defect size and the enlargement has taken into consideration as initial defect width $2a_i$ for the fracture mechanics evaluation with regard to the further creep crack development. At the edge of this overall defect local creep damage (cavities at grain boundaries) embedded in the cleavage fracture surface can be observed. The UT indication appertaining is plotted versus the time of the creep

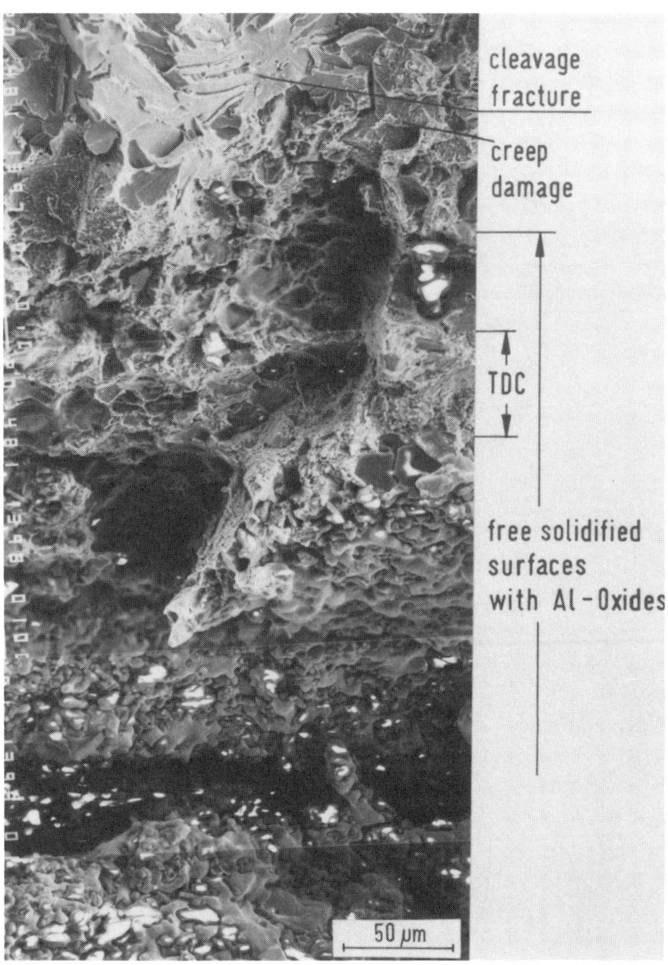


Fig. 8 - Defect with transcrystalline ductile crack and creep damage (specimen UA).

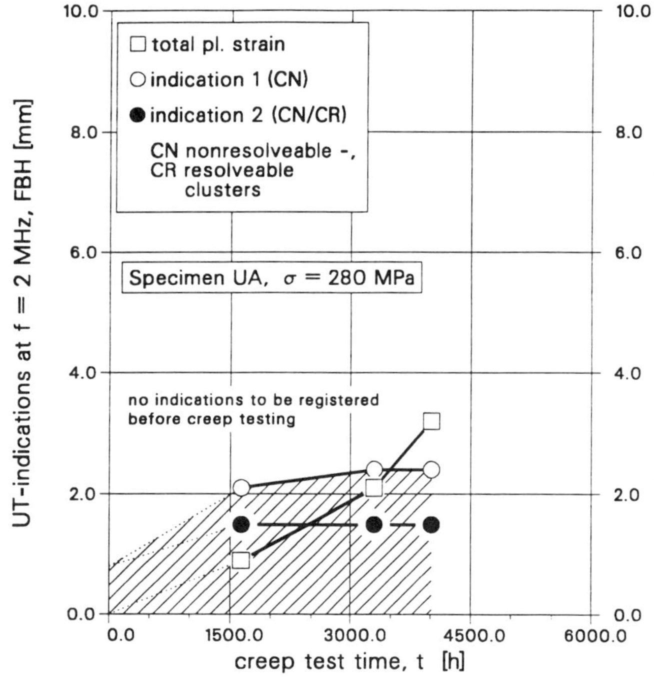


Fig. 9 - UT-indications and total plastic strain versus time (specimen UA).

test in Figure 9. At the UT conducted before the beginning of the creep test no indications exceeding the limit ≥ 1 mm FBH was found. After a loading time of 1 600 h ($\varepsilon_p$ = 0.9 %) an indication with 2.1 mm FBH was observed. The increase of the amount up to 2.4 mm until the end of the creep test seems to be negligible. The indication at 1 600 h can be traced back to the gaping of defects in connection with partially present TDC.

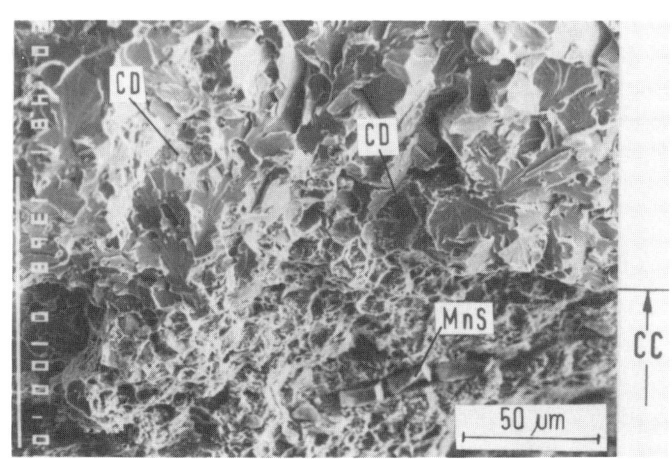


Fig. 10 - Creep damage in the cleavage fracture surface in front of a creep crack (specimen UB).

An example for a defect with inclusions and free solidified surfaces that showed a distinct creep crack (specimen UB, $\sigma$ = 280 MPa, t = 5 026 h) is exhibited in Figure 10. The defect width determined after the creep test was approx. 10 mm. With an UT-indication of 4.8 mm FBH it has already been determined after the test U III before starting the creep test. In the cleavage fracture surface intercrystalline creep damage can be seen.

Ongoing fractographic investigations of the long-term loaded specimen TA ($\sigma$ = 262 MPa, t = 24 000 h) show a distinctive creep crack growth initiated at a defect with free solidified surfaces.

MANAGESE SULPHIDES AT MEDIUM AND LOWER STRESSES - The original UT-indications in specimen K1 ($\sigma$ = 262 MPa, t = 4 824 h) were identified as Al- and Ca-inclusions in a confined area of approx. 50 mm diameter around the original turbine shaft axis. During the creep test additional UT indications were observed, which increased partly due to the creep stressing. These indications were identified as a lot of MnS-bands running parallel to the rotor shaft axis. Strips with TDC appeared within these MnS-bands depending on the density of the included MnS.

The total surface of such linked up separations is used again as initial defect size for the fracture--mechancis evaluation as already mentioned. The order of magnitude of the stress intensity of such an

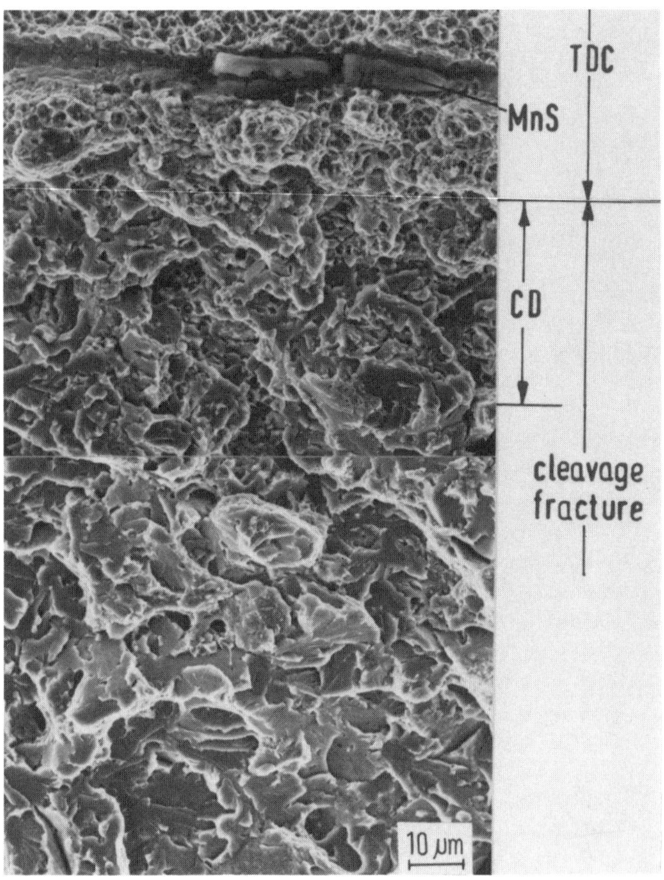


Fig. 11 - Creep damage in front of a transcystalline ductile crack (specimen K1).

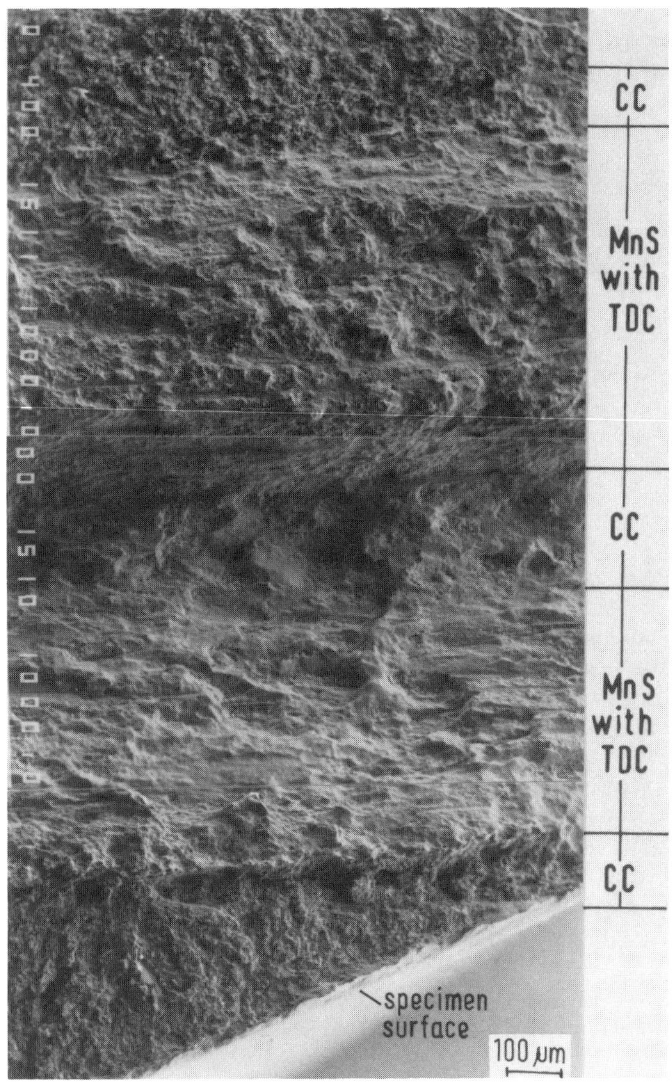


Fig. 12 - Creep cracks at MnS-bands with TDC (specimen K1).

overall defect may lead to creep damage and creep crack initiation in the course of creep loading. At the specimen K1 intercrystalline creep damage was found at the edge of MnS-bands with TDC, cf. Figure 11. Creep cracks were detected in the case where two broad bands including MnS were so densely next to each other that reciprocal influence enhanced the stress intensity between them, see Figure 12.

## ASSESSMENT OF DEFECT BEHAVIOUR

In order to delineate creep crack initiation and growth occurring at small defects the LEFM stress intensity factor K is applied. This parameter proved to be appropriate due to the fact that the defects will cause only low order of magnitude of the stress intensity K compared with the nominal stress of the ligament. In that case a ligament damage can generally be assumed, i.e. the nominal stress is governing the creep failure of the cross-section. The presence of this failure mode is based on the following observations:

- the specimens showed a relatively high creep deformation
- the (extrapolated) rupture times of specimens with defect areas are in coincidence with those specimens without defects
- only small amounts of creep crack growth could be observed up to now.

Obviously, the defects only generate low peaks of stress intensity in the nominal stress field of the specimen ligament. In the following the behaviour of specimens with defects will be compared with results of creep-crack growth investigations at precracked CT- or DENT- specimens (3,4).

The determination of the stress intensity at defects was based on (5) and on relations proved in the investigations of (2). In Figure 13 the calculated maximum K-values for the particular specimens are shown versus the loading time in the creep test. The defect types and findings (with and without creep crack initiation, creep crack growth) are indicated by different symbols. These findings were compared with results

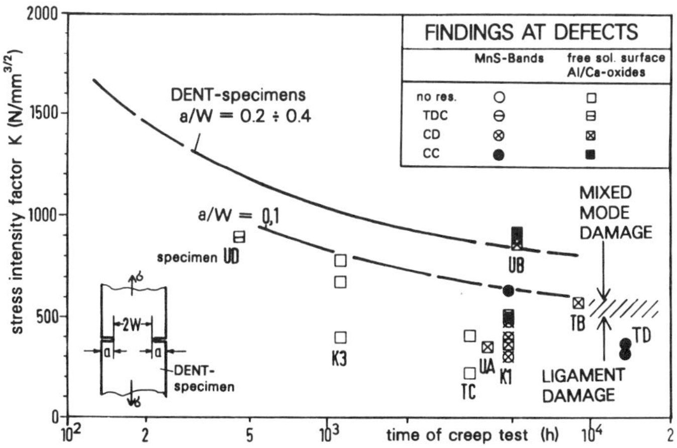


Fig. 13 - Stress intensity at defects versus creep test time with indicated type and finding.

of creep crack growth tests on large sized DENT-specimens with different initial crack-length relations a/W (3). For these specimens creep crack initiation was established at creep crack length of 0.1 mm. In order to compare the exhibited crack initiation curves of DENT-specimens with the results of specimens with defects, it has to be taken into account that the latter ones have very low ratios of a/W. Therefore, only larger defects with high K-values, such as specimen UB, show a good agreement with the behaviour of DENT-specimens. In these cases a mixed mode damage occurred, i.e. also crack tip damage (intercrystalline creep crack growth) besides ligament damage can be observed.

In specimens with very small defects, such as specimen TD ligament damage is dominating. Obviously, the application of K is not relevant in describing the global behaviour of the specimens. Nevertheless, it has to be considered that at lower nominal stress levels the damage mode could change due to the relative increase in the stress intensity at defect tip in relation to the nominal stress field in the ligament.

To describe the creep-crack initiation behaviour the two-criteria-diagram (6) was used (Figure 14). The first criterion plotted on the ordinate is a minimum-to-maximum stress ratio, $R_\sigma$, which is a measure of a time-dependent strength parameter, i.e. the nominal stress, $\sigma_n$, of the part

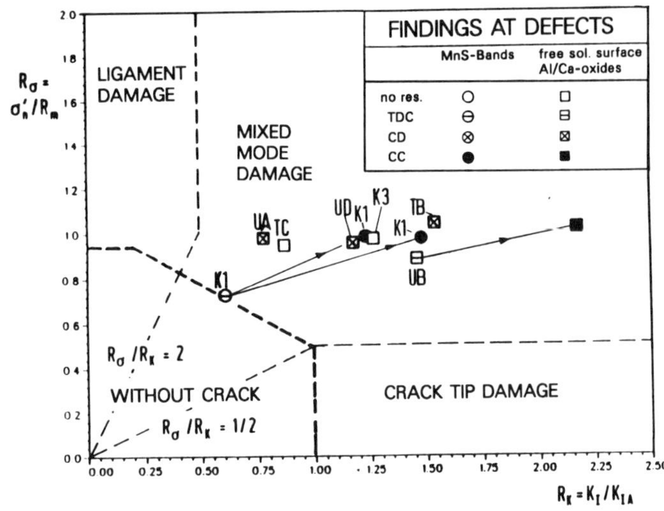


Fig. 14 - Two-criteria-diagram with results for evaluated defects.

under investigation in relation to the creep rupture strength, $R_m$, of a smooth small-scale reference tensile specimen at the corresponding temperatures and loading periods. The second criterion plotted on the abscissa is a stress intensity ratio, $R_k$, which is a measure of a time-dependent stress intensity factor, i.e. the stress intensity factor, $K_I$, of the part under investigation in relation to the stress intensity factor, $K_{IA}$, for creep crack initiation of a small-scale reference CT-specimen at the corresponding temperature and loading period.

In the two-criteria-diagram, the ratio $R_\sigma / R_K$ characterizes various types of damage. The following four areas, designated in Figure 14 are of interest for engineering application: 1) no crack growth, 2) ligament damage, 3) crack tip damage and 4) mixed mode damage.

The plotted points of the specimens with defects are arranged in the area of mixed mode damage. Evaluating the results it has to be taken into account that

- defects are at least in the initial state not really crack-like
- enlargement of the defect size by TDC
- the initial size of the defects is very small compared with the initial crack size of precracked specimens.

Therefore, a shift of the results from the specimens with defects to the left of the diagram is possible. Nevertheless, the temporary results of the evaluation show that the prediction of this method is conservative.

## EVALUATION OF THE ULTRASONIC FINDINGS

The UT results within the scope of these investigations enhanced the present knowledge on the comparison of UT-findings with the real natural defects (7). It has been confirmed that the equivalent diameter of the true defect area may be larger up to factor 3 than the diameter of the FBH.

The determination of the defects width is most important to the fracture-mechanics assessment in the case of longitudinal defects in shafts. The fact that the true max. defect width has never been larger than the appertaining FBH-diameter determined with 4 MHz is shown in Figure 15 according to (7). In the case of a test frequency of 2 MHz the true defects width is not greater than 80 % of the FBH diameter. A comparison with the defects investigated confirms this correlation.

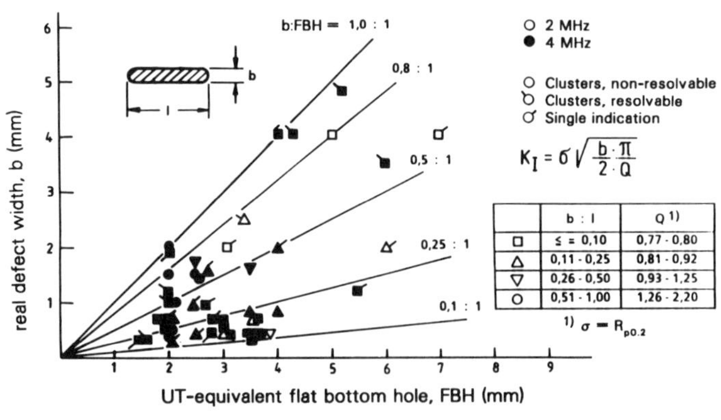


Fig. 15 - Comparison of UT-results and the real width of defects (7).

## DISCUSSION AND CONCLUSIONS

TYPES AND BEHAVIOUR OF DEFECTS - The metallographic and fractographic investigations showed that the natural

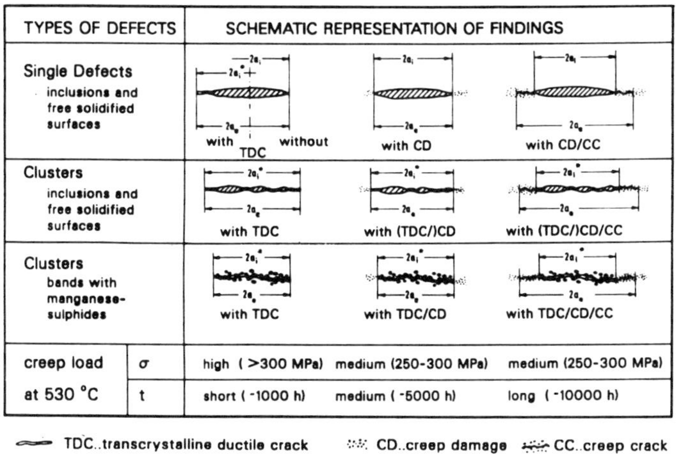


Fig. 16 - Types of defects and findings (2a = defect width).

defects can be classified schematically according to Figure 16. The defects were mainly inclusions of aluminium and calcium oxides and free solidificated surfaces which were more or less densely covered with inclusions. Furthermore, distinctive bands with needle-shaped manganese sulphides were found.

It was proved that a relatively high stress (short exposure time) under creep conditions generates transcrystalline ductile cracks (TDC) at the front of larger defects. In the matrix around clusters TDC can also occur at lower nominal stresses due to locally increased stresses which depend on size and density of the defects. This crack mode can be additionally promoted by a reduced deformation capacity in the areas with defects.

At the edges of defects and TDC intercrystalline creep damages may occur at decreasing nominal stresses and increasing times leading to intercrystalline creep cracks in the course of time.

ASSESSMENT OF DEFECT BEHAVIOUR - The fractografical examinations proved that creep crack initiation occurs at defects if the following conditions are met:

- larger defect size and low nominal stress
- enlargement of defect size by link-up of single small defects/ clusters as a result of TDC and also under low nominal stress.

The observed failure mode can generally be described as ligament damage due to the small defect sizes and the relatively high deformation. However, in the case of creep crack initiation and growth the LEFM parameter K is a suitable tool describing the defect behaviour.

EVALUATION OF UT-INDICATIONS - The comparison of fractographic investigations and UT with regard to defect size and location showed an acceptable accordance. It is possible to characterize the behaviour of specimens in a conservative manner on the basis of evaluating the UT-indications by LEFM.

## REFERENCES

1- Maidorn, C. and D. Blind, Nuclear Engin. a. Design 84, No. 2, p 285 - 296 (1985)
2- Berger, C., J. Foth, K. H. Mayer, W. Oberparleiter and R. B. Scarlin, VGB Kraftwerkstechnik 67, p177-190 (1987)
3- Granacher, J., R. Tscheuschner, K. Maile and W. Eckert, Creep-crack Behaviour of Typical Power Plant Steels, Steel Research 11, p514-521 (1989)
4- Ewald, J., K.-H. Keienburg and K. Maile, Nuclear Engin. a. Design 87, p 389 - 398 (1985)
5- Murakami, Y. et al, Stress Intensity Factors Handbook, Pergamon Press, Oxford (1987)
6- Ewald, J., K. Maile and R. Tscheuschner, 13. MPA-Seminar, Stuttgart, FRG, 8./9. Oct.1987
7- Mayer, K.-H., C. Berger, G. Gnirß, D. Heinrich and W. Prestel, Nuclear Engin. a. Design, to be published.

Aknowledgement: The investigations presented in this paper were sponsored by Forschungsvereinigung Verbrennungskraftmaschinen e.V. (FVV-Nr. 357) and Arbeitsgemeinschaft industrieller Forschungsvereinigungen (AIF-Nr. 6519).

*Proceedings of the Fifth International Conference on Creep of Materials, Lake Buena Vista, Florida, USA, 18-21 May 1992*

# Development of High Strength Ferritic Heat Resistant Steels

**X.Y. Liu**
Northeast University of Technology
Shenyang, China

**T. Fujita**
The University of Tokyo
Tokyo, Japan

## Abstract

Extensive studies have been carried out to improve the elevated-temperature strength of 9-12% Cr heat resistant steels. As a result, a new ferritic steel, which showed extrapolated $10^5$ hour creep rupture strengths of as high as 15-20 kgf/mm$^2$ at 600°C, and 7-10 kgf/mm$^2$ at 650°C, has been successfully developed. This paper describes a part of the work, with an emphasis on the effects of Mo and W contents on the creep rupture properties.

Increase in Mo contents to over 1.5% improved the creep rupture strength at 550°C, but reversely impaired the long-term creep resistance at 650°C and 700°C. It was revealed that the decreased creep resistance of the high-Mo containing steels in the long-term side was due mainly to the massive formation of the $M_6C$ type carbides which reduced both the amount of dissolved Mo in matrix and the amount of fine $M_{23}C_6$ type carbides, and coarsened at a relatively high rate.

The substitution of W for a part of Mo suppressed the formation of $M_6C$ carbides and stabilized the microstructure, and thus improved the long-term creep rupture strength. The formation of the $Fe_2$(Mo,W) type intermetallics in the W-containing steels was also considered a beneficial factor to creep rupture resistance in the long-term side.

In the case of the combined addition of Mo and W, the optimum Mo and W contents were in the range of 1.2 to 1.6% in terms of the Mo equivalent (Mo+1/2W in mass percentage).

CONSTANT EFFORTS have been made to improve the thermal efficiency of fossil-fired power plants through increasing the inlet steam temperature and pressure. The development programs of modern power plants operating at ultra-super critical steam conditions (1) in recent years have raised an urgent demand for the development of new turbine rotor materials, because the currently used Cr-Mo-V steels can not meet the requirements for elevated-temperature strength and the oxidation/corrosion resistance. Austenitic steels, on the other hand, possess the shortcomings of having greater thermal expansion coefficients and poor thermal conductivity, and thus undesirable for the intended application.

The development interests, therefore, have been devoted to the ferritic steels of the 9-12% Cr grades which lie between the Cr-Mo-V and the austenitic steels. However, when used at temperatures as high as 600°C to 650°C, the elevated-temperature strength of the existing grades is insufficient and needs further improvement. As a part of the efforts made to develop new ferritic heat resistant steels with improved creep rupture strength and reasonable toughness, this study was carried out to optimize the Mo and W contents in 9-12% Cr ferritic heat resistant steels.

## Experimental Procedures

The materials used were either from the laboratory melted ingots weighing 100 kg or the industrially produced ingots weighing 2 tons. The ingots were forged in the same ratio as that of commercial turbine rotors at 1150°C-950°C. Since the cooling rate in the center part of a turbine

Table I. Chemical Compositions (in mass %)

| Heat | C | Si | Mn | Cr | Ni | Mo | W | V | Nb | N |
|---|---|---|---|---|---|---|---|---|---|---|
| T0 | 0.15 | 0.05 | 0.44 | 8.96 | 0.47 | 1.49 | ------ | 0.15 | 0.061 | 0.0335 |
| T1 | 0.15 | 0.05 | 0.45 | 8.76 | 0.49 | 1.78 | ------ | 0.14 | 0.063 | 0.0330 |
| T2 | 0.15 | 0.05 | 0.40 | 8.89 | 0.47 | 1.34 | 0.61 | 0.16 | 0.059 | 0.0336 |
| T3 | 0.15 | 0.05 | 0.43 | 8.98 | 0.48 | 0.69 | 1.47 | 0.16 | 0.061 | 0.0362 |

rotor with 1000 mm in diameter is about 100°C/h when oil quenched, and the cooling rate after austenitization has a substantial effect on the creep rupture strength and room temperature toughness, the specimens were cooled down to 600°C in a rate of 100°C/h and then air cooled after austenitized at 1100°C for 5 hours. Tempering was carried out at 680°C for 1 hour. The creep rupture test was performed at 550°C through 700°C. Toughness of the as-tempered and further aged specimens were evaluated through Charpy impact tests using V-notched specimens. Microstructures and precipitation characteristics were investigated by optical and transmission electron microscopy, and X-ray diffraction of the electrolytically extracted residues.

## Results and Discussions

**Effects of Mo content on mechanical properties of 9% Cr steels.** Among all the ferritic heat resistant steels, the modified TAF steel (2) had been the strongest. An important feature of the Modified TAF steel that differentiated it from the other 12% Cr grades was that it contained 1.5% Mo (in mass percentage, and the same hereafter if without specification) which is much higher than the amounts up to 1.0% contained in similar grades. A subsequent investigation by the authors (3) confirmed that the increase in Mo content from 1% to 1.5% did result in important improvement in the creep rupture strength. It was, therefore, felt necessary to investigate further the effects of Mo in a higher content range.

Two heats, with Mo contents of 1.5% and 1.8% respectively, and designated as T0 and T1 in Table I, were prepared by the electro-slag-remelting technique. The Cr content was reduced to 9% to avoid the formation of delta ferrite which in an earlier work (4) was supposed to have led to the decreased creep resistance in a 12% Cr steel containing 2% Mo. The absence of delta-ferrite in Heats T0 and T1 was confirmed by optical microscopy investigation into the as-tempered specimens.

The creep rupture data showed that increasing Mo content resulted in improved creep rupture strength at 550°C, but the beneficial effects disappeared at 600°C. At higher temperatures of 650°C and 700°C, the increase in Mo content reversely impaired the long-term creep rupture strength. This result agreed well with the conclusion reached by Takahashi et al. (4) that the optimum molybdenum content in the 12% Cr steels was about 1.5%.

As both Heat T0 and Heat T1 were free of delta-ferrite, the decreas in creep rupture strength in the 1.8% Mo steel could not have been caused by delta ferrite that was found detrimental to creep resistance (4). To clarify the reasons, the precipitation characteristics and microstructures have been investigated.

X-ray diffraction of the extracted residues revealed that the major carbide was $M_{23}C_6$ for all the as-tempered and the subsequently crept specimens. A small amount of Nb(C,N) also existed in all the cases, but $M_6C$ was detected only after long-term creep. It became clear that the variation in Mo content resulted in different amounts and different precipitation kinetics of $M_6C$ type carbides. The more the Mo content, the more the relative amounts of $M_6C$ in residues, and the earlier the formation of $M_6C$.

Transmission electron microscopy investigation further revealed that the $M_6C$ type carbides usually nucleated on prior austenite grain boundaries, and showed a higher coarsening rate than $M_{23}C_6$. For instance, the carbides grown to about 1 μm after soaked at 650°C for about 1500 hours, in contrast to the maximum of 0.3 to 0.4 μm of the $M_{23}C_6$ carbides. Such a large particle size was obviously less effective in dispersion hardening. Furthermore, it was observed that in the vicinity of the large $M_6C$ particles, there existed precipitate-free-zones where dislocation microstructures recovered more rapidly than the other areas.

It has not been clear whether $M_6C$ nucleated separately from the formerly existing $M_{23}C_6$ or transformed directly from $M_{23}C_6$, but the formation of $M_6C$ was expected to be at the expense of $M_{23}C_6$, because the carbon necessary for the formation of $M_6C$ could not be obtained without the dissolution of $M_{23}C_6$ carbides. The formation of the precipitate-free-zones at around large $M_6C$ carbides was believed to be the results of the redissolving of the $M_{23}C_6$ type carbide during the growth of the $M_6C$ carbides.

Mo has been known to increase creep resistance by both solid solution hardening and stabilizing the $M_{23}C_6$ type carbides. To function in this way, it is necessary for Mo to exist in solid solution as well as in the $M_{23}C_6$ type carbides. Quantitative analysis of the extracted residues indicated that when $M_6C$ was massively formed, a large quantity of Mo was consumed because $M_6C$ contains more percentage of Mo than $M_{23}C_6$ in a unit weight. Thus, the formation of $M_6C$ not only reduced the fine precipitates of $M_{23}C_6$ type, but also reduced the dissolved amount of Mo. It can be concluded that the drop in creep rupture strength of the 1.8% Mo steel on the long-term side was mainly caused by the massive formation and the rapid coarsening of the $M_6C$ type carbides.

**Effects of W contents on mechanical properties in 9% Cr steels.** From the above results and discussions, it can be expected that one way to improve the long-term creep rupture strength is to avoid the formation of the $M_6C$ type carbides. W has been known to strengthen steels in a similar way to Mo, but the isothermal sections of Fe-C-Mo and Fe-C-W phase diagrams at 600°C indicated that W had a weaker $M_6C$ formation tendency than Mo at the composition ranges of interest. Therefore, the substitution of W for a part of Mo was chosen to improve the elevated temperature strength. Two additional heats with different Mo and W contents and designated as T2 and T3 in Table I, were prepared and tested in the same conditions as that of Heats T0 and T1.

A comparison of the creep rupture behaviors of the Heats T2 and T3 with T1 was shown in Figure 1. The substitution of W for a part of Mo gave no change to the creep rupture strength at 550°C and 600°C, but the long-term strength at 650°C and 700°C was significantly improved. It could also be seen that the higher the W content, the higher the

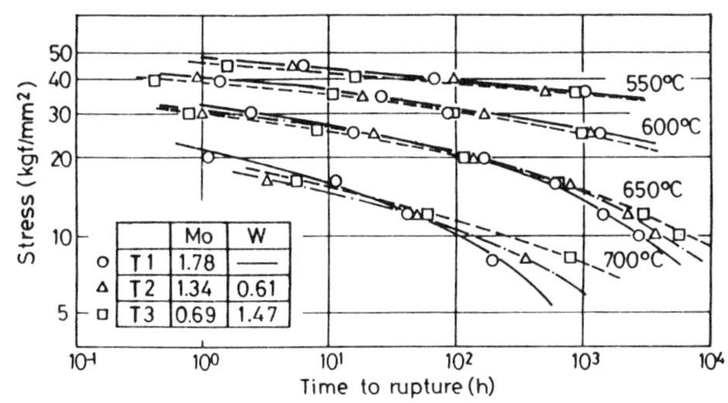


Fig.1. Effects of W content on creep rupture properties of a 9% Cr steel.

long-term creep rupture strength. The higher strength on the long-term side of 650°C and 700°C would lead to the extrapolation of higher $10^5$ hour rupture strengths at 600°C.

Identified precipitates by X-ray diffraction of the extracted residues were shown in Table II. Substitution of W for a part of Mo did not change the main carbide of the $M_{23}C_6$ type and the minor

Table II. Identified Precipitates by X-ray Diffraction.

| Heat | As Tempered | 600°C 30kgf/mm² | 600°C 25kgf/mm² | 650°C 20kgf/mm² | 650°C 12kgf/mm² | 650°C 10kgf/mm² | 700°C 8kgf/mm² |
|---|---|---|---|---|---|---|---|
| T1 | $M_{23}C_6$<br>Nb(C,N) | $M_{23}C_6$<br>Nb(C,N)<br><br>(87.3) | $M_{23}C_6$<br>Nb(C,N)<br><br>(1430) | $M_{23}C_6$<br>$M_6C$<br><br>(173) | $M_{23}C_6$<br>$M_6C$<br>Nb(C,N)<br>(1488) | $M_{23}C_6$<br>$M_6C$<br>Nb(C,N)<br>(2832) | $M_{23}C_6$<br>Nb(C,N)<br>Nb(C,N)<br>(200) |
| T2 | $M_{23}C_6$<br>Nb(C,N) | $M_{23}C_6$<br>Nb(C,N)<br><br>(164) | $M_{23}C_6$<br>Nb(C,N)<br><br>( 1203) | $M_{23}C_6$<br>Nb(C,N)<br><br>(153) | $M_{23}C_6$<br>M6C<br><br>(2232) | $M_{23}C_6$<br>M6C<br>Nb(C,N)<br>$Fe_2$(Mo,W)<br>(3790) | $M_{23}C_6$<br>Nb(C,N)<br>$Fe_2$(Mo,W)<br>Nb(C,N)<br>(548) |
| T3 | $M_{23}C_6$<br>Nb(C,N) | $M_{23}C_6$<br>Nb(C,N)<br><br>(96.7) | $M_{23}C_6$<br>Nb(C,N)<br><br>(1007) | $M_{23}C_6$<br>Nb(C,N)<br><br>(122) | $M_{23}C_6$<br>$Fe_2$(Mo,W)<br><br>(3000) | $M_{23}C_6$<br>$Fe_2$(Mo,W)<br>Nb(C,N)<br>(5880) | $M_{23}C_6$<br>Nb(C,N)<br>Nb(C,N)<br>(805) |

( ): Rupture life in hour

Table III. Chemical Compositions (in mass %)

| Heat | C | Si | Mn | Cr | Ni | Mo | W | V | Nb | N | Moeq. |
|---|---|---|---|---|---|---|---|---|---|---|---|
| S1 | 0.14 | 0.07 | 0.47 | 10.48 | 0.70 | 0.29 | 1.78 | 0.20 | 0.054 | 0.055 | 1.18 |
| S2 | 0.13 | 0.05 | 0.49 | 10.30 | 0.70 | 0.49 | 1.78 | 0.20 | 0.056 | 0.042 | 1.38 |
| S3 | 0.13 | 0.06 | 0.48 | 10.44 | 0.70 | 0.68 | 1.79 | 0.20 | 0.055 | 0.052 | 1.58 |
| S4 | 0.13 | 0.06 | 0.47 | 10.45 | 0.70 | 0.29 | 2.17 | 0.20 | 0.055 | 0.052 | 1.38 |
| S5 | 0.12 | 0.07 | 0.46 | 10.52 | 0.70 | 0.69 | 1.81 | 0.20 | 0.054 | 0.059 | 1.60 |
| S6 | 0.13 | 0.07 | 0.46 | 10.40 | 0.70 | 0.69 | 1.79 | 0.20 | 0.054 | 0.052 | 1.59 |
| S7 | 0.13 | 0.06 | 0.48 | 10.51 | 0.70 | 0.48 | 1.44 | 0.20 | 0.058 | 0.056 | 1.20 |

carbonitride of the Nb(C,N) type, but it changed the precipitation characteristics of $M_6C$ on the long-term sides at 650°C and 700°C. That is, the substitution of W for a part of Mo prohibited the formation of $M_6C$ carbides, and accelerated the precipitation of Laves phase of the $Fe_2(Mo,W)$ type at the same time. The examination of microstructures also revealed that in W containing steels the precipitates remained relatively fine and the dislocation structures were less recovered even after long time creep. It was believed that these changes in precipitates and microstructures accounted for the improvement in long-term creep rupture strength.

**Optimization of Mo and W content in a 10.5% Cr steel.** Now that substituting W for a part of Mo is beneficial to the long-term creep rupture strength, it is necessary to further optimize the Mo and W contents. According to earlier research by the authors (5), Cr has also important effects on creep resistance, and the optimum Cr content for the steel in consideration was in the range of 10.5 to 11.5%. The investigation into the optimum amounts of Mo and W was carried for a 10.5% Cr steel.

The compositions of the specimens used were presented in Table III. In Heats S1 through S7, the Mo and W contents were so designed that W contents in all the seven heats were more than that of Mo, and the Mo equivalent (Mo+1/2W in mass percentage) varied in the range of 1.2-1.6%. All the heats were prepared by vacuum-induction-melting and cast into 100 kg ingots. The heat treatment conditions were the same as those of heats T0 through T3. It was confirmed that all the heats were free from delta-ferrite.

The creep rupture curves for heat S1 through S7 were presented in Figures 2 through 4. From Figure 2 it could be seen that when the Mo equivalent was kept constant at about 1.2%, the heat with higher W content showed higher long-term creep rupture strength at 700°C, although no apparent difference in creep rupture strength could be observed at 600°C and 650°C within the time periods tested. Figures 3 and 4 indicated that increasing in Mo or W contents while keeping other gradients constant improved the creep rupture strength at 600°C, but had no obvious

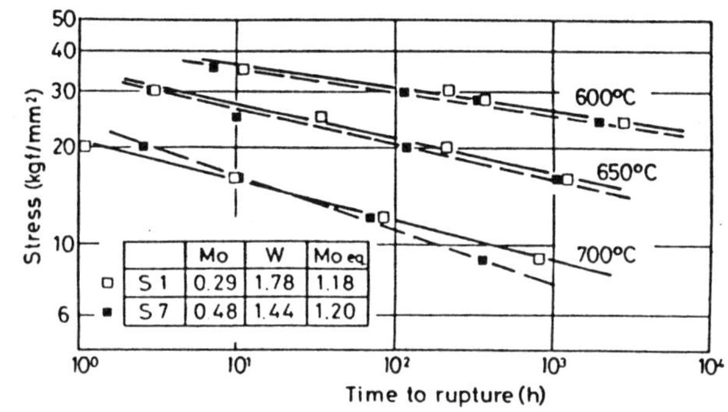


Fig.2. Comparison of the effects of Mo and W in the same Mo equivalent.

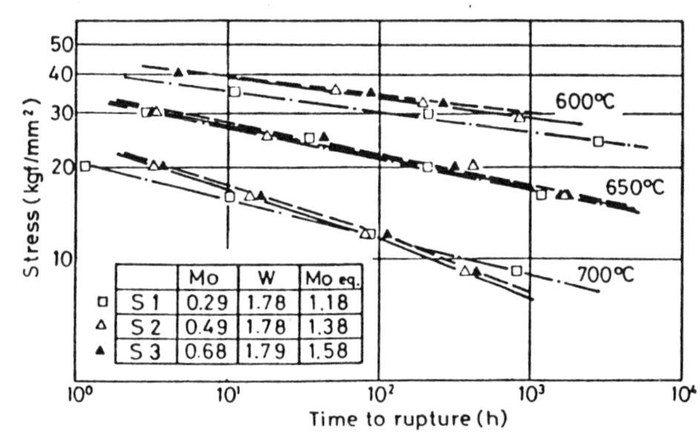


Fig.3. Effects of Mo content on creep rupture strength.

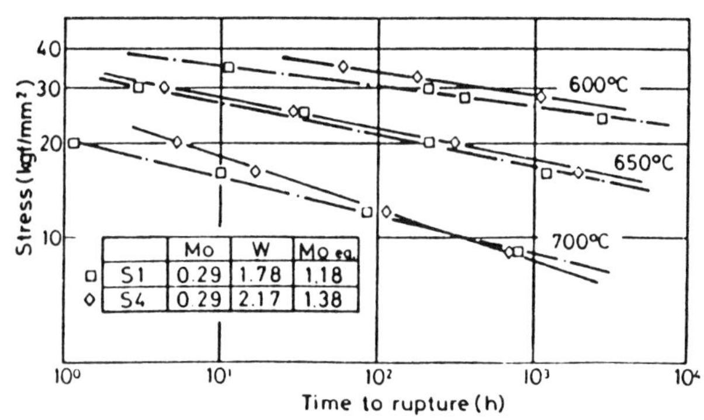


Fig.4. Effects of W content on creep rupture strength.

effects at 650°C. On the long-term side of 700°C, however, the heats with lower Mo or W contents showed higher creep rupture strength. These results indicated that excessive addition of Mo or W was not desirable for long-term service, and the optimum Mo and W contents were in the range of 1.2-1.6% in terms of the Mo equivalent.

The master rupture curves of the steels as a function of the Larson-Miller parameter were shown in Figure 5. By the substitution of W for a part of Mo and the optimization of the contents of other alloying elements, excellent $10^5$ hour rupture strengths were obtained at 600°C through 650°C. The extrapolated values were 15-20 $kgf/mm^2$ at 600°C and 7-10 $kgf/mm^2$ at 650°C. A further comparison in $10^5$ hour rupture strength of the steel developed with those of conventional ferritic steels was presented in Figure 6. It could be seen in this figure that the newly developed steels were much stronger than the existing ones.

Charpy impact tests showed that the steels

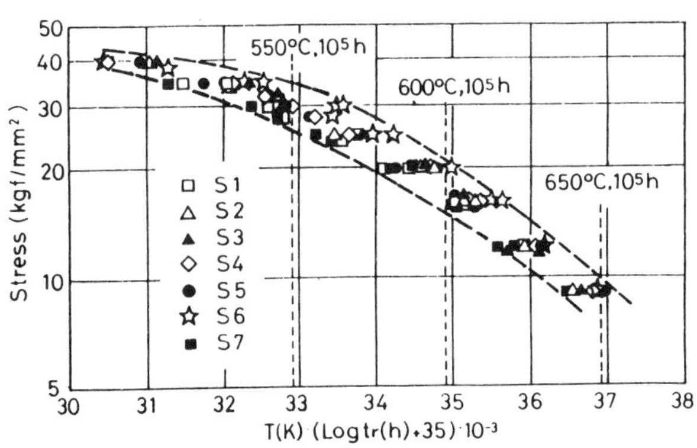


Fig.5. Master rupture curves of Steels S1-S7.

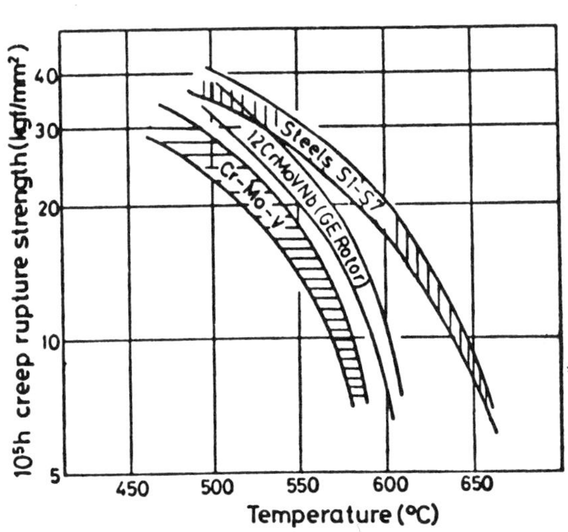


Fig.6. Comparison of $10^5$ hour creep rupture strength of the steels developed with those of conventional ferritic heat resistant steels

developed also possessed reasonable toughness. The impact values of the steels in the as-tempered conditions were all higher than 16 $kgfm/cm^2$, and higher than 3 $kgfm/cm^2$ even after 10000 hour soaking to 600°C.

## Conclusions

Mo and W are two of the most important alloying elements in heat resistant steels that bring about solid solution hardening and enhance precipitation hardening. Increasing Mo contents to over 1.5% improved the creep rupture strength at 550°C, but reversely impaired the long-term creep resistance at 650°C and 700°C. This was because the excessive addition of Mo resulted in the massive formation of $M_6C$ type carbides on long-term soaking to high-temperatures which consumed large quantity of Mo and coarsened at a relatively high rate.

The substitution of W for a part of Mo suppressed the formation of $M_6C$ and stabilized the microstructure, and thus improved the long-term creep rupture strength. The formation of the $Fe_2(Mo,W)$ type intermetallics in the W-containing steels was also considered a beneficial factor for the improvement of creep resistance on the long-term side.

In the case of the combined addition of Mo and W, the excessive amounts of Mo or W would also lead to the decrease in long-term creep rupture strength. The optimum Mo and W contents were in the range of 1.2 to 1.6% in terms of Mo equivalent.

By addition of W and further optimization of the compositions, new 12% Cr steels with excellent long-term rupture strength and reasonable toughness have been developed. The extrapolated $10^5$ hour rupture strengths of the newly developed steels were as high as 15-20 $kgf/mm^2$ at 600°C and 7-10 $kgf/mm^2$ at 650°C.

## References


1. K. Kougami, and H. Isaka, J. Iron Steel Inst. Japan, 76, 1990, 1043-52.
2. I. M. Park, R. Watanabe, and T. Fujita, J. Iron Steel Inst. Japan, 66, 1980, 82-91.
3. X. Y. Liu, T. Tsuchiyama, and T. Fujita, J. Iron Steel Inst. Japan, 69, 1983, S1257.
4. N. Takahashi, and T. Fujita, J. Iron Steel Inst. Japan, 60, 1974, 1506-18.
5. X. Y. Liu, and T. Fujita, ISIJ International, 29, 1989, 680-686

Proceedings of the Fifth International Conference on Creep of Materials, Lake Buena Vista, Florida, USA, 18-21 May 1992

# A CDM Motivated Safety Factor for Creeping Structures in Plane Stress Conditions

**M. Chrzanowski**
Technical University of Cracow
Krakow, Poland

**J. Madej**
Technical University of Lodz
Bielsko-Biala, Poland

ABSTRACT

Two different definitions of exertion factors have been used to analyze both time- and load-dependent material safety in creep conditions. The analysis performed made use of Continuous Damage Mechanics methods, which allowed to deal in a similar manner both phenomena: creep induced deterioration, and time-independent embrittlement leading to instantaneous failure. As a result, the curves in the space of exploitation time and applied constant stress have been found, to define areas in which one of the above phenomena is prevailing.

SAFETY EVALUATION WHICH IS A COMMON practice in design process has its special importance in creep conditions. High temperatures (in a homologous scale) which cause that time-dependent material properties come to play an essential role in structural behavior role are often associated with energetic applications which deserve high reliability (cf. [1]). On the other hand, the analysis of material and structural integrity becomes more complex, as irreversibility of phenomena observed has to be taken into account. This applies not only to material deformability but also to deterioration process which ultimately leads to a failure. Therefore classical stress-strain analysis has to be replaced by that which describes time dependence of deformations and defects formation and growth. This can be done by using one of so called coupled theory which take into account both effects (cf. e.g. [2]). In this situation the material exertion depends on a distance from actual state to ultimate one in stress-time space. The safety factor (which can be viewed as the exertion reciprocal) which in time independent analysis is solely related to the stress space, has to be replaced by a pair of two factors: one connected with applied stress, the other - with exploitation time. The analysis which takes into account both factors are given below. What makes this analysis even more complex is fact that safety evaluation only past loading stress-time history, but also on a forecasting of a future behavior. In any instant of time both stress and exploitation time can grow independently, and some assumptions have to be made according to the future loading path. In the present analysis it is assumed that these can grow independently, i.e. in any instant of stress-time history stress can grow at constant exploitation time, or the latter one can grow at constant stress. This an obvious simplification which makes the analysis more digestible. A generalization on the case when both stress and exploitation can grow may be easily made.

The ultimate goal of present analysis is to find in the stress-time space a path along which both stress- and time-dependent exertions (and hence safety factors as well) are equal. It is perhaps the most desirable structure behavior if future loading history is nonpredictable, and it allow to optimize a structure with respect to its safety.

## QUALITATIVE ANALYSIS

Let us define two exertion factors:

Time dependent exertion factor:

$$W_t = t_e / t^* \quad \text{at } \sigma_e = \text{const.} \qquad (1)$$

and stress dependent exertion factor:

$$W_\sigma = \sigma_e / \sigma^* \quad \text{at } t_e = \text{const.}, \qquad (2)$$

where:

- $t_e$ is exploitation time,
- $t^*$ is time to failure at constant $\sigma_e$
- $\sigma_e$ is applied stress
- $\sigma^*$ is stress to cause failure at constant $t_e$,

and stress $\sigma$ stands for any stress representation through stress tensors or its invariants (in the simplest case of uniaxial state of stress it is represented by a single normal stress $\sigma$).

The classical exertion factor W, employed in time independent analysis can be dealt with as a special case of time dependent exertion factor $W_\sigma$ when $t_e = 0$, and $\sigma^* = R$ is the ultimate strength:

$$W = \sigma_e / R \quad \text{at } t_e = 0\,. \qquad (3)$$

As $\sigma^* \le R$ for $t \ge 0$, there is always $W_\sigma \ge W$, what is a result of ultimate strength reduction due to progressive material deterioration.

All variables defined above are illustrated by Fig.1, which shows also the curves $t^* = t^*(\sigma_e)$ and $\sigma^* = \sigma^*(t_e, \sigma_e\text{=const.})$. In fact this figure represents only a projection of a surface given by $\sigma^* = \sigma^*(t_e, \sigma_e)$ on the $\sigma^* - t^*$ plane. This surface is schematically sketched in Fig.2 in $\sigma^* - \sigma_e - t_e$ space.

It is seen that the curves RB', RC',

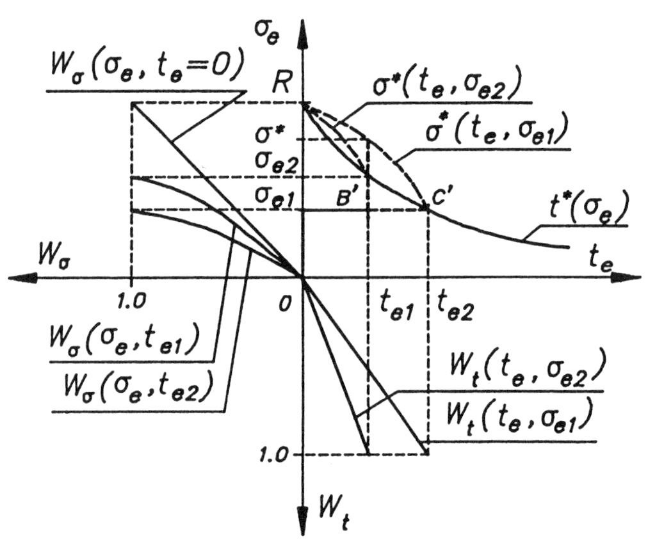


Fig.1. Variable definitions

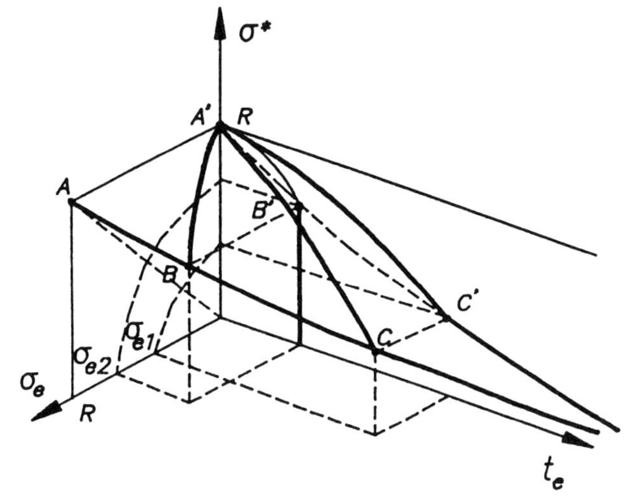


Fig.2. Failure surface in $\sigma^*$, $\sigma_e$, and $\tau_e$ space

and RB'C' shown in Fig.1 are projections of the curves RB, RC, and ABC sketched in Fig.2. In a similar way one can consider projection of these curves on $\sigma^* - \sigma_e$ plane to obtain the residual strength for constant $t_e$.

Now, with the help of Fig.1, one can easily give an interpretation of exertion definitions given by Eqs 1 and 2. As an example two values of $\sigma_e$ namely $\sigma_{e1}$ and $\sigma_{e2}$ with corresponding $t^*$ equal to $t_{e1}$ and $t_{e2}$ have been chosen to illustrate $W_\sigma$ and $W_t$ dependence on $\sigma_e$ and $t_e$. Again, both exertion factors can be analyzed as a function of two variables $\sigma_e$ and $t_e$, and curves of $W_\sigma(\sigma_e)$ and $W_t(t_e)$ shown in Fig.1 can be viewed as projections of corresponding spatial curves on $W_\sigma - \sigma_e$ and $W_t - t_e$ planes, respectively.

It is the main goal of present analysis to describe functions: $W_\sigma(\sigma_e, t_e)$ and $W_t(\sigma_e, t_e)$, and to find out the curve $f(\sigma_e, t_e)=0$ which satisfies the condition:

$$W_\sigma = W_t \ . \tag{4}$$

## QUANTITATIVE ANALYSIS

As a stress was chosen as independent exertion parameter, and material behavior is considered only, there is convenient to use stress governed damage growth law used in [3]:

$$\frac{d\omega}{dt} = \mathbf{A}_o \left(\frac{\sigma_{eq1}}{1 - \omega}\right)^{m_o} \frac{d\sigma_{eq1}}{dt} + \mathbf{A}\left(\frac{\sigma_{eq2}}{1 - \omega}\right)^{m} \tag{5}$$

which describes both: time independent deterioration (first term on the right hand side of the above equation) and time dependent one (second term). The following notations have been used here: $\omega$ is a dimensionless damage parameter ($0 \le \omega \le 1$), t is time, $\mathbf{A}_o$, $\mathbf{A}$ and $m_o$, m are material constants, and $\sigma_{eq1}$ and $\sigma_{eq2}$ are equivalent stresses which in general can be given by a formula proposed in [4]:

$$\sigma_{eq} = \alpha\, \sigma_1 + (1-\alpha)\sigma_{eff} \ , \tag{6}$$

where $\sigma_1$ is a positive principal stress, $\sigma_{eff}$ stands for von Mises effective stress, and $\alpha$ ($0 \le \alpha \le 1$) is a parameter responsible for interaction between the two.

In Eq.5 the notation $\sigma_{eq1}$ and $\sigma_{eq2}$ was used to emphasize that different mechanisms can be responsible for time dependent and independent deterioration processes, and represented by two different values of $\alpha$.

For the uniaxial state of stress there is $\sigma_{eff} = \sigma_1$ and $\sigma_{eq1} = \sigma_{eq2} = \sigma_1$, and Eq. 5 becomes:

$$\frac{d\omega}{dt} = \mathbf{A}_o\left[\frac{\sigma}{1-\omega}\right]^{m_o}\frac{d\sigma}{dt} + \mathbf{A}\left[\frac{\sigma}{1-\omega}\right]^{m}, \tag{7}$$

where $\sigma_1 = \sigma$ was put for a better readability.

With dimensionless variables defined as follows:

stress: $s = \sigma/R \quad (0 \le s \le 1)$,

time: $\tau = t\,\mathbf{A}\,R^m$, and $\mathbf{A}_o = 1/\,R^{m_o+1}$,

this equation can be rewritten as:

$$\frac{d\omega}{dt} = \left[\frac{s}{1-\omega}\right]^{m_o}\frac{ds}{dt} + \frac{1}{m+1}\left[\frac{s}{1-\omega}\right]^{m}. \tag{8}$$

With $\tau_e$, $\tau^*$ and $s_e$, $s^*$ denoting dimensionless quantities which appeared in Eqs 1-3, the exertion factors are:

$$W_t = \tau_e / \tau^* \quad \text{at } s_e = \text{const.} \tag{9}$$

$$W_\sigma = s_e / s^* \quad \text{at } \tau_e = \text{const.}, \tag{10}$$

and

$$W = s_e \quad \text{at } \tau_e = 0\,. \tag{11}$$

To find the functions $\tau^*(s_e)$ and $s^*(s_e,\tau_e)$, and consecutively functions $W_\sigma(s_e,\tau_e)$ and $W_t(s_e,\tau_e)$, the Eq. 8 will be integrated for the following loading history:

1: $0 \le s \le s_e$ for $\tau = 0$,

2: $0 \le \tau \le \tau_e$ for $s = s_e$, with $\tau_e = \tau^*$ as a special case, and

3: $s_e \le s \le s^*$ for $\tau = \tau_e$.

A straightforward calculation yields:

For stage 1:

$$\omega\,(s_e, \tau{=}0) \overset{df}{=\!=} \omega_1 = 1 - \left[1 - s_e^{\,m_o+1}\right]^{\frac{1}{m_o+1}} \tag{12}$$

with an obvious limiting case:

$$\omega_1 = 1 \text{ for } s_e = 1 \text{ (i.e. } \sigma_e{=}R).$$

For stage 2 the integration of Eq. 8 gives:

$$\omega(s_e, \tau{=}\tau_e) \overset{df}{=\!=} \omega_2 = 1 - \left[\left(1 - s_e^{\,m_o+1}\right)^{\frac{m+1}{m_o+1}} - \tau_e s_e^m\right]^{\frac{1}{m+1}}. \tag{13}$$

For a special case $\tau_e = \tau^*$, $\omega_2 = 1$, and we have:

$$\tau^* = \frac{1}{s_e^m}\left[1 - s_e^{\,m_o+1}\right]^{\frac{m+1}{m_o+1}}. \tag{14}$$

Finally, for stage 3 integration yields the formula for stress to failure after repeated loading (ultimate residual strength):

$$s^* = \left[s_e^{\,m_o+1} + \left(1 - s_e^{\,m_o+1}\right)^{\frac{m+1}{m_o+1}} - \tau_e s_e^m\right]^{\frac{1}{m+1}}. \tag{15}$$

Now, Eqs 14 and 15 can be used to define functions $W_\sigma(s_e,\tau_e)$ and $W_t(s_e,\tau_e)$ according to Eqs 9 and 10:

$$W_t = s_e^m \tau_e \left[ 1 - s_e^{m_o+1} \right]^{-\frac{m+1}{m_o+1}} , \quad (16)$$

$$W_\sigma = s_e \left[ s_e^{m_o+1} + \left(1 - s_e^{m_o+1}\right)^{\frac{m+1}{m_o+1}} - \tau_e s_e^m \right]^{-\frac{1}{m+1}} . \quad (17)$$

## EXAMPLE

Let us consider the special case of $m=m_o$ (for most metals is $m_o>m$).
Eqs 14 and 15 reduce to:

$$\tau^* = \frac{1}{s_e^m} \left[ 1 - s_e^{m+1} \right] , \quad (18)$$

and

$$s^* = \left[ 1 - \tau_e s_e^m \right]^{\frac{1}{m+1}} . \quad (19)$$

In Fig.3 the curves $\tau^*(s_e)$ and $s^*(\tau_e; s_e=\text{const.})$ are shown. The projections of $s^*$ surface for constant values of $\tau_e$ on $s^* - s_e$ plane are shown in Fig.4. The exertion factors are:

$$W_t = s_e^m \tau_e \left[ 1 - s_e^{m+1} \right]^{-1} , \quad (20)$$

$$W_\sigma = s_e \left[ 1 - \tau_e s_e^m \right]^{-\frac{1}{m+1}} . \quad (21)$$

The condition (Eq.4) yields an implicit expression for $s_e - \tau_e$ dependence such that $W_t = W_\sigma$:

$$s_e^{m-1} \tau_e \left[ 1 - \tau_e s_e^m \right]^{\frac{1}{m+1}} = 1 - s_e^{m+1} . \quad (22)$$

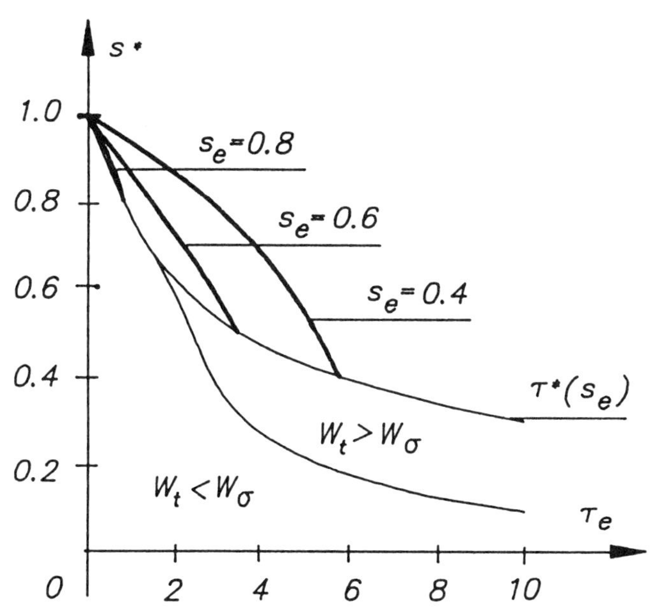


Fig.3. Failure curves for constant $s_e$

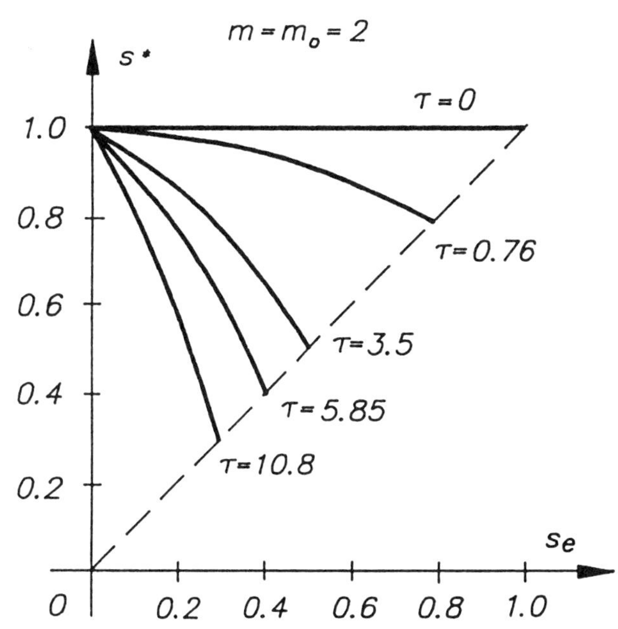


Fig.4. Failure curves for constant $\tau_e$

The corresponding curve is shown in Fig.3, separating fields for which there is $W_\sigma < W_t$ and $W_\sigma > W_t$.

## SUMMARY

The attention to safety evaluation in creep conditions has been drawn by defining two types of exertion factors: stress and time dependent. In any instant of loading history both have to be calculated, as residual strength reduction can endanger stucture's integrity more than deterioration related to creep process. In the analysis performed a simplified time-history loading was considered. But even in this case the importance of full analysis of damage development was demonstrated. It indicates that similar analysis may be of high importance when more complex loading history is prescribed.

## ACKNOWLEDGEMENT

The work reported in the paper has been performed partially under project no. 3/0917/91/01 granted by KBN (Polish Committee for Scientific Research).

## REFERENCES

1 -Proceedings of SMiRT-11 Conference, vol.L, (1991)

2 -Chrzanowski,M., Bull.Pol.Ac.Sc., Ser. sc.techn., vol.20,3,(1972)

3 -Chrzanowski,M., Int. J. Mech.Sci., 18, 69-73,(1976)

4 -Hayhurst,D.R., Dimmer,P.R., Morrison, C.J., Phil.Trans.R.Soc.London, A311, 103-129, (1984)

Proceedings of the Fifth International Conference on Creep of Materials, Lake Buena Vista, Florida, USA, 18-21 May 1992

# Dutch Developments in Design and Behavior of Creep Loaded Structures

**C.F. Etienne, J.H. Heerings, and J.C. van Wortel**
TNO Industrial Research
Apeldoorn, The Netherlands

**L.B. Dufour and G.M. van Dijk**
KEMA
Arnhem, The Netherlands

**J.H. Fokkens**
Netherlands Energy
Research Foundation
Petten, The Netherlands

## ABSTRACT

In the N.I.L.-project "Creep of Installations" several aspects with regard to creep loaded structures were investigated: optimization of design, repair welding and inspection techniques. Besides a damage inventory has been performed.
The optimization of design was focused on nozzle-sphere and nozzle-cylinder connections. It was shown that the lifetime is dependent on the reinforcement strategy and chosen design concept. A reinforcement of both the sphere/cylinder and the nozzle is most effective.
The investigations have confirmed that lifetime of creep damaged components can be increased considerably by local weld repair. A weld repair procedure is proposed. A heat treatment after welding seems to be superfluously.
For assessment of creep deformation a monitoring-system has been developed. By numerical speckle pattern analysis the deformation of components may be monitored on spots provided with a local clad layer of platinum. This method has good prospects for practical application. Further development on practice oriented components still has to be carried out.
From the damage-inventory it is concluded that only a limited number of premature damages in components for electrical power generation have occurred. In case of premature damage external forces and moments were a major cause of failure.

IN THE NETHERLANDS a three years' research project, aimed at the optimization of design, repair of creep loaded structures and inspection, was recently completed [1]. The investigations were directed to ferritic materials for operation up till 600°C. The project has been organized by the N.I.L. (Netherlands Institute of Welding) and was carried out by various research institutes in cooperation with industry and government.

In the project the following subjects were dealt with:
- optimization of design
- repair welds in thick-walled headers
- creep damage assessment by non destructive testing
- damage inventory

## OPTIMIZATION OF DESIGN

The present design rules in the Netherlands [2] are only based on the elastic behaviour of a structure. At the design stage the creep behaviour is taken into account indirectly, viz. at the determination of allowable design stress. Once this has been done the normal elastic relations are applied to structures under creep conditions. Particularly for creep only some general remarks are given about geometrical discontinuities, which should be manufactured as smooth as possible. Creep damage in high temperature installations is usually concentrated at a limited number of structural configurations, such as nozzle-sphere and nozzle-cylinder connections [3]. Besides it is often necessary to replace a complete component, even though the limits for creep damage are only exceeded very locally. Therefore it appears that the present design rules are not optimal for geometries with stress concentrations.

In this project the research was aimed at the design optimization of creep loaded nozzle-sphere and nozzle-cylinder connections. For these components the following questions had to be answered:
- what is an optimal geometry for creep?
- what is a simple design method to define the optimal geometry?

- which recommendations for adaption of the Dutch design rules can be given?

Initially a literature survey has been performed, in which special attention was paid to the design rules, the limit load and the creep behaviour of nozzle-sphere and nozzle-cylinder connections.

Many references were found concerning the elastic and limit load behaviour of thin-walled nozzle-sphere and nozzle-cylinder connections with a small diameter ratio [4,5,6,7]. Very few references [8,9,10,11,12] were found about the creep behaviour of thick-walled nozzle-sphere and nozzle-cylinder connections with a large diameter ratio. Also no special design rules incorporating creep behaviour could be found. The applied design rules for creep loaded structures are based on elastic behaviour with a correction for initial plasticity.

From this literature survey it has been postulated that a structure with an optimal limit load should also be optimal with regard to creep. If this assumption can be proven to be valid it will be possible to use limit load data available in the literature for the design of creep loaded structures. Therefore numerical analyses have been performed on the nozzle-sphere (thin- and thick-walled) and the nozzle-cylinder (thin-walled) connection. Most attention was paid to the thick-walled nozzle-sphere connection.

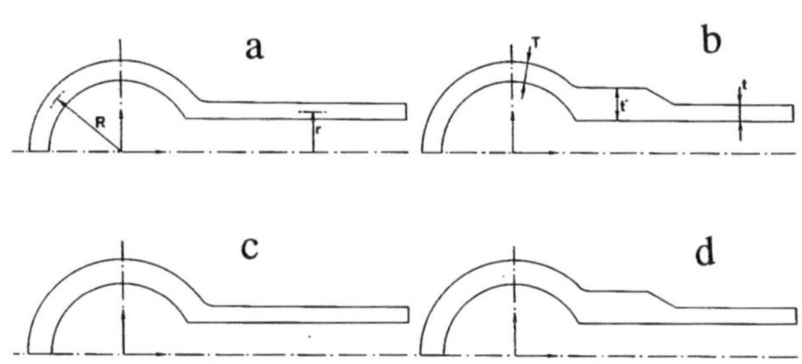


a: unreinforced
b: nozzle reinforcment
c: sphere reinforcment
d: total reinforcement

Fig.1 Reinforcement strategies for thick-walled nozzle-sphere connections.

For the various geometries the influence of different reinforcement strategies, as illustrated in figure 1, has been investigated. For each strategy a reinforcement factor can be defined as the ratio between the reinforced and unreinforced wall thickness. The influence of reinforcement has been evaluated with three different concepts:

- calculation of stresses and strains by creep analysis with the finite element method
- calculation of the limit load with the finite element method
- the Dutch design rules.

For each concept an optimum of the reinforcement factor can be determined. For the creep analysis the optimum is reached when the maximum equivalent creep strain in the connection is equal to the equivalent creep strain in the undisturbed, unreinforced parts. The influence of reinforcement strategy and reinforcement factor according to creep analysis is illustrated in figure 2.

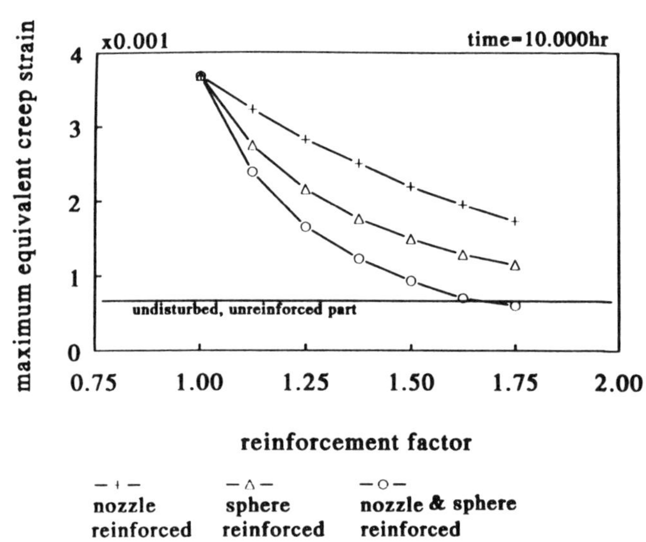


Fig.2 Creep analysis of the thick-walled nozzle-sphere connections.

Besides the influence on lifetime has been calculated for the thick walled nozzle-sphere connection, based on the strain criteria, particularly the 1% strain criterium, from ASME Code Case N47. Results of these calculations are presented in figure 3. Based on the optimal reinforcement factors according to the three different concepts, the corresponding lifetime can be deduced. The lifetime according to the creep analysis appears to be 250,000 hr, while the lifetime is shorter for the other two concepts, viz. the limit load approach and the Dutch design code. Nevertheless the last two concepts approach the creep analysis reasonably.

Although the extent of the numerical analyses and configurations is too limited for general conclusions, the following tendencies have been shown:

- a reinforcement of the nozzle and the cylinder or

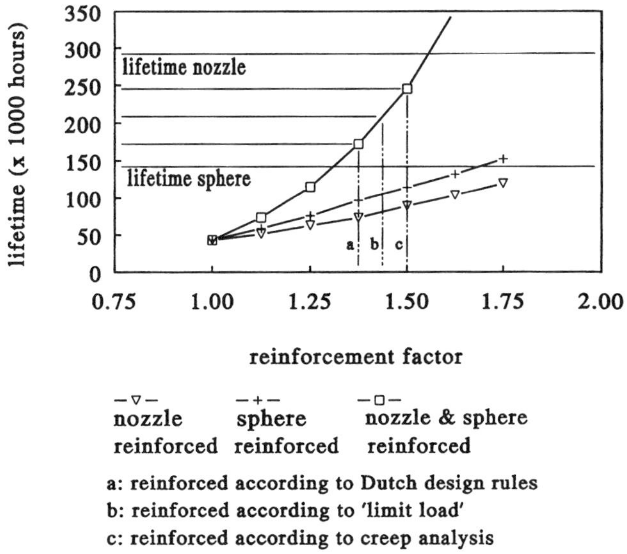


Fig.3 Lifetime of thick-walled nozzle-sphere connections as a function of reinforcement strategy.

sphere is much more effective than a reinforcement of the nozzle only;

- a local reinforcement of a sphere only leads to a 'creep strain' concentration at the wall thickness transition. A local reinforcement in a cylinder does not lead to a 'creep strain' concentration at the wall thickness transition;
- connections reinforced according to the Dutch design rules have a lower lifetime than the undisturbed parts of the connection;
- connections reinforced according to the limit load have a lower lifetime than the undisturbed parts, but higher than with the Dutch design rules;
- an evaluation of a nozzle-sphere connection according to the Dutch design rules should rather be based on the weakest of the component parts, instead of on the sphere only.

The numerical analysis for the nozzle-cylinder connections demonstrated the same tendency. This will be demonstrated in more detail in the next three years project: Life Time of High Temperature Installations, which has started at the end of 1990. In this project attention will also be paid to the behaviour of creep loaded welds.

## REPAIR WELDS IN THICK-WALLED HEADERS

Weld form a potential risk for premature failure of creep loaded structures [3,13]. By repairing these structures they can be used for longer times. The optimal situation should be achieved if the lifetime of the undisturbed material could be reached.

With respect to weldability and residual service life a programme has been performed on the consequences of repairing cracked locations in ½Cr-½Mo-¼V, 1Cr-½Mo and 2¼Cr-1Mo headers with wall thicknesses of 25, 35 and 70 mm respectively.

The headers had been used in fossil fired power plants for 40,000-90,000 hr and were replaced because cracking had been observed in the saddlepoint of the nozzle-cylinder connection.

In the first instance full restraint SMAW and GTAW circumferential testwelds have been welded in the cylindrical uncracked parts of the headers. For both welding methods 4 procedures have been used, viz. with and without preheating and post weld heat treatment (PWHT).

The aim of this part of the investigation was to qualify the optimal welding method/procedure with respect to both weldability and residual service life. Therefore, isostress creep tests have been performed on crossweld specimens at 675, 650 and 625°C with a stress level of 60 MPa (½Cr-½Mo-¼V) and 45 Mpa (1Cr-½Mo and 2¼Cr-1Mo); rupture times varied from 100-3000 hours.

The testwelds always fractured in the fine grained HAZ, while at other locations no cavities or cracks could be detected. The creep rupture strength of the GTAW weld was on average somewhat better than for the SMAW welds. However, the influence of preheating and PWHT on the creep rupture strength of the weldments was not unambiguous. The qualification with respect to weldability was done by magnetic particle and ultrasonic inspection after welding. In no case cracks were detected.

The extrapolated time to rupture (residual service life) of the testwelds was always >80,000hr, even in the conditions where no preheating and/or no PWHT were carried out, see figure 4.

In addition to the investigation on test welds real repair of cracked headers have been performed. The cracked headers have been repaired with the SMAW process, because of the flexibility of this process. For comparison a GTAW repair weld has been performed. Although the testwelds never showed cracks, it was decided to repair the cracked headers in the preheated condition to exclude the risk of cold cracking. A PWHT was omitted, but for comparison one 2¼Cr-1Mo header was post weld heat treated at 680°C.
Figure 5 shows a cross sections of a repaired location in 1Cr-½Mo.

The creep tests on the repaired parts have been performed at 610-620°C with the same stress levels as the testwelds. In the cross weld specimens the following

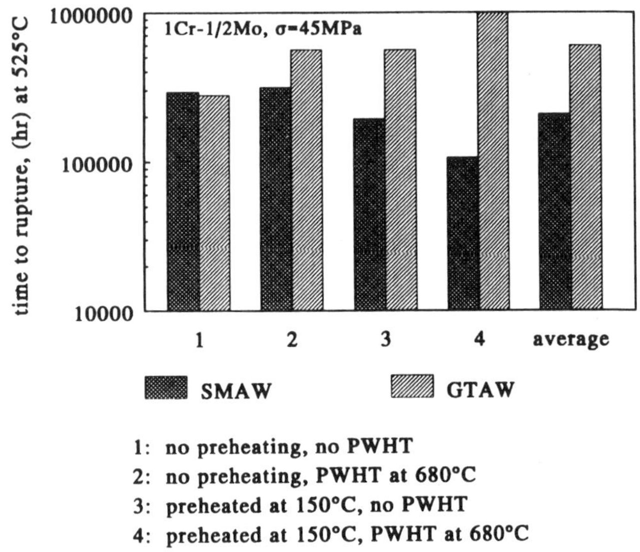


Fig.4 Extrapolated time to rupture of 1Cr-½Mo test-welds at 525°C.

zones were always included: used base metal, new weld metal and used weld metal. The fracture location of the creep specimens was never in the base material. Usually it was situated in one of the fine grained heat affected zones.

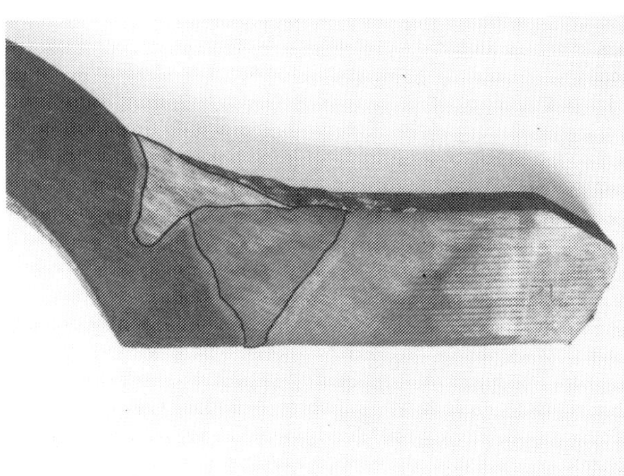

Figure 5 Cross section of a SMAW repair weld in 1Cr-½Mo.

In the GTAW repair weld the fracture location was in the weld metal of the repair weld.

The time to rupture of the repaired headers was equal or lower than for the testwelds. The drop was most pronounced for the GTAW repair weld.

PWHT had a slightly negative influence on the time to rupture. Nevertheless the extrapolated residual lifetime of the headers was always more than 4 years.

It was concluded that the residual lifetime of locally repaired headers can be increased with at least several years and that no extreme care has to be taken into account with respect to the weldability. It proved to be essential to grind out the crack sufficiently wide. There is no unambiguous difference between SMAW and GTAW. A post weld heat treatment does not seem necessary.

Some unclearness, however, still exists with respect to the unexpected behaviour of the GTAW repair weld and the ductility of repaired headers when no PWHT has been carried out. After clarifying these aspects in the aforementioned new project, a definitive statement about the optimal repairing procedure can be made.

## CREEP DAMAGE ASSESSMENT BY NON DESTRUCTIVE TESTING

Current practice with regard to non destructive creep damage assessment is based on surface replication, dimensional control, magnetic particle inspection, liquid penetrant inspection and visual inspection. Capacitive strain gauges offer a good possibility to monitor the overall creep deformation, but practical application is still limited [14]. There remains a need for methods to detect creep damage at an earlier stage, however, in order to allow a more reliable judgement on the necessity to repair or to replace. In this program the potential for enhancement of damage assessment methods has been investigated and developed.

Firstly, a literature survey was accomplished to establish potentials for an improvement of non destructive testing techniques compared to the current practice. Secondly an experimental programme was carried out to investigate or develop some candidate methods with a potential for improved detection: ultrasonics and optical deformation measurement.

**Ultrasonics.** Ultrasonic detection of subsurface creep damage was investigated by using two different techniques:
-ultrasonic backreflection
-ultrasonic transmission of Raleigh waves.
Measurements using both advanced ultrasonic techniques were made on specimens with either service induced creep damage or specimens with creep damage induced during long-term temperature/stress exposure in a laboratory set-up. Effects observed appeared to be marginal and ambiguous. For practical application(s) both ultrasonic techniques don't appear to exhibit a significant potential for creep detection and damage assessment at present.

**Optical deformation measurement.** This technique makes use of a 'speckle image', which is an optical fingerprint of the surface of a restricted area (1 $cm^2$). Surface deformation patterns can be derived from comparison of the images before and after deformation of this area. Figure 6 illustrates the experimental set-up of this technique.

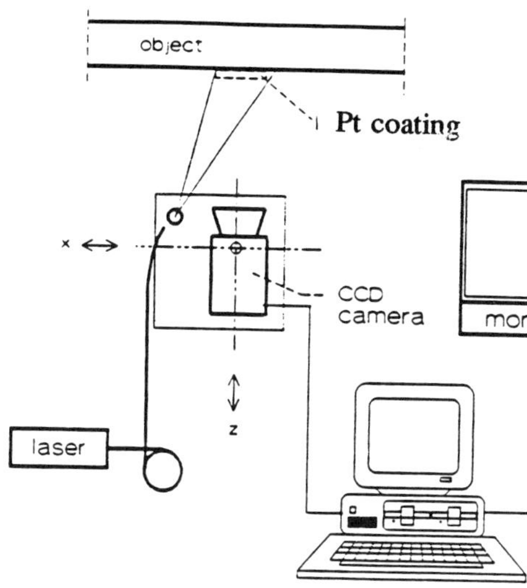


Figure 6 Experimental set-up for optical deformation measurement.

In the project the technique for optical deformation monitoring was investigated and developed. The first development was focused on the selection of a coating-layer and a coating technique to provide an optically stable spot on the outer surface of a component to be monitored. The coating must follow the deformation of the substrate of the component while its surface roughness should remain unaffected by corrosion or oxidation. The technique of explosive welding a platinum layer seems to satisfy the requirements and is used so far.

The second development was directed to the numerical automation of the optical measuring technique to quantify the speckle image on the coated area. The needed software has been realized. A measurement provides a quantitative strain distribution in 2 orthogonal directions on the selected area. In order to test the system, measurements were made on specimens with artificially induced deformations both uniformly and locally. The accuracy of the strain measurement was found to be 5‰ for local strain (2mm spacing) and 1‰ for uniform area strain (10mm spacing).

Exposure tests have been performed at 550°C to verify if the coating fulfils its requirements. The correlation coefficient of the measurements has been determined periodically. Its value remains above the required threshold value, see fig. 7.

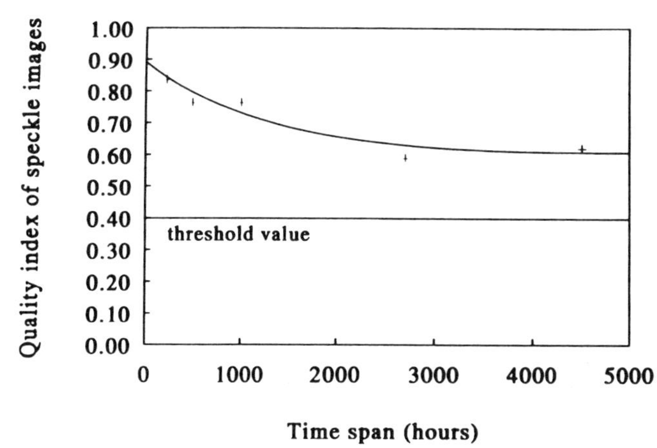


Figure 7 Correlation coefficient as function of exposure time at 550°C.

For monitoring in practice, the measuring device should be installed in a plant periodically during shutdown. Therefore there is a need to replace the device on the same spot. In order to be able to replace properly an optical system has been developed and tolerances for replacement have been defined.

In this project the optical deformation technique has proven to have significant potential for creep damage assessment: local strains can be measured with high resolution. However, further development efforts are needed. In the aforementioned new project a portable device will be designed and manufactured and further testing will be executed on practice oriented situations and components.

A patent application with respect the optical deformation technique has been filed (Nº 9200087 of Dutch Patent Office) by KEMA.

## DAMAGE INVENTORY

The tentative rule T0102V of the Dutch authority Stoomwezen [2] states that additional examinations of components should be performed as soon as the calculated usage factor due to steady-state creep loading exceeds 0.6. This rule is applicable as long as the number of start-stop cycles is less than 1,000 during the total operating time. The additional examination consists of magnetic particles inspection and surface replication on the outside of components. In the project a damage inventory has been carried out to check whether amendments of rule T0102V are necessary.

Users of high-temperature installations have been

approached to provide information about damage cases. The available damage data have been systematically ordered on data forms. Damages of components with diameters less than 100 mm have been excluded from the inventory. Ninety percent of the total amount of 106 damages originated from public electricity plants and almost all damages concerned ferritic materials. For evaluation various data sets were prepared ,like:

- damages in relation to component types
- damages in relation to material types
- damages in relation to component and material types
- damages in relation to cause and component type
- damages in relation to operating time
- damages in relation to usage factor and material type

The inventory leads to the following percentages of damaged components in electricity generating plants, see table 1.

Table 1 Classification of 106 damages over types of component.

| Type of component | Number of damaged components | Percentage of damaged components |
|---|---|---|
| straight pipe | 11 | 0.03 *) |
| header | 45 | 5 |
| T-piece | 22 | 3 |
| spherical head | 9 | 2.5 |
| pipe bend | 7 | 0.2 |
| other | 12 | 1.5 |

*) considering damage of more than 1 meter pipe

It is clear that headers and T-pieces are damaged most. Pipe bend damages appear to occur only in a relatively few cases. Approximately 90% of all damages occurred in welds. No specific material is damaged preferentially.

Fifteen percent of the damages occurred at usage factors less than 0.6. In 62.5% of these cases external pipe forces played a role. Therefore it is concluded that there are no reasons for adjusting the factor 0.6. Nevertheless it is advisable to incorporate primary stresses due to external forces/moments into the calculation of usage factors. An alternative would be to inspect suspected areas, especially if the number of start-stop cycles is relatively high.

The inventory did not reveal damages which were initiated at the inside of components. This implies that surface replication is effective to detect creep damage. Corrosion did not play a role in the damages.

For the time being, there are no reasons for adjusting the number of start-stop cycles in rule T0102V, although there is a very slight indication that the number of cold starts do play a role in damaging headers.

## CONCLUSIONS

The Dutch project of the Netherlands Institute of Welding has yielded some contributions to the major aspects concerning optimization of creep loaded structures.

The conclusions for the relevant aspects can be summarized as follows:

With respect to optimal design of critical component details it could be concluded that creep calculations are essential for real optimization. With simplified calculations (limit load approach or design according to the Dutch code) a reasonable design can be achieved, but the lifetime will be shorter (up to a factor of 2). This has been demonstrated for nozzle-sphere connections, but on the basis of preliminary results for nozzle-cylinder connections (T-pieces) it might be anticipated upon a more general validity.

With respect to weld repair it could be concluded that repair of locally damaged areas in T-pieces offers the possibility to enlarge the lifetime considerably (several years). The following procedure is recommended for an optimal repair of thick-walled connections:

- completely grinding out of existing cracks (not only for sufficient depth, but for certain also for sufficient width),
- for symmetrical parts it is beneficial to grind also the area located diametrically opposite, e.g. both saddle-points of T-piece,
- final check on non-attendance of rests of cracks by means of replicas,
- preheating of material (≈200°C)
- repairing by GTAW (TIG) or by SMAW (electrode),
- a post weld heat treatment seems superfluously (at least from point of view of rest life).

With respect to an early detection of creep damage a special NDT-technique could be developed which permits to measure with interferometry the deformation of an explosive cladded platinum layer. The prospect of this method is proven to be very prospective. The method allows not only the measurement of overall deformation but also the determination of local deformation in e.g. the HAZ of a weldment or a curved area. Experience under practical conditions is necessary before a definite conclusion on the practicability of the method can be drawn and the applied patent can be commercialized.

From an inventory of creep damages in components for electricity generating plants it is found that 90% of 106 cases of damage occurred in weldments. A relatively small amount of premature failures has been found (15%), which were mostly related to the presence of external forces or moments. Therefore it is recommended to avoid undesirable system stresses as far as possible or to take these into account in the design stage.

## ACKNOWLEDGMENTS

The authors would like to acknowledge the support by the Netherlands Institute of Welding and the financial support by the sponsors of the project, particularly the Dutch Ministry of Economic Affairs. The co-authors express their thanks to mr. C.F. Etienne for the overall project management and the many useful discussions.

## REFERENCES

[1] Final report "Creep of Installations", Netherlands Institute of Welding, The Hague, march 1991 [in Dutch]

[2] Rules for pressure vessels, Dutch licensing authority Stoomwezen, Staatsuitgeverij, 1986.

[3] C.F. Etienne, J. Prij, Int. Conference "Life assessment and Extension", The Hague, june 1988, p.III 38-47

[4] Robonsin, M., Kirk, A. and Gill, S.S., An Experimental Investigation into the Plastic Behaviour of Oblique Flush Nozzles in Spherical Pressure Vessels.
Int. J. Mech. Sci., 13, 41-61, (1971)

[5] Robonsin, M. and Gill, S.S., A Lower Bound to the limit Pressure of a Flush Oblique Cylindrical Branch in a spherical Vessel.
Int. J. Mech. Sci., 14, 579-601, (1972).

[6] Robonsin, M. and Gill, S.S., Limit Analysis of Flush Oblique Cylindrical Nozzles in Spherical Pressure Vessels. Part I: A Parametric Survey of Results.
Int. J. Pres. Ves. and Piping, 1, 199-231, (1973).

[7] Robonsin, M. and Gill, S.S., Limit Analysis of Flush Radial and Oblique Cylindrical Nozzles in Spherical Pressure Vessels. Part 2: Application of Results in a Design Procedure.
Int. J. Pres. Ves. and Piping, 1, 233-244, (1973).

[8] Creep analysis of axisymmetric bodies using finite elements. G.A. Greenbaum; M.F. Rubinstein; Nuclear Engineering and Design; Vol. 7; page 379-397; April 1968.

[9] Creep behaviour of a sphere-nozzle connection of 1Cr-½Mo; M.J.J.M. Leemans; Netherlands Institute of Welding; NIL-RLD 83-69; june 1983 [in Dutch].

[10] Assessment of creep deformation in a T-piece of 1Cr-½Mo; M.J.J.M. Leemans; Netherlands Institute of Welding; NIL-RLD 85-06; january 1984 [in Dutch].

[11] Creep in T-pieces and sphere-nozzle connections; J. van den Eikhoff; Netherlands Institute of Welding; NIL-RLD 85-09; october 1984 [in Dutch].

[12] Final report, part C; Summary of creep and lifetime calculations for components in the framework of the N.I.L. Project Residual Lifetime; J. van den Eikhoff; J. Prij; Netherlands Institute of Welding; NIL-RLD 86-01; december 1985 [in Dutch].

[13] C.H.A. Townly, Int. Conference "High Temperature Structural Design", Venice, october 1990.

[14] M. Henrichs, P. Hoffstötter, H.H. Reichel, H. Williams, E. Winschuh, VGB-Conference "Werkstoffe und Schweißtechnik in Kraftwerk", Essen, january 1991.

*Proceedings of the Fifth International Conference on Creep of Materials, Lake Buena Vista, Florida, USA, 18-21 May 1992*

# Lifetime Evaluation of Creeping Structures

**A. Bodnar, M. Chrzanowski, and P. Latus**
Technical University of Cracow
Krakow, Poland

ABSTRACT

The paper gives an evaluation of two groups of parameters upon time-to-failure of structures in creep conditions. The first group is related to material characteristics (constants and functions) and therefore the analysis of these is referred to as material properties sensitivity. The other describes structure shape and loading and corresponding analysis is called boundary conditions sensitivity.

Among numerous material parameters those have been chosen which are related directly to fracture process: mode of failure and influence of compressive stresses. The analysis has been performed for thin plates, rather complicated, but important case for engineering practice.

In contrast to that, boundary conditions sensitivity was studied for a rather simple structure which layout allowed to systematize the analysis. Material properties were fixed, except one free dimensionless parameter.

In both cases analysis was aimed in determining the time to First Crack Appearance (FCA) and the place where this crack (FC) appeared.

LIFE-TIME EVALUATION OF CREEPING STRUCTURES can be based on two criteria: allowance of an extreme deformability of structure's member, or structure's integrity which may be violated due to progressive process of material deterioration. The latter is often considered a crucial one, as it may occur at small strains and therefore at small deformation, with no warning on coming critical state. The branch of fracture mechanics, often referred to as Continuous Damage Mechanics [1], is dealing with this phenomenon. It was originated already in late 50-ties [2,3], and after some twenty years of proliferation spread over the whole community of engineers working in the field [4-7]. The paper deals with this type of limiting stage of a structure.

The nature of deterioration process is very complex, and its description for metals can be found elsewhere [8]. It is reflected by constitutive equations, which describe both material deformability and evolution of material variable responsible for deterioration. The latter should reflect its anisotropic nature through appropriate representation [9] and limit state yielding criterion [10]. In earlier works [11,12] authors examined the influence of material deterioration sensitivity to the invariants of the stress state and influence of compressive stresses upon time to failure. Parallel, the works were carried out to examine the influence of boundary conditions on redistribution of fractured zones [13,14]. In fact, the fracture process which initiates in a given material point where failure conditions have been satisfied, develops throughout a structure. In the case of a beam structures in creep conditions three stages can be identified: time to First Crack Appearance (FCA), time to the Brittle Hinge Formation (BHF) and time to the Structure Total Collapse (STC) (cf.[13]). Interrelations between these stages depend on loading conditions and structure layout, i.e. on boundary conditions.

In the present paper both effects, i.e. material deterioration and structural failure have been studied separately. First, material behaviour has been examined, with a plate bending as an example of case study, followed by analysis of structural behaviour of a simple bar structure.

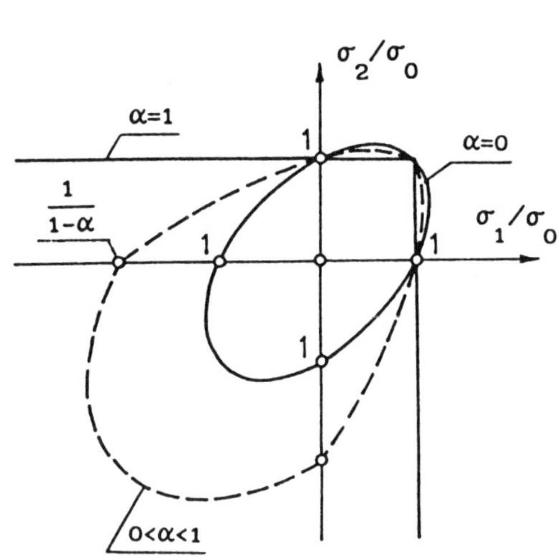


Fig.1 - Isochronous failure curves for different modes of fracture (*unilateral* material)

## MATERIAL PROPERTIES SENSITIVITY

**Constitutive equations.** The characteristic feature of structural components which operate in creep conditions is a gradual deterioration which in metals is caused by nucleation and growth of micro-defects such as fissures and voids at grain boundaries. This phenomenon of material weakening manifests by stress redistribution, strain rate increase, and formation of macrocracks, which terminates the process. A proper description of this process through the choice of adequate constitutive equations should be simple enough to allow numerical analysis, and complex to cover all above mentioned phenomena. Following [12] we will make use of the following set of constitutive equations:

$$\varepsilon_{ij} = \varepsilon^{e}_{ij} + \varepsilon^{c}_{ij} \tag{1}$$

$$\varepsilon^{e}_{ij} = C_{ijkl}\sigma_{kl} \tag{2}$$

$$\frac{\partial \varepsilon^{c}_{ij}}{\partial t} = \gamma\left(\frac{\sigma_e}{1-\omega}\right)^{n}\frac{\partial \sigma_e}{\partial \sigma_{ij}} \tag{3}$$

$$\frac{\partial \omega}{\partial t} = A\left[\alpha max\left\{\frac{\langle\sigma_i\rangle}{1-\omega}, \frac{\langle-\sigma_j\rangle}{1-h\omega}\right\}+(1-\alpha)\frac{\sigma_e}{1-\omega}\right]^{m} \tag{4}$$

where $\varepsilon_{ij}$ and $\sigma_{ij}$ stands for strain and stress components, respectively, superscripts $^e$ and $^c$ correspond to elastic and creep strain components, $\sigma_e=(3s_{ij}s_{ij}/2)^{1/2}$ is effective stress ($s_{ij}$ is deviatoric stress tensor). Deterioration process is described by a single state parameter $\omega$, and t denotes time. Material properties are represented by the following matrices and constants: $C_{ijkl}$ - elastic constants matrix, $\gamma$ and $n$ - steady state creep constants, $A$ and $m$ - damage evolution law constants, $\alpha$ and $h$ - material constants reflecting failure mode.

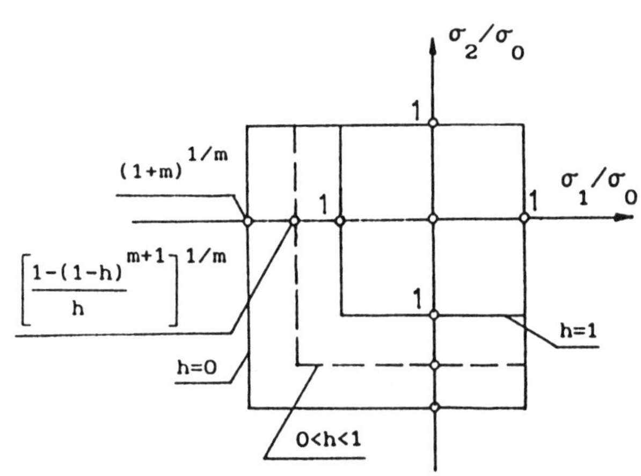


Fig.2-Isochronous failure curves for *non-unilateral* materials and α=1 (*brittle* material)

These equations deserve some explanations: Eq.1 shows that non-stationary creep can be described, which allows stress redistribution to be taken into account. Elastic strain is given by Eq.2, whereas Eq.3 is a general form of nonlinear, steady-state associated creep flow of Norton-Bailey type, coupled with deterioration process through term $(1-\omega)$. In Eq. 4 the following notation is used: $i\neq j=1,2$ designate principal stresses, and $\langle\sigma\rangle=\sigma$ for $\sigma>0$ and $\langle\sigma\rangle=0$ for $\sigma\leq 0$. Parameter $\alpha$ determines the mode of failure, with $\alpha=1$ corresponding to *brittle* fracture, and $\alpha=0$ - to the *ductile* type, as shown in Fig.1. In this figure (as well as in all others) $\sigma_0$ is the reference stress. Intermediate values of $\alpha$ describe a mixed mode of failure. Parameter $h$ is responsible for negative principal stress influence upon deterioration process [15]: its presence in Eq.4 defines a material as *non-unilateral* with an extreme value of $h=1$ defining a *bilateral* material, i.e. material equally sensitive to positive and negative stresses. A material described by a simplified version of Eq.4:

$$\frac{\partial\omega}{\partial t} = A\left[\alpha\,\frac{\sigma_1}{1-\omega} + (1-\alpha)\frac{\sigma_e}{1-\omega}\right]^m \qquad (5)$$

in which $\sigma_1$ is positive principal stress and the effect of negative principal stress is included through $\sigma_e$ only, will be referred to as *unilateral* one.

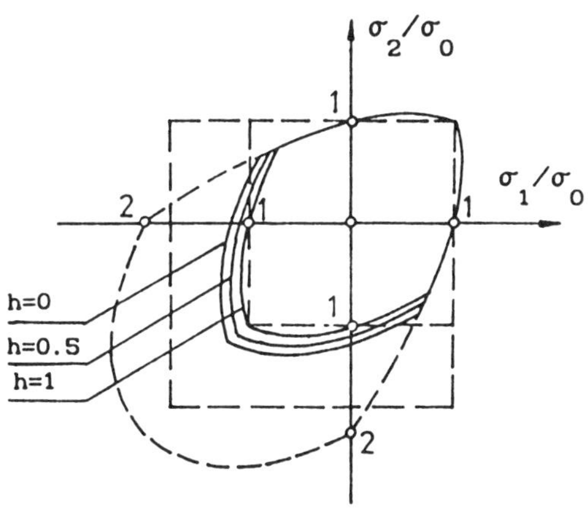


Fig.3 - Isochronous failure curves for combined theory (Eq.4)

Material behaviour which takes into account both effects is reflected by Fig.3, which corresponds to Eq.4 for $\alpha=0.5$ and $m=2$. Next section gives results of plate analysis according to Eqs (1-4), with the determination of time and place of FCA as a main goal.

**Example: plate in bending**. For two types of plate edges support: clamped and simply supported one, the numerical calculations using FEM have been performed, making use of Eqs (1-4). The plates were square in shape (1.0m x 1.0m) and of the thickness of 0.01m. The load of 0.2 MPa was uniformly distributed over the upper plate surface, and material constants were assumed as for a Ti-6Al-2Cr-2Mo titanium alloy at 675 K.

The computer code used in the calculations employed rectangular finite elements with twelve degrees of freedom; the Hermitian interpolation formulae for shape functions were used, and each quadrant of plate was divided into 256 elements. The details of numerical procedure can be found in [16]. As the main goals of the analysis were: the time to FCA, denoted as $t_I$, and a FC location (defined as $\omega=1$), these are given by Table I, and Figs 4a and 4b for clamped plate and by Table II, and Figs 5a and 5b for simply supported one.

Table I. Time to FCA for clamped plate (in $10^5$ hrs)

| $\alpha$ | Uni-lateral | Non-unilateral | | Bilateral |
|---|---|---|---|---|
| | | h=0 | h=0.5 | h=1 |
| 0 | 2.418 | 2.418 | 2.418 | 2.418 |
| 0.5 | 2.173 | 1.301 | 1.186 | 0.932 |
| 1 | 0.890 | 0.590 | 0.531 | 0.382 |

Table II.Time to FCA for simply supported plate (in $10^5$ hrs)

| $\alpha$ | Uni-lateral | Non-unilateral | | Bilateral |
|---|---|---|---|---|
| | | h=0 | h=0.5 | h=1 |
| 0 | 0.224 | 0.224 | 0.224 | 0.224 |
| 0.5 | 0.300 | 0.199 | 0.189 | 0.170 |
| 1 | 0.211 | 0.153 | 0.144 | 0.121 |

In Figs 6a, 6b and 7a, 7b the time evolution of damage parameter in the point of FCA is shown. In all cases figures numbered with "a" refer to $\alpha=0.5$ and these with "b" to $\alpha=1$. The results

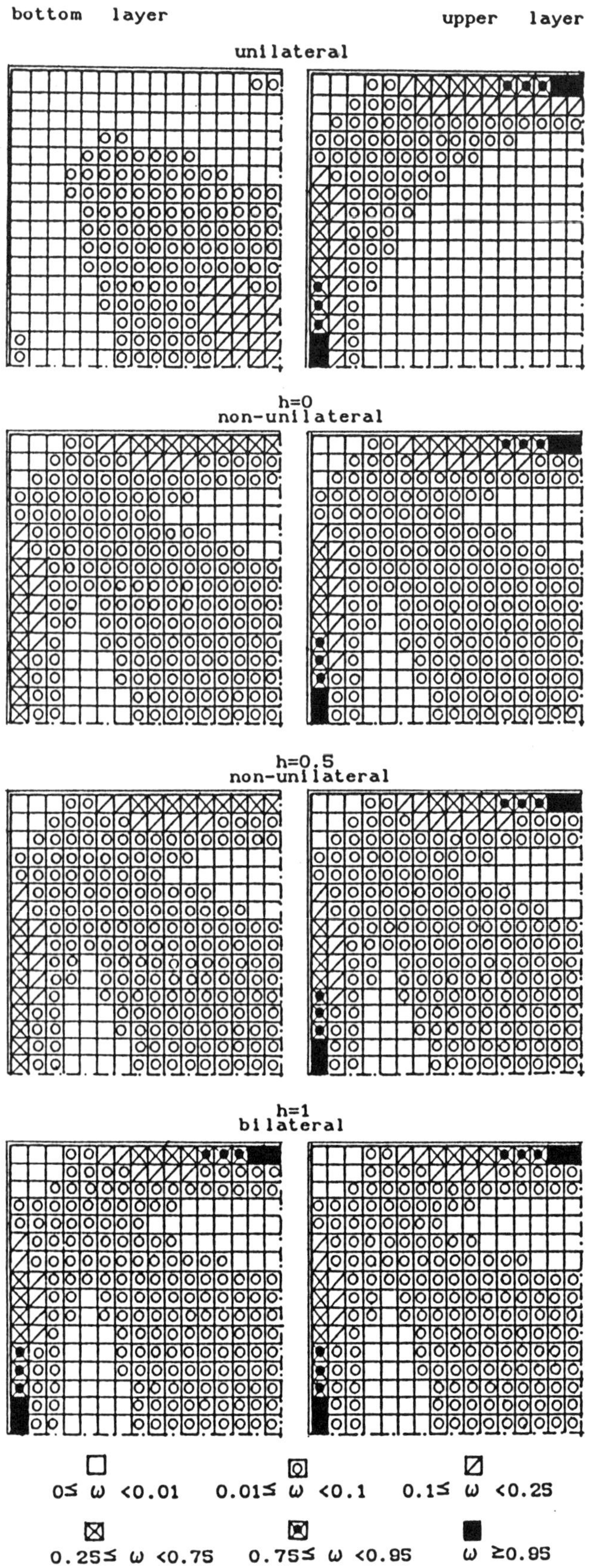


Fig.4a - Damage parameter distribution at time of FCA for clamped plate, $\alpha$=0.5

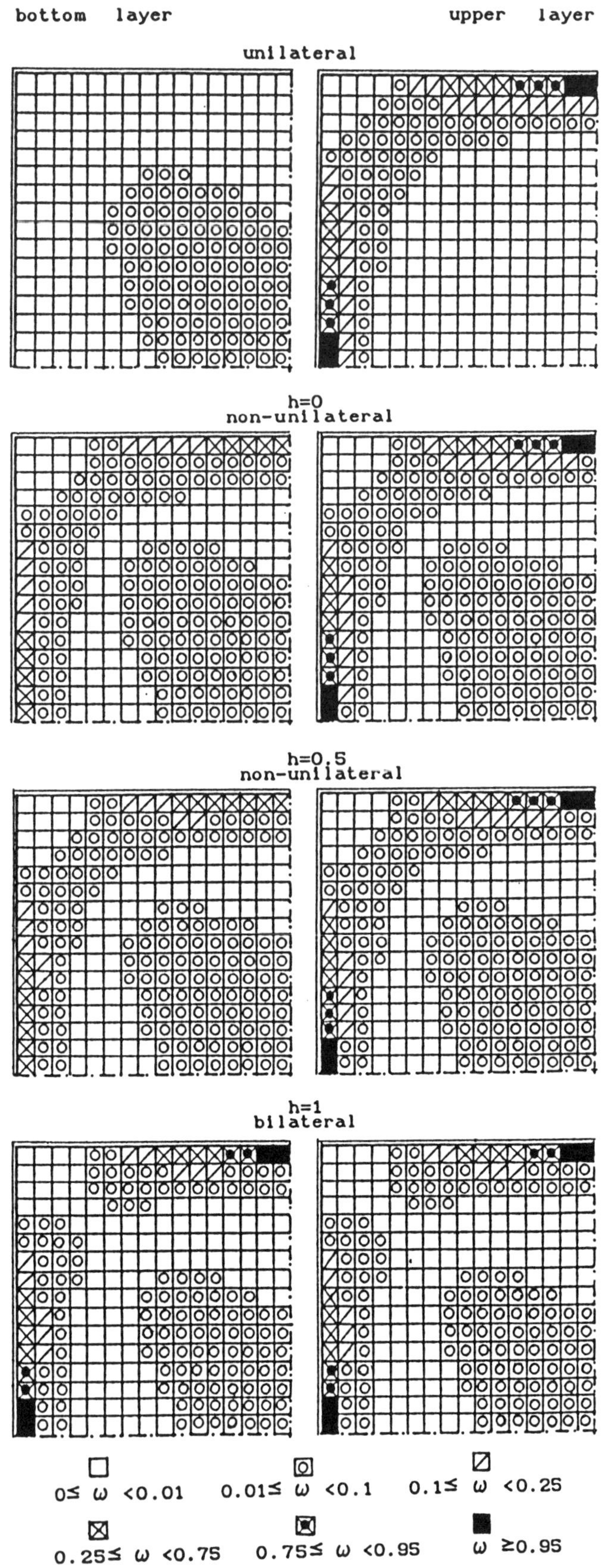


Fig.4b - Damage parameter distribution at time of FCA for clamped plate, $\alpha$=1

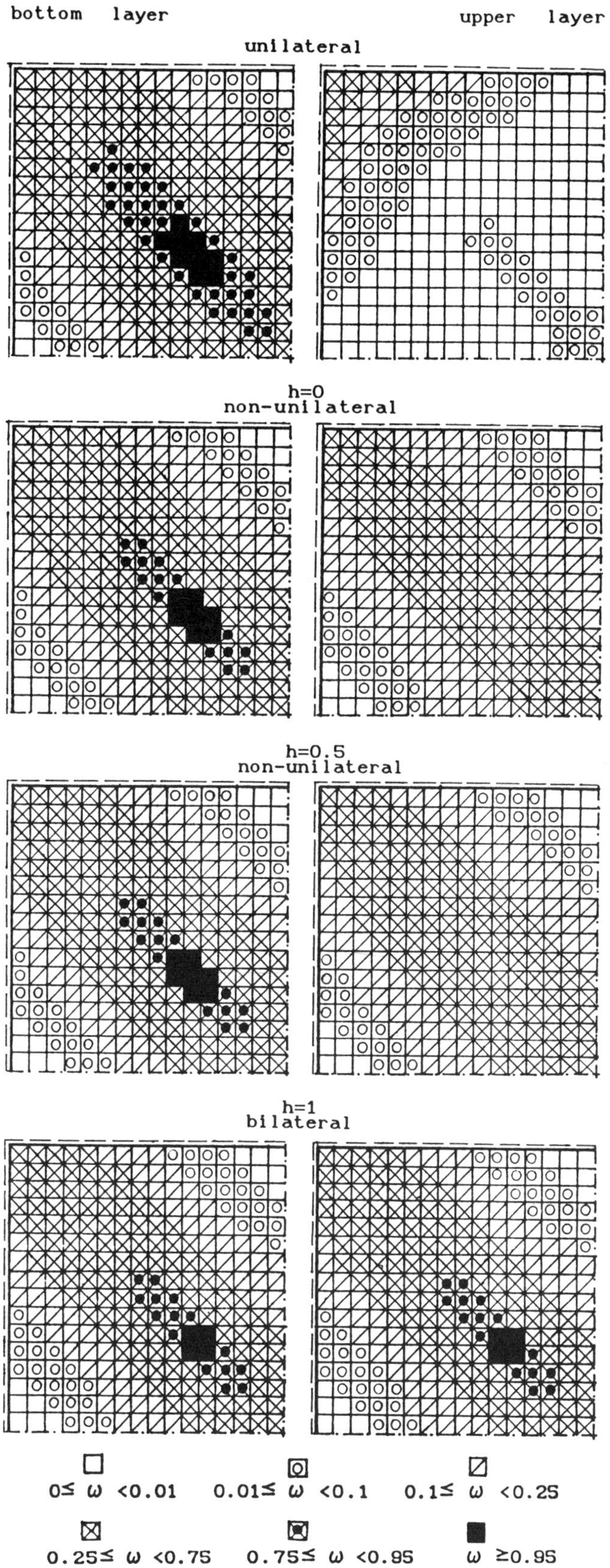


Fig. 5a - Damage parameter distribution at time of FCA for simply supported plate, $\alpha$=0.5

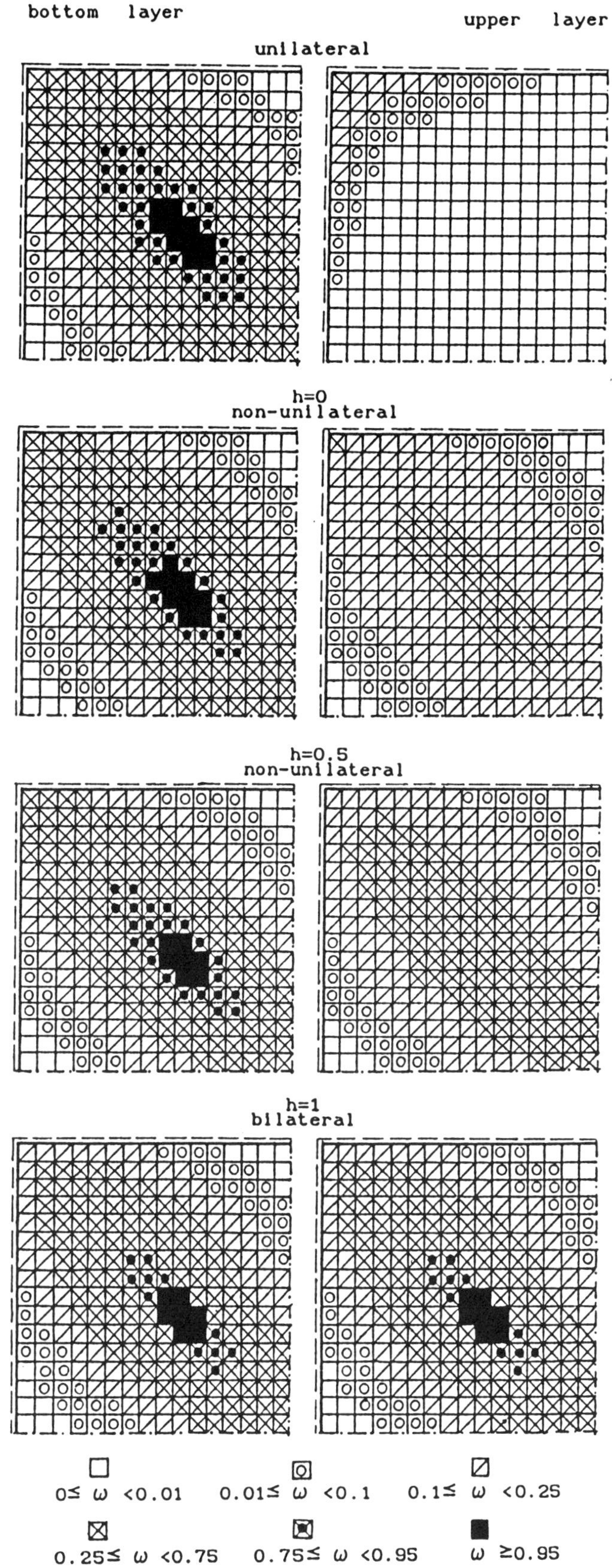


Fig. 5b - Damage parameter distribution at time of FCA for simply supported plate, $\alpha$=1

for three values of negative stresses influence parameter *h* (0, 0.5, and 1 - *bilateral*) are shown, as well as these for an *unilateral* material.

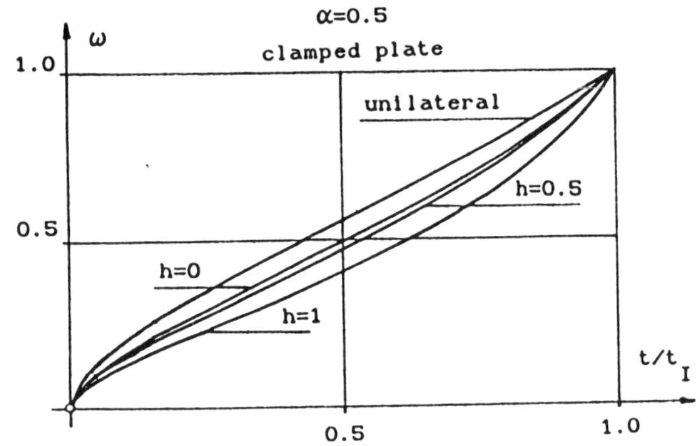


Fig.6a - Time dependence of damage parameter at the point of FCA for clamped plate, $\alpha$=0.5

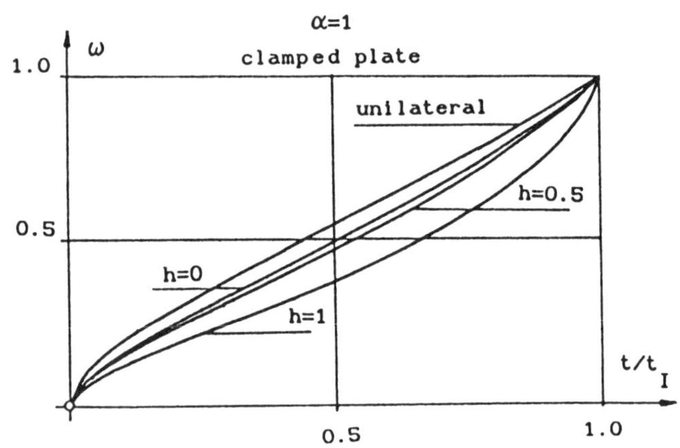


Fig.6b - Time dependence of damage parameter at the point of FCA for clamped plate, $\alpha$=1

**Conclusions.** Tables I and II show that the time to FCA is affected by both mode of failure (expressed by value of $\alpha$) and sensitivity to compressive stresses (through value of *h*). The influence of $\alpha$ is not as much pronounced as that of *h*. In the first case the reduction of time to FCA is strongly dependent on boundary conditions; for clamped plate the reduction with $\alpha$ changing from 0 to 1 is between 2.72 for *unilateral* material, and 6.33 for *h*=1, whereas for simply supported plate corresponding reduction factors are: 1.06 and 1.85. However, the general tendency is the same: higher the value of $\alpha$ (i.e. more important is the influence of principal stress $\sigma_1$ on damage accumulation), lower the value of time to FCA.

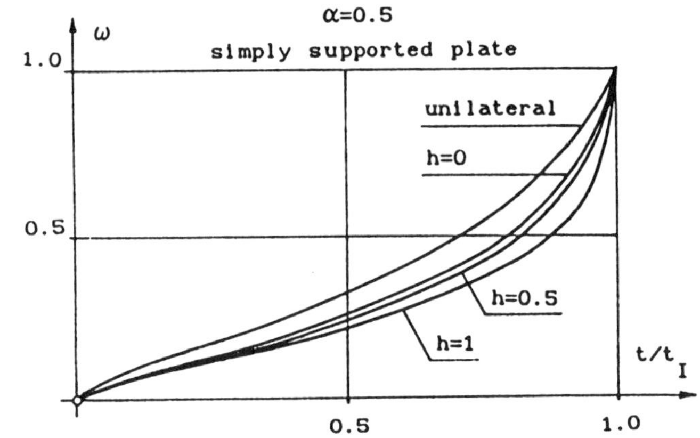


Fig.7a - Time dependence of damage parameter at the point of FCA for simply supported plate, $\alpha$=0.5

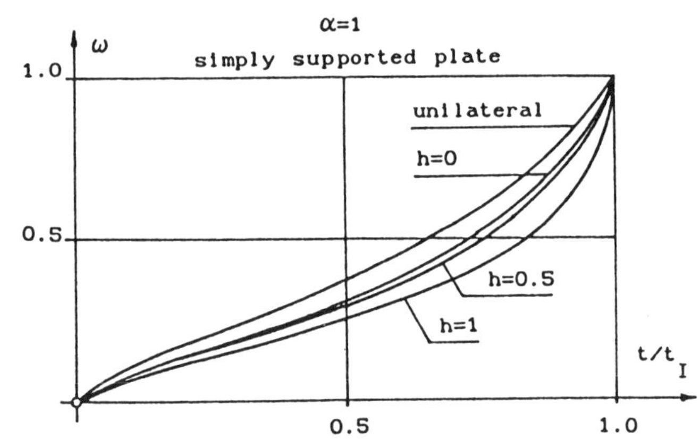


Fig.7b - Time dependence of damage parameter at the point of FCA for simply supported plate, $\alpha$=1

The effect of compressive stresses upon $t_I$ is not so much influenced by boundary conditions and ratios of $t_I$ vary between 2.33 for clamped plate to 1.76 for simply supported one. It appeared to be practically independent of mode of the failure (parameter $\alpha$), but , of course, compressive stresses reduce considerably the time to FCA, if material exhibits such a sensitivity.

Location of FC is practically insensitive to both: mode of failure on compressive stresses dependence, but boundary conditions play a decisive role (cf. Figs 4a,5a and 4b,5b).

The inspection of time dependence of damage parameter evolution (Figs 6 and 7) exhibits a limited influence of both $\alpha$

and $h$. It may be observed, however, that higher degree of redundancy (clamped plate, Figs 6a and 6b) makes the process of damage growth almost linear, even with high exponent in damage evolution law ($m$=5.79 was used in this example). This is an obvious effect of stress redistribution, which is less pronounced in the case of lower redundancy (simply supported plate, Figs 7a and 7b).

## BOUNDARY CONDITIONS SENSITIVITY

**Stages of deterioration process.** The influence of one of the elements of boundary conditions (kinematic boundary conditions) was striking even in the previous analysis which was meant mostly for material aspects analysis. The essential influence of two other elements of boundary conditions (structure shape and loading) may be expected, too. It does not undergo systematic analysis easily, as it can not be expressed through simple parameters. More important seems to be another feature of deterioration process, which is its progression through three characteristic stages: the period of first crack formation terminated by FCA at time denoted as $t_I$, period of main crack propagation through structure element thickness (BHF) at time $t_{II}$, and finally the period when sufficient number of "*brittle hinges*" are formed to cause structure total collapse (STC) at time $t_{III}$. This idea of splitting the life-time of a structure into three stages were pursued in [13] and applied to a statically indeterminate beam (of redundancy equal to 1). In the present paper the analysis will be carried out for a truss structure. Due to stress uniformity in each truss member the coincidence $t_I \equiv t_{II}$ occurs, and analysis can be limited only to the two characteristic times $t_I$ and $t_{III}$ (the difference between these being denoted by $\Delta t$).

**Example: three bar truss.** A simple truss consisting of three bars shown in Fig.8 was analyzed using the set of constitutive equations given by Eqs. 1-4

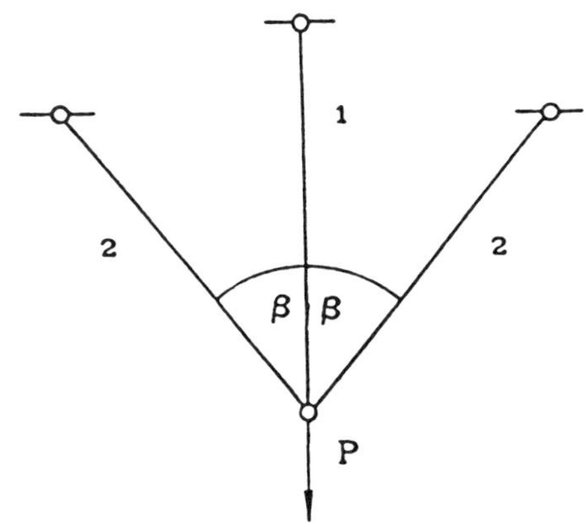


Fig.8 - A three bar truss

for an *unilateral* material and $\alpha$=1. For the sake of simplicity a nondimensional variables were introduced according to the definitions:

stresses in truss bars:

$$s_k = \frac{\sigma_k}{\sigma_*} \tag{6}$$

strain in truss bars:

$$e_k = \frac{\varepsilon_k}{\varepsilon_*} \tag{7}$$

time:

$$\tau = \frac{t}{t_K} \tag{8}$$

where $\sigma_*$ and $\varepsilon_*$ are reference stress and strain, respectively, $t_K=1/A(m+1)\sigma_*^m$ is reference time to failure, and k=1,2,3 corresponds to individual bars. With these denotations the basic set of equations takes form:

$$e_k = e_k^e + e_k^c \tag{9}$$

$$e_k^e = s_k^e \tag{10}$$

$$\frac{\partial e_k^c}{\partial \tau} = \frac{L}{(m+1)} \left[\frac{s_k}{1-\omega_k}\right]^n \tag{11}$$

$$\frac{\partial \omega_k}{\partial \tau} = \frac{1}{(m+1)} \left[\frac{s_k}{1-\omega_k}\right]^m \tag{12}$$

where **L** is a dimensionless material constant which for a material considered in this paper is equal to 7.5, and has to be completed with an appropriate displacement compatibility equation. In what follows the truss geometry will be solely determined by an angle $\beta$ between bars by assuming symmetry of truss geometry (bars' cross section areas and lengths are indentical).

The results of numerical calculations are summarized in Figs 9 through 12. As a reference solution the simplest possible theory which does not take into account stress redistribution (by neglecting elastic stresses and coupling between strain and damage) was used; corresponding curves are given in Figs 9-12 by dashed lines.

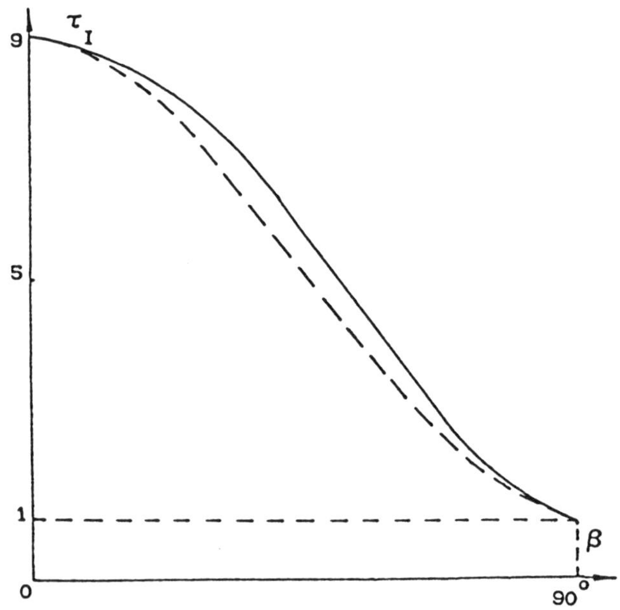


Fig.9 - Time to the FCA as function of truss angle $\beta$

**Conclusions**. Figs 9-11 show values of time $\tau_I$ to FCA , time increment $\Delta\tau$ to STC, and $\Delta\tau$ relative to $\tau_I$, respectively, all as functions of angle $\beta$. There is strong reduction of $\tau_I$ with $\beta$, as side bars initially are carrying less load for higher value of $\beta$. The area between dashed curve and that shown by full line represents the influence of stress redistribution due to nonstationary creep.

It is interesting to notice a significant difference between exact solution (according to the assumptions made) and that simplified one. This difference, caused by spatial

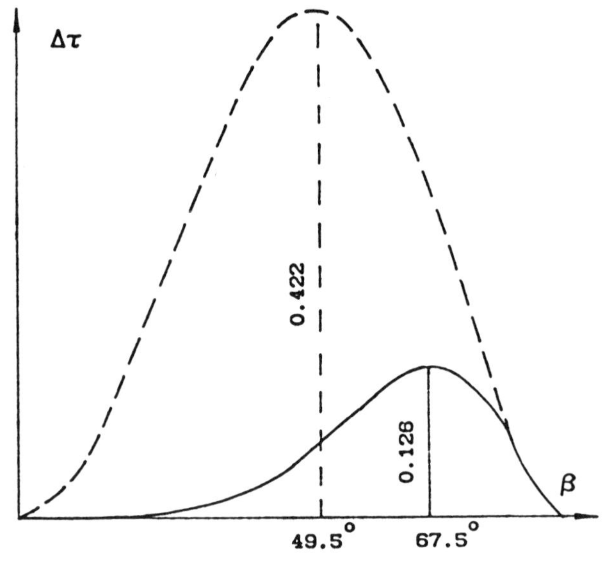


Fig.10 - Time interval $\Delta\tau$ as function of truss angle $\beta$

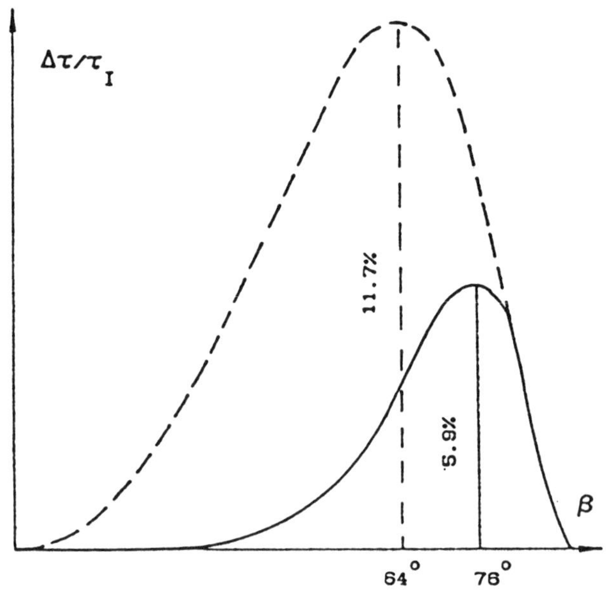


Fig.11 - Relative time $\Delta\tau/\tau_I$ as function of truss angle $\beta$

redistribution of internal forces, is manifested by the different values of the angle $\beta$ for which maximum of a given variable is reached, as well as by the value of that variable.

Both $\Delta\tau$ and $\Delta\tau/\tau_I$ curves have maxima, which occurs at different angles $\beta$ for simplified (dashed line) and full theory (solid line). The relative values of these maxima are rather small (5.9% for full analysis, and 11.7% for simplified

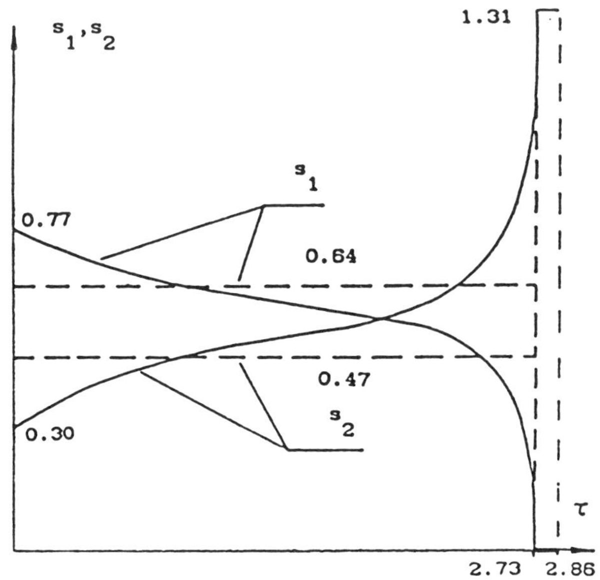


Fig.12 - Internal forces redistribution for truss angle $\beta$=67.5°as a function of time

one) because of low structure redundancy (equal 1 for analyzed structure). Nevertheless, one can think of optimization of structures with respect to $\Delta t$, or $\Delta t/t_I$.

The above analysis falls into the frame of a general one performed in paper [17]. A bundle of fibers of different material properties were studied there to define different types of structural behaviour (immediate versus delayed, and instantaneous versus gradual). The present analysis falls into delayed gradual class, but it is enhanced by a new parameter which takes into account spatial redistribution of internal forces.

SUMMARY

Two groups of parameters which may influence the structural behaviour with respect to the time-to-failure of a structure were distinguished: one was referred to as material behaviour and was represented by different mechanisms of deterioration process, whereas the other one was purely geometry and load dependent. It was shown that both play significant role in evaluation of the time to FCA, as well as in a STC as a consequence of formation of a collapse mechanism. The combination of the two effects should be taken into account if an accurate life-time evaluation is essential for structure safe operation.

ACKNOWLEDGEMENT

The work reported in the paper has been performed partially under project no. 3/0917/91/01 granted by KBN (Polish Committee for Scientific Research).

REFERENCES

1-Janson,J.,Hult,J.,J.mech.appl.,vol.1., 1,69-84 (1977)
2-Kachanov,L.M., Izv.AN SSSR, OTN, 26-31 (1958)
3-Rabotnov,Yu.N.,"Vopr.Prochn.M.Konstr." Izv.AN SSSR, Moskva, 5-7 (1959)
4-Penny,R.K.,Met. Mater.,278-283 (1974)
5-Chaboche,J.L., Nucl.Eng.Design,vol.64, 233-247 (1981)
6-Krajcinovic,D., Appl.Mech.Rev.,vol.37, 1-6 (1984)
7-Lemaitre,J., Nucl.Eng.Design,vol.80, 233-245 (1984)
8-Garofalo,F.,"Fundamentals of Creep and Creep Rupture in Metals", MacMillan, N.Y. (1965)
9-Betten,J., J.Mec. teor. appl.,vol.2,1, 13-32 (1983)
10-Hayhurst,D.R.,Dimmer,P.R.,Morrison,C., Phil.Trans.R.Soc.London, A311, 103-121 (1984)
11-Bodnar,A., SMiRT 11 Trans., vol.L, 391-396 (1991)
12-Bodnar,A.,Chrzanowski,M., "Creep in Structures", Proc.IUTAM Symp.,287-293, (1991)
13-Chrzanowski,M., Latus,P., Mech. Teor. Stos.,vol.28,1-2,45-54 (1990)
14-Chrzanowski,M., Madej,J., Latus,P., SMiRT 11 Trans.,vol.L, 355-365 (1991)
15-Lemaitre,J., CISM Course on Damage Mechanics, 13-16 Sept. (1986)
16-Bodnar,A.,Chrzanowski,M., Mech. Komp., vol.9, 101-115 (1989)
17-Chrzanowski,M., Hult,J., Eng. Fract. Mech.,vol.28,5/6, 681-688 (1987)

*Proceedings of the Fifth International Conference on Creep of Materials, Lake Buena Vista, Florida, USA, 18-21 May 1992*

# An Expert System in the Domain of Life Assessment of High Temperature Power Plant Components

**K. Kussmaul, A. Jovanovic, and K. Maile**
Staatliche Materialprufungsanstalt
Stuttgart, Germany

**H. Kautz**
Grosskraftwerk Mannheim AG
Stuttgart, Germany

**Abstract**

The paper presents background and first results in development of a knowledge based (expert) system aimed to provide technical support to engineers performing assessment and management of remaining life of the high temperature pressurized components (mainly piping) in power plants. The system (ESR – Expert System for Remaining life assessment), is being developed at MPA Stuttgart (FR Germany), with support and under sponsorship of nine European utilities, and it is envisaged as a way to summarize, structure and make re–usable human expert experience gathered in current practice. The system includes parts providing assistance when dealing with: a) damage analysis, piping analysis and assessment; b) recommendations, standards, regulatory requirements and case histories; and c) engineering calculations and data storage and retrieval in/from data bases. The system runs on a personal computer.

## 1. INTRODUCTION

Much of the recent research effort in Europe, USA and in Japan has been devoted to expert (or knowledge–based) systems applied in the fields of power plant and structural engineering.. These systems deal typically with different aspects of operation, maintenance and analysis, and they are applied to various systems and components in power plants (e.g. turbine rotors, blades and casings, boiler tubes, piping, etc.).

From the "application domain related" point of view most of the systems can be categorized into two main classes:

(1) analysis–oriented expert systems that diagnose a failure mechanism or the root cause of a failure, or systems that assess the remaining useful life of a component and provide guidance on run/repair/replace decisions; and

(2) expert systems that help control or diagnose problems with process variables such as water and steam chemistry, that affect the performance and reliability of high–temperature components (including the systems addressing general and specific corrosion problems).

Several characteristic American and European expert systems in this domain are presented more in detailed in the work of Jovanovic and Gehl (1991).

The ESR project of MPA Stuttgart (ESR – Expert System for Remaining life assessment, or Expertensystem für Schädigungsanalyse und Restlebensdauerermittlung) addresses the problem related to possible failures of high temperature pressurized components (piping, pipework subcomponents, etc., in power and other industrial plants. The development of the system is sponsored by nine electric utilities from Austria, Belgium, Germany, Finland and Portugal. Companies from some other countries expressed their interest to join the group, too. Further support is expected also from the Association of German Large Electric Utility Companies (VGB) and from the European Community (SPRINT and BRITE programs). The system is envisaged as a way to summarize, structure and make reusable human expert experience gathered in current practice, in an engineering environment where the regulatory requirements are numerous and tough and where the tradition to rely on deterministic and numerical calculations is very strong. The major generic requirements imposed by the utilities are:

- improve safety and availability of piping systems
- support and help personnel in performing usual maintenance tasks
- support and help personnel when dealing with specific sub–problems
- facilitate search and use of necessary documentation
- save and re–use knowledge of human experts (personnel)
- preserve experience and knowledge from manufacturing/construction phases
- perform strategic analyses and analyses of case studies (will be implemented only in the future versions of ESR).
- evaluate better single aspects of new cases and compare them with stored ones
- use both "plant specific" (e.g. NDE – non–destructive examination results) and "plant neutral" data (e.g. regulatory requirements, standards, etc.).

The first phase of the ESR project has provided an "in–breadth" answer to many of the above requirements. An "in–depth" answer will be provided in the second phase of the project (1992–94).

## 2. THE ENGINEERING PROBLEM TO BE SOLVED

Life assessment of high temperature pressurized components, as performed nowadays, is based mainly on:

- numerical calculation of the consumed creep and fatigue life (e.g. the German code TRD 508 – Zusätzliche Prüfungen an Bauteilen, berechnet mit zeitabhängigen Festigkeitskennwerten)
- destructive (material) testing, nondestructive examinations and strain measurements, e.g. the VGB–Richtlinie (Guideline) – Wiederkehrende Prüfungen an Rohrleitungsanlagen in fossilbefeuerten Wärmekraftwerken – VGB–R 509 L, and
- human expertise and experience involved in performing the preceding steps and in analysis of their results.

Improving methods and procedures for the assessment and management of remaining life of the target components is therefore extremely important (Den Haag, 1988), but the development and application of new, more sophisticated numerical engineering methods and tools is often coupled with severe limitations. The limitations are in most cases imposed by unavailability of material data, time, resources and highly qualified personnel which is required for a sophisticated analysis.

Furthermore, practical experience and case studies (Garleff, Dooley, Kautz, van Liere and Plate, 1988; Kautz and Zürn, 1987) confirm importance of various factors like uncertainty (scattering) in material properties, uncertainties related to oscillations of operational parameters like pressure and/or temperature in particular, etc.

Human expertise of different specialists is essential for the assessment, but is often unavailable at the plants in the very moment when it is needed. In its absence, the plant engineers, are thus often facing the question what to do with the component and/or plant (e.g. shut down and re–inspect, reduce load, etc.).

## 3. THE EXPERT SYSTEM BASED SOLUTION

Practical experience (Jovanovic et al., 1989; Jovanovic, Bogaerts,1991) with expert systems has clearly shown that conventional (production rule based) expert systems alone can deal successfully only with a very limited range of practical problems in the above described domain. In addition the conventional systems are almost always interactive (dialog–based) and as such they often tend to "block" the dialog at the moment when the user is not sure what to answer the system, either because he needs an explanation of the question or because he is asked to provide (to the system) some additional information which is currently not known/available.

A typical example of these cases are questions of the type "please input the value of the factor XY given in standards YZ". If the required value in the standard YZ is given (e. g.) through a nomogram, table, etc., and if the user does not have the paper–copy of the standard YZ in front of him and if he does not know the required value by heart, he will have to stop the session and start searching the information "manually", e. g. in some library. A more user–friendly conventional system would presumably also add that factor XY can be found in document YZ, but that would probably be the end. On the other hand, even when the user finally finds the required paper–copy, it can still be necessary to (e. g.) interpolate values or perform some other intermediate calculations. Only when the user has done all this he will be able to continue the session. But in practice this would frequently mean that he will have to restart the session from the very beginning.

Possible answer to this and similar questions is to provide, by means of integration, all the tools necessary to avoid the above mentioned "blocking" of the dialog (e. g. numerics including finite elements, databases, etc.). This idea is the baseline of the MPA's approach called "KISS" – <u>K</u>nowledge–based <u>I</u>ntegrated <u>S</u>oftware <u>S</u>ystems, or <u>K</u>nowledge–based Intelligent <u>I</u>ntegrated <u>S</u>oftware <u>S</u>ystems (Jovanovic, Maile, 1990).

KISS allows different "weighting" of single elements (depending on the characteristics of a given application). So for instance, an application aimed to provide consultance related to codes and standards will have a stronger hypertext component and, e.g., weaker database component than an application providing an intelligent engineering prediction based on a large number of data. Differences in "weighting" can thus result in (e.g.) Intelligent Databases, Intelligent Hypermedia Systems KBSs with Hypermedia Support, Numerical programs + Intelligent Front–Ends, etc. (Parsaye et al. 1989; Wallmannsberger, 1990; Bielawski, Lewand, 1991).

The approach can be illustrated by an "ESR–related" example. The typical "domain question" would be "what to do with an old pressurized component in a power plant?" The component may be dangerous, people who built it may be long gone, a lot of data is missing, documentation is in a poor state, ... still, the new component may be very expensive, its delivery time long, and, after all, why should one replace it if it still works, or can continue to work after a minor repair!. The engineer's answer would then contain the following steps:

a) Retrieve necessary data / information regarding
- material
- operational history
- loads
- applicable standards, codes, etc.,

**KISS–support:** <u>"conventional" databases</u>.

b) Retrieve necessary background data / information <u>(the "full text"!)</u> regarding
- applicable standards, codes, etc.
- similar cases (case studies)
- relevant literature

**KISS–support:** <u>hypermedia databases</u>.

c) Make necessary calculations / analyses (e.g. finite element ones)

**KISS–support:** <u>numerical programs</u>

d) Recommend necessary further actions involving (a), (b) or (c), ask for missing information / data

**KISS–support:** <u>knowledge–based systems</u>

d) Make all necessary warnings recommend solutions and support the decision

**KISS–support:** <u>knowledge–based systems</u>

e) Communicate with the user and record the work in the most appropriate way
**KISS–support:** <u>knowledge–based systems and hypermedia</u>

When developing each of these KISS–parts one regularly meets the same dilemma: to program these parts in the expert system shell (and thus avoid the compatibility problems) or to integrate off–the–shelf solutions (and thus avoid the problem of developing in unoptimized environment). E.g. one can program a database also in Lisp but at what cost and with what performance. There is also regularly the need to optimize the system ("a system preferably as big as possible on a machine as small as possible") and this objective can practically be reached by using, i.e. integrating the software already optimized for particular tasks. The second option, to integrate off–the–shelf solutions, is the approach adopted by MPA within KISS.

A KISS can thus help to integrate two basic elements of each engineering decision (Fig. 1)

a) knowledge AND
b) information.

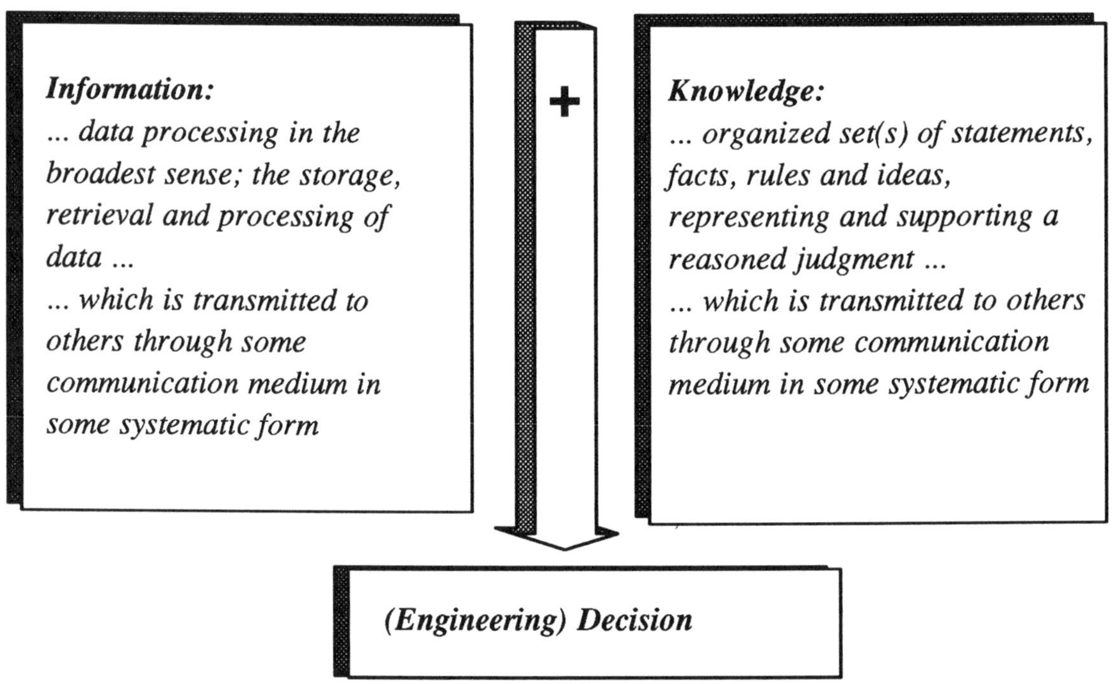


Fig. 1 Two basic elements of each engineering decision

ESR System is based on KISS and designed as an engineering "tool box" structured on two basic levels (Fig. 2):

- the level containing the expert systems shell and the hypermedia based modules,
- the level of ESR Utilities (e.g. data bases, spreadsheets, etc.).

The system is built on top of commercially available software development toolkits (expert system shell KAPPA, hypermedia tools like GUIDE and Toolbook, etc.).

The object oriented programming (OOP) appears both at the level of overall ESR architecture (each part of ESR is an object exchanging messages with other ones) and at the level of its single parts. The architecture allows to introduce new modules or to reorganize the existing ones any time.

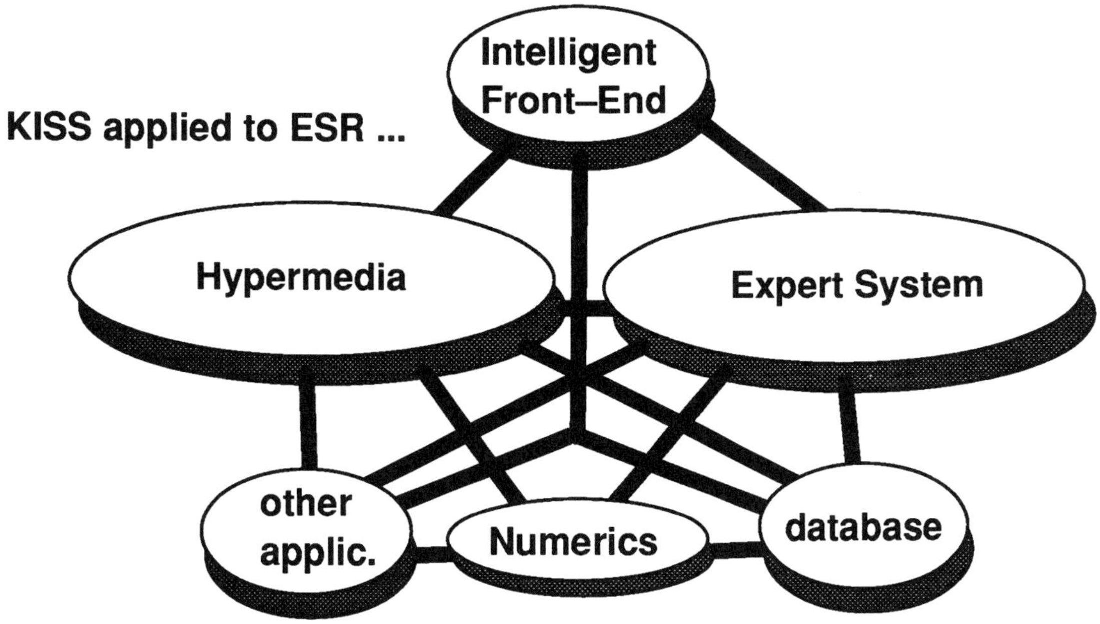


Fig. 2 – Structure of ESR

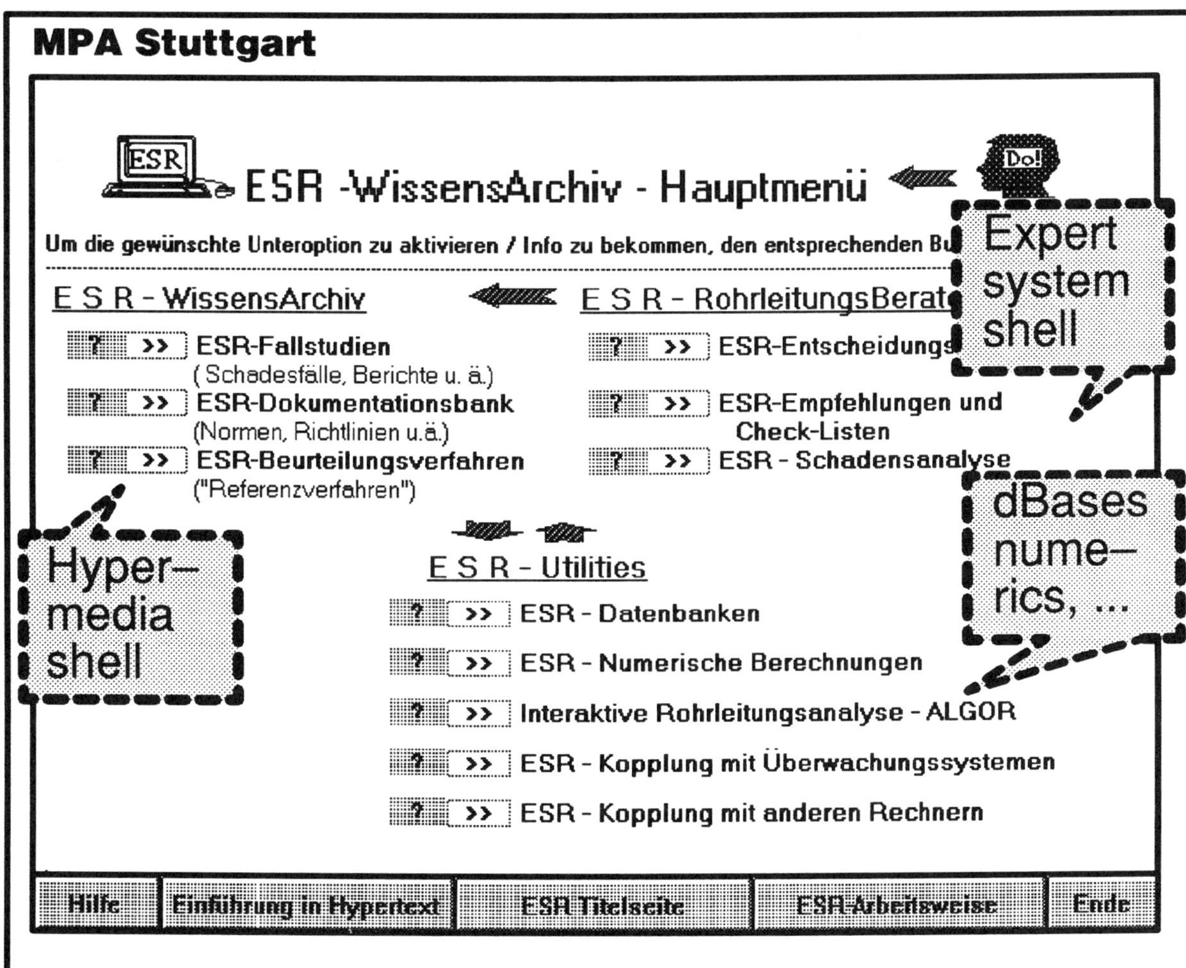


**ESR – System** (screen printout)

The integrated multitasking and communication among modules/parts is performed mainly via DDE – Dynamic Data Exchange and DLL – Dynamic Link Libraries.

## 4. MODULES OF THE ESR–SYSTEM

**The Reference Procedure** (Jovanovic, 1989; Jovanovic and De Witte, 1989; also Assessment Procedure, "Beurteilungsverfahren", "Referenzverfahren") lies in the background of the ESR system . It has been envisaged as a "procedural summary of the domain engineering knowledge", i.e. as a summary of good practice, recommendations, guidelines, standard regulatory requirements and human expertise. Partly in the form of check lists, the procedure identifies and explains steps in the process of assessment. Starting from very "compressed" representation, one can proceed to the desired level of explanation simply by expanding single step and substeps in the procedure. The Procedure contains also the natural language structured, easy manageable set of heuristic rules and corresponding representation of the "raw" domain knowledge acquired.

**Documentation Bank** ("Dokumentationsbank") is probably the most "conventional" hypertext application in ESR. It contains the frequently used codes, standards and other documents (Figure 3) and allows multilevel explanation of the documents. The example in Figure 4 shows how the explanations (from German boiler code TRD 300: Festigkeitsberechnung von Dampfkesseln) are conveyed to the reader of the original document (German guidelines: VGB Vereinbarung – section 2

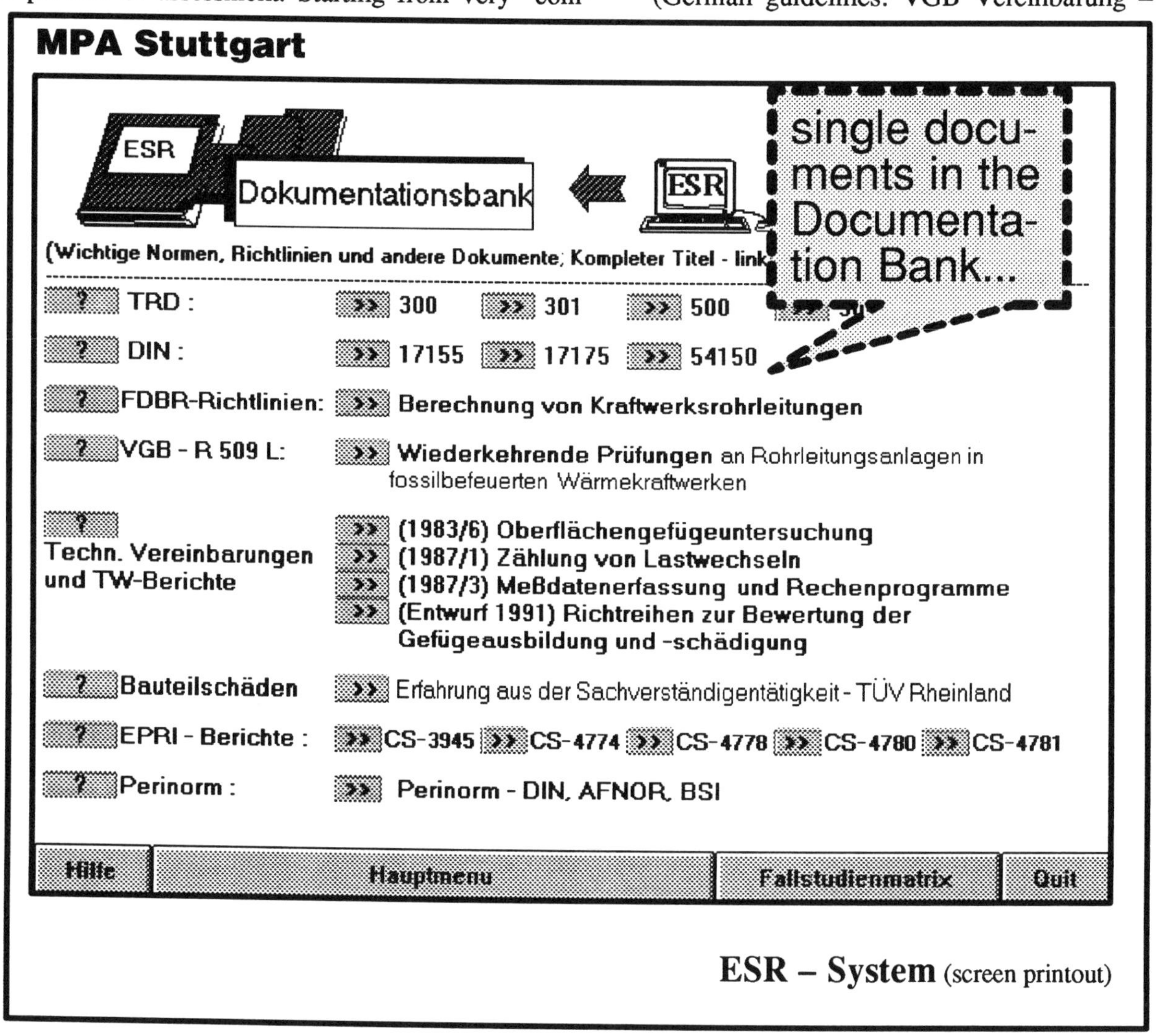


Fig. 3 – The ESR Documentation Bank

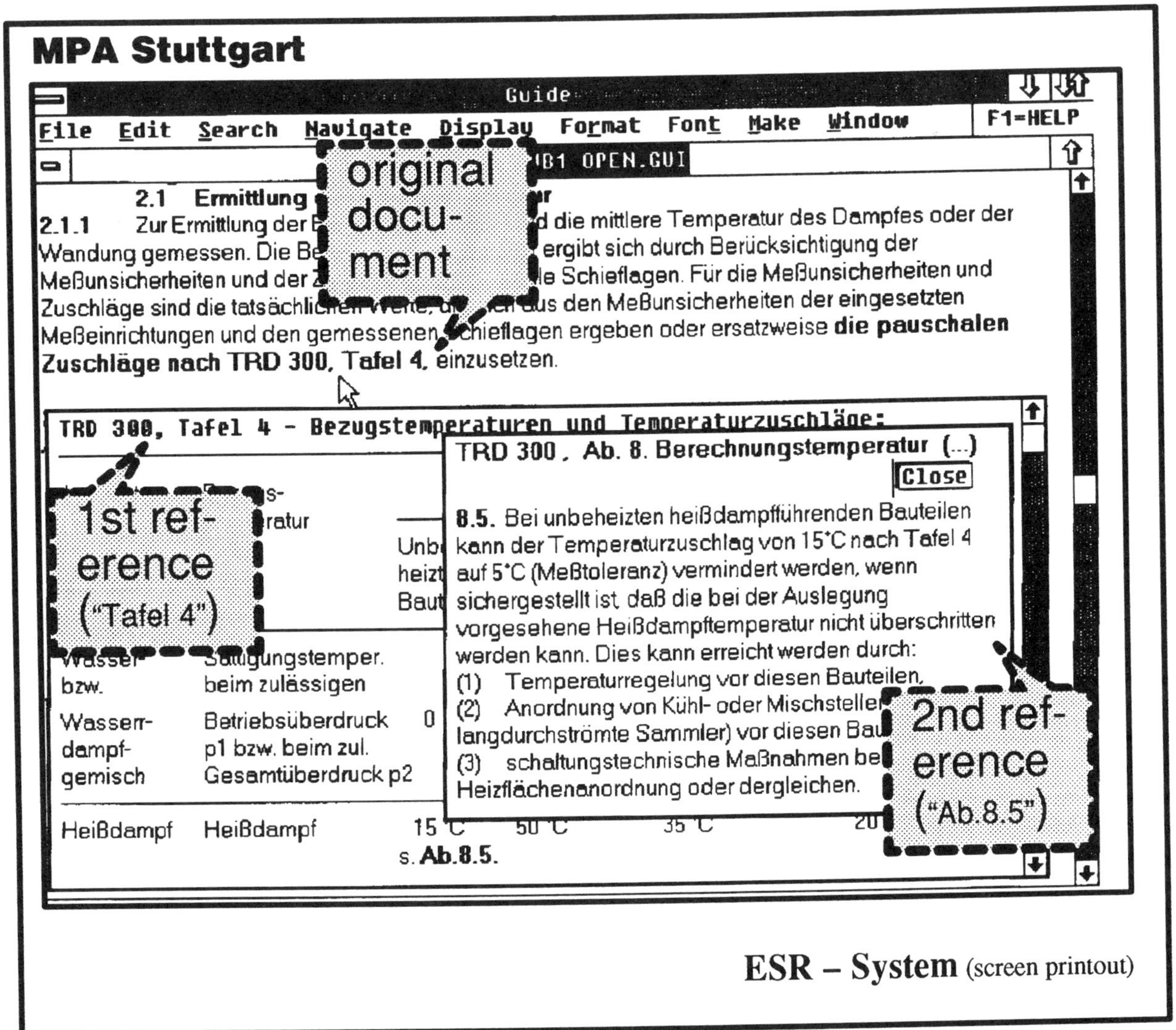


Fig. 4 – A hypertext expansion of one document from the Documentation Bank with the reference to two parts of another one (TRD 300)

"Allgemeine Anforderungen"). Through the Documentation Bank the user has also access to the CD–ROM based information base Perinorm containing German, French and British standards (DIN, AFNOR and BSI) and the corresponding cross references.

**Case Studies** ("Fallstudien"): In order to provide a sound technical base to ESR it has been necessary to identify and analyze a series of case studies. The case studies have provided a source of further evidence and experience in ESR. The ESR matrix of cases, although not yet completely "filled" contains typical combinations of component types, materials and types of case studies (remaining life assessment and damage analysis). Hypermedia allow the incorporation of full text and images (Figure 5). A particular way of compressing gray images has been developed and applied in order to reduce the amount of computer memory necessary for storing of gray images (material micro structure, fracture surfaces, etc.). The system helps the user to retrieve those stored case histories which are most relevant for the currently analyzed case.

The current version of ESR contains several expert system shell based modules, covering three basic aspects (Fig. 2):

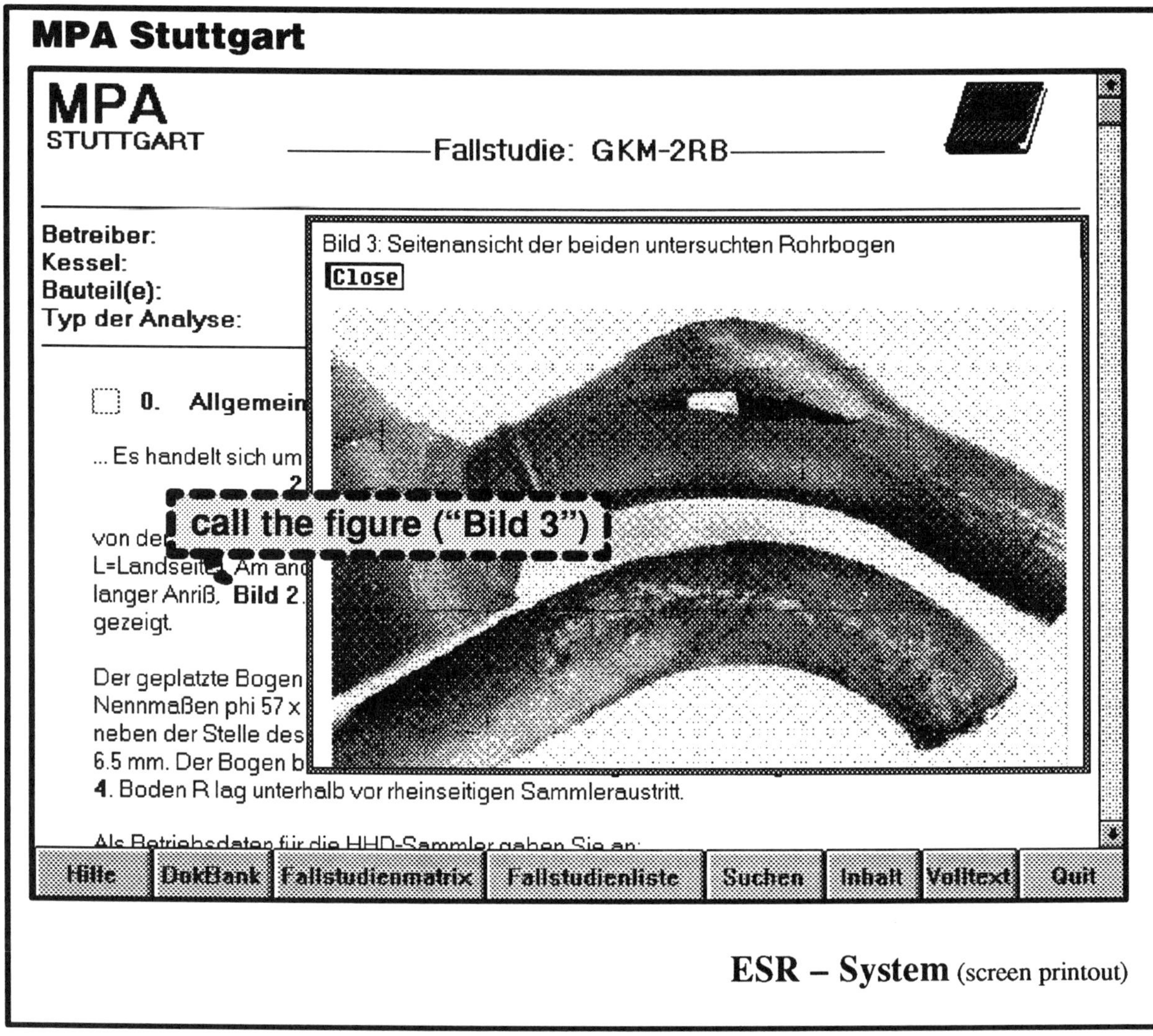


Fig. 5 – A hypermedia expansion of a case study in ESR (calling an image/figure from the text)

a) Decision making according to VGB guidelines (decision aid for making the "(final) disposition decision" like *replace, shutdown, repair, use_under_reduced_load*, etc.). This decision is based partly on the regulatory guidelines (VGB), partly on the experience and heuristic knowledge incorporated into the expert system.

b) Recommendation during the annual inspection (revision)

c) Damage analysis

Further development of the ESR system will be devoted mainly to these modules.

**Data Bases and Numerics – ESR Utilities:** These modules offer assistance when:

- searching and retrieving of data
- performing numerical calculi according to TRD standards
- performing piping and component stress analyses.

Currently, there are four data bases in ESR: one for material properties (DIN standards), one for data regarding power plants and single units, one for data regarding the components, and one including their operational history (hours, pressures and temperatures). The bases include both "plant specific" and/or "utility specific" data.

Oversimplification of conventional (numeric) programs was common practice in many expert system applications in the past. This has been avoided in the ESR System where conventional (in the sense of applied software tools), but very advanced (in the sense of the built–in capabilities) computer codes (like ALGOR for piping analysis) have been integrated, as well as a set of standard numerical programs for engineering calculi, like those of the boiler code TRD 508 and TRD 300.

## 5. USE AND VERIFICATION OF ESR–SYSTEM

Once the system has been started there are two ways to continue: either the user knows already which part of ESR will match his needs in a given case and uses this part correspondingly, or the user "allows" the ESR to guide him, through the dialogue.

The system communicates with the user via menus, dialogue, active images, etc., and virtually no knowledge about computers, expert systems and/or programming is needed. Most domain engineers (working normally in fields of design, material science, etc.) have been able to use the system after some initial explanations. Using all the the existing and yet to be developed features of the system will however, require somewhat longer acquaintance times.

First experience with test installations in power plants have clearly shown how complex the overall problem of the remaining life assessment is, and how many aspects does it have, but it has also indicated that the approach adopted in ESR is an adequate one for a whole series of sub–problems in this field.

Quality of decision and advice, the human–computer interaction, the system efficiency and the cost effectiveness will be imposed as the main verification/evaluation criteria.

## 6. CONCLUSIONS

ESR is an open knowledge based software system, both in the terms of the domain and in terms of software. In terms of software, the system has been designed as an open one in order to fit and profit out of the current developments on the market.

Such an "open character" of the system implies the need for continuous maintenance. The maintenance is provided by MPA Stuttgart for the "reference version" of ESR. In addition, the system is open both for "plant specific" and for "utility specific" data.

ESR exploits fully all the capabilities of 386–processor based PCs. Excellent speed, user friendliness and high degree of software transparency are maintained even when dealing with the most complex tasks.

Running on a standard PC/ DOS platform ESR can be easily integrated in virtually any existing EDP system, both from software and from the hardware point of view. No LAN version of ESR is available at present, but external data bases on LAN (Decnet–PCSA in the case of MPA) or other networks (EAN, Datex–P) can be reached.

The development of the ESR system and its future deployment in power plants should help in achieving a series of economic and technical benefits. Improved availability of systems and plants, shorter and better utilized maintenance periods, reduced costs of scheduled inspections due to the optimized inspection strategy– reduced costs of daily operation (specialists called only when necessary), reduced unplanned costs, or improved possibilities for the life extension of the plants, are only some of the goals to be mentioned. ESR alone, in particular at this stage of development, will not bring all these benefits, but the first practical experience in this sense is very encouraging.

***Acknowledgments:*** *The development of the ESR system has involved, in various forms of cooperation, a number of institutions and persons. To all of them and in particular to the colleagues who performed some pats of the programming (Messrs. Friemann, Helf and Vrhovac of MPA, Mr. Arents of KU Leuven, Belgium) and to Messrs. Bulens, De Witte, Rantala and other colleagues from nine utility companies sponsoring the project (Badenwerk, EVS, GKM, Neckarwerke and TWS – Germany, IVO – Finland, ISQ – Portugal, SGP–VA – Austria and Laborelec – Belgium), authors express their gratitude for their efforts, support and help.*

## REFERENCES


Bielawski, L. and Lewand, R. (1991). Intelligent systems design: Integrating expert systems, hypermedia, and database technologies. John Wiley & Sons Inc., New York, Chichester, Brisbane, Toronto, Singapore, 302 pp.

Den Haag (1988): Proceedings of the International Conference "Life Assessment and Extension". June 13 to 15, 1988, The Hague, The Netherlands

Garleff, Dooley, Kautz, van Liere and Plate (1988). Lebensdauerverlängerung – ein komplexes Planungsziel. VGB Kraftwerkstechnik, 69. Jahrgang, Heft 12, Dez. 1988

Jovanovic, A. (1989). ESR project: Expert system aided assessment and management of remaining life of pressurized high temperature components (in Expert System in Structural Safety Assessment. Jovanovic, A., Kussmaul, K.F., Lucia; A.C., Bonissone, P., Eds.), Springer–Verlag

Jovanovic, A., Kussmaul, K.F., Lucia; A.C., Bonissone, P., Eds. (1989). Expert system in structural safety assessment, – Lect. Notes in Engineering, vol. 53, Springer–Verlag

Jovanovic, A., De Witte, M. (1989). Development of a reference procedure used in expert system assisted life assessment of high temperature components, Proc. of the 15th Seminar, MPA Stuttgart, FR Germany, October

Jovanovic, A., Maile, K. (1990). ESR–System – Expertensystem für Schädigungsanalyse und Restlebensdauerermittlung / Expert System for Remaining life assessment, Zwischenbericht – Interim report (Auftr. Nr. 753245), MPA Stuttgart, September 1990

Kautz, H.R., Zürn, H.E.D. (1987). Das Verhalten von Schweisskonstruktionen im Kriechbereich. DVS–Berichte, Band 115 (1987), DVS Verlag, Düsseldorf

Jovanovic, A., Bogaerts, W. (1991). Hybrid knowledge–based and hypermedia systems for engineering applications, Avignon '91 Conference Expert Systems and their Applications (vol. Tutorial Nr. 13), Avignon, May 27–31, 1991

Jovanovic, A., Gehl, S. (1991). Some expert systems for power plant components in Europe and USA, Proc. of the SMiRT 11 Post Conference Seminar Nr. 13 "Expert Systems and AI Applications in the Power Generation Industry", Hakone (Japan), Aug. 26–28, 1991

Parsaye, K., Chignell, M., Khoshafian, S. and Wong, H. (1989). Intelligent databases: Object–oriented, deductive hypermedia technologies. John Wiley & Sons Inc., New York, Chichester, Brisbane, Toronto, Singapore, 479 pp.

Wallmannsberger, J. (1990). Open hypertext environments: towards the integrated information workplace. In Proceedings of the 14th International Online Information Meeting (London, December 11–13) (Ed. Learned Information Ltd.). Learned Information Ltd., Oxford, New Jersey, pp. 361–368.

*Proceedings of the Fifth International Conference on Creep of Materials, Lake Buena Vista, Florida, USA, 18-21 May 1992*

# Application of Creep Swelling Data in Life Assessment of High Temperature Piping

**F.V. Ellis**
Tordonato Energy Consultants, Inc.
Chattanooga, Tennessee

**B.W. Roberts, M.F. Steakley, and S.O. Hilton**
Tennessee Valley Authority
Chattanooga, Tennessee

## ABSTRACT

In the fall of 1990, TVA experienced a loss of differential pressure across the main steam flow-indicating nozzle at the Gallatin Unit 3 fossil plant. Subsequent investigation revealed that diametral swelling of approximately 5% had occurred in the flow nozzle region piping, leading to disengagement of the nozzle retaining pins. Widespread evidence of creep cracking at the through-wall penetrations was also found. This paper will present an overview of the findings for this inspection, an analysis of the pipe swelling, results of other flow-nozzle inspections, considerations in development of an acceptance limit for pipe deformation, and the use of strain measurements in a steamline assessment program.

THE TENNESSEE VALLEY AUTHORITY operates 59 fossil units that were built over a time span from the mid 1950's to the mid 1970's. The newest plants are approaching 110,000 hours operating time while the older plants have logged 270,000 hours. The materials of construction of the high temperature piping is 1-1/4Cr-1/2Mo-Si and 2-1/4Cr-1Mo steel. Nominal design temperatures are either 540C or 566C.

A systematic steamline inspection program was initiated in 1978 with the objectives of maintaining unit reliability and insuring personnel safety. The routine inspection methods utilized by TVA included traditional NDE techniques (MT, UT, PT), metallurgical techniques (hardness, plastic replication), and dimensional (pi tape diameter and UT thicknesses). The results of the inspections are tracked with a computer program to identify the areas not yet inspected and those areas requiring re-inspection.

This program includes inspection of all the main and high temperature reheat steam piping system components in order of priority. Selected welds and/or components are inspected at each regularly scheduled outage. The initial priority order was based on TVA's and the utility industry's experience and is shown below in descending order of priority:

- Wye blocks and the adjacent welds
- Tee blocks and the adjacent welds
- Welds at stop valves
- Welds at outlet headers
- Welds at bends
- Welds in straight pipe

The priority is changed when events warrant a reordering of the items to address an immediate area of concern. The first example of such reordering of priority was the extensive inspection effort initiated by TVA following the reheat steam line failures at Southern California Edison's Mohave Station and Detroit Edison's Monroe Station. The second reordering occurred during operation of one of the fossil plants in 1990 when the differential pressure indication from the main steam line flow nozzle was lost. The flow nozzle was located in a vertical section of piping. It rests on a ledge and is secured by four mechanical pins. Subsequent investigation revealed that the pipe had swollen to such an extent that the flow pins were disengaged and the nozzle fell, eventually lodging in the next girth weld backing ring. Diameters measured by pi tape in the flow nozzle region varied along the length with a peak diameter swelling of five percent. The extent of swelling led to two actions. First, flow nozzle inspections were performed at other plants on a priority basis. Second, renewed interest was given to the utilization of pi tape measurements which had been gathered systematically over a large number of years. The current paper presents an overview of the flow-nozzle failure investigation and steamline assessment by strain measurement.

## Pipe Swelling in Flow Nozzle Region

A comprehensive review of the flow-nozzle swelling including the repair, restrictions on operation and planned replacement are given elsewhere (1) and are only summarized herein. Gallatin is a four unit, fossil-fired plant with Combustion Engineering boilers producing steam for Westinghouse turbo-generators. Units 3 and 4 are rated at 275 MW with design steam conditions of 566C/14.8 MPa main steam and 566C/3.3 MPa reheat steam. The maximum continuous steam flow rating is 889,000 kg of steam per hour.

On September 5, 1990 the main steam flow indicating differential pressure signal on TVA's Gallatin Unit 3 went to zero while the unit was at full load. An investigation by the plant maintenance personnel determined that the problem was not an instrument malfunction but a loss of differential pressure at the source.

The source of the main steam flow signal was an internal restricting device or flow nozzle shown in Figure 1 and was located above the main floor in a vertical section of straight pipe. The nozzle was retained in place by a locating ring and by four pins protruding through the pipe wall into the nozzle. The locating ring was fillet welded to the inside of the pipe, and the pins were seal welded on the pipe O.D. There were two pressure taps. They were located one above and one below the flow nozzle. The signal from these pressure differential sensors were used for feedback to control the automatic boiler firing system, which was especially critical during start-up.

During its original manufacture, this section of steam pipe received an additional machine turning operation on the inside surface to ensure a uniform, accurate diameter both upstream and downstream from the nozzle. This machining was performed for a distance of approximately 3 meters and caused this section to be very close to the minimum wall for most of this length. Following the locating of the flow nozzle by video probe through a steam sampling penetration, the insulation from a section of pipe approximately 24 meters long was removed to allow for inspection and restraining of the pipe. Restraining was required to prevent the loss of cold spring and to maintain the alignment for refitting the pipe section.

Inspection found cracks at the 4 seal welds for the nozzle retaining pins, the 2 pressure tap welds, 2 of the 3 gamma plug welds, some of the hanger lug attachment welds, and the 2 steam sampling plug welds. The steam sampling plug welds had been removed earlier to allow for the inspection inside the pipe and were found to be severely cracked. It appeared that the plugs would have leaked or blown out at some time in the near future. The hole left by removing the steam sampling plug was visibly distorted. The hole dimensions were measured using a telescoping gauge and micrometer in both the longitudinal and circumferential directions, and the percent swelling was calculated based on the original dimensions of the threaded tapped hole. The calculated swelling percentages are given in Table 1:

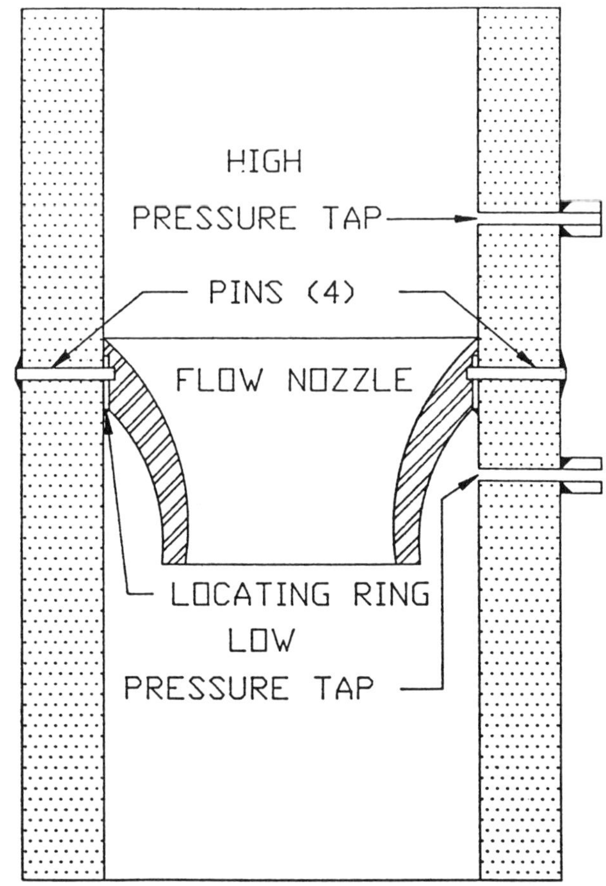


Fig. 1. Cross section drawing of flow nozzle region showing flow nozzle, pressure taps and pins.

Table 1. Steam Sampling Penetration Swelling.

| | Swelling - % | |
|---|---|---|
| | Longitudinal | Circumferential |
| Near Inside | 4.3 | 14.2 |
| Near Outside | 7.1 | 14.2 |

The swelling in the longitudinal direction would not be expected for internal pressure loading alone and indicates the possible influence of bending moments. The 14% circumferential swelling for the pipe penetration was approximately three times the measured strain of 5% for the pipe body.

After the section of pipe was removed, the flow nozzle and locating ring were lifted out of the pipe end. The ring is shown in Figure 2. The edge of the ring welded to the pipe inside surface was expanded by the swelling of the pipe. The other edge, being free, remained near the original diameter. This is the most dramatic evidence of the magnitude of change of pipe inside diameter.

Fig. 2. Photograph of flow nozzle following removal.

The pipe at the retaining pin location had an obvious bulge. A less severe second bulge was just above the first. Pi tape diameter readings were taken to determine the extent of swelling on the lower bulge. For comparison, measurements were taken just above the girth weld, just below the nozzle retaining pins, and about 1 meter below the girth weld on the next spool piece. Even though no original measurements were available, sufficient accuracy was obtained by using the nominal pipe diameter from the piping contractors spool piece drawings or the maximum allowed for that nominal size increased to the maximum allowed by the ASTM piping specification. The pipe swelling given in Table 2 was calculated assuming that the original diameter was at the maximum allowed by the original piping specification.

Table 2. Pipe Swelling Data.

| Measurement Position | Pipe Swelling - % |
|---|---|
| Girth Weld | 3.4 |
| Retaining Pins | 5.0 |
| 1 meter Below Girth Weld | 2.4 |

The ultrasonic thickness measurements in the swollen area corresponded well with a calculated thickness reduction caused by a 5.6% increase in outside diameter while maintaining a constant cross sectional area. Calculations show that under this assumption, the inside diameter increased more than the outside diameter.

Because of the severity of the pipe swelling and cracked welds found, it was decided to examine the entire main steam piping system. This included removing the asbestos insulation, cleaning all welds for inspection and performing NDT inspections of all welds. The usual steam line inspections were performed with the addition of extra replicas at each girth weld.

## Steam Line Inspection Results

The inspection of the entire steam piping system found several areas of concern. Figure 3 is an isometric weld map showing the relative main steam piping configuration and notes on the inspection results.

The results of the replica evaluation found that 4 welds contained oxide filled creep cracks. Microcracking was found in 5 girth welds, and interconnected cavitated inclusions were found in the weld metal of 1 weld and the base metal adjacent to 7 welds. The indication orientations were all in the axial direction of the pipe and were the result of long term creep from pressure induced stresses.

Brinell hardness values were obtained using an Equotip portable hardness tester at each replica location in the base metal, weld metal, and in the heat affected zone (HAZ). The minimum hardness reading found was 106 HRB in the base metal adjacent to the girth weld just below the flow nozzle. Other base metal hardness readings ranged from 110 HRB to 140 HRB and averaged 121 HRB. The weld metal hardnesses ranged from 110 HRB to 184 HRB and averaged 145 HRB.

A portable energy dispersive x-ray alloy analyzer was used to verify the correct alloy for all weld and base metals. All the piping system weld and base metals were confirmed to be 2-1/4Cr-1Mo steel.

The magnetic particle inspection found cracks in many piping attachments. Cracks were found in hanger lug welds, stanchion support pipes, and small diameter through-wall penetrations such as thermocouple wells, branch connections, and radiographic inspection plugs.

Thickness measurements were taken at several locations around each girth weld. The original piping supplied by NAVCO was specified to be .5842 meter outside diameter and .4016 meter inside diameter, forged and bored pipe in the straight sections. The sections for bending were .5969 meter in diameter. The ASME B31.1 minimum wall thickness was 86.9 mm using the design conditions of 567C and 14.8 MPa and a zero corrosion

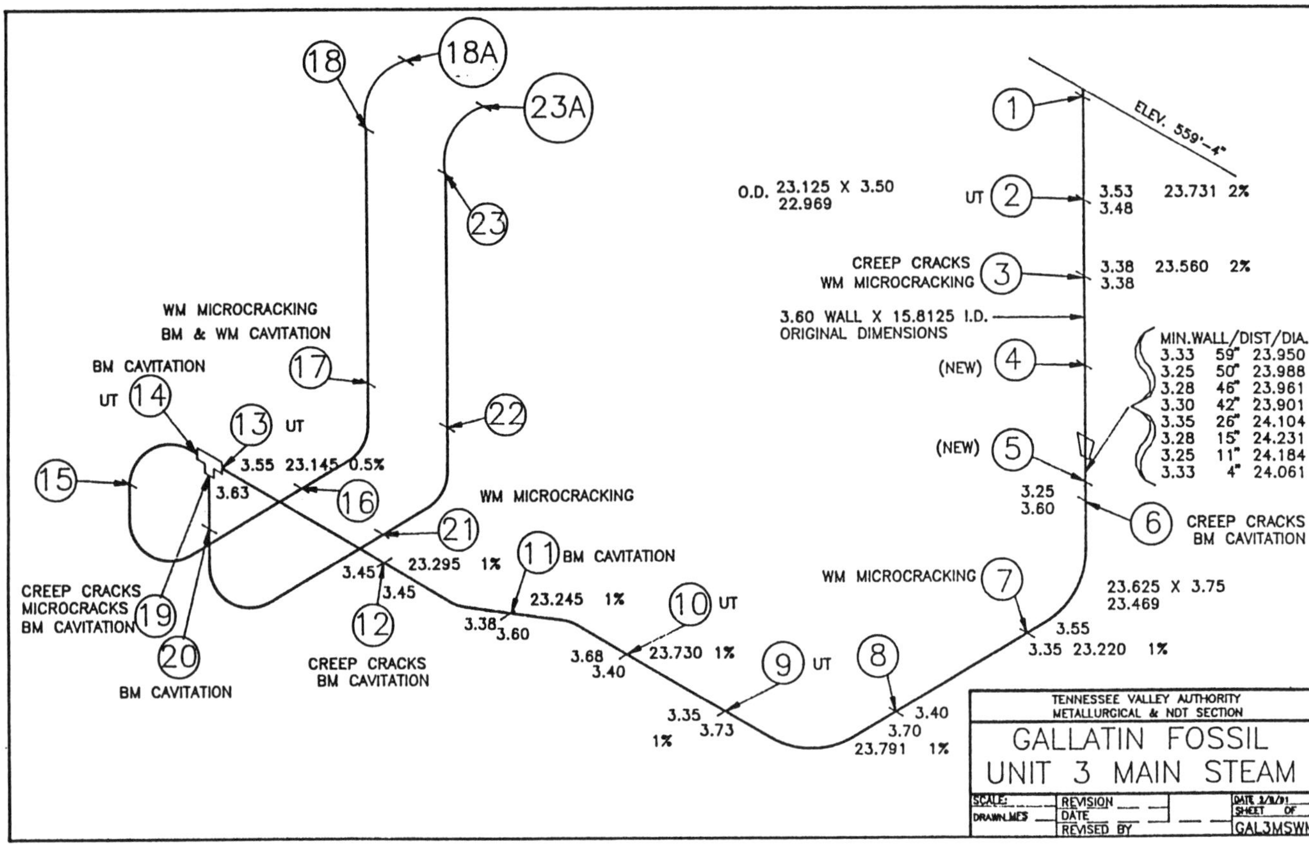


Fig. 3. Isometric main steamline weld map showing steamline inspection results.

allowance. A corrosion allowance of 1.6 mm would make the minimum wall 88.5 mm. Thickness readings below this were found in the area of swelling at the flow nozzle and at the 6:00 position at several butt welds in a horizontal pipe run.

As detailed in Figure 3, the pi tape measurements for piping sections other than the swollen area at the flow nozzle, showed pipe swelling ranging from 1% to 2% in the nominally .584 meter diameter sections. For the .419 meter diameter piping, the indicated strains were less than 1%.

## Analysis of Pipe Swelling

The analysis of the swelling of the Gallatin Unit 3 main steam piping was performed using two different methods. The first comparison made was the apparent swelling strain to the uniaxial creep strain data and tertiary creep equation predictions while the second comparison was a life fraction calculation using the design stress and operating hours to the equivalent time for stress rupture failure. Both of these methods indicated that service life of the main steam piping was consumed beyond half life.

**Strain Based**. The graph in Figure 4 was used in the evaluation of the pipe swelling data. This graph was developed using published creep rupture data for 2-1/4Cr-1Mo annealed steel published by the Institution of Mechanical Engineers in England (2). This data indicates that approximately half of the life is consumed at 2% creep strain and two-thirds of life is consumed at 5% creep strain. This life fraction value at 2% strain is equal to that found using Cane's tertiary creep equation (3) with values typical for 2-1/4Cr-1Mo steel of rupture elongation of 30 percent and Monkman-Grant constant of 3 percent. Using Cane's tertiary creep equation with appropriate constants, a life fraction of .84 would be calculated at 5% strain. Another creep equation based on a linear relationship between log strain rate and linear strain in the tertiary creep regime (4) yields similar results to Cane's equation. This equation can be written as

$$\epsilon = -A \ln (C - Ct/t_r) \qquad (1)$$

where

$\epsilon$ is the creep strain
A is a constant similar to the Monkman-Grant Constant
C is a constant
t is time
and $t_r$ is the rupture time.

Using C equals 1 and A equals 3%, Equation 1 predicts life fractions of .5 and .81 at creep strains of 2% and 5% respectively. The tertiary creep equations are compared to the Institution of Mechanical Engineers data in Figure 4.

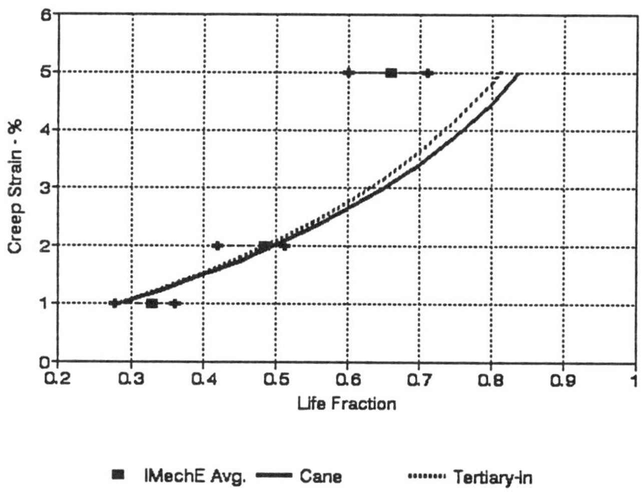


Fig. 4. Creep strain versus life fraction relationships for 2-1/4Cr-1Mo steel.

The magnitude of swelling was an important factor in the condition assessment process. The swelling in the bulged area was very high at 5% and was an immediate concern. Other areas were found to be swollen about 2%. Based in part on the considerations presented later, TVA currently uses the 2% swelling value as an indicator that partial or entire replacement and/or operational changes should be considered.

**Stress Based**. Equation 2 was used to calculate the time to failure, and subsequently, the life fraction consumed. The equation constants were derived from Larson-Miller curve fits to data from the USA, UK, and Japan. The calculated time to failure was then divided by the number of operating hours accumulated by the particular unit to give the life fraction.

$$\log t_{min} = A_0 + A_1/T + A_2 (\log S + 1.65\ SEE)/T + A_3 10^{(\log S + 1.65\ SEE)}/T \qquad (2)$$

where

$t_{min}$ is the minimum time to failure in hours
T is absolute temperature in Rankine (C = (F-32)/1.8)
S is the stress in KSI (1 KSI = 6.895 MPa)
SEE is the Standard Error of Estimate in log stress
$A_0$ = -18.2
$A_1$ = 39993
$A_2$ = -3912
$A_3$ = -159
SEE = 0.049444

Equation 2 obtains an estimate of the minimum rupture time by using the SEE in log stress not in log time. For isothermal log stress log rupture time curves with negative curvature, this method yields prediction for extrapolations to lower stress which are not as pessimistic as those obtained using an SEE in log time. Using Equation 2, the calculated minimum rupture life was 248,000 hours for the .584 meter diameter pipe and based on an operating time of 221,000 hours, the calculated life fraction was 0.89.

Although the predictions of Equation 2 may be pessimistic due to oxidation effects on rupture life for small diameter specimens tested in air, it should be noted that Equation 2 is consistent with the ASME Code Allowable stresses and calculations using this equations lead to the expectation of creep life problems in the 200,000 to 300,000 hour time period for 2-1/4Cr-1Mo operating at 566C and code stress of 40 MPa. The same material operating at 538C and at the code stress (53.8 MPa) is not likely to encounter the same problem until 300,000 to 400,000 hours of operation.

## Results of Inspections on Other Units

In order to prioritize the planned flow nozzle inspections, life fraction calculations were performed on the 59 main steam and 29 hot reheat piping systems. Minimum stress rupture properties, consistent with the most recent ASME Code assessment of elevated temperature strength, were used in the analysis. Partial vindication of the technique was contained in two observations. First, the plant with the swollen flow nozzle had the highest life fraction consumed - 0.89 versus 0.82 for the next highest plant. Second, a plot of the maximum diametral strain from TVA's data base versus life fraction consumed, while displaying substantial scatter, showed a reasonable correlation between the two variables.

The main steam flow-nozzle location of all other twenty-eight 566C units were scheduled to be inspected for swelling. To date, 12 of these inspections are complete. The inspections did not reveal any other units with swelling to the extent found on Gallatin Unit 3; however, the maximum swelling found was 2.2% on Gallatin Unit 4, the sister unit to Gallatin Unit 3. This unit's entire main steamline was inspected due to the extent of swelling found. The results were similar to those found on Unit 3, and the main steam piping is scheduled for replacement during the fall of 1992. The remaining units ranged from 1% to 1.9% with an average swelling of 1.29%. While these inspections found less-serious swelling, 2 large cracks were found on one unit. The cracks were located at girth welds and were apparently caused by bending stresses. These were repaired.

On 6 of the 12 units inspected, the high temperature reheat flow-nozzle region was also inspected. 1.85% swelling was found on one unit and this section will be replaced. The other five showed about 1% swelling.

## Pipe Swelling Limit

The use of creep strain measurements for remaining life assessment of boiler components has been recently reviewed by Mlynarski and Taler (5). There were three approaches discussed including the Monkman-Grant relation, the tertiary creep equation suggested by Cane (3) and the use of an exponential primary plus steady state creep equation with a maximum inelastic strain limit. Values for this maximum creep strain of 1 and 2 percent have been typically used depending on the degree of conservatism desired and guidance from possible regulatory guidelines. The same problem arises in the current study because in order to interpret the flow nozzle inspection data, it was necessary to establish an acceptable upper strain limit for continued operation within design conditions. The approach taken included analysis of the uniaxial creep ductility for Cr-Mo steels supplemented by engineering judgement and service experience.

The analysis of the uniaxial creep rupture data is relatively straightforward. The data base used was developed by the National Research Institute of Japan (6,7) and has long term data as well as multiple heats of steel and several products forms - plates and tubes.

The selection of the regression model was based on observation, prior results, and need. First, it was observed that the isothermal plots of log rupture elongation versus log rupture time given in the NRIM data compilation developed a nearly linear relation although possessing a high degree of scatter. Second, Viswanathan and Fardo (8) have shown good correlation for the log of average rate of elongation, i.e. total elongation/rupture time, and the log of the rupture time for a single heat of 1-1/4Cr-1/2Mo steel. Their relationship can be rewritten to be equal of the simple log-log relation illustrated by the NRIM data. Lastly, the purpose of the analysis was to aid in setting a limit on elongation at typical service times and not the correlation of elongation with operating stress and temperature.

Figure 5 shows the calculated and observed rupture elongation versus rupture time on a log-log plot for 2-1/4Cr-1Mo tube and plate material at 600C. The relation is written as

$$\log e_r = 2.3567 - .21346 \log t_r \qquad (3)$$

where

$e_r$ is the rupture elongation in percent
and $t_r$ is the rupture time in hours.

The $R^2$ values for this equation was only 0.57 but the standard error of the estimate, SEE, was 0.10. The minimum shown in Figure 5 was estimated by subtracting 1.65 SEE from the mean. The residuals in log $e_r$ (observed minus calculated) were randomly distributed around zero with no systematic trend towards under or overprediction at long times indicating reasonable extrapolation would be expected. Similar results in terms of slope were obtained using the combined tube and plate data at 550C but the data lay slightly above the 600C data. This would suggest that a model with a temperature dependent intercept term may be a good description.

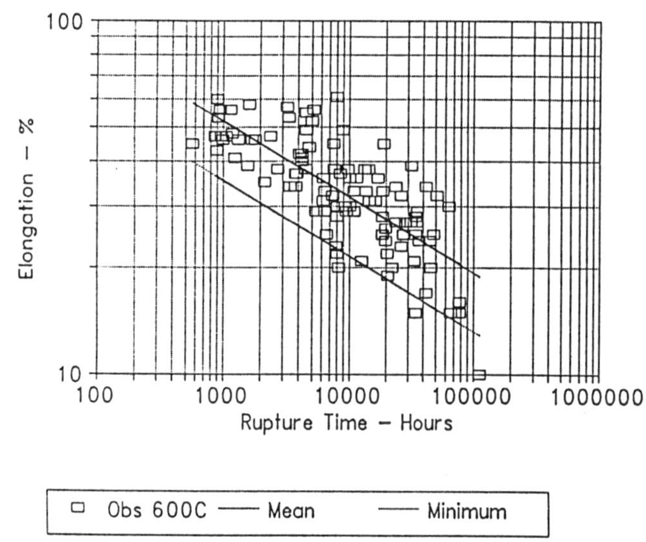


Fig. 5. Comparison of calculated and observed rupture elongation versus rupture time data for 2-1/4Cr-1Mo steel at 600C.

Figure 6 shows the comparison of observed and calculated rupture elongation versus rupture time data for 2-1/4Cr-1Mo tube and plate material at 550C and 600C using the temperature dependent model written as

$$\log e_r = 1.7587 + 580/T - .2345 \log t_r \qquad (4)$$

where

$e_r$ is the rupture elongation in percent
$t_r$ is the rupture time in hours
and T is the absolute temperature in Kelvin.

There were 127 data points used in this regression with the SEE equal 0.094 and an $R^2$ value of 0.69. This model calculates minimum uniaxial rupture ductilities greater than 10 percent for typical steamline operating temperature of 538C and 566C and out to service times of approximately 300,000 hours.

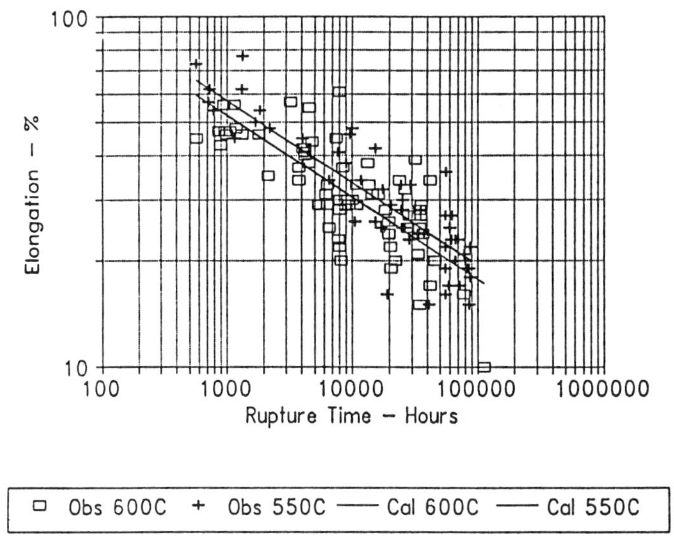


Fig. 6. Comparison of calculated and observed rupture elongation versus rupture time data for 2-1/4Cr-1Mo steel at 550C and 600C.

Engineering judgement is required to set realistic strain limits due to considerations of the multiaxial stress state, uniform versus total elongation, and cracking associated with welds or penetrations. The multiaxial stress state typical of a creeping internally pressurized cylinder tend to decrease the uniaxial based strain limit in two ways. First, the ductility of the material is reduced for multiaxial stresses (3,5) compared to uniaxial stress and depends on the degree of triaxiality of the stress. Second, the effective strain rates are higher in the interior of a pressurized cylinder undergoing steady state creep that at the pipe O.D. where the strain measurements are made (5). Another factor that makes the uniaxial data analysis nonconservative was the use of total elongation rather than the .2% offset tertiary strain which would be more representative of uniform deformation. The final and most important consideration in setting a realistic strain limit was cracking and failure at penetrations and welds. The expected creep damage mechanism in ferritic base material is structural coarsening leading to strain accumulation (9). Therefore, strain assessment methods are considered appropriate for remaining life estimates of thick section components (9,10). The ASME Code stress used in the stress based life fraction calculation should be applicable to the rupture of an unpenetrated pipe. To account for the stress concentration due to the penetrations, a circumferential ligament efficiency factor is often used in life assessment. Full scale header creep tests indicate that higher hoop strains (by approximately a factor of 2) are measured in the ligament region of the header compared to the unpenetrated pipe regions (11). In another test, an accelerated creep test on a service exposed header failed by steam leakage at a stub tube penetration at a hoop strain of approximately 6% in the header body (12). A similar failure also occurred at a radiograph plug in a 1-1/4Cr-1/2Mo main steamline pipe with approximately 5% strain in pipe body (9). While there is no consensus on the definition of "failure", it would appear that an upper bound on strain accumulation in the pipe body material is a reasonable approach. By the time a homogeneous strain of from approximately 2 to 5 percent is obtained in the pipe, extensive cracking and leaks would have occurred in penetrated, welded or strain concentrated regions.

## Strain Assessment Based on Diameter Measurement

Techniques for measuring creep strain accumulation in piping base material include high temperature capacitance strain gages, welded strain pips or pins, (5) strain grids (13) and pi-tape readings. For detection of gross swelling or comparisons of trends in an overall piping system, pi-tape readings are considered adequate and have a cost savings relative to the other methods.

Nevertheless, there are several concerns with the use of diametrical readings obtained by the use of pi-tapes for creep swelling measurements. A major concern regards the lack of specific information on the as-built dimensions of the steamline piping in most instances. Two other effects which influence the measured diameters are the presence or absence (ground or grit blasted removal) of air side oxide scale and the possible non-uniformity in strain around the pipe circumference due to piping system stresses. The effect of system stress (applied bending moment) on circumferential variations in creep strain accumulation were addressed by Wells, Lange and Roberts (14) and the other two issues are discussed here.

There were measured diameter and wall thickness values for the main steamline piping for five plants. The piping was specified as a .3048 meter inside diameter and 63.5 mm wall thickness yielding a nominal .4318 meter outside diameter. Because of the magnitude and fact that the tolerances were specified on the wall thickness and inside diameter, the actual outside diameter would be expected to exceed the nominal diameter plus permissible variations suggested by the ASTM specifications for alloy steel pipe.

Figure 7 shows the histogram for the as-built diameter for the main steam piping from these five plants. The average diameter was 436.49 mm and the standard deviation was 0.40 mm. The X axis of the histogram has been plotted in deviations in units of standard deviation i.e. (measured minus average) divided by standard deviations, so the observed can be compared to that expected for the Gaussian distribution function. Except for the larger number of observations near mean than predicted by Gaussian, the diameter data is reasonably represented by the Gaussian distribution function.

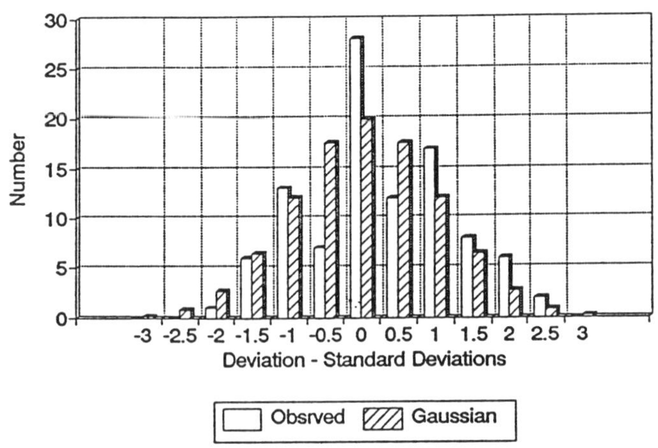


Fig. 7. Histogram for as-built diameter of main steamline piping compared to Gaussian distribution.

In addition to pipe diameter increases in service caused by creep swelling, there is an increase due to the high temperature oxide scale growing with time at temperature. Inspections have been traditionally performed on the pipe areas in the vicinity of girth butt welds. The air side oxide scale has been removed in these areas. The air side scale is similar to the steamside scale both in kinetics and morphology. The scale consists of a dense inner portion termed penetration next to the metal and a porous outer layer. The interface between the two marks the position of the original outside diameter position and therefore should be used in strain calculations. Thus, to use pi-tape readings on descaled pipe, a correction factor should be added to the measured diameter to account for the loss of the penetration portion of the scale. Equation 5 gives a scale correction relation

$$D = D_u + 2X_p\,(t,T) \qquad (5)$$

where

D is the current pipe outside diameter
$D_u$ is the unscaled pipe diameter
and $X_p$ is the depth of the air scale penetration as a function of time, t, and temperature, T.

Correlations for total steam scale thickness have been reviewed by Viswanathan, Foulds and Roberts (15). The linear steamside scale growth kinetics relation for 2-1/4Cr-1Mo steel would calculate a total scale thickness of approximately 750 micrometers and 350 micrometers after 150,000 hours of service at 566C and 538C, respectively. Although the magnitude of the scale correction is not large, failure to apply the correction would result in a systematic error in calculated strain. Assuming the air side and steamside scale growth rates are similar, the linear kinetics cited above could be used to obtain estimates of the total scale thickness which could then be multiplied by a factor to obtain estimates for the penetration depth to be used in Equation 5.

## Conclusions

Based on the results of this study, the following conclusions were reached:

- The pipe swelling in the flow nozzle region was attributed to the pipe section wall thickness being manufactured close to the minimum, the creep properties falling very close to the bottom of the stress rupture scatter band and probable bending moment effects.
- An analysis of the uniaxial ductility for the 2-1/4Cr-1Mo steel using a temperature dependent relationship between the log of the rupture strain and the log of the rupture time, indicates that in the time and temperature regime of interest, total elongation at rupture should exceed ten percent.
- Based on considerations regarding accuracy of strain measurement, multiaxial stress states, and cracking associated with welds or penetrations, TVA currently uses a 2% swelling value as an indicator that partial or entire replacement and/or operational changes should be considered.
- Statistical analysis performed on a limited data base indicate that the as-built diameter measurements show reasonable conformity to the Gaussian distribution function.
- Pi tape diameter measurements, by indicating the creep strain, are a valuable tool in the analysis of steam lines. A method to correct diameter measurements for the removal of the air side scale was developed. Regarding dimensions, the calculations of swelling can be made using the original nominal specification diameter even if original exact measurements are not available. The diameter tolerance of the original piping specification can be used to estimate the range of possible swelling. Of course this is a more difficult task where the original pipe dimensions were based on an inside diameter and minimum wall thickness. However, sufficient accuracy is still available to indicate gross creep swelling, especially over an entire steam piping system.
- After 200,000 hours of operation at 566C, a comprehensive inspection is essential for the continued safe operation of steam piping.

## References


1. M. F. Steakley, B. W. Roberts, S. O. Hilton, "Long Term Creep Deformation of Main Steam Piping Operating at 1050°F," Third EPRI Conference on Fossil Plant Inspections, August 13-15, 1991, Baltimore, Maryland.
2. Institution of Mechanical Engineers. The Creep of Steels Working Party. High Temperature Design Data for Ferritic Pressure Vessel Steels. Mechanical Engineering Publications Ltd. London, England.
3. B. J. Cane, "Remaining Creep Life Estimation by Strain Assessment on Plant," Int. J. Pressure Vessels and Piping, Vol. 10, 1982, pp. 11-30.
4. F. V. Ellis and J. E. Bynum, "Creep and Tensile Properties of SA508 Class 3 Forging Material," Damage Assessment, Reliability, and Life Prediction of Power Plant Components, ASME, NY, NY, 1990, pp.37-46.
5. F. Mlynarski and J. Taler, "Residual Life Assessment for Boiler Pressurized Components Based on Measurements of Creep Strains," J. of Pressure Vessel Technology, Vol. 110, 1988, pp. 308-313.
6. "Data Sheets on the Elevated-Temperature Properties of 2.25Cr-1Mo Steel for Boiler and Heat Exchange Seamless Tubes," National Research Institute for Metals, Tokyo, Japan, 1986.
7. "Data Sheets on the Elevated Temperature Properties of Normalized and Tempered 2.25Cr-1Mo Steel Plates for Pressure Vessels," National Research Institute for Metals, Tokyo Japan, 1980.
8. R. Viswanathan and R. D. Fardo, "Parametric Techniques for Extrapolating Rupture Ductility," Ductility and Toughness Considerations in Elevated Temperature Service, ASME, NY, NY, 1978, pp. 11-19.
9. F. V. Ellis, J. F. Henry, B. W. Roberts, M. C. Askins, M. S. Shammas, and B. J. Cane, "Remaining Life Estimation of Boiler Pressure Parts, Vol. 1: Identification of Relevant Damage Mechanisms," EPRI CS-5588, Vol. 1, Electric Power Research Institute, Palo Alto, CA, 1988.
10. J. D. Parker and D. Sidey, "Residual Life Assessment of Boiler Components," Metals Forum, Vol. 9, No. 1 and 2, 1986, pp. 78-89.
11. J. K. Hepworth and J. A. Williams, "The Assessment of Remnant Creep Life Estimation Methods Using Full Size Component Tests," Central Electricity Generating board, CEGB Report TPRD/M/1619/R87, March, 1987, CEGB, Marchwood, U.K.
12. R. V. Day, Rowley, T and J. A. Williams, "Examination of the Deformation and Failure Modes of an Overheated 1Cr-1/2Mo West Thurnack Boiler Header Assembly of Complex Form," CEGB Report R/M/R277, 1979, CEGB, London, U.K.
13. C. Eggeler, C. Millward, B. J. Cane, "A Remnant Life Assessment Technique Based on Local Strain Measurements Using Replication," Fossil Power Plant Rehabilitation, ASM International, Materials Park, OH, 1989, pp. 13-18.
14. C. H. Wells, C. H. Lange and B. W. Roberts, "Creep Analysis of a Thick-Walled Pipe Under Bending and Internal Pressure," presented at Fifth International Creep Conference on Creep of Materials, Orlando, FL, 1992.
15. R. Viswanathan, J. R. Foulds, and D. I. Roberts, "Methods for Estimating the Temperature of Reheater and Superheater Tues in Fossil Boilers," Presented at EPRI Conference on Boiler Tube Failures in Fossil Plants, Atlanta, GA, Nov. 1987.

Proceedings of the Fifth International Conference on Creep of Materials, Lake Buena Vista, Florida, USA, 18-21 May 1992

# Study on Remaining Life of 12CrMo, 15CrMo Steam Mains Operating Beyond Design Life—Effected by Creep-Fatigue Interaction

**W.F. Ven, K.Y. Jun, L.Y. Jun, S.G. Gon, and L.S. Hua**
Thermal Power Engineering Research Institute
Ministry of Energy
Xian, China

ABSTRACT

Based on extensive investigations and experiments, this paper is de voted to the study of remanant life of 12CrMo, 15CrMo steam mains used for power plant peak load adjustment. Such steam mains are under the effect of creep-fatigue interaction. Samples were tacken from steam mains which have been working at high temperature for(2.2 - 2.5) x $10^5$ hours. The creep-fatigue interaction tests were conducted at high temperature and under controlled load conditions with dwell time. Metallographic examinations were carried out to determine the extent of creep and fatigue damages and the deterioration of microstructure. As a direct result of this study, more resonable start-up and shutdown procedures have been proposed and various failure criterion suggested.

WITH the shift of power consumption pattern in China, there has been a greater need for peak load adjustment in power plant. Generating units used for regular power supply in the past are now being used as stand-bys for the peak load adjustment, which feature frequent start-up and shutdowns. Therefore, it is urgent to find out a way to evaluate the extent of life consumption of the steam mains of the high temperature high pressure generating units running beyond their design lives and working under frequent start-up and shutdown conditions. The solution to this problem will contribute to the safe running of large number of such generating units under new working conditions in our country.

As far as the steam mains of the main power supply generating units are concerned, their life expectancy is determined chiefly by creep consumption, as the number of start-up and shutdown is small and the load of the units remains constant or with little fluctuation. But it is another picture for the stand-by units: they have to be frequently started up and shut down and the load fluctuates significantly. In these circumstances, the effect of stress variation, i.e., the effect of fatigue, must be tacken into consideration. And the life consumption becomes the consumption under the interaction of creep and fatigue. More precisely, it is the interaction of fatigue caused by frequent start-up and shutdown and load variation and the creep caused by in the period of stable running.

The damage mode caused by creep life consumption alone differs from that caused by the interaction of creep and fatigue. The former is mainly damages of creep voids at the grain boundaries, such as creep voids and creep cracks. The latter is a mixture of transgranular and grain boundary damages with the following forms of interaction:

FCF: Fatigue crack initiation accompanied by creep crack initiation, the fatigue cracks propagate preferentially and lead to the final fracture.

CFF: Creep crack initiation accompanied by fatigue crack initiation, the propagation of fatigue cracks leads to the final fracture.

CFC: Creep crack initiation accompanied by fatigue crack initiation, the propagation of creep crack leads to the final fracture.

The above forms of combination reflect not only just the interaction but the extent of influence each element exerts, i.e., whether the chief damage mode being the creep damage or the fatigue damage.

The difference in damage mode leads to difference in life consumption and in the method of life predication. With the above points in mind, we decide the purpose of this article is to:

(1) determine the life consumption of the 12CrMo, 15CrMo staem mains running beyond their design lives and working under the interaction of creep and fatigue and to propose a reasonable start-up and shutdown procedures for the power plant peak load adjustment.

(2) study the laws of deformation and the microstructure change of the 12CrMo, 15CrMo steam mains running beyond their design lives and working under the interaction of creep and fatigue and to suggest failure criteria for them.

## TEST MATERIAL AND TEST METHOD

TEST MATERIAL. The test material for the creep and fatigue life study was taken from 12CrMo, 15CrMo steam mains running beyond their design life. The related parameters and their chemical composition are listed in table 1 and 2 respectively.

POWER PLANT WORKING CONDITION INVESTIGATION. A thorough investigation was made on the generating unit's start-up and shutdown patterns and load variation during 24 hours. The investigation results formed the basis for deciding the test load spectrum of the creep-fatigue interaction tests. Fig 1, 2 and 3 are curves of start-up, shutdown and load variation of the generating unit respectively.

DECIDING LOAD SPECTRUM WAVE FORM. From the investigation it is known that the major load variation occurs during the start-up and shutdown operation. Therefore, the load spectrum of the analogical creep-fatigue interaction test must be in trapezoid wave form, which corresponds to the following procedures:

Start-up (loading)→Stable running or with little load variation (load dwell)→Shutdown (unloading)

The above procedure is repeated and is shown in Fig 4.

While determing the test load spectrum, the nature of the cycles, that is whether it is stress controlled or strain controlled, must also be decided, and as well as the following parameters: the average stress $\sigma_m$, the upper load $P_u$, the lower load $P_l$, the dwell time for the upper and lower load $t_u$ and $t_l$ and the rate of loading and unloading. We have the following consideration in these respects:

a. Stress control: in the load dwell period, the load remains constant; stress caused by the internal pressure exists in the steam mains, all the time during the running state, leading to creep deformation. Influenced further by the start-up and shutdown operation, steam mains are under the interaction of creep and fatigue. Strain control: the total strain keeps constant; steam mains are under the interaction of relaxation and fatigue. In our case stress control applies more than strain control.

b. Load dwell time $t_u = 5$ min (during pre-test, it was 10 minutes or more); $t_l = 2$ min (during pre-test it was 5 minutes or more). Creep damage occurs when tensile stress is present; therefore, if we take $t_u > t_l$ so that the creep deformation can fully develop and yet much test time can be spared.

c. The average stress: $\sigma_m = (\sigma_u + \sigma_l)/2$, in our tests, $\sigma_m$ never equalled zero.

d. The rate of loading and unloading was varied to analog the different speed of start-up and shutdown. In our test, we kept the loading and unloading time fixed to 3.2 sec, changing the magnitude of $P_u$ and $P_l$ to obtain the different speed of loading and unloading.

e. The total number of cycles was 2500. From the practical point of view, it represents 48 years of running (420,000 hours) for start-up and shutdown in cold condition once every week, which is more than enough in our cases. If twice every week, which is rather frequent, it represents 24 years of running.

TEST SCHEME. According to the purpose of the study, every section of the steam mains should undergo the following tests:

a. Creep-fatigue interaction test at the same temperature but under different load levels.

b. Creep-fatigue interaction test under the same load but at diffrent temperatures.

c. Creep-fatigue interaction test under different average stress and stress amplitude.

d. Pure fatigue test under the condition of $t_u = t_l = 0$ to contrast the creep-fatigue interaction test.

CREEP FATIGUE INTERACTION TEST AND PURE FATIGUE TEST. Table 3 shows the test parameters. The stress and strain values were measured while the repetition cycles were in progress.

METALLOGRAPHIC STUDY. Samples having been undergone the creep-fatigue test were again taken for metallographic studies to determine their microstructural change and the extent of creep damage and fatigue damage (both qualitative and quantitative).

## TEST RESULTS

The microstructures of sample number series 894 and 895 are shown in Fig 5 and 6 respectively.

CREEP-FATIGUE INTERACTION TEST RESULTS. The relationship between the maximum strain $\varepsilon_{max}$ and the number of cycles N in the creep-fatigue test for the sample series 894 and 895 is shown in Fig 7 and 8 respectively.

METALLOGRAPHIC EXAMINATION OF THE CREEP-FATIGUE INTERACTION SAMPLE. The samples were examined under the TAS Plus instrument to measure carbide average radius and to make sure if there were creep voids and the extent of creep and fatigue damages. The results are listed in table 4.

## DISCUSSION OF THE RESULTS

THE OPTIMIZATION OF THE START UP AND SHUTDOWN PROCEDURES FOR THE GENERATING UNIT. When the generating unit starts up in the cold state, the medium (steam) temperature inside the steam mains rises gradually. Because of heat conduction, the temperature distribution along the pipe section varies with time; as a result, thermal stress is induced. Fig 9 schematically shows, for a straight pipe with outer wall thermal insulation, the change of inner and outer wall temperatures and the induced thermal stress versus time during the start up process. In the figure, $\sigma_w$ represent membrane stress; $\theta_i$, inner wall temperature; $\theta_a$, outer wall temperature; $\Delta\theta(t)$, temperature difference.

Let r represents the coordinate of any point in the wall in the radial direction, the thermal stress

at that point can be expressed as in Eq. 1., 2. and 3.
The axial stress:

$$\sigma_{z\theta} = \frac{\alpha E}{1-\gamma}(\Delta\theta_m - \Delta\theta) \quad (1)$$

The circumferential stress:

$$\sigma_{\varphi\theta} = \frac{\alpha E}{1-\gamma}\left(\frac{r^2+r_i^2}{2r^2}\Delta\theta_m + \frac{1}{r^2}\int_{r_i}^{r}\Delta\theta r dr - \Delta\theta\right) \quad (2)$$

The radial stress:

$$\sigma_{r\theta} = \frac{\alpha E}{1-\gamma}\left(\frac{r^2-r_i^2}{2r^2}\Delta\theta_m - \frac{1}{r^2}\int_{r_i}^{r}\Delta\theta r dr\right) \quad (3)$$

Where:

$\alpha$ - coefficient of expansion
$E$ - modulus of elasticity
$\gamma$ - Poisson's ratio
$\Delta\theta_m$ - average temperature difference

The pipe is cylindrical in shape; the temperature distribution inside the wall is shown in Fig 10.

Calculated on the basis of the volume of the cylindrical body, the average temperature difference $\Delta\theta_m$ can be expressed as:

$$\Delta\theta_m = \frac{2}{r_a^2 - r_i^2}\int_{r_i}^{r_a}\Delta\theta(r) r dr \quad (4)$$

Where $\Delta\theta(r)$ is the temperature difference between any two points inside the temperature field under the metastable state, which, in turn, can be expressed as follows:

$$\Delta\theta(r) = \theta(r) - \theta_i = \frac{V_\theta}{2a}\left(\frac{1}{2}(r^2 - r_i^2) - r_a^2 \ln\frac{r}{r_i}\right) \quad (5)$$

Where $V_\theta$ is the inner wall temperature changing rate, which can be expressed as:

$$V_\theta = \frac{\partial\theta_i}{\partial t} \quad (6)$$

Where t is the time, and a the temperature conduction coefficient:

$$a = \frac{\lambda}{C\rho} \quad (7)$$

Where:

$\lambda$ - heat conduction coefficient
$C$ - specific heat
$\rho$ - density of material

Substitute Eq. 5. into Eq. 4, we have:

$$\Delta\theta_m = \frac{V_\theta}{8a}\left(3r_a^2 - r_i^2 - \frac{4r_a^4}{r_a^2 - r_i^2}\ln\frac{r_a}{r_i}\right) \quad (8)$$

Replace $u = r_a/r_i$ into the above equation and simplify:

$$\Delta\theta_m = \frac{V_\theta S^2}{a}\psi_{zi} \quad (9)$$

Where:

$S$ - wall thickness

$$\psi_{zi} = \frac{1}{8}\,\frac{(u^2-1)(3u^2-1)-4u^4 \ln u}{(u^2-1)(u-1)^2} \quad (10)$$

Similarly, replace the average temperature difference based on sylinder volume by the total average temperature difference $\Delta\theta_a$ and we have:

$$\Delta\theta_a = \frac{V_\theta}{2a}\left[\frac{1}{2}(r_a^2 - r_i^2) - r^2 \ln\frac{r_a}{r_i}\right] \quad (11)$$

$$\Delta\theta_m - \Delta\theta_a = \frac{V_\theta S^2}{a}\psi_{za} \quad (12)$$

Where:

$$\psi_{za} = \frac{1}{8}\,\frac{(u^4-1)-4u^2 \ln u}{(u^2-1)(u-1)^2} \quad (13)$$

By analyzing, it can be seen that the maximum thermal stress occurs on the inner and outer surfaces of the pipe. Substitute Eq. 9 and 13. into Eq. 1., 2. and 3; the maximum thermal stress can be expressed as:

$$\sigma_{zi} = \sigma_{\varphi i} = \frac{\alpha E}{1-\gamma}\cdot\frac{V_\theta S^2}{a}\cdot\psi_{zi} \quad (14)$$

$$\sigma_{za} = \sigma_{\varphi a} = \frac{\alpha E}{1-\gamma}\cdot\frac{V_\theta S^2}{a}\cdot\psi_{za} \quad (15)$$

$$\sigma_{ri}=\sigma_{ra} = 0 \qquad (16)$$

The calculation of thermal stress and internal pressure stress of the steam mains is based on the steam temperature and pressure changing data measured during the start-up process. The circumferential stress $\sigma_{\psi}^{p}$ induced by the internal pressure can be calculated by the following equation:

$$\sigma_{\psi}^{p}= P\left[\left(\frac{d_a}{d_i}\right)^2+ 1\right] / \left(\frac{d_a}{d_i}\right)^2-1 \qquad (17)$$

Where:

P - steam pressure

$d_a$- outside diameter

$d_i$- inside diameter

The inner wall surface thermal stress $\sigma_{zi}$, $\sigma_{\psi i}$ and outer wall surface stress $\sigma_{za}$, $\sigma_{\psi a}$ can be calculated by the Eq. 14 and 15. respectively. The calculated inner wall stress change during the start-up and shutdown operation is presented in its simplied form in Fig 11. In the figure, curve 1 represents thermal stress: curve 2, the internal pressure stress; and curve 3, the composite stress.

From the figure, we know that the most important stress which causes the inner wall fatigue is the compressive stress at point A in Fig 11. On the outer wall surface, the stress is tensile cyclic in nature. As cyclic plastic deformation is the main factor leading to fatigue damage, from the point of view of start-up and shutdown, the inner wall surface is the main region where fatigue failure is likely to occur.

From the figure, the change of stress during the start-up operation is as follows: Before it is started, both the internal pressure and the thermal stress are equal zero. As the steam temperature rises, the thermal stress increases at the same rate as that of the temperature increase. At this moment, the internal pressure stress is low as the pressure is low, and the composite stress is compressive. Then, the temperature increases at the rate of 1°C per minute. After half an hour, the internal pressure stress exceeds the thermal stress, which corresponds to point B in Fig 11. Thereafter, the internal pressure stress greatly exceed the thermal stress. When the stable state is reached, the stress is at its maximum. During the shutdown operation, the temperature falls steadily; the thermal stress quickly arrives at a certain value (tensile stress). The internal pressure stress falls as the pressure decreases (curve 2 in the figure). The composite stress is shown as the curve 3.

The stress curves in Fig 11 are consistent with the test load spectrum, as both have the following features:

(1) The stress curves and the load spectrum are similar in shape and consistent in direction; both are asymmetrical cycles (refer to Fig 4.).

(2) During the load dwell period the stress keeps constant.

The magnitude of the composite stress and its rate of change are determined by the internal pressure and the rate of temperature rise. In our test, the time for loading and unloading was fixed; the levels of the upper and lower loads were varied to analog the range of change. Moreover, as the stress at the upper and lower load dwell period in our test was as 1.5 to 3 times high as the stress actually happens, the rate of temperature change during loading and unloading in the test was much faster than the actual and the high test temperature used, all these would lead to the quickened steps for the creep damage and fatigue damage in the one hand, and the consideration of stress concentration in the other hand.

PROPOSAL FOR THE START-UP AND SHUTDOWN PROCEDURES FOR THE GENERTING UNIT. According to the metallographic examinations, there never discovered any creep damages and fatigue damages during the 2,500 test cycles under the much higher rate of thermal stress change and stress magnitude three times higher than the actual (refer to table 4). It indicates that the steam mains, although having been undergone long hours of running (252,000 hours for the 12CrMo, 15CrMo steam mains) are able to endure further 2,500 times of start-up and shutdown operation at a faster speed than the actual without inflicting any creep or fatigue damages.

The proposed fast start-up procedure is listed in table5, which can reduce start-up time and improve the utilization efficiency of the generating unit.

In the shutdown operation, as there is no pressure stress and if the generating unit allows, it can be shutdown as fast as possible without any risk of fatigue damage.

CREEP-FATIGUE LIFE PREDICATION. In summary, we have conducted a series of tests on the 12CrMo, 15CrMo steam mains, which have been running for 252,000 hours. From the test results, we have the following observations:

(1) Comparing the $\varepsilon_{max}$ - N curves of those having dwell time with those having no dwell time, though they are under the same temperature, $\sigma_m$ and $\sigma_a$ conditions, they have different rate of deformation. While in the pure fatigue test, the deformation almost keeps constant (894-6 in Fig 7 and 895-9 in Fig 8) during the 2,500 test cycles and the magnitude of the deformation is small (0.06% for 894-6 and merely 0.072% for 895-9). For those with dwell time, for example, 894-17 in Fig 7, the deformation is about 1.4%. The difference between them, 1.34%, can be seen as the portion of the creep deformation inflicted during the dwell period.

The lower the temperature and the smaller the $\sigma_m$, the smaller is the deformation. That is consistent with the relationship between the creep deformation and the temperature and stress.

As no fatigue damage has ever been discovered in both the creep-fatigue test and the pure fatigue test in the above stress and temperature ranges, it can be concluded that as long as the temperature and stress amplitudes do not exceed the test range, the mechanism of creep life consumption is the main controlling mechanism of the life consumption under the interaction of creep and fatigue in the 2,500 test cycles. Therefore, the creep life evaluation method can be used to predicate the life expectancy under the interaction of creep and fatigue.

(2) The metallographic examination results show:

a) Comparing the extent of pearlite spheroidization, there has been no significant deterioration of the test samples after the test.

b) The data of the measured average carbide radius (table 4) show that even just for 300 hours of the creep-fatigue interaction test, the enlargement of the average radius has been significant. It means that the frequent start-up shutdown operation will promote the speed up of the microstructural change or the aging of the 12CrMo, 15CrMo steam mains.

(3) From the fact that no creep damage and fatigue damage have ever been discovered in all the test samples listed in table 4, it can be expected that as long as the start-up and shutdown procedures proposed in this article are observed, no damages of any kinds will occur for 2,500 times of start-up and shutdown operations, although the average carbide radius will become larger.

FAILURE CRITERIA. In considering the failure criteria for steam mains under the interaction of creep and fatigue, it was previously thought that the effect of fatigue must be taken into account. However, as the test results show, for the 12CrMo, 15CrMo steam mains which have already been running for 252,000 hours, the creep life failure criteria alone can be used in the next 2,500 times of start-up and shutdown operations, without considering the effect of fatigue.

CREEP DEFORMATION CRITERIA. In the evaluation of creep life expectancy, it has been thought that the 1% creep deformation specified in the "Metal Technology Monitoring Regulation" could be used as the failure criteria for creep life consumption, although, at the same time, it has been thought to be conservative.

In our test, it has been proved that, in the 2,500 number of times of load dwell cycles, the creep deformation amounted to 1.34% (sample No. 894-6) without inflicting any creep or fatigue damages. If we add the 0.73% creep deformation accumulated during the past long hours of running, the total deformation reached 2.07% and it could still be running safely.

Therefore, to evaluate the life consumption under the interaction of creep and fatigue, the 1% creep deformation criterion can be used as a failure criterion for the time being as well. And, still, it is a conservative criterion.

AVERAGE CARBIDE RADIUS FAILURE CRITERION. The test results show that the enlargement in carbide radius is basically in agreement with the following dynamic equation:

$$r_t^3 - r_o^3 = kt \qquad (18)$$

Where:

$r_t$ - average carbide radius after t hours of running

$r_o$ - average carbide radius before the steam mains is put into operation

k - the rate of carbide enlargement

Value k reflects the speed of carbide coalescing of the material. Under the same stress conditions, we have the following relationship:

$$k = k_o e^{RT}$$

or:

$$\ln k = \ln k_o + RT$$

Where:

$k_o$ - speed constant (carbide enlargement

T - temperature

R - constant

From the data in table 4, we have:

For 15CrMo grade of steel:

$$\ln k = 0.0271T - 61.6752 \qquad (19)$$

For 12CrMo grade of steel:

$$\ln k = 0.0567T - 89.1659 \qquad (20)$$

If the operation temperature T is known, the value k can be calculated by using Eq. 19 or 20. And substitute the value k into Eq. 18, the value $r_t$ can be determined. If there is a failure criterion for value $r_t$, the remanant life expectancy can be predicated.

## CONCLUSIONS

SUMMARY OF TEST DETAILS. Samples were taken from the 15CrMo, 12CrMo steam mains which have already been running for 252,000 hours and were tested on a creep-fatigue interaction test machine to study the effect of the creep-fatigue interaction. The test load spectrum was based on a broad investigation of the actual working conditions of the generating unit and on extensive experiments, which was very similar with the actual conditions and had wider ranges for the variation of parameters. Data obtained from the test enable us to make very reliable predications on the remanant life expectancy of the steam mains working beyond their design lives and under the interaction of creep and fatigue. As a direct result, more reasonable start-up and shutdown procedures were proposed for the generating unit.

The 2,500 number of times

of stress cycles used in the test were more than enough for our purpose of life expectancy evaluation.

MAIN FACTOR OF FATIGUE FAILURE. For the life consumption under the interaction of creep and fatigue, the main factor causing fatigue failure to occur is the cycles of plastic deformation. During the start-up of the steam mains, the combination of thermal stress and internal pressure stress induces alternative compress-tensile stress cycles on the inner wall of the steam mains, causing fatigue failure to occur. As the maximum stress is chiefly determined by the internal pressure stress, the magnitude of the alternative stress on the inner wall surface, $\Delta\sigma_i$, is chiefly determined by the negative value stress, i.e., the sum of:

$$\sigma^T_{\psi i} + \sigma^p_{\psi i}$$

The more negative the value is, the more severe is the extent of fatigue damage.

The proposed start-up procedures, derived from an analysis of the test results, are able to limit the internal pressure stress in a certain range, thus reducing the risk of fatigue failure. The enforcement of the proposed start-up and shutdown procedures will not only lead to the reducing the start-up and shutdown time and the improvement in overall utilization of the generating unit, but guarantee the safe running of the steam mains in the next 2,500 number of times of stress cycles as well, without inflicting any creep or fatigue damages.

CREEP-FATIGUE LIFE EXPECTANCY EVALUATION. The test results have shown that, during the 2,500 number of times of cycles, as long as the magnitude of the temperature and the stress did not exceed the test ranges, the creep life consumption mechanism alone was the controlling mechanism for life consumption under the interaction of creep and fatigue. Therefore, the method of creep life evaluation can be used to evaluate the life expectancy under the interaction of creep and fatigue.

CREEP DEFORMATION FAILURE CRITERION. As discussed above, the creep life failure criterion alone can be used to evaluate the life expectancy under the interaction of creep and fatigue. It is suggested that, for the time being, the creep deformation failure criterion, 1% creep deformation , can be used as a criterion for life consumption under the interaction of creep and fatigue.

AVERAGE CARBIDE RADIUS FAILURE CRITERION. The test results indicate that the only bad effect caused by the 2,500 number of times of cycles of the test was the enlargement of the average carbide radius. The test has proved that, for 15CrMo steam main material which has already been running for 252,000 hours, the equation:

$$\ln k = 0.0271T - 61.6572$$

is valid; for the 12CrMo steam main material,

$$\ln k = 0.0567T - 89.1659$$

is valid. If the working temperature T is known, the failure criterion for average carbide radius can be determined from the above equations.

REFERENCES


1- Wu Fee Ven, "A Study of Life Predication Method for Pearlitic Steel Main Steam Pipes in Power Plant in China", International Conference of Advanced Method in Life Predication, Albany (1983)
2- Wu Fee Ven, "Life Expectancy of Power Plant Steam Mains in Abroad", Electric Power Industry Publishing Co., Beijing (1981)
3- Wu Fee Ven, "Behaviour of Heat Resistant Steel in Power Plant", Electric Power Industry Publishing Co., Beijing (1979)
4- Xu Zhi Lun, "Elasticity Mechanics", People's Education

Publishing Co., Beijing (1982)
5- "Alloy Steel Handbook, Vol. 4. Heat Resistant Steel", Metallurgical Industry Publishing Co., Beijing (1983)
6- Bela I. Sandor, "Fundamentals of Cyclic Stress and Strain", The University of Wisconsin Press (1972)

Table 1. Related parameter of the test samples

| Sample No. | Steel grade | Nominal size (mm) | Measured thickness (mm) | Running parameter | No. of start up shutdown | Accumulated running time ( h ) |
|---|---|---|---|---|---|---|
| 894 | 15CrMo | 273x26 | 30 | 510°C/9.8MPa | 134 | 252000 |
| 895 | 12CrMo | 273x26 | 28 - 29 | 510°C/9.8MPa | 133 | 252000 |

Table 2. Chemical analysis of steam mains

| steel grade | chemical composition (%) | | | | | |
|---|---|---|---|---|---|---|
| | C | Si | Mn | S | P | Cr | Mo |
| 15CrMo | 0.13 | 0.22 | 0.57 | 0.023 | 0.014 | 1.01 | 0.43 |
| 12CrMo | 0.13 | 0.25 | 0.57 | 0.020 | 0.015 | 0.59 | 0.45 |

Table 3. The creep-fatigue interaction test list

| Sample No. | Test temp. (°C) | Loading rate (kN/sec) | Load(kN)Dwell (min) $P_u$ | $P_1$ | $T_u$ | $T_1$ | No. of cycles (N) |
|---|---|---|---|---|---|---|---|
| 894-2 | 510 | 7.73 | 12.36 | -12.36 | 10 | 5 | 208 |
| -4 | 510 | 7.73 | 12.36 | -12.36 | 10 | 5 | 422 |
| -14 | 510 | 7.88 | 12.6 | -12.6 | 10 | 5 | 459 |
| -3 | 540 | 3.73 | 6.29 | -5.64 | 5 | 2 | 2091 |
| -12 | 540 | 3.73 | 5.64 | -6.29 | 5 | 2 | 2500 |
| -1 | 560 | 3.73 | 6.29 | -5.64 | 5 | 2 | 2502 |
| -15 | 570 | 3.73 | 6.29 | -5.64 | 5 | 2 | 2692 |
| -17 | 570 | 4.42 | 7.86 | -6.28 | 5 | 2 | 2500 |
| -10 | 570 | 4.91 | 9.42 | -6.28 | 5 | 2 | 2500 |
| -6* | 570 | 4.42 | 7.85 | -6.28 | 0 | 0 | 5000 |
| 895-12 | 540 | 3.73 | 6.29 | -5.64 | 5 | 2 | 2500 |
| -5 | 540 | 3.73 | 6.29 | -5.64 | 5 | 2 | 651 |
| -4 | 570 | 3.73 | 6.29 | -5.64 | 5 | 2 | 1168 |
| -2 | 570 | 3.73 | 6.28 | -5.64 | 5 | 2 | 2514 |
| -11 | 570 | 4.91 | 9.42 | -6.28 | 5 | 2 | 2504 |
| -9* | 570 | 4.42 | 7.85 | -6.28 | 0 | 0 | 5004 |

* pure fatigue

Table 4. Metallographic examination results after creep-fatigue interaction test

| Sample No. | Test temp. (°C) | Average stress (MPa) | Stress amplitude (MPa) | Average carbide radius (um) | Creep Void | Fatigue damage |
|---|---|---|---|---|---|---|
| 894 | / | / | / | 0.212 | No | No |
| 894-3 | 540 | 4.14 | 151.9 | 0.221 | No | No |
| 894-12 | 540 | -4.14 | 151.9 | 0.226 | No | No |
| 894-1 | 560 | 4.14 | 151.9 | 0.232 | No | No |
| 894-15 | 570 | 4.14 | 151.9 | 0.238 | No | No |
| 894-17 | 570 | 10.06 | 180.0 | 0.224 | No | No |
| 894-6** | 570 | 10.06 | 180.0 | 0.207 | No | No |
| 895 | / | / | / | 0.186 | No | No |
| 895-12 | 540 | 4.14 | 151.9 | 0.202 | No | No |
| 895-4* | 570 | 4.14 | 151.9 | 0.198 | No | No |
| 895-2 | 570 | 4.14 | 151.9 | 0.210 | No | No |
| 895-9** | 570 | 10.06 | 180.0 | 0.226 | No | No |

Note: * Cycles times 1168
**Pure fatigue test with no load dwell period

Table 5. The proposed fast start-up scheme

| Pressure range MPa \ Rate of Temp. rise °C/min \ Scheme | I | II | III |
|---|---|---|---|
| 0 - 0.4 MPa | 0.5 | 1.0 | 1.0 |
| 0.4 - 0.5 MPa | 1.0 | 1.0 | 1.0 |
| 0.6 - 1.2 MPa | 2.0 | 2.0 | 2.0 |
| 1.2 - 9.4 Mpa | 3.0 | 3.0 | 4.0 |
| Total strat-up time (h) | 3.17 | 3.08 | 2.75 |
| Per cent of time sparing | 21% | 23% | 31% |

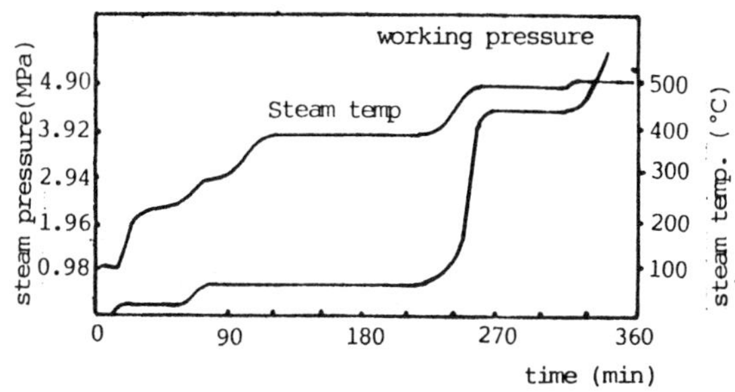


Fig. 1. Generating unit start-up curves

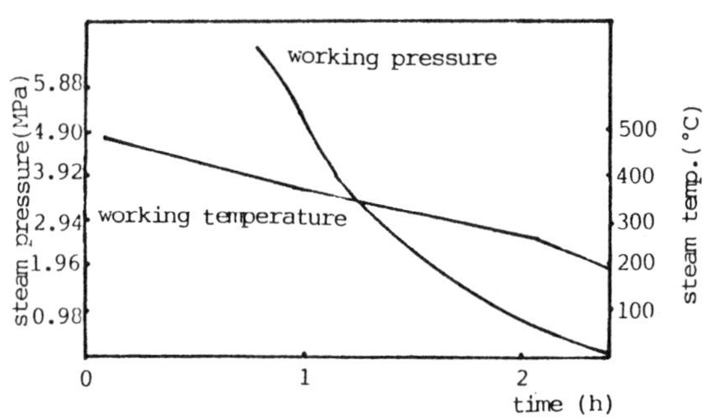


Fig. 2. Generating unit shutdown curves

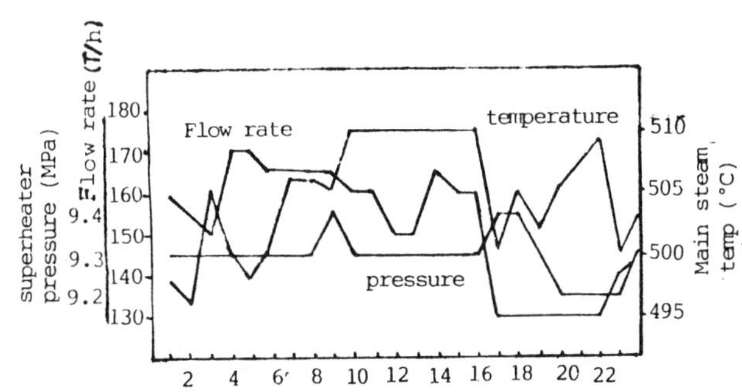


Fig. 3. Load variation during 24 hours

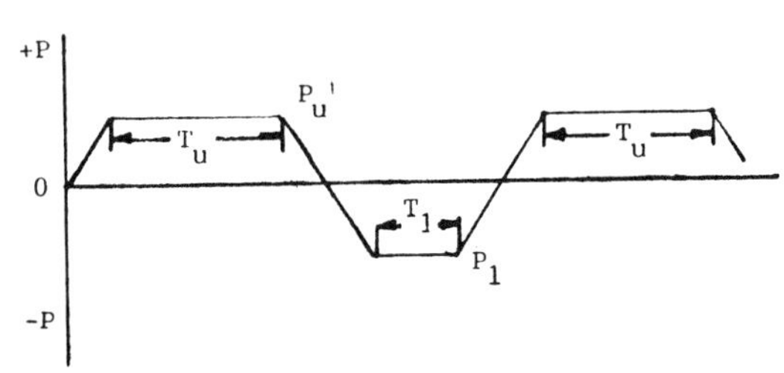


Fig. 4. Load spectrum of the creep-fatigue interaction tests

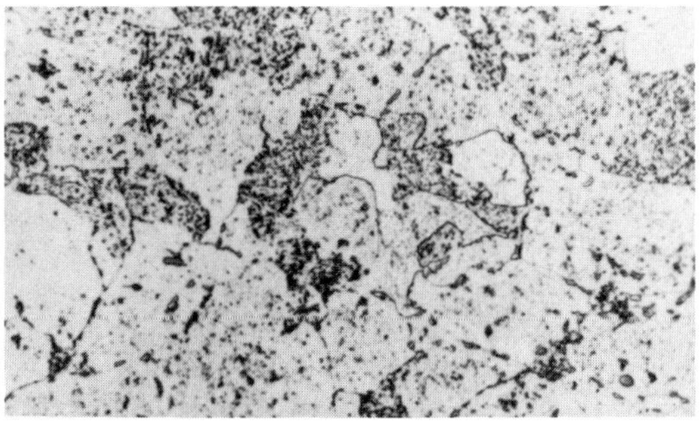

Fig. 5. Microstructure of 15CrMo steam mains after having been running for 252,000 hours 500X

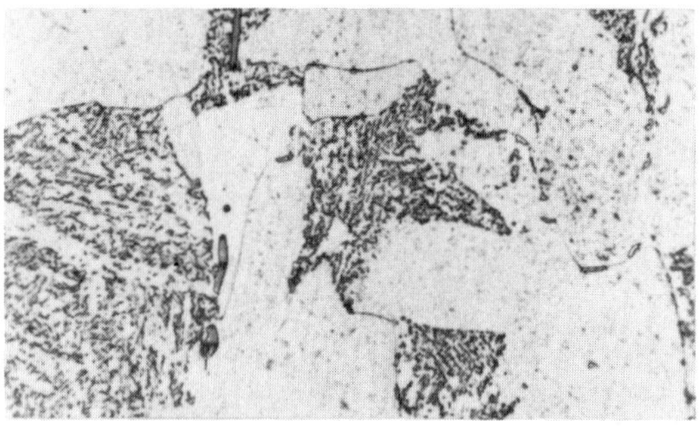

Fig. 6. Microstructure of 12CrMo steam mains after having been running for 252,000 hours 500X

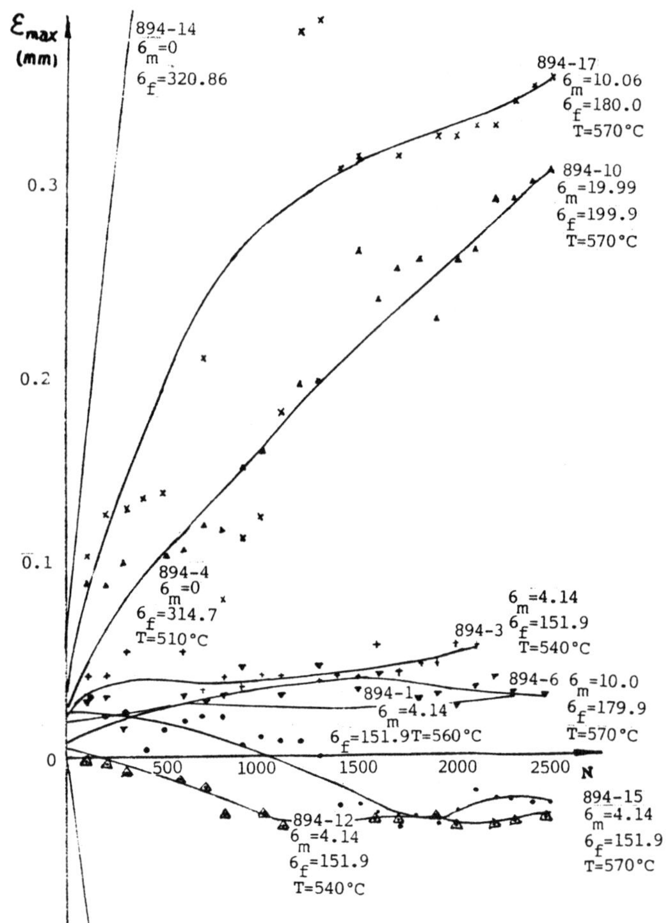


Fig. 7. The $\varepsilon_{max}$ -N curves under the creep-fatigue interaction of the 15CrMo steam mains after 252,000 hours of running

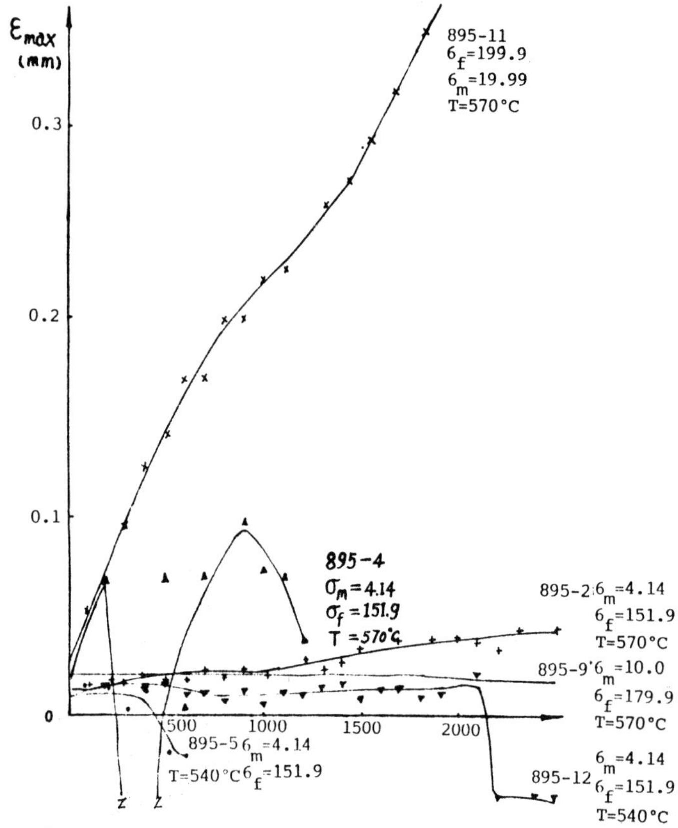


Fig. 8. The $\varepsilon_{max}$ -N curves under the creep-fatigue interaction of the 12CrMo steam mains after 252,000 hours of running

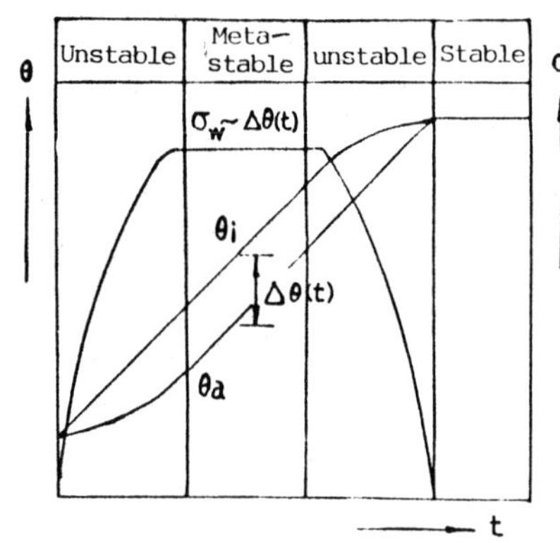


Fig. 9. Schematic temperature thermal stress versus time diagram for thermal insulated pipes during start-up process

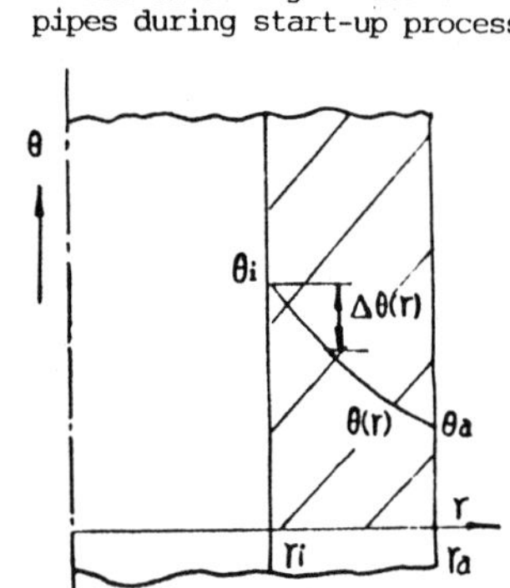


Fig. 10. The temperature distribution inside the cylinder wall

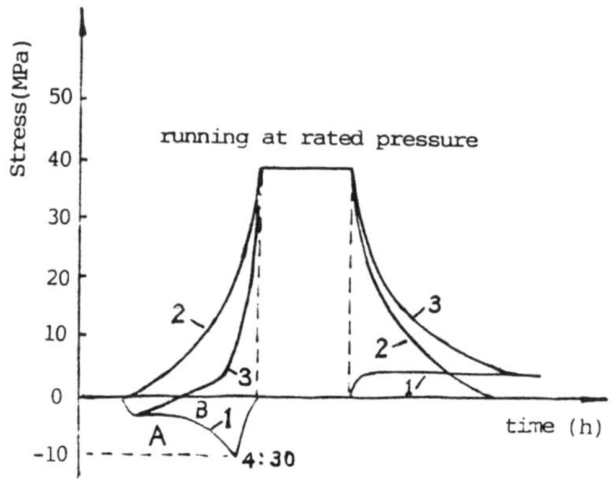


Fig. 11. Inner wall thermal stress, internal pressure and composite stress versus time during start-up and shutdown

*Proceedings of the Fifth International Conference on Creep of Materials, Lake Buena Vista, Florida, USA, 18-21 May 1992*

# A New Method for Predicting the Life of Furnace Tube Operated at High Temperature

**F. Shen and X. Lin**
Nanjing Institute of Chemical Technology
Nanjing, China

## ABSTRACT

This paper describes a new numerical method developed for computer simulation of stresses in furnace tubes used in the synthetic ammonia plant. The method, by introducing the 'damage–coupled mesh' into BEM (boundary element method) analysis and considering the coupling between strain and damage due to the fact that the damage decreases the rigidity and the strength of materials, predicts the damage to the tube material and thus provides an assessment of both tube design life and residual life of tubes that have been subjected to service exposure. The analysis procedure has been incorporated into a computer code (ABEMC), which has been applied to an practice example. The calculation result is efficient in creep damage analysis and residual life predicting of furnace tubes. In addition, an experimental measurement approach for creep damage called natural frequency method (NFM) has been studied which determines the critical damage factor to be essential to the numerical analysis for furnace tube. The equipment and technique of the approach are introduced.

**THE REFORMER FURNACE** in which there are a great number of vertical reformer tubes containing the catalyst is an important equipment for the production of hydrogen in the synthetic ammonia plant. In the process of the steam reforming of hydrocarbons, these tubes service at the temperature of about $900^0$C. The typical internal operating pressure is around 3.43 MPa. The heat must be offered outside of tube by firing because the overall reaction is endothermal. The heat flux gradient through tube wall and the temperature difference between both sides of the wall is up to $100^0$C. These tubes serve the harsh conditions. In addition they are under the influence of thermal fluctuation and mechanical changing load due to startup and shutdown. These tubes often fail by creep stress rupture cracking. Fig.1 shows the type of creep damage seen in tubes from reformer furnace at end–of–life conditions.

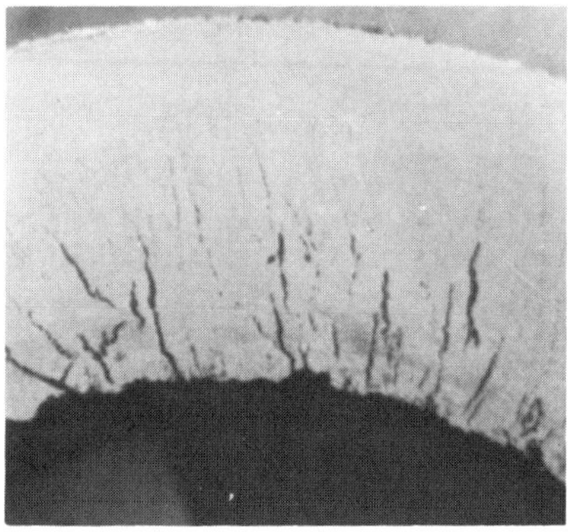

Fig.1 Typical creep stress rupture failure.

Traditional life prediction methods involve the calculation of stress using the mean diameter equation based upon the design pressure for steady–state operating conditions. the computed stress then is used to calculate the stress rupture life of the furnace tube. The maximum outside wall temperature is used, in association with the minimum rupture strength of $10^5$ hours at this temperature. This approach is simple and straightforward, but it often predicts a life of 20 or even 100 years. In contrast, actual experience may often show a life of only 5 to 15 years. This shows that the method for design of furnace tube is of shortfall. So that we find proper furnace tube design and operating are essential to obtaining long–term reliable performance from furnace tubes.

BEM is an effective tool in the structure analysis. The authors have used the method to be satisfy for creep analysis (1). Classical methods of predicting the life of a component are generally based on fracture mechanics concepts. For HK–40 tubes, where multi–cracks distributed along the radial direction (Fig.1) are present , such classical approaches lead to much difficulties. In view of the damage of furnace tubes, a new numerical method for predicting the life of a structure is developed in this paper through combination of CDM (continuum damage mechanics) and BEM. In order to give proof of effectiveness for the numerical results, a new method called natural frequency method (NFM) is introduced to measure the damage of the furnace tube.

## FUNDAMENTAL EQUATION OF CREEP ANALYSIS

Structure geometries and subjected loads of furnace tube are both axisymmetric. Therefore the creep problem of furnace tubes can be treated axisymmetrically.

According to the constitutional equation of incremental mode, when the body forces are ignored, quasi–stationary state equations are

$$\frac{\partial \dot{\sigma}_{rr}}{\partial r} + \frac{\partial \dot{\sigma}_{rz}}{\partial z} + \frac{(\dot{\sigma}_{rr} - \dot{\sigma}_{\theta\theta})}{r} = 0 \qquad (1.a)$$

$$\frac{\partial \dot{\sigma}_{zr}}{\partial r} + \frac{\partial \dot{\sigma}_{zz}}{\partial z} + \frac{\dot{\sigma}_{zr}}{r} = 0 \qquad (1.b)$$

Total strain rate can be written (plastic strain is ignored as the tubes are always considered to be in the range of elasticity)

$$\dot{\varepsilon}_{ij} = \dot{\varepsilon}^{e}_{ij} + \dot{\varepsilon}^{c}_{ij} (i,j = r.z) \qquad (2)$$

If the strain is small, the geometrical equation can be written as

$$\dot{\varepsilon}_{ij} = \frac{1}{2}(\dot{u}_{i,j} + \dot{u}_{j,i})(i,j = r.z) \qquad (3.a)$$

$$\dot{\varepsilon}_{\theta\theta} = \frac{\dot{u}_r}{r} \qquad (3.b)$$

Where $\dot{\varepsilon}^{e}_{ij}$ is the elastic strain rate and $\dot{\varepsilon}^{c}_{ij}$ is the creep strain rate.

Constitutive equation of stress–strain is

$$\dot{\sigma}_{ij} = 2\mu(\dot{\varepsilon}_{ij} - \dot{\varepsilon}^{c}_{ij}) + \frac{2\mu v}{1-2v}(\dot{\varepsilon}_{kk} - \dot{\varepsilon}^{c}_{kk})\delta_{ij} \qquad (4)$$

Where $\mu$ is the shear modulus and $v$ is the Possion's ration.

## THE BEM EQUATIONS FOR CREEP ANALYSIS

For axisymmetric creep the boundary integral equation is

$$C_{ij}(P)\dot{u}_j(P) + 2\pi\int_{\Gamma} T_{ij}(P,Q)\dot{u}_j(Q)r_Q d\Gamma$$

$$= 2\pi\int_{\Gamma} U_{ij}(P,Q)\dot{t}_j(Q)r_Q d\Gamma$$

$$+ 2\pi\int_{\Gamma} \Sigma_{i\alpha\beta}(P,q)\dot{\varepsilon}^{c}_{\alpha\beta}(q)r_q dA$$

$$(i,j = r.z;\alpha,\beta = r.z.\theta) \qquad (5)$$

where $\Gamma$ is the boundary of region A (the

axisymmetric meridian plane), $\dot{u}_j$ and $\dot{t}_j$ are the rates of displacement and transaction, respectively. $U_{ij}$, $T_{ij}$ and $\Sigma_{i\alpha\beta}$ are the kernels function of displacement, transaction and stress, respectively.

The expressions of $U_{ij}$, $T_{ij}$ and $\Sigma_{i\alpha\beta}$ are given in the appendix of reference (2).

For the nonlinear problems, it is necessary to know the stresses of the boundary and internal point. the stress of internal point is given by

$$\dot\sigma_{ij}(p) = 2\pi\int_\Gamma D_{ijk}(p,Q)\,\dot t_k(Q)\,r_Q\,d\Gamma$$

$$- 2\pi\int_\Gamma S_{ijk}(p,Q)\,\dot u_k(Q)\,r_Q\,d\Gamma$$

$$+ 2\pi\int_\Gamma B_{ij\alpha\beta}(p,q)\,\dot\varepsilon^c_{\alpha\beta}(q)\,r_q\,dA$$

$$+ h_{ij}(\dot\varepsilon^c_{\alpha\beta}(p)) \qquad (6)$$

The expressions of kernels $D_{ijk}$, $S_{ijk}$ and $B_{ij\alpha\beta}$ are given in the appendix of reference (2). For the point of the boundary, rates of stress can be determined by changing p (internal point) into P (boundary point), thus the strong singular integrals must be dealt with. This can be solved because the rates of displacement and traction were already known. In Fig.2, on a local coordinate system.

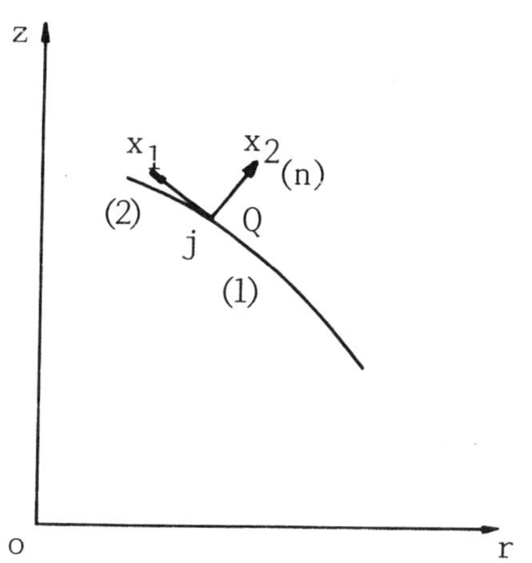


Fig.2 A local coordinate system.

$$\dot\varepsilon_{11} = \frac{\dot u_1}{x_1} \qquad \dot\varepsilon_{\theta\theta} = \frac{\dot u_r}{r}$$

$$\dot\sigma_{12} = \dot t_1 \qquad \dot\sigma_{22} = \dot t_2$$

from eq. 4 we have

$$\dot\varepsilon_{11} = \frac{1}{E}[\dot\sigma_{11} - \nu(\dot\sigma_{22} + \dot\sigma_{\theta\theta})] + \dot\varepsilon^c_{11}$$

$$\dot\varepsilon_{\theta\theta} = \frac{1}{E}[\dot\sigma_{\theta\theta} - \nu(\dot\sigma_{22} + \dot\sigma_{11})] + \dot\varepsilon^c_{11}$$

combine the two equations

$$\dot\sigma_{11} = \frac{\nu}{1-\nu}\dot\sigma_{22} + \frac{E}{1-\nu^2}(\dot\varepsilon_{11} + \nu\dot\varepsilon_{\theta\theta}) - \frac{E}{1-\nu^2}(\dot\varepsilon^c_{11} + \nu\dot\varepsilon^c_{\theta\theta}) \qquad (7.a)$$

$$\dot\sigma_{\theta\theta} = \frac{\nu}{1-\nu}\dot\sigma_{22} + \frac{E}{1-\nu^2}(\nu\dot\varepsilon_{11} + \dot\varepsilon_{\theta\theta}) - \frac{E}{1-\nu^2}(\nu\dot\varepsilon^c_{11} + \dot\varepsilon^c_{\theta\theta}) \qquad (7.b)$$

In Eq.7 and Eq.8, $\dot\varepsilon_{11}$ is the only unknown term. In Fig.2, we take a unit j which includes the desired point where $L_j$ is the length of that unit. The following equation can be obtained after a series of operations (2)

$$\dot\varepsilon_{11} = \frac{(\dot u_r^{j1} n_z - \dot u_z^{j1} n_r - \dot u_r^{j2} n_z + \dot u_z^{j2} n_r)}{L_j} \qquad (8)$$

$$\begin{Bmatrix} \dot t_1 \\ \dot t_2 \end{Bmatrix} = \begin{bmatrix} -n_z & n_r \\ n_r & n_z \end{bmatrix} \begin{Bmatrix} \dot t_r \\ \dot t_z \end{Bmatrix} \qquad (9)$$

And

$$\dot\varepsilon^c_{11} = \dot\varepsilon^c_{rr} n_z^2 + \dot\varepsilon^c_{zz} n_r^2 - \dot\varepsilon^c_{rz} n_r n_z \qquad (10)$$

$\dot\sigma_{11}$ and $\dot\sigma_\theta$ can be obtained from eq. 7 and 8. other rates of stress can be determined by using the following three equations.

$$\dot\sigma_{rr} = \dot\sigma_{22} n_r^2 + \dot\sigma_{11} n_z^2 - \dot\sigma_{12} n_r n_z \qquad (11)$$

$$\dot\sigma_{zz} = \dot\sigma_{11} + \dot\sigma_{22} - \dot\sigma_{rr} \qquad (12)$$

$$\dot\sigma_{rz} = (\dot\sigma_{22} - \dot\sigma_{11}) n_r n_z + \dot\sigma_{12}(n_r^2 - n_z^2) \qquad (13)$$

## BEM MODE FOR CREEP DAMAGE

Because of the nonhomogeneous of stress and the existence of the initial damage on the section of a component, the damage will certainly be nonhomogeneous under the condition of high temperature creep and the weak part will reach the tertiary stage of creep quickly. A typical creep curve includes three stages. The primary stage results from the change of the dislocation density , the second stage from diffusion and equilibrium of dislocation and the third from the unrecoverable damage . In this paper, the primary stage is not considered due to the difficulty of obtaining the necessary fitting data and it does not play a leading role .

Hayhurst has proposed a set of comprehensive equations for creep and evolution of damage under the multi–axial stress , the basic forms are

(1) for materials prior to the damage stage (the second stage)

$$\dot{\varepsilon}_{ij}^{c} = \frac{3}{2} B \bar{\sigma}^{n-1} S_{ij} \tag{14}$$

(2) for materials in the damage stage (the third stage)

$$\dot{\varepsilon}_{ij}^{c} = \frac{3}{2} B \frac{\bar{\sigma}^{n-1}}{(1-D)^{n}} S_{ij} \tag{15}$$

$$\dot{D} = H\left(\frac{\varphi(\sigma)}{1-D}\right)^{q} \tag{16}$$

where $\bar{\sigma}$ is the effective stress, $S_{ij}$ is the stress deviator tension. $S_{ij} = \sigma_{ij} - \delta_{ij}\sigma_{ij}$, where $\sigma_{ij}$ is the stress tension and $\delta_{ij}$ is the Kronecter delta. $\varphi(\sigma)$ is the equivalent stress for determining the rule of material failure and can be defined by following linear combination

$$\varphi(\sigma) = \alpha\sigma_1 + \beta\bar{\sigma} + \gamma J_1 \tag{17}$$

where $\alpha$、 $\beta$ and $\gamma$ are constants and $\alpha+\beta+\gamma=1$. $\sigma_1$ is the maximum principle stress. $J_1$ is the first constant variable and $J_1 = \sigma_1+\sigma_2+\sigma_3 = 3\sigma_m$. Function $\varphi(\sigma)$ is called instantaneous plane and its detailed description is shown in (4). The proposal of this theory established the foundation of practical problem solving. When $\alpha=1$, $\beta=0$ and $\gamma=0$, the instantaneous plane is determined by the maximum principle stress. When $\alpha=0$, $\beta=1$ and $\gamma=0$, the instantaneous plane is determined by the effective stress. In most cases, $\gamma$ is so small that we consider $\gamma=0$, thus the eq. 17 becomes

$$\varphi(\sigma) = \alpha\sigma_1 + (1-\alpha)\bar{\sigma} \tag{18}$$

## NATURAL FREQUENCY METHOD

The strain equivalent law considered the coupling between the creep damage and visco–plasticity allows to evaluate the damage by means of its influence on the elasticity modulus(4)

$$\varepsilon_e = \frac{\sigma}{E(1-D)} = \frac{\sigma}{\tilde{E}} \tag{19}$$

We see that the effective elasticity modulus $\tilde{E} = E(1-d)$ decreases as D increase. If Young's modulus E is known from a test on the non–damageed material and if $\tilde{E}$ is measured in the damaged state, the corresponding value of damage is

$$D = 1 - \frac{\tilde{E}}{E} \tag{20}$$

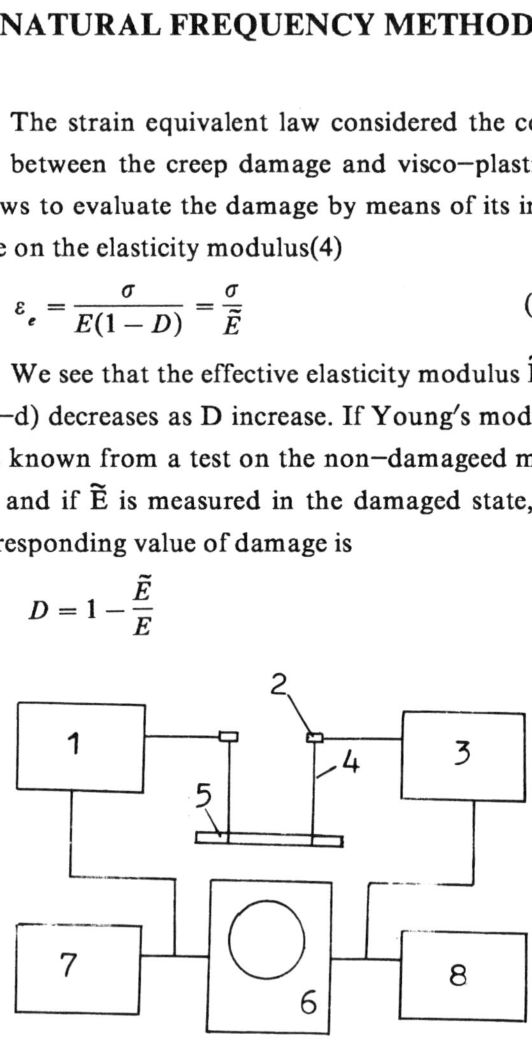


1. Signal generator 2. Transducer 3. Amplifier 4. Copper wire 5. Specimen 6. Oscillograph 7. Frequency teller 8. Indicating instrument

Fig.3 The equipment diagram of the NFM.

Here, this paper applied NFM to measure the initial and actual damaged elasticity modulus. Fig.3 illustrates the diagram of the equipment of the NFM.

Tab. 1

| No. | Length (mm) | Diameter (mm) | Mass (g) | Test Time (hrs) |
|---|---|---|---|---|
| Y1 | 139.88 | 6.025 | 31.1476 | 0 |
| Y2 | 140.06 | 6.023 | 31.1265 | 24.75 |
| Y3 | 139.99 | 6.023 | 31.1065 | 67.25 |
| Y4 | 140.04 | 6.002 | 30.943 | 80* |

* Rupture Time $t_f$.

Tab.2

| No. | Response Frequency (Hz) | Elasticity Modulus (MPa) | Damage |
|---|---|---|---|
| Y1 | 1164.9 | 141292 | 0.2411 |
| Y2 | 1150.3 | 138099 | 0.2583 |
| Y3 | 1107.4 | 127716 | 0.3139 |
| Y4 | 1099. | 127021 | 0.3178 |

Considering a bar with homogeneous section and free tow–ends (to see Fig.3), when response occurs, the relationship between the basic natural frequency and the elasticity modulus is

$$E = 1.606K\frac{ml^3}{d^4}f^2(Pa) \tag{21}$$

Where m, l and d are the mass, length and diameter, respectively. f is the basic natural frequency and K is a correct factor. An example of damage measurement on furnace tubes has been carried out.(6) The dimension of specimens from tubes served for 69000 hrs is shown in Tab. 1. Three endurance life tests were performed from specimens Y2, Y3 and Y4. Tab.2 illustrates the result from natural frequency measurement. The variation of creep damage as calculated from the measured frequency and eq.21 is shown in Tab.2.

## NUMERICAL CALCULATION

**Dispersion Boundary and Internal Region.** Region A on the meridian plane and boundary $\Gamma$ dispersed several quadrilateral cells and boundary elements, respectively. From Eq.5 , we have

$$\mathbf{H}\{\dot{u}\} = \mathbf{G}\{\dot{t}\} + \mathbf{Q}\{\dot{\varepsilon}^c\} \tag{22}$$

Stress derived is

$$\{\dot{\sigma}\} = \mathbf{D}\{\dot{t}\} - \mathbf{S}\{\dot{u}\} + \mathbf{R}\{\dot{\varepsilon}^c\} \tag{23}$$

**Determination of Iterative Mode and Time Step.** In solving creep problems, two kinds of initial value problems may be derived from the principle equation, one concerns the initial value of stress, the other the initial value of strain. The later is used in this paper because its results are considered to be more convincible ones.

In computing creep strain, the Eular reiterative formula is applied

$$\varepsilon^c_{ij}(n+1) = \varepsilon^c_{ij}(n) + \Delta t_{n+1}$$

$$[(1-\theta)\dot{\varepsilon}^c_{ij} + \theta\dot{\varepsilon}^c_{ij}] \tag{24}$$

where n is the nth reiteration. $\Delta t_{n+1} = t_{n+1} - t_n$ is the step to time. $\theta$ is the coefficient of the reiterative mode and the field is $0 < \theta < 1$. It has been proved that it will be convergent when $\theta > 0.5$, therefore in the paper we use $\theta = 0.5$.

The first rule to select the magnitude of the time step consists in limiting the maximum amount of incremental creep strain to a fraction of the total accumulated strain according to

$$\Delta t_n = \tau \left[ \frac{\varepsilon^c_{ii}(n)}{\dot{\varepsilon}^c_{ii}} \right]^{\frac{1}{2}} \tag{25}$$

where is $\tau$ a pre–specified value and the minimum value of $\Delta t_n$ is taken with respect to all integrating in the domain. The second rule limits the change in the time step length between any two intervals, i.e.

$$\Delta t_{n+1} \leqslant k \Delta t_n \tag{26}$$

where k is a given constant. Numerical experience to date indicates that suitable values for $\tau$ and k are in the range $0.01 < \tau < 0.15$ and $0.5 < k < 1.5$ respectively, although there are no fixed criteria for their specification.

**Singularities Associated With The Axisymmetric Kernels .** The integration of the axisymmetric kernel function requires special attention in the following circumstances :

( 1 ) when the radial co–ordinate of the source P falls on the origin ($R_P = 0$),

( 2 ) in the displacement rate equation when the source point P and the field point Q coincide, and

( 3 ) in the stress (or strain) rate equation when the source point p and the integration q (field point) coincide.

In the first case when the source point P falls on the origin, the axisymmetric kernel functions are singular. This problem can be resolved by moving the point a small radial distance from the origin. This is a common practice also employed in the finite element analysis of axisymmetric problems . In the equations for stress and strain rates, experience shows that the minimum distance should be 0.5 per cent of the difference between the maximum and minimum z co–ordinate at the origin. For the displacement rate equations this distance can be appreciably less. Alternatively, a separate limiting from for the displacement rate at the origin can be derived (2).

The second case concerns the integration of the displacement rate equation when a field point coincides with source point. The integrations of the $U_{ij}$ and $\Sigma_{i\alpha\beta}$ kernels are weakly singular in this circumstance and can be integrated numerically using element subdivisions near the singular point(1). On the other hand, the integration of the $T_{ij}$ kernel of the displacement rate equation is strongly singular and must be integrated as a Lebesgue integral over the singular point. This integral can be decomposed into a free term and cauchy principal–value integral but accurate numerical is still difficult. When constructing the equations for the boundary system, the singular integration involving the $T_{ij}$ kernel can be circumvented by using an alternative method(1,2), analogous to the 'rigid body' displacement technique(2). This procedure known as the 'inflation mode for axisymmetry' assumes two independent, admissible displacement fields that are applied to each equation, allowing the coefficient corresponding to singular point (node) to be determined.

The final case involving the stress (or strain) rate calculation is handled in two ways, depending on whether the source point falls on the boundary or in the interior of domain. For a point on the boundary, all kernel functions of the stress rate equation are all strongly singular and difficult to integrate numerically. However, the stress rate on the boundary can be obtained from the boundary traction and displacement rates without any integrations, as it is described above by the authors. When the source point lies in the interior of the body, the $D_{ijk}$ and $S_{ijk}$ kernels of the stress rate equations are well behaved and require no special attention. However, the $B_{ij\alpha\beta}$ kernel is strongly singular and must be integrated as a Lebesgue integral over the singular point. Once again, this type of integral can be decomposed into a free term (given in (2)) and a cauchy principal–value integral. Although difficulty and expensive, reasonable accuracy can be obtained when this integral is numerically integrated using technique described by Banerjee and Raveendra (7).

**Treatment of the Failed Cell.** When damage in a cell has exceed $D_{cr}$, the cell is considered to be failure. In the present work the failed cell is canceled from the structure. Thus the geometric shape and boundary condition will be changed. The overall coefficient matrix are reformulated with the failed cell omitted. So that a movable boundary problem is required to treat.(2)

According to above way to treat problem. The authors have developed the BEM program ABEMC for analysis of creep and creep damage.

## PREDICTION OF CREEP DAMAGE IN FURNACE TUBE

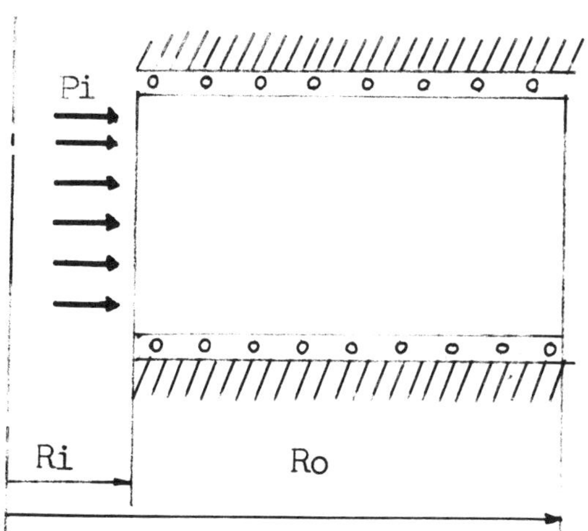


Fig.4 A creep analysis model for furnace tube.

**Computation Model.** For numerical analysis of creep problem, a great number of iterative generally must be required. For example, even for a simple tube segment of twenty nodes , its CPU time for completing the calculation by FEM analysis is about thirty hours on IBM PC / AT computer. So it is very difficult to analyse the global furnace tube ten metres in length in microcomputer. In order to overcome it, the authors use a method called 'section method' presented by Wu Guanghua to fulfill the calculation whose validity has been proved by (8). The method can save the memory capacity of the computer. The model applied to creep problem analysis is shown in Fig.4.

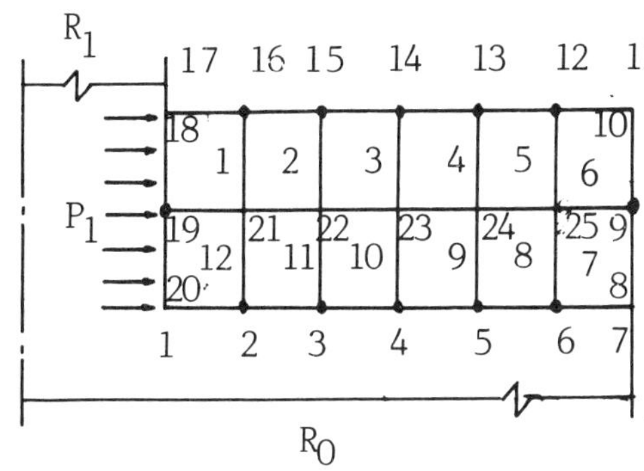


Fig.5 BEM mesh and element for furnace tube.

The inner and outer tube surfaces are at prescribed potentials $T_i$ and $T_o$ that are known to be input. The analytical solution of the purely radial of potential is

$$T = T_i + \frac{ln\left(\frac{r}{r_i}\right)}{ln\left(\frac{r_o}{r_i}\right)}(T_o - T_i) \qquad (27)$$

where T and r are temperature and radial, respectively. The subscribe i and o express the inner and outer tube surfaces, respectively.

Fig.5 illustrates BEM mesh and element for furnace tube.

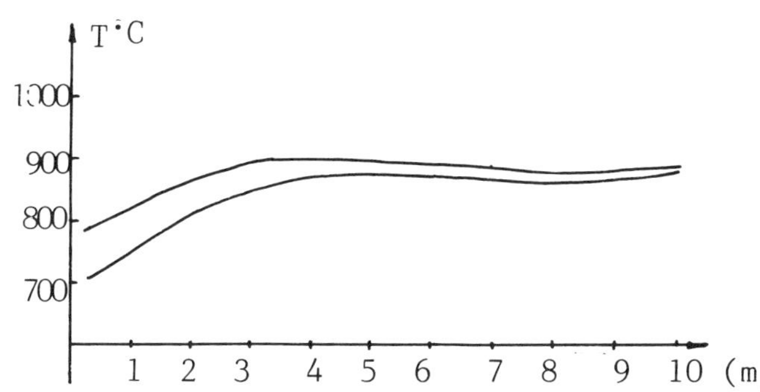


Fig.6 Temperature distribution of tube.

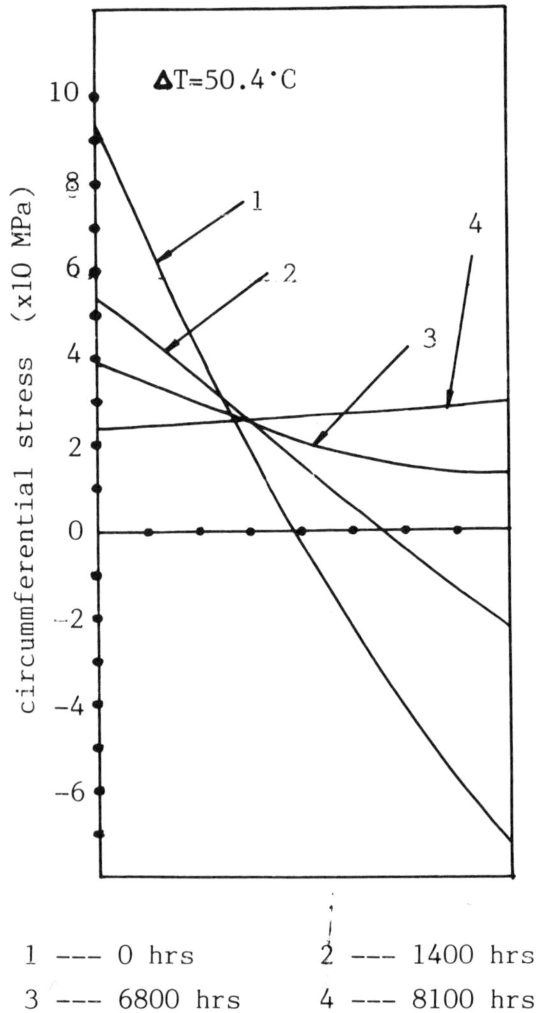


Fig.7 Hoop stress distribution history through wall of tube.

**Numerical Example.** As an example, a typical reformer tube of HK−40 alloy is calculated. Its temperature distribution along length direction is shown in Fig.6. Through creep rupture test of some specimens, the material constants n , B , H and q are obtained. The critical damage factor $D_c = 0.3178$ has been determined by the FEM presented by authors (6). The other known parameter employed are : $P_i =$ 3.42 MPa (internal pressure), $r_i =$ 74.4 mm and $r_o =$ 104.1 mm. Hoop stress distribution history through wall of tube is illustrated in Fig.7. The initial state of stress at vicinity of inner wall of the tube is considerably higher than that near the outer wall of the tube. This is caused by the internal pressure and thermal fluctuation interaction. Compressive stress occurs on the outer wall surface and tensile stress occurs on the inner wall surface. As operating time proceeds, these stresses are redistributed due to relaxation of stress brought on by creep deformation.

Fig.8 shows the calculated creep damage as it is distributed through the wall of the tube. It is seen that the area near the inner wall is much damaged. This agrees well with the fact as shown in Fig.1.

The residual life furnace tube served for 69000 hours predicted by the method presented here is 162175 hours. As some influence factors, such as standup and shutdown, are not considered, the result is more than the real life, but it is less than the results predicted by (8) and L−M method. The CPU time is only 18 hours and considerably less than the FEM analysis (8) without damaging the accuracy.

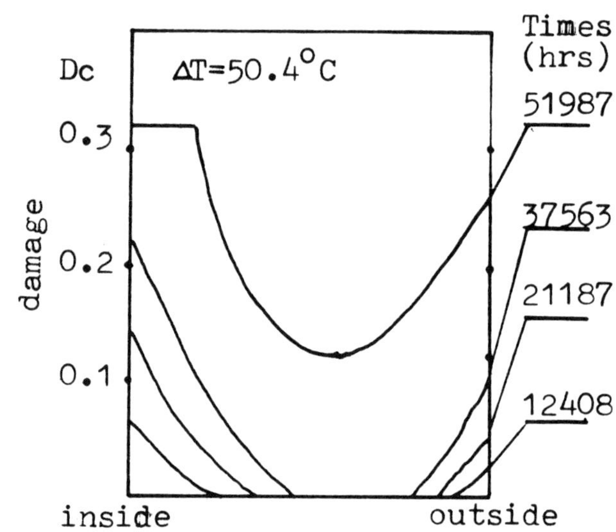


Fig.8 Predicted creep damage distribution history through wall of tube.

## CONCLUSION

1 ) The basic formula on axisymmetric creep problem has been proposed in the paper. By introducing the 'damage−coupled mesh' into BEM analysis, the authors have established a new model suita-

ble for boundary element and internal cell damage analysis. A step further is to take into account the coupling between strain and damage due to the fact damage decreases the rigidity and the strength of materials. Then, by means of coupled constitutive equations, strain and damage fields are calculated in one step.

2 ) In numerical implementation, the iterative form based on initial strain method and controlling criterion of time step are achieved. The singularities with these kernels are discussed and the techniques are described to calculate the coefficients associated with the singular node. The computation software of BEM for axisymmetric problem (ABEMC) has been developed.

3 ) The method presented here has been applied to an practice example. The calculation result that ABEMC software compiled is efficient in creep damage analysis and residual life predicting of furnace tubes. Its advantage over finite element method lie in less computation without damaging the accuracy.

## REFERENCES


1.Lin Xiang and Cai Ruiyin, Chemical Engineering and Machinery (In Chinese), 38 (1992) (to press)

2.Lin Xiang, Thesis for the M.S., Nanjing Institute of Chemical Technology (1991)

3.Hayhurst, D. R. J., Mech. Phs. Solids, 20 (6), 381–390 (1972)

4.Wu Hongyao, "Damage Mechanics", National Defense Industry Publishing House, Beijing (1990)

5.Boyle, J. T. and Spence, J., "Stress Analysis for Creep ", England (1983)

6.Lin Xiang , Chou Jinshi and Shen Fuzhong , Chemical Engineering and Machinery (In Chinese) , 38 (1992) (to press)

7.Banerjee, P. K. and Raveendra, S. T., Int. Numer. Methods Eng., 23, 985–1002 (1986)

8.Wu Guanghua, Thesis for the M.S., Nanjing Institute of Chemical Technology (1989)

*Proceedings of the Fifth International Conference on Creep of Materials, Lake Buena Vista, Florida, USA, 18-21 May 1992*

# Creep-Life Prediction of Thin-Walled Zirconium Alloy Tubings

**K.L. Murty**
North Carolina State University
Raleigh, North Carolina

## Abstract

The paper summarizes methodologies used in predicting the dimensional changes of anisotropic textured zirconium alloys starting from a quantitative description of the grain orientations or the textures through the crystallite orientation distribution functions (CODF) coupled with appropriate plasticity models based on slip behavior(s) of low c/a-hcp single crystals. Anisotropic creep behavior is characterized using biaxial creep through creep-locus while the temperature and stress dependences of the steady-state creep-rate were modelled on the climb of the edge dislocations as the underlying creep mechanism. The transient and steady-state creep behaviors are considered and thus the creep strain at any temperature, stress and time could be predicted. The superimposed effects of neutron irradiation are included to model the creep characteristics of fuel rods in light water reactors and the model predictions correlated with the post-irradiation examination data of Zircaloy cladding tubes in Oconee reactor, once the stress-free radiation-growth which itself is anisotropic, is taken into account.

ZIRCONIUM ALLOYS are used in light water reactors primarily as thin-walled tubings to prevent coolant-fuel contact and to retain fission gases. The structural integrity of these materials is of prime consideration to avoid leakage of fission gases into the coolant and thus a knowledge of the dimensional changes of these cladding tubes in-reactor is of utmost significance for the safe operation of the reactors. In service, these thin-walled tubes are subject to complex multiaxial loading. Principal factors contributing to the stresses on Zircaloy cladding are the external pressure of the coolant, flow-induced vibration, internal pressure due to released fission gases, pellet cladding interaction and spacer grid forces due to differential growth leading to fuel rod bow.

These alloys have hcp crystal structure with low c/a ratio at and below the reactor operating temperatures and exhibit preferred orientations or textures which are dependent on the specific thermomechanical treatments during the tube reduction processes. The anisotropic mechanical properties in turn affect their *in-service* behavior such as in-pile creep-down as well as stress-free radiation-growth of the cladding tubes. X-ray diffraction techniques are used to determine the textures of cold-worked stress-relieved [CWSR] and recrystallized [$R_X$] Zircaloy cladding tubes from which the crystallite orientation distribution functions (CODF's) are derived (1). Plasticity models are employed along with these CODF's to predict the anisotropic creep behavior which correlated with experimental creep results evaluated using *in-situ* axial and diametral strain measurements at varied stress ratios at constant temperature. The experimental and predicted creep loci are fit to the Hill's phenomenological equation for the generalized stress.

The temperature and the stress variations of the steady-state creep-rate were derived from uniaxial tests which correlated with the predictions based on creep due to the climb of the edge dislocations. Extensions of the model to transient creep lead to a single unified equation (2), in terms of the temperature and stress compensated steady-state creep-rate, that can be used to predict the creep strain at any given time, temperature and stress. However, these models did not consider the strain transients following stress changes (3). In addition, the high-stress exponential-stress dependence of the creep-rate noted in laboratory creep tests (4) was assumed to be applicable at the reactor operating conditions which correspond to relatively low stresses where power-law creep is more appropriate (5); these relevant extensions are considered here.

Post irradiation examination of the experimental fuel pin samples in the Oconee reactor revealed that neutron irradiation does not produce any significant changes in the creep anisotropy of the CWSR Zircaloy when radiation growth, which itself is anisotropic, is taken into account (2,3). These findings, in conjunction with appropriate constitutive equations for thermal and radiation creep along with radiation growth, make it possible to develop quantitative procedures to predict the in-service dimensional changes of the fuel rods such as the clad creep down and the axial growth. We demonstrate the applicability of our CODF-CREEP model in predicting the in-reactor dimensional changes of CWSR cladding and these effects following recrystallization are briefly discussed.

## Materials

Zirconium alloyed with Sn and Fe, known as Zircaloys, are used in light water reactors to clad radioactive $UO_2$ and we consider here Zircaloy-4 which is used in pressurized water reactors (PWRs). The alloy is used as thin-walled tubing which is suitable for biaxial creep testing by internal pressurization with superimposed axial loading. The experimental tubing material was supplied in the cold-worked stress-relieved (CWSR) condition, and contained 1.5%Sn, 0.21%Fe and 0.10%Cr by weight. The biaxial creep characteristics in terms of creep anisotropy are sensitive functions of the amounts of cold work and the degree of recrystallization (3). We describe here the crystallographic textures and CODFs of this material followed by the formulation of creep anisotropy, the biaxial creep experiments, model creep equations and finally the application to prediction of in-reactor clad creep down.

## Crystallographic Texture and CODF

Crystallographic pole-figures are commonly used in the description of preferred orientations. Each pole-figure represents the spatial distribution of a given crystallographic plane within the polycrystal as portrayed in stereographic coordinates. The inverse pole figures were generated along the 3 orthogonal directions (r, θ, z) of the tubing while direct pole figures were obtained using appropriate samples, and the basal (00.2), prism (10.0) and pyramidal (10.2) pole figures are considered. The basal and prism texture pole figures for the CWSR Zircaloy-4 are included in Fig.1 which clearly reveal the typical bimodal basal pole intensity distributions around the tube thickness direction in the transverse [r-θ] plane as well as the clustering of the prism (10.0) poles around the axial (z) direction.

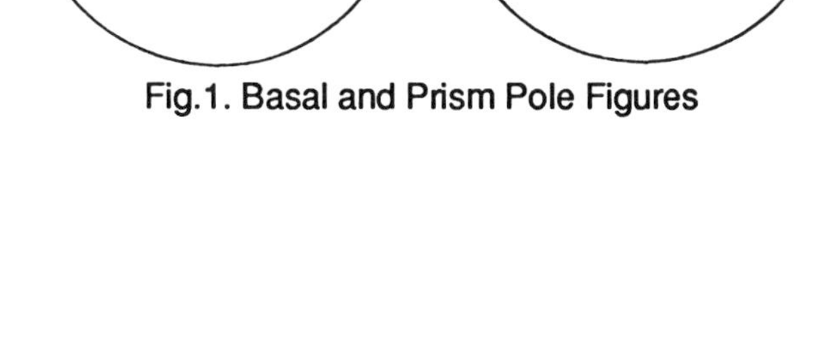


Fig.1. Basal and Prism Pole Figures

Quantitative description of the textures is possible through an evaluation of a distribution function representing the probability that a crystallite [grain] lies within a certain range of orientations with respect to processing or any specified macroscopic coordinates (6). The probability that a crystallite has an orientation θ,ψ,φ with respect to the specimen axes is expressed as a series of generalized spherical harmonics,

$$\omega(\theta,\psi,\phi) = \sum_{l=0}^{\infty}\sum_{m=-l}^{l}\sum_{n=-l}^{l} W_{lmn} Z_{lmn}(\cos\theta)\, e^{-im\psi}\, e^{-im\phi} . \quad (1a)$$

where $W_{lmn}$ are the appropriate series coefficients and $Z_{lmn}$ are the augmented Jacobi polynomials. This probability function $\omega(\theta,\psi,\phi)$ is known as the crystallite orientation distribution function [CODF] and $(\theta,\psi,\phi)$ are the Euler angles connecting coordinate axes embedded in the crystallite with axes coincident with specimen and the principal working directions of the material. Input for the series method consists of data from a set of pole figures, the intensities of which are also represented by a series of spherical harmonics. Thus if sufficient sets of pole-figure data are available, the system of linear equations can be solved for any order of the W coefficients. Due to physical and experimental constraints, usually three to five pole-figures are readily determinable; thus the $W_{lmn}$ values are typically evaluated to an order of $l = 20$ in cubic and hexagonal polycrystals. This order is sufficiently high for most purposes to represent the CODF reliably and, in addition, experimental errors inherent in the measurement of pole-figures mask the truncation error corresponding to this or lower cutoff level. For hexagonal symmetry, l and m are restricted to even values and n to integer multiples of 6 in the series expansion for CODF. With these symmetry restrictions a truncation level of $l = 16$ is reached with three pole-figures.

Physical and mechanical properties of single crystals are anisotropic and this extends to the bulk properties of the polycrystals whenever preferred orientations are present. To the first order, the bulk property is related to the microscopic property, $\rho(\theta,\psi,\phi)$, by statistical averaging using the CODF. This approach neglects interactions between crystallites in the aggregate. The average property of the polycrystalline aggregate, $\langle\rho(\theta,\psi,\phi)\rangle$, is then given by

$$\langle\rho(\theta,\psi,\phi)\rangle = \int_0^{2\pi}\int_0^{2\pi}\int_{-1}^{1} \rho(\theta,\psi,\phi)\, \omega(\theta,\psi,\phi)\, d(\cos\theta)\, d\psi\, d\phi . \quad (1b)$$

## Thermal Creep - Experimental

Thin-walled tubing facilitates the achievement of multiaxial stresses by combining internal pressurization with tensile loading; we will confine here to the positive quadrant and thus the strength differential effects would not be considered here. The biaxial creep system comprises of a lever arm creep machine, ATS@ model-2330 fitted with a custom-designed furnace with windows for the diametral strain measurement using a laser telemetric extensometer, Lasermike@ model 60-30 (6,9). In addition, the machine is equipped with an LVDT extensometer for monitoring axial strains. The internal pressurization was accomplished with a sustained gas pressurization system capable of obtaining pressures in excess of 70 MPa. The machine is fitted with pressure fittings suitable for testing tubings of 8-15mmOD while the LVDT and laser extensometers cover a range to 25 mm OD. Thus both the axial and hoop strains are continuously monitored as a function of the applied biaxial stresses [$\sigma_\theta$ and $\sigma_z$ with $\sigma_r$ neglected using thin wall approximation]. An IBM@ XT micro-computer is dedicated to the machine for on-line data acquisition and analyses.

The hoop and axial stresses for an internal pressure (P) and axial load (W) are calculated from the following:

$$\sigma_\theta = \frac{P\,a^2}{b^2 - a^2}\left(1 + \frac{b^2}{a^2}\right) \quad \text{and}$$

$$\sigma_z = \frac{P\,a^2}{b^2 - a^2} + \frac{W}{\pi\,(b^2 - a^2)}, \tag{2}$$

where a and b are the inner and outer radii of the tubing.

The pressure and load were varied to obtain stress ratios, $\alpha$ (= $\sigma_\theta/\sigma_z$), from 0 to 2 with 5 intermediate values. The anisotropic creep behavior was thus characterized using biaxial creep data from which the generalized stresses and strain-rates could be related through the derived anisotropy parameters.

The temperature and stress dependences of the steady-state creep-rates were studied using tensile creep tests on tubing samples of about 10.2cm gauge length under axial load with no internal pressure ($\alpha$ = 0). These test results were used to develop the primary creep model for predicting the creep strain as well as the transients in creep strain following stress changes.

## Thermal Creep - Results

The temperature dependence of the steady-state creep-rate yielded an activation energy for creep of 268±7 kJ/mole which agreed with that for self-diffusion of 260 kJ/mole (7). The close agreement of the activation energies for creep and self-diffusion suggests a diffusion controlled creep mechanism. Steady-state creep-rate is then given by,

$$\dot{\varepsilon} = \frac{DEb}{kT}\, f\!\left(\frac{\sigma}{E}\right), \tag{3}$$

where $f(\sigma/E) = A\,(\sigma/E)^n$ for $\sigma \le 10^{-3}$ E and $A_1\, e^{B(\sigma/E)}$ for $\sigma \ge 10^{-3}$ E. In this equation, E is the elastic modulus and D is the lattice diffusion. At high stresses, the creep-rates are correlated to the applied stresses through a semi-log relation with an activation area of 19±3 $b^2$. The equivalence of the activation energy for creep with that for diffusion combined with this value for the activation area points to the fact that the climb of the edge dislocations is the rate controlling mechanism [8]. Power-law stress dependence with an exponent of 5.5 is noted at lower stresses and thus the entire set of data are describable using a sinh-function:

$$\dot{\varepsilon} = A_2\, D\,\{\sinh(B_2 \frac{\sigma}{E})\}^n \tag{4}$$

where $A_2$, $B_2$ and n are constants whose values were determined to be 1.9x$10^9$, 500 and 5.5±0.2 respectively. Figure 2 is a double-log plot of strain-rate vs sinh(5x$10^{-5}\sigma$) at a constant temperature of 385°C illustrating this behavior.

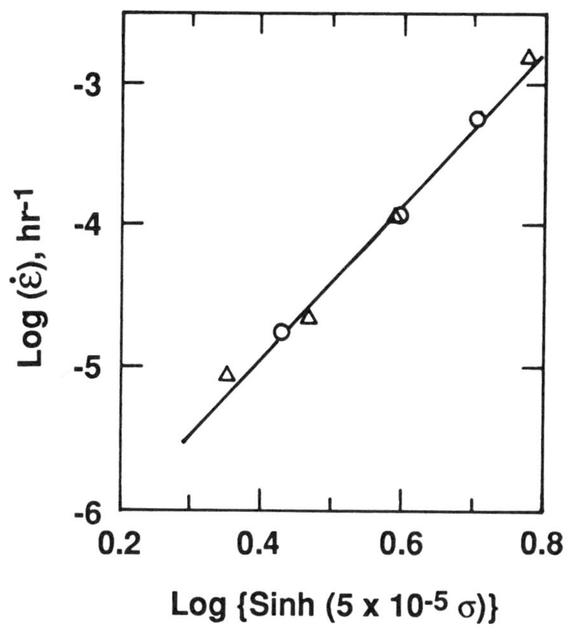


Fig.2. Stress Dependence of Steady-State Creep-Rate

In predicting the creep strain at any given time following exposure to a specified stress and temperature, one needs to consider the primary creep as well; indeed, in *real* situations, structures usually accumulate deformations which fall in the primary creep regime. In general, an Avrami type equation is used in describing the total creep curve comprising both the steady-state and primary stages,

$$\varepsilon_p = \dot{\varepsilon}_s t + A'\,(1 - e^{-B'\,\dot{\varepsilon}_s t}), \tag{5a}$$

where $\varepsilon_p$ is the time-dependent plastic strain and A' is a material constant, a function of the initial creep-rate (at t=0) and the steady-state value and characterizes the extent of the primary creep strain, while B' is related to the ratio of the initial creep-rate to the steady-state value

(9). This model implies that the initial creep-rate is proportional to the steady-state value and is independent of the test conditions such as stress and temperature. While this behavior has been shown to be valid for many metals, there exist as many examples where this equality is seen to be invalid; this model however, predicts a finite initial creep-rate as is often observed.

Although this above equation (5a) was seen to describe the creep data for zirconium alloys reasonably well (9), we found that the following alternate expression gives a better correlation,

$$\varepsilon_p = \dot{\varepsilon}_s t + \frac{A\dot{\varepsilon}_s t}{1 + AB\dot{\varepsilon}_s t} . \qquad (5b)$$

Here, A and B are material constants related to the ratio of the initial creep-rate to the steady-state value and inverse of the transient creep strain respectively,

$$A = \frac{\dot{\varepsilon}_i}{\dot{\varepsilon}_s} \quad \text{and} \quad B = \frac{1}{\varepsilon_{tr}} \qquad (5c)$$

where $\varepsilon_{tr}$ is the extent of the primary creep. The temperature and stress dependences are built into the steady-state creep-rate equation (Eq.4). Figure 3 is a plot of the creep strain versus temperature and stress compensated time and demonstrates the applicability of the model equations with the following parameters,

$$A = 10 \quad \text{and} \quad B =4\ 75. \qquad (5d)$$

One could thus use the above equations 4 and 5 along with either the time-hardening or the strain-hardening rule to evaluate the creep strain at any time for a given stress and temperature. However, these models do not consider the transients in creep that arise when sudden changes in the stress levels are encountered.

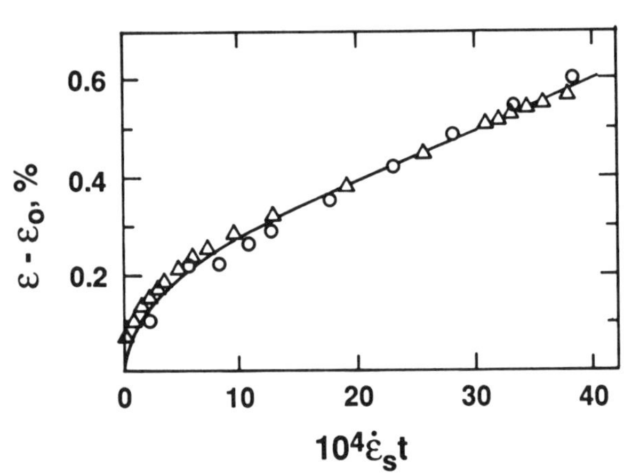


Fig.3. Experimental Creep Results (o) Compared with Model Prediction (--) based on Eq.5b

The strain transients immediately following load or stress changes were described by Murty and Yoon (3) through Hart's anelastic strain model. Accordingly, strain at any given instant is given by

$$\varepsilon = \varepsilon_E + \varepsilon_p + \varepsilon_a , \qquad (6a)$$

where the subscripts E and a stand for instantaneous elastic and time-dependent anelastic recoverable strains. A sudden change in stress results in short-term anelastic deformation in addition to time dependent plastic strain. An anelastic strain is *time-dependent recoverable* deformation in contrast to the permanent plastic deformation which is not recoverable whereas elastic component is recoverable but instantaneous (time-independent). The total anelastic strain change after each load variation is thus completely reversible upon reversing the sign of the load change. This anelastic component of the total strain reaches a saturation value that is determined by the new stress level. Accordingly (3), the anelastic strain ($\varepsilon_a$) is given by,

$$\varepsilon_a = \frac{\sigma_s}{\mu} - \{[\frac{\sigma_s}{\mu} - \frac{\sigma_o}{\mu}]^{M-1} + (M-1)\,\alpha^*\,(t - t_o)\}^{-1/(M-1)}. \qquad (6b)$$

In this equation, $t_o$ is the time at which the stress was changed from $\sigma_o$ to $\sigma_s$, M and $\alpha^*$ are material constants in Hart's equation of state, and $\mu$ is the anelastic modulus (10). While a value of 8 for M was shown to be valid for many metals (10), the validity in Zircaloy-4 was demonstrated by Huang et al (11) and Murty and Yoon (3). $\alpha^*$ was determined by Huang et al (11) using load relaxation tests on Zircaloy cladding while Murty and Yoon (3) evaluated $\mu$ for the same material. The following parameters were, therefore, selected for input to the equation 6b :

$$M = 8, \quad \alpha^* = 2.52\text{x}10^{22}, \quad \mu = 17 \text{ x } 10^6 \text{ psi.} \qquad (6c)$$

Before applying these model equations to the experimental data, it should be pointed out that the existence of the universal master curve describing creep at different test conditions (stresses and temperatures) points to the validity of the strain-hardening rule implying that the microstructures developed at any specific strain level are identical irrespective of the loading conditions. Thus the strain-hardening rule is used in modelling the strain transients. Figures 4a and 4b depict the experimental results. The model predictions correctly revealed the negative creep following load drops followed by accelerating or inverse primary creep till the new steady-state rate is reached. The relatively small deviations noted here seem to stem from the steady-state creep model rather than the transient. The model accurately predicted the short normal transients following the stress increases. It is believed that an optimization of the parameters in the model equations will yield far better agreement and an extended test matrix is under consideration.

## Anisotropic Deformation and Creep Formulation

Hill proposed a modified Von-Mises yield criterion for an anisotropic material with the applied stresses coincident with the axes of anisotropy. Following the later modifications by Backofen and others, a generalized stress, $\sigma_g$, is defined as the uniaxial stress along the axial direction so that the yield criterion reads as follows:

$$\sigma_g=\sqrt{\frac{R(\sigma_r - \sigma_\theta)^2 + RP(\sigma_\theta - \sigma_z)^2 + P(\sigma_z - \sigma_r)^2}{P(R+1)}}, \quad (7)$$

where $\sigma_r$, $\sigma_\theta$ and $\sigma_z$ are the stresses along the radial, hoop and axial directions of the tube, and R and P are the coefficients of anisotropy. Using the Prandtl-Reuss energy balance in conjunction with the above yield criterion, the strain increments or strain-rates along the three orthogonal directions are related to the respective stresses,

$$\begin{bmatrix} \dot{\varepsilon}_r \\ \dot{\varepsilon}_\theta \\ \dot{\varepsilon}_z \end{bmatrix} = \frac{\dot{\varepsilon}_g}{P(R+1)} \begin{bmatrix} R+P & -R & -P \\ -R & R(1+P) & -RP \\ -P & -RP & (R+1)P \end{bmatrix} \begin{bmatrix} \sigma_r \\ \sigma_\theta \\ \sigma_z \end{bmatrix}, \quad (8)$$

where $\dot{\varepsilon}_g$ is the generalized strain rate corresponding to the generalized stress $\sigma_g$, and R and P are the anisotropy coefficients. These above equations define the yield and flow loci, and the corresponding creep locus is referred to at a constant energy dissipation-rate, $\dot{W}(= \sigma_{ij}\,\dot{\varepsilon}_{ij} = \sigma_g\,\dot{\varepsilon}_g)$.

## Biaxial Creep - Experimental Results

As noted earlier, the creep machine described before facilitates tests under biaxial loading with internal pressurization superimposed on axial loading from which the steady-state creep-rates along the hoop ($\theta$) and axial (z) directions of the tubing are evaluated at varied stress levels under different stress ratios ($\alpha = \sigma_\theta/\sigma_z$) from which a creep locus can be constructed at a specified dissipation energy rate [W]. Since power-law creep is obeyed in the range of stresses studied, a double-log plot of dissipation energy rate versus generalized stress or any stress component will result in a straight line, as shown in Fig.5. A creep locus is referred to the relation between the principal component stresses at equivalent dissipation energy rate and Fig.6 shows such a locus obtained at a constant energy release rate of 8 J/m$^3$-s.

It is interesting to note that the effect of recrystallization heat treatment results in a shift of the creep resistance from low [$\leq$0.75] $\alpha$ [$=\sigma_\theta/\sigma_z$] values [or high $\sigma_z$] for CWSR to relatively high $\alpha$ values corresponding to internal pressurization [$\alpha$ =2] (2). This is distinctly noted in the strain-rate ratios under equibiaxial loading [$\sigma_\theta = \sigma_z$, $\alpha$=1] :

$$\left[\frac{\dot{\varepsilon}_\theta}{\dot{\varepsilon}_z}\right]_{1:1} = \begin{cases} 3.2 \text{ for CWSR} \\ 0.4 \text{ for } R_x \end{cases} . \quad (9)$$

Thus the relatively weak hoop direction for CWSR became stronger following recrystallization anneal, and illustrates the profound effect of heat treatment on creep anisotropy of Zircaloys. We also note that the deviation of the ratio from unity is indicative of the extensive creep anisotropy of these materials that exists even after complete recrystallization.

By fitting these biaxial creep results to the Hill's equation, we obtained the following anisotropy parameters for the stress-relief annealed Zircaloy-4:

$$R = 1.5 \quad \text{and} \quad P = 0.4, \quad (10)$$

which are quite different from those derived from ambient strength data (1). This indicates that the micromechanism controlling deformation has distinct influence on the mechanical anisotropy of these materials in addition to the stress and temperature dependencies.

## CODF-CREEP Model & Predictions

Plasticity models are applied in conjunction with CODF to predict creep anisotropy for the materials and textures discussed earlier. Explicit dependence of anisotropic deformation and creep on preferred orientation can be obtained from matter tensorial descriptions of crystal plastic theories. Once the orientation distribution of crystallites in the polycrystalline materials is evaluated and the appropriate CODF is known, the crystallite behavior in each orientation can be weighted by the volume fraction of crystallites in that orientation so that the behavior of the polycrystalline aggregate can be predicted. Using information on the probable operating slip modes in the material under investigation, the deformation behavior of the individual crystallites as well as the polycrystalline aggregate can be evaluated.

Rigid body models such as Sachs, Taylor and Bishop-Hill are used to model the low temperature deformation behavior where specific slip systems are postulated to operate. Lower-bound estimates satisfy stress equilibrium but do not include considerations of strain compatibility among the crystallites, and these uniform stress lower-bound models are considered appropriate for creep conditions. Power-law creep is applicable for many materials, including hexagonal close packed and accordingly,

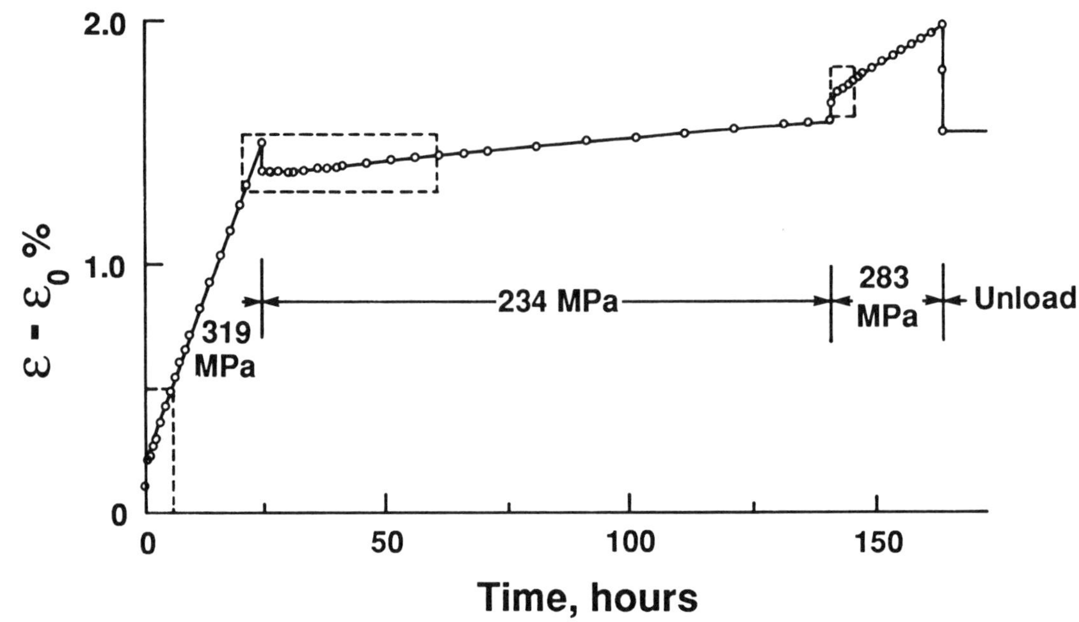


Fig.4a. Creep Curve Depicting the Effect of Stress Changes on Strain Transients
Boxed portions are enlarged in Fig. 4b along with model predictions.

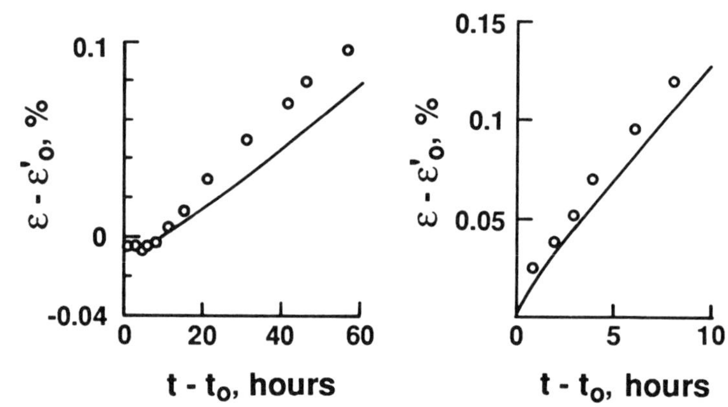


Fig.4b. Transient Creep Regions following Stress
Changes along with Model Predictions

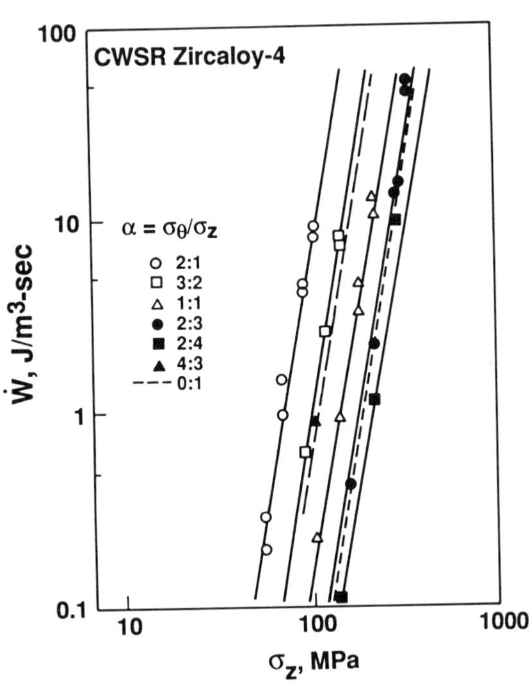


Fig.5. Double-Log Plot of Dissipation Energy Rate
versus Generalized Stress

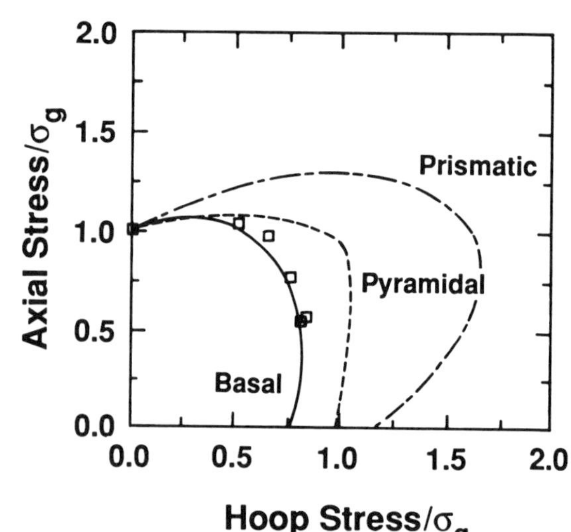


Fig.6. Experimental Creep Locus Compared with
Model ( CODF-CREEP) Predictions

$$\dot{\gamma}^{k} = A\left(\frac{\tau^{k}}{\tau_{o}^{k}}\right)^{n} \quad , \tag{11}$$

where $\tau_o$ is the reference shear stress on the $k^{th}$ slip system, n the stress exponent and C the reference strain-rate which depends on the test temperature. The stress exponent is taken to be 5.5 for the creep behavior as per the equation 4. The total strain-rate in the crystal is expressed as the sum of the contributions from each of the active slip systems,

$$\dot{\varepsilon}_{ij}^{c} = \sum_{kl} \mu_{ij}^{kl} \dot{\gamma}_{ij}^{kl} \quad , \tag{12}$$

where $\mu_{ij}$ is the geometric tensor connecting slip coordinates with crystal coordinates. The shear stress is related to the crystallite stress:

$$\tau^{k} = \sigma_{ij}^{c} \mu_{ij}^{k} \quad , \tag{13}$$

where the superscript k specifies the slip system being considered.

In the lower-bound method which is commonly used in creep, the stress state in each crystallite is regarded identical to the applied macroscopic stress state and the creep-rate is calculated using the constitutive equation. The bulk creep-rate is then obtained by averaging the creep-rate of the individual grains weighted by CODF using equation 1b.

The CODFs derived from the texture pole figures described earlier were used to obtain the model predictions which were also appropriately scaled to the same rate of dissipation (12). The models were applied to evaluate the *dominant contributions* of the basal ({00.1}<11.0>), prism ({10.0}<11.0>) and pyramidal ({11.2}<11.3>) slip systems; both the upper and lower bounding schemes yielded essentially identical creep loci. We note (Fig.6) that the model based on basal slip dominance yielded the best correlation with the experimental results whereas the low c/a ratio materials such as Zr and Ti are believed to exhibit prism slip. No *direct* experimental evidence exists to corroborate the observed correlation. The large amount of cold-work in the CWSR material might have hardened the operating prism slip planes so much that the secondary slip systems such as the basal slip become dominant under the test conditions. This observation is corroborated by the fact that the biaxial creep in recrystallized material is indeed correlatable with the predictions based on prism slip dominance (12). The compliance factors in the plasticity equations were adjusted to yield the following generalized stresses and strain-rates for the experimental material :

$$\sigma_g = 220 \text{ MPa}, \qquad \dot{\varepsilon}_g = 3.8\text{x}10^{-8} \text{ sec}^{-1}. \tag{14}$$

It should be noted that in these models, all of the slip systems are considered to operate simultaneously; however, the contribution of any specified slip system is emphasized by the appropriate choice of the reference stresses.

Once the anisotropic creep behavior is modelled using CODF and crystal plasticity, the appropriate anisotropy parameters can be evaluated. The model predictions as noted above are fit to the modified Hill's equation for the generalized stress with R and P values of 1.55 and 0.45 which are, as expected, similar to those reported above (Eq.10). It is to be noted that the derived R and P values agree extremely well with the experimental strain-rate ratios :

$$\left(\frac{\dot{\varepsilon}_{\theta}}{\dot{\varepsilon}_{z}}\right)_{1:1} = 3.2 \text{ while } \frac{R}{P} = 3.44 \quad ; \tag{15}$$

such a correlation attests to the validity of the models used here.

## Radiation Creep and Growth

In service, zirconium alloys experience stress-free radiation growth mainly due to their inherent crystalline anisotropy as well as radiation-induced and/or radiation-enhanced creep under applied stress in addition to thermal creep and elastic deformation. In general, the axial growth is directly proportional to the flux while the change in the diameter is negligible. The post-irradiation examination [PIE] of the experimental *stress-free* tubing samples in Oconee reactor yielded the following relation for the axial growth (2,6),

$$\varepsilon_{z_{sf}} = 4.9\text{x}10^{-25} \phi \, t \tag{16}$$

where the subscript sf indicates stress-free radiation growth and $\phi$ the fast (>1MeV) neutron flux in n/cm$^2$-s. Negligible changes were noted in the diameter of the unstressed clad [growth] samples up to the fluence considered (end of the 3rd cycle).

Irradiation creep-rate is modelled with linear stress and flux dependencies. Thus the radiation induced hoop creep strain is given by,

$$\varepsilon_{irr} = B \, \phi^{p} \sigma t \tag{17}$$

where $\phi$ is the neutron flux (>1 MeV), t the time of irradiation, B and p ($\approx$1) are material constants. The total creep-rate is given by the addition of the thermal and radiation components.

## Prediction of In-Reactor Clad Creep-Down

We can now summarize the application to in-pile creep predictions for CWSR cladding by referring to the above formulations for thermal creep, radiation growth and irradiation creep along with the creep anisotropy predictable from CODF-CREEP model. Using the post-irradiation examination data on well controlled cladding tubes [both stressed and unstressed], Clevinger et al (13) and Wilson et al (14) noted that radiation *does not* alter the creep anisotropy for CWSR Zircaloys and thus the CODF-CREEP textural model can be used to derive the anisotropy parameters in-pile just as in the thermal case described above. On the contrary, it is seen that the creep anisotropy of the recrystallized material is quite sensitive to the radiation exposure and radiation-induced microstructural modifications (2,12).

With the various constants determined and optimized from laboratory and PIE results along with the creep anisotropy parameters derived from CODF, it is thus possible to make predictions of the dimensional changes of the fuel rods in-service. As noted earlier, the thermal creep is modelled using strain-hardening rule with the appropriate equations developed for steady-state creep-rate, creep strain under constant stress and the strain transients during load changes with anelastic strain. These equations are input into the code as generalized stress and strain components, while the anisotropy is incorporated with the parameters derived using CODF-CREEP model for the textures of the specific tubing materials used in the Oconee reactor.

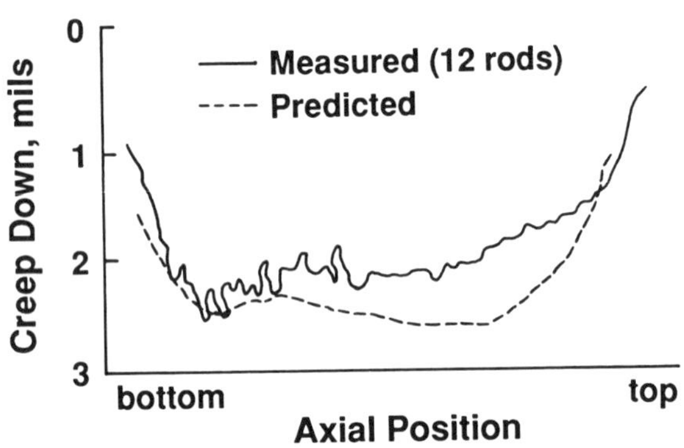


Fig.7. In-Reactor Creep-Down of Fuel Rods from Oconee Reactor Compared with Model Predictions

Figure 7 shows the axial position dependence of the decrease in the diameter known as creep-down which clearly indicates extensive deformation at the center of the tubing. This is due to the fact that both the neutron radiation flux and temperature are maximum at these locations in comparison to the top and bottom positions. The model predictions yield good correlations exactly as found earlier (2) where relatively simple models were used which did not consider the transients following load changes as well as the transition from the exponential stress dependence of the steady-state creep-rate to the power-law. This is a reflection of the fact that in this particular case, radiation creep is dominant compared to the thermal creep. The over prediction of the creep-down at the top of the rod was attributed to two-point contact between the cladding and the fuel stack thereby retarding further creep-down.

## Acknowledgements

This work is supported by the National Science Foundation through the grant DMR-9105178.

## References


1. Beauregard, R.J., G.S. Clevinger and K.L.Murty, "Effect of Annealing Temperature on the Mechanical Properties of Zircaloy-4 Cladding," in Proceedings of the 4th International Conference on Structural Mechanics in Reactor Technology, SMiRT-4, San Francisco, 1977, paper C3/5.
2. Murty, K.L., "Biaxial Creep Behavior of Textured Zircaloy Tubing," Journal of Metals (1992) in print.
3. Murty, K.L. and K.K. Yoon, Scripta Met. 13, 299-302 (1979)
4. Murty, K.L., "Anisotropic Biaxial Creep of Zircaloy Tubing," in Proceedings of the Fourth International Conference, The Institute of Metals, Swansea, 877-886 (1990)
5. Ravi, J, S.T. Mahmood and K.L. Murty, "High Temperature Creep of *High Temperature Creep of Nb-Modified Zircaloy,* Proceedings of 11th International Conference on Structural Mechanics in Reactor Technology (SMiRT-11), Tokyo, Japan (1991) paper C03/1.
6. Murty, K.L., "Technological Applications of Crystallographic Textures of Zirconium Alloys" *Materials Forum,* Vol.15, 217-230 (1991)
7. Lyashenko,V.S., Bykor, V.N. and Pavlinov, L.V., Phys. Met. Metal., Vol.8, 40-45 (1960)
8. Murty, K.L. M. Gold and A.L. Ruoff, J. Appl. Phys., Vol.41, 4917-4927 (1970)
9. Murty, K.L., Trans. Indian Institute of Metals, Vol.39, 357-368 (1986)
10. Hart, E.W., J. Eng. Materials and Technology, Vol.98H, 193-207 (1976)
11. Huang, F.H., G.P. Sabol, S.G. McDonald and Che-Yu Li, J. Nucl. Mater., Vol.79, 214 (1979).
12. Murty, K.L. and S.T. Mahmood, in *'Zirconium in the Nuclear Industry :9th International Symposium', ASTM STP 1132,* American Society for Testing and Materials, Philadelphia, 1991, pp.198-217.
13. Clevinger, G.S., B.L. Adams, and K.L. Murty, "Analysis of Irradiation Growth and Multi-Axial Deformation Behavior of Nuclear Fuel Cladding," in *'Zirconium in the Nuclear Industry (Fourth Conference)', ASTM STP 681,* American Society for Testing and Materials, Philadelphia, 1979, pp.189-201.
14. Wilson, H.W., H.J. Hsieh, K.K. Yoon and G.S. Clevinger, "Development and Correlation of Cladding Creep and Irradiation Growth Models," in Proceedings of IAEA Specialists Meeting on *Water Reactor Fuel Element Performance Computer Modelling*; Blackpool, UK (1980)

*Proceedings of the Fifth International Conference on Creep of Materials, Lake Buena Vista, Florida, USA, 18-21 May 1992*

# Development of Constitutive Relations for Creep of Composites Using an Indentation Technique

**M.G. Sharma and L. Fang**
Pennsylvania State University
University Park, Pennsylvania

## Abstract

Creep behavior of two metal matrix composites and a polymer matrix composite are studied using an indentation technique. The materials tested are aluminum oxide particles in aluminum, silicon carbide whiskers in aluminum and chopped glass fiber in Nylon. Sheet specimens were placed on a rigid base and subjected to a step loading by an indenter with a spherical tip. The resulting vertical deformation of the indenter with time was measured. A material parameter identification model was developed to back calculate constitutive parameters for all the three composites tested.

**In The Measurement Of Mechanical Properties Of Composites** It is desirable to have recourse to a simple and convenient method which requires no elaborate specimen preparation and gripping and which involves relatively simple test apparatus. The indentation technique fits the above description and is used extensively to measure various types of mechanical properties in the past. The indentation technique has been used to obtain stress-strain relations for aluminum (1) The dynamic indentation method has been used effectively to determine the high strain rate and flow behavior of ductile materials (2). Fracture toughness measurements of ceramic materials were made using an indentation technique (3,4). The need for a technique for the determination of mechanical properties of thin films has given impetus to the development of microhardness, and submicron indentation devices (5). In this paper creep properties of two metal matrix composites and a polymer matrix composite are studied using the indentation technique. The metal matrix composite displayed both primary and secondary creep response where as the polymer matrix composite was found to be linear viscoelastic in nature.

## Experimental Investigation

**Apparatus** A schematic of the indentation test apparatus for the evaluation of creep properties of composite materials is shown in Figure 1. The apparatus uses the specimens in the form of a thin film or a thick plate, placed on a massive substrate or bonded to it. For eliminating or reducing the influence of substrate properties on the measured properties of test material, it is necessary to have the indentation depths much smaller than the thickness of the sample. The specimen is indented by an indenter of spherical or conical tip. The load and the indentation depths were measured by a load cell and a linear variable differential transformer (LVDT) respectively. The specimen was enclosed in a thermal chamber in which any desirable temperature within a range of 25-500°C could be maintained during the test. The step loading needed in the creep test was obtained by operating a Bellofram load cylinder connected to a compressed air supply via a solenoid valve. The solenoid valve in turn is controlled by a pulsing unit which can be set to produce a load pulse of a desirable duration.

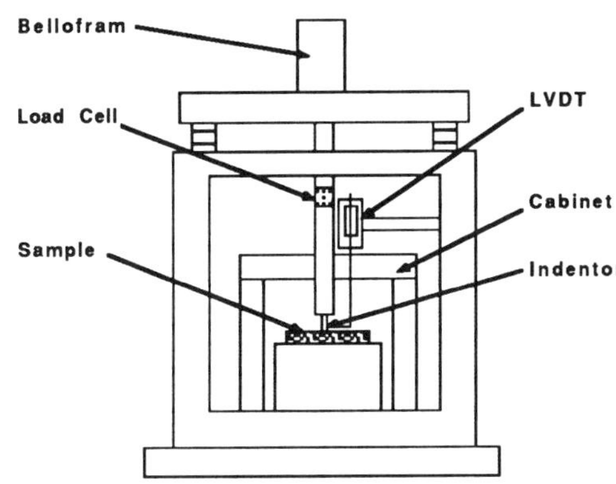


Figure 1. Schematic Diagram of the Indentation Setup

**Experimental Procedure and Results.** Creep behavior of three composites were studied using the indentation technique. These were aluminum oxide particles in aluminum (15 percent $Al_2O_3$ particles), silicon carbide whiskers in aluminum (20 percent SiC whiskers) and chopped glass fibers in Nylon (20 percent chopped glass fibers). Sheet specimens (0.0254 m. (1 in.) length, 0.0127 m. (.5 in.) wide, 0.0032 m. (0.125 in.) thick) of these three materials were placed on a rigid base and indented by a 0.0064 m.(0.25 in.) diameter cylindrical rod with a spherical tip. Creep tests were conducted by subjecting specimens to step loading of various magnitudes and the measurement of vertical deformation of the indenter with time. Figure 2 shows the displacement versus time curves obtained from the experimental data for the aluminum oxide particles in aluminum ($Al_2O_3$ (P)/Al) for three loads and a temperature of 197°C (387°F). The displacement versus time curves for silicon carbide whiskers in aluminum matrix (SiC(W)/Al) for various loads and a temperature of 94°C (201°F) are shown in Figure 3. In the same figure are shown creep curves for the aluminum matrix materials for comparison. In addition to the above matrix material, creep behavior of Glass/Nylon composite was studied for various temperatures. Figure 4 shows a typical load versus time curve for a load of 201N (45 lb). and a temperature of 81°C (178°F).

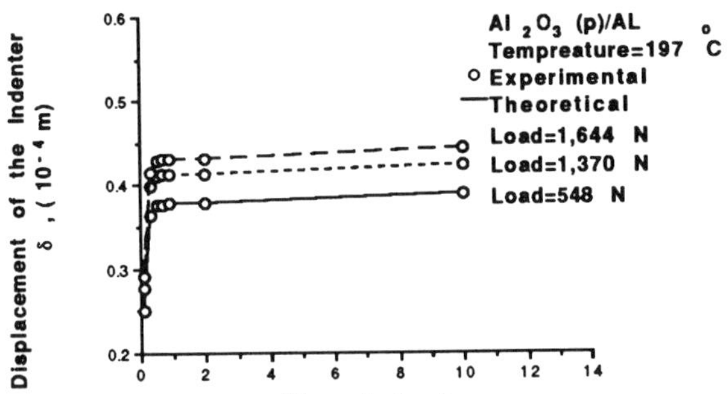

Figure 2. Indenter Displacement Versus Time Curves for $Al_2O_3$ (p)/Al Composite

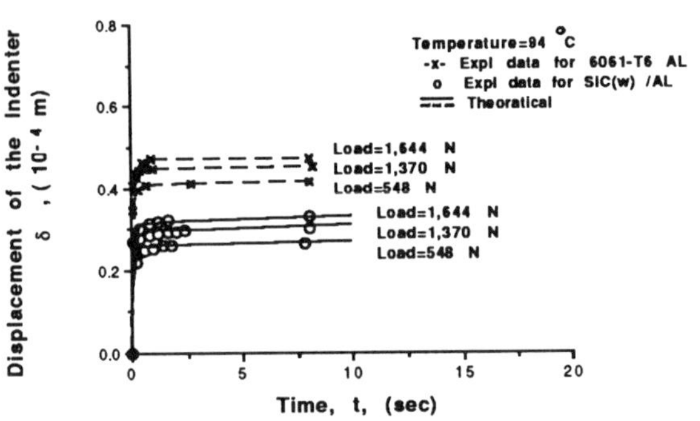

Figure 3. Indenter Displacement Versus Time Curves for SiC(w)/Al Composite and Aluminum Matrix

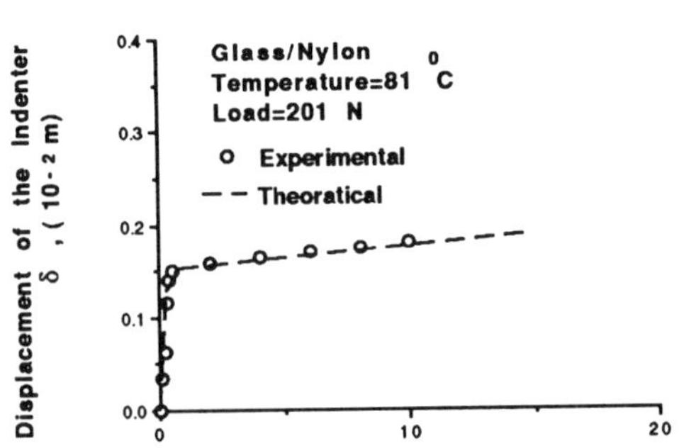

Figure 4. Indenter Displacement Versus Time Curve for Glass/Nylon Composite

# Theoretical Considerations

Creep behavior of the metal matrix composites tested was modeled according to a multiaxial theory of creep deformation for particle strengthened composites developed by Zhu and Weng [6]. The constitutive relations based upon this theory for transient and steady state creep are.

Transient Creep

$$\dot{\epsilon}_t^* = b\left\{ d\left[1-\exp\left\{-p\left(\pi^{\frac{1}{2}}/c_1^{\frac{1}{2}}-2\right)^2 r\right\}\right]\sigma^{*n}_{-}\,\epsilon^*\right\} \tag{1}$$

Steady State Creep

$$\dot{\epsilon}_s^* = a\left\{1-\exp\left[-p\left(\pi^{\frac{1}{2}}/c_1^{\frac{1}{2}}-2\right)^2 r\right]\right\}\sigma^{*n} \tag{2}$$

where $c_1$ = volume concentration of the inclusions
r = particle size of inclusions
p = material constant of the matrix
a, b, c, d and n are the material constants of the matrix

$$\sigma^* = \left(\frac{3}{2}Sij\ \ Sij\right)^{\frac{1}{2}}$$

$$\epsilon^* = \left(\frac{2}{3}\epsilon ij\ \epsilon ij\right)^{\frac{1}{2}}$$

Sij = deviatoric stress tensor
and $\epsilon ij$ = strain tensor

Using Eq. 1 and Eq. 2 in a material parameter parameter identification model which used a finite element submodel in conjunction with Conmin optimization scheme [7] constitutive parameters were evaluated. The finite element submodel analyses the stresses and strains in a narrow plate resting on a rigid semi infinite medium (see Figure 5). In figure 6 is given the flow chart for the material parameter identification model.

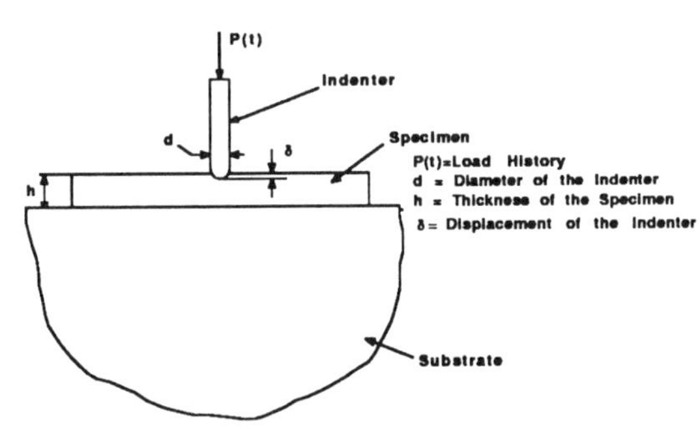

Figure 5. Plate on a Rigid Semi Infinite Medium

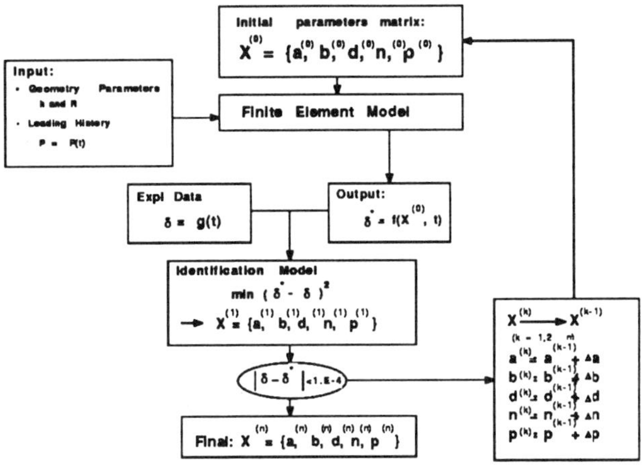

Figure 6. Flowchart for Prediction of Constitutive Parameters

For the range of loads and temperature the indentation test data on glass/Nylon composite material indicated that the material is linear viscoelastic in nature. The elastic analysis of a narrow plate on a semi infinite medium subjected to a contact load developed elsewhere (8) was modified to obtain the solution for a viscoelastic plate resting on semi infinite elastic medium. Based upon this analysis the expression for shear creep compliance can be shown to be

$$J(t) = \frac{16\sqrt{R}}{3P_o} f(R,\delta,h)[\delta(t)]^{\frac{3}{2}} \tag{3}$$

where $P_o$ = step load applied to the specimen by the indenter
$R$ = radius of the indentor
$h$ = thickness of the sample
$\delta$ = vertical displacement of the indentor
$J(t)$ = shear creep compliance of the composite
$f$ = a function of the radius and the depth of the indenter and the thickness of the sample

The function f is expressed as an integral and is evaluated numerically (8). The function f tends to 1 when the thickness of the sample becomes large compared to the indention radius and the depth of indenter. When the function f becomes one, the expression for the creep compliance reduces to the solution for a linear semi infinite linear viscoelastic medium.(9)

## Analysis and Discussion of Results

The material parameter identification model was used to determine constitutive parameters based upon equation 1 and Eq. 2. In table 1 are given the constitutive parameters that characterize the creep properties of $Al_2O_3(P)/Al$ and SiC(W)/Al metal matrix composites. Figures 2 and 3 show the indenter depth variation with time for the metal matrix composites evaluated from Eq. 1 and Eq. 2 and the constants given in table 1 The above curves fit the experimental data rather well. equation. 3 was used to evaluate the creep compliance versus time curves for glass/Nylon Composite for four temperatures (28, 38, 49 and 81°C) as shown in Figure 7. The time-

**Table 1: Material Parameters for Composites**

| | a | b | d | n | p |
|---|---|---|---|---|---|
| SiC(W)/AL | 3.4E-7 | 9.582 | 6.5E-5 | 0.3896 | 0.135 |
| $Al_2O_3$(p)/AL | 2.7E-7 | 11.273 | 8.6E-5 | 0.4130 | 0.257 |

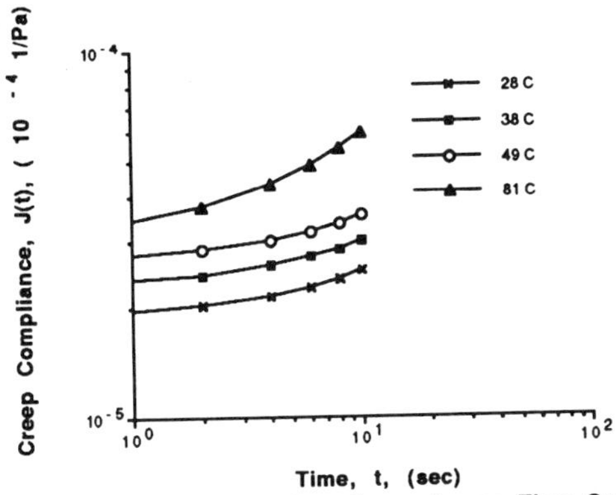

Figure 7. Creep Compliance Versus Time Curves for Glass/Nylon Composite

temperature shift hypothesis was then used to construct master creep compliance versus reduced time curve for this material (see figure 8). The master creep compliance curve represents the creep behavior for an extended time scale for a selected reference temperature. The shift factor that was used to construct the master curve is a function of temperature and was found to conform to the well known WLF equation (see Figure 9). If indentation technique is to be used effectively in the evaluation of constitutive parameters, the size of the indentor and the indentation depth in relationship to the thickness of the sample determine whether the substrate (see figure 5) properties influence the measurement of

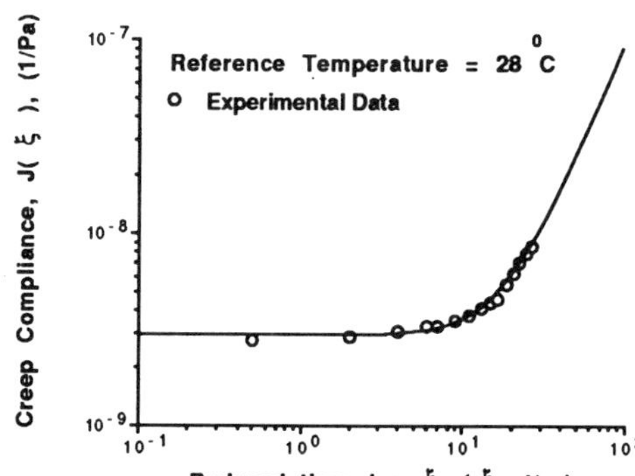

Figure 8. Master Creep Compliance Curve for Glass/Nylon Composite

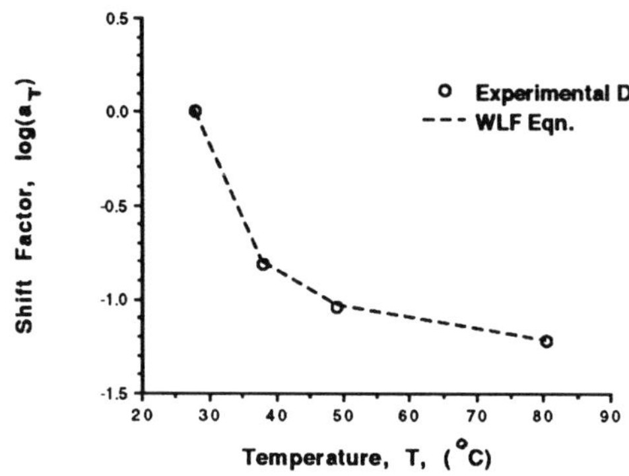


Figure 9. Shift Factor for Galss/Nylon Composite

properties of test material. Figure 10 shows that for the indentation depth and the size of the indenter used creep properties of glass/Nylon is not influenced by the substrate properties.

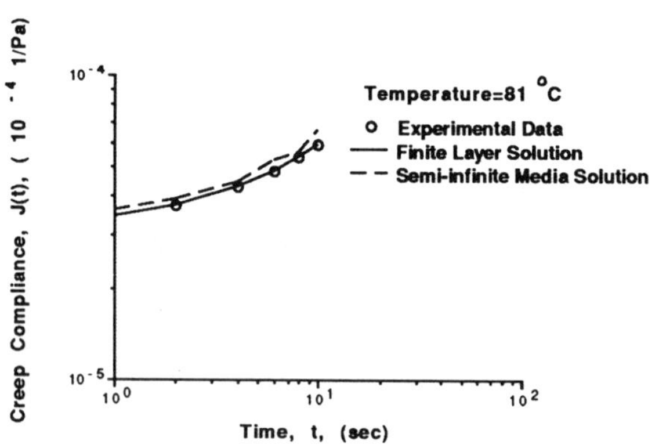


Figure 10. Creep Compliance for Glass/Nylon Composite

## Conclusions

The identation technique is a convenient and simple method for the evaluation of creep properties of metal matrix and polymer matrix composites. By proper selection of the size of the indentor and the indentation depth, the method can be effectively used for the measurement of creep properties for very thin films bonded to a substrate or a thicker plate specimen placed on a substrate and subjected a step load by an indenter.

## References


1 - Robinson, W. H. and S. D. Truman, J. Mat. Sci. 12, 1961-1965 (1977)
2 - Tirupataiah, Y. and G. Sundararajan, J. Mech. Phys. Solids, 39, 243-271 (1991)
3 - Anstis, G. R., P. Chantikul, B. R. Lawn, and D.B. Marshall, J. Am. Ceram. Soci., 64, 533-538 (1981)
4 - Marshall, D.B., and B.R. Lawn, "Micro Indentation Technique in Material Science and Engineering", ASTM STP 889, P.J. Blau and B.R. Lawn, Eds., ASTM, Phila. pp. 26-46 (1986)
5 - Doermer, M.F, and W.D. Nix, J. Mat. Res. 1, 601-609 (1986)
6 - Zhu, S.G. and G.J. Weng, J. Eng. Mat. and Tech 111 99-105, (1989)
7 - Vanderplaats, G.N. and F. Moses, Computers and Structures 3, 739-755 (1973)
8 - Yu, H.Y., S.C. Sanday and B.B. Rath., J. Mech. Phys. Solids 38, 745-764 (1990)
9 - Lee, E.H. and J.R.M. Radok, J. App. Mech 27, 438 (1960)
10 - Ferry, J.D., "Viscoelastic Properties of Polymers" 2nd Ed., Wiley, NY (1970)

*Proceedings of the Fifth International Conference on Creep of Materials, Lake Buena Vista, Florida, USA, 18-21 May 1992*

# Application of Quantitative Fractography to Creep Studies

**A.M. Gokhale and S. Mishra**
School of Materials Engineering
Georgia Institute of Technology
Atlanta, Georgia

## ABSTRACT

The fracture surface topography is expected to depend on the microstructural features actively involved in the failure process and the descriptors of the failure processes responsible for the fracture. In this investigation, an attempt has been made to develop detailed quantitative correlations among the descriptors of the fracture surface topography such as surface roughness, fractal dimension (measure of tortuosity) and fraction overlap (measure of reentrancies), and the amount of preexisting damage (voids), magnitude of applied stress during creep, and the type of failure (tensile or creep). An important feature of the present work is that the fracture surface topography is quantified by using the parameter that are estimated using the assumption-free and unbiased stereological methods developed very recently. It is observed that the fracture surfaces become rougher, more tortuous, and have more reentrant regions as the amount of preexisting damage increases. Creep fracture surfaces have significantly higher overlaps (reentrancies) than the tensile fracture surfaces (higher by a factor of two). Among the creep fracture surfaces, lower stress leads to fracture surface that is rougher, more tortuous and has more reentrant regions.

## INTRODUCTION

Deformation and fracture processes are responsible for the generation of fracture surface. Thus, the geometry and topography of the fracture surface may contain the information concerning the processes responsible for failure. The present investigation is an 'exploratory' study to determine whether the topography of the fracture surface, as characterized by its important global geometric attributes, changes significantly and systematically with the following:

(1) type of failure process (creep or tensile)
(2) the amount of microstructural damage (volume fraction of voids
(3) applied stress responsible for failure (normal stress during creep test)

The important aspect of the present investigation is that the global attributes of the fracture surface are experimentally measured by using the stereological techniques that are unbiased and assumption-free. Thus, the resulting correlations with the deformation and fracture process descriptors are expected to be 'real'. The fracture surface roughness, the fracture surface fractal dimension, and the fraction of fracture surface area having overlaps, are the three parameters stereologically measured in this investigation. The vertical section profilometry was utilized to measure the digitized co-ordinates of the points on the vertical section fracture profiles. These data were obtained by using the digital image analysis. The fracture surfaces were generated by using the creep as well as the tensile tests; a number of tensile tests were performed on interrupted creep test specimens in order to vary the amount of microstructural damage (voids) systematically. The quantitative fractographic measurements were performed on the fracture surfaces of fourteen specimens. To the best knowledge of the authors, these are the first quantitative fractographic measurements on the creep fracture surfaces, and the tensile fractures of interrupted creep specimens. The results clearly show that the topography of fracture surfaces is quite sensitive the type of failure (tensile or creep), amount of preexisting damage, and, in the case of creep fractures, the applied tensile stress. Thus, the quantitative fractographic measurements may provide useful input for failure analysis, and to obtain the information concerning the deformation and fracture processes that generate the fracture surface.

## MATERIALS AND EXPERIMENTAL PROCEDURE

The experiments were performed on the high purity oxygen free electrolytic copper (OFE, ASTM F 68-77). The creep tests were conducted on the SATEC dead weight creep machines. All the tests were conducted at 400°C, and in the high purity nitrogen atmosphere. The experimental program involved creep tests on fourteen specimens: nine interrupted creep tests at 41.4 MPa uniaxial stress level for the time periods ranging from 10% to 90% of creep life, three creep tests to failure at the stress levels of 34.5, 37.9 and 41.4 MPa, two interrupted creep tests at 34.5 and 37.9 MPa stress for the time period of 10% of the creep life. The interrupted creep tests generated a series of specimens having different amounts of creep cavities or voids, i.e., damage. The tensile test specimens were machined from these interrupted creep test specimens, and a standard tensile test was carried out at the strain rate of 0.1% per sec at room temperature, to generate the tensile fracture surfaces originating from the specimens having different amounts of microstructural damage. In addition, a tensile test was also performed on an uncrept (zero damage) specimen. Further, tensile specimens were also machined from one half of the creep failure specimens, and similar tensile test was performed, to generate the fracture surfaces originating from the specimens having the same damage as creep failure specimens but a different mode of failure. The details of experimental procedures are described elsewhere[1].

## FRACTOGRAPHY

The fracture surfaces were plated with electroless nickel, and then they were electroplated with a thick layer of copper. The specimens were sectioned along a plane parallel to the tensile axis and through the center of the cross section. The standard metallographic polishing techniques were utilized. Figures 1 to 6 illustrate the fracture profiles obtained in this manner.

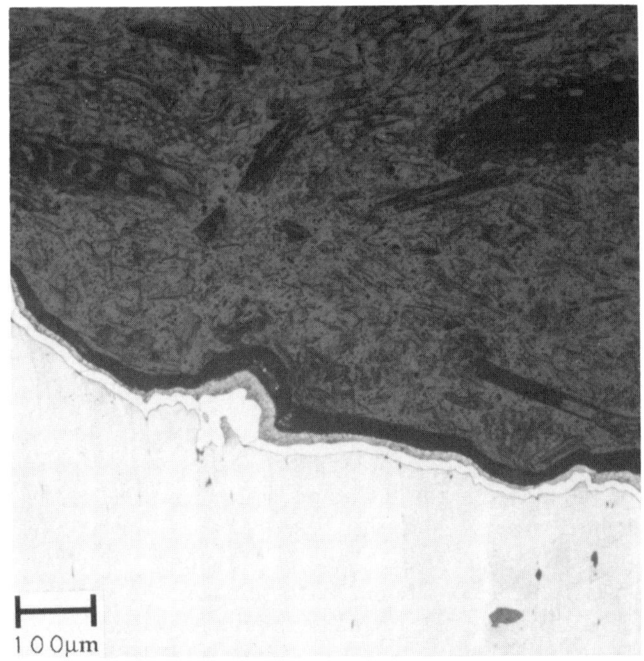


Fig. 1 - (Tensile Fracture) Fracture Profile of Specimen Tested in Tension.

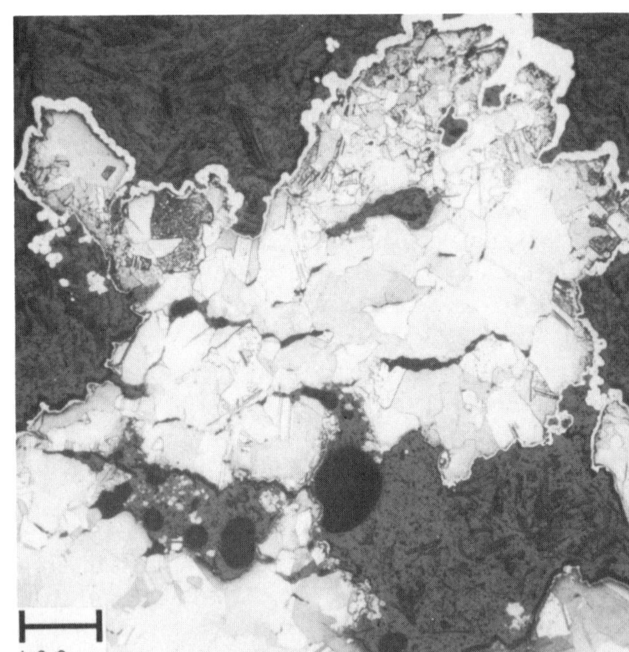


Fig. 2 - (Creep Rupture) Fracture Profile of Specimen Crept to Failure at 41.4 MPa Stress Level.

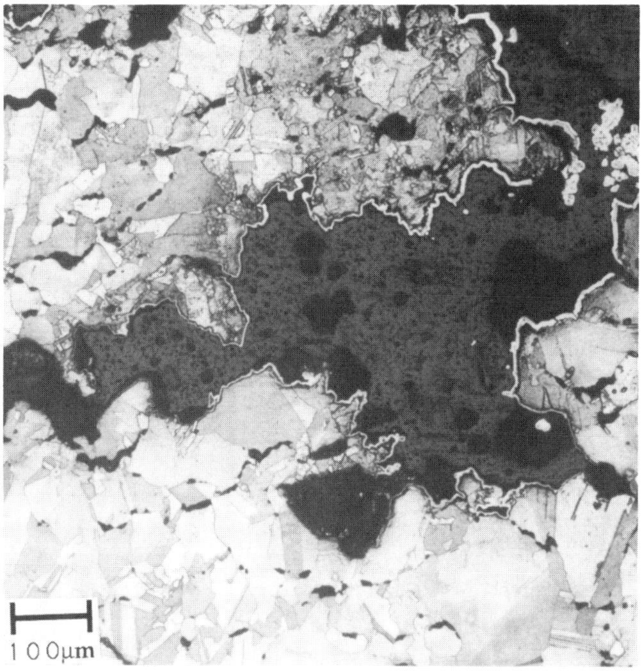


Fig. 3 - (Creep Rupture) Fracture Profile of Specimen Crept to Failure at a 34.5 MPa Stress Level.

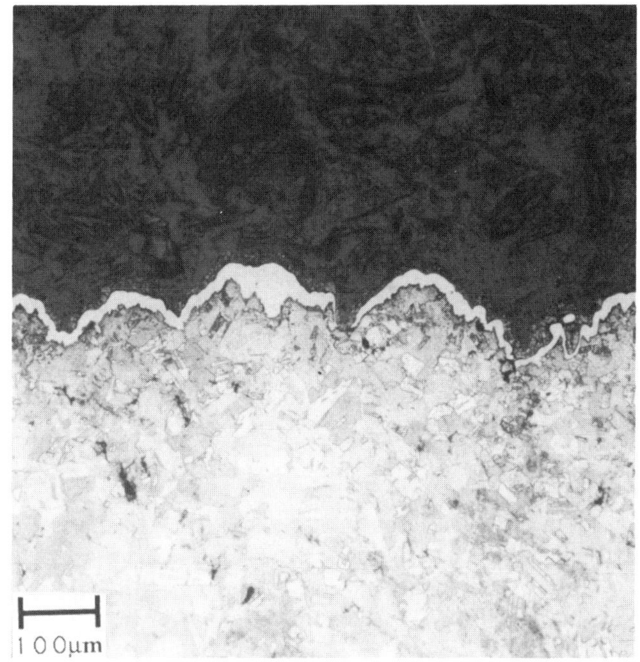


Fig. 5 - (Tensile Fracture) Fracture Profile of Specimen Crept to 10% Creep Life at 41.4 MPa Stress Level.

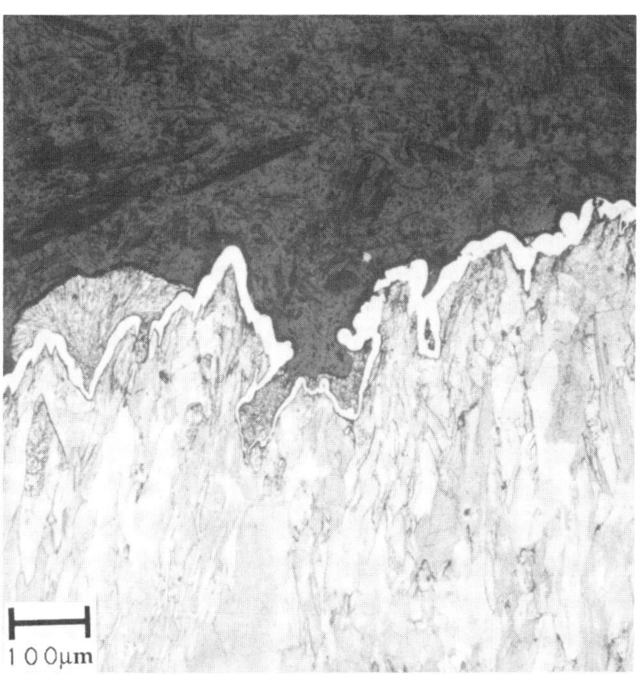


Fig. 4 - (Tensile Fracture) Fracture Profile of Specimen Crept to 90% Creep Life at 41.4 MPa Stress Level.

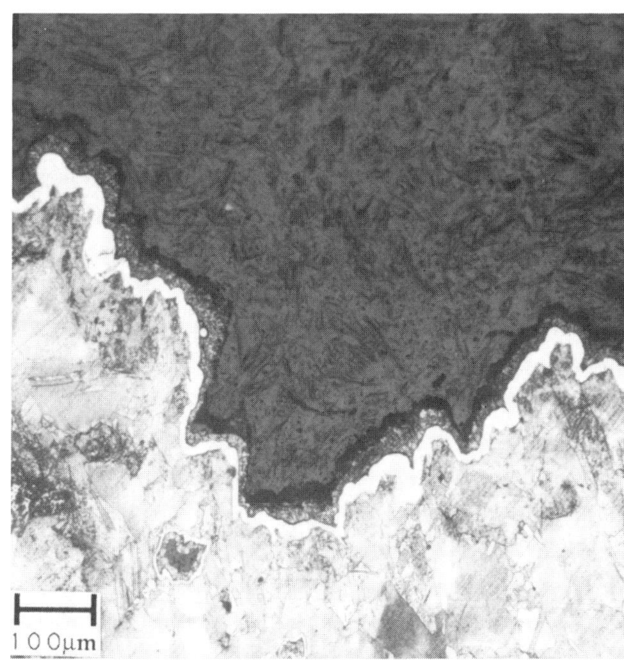


Fig. 6 - (Tensile Fracture) Fracture Profile of Specimen Crept to 100% Creep Life at 41.4 MPa Stress Level.

Inspections of these fractographs leads to the following qualitative observations.
(1) Creep fracture surfaces are rougher, more tortuous, and have very high degree of overlaps (reentrancies) as compared to tensile fracture surfaces (compare Figures 1 and 2).
(2) The creep fracture surface generated at a lower stress level is more tortuous, rougher, and has more overlaps than the one generated at the higher stress level (compare Figures 2 and 3).
(3) As the amount of preexisting damage increases, the fracture surfaces become more tortuous, rougher, and they have more overlaps (compare Figures 4 and 5).
(4) For the same amount of damage, the creep process generates a fracture surface that is more rough and tortuous as compared to the tensile fracture surface (compare Figures 2 and 6).

The above qualitative observations show that the nature and the topography of fracture surface are sensitive to the type of failure (tensile or creep), the amount of preexisting damage, and the magnitude of stress. Thus, it is of interest to quantify these attributes of the fracture surface and to develop quantitative correlations with the amount of damage and the nature of the failure process.

## QUANTITATIVE FRACTOGRAPHY

The fracture profiles were digitized by using an image analyzer. The digitization was carried out at the magnification of 200X, and sufficient length of a profile was scanned to obtain about 10,000 digitized (X,Y) co-ordinates from each profile; the Y-axis was oriented parallel to the stress axis. The coordinates were sampled at an interval of about 1 μm. This data was transferred to the VAX computer to calculate the following parameters.
(1) Profile roughness parameter $R_L$ defined (2,3) as the ratio of the fracture profile length and its apparent projected length on the x-axis. The rougher the profile, higher is the value $R_L$;$R_L = 1$ for a straight profile parallel to x-axis. In general, $R_L$ can be anywhere from 1 to ∞.

(2) Profile fractal dimension, $D_L$, which is the quantitative measure of the tortuosity of the profile. $D_L$ can be calculated from digitized coordinates data; the details are given elsewhere [4].
(3) The fracture overlap, $O_L$ which is defined as the ratio of the total lengths of the profile segments that do not contribute to the apparent projected length on the X-axis, and the total fracture profile length. $O_L$ is the tensile fracture surfaces.

(3) The fraction overlap, $O_L$ which is defined as the ratio of the total lengths of the profile segments that do not contribute to the apparent projected length on the X-axis, and the total fracture profile length. $O_L$ is the quantitative measure of the reentrancies on the fracture profile.

The fracture surface roughness $R_S$, defined as the ratio of the true area of the fracture surface and its apparent projected area on the plane perpendicular to the vertical axis (in this case tensile axis), is related to the measured profile roughness parameter $R_L$ by the following assumption-free and unbiased, general stereological relationship[5].

$$R_S = \overline{R_L \Psi} \tag{1}$$

where, the 'bar' represents the averaging over all the vertical sections profiles, and $\Psi$ is the profile structure factor that can be calculated from the digitized profile data [8]. Similarly, the fracture <u>surface</u> fractal dimension $D_S$ can be calculated from the profile fractal dimension $D_L$, without any assumptions[7]. The fraction of fracture surface overlap can be also calculated from the digitized profile data, by using equation (1). Figure 7 shows the plot of the fracture surface roughness $R_S$ vs. the volume fraction of the cavities (damage).
Figure 8 illustrates the variation of the fracture surface fractal dimension with the volume fraction of cavities (damage). Figure 9 shows the variation of the fraction overlap with the extent of damage, i.e., cavity volume fraction.

## RESULTS AND DISCUSSION

Inspection of the trends present in the Figures 7 to 9 and the qualitative observations from the fractographs shown in Figures 1 to 6 lead to the following.

(1) Figure 7 clearly shows quantitative correlation between the roughness of the fracture surface and the amount of microstructural damage (cavity volume fraction). The surface roughness increases monotonically with the amount of the damage. For the tensile fraction surfaces generated from the tensile test specimens having preexisting damage, there is one to one single valued correlation between the surface roughness and cavity volume fraction. Interestingly enough, for the same amount of damage, the creep fracture surfaces have higher roughness than the tensile fracture surfaces.

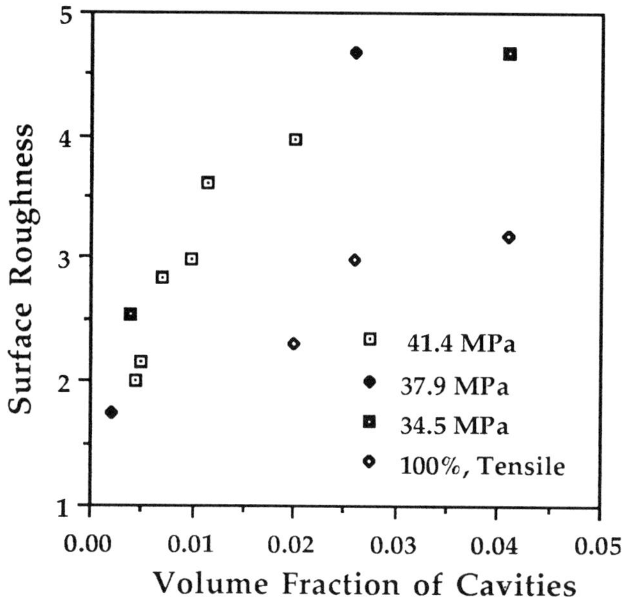


Fig. 7 - Plot of Fracture Surface Roughness ($R_S$) versus Volume Fraction of Cavities $(V_V)^C$.

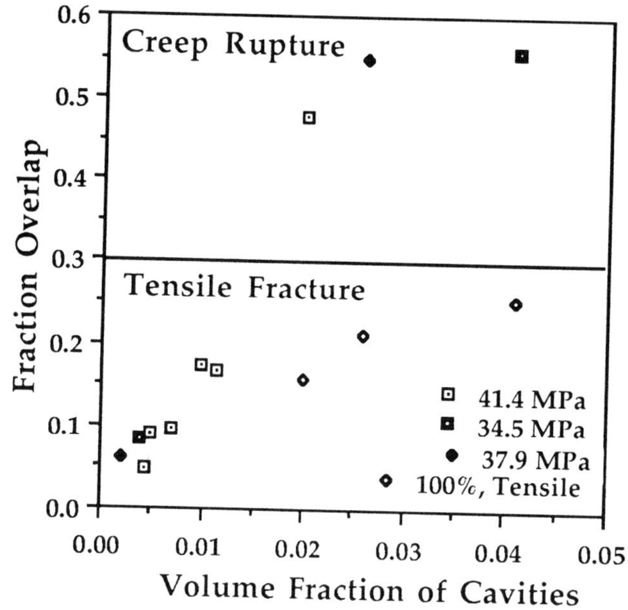


Fig. 9 - Plot of Fraction of Overlap ($O_L$) versus Volume Fraction of Cavities.

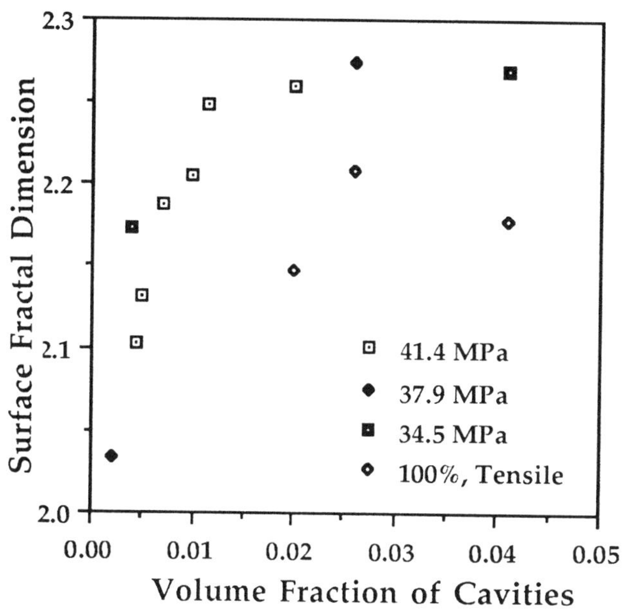


Fig. 8 - Plot of Fracture Surface Fractal Dimensions ($D_S$) versus Volume Fraction of Cavities $(V_V)^C$.

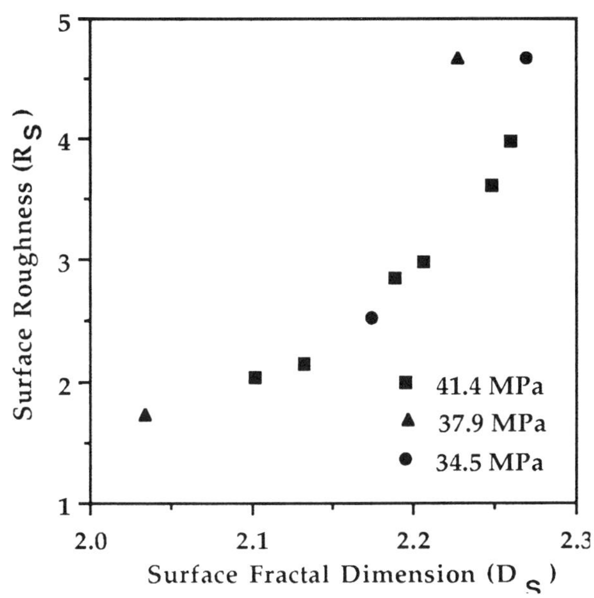


Fig. 10 - Plot of Fracture Surface Roughness ($R_S$) versus Fracture Surface Fractal Dimension ($D_S$).

Further, among the creep fracture surfaces, lower stress leads to higher quantitative value of the fracture surface roughness.

(2) Figure 8 shows a similar trend in the variation of the fracture surface fractal dimension and the cavity volume fraction. Note that the fractal dimension quantifies the tortuosity of the fracture surface. The fractal dimension increases with the amount of preexisting damage in the tensile specimens. For creep fracture surfaces, the fractal dimension (and hence the tortuosity) increases with the decrease in the applied stress. Further, the creep fracture surfaces are more tortuous than the tensile fracture surfaces.
(3) The fraction overlap (Figure 9) shows some very interesting variations. Although, the fraction overlap increases with the amount of preexisting damage in the tensile specimens, the maximum value of fraction overlap for the tensile fracture surfaces is about 0.25. On the other hand, the minimum value of the fraction overlap for the creep fracture surfaces is about 0.48. Thus, the creep fracture surfaces have at least two times as many overlaps and reentrant regions as compared to the tensile fracture surfaces (even when the amount of damage is the same)! This is a very important observation. The fraction overlap measurement may provide useful input for failure analysis.

A plot of the fracture surface roughness $R_S$ versus the fracture surface fracture dimension $D_S$ is shown in Figure 10. The data indicates that the rougher fracture surfaces have the higher tortuosity. This is not always true; in some composites [7], as the amount of damage increases, the fractal dimension remains the same but the surface roughness increases!

## CONCLUSIONS

The topography of the fracture surfaces can be quantified through the parameters such as the surface roughness parameter, fractal dimension and fraction overlap. These attributes can be experimentally measured without involving any assumption or bias; the measurement are practically feasible via digital image analysis. All the three parameters, surface roughness, fractal dimension, and fraction overlap increase systematically with the amount of damage, i.e., the fracture surfaces of the metal having more damage are rougher, more tortuous, and have more reentrancies. The creep fracture surfaces have much higher fraction overlaps (reentrant regions) than the tensile fracture surfaces; the fraction overlap increases almost by a factor of two. These observations may provide useful input for failure analysis. Among the creep fracture surfaces, the roughness, tortuosity and fraction overlap increase as the applied stress decreases.

## ACKNOWLEDGEMENTS

This work is supported by NSF sponsored project DMR-9013098 entitles "Quantitative Analysis of the Fracture Surfaces Using Stereological Methods"; Dr. B. McDonald is the project monitor. The financial support is gratefully acknowledged.

## REFERENCES

1. Mishra, S., M.S. Thesis, Georgia Institute of Technology, 1992.
2. Underwood, E.E., and Banerji, K., "Quantitative Fractogrpahy" in ASM Metals Handbook, 9th Ed., Vol. 12, 193-210.
3. Pickens, J.R., and Gurland, J., 4th Int. Congr for Stereology, E.E. Underwood, R. deWit, and G.A. Moore eds, NBS Spec. Publ. 431, Gaithersburg, MD, 1976, pp 269-72.
4. Banerji, K., Metall. Trans. A, 1988, Vol. 19, pp. 264-71.
5. Gokhale, A.M., Underwood, E.E., Metall. Trans. 21A, pp 1193-99 (1990).
6. Gokhale, A.M., and Drury, W.J., Metall. Trans. 21A, 1201-07 (1990).
7. Gokhale, A.M., and Drury, W.J., "Metallography - 75 Years Later", ASTM Spec. Tech. Pub., in press.

Proceedings of the Fifth International Conference on Creep of Materials, Lake Buena Vista, Florida, USA, 18-21 May 1992

# Mechanism of High Temperature Brittle Intergranular Cracking in High Strength Nickel Alloys

**J. Kameda and A.J. Bevolo**
Ames Laboratory
Iowa State University
Ames, Iowa

## Abstract

The mechanism of high temperature (823-923 K) brittle intergranular cracking has been studied in several γ' strengthening nickel alloys undoped and doped with S, Zr and/or B. Alloys with low and high S showed high susceptibility to high temperature brittle intergranular cracking. The addition of B in S-doped alloys more substantially increased the brittle cracking resistance than that of Zr. The alloy with high S exhibited the highest intergranular S segregation but the S segregation did not reach very high levels because of the formation of Ti rich sulfides. The presence of Zr mostly scavenged S by forming sulfides in the alloy with high S and Zr. A high amount of segregated B at grain boundaries resulted in a reduction in S segregation in the alloy with high S & B. The intergranular segregation of S and B gives rise to global embrittling and toughening effects, respectively. The alloys with low S and high S & B had more S segregation on high temperature fracture surfaces, which was much greater than at grain boundaries, compared with the alloys with high S and high S & Zr. A remarkable S enrichment was observed near sulfides at stressed grain boundaries in the S-doped alloys. It is proposed that diffusive S fluxes from crack surfaces and near sulfides at stressed grain boundaries to the crack tip induce local embrittlement. The composition and temperature dependence of the brittle cracking behavior is elucidated in terms of the global and local embrittling effects.

HIGH TEMPERATURE CRACKING frequently occurs along grain boundaries. In low to medium strength materials, high temperature intergranular cracking results from the cavity formation. It has been shown [1,2] that the nucleation and growth of cavities are controlled by the condensation of vacancies at particles and/or grain boundary sliding, and the self diffusion at grain boundaries or on cavity surfaces, respectively. On the other hand, it is possible in high strength materials that high temperature cracking exhibits a brittle feature which is not associated with cavitation [3-6]. Although high temperature brittle intergranular cracking is known to be related to the dynamic segregation of embrittling solute like S, the segregation mechanism near the high temperature crack tip which controls brittle cracking remains still controversial [7-10].

It is well recognized in nickel base superalloys [6,11,12] that the presence of residual impurities such as S deteriorates the high temperature fracture properties while the addition of Zr and/or B improves them. Since nickel base superalloys generally exhibit mixed modes of intergranular cavitation and brittle cracking, however, the composition effect on the high temperature brittle cracking behavior in high strength nickel alloys has not been fully understood [6,12].

The present paper attempts to summarize recent results of high temperature brittle intergranular cracking and solute segregation behavior in high strength nickel alloys undoped and doped with S, Zr and/or B, which are presented in detail elsewhere [13,14]. The high temperature cracking behavior was characterized in terms of the crack growth rate, da/dt, vs. applied stress intensity, K, relationship. Solute segregation at grain boundaries and on high temperature intergranular fracture surfaces was examined using scanning Auger microscopy (SAM). The mechanism for high temperature brittle intergranular cracking is qualitatively proposed.

## Experimental

The materials used in this study were five γ' strengthening nickel alloys undoped and doped with S, Zr and/or B. The nickel alloys were prepared using a vacuum drop casting method. The chemical composition of the alloys is shown in Table I. The alloys with low S & Al, low S, high S, high S & Zr, and high S & B were designated as LSA, LS, HS, HSZ and HSB, respectively. All the alloys except the LSA alloy contained the same content of Al, and the LSA and LS alloys were undoped with S. The conditions of processing, and recrystallization and aging treatments of the alloys are shown in Ref. [13].

High temperature (823-923 K) cracking behavior was examined in vacuum (1.4 x $10^{-5}$ Pa) under a constant loading condition by measuring the da/dt vs. applied K relationship. The da/dt was determined using a DC potential method. High temperature (923 K) stress relaxation behavior in the various alloys was investigated by smooth compressive specimen tests under a constant displacement condition. The geometry of specimens used in this study and the detailed experimental

Table I. Chemical composition of several nickel alloys (wt. %).

| | Ti | Al | S | Zr | B | C | P | O | N |
|---|---|---|---|---|---|---|---|---|---|
| Low S & Al (LSA) | 5.65 | 3.47 | 0.0030 | 0.0006 | <0.0027 | 0.0062 | <0.0001 | 0.0039 | 0.0004 |
| Low S (LS) | 5.95 | 5.20 | 0.0031 | 0.0007 | <0.0028 | 0.0059 | <0.0001 | 0.0038 | 0.0002 |
| High S (HS) | 5.70 | 5.14 | 0.0105 | 0.0007 | <0.0028 | 0.0060 | <0.0001 | 0.0023 | 0.0003 |
| High S & Zr (HSZ) | 5.80 | 5.46 | 0.0107 | 0.2300 | 0.0030 | 0.0060 | <0.0001 | 0.0022 | 0.0001 |
| High S & B (HSB) | 5.94 | 5.14 | 0.0108 | 0.0006 | 0.0410 | 0.0057 | <0.0001 | 0.0010 | 0.0002 |

method of measuring the crack growth rate and stress relaxation are described in Ref. [13].

Solute segregation at grain boundaries and on high temperature fracture surfaces was analyzed using SAM. SAM specimens were extracted from compact tension (CT) specimens load-interrupted during the high temperature crack growth experiment so as to contain the crack tip. Notched SAM specimens also were prepared from the portion of the load-interrupted specimen not affected by the applied stress. The SAM specimens of the LSA and HSB alloys were hydrogenated in order to promote intergranular embrittlement. All the SAM specimens were fractured at about 100 K in an ultra high vacuum Auger chamber (2.7 x $10^{-8}$ Pa). Selected area and line profile SAM analyses on contiguous low and high temperature intergranular fracture areas were carried out using a cylindrical mirror analyzer of Physical Electrics model 600 operated at 10 keV. The detailed SAM analysis is presented in Ref. [14].

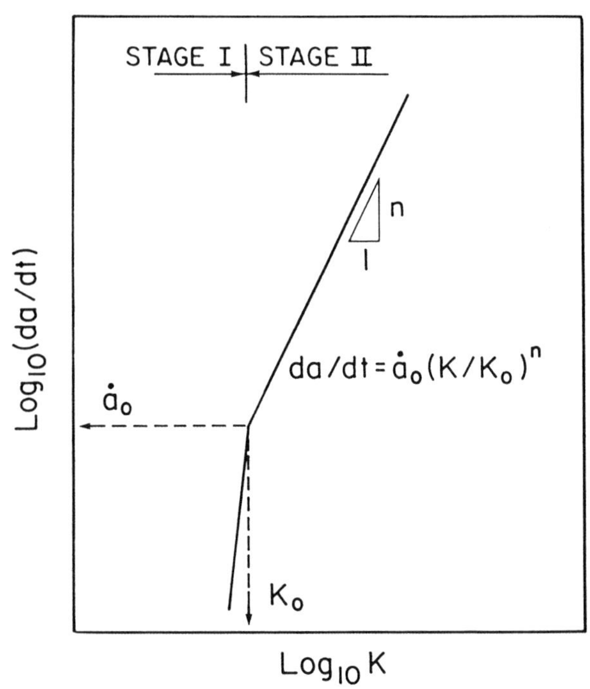


Fig. 1. Schematic representation of high temperature cracking behavior.

## High Temperature Crack Growth Behavior

As schematically illustrated in Fig. 1, the high temperature cracking behavior in all the alloys was found to be divided into two regimes; at lower applied K levels, da/dt very steeply depends on the applied K (stage I) and above the critical applied K, $K_o$, da/dt more weakly increases with increasing applied K (stage II) [13]. The stage II cracking behavior can be characterized by

$$\frac{da}{dt} = \dot{a}_o\left(\frac{K}{K_o}\right)^n \qquad (1)$$

where $K_o$ and $\dot{a}_o$ represent a transition from the crack nucleation to growth controlling process and n indicates the stress dependence of da/dt.

Figure 2 shows the variations of $K_o$ and n with the alloy composition and test temperature. Since the Al content did not affect the cracking behavior [13], the result of undoped alloys was represented by that of the LSA alloy. The alloy

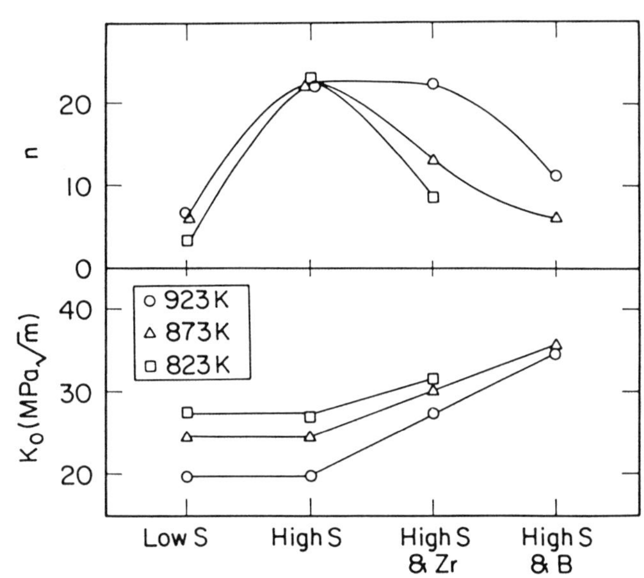


Fig. 2 Variation of $K_o$ and n with alloy composition and test temperature.

composition and temperature produced quite different effects on $K_0$ and n. The LSA and HS showed the lowest $K_0$ among all the alloys and therefore the bulk content of S did not influence the value of $K_0$. The addition of Zr or B in S-doped alloys increased $K_0$. Higher values of $K_0$ were observed in the HSB alloy than in the HSZ alloy. The LSA and HS alloys showed stronger temperature dependence of $K_0$ than the HSZ and HSB alloys. On the other hand, increasing the bulk S content drastically increased n. In contrast to Zr doping, the addition of B in S-doped alloys decreased more greatly n, which almost reached the level in the LSA alloy. The HSZ alloy exhibited the strongest temperature dependence of n while the HS alloy had no temperature dependence. The composition dependence of $\dot{a}_0$ was found similar to that of $K_0$ [13]. The temperature dependence of $\dot{a}_0$ is discussed later.

As shown in Fig. 3, high temperature fracture surfaces in all the alloys exhibited mostly a brittle intergranular feature which is not associated with cavitation. However, smoother fracture surfaces were observed in the LSA alloy compared with in the other alloys [13]. The different fracture morphology is reflected by the grain boundary feature which is affected by the size of γ' precipitates [13]. Moreover, the LSA alloy exhibited occasional striations probably caused by crack arrest in all the specimens and partial intergranular cavitation in specimens tested at lower temperatures and higher da/dt [13,14]. It was also found [14] that in load-interrupted specimens the high temperature crack tip is often located in the mid part of grain boundary facets in the LSA and HSB alloys while cracks are mostly arrested at the triple junction of grain boundaries in the HS and HSZ alloys. These results imply that different microcracking processes might occur depending on the alloy composition. This point is discussed later. Note that the HS and HSZ alloys showed intergranular fracture partly associated with ductile tearing even at 300 K [15].

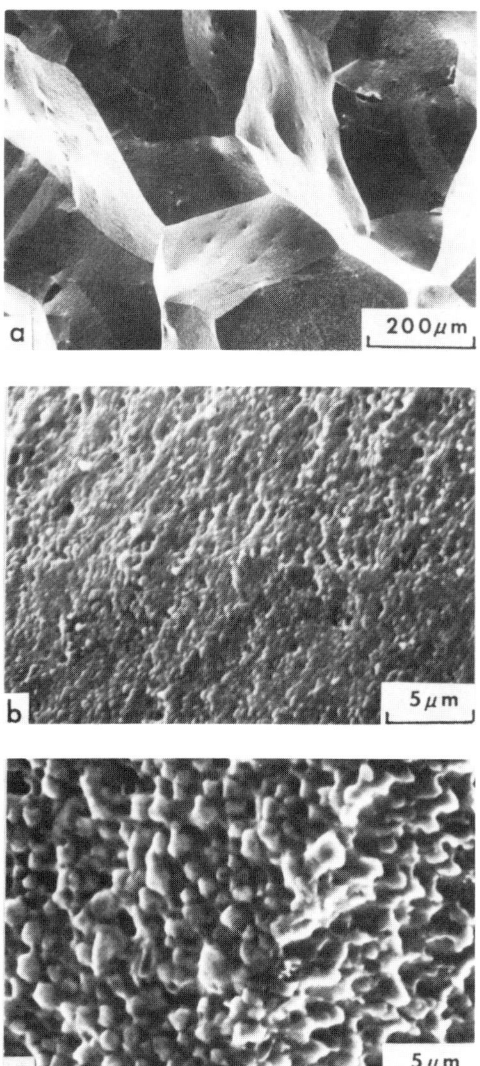


Fig. 3. Scanning electron micrographs on high temperature (923 K) fracture surfaces in alloys with (a) & (b) low S & Al and (c) high S.

High temperature brittle intergranular cracking is considered to be controlled by stress, as opposed to cavitation which is driven by strain. From a mechanical point of view, high temperature brittle cracking more readily occurs in materials with high stress relaxation resistance because high stress intensification can be maintained ahead of the high temperature crack tip. Therefore, the stress relaxation behavior, which might be related to the brittle cracking, was investigated in the various alloys. Figure 4 shows the result of high temperature (923 K) stress relaxation in the various alloys. The applied stress was more readily relaxed as follows: the LSA alloy > the LS alloy > the HS alloy > the HSB alloy > the HSZ alloy. It is clear that the composition dependence of brittle intergranular cracking is not highly correlated with that of stress relaxation. The reason is discussed later.

Several research groups [9,10,16] have attempted to discuss the rate process controlling high temperature brittle cracking in ferritic alloy steels in terms of the Arrhenius plot of da/dt at given applied K levels. The estimated activation energies were found to be greater than those for S and self diffusion in lattice. The estimate of unreasonably high activation energies is ascribed to the fact that the local stress and S enrichment near the crack tip, which control the brittle cracking behavior, change depending on the da/dt and temperature even at the same applied K level [13]. Since the quantity of $\dot{a}_0$ represents a specific cracking behavior

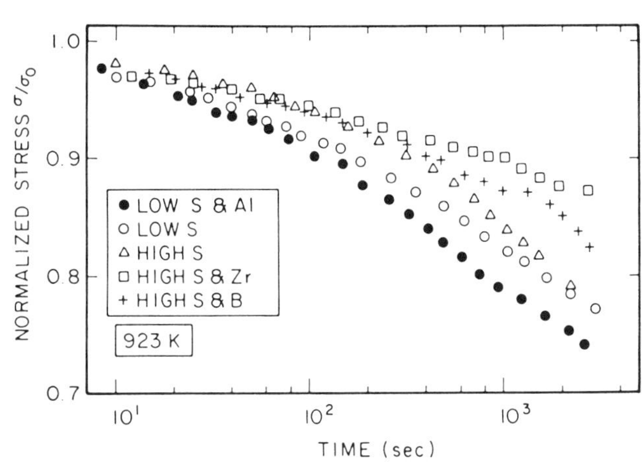


Fig. 4. Variation of stress normalized by initial applied stress, $\sigma_0$, with relaxation time at 923 K in various alloys.

corresponding to the transition from the crack nucleation to growth controlling process, the Arrhenius plot of $\dot{a}_0$ provides a more reasonable estimate of the activation energy for brittle cracking (Fig. 5). It has been shown [13] that the estimated activation energies varying from 64 to 103 kJ/mol are in close agreement with those for S diffusion at grain boundaries and on crack surfaces.

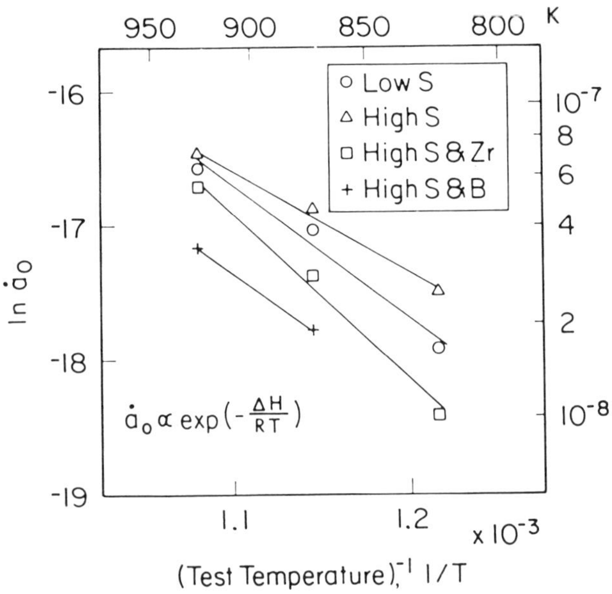


Fig. 5. Arrhenius relationship between $\dot{a}_0$ and test temperature in various alloys.

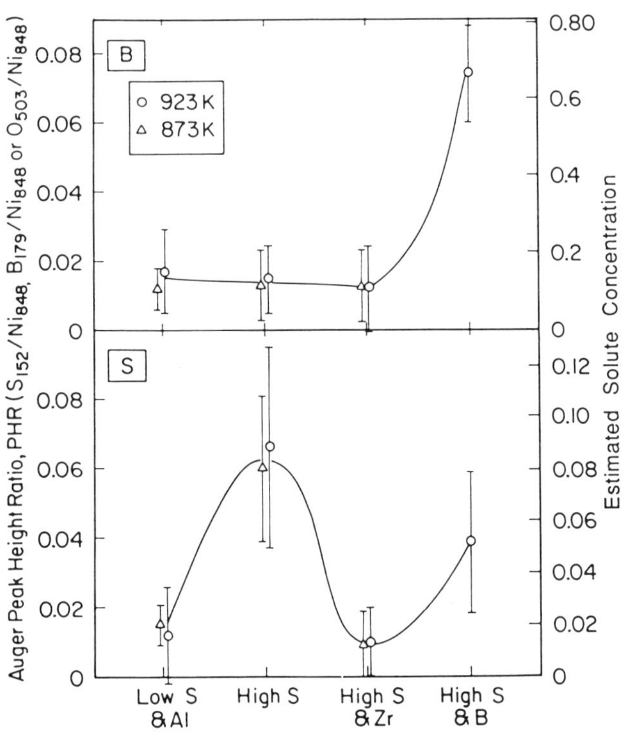


Fig. 6. Variation of Auger peak height ratio (PHR) and estimated concentration of S and B at unstressed grain boundaries with alloy composition and test temperature.

## Solute Segregation Analysis

In order to examine the effect of solute segregation on the high temperature brittle cracking behavior, SAM analyses were performed at unstressed grain boundaries. The SAM analysis at unstressed grain boundaries indicated the segregation of S, O and B giving rise to global embrittling and toughening effects. Figure 6 shows the variations of the Auger peak height ratio (PHR) of S & B to Ni and the estimated solute concentration with the alloy composition and temperature. The HS alloy showed the highest S segregation. However, the S segregation did not reach very high levels because of the formation of Ti rich sulfides [14]. The addition of Zr in S-doped alloys decreased S segregation to the level in the LSA alloy. The formation of Zr rich sulfides resulted in scavenging nearly all of the S in the HSZ alloy [14]. Intergranular B segregation was substantially found in the HSB alloy and the site competition by segregated B caused S segregation to decrease partly. All the alloys showed the same amount of segregated O. Thus the composition dependence of brittle cracking can not be explained by O segregation-induced embrittlement [14]. The test temperature did not affect solute segregation at unstressed grain boundaries.

To clarify the mechanism of brittle intergranular cracking, SAM analyses of solute segregation near the high temperature crack tip were further required. Figure 7 shows the line profiles of adsorbed O and segregated S on contiguous low and high temperature intergranular fracture surfaces in LSA alloy specimens load-interrupted at da/dt = 3 x $10^{-7}$ m/s. The build up of O on high temperature fracture surfaces is considered to be achieved mostly during the exposure of load-interrupted specimens to room temperature air because free surfaces were found not to absorb O much under a simulated mechanical testing condition* [15]. High temperature intergranular fracture surfaces exhibited much higher S segregation than grain boundaries and a peak of segregated S

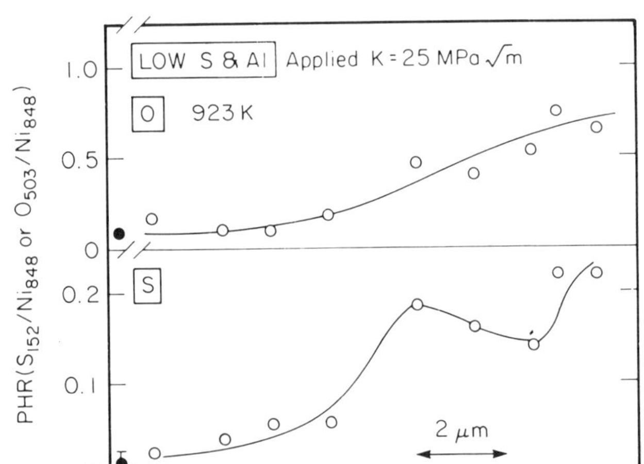


Fig. 7. Line profiles of adsorbed O and segregated S on contiguous low and high temperature fracture surfaces indicated by LT and HT, respectively, in alloy with low S & Al. Closed points indicate the result at unstressed grain boundaries.

*The oxygen partial pressure was less than 1.4 x $10^{-6}$ Pa was observed in a few μm distance away from the crack tip in all the alloys. Argon sputtering experiments indicated that S is segregated in a monolayer of high temperature fracture surfaces [14]. The effect of alloy composition on S segregation on high temperature fracture surfaces (at constant da/dt = 3 x $10^{-7}$ m/s but different applied K) was found to be different from that at grain boundaries [14]. The S segregation on high temperature fracture surfaces increased in the following order: the HSB alloy > LSA alloy > the HSZ alloy > the HS alloy. The amount of segregated S on high temperature fracture surfaces increased with increasing temperature and B enrichment on high temperature fracture surfaces was not observed in the HSB alloy [15].

Although the stress intensification ahead of the crack tip slightly enhanced S segregation at grain boundaries in the LSA alloy (Fig. 7), no significant effect of applied stress on intergranular S segregation was observed in the other alloys. This result is not in agreement with that observed in ferritic steels [7,8,17,18]. The disagreement of the stress effect might be related to the difference in grain boundary structure between nickel and iron. As shown in Fig. 8, however, a remarkable enrichment of S was found near sulfides at stressed grain boundaries located just ahead of the crack tip.

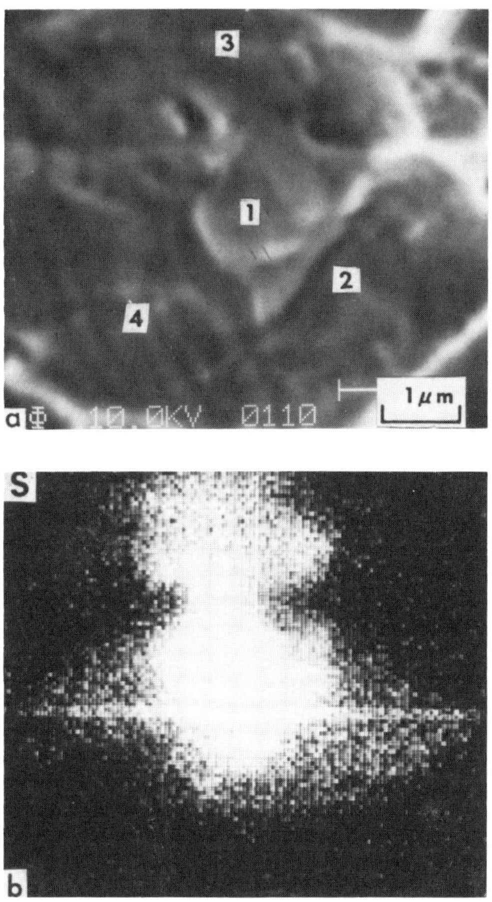


| Spot | $S_{152}$ & $Zr_{147}/Ni_{848}$ |
|---|---|
| 1. | 6.14 |
| 2. | 0.60 |
| 3. | 0.67 |
| 4. | 0.07 |

Fig. 8. Scanning electron micrograph of Zr rich sulfide at stressed grain boundaries and Auger map (bright image) for S & Zr along with PHR at different spots in alloy with high S & Zr. The presence of Zr was identified by measuring a high energy peak (1845 keV) of Zr.

## Mechanism of Local S Enrichment and High Temperature Brittle Intergranular Cracking

Based on the activation energies estimated for brittle cracking and SAM result, an attempt is made to develop a qualitative model for S segregation-induced high temperature brittle intergranular cracking. The sequence of S segregation near the crack tip during the heating and/or loading of specimens is shown in Fig. 9. Precrack surfaces are fully covered with oxides and/or adsorbed O before heating the CT specimen (Fig. 9a). From the SAM observation of S segregation on oxidized free surfaces under a simulated mechanical testing condition [14], it is expected that S segregation on the precrack surface increases with increasing heating time by replacing O which dissolves into the grain interior before applying stress (Fig. 9b). When stresses are applied, diffusive S fluxes from precrack or high temperature

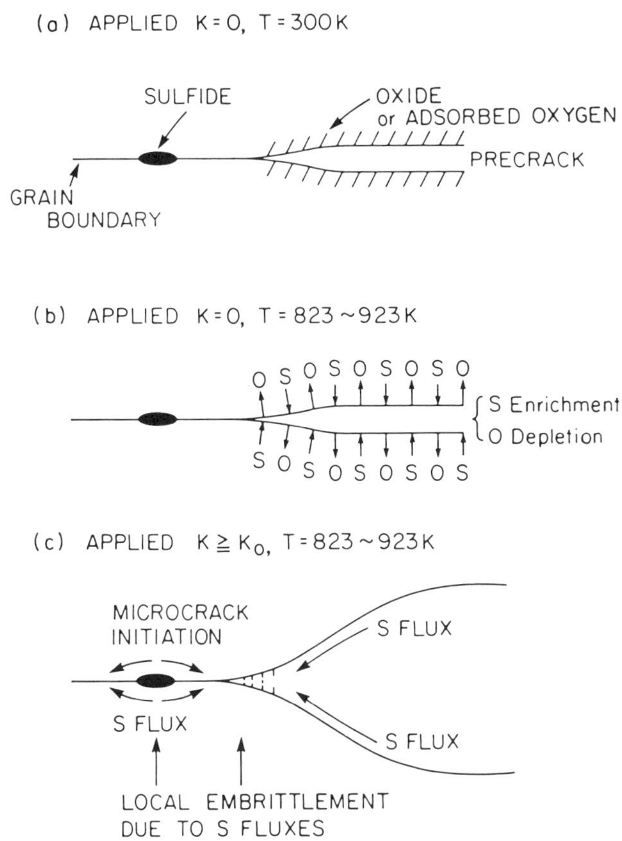


Fig. 9. Schematic illustration of variation of adsorbed O and segregated S on precrack surfaces and near sulfides and local embrittlement process: (a) precrack surfaces covered with oxide or adsorbed O at applied K = 0 and T = 300 K, (b) S enrichment and O depletion on precrack surfaces at applied K = 0 and T = 824-923 K and (c) diffusive S fluxes from crack surfaces and near sulfides facilitating local decohesion at applied $K \geq K_o$ and T = 823-923 K.

crack surfaces to the crack tip are assumed to be driven by the difference in S concentration between the crack surface and crack tip region, and the reduction in chemical potentials at the bond stretching region (Fig. 9c). Vacancy fluxes occurring from the crack tip to crack surfaces assist rapid S migration on crack surfaces via vacancy exchange mechanism. Moreover, S fluxes are released from sulfides at stressed grain boundaries which is probably attributed to the annihilation flux of vacancies to the sulfide interface. The S fluxes from crack surfaces and near sulfides to the crack tip facilitates locally intergranular decohesion which controls high temperature brittle cracking (Fig. 9c). It is also possible that sulfides accompanied by S enriched regions become nuclei of microcracks and the coalescence of several microcracks causes macrocrack extension.

The proposed model is somewhat analogous to one presented by Shin and McMahon [9]. In their model, it is instead assumed that sulfides on high temperature crack surfaces which dissolve into the grain interior play a major role in inducing brittle cracking. Hippsley et al. [7] have proposed a drift model on the assumption that S enrichment near the crack tip is achieved via lattice S diffusion assisted by the local stress gradient. The drift model indicated a broad depth profile of segregated S near the crack tip. However, the assumption and result of the drift model is not consistent with the present experimental findings of the activation energy for brittle cracking lower than for lattice S diffusion and monolayer S coverage on high temperature fracture surfaces.

Since the susceptibility of the alloys to high temperature brittle intergranular cracking was not highly correlated with the stress relaxation behavior in the smooth compressive test, the dislocation activity near the crack tip where S segregation is accumulated appears to be different from in the grain interior. Rice and Thomson [19] have shown that an energy barrier for the formation of dislocation loops at the crack tip increases with decreasing surface energy. It is most likely that the creation of dislocations at the crack tip is substantially hindered by reducing the surface energy due to the S enrichment during the high temperature cracking process. Thus the S-induced embrittling effect controls more strongly the high temperature brittle intergranular cracking than the local plastic flow.

While high temperature cracks grow in a subcritical manner, two different types of microcracking probably occur, i.e., intermittently rapid and continuously slow growth of cracks [7,20,21]. The preferential occurrence of the microcracking processes depends on the relative magnitude of the global and local embrittling effects. The macrocrack growth rate controlled by the rapid microcracking is related proportionately to the unit crack jumping distance (changing from the striation spacing to the grain boundary facet size) and inversely to the incubation time required for the build up of S near the crack tip. Increasing the global embrittling effect facilitates the intermittently rapid crack growth by increasing the unit jumping distance and decreasing the incubation time. In materials with weak global embrittling effects, on the other hand, cracks can grow slowly by maintaining the embrittling regime induced by S fluxes from crack surfaces to the crack tip because diffusive S fluxes have stronger ability to decrease the bond strength than nondiffusive ones [22,23].

Finally, the effects of alloy composition and temperature on n and $K_o$ (Fig. 2) are discussed in terms of the global and local embrittling effects which are summarized in Table II. The HS and HSZ alloys exhibiting intergranular fracture even at 300 K have greater global embrittlement. In these alloys, it is possible that rapid growth of cracks intermittently occurs at the stage II. The high amount of segregated S at grain boundaries produces the highest value of n independent of the temperature in the HS alloy (Fig. 6). In the HSZ alloy, S fluxes near Zr rich sulfides at stressed grain boundaries control the brittle cracking behavior by facilitating the formation of microcracks at sulfides and the temperature dependence of n becomes strong (Fig. 8). On the contrary, the LSA and HSB alloys possess a weak global embrittling effect due to small amounts of segregated S and sulfides, and grain boundary toughening by B segregation, respectively (Fig. 6). Diffusive S fluxes from crack surfaces is most likely to induce slow crack growth and the value of n is lowered in the LSA and HSB alloys.

Table II. Relative magnitude of global vs. local embrittlement (GE vs. LE) controlling brittle cracking in various nickel alloys.

| Alloy | GE | | LE | |
|---|---|---|---|---|
| | gb | sp | ct | sp |
| LSA | weak | very weak | medium-strong | very weak |
| HS | strong | weak | weak | strong |
| HSZ | weak | medium | weak | strong |
| HSB | weak | weak | low-medium | low-medium |

gb: weakening at grain boundaries
sp: weakening near sulfide particles
ct: weakening near crack tip

With decreasing da/dt, higher S fluxes which produce a great local embrittling effect can be maintained because of an increase in the diffusion time of S near the crack tip. Then the LSA and HS alloys show the lowest value and strong temperature dependence of $K_o$. However, the value of of $K_o$ in the HS alloy is not lowered below that found in the LSA alloy. This is partly because a critical stress might be required for intensifying S fluxes from crack surfaces to the crack tip [24]. The S scavenging effect of Zr and toughening effect by segregated B appears to become more effective with decreasing da/dt, and high values of $K_o$ are required for brittle cracking in the HSZ and HSB alloys. Increasing the applied K overwhelms the thermal effect for increasing S fluxes. Thus the HSZ and HSB alloys show weak temperature dependence of $K_o$.

## Conclusion

The mechanism for high temperature brittle intergranular cracking in high strength nickel alloys undoped and doped with S, Zr and/or B has been studied in light of the influence of solute segregation. High temperature brittle cracking in alloys with low and high S occurred at lower applied K levels than in alloys with high S & Zr or B. The activation energies for brittle cracking were closely related to those for S diffusion at grain boundaries and on crack surfaces. The alloy with high S exhibited the highest S segregation. The addition of Zr in S-doped alloys mostly scavenged S by forming Zr rich sulfides. The alloy with high S & B had substantially high intergranular B segregation. High temperature intergranular fracture surfaces depicted more strongly S segregation than grain boundaries. A remarkable S enrichment was observed near sulfides at stressed grain boundaries. It is proposed that S fluxes from crack surfaces and near sulfides at stressed grain boundaries to the crack tip induces local embrittlement. The composition and temperature dependence of the brittle cracking behavior is elucidated in terms of the proposed model.

## Acknowledgment

Ames Laboratory is operated for the U. S. Department of Energy by Iowa State University under contract no. W-7405-ENG-82. This work was supported by the Office of Basic Energy Sciences, Division of Materials Sciences.

## References

1. "Proc. Third Int. Con. on Creep and Fracture of Engineering Materials and Structures" (edited by B. Wilshire and R. W. Evans), The Inst. of Metals, London (1987).
2. "Perspective in Creep Fracture" (edited by M. F. Ashby and L. M. Brown), Pergamon Press Ltd., Oxford (1983).
3. Ritter, J. C. and R. McPherson, Metal Technol., 1, 506 (1974).
4. McMahon, C. J., Jr., R. J. Dobbs and D. H. Gentner, Mater. Sci. Eng., 37, 179 (1979).
5. Hippsley, C. A., J. F. Knott and B. C. Edward, Acta Metall., 28, 869 (1980).
6. Floreen, S. and J. M. Davidson, Metall. Trans. A, 14A, 895 (1984).
7. Hippsley, C. A., H. Rauh and R. Bullough, Acta Metall., 32, 1381 (1984).
8. Hippsley, C. A., Acta Metall., 35, 2399 (1987).
9. Shin, J. and C. J. McMahon, Jr., Acta Metall., 32, 1535 (1984).
10. Lewandowski, J. J., C. A. Hippsley and J. F. Knott, Acta Metall., 35, 2081 (1987).
11. Holt, R. T. and W. Wallace, Int. Metals Rev., 21, 1 (1976).
12. Decker, R. F. and J. W. Freeman, Trans. Metall. Soc. of A.I.M.E., 218, 277 (1960).
13. Kameda, J., submitted to Acta Metall. et Mater.
14. Kameda, J. and A. J. Bevolo, submitted to Acta Metall. et Mater.
15. Kameda, J. and A. J. Bevolo, unpublished work at Ames Laboratory (1991).
16. Hippsley, C. A., Metal Sci. Technol., 1, 475 (1985).
17. Lewandowski, J. J., C. A. Hippsley, M. B. D. Ellis and J. F. Knott, Acta Metall., 35, 593 (1987).
18. Misra, R. D. K. and T. V. Balasubramanian, Acta Metall., 38, 1263 (1990).
19. Rice, J. R. and R. Thomson, Phil. Mag., 29, 73 (1974).
20. Kameda, J., Acta Metall., 34 1721 (1986).
21. Kameda, J. and M. L. Jokl, Scripta Metall., 16, 325 (1982).
22. Rice, J. R., "Effect of Hydrogen on Behavior of Materials" (edited by I. M. Bernstein) p. 445, T.M.S.-A.I.M.E., New York (1976).
23. Hirth, J. P. and J. R. Rice, Metall. Trans. A 11A, 1501 (1980).
24. Kameda, J. and C. J. McMahon, Jr., Metall. Trans. A, 14A, 903 (1983).

*Proceedings of the Fifth International Conference on Creep of Materials, Lake Buena Vista, Florida, USA, 18-21 May 1992*

# Creep Behavior of a Tungsten-3.6w/o Rhenium-0.33w/o Zirconium Carbide at Ultrahigh Temperatures

**A. Luo and D.L. Jacobson**
Department of Chemical, Bio, and Materials Engineering
Arizona State University
Tempe, Arizona

## Abstract

The creep behavior of an arc-melted tungsten-3.6w/o rhenium-0.33w/o zirconium carbide was evaluated at ultrahigh temperatures (1900-2400 K) under 15-80 MPa stresses in a vacuum of better than $1.3x10^{-6}$ Pa ($1.0x10^{-8}$ torr). The effects of temperature and stress on the steady-state creep rate were determined and the stress exponent and activation energy for creep deformation were measured. The present study showed that ZrC particle was very effective in strengthening tungsten at high temperatures. The temperature-compensated creep rate of a W-3.6Re-0.33ZrC at temperatures above 0.5 $T_m$ was about one order lower than that of a W-5Re, and was about two orders lower than that of pure W tested under the identical condition. The creep behavior of a W-3.6Re-0.33ZrC was also correlated to its microstructure developed during high temperature deformation. Transmission electron microscopy (TEM) micrographs of crept W-3.6Re-0.33ZrC samples showed that the high creep-resistance of this alloy was associated with the presence of ZrC particles which retarded the movements of dislocations and subgrain boundaries, resulting in both direct particle strengthening and indirect particle strengthening.

TUNGSTEN possesses an unique thermionic properties and the highest melting temperature among all metals, and thus is considered as an ideal material for the emitter of thermionic energy converter. Pure tungsten is subject to brittle fracture at room temperature, and as a result, the use of tungsten has been severely limited due to the difficulties in fabrication. An effective way to lower the ductile-brittle transition temperature of tungsten is to alloy with rhenium at a composition of about 4.0 pct Re (1-7). Previous studies showed that solid-solution softened W-Re alloys could be ductile at room temperature if the fabrication history was carefully controlled. On the other hand, dispersion with stable second-phase particles is an effective way to strengthen tungsten at temperatures above 0.5 $T_m$ (melting temperature of tungsten in Kelvin). In studying the mechanical properties of arc-melted W-Re-Hf-C alloys, Klopp and Witzke (8) found that the high-temperature strengthening effect of second-phase particles in tungsten could be combined with the room-temperature ductilizing effect of rhenium without detrimental interaction. It is therefore interesting to develop particle strengthened W-Re alloys for the space power applications. Zirconium carbide (ZrC), a compound with very high melting temperature and formation energy, seems to be a potential high temperature strengthener in tungsten. But so far the effect of ZrC on the high temperature mechanical properties of dilute W-Re has not been studied.

The objectives of the present study were: 1) to study the creep behavior of a W-3.6Re-0.33ZrC in detail by determining the effects of stress, temperature, and microstructure on creep rate; 2) to compare the creep strength and creep rate of a W-3.6Re-0.33ZrC with respect to pure W and W-5Re, and 3) to find the strengthening mechanism of dispersed ZrC particles in tungsten at ultrahigh temperatures.

## Experimental Procedures

The material used for the present study was an arc-melted W-3.6w/o Re-0.33w/o ZrC alloy in the form of a 25.4 mm diameter bar. Flat-plate creep specimens with a gage section of 12.7 mm x 3.175 mm x 0.8 mm were prepared by an electric discharge machine (EDM). The specimens so obtained were first mechanically polished with 400 and 600 grit emery papers, and then chemically polished in a 10% NaOH solution to ensure a smooth surface condition. All creep specimens were annealed at 2473 K for 2 hours to produce a stable grain size prior to testing. The average grain size, as determined by the intercept method, was 110±5 micron.

Specimens were tested in tension by a step-load method in a vacuum of better than $1.3x10^{-6}$ Pa ($1.0x10^{-8}$ torr) on a custom-built UHV creep testing facility interfering with a computer data recording and analysis system. No contamination occurred during annealing and testing. Specimen was tested in a stress range of 15 MPa to 80 MPa in a temperature range of 1900-2400 K. The load was applied to the specimen through a bellows, and the stresses were maintained to ±1 pct of the selected value by making periodical corrections for the changes in specimen cross-sectional area during deformation and for the changes in the bellows spring force due to the extension of the load column. Elongation was measured with a linear variable differential transducer (LVDT) by which it was possible to measure an elongation of 1 micron. Testing temperature was measured by an optical pyrometer which had been calibrated in advance with a standard filament. The maximum uncertainty of the measured temperature was within 5 K in the entire temperature range employed in the present study. There was no measurable temperature gradient along the specimen gage length.

The dislocation substructure of deformed specimens was studied by TEM. Coupons taken from the gage section of post-test creep specimens were first thinned by electropolishing in a 2.0 pct NaOH solution at a cell potential of 15 v, and then examined with a JEOL 2000FX high-resolution transmission electron microscope operating at 200 kv.

## Results

The normal three-stage creep was observed in the step-load creep tests of this alloy, i.e. primary, steady-state, and tertiary creeps. Figure 1 shows the typical strain-time creep curves of the step-load creep specimens tested at 1900 K and 2200 K, respectively. These two curves show that upon initial loading a primary creep region with decreasing creep rate occurred, followed by a steady state creep region with a constant creep rate. After the steady state creep rate was measured at a given stress, the stress was abruptly changed to a new level and the primary creep took place again. Eventually tertiary creep occurred with necking in the gage section followed by creep rupture. The steady state creep, which was of primary interest in the present study, was relatively long for most of the test conditions employed in the present study. The effect of stress and temperature on the steady state creep behavior of a W-3.6Re-0.33ZrC alloy can be obtained by analyzing these creep curves.

Figure 2 shows the stress-dependence of the steady state creep rate of a W-3.6Re-0.33ZrC at various temperatures obtained in step-load creep tests. Straight parallel lines were obtained in this log-log plot, which suggested that the steady state creep rate of this alloy had a power-law relationship with the applied stress as following:

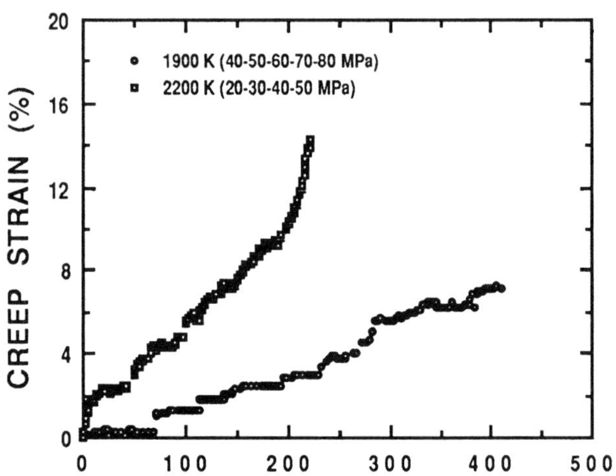


Fig. 1 - Typical strain-time creep curves of the step-load creep tested W-3.6Re-0.33ZrC.

$$\dot{\varepsilon} \propto \sigma^n \tag{1}$$

where $\dot{\varepsilon}$ is the steady state creep rate, $\sigma$ is applied stress, and **n** is the stress exponent for creep deformation, which is an indication of the stress sensitivity of the steady state creep rate. The value of **n** at given temperature can be obtained by measuring the slope of each straight line in Figure 2. The values of **n** obtained in a W-3.6Re-0.33ZrC ranged from 5.2 to 5.8 at temperatures from 1900 to 2400 K, indicating that the stress exponent of this alloy was not sensitive to the testing temperature. A least squares approximation of the creep data showed that the **n** value of this alloy was about 5.4.

The steady state creep rate of a metallic material at a constant temperature can also be expressed as a function of the normalized stress (9), i.e.

$$\dot{\varepsilon} \propto \left(\frac{\sigma}{E}\right)^n \tag{2}$$

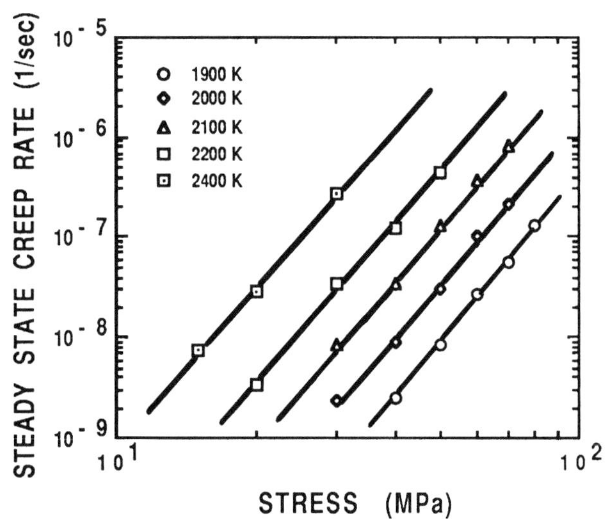


Fig. 2 - Stress-dependence of the steady state creep rate of a W-3.6Re-0.33ZrC.

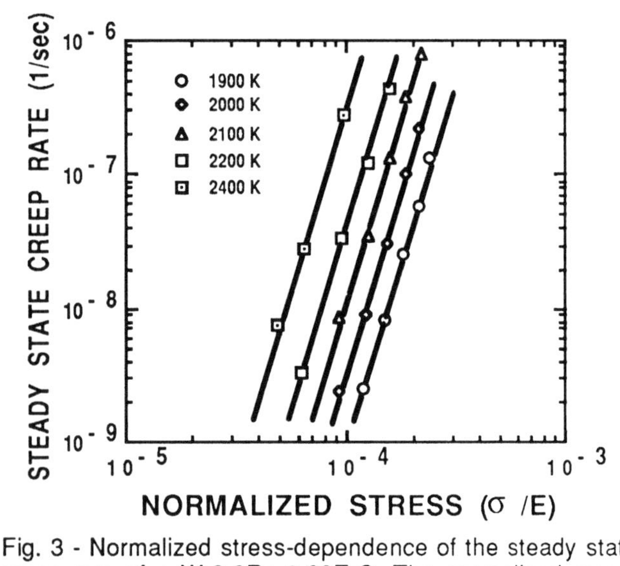


Fig. 3 - Normalized stress-dependence of the steady state creep rate of a W-3.6Re-0.33ZrC. The normalized stress exponent for creep was also determined to be about 5.4.

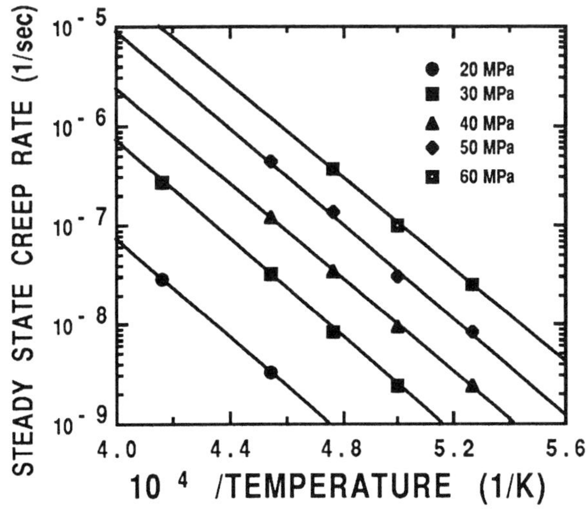


Fig. 4 - Effect of temperature on the steady state creep rate of a W-3.6Re-0.33ZrC.

where E is the Young's modulus of matrix at given temperature. If we normalize the applied stress by the Young's modulus of polycrystalline tungsten reported by Lowrie and Gonas (10), assuming that the addition of Re has little effect on the modulus, there still exists a linear relationship between creep rate and normalized stress on a logarithmical scale as shown in Figure 3. The normalized stress exponent for creep of a W-3.6Re-0.33ZrC was also found to be approximately 5.4.

The effect of temperature on the steady-state creep rate of a W-3.6Re-0.33ZrC alloy is shown in Figure 4. This figure is a plot of steady state creep rate (in log scale) versus the reciprocal temperature. The apparent activation energy for creep deformation of this alloy was determined experimentally at constant stress by applying the following expression (11):

$$\dot{\varepsilon} \; \alpha \; \exp(-\frac{Q_c}{RT}) \quad (3)$$

where $Q_c$ is the apparent activation energy, R and T have their usual meanings. The slope of each straight line in Figure 4 was measured and it was found that in temperature region of 1900 to 2400 K the apparent activation energies of this alloy ranged from 104 to 112 Kcal/mol with an average value of 109 Kcal/mol.

From the results obtained in the present study, the stress- and temperature-dependence of the steady state creep rate of a W-3.6Re-0.33ZrC alloy can be described by the following equation

$$\dot{\varepsilon} = A \bullet (\frac{\sigma}{E})^n \bullet \exp(-\frac{Q_c}{RT}) \quad (4)$$

where A is a constant and other parameters have been defined previously. This equation is in accordance with the equations for steady state creep rate of common metallic materials proposed by Mukherjee, Bird, and Dorn (12) and by Weertman (13). By substituting the known values of **n** and $Q_c$ into Eq. 4, the value of constant A was determined to be about $1.1 \times 10^{25}$/sec. Therefore, Eq. 4 becomes

$$\dot{\varepsilon} = 1.1 \bullet 10^{25} \bullet (\frac{\sigma}{E})^{5.4} \bullet \exp(-\frac{109{,}000}{RT}) \quad (5)$$

This equation allows us to predict the steady state creep rate of a W-3.6Re-0.33ZrC alloy at a given stress of σ and a temperature of T (1900 K≤T≤2400K) as long as the power-law relationship is valid. According to this equation, a thermionic emitter made by a W-3.6Re-0.33ZrC alloy under a load of 10.4 MPa (1,500 psi) at 2100 K would have a steady state creep rate of $3 \times 10^{-11}$/sec, i.e., it would take about 10 years for such an emitter operating under these conditions to reach a strain of 1.0 pct. Therefore, W-Re-ZrC alloy seems to be a very promising material for the space nuclear power applications

## Discussion

In order to characterize the strengthening effect of ZrC particles in tungsten at high temperatures, we may compare the creep properties of a W-3.6Re-0.33ZrC with those of pure W and W-Re alloy. Figure 5 shows the creep strength of a W-3.6Re-0.33ZrC with pure W (14) and a W-5Re (15) for a constant creep rate of $10^{-6}$/sec. It is very clear that ZrC particles had a very significant strengthening effect on the W-Re matrix. At 1900 K, the creep strength of a W-3.6Re-0.33ZrC is about 44 MPa greater than that of a W-5Re. However, the strengthening effect of ZrC particles decreases as temperature increases. At 2200 K, the strength advantage of a W-3.6Re-0.33ZrC over a W-5Re is reduced to about 18 MPa. If we extrapolate the creep strength of a W-5Re to higher temperatures, we will see that both W-3.6Re-0.33ZrC and W-5Re alloys have almost the same creep strength at about 2500 K. That means ZrC particles lose all the strengthening effect on the W-Re matrix at 2500 K and higher temperatures.

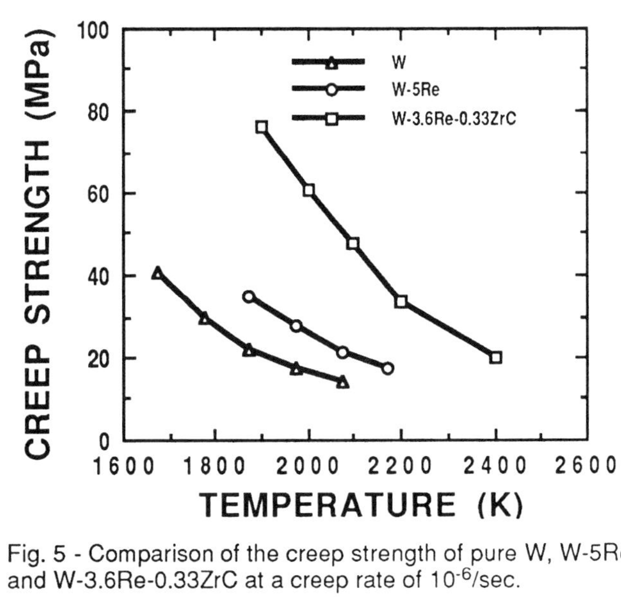


Fig. 5 - Comparison of the creep strength of pure W, W-5Re, and W-3.6Re-0.33ZrC at a creep rate of $10^{-6}$/sec.

One of the most often used parameters for the comparison of creep properties of different materials is the temperature-compensated creep rate, Z, so called Zener-Holloman parameter (16) which is calculated from the relation

$$Z = \dot{\varepsilon} \cdot \exp(-\frac{Q_c}{RT}) \quad (6)$$

By using such a parameter, the creep property of a W-3.6Re-0.33ZrC can be compared with those of pure W (14) and a W-5Re (15) in a temperature-compensated creep rate vs. normalized stress plot. In Figure 6, the Z values of these three materials were plotted logarithmically against the normalized stress. This figure clearly shows the strength advantage of a W-3.6Re-0.33ZrC over pure W and a W-5Re in a wide range of testing temperature and applied stress. At any given stress, the temperature-compensated creep rate of a W-3.6Re-0.33ZrC is about one order lower than that of a W-5Re, and is about two orders lower than that of pure W. The superior creep-resistance of a W-3.6Re-0.33ZrC over pure W and a W-5Re at ultrahigh temperatures is apparently due to the second-phase particle strengthening of ZrC in tungsten matrix.

Typical transmission electron micrographs of crept W-3.6Re-0.33ZrC specimens are shown in Figures 7 and 8. The interaction between dislocations and ZrC particles is clearly seen in Figure 7. This figure reveals that ZrC particles tend to retard the movement of dislocations. The dislocations were bent and constrained by ZrC particles, indicating that the motion of the individual dislocation is impeded by ZrC particles. Figure 8 shows the dislocation substructure of specimen crept at 2200 K. It reveals some interesting characteristic creep microstructures of this alloy. Evidently, subgrains were formed and dislocation tangles were no longer existed. Subboundaries pinned by spherical ZrC particles usually consist of either dislocation networks or dislocation walls (simple tilt boundaries). Zirconium carbide particles were located in the subboundaries as well as the subgrain interior. Dislocation density in the subgrain interior is usually very low, and the dislocations in the subgrain interior are always seen to be pinned by ZrC particles. In the examination of the subgrain structure, it was also found that the subgrain size was normally smaller near the grain boundary. In other words, the density of the subgrains is higher near grain boundaries.

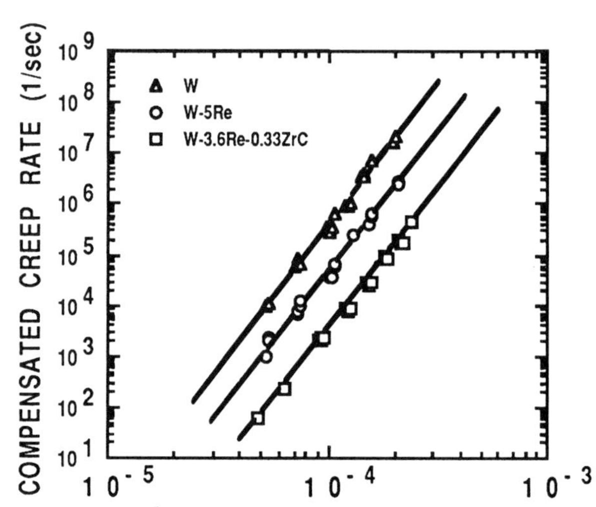


Fig. 6 - Creep rate comparison for pure W, W-5Re, and W-3.6Re-0.33ZrC. The temperature-compensated creep rate of a W-3.6Re-0.33ZrC is about one order lower than that of a W-5Re, and about two orders lower than that of pure W.

High density grain boundary dislocations (GBDs) are also observed in Figure 8.

From above microstructural examinations, it can be concluded that the interactions between ZrC particles and dislocations and the formation of subgrains are the two typical substructural features of crept W-3.6Re-0.33ZrC specimens. The strengthening effect of dispersed ZrC particles in tungsten matrix is therefore attributed to both of the direct and indirect particle strengthening. Direct particle strengthening is due to the immobilization of dislocations by direct interaction with ZrC particles. While, indirect particle strengthening is due to the stabilization effect of ZrC particles on grains and subgrains that in turn reduces the distance that mobile dislocations can move before being immobilized at subboundaries or grain boundaries.

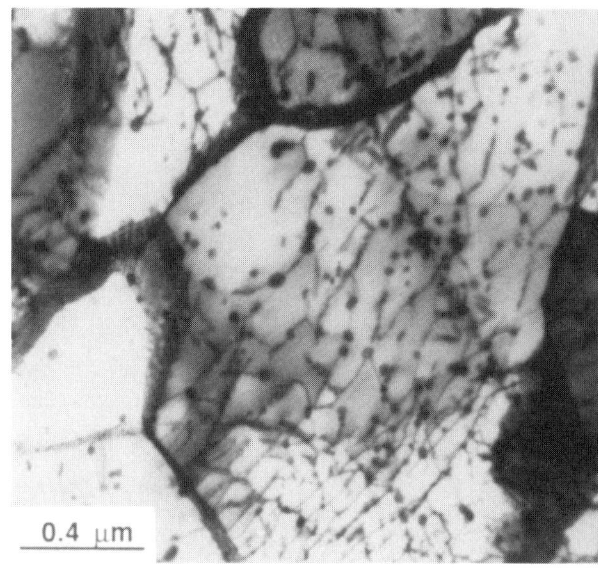


Fig. 7 - Dislocation substructure in a W-3.6Re-0.33ZrC specimen crept at 1900 K.

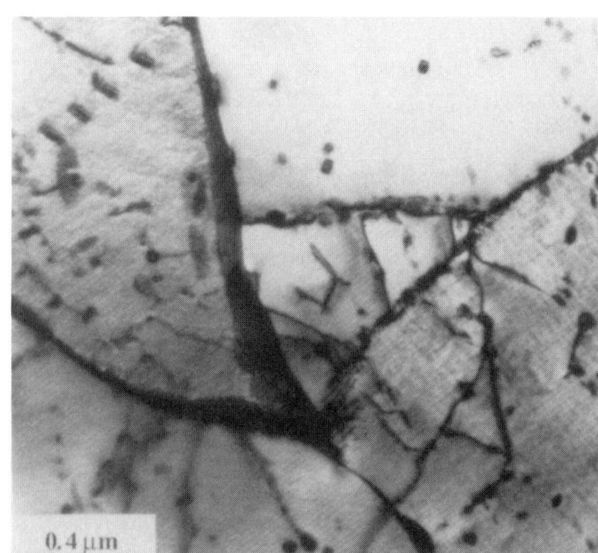


Fig. 8 - Subgrain structure near a grain boundary in a W-3.6Re-0.33ZrC specimen crept at 2200 K.

It was shown in Figure 5 that the strengthening effect of ZrC particles decreases as temperature increases. This was most likely due to the particle coarsening at high temperatures. Comparing the ZrC particle size in Figure 7 with that in Figure 8, it is clear that the particle size increases with increasing temperature. The driving force for precipitate growth is the reduction of total surface energy of precipitates. A theory of diffusion controlled coarsening, in which large precipitates grow at the expense of small ones, had been derived by Lifshitz and Slyozov (17) and by Wagner (18). The LSW theory predicts that the average particle radius is an increasing function of time and temperature. Therefore, it could be concluded that the strengthening effect of ZrC particle decreases with increasing particle size. This is because the total number of the particles available in the matrix decreases rapidly with increasing particle size so that the particle growth reduces not only the pinning effect of particles on dislocations but also the retarding effect of particles on the growth of grains and subgrains. Therefore, both direct particle strengthening effect (interaction of dislocations with particles) and the indirect particle strengthening effect (stabilization of grain and subgrain sizes) decreases with increasing temperature.

## Conclusion

The following conclusions can be drawn from the present study on the creep behavior of an arc-melted W-3.6Re-0.33ZrC alloy.

1) The stress exponent for the steady state creep of a W-3.6Re-0.33ZrC alloy is shown to be about 5.4, and is not sensitive to the testing temperature.

(2) At temperatures from 1900 to 2400 K, the activation energy for creep deformation of this alloy is about 109 Kcal/mol, and is also not sensitive to the applied stress in the stress range employed in the present study.

(3) The temperature-compensated creep rate of a W-3.6Re-0.33ZrC is about one order lower than that of a W-5Re, and is about 2 order lower than that of pure W at temperatures above 0.5 $T_m$.

4) The high creep-resistance of this alloy is associated with the presence of ZrC particles which retard the movements of dislocations and subgrain boundaries, resulting in both direct particle strengthening (interaction of dislocations with ZrC particles) and indirect particle strengthening (stabilization of grain and subgrain sizes).

5) ZrC strengthening effect decreases with increasing temperature due to the particle coarsening.

## Acknowledgment

This work was supported by the Wright Research and Development Center (WRDC), WPAFB, Ohio, under contract No. F-33615-91-C-2109. The contract manager is Mr. Steven Adams. The authors would like to thank Mr. J. Clark for the use of the analytical electron microprobe, which was purchased with the aid of NSF grant EAR 8408163, in the Chemistry Department of Arizona State University. The authors would also like to thank Mr.D.R. Bosch for his comments on and assistance in preparing this manuscript.

## References

1 - C. T. Sims and R. I. Jaffee, *Trans. ASM*, 52, 929-937 (1960).
2 - L. Ratliff, D. J. Maykuth, H. R. Ogden and R. I. Jaffee, *Trans. AIME*, 230, 490-500 (1964).
3 - W. D. Klopp, W. R. Witzke, and P. L. Raffo, NASA Tech. Note D-3483, Lewis Research Center, Cleveland, Ohio (1966).
4 - P. L. Raffo, NASA Tech. Note D-4567, Lewis Research Center, Cleveland, Ohio (1968).
5 - P. L. Raffo, *J. Less-Common Metals*, 17, 133-149 (1969).
6 - M. Garfinkle, W. R. Witzke and W.D. Klopp, *Trans. AIME*, 245, 303-315 (1969).
7 - J. R. Stephens and W. R. Witzke, *J. Less-Common Metals*, 23. 325-335 (1971).
8 - W.D. Klopp and W.R. Witzke, NASA TN D-5348, Lewis Research Center, Cleveland, Ohio (1969).
9 - O.D. Sherby, *Acta Metall.*, 10, 135-143 (1962).
10 - K. Lowrie and A.M. Gonas, J. Appl. Phys., 36, 2189-2196 (1965).
11 - C. R. Barrett, A. J. Ardell, and O. D. Sherby, *Trans. AIME*, 230, 200-212 (1964).
12 - A. K. Mukherjee, J. E. Bird, and J. E. Dorn, *Trans. ASM*, 62, 155-163 (1969).
13 - J. Weertman, *J. Appl. Phys.*, 28, 362-370 (1957).
14 - R. R. Vandervoort and W. L. Barmore, p.108, Proc. 6th Plansee Seminar, Metallwerk Plansee Ag., Reutte-Tyrol (1969)
15 - R.R. Vandervoort, *Metall. Trans.*, 1, 857-864 (1970).
16 - C. Zener and J. H. Hollomon, *J. Appl. Phys.*, 15, 22-34 (1944).
17 - I. M. Lifshitz and V. V. Slyozov, *J. Phys. Chem. Solids*, 19, 35-49 (1961).
18 - C. Wagner, *Z. Elektrochem.*, 65, 581-590 (1961).

Proceedings of the Fifth International Conference on Creep of Materials, Lake Buena Vista, Florida, USA, 18-21 May 1992

# Creep, Cyclic Creep and Fatigue of a Rotor Steel at Elevated Temperature

**S.J. Zhu, Z.A. Yang, and Z.G. Wang**
Institute of Metal Research
Shenyang, China

## ABSTRACT

Effect of stress ratio(from 1 to −1) on strain rate and rupture time of 34CrMo rotor steel were investigated at constant maximum stresses and at 550℃. The minimum strain rates decrease with an increase in stress ratio up to zero and increase with an increase in stress ratio beyond zero. The rupture time behavior was explained by microscopic analysis of dislocation structures. Although the dislocation cell sizes are not dependent on stress ratio, the dislocation structures in cell walls are dependent on stress ratio.

THE ROTORS IN STEAM TURBINE are important components subjected to creep / fatigue interaction. Therefore, there are great interests in creep, cyclic creep and fatigue behaviour of rotor materials at high temperature. Cyclic stress component superimposed onto the sustained stress component can either accelerate or decelerate the creep deformation in dependence on material, temperature and stress amplitude. For example, cyclic creep acceleration was observed at room temperature and cyclic creep deceleration was observed at 550℃ in Cr−Mo−V steel (1).

The effect of the magnitude of the cyclic stress component on the life time was often found to be non−monotonous. Results of Paulson et al (2) on Udimet 115 at 760℃ show that for a constant maximum stress in stress cycle the dependence of the life time on the stress ratio R exhibits a maximum at about $R=0.3$. Thus, a small cyclic stress component has a beneficial effect in comparison with pure creep, while a large cyclic stress component may have a detrimental effect. No corresponding data on the R−dependence of the cyclic creep strain rate are available.

The aim of this paper is to present experimental results on the interaction of fatigue with creep obtained on 34CrMo rotor steel at 550℃. The effect of stress ratio on both deformation and fracture was investigated by mechanical testing and micro analysing.

## EXPERIMENTAL PROCEDURES

The material used in this investigation was a new rotor steel. The chemical composition of the steel (in wt%) was 0.38C, 1.06Cr, 0.54Mn, 0.28Si, 0.019S, <0.1P, 0.23Mo, 0.09Ni and the balance Ni. The rotor was quenched into oil from 860℃ and then tempered at 580℃. The specimens for mechanical properties were machined from outside wall along axial direction with a gauge length of 11 mm and a diameter of 7.6 mm. The microstructures of the specimens consist of sorbite and ferrite.

The tests of creep, cyclic creep and fatigue were conducted in mechanical−servo Mayes ESM−100 testing machine in air. The cyclic stress frequency was 0.5 Hz with wave form of triangle. The strain was measured with a linear variable differential transducer and continuously monitored on a strip chart recorder. The temperature for testing was 550℃ which was maintained within ± 2℃. The ratios of the minimum stress to the maximum stress were from −1 to 1 and the maximum stresses were from 385 to 410 MPa. The internal stress for both static and cyclic creep was measured by stress dip test method.

Several specimens were taken to a certain strain and then cooled rapidly under load to preserve the substructure formed during creep, cyclic creep and fatigue. The gauge sections of the interrupted specimens were cut into slices along the bar axis using a electric discharge machine. The slices were mounted and reduced to about 100 $\mu$m by mechanical polishing, and then electrolytically thinned in a double jet. The dislocation structures in the slices for creep, cyclic creep and fatigue were observed by Philips EM420 transmission electron microscopy(TEM). The fracture surfaces were observed by Cambridge S360 scanning electron microscopy(SEM).

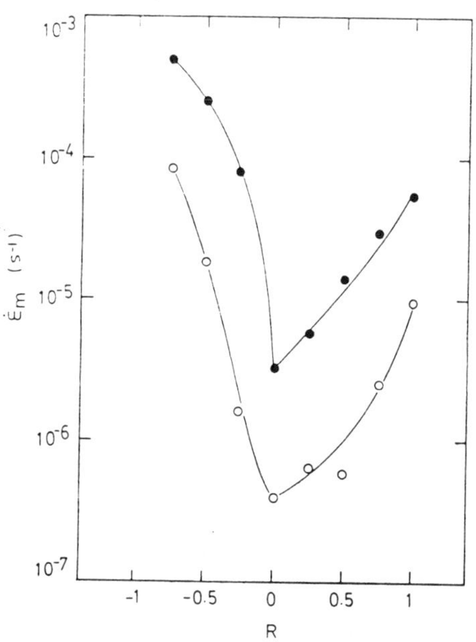


Fig.1—Minimum strain rate versus stress ratio in 34CrMo rotor steel at 550℃ (○, maximum stress of 385 MPa; ●, maximum stress of 410MPa)

## RESULTS AND DISCUSSION

The dependence of the minimum creep strain rate on the R−ratio is shown in Fig.1. It can be seen that the creep strain rate decreases with increasing value of R up to R = 0, while the creep strain rate increases with increasing value of R beyondR = 0. Correspondingly, the dependence of the life time on the R−ratio exhibits a maximum at R = 0, as shown in Fig.2. It can be seen that the effect of cyclic stress is strong. Both the creep strain rate and the life time vary by two orders of magnitude in the investigated range of R−values.

Figs.1 and 2 suggest that there is a correlation between the life time and the creep strain rate. Fig.3 shows this very close correlation in the log−log plot. This means that Monkman−Grant relation can be used for cyclic creep for the present material.

Dislocation structures of specimens interrupted during creep, cyclic creep and fatigue are shown in Fig.4. The dislocation structures are cell structures. The cell size is not measurably dependent on the R−ratio, being about 0.5$\mu$m.

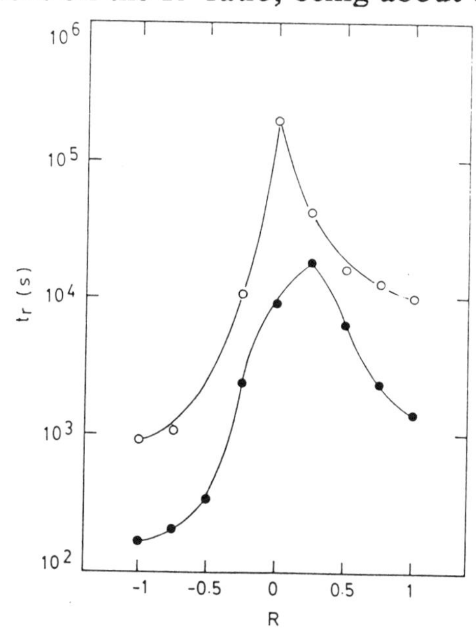


Fig.2—Rupture time versus stress ratio in 34CrMo rotor steel at 550℃ (○, maximum stress of 385 MPa, ●, maximum stress of 410MPa)

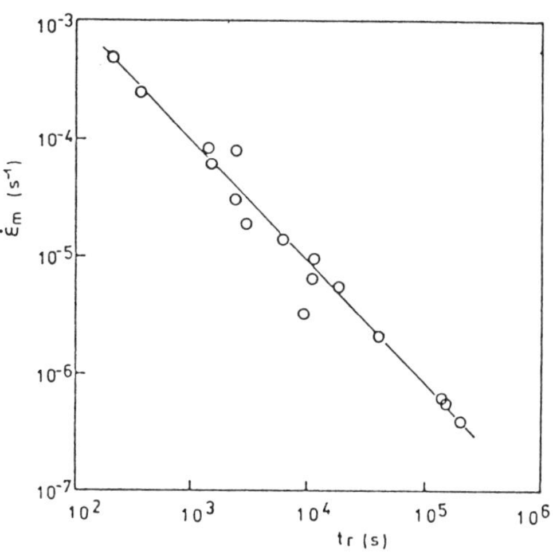


Fig.3—Minimum strain rate versus rupture time in 34CrMo rotor steel in different stress ratios and at 550℃

The only effect of the R—ratio on dislocation structures lies in the sharpeness of the cell walls. The cell walls for R = 0 are more diffuse. There are dislocation networks for R = 1, shown in Fig.5, which is a typical feature for creep at high temperature. This suggests that the details of the dislocation configuration in the cell walls play an important role in the deformation process.

Maghrabi (3) pointed out that the back stress determining the easiness of plastic deformation is thus critically dependent not only on the cell size and the volume fraction of the cell walls, but also on the details of the dislocation arrangement within the walls. The less equilibrium the dislocation configuration in the walls, the higher the back stress. In reference 4 , the internal stresses deflecting long range back stress were measured by stress dip test method for both static and cyclic creep. The internal stress for cyclic creep was higher than that for static creep. This is consistent with the less equilibrium dislocation configuration in the walls for cyclic creep (3). For cyclic creep, the more diffuse dislocation structures in cell walls

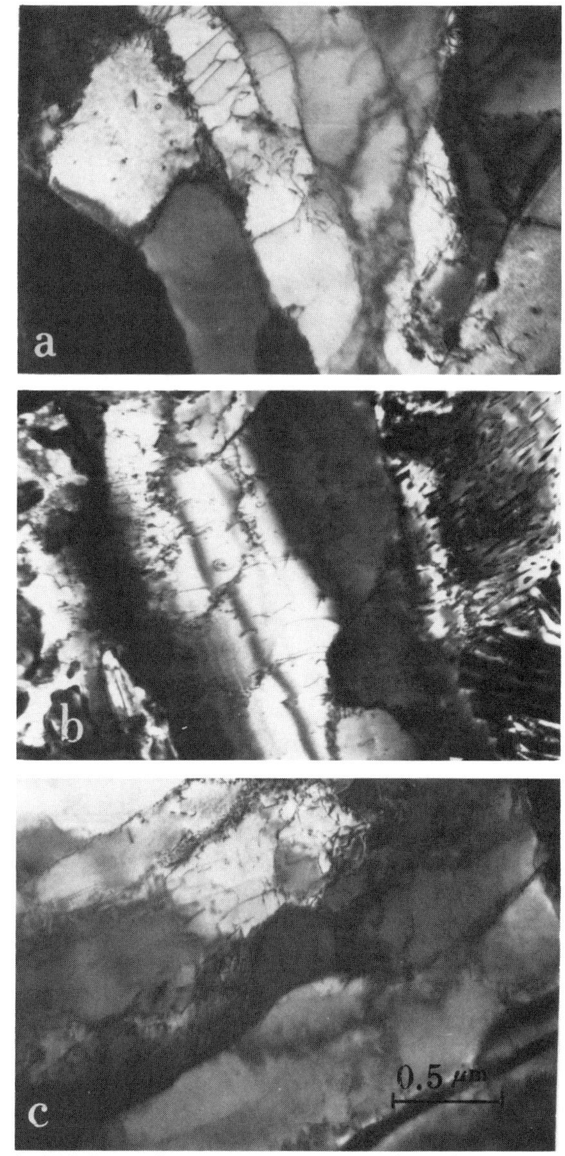


Fig.4—Dislocation structures in interrupted specimens of 34CrMo rotor steel at 550℃ and at 410 MPa: a)R = −1; b) R = 0; c) R = 1

produce higher internal stress i.e. lower effective stress (applied stress minus internal stress). As a result, creep strain rate decreases with decreasing value of R from R = 1 to R = 0. The compressive stress increases with decreasing value of R from R = 0 to R = −1 at constant maximum stress. The dislocation cell size may be determined by the maximum stress. The dislocation structures in cell walls become denser

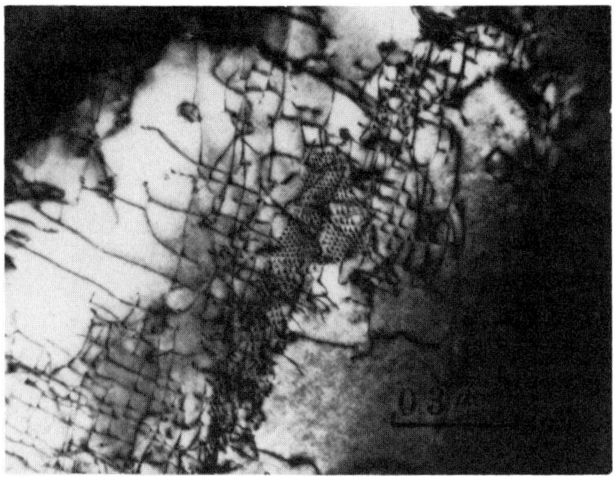

Fig.5–Dislocation networks in cell walls in crept specimen of 34CrMo rotor steel at 550℃ and at 410 MPa

and sharper with decreasing R–ratio. This is almost the came with Lukas′ results (5) on steels.

For pure creep and for tests with small cyclic stress components the final fractue is of ductile type and preceded by necking. In tests with higher cyclic stress components transcrystalline fatigue crack propagation procedes necking and final fracture. Fig.6 shows the fracture surfaces for $R=1, 0, -1$. It can be seen that there are dimples for $R=1$ and there are fatigue striations for $R=0$ and $R=-1$. It should be mentioned that it is also dependent on the roughness of the specimen surface if fatigue striations can be seen. Even for $R=-1$ there are positive creep strain since tensile creep strain is higher than compressive creep strain. This phenomenon is usually not observed for fatigue at room temperature. The relation of roughness of fracture surfaces with properties will be studied in the future.

## CONCLUSION

The deformation and fracture behaviour of 34CrMo rotor steel were studied at 550℃ and in condition of different stress ratios with constant maximum stress. The creep strain rates are well correlated with the rupture time. Monkman –Grant relation seems suitable for cyclic creep. The creep strain rate is lowest when stress ratio is close to zero. This can be explained by dislocation structures. The dislocation cell size is almost constant with the increase of stress ratio. However, the dislocation structures in cell wall change with stress ratio, which play the main role for the mechanical property.

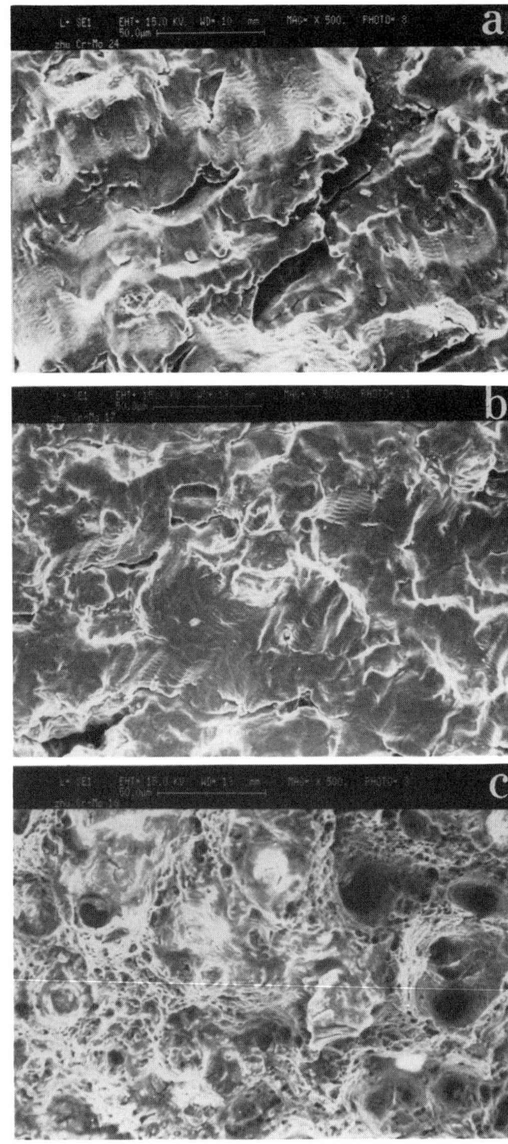


Fig.6–Fracture surface of 34CrMo rotor steel at 550℃ and at 410 MPa: a) $R=-1$; b) $R=0$; c) $R=1$

## ACKNOWLEDGEMENT

We are grateful for the finicial support provided by the National Natural Science Foundation of China under Grant No. 59071020.

## REFERENCES

1– Wang Z.G., K. Rahke and C. Laird: Fatigue Fract. Engng. Mats. Struct., 9, 219–230(1986)

2– Paulson, L.G. Fritzmeier and J.K. Tien: Metall. Trans., 14A, 727–731(1983)

3– H. Mughrabi, Acta Metall., 9, 1367–1379(1983)

4– S.J. Zhu, Z.A. Yang and Z.G. Wang, Submitted to Chin. J. Met. Sci. Technol. 1992

5– P. Lukas, "Fatigue 90", P1725, Materials and Component Engineering Publications Ltd, Birmingham(1991).

*Proceedings of the Fifth International Conference on Creep of Materials, Lake Buena Vista, Florida, USA, 18-21 May 1992*

# Creep Damage of Hastelloy XR at Very High Temperatures in Simulated Helium Gas

**Y. Nakasone and T. Tanabe**
National Research Institute
for Metals
Tsukuba Laboratories
Ibaraki, Japan

**H. Tsuji and H. Nakajima**
Japan Atomic Energy
Research Institute
Tokai Research Establishment
Ibaraki, Japan

**T. Ohba and K. Yagi**
National Research Institute
for Metals
Tokyo, Japan

## ABSTRACT

The present paper investigates early-stage-creep damage of Hastelloy XR and XR-II alloys, modified versions of Hastelloy X alloy, which have been developed in Japan as most promising candidate structural alloys for Japanese high-temperature gas-cooled reactors (HTGRs). Creep tests were made on Hastelloy XR forging, tube and XR-II tube at 1,123 to 1,273 K in a simulated HTGR helium gas environment. The tests were interrupted at different strain levels of up to 5 % in order to evaluate creep damage via intergranular voids. The void size along a grain boundary and the *A*-parameter, the ratio of the number of damaged grain boundaries, on which one or more voids are found, to that of the total grain boundaries observed are used to evaluate creep damage. Although the *A*-parameter is usually treated as a deterministic parameter, the values of the parameter show wide variations and follow the Weibull distribution, reflecting spatial randomness of the voids. The void sizes along grain boundaries, on the other hand, follow the log-normal distribution. The mean void size $d_m$ and the mean value of the *A*-parameter $A_m$ are plotted against the Larson-Miller parameter $P_L'$ in terms of interruption life, or the time to reach a given interruption creep strain. The resultant $d_m$ vs. $P_L'$ and $A_m$ vs. $P_L'$ diagrams show that Hastelloy XR forging has suffered more damage than Hastelloy XR tube; nevertheless, the forging has longer interruption life. This result indicates that grains may have been deformed more easily in Hastelloy XR in the form of tube than in the form of forging. The diagrams also imply that the addition of boron has suppressed the nucleation as well as the growth of voids and thus has brought about longer interruption life of Hastelloy XR-II tube.

**IN THE COURSE** of Japanese R & D of high-temperature gas-cooled reactors (HTGRs), Hastelloy XR and XR-II alloys have been developed in Japan by modifying the conventional Hastelloy X alloy. Hastelloy XR alloy has gained better corrosion resistance against helium gas by controlling the contents of manganese (Mn) and silicon (Si) in the optimum ranges, and also by reducing the contents of aluminum (Al) and titanium (Ti). Moreover, by diminishing cobalt (Co) content, the alloy can lower induced radioactivity for nuclear engineering application purposes (1). The addition of boron (B) to Hastelloy XR alloy has brought Hastelloy XR-II alloy which has achieved higher high-temperature creep strength than the other two alloys (2,3). At present, they are the most promising candidate structural alloys for the intermediate heat exchangers (IHXs) in Japanese high-temperature gas-cooled reactors. In fact, Hastelloy XR alloy is being used for the structural components in the IHXs in a Japanese high-temperature engineering test reactor (HTTR) which is now under construction in Japan (4).

The present paper investigates the effect of product form and that of boron-addition on early-stage creep damage in these modified Hastelloy X alloys at a very high temperature range of 1,123 to 1,273 K in a simulated HTGR helium gas environment (*He 2'*). Creep damage is evaluated quantitatively via the sizes of intergranular voids along grain boundaries and the *A*-parameter (5). The *A*-parameter is defined as the ratio of the number of damaged grain boundaries, on which one or more intergranular voids are found, to that of the total grain boundaries observed. Although this parameter is usually treated as a deterministic parameter (5), the values of the parameter show wide variations in the present experiments and follow the Weibull distribution. For this reason, statistical treatment is applied to the void sizes and the *A*-parameter in this study. The bulk carbon analysis of the specimens were also made after the creep tests in order to investigate the carburization behavior of the materials tested in the helium gas environment.

## EXPERIMENTAL PROCEDURES

**Materials Tested and Specimen Geometry.** Materials tested in this study are Hastelloy XR forging, tube and XR-II tube whose chemical compositions are listed in Table I. As indicated in the table, Hastelloy XR forging and tube have the same composition. The content of Co in Hastelloy XR is reduced to about one tenth of that in the conventional Hastelloy X. In addition to this modification, Hastelloy XR-II contains 40 mass ppm of B in order to improve creep strength. Prior to specimen fabrication,

Table I. Chemical compositions of Hastelloy XR forging, tube and XR-II tube tested in this study (mass%).

| Alloy | C | Cr | Mn | Si | Co | Mo | Fe | W | Al | Ti | P | S | B | N | Ni |
|---|---|---|---|---|---|---|---|---|---|---|---|---|---|---|---|
| Hastelloy XR | 0.07 | 21.98 | 0.95 | 0.33 | 0.03 | 9.11 | 18.61 | 0.49 | 0.02 | 0.01 | <0.001 | <0.001 | 0.0002 | 0.007 | Bal. |
| Hastelloy XR-II | 0.07 | 21.96 | 0.87 | 0.27 | 0.12 | 8.95 | 18.33 | 0.46 | 0.03 | 0.05 | 0.005 | 0.001 | 0.004 | - | Bal. |

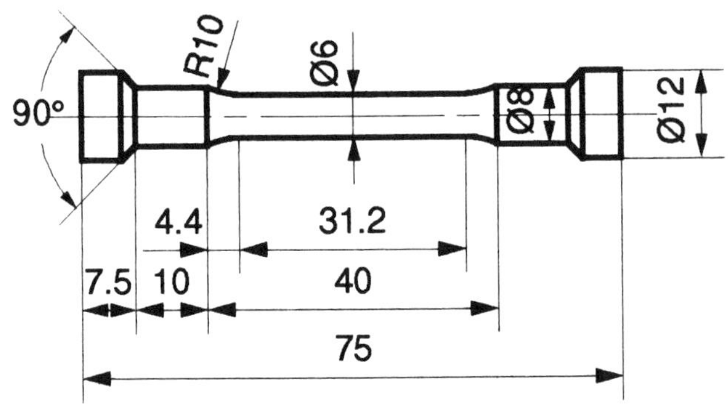


Fig. 1 - Specimen geometry used in this study (unit: mm).

Table II. Chemical analysis of helium gas *He 2'* (mass ppm).

| $H_2$ | $CH_4$ | CO | $CO_2$ | $H_2O$ | O | N | He |
|---|---|---|---|---|---|---|---|
| 300 | 15 | 100 | 1 | 3 | N.D.* | N.D.* | Bal. |

* N.D.: Not Detected.

Table III. Matrix of the present test conditions.

| Alloy | Temp. *T*, K | Stress σ, MPa | Interruption strain $\varepsilon_{int}$, % |
|---|---|---|---|
| Hastelloy XR forging | 1,123 | 31.4 | 1.0, 2.0, 5.0 |
| | 1,173 | 22.6 | 1.0, 1.7, 2.0, 5.0 |
| | 1,223 | 14.7 | 1.0, 2.0, 3.0, 5.0 |
| | 1,273 | 9.8 | 1.0, 2.0, 5.0 |
| Hastelloy XR tube | 1,123 | 31.4 | 1.0, 2.0, 5.0 |
| | 1,223 | 14.7 | 1.0, 1.6, 1.7, 2.0, 3.0, 5.0 |
| Hastelloy XR-II tube | 1,123 | 31.4 | 0.7 |
| | 1,173 | 22.6 | 0.3, 0.7, 1.5 |
| | 1,223 | 19.6 | 1.0, 2.0, 5.0 |
| | | 14.7 | 1.5 |

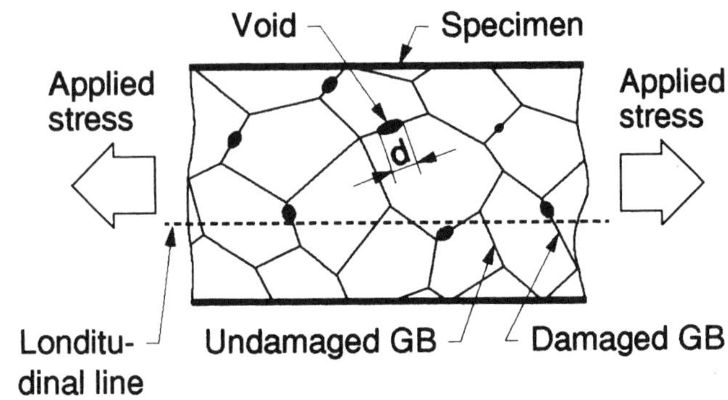


*A*-parameter: $A = n_D / n_T = n_D / (n_U + n_D)$
where $n_D$ : no. of damaged GBs,
$n_T$ : no. of the total GBs observed,
and $n_U$: no. of undamaged GBs

Fig. 2 - Schematic illustration explaining the *A*-parameter.

solution-treatment was given to the three materials; i.e., both Hastelloy XR forging and tube were hold at 1,453K for one hour followed by water quenching, and Hastelloy XR-II tube at 1,463 K for one hour followed by water-spray quenching. The ASTM micro-grain size number (6) is 2.5 in Hastelloy XR forging and tube, and 3.0 in Hastelloy XR-II tube, respectively.

Specimens used in this study are of unnotched-round-bar type, having a diameter of 6 mm and a gage length of about 31 mm as illustrated in Fig. 1.

**Interruption Creep Tests.** Creep tests of the present materials were made by using conventional lever-type creep testers attached to a helium-gas recirculating system in a temperature range of 1,123 to 1,273 K in an impure helium gas environment called *He 2'* (7). This temperature range corresponds to the upper limit of the operating temperature of the IHXs in the HTTR (8). *He 2'* is one of the impure helium gases proposed for the gaseous coolant to be used in Japanese R & D of structural materials for HTGR applications (7). The chemical composition of the present helium gas, which is listed in Table II, had been well-controlled throughout the tests by the helium-gas recirculating system (7).

Creep tests were interrupted at different creep strain levels of up to 5 % in order to evaluate early-stage creep damage via the void sizes along grain boundaries and the *A*-parameter. Table III lists the values of test temperature, applied stress level and interruption strain $\varepsilon_{int}$, or the strain level at which the creep tests were interrupted.

After interruption creep tests at different creep strain levels had been terminated, central portions of the specimens having length longer than 10 mm were cut into two semi-cylinders; the longitudinal sections of 6 mm by 10 mm or longer were used for optical-microscopic observation of creep voids. Creep void sizes along grain boundaries and the *A*-parameter within the longitudinal section were measured through an optical microscope at a magnification of x 400. All the voids observed in this study were intergranular voids.

The *A*-parameter has been proposed as one of quantitative parameters for evaluating creep damage, and is defined as the ratio of the number of damaged grain boundaries, on which one or more voids are found, to that of the total grain boundaries observed (5). Although this parameter is usually treated as a deterministic parameter, the values show wide variations in this study. Thus, for the statistical analysis of the *A*-parameter, a set of twenty values of the parameter were measured for each specimen by setting twenty lines lying parallel to the applied stress axis at random in the longitudinal section as illustrated in Fig.2.

Bulk carbon contents of the specimens before and after the tests were also analyzed for investigating the carburization behavior of the present materials by using a carbon analyzer.

## RESULTS AND DISCUSSION

**Relationship between Interruption Strain and Interruption Life.** Figure 3 shows the relationship between interruption strain $\varepsilon_{int}$ and interruption life $t_{int}$, or time to reach a given interruption strain level. Although the data scatter a little, the interruption life $t_{int}$ is the longest in Hastelloy XR-II tube, and the shortest in Hastelloy XR tube in respective conditions. As shown in Table I, Hastelloy XR forging and tube have the same composition; nevertheless, $t_{int}$ is longer in the forging than in the tube. This is considered as one of the effects that may have been brought about by manufacturing process.

**Relationship between Applied Stress and the Larson-Miller Parameter.** In Fig. 4, log applied stress is plotted against the Larson-Miller parameter $P_L'$ expressed in terms of the interruption life $t_{int}$ for $\varepsilon_{int}$ = 1, 2 and 5 %; i.e.,

$$P_L' = T(C_L + \log_{10} t_{int}) \qquad \text{... (1)}$$

where $T$ is the test temperature, and $C_L$ some constant. The Larson-Miller parameter is introduced in order to compare the data obtained at the different test temperatures in this study via the trade-off of temperature for time (9).

The solid line labeled "rupture" in Fig. 4 is the master creep rupture curve, or the log applied stress vs. Larson-Miller parameter $P_L$ expressed in terms of the rupture life $t_r$ obtained for the present materials (10); i.e.,

$$P_L = T(C_L + \log_{10} t_r). \qquad \text{... (2)}$$

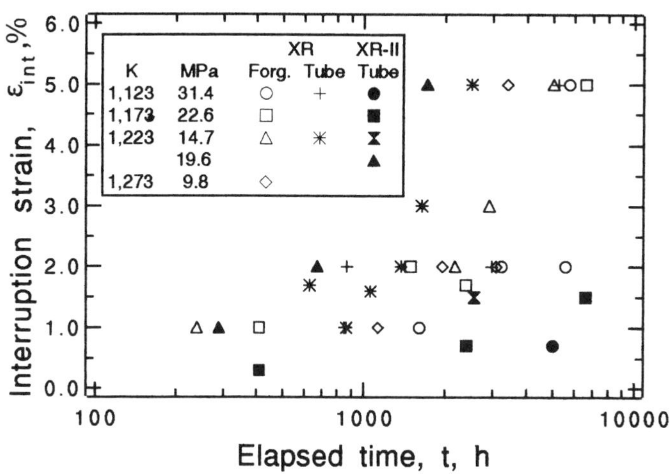


Fig. 3 - Interruption strain $\varepsilon_{int}$ vs. elapsed time $t$ diagrams of Hastelloy XR and XR-II alloys.

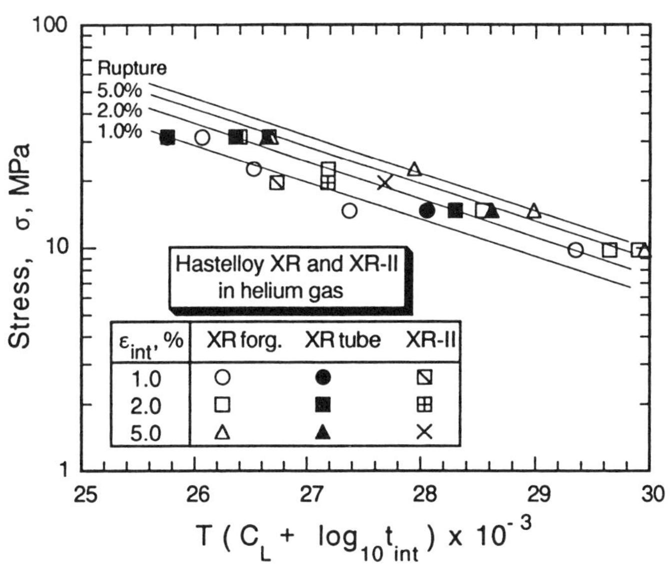


Fig. 4 - Log applied stress σ vs. Larson-Miller parameter $P_L'$ diagrams of Hastelloy XR and HR-II alloys; $C_L$=20 for Hastelloy XR forging and tube, and $C_L$=19.4 for XR-II tube.

All of the creep rupture curves obtained for the present three materials in a temperature range from 1,123 to 1,273 K fit well on this master curve (10). The constant $C_L$ in the Larson-Miller parameter is 20 for Hastelloy XR forging and tube, whereas 19.4 for Hastelloy XR-II. The same values of $C_L$ as those in Eq. 1 were used to obtain the log applied stress vs. $P_L$ relations; thus, it is concluded that the addition of 40 mass ppm of B has brought about longer interruption life $t_{int}$ as well as longer creep rupture life $t_r$ in Hastelloy XR-II.

**Statistical distributions of Intergranular Void Size and the *A*-parameter.** Figure 5 shows cumulative probability $F$ vs. void size along grain boundary $d$ plotted on the log-normal probability paper. The void size $d$ can be regarded to follow the log-normal distribution in all the cases tested in this study. The present interruption creep tests show a tendency that the density of the voids observed on the longitudinal cross section is low in Hastelloy XR-II, whereas high in Hastelloy XR forging in respective conditions. The maximum void density is about 1630 per $cm^2$ in Hastelloy XR forging for $\varepsilon_{int}$ = 5 % at 1,223 K, whereas the minimum

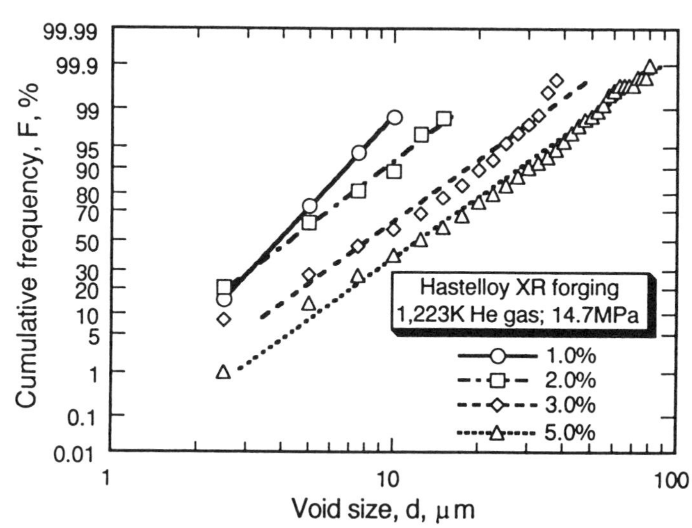


Fig. 5 - Statistical distributions of intergranular void size at different interruption strain levels in Hastelloy XR forging crept at 14.7 MPa in the 1,223 K *He 2'* environment plotted on the log-normal probability paper.

one is about 65 per $cm^2$ in Hastelloy XR-II tube for $\varepsilon_{int} = 0.7$ % at 1,123 K. The maximum void size is 79 μm, larger than half the average grain size, in Hastelloy XR forging for $\varepsilon_{int} =$ 5 % at 1,223 K. The values of the correlation coefficient between the void sizes along grain boundaries and the aspect ratios of the voids, i.e., the ratios of the void sizes perpendicular to grain boundaries to those along the boundaries, were also calculated. The results reveal that all the values obtained are negative and their magnitudes are smaller than 0.7, indicating that the larger the voids, the flatter they tend to become; however, the tendency is not so strong.

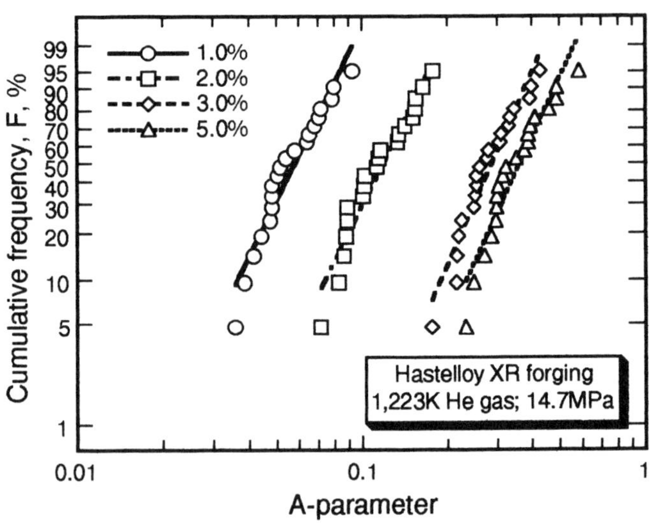


Fig. 6 - Statistical distributions of the *A*-parameter at different interruption strain levels in Hastelloy XR forging crept at 14.7 MPa in the 1,223 K *He 2'* environment plotted on the Weibull probability paper.

Figure 6 shows cumulative probability $F$ vs. the $A$-parameter plotted on the log-normal probability paper. The $A$-parameter can be regarded to follow the Weibull distribution in all the cases tested in this study. Although this parameter is usually treated as a deterministic parameter, the values of the parameter show wide variations in this study; i.e., the coefficient of variation, or the ratio of the standard deviation to the mean of void sizes, is 20 to 86 %. This result implies spatial randomness of the voids. The voids tend to be initiated at grain boundaries unevenly rather than evenly distributed in a crept specimen as in the case of multiple small cracks in fatigue (11, 12).

**Mean Void Size and Mean of the *A*-parameter vs. Larson-Miller Parameter Diagrams.** The mean void size $d_m$ and the mean value of the $A$-parameter $A_m$ were calculated in the statistical analysis described above and plotted against the Larson-Miller parameter $P_L'$. The resultant $d_m$ vs. $P_L'$ and $A_m$ vs. $P_L'$ diagrams are shown in Figs. 7 and 8, respectively. These figures show that the data seem to be divided into three groups in accordance with applied stress level, and that lower applied stress shifts the relations rightwards; i.e., the lower the applied stress level, the less the early creep damage for the same value of $P_L'$. At lower stress levels, however, the creep damage becomes larger for a given value of $\varepsilon_{int}$. In other words, intergranular voids may have grown larger in longer elapsed time span, being exposed to the high-temperature helium gas environment. Hastelloy XR forging has suffered more damage than Hastelloy XR tube; nevertheless, the forging has longer interruption life as shown previously in Fig. 2. This result implies that grains can be deformed more easily in Hastelloy XR in the form of tube than in the form of forging.

The data points in Figs. 7 and 8 also show less creep damage in Hastelloy XR-II tube than in Hastelloy XR forging and tube. Taking the smaller value of $C_L$ for Hastelloy XR-II

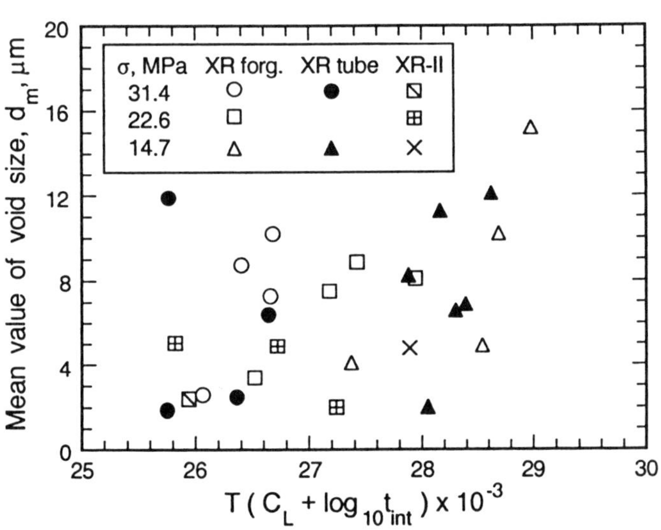


Fig. 7 - Mean void size $d_m$ vs. Larson-Miller parameter $P_L'$ diagrams of Hastelloy XR and XR-II alloys crept at very high temperatures in the *He 2'* gas environment; $C_L$=20 for Hastelloy XR forging and tube, and $C_L$=19.4 for XR-II tube.

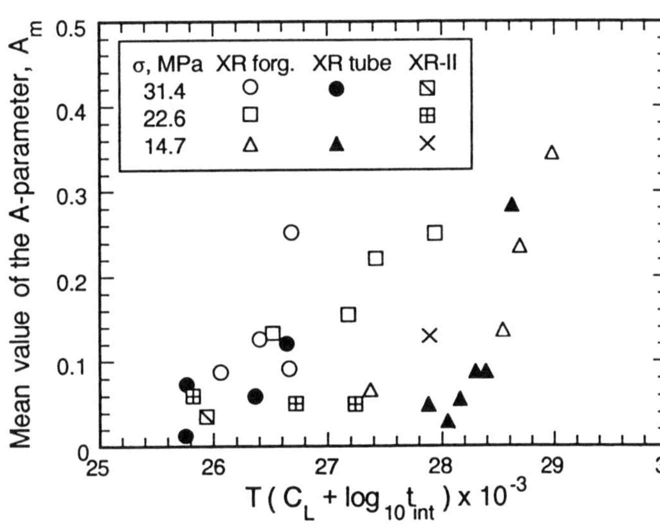


Fig. 8 - Mean of the *A*-parameter $A_m$ vs. Larson-Miller parameter $P_L'$ diagrams of Hastelloy XR and XR-II alloys crept at very high temperatures in the *He 2'* gas environment; $C_L$=20 for Hastelloy XR forging and tube, and $C_L$=19.4 for XR-II tube.

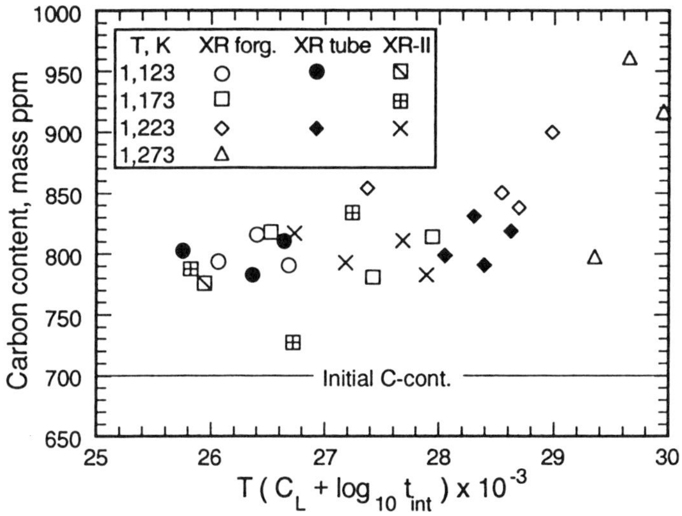


Fig. 9 - Bulk carbon content vs. Larson-Miller parameter $P_L'$ diagrams of Hastelloy XR and XR-II alloys after interruption creep tests at very high temperatures in the *He 2'* gas environment; $C_L$=20 for Hastelloy XR forging and tube, and $C_L$=19.4 for XR-II tube.

into account, it is concluded that the addition of 40 mass ppm of B has suppressed the nucleation as well as the growth of intergranular voids and thus has brought about longer interruption life of Hastelloy XR-II tube.

**Bulk Carbon Analysis of the present Materials.** Figure 9 shows the relations between the bulk carbon contents in the specimens after the interruption creep tests and the Larson-Miller parameter $P_L'$. Carburization occurred in all the three materials tested in the present *He 2'* gas environment. Increase in bulk carbon content due to carburization keeps a constant value of about 100 mass ppm for the values of $P_L'$ up to 28. Above $P_L'$ = 28, however, it suddenly becomes larger with increasing value of $P_L'$. The acceleration of carburization may have needed some duration of time. The maximum increase in the bulk carbon content is about 260 mass ppm. Some of the carbon is considered to be transformed into deleterious carbides like $M_6C$ and $M_{12}C$ (3). Figure 9 depicts that there is little difference in the carburization behavior with respect to material; nevertheless, it is reported that, in Hastelloy XR-II, the addition of B brings about the formation of the beneficial carbide $M_{23}C_6$ to decrease the amount of the deleterious carbides instead (3). This effect may bring about less creep damage in the form of intergranular voids and thus higher creep strength in Hastelloy XR-II tube. The bulk carbon content vs. elapsed time relation does not show such a systematic variation probably due to no trading-off of temperature for time.

## CONCLUSION

The present paper investigates early-stage creep damage of Hastelloy XR and XR-II alloys, modified versions of Hastelloy X alloy, the most promising candidate structural alloys for Japanese high-temperature gas-cooled reactors (HTGRs). Creep tests were made on Hastelloy XR forging, tube and XR-II tube in a temperature range from 1,123 to 1,273 K in a simulated HTGR helium gas environment. The tests were interrupted at different strain levels of up to 5 % in order to evaluate creep damage via intergranular voids. The void size along a grain boundary and the *A*-parameter are used to evaluate creep damage. Although the *A*-parameter is usually treated as a deterministic parameter, the values of the parameter show wide variations and follow the Weibull distribution, reflecting spatial randomness of the voids. The void sizes along grain boundaries, on the other hand, follow the log-normal distribution. The mean void size $d_m$ and the mean value of the *A*-parameter $A_m$ were plotted against the Larson-Miller parameter $P_L'$ in terms of interruption life. The resultant $d_m$ vs. $P_L'$ and $A_m$ vs. $P_L'$ diagrams reveal that Hastelloy XR forging has suffered more damage than Hastelloy XR tube; nevertheless, the forging has longer interruption life. This result indicates that grains may have been deformed more easily in Hastelloy XR in the form of tube than in the form of forging. The diagrams also imply that the addition of boron has suppressed the nucleation as well as the growth of voids and thus has brought about longer interruption life of Hastelloy XR-II tube.

## ACKNOWLEDGEMENT

The present research was sponsored in part by the Japan Atomic Energy Research Institute under contract No. H2-3. The authors would like to express their deep gratitude to Mr. K. Hiwatari, a student at Meisei University, for his assistance.

## REFERENCES

1- Shindo, M. and T. Kondo, in Proc. Conf. on Gas Cooled Reactor Today, vol. 2, Brit. Nucl. Soc., London, 179-185 (1982).
2- Kondo, T., K. Watanabe, K. Sato, T. Nakanishi, K. Sahira, H. Tsuji, Y. Kuroda, T. Tsukada and K. Ozawa, in JAERI-M, Japan Atomic Energy Research Institute, No. 86-003 (1986) (in Japanese).
3- Tanabe, T., F. Abe, Y. Sakai and M. Okada, *Trans. the Iron and Steel Inst. of Japan, 26*, 968-976 (1986).
4- Executive Committee of Technical Meeting on HTGRs, in JAERI-M, Japan Atomic Energy Research Institute, No. 90-078 (1990) (in Japanese).
5- Askins, M.C. *et al.*, EPRI PR2253-1 (1986).
6- ASTM Std. E112-82, in "1983 Annual Book of ASTM Standards Vol. 03.03," pp. 121-155, ASTM, Philadelphia (1983).
7- Tanabe, T., Y.Sakai, T. Shikama, M. Fujitsuka, H. Yoshida and R. Watanabe, *Nucl. Tech., 66,* 260-272 (1984).
8- Japan Atomic Energy Research Institute, "Present Status of HTGR Research & Development", Japan Atomic Energy Research Institute (1990) (in Japanese).
9- Dieter, G.E., "Mechanical Metallurgy", 3rd ed., pp. 461-465, McGraw-Hill Book Co., New York (1986).
10- Nakasone, Y., H. Tsuji, H. Nakajima and T. Tanabe, to be presented at the Winter Annual Meeting of JSME.
11- Kitagawa, H., Y. Nakasone and S. Miyashita, in "ASTM STP 811," pp. 233-263, ASTM, Philadelphia (1983).
12- Kitagawa, H. and Y. Nakasone, *J. Society Mat. Sci., Japan, 33-364*, 14-20 (1984) (in Japanese).

*Proceedings of the Fifth International Conference on Creep of Materials, Lake Buena Vista, Florida, USA, 18-21 May 1992*

# Creep Collapse of Heat Transfer Tubes Under External Pressure

**Y. Kaji and Y. Miyamoto**
Japan Atomic Energy Research Institute
Ibaraki-ken, Japan

**I. Ioka**
Science and Technology Agency
Tokyo, Japan

**I. Nishiguchi**
Kanagawa Institute of Technology
Kanagawa-ken, Japan

## ABSTRACT

The effect of the initial ovality, wall thickness and pressure level on the collapse time of the heat transfer tube subjected to external pressure at elevated temperature were examined experimentally and analytically. The creep deformation of the tube obtained by the ABAQUS code was a good approximation. The predictions of the creep collapse time were in fairly good agreement with the experimental results. The simplified methods predicted conservative deformation behaviors compared with the ABAQUS solution. Though many cracks initiated by creep collapse were observed on the outer surface of a few test tubes, the cracks did not pass through the tube thickness and the leak tightness was maintained despite a collapse for all tubes tested.

EXPERIMENTAL AND THEORETICAL INVESTIGATIONS were carried out on heat transfer tubes of an intermediate heat exchanger (IHX). The IHX is a helium/helium heat exchanger of the High Temperature engineering Test Reactor (HTTR) which is a high temperature gas–cooled reactor being under construction in Japan.

In the IHX, if the secondary gas were lost in an accident, the tubes would be subjected to the external primary gas of 4MPa at high temperature and the possibility of the creep collapse would occur. Thus, it is indispensable to evaluate the collapse time of the tubes and confirm a calculated method to predict the creep collapse behavior of the tubes under external pressure.

Regarding theories of creep collapse of a tube under external pressure, simplified methods were proposed by Hoff et al. [1], Bargmann [2,3] and Nishiguchi et al. [4]. The Hoff's method is based on the sandwich model and elasticity during deformation is ignored in it. The Bargmann's method was modified to include elasticity throughout the deformation. The creep constitutive relations in the above methods, however, are limited to Norton–Bailey type. In the simplified method proposed by Nishiguchi et al., the restriction on creep constitutive equation is removed and elasticity during deformation is taken into account.

In this paper, the creep collapse tests were performed to examine the effect of the initial ovality, wall thickness and pressure level on the creep collapse time of the tube subjected to external pressure at elevated temperature [5]. The results were compared with the analyzed ones by a general purpose FEM–code ABAQUS [6] and those of simplified methods by Nishiguchi et al.. Moreover, inspection of the collapsed tubes was carried out because leak tightness of a primary coolant gas is the prime concern for the accident.

## EXPERIMENTAL METHOD

Material was a nickel–based superalloy Hastelloy XR which used to be for the IHX. Chemical composition and mechanical properties are shown in Table 1. Hastelloy XR is a modified version of Hastelloy X, and it is known to have substantially improved creep strength and corrosion resistance in a simulated High Temperature Gas–cooled Reactor (HTGR) helium environment as compared with Hastelloy X. The grain size of material were ASTM No.2–2.5.

In creep buckling tests, the dimensions of the test tubes were 31.8mm in outer diameter, 400mm in length, and 2mm to 4mm in thickness. Some of the test tubes were machined from the forged material so as to possess accurate quasi–elliptical cross section. One of the hot extruded tubes was intentionally deformed since the

Table 1. Chemical compositions and mechanical properties of Hastelloy XR.

Chemical compositions (wt%)

| Product form | C | Mn | Si | Cr | Co | Mo | W | Fe | Ni |
|---|---|---|---|---|---|---|---|---|---|
| Forged | 0.07 | 0.94 | 0.33 | 21.93 | 0.02 | 8.96 | 0.48 | 18.21 | Bal. |
| Extruded | 0.07 | 0.95 | 0.35 | 21.83 | 0.02 | 9.02 | 0.49 | 18.26 | Bal. |

Mechanical properties at 900°C

| Product form | Tensile strength (MPa) | 0.2% proof strength (MPa) | Creep rupture test | | |
|---|---|---|---|---|---|
| | | | Stress (MPa) | Time (h) | Elongation (%) |
| Forged | 203.8(700.7) | 155.8(319.5) | 49.0 | 146 | 26 |
| Extruded | 237.2(696.8) | 170.5(310.7) | 49.0 | 138 | 42 |

( ) : Strength at room temperature.

helical hot extruded tubes would be used in the IHX. The ovality of the test tube was defined by the following equation,

$$\phi = \frac{R_{max} - R_{min}}{R} \times 100 \qquad (1)$$

where $R_{max}$ and $R_{min}$ are the maximum and the minimum outer radius, respectively; R(=15.9mm) is the mean outer radius.

In order to study the initial geometric imperfection in the shape of cross sections, outer radii and wall thickness were measured at every 45° of circumferential angle of eight distinct cross sections. Residual stress in the test tube caused by set-up was released before experiments. The initial ovality of the test tubes was approximately 0% and 6%.

The cross section of the test apparatus is shown in Fig. 1. The test apparatus can be operated at a temperature up to 1100°C and pressure up to 6.5MPa. The upper end of the test tube is fixed, while the lower end is sealed by O-Ring to allow axial elongation. External pressure on the test tube is controlled to within ± 0.02MPa. The temperature is controlled to within ± 5°C at the above-mentioned eight positions along the test tube. The apparatus is designed to measure the creep collapse time by detecting sudden rise in the internal pressure of the test tube.

In the test, the external pressure of 5.0 or 6.0MPa was imposed on the test tube at 950°C. The creep collapse test was interrupted halfway to measure the ovality and thickness of the test tubes at room temperature.

For most of the test tubes the leak test was carried out after creep collapse test. Internal pressure of 6.0MPa was imposed on the test tubes by helium gas fed from gas cylinder at room temperature. The change in the pressure was measured using pressure gauge with a sensitivity of 0.01MPa and was continuously monitored for half an hour by recorder.

The outer and inner surfaces of the test tubes were observed by liquid penetrant examination to check the initiation of cracks. For the cracks detected on the surface of the test tube, observation of the cross section was performed by means of a microscope to measure the depth of the cracks.

## EXPERIMENTAL RESULTS

The experimental results of creep collapse tests are shown in Table 2. Fig. 2 shows relationship between collapse time and thickness of the test tube. It is found that the collapse time becomes shorter as the initial ovality increases and the wall thickness of the test tubes decreases.

Experiments of Nos. CR-4, CR-6 and CR-7 were carried out under identical conditions to examine the reproducibility of test results. No. CR-4 tube was machined from the hot extruded material and Nos. CR-6 and CR-7 were machined from forged one. Good reproducibility was observed as shown in Table 2. Therefore, it is found that the effect of the manufacturing method on the creep collapse behavior of the test tube is almost negligible.

The photograph of the test tube, 3.5mm in wall thickness and 6% in initial ovality, after creep collapse is shown in Fig. 3. The shape of cross section showed two-lobed deformation mode. Others also showed the same deformation mode. Maximum deformation of radius was recorded at the position where initial out-of-roundness in radius was largest.

In the leak tests, the pressure change was not observed in all collapsed test tubes. Many cracks on the outer surface of a few test tubes were observed by liquid penetrant examination while the crack on the inner surface was not observed. As an example, the cross-sectional view of the collapsed test tube, 3.89mm in average wall thickness and 6.16% in average initial ovality under 5MPa is shown in Fig. 4. In the outer surface, the cracks and voids are mainly

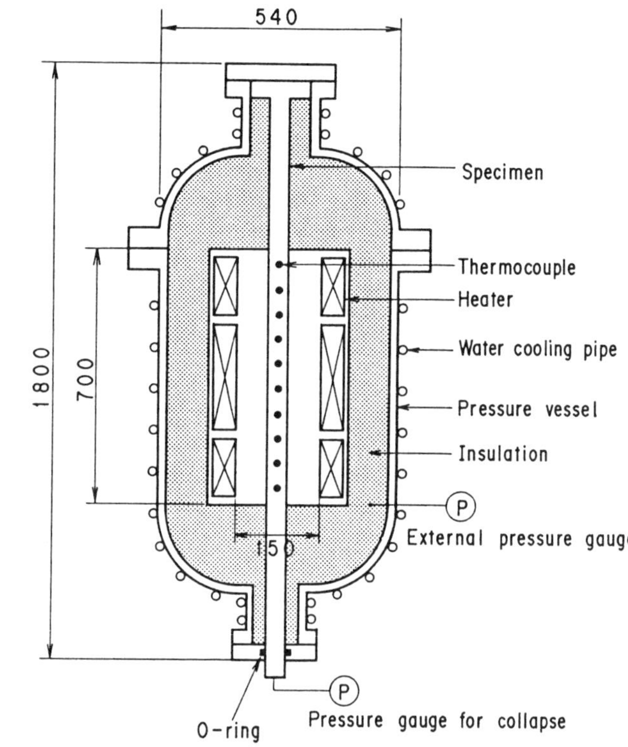


Fig. 1 Cross section of test apparatus

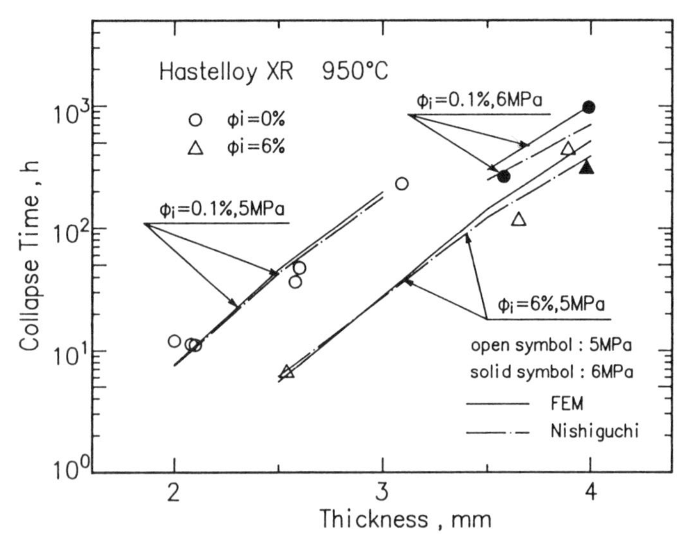


Fig. 2 Relationship between collapse time and thickness of the test tube.

Table 2. Experimental and calculated times to collapse.

| No. | Temp. (°C) | External pressure (MPa) | Experimental conditions Thickness (mm) | Experimental conditions Initial ovality (%) | Analytical conditions Thickness (mm) | Analytical conditions Initial ovality (%) | Time to collapse Exp. (h) | Time to collapse FEM (h) | Time to collapse Nishiguchi et al. (h) |
|---|---|---|---|---|---|---|---|---|---|
| CR-1 | 950 | 6.0 | 3.58 (3.47) | 0.01 (0.03) | 3.5 | 0.1 | 263.6 | 298.8 | 246.6 |
| CR-2 | 950 | 5.0 | 3.65 (3.52) | 6.36 (6.42) | 3.5 | 6.0 | 114.0 | 142.3 | 121.8 |
| CR-3 | 950 | 6.0 | 3.99 (3.87) | 0.00 (0.03) | 4.0 | 0.1 | 973.1 | 1010.0 | 704.6 |
| CR-4 | 950 | 5.0 | 2.00 (1.87) | 0.06 (0.31) | 2.0 | 0.1 | 12.0 | 7.7 | 7.5 |
| CR-5 | 950 | 5.0 | 3.89 (3.77) | 6.16 (6.82) | 4.0 | 6.0 | 432.0 | 515.0 | 387.6 |
| CR-6 | 950 | 5.0 | 2.10 (2.01) | 0.03 (0.13) | 2.0 | 0.1 | 11.1 | 7.7 | 7.5 |
| CR-7 | 950 | 5.0 | 2.08 (1.96) | 0.04 (0.19) | 2.0 | 0.1 | 11.3 | 7.7 | 7.5 |
| CR-8 | 950 | 5.0 | 2.58 (2.54) | 0.03 (0.16) | 2.5 | 0.1 | 36.5 | 45.9 | 42.8 |
| CR-9 | 950 | 5.0 | 2.60 (2.52) | 0.07 (0.16) | 2.5 | 0.1 | 47.4 | 45.9 | 42.8 |
| CR-10 | 950 | 5.0 | 3.09 (3.01) | 0.02 (0.05) | 3.0 | 0.1 | 230.1 | 197.9 | 177.9 |
| CR-11 | 950 | 5.0 | 2.54 (2.33) | 5.44 (6.04) | 2.5 | 6.0 | 6.5 | 5.5 | 6.1 |
| CR-12 | 950 | 6.0 | 3.98 (3.92) | 5.59 (5.88) | 4.0 | 6.0 | 303.2 | 211.0 | 159.3 |

( ) Minimum value in thickness, maximum value in initial ovality

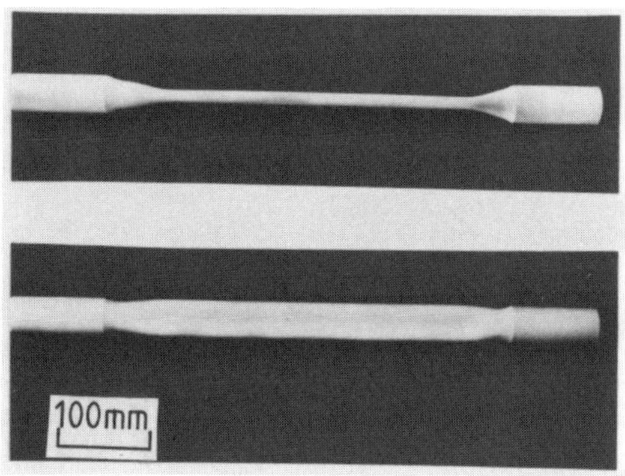


(a) appearance

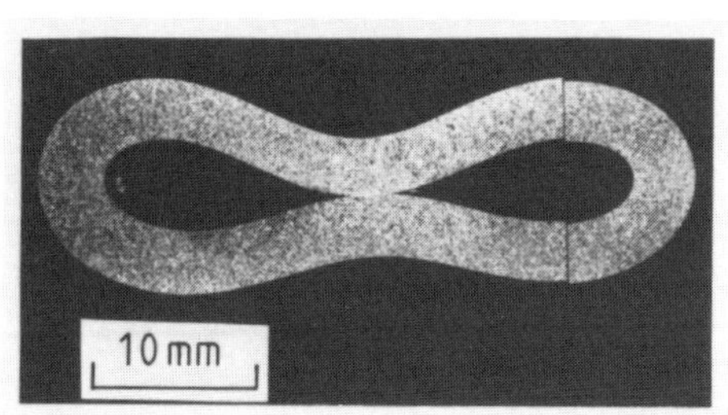


(b) cross section

Fig. 3 Test tube after creep collapse (a)appearance and (b)cross section. 3.5mm in wall thickness and 6% in initial ovality.

observed to have grown along the grain boundary. The precipitations were also found in the grain boundary and the grain of the collapsed tube. The metallurgical structure was typical in the creep fracture. The crack depth was less than 0.6mm. And the cracks did not pass through the wall of the test tube. It is concluded that the leak tightness was maintained despite a collapse of the tube.

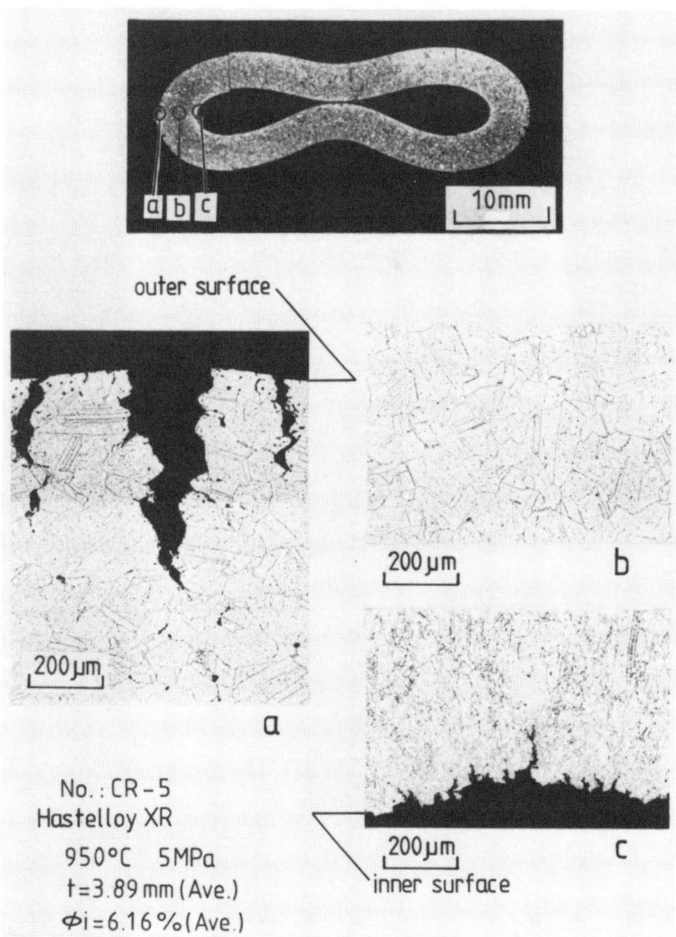


Fig. 4 Cross-sectional views of the collapsed tube for 3.89mm(Ave.) in wall thickness and 6.16%(Ave.) in initial ovality under 5MPa.

## ANALYTICAL METHOD

**Material properties.** Constant stress creep tests were done by a balanced weight inside an electric–heated furnace at 950°C in air. Tests were carried out under the assumption of invariable volume within the gauge length measured by a extensometer. A next Norton type creep constitutive equation [7] was established for the further analyses of creep deformation.

$$\dot{\varepsilon}_c = 3.3025\text{x}10^{-11}\sigma^{4.8458} \quad (2)$$

where $\dot{\varepsilon}_c$ is creep strain rate (1/h), $\sigma$ is applied stress(MPa).

Young's modulus at 950°C was taken to be E=1.285x$10^5$ MPa. And elastic and inelastic Poison's ratio were taken to be $\nu^e$=0.3 and $\nu^{in}$=0.5, respectively.

**FEM analysis by ABAQUS.** The creep collapse processes were analyzed by the FEM using a general purpose code ABAQUS (version 4.8) in which the update Lagrangian formulation is adopted and the change of the direction of external pressure caused by the deformation of the outer surface is taken into account. In the formulation, the Jaumann rate of the Kirchhoff stress is used as the objective stress rate for the constitutive equation. Also the additive decomposition of the rate of deformation into elastic and creep components is assumed.

The first quadrant of the cross section of a tube was modeled by two dimensional 8–node isoparametric elements and plane strain condition was enforced. Initial out–of–roundness was assumed to be quasi–elliptical. Numbers of elements and nodes were 54 and 205, respectively. A finite element model is shown in Fig. 5 with deformed shapes. The calculations were performed on a FACOM VP–2600 supercomputer in JAERI.

**Simplified method by Nishiguchi et al.[4].** Consider a long straight tube which is subjected to external pressure p. The shape of the median line of cross section of the tube is assumed to be quasi–elliptical at any time

$$R(\theta,t) = R_m(1+\alpha\cos 2\theta) \quad (3)$$

where R and θ are polar coordinates as shown in Fig. 6 , $R_m$ is the average radius of the tube, α is a time–dependent parameter and referred to as the shape factor; the shape of the median line is

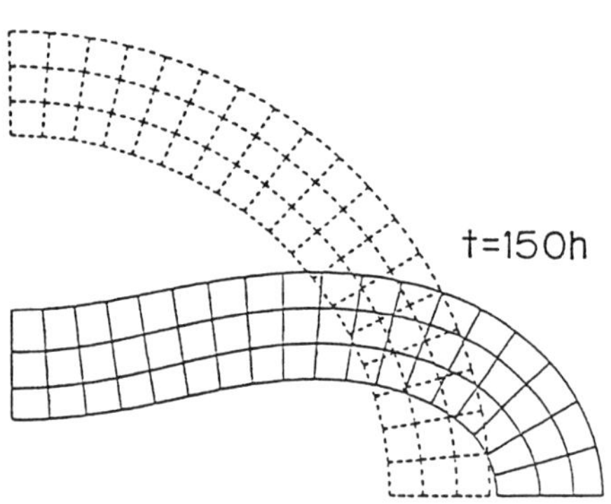


Fig. 5 Results of FEM for 3.5mm in wall thickness and 6% in initial ovality. Undeformed mesh (dashed lines) and deformed mesh (solid lines) are shown.

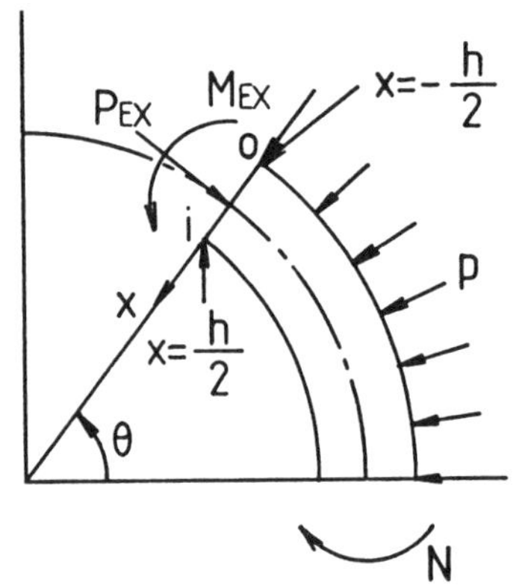


Fig. 6 Forces and moments acting on a tube

determined solely by the single variable α throughout the deformation.

As shown by Hoff et al.[1], the curvature κ of the median line expressed by equation (3) is for a small value of α

$$\kappa = (1+3\alpha\cos 2\theta)/R_m \quad (4)$$

which is linear with respect to α.

Hoff et al.[1] also derives the expressions for the bending moment $M_{EX}$ and the compressive force $P_{EX}$ per unit length of the tube by the external pressure p.

Differentiation of these equations with respect to time gives

$$\dot{M}_{EX} = (\dot{\alpha}p+\alpha\dot{p})R_m^{\ 2}\cos 2\theta \quad (5)$$

$$\dot{P}_{EX} = \dot{p}R_m(1+\alpha\cos 2\theta)+pR_m\dot{\alpha}\cos 2\theta \quad (6)$$

where the terms of $\dot{p}$ are kept for generality.

The time derivative of the bending moment and the compressive force can be alternatively expressed by the stress rate distribution in the cross section of the tube wall

$$\dot{M}_{IN} = \int_{-h/2}^{+h/2} \dot{\sigma}x\, dx \quad (7)$$

$$\dot{P}_{IN} = \int_{-h/2}^{+h/2} \dot{\sigma}\, dx \quad (8)$$

where $M_{IN}$ is the bending moment and $P_{IN}$ is the compressive force per unit tube length both calculated from the distribution of the compressive normal stress σ in the circumferential direction in the tube wall, and x is the coordinate in the radial direction shown in Fig. 6.

The rate $\dot{\sigma}$ of the circumferential stress in equations (7) and (8) is connected with the compressive strain rate in the circumferential direction, $\dot{\varepsilon}$, by the following constitutive equation:

$$\dot{\varepsilon} = \dot{\varepsilon}^e + \dot{\varepsilon}^{in} \quad (9)$$

where

$$\dot{\varepsilon}^e = (1/E')\dot{\sigma} \tag{10}$$

$$\dot{\varepsilon}^{in} = C_\sigma^{in'}\sigma + C_\eta^{in'}\eta \tag{11}$$

where $\eta$ is the compressive stress in the axial direction, E' is the modified Young modulus, $C_\sigma^{in'}$ and $C_\eta^{in'}$ denote the modified inelastic compliances for $\sigma$ and $\eta$, respectively [4].

Assuming the circumferential strain distribution in the tube wall is linear, $\dot{\varepsilon}$ in equation (9) can be expressed as

$$\dot{\varepsilon} = \frac{1}{2}(\dot{\varepsilon}_i + \dot{\varepsilon}_o) + \frac{x}{h}(\dot{\varepsilon}_i - \dot{\varepsilon}_o) \tag{12}$$

where $\dot{\varepsilon}_i$ and $\dot{\varepsilon}_o$ are the circumferential strain rate at the inner and outer surface of the tube, respectively.

From equations (9) and (12), we have an expression for $\sigma$, by which equations (7) and (8) can be rewritten as

$$\dot{M}_{IN} = \frac{E'h^2}{12}(\dot{\varepsilon}_i - \dot{\varepsilon}_o) - E'\int_{-h/2}^{+h/2} \dot{\varepsilon}^{in} x dx \tag{13}$$

$$\dot{P}_{IN} = \frac{E'h}{2}(\dot{\varepsilon}_i + \dot{\varepsilon}_o) - E'\int_{-h/2}^{+h/2} \dot{\varepsilon}^{in} dx \tag{14}$$

In the present context, $\dot{\varepsilon}_i - \dot{\varepsilon}_o$ in equation (13) can be related with $\kappa$ by

$$\dot{\varepsilon}_i - \dot{\varepsilon}_o = h\dot{\kappa} = (3\dot{\alpha}h/R_m)\cos 2\theta \tag{15}$$

Using the foregoing relation, equation (13) can be rewritten as

$$\dot{M}_{IN} = \frac{E'h^3}{4R_m}\dot{\alpha}\cos 2\theta - E'\int_{-h/2}^{+h/2} \dot{\varepsilon}^{in} x dx \tag{16}$$

Though $\dot{M}_{IN}$ should be equal to $\dot{M}_{EX}$ for all $\theta$ throughout the course of the deformation, the exact coincidence can not be established due to the approximations included in the formulation. By an inspection of equation (5) it is found that the absolute value of $\dot{M}_{EX}$ decreases from its maximum value at $\theta=0$ to zero at $\theta=\pi/4$. A compromise suggested by this nature may be

$$\int_0^{\pi/4} \dot{M}_{EX}\, d\theta = \int_0^{\pi/4} \dot{M}_{IN}\, d\theta \tag{17}$$

Therefore, from equations (5) and (16), a growth law for $\alpha$ is established as

$$\dot{\alpha} = \frac{\dfrac{8R_m}{h^3}\displaystyle\int_0^{\pi/4}\int_{-h/2}^{+h/2} \dot{\varepsilon}^{in} x dx d\theta + \frac{\dot{p}}{p_c}\alpha}{1 - p/p_c} \tag{18}$$

where $p_c$ is the elastic critical pressure given by Timoshenko and Gere [8].

On the other hand, from equations (6), (14) and (15), we have the expressions for $\dot{\varepsilon}_i$ and $\dot{\varepsilon}_o$ as

$$\dot{\varepsilon}_i = \frac{\dot{p}R_m}{E'h} + \{(\frac{pR_m}{E'h} + \frac{3h}{2R_m})\dot{\alpha} + \frac{R_m\dot{p}}{E'h}\alpha\}\cos 2\theta + \frac{1}{h}\int_{-h/2}^{+h/2} \dot{\varepsilon}^{in} dx \tag{19}$$

$$\dot{\varepsilon}_o = \frac{\dot{p}R_m}{E'h} + \{(\frac{pR_m}{E'h} - \frac{3h}{2R_m})\dot{\alpha} + \frac{R_m\dot{p}}{E'h}\alpha\}\cos 2\theta + \frac{1}{h}\int_{-h/2}^{+h/2} \dot{\varepsilon}^{in} dx \tag{20}$$

A growth rate of $\eta$ in equation (11) is given as

$$\dot{\eta} = \nu^e\dot{\sigma} - (E/E^{in})(\eta - \nu^{in}\sigma) \tag{21}$$

Based on the foregoing formulas, evolution of $\alpha$ can be obtained straightforwardly. Having known the stress field and the states of internal variables, if any, which determine $\dot{\varepsilon}^{in}$, $\dot{\alpha}$ is obtained from equation (18); from which the strain rate field is computed from equation (19), (20) and (12), which determines $\dot{\varepsilon}$ of the left-hand side of equation (9). The stress rate field is determined by solving equations (9) and (21) for $\dot{\sigma}$ and $\dot{\eta}$. Finally, the stress field and internal variables are updated for the next step of computation.

## DISCUSSION OF ANALYTICAL AND EXPERIMENTAL RESULTS

Using the Norton type creep constitutive equation (2), the creep collapse time analyzed by the FEM code and the simplified method proposed by Nishiguchi et al. are shown in Table 2 and Fig. 2. As can be seen, the FEM and simplified method predict almost conservative creep collapse time compared with the experimental results. And for a thinner tube, the difference between solutions by FEM and simplified theories is reduced.

As an example, the calculated ovality curve over time for 3.5mm in wall thickness and 0.1% in initial ovality under 6MPa is shown in Fig. 8. Open circles denote the experimental results for 3.47mm in minimum wall thickness and 0.03% in maximum initial ovality. It is found from the experiment the change of ovality is very little for most of the time to collapse, then a very rapid increase of ovality is observed at the final stage. As for the time to collapse, the simplified method gives a reasonable prediction for the experimental result. As can be seen, the simplified method predicts shorter creep collapse time than the FEM.

In addition to the foregoing example, a thinner tube of 2mm in wall thickness and 0.1% in initial ovality was analyzed. The results are shown in Fig. 9. The times to collapse of experiments of Nos. CR-4, CR-6 and CR-7 are shown in this figure. It is found that the FEM and the simplified method provide conservative prediction for the time to collapse as compared with the experimental results. For a thinner tube, the creep deformation behavior predicted by the simplified method is in fairly good agreement with the results of the FEM.

In the creep collapse analyses, the number of degrees of freedom is one since the simplified method assumes merely the quasi-elliptical cross section with a single parameter for the displacement field whose growth is governed by an ordinary differential equation. However, the FEM code needs far more computation time for the finite deformation with greater degrees of freedom.

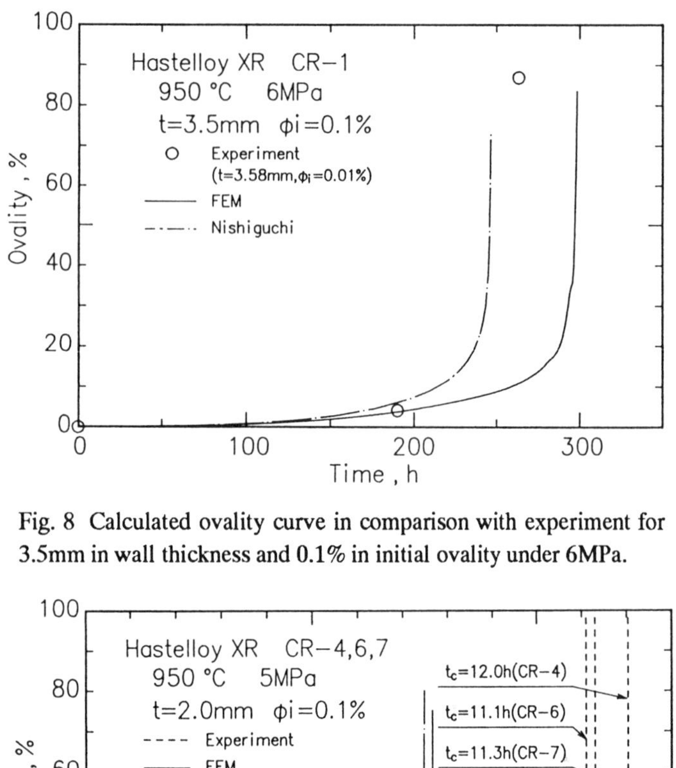


Fig. 8 Calculated ovality curve in comparison with experiment for 3.5mm in wall thickness and 0.1% in initial ovality under 6MPa.

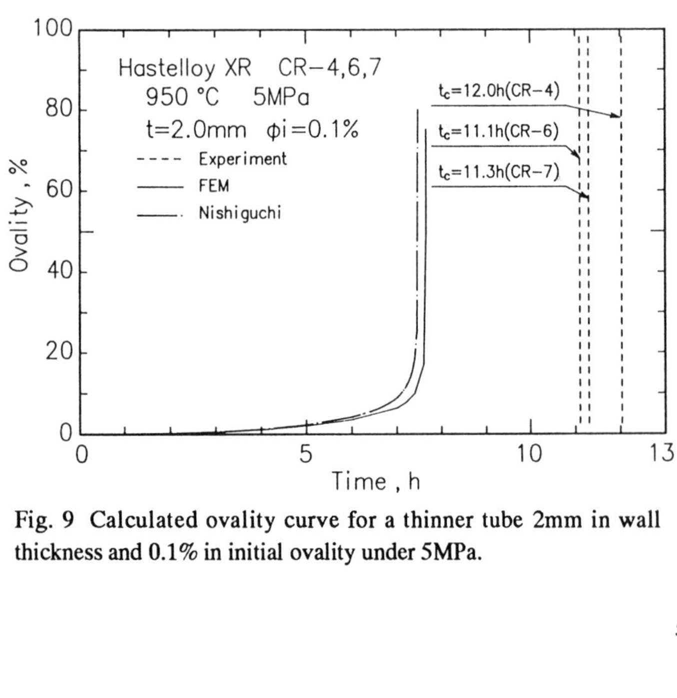


Fig. 9 Calculated ovality curve for a thinner tube 2mm in wall thickness and 0.1% in initial ovality under 5MPa.

## CONCLUSIONS

The creep collapse tests and analyses of the heat transfer tubes subjected to constant external pressures under constant temperature were performed.

Analyses have shown that both the simplified method proposed by Nishiguchi et al. and the FEM reasonably predict the creep collapse as compared with experimental results. The simplified method provides satisfactory and more importantly conservative prediction for collapse time of tube under external pressure compared with the FEM.

Though many cracks were observed on the outer surface of a few test tubes initiated by creep collapse, the cracks did not pass through the tube thickness. All the test tubes which carried out the leak test have shown that the leak tightness was maintained irrespective of the cracks caused by creep collapse.

## ACKNOWLEDGEMENT

The authors are grateful to Mr. Y. Muto, JAERI, for his helpful suggestions and comments. Thanks are also extended to Mr. K. Matsumoto, JAERI, and Mr. T. Yamamura, IHI Co., for help in conducting the experiments.

## REFERENCES

1. Hoff, N.J., Jahsman, W.E. and Nachbar, W., Journal of the Aero/Space Sciences, 663-669(1959).
2. Bargmann, H., AIAA J., 327-329(1972).
3. Bargmann, H., Nuclear Engineering and Design, 22, 51-62(1972).
4. Nishiguchi, I., Kaji, Y., Ioka, I. and Yamamura, T., Journal of Pressure Vessel Technology, 112(3), 233-239(1990).
5. Ioka, I., Kaji, Y., Miyamoto, Y. and Nishiguchi, I., SMiRT 11 Transactions Vol.L, L03/4(1991).
6. Hibbitt, Karlsson & Sorensen, Inc., ABAQUS Theory Manual Version 4.8(1989).
7. Norton, F. H., "Creep of Steel at High Temperatures", McGrow Hill, New York,(1929).
8. Timoshenko, S. and Gere, M., "Theory of Elastic Stability", Second Edition, McGraw-Hill(1963).

*Proceedings of the Fifth International Conference on Creep of Materials, Lake Buena Vista, Florida, USA, 18-21 May 1992*

# Development of a Lifetime Evaluation Method for 2.25Cr-1Mo Heat Transfer Tube Serving for Operating Boiler Plants

**M. Nakashiro and S. Kihara**
Ishikawajima-Harima Heavy Industries Company, Ltd.
Tokyo, Japan

## 1. Introduction

In Japan, many pieces of thermal power plant equipment have exceeded service life of 100,000 hours, and aging is proceeding further. Among the heat resisting parts of boilers, damage due to creep and corrosion —especially the kinds that are caused by high temperature oxidation and steam oxidation— is particularly heavy for the superheater tubes and the reheater tubes that have to keep operating in the highest domain of service temperatures. Since a significant part in the current practice of boiler maintenance is expended in finding and replacing those boiler tubes that appear to have sustained large metal loss by corrosion or large creep damage due to overheat, developing of a method of accurately evaluating the degree of damage accumulated, hence assessing the residual service lifetime, is urgently called for.

Now, the commonly adopted method to evaluate the life of a heat transfer tube is to perform the rupture test on it. We at IHI, too, have examined some fifty utility boilers in the past five years; as for 2.25Cr-1Mo steel tubes (ASME T22), in particular, nearly as many as 1,000 creep specimens have been tested. One trouble with this method is the fact that, as shown in Figure 1, even though the test results indicate clear decrease of creep strength for high stresses, what they predict for the actual operation stress is a strength that is not much different from virgin tube's, all regardless the time the sample tube has expended in actual service. We consider this discrepancy as arising from our inability to assess the comparatively small creep damages —the kinds that are encountered in real services— accurately enough.

It is to overcome this difficulty that we at IHI have devised a transmission electron microscopy

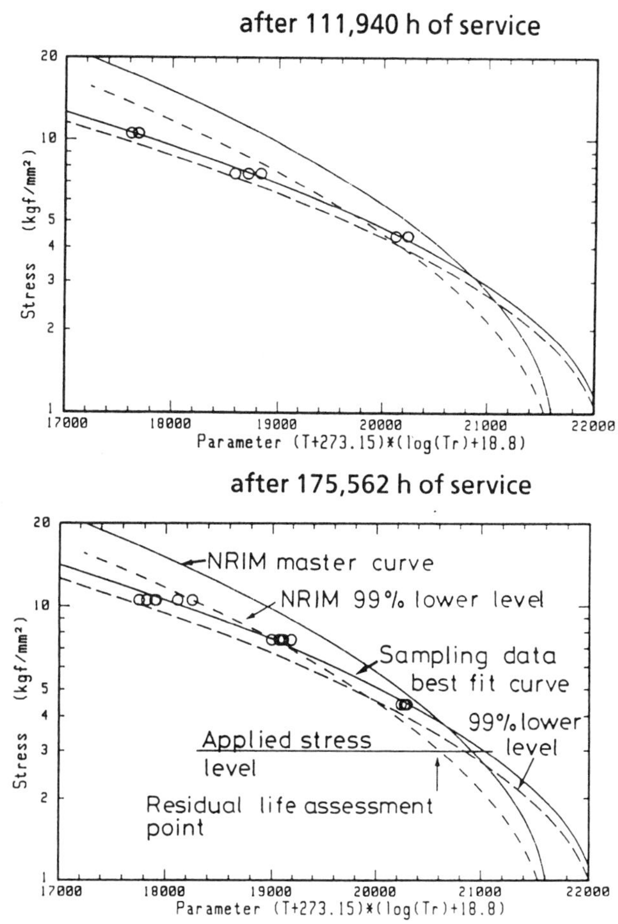


Fig. 1 Stress versus L.M. parameter curves of super-heater tube (2.25Cr-1Mo steel) after 111,940 h and 175,562 h of services at operating plants.

(TEM) method of evaluating the degree of creep damage(1) and a method of correcting the NRIM creep data sheet(2) for the effect of high temperature oxidation so as to estimate the true creep strength for the real operation conditions(3). Owing to these studies, we have been able to clarify, partially, at least, the behaviors of heat transfer tubes being damaged in real operation conditions(4).

In this paper, then, we intend to present a simulation-cum-evaluation method of the process of heat transfer damage that we have developed on these studies and to discuss the efficacy of this method as examined against the NRIM creep data sheet.

## 2. The Method of Evaluation

For analysis, which has been formulated as a computer program, we use the tube dimensions (outer diameter and wall thickness), the operation temperature, the steam pressure, the heat absorption rate for the operation conditions, and the heat conductivity of the steam scale as the design data. Next, we enter the time expended in service (i.e., the operation time), the tube dimensions (outer diameter and wall thickness), and the mean operation temperature, all as acquired on the latest inspection, as the monitor data. Here, we determine the mean operation temperature on the basis of the operation time and the observed microstructure referring to the standard microstructures; even when the mean operation temperature differs significantly from the design temperature because of overheating and such, we take the mean temperature as determined this way for the input data.

Then we calculate the correction factors for the rates of high temperature oxidation and steam oxidation from the tube dimensions determined on inspection, and assuming the oxidation to proceed at the rate so corrected even after the inspection, we calculate the amount of scale formed in the respective oxidation processes. This allows us to calculate the rise in the metal temperature due to formation of the steam scale.

Appreciation of the results of analysis for creep damage is done by referring to both the NRIM creep data sheet as published(2) and the same but corrected for oxidation that has taken place during the test (to be called "the corrected data sheet" hereinafter)(3): for either, two sorts of evaluation curves, one for the mean values and the other for the 99% lower limits, are drawn on which to determined the extent of creep damage per unit time. We then calculate the total creep damage fraction, Dec, according to the linear damage accumulation law, and decide on a rupture life as that time at which the sum of Dec has reached unity. Finally, we check these calculated damage data against the creep ruptured tests and the creep damage index, Nsb, determined from TEM observations at the same time.

## 3. Results of Analysis and Discussion

### 3.1 Rate of Scale Growth on Fire Side and on Steam Side

Figure 2 presents the decrease of creep specimen diameter, $D_0$–D, i.e., the metal loss, obtained in various tests in terms of the Larson-Miller parameter (the "L.M. parameter")(3, 5). Here, the diametric loss is to represent the amount of the oxide scale formed in high temperature oxidation, because such a scale is easy to exfoliate away.

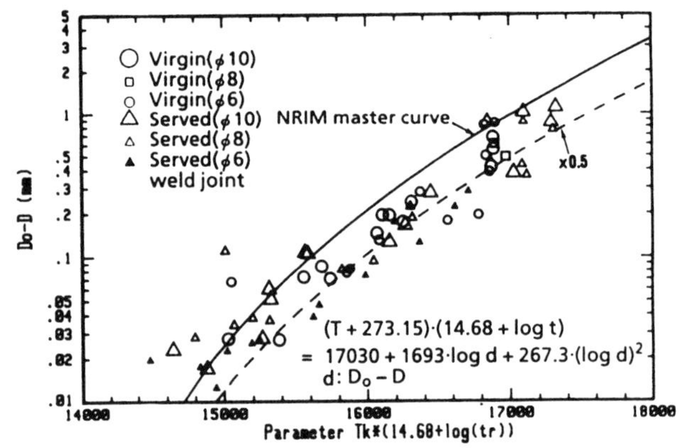


Fig. 2 Relationship between the specimen diameter reduction by oxide scale and L.M. parameter for 2.25Cr-1Mo steel in air atmosphere test.

It will be seen that one single regression curve (Eq.(1)) represents all the oxidation metal losses, whether the tube is virgin or served in an operating plant for a while. Therefore, not only the degree of oxidation can be calculated readily by specifying the service temperature and the operation time, but also the rate of metal loss can be computed out by multiplying the oxidation loss with a correction factor.

$$T_K \cdot (14.68 + \log t) = 17030 + 1693 \cdot (\log d) + 267.3 \cdot (\log d)^2 \quad (2)$$

where d: $D_0 - D$,
$T_K$: Temperature (°K)
t: Operation time

Since the steam scale is highly cohesive, on the other hand, its growth rate can be determined directly from the thickness, as done by Rehn in his EPRI report(6), which is shown in Figure 3. The evaluation formula given below as Eq. (2) for the inside scale thickness and Eq. (3) for the total scale thickness are those that have been developed on the mean values of respective data bands:

$$\log d_{IN} = 2.15 \times 10^{-4} \cdot P_{ar} - 5.978 \quad (2)$$

and $\log d_T = 2.11\times10^{-4}\cdot P_{ar} - 5.503,$ (3)

where $d_{IN}$: the thickness of the inside scale ($\mu$m), which is taken as equal to the decrement of the inside tube wall thickness i.e., the metal loss ($\mu$m),

$d_T$: the total scale thickness ($\mu$m),

$P_{ar}$: the L.M. parameter, which is $(T+491.67)\times(20+\log t)$,

t: the operation time (h),

and T: the operation temperature (°F).

Here, the rate of steam oxidation was modified by multiplying a correction factor as for the case of high temperature oxidation, and the rise in the metal temperature on steam oxidation was computed on $d_T$ as will be discussed in next item.

**3.2 Estimation of Mean Operation Temperature and of Rise in Metal Temperature due to Formation of Steam Scale**

The Cr-Mo steels, of which the 2.25Cr-1Mo steel of this study is one, undergo microstructural changes on prolonged service at high temperatures, and these changes can be systematically arranged in terms of the L.M. parameter. Figure 4 shows the standard microstructure album for 2.25Cr-1Mo steel. Sample's microstructure is identified with one of these photomicrographs, whose heating time and temperature give a L.M. parameter, from which the mean operation temperatures is computed back, assuming equivalency of the two L.M. parameters, from the actual operation time as following equation.

$$P=(T+273.15)\cdot(\log t+18.8), \quad (4)$$

where t is the operation time (h), and T, the heating temperature (°C).

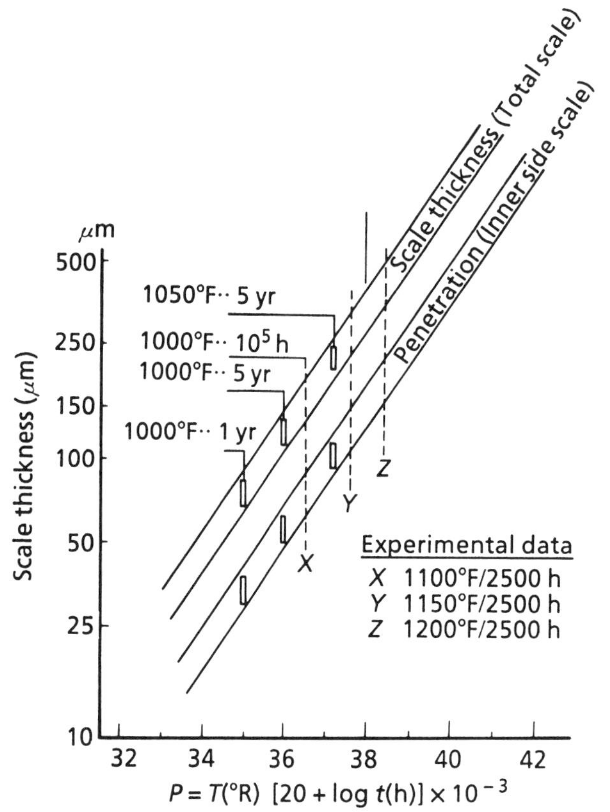


Fig. 3 Steam-side oxidation growing thickness for 1~3% Cr steel. (EPRI Data)

The temperatures estimated from the microstructure observed for test pieces after creep testing and their testing time as in Figure 4 are

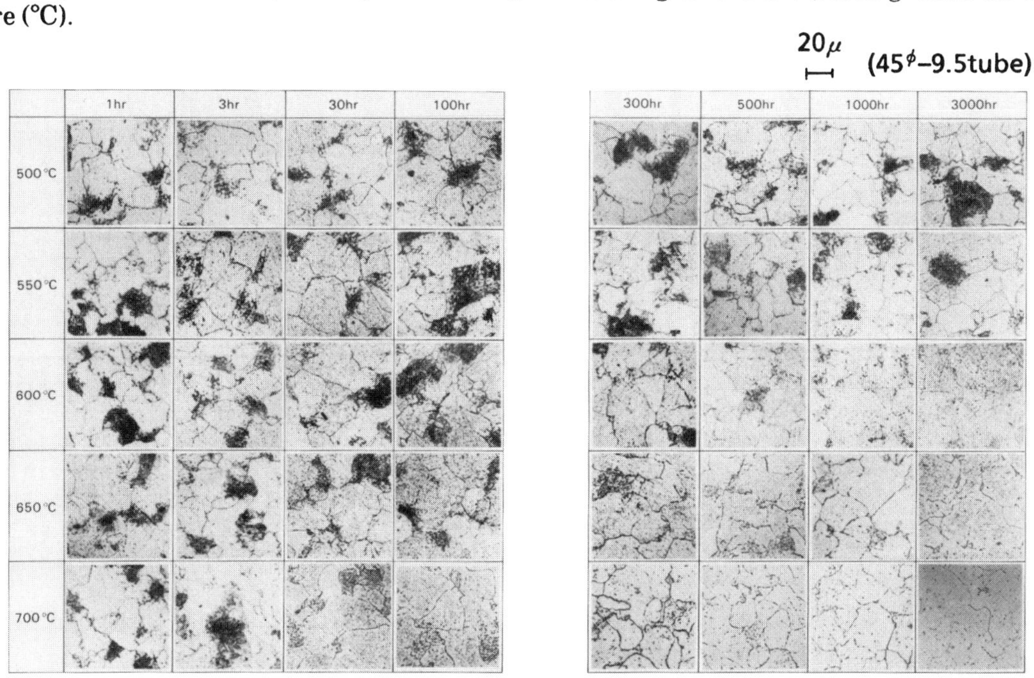


Fig. 4 Variation of microstructure of 2.25Cr-1Mo steel (T-22) with long time heating

compared to the actual testing temperature in Figure 5. Since the test pieces were all those that were subjected to stress-acceleration tests with creep times less than 3000 h, a tendency to slightly overestimate the temperature may be seen; yet the agreement should be considered satisfactory enough. Since the operation stress in actual service is much lower and the operation time much longer, however, the precision will undoubtedly be better for real application.

The method of calculating the increase in the metal temperature due to formation of oxide scale is schematically illustrated in Figure 6. For a heat absorption rate q, which is known for any one of the heat transfer tubes from the design conditions, and the heat conductivity of steam scale $\lambda_S$, for which a value of 1.5kcal/m·h·°C has been recommended for boiler environments(7), the temperature rise $\Delta T_S$ is given by:

$$\Delta T_s = q/(\lambda_s/t_s) = (q/\lambda_s)\cdot t_s, \qquad (5)$$

where $t_s$ is the thickness of the steam scale (m).

### 3.3 Evaluation of Creep Damage

Figure 7 shows the NRIM creep data sheet as rearranged in terms of the L.M. parameter. Since the original data sheet includes many cases that exceeded 100,000 h of creep time, and since the specimens are rather small —only 6mm in diameter—, the data on the high temperature, long term side are subject to appreciable degradation of strength due to oxidation. This is the main reason why the 99% lower limit creep strength regression curve intersects with the ASME allowable design strength curve in low stress domain.

On correcting the data for oxidation by Eq. (1), however, we obtain Figure 8. It will be noted that both the mean value regression curve and the lower limit regression curve drawn on the corrected NRIM data now run approximately in parallel to the ASME allowable design stress curve. From this, we conclude that the ASME allowable strengths can be taken as the standard values for the cases where the effect of oxidation needs not to be considered. That is to say, the creep damage can well be evaluated on the creep strength curve of Figure 8, provided that the effect of oxidation incurred during the service be assessed separately through Eq. (1); the creep damage thus determined should afford accurate enough prediction of lifetime.

### 3.4 Residual Lifetime Estimation on Nsb Value

According to the TEM observation method of assessing the creep damage we have devised(1), the damage a steel has sustained through its service up to the time of testing is represented by the Nsb value, which is a ratio of the number of grains that have developed subgrains within themselves as shown in Figure 9 to the total number of grains inspected. The notion of underlying this method is schematically depicted in Figure 10.

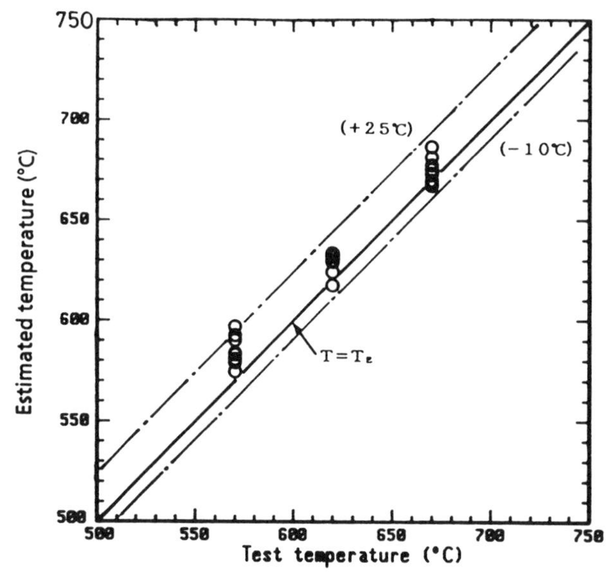


Fig. 5 Relationship between the test temperature and the estimated temperature.

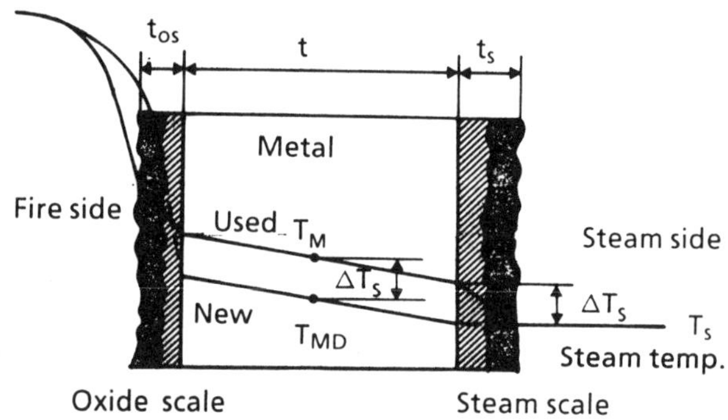


$t_{os}$: Oxide scale thickness (mm)
t : Metal thickness (mm)
$t_s$ : Steam scale thickness (mm)
q : Heat absorption rate
ex. { Super-heater 40000 (kcal/m$^2$h)
  Re-heater 29000 (kcal/m$^2$h)
$\lambda_s$ : Heat conductivity of steam scale
$\lambda_s$ = 1.5 (kcal/m h °C)

$\Delta T_s: T_M - T_{MD}$

$$\Delta T_s = \frac{q}{\lambda_s/t_s} = \frac{q}{\lambda_s}\cdot t_s$$

$T_s$ : Steam temp.
$T_{MD}$ : Design metal temp.
$T_M$ : Operating metal temp.
$\Delta T_s$ : Increase in metal temp.

Fig. 6 Estimation method of increase in metal temperature due to formation of steam scale.

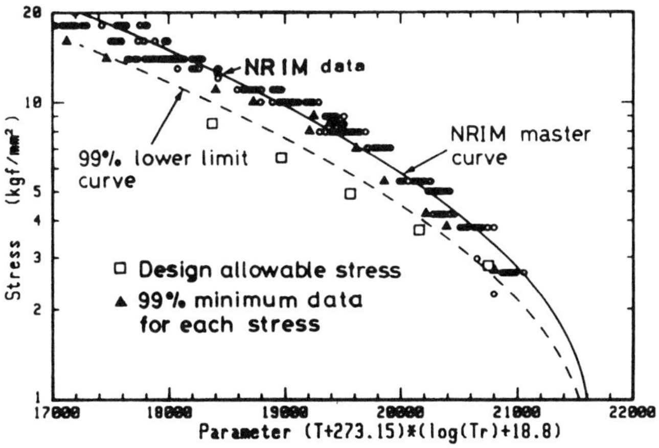


Fig. 7 Stress versus L.M. parameter relations of STBA24 in NRIM creep data sheet and ASME design allowable stresses.

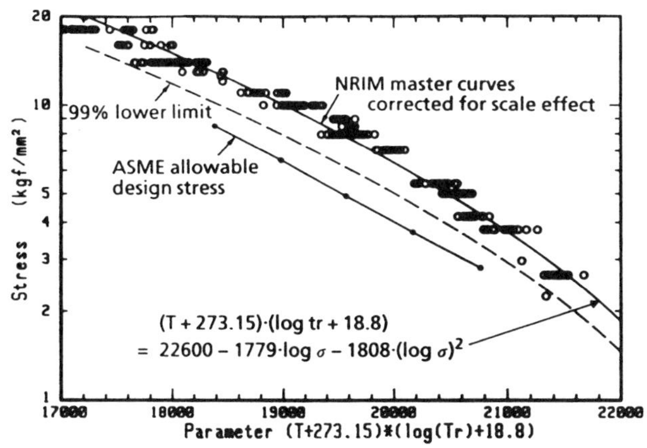


Fig. 8 Stress versus L.M. parameter curves and creep data of STBA24(NRIM) for calculating oxide scale effect.

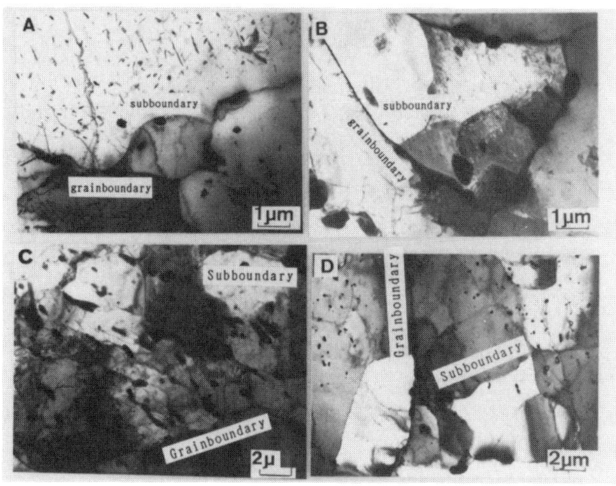


Fig. 9 TEM microstructures of creep damaged specimen.

A : Series 2 650°C 29.4MPa 500.5h t/tr = 0.08
B : Series 1 670°C 26.5MPa 764.0h t/tr = 0.25
C : Series 1 670°C 26.5MPa 1556.4h t/tr = 0.52
D : Series 1 670°C 26.5MPa 2257.8h t/tr = 0.75

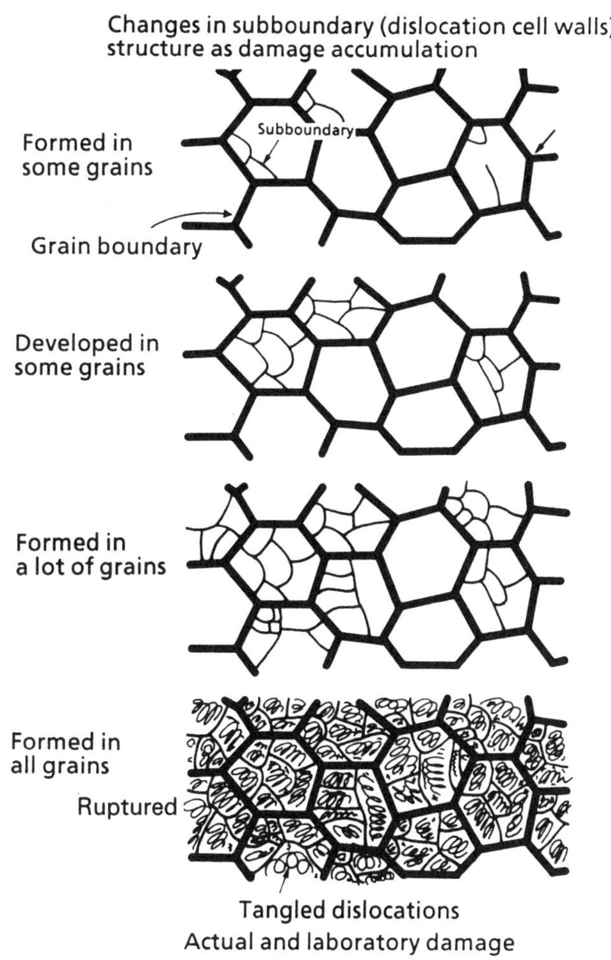


Fig. 10 Schematic diagram of microstructure changes with increasing creep damage.

The relations among Nsb value, creep damage fraction, and typical creep curve are shown in Figure 11. As may be seen in Figure 11, the Nsb value increases steadily even in the steady creep stage where the creep strain is very small. This allows us to assess the accumulated damage at a high degree of accuracy by simply determining Nsb for samples taken on inspection.

**3.5 Method of Evaluating the Accumulated Creep Damage**

Based on the methods of assessing the creep damage presented above, we have developed a computer program to evaluate the accumulated creep damage, hence the residual service life. The calculation flow chart is shown in Figure 12 with the input data and stipulations for computation.

First, we determine the correction factor K for growth rates of the high temperature oxide scale and the steam scale for the operation temperature which has been checked against the results obtained with the inspection samples. Then the scale thickness, tube wall thickness, metal temperature, and accumulated creep damage are calculated using the K value determined above until the accumulated creep damage fraction has become unity. These quantities are plotted as a function of time so as to assess the residual lifetime. The calculated damages are checked against the observations acquired on each inspection, particularly in terms of the Nsb value and the creep damage deduced from the rupture test results, to decide the reliability of the sample tubes damage accumulation curve. Then, prediction of residual service lifetime and determination of the next inspection time and the time of tube replacement can be made on this curve with confidence.

### 3.6 Results of Calculation

Figure 13 presents the case of superheater (SH) tube. Since SHs operate under a high steam pressure, their tubes' walls are rather thick. Therefore, the stress does not change much even when a considerable metal loss has been incurred by high temperature oxidation. For this reason, the accumulated creep damage fraction Dec can attain unity before the metal loss has reached the design allowance usually, meaning that evaluation of creep damage is more important for this kind of heat transfer tube.

The case of reheater (RH) tube is shown in Figure 14. Since the steam pressure is low in RHs — their heat transfer tubes are of thinner walled—, the rise in stress due to metal loss is large. The RH life time is shorter than SH. Also, it is seen that the wall thickness can decrease to below the design allowance before the Dec has become unity. Therefore, measurement of the wall thickness is usually more important for this kind of heat transfer tubes, because their replacement time is determined by attainment of the design allowance thickness.

Next, we evaluate Dec's and Dnsb's for various heat transfer tubes, some of which have failed in actual serving by erosion or over heating. Dnsb is creep damage ratio by evaluating from Nsb-t/tr master curve in Figure 11. Figure 15 presents the relationship between Dnsb and Dec, which is calculated the creep damage with correcting creep strength for oxide scale effect. Since Dnsb are almost located between mean and minimum Dec values, these calculation methods are considered to be reasonableness. Figure 16 presents the calculated results with NRIM creep data sheet. These evaluation methods are not to be used because the evaluate Dec greater than Dnsb or actually tube conditions. Figure 17 shows the correcting creep damage curves to get the same value of Dnsb and Dec at maintenance and inspection time, and each correcting curves of mean and minimum are almost same curves. Therefore, the correcting curve is the most accuracy life evaluation curve at the maintenance time.

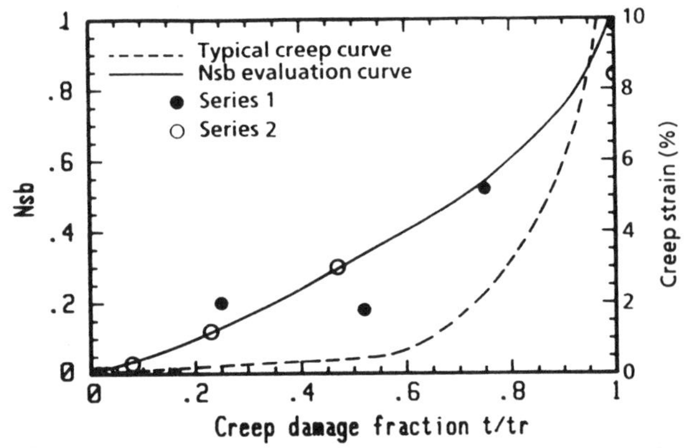


Fig. 11 Variations in Nsb and creep strain with creep damage ratio.
(Nsb: Ratio of subboundary-containing grains to all inspected grains.
Series 1; 670°C, 26.7MPa, tr=3018 h,
Series 2; 650°C, 29.4MPa, tr=6607 h)

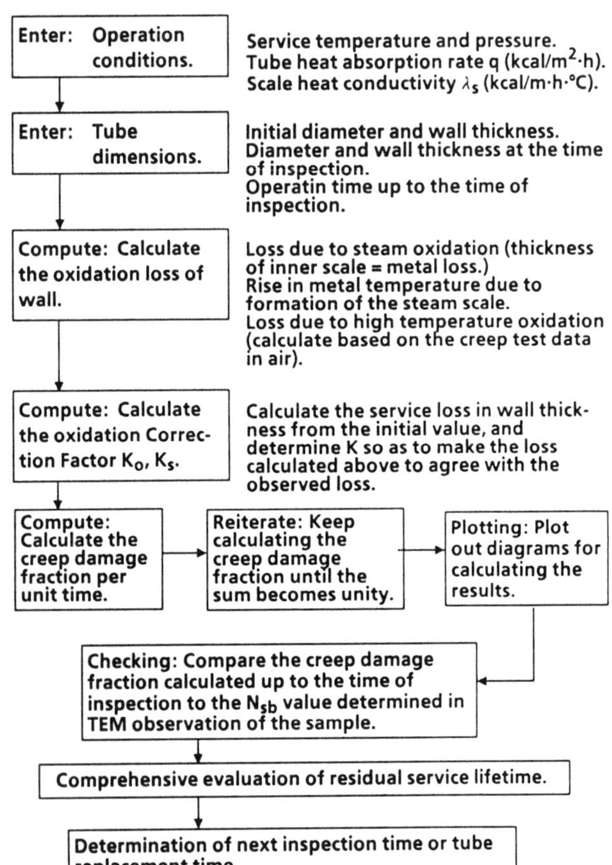


Fig. 12 Flow chart for evaluating creep damage accumulated in boiler heat transfer tubes

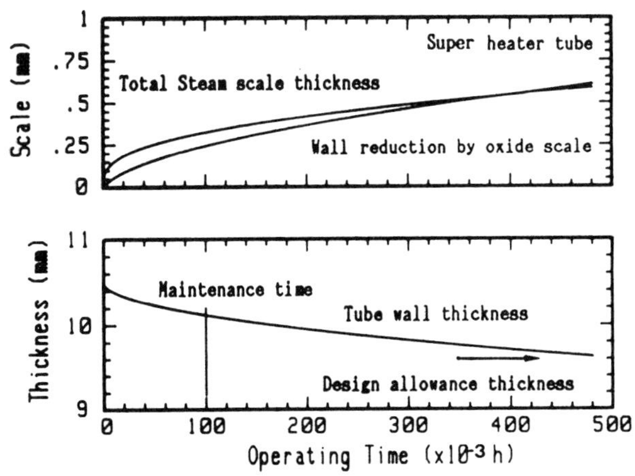


Variation of tube thickness and oxide and steam scale thickness with operating time for super-heater tube.

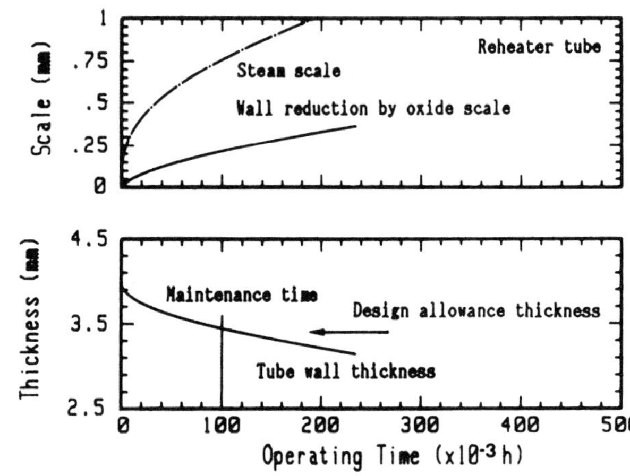


Variation of tube thickness and oxide and steam scale thickness with operating time for reheater tube.

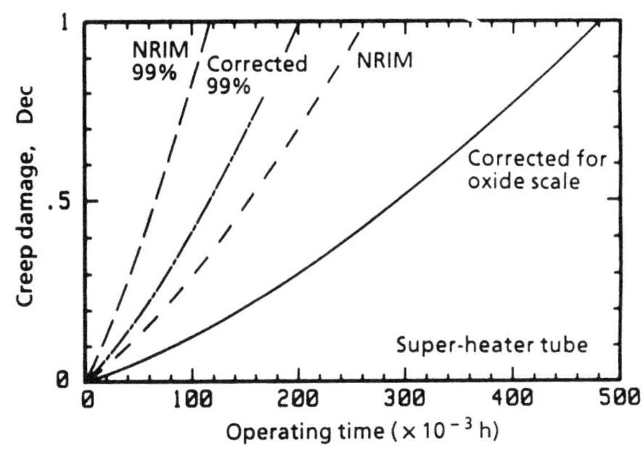


Fig. 13 Creep damage variation with operating time by NRIM creep data sheet, 99% minimum creep strength and corrected for oxide scale for super-heater tube.

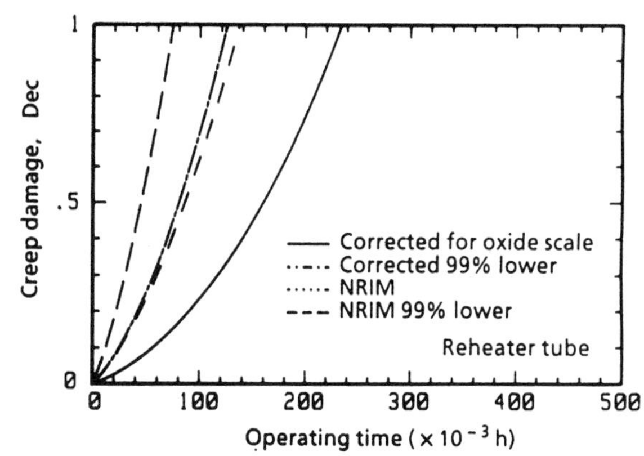


Fig. 14 Creep damage variation with operating time by NRIM creep data sheet, 99% minimum creep strength and corrected for oxide scale for reheater tube.

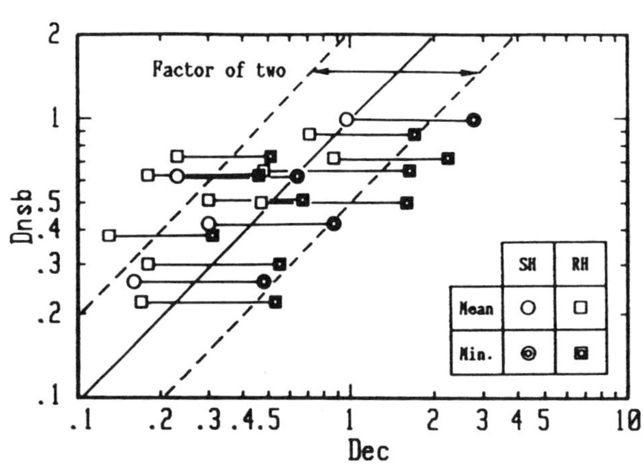


Fig. 15 Relationship between Dec and Dnsb from correcting data.
(Dnsb: Evaluation damage from Nsb master curve)

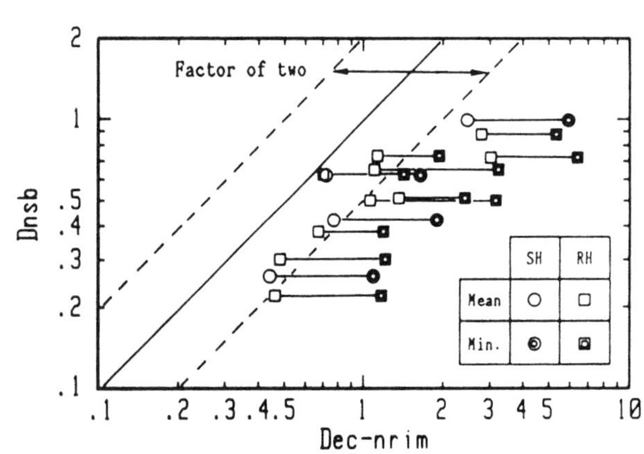


Fig. 16 Relationship between Dec-nrim and Dnsb from NRIM creep data sheet.
(Dnsb: Evaluation damage from Nsb master curve)

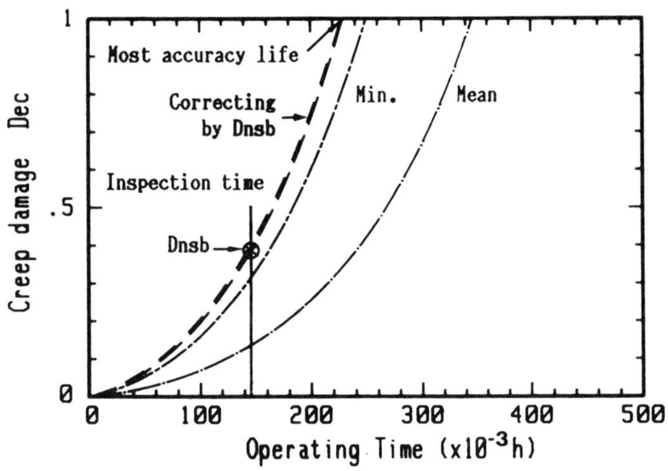


Fig. 17 Creep damage variation with operating time by correcting of oxide scale effection and 99% minimum strength.

## 4. Conclusions

(1) To evaluate the creep damage incurred in a boiler heat transfer tube, the use of the creep strength regression curve of the NRIM data sheet curves is the most proper, provided the strengths be corrected for formation of oxide scale.

(2) The computation algorithm we have proposed to calculate the accumulated damage fractin Dec on two correction factors Ko and Ks is able to calculate the Dec as a function of time with reasonable accuracy, provided the operation conditions of the tube be given correctly in terms of the data acquired on regular inspection, i.e., the tube dimensions, thickness of scale, microstructure, mean operating temperature, and the Dnsb value as determined through TEM observation as the ratio of the number of subboundary-containing grains to the total number of grains inspected.

(3) The Dnsb value and the Dec value as calculated above agree each other quite well, affording reliable estimation of residual lifetime by either.

(4) For the superheater tubes, rupture is expected to take place when the Dec value attains unity before the design allowance thickness is reached due to metal loss on oxidation, so that determination of Dec should be more important.

(5) For the reheater tubes, the design allowance thickness will be reached before Dec becomes unity, so that care should be paid to the wall thickness.

## References


1- K.Murakami, S.Kihara, M.Nakashiro, Y.Nakagawa, Ishikawajima-Harima Engineering Review, Vol. 31. No.5, Sep. (1991) pp.353-358.

2- *NRIM Creep Data Sheet No. 3B*, National Research Institute for Metals (Tokyo, 1986).

3- M.Nakashiro, S.Kihara, F.Kishimoto and T.Fujimori: ISIJ International Vol.30 No.10 (1990) pp.283-828.

4- M.Nakashiro, K.Murakami, S.Kihara, *ICM-6*, **2** WS6b5, 131 (1991).

5- T.Kaneko, H.Hongho, N.Nagashima, Y.Monma, and C.Tanaka, *CAMP ISIJ*, 1, 900 (1988).

6- I.M.Rehn; Corrosion Problems in Coal-Fired Boiler Superheater and Reheater Tubes: Steam-Side Oxidation and Exfoliation, EPRI CS 1811 Project 644-1 (1981).

7- CRIEPI REPORT 277018 Dec. (1977).

*Proceedings of the Fifth International Conference on Creep of Materials, Lake Buena Vista, Florida, USA, 18-21 May 1992*

# Fractography of Notched Specimens of Wrought Superalloys Under Static Creep and Cyclic Creep Conditions

**Z. Guanghua**
Luoyang Ship Material Research Institute
Luoyang Henan, China

## Abstract

The fractography of notched bar static creep specimens and notched bar cyclic creep specimens of three wrought supperalloys, i. e. , GH33, GH37 and GH49, are made in this article. The results indicate that for GH33 alloy, in comparison with the fracture surfaces of notched bar static creep specimens, on the fracture surfaces of notched bar cyclic creep specimens there were a large number of slip lines and fatigue striations besides creep cavities in addition, subcracks increased obviously and chair-like carbides obviously grew along grain boundaries and were interconnected with each other resulting in the formation of cracks, and trifurcate cracks, originally existing under the notched bar static creep condition, disappeared. For GH37 alloy, there were quantities of creep cavities on both fractures. In comparison with the fractures of notched bar static creep specimens, on the fractures of notched bar cyclic creep specimens creep cavities grew obviously, grain boundary oxidation was aggravated and subcracks increased, but fatigue striations decreased obviously. For GH49 alloy, the creep cavities, grain boundary oxidation and subcracks on the fracture surfaces of notched bar cyclic creep specimens were much more serious than those on the fracture surfaces of notched bar static creep specimens, but there was no fatigue striation on the former except that there was a small amount of slip lines in a few zones. Based on these results, the substance of the considerable decrease of the creep rupture lives of notchedspecimensbycyclicloading is also discussed in this article.

IT IS WELL KNOWN that cyclic loading has an obvious influence on the creep rupture lives of materials under a controlled load. This fact indicates that it is unreasonable to design a high temperature structural element working under a condition of variable temperature and loading by employing conventional static creep rupture life data. However how to explain this fact is still a problem to be studied deeply. There phenomena have once been attributed to a materials cyclic softening or hardening. But in general, an obvious cyclic softening or hardening would not occur because a specimen can be elongaled freely during its load-holding time, and the amount of creep would be changed with cyclic loading. Based on our previous work, it was indicated that this was a more complex mechanical-metallurgical pheomenon, so that it is difficult to explain by cyclic softening or cyclic hardening. It is believed that cyclic loading has an influence on creep rupture lives mainly because of its effects on the creep mechanism itself, and at ydifferent testing temperatures and stresses cyclic loading would exert an influence on creep deformation and rupture rate in different forms. However the mechanics phenomena existing in notched specimens under cyclic creep conditions are difficult to explain completely by this point of view. For example, for GH49 alloy (in comparison with smooth bar static creep condition) under smooth bar cyclic creep condition, the rupture life is decreased, however under notched bar cyclic creep condition, the rupture life is increased. Form this it seems that the substance of the phenomenon mentioned above might be not a cyclic strengthening but a notch strengthening effect.

In order to make certain the substance of the effect of cyclic loadingoncreeprupturelives, observations and analyses of fractures of rupture specimens in notched bar static creep tests and notched bar cyclic creep tests were made respectively.

## Experiments

GH33, GH37 and GH49 alloys used for the work are solution treated aging hardened Ni-base superalloys. The chemical compositions, heat treatments, testing parameters , mechanical properties and results were shown in Reference(1)。

After the fractures were observed under a reading microscope ,half the fractures were taken to prepare ordinary metallugraphic samples and the other was observed under a JSM—35c Mode scanning electro microscope. The ruptured specimens include those in notched bar static creep tests and notched bar cyclic creep tests.

## Results

GH33 alloy Macroscopically, either the fractures of notched bar static creep specimens or those of notched cyclic creep specimens had an unever appearance ,but most of the former exhibited a black colour and the latter blue one , which indicated that the transient yuptured zone of the former was far smaller than that of the latter. The examinations under a convinient microscope and scanning electron microscope shown that either in the the notched bar static creep tests or in the notched bar cyclic creep tests a ruptured specimen exhibited an intergranular fracture . But an obvious trifurcate crack existed on the fracture surfaces of ruptured specimens in notched bar static creep tests, shown in Fig. 1, with less subcracks found . On the other hand , on the fracture surfaces of ruptured specimens in notched bar cyclic creep tests, there were few wedge-shaped cracks found with subcreaks increased and boundary damage being serious ,as shown in Fig. 2. In addition , there was a large amount of slip lines and fatigue striations, shown in Fig. 3.

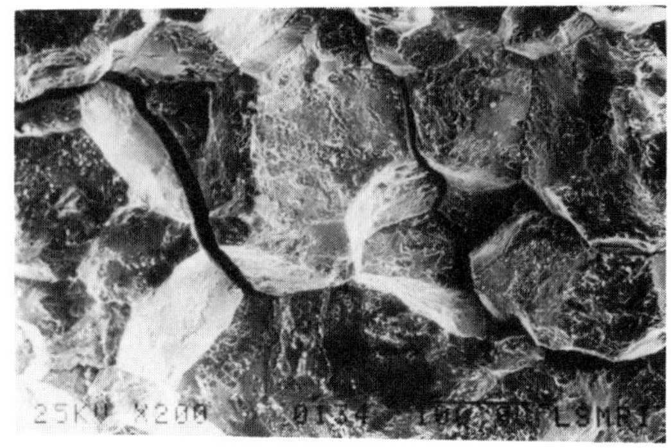

**Fig. 1 Fractograph of a ruptured specimen of GH 33 alloy in a notched bar static creep test (200X) (Testing temperature: 700℃ Stress: 431 MPa rupture life: 355h)**

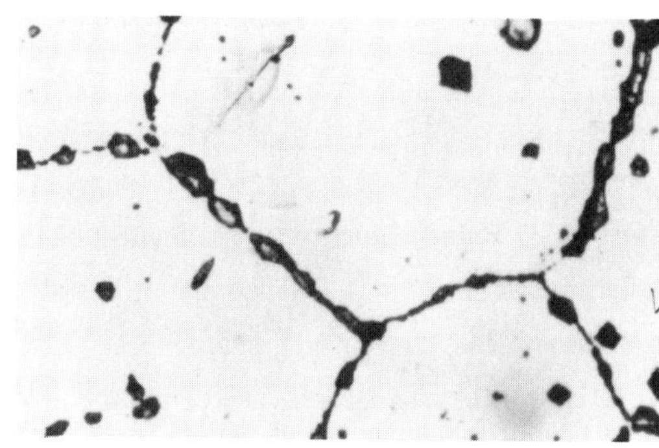

**Fig. 2 Microfractograph of a ruptured specimen of GH 33 alloy in a notched bar cyclic creep test. (340X) (Testing temperature : 700℃, stress: 383MPa, hold time : 0. 5h, Rupture life : 174h)**

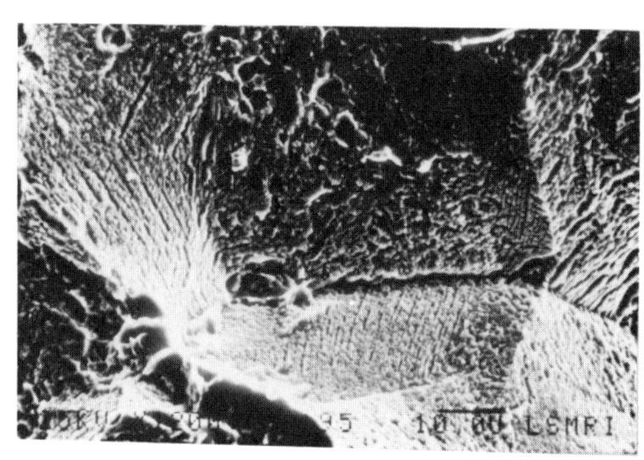

**Fig. 3 Scanning electron microfractograph of a ruptured specimen of GH 33 alloy in a notched bar cyclic creep test . (1200X) (Testing temperature : 700℃; stress: 383 MPa; hold time : 0. 5h; Rupture life : 174h)**

GH37 alloy: Macroscopically, there was no significant difference in fracture appearance between the ruptured specimens in notched bar static creep tests and those in notched bar cyclic creep tests, both showing an uneven fracture appearance, and most of the fractures exhibited a black colour with the transient ruptured zone being smaller and an intergranular fracture being predominent. The examinations under a microscope and scanning electron microscope indicated that there was a significant difference between the two mentioned above. Making a comparison between the fractures of ruptured specimens in notched bar static creep tests and those in notched bar cyclic creep tests, it can be seen that on the latter, subcracks were obviously increased with a more serious damage and obvious creep cavity growth, shown in Fig. 4 and Fig. 5 respectively. In addition, there were fatigue striations in local zones, shown in Fig. 6; In particular, fatigue striations coexisted with a bright oxidation zone at anglar grains, shown in Fig. 7.

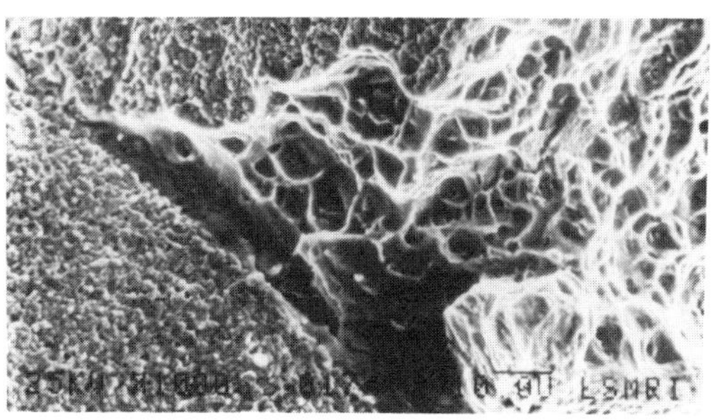

**Fig. 4 Scanning electron microfractograph of a ruptured specimen of GH37 alloy in a notched bar static creep test. (1000X) (Testing temperature: 800℃, Stress: 294MPa, Rupture life: 61h)**

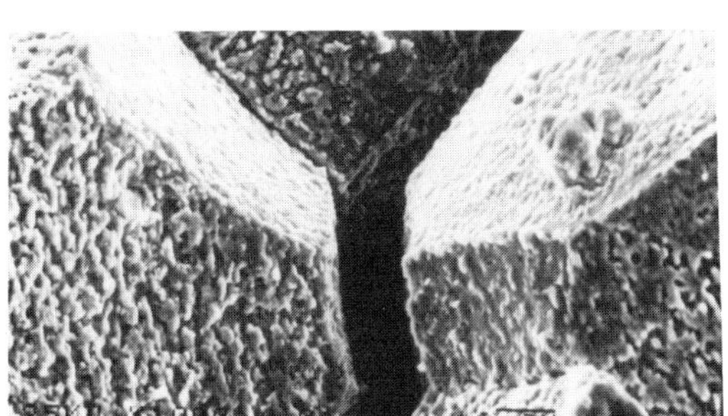

**Fig. 5 Scanning electron microfractograph of a ruptured specimen of GH37 alloy in notched bar cyclic creep test. (1000X) (Testing temperature: 800℃, Stress: 294MPa, hold time: 0. 5h, Rupure life: 20h)**

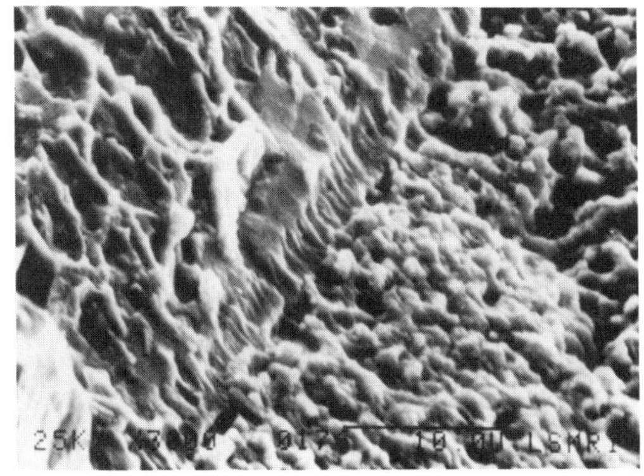

**Fig. 6 Scanning electron microfractograph of a ruptured specimen of GH37 alloy in a notched bar cyclic creep test. (3000X) (Testing temperature: 800℃, Stress: 196MPa, hold time: 0. 5h Rupture life: 121h)**

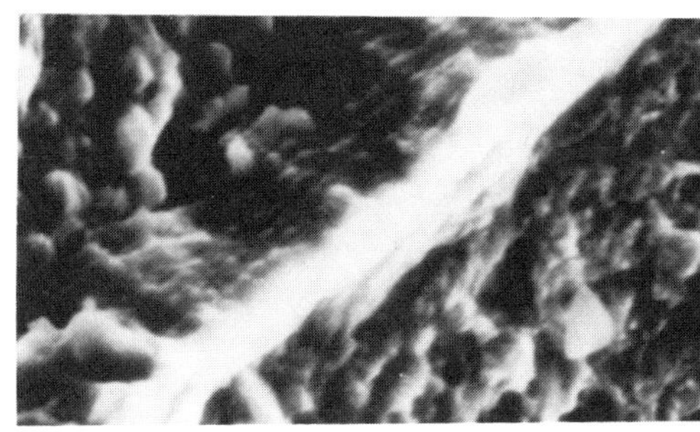

**Fig. 7 Scanning electron microfractograph of a ruptured specimen of GH37 alloy in notched bar cyclic creep test. (2000X) (Testing temperature: 800℃, Stress: 294MPa, hold time: 0. 5h Rupture life: 20h)**

GH49 alloy: Macroscopically, there was no significant difference in fracture appearance between the ruptured specimens under a notched bar static creep condition and those under a notched bar cyclic creep condition, both showing black-colored and relatively even fracture. The microscopic examinations indicated that under the two conditions, an intergranular fracture was predominent with subcracks existing, but under the notched bar cyclic creep condition more serious subcracks and grain boundary oxidation existed on fracture surfaces (shown in Fig. 8 and Fig. 9 respectively) with a more obvious creep cavity growth, shown in Fig. 10, and there was no fatigue striation found. However in this case, there were a few unnoticable slip lines found only in a few locations.

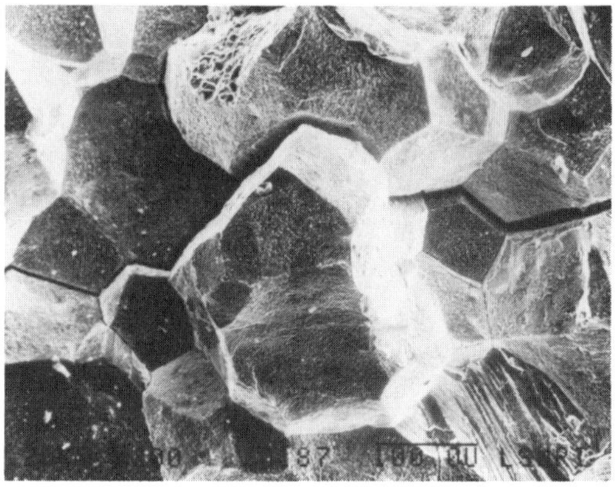

**Fig. 8 Microfractograph of a ruptured specimen of GH49 alloy in a notched bar cyclic creep test . (200X) (Testing temperature ; 820℃, Strss; 392MPa, hold time ; 0. 5h, Rupture life ; 282h)**

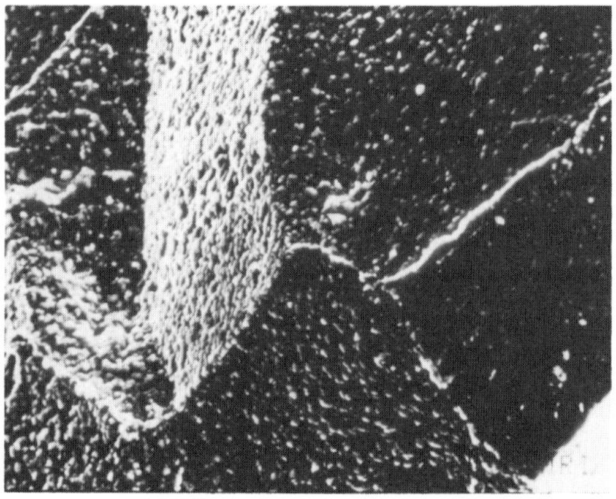

**Fig. 9 Scanning electron microfractograph of a ruptured specimen of GH 49 alloy in a notched bar static creep test. (1000X) (Testing temperature; 820℃, stress; 392MPa, Rupture life; 1408h)**

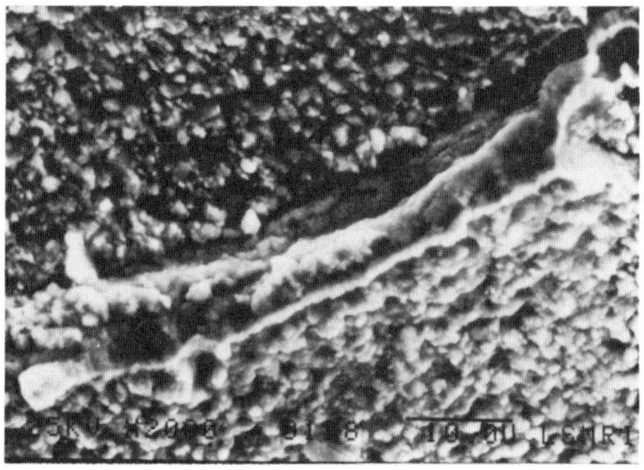

**Fig. 10 Scanning electron microfractogratograph of a rupture specimnn of GH49 alloy in a notched bar cyclic creep test (2000X) (Testing temperature; 820℃, Stress ; 392MPa, hold time ; 0. 5h, Rupture life; 282h)**

# Discussions

For GH33 alloy , there were not only creep cavities but also more fatigue striations on the fracture surfaces of ruptured specimens in notched bar cyclic creep tests, morcover creep cavities coexisted with fatigue striations on some fracture surfaces, as shown in Fig. 3. This is different either from the case in notched bar static creep tests or from the case in smooth bar cyclic creep tests. At the same test temperature and stress, no fatigue striation was found on the fracture surfaces of ruptured specimens in smooth bar cyclic creep tests and the only phenomenon found in this case was that the slip lines were increased compared with those on the fracture surfaces of ruptured specimens in smooth bar static creep tasts(2). This difference mentioned above may be associated with the difference in stress effect between smooth specimens and notched specimens under cyclic loading conditions. The stress used in this work is in the range from 353 to 431 MPa, and the material' s yield strength at testing temperature is 581 MPa. The testing stress in a smooth specimen was much lower than material' s yield strength, but in a notched specimen , the actual stress level at its notched root was 2～3 times higher than the testing stress due to stress concenteration(3), that is , it may be up to 706～1293 MPa, being far beyand the material' s yield strength, and the plastic deformation at local zones would be , in turn a precondition for fatigue damage.

The serious grain boundary damage, shown in Fig. 2, is a remarkable phenomenon. This is considered to be attributed to the carbide aggregation, growth and then cracking at grain boundaries, consequently providing a passage for the propergation of intergranular cracks. During a static creep rupture test, it was found that stress aging would promote the $M_{23}C_6$ carbide growth at grain boundaries in the alloy, and then it remains to be proved whether notch effect plus cyclic loading factor would further accelerate the $M_{23}C_6$ growth, and whether the accelerated growth could be attributed to the increase of actual stress due to notch factor or to the action of a certain mechanism due to cyclic loading factor or to the action of the two factors above.

It could be associated with the unloading occuring during cyclic loading that on the fracture surfaces of ruptured specimens in notched bar cyclic creep tests subcracks increased and trifurcate cracks disappeared. This is because the stress concenteration could be relaxed and slip dispersed during unloading and its holding time.

Under notch bar cyclic creep conditions ,the microfratigraph of ruptured specimens of GH37 alloy was obviously different from that of GH33 alloy ,for GH37 alloy ,on the intergranular fracture an obvious growth trace of creep cavity was found and though fatidue striations were much less, they coexisted with creep cavities and oxidation zone (Fig. 7), in comparison with the case of GH33 alloy. This phenomenon indicates that cyclic loading is an important factor accelerating the grain boundary oxidation and creep cavity growth.

Northwood in the inverstigation on Nimonic 90 alloy found that the oxide penetration depth along grain boundaries was increased by cyclic loading and a deep penetera-tion in local zones occured at temperatures over 704℃ (4). During the measurements of the effcets of fatigue deformation on the oxidation rate of ferro alloys , it was found that as the fatigue deformation was more than the damage deformation of oxides ,even if the test stress was much lower than material' s fatigue limit, a linesr increase in the oxidation rate occurred(5). GH37 alloy containing lower Cr content was tested at a higher temperature , so that it would be subjected to a more critical oxidation. Furthermore for a notched specimen the actual stress level at its notched root was initially much higher than the material' s fatigue limit, so that its oxidation rate under a notched bar cyclic creep condition would be higher notonly than that under a notched bar static creep condition but also than that under a smooth bar cyclic creep condition. Oxidation would reduce material' s ductility and embrittle alloys. So as to play an important role in the acceleration of the initiation and propergation of intergranular cracks.

On the other hand through making a comparison between Fig. 5 and Fig. 7, it can be seen that at the same temperature and stress the rupture life under the notched bar cyclic creep condition is only 1/3 of that under the notched bar static creep condition, but the creep cavities in the former case were obviously larger than those in the latter case. This indicate that cyclic loading does accelerate creep cavity grown. The previous work indicated that cyclic loading could produce a large amount of vacancies and be of benefit to the interconnection of vacancies(6), and the cracks caused by vacancy condensation could grow on the surfaces of cavities(7). Wells though that the intergranular cracks perpendicular to the tensile stress axis would grow mainly during the vacancy condensation (8). Masatern ohnami in the investigation on the creep-fatigue interaction in notched specimens found that there were a number of cracks appearing at notch root and those cracks wound be interconnected with each other around the notch root in the early stage of crack propergation, with only a sharp macro circular crack propergation insidewards in the direction perpendicular to the specimen' s axix (9). It is obvious that this special three-dimentional stress at the notch root would bring about a favourable condition for more intergranular cracks perperndicular to the principle stress.

It seems that in terms of the effect of cyclic loading on the creep rupture life of a notched cpecimen, a direct damage due to fatigue would not be an important factor, and the most important effective factor would be that cyclic loading in combination with notch effect could aggravate grain boundary oxidation ,and promote intergranular embrittlement and accelerate the creep cavity growth on intergranularsurfaces. Under a notched a notched bar cyclic creep condition, GH49 alloy' s fractures are much similar to GH37 alloys. On the former the aggavations of creep cavity growth and grain boundary oxidation are still very outstanding ,but the direct fatigue damage was weakened only because of the further increase of testing temperture. Making a comparison between Fig. 9 and Fig. 10, it can be found that the creep cavity size and width of grain boundary oxidation zone on the fracture surface under a notched bar cyclic creep condition are much larger than those under a notched bar static creep condition ,though the creep rupture life under the former condition is only 1/5 of that under the latter condition.

## Conclusions

1. The fracture of ruptured specimens of GH33 alloy in notched bar cyclic creep tests are intergranular fracture and the main damage forms are creep , fatigue and fatigue-creep interaction. the combined effects of notch with cyclic

loading could accelerate the aggregation and growth of the chair-like carbides at grain boundaries and cracking , resulting in a passage for the propergation of intergranular cracks. It could be an important damage mechenism.

2. An intergranular rupture is predominent for both GH37 alloy and GH49 alloy. The dual action of notch and cyclic loading would aggravate grain boundary oxidation, embrittlement and interganular creep cavity growth , which would be the important damage forms for these alloys under a notched bar cyclic creep condition.

3. At a lower temperature and higher stress, the direct fatigue damage caused by the dual action of notch and cyclic loading would be imporant ; and at a higher temperature and lower stress, the indirect danage carsed by the dual action mentioned above through acceleration the grain boundary oxidation and creep cavity growth would be important .

## Reference


1-Zhou Guang hua, Development and Application of Materials, 5, 19-24(1991)

2-Zhou Guang hua, Metallurgical Analysis and Testing , 5, 1-7(1984)

3-D. Brock, "Engineer Fracture Mechanics Base", Beijing Science Publisher first edition, 47-49(1980)

4-E. Northwood , R. S. Smith and N. stephenson, Memorandum, No. M. 325, National Gas Turbine Establishment, 1-35(1959)

5-"Superalloys for Marine Gas Turbines", shipbuilding Materials Translation Colletion, 15, 164(1978)

6-R. R. Sktiton, philosophical Magazine, 14, 563-672 (1966)

7-Garfalo, Frank, Fundamentals of Creep and Creep-fraceure in Metals, Macmillan New york, 241(1965)

8-S. S. Manson, "Metal Fatigue Damage", National Defence Industry Publisher 65-128(1978).

9-Masao Sakane, Masutern Ohnami, Joural of Engineering Materials and Technolohy, 108, 281(1980)

*Proceedings of the Fifth International Conference on Creep of Materials, Lake Buena Vista, Florida, USA, 18-21 May 1992*

# Creep Deformation and Fracture Behavior of a Type 316LN Stainless Steel

**M.D. Mathew, G. Sasikala, K.B.S. Rao, and S.L. Mannan**
Materials Development Division
Indira Gandhi Centre for Atomic Research
Tamilnadu, India

**ABSTRACT**

This paper discusses experimental results related to the creep properties of a nitrogen alloyed low carbon nuclear grade type 316 (316LN) stainless steel (SS) at temperatures of 873 and 923 K in the stress range 145-335 MPa. The creep properties of type 316LN steel are compared with the results obtained on a nuclear grade of type 316 SS. Both these alloys possessed identical grain size and similar chemical composition with respect to the major alloying elements. Type 316LN SS exhibited a higher rupture life, a lower steady state creep rate and generally a higher rupture ductility at all the test conditions. The improvement in creep properties has been attributed to the combined influence of precipitation strengthening by fine intra and intergranular carbides and strengthening arising from nitrogen possibly due to the formation of a complex of nitrogen and substitutional elements. The micromechanisms governing the creep deformation and fracture are discussed in terms of the specific roles played by the interstitial elements and the precipitation behaviour.

**TYPE 316 STAINLESS STEEL** finds widespread application as a material for structural components of fast breeder reactors in view of its good high temperature mechanical properties, compatibility with the sodium coolant and adequate weldability. However, this material has relatively poor resistance to intergranular stress corrosion cracking (IGSCC) in chloride and caustic environments. Failures due to IGSCC have been experienced in the heat affected zone of type 316 SS welds exposed to marine environments due to sensitization and residual stresses introduced during welding. To avoid this problem, low carbon versions of type 316 SS with additions of nitrogen to ensure mechanical strength equivalent to stainless steels with normal carbon content, are generally preferred.

Type 316 SS with carbon content in the range 0.02-0.03 wt% and alloyed with nitrogen (0.06 - 0.08 wt%) called type 316LN SS has been selected as the material for high temperature structural components of the proposed Prototype Fast Breeder Reactor. Fast reactor components operating in the creep range are designed on the basis of ASME nuclear code case N-47 (1). Since design data for this grade of material is not available in the ASME nuclear code, the components are being designed based on data for type 316 SS. This paper presents a comparative study of the creep properties of nuclear grades of type 316 and type 316LN stainless steels at 873 and 923 K. The micromechanisms governing the creep deformation and fracture are discussed in terms of the specific roles played by the interstitial elements and the precipitation behaviour.

## EXPERIMENTAL

The chemical compositions of type 316LN and type 316 stainless steels employed in this study are given in Table I. Type 316LN SS was obtained in the form of 2500x1500x20 mm plates in the mill-annealed condition. Bars were machined from the plates and were given a solution treatment at 1373 K for one hour and then water-quenched in order to have both carbon and nitrogen in solid solution. This heat treatment yielded an average grain size of 0.088 mm. Constant load creep tests have been conducted in air in single lever machines using round specimens of 10 mm gauge diameter and 50 mm gauge length. Loading axis of the creep specimens was parallel to the rolling direction of the plate. Creep tests were carried out at stresses in the range 215-335 MPa at 873 K and 115-245 MPa at 923 K. Optical, scanning electron and transmission electron microscopy were used to evaluate the precipitation, damage and fracture behaviour of the materials.

Table I. Chemical composition(wt%) and grain size of type 316 and 316LN stainless steels

| Element | 316 SS | 316LN SS |
|---|---|---|
| C | 0.0485 | 0.023 |
| Mn | 1.76 | 1.65 |
| Si | 0.295 | 0.29 |
| S | 0.017 | 0.003 |
| P | 0.028 | 0.024 |
| Ni | 12.34 | 12.21 |
| Cr | 16.1 | 17.12 |
| Mo | 2.085 | 2.31 |
| Cu | 0.225 | 0.10 |
| B,ppm | 5 | 12 |
| N | 0.031 | 0.086 |
| grain size(mm) | 0.070 | 0.088 |

## RESULTS

The creep curves of type 316LN SS were characterised by well defined primary, secondary and tertiary stages at all the stress levels. The log-log variation of steady state creep rate ($\dot{\epsilon}_s$) with applied stress($\sigma$) for 316LN and 316 stainless steels at 873 K and 923 K is shown in Fig.1. The steady state creep rate of type 316LN SS was found to be lower than that of type 316 SS at all the test conditions. The relation between $\dot{\epsilon}_s$ and $\sigma$ could be

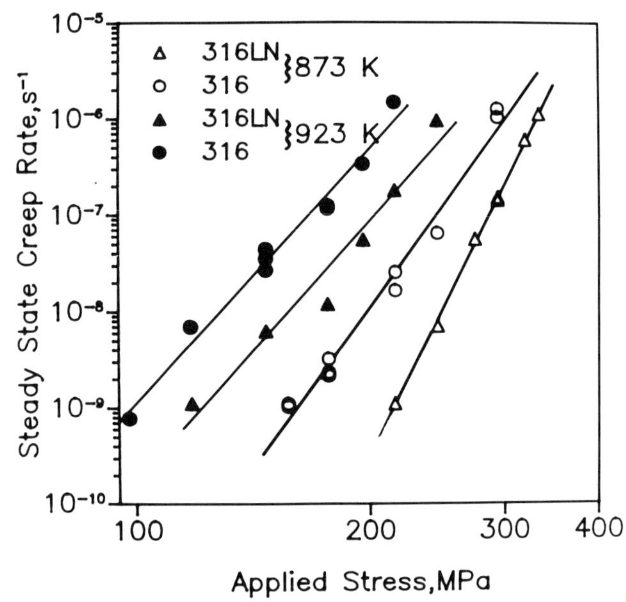


Fig.1. Variation of steady state creep rate with applied stress at 873 and 923 K

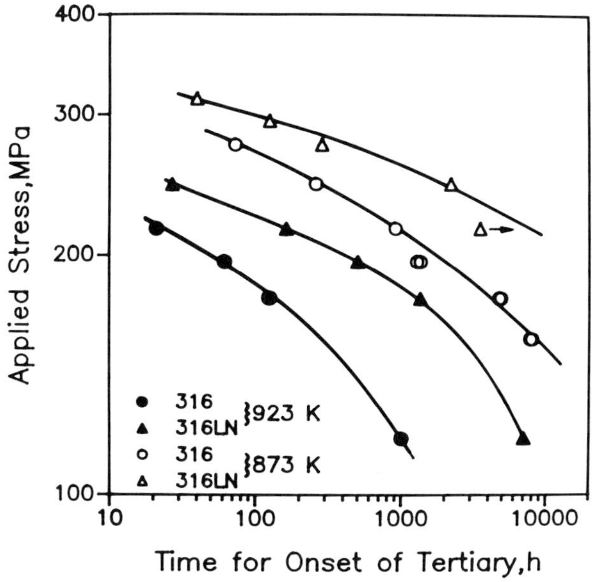


Fig.2. Variation of time for onset of tertiary creep with applied stress

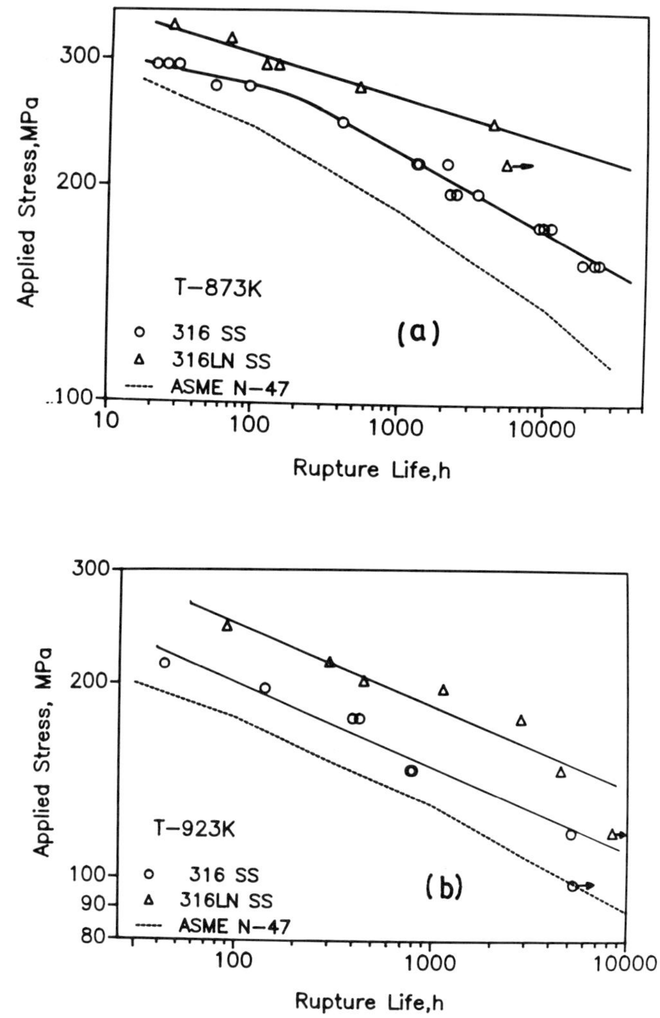


Fig.3. Variation of rupture life with applied stress and comparison with ASME N-47 curve at (a)873 and (b)923 K.

described by a power law relation i.e. $\dot{\epsilon}_s = A\sigma^n$ where n is the stress exponent and A is a material constant. A value of n=9 has been obtained at 923 K for both the steels while at 873 K, values of n=10 and n=16 have been obtained for 316 and 316LN steels respectively. The high value of n (i.e. n 5) suggests that climb controlled dislocation creep mechanism is operative in both the steels. Fig.2 shows the variation of time for the onset of tertiary creep ($t_t$) with applied stress. At all the test conditions $t_t$, was more for type 316LN SS as compared to type 316 SS. The variation of rupture life ($t_r$) with applied stress is shown in Figs. 3(a) and 3(b) for temperatures of 873 and 923 K respectively. Rupture life of both the materials increased linearly with decrease in applied stress (on log -log scale). The slope at 873 K was higher for 316 SS while at 923 K slopes of the log-log plots were identical for both the materials. At both temperatures, type 316LN SS showed significantly higher rupture life than type 316 SS. The rupture strengths of both the materials were higher than the ASME values for the expected minimum stress to rupture (Figs.3(a) and 3(b)).

For the two steels, rupture ductility (expressed as percentage rupture elongation) varied as a complex function of applied stress and temperature. At 873 K, type 316LN SS exhibited higher rupture ductility at all the stress levels (Fig.4a). The variation of ductility with stress showed a minimum in the case of type

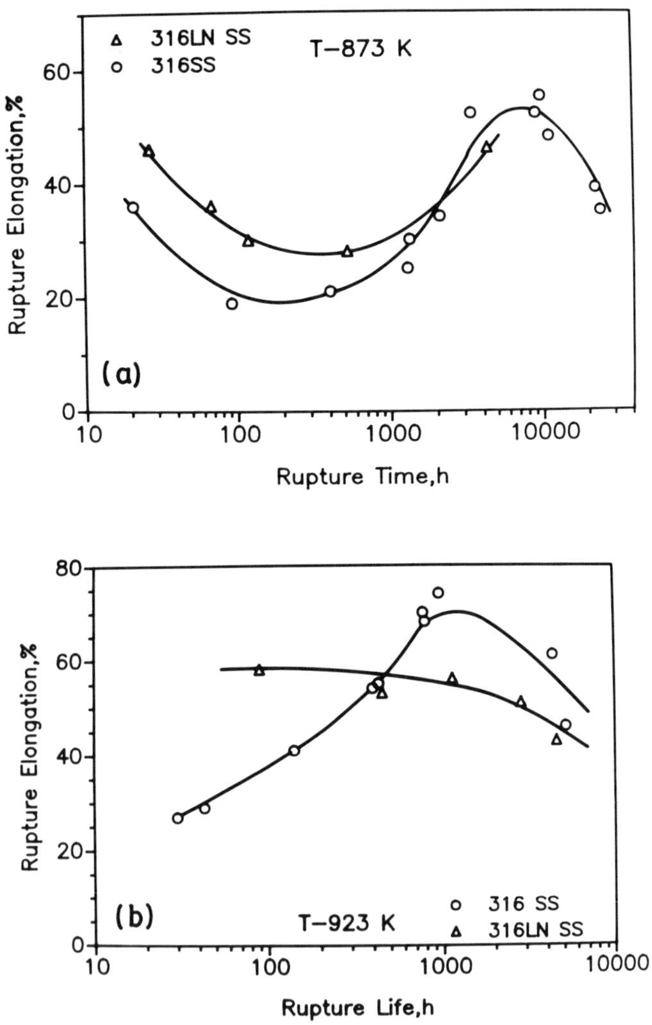


Fig.4. Comparison of the rupture elongation of type 316 and 316LN stainless steels at (a)873 and (b)923 K

316LN SS. In comparison, over the stress range studied, type 316 SS showed a minimum in ductility at intermediate stress levels and a peak at low stress levels. At 923 K, type 316LN SS showed better rupture ductility only at high stress levels (Fig.4(b)). The ductility of 316LN steel showed a gradual decrease with decrease in applied stress whereas a peak in ductility was shown by type 316 SS.

## DISCUSSION

The average grain size of type 316LN SS (0.088 mm) is slightly larger than that of the reference heat of type 316 SS(0.07 mm). Extensive studies carried out by Mannan and Rodriguez(2,3) on influence of grain size on the creep properties of type 316 SS have indicated that a variation in grain size between 0.07 and 0.088 mm will not have any significant effect on the creep properties at 873 and 923 K. As seen from Table I, the amounts of various alloying and residual elements except carbon, nitrogen and boron are comparable in the two grades of steel. Hence the observed differences between the creep properties of type 316 and 316LN SS can be primarily associated with the differences in the interstitial contents in these materials.

**RUPTURE STRENGTH.** Studies carried out by Grover and Wickens (4) on the effects of interstitial elements carbon (0.04 - 0.05%), nitrogen (0.03 - 0.05%) and boron (1-50 ppm) on the long term creep rupture properties of type 316 SS in the temperature range 823 - 973 K have shown an increase in the rupture life with increasing amounts of interstitials. The increase in rupture life was attributed to an increase in the time to onset of tertiary creep since the steady state creep rate did not change significantly by varying the carbon, nitrogen or boron content. In the present investigation, increases in the amount of nitrogen from 0.031 to 0.086 wt% as well as of boron from 5 ppm to 12 ppm, in spite of a decrease in the carbon content from 0.0485 to 0.023 wt%, have led to an increase in the time to onset of tertiary creep and a decrease in the steady state creep rate. Rupture life increased linearly with increase in time to the onset of tertiary creep (on log-log scale) and was independent of test temperature and the type of steel as shown in Fig.5. Similarly, rupture life was found to increase linearly with decrease in the steady state creep rate (on log-log scale) and was again independent of test

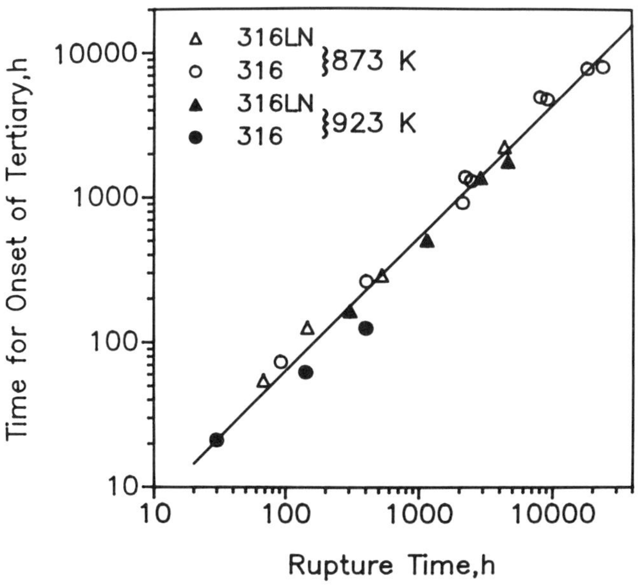


Fig.5.Variation of time for onset of tertiary creep with rupture time.

temperature and the type of steel (Fig.6). The better creep rupture life could hence be correlated with the lower steady state creep rate (Fig.1) and increased time to the onset of tertiary creep(Fig.2) of type 316LN SS as compared to type 316 SS.

It may be noted that Fig.6 is a graphical representation of the well known Monkman - Grant relationship. Validity of this relationship implies that the creep fracture process is controlled by the creep deformation of the material (5). The Monkman-Grant equation has been analysed independently for type 316 SS using data at 873 and 923 K and a relation of the form $\dot{\epsilon}_s = 0.38\ t_r^{-1.08}$ has been obtained. The rupture life of type 316LN SS has been estimated using the steady state creep rate for type 316 LN SS and the above relation for type 316 SS. Table II gives a comparison between the experimental and the estimated rupture lives of type 316LN SS. This evaluation establishes the applicability of Monkman-Grant relationship for predicting the rupture life of type 316LN SS by knowing the steady state creep rate. Therefore the data base on the creep rupture properties of type type 316 SS could be used to predict reasonably well the

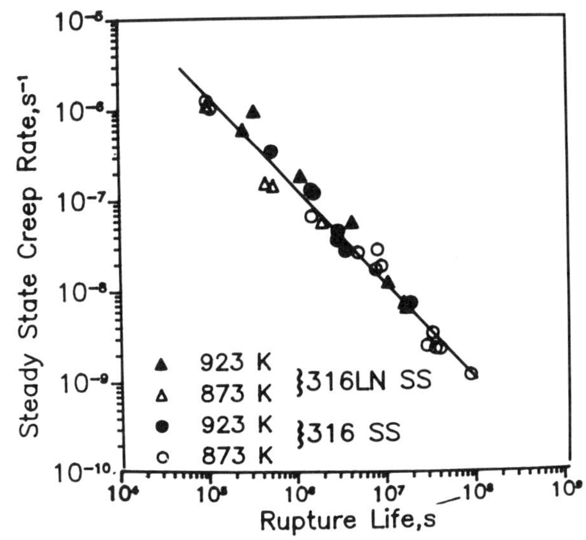


Fig.6. Monkman-Grant plot showing variation of steady state creep rate with applied stress.

life of type 316LN SS without carrying out tests upto rupture.

The creep deformation processes in type 316 SS are known to be influenced by solid solution strengthening by interstitial and substitutional elements and precipitation strengthening by carbides (6,7). Precipitation of intragranular carbides inhibits dislocation motion and also helps in stabilising a fine network of dislocations; finer and more closely spaced precipitates contribute to lower creep rates(6). Fig 7(a) is a transmission electron micrograph showing carbides precipitated on matrix dislocations in type 316 SS creep tested at 873 K for 22000 hours. At low stress levels, i.e. longer creep exposure times, coarsening of carbides and precipitation of intermetallic sigma phase on grain boundaries would occur leading to a decrease in grain boundary as well as matrix strengthening. Figs. 7(b) and 7(c) show the presence of sigma phase particles on grain boundaries in type 316 SS ruptured at 873 K after

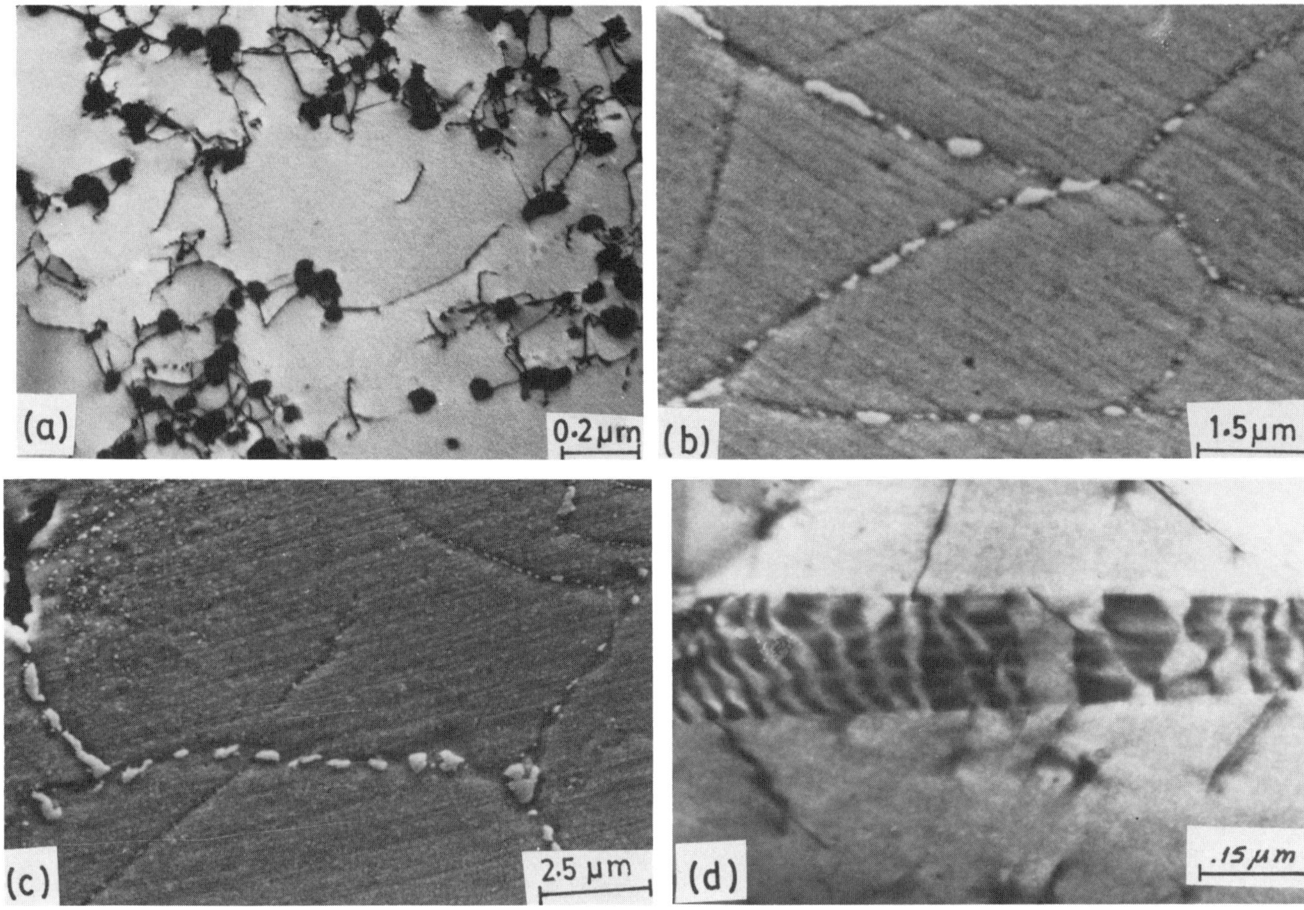


Fig.7. Transmission electron micrograph showing (a) uniform distribution of dislocations and matrix carbides in type 316 SS at 873 K, $t_\gamma$ -22000 hours and (d) planar slip bands in type 316LN SS at 873 K, $t_\gamma$ -120 hours. Optical micrographs showing sigma phase particles on grain boundaries in type 316 SS at (b)873 K, $t_\gamma$ -10000 hours and (c)923 K, $t_\gamma$ -5000 hours

10000 hours and at 923K after 5000 hours respectively. The size of the sigma phase particles at 923 K was nearly twice of that at 873 K although the test time was less by fifty percent. Thus the benefits of solid solution strengthening and precipitation strengthening are lost appreciably at higher temperatures and lower stress levels i.e. at long test durations.

The role of nitrogen on the deformation behaviour of austenitic stainless steels is not clearly understood although it has been generally attributed to one or more of the following effects (6,8-15), namely (i) solid solution strengthening, (ii) precipitation strengthening by nitrides, (iii) decrease in stacking fault energy and (iv) formation of a complex of nitrogen and substitutional elements like chromium and molybdenum. The presence of nitrogen is also known to delay the rate of coarsening of $M_{23}C_6$ carbides in austenitic stainless steels since the diffusivity of carbon and chromium atoms is reduced by the presence of nitrogen(15). Besides, in

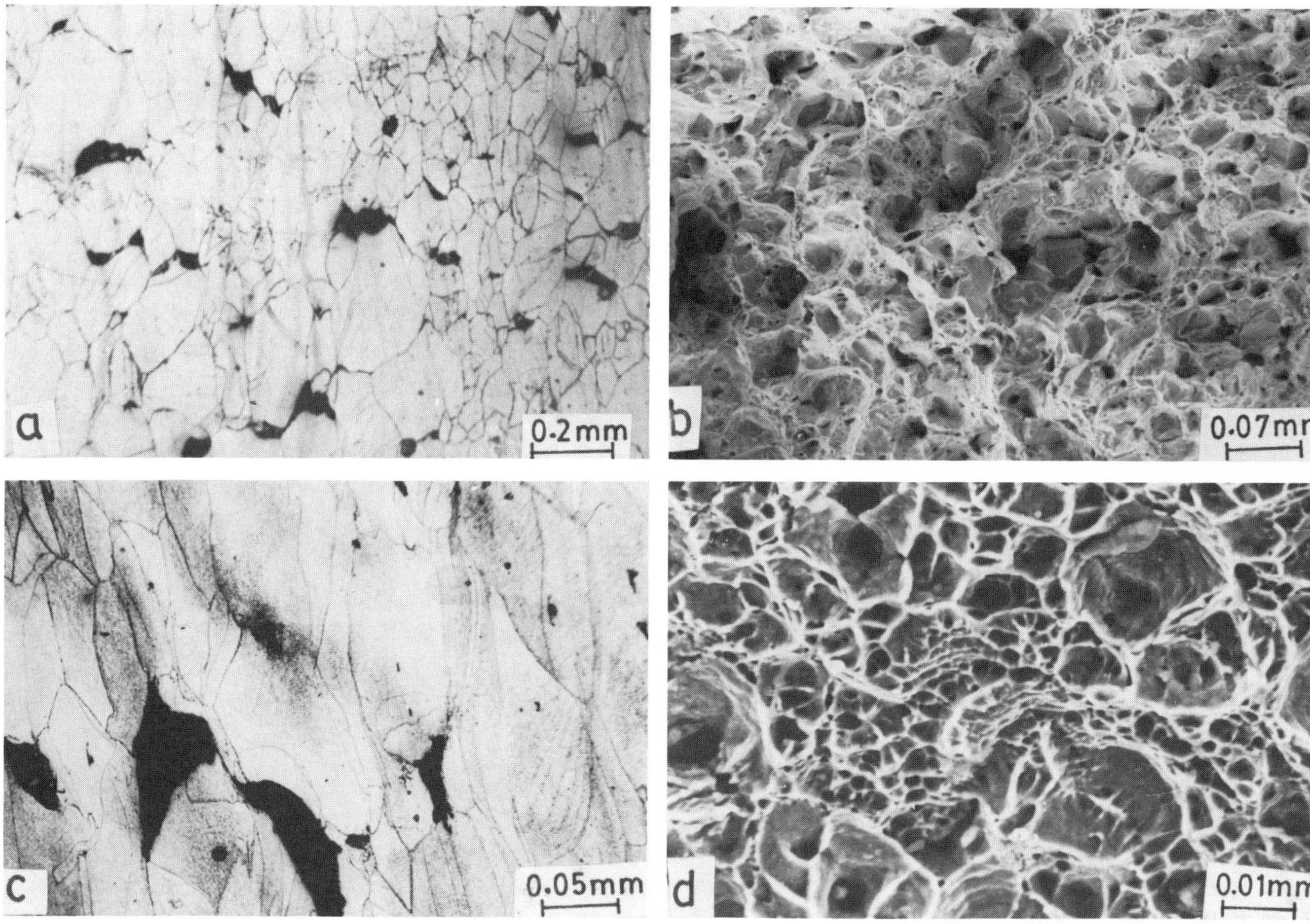


Fig.8.Optical micrographs showing creep damage in (a) type 316LN SS at 873 K, σ-275 MPa and (c) type 316L SS at 873 K, σ-195 MPa. Scanning electron micrographs showing (b) typical intergranular fracture surface in type 316 SS at 873 K, σ-245 MPa and (d) typical ductile fracture surface in type 316 SS at 873 K, σ-196 MPa.

type 316LN SS, because of its lower carbon content(Table I), the incubation period for precipitation of carbides itself is large as compared with that in type 316 SS. Transmission electron microscopic studies showed that under similar temperature and test durations, the carbide precipitates in type 316LN SS were much finer in size.

The density of dislocations in creep tested type 316LN steel was appreciably lower than in type 316 SS. Moreover, evidences of planar slip were also seen in the creep tested nitrogen alloyed steel (Fig.7(d)) whereas a uniform distribution of dislocations associated with carbide particles was generally observed in the nitrogen lean material(Fig.7(a)). It is known that low stacking fault energy promotes concentration of slip in planar glide bands. The presence of easily shearable precipitates of short range order, short range clustering and large grain size also favour planar slip(16,17). Planar slip during creep deformation has not been widely observed in nitrogen bearing austenitic stainless steels, although Alden and Aronsson(18) have reported the presence of rows of dislocations confined to the slip plane in type 316 SS containing 0.17 % nitrogen. Type 316LN SS used in this study has shown well developed planar slip bands under

Table II. A comparison between the experimental(Exp.) and the predicted rupture lives of type 316LN SS

| Temp.,K | Stress,MPa | Rupture Life,h Exp. | Predict |
|---|---|---|---|
| 873 | 333 | 27 | 38 |
| | 314 | 137 | 177 |
| | 275 | 525 | 616 |
| | 245 | 4297 | 4173 |
| 923 | 245 | 90 | 45 |
| | 216 | 304 | 205 |
| | 196 | 1138 | 616 |
| | 176 | 2845 | 2530 |
| | 147 | 4550 | 4602 |

strain controlled fatigue at 873 K due to dynamic strain ageing effects(19). It is suggested that the lower creep rate and higher rupture life of type 316LN SS observed in the present investigation could be attributed to the combined influence of precipitation strengthening by fine carbides and strengthening by nitrogen possibly due to the formation of a complex of nitrogen with substitutional elements.

**RUPTURE DUCTILITY.** The contributions to creep rupture ductility comes from intragranular deformation and grain boundary sliding with matrix deformation accounting for a major share of the ductility. A strong grain boundary is necessary in order to minimise the intergranular damage and increase intragranular deformation. The influence of applied stress and temperature on the intergranular precipitation behaviour and development of creep damage in nitrogen alloyed and nitrogen lean type 316 SS has been reported earlier(7).It was observed that absence of intergranular precipitation facilitates grain boundary sliding leading to the formation of sharp cracks on grain boundaries (Fig.8(a)) resulting in intergranular failure (Fig.8(b)). However, when fine carbides precipitate on the grain boundaries and strengthen against sliding, creep cracks are few in number and are elongated in the direction of applied stress (Fig.8(c)). Large grain deformation and ductile fracture have then been obtained (Fig.8(d)). The fall in ductility observed in type 316 SS at low stress levels (Figs.4(a) and 4(b)) was found to be due to the formation of coarse sigma particles as shown in Figs.7(b) and 7(c). The higher rupture ductility of type 316LN SS in the present investigation is believed to be due to the lesser amount of creep damage that occurs in the material.

## CONCLUSIONS

Type 316LN SS has a lower steady state creep rate, higher rupture strength and generally higher rupture ductility compared to type 316 SS. Uniform distribution of dislocations has been observed in type 316 SS whereas arrays of dislocations confined to planar slip bands have been observed in the case of type 316LN SS. The improvement in the creep properties of nitrogen alloyed steel is attributed to strengthening due to precipitation of finer carbides and possibly due to the formation of a complex of nitrogen with interstitial atoms.

## ACKNOWLEDGMENT

The authors wish to thank Dr. Placid Rodriguez, Head, Metallurgy and Materials Programme for his keen interest in this work and for many useful discussions. Experimental assistance provided by Miss. S. Latha and Mrs. K.S. Chandravathi is gratefully acknowledged.

## REFERENCES

1. ASME Boiler and Pressure Vessel Code, American Society of Mechanical Engineers, New York,Code Case N-47 (1989).
2. Mannan,S.L. and P. Rodriguez,in

Proceedings Sixth Int. Conf. on Fracture, Pergamon Press, Oxford, 3, 2303-2309(1984).
3. Mannan,S.L. and P. Rodriguez, Metal Science, 17, 63-69(1983).
4. Grover,H.K. and A. Wickens, in Proceedings Int. Conf. on Mechanical Behaviour and Nuclear Applications of Stainless Steel at Elevated Temperatures,Varese, 81-87 (1982).
5. Nix,W.D. and J.C.Gibeling, "Flow and Fracture at Elevated Temperatures",p 1, American Society of Metals, Metals Park, Ohio (1984).
6. Morris,D.G. and D.R. Harries, Metal Science,12, 525-531(1978).
7. Mathew, M.D., G. Sasikala, K. Bhanu Sankara Rao and S.L. Mannan, Mater. Sci. Eng.(1991),(in press).
8. Soussen, A., S. Degallaix and T. Magnin,Mater. Sci. Eng., A142, 169-176(1991).
9. Byrnes, M.L.G., M.Grujicic and W.S.Owen, Acta Metall., 35, 1853-1862(1987).
10. Wahlberg, G.,U.Rolander and H.O. Andren in Proceedings Int.Conf. on High Nitrogen Steels, Lille,France, 18-20 May,1988, p.163.
11. Matsuo,T., N.Morioka, S.Kaise, M.Kikuchi and R.Tanaka in Proceedings Int.Conf. High Nitrogen Steels, Lille, France, 18-20 May,1988, p.213.
12. Stoltz,R.E. and J.B. Vander Sander, Met. Trans., 11A, 1033-1037 (1980).
13. Sobotka, J., M.Liska, M.Sobotkova, M.Tomasova, V.Vodarek and V.Bina in Proceedings Int.Conf.High Nitrogen Steels, Lille, France,18-20 May,1988, p.363.
14. Nakazawa,T., H.Abo, M.Taniano, H.Komatsu, T.Nishida and M.Tashimo, in Proceedings Sixth Int. Conf. on Pressure Vessel Technology, Beijing, 11-15 Sept.,1988, p.1041.
15. Lai, J.K.L., Mater. Sci. Eng.,61, 101-109(1983).
16. Gerold,V. and H.P. Karnthaler, Acta Met.37, 2177-2183(1989).
17. Bressers,J. and M. Steen, Int. Journal of Pressure Vessel and Piping, 47, 217-245(1991).
18. Alden,G. and B. Aronsson, in Proceedings of a Meeting on Creep Strength in Steel and High Temperature Alloys, Sheffield,1972, p.67
19. Srinivasan, V.S., R. Sandhya, K. Bhanu Sankara Rao and S.L. Mannan,Int. J. Fatigue,13, 471-478(1991).

*Proceedings of the Fifth International Conference on Creep of Materials, Lake Buena Vista, Florida, USA, 18-21 May 1992*

# Validation of a Viscoplastic Model in Creep Regime

**B. Riou**
Franatome
Lyon, France

**N. Waeckel**
EDF Septen
Lyon, France

## ABSTRACT

A U-bend specimen subjected to loads equivalent to those of LMFBR's structures and designed to study simultaneously creep under tension and compression is analysed.

The inelastic calculation and life evaluation are carried out with the viscoplastic model and the continuum damage model proposed by ONERA (Office National d'Etudes et de Recherches Aérospatiales - France).

This paper presents the geometry of the specimen, the loads applied and the results of the inelastic analysis. A comparison with experimental data and with the results of the RCC-MR(1) elastic analysis enables to evaluate the accuracy of such an inelastic approach and the benefit brought when compared with traditional simpler analyses.

LMFBR's STRUCTURES are highly sollicitated, on one hand with fatigue due to thermal transients and on the other hand with creep due to hold periods at high temperature.

The design of these structures is usually based on elastic analyses. However, some of them require elastoplastic calculations when damages are beyond or close to the allowable limits. Yet, both approaches are known to be overconservative and, in some cases, more accurate analyses become necessary.

Thus, NOVATOME has been led to take an interest in more elaborated models and in particular in the viscoplastic and the continuum damage models proposed by ONERA. The identification of the parameters of these models has already been performed by NOVATOME and their present purpose is to validate these identifications on multiaxial specimens.

## PRESENTATION OF EXPERIMENTS

The experiments were carried out in CEA-SACLAY (2) on 15 U-bend specimens at 600 °C. These specimens were subjected either to pure fatigue or creep-fatigue loads. During the tests, one end of the specimen is fixed whereas a displacement is applied to the other.

The geometry and the loads are described in figure 1. Table I summarizes the set of tests and the corresponding instrumentation.

The material considered is the stainless steel 316 L(N). The test chosen for the calculation is the creep-fatigue test with an hold period $T_M$ = 30 min (tests 5b, 7b, 8b, 9b).

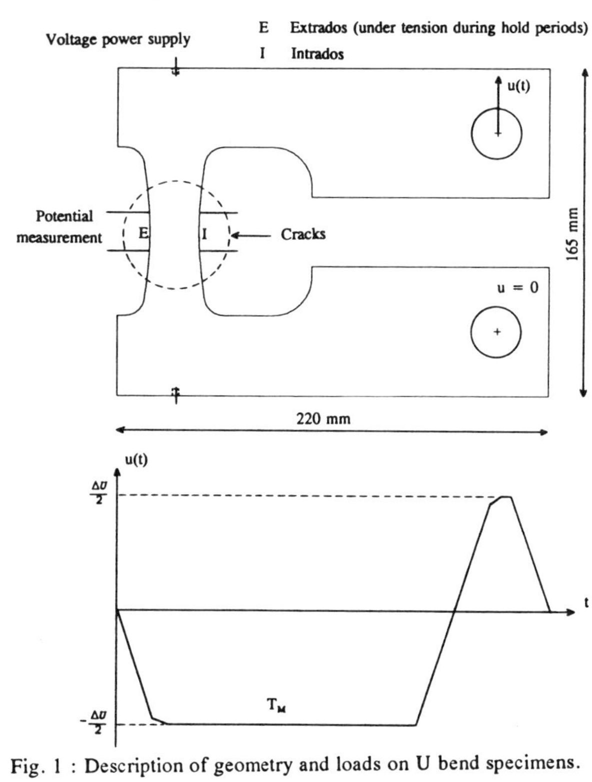


Fig. 1 : Description of geometry and loads on U bend specimens.

Table I. Summary of experiments.

| Test number | Steel plate | $T_M$ (min) | Frequency (Hz) | Instrumentation and measurements | Δu (mm) |
|---|---|---|---|---|---|
| 1 | DAS | 0 | 0.2 | (2)(3)(4) | 6.74 |
| 2* | DAS | 0 | 0.2 | (2)(3) | 7.24 |
| 3 | DAS | 15 | / | (2)(3) | 7.24 |
| 4 | DAS | 15(1) | / | (2)(3) | 7.24 |
| 5 | DAS | 15(1) | / | (2)(3) | 10.00 |
| 1b | SR | 0 | 0.03 | (2)(3) | 7.24 |
| 2b* | SR | 0 | 0.03 | (2)(3) | 7.24 |
| 3b | SR | 0 | 0.03 | (2)(3) | 7.24 |
| 4b* | SR | 15 | / | (2) | 7.24 |
| 5b | SR | 30 | / | (2) | 7.24 |
| 6b | SR | 60 | / | (2) | 7.24 |
| 7b* | SR | 30 | / | (2) | 7.24 |
| 8b* | SR | 30 | / | (2) | 7.24 |
| 9b* | SR | 30 | / | (2)(4) | 7.24 |
| 10b* | SR | 60 | / | (2) | 7.24 |

(1) with fatigue blocs,
(2) measurement of the strength, the applied displacement and electric monitoring of cracks,
(3) measurement of longitudinal strains (extrados and intrados),
(4) hardness testing

* test stopped before collapse of the specimen

## DESCRIPTION OF THE VISCOPLASTIC MODEL

The viscoplastic model proposed by J.L. CHABOCHE (ONERA) is a unified model based on the concept of the viscous stress, with non linear kinematic hardening, isotropic hardening, plastic strain memorization and time recovery on kinematic and isotropic hardening.

The complete set of constitutive equations of the viscoplastic model is presented in table II.

Table II. Equations of the viscoplastic model

Kinematic hardening

$$\chi = \chi_1 + \chi_2$$

$$(\dot{\chi}_i)_{hard} = \frac{2}{3} C_i \dot{\varepsilon}^p - \gamma_i \chi_i \dot{p}$$

$$\dot{\gamma}_i = \gamma_i^{\circ}[a_\infty + (1 - a_\infty) e^{-bp}]$$

Time-recovery for kinematic hardening

$$\dot{\chi}_i = (\dot{\chi}_i)_{hard} - \gamma_{x_i} [J(\chi_i)]^{m_i - 1} \chi_i$$

Isotropic hardening

$$(\dot{R})_{hard} = b(Q - R)\dot{p}$$

Time-recovery for isotropic hardening

$$\dot{R} = (\dot{R})_{hard} + \gamma_r (Q_r - R)^{m_r} \, sign\,(Q_r - R)$$

$$Q_r = Q - Q_r^{*}\left[1 - \left(\frac{Q_M - Q}{Q_M}\right)^2\right]$$

Table II. Equations of the viscoplastic model (continued)

Plastic strain memorization

$$Q = Q_o + (Q_M - Q_o)(1 - e^{-2\mu q})$$

$$F = \frac{2}{3} J(\varepsilon^p - \xi) - q = 0$$

$$\dot{q} = \eta\, H(F) \langle n : n^* \rangle \dot{p}$$

$$\xi = \sqrt{\frac{3}{2}} (1 - \eta)\, H(F) \langle n : n^* \rangle n^* \dot{p}$$

$$n = \sqrt{\frac{3}{2}} \frac{\sigma' - \chi'}{J(\sigma - \chi)} \; ; \; n^* = \sqrt{\frac{3}{2}} \frac{\varepsilon^p - \xi}{J(\varepsilon^p - \xi)}$$

Viscoplastic strain rate

$$\dot{\varepsilon}^p = \frac{3}{2}\dot{p} \frac{\sigma' - \chi'}{J(\sigma - \chi)}$$

$$\dot{p} = \left\langle \frac{\sigma_v}{K} \right\rangle^n e^{\alpha \left\langle \frac{\sigma_v}{K} \right\rangle^{n}}$$

Viscoplastic potential

$$\Omega = \frac{K}{n + 1} \left\langle \frac{\sigma_v}{K} \right\rangle^{n+1}$$

$$\sigma_v = J(\sigma - \chi) - \alpha_R R - k$$

$$\sigma_v \geq 0$$

Viscosity

$$K(R) = K_o + \alpha_K R$$

The parameters of this model have been identified on the "best fit" curves of the stainless steel 316 L(N) (3). The viscoplastic identification at 600 °C is shown in table III.

Table III. Viscoplastic identification of the stainless steel 316 L(N) at 600 °C.

| MATERIAL PARAMETERS | | 600 °C |
|---|---|---|
| E | (MPa) | 145 000 |
| $\nu$ | | 0.3 |
| k | (MPa) | 35 |
| $\alpha_R$ | | 1 |
| b | | 12 |
| $Q_o$ | (MPa) | 40 |
| $Q_M$ | (MPa) | 460 |
| $\mu$ | | 19 |
| $\eta$ | | 0.04 |
| $\gamma_r$ | ($MPa^{-m_r}s^{-1}$) | $2.10^{-7}$ |
| $Q_r$* | (MPa) | 200 |
| $m_r$ | | 2 |
| $c_1$ | (MPa) | 65 000 |
| $\gamma_1^0$ | | 1 300 |
| $c_2$ | (MPa) | 1 950 |
| $\gamma_2^0$ | | 50 |
| $a_\infty$ | | 0.5 |
| $\gamma_{x1}$ | ($MPa^{-m_1}s^{-1}$) | $10^{-12}$ |
| $m_1$ | | 4 |
| $\gamma_{x2}$ | ($MPa^{-m_2}s^{-1}$) | $2.10^{-13}$ |
| $m_2$ | | 4 |
| $k_o$ | ($MPas^{1/n}$) | 70 |
| $\alpha_k$ | | 1 |
| n | | 24 |
| $\alpha$ | | 0 |

## VISCOPLASTIC CALCULATION

The mechanical calculation is carried out with the SYSTUS code. The mesh of the structure which consists of 152 quadratic 8 noded elements is shown in figure 2.

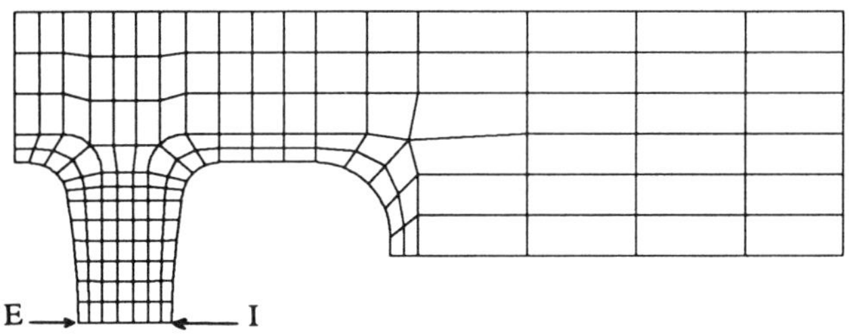


Fig. 2 : Mesh of the specimen

25 cycles have been computed before reaching the stabilized cycle. Figure 3 shows the cyclic behaviour at the intrados (I) and extrados (E) with $\sigma_{yy} = f(\varepsilon_{yy})$ curves.

The most loaded part of the structure is the intrados zone (I) where the equivalent stress $\bar{\sigma} = 286$ MPa and the equivalent strain range $\overline{\Delta\varepsilon} = 0.68$ %. These values are to be compared with $\bar{\sigma} = 276$ MPa and $\overline{\Delta\varepsilon} = 0.64$ % in the extrados zone (E). It is also to be noted that the stress field in those regions is uniaxial. Figure 4 shows the $\sigma_{yy}$ stress map at the end of the hold period.

a/

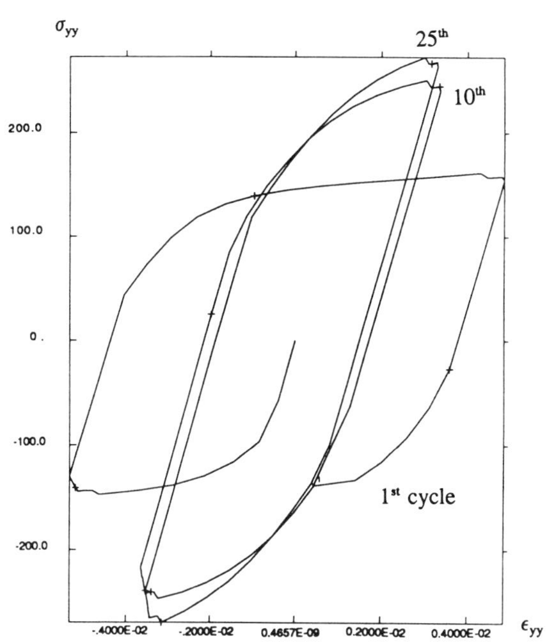


b/

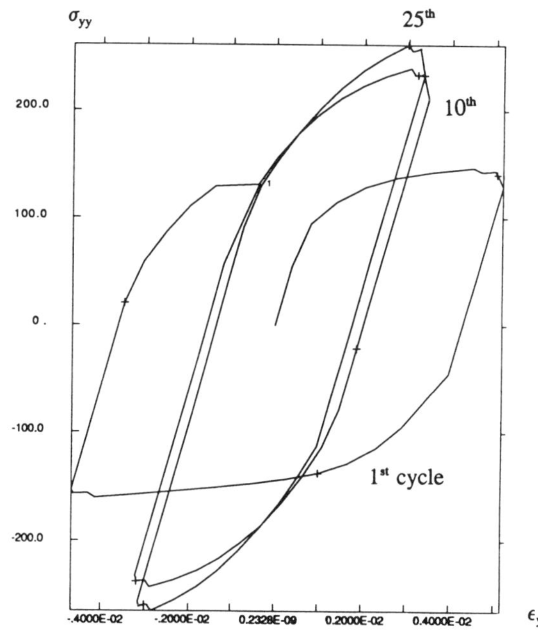


Fig. 3 : Cyclic behaviour at the intrados a/ and extrados b/.

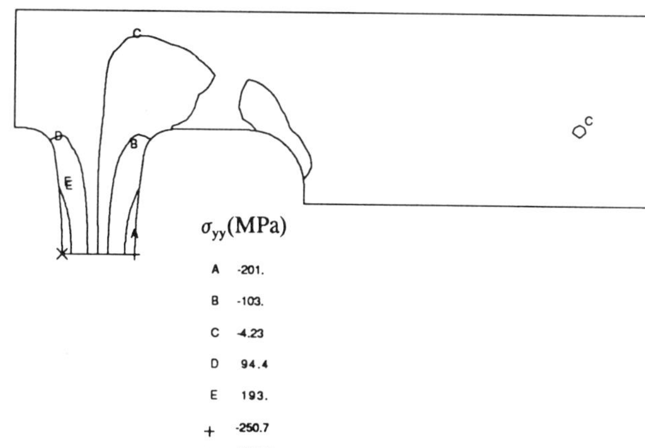


Fig. 4 : $\sigma_{yy}$ stress map at the end of the hold period.

## INELASTIC CREEP-FATIGUE ANALYSIS

The creep-fatigue analysis is performed with a post-processor of SYSTUS code which computes the number of cycles to failure, using the stress evolution of a cycle to which a structure is submitted.

The damage model is the continuum damage model (4) proposed by J.L. CHABOCHE (ONERA). The parameters of this model have been previously determined for the stainless steel 316 L(N) (5). The set of parameters at 600 °C is shown in table IV.

Table IV. Continuum damage model identification of the stainless steel 316 L(N) at 600 °C.

| MATERIAL PARAMETERS | | 600 °C |
|---|---|---|
| $\beta$ | | 4.438 |
| b | $(MPa^{-1})$ | 0.00167 |
| M | (0) (MPa) | 1 473 |
| $\sigma_l$ | (0) (MPa) | 150 |
| $\sigma_u$ | (MPa) | 600 |
| A | $(MPa.S^{1/r})$ | 3 207 |
| r | | 6.825 |
| $\gamma$ | | 0 |
| $\delta$ | | 0 or 0.133 |
| k | | 5 |
| a | | 0.2 |

Creep is taken into account considering the equivalent stress :

$$\chi(\underset{\sim}{\sigma}) = (1-\gamma-\delta)\,\bar{\sigma} + \delta\, tr(\underset{\sim}{\sigma}) + \gamma\, Max(\sigma_1, \sigma_2, \sigma_3),$$

where $\underset{\sim}{\sigma}$ is the stress tensor, $\bar{\sigma}$ the VON MISES equivalent stress and $\sigma_1$, $\sigma_2$, $\sigma_3$ are the principal stresses.

The model is therefore able to distinguish states under tension from states under compression.

The life assessment using the continuum damage model leads to the following results :

| | Intrados | Extrados |
|---|---|---|
| CONTINUUM DAMAGE MODEL "BEST FIT" | (1) Nr = 648 | Nr = 803 |
| | (2) Nr = 1 063 | |
| | (3) Nr = 3 106 | |

(1) Assuming no difference between creep under tension and creep under compression ($\delta$ = 0).

(2) Creep under compression according to RCC-MR addendum 1987 ($\delta$ = 0.133).

(3) Assuming no damage with creep under compression.

Nr Number of cycles to failure.

Fig 5 shows the iso-damage map (resulting from the stabilized cycle) in the central part of the specimen in the case when creep under compression is taken into account according to RCC-MR addendum 1987.

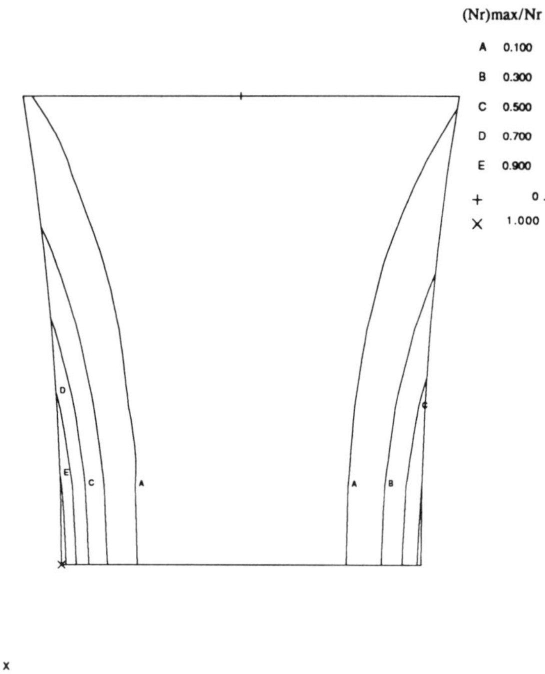


Fig. 5 : Iso-damage map.

A comparison is made with the inelastic procedure of the RCC-MR as described in RB 3262.22. This analysis includes the latest results on fatigue curves of the stainless steel 316 L(N) provided by the Design and Construction Rules Committee for the EFR (European Fast Reactor) project.

The results of the "best fit" analysis written in term of cycles to failure are as follows :

| | Intrados | Extrados |
|---|---|---|
| RCC-MR INELASTIC "BEST FIT" | (1) Nr = 582 | Nr = 781 |
| | (2) Nr = 1 243 | |
| | (3) Nr = 4 240 | |

Nota : "Best fit" means without the coefficients respectively 2 and 20 for the fatigue curves and 0.9 for the equivalent stress with creep damage calculated on the average $S_R$-curve.

Both analyses lead therefore to close results.

An elastic analysis has been performed in order to appreciate the benefit brought by inelastic analyses. Figure 6 shows the VON MISES stress map at one extremity of the load cycle.

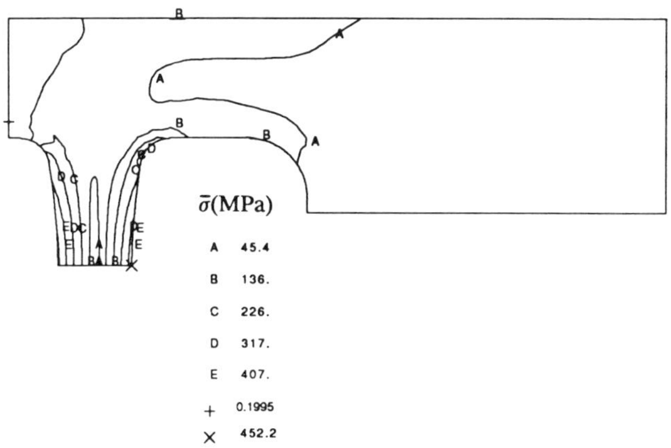


Fig.6 : VON MISES stress at one extremity of the load cycle (elastic calculation).

The values of stress ranges in the intrados and extrados region are respectively $\overline{\Delta\sigma} = 905$ MPa and $\overline{\Delta\sigma} = 860$ MPa. The elastic creep-fatigue analysis leads therefore to the following results :

| | Intrados | Extrados |
|---|---|---|
| RCC-MR ELASTIC "BEST FIT" | Nr = 241 | Nr = 292 |

This table shows that the benefit brought in this case by the use of a full inelastic analysis is a factor 3.2 on the number of cycles to failure in comparison with the result of an elastic analysis.

## COMPARISON WITH EXPERIMENTS

The experimental results available consist in the evolution of the strength versus time and the damage data coming from destructive tests and electric monitoring of cracks.

The strength range at stabilized cycle calculated with the viscoplastic model is $\Delta F = 9\ 260$ N. This value is very close to the average found on four tests $(\Delta F)_{exp} = 9\ 175$ N. Concerning strength relaxation during hold period, the calculation overestimates the relaxation of the structure. It gives $(\Delta F)_{relax} = 454$ N which is to be compared with a measurement of 350 N at the stabilized cycle. This difference may be explained by a slight underestimation of the hardening of the material.

As far as damage assessment is concerned, table V shows the results derived from destructive tests and crack monitoring. The numbers of cycles given in this table are higher than those calculated with the continuum damage model and this means that this model leads either to slightly conservative results or to results corresponding to an early stage of the damage.

Table V. Damage results deduced from experiments

| Test | 5b | 7b | 8b | 9b |
|---|---|---|---|---|
| $N_f$ | 2 820 | 1 930* | 2 315* | 1 448* |
| $N_R$ (5 %) | 2 740 | / | / | / |
| $N_i$ | 1 750 | 1 400 | 1 525 | / |
| Cycles corresponding to a = 0.1 mm | intrados : 1 280<br>extrados : 1 365 | | | |
| Cycles corresponding to a = 0.4 mm | intrados : 1 910<br>extrados : 1 730 | | | |
| Cycles corresponding to a = 0.8 mm | intrados : 2 370<br>extrados : 1 920 | | | |

(*) test interrupted before failure.
$N_f$ Number of cycles at the end of experiments.
$N_R$(5 %) Number of cycles corresponding to 5 % of loss of strength range.
$N_i$ Number of cycles deduced from electric monitoring of cracks.
a Crack depth.

It is also interesting to note that the number of cycles necessary to reach significant crack depth values are higher in the intrados region (compression during hold period) than in the extrados region whereas the latter is submitted to a higher stress level. This trend is only reversed for very low crack depth values for which the difference of the corresponding numbers of cycles is not significant. This enables to conclude that crack growth in region with creep under compression is slower than in region with creep under traction. It is therefore justified to take into account the identification of the continuum damage model using the less damaging effect of creep under compression.

Figures 7 and 8 show characteristic crack growth mechanisms in the intrados and extrados regions for creep-fatigue tests. It is to be noted that crack growth is intragranular in the region under compression during hold periods and intergranular in the region under tension. Damage mechanisms are therefore totally different and confirm that creep under tension and compression have to be considered separately.

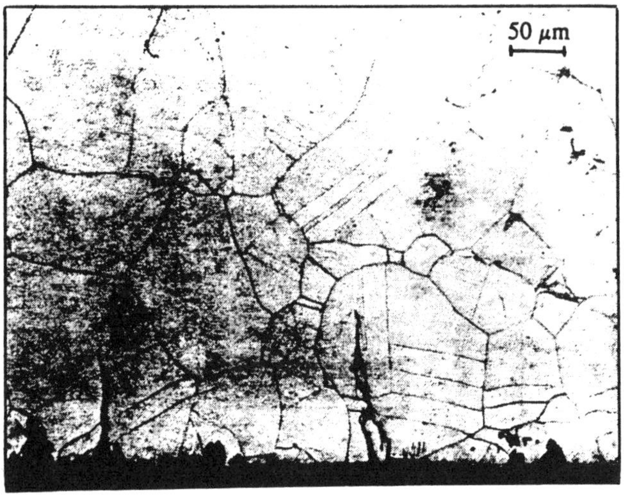


Fig.7 : Intragranular crack growth in the intrados region (compression during hold period).

Fig.8 : Intergranular crack growth in the extrados region (tension during hold period).

## CONCLUSIONS

The inelastic calculation and creep fatigue analysis of a U-bend specimen have been performed and have shown the applicability of the viscoplastic model and the continuum damage model proposed by ONERA to deal with LMFBR's structures.

The comparison with experiments shows that the results of the calculation are close to experimental results even if life assessment seems slightly conservative. However, the benefit brought in this case by the full inelastic analysis when compared with the elastic analysis reaches a factor 3.2 on the number of cycles to failure. Such a factor may justify in some cases the use of inelastic analyses even if they imply an important increase of the cost.

The study confirms finally the less damaging effect of creep under compression and the necessity to take it into account in design calculations.

## ACKNOWLEDGMENT

The authors wish to thank the CEA-SACLAY and the DAS for providing the results and comments on the U-bend tests.

## REFERENCES

1. Design and construction rules for mechanical components of FBR Nuclear Islands. RCC-MR AFCEN June 1985 Edition and addendum 1987.

2. Y. Lejeail, Thèse de Docteur de l'Ecole Nationale Supérieure des Mines de Paris (1991).

3. Rive D., C. Escaravage, B. Riou and M.T. Cabrillat, SMIRT conference, Tokyo, Japan, Aug.1991, L06/1.

4. Chaboche J.L., Nuclear Engineering and Design 105, 19-33 (1987).

5. Riou B., M. Spérandio, C. Poëtte and N. Waeckel, SMIRT conference, Tokyo, Japan, Aug.1991, E06/2.

*Proceedings of the Fifth International Conference on Creep of Materials, Lake Buena Vista, Florida, USA, 18-21 May 1992*

# High-Temperature Creep Phenomena and Their Associated Superplasticity

**B.V. Gorev, A.A. Ratnichkin, and O.V. Sosnin**
Lavrentyev Institute of Hydrodynamics
Russian Academy of Sciences
Novosibirsk, Russia

ABSTRACT

The strain processes at high temperatures ( T>0.6 of absolute melting temperature) and low strain rates ($\dot{\varepsilon} < 10^{-3}$ $s^{-1}$) are considered. It has been shown that over a wide temperature-rate range a material behaves as a thermoviscoelastic medium, timeless plastic strains do not develop, $\varepsilon^p = 0$, a main contribution into a total strain is made by creep strain $\varepsilon^c$, tensor-linear relation between creep strain rate and stresses is confirmed by experiments over a wide range of continuously varying loads and temperatures. Analogous temperature-load restrictions as basic conditions are inherent to superplasticity. When widening a temperature-rate range and ignoring structure factors, superplastic flows are not obtained, however, in terms of continuum mechanics, general regularities of strain remain valid. In this sense, strain processes with the above-mentioned restrictions only in T and $\dot{\varepsilon}$ varying over a wide range are defined as the regimes close to superplastic ones.

IN MECHANICS OF DEFORMABLE SOLID BODY the processes of material deformation are divided traditionally into two classes. They are:

- quasistatic processes with strain rates of the order of several percent per second and less, $\dot{\varepsilon} < 10^{-1} s^{-1}$, and
- dynamic processes occurring at strain rates of the order of hundreds percent per second and higher, $\dot{\varepsilon} > 1$ $s^{-1}$.

At moderate temperatures and in time-limited loading ranges, dividing all the types of strain into the above two classes turned out to be sufficient to solve most of applied problems. However, high-temperature processes with low, about $\dot{\varepsilon} < 10^{-3} s^{-1}$, strain rates have attracted wide attention recently. In such temperature-rate loading regimes, the strain process differs both qualitatively and quantitatively from the two above-mentioned ones.

It is known that at high temperatures, creep strain makes the greatest contribution into a total value of irreversible deformation. It should be noted that for most materials the creep occurs without the first stage of hardening and its third stage is insignificant. Ignoring the latter, one may consider that high-temperature creep occurs at constant strain rate corresponding to the second stage of creep. On the basis of this assumption, one may suggest that for any elevated temperature there exists such value of strain rate $\dot{\varepsilon}_0$ that at $\dot{\varepsilon} \leqslant \dot{\varepsilon}_0$ the loading process falls inside the elestic limits. This means that during strain, the so-called instantaneous plastic deformations (i.e. irreversible deformations associated with the loading parameter rather than real physical time) are not observed and the material behaves as a thermoviscous elastic medium. It is evident that such strain processes are to be distinguished into a particular, third, class of processes, which in terms of mechanics of deformable solid body are directly referred to phenomenology of superplasticity.

The phenomenon of superplasticity (SP-flow) is usually associated with three conditions, such as (i) a definite physical state of a material or the instant of transformation from one state into the other one, (ii) rather high (T > 0.6 of absolute melting temperature) temperature level with a narrow temperature range, (iii) low ( $\dot{\varepsilon} < 10^{-3}\ s^{-1}$) strain rate over a rather narrow rate range.

The first condition is associated with physics of metals and is difficult to describe within the framework of continuum mechanics. Varying any of the above conditions leads to the change, sometimes rather essential, in the character of SP-flow. Dependent on the degree of validity of these conditions, three types of SP-flows have been distinguished:

1. Superplasticity state. All the above-mentioned conditions responsible for the SP-flow formation are valid. The strain values may be as high as hundreds of dozens per cent.

2. Superplasticity regime. An ordinary structural material is considered, deformation occurs within a narrow temperature-rate range. Thus, the restriction imposed by the first condition associated with a physical state of a material has been eliminated. The strain values may be as high as several hundred per cent.

3. Near-superplastic regime. An ordinary structural material is analysed, strain occurs over a wide temperature-rate range, temperature being below bounded, T > 0.6 of absolute melting temperature, and strain rate being upper bounded, $\dot{\varepsilon} \leqslant \dot{\varepsilon}_0(T)$ . Thus, the three conditions responsible for the SP-flow formation have been removed or weakened. The strain values are as high as hundreds per cent.

It is not difficult to see that the near-superplastic regime just represents the third class of strain under which the materials behave as thermoviscoelastic media. Some peculiarities of SP-flow inherent to the state or regime of superplasticity probably do not manifest themselves in the near-superplastic regimes, however, the main regularities of strain in terms of continuum mechanics are likely to be the same for the three classes.

In applied problems of mechanics of deformable solid body, a total deformation is assumed to be a sum of four components: $\varepsilon=\varepsilon^e+ \varepsilon^T+ \varepsilon^p+ \varepsilon^c$ , where $\varepsilon^e$ and $\varepsilon^T$ are the reversible elastic and temperature strains, $\varepsilon^p$ and $\varepsilon^c$ are the irreversible plastic and creep strains. In near-superplastic regimes ( $\varepsilon^p \equiv 0$ ), reversible components account for no more than 1 % of total deformation, thus, creep strain occurs to be most contributive. Consequently, regularities of material deformation in near-superplastic regimes should be studied within the framework of high-temperature creep theory. In this paper, the deformation processes are considered on the basis of Al, Ti and Fe alloys, a system of governing equations is suggested, and the methods of determining characteristics of the material on the basis of one-dimensional processes are developed. Using these relations, the deformation process is described over a wide range of continuously varying loadings and temperatures.

## SOME PECULIARITIES OF HIGH-TEMPERATURE SLOW DEFORMATION OF MATERIALS

The structural commercial alloys, used in the experiments, were not subject to preliminary thermomechanical treatment. The specimens were made of the rod 45 mm in dia. It was longitudinally cut into two parts to produce the specimens to be subject to tension and compression. The specimens to be subject to compression were 10 mm in dia., the working part length l being 40 mm and the total length L being 70 mm. The two types of specimens were used in the experiments: (i) $d_1$ = 10 mm, $l_1$ = 50 mm and $L_1$ = 130 mm, for low-intensive tension at moderate temperatures; (ii) $d_2$ = 7 mm, $l_2$ = 32 mm, $L_2$ = 64 mm for higher temperatures.

The tubular specimens with outer diameter $D_H$ = 21 mm, inner diameter $d_{in}$ = 18 mm, l = 35 mm and L = 110 mm were used in tensile-torsion stress experiments, the initial rod being not longitudinally cut into two parts. All the experiments were carried out in air, the specimens were not provided with protecting coatings, and a total heating and loading duration was not more than 4 hours.

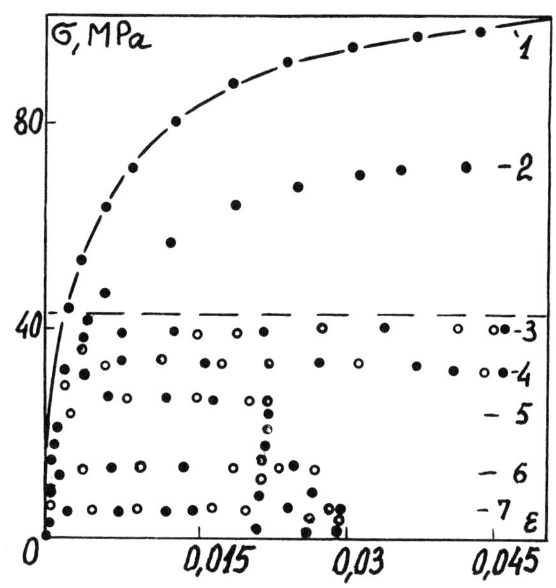


Fig.1. The $\sigma - \varepsilon$ diagram at T = 450 °C for Al alloy with different strain rates.

Figure 1 displays the $\sigma$-$\varepsilon$ diagrams for Al alloy at T = 450°C with different mean strain rates:

(1) $\dot{\varepsilon}_1 \approx 3 \cdot 10^{-2}\ s^{-1}$,

(2) $\dot{\varepsilon}_2 \approx 1 \cdot 10^{-2}\ s^{-1}$,

(3) $\dot{\varepsilon}_3 \approx 7\cdot10^{-3}\ s^{-1}$,

(4) $\dot{\varepsilon}_4 \approx 5\cdot10^{-3}\ s^{-1}$,

(5) $\dot{\varepsilon}_5 \approx 2\cdot10^{-3}\ s^{-1}$,

(6) $\dot{\varepsilon}_6 \approx 4\cdot10^{-4}\ s^{-1}$,

(7) $\dot{\varepsilon}_7 \approx 2\cdot10^{-5}\ s^{-1}$.

The last three diagrams present the results for the same repeatedly loaded specimens. It is evident that they are absolutely identical, i.e. no hardening associated with the presence of plastic strains, takes place.

Figure 2 presents analogous results for Ti alloy at T=800°C:

(1) $\dot{\varepsilon}_1 \approx 1\cdot10^{-1}\ s^{-1}$,

(2) $\dot{\varepsilon}_2 \approx 1\cdot10^{-2}\ s^{-1}$,

(3) $\dot{\varepsilon}_3 \approx 2\cdot10^{-3}\ s^{-1}$,

(4) $\dot{\varepsilon}_4 \approx 7\cdot10^{-4}\ s^{-1}$,

(5) $\dot{\varepsilon}_5 \approx 3\cdot10^{-4}\ s^{-1}$,

(6) $\dot{\varepsilon}_6 \approx 3.5\cdot10^{-5}\ s^{-1}$.

For Fe alloy at T = 800°C the data are as follows:

(1) $\dot{\varepsilon}_1 \approx 3\cdot10^{-2}\ s^{-1}$,

(2) $\dot{\varepsilon}_2 \approx 1\cdot10^{-2}\ s^{-1}$,

(3) $\dot{\varepsilon}_3 \approx 7\cdot10^{-5}\ s^{-1}$,

(4) $\dot{\varepsilon}_4 \approx 1.5\cdot10^{-5}\ s^{-1}$,

(5) $\dot{\varepsilon}_5 \approx 3\cdot10^{-6}\ s^{-1}$.

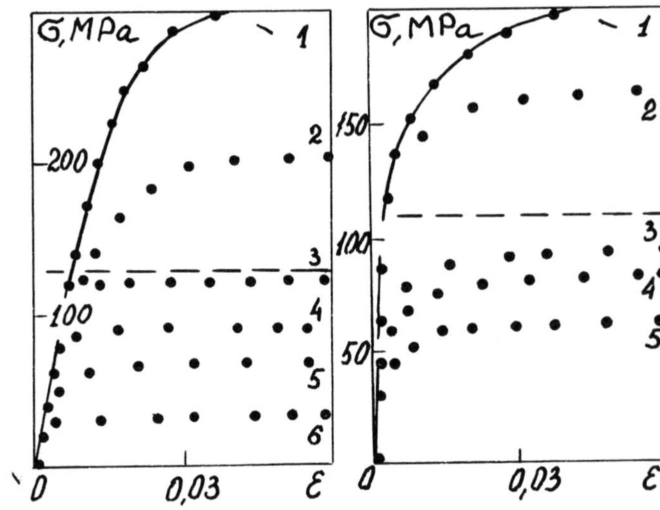


Fig.2. The $\sigma$ - $\varepsilon$ diagram at T = 800°C for Ti and Fe alloys with different strain rates.

The specimens were repeatedly loaded again with slow strain rates, however, the diagrams turned out to be identical too. The diagrams for higher strain rates are beyond the elastic limit and do not possess the repetition property.

Thus, the diagram 1 may be assumed to be the so-called instantaneous elastic-plastic diagram. For all the subsequent diagrams obtained at lower strain rates, total strain value comprises the component dependent on real physical time, namely, creep strain. The dashed line denotes the stress level $\sigma_e$ , above which in the material there appear plastic strains independent of real physical time. The diagrams, placed below this level, do not demonstrate instantaneous plastic strains; they quickly tend to a horizontal asymptote, from where it follows that in such regimes the creep is not accompanied by hardening and the creep rate $\eta$ is a function of current stress and temperature only, $\eta = \varphi(\sigma,T)$. Actually let the strain rate be $\dot{\varepsilon} = V$. Then within the domain $\sigma < \sigma_e$ from $\varepsilon = \varepsilon^e+\varepsilon^T+\varepsilon^c$, where $\varepsilon^e = \sigma/E$ is the elastic component, $\varepsilon^T$ is the temperature component, $\varepsilon^c$ is the creep strain component, T = const., it follows that $V = \frac{1}{E}\frac{d\sigma}{dt} + \eta$.

Substituting $\frac{d\sigma}{dt} = \frac{d\sigma}{d\varepsilon}\cdot V$, we obtain $\frac{d\sigma}{d\varepsilon} = E(1 - \varphi(\sigma,T)/V)$ and when reaching the stress level at which the creep rate $\eta = \varphi(\sigma,T)$ is comparable to the total strain rate $\dot{\varepsilon} = V$, the diagram $\sigma - \varepsilon$ tends to a horizontal asymptote. From here it follows that an upper bound of strain rate $\dot{\varepsilon}_o$ may be found by creep experiment at $\sigma = \sigma_e$ and then by determining creep strain rate. Thus, for Ti alloy at $T = 800^\circ C$, $\sigma_e = 110$ MPa, $\eta_o = 1.5\cdot10^{-3}\ s^{-1}$. The $\sigma - \varepsilon$ diagrams obtained at $\dot{\varepsilon} \leqslant 1.5\cdot10^{-3}\ s^{-1}$ are within the elastic region.

If the parameters $q_i$ showing the effects of hardening or softening should be taken into account when building the plot of $\eta$ vs $\sigma$ and T, there are no horizontal asymptotes in the diagrams. When the process is accompanied by hardening, the diagrams are qualitatively similar to diagram 1, however, the inclination to the horizontal axis is less pronounced. If the process is accompanied by softening (the presence of the third creep stage), there appears the falling branch of the diagram. The absence of the parameters $q_i$ in $\eta = \varphi(\sigma,T)$ is also confirmed by the behaviour of cyclic diagrams (Figs.1 and 2): 10 to 20 loading cycles were carried out at different temperatures and strain rates, and the results were the same, namely, the material did not "remember" the prehistory of the process.

Thus, if the above-mentioned restrictions on temperature and strain rate are taken into account when studying the process, in the total strain value $\varepsilon^p \equiv 0$, and the irreversible component $(\varepsilon^e + \varepsilon^T)$ is negligible as compared to $\varepsilon^c$. Thus, the whole process of strain may be described within the framework of the regularities of high-temperature creep and, if necessary, by summing strains $\varepsilon^c$ and reversible components, each being counted independently, according to special laws.

## CHARACTERISTICS OF HIGH-TEMPERATURE CREEP UNDER UNIAXIAL LOADING

To determine creep characteristics, a series of experiments on tension-compression of cylindrical specimens and torsion of tubular ones were carried out at fixed stresses and temperatures over the ranges corresponding to the regimes close to superplastic ones. During the experiment, the cross-sectional area of the specimen was re-computed after each 0.5 % of axial deformation taking into account incompressibility of the material, and the loading value was corrected in order to hold stress values constant. Some experiments with constant $\sigma$ and T were carried out up to a complete fracture of the specimen. In some of them $\sigma$ and T were changed step by step until the process transited into the third stage of creep ( the laws of creep at the third stage were not considered). In both cases the creep strain rates coincide, $\sigma$ and T being the same, i.e. the existence of dependences of the type $\eta = \varphi\ (\sigma, T)$ was confirmed experimentally for the materials under consideration.

It is assumed for a spatially stressed state that the same dependence relates the equivalent strain rate

$$\eta_i = (\tfrac{2}{3}\eta^o_{kl}\eta^o_{kl})^{1/2}$$

to the equivalent stress

$$\sigma_i = (\tfrac{3}{2}\sigma^o_{kl}\sigma^o_{kl})^{1/2},$$

where $\eta^o_{kl} = \eta_{kl} - \frac{1}{3}\ \delta_{kl}\eta_{nn}$, $\sigma^o_{kl} = \sigma_{kl} - \frac{1}{3}\ \delta_{kl}\sigma_{nn}$ are the deviators of strain rate and stress tensors. Assuming the material incompressibility to be plastic, we have $\eta^o_{kl} = \eta_{kl}$. A special form of this dependence is as follows:

$$\eta_i = exp\ [f(\sigma_i, T)] \qquad (1)$$

Besides, it is assumed that the creep rate tensor and stress tensor are quasi-linearly related by

$$\eta_{kl} = \lambda\ \sigma^o_{kl} \qquad (2)$$

By multiplying and summarizing indices, we get $\lambda = 3\eta_i/2\sigma_i$ and taking into account (1), we obtain

$$\eta_{kl} = (3\sigma^o_{kl}/2\sigma_i)\ exp[f(\sigma_i, T)] \qquad (3)$$

Figures 3a,b,c represent the experimental dependence of $ln\ \eta_i$ on $\sigma_i$ , respectively , for Al, Fe and Ti alloys.

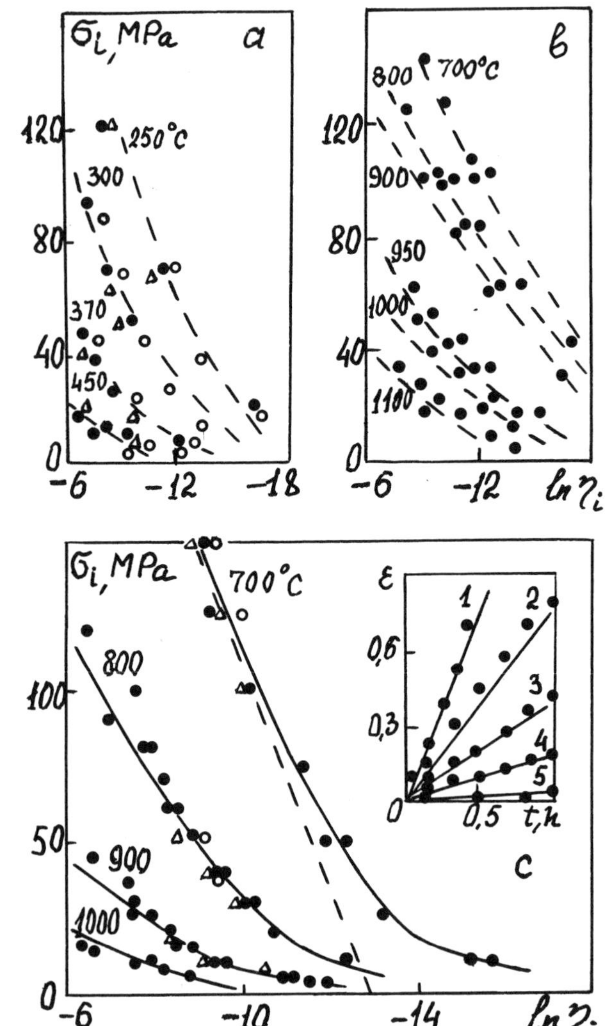


Fig.3. The creep diagram $ln\ \eta_i$ - $\sigma_i$ (T) for Al, Fe and Ti alloys.

Here the experimental results on uniaxial tension are denoted by dark dots, on compression by light ones and on torsion of thin-walled tubes by triangles. Dimension $\sigma_i$ - MPa, $\eta_i$ - $s^{-1}$. It is evident that the experimental points at any fixed temperature do not lie on a straight line. There is no difference in creep rates upon tension, compression and torsion, i.e. the materials behave as isotropic media, although at lower temperatures anisotropy of these materials becomes more pronounced.

Now let us consider in more detail the experimental results on Ti alloys. For Al and Fe alloys the results turned out to be qualitatively analogous.

For each fixed value of temperature, the right-hand side of the dependence $ln\ \eta_i = f(\sigma_i)$ may be approximated by the polynomial

$$f(\sigma_i) = \sum_{k=-n}^{m} a_k \sigma_i^k$$

In its turn, each coefficient $a_k$ is assumed to be dependent on temperature T and may be approximated by an analogous polynomial. The final form of the expression is

$$f(\sigma_i) = A + B\,\sigma_i + C\,\frac{1}{\sigma_i}, \qquad (4)$$

$$A = a_o + a_1 T + a_2 T^2,$$

$$B = b_o + b_1 T + b_2 T^2,$$

$$C = c_o + c_1 T + c_2 T^2,$$

with the numerical values of the coefficients

$a_o = -52.5, \quad a_1 = 8.750 \cdot 10^{-2},$

$b_o = 0.792, \quad b_1 = -2.175 \cdot 10^{-3},$

$c_o = -255.4, \quad c_1 = 48.95 \cdot 10^{-2},$

$a_2 = -4.46 \cdot 10^{-5},$

$b_2 = 1.55 \cdot 10^{-6},$

$c_2 = -2.35 \cdot 10^{-4},$

where T is measured in centigrade degrees, $\sigma_i$ in MPa and $\eta_i$ in $s^{-1}$. It is seen that $f(\sigma_i)$ in (4) consists of a linear part (it is shown by dashed line for 700°C in Fig. 3c) and a hyperbolic one. The latter makes a significant contribution if only the values of $\sigma_i$ are low. Hence, as far as the processing of experimental data is concerned, it is recommended first of all to determine a linear part $f(\sigma_i)$, as in ordinary engineering creep theories at high stress levels, ignoring a hyperbolic part, and then to find the coefficient C by the substitution method at small values of $\sigma_i$.

Solid lines in Fig.3c demonstrate the diagrams calculated from (1) with the above-mentioned coefficients. Presented in the right upper corner of Fig.3c are the results of some experiments in the form $\varepsilon^c = \varepsilon(t)$ at fixed values of $\sigma$ and T:

1. T = 800°C, $\sigma$ = 80 MPa,
2. T = 1000°C, $\sigma$ = 7.5 MPa,
3. T = 800°C, $\sigma$ = 50 MPa,
4. T = 950°C, $\sigma$ = 4 MPa,
5. T = 700°C, $\sigma$ = 75 MPa.

The dotted line shows the experimental values of $\varepsilon^c = ln(l/l_0)$, where $l = l(t)$ is the current working part length of the specimens and the solid line illustrates the results calculated from (1).

As seen from the diagram, the first unsteady creep stage is absent for the temperature range $700^\circ C \leqslant T \leqslant 1000^\circ C$, and the process transforms into the second stage with constant strain rate. Transformation into the third stage of creep depends on temperature: at $T = 700^\circ C$ the third stage starts when $\varepsilon^c$ accounts for 60 % and is comparable with the second one. As temperature increases, duration and value of strains $\varepsilon^c$ at the third stage decreases as compared with the second one and at $T = 1000^\circ C$ it completely disappears. At $T = 600^\circ C$ and below, all the three stages of creep are pronounced. Thus, in the regimes close to superplastic ones, the Ti-alloy behaves as an isotropic thermoviscoelastic medium.

## CREEP UNDER VARIABLE LOADS AND TEMPERATURES

The reliability of the proposed approaches has been varified by the description of strain of the specimens over a wide range of loadings and temperatures.

Figure 4 represents the experimental results on tension of a specimen for stresses and temperatures arbitrarily varying in time, $\sigma = \sigma(t)$ and $T = t(t)$. Loads and temperatures were varied manually, and the data were recorded each 5 minutes. The corresponding plots of $\sigma = \sigma(t)$ and $T = T(t)$ are shown in the left part of Figure 4. The dotted line in the right part shows the experimentally measured strain $\varepsilon = ln(l/l_0)$, and the solid line illustrates the values of strain calculated from

$$\varepsilon = \frac{\sigma}{E} + \alpha(T-T_0) + \int_0^t exp[f(\sigma,T)]\, dt . \quad (5)$$

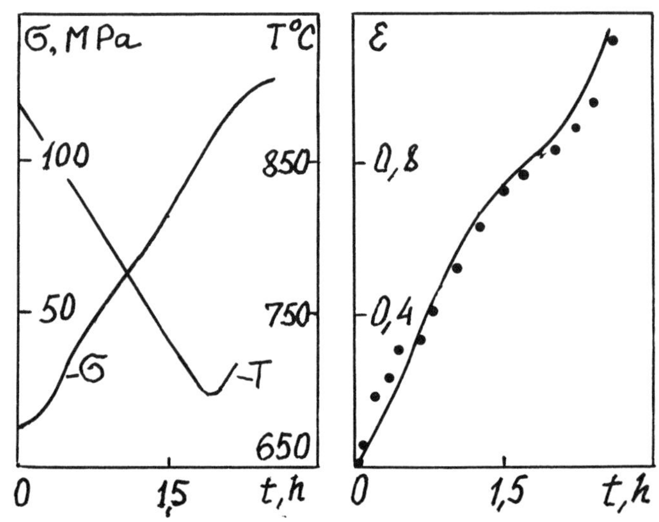


Fig.4. Creep diagram for Ti alloy at $\sigma = \sigma(t)$ and $T = T(t)$.

In the course of calculations it was assumed that E =20 GPa and $\alpha = 2 \cdot 10^{-5}$ grad$^{-1}$ over the whole temperature range. As was mentioned above, a total value of the first two summands of strain is about 1 %. This value is too low as compared to a total strain, and there is no sense to assume E and $\alpha$ to be dependent on temperature. The creep strain is determined from (1) with the coefficients found from the stationary experiments. On the whole, the experimental and calculated values are in satisfactory agreement.

Shown in Fig.5 are the experimental results on tension and torsion of tubular specimens at constant temperature. According to (1), i.e. the hypothesis of a unique curve, constant equivalent stress $\sigma_i$ = const should correspond to a constant value of equilibrium creep strain rate $\eta_i$ = const. According to (2), i.e. the hypothesis of similarity of deviators of the corresponding tensors, $\frac{\Delta\gamma}{\Delta\varepsilon} = 3\frac{\tau}{\sigma}$ should be valid, where $\gamma^2 = (2\varepsilon_{12})^2 + (2\varepsilon_{23})^2$ is the shear strain. In the course of experiments, the loads were changed step-by-step, the values of $\sigma$ and $\tau$ being constant at each step. The strains were defined from the dependences

$$\varepsilon = ln\, 1/l_0 , \quad \gamma = \sum_{k=1}^{n} \Delta\gamma_k ,$$

$$\Delta\gamma_k = \Delta\psi_k\, r_0 / l_k ,$$

where $r_0$ is the mean radius of a tube, $l_k$ is the current working part length, $\Delta\psi_k$ is the angle of twist. The equivalent stresses and strains are respectively, $\sigma_i = (\sigma^2 + 3\tau^2)^{1/2}$ and

$$\varepsilon_i = (\varepsilon^2 + \frac{1}{3}\gamma^2)^{1/2}.$$

At 800°C the process is four-stage, the designations used for the first and second diagrams are the same, dark points correspond to a uniaxial loading, and the results of the previous series of experiments are used. The solid lines in the second diagram were calculated from (1), and the figures near the lines correspond to the values of $\sigma_i$ (MPa). The dots in the third diagram show the values of the ratios $\Delta\gamma/\Delta\varepsilon$ measured each 5 minutes, the solid lines demonstrate the ratios $3\tau/\sigma$ at the corresponding stage of loading. On the whole, the both hypotheses (1) and (2) have been satisfactorily confirmed by experiment.

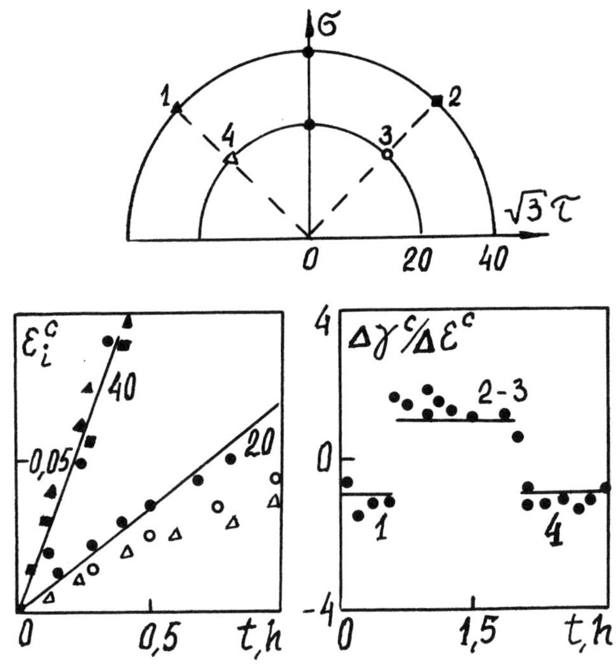


Fig.5. An experimental verification of the hypotheses of a unique curve and of deviator similarity.

The experimental results on determination of axial $\varepsilon$ and shear $\gamma$ strains of a tubular specimen at arbitrary temperatures $T = T(t)$ and loads $\sigma = \sigma(t)$ , $\tau = \tau(t)$ are presented in Fig.6. As in uniaxial experiments, temperature and load were varied manually, the data being recorded each 5 minutes. The data on axial and angular displacements were simultaneously recorded. The plots $T = T(t)$, $\sigma = \sigma(t)$, $\tau = \tau(t)$ and $\sigma_i = \sigma_i(t)$ are presented in the first four diagrams, in the last two diagrams the points show the experimental values of axial $\varepsilon = \varepsilon(t)$ and shear $\gamma = \gamma(t)$ strains of a tubular specimen. Solid lines illustrate the calculation by numerical integration of dependence (3) with adding elastic and temperature components, as in (5). Similar to uniaxial experiments, the experimental and theoretical values of $\varepsilon$ and $\gamma$ turned out to be in rather satisfactory agreement. At strains $\varepsilon_i$ of the order of 20 % the experiment was terminated because of stability loss of the tubular specimen.

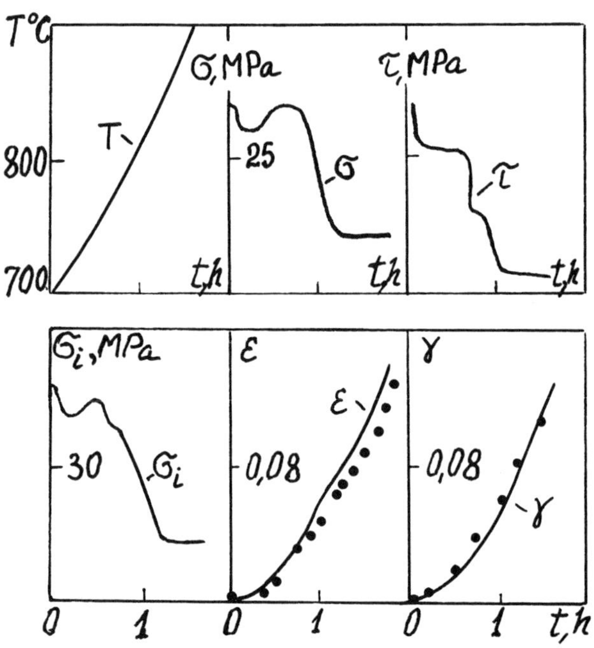


Fig.6. Creep diagrams for Ti alloys at $\sigma = \sigma(t)$, $\tau = \tau(t)$, $T=T(t)$.

CONCLUSION

Analysis of the experimental results taking into account analogous data on Al- and Fe-alloys [1,2] demonstrates that in the strain regimes close to superplasticity, the material behaves as an ideal thermoviscoelastic medium, "instantaneous" plastic deformations do not occur and creep strains make the greatest contribution into total deformation. The system of governing creep equations is thereby of very simple and widespread form.: strain rate tensor and stress one are coaxial, their deviators are similar, the second invariant of strain rate tensor deviator $\eta_i$ is a function of only the second stress deviator invariant $\sigma_i$ and of temperatures without any hardening-softening parameters. The deformation process is not divided into an active part and a passive one. All the abovementioned factors noticeably simplify the construction of corresponding dependences and their usage in applied problems.

REFERENCES


1. Gorev B.V., Ratnichkin A.A., Sosnin O.V. Zhurnal problemy prochnosti, Kiev, 11, 36-47 (1987) (in Russian).
2. Sosnin O.V., Gorev B.V. & Ratnichkin A.A. In: Problemy nelineinoi mekhaniki deformiruemogo tverdogo tela, Sverdlovsk, p. 41-52 (1990) (in Russian).

*Proceedings of the Fifth International Conference on Creep of Materials, Lake Buena Vista, Florida, USA, 18-21 May 1992*

# Influence of Orientation on the Fatigue and Creep Behavior of a Single Crystal Nickel Base Superalloy

**D.J. Smith, L. Shu-Xin, and G.E. Ellison**
Department of Mechanical Engineering
University of Bristol
Bristol, England

## ABSTRACT

The absence of grain boundaries in single crystal nickel base superalloys leads to material anisotropy which is dependent on crystal orientation. An experimental study was carried out on the single crystal superalloy SRR99 to examine the mechanical behaviour for different orientations. Smooth bar specimens tested at 950°C were subjected to fully reversed strain controlled fatigue-creep cycles for various strain ranges with creep dwells at peak tensile and compressive dwells. It was observed that both the cyclic stress-strain response and fatigue life exhibited an orientation dependence. This dependence is found to be related primarily to the anisotropic based on the elastic properties of the single crystal superalloy. The elastic modulus for the [111] orientation is larger compared with, for example [001]. Consequently for tests at the same total strain range [111] specimens exhibit a higher stress range and also a high inelastic strain range; thereby leading to shorter fatigue lives. It is shown that the orientation dependence of the mechanical behaviour and fatigue life can be described in terms of an orientation function.

IN GAS TURBINES the high temperatures involved have necessitated the development of creep resistant nickel base superalloys in single crystal as well as in columnar grain and conventional equiaxed form. The alloys used are designed to have high creep strength at temperatures as high as 0.8 $T_m$ (where $T_m$ is the melting temperature in Kelvin) but must also have adequate oxidation and not corrosion resistance. The desire for increased turbine entry temperatures and compression ratios to increase engine efficiency and performance has led to investigations into two major properties of the alloys - corrosion/oxidation resistance and creep-fatigue interaction. Superior high temperature properties of single crystal over the conventional polycrystalline nickel base superalloys are attributed to the absence of grain boundaries in single crystals. However, the absence of grain boundaries in single crystal alloys also leads to material anisotropy which produces material response which is dependent on crystal orientation.

A fundamental requirement for the successful application of single crystal turbine materials is to provide a clear understanding of the influence of orientation on the mechanical behaviour. Previous studies (1-3) have shown that the monotonic deformation response of single crystal nickel base superalloys is significantly dependent on orientation. Recent work by Gabb et al (4) examined the influence of orientation on the material response of single crystal René N4 under cyclic loading, at temperatures where there was very little creep deformation. Fatigue-creep behaviour of single crystal PWA 1480 has been studied (5) at temperatures where creep is significant, but the influence of orientation was not investigated.

In this paper we present the results of an investigation into the orientation dependence of the single crystal nickel base superalloy SRR99 when subjected to combined creep and fatigue loading at 950°C. Experimental results in terms of deformation response and fatigue-creep life are examined in relation to the crystal orientation.

## EXPERIMENTS

Single crystal SRR99 is a high $\gamma^1$ volume fraction nickel base superalloy, with a chemical composition as shown in Table 1. The material was provided in the form of round bar specimens in the fully heat treated condition i.e. solution heat treated together with a 1 hour dummy aluminising treatment at 1100°C followed by ageing for 16 hours at 870°C. The crystal orientation of each specimen was obtained by using the Laue back reflection X-ray technique.

Button ended round bar specimens with 8.5mm diameter and 12mm gauge length were used to perform fatigue - creep tests at 950°C. The specimens were heated by a 10 KW RF induction coil heater whilst temperature control was achieved by using a Pt/Pt 13% Rh thermocouple attached to each specimen in the centre of the gauge length coupled to a Eurotherm 3-term controller. Two further thermocouples were mounted at the ends of the gauge length to monitor temperature and temperature gradients so that the specimen gauge length is controlled within ± 3°C. The tension-compression rig, used in the present work, was a servo-hydraulic test machine with loads measured by a strain-gauged load cell in the machine crosshead. Strain was measured using a MTS high temperature side contacting extensometer consisting of two ceramic rods pushed by spring pressure radially onto the gauge length of the specimen.

All tests were performed under fully reversed total strain control at a strain rate of $0.004s^{-1}$, see Figure 1. In fatigue-creep tests, constant strain dwells with a duration of 2 minutes at peak strain either in tension, compression or both were employed. In all tests failure was defined as specimen rupture or a reduction of the stabilised maximum cyclic stress range by 30 per cent. Elastic moduli were measured at room temperature (20°C) and 950°C. The results in this paper have been normalised; total stress range is normalised with respect to the critical resolved shear stress (CRSS) of SRR99 at 950°C, and total strain range normalised with respect to the total strain corresponding to the critical resolved shear stress.

**Table 1. Alloy Composition of SRR99**

| Elements | Ni | C | Cr | W | Al | Co | Ta | Ti |
|---|---|---|---|---|---|---|---|---|
| Wt% | Bal. | 0.015 | 8.5 | 9.5 | 5.5 | 5.0 | 2.8 | 2.2 |

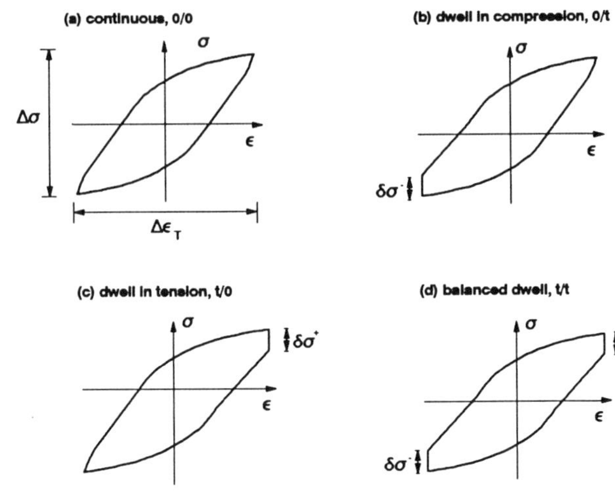


**Fig.1. Schematic of Hysteresis Loops of Different Tests**

## RESULTS

**Cyclic Mechanical Response.** The normalised cyclic stress range- total strain range behaviour for all tests under various loading conditions are shown in Figure 2. Illustrated in Figure 2 (and subsequently figures 3 to 5) are the orientations of representative specimens in terms of the orientation parameter $A_{[hkl]}$ (described later) which varies from 0 at [001] to 1/3 at [111]. The results indicate that the behaviour is highly orientation dependent. At the same total strain range, specimens with a crystal orientations near [111] exhibit higher stress responses than for specimens with orientations near [001].

Figure 3 shows the normalised stress response for four specimens with different crystal orientations subjected to the same cyclic strain range. For the specimen subjected to continuous cycling (with no strain dwells) the material is essentially cyclically stable throughout the life of the specimen. For tests with a dwell in tension or compression alone the material initially cyclically softens within the first five per cent of the life. Thereafter the material is cyclically stable. In contrast, the cyclic test with strain dwells at both peak tensile and compressive strain, progressively cyclically softens, with the total stress range reducing rapidly during the first five per cent of life.

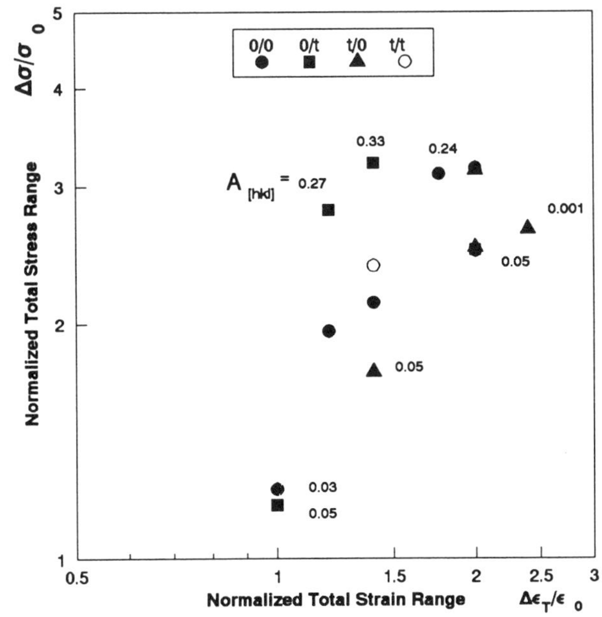


Fig.2. Cyclic Stress-Strain Relationship

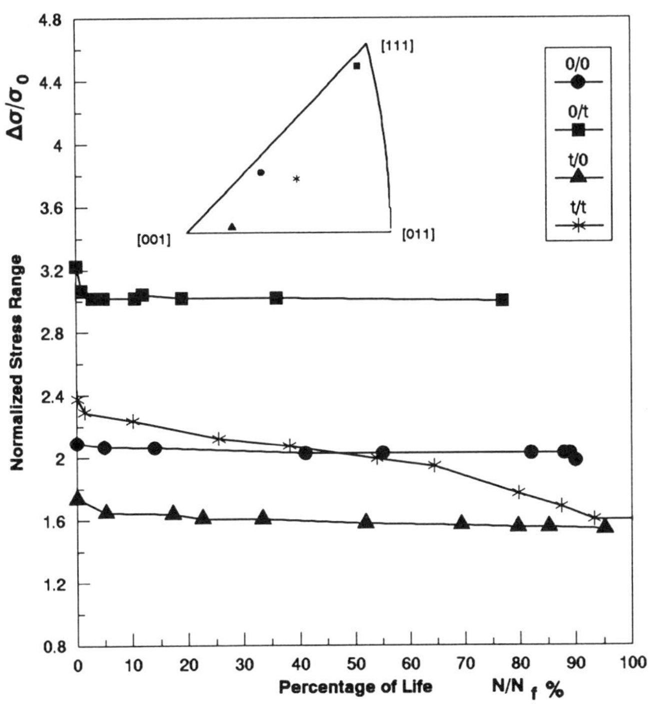


Fig.3. Cyclic Stress Response

The total extent of stress relaxation $\delta\sigma$, normalised with respect to the CRSS at 950°C, during two minute strain dwells for the various test cycles is shown in Fig 4 as a function of half the normalised strain range. It was found that stress relaxation, whether for tensile or compressive dwell, occurred rapidly during the first five per cent of life, with the extent of total stress relaxation remaining almost constant in the following dwell periods. The orientation dependence of the total stress relaxation response is evident in Fig 4. The amount of stress relaxation increases as the crystal orientation moves away from [001].

**Fatigue Life**. The fatigue life for various normalised total strain ranges and cycle types are shown in Figure 5. It is evident that the fatigue life is highly orientation dependent with the [001] orientation for a given total strain range and cycle type providing the longest fatigue life.

The ratio of the lives of tests with compressive dwells to the lives of continuous tests was always less than unity. In constrast the life ratio for tests with dwells in tension compared with continuous tests was always greater than unity. These ratios are relevant to tests with similar orientations and total strain ranges.

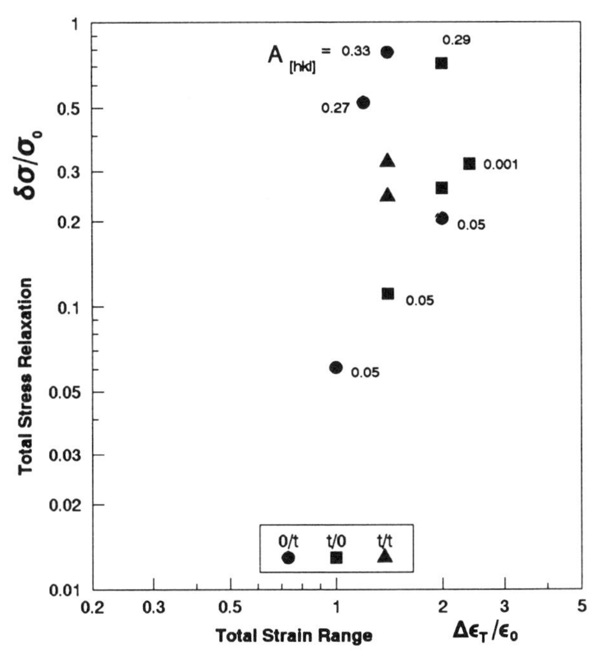


Fig.4. Total Stress Relaxation As Function of Total Strain Range

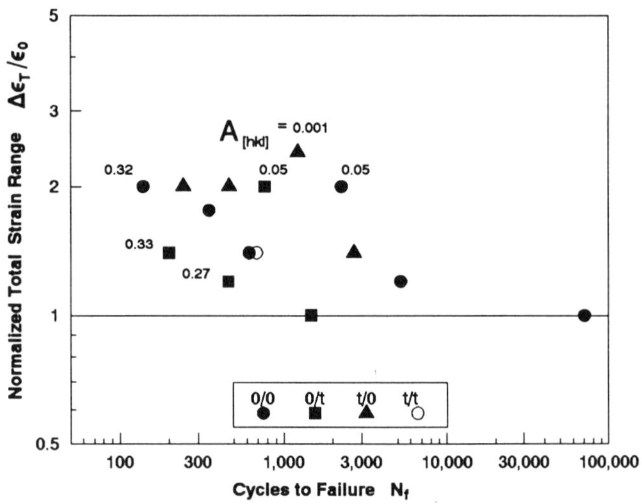


**Fig.5. Fatigue-Creep Life**

## DISCUSSION

The experimental results for SRR99 at 950°C indicate that the fatigue-creep mechanical response and fatigue life are strongly dependent on orientation. The orientation dependence of the elastic moduli of single crystals is well documented (6), but information on the influence of orientation on combined elastic, plastic and creep behaviour is limited. Here, for simplicity we assume that the influence of orientation on the total mechanical response is similar to that for the elastic response. However, we recognise that the orientation dependence of plastic deformation will be different to that for elastic behaviour. Nevertheless, when the plastic strains are small, as is the case for the present study, the behaviour will be dominated by elastic response.

Classical anisotropic elasticity theory provides an analysis for the determination of the elastic moduli in any given direction and relative to the loading direction assuming that all elements of the stiffness (or compliance) matrix are known. For crystals with cubic symmetry this requires information for $E_{[001]}$ $(=1/S_{11})$, $\nu$ $(=-S_{12}E)$ and $G_{[001]}$ $(=1/2S_{44})$ at the temperature of interest. The elastic modulus $E_{[hkl]}$ for a known orientation, described by the direction ratios, *h,k,l,* of the plane normal is given by

$$E_{[hkl]} = E_{[001]}/(1 - D\,A_{[hkl]}) \quad (1)$$

where $E_{001}$ is the elastic modulus in the [001] orientation, and D is given by

$$D = 2 + 2\upsilon - (E_{[001]}/G_{[001]}) \quad (2)$$

with

$$A_{[hkl]} = \frac{h^2k^2 + k^2l^2 + l^2h^2}{(h^2 + k^2 + l^2)^2} \quad (3)$$

Since there are no available data for the three elastic constants for SRR99 at elevated temperature, the uniaxial elastic modulus $E_{[hkl]}$ measured in each test at 950°C were correlated with $A_{[hkl]}$, Eq.1, yielding the constant *D*, Eq.2. Further results, reported in (7), have shown that *D* is weakly dependent on temperature, and at temperatures between 20°C and 950°C the ratio of the elastic moduli to the modulus at [001] given by

$$\frac{E_{[hkl]}}{E_{[001]}} = f(A_{[hkl]}) = \left\{1 - 1.8\,A_{[hkl]}\right\}^{-1} \quad (4)$$

Figures 6 to 9 illustrate the experimental results shown in Figures 2 to 5 but subsequently modified using Eq.4. The normalised cyclic stress range is now related to the orientation modified normalised cyclic strain range so that the cyclic stress strain response at 950°C can be expressed as

$$\frac{\Delta\varepsilon_T\, f(A_{[hkl]})}{\varepsilon_0} = \frac{\Delta\sigma}{\sigma_0 E_{[001]}} + B\left(\frac{\Delta\sigma}{\sigma_0}\right)^n \quad (5)$$

where B and n are material constants.

Figure 6 illustrates that there are two distinct deformation regimes - elastic and plastic - that are described by Eq.5, which are approximately independent of orientation. For normalised stress ranges and orientation modified normalised total strain ranges less than about 3 and 2.5 respectively the deformation response is predominately elastic.

The normalised total stress range shown in Fig.3, for different loading conditions and at the same normalised strain range of 1.4, has been modified to take account of the orientation of each specimen. The results are shown in Figure 7. Since the normalised total stress range is less than about 3 it is expected that the predominant deformation mode is elastic, and correspondingly the normalised curves shown in Figure 7 are similar. An important feature of the modified curves is that continuous cyclic (o/o) and balanced dwell (t/t) tests exhibit higher stress ranges in the first few cycles. In contrast, the unbalanced dwell tests (t/o and o/t) show reduced stress ranges in the initial stages. These features are similar to those observed by Plumbridge and Ellison (8) for MAR-M002, a polycrystalline nickel base superalloy.

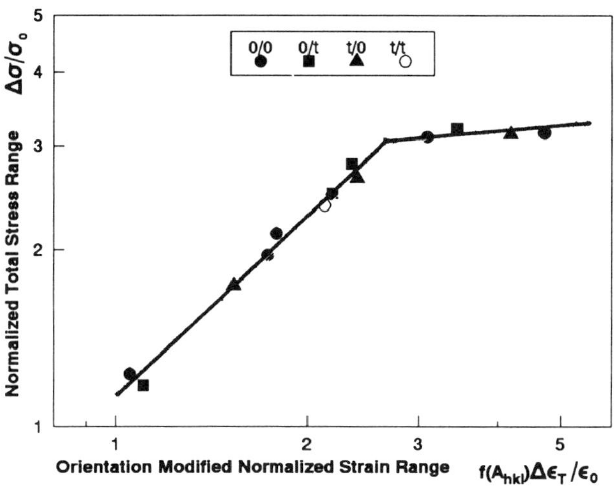


**Fig.6. Orientation Modified Cyclic Stress-Strain Relationship**

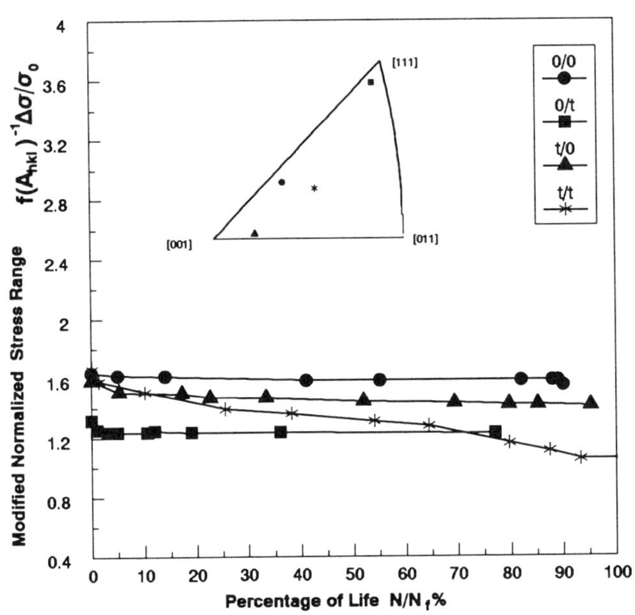


**Fig.7. Orientation Modified Total Stress Range Response**

Stress relaxation during a dwell at constant strain is a consequence of the recovery of elastic deformation. The extent of stress relaxation for a given strain range shown in Figure 4 has been modified in Figure 8 to take into account the orientation of each specimen. The relationship between the total amount of stress relaxation and the total strain range can be expressed as

$$\frac{\delta\sigma}{\sigma_0} = H \left\{ \frac{\Delta\varepsilon_T}{\varepsilon_o} f(A_{[hkl]}) \right\}^m \qquad (6)$$

where H and m are material constants. This equation appears to be applicable to stress relaxation during either tensile or compressive strain dwell.

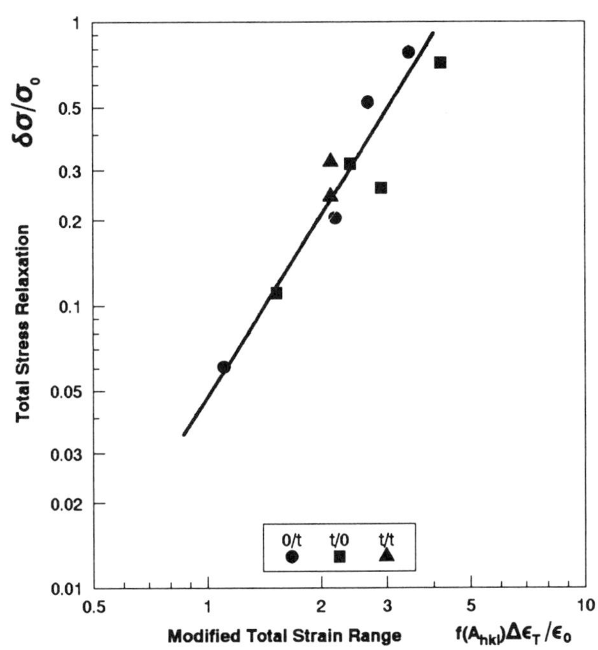


**Fig.8. Total Stress Relaxation As Function of Orientation Modified Total Strain Range**

The fatigue life of SRR99 for orientation modified total strain ranges less than 2.5 will be dominated by the elastic response. Fig 9 illustrates the results shown in Fig 5 where the total strain range is modified by the orientation parameter f(Ahkl). The scatter shown in Fig 9 is now much less than in Fig 5. At low strain ranges there is some evidence that tests with dwells in compression have shorter lives than continuously cycled tests. At high strain ranges the distinction in life between tests with and without dwells is not so clear. This may in part be due to the imprecise evaluation of the influence of orientation on combined elastic and plastic deformation.

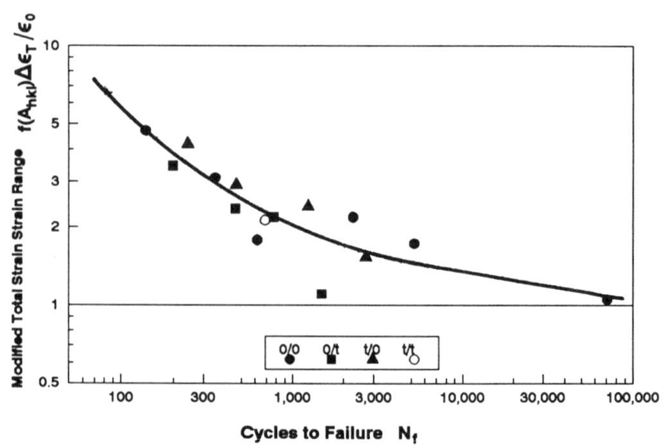


Fig.9. Orientation Modified Fatigue-Creep Life

## CONCLUSIONS

The single crystal nickel base superalloy SRR99 has been subjected to combined fatigue and creep loading at 950°C to study the influence of orientation of the subsequent deformation and fatigue-creep life. Specimens with orientation near [111] exhibit higher cyclic stress ranges, larger degrees of stress relaxation and high inelastic strain ranges.

The orientation dependence of the deformation properties and fatigue life is found to be related mainly to the elastic anisotropic properties of the single crystal and can be reasonably described in terms of an orientation function.

## ACKNOWLEDGEMENTS


This work was funded by Rolls Royce plc and their support is gratefully acknowledged. The interpretation of the results and the views expressed in this paper are those of the authors and do not necessarily represent those of Rolls Royce.


## REFERENCES


1- Copley,S.M., B.H.Kear and G.M.Rowe, Mater Sci. Eng. 10, 87-91 (1972)

2- Miner R.V., T.P. Gabb, J.Gayda and K.J. Hemker, Metall. Trans. A 17A, 507-512 (1986)

3- Shah, D.M. and D.N. Duhl, in Superalloys, Proc. 5th Int. Symposium on Superalloys, M. Gell et al (eds.) AIME Warrendale PA 1984, 105-114.

4- Gabb, T.P., Gayda, J., and Miner R.V., Metall. Trans. A, 17A, 497-505 (1986)

5- Miner, R.V., Gayda, J. and Hebsar, G. in Low Cycle Fatigue and Life Prediction, ASTM STP 770, C. Amzallag, B.N. Leis and P. Rabbe, Eds., ASTM 1982 105-134.

6- Zener,C. Elasticity and Anelasticity of Metals p7 The University of Chicago Press, Chicago, (1965).

7- Li S X, D.J. Smith and E.G. Ellison. unpublished research, Dept. of Mechanical Engineering, University of Bristol, 1991.

8- Plumbridge W.J. and E.G. Ellison. Fatigue Fracture Engng. Mater. & Stuct. Vol.4, pp.373-389, (1991).

*Proceedings of the Fifth International Conference on Creep of Materials, Lake Buena Vista, Florida, USA, 18-21 May 1992*

# Mechanistic Creep Modeling for Weldment Life Prediction

**A.J. Tack and J.M. Brear**
ERA Technology Limited
Surrey, United Kingdom

**F.J. Seco**
Inasmet
San Sebastian, Spain

## ABSTRACT

This paper describes remanent creep life modelling work performed in two collaborative projects within the Commission of the European Communities BRITE programme. The participants include European power utilities, petrochemical industries and research organisations. Both projects address condition assessment, performance evaluation and life prediction of low alloy ferritic steel pressure parts for elevated temperature service. The former concentrates on base material and the latter on weldments. The evaluation approaches are centred on a simple mechanistic creep model which describes the dominant degradation process - microstructural coarsening and creep cavitation - in terms of parameters readily obtainable by non-destructive, on-site condition monitoring techniques. In its original form (5) the model covered secondary and tertiary creep processes controlled by thermal softening and cavitation. These projects have enabled further development of the model for base material and allowed its application to weldments.

LOW ALLOY FERRITIC STEELS are routinely used in the power generation and process industries for high temperature steam and process pipework. With much of the World's existing plant approaching, or now operating beyond, its original design life there is considerable interest in the development of techniques for assessing the remanent life of critical components such that maintenance and repair schedules can be planned to ensure optimum plant availability, and hence, maximised economic efficiency.

Pressure vessels exposed to high temperature conditions under which creep and other time dependent processes occur, will suffer degradation of their properties over periods of extended operation. This damage subsequently leads to cracking and failure. It is essential, during the operation of high temperature components, that the necessary safety requirements are met in addition to gaining the maximum economic benefit from extended operating lives. Two approaches are generally recognised in predicting the remanent lives of components. These are: i) operational condition monitoring which facilitates a preliminary estimate of remanent life based on operational parameters, simple stress calculations and the use of standard lower bound material data in association with simple damage summation rules and ii) contact with the component during inspection intervals involving the application of methodologies based on metallography, micro-sampling, strain measurement, post-exposure creep rupture testing and more detailed stress analysis.

In general, however, the preliminary approach is subject to many uncertainties arising from, in particular, the scatter and extrapolation of standard material design data. In addition, since long term data are obtained on relatively small test specimens oxidation is significant and will lead to pessimistic predictions when applied to thick section components. Consequently, although the application of the preliminary approach is an initial stage often necessary to obtain insurance certification for continued operation, it does not, in general, provide a realistic estimate of the remanent life on which to base life extension strategies or to plan for component replacement.

The second approach allows more accurate remanent life predictions to be made, not least because it is not reliant on standard materials data and their associated uncertainties. It is important, therefore, that procedural routes utilising techniques within this approach are developed and implemented to improve remanent life predictions and define future inspection requirements.

To determine remanent life by component examinations or sampling requires predictive models for damage evolution and failure which relate a relevant damage feature or property to the life fraction consumed. Many of these features and properties have been reviewed (1).

Creep damage leading to failure in pressure vessels constructed from low alloy steel results from: i) structural degradation leading to a continuous reduction in creep strength during service exposure (2,3,4) and ii) intergranular creep cavitation. Both processes normally occur simultaneously; the prevalence of either is determined by the initial structural state and purity, conditions of stress and temperature.

In a previous paper (5) a mechanistic model for predicting creep life was presented which took account of both structural coarsening and creep cavitation. The cavitation was incorporated using a Kachanov approach whilst structural damage was modelled in terms of interparticle spacing. The marked heterogeneous nature of particle dispersions in low alloy ferritic steels gives rise to difficulties when trying to obtain a non-subjective representative measure of the interparticle spacing. Consequently a non-subjective approach using hardness change measurements was proposed for use in conjunction with the mechanistic model for creep life prediction.

The current paper describes how recent data have allowed the original model to be extended to include the influence of primary creep and strain softening and then by normalising the applied stress with respect to the yield strength, estimated from the hardness via an Orowan type relation goes on to illustrate how a single set of parameters developed for a range of steel types can be used to describe the creep response of all heats to that specification. In a further advance cross-weld data have been analysed using the mechanistic model. This has facilitated the formulation of a weldment predictor diagram which can be used to determine where, within a weldment, failure is most likely to occur.

## MATERIALS

Various heats of new and ex-service 2.25Cr1Mo steels were procured for this work. The material conditions employed are summarised in Table I.

Table I. 2.25Cr1Mo experimental materials

| Material Heat | Type | History/treatment |
|---|---|---|
| 2CM/V1 | New | Welded as received |
| 2CM/V2 | New | Welded as received |
| 2CM/A1 | Aged 2CM/V1 | 500 hrs at 640°C prior to welding |
| 2CM/V1/A3 | Aged 2CM/V1 | 3000 hrs at 640°C prior to welding |
| 2CM/V2/A1 | Aged 2CM/V2 | 500 hrs at 710°C. No weld |
| 2CM/U1 | Ex-service | 11650 hrs at 540°C prior to welding |
| 2CM/U2 | Ex-service | 170600 hrs at 550-575°C prior to welding |
| 2CM/U3 | Ex-service | 17650 hrs at 550-575°C prior to welding |

All the materials were metallographically characterised to provide a quantitative record of the pre-weld repair condition. For the weldment study a number of the new and ex-service pipe segments were sectioned and repair welded using current procedural practice and matching

2.25Cr1Mo filler metal. The new weldments were then fully characterised metallographically. In addition sections of the new material were aged prior to repair welding to simulate service exposure.

## THE CREEP MODEL

The model is based on a Norton creep law, with a threshold stress term, $\sigma_0$, which has its physical origin in the interaction between moving dislocations and precipitate particles:

$$\dot{\varepsilon} = A(\sigma - \sigma_0)^n \tag{1}$$

where $\sigma$ is the applied stress, $\dot{\varepsilon}$ the creep rate and $A$ and $n$ are constants. The threshold stress is inversely proportional to the interparticle spacing, $\lambda$, but since this is problematic to measure directly, a relationship between threshold stress and hardness, $H$, is commonly accepted:

$$\sigma_0 \sim \frac{1}{\lambda} \sim (H - H_{ss}) \tag{2}$$

where $H_{ss}$ is the contribution to the hardness due to solid solution strengthening. As the precipitates coarsen with time, $t$, at temperature, the hardness and threshold stress decrease:

$$\sigma_0 \sim (H - H_{ss}) \sim \frac{1}{\lambda} \sim t^{-\frac{1}{3}} \tag{3}$$

The model had reached this stage of development prior to the commencement of the current work (5,6). The immediate aims here were to calibrate and refine it where appropriate using the body of test data available and then apply it to the prediction of test piece and component behaviour.

The first refinement of the model involved the incorporation of a strain softening term. Examination of the hardnesses measured from both the gauge length of the test specimens and the section contained within the grips of the creep machine indicated a semi-empirical strain softening relationship of the form:

$$\frac{H_t - H_{t,\varepsilon}}{H_t - H_{ss}} = a + b\ln(\varepsilon) \tag{4}$$

where $H_t$ is the hardness due to thermal softening alone and $H_{t,\varepsilon}$ is the result of thermal and strain softening. The relationship was calibrated on the available data and incorporated in the creep model. Whilst this approach is not fully rigorous, it does generate a better description of the tertiary creep curve shape for the materials investigated.

Consideration of the creep data showed that primary creep cannot be ignored, as initially assumed. Whilst primary strains are quite small, the duration of primary creep is considerable (between ten and fifty percent of life). Accordingly a method of incorporating this effect was required. The basic creep law represents the threshold stress existing under steady state flow, where dislocation movement is limited by the need to climb over precipitate particles. On initial loading, however, the dislocations are largely free of the precipitates and some time passes before the steady state is reached. This can be represented by rewriting Eq. 1 as:

$$\dot{\varepsilon} = A(\sigma - k\sigma_o)^n \tag{5}$$

with:

$$k = 1 - \exp\left(-\frac{t}{\tau_p}\right) \quad (6)$$

where $\tau_p$ is the primary time constant. This modification facilitates a much improved definition of the initial portion of the creep curve. An analogous form of the relationship using the ratio $\frac{\varepsilon}{\varepsilon_p}$ is also possible.

During the initial calibration of the creep model, the heats of material were treated separately. It transpired, however, that the data could be merged by a simple modification to the model, reflecting the relative flow strengths of the materials. Thus:

$$\dot{\varepsilon} = A\left(\frac{\sigma - k\sigma_0}{\sigma_y}\right)^n \quad (7)$$

where $\sigma_y$ is the flow stress given by:

$$\sigma_y = KH_{t,\varepsilon} \quad (8)$$

by analogy with the existing relation (5,6):

$$\sigma_0 = K\alpha^{/}(H_{t,\varepsilon} - H_{ss}) \quad (9)$$

where $\alpha^{/}$ is a geometric term.

There are clearly further, less important effects which influence the fit of the model to the observed data, however, it is considered that the above represents a robust and straightforward approach to predicting the creep behaviour of this class of material. Figure 1 shows predictions obtained using the model. Data obtained from parent material specimens of 2CM/V2 (hardness = 161 Vickers) tested at 680°C and a stress of 40MPa are plotted together with a series of curves showing the development of the model. Curve a) represents the model in its original form, with only thermal softening included. Curves b) and c) show the effects of adding strain softening and primary creep individually, whilst curve d) shows the combined effect and represents the model in its final state. The agreement with the experimental data is good. Finally curve e) shows the predicted and observed behaviour of the same heat in the overaged, 2CM/V2/A1, condition (hardness = 117 Vickers).

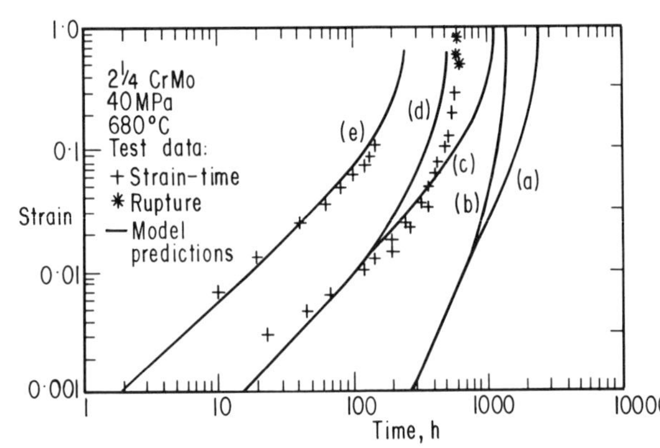


Fig. 1 - Creep model predictions illustrating the refinements to the model

Figure 2 shows the application of the model to life prediction, in combination with photogrammetric strain measurements. Curve a) represents the modelled behaviour of a specimen taken to approximately 50 percent of its life at 610° C and 70 MPa. The specimen test temperature was then raised to 635 °C and the test continued to failure. The model prediction, b) and the observed behaviour are in excellent agreement.

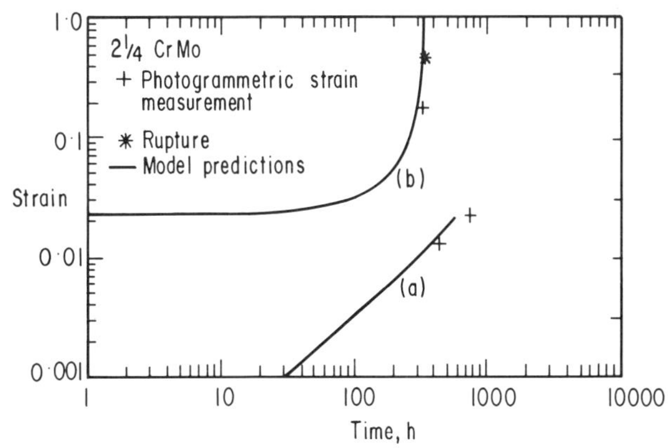


Fig. 2 - Prediction of life based on photogrammetric strain measurement

## WELDMENT MODELLING

Work has centred on the evaluation of creep rupture data obtained from parent and cross-weld specimens extracted from the experimental materials. Two specimen geometries were employed; small 15mm diameter cylindrical test pieces and large 60 x 60mm square section specimens. It was observed that the small specimens can be treated as a simple series of parent, heat affected zone and weld metal units with no complications arising from constraint effects. In the large specimens however, constraint effects appear to predominate, leading to localised failures in the weaker zones.

The analysis of the small diameter specimen data was carried out using a similar approach to that set out in the previous section. In this case, however, the model is modified to describe the rupture response of the materials and not the shape of the creep curve. As before the starting point is a formulation of the Norton Law:

$$\dot{\varepsilon} = \dot{\varepsilon}_0 \left( \frac{\sigma_e}{\sigma_y} \right)^n \quad (10)$$

where $\dot{\varepsilon}_0$ is a temperature dependent creep rate, $\sigma_e$ the effective stress for creep and $\sigma_y$ a reference flow stress. For precipitate strengthened materials the effective stress for creep can be expressed as

$$\sigma_e = \sigma_g - \sigma_0 \quad (11)$$

where $\sigma_g$ is the stress acting within the grains. Thus $\sigma_g$ is the applied stress corrected for stress concentration effects, cavitation and overall deformation, and $\sigma_0$ is related to the shear modulus, $\mu$ , the Burgers vector, $b$ and the interparticle spacing, $\lambda$ , by:

$$\sigma_0 = \frac{\alpha' \mu b}{\lambda} \quad (12)$$

It has previously been shown that $\sigma_0$ may be related to the hardness of a material using Eq. 9. This assumes that the overall hardness, $H$, is made up of two components, $H_{ss}$ being the contribution due to solid solution strengthening with the balance due to precipitate strengthening. The constant, $K$, defines the relationship between hardness and tensile strength.

Traditionally the flow stress for materials of this class has been equated with the Orowan stress:

$$\sigma_y = \frac{\mu b}{\lambda} \quad (13)$$

However, this implies a zero flow resistance for precipitate free (ie overaged) material, which is clearly not the case, and the observed relation:

$$\sigma_y = KH \quad (14)$$

leads to formal inconsistencies between Eq. 13 and Eqs. 11,12,14. These problems may be overcome by rewriting Eq. 13 as:

$$\sigma_y = \sigma_{f(ss)} + \frac{\mu b}{\lambda} \quad (15)$$

where $\sigma_{f(ss)}$ is the solid solution (plus other internal friction effects) contribution to the flow resistance. Combining the above leads to:

$$\dot{\varepsilon} = \dot{\varepsilon}_0 \left( \frac{\sigma_g - \sigma_0}{\sigma_y} \right)^n \quad (16)$$

or:

$$\dot{\varepsilon} = \dot{\varepsilon}_0 \left( \frac{\sigma_g - K\alpha'(H - H_{ss})}{KH} \right)^n \quad (17)$$

This creep rate will not be constant but will change as the hardness evolves with time at temperature and possibly with strain, and will increase as $\sigma_g$ increases with strain and the accumulation of any grain boundary cavitation. Generally, the minimum creep rate, $\dot{\varepsilon}_m$, will be obtained by substituting initial values of $H_i$ and $\sigma_{gi}$ for $H$ and $\sigma_g$.

Assuming a Monkman-Grant relationship:

$$\dot{\varepsilon}_m t_r = \varepsilon_s \quad (18)$$

the rupture life, $t_r$, is given by:

$$t_r = \frac{\varepsilon_s}{\dot{\varepsilon}_0} \left( \frac{\sigma_{gi} - K\alpha'(H_i - H_{ss})}{KH_i} \right)^{-n} \quad (19)$$

Preliminary trials with the experimental results showed that simpler models (obtained by neglecting $\sigma_0$ or treating $\sigma_y$ as hardness independent) were not adequate for describing the data. Accordingly Eq. 19 was fitted to the rupture lives of the parent and weld metal for the experimental materials. The results are illustrated in Figure 3. Values for the individual terms were determined as follows:

$\frac{\varepsilon_s}{\dot{\varepsilon}_0}$ = Single values obtained by fitting the data

$\frac{\varepsilon_s}{\dot{\varepsilon}_0}$ (parent) = 0.08516 h

$\frac{\varepsilon_s}{\dot{\varepsilon}_0}$ (weld) = 0.01055 h

$n$ = Single value obtained by fitting. $n$ = 4.7

$K$ = Consideration of tensile data available in the project and other information indicates a value of $K$ = 1.56 MPa/HV.

$\alpha'$ = Obtained from the literature (7,8). $\alpha'$ = 0.32.

$H_{ss}$ = From previous work (6) a value was assumed for parent material of $H_{ss}$ = 100 HV and fitting the weld metal data yielded a value of $H_{ss}$ = 155 HV.

$\sigma_{gi}$ = The applied stress. No cavitation was observed and so no correction for strain is required in the rupture life equation.

$H_i$ = The measured macro hardness. No variation with time, temperature or strain is required in the rupture life equation (nor is any significant softening observed for the materials studied here).

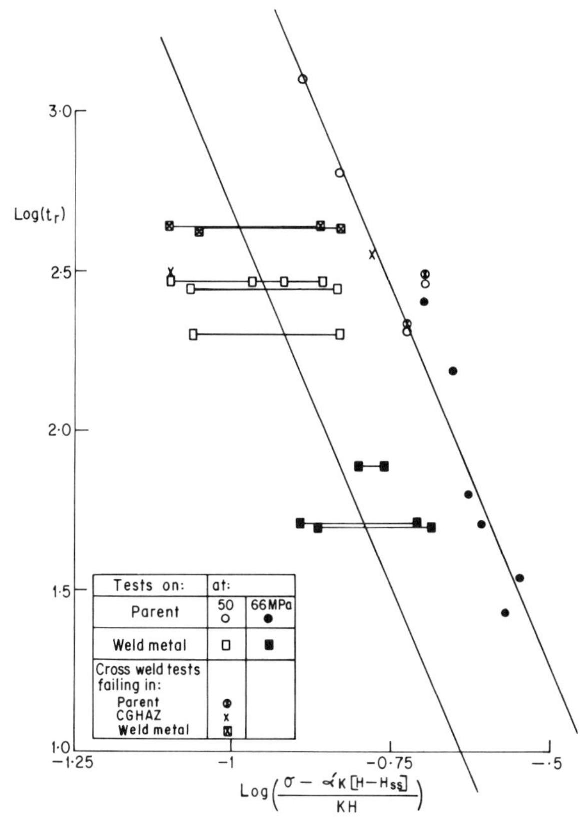


Fig. 3 - Creep rupture data correlated using the hardness based creep model (tie bars link points where two hardness data points were available)

Considerable scatter is seen for the weld metal data reflecting the local variability in hardness in each weld. The parent material data, however, are well correlated and the effects of hardness at constant stress and of stress at constant hardness are closely comparable, confirming the suitability of the general form adopted for the model.

The rupture data arising from the small specimen cross-weld tests have been added to Figure 3 for comparison. It appears that there are no significant constraint effects and that the rupture lives lie within the bands of behaviour appropriate to the material in which failure occurred. Broadly the behaviour of the coarse-grained heat affected zone material is comparable to that of the weld metal.

The rupture life equations for the parent and weld materials can then be used to define a line giving the relative hardnesses required to produce equal creep strengths and equal rupture lives. Such a line for an applied stress of 50 MPa is shown in Figure 4.

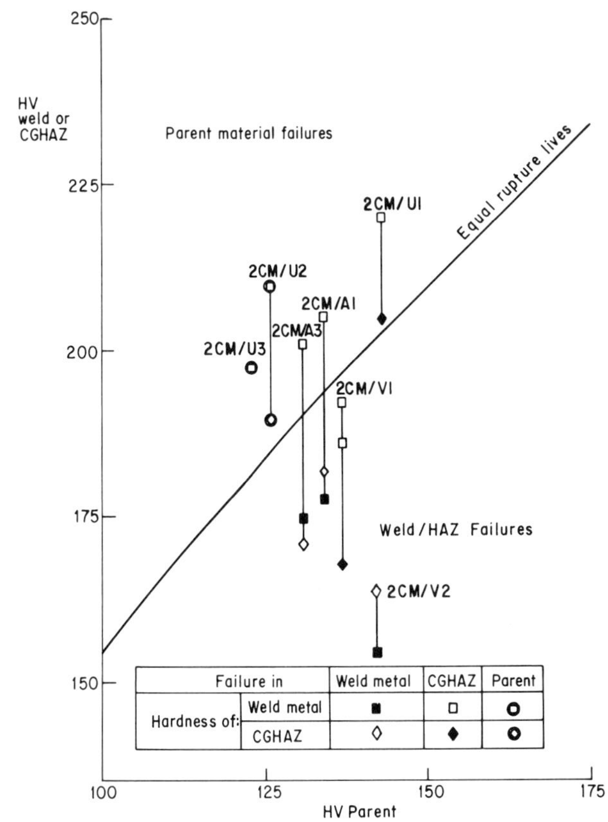


Fig. 4 - Rupture data plotted in terms of constituent microstructural hardnesses and failure location. The parent/weld and HAZ equal rupture life line is shown.

Points defining an individual weldment may be plotted on this graph and from their position the failure location may be predicted. Points above the line indicate that the weld metal is stronger than the parent metal and therefore failure in the parent material is to be expected. The converse applies below the line. A summary of the data is given below:

2CM/V1 The weld and HAZ hardnesses lie below the line. Failure occurs as expected, in the weaker HAZ.

2CM/V2 The weld and HAZ hardnesses lie below the line. Failure in this case occurs in the weaker weld metal.

2CM/A1 The weld metal hardnesses span the line and extend below the HAZ hardnesses. Failure occurred in (probably the weakest part of) the weld metal.

2CM/A3 The weld metal hardnesses span the line but lie above the The HAZ hardness. Failure occurred in the weld metal although it would have been expected in the HAZ. However, the error is small in relation to the the scatter of the input data.

2CM/U1 The weld metal hardness is well above the line and the HAZ slightly above. Failure occurred in the HAZ whilst it would have been expected in the parent material. However, the error is small compared with the scatter in the input data.

2CM/U2 and 2CM/U3 For both of these materials the weld metal and HAZ hardnesses lie above the line. Failures occurred in the parent material as expected.

## CONCLUSION

Significant refinements to an existing mechanistic creep model have been made, allowing firstly, the effects of primary creep and strain softening to be incorporated in an expression defining the creep response of a range of 2.25Cr1Mo steels, and secondly, the extension of the model to predict failure locations in weldments. The model has a clear application to high temperature creep life assessment, however, before the model can be applied with confidence to weldments in thick section components further work is required to include the effects of stress redistribution and alternative failure modes.

## ACKNOWLEDGMENTS

This paper has been produced using the results of two Commission of the European Communities funded BRITE programmes: P-1290 'Condition Assessment of Industrial Pressure Parts Operating at Elevated Temperatures' and P-2341 'Performance and Reliability Evaluation of Weldments in Elevated Temperature Service.' The partners involved in these programmes are: ERA, ENEL-CRTN, Shell, Tecnatom, Inasmet, ICI, National Power, Powergen and Nuclear Electric. The authors would like to thank Dr Davide D'Angelo of ENEL-CRTN and Mr Angel Perez Sainz of the CEC in particular, for their helpful suggestions and encouragement concerning this work.

This paper is published with the permission of ERA Technology Ltd.

## REFERENCES

1 - Cane, B. J., High Temperature Technology , Vol 1, 1983, p. 4.

2 - Williams, K. R. and Cane, B. J. Materials Science and Engineering, Vol 38, 1979, p. 199.

3 - Burt, H., Dennison, J. P. and Wilshire, B. Met Science, Vol 13, No. 5, May 1979, p. 295.

4 - Tipler, H. R. and Peck, M. S., 'COST 50 materials for gas turbines, UK 17.' NPL Report ref DMA(A) 33, National Physical Laboratory, Teddington, UK, June 1981.

5 - Cane, B. J., Aplin, P. F. and Brear, J. M., 4th Int. Conf. on Creep, JSME, I.Mech.E., ASME, ASTM. Tokyo, April 14-18, 1986. p. 447.

6 - Cane, B. J., Brear, J. M. and Aplin, P. F., 3rd Int. Conf. on Creep and Fracture of Engineering Materials and Structures, Inst. Metals, Swansea, April 5-10, 1987. p. 853.

7 - Foreman, A. J. E. and Makin, M. J., Phil. Mag., No. 14, 1966. p. 911.

8 - Shewfelt, R. S. W. and Brown, L. M., Phil Mag., Vol 35, No. 4, 1977. p. 945.

*Proceedings of the Fifth International Conference on Creep of Materials, Lake Buena Vista, Florida, USA, 18-21 May 1992*

# Effect of Calcium Treatment, Sulfur Content, and Specimen Orientation on Creep Resistance of AISI 4140 Steel

**U. Muralidharan**
Technology Center
The Timken Company
Canton, Ohio

## Abstract

AISI 4140- type steel was studied to evaluate the effects of calcium treatment, sulphur levels, and specimen orientation on high temperature mechanical properties including creep resistance. Hot tensile tests were performed at 427 C (800 F), 482 C (900 F) and 538 C (1000 F). Creep tests were conducted at 482 C and 538 C. The creep resistance is presented in terms of steady state creep rates and the times required to deform to a given strain.

The calcium treated steels showed at least equivalent or slightly better creep resistance relative to non-Calcium treated steels. There was a slight reduction in creep resistance in the transverse direction relative to the resistance in the longitudinal direction.

CALCIUM TREATMENT is performed on steels to obtain improved machinability [1] and improved transverse mechanical properties. Machinability is enhanced at higher sulphur levels by the modification of inclusion morphology (aluminum oxides get encapsulated by calcium sulphides) and chemistry. Calcium treating is used to lower the oxygen and sulphur content of alloy steels which reduces the number of inclusions present. Calcium treatment also influences sulphide shape. (sulphides are made globular as opposed to elongated). This results in improved transverse toughness [2].

There was a concern in the power generation industry regarding the use of calcium treated 4140 type steels in their high temperature fasteners. As a consequence, calcium treated steels are not recommended for high temperature fastener applications. In these applications it is not the creep rupture life that is important in the design but rather the resistance to creep deformation itself. This is because the bolting stresses relax, due to time-dependent plastic deformation and creep of the bolt, which in turn may lead to loss of clamping and subsequent leakage in the pressure vessel [3]. Literature was searched extensively, but no relevant research could be found which addresses the issues and concerns dealt with in this paper.

An extensive test program was conducted to answer some of the concerns in using calcium treated steels at moderately elevated temperatures up to 538 C (1000 F). Elevated temperature mechanical properties, including creep resistance, were studied. The primary objectives of this test program were to evaluate the effects of :

1) calcium treatment at 0.018 wt% sulphur level;

2) sulphur levels in the range of 0.0035 wt% - 0.038 wt% in the calcium treated steel;

3) orientation ( longitudinal or transverse ) for the calcium treated steels.

The results of this test program are presented and discussed in this paper.

## Materials, Heat Treatment and Experimental Procedures

Three heats of Ca treated 4140 steel containing " low ", " nominal ", and " high" sulphur levels of 0.0035 wt %, 0.018 wt. % and 0.038 wt. % , respectively, were tested. In addition, non-calcium treated 4140 steel containing nominal sulphur of 0.018 wt.% was selected for this study.

The chemistries of the steels tested are shown in Table 1. While selecting the heats, the composition of elements (other than S) are kept as close to the nominal composition of the non-Ca treated heat as possible (especially P, Cu, Al, and O). Effort was also made to keep the ratio of Al to Ca constant among the heats. This was done so that the inclusion modification from stringer to globular would be directly controlled by the sulphur level. The following heats of 4140 steel were selected for the study.

Heat CLS --------- 82.6 mm diameter bar.
Heat CAS ---------- 152.4 mm diameter bar.
Heat CHS-----------120.7 mm diameter bar.
Heat NCT------------279.4 mm OD x 79.4 mm wall tube.

The hot tensile samples were 9 mm in gage diameter and 44.5 mm in gage length. The blanks for these samples were cut in both longitudinal (parallel to rolling direction) and transverse (perpendicular to rolling direction) orientation except for Heats CLS and NCT for which no transverse samples could be obtained because of product size limitations.

The creep rupture samples were 6.4 mm in gage diameter with the geometry shown in Figure 1. The blanks were cut from the above heats in both longitudinal and transverse orientations except for Heat NCT, for which only longitudinal samples could be obtained.

They were heat treated as follows.

Normalize at 843 C (1550 F), Air Cool
Austenitize at 843 C (1550 F) - 1 hour, Oil Quench
Temper at 593 C (1100 F) - 2 hours.

Hot Tensile tests were performed on these steels to determine the baseline properties prior to conducting creep rupture tests. These tests were performed at 427 C (800 F), 482 C (900 F), 538 C (1000 F). The strain rates involved in the testing range were $5 \times 10^{-5}$ per second (up to yield) and $5 \times 10^{-4}$ per second (after yield).

Constant load creep rupture tests were performed at six different stress levels at 482 C (900 F) and 538 C (1000 F) in an air environment. Three chrcmel- alumel thremocouples were bound to each specimen at three locations equally spread across the 25.4 mm (1.0 inch) gage length. The central thermocouple was used for feedback of the closed loop control of the furnace. The top and bottom thermocouples were used for monitoring and also for adjusting the heating of the three zones of the furnace to maintain the longitudinal temperature gradient to within 1.1 C ( 2 F). The test specimen was loaded uniaxially by a lever arm loading mechanism. A high temperature extensometer with knife edges positively seated on the grooves of the specimen was used to monitor the deformation of the sample across the 25.4 mm (1.0 inch) gage. The procedures conformed to ASTM E 139- 83 standards [4].

## Results

### Hot Tensile Data:

The strain rates in this test are about an order of magnitude higher than the steady state creep rates observed for the highest stress level tested. A summary of all the data collected is shown in Table 2. The data are compared in Figure 2.

There was a consistent reduction in ductility in the transverse orientation which was more pronounced for steels with 0.018 wt% sulphur at 427 C and 482 C (800 and 900 F). In general, the strength remained unchanged, whereas the ductility decreased with increasing sulphur levels. The calcium treated steel had superior longitudinal hot tensile properties at 538 C (1000 F). The hot tensile properties were identical for both Ca-treated and non-Ca-treated steels at 427 C and 482 C.

### Creep Data:

Creep tests were conducted at only two temperatures, 482 C and 538 C. The corresponding homologous temperatures, ratio of temperature to melting point in absolute scale, are 0.465 and 0.5 respectively. This was due to the extremely long test times required to run tests at 427 C. Tests were done at four to six different stress levels.

The summary of all the creep rupture tests conducted is shown in Tables 3 (a) through 3 (d). The steady state creep rates (the slope of the steady state portion of the creep curve), the rupture times, and the area reduction (measure of creep ductility) are also shown in the tables.

The steady state creep rates plotted against applied stress on log-log coordinates show a straight line relationship (Figure 3), indicating a power-law relationship between the two variables. The power-law relation- ships obtained using least-squares procedure are also shown in these figures. This equation is actually a simplified form of the Arhenius equation for a given temperature [5]. The value of stress exponent varied between 9.0 and 9.5 for 482 C and 5.0 and 5.5 for 538 C. A value of 5.0 has been observed for pure metals and 11.0 for 316 stainless steel at 704 C (1300 F) [5].

### Comparisons of Creep Resistance:

#### Effect of Orientation:

There was a slight reduction in creep resistance in the transverse direction relative to the resistance in the longitudinal direction at both 482 C and 538 C.

**Effect of Sulphur Level:**
The sulphur level (0.003 to 0.038 wt %) in calcium treated steel did not significantly affect the creep properties at 482 C and 538 C at all stress levels tested.

**Effect of Calcium Treatment:**
The calcium treatment does not cause a reduction in the creep resistance of 4140 steel at 482 C and 538 C (Figure 3). The steady state creep rates of the calcium treated steel are lower than that of non-calcium treated steels indicating the superior creep resistance of calcium treated steels for the 0.018 wt% sulphur level.

The creep resistance of the steel can also be compared in terms of time required to reach a given deformation level (e.g. 2% creep strain). This is shown in Figure 4. It is clear that the calcium treated steel takes longer to achieve the same strain at a given applied stress. The same information is also presented in the more familiar "Master Curve" format using the Larson-Miller Parameter in Figure 5.

## Discussion

The reasons for the difference in creep behavior of the steels were explored. The longitudinal sectional samples were carefully polished and the inclusion contents were analyzed using quantitative metallography (QTM). A total of 500 fields (with each field size being 275 um x 300 um) were analyzed using a magnification of 200X. Both the nominal sulphur Ca-treated heat (CAS) and the non-Ca Treated heat with nominal sulphur (NCT) showed an average inclusion lengths of 9 to 11 microns and an overall area percent of inclusions in the range of 0.16 - 0.20 %. The overall inclusion contents of these steels (size distribution and shape) were essentially the same and could not have contributed to the differences in creep behavior.

Another hypothesis was made that the coherence of the inclusions with the steel matrix could affect the deformation under creep. The duplex type inclusions, CaS covered oxides, would be more coherent with the steel matrix than the aluminum oxide inclusions. Some critical experiments were conducted to understand the differences in creep behavior. The creep rupture tests were rerun at the same stress level of 386 MPa and 482 C temperature. But this time, the tests were not run to rupture. Instead, the creep deformation was accumulated in each of the tests to 50.0 hours and, subsequently, the test specimens were ruptured in liquid nitrogen at the highest strain rates possible in the tensile test machine. The fracture surfaces were examined for any void formation or cracking near inclusions in these tests, but none were found. It became clear that at moderately elevated temperatures, the void formation or cracking near inclusion-matrix interfaces do not occur in the early stages of creep deformation [6].

Lower residual stresses develop in the matrix adjacent to duplex type inclusions for the calcium treated steel due to encapsulation of oxides by sulphide envelopes relative to non-calcium treated steel in which aluminum oxides tend to remain free in the matrix [7]. This occurs during solidification, according to Brooksbank and Andrew, due to the smaller difference in thermal expansion coefficients between the calcium sulphide and iron matrix compared to the difference in thermal expansion coefficients between aluminum oxide and iron matrix. Hence, calcium treated steel would develop lower local stresses around the inclusions which apparently leads to lower creep deformation rates for this steel relative to non-calcium treated steel as shown in Figures 3 and 4.

The observed difference could also be due to statistical variation from heat to heat. This conclusion was further corroborated by Prager's work on 2-1/4 Cr-1 Mo- steels in which the calcium treatment (which was used for sulphide shape control) was not found to cause significant difference in creep resistance relative to the non-calcium treated steels [8].

## Summary

1. Based on the data obtained at 482 C and 538 C the calcium treated 4140 steel provides slightly better creep resistance at all stress levels than the non-Ca treated steel. This can be expected even at temperatures lower than 482 C.

2. If the calcium treatment is desired in these steels either to achieve i) improved machinability at higher sulphur levels or, ii) improved transverse mechanical properties (including toughness) through sulphide shape control at low sulphur levels, this can be done without concern for creep resistance.

3. A sulphur level in the range of 0.0035 to 0.38 wt% in calcium treated steel does not affect creep properties at 482 C and 538 C at all stress levels.

4. There was only a slight reduction in creep resistance in the transverse orientation relative to longitudinal orientation. However, there was a consistent reduction in hot tensile ductility in the transverse direction which is more pronounced for steels with 0.018 wt% Sulphur at 427 C and 482 C.

## Acknowledgments

The author would like to thank The Timken Company for permission to publish this paper.

I am grateful to Mr. R. Schoone for skillfully conducting the experiments. Sincere appreciation is expressed to Mr. P. Kavadas for preparing the tables and figures.

Stimulating and useful discussions with Dr. R. Fuquen, Mr. C. V. Darragh and Dr. R. H. Klundt are gratefully acknowledged.

## References

[1] Tipnis, V. A.," Calcium Treatment for Improved Machinability," Sulphide Inclusions in Steel, ASM, p. 480 (1975).

[2] Hetzner, D. W. and Pint, B.A.," Sulphur Content, Inclusion Chemistry, and Inclusion Size Distribution in Calcium Treated 4140 Steel," Inclusions and Their Influence on Material Behavior, Symposium Proceeding, Edr. Ravi Rungta, ASM, p. 35 - 48 (1988).

[3] Finnie, I. and Heller, W. R., "Creep of Engineering Materials", p. 129 - 134, McGraw-Hill (1959).

[4] "Standard Practice for Conducting Creep, Creep-Rupture, and Stress-Rupture Tests of Metallic Materials", E 139-83, Annual Book of ASTM Standards, p. 306 - 320, (1983).

[5] Manson, S. S., Muralidharan, U. and Halford, G. R.," Tensile and Compressive Constitutive Response of 316 Stainless Steel at Elevated Temperatures," NASA Technical Memorandum 83506, (1982).

[6] Ashby, M. F. and Dyson, B. F.,"Creep Damage Mechanics and Micromechanisms," Proceeding of ICF 6, Vol. 1, p.3 (1984).

[7] Brooksbank, D. and Andrew, K. W.," Stresses associated with Duplex Oxide - Silphide Inclusions in steel," Journal of The Iron and Steel Institute, p. 582 - 586 (1970).

[8] Private communication with Mr. Martin Prager of Metals Properties Council, ASME, New York.

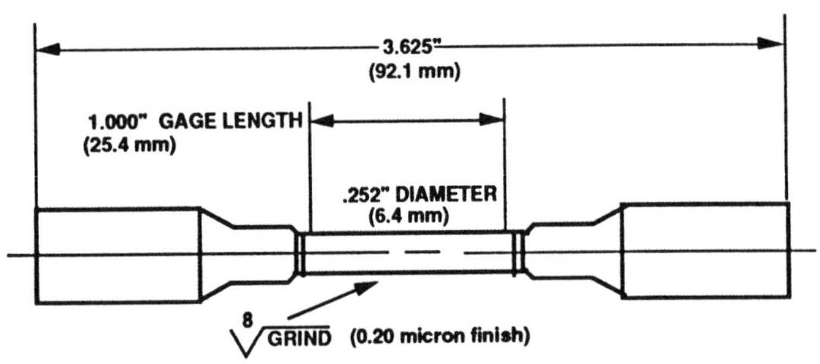


**Figure 1. Creep Rupture Specimen.**

**Table 1**
**Chemistry of Steels Tested**

| | Steel | C | Mn | Cr | Ni | Mo | P | Si | S | Cu | Al | V | Ti | Ca | Oxygen (ppm) | Nitrogen (ppm) |
|---|---|---|---|---|---|---|---|---|---|---|---|---|---|---|---|---|
| Calcium Treated | Heat CLS | .41 | .91 | 1.00 | .11 | .21 | .012 | .27 | .0035 | .10 | .032 | .005 | .002 | .0022 | 7 | 98 |
| | Heat CAS | .41 | .89 | 1.00 | .13 | .21 | .009 | .31 | .018 | .11 | .021 | .003 | .003 | .0022 | 14 | 81 |
| | Heat CHS | .42 | .93 | 1.03 | .18 | .21 | .009 | .22 | .038 | .13 | .038 | .006 | .006 | .0019 | 4 | 82 |
| Non-Calcium Treated | Heat NCT | .44 | .96 | 1.07 | .13 | .18 | .008 | .27 | .018 | .09 | .027 | .01 | .007 | .0002 | 10 | 117 |

## Table 2

## Summary of Hot Tensile Data

| Code | Orientation | Temp (C) | Yield Strength (MPa) | Ultimate Tensile Strength (MPa) | % Elongation | % Reduction in Area |
|---|---|---|---|---|---|---|
| CLS | LONGITUDINAL | 427 | 676.0 | 811.0 | 23.2 | 78.0 |
| CLS | LONGITUDINAL | 482 | 607.0 | 700.7 | 27.1 | 83.1 |
| CLS | LONGITUDINAL | 538 | 522.4 | 598.6 | 29.2 | 86.4 |
| CAS | LONGITUDINAL | 427 | 688.2 | 819.0 | 23.7 | 76.2 |
| CAS | LONGITUDINAL | 482 | 617.2 | 707.6 | 25.0 | 79.5 |
| CAS | LONGITUDINAL | 538 | 550.0 | 630.3 | 29.3 | 85.4 |
| CAS | TRANSVERSE | 427 | 638.3 | 764.1 | 13.2 | 37.9 |
| CAS | TRANSVERSE | 482 | 581.0 | 660.7 | 11.7 | 38.3 |
| CAS | TRANSVERSE | 538 | 578.0 | 580.7 | 21.4 | 55.1 |
| CHS | LONGITUDINAL | 427 | 667.2 | 796.5 | 19.3 | 73.3 |
| CHS | LONGITUDINAL | 482 | 606.2 | 669.0 | 16.0 | 45.0 |
| CHS | LONGITUDINAL | 538 | 546.2 | 598.6 | 24.6 | 84.2 |
| CHS | TRANSVERSE | 427 | 649.3 | 744.8 | 11.7 | 27.6 |
| CHS | TRANSVERSE | 482 | 606.0 | 669.3 | 14.3 | 37.7 |
| CHS | TRANSVERSE | 538 | 576.2 | 593.1 | 21.0 | 45.8 |
| NCT | LONGITUDINAL | 427 | 593.1 | 775.2 | 25.4 | 79.6 |
| NCT | LONGITUDINAL | 482 | 572.4 | 689.0 | 27.5 | 82.0 |
| NCT | LONGITUDINAL | 538 | 347.0 | 503.4 | 29.6 | 86.6 |

CLS = Ca Treated Low Sulfur
CHS = Ca Treated High Sulfur
CAS = Ca Treated Average Sulfur
NCT = Non - Ca Treated

## Table 3 (a)
## Summary of Creep Rupture Data

**Calcium Treated Steel with .004 wt% Sulfur**

| Temp (C) | Stress Level (MPa) | Steady State Creep Rate X $10^{-8}$ $SEC^{-1}$ | Rupture Time, Hrs | Reduction in Area, % | Elongation, % |
|---|---|---|---|---|---|
| | | | LONGITUDINAL | | |
| 482 | 331.0 | 4.2 | 1114.0 | 85.0 | 41.4 |
| 482 | 386.0 | 12.5 | 283.2 | 83.0 | --- |
| 482 | 441.0 | 39.7 | 68.6 | 82.4 | 58.0 |
| 482 | 483.0 | 163.0 | 21.7 | 82.0 | 30.0 |
| 482 | 524.0 | 313.0 | 6.7 | 82.0 | 49.0 |
| 538 | 172.0 | 8.7 | 437.1 | 91.4 | 78.2 |
| 538 | 248.0 | 68.0 | 79.2 | 91.0 | 61.0 |
| 538 | 290.0 | 129.0 | 29.5 | 91.7 | 54.0 |
| 538 | 331.0 | 439.0 | 10.6 | 97.0 | 38.0 |
| | | | TRANSVERSE | | |
| 482 | 331.0 | 6.0 | 755.8 | 71.9 | 34.5 |
| 482 | 441.0 | 61.1 | 40.5 | 66.5 | 26.0 |
| 482 | 483.0 | 148.0 | 15.2 | 65.5 | 24.3 |
| 482 | 524.0 | 366.0 | 4.75 | 67.7 | 22.5 |
| 538 | 172.0 | 14.1 | 392.5 | 82.5 | 58.5 |
| 538 | 248.0 | 94.1 | 58.8 | 83.0 | 51.7 |
| 538 | 331.0 | 397.0 | 8.5 | 78.5 | 40.5 |

## Table 3 (b)
## Summary of Creep Rupture Data

**Calcium Treated Steel with .018 wt% Sulfur**

| Temp (C) | Stress Level (MPa) | Steady State Creep Rate X $10^{-8}$ $SEC^{-1}$ | Rupture Time, Hrs | Reduction in Area, % | Elongation, % |
|---|---|---|---|---|---|
| | | | LONGITUDINAL | | |
| 482 | 290.0 | 1.8 | 2878.9 | 58.5 | 39.0 |
| 482 | 331.0 | 3.12 | 987.9 | 72.6 | 40.0 |
| 482 | 386.0 | 10.33 | 308.2 | 76.8 | 44.0 |
| 482 | 483.0 | 63.50 | 45.3 | 78.5 | 56.0 |
| 482 | 524.0 | 167.0 | 14.4 | 58.5 | 39.0 |
| 538 | 172.0 | 10.9 | 435.2 | 77.4 | 80.0 |
| 538 | 207.0 | 15.7 | 178.0 | ------ | 61.5 |
| 538 | 248.0 | 60.5 | 80.4 | 79.2 | 74.0 |
| 538 | 290.0 | 139.0 | 13.9 | 80.0 | 42.0 |
| 538 | 332.0 | 319.5 | 10.3 | 84.0 | ---- |
| | | | TRANSVERSE | | |
| 482 | 331.0 | 2.75 | 1396.6 | 74.1 | 35.4 |
| 482 | 386.0 | 11.50 | 250.5 | 48.0 | 24.3 |
| 482 | 483.0 | 113.0 | 19.6 | 39.6 | 17.4 |
| 482 | 524.0 | 231.0 | 8.5 | 44.5 | 38.0 |
| 538 | 172.0 | 8.7 | 409.0 | 44.9 | 41.4 |
| 538 | 207.0 | 42.0 | 165.4 | 57.0 | 50.0 |
| 538 | 290.0 | 222.0 | 23.4 | 58.9 | 40.5 |
| 538 | 248.0 | 83.3 | 60.7 | 48.5 | 35.4 |
| 538 | 331.0 | 347.0 | 10.8 | 57.5 | 42.2 |

## Table 3 (c)
## Summary of Creep Rupture Data

**Calcium Treated Steel with .038 wt% Sulfur**

| Temp (C) | Stress Level (MPa) | Steady State Creep Rate $X 10^{-8} SEC^{-1}$ | Rupture Time, Hrs | Reduction in Area, % | Elongation, % |
|---|---|---|---|---|---|
| | | | LONGITUDINAL | | |
| 482 | 331.0 | 4.7 | 1059.8 | 68.7 | 40.5 |
| 482 | 386.0 | 11.9 | 307.8 | 74.6 | 54.0 |
| 482 | 441.0 | 55.6 | 52.0 | 78.0 | 35.5 |
| 482 | 483.0 | 108.5 | 24.8 | --- | 29.0 |
| 482 | 524.0 | 421.3 | 4.9 | --- | 32.0 |
| 538 | 172.0 | 11.4 | 384.8 | 47.3 | 41.0 |
| 538 | 207.0 | 18.9 | 194.7 | --- | 47.5 |
| 538 | 248.0 | 78.1 | 72.0 | 67.5 | 43.0 |
| 538 | 290.0 | 143.0 | 25.0 | 76.0 | 50.0 |
| 538 | 331.0 | 385.0 | 9.8 | --- | 52.5 |
| | | | TRANSVERSE | | |
| 482 | 386.0 | 14.5 | 241.6 | 55.5 | 31.1 |
| 482 | 441.0 | 39.7 | 67.5 | 55.2 | 28.5 |
| 482 | 524.0 | 394.0 | 6.2 | 51.0 | 20.0 |
| 538 | 172.0 | 13.9 | 392.8 | 32.0 | 32.8 |
| 538 | 207.0 | 30.6 | 157.0 | 44.7 | 45.7 |
| 538 | 248.0 | 70.1 | 66.9 | 49.2 | 46.5 |
| 538 | 290.0 | 181.0 | 22.6 | 51.4 | 43.1 |
| 538 | 331.0 | 441.0 | 8.7 | 65.6 | 41.4 |

## Table 3 (d)
## Summary of Creep Rupture Data

**Non-Calcium Treated Steel with .018 wt% Sulfur**

| Temp (C) | Stress Level (MPa) | Steady State Creep Rate $X 10^{-8} SEC^{-1}$ | Rupture Time, Hrs | Reduction in Area, % | Elongation, % |
|---|---|---|---|---|---|
| | | | LONGITUDINAL | | |
| 482 | 290.0 | 2.6 | 2030.5 | 53.0 | 32.0 |
| 482 | 331.0 | 6.5 | 761.0 | 67.1 | 34.5 |
| 482 | 386.0 | 28.6 | 168.3 | 75.9 | 38.8 |
| 482 | 441.0 | 109.0 | 48.1 | 78.3 | 37.1 |
| 482 | 483.0 | 206.5 | 15.7 | 79.6 | 30.0 |
| 482 | 524.0 | 430.0 | 6.2 | 81.7 | 32.8 |
| 538 | 138.0 | 6.1 | 793.4 | 50.5 | 25.0 |
| 538 | 172.0 | 20.9 | 292.0 | 55.8 | 45.7 |
| 538 | 207.0 | 49.7 | 142.0 | 57.4 | 68.0 |
| 538 | 248.0 | 110.0 | 51.8 | 68.3 | 61.9 |
| 538 | 290.0 | 296.0 | 17.8 | 75.5 | 46.5 |
| 538 | 331.0 | 625.0 | 7.3 | 81.5 | 55.0 |

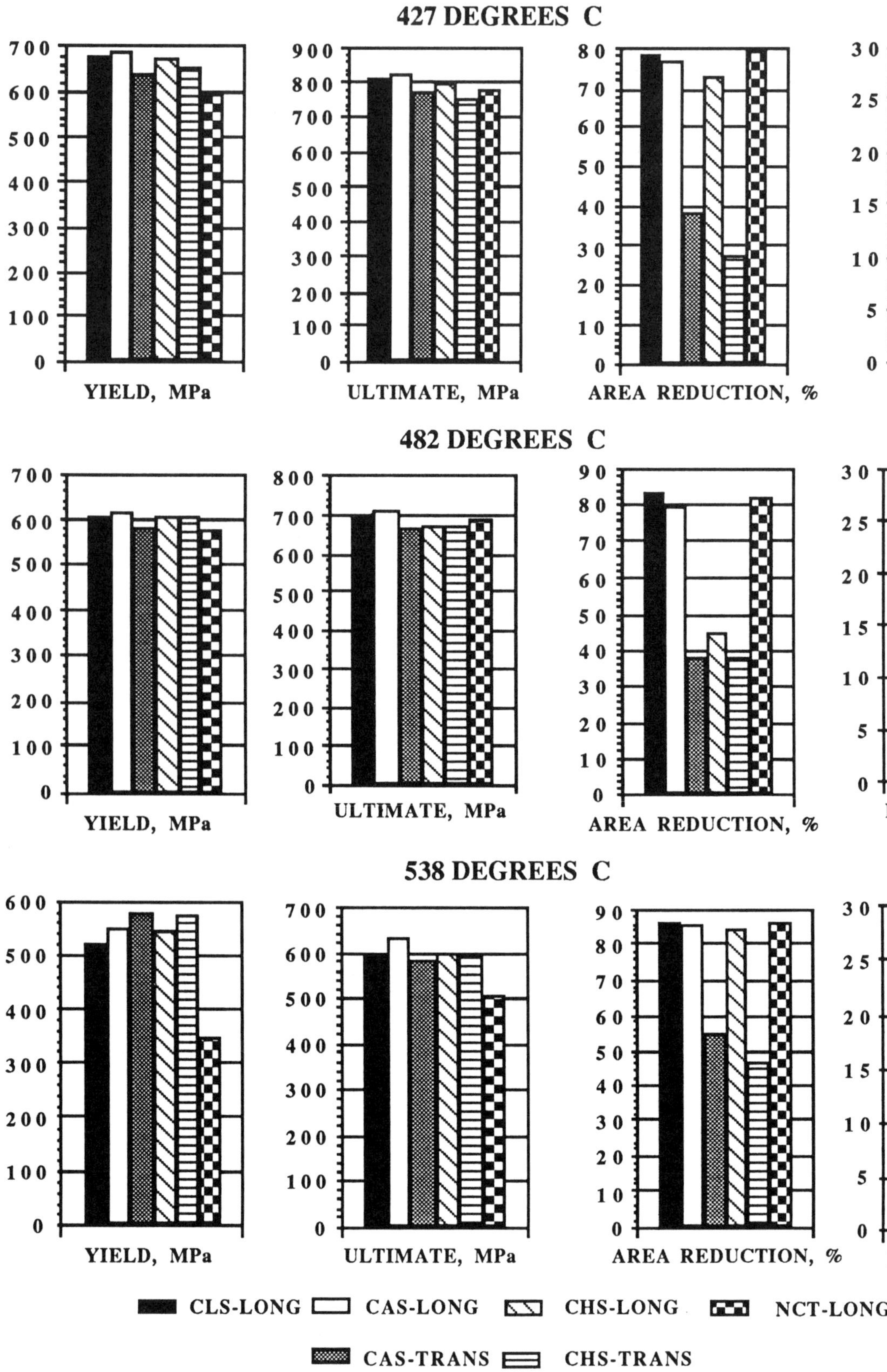


**Figure 2. Comparison of Hot Tensile Properties of Calcium and Non- Calcium Treated 4140 Steels.**

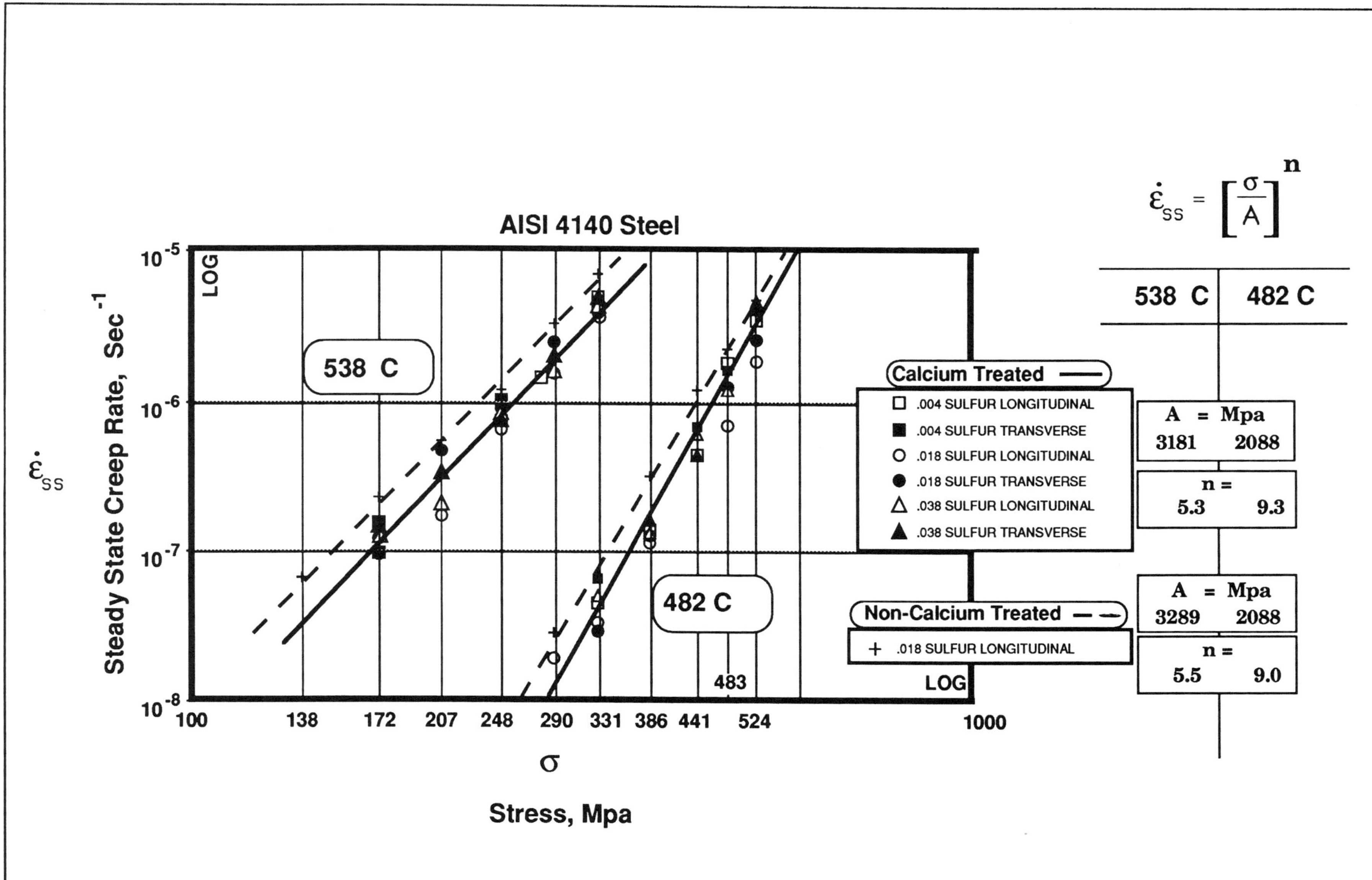


Figure 3. Comparison of Steady State Creep Rates for Calcium and Non-Calcium Treated Steels.

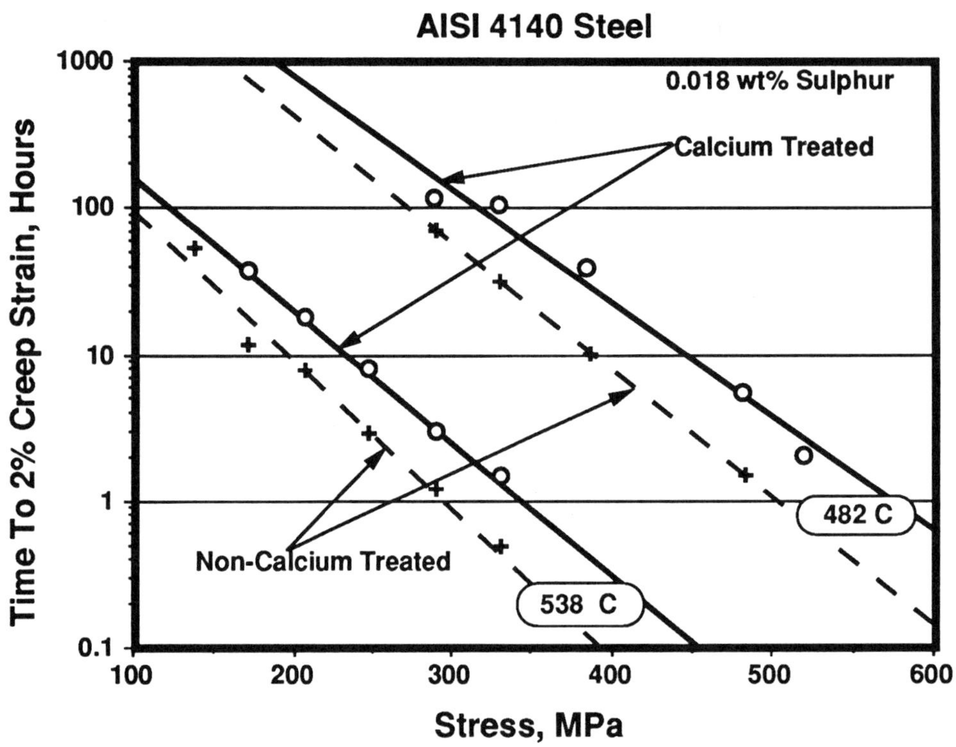


**Figure 4. Effect of Calcium Treatment On Times to 2% Creep Strain.**

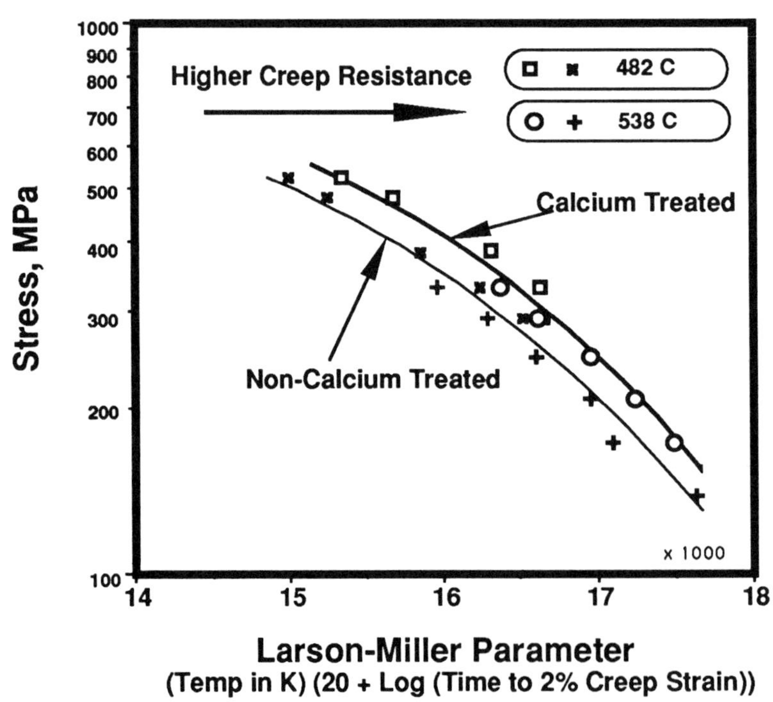


**Figure 5. Comparison of Creep Resistance in terms of Larson-Miller Parameter.**